Problem Books in Mathematics

Series Editor:
Peter Winkler
Department of Mathematics
Dartmouth College
Hanover, NH 03755
USA

More information about this series at http://www.springer.com/series/714

Vladimir V. Tkachuk

A *Cp*-Theory Problem Book

Compactness in Function Spaces

 Springer

Vladimir V. Tkachuk
Departamento de Matematicas
Universidad Autonoma Metropolitana
Iztapalapa, Mexico

ISSN 0941-3502   ISSN 2197-8506   (electronic)
Problem Books in Mathematics
ISBN 978-3-319-36536-7   ISBN 978-3-319-16092-4   (eBook)
DOI 10.1007/978-3-319-16092-4

Mathematics Subject Classification: Primary 54C35, Secondary 46E10

Springer Cham Heidelberg New York Dordrecht London
© Springer International Publishing Switzerland 2015
Softcover reprint of the hardcover 1st edition 2015

Printed on acid-free paper

Springer International Publishing AG Switzerland is part of Springer Science+Business Media (www.springer.com)

Preface

This is the third volume of the series of books of problems in C_p-theory entitled "A C_p-Theory Problem Book", i.e., this book is a continuation of the two volumes subtitled *Topological and Function Spaces* and *Special Features of Function Spaces*. The series was conceived as an introduction to C_p-theory with the hope that each volume could also be used as a reference guide for specialists.

The first volume provides a self-contained introduction to general topology and C_p-theory and contains some highly non-trivial state-of-the-art results. For example, Section 1.4 presents Shapirovsky's theorem on the existence of a point-countable π-base in any compact space of countable tightness and Section 1.5 brings the reader to the frontier of the modern knowledge about realcompactness in the context of function spaces.

The second volume covers a wide variety of topics in C_p-theory and general topology at the professional level, bringing the reader to the frontiers of modern research. It presents, among other things, a self-contained introduction to Advanced Set Theory and Descriptive Set Theory, providing a basis for working with most popular axioms independent of ZFC.

This present volume basically deals with compactness and its generalizations in the context of function spaces. It continues dealing with topology and C_p-theory at a professional level. The main objective is to develop from scratch the theory of compact spaces most used in Functional Analysis, i.e., Corson compacta, Eberlein compacta, and Gul'ko compacta.

In Section 1.1 of Chapter 1, we build up the necessary background presenting the basic results on spaces $C_p(X)$ when X has a compact-like property. In this section, the reader will find the classical theorem of Grothendieck, a very deep theorem of Reznichenko on ω-monolithity, under MA+¬CH, of a compact space X if $C_p(X)$ is Lindelöf, as well as the results of Okunev and Tamano on non-productivity of the Lindelöf property in spaces $C_p(X)$.

The main material of this volume is placed in Sections 1.2–1.4 of Chapter 1. Here we undertake a reasonably complete and up-to-date development of the theory of Corson, Gul'ko, and Eberlein compacta. Section 1.5 develops the theory of splittable spaces and gives far-reaching applications of extension operators in both C_p-theory and general topology.

We use all topological methods developed in the first two volumes, so we refer to their problems and solutions when necessary. Of course, the author did his best to keep *every* solution as independent as possible, so a short argument could be repeated several times in different places.

The author wants to emphasize that if a postgraduate student mastered the material of the first two volumes, it will be more than sufficient to understand every problem and solution of this book. However, for a concrete topic much less might be needed. Finally, let me outline some points which show the potential usefulness of the present work.

- the only background needed is some knowledge of set theory and real numbers; any reasonable course in calculus covers everything needed to understand this book;
- the student can learn all of general topology required without recurring to any textbook or papers; the amount of general topology is strictly minimal and is presented in such a way that the student works with the spaces $C_p(X)$ from the very beginning;
- what is said in the previous paragraph is true as well if a mathematician working outside of topology (in functional analysis, for example) wants to use results or methods of C_p-theory; he (or she) will find them easily in a concentrated form or with full proofs if there is such a need;
- the material we present here is up to date and brings the reader to the frontier of knowledge in a reasonable number of important areas of C_p-theory;
- this book seems to be the first self-contained introduction to C_p-theory. Although there is an excellent textbook written by Arhangel'skii (1992a), it heavily depends on the reader's good knowledge of general topology.

Mexico City, Mexico   Vladimir V. Tkachuk

Contents

Detailed Summary of Exercises

1.1. The Spaces $C_p(X)$ for Compact and Compact-like X.

1.2 Corson Compact Spaces

1.3 More of Lindelöf Σ-Property. Gul'ko Compact Spaces

1.4 Eberlein Compact Spaces

1.5 Special Embeddings and Extension Operators

Introduction

The term "C_p-theory" was invented to abbreviate the phrase "The theory of function spaces endowed with the topology of pointwise convergence". The credit for the creation of C_p-theory must undoubtedly be given to Alexander Vladimirovich Arhangel'skii. The author is proud to say that Arhangel'skii also was the person who taught him general topology and directed his PhD thesis. Arhangel'skii was the first to understand the need to unify and classify a bulk of heterogeneous results from topological algebra, functional analysis, and general topology. He was the first to obtain crucial results that made this unification possible. He was also the first to formulate a critical mass of open problems which showed this theory's huge potential for development.

Later, many mathematicians worked hard to give C_p-theory the elegance and beauty it boasts nowadays. The author hopes that the work he presents for the reader's judgement will help to attract more people to this area of mathematics.

The main text of this volume consists of 500 statements formulated as problems; it constitutes Chapter 1. These statements provide a gradual development of many popular topics of C_p-theory to bring the reader to the frontier of the present-day knowledge. A complete solution is given to every problem of the main text.

The material of Chapter 1 is divided into five sections with 100 problems in each one. The sections start with an introductory part where the definitions and concepts to be used are given. The introductory part of any section *never exceeds two pages and covers everything that was not defined previously.* Whenever possible, we try to save the reader the effort of ploughing through various sections, chapters, and volumes, so we give the relevant definitions in the current section not caring much about possible repetitions.

Chapter 1 ends with some bibliographical notes to give the most important references related to its results. The selection of references is made according to the author's preferences and by no means can be considered complete. However, a complete list of contributors to the material of Chapter 1 can be found in our bibliography of 400 items. It is my pleasant duty to acknowledge that I consulted the paper of Arhangel'skii (1998a) to include quite a few of its 375 references in my bibliography.

Sometimes, as we formulate a problem, we use without reference definitions and constructions introduced in other problems. The general rule is to try to find the relevant definition *not more than ten problems before.*

The first section of Chapter 1 deals with the spaces $C_p(X)$ for a compact X. It includes the classical theorem of Grothendieck, a deep result of Reznichenko on ω-monolithity, under MA$+\neg$CH, of any compact space X such that $C_p(X)$ is Lindelöf, as well as the examples of Okunev and Tamano on non-productivity of the Lindelöf property in function spaces.

Sections 1.2–1.4 of Chapter 1 represent the core of this volume; they constitute a development of the theory of Corson, Gul'ko, and Eberlein compacta. These classes of compact spaces are of utmost importance not only in topology but also in functional analysis. These sections include all classical results (characterizations via families of retractions, categorical and topological properties of these compact spaces, etc.) together with comparatively new material on Sokolov spaces and a radical generalization of a cornerstone result of Gul'ko on the Corson property of any compact space X whose $C_p(X)$ is Lindelöf Σ.

It is worth mentioning that Section 1.3 features a detailed description and proof of the properties of a famous example of Reznichenko of a Talagrand compact space X such that $X = \beta(X \setminus \{a\})$ for some point $a \in X$. This example has never been published by the author; since it disproves quite a few conjectures in both topology and functional analysis, the respective paper circulated as a preprint for more than 20 years before the example was described in a monograph on Functional Analysis.

Every Section of Chapter 1 has numerous topics that are developed up to the frontier of the present-day knowledge. In particular, Section 1.2 introduces the reader to the technique of adequate sets for constructing Corson compact spaces, Section 1.3 contains the results on domination of $C_p(X)$ by second countable spaces, and Section 1.4 gives a Gruenhage's characterization of Eberlein compactness of X in terms of a covering property of $(X \times X) \setminus \Delta$.

The complete solutions of all problems of Chapter 1 are given in Chapter 2. Chapter 3 begins with a selection of 100 statements which were proved as auxiliary facts in the solutions of the problems of the main text. This material is split into 8 sections to classify the respective results and make them easier to find. Chapter 4 consists of 100 open problems presented in 8 sections with the same idea: to classify this bulk of problems and make the reader's work easier.

Chapter 4 also witnesses an essential difference between the organization of our text and the book of Arhangel'skii and Ponomarev (1974): *we never put unsolved problems in the main text as is done in their book.* All problems formulated in Chapter 1 are given complete solutions in Chapter 2, and the unsolved ones are presented in Chapter 4.

There is little to explain about how to use this book as a reference guide. In this case, the methodology is not that important and the only thing the reader wants is to find the results he (or she) needs as fast as possible. To help with this, the titles of chapters and sections give the first approximation. To better see the material of a chapter, one can consult the second part of the Contents section where a detailed summary is given; it is supposed to cover all topics presented in each section. Besides, the index can also be used to find necessary material.

To sum up the main text, I believe that the coverage of C_p-theory will be reasonably complete and many of the topics can be used by postgraduate students who want to specialize in C_p-theory. Formally, this book can also be used as an introduction to general topology. However, it would be a somewhat biased introduction, because the emphasis is always given to C_p-spaces and the topics are only developed when they have some applications in C_p-theory.

To conclude, let me quote an old saying which states that the best way for one to learn a theorem is to prove it oneself. This text provides a possibility to do this. If the reader's wish is to read the proofs, there they are concentrated immediately after the main text.

Chapter 1
Behavior of Compactness in Function Spaces

The reader who has found his (or her) way through the first thousand problems of this book is fully prepared to enjoy working professionally in C_p-theory. Such a work implies choosing a topic, reading the papers with the most recent progress thereon and attacking the unsolved problems. Now, the first two steps are possible without doing heavy library work, because Chapter 1 provides information on the latest advances in all areas of C_p-theory, where compactness is concerned. Here, many ideas, results and constructions came from functional analysis giving a special flavor to this part of C_p-theory, but at the same time making it more difficult to master. I must warn the reader that most topics, outlined in the forthcoming bulk of 500 problems, constitute the material of important research papers—in many cases very difficult ones. The proofs and solutions, given in Chapter 2, are complete, but sometimes they require a very high level of understanding of the matter. The reader should not be discouraged if some proofs seem to be unfathomable. We still introduce new themes in general topology and formulate, after a due preparation, some non-trivial results which might be later used in C_p-theory.

Section 1.1 contains general facts on "nice" behavior of $C_p(X)$ and its subspaces when X has some compactness-like property. There are two statements which deserve to be called the principal ones: Arhangel'skii's theorem on compact subspaces of $C_p(X)$ for X Lindelöf, under PFA (Problem 089) and Okunev's theorem on Lindelöf subspaces of $C_p(X)$ for a compact separable space X, under MA+¬CH (Problem 098).

Section 1.2 deals with Corson compact spaces and their applications. The most important results include a theorem of Gul'ko, Michael and Rudin which states that any continuous image of a Corson compact space is Corson compact (Problem 151), Sokolov's example of a Corson compact space which is not Gul'ko compact (Problem 175) and Sokolov's theorem on Lindelöf property in iterated function spaces of a Corson compact space (Problems 160 and 162).

In Section 1.3 we present the latest achievements in the exploration of Lindelöf Σ-property in X and $C_p(X)$. Many results of this section are outstanding. We would

© Springer International Publishing Switzerland 2015

V.V. Tkachuk, *A Cp-Theory Problem Book*, Problem Books in Mathematics, DOI 10.1007/978-3-319-16092-4_1

like to mention Gul'ko's theorem which says that every Gul'ko compact is Corson compact (Problem 285), Okunev's theorems on Lindelöf Σ-property in iterated function spaces (Problems 218 and 219), Leiderman's theorem on existence of dense metrizable subspaces of Gul'ko compact spaces (Problem 293) and Reznichenko's example of a Gul'ko compact space which is not Preiss–Simon (Problem 222).

Section 1.4 outlines the main topics in the theory of Eberlein compact spaces. Here, the bright results are also numerous. It is worth mentioning a theorem of Amir and Lindenstrauss on embeddings of Eberlein compact spaces in $\Sigma_*(A)$ (Problem 322), Rosenthal's characterization of Eberlein compact spaces (Problem 324), a theorem of Benyamini, Rudin, Wage and Gul'ko on continuous images of Eberlein compact spaces (Problem 337), Gruenhage's characterization of Eberlein compact spaces (Problem 364) and Grothendieck's theorem on equivalence of the original definition of Eberlein compact spaces to the topological one (Problem 400).

Section 1.5 is devoted to the study of splittable spaces and embeddings which admit nice extension operators. It has two main results: Arhangel'skii and Shakhmatov's theorem on pseudocompact splittable spaces (Problem 417) and a theorem of Arhangel'skii and Choban on compactness of every t-extral space (Problem 475).

1.1   The Spaces C$_p$(X) for Compact and Compact-Like X

Given an infinite cardinal κ, call a space X κ-*monolithic* if, for any $Y \subset X$ with $|Y| \leq \kappa$, we have $nw(\overline{Y}) \leq \kappa$. A space is *monolithic* if it is κ-monolithic for any κ. A space X is zero-dimensional if it has a base whose elements are clopen. The space X is *strongly zero-dimensional (this is denoted by* $\dim X = 0$*)* if any finite open cover of X has a disjoint refinement. If we have topologies τ and μ on a set Z then τ is *stronger than* μ if $\tau \supset \mu$. If $\tau \subset \mu$ then τ is said to be *weaker than* μ.

Given a space X and $A \subset C(X)$, denote by $\mathrm{cl}_u(A)$ the set $\{f \in C(X) :$ there exists a sequence $\{f_n : n \in \omega\} \subset A$ such that $f_n \rightrightarrows f\}$. If X is a space, call a set $A \subset C(X)$ *an algebra*, if A contains all constant functions and $f + g \in A$, $f \cdot g \in A$ whenever $f, g \in A$. If we have spaces X and Y say that a set $A \subset C(X, Y)$ *separates the points of* X if, for any distinct $x, y \in X$, there is $f \in A$ with $f(x) \neq f(y)$. The set A *separates the points and the closed subsets of* X if, for any closed $F \subset X$ and any $x \in X \backslash F$ there is $f \in A$ such that $f(x) \notin \overline{f(F)}$.

The space $\mathbb{D}$ is the two-point set $\{0, 1\}$ endowed with the discrete topology. If T is a set and $S \subset T$ then $\chi_S^T : T \to \mathbb{D}$ is *the characteristic function of* S *in* T defined by $\chi_S^T(x) = 1$ for all $x \in S$ and $\chi_S^T(x) = 0$ whenever $x \in T \backslash S$. If the set T is clear we write χ_S instead of χ_S^T. A map $f : X \to Y$ is *finite-to-one* if $f^{-1}(y)$ is finite (maybe empty) for any $y \in Y$.

If X is a space and $A, B \subset C_p(X)$ let $\mathrm{MIN}(A, B) = \{\min(f, g) : f \in A, g \in B\}$ and $\mathrm{MAX}(A, B) = \{\max(f, g) : f \in A, g \in B\}$. For any $n \in \mathbb{N}$ consider the set $G_n(A) = \{af + bg : a, b \in [-n, n], f, g \in A\}$. Given $Y \subset C_p(X)$ we let $\mathcal{S}_1(Y) = \{Y\}$. If we have $\mathcal{S}_k(Y)$ for some $k \in \mathbb{N}$, let $\mathcal{S}_{k+1}(Y) = \{\mathrm{MIN}(A, B) : A, B \in \mathcal{S}_k(Y)\} \cup \{\mathrm{MAX}(A, B) : A, B \in \mathcal{S}_k(Y)\} \cup \{G_n(A) : A \in \mathcal{S}_k(Y), n \in \mathbb{N}\}$. This defines a family $\mathcal{S}_n(Y)$ for every $n \in \mathbb{N}$; let $\mathcal{S}(Y) = \bigcup \{\mathcal{S}_n(Y) : n \in \mathbb{N}\}$.

The expression $X \simeq Y$ says that the spaces X and Y are homeomorphic. If $\mathcal{P}$ is a topological property then $\vdash \mathcal{P}$ is to be read "has $\mathcal{P}$". For example, $X \vdash \mathcal{P}$ says that a space X has the property $\mathcal{P}$. For a space X, the class $\mathcal{E}(X)$ consists of all continuous images of products $X \times K$, where K is a compact space. A class $\mathcal{P}$ of topological spaces is called k-*directed* if it is finitely productive (i.e., $X, Y \in \mathcal{P} \Longrightarrow X \times Y \in \mathcal{P}$) and $X \in \mathcal{P}$ implies that $\mathcal{E}(X) \subset \mathcal{P}$. A k-directed class $\mathcal{P}$ is *sk-directed* if $\mathcal{P}$ is closed-hereditary, i.e., if $X \in \mathcal{P}$ then $Y \in \mathcal{P}$ for any closed $Y \subset X$. A property (or a class of spaces) $\mathcal{Q}$ is *weakly k-directed* if any metrizable compact space has $\mathcal{Q}$ (belongs to $\mathcal{Q}$) and $\mathcal{Q}$ is preserved (invariant) under continuous images and finite products.

Given a space X and an infinite cardinal κ, the space $o_\kappa(X) = (X \times D(\kappa))^\kappa$ is called *the* κ-*hull of* X. The space X is called κ-*invariant* if $X \simeq o_\kappa(X)$. A class $\mathcal{P}$ is called κ-*perfect* if, for every $X \in \mathcal{P}$, we have $o_\kappa(X) \in \mathcal{P}$, $\mathcal{E}(X) \subset \mathcal{P}$ and $Y \in \mathcal{P}$ for any closed $Y \subset X$. If $\mathcal{P}$ is a class of spaces, then $\mathcal{P}_\sigma$ consists of the spaces representable as a countable union of elements of $\mathcal{P}$. The class $\mathcal{P}_\delta$ contains the spaces which are countable intersections of elements of $\mathcal{P}$ in some larger space. More formally, $X \in \mathcal{P}_\sigma$ if $X = \bigcup \{X_n : n \in \omega\}$ where each $X_n \in \mathcal{P}$. Analogously, $X \in \mathcal{P}_\delta$ if there exists a space Y and $Y_n \subset Y$ such that $Y_n \in \mathcal{P}$ for all $n \in \omega$ and $\bigcap \{Y_n : n \in \omega\} \simeq X$. Then $\mathcal{P}_{\sigma\delta} = (\mathcal{P}_\sigma)_\delta$.

A space is called *k-separable* if it has a dense σ-compact subspace; the space X is *Hurewicz* if, for any sequence $\{\mathcal{U}_n : n \in \omega\}$ of open covers of X, we can choose, for each $n \in \omega$, a finite $\mathcal{V}_n \subset \mathcal{U}_n$ such that $\bigcup\{\mathcal{V}_n : n \in \omega\}$ is a cover of X. We say that X is *an Eberlein–Grothendieck space (or EG-space)* if X embeds into $C_p(Y)$ for some compact space Y. A space X is *radial* if, for any $A \subset X$ and any $x \in \overline{A}\backslash A$, there exists a regular cardinal κ and a transfinite sequence $S = \{x_\alpha : \alpha < \kappa\} \subset A$ such that $S \rightarrow x$ in the sense that, for any open $U \ni x$, there is $\alpha < \kappa$ such that, for each $\beta \geq \alpha$, we have $x_\beta \in U$. The space X is *pseudoradial* if $A \subset X$ and $A \neq \overline{A}$ implies that there is a regular cardinal κ and a transfinite sequence $S = \{x_\alpha : \alpha < \kappa\} \subset A$ such that $S \rightarrow x \notin A$. A subset A of a space X is called *bounded* if every $f \in C(X)$ is bounded on A, i.e., there exists $N \in \mathbb{R}$ such that $|f(x)| \leq N$ for all $x \in A$. A family $\mathcal{U}$ is an *ω-cover of a set A*, if, for any finite $B \subset A$, there is $U \in \mathcal{U}$ such that $B \subset U$. A *Luzin space* is any uncountable space without isolated points in which every nowhere dense subset is countable. An *analytic space* is a continuous image of the space $\mathbb{P}$ of irrational numbers. Given a space X, let $vet(X) \leq \kappa$ if, for any $x \in X$ and any family $\{A_\alpha : \alpha < \kappa\} \subset \exp X$ with $x \in \bigcap\{\overline{A_\alpha} : \alpha < \kappa\}$, we can choose, for each $\alpha < \kappa$, a finite $B_\alpha \subset A_\alpha$ such that $x \in \overline{\bigcup\{B_\alpha : \alpha < \kappa\}}$. The cardinal $vet(X) = \min\{\kappa \geq \omega : vet(X) \leq \kappa\}$ is called *fan tightness of the space X*.

It is also important to mention that the wizards of set theory invented an axiom which is called PFA (from Proper Forcing Axiom). Any reasonably comprehensive formulation of PFA is outside of the reach of this book. However, any professional topologist must know that it exists and that it is consistent with the usual system of axioms (referred to as ZFC) of set theory provided there exist some large cardinals. I am not going to give a rigorous definition of large cardinals because we won't need them here. However, we must be aware of the fact that it is absolutely evident for everybody (who knows what they are!) that their existence is consistent with ZFC notwithstanding that this consistence is not proved yet. So, it is already a common practice to use nice topological consequences of PFA. The one we will need is the following statement: "If X is a compact space of weight ω_1 and $t(X) > \omega$ then $\omega_1 + 1$ embeds in X".

Given a space X a continuous map $r : X \rightarrow X$ is called *a retraction* if $r \circ r = r$. If f is a function then $\mathrm{dom}(f)$ is its domain; given a function g the expression $f \subset g$ says that $\mathrm{dom}(f) \subset \mathrm{dom}(g)$ and $g|\mathrm{dom}(f) = f$. If we have a set of functions $\{f_i : i \in I\}$ such that $f_i|(\mathrm{dom}(f_i) \cap \mathrm{dom}(f_j)) = f_j|(\mathrm{dom}(f_i) \cap \mathrm{dom}(f_j))$ for any indices $i, j \in I$ then we can define a function f with $\mathrm{dom}(f) = \bigcup_{i \in I} \mathrm{dom}(f_i)$ as follows: given any $x \in \mathrm{dom}(f)$, find any $i \in I$ with $x \in \mathrm{dom}(f_i)$ and let $f(x) = f_i(x)$. It is easy to check that the value of f at x does not depend on the choice of i so we have consistently defined a function f which will be denoted by $\bigcup\{f_i : i \in I\}$.

001. Prove that, if X is a normal space and $\dim X = 0$ then $\dim \beta X = 0$ and hence βX is zero-dimensional.

002. Let X be a zero-dimensional compact space. Suppose that Y is second countable and $f : X \to Y$ is a continuous onto map. Prove that there exists a compact metrizable zero-dimensional space Z and continuous onto maps $g : X \to Z$ and $h : Z \to Y$ such that $f = h \circ g$.

003. Prove that there exists a continuous map $k : \mathbb{K} \to \mathbb{I}$ such that, for any compact zero-dimensional space X and any continuous map $f : X \to \mathbb{I}$, there exists a continuous map $g_f : X \to \mathbb{K}$ such that $f = k \circ g_f$.

004. Prove that, for any zero-dimensional compact X, the space $C_p(X, \mathbb{I})$ is a continuous image of $C_p(X, \mathbb{D}^\omega)$.

005. Given a countably infinite space X prove that the following conditions are equivalent:

   (i) $C_p(X, \mathbb{D})$ is countable;
   (ii) $C_p(X, \mathbb{D}) \simeq \mathbb{Q}$;
   (iii) X is compact.

006. For an arbitrary space X prove that

   (i) for any $P \subset C_p(X)$ there is an algebra $A(P) \subset C_p(X)$ such that $P \subset A(P)$ and $A(P)$ is *minimal* in the sense that, for any algebra $A \subset C_p(X)$, if $P \subset A$ then $A(P) \subset A$;
   (ii) $A(P)$ is a countable union of continuous images of spaces which belong to $\mathcal{H}(P) = \{P^m \times K$ for some $m \in \mathbb{N}$ and metrizable compact $K\}$.
   (iii) if $\mathcal{Q}$ is a weakly k-directed property and $P \vdash \mathcal{Q}$ then $A(P) \vdash \mathcal{Q}_\sigma$, i.e., $A(P)$ is a countable union of spaces with the property $\mathcal{Q}$;

007. Given a compact space X suppose that $A \subset C_p(X)$ is an algebra. Prove that both $\overline{A}$ and $\mathrm{cl}_u(A)$ are algebras in $C_p(X)$.

008. Let X be a compact space. Suppose that $A \subset C_p(X)$ separates the points of X, contains the constant functions and has the following property: for each $f, g \in A$ and $a, b \in \mathbb{R}$ we have $af + bg \in A$, $\max(f, g) \in A$, $\min(f, g) \in A$. Prove that every $f \in C_p(X)$ is a uniform limit of some sequence from A.

009. Let X be a compact space and suppose that $Y \subset C_p(X)$ separates the points of X. Prove that

   (i) for any algebra $A \subset C_p(X)$ with $Y \subset A$, we have $\mathrm{cl}_u(A) = C_p(X)$;
   (ii) if Y contains a non-zero constant function then $\mathrm{cl}_u(\bigcup \mathcal{S}(Y)) = C_p(X)$.

010. For a space X, suppose that $Y \subset C_p(X)$ and $\mathrm{cl}_u(Y) = C_p(X)$. Prove that $C_p(X) \in (\mathcal{E}(Y))_\delta$.

011. Prove that every k-directed non-empty class is weakly k-directed. Give an example of a weakly k-directed class which is not k-directed.

012. Prove that any class $\mathcal{K} \in \{$compact spaces, σ-compact spaces, k-separable spaces$\}$ is k-directed. How about the class of countably compact spaces?

013. Let $\mathcal{P}$ be a weakly k-directed class. Prove that, for any $Y \subset C_p(X)$ such that $Y \in \mathcal{P}$, we have $\mathcal{S}(Y) \subset \mathcal{P}$.

014. Given a k-directed class $\mathcal{Q}$ and a compact space X suppose that some set $Y \subset C_p(X)$ separates the points of X and $Y \in \mathcal{Q}$. Prove that $C_p(X) \in \mathcal{Q}_{\sigma\delta}$, i.e., there is a space Z such that $C_p(X) \subset Z$ and $C_p(X) = \bigcap\{C_n : n \in \omega\}$ where every $C_n \subset Z$ is a countable union of spaces with the property $\mathcal{Q}$.

015. For a compact space X suppose that $Y \subset C_p(X)$ separates the points of X. Prove that there exists a compact space K and a closed subspace $F \subset o_\omega(Y) \times K$ such that $C_p(X)$ is a continuous image of F.

016. Prove that, for any compact space X, there exists a compact space K and a closed subspace $F \subset (C_p(X))^\omega \times K$ such that $C_p(X^\omega)$ is a continuous image of F.

017. Let X be a compact space such that $(C_p(X))^\omega$ is Lindelöf. Show that $C_p(X^\omega)$ is Lindelöf. As a consequence, $C_p(X^n)$ is Lindelöf for each $n \in \mathbb{N}$.

018. Assume that X is compact and $\mathcal{P}$ is an ω-perfect class. Prove that it follows from $C_p(X) \in \mathcal{P}$ that $C_p(X^\omega) \in \mathcal{P}$.

019. Let $\mathcal{P}$ be an ω-perfect class of spaces. Prove that the following properties are equivalent for any compact X:

  (i) the space $C_p(X)$ belongs to $\mathcal{P}$;
  (ii) there exists $Y \subset C_p(X)$ such that Y is dense in $C_p(X)$ and $Y \in \mathcal{P}$;
  (iii) there exists $Y \subset C_p(X)$ which separates the points of X and belongs to $\mathcal{P}$;
  (iv) the space X embeds into $C_p(Z)$ for some $Z \in \mathcal{P}$.

020. Prove that the class $L(\Sigma)$ of Lindelöf Σ-spaces is ω-perfect. As a consequence, for any compact X, the following properties are equivalent:

  (i) the space $C_p(X)$ is Lindelöf Σ;
  (ii) there exists $Y \subset C_p(X)$ such that Y is dense in $C_p(X)$ and $Y \in L(\Sigma)$;
  (iii) there exists $Y \subset C_p(X)$ which separates the points of X and belongs to $L(\Sigma)$;
  (iv) the space X embeds into $C_p(Y)$ for some Lindelöf Σ-space Y.

021. Let X be a compact space such that $C_p(X)$ is Lindelöf Σ. Show that $C_p(X^\omega)$ is a Lindelöf Σ-space and so is $C_p(X^n)$ for each $n \in \mathbb{N}$.

022. Prove that the class $K(\mathcal{A})$ of K-analytic spaces is ω-perfect. Thus, for any compact X, the following properties are equivalent:

  (i) the space $C_p(X)$ is K-analytic;
  (ii) there exists $Y \subset C_p(X)$ such that Y is dense in $C_p(X)$ and $Y \in K(\mathcal{A})$;
  (iii) there exists $Y \subset C_p(X)$ which separates the points of X and belongs to $K(\mathcal{A})$;
  (iv) the space X embeds into $C_p(Y)$ for some K-analytic space Y.

023. Let X be a compact space such that $C_p(X)$ is K-analytic. Show that $C_p(X^\omega)$ is a K-analytic space and so is $C_p(X^n)$ for each $n \in \mathbb{N}$.

024. Observe that any K-analytic space is Lindelöf Σ. Give an example of a space X such that $C_p(X)$ is Lindelöf Σ but not K-analytic.

025. Give an example of X such that $C_p(X)$ is K-analytic but not $K_{\sigma\delta}$.

026. Let X be a Lindelöf Σ-space. Prove that $C_p(X)$ is normal if and only if $C_p(X)$ is Lindelöf. In particular, if X is compact then $C_p(X)$ is normal if and only if it is Lindelöf.

027. Suppose that X is a Lindelöf Σ-space such that $C_p(X) \backslash \{f\}$ is normal for some $f \in C_p(X)$. Prove that X is separable. In particular, if X is ω-monolithic and $C_p(X) \backslash \{f\}$ is normal for some $f \in C_p(X)$ then X has a countable network.

028. Let X and $C_p(X)$ be Lindelöf Σ-spaces and suppose that $C_p(X) \backslash \{f\}$ is normal for some $f \in C_p(X)$. Prove that X has a countable network.

029. Let M_t be a separable metrizable space for all $t \in T$. Suppose that Y is dense in $M = \prod \{M_t : t \in T\}$ and Z is a continuous image of Y. Prove that, if $Z \times Z$ is normal then $ext(Z) = \omega$ and hence Z is collectionwise normal.

030. Prove that, for any infinite zero-dimensional compact space X, there exists a closed $F \subset C_p(X, \mathbb{D}^\omega) \subset C_p(X)$ which maps continuously onto $(C_p(X))^\omega$.

031. Prove that, for any infinite zero-dimensional compact space X, there exists a closed $F \subset C_p(X, \mathbb{D}^\omega) \subset C_p(X)$ which maps continuously onto $C_p(X^\omega)$.

032. Prove that the following conditions are equivalent for an arbitrary zero-dimensional compact X:

    (i) $C_p(X, \mathbb{D}^\omega)$ is normal;
    (ii) $C_p(X, \mathbb{I})$ is normal;
    (iii) $C_p(X)$ is normal;
    (iv) $C_p(X)$ is Lindelöf;
    (v) $(C_p(X))^\omega$ is Lindelöf;
    (vi) $C_p(X^\omega)$ is Lindelöf.

033. Observe that $C_p(X)$ is monolithic for any compact X. Using this fact prove that, for any compact space X, each compact subspace $Y \subset C_p(X)$ is a Fréchet–Urysohn space.

034. Prove that, for any metrizable space M, there is a compact space K such that M embeds in $C_p(K)$.

035. Prove that the following conditions are equivalent for any compact X:

    (i) there is a compact $K \subset C_p(X)$ which separates the points of X;
    (ii) there is a σ-compact $Y \subset C_p(X)$ which separates the points of X;
    (iii) there is a σ-compact $Z \subset C_p(X)$ which is dense in $C_p(X)$;
    (iv) X embeds into $C_p(K)$ for some compact K;
    (v) X embeds into $C_p(Y)$ for some σ-compact Y.

036. Suppose that X is compact and embeds into $C_p(Y)$ for some compact Y. Prove that it is possible to embed X into $C_p(Z)$ for some Fréchet–Urysohn compact space Z.

037. Give an example of a compact space X embeddable into $C_p(Y)$ for some compact Y but not embeddable into $C_p(Z)$ for any compact first countable space Z.

038. Suppose that X embeds into $C_p(Y)$ for some compact Y. Prove that it is possible to embed X into $C_p(Z)$ for some zero-dimensional compact space Z.

039. Suppose that X embeds into $C_p(Y)$ for some countably compact Y. Prove that it is possible to embed X into $C_p(Z)$ for some zero-dimensional countably compact space Z.

040. Give an example of a space Y which embeds in $C_p(X)$ for a pseudocompact space X but does not embed in $C_p(Z)$ for any countably compact Z.

041. Prove that a countable space Y embeds into $C_p(X)$ for some pseudocompact space X if and only if Y embeds into $C_p(Z)$ for some compact metrizable space Z.

042. Give an example of a space Y which embeds into $C_p(X)$ for a countably compact space X but does not embed into $C_p(Z)$ for a compact space Z.

043. Let $\xi \in \beta\omega \setminus \omega$. Prove that the countable space $\omega_\xi = \omega \cup \{\xi\}$, considered with the topology inherited from $\beta\omega$, does not embed into $C_p(X)$ for a pseudocompact X.

044. (Grothendieck's theorem). Suppose that X is a countably compact space and $B \subset C_p(X)$ is a bounded subset of $C_p(X)$. Prove that $\overline{B}$ is compact. In particular, the closure of any pseudocompact subspace of $C_p(X)$ is compact.

045. Prove that there exists a pseudocompact space X for which there is a closed pseudocompact $Y \subset C_p(X)$ which is not countably compact.

046. Let X be a σ-compact space. Prove that any countably compact subspace of $C_p(X)$ is compact.

047. Let X be a space and suppose that there is a point $x_0 \in X$ such that $\psi(x_0, X) = \omega$ and $x_0 \notin \overline{A}$ for any countable $A \subset X$. Prove that there is an infinite closed discrete $B \subset C_p(X)$ such that B is bounded in $C_p(X)$.

048. Prove that there exists a σ-compact space X such that $C_p(X)$ contains an infinite closed discrete subspace which is bounded in $C_p(X)$.

049. Prove that there exists a σ-compact space X such that $C_p(X)$ does not embed as a closed subspace into $C_p(Y)$ for any countably compact space Y.

050. Given a metric space (M, ρ) say that a family $\mathcal{U} \subset \exp M \setminus \{\emptyset\}$ is ρ-*vanishing* if $\operatorname{diam}_\rho(U) < \infty$ for any $U \in \mathcal{U}$ and the diameters of the elements of $\mathcal{U}$ converge to zero, i.e., the family $\{U \in \mathcal{U} : \operatorname{diam}_\rho(U) \geq \varepsilon\}$ is finite for any $\varepsilon > 0$. Prove that, for any separable metrizable X, the following conditions are equivalent:

  (i) X is a Hurewicz space;
 (ii) for any metric ρ which generates the topology of X, there is a ρ-vanishing family $\mathcal{U} \subset \tau(X)$ such that $\bigcup \mathcal{U} = X$;
(iii) for any metric ρ which generates the topology of X, there exists a ρ-vanishing base $\mathcal{B}$ of the space X;

(iv) there exists a metric ρ which generates the topology of X, such that, for any base $\mathcal{B}$ of the space X, there is a ρ-vanishing family $\mathcal{U} \subset \mathcal{B}$ for which $\bigcup \mathcal{U} = X$;

(v) for any metric ρ which generates the topology of X and any base $\mathcal{B}$ of the space X there is ρ-vanishing family $\mathcal{B}' \subset \mathcal{B}$ such that $\mathcal{B}'$ is also a base of X;

(vi) every base of X contains a family which is a locally finite cover of X.

051. Prove that X^ω is a Hurewicz space if and only if X is compact.

052. Prove that any separable Luzin space is a Hurewicz space.

053. Prove that any Hurewicz analytic space is σ-compact.

054. Give an example of a Hurewicz space which is not σ-compact.

055. Prove that, under CH, there exists a Hurewicz space whose square is not normal.

056. Prove that X^n is a Hurewicz space for every $n \in \mathbb{N}$, if and only if, for any sequence $\{\gamma_k : k \in \omega\}$ of open ω-covers of the space X, we can choose, for each $k \in \omega$, a finite $\mu_k \subset \gamma_k$ such that the family $\bigcup\{\mu_k : k \in \omega\}$ is an ω-cover of X.

057. Let X be any space. Prove that X^n is a Hurewicz space for all $n \in \mathbb{N}$ if and only if $vet(C_p(X)) \le \omega$.

058. Prove that if $C_p(X)$ is Fréchet–Urysohn then $vet(C_p(X)) \le \omega$.

059. Prove that, under MA+¬CH, there exists a second countable space X such that X^n is a Hurewicz space for each natural n, while X is not σ-compact.

060. Say that a space is *subsequential* if it embeds in a sequential space. Prove that every sequential space has countable tightness and hence each subsequential space also has countable tightness.

061. For any point $\xi \in \beta\omega \backslash \omega$ prove that the countable space $\omega \cup \{\xi\}$ is not subsequential.

062. Prove that $C_p(\mathbb{I})$ is not subsequential.

063. Prove that the following are equivalent for any pseudocompact X:

(i) $C_p(X)$ is a Fréchet–Urysohn space;

(ii) $C_p(X)$ embeds in a sequential space;

(iii) X is compact and scattered.

064. Prove that radiality is a hereditary property; show that pseudoradiality is closed-hereditary. Give an example showing that pseudoradiality is not hereditary.

065. Prove that any quotient (pseudo-open) image of a pseudoradial (radial) space is a pseudoradial (radial) space.

066. Prove that any radial space of countable tightness is Fréchet–Urysohn.

067. Prove that a space is radial (pseudoradial) if and only if it is a pseudo-open (quotient) image of a linearly ordered space.

068. Prove that any radial space of countable spread is Fréchet–Urysohn.

069. Prove that any radial dyadic space is metrizable.

070. Prove that $\beta\omega \backslash \omega$ is not pseudoradial.

071. Prove that $\mathbb{D}^{\omega_1}$ is not pseudoradial under CH and pseudoradial under MA+¬CH.

072. Prove that it is independent of ZFC whether every dyadic pseudoradial space is metrizable.

073. Prove that, for any space X, the space $C_p(X)$ is radial if and only if it is Fréchet–Urysohn.

074. An uncountable cardinal κ is called ω-*inaccessible* if $\lambda^\omega < \kappa$ for any cardinal $\lambda < \kappa$. Recall that, if ξ is an ordinal then cf(ξ) = min$\{|A| \,:\, A$ is a cofinal subset of $\xi\}$. Prove that, for an infinite ordinal ξ, the space $C_p(\xi)$ is pseudoradial if and only if either cf(ξ) $\leq \omega$ or ξ is an ω-inaccessible regular cardinal (here, as usual, ξ is considered with its interval topology). Observe that ω-inaccessible regular cardinals exist in ZFC and hence there exist spaces X such that $C_p(X)$ is pseudoradial but not radial.

075. Let X be a compact space. Prove that, if $C_p(X)$ is pseudoradial then it is Fréchet–Urysohn (and hence X is scattered).

076. Let X be any space such that $C_p(X, \mathbb{D}) \times \omega^\omega$ is not Lindelöf. Prove that the space $C_p(X, \mathbb{D}^\omega)$ is not Lindelöf.

077. Suppose that X is a compact space such that a countable set $M \subset X$ is open and dense in X. Assume also that the set of isolated points of $Y = X \backslash M$ is uncountable and dense in Y. Prove that $ext(C_p(X, \mathbb{D}) \times \omega^\omega) > \omega$.

078. Suppose that X is a compact space such that a countable set $M \subset X$ is open and dense in X. Assume also that the set I of isolated points of $Y = X \backslash M$ is uncountable and dense in Y; let $F = Y \backslash I$. Prove that, under MA+¬CH, any uncountable subset of the set $E = \{f \in C_p(X, \mathbb{D}) : f(F) = \{0\}\}$ contains an uncountable set D which is closed and discrete in $C_p(X, \mathbb{D})$.

079. Let X be a compact space of weight ω_1 in which we have a countable dense set L and a nowhere dense closed non-empty set F. Assuming MA+¬CH prove that there exists $M \subset L$ such that $\overline{M} \backslash M = F$ and all points of M are isolated in the space $M \cup F$.

080. Prove that, under MA+¬CH, if X is a compact space such that $C_p(X)$ is normal, then X is Fréchet–Urysohn, ω-monolithic and has a dense set of points of countable character.

081. Assume MA+¬CH. Show that, if a compact space X has the Souslin property and $C_p(X)$ is normal then X is metrizable.

082. Prove that $w(X) = l(C_p(X))$ for any linearly orderable compact space X. In particular, if $C_p(X)$ is Lindelöf then X is metrizable.

083. Given an infinite compact space X prove that we have $|\overline{Y}| \leq 2^{l(Y) \cdot c(X)}$ for any $Y \subset C_p(X)$.

084. Suppose that X is a compact space with the Souslin property and $C_p(X)$ has a dense Lindelöf subspace. Prove that $w(X) \leq |C_p(X)| \leq 2^\omega$.

085. Prove that, for any uncountable regular cardinal κ, if $Z \subset C_p(\kappa + 1)$ separates the points of $\kappa + 1$ then $l(Z) \geq \kappa$.

086. Prove that, if X is a dyadic space and $Y \subset C_p(X)$ then $nw(Y) = l(Y)$. In particular, any Lindelöf subspace of $C_p(X)$ has a countable network.

087. Prove that, if X is a dyadic space and $C_p(X)$ has a dense Lindelöf subspace then X is metrizable.

088. Given a space X suppose that $K \subset C_p(X)$ is a compact space of uncountable tightness. Show that there exists a closed $X_1 \subset X$ such that $C_p(X_1)$ contains a compact subspace of weight and tightness ω_1.

089. Prove that the axiom PFA implies that, for any Lindelöf space X and any compact $K \subset C_p(X)$, we have $t(K) \leq \omega$.

090. Given a space X and a set $A \subset X$ denote by τ_A the topology on X generated by the family $\tau(X) \cup \exp(X \setminus A)$ as a subbase; let $X[A] = (X, \tau_A)$. In other words, the space $X[A]$ is constructed by declaring isolated all points of $X \setminus A$ and keeping the same topology at the points of A. Prove that, for any uncountable Polish space M and $A \subset M$ the following conditions are equivalent:

  (i) the space $(M[A])^\omega$ is Lindelöf;
  (ii) if $\mathcal{F}$ is a countable family of finite-to-one continuous maps from the Cantor set $\mathbb{K}$ to M then $\bigcap \{f^{-1}(A) : f \in \mathcal{F}\} \neq \emptyset$;
  (iii) if $\mathcal{F}$ is a countable family of injective continuous maps from the Cantor set $\mathbb{K}$ to M then $\bigcap \{f^{-1}(A) : f \in \mathcal{F}\} \neq \emptyset$.

  Deduce from this fact that, for any uncountable Polish space M there is a disjoint family $\{A_\alpha : \alpha < \mathfrak{c}\}$ of subsets of M such that $(M[A_\alpha])^\omega$ is Lindelöf for any $\alpha < \mathfrak{c}$.

091. Given a space X and a set $A \subset X$ denote by τ_A the topology on X generated by the family $\tau(X) \cup \exp(X \setminus A)$ as a subbase; let $X[A] = (X, \tau_A)$. Prove that, if M is a Polish space, $A \subset M$ and $n \in \mathbb{N}$ then the following conditions are equivalent:

  (i) the space $(M[A])^n$ is Lindelöf;
  (ii) if $\mathcal{F}$ is a family of finite-to-one continuous maps from the Cantor set $\mathbb{K}$ to M and $|\mathcal{F}| \leq n$ then $\bigcap \{f^{-1}(A) : f \in \mathcal{F}\} \neq \emptyset$;
  (iii) if $\mathcal{F}$ is a family of injective continuous maps from the Cantor set $\mathbb{K}$ to M and $|\mathcal{F}| \leq n$ then $\bigcap \{f^{-1}(A) : f \in \mathcal{F}\} \neq \emptyset$.

  Deduce from this fact that, for any uncountable Polish space M there is a disjoint family $\{A_\alpha : \alpha < \mathfrak{c}\}$ of subsets of M such that for every $\alpha < \mathfrak{c}$ the space $(M[A_\alpha])^k$ is Lindelöf for any $k \in \mathbb{N}$ while $(M[A_\alpha])^\omega$ is not Lindelöf.

092. Suppose that $\mathcal{P}$ is an sk-directed class of spaces and $Y \in \mathcal{P}$. Prove that if $X \subset C_p(Y)$ and the set of non-isolated points of X is σ-compact then $C_p^*(X)$ belongs to the class $\mathcal{P}_{\sigma\delta}$.

093. Prove that there exist separable, scattered σ-compact spaces X and Y such that both $(C_p(X))^\omega$ and $(C_p(Y))^\omega$ are Lindelöf while $C_p(X) \times C_p(Y)$ is not normal and contains a closed discrete set of cardinality $\mathfrak{c}$.

094. Show that there is a separable scattered σ-compact space X and a countable space M such that the space $(C_p(X))^\omega$ is Lindelöf while we have the equality $ext(C_p(X) \times C_p(M)) = \mathfrak{c}$ and $C_p(X) \times C_p(M)$ is not normal.

095. Prove that, under CH, there exists a separable scattered compact space X such that $(C_p(X, \mathbb{D}))^n$ is Lindelöf for any natural n, while $(C_p(X, \mathbb{D}))^\omega$ is not Lindelöf.

096. Prove that there is a scattered, separable, zero-dimensional σ-compact space X with $(C_p(X, \mathbb{D}))^n$ Lindelöf for each natural n, while $(C_p(X, \mathbb{D}))^\omega$ is not Lindelöf.

097. Assume MA+¬CH. Let X be a space with $l^*(X) = \omega$. Prove that any separable compact subspace of $C_p(X)$ is metrizable.

098. Assume MA+¬CH. Let X be a separable compact space. Prove that, for any $Y \subset C_p(X)$ with $l^*(Y) = \omega$, we have $nw(Y) = \omega$.

099. Prove that there exists a separable σ-compact space X such that $(C_p(X))^\omega$ is Lindelöf and $s(X) > \omega$.

100. Assume MA+¬CH. Prove that there is a separable σ-compact space X such that $C_p(X)$ does not embed into $C_p(Y)$ for a separable compact space Y.

1.2 Corson Compact Spaces

All spaces are assumed to be Tychonoff. For a product $X = \prod\{X_t : t \in T\}$ of the spaces X_t and $x \in X$, let $\Sigma(X, x) = \{y \in X : |\{t \in T : y(t) \neq x(t)\}| \leq \omega\}$. The space $\Sigma(X, x)$ is called the Σ-*product of spaces* $\{X_t : t \in T\}$ *with the center* x. Again, if $x \in X = \prod\{X_t : t \in T\}$, let $\sigma(X, x) = \{y \in X : |\{t \in T : y(t) \neq x(t)\}| < \omega\}$. The space $\sigma(X, x)$ is called the σ-*product of spaces* $\{X_t : t \in T\}$ *with the center* x. If some statement about Σ-products or σ-products is made with no center specified, then this statement holds (or must be proved) for an arbitrary center. The symbols $\Sigma(A)$ and $\sigma(A)$ are reserved for the respective Σ- and σ-products of real lines with the center zero, i.e., $\Sigma(A) = \{x \in \mathbb{R}^A : |\{a \in A : x(a) \neq 0\}| \leq \omega\}$ and $\sigma(A) = \{x \in \mathbb{R}^A : |\{a \in A : x(a) \neq 0\}| < \omega\}$. Now, $\Sigma_*(A) = \{x \in \mathbb{R}^A :$ for any $\varepsilon > 0$ the set $\{a \in A : |x(a)| \geq \varepsilon\}$ is finite$\}$. A family $\mathcal{U} \subset \exp A$ is called ω-*continuous* if $\bigcup\{U_n : n \in \omega\} \in \mathcal{U}$ whenever $U_n \in \mathcal{U}$ and $U_n \subset U_{n+1}$ for all $n \in \omega$. The family $\mathcal{U}$ is ω-*cofinal* if, for every countable $B \subset A$, there exist $U \in \mathcal{U}$ such that $B \subset U$. If $B \subset A$ and $x \in \Sigma(A)$, let $r_B(x)(a) = x(a)$ if $a \in B$ and $r_B(x)(a) = 0$ otherwise. Clearly, $r_B : \Sigma(A) \to \Sigma(A)$ is a continuous map. Call a set $Y \subset \Sigma(A)$ *invariant* if the family $\{B \subset A : r_B(Y) \subset Y\}$ is ω-continuous and ω-cofinal.

A compact space X is *Corson compact* if it embeds into $\Sigma(A)$ for some A. Given a space X and an infinite cardinal κ, the space $(X \times D(\kappa))^\kappa$ is called *the* κ-*hull of* X and is denoted by $o_\kappa(X)$. The space X is called κ-*invariant* if $X \simeq o_\kappa(X)$. For a space X, the class $\mathcal{E}(X)$ consists of all continuous images of products $X \times K$, where K is a compact space. Define a class $\mathcal{P}$ to be κ-*perfect* if, for any $X \in \mathcal{P}$, we have $o_\kappa(X) \in \mathcal{P}$, $\mathcal{E}(X) \in \mathcal{P}$ and $Y \in \mathcal{P}$ for any closed $Y \subset X$.

Let T be an infinite set. An arbitrary family $\mathcal{A} \subset \exp T$ is called *adequate* if $\bigcup \mathcal{A} = T$, $\exp A \subset \mathcal{A}$ for any $A \in \mathcal{A}$, and $A \in \mathcal{A}$ whenever all finite subsets of A belong to $\mathcal{A}$. Given $A \subset T$, let $\chi_A(t) = 1$ if $t \in A$ and $\chi_A(t) = 0$ if $t \notin A$. The map $\chi_A : T \to \{0, 1\}$ is called *the characteristic function of* A *in the set* T. The symbol $\mathbb{D}$ denotes the two-point discrete space $\{0, 1\}$. If we have a set T and an adequate family $\mathcal{A}$ on T let $K_\mathcal{A} = \{\chi_A \in \mathbb{D}^T : A \in \mathcal{A}\}$. Another object associated with $\mathcal{A}$ is the space $T_\mathcal{A}^*$ whose underlying set is $T \cup \{\xi\}$, where $\xi \notin T$, all points of T are isolated in $T_\mathcal{A}^*$ and the basic neighbourhoods of ξ are the complements of finite unions of elements of $\mathcal{A}$. A subspace $X \subset \mathbb{D}^T$ is called *adequate* if $X = K_\mathcal{A}$ for some adequate family $\mathcal{A}$ on T.

Given an uncountable cardinal κ, a space X belongs to $\mathcal{M}(\kappa)$ if there exists a compact K such that X is a continuous image of a closed subset of $L(\kappa)^\omega \times K$. The space X is called *primarily Lindelöf* if X is a continuous image of a closed subspace of $L(\kappa)^\omega$ for some uncountable cardinal κ.

Suppose that S_α is homeomorphic to $\omega + 1$ for each $\alpha < \kappa$. In the space $S = \bigoplus\{S_\alpha : \alpha < \kappa\}$ let F be the set of non-isolated points. Introduce a topology τ on the set $S/F = \{F\} \cup (S \backslash F)$ declaring the points of $S \backslash F$ isolated and defining the local base at F as the family of all sets $\{F\} \cup (U \backslash F)$ where U is an open set (in S) which contains F. The space $V(\kappa) = (S/F, \tau)$ is called the *Fréchet–Urysohn κ-fan*. A family $\mathcal{U} \subset \exp X$ is *T_0-separating in X* if, for any distinct $x, y \in X$, there exists $U \in \mathcal{U}$ such that $|U \cap \{x, y\}| = 1$. A continuous surjective map $f : X \to Y$ is *irreducible* if, for any closed $F \subset X$ with $F \neq X$, we have $f(F) \neq Y$.

Let X be a space. Denote by $AD(X)$ the set $X \times \{0, 1\}$. Given $x \in X$, let $u_0(x) = (x, 0)$ and $u_1(x) = (x, 1)$. Thus, $AD(X) = u_0(X) \cup u_1(X)$. Declare the points of $u_1(X)$ isolated. Now, if $z = (x, 0) \in AD(X)$ then the base at z is formed by the sets $u_0(V) \cup (u_1(V) \backslash \{u_1(x)\})$ where V runs over the open neighbourhoods of x. The space $AD(X)$, with the topology thus defined, is called *the Alexandroff double of the space X*. Recall that, if we have a map $f : X \to Y$ then the map $f^n : X^n \to Y^n$ is defined by $f^n(x) = (f(x_1), \ldots, f(x_n))$ for any $x = (x_1, \ldots, x_n) \in X^n$. A space X is called *Sokolov space*, if, for any family $\{F_n : n \in \mathbb{N}\}$ such that F_n is a closed subset of X^n for each $n \in \mathbb{N}$, there exists a continuous map $f : X \to X$ such that $nw(f(X)) \leq \omega$ and $f^n(F_n) \subset F_n$ for all $n \in \mathbb{N}$. Given a space X, we let $C_{p,0}(X) = X$ and $C_{p,n+1}(X) = C_p(C_{p,n}(X))$ for all $n \in \omega$. The spaces $C_{p,n}(X)$ are called *iterated function spaces of X*.

In this section, we make use of a two-player game with complete information introduced by G. Gruenhage. In this game (which we call the Gruenhage game or W-game), there are two players (the concept of "player" is considered axiomatic), who play a game of ω moves on a space X at a fixed set $H \subset X$. The first player is called OP (for "open") and the second one's name is PT (for "point"). The n-th move of OP consists in choosing an open set $U_n \supset H$. The player PT responds by choosing a point $x_n \in U_n$. After ω moves have been made, the sequence $\mathcal{P} = \{(U_n, x_n) : n \in \mathbb{N}\}$ is called a *play of the game*; for any $n \in \mathbb{N}$, the set $\{U_1, x_1, \ldots, U_n, x_n\}$ is called *an initial segment of the play $\mathcal{P}$*.

Now, if $\mathcal{P} = \{(U_n, x_n) : n \in \mathbb{N}\}$ is a play in the W-game at H in the space X then the set $\{x_n : n \in \mathbb{N}\}$ is taken into consideration to determine who won the game. If $x_n \to H$ in the sense that any open $U \supset H$ contains all but finitely many points x_n, then OP wins. If not, then PT is the winner.

A *strategy for the player OP* is any function s, whose domain is the family $\mathrm{dom}(s) = \{\emptyset\} \cup \{F : F = (U_1, x_1, \ldots U_n, x_n), n \in \mathbb{N}, H \subset U_i \in \tau(X)$ and $x_i \in U_i$ for all $i \leq n\}$ and $s(F)$ is an open set containing H for any $F \in \mathrm{dom}(s)$. If $P = \{(U_n, x_n) : n \in \mathbb{N}\}$ is a play, we say that *OP applied the strategy s in P*, if $U_1 = s(\emptyset)$ and, for any $n \geq 2$, we have $U_n = s(U_1, x_1, \ldots, U_{n-1}, x_{n-1})$. The strategy s is called *winning* if any play in which s is applied, is favorable for the player OP, i.e., OP wins in every play where he/she applies the strategy s. A set $H \subset X$ is *a W-set (or has the W-property)* if OP has a winning strategy in the game on X with the fixed set H. If every point of X is a W-set, X is called *W-space (or a space with the W-property)*. A family $\mathcal{U}$ of subsets of X is called *point-countable* if, for any $x \in X$, the family $\{U \in \mathcal{U} : x \in U\}$ is countable. A space X is *metalindelöf* if any open cover of X has a point-countable open refinement. A space X is *metacompact* if any open cover of X has a point-finite open refinement.

101. Let M_t be a metrizable space for each $t \in T$. For an arbitrary point $a \in M = \prod\{M_t : t \in T\}$, prove that $\Sigma(M, a)$ is a Fréchet–Urysohn space. In particular, $\Sigma(A)$ is a Fréchet–Urysohn space for any A.

102. Let M_t be a metrizable space for each $t \in T$. For an arbitrary point $a \in M = \prod\{M_t : t \in T\}$, prove that $\Sigma(M, a)$ is a collectionwise normal space. In particular, $\Sigma(A)$ is a collectionwise normal space for any A.

103. Let M_t be a second countable space for each $t \in T$. For an arbitrary point $a \in M = \prod\{M_t : t \in T\}$, prove that $ext(\Sigma(M, a)) \leq \omega$. In particular, $ext(\Sigma(A)) = \omega$ for any set A.

104. Let M_t be a second countable space for any $t \in T$. Take any point $a \in M = \prod\{M_t : t \in T\}$. Prove that, if a compact space X is a continuous image of a dense subspace of $\Sigma(M, a)$ then X is metrizable. In particular, if a compact X is a continuous image of $\sigma(M, a)$ or $\Sigma(M, a)$ then X is metrizable.

105. Prove that, if $|A| = \kappa \geq \omega$ then the space $\Sigma_*(A)$ is homeomorphic to $C_p(A(\kappa))$.

106. Prove that, if $|A| = \kappa > \omega$ then the space $\Sigma(A)$ is homeomorphic to $C_p(L(\kappa))$.

107. Prove that, for any κ, there is a compact subspace of $C_p(A(\kappa))$ which separates the points of $A(\kappa)$. As a consequence, $C_p(A(\kappa))$ and $\Sigma_*(\kappa)$ are $K_{\sigma\delta}$-spaces and hence Lindelöf Σ-spaces.

108. Prove that $\sigma(A)$ is a σ-compact space (and hence a Lindelöf Σ-space) for any A.

109. Prove that, for any uncountable set A, there is a closed countably compact non-compact subspace in $\Sigma(A)$ and hence $\Sigma(A)$ is not realcompact.

110. Prove that, for any infinite A, every pseudocompact subspace of $\Sigma_*(A)$ is compact.

111. Prove that any metrizable space M embeds in $\Sigma_*(A)$ for some A.

112. Observe that any pseudocompact continuous image of $\Sigma_*(A)$ is compact and metrizable for any infinite A. Give an example of a countably compact non-compact space which is a continuous image of $\Sigma(\omega_1)$.

113. Prove that, for any uncountable A, the space $\Sigma(A)$ is not embeddable into $\Sigma_*(B)$ for any set B.

114. Prove that, for any uncountable A, the space $\Sigma_*(A)$ is not embeddable into $\sigma(B)$ for any set B.

115. Prove that, for any uncountable A, neither of the spaces $\Sigma(A)$ and $\Sigma_*(A)$ maps continuously onto the other.

116. Prove that, for any A, the space $\Sigma(A)$ embeds in a countably compact Fréchet–Urysohn space.

117. Show that, if A is an uncountable set, then $\Sigma_*(A)$ cannot be embedded in a σ-compact space of countable tightness. In particular, neither $\Sigma(A)$ nor $\Sigma_*(A)$ are embeddable in a compact space of countable tightness if $|A| > \omega$.

118. Let X be a compact space. Prove that X is Corson compact if and only if X has a point-countable T_0-separating family of open F_σ-sets. Deduce from this fact that any metrizable compact space is Corson compact.

119. Let M_t be a second countable space for any $t \in T$. Prove that, for any point $a \in M = \prod\{M_t : t \in T\}$, every compact subset of $\Sigma(M, a)$ is Corson compact.

120. Prove that any Corson compact space is monolithic, Fréchet–Urysohn and has a dense set of points of countable character. As a consequence, $\omega_1 + 1$ is not Corson compact.

121. Prove that $d(X) = w(X)$ for any Corson compact space. Thus, the two arrows space is not Corson compact.

122. Let X be a Corson compact space such that $C_p(X)\backslash\{f\}$ is normal for some $f \in C_p(X)$. Prove that X is metrizable. In particular, if $C_p(X)$ is hereditarily normal, then X is metrizable.

123. Prove that any linearly ordered and any dyadic Corson compact space is metrizable.

124. Let X be a Corson compact space. Prove that the Alexandroff double $AD(X)$ is also Corson compact. In particular, $AD(X)$ is Corson compact for any metrizable compact X.

125. Let X_t be a Corson compact space for any $t \in T$. Prove that the one-point compactification of the space $\bigoplus\{X_t : t \in T\}$ is also Corson compact.

126. Prove that, under CH, there exists a compact space of countable spread which is not perfectly normal.

127. Let X be a Corson compact space such that $s(X) = \omega$. Prove that X is perfectly normal.

128. Let X be an ω-monolithic compact space such that $s(C_p(X)) = \omega$. Prove that X is metrizable. In particular, a Corson compact space X is metrizable whenever $s(C_p(X)) = \omega$.

129. Let X be a compact space of countable tightness. Prove that X maps irreducibly onto a Corson compact space.

130. Given spaces X and Y assume that there exists a closed continuous irreducible onto map $f : X \to Y$. Prove that $d(X) = d(Y)$ and $c(X) = c(Y)$.

131. Prove that, under the Jensen's axiom ($\diamondsuit$), there is a perfectly normal non-metrizable Corson compact space X. Therefore, under $\diamondsuit$, a Corson compact space X need not be metrizable if $c(X) = \omega$.

132. Prove that any Corson compact space X, with ω_1 precaliber of X, is metrizable.

133. Assuming MA+¬CH, prove that any Corson compact space X, with $c(X) = \omega$, is metrizable.

134. Prove that a compact space X can fail to be Corson compact being a countable union of Corson compact spaces.

135. Prove that there exists a compact space X which is not Corson compact being a union of three metrizable subspaces.

136. Suppose that X is compact and X^ω is a countable union of Corson compact subspaces. Prove that X is Corson compact.

137. Prove that any countable product of Corson compact spaces is Corson compact. In particular, X^ω is Corson compact whenever X is Corson compact.

138. Let X be a Corson compact space. Prove that X has a dense metrizable subspace if and only if it has a σ-disjoint π-base.

139. Prove that $\mathcal{M}(\kappa)$ is an ω-perfect class for any κ.

140. Prove that for any Corson compact space X the space $C_p(X)$ belongs to $\mathcal{M}(\kappa)$ for some uncountable κ.

141. Prove that if κ is an uncountable cardinal and $Y \in \mathcal{M}(\kappa)$ then Y^ω is Lindelöf. In particular, $(C_p(X))^\omega$ is Lindelöf for any Corson compact space X.

142. Prove that any countable union of primarily Lindelöf spaces is a primarily Lindelöf space.

143. Prove that any countable product of primarily Lindelöf spaces is a primarily Lindelöf space.

144. Prove that any continuous image as well as any closed subspace of a primarily Lindelöf space is a primarily Lindelöf space.

145. Prove that any countable intersection of primarily Lindelöf spaces is a primarily Lindelöf space.

146. Prove that primarily Lindelöf spaces form a weakly k-directed class.

147. Given a space X let $r : X \to X$ be a retraction. For any $f \in C_p(X)$ let $r_1(f) = f \circ r$. Prove that $r_1 : C_p(X) \to C_p(X)$ is also a retraction.

148. Given an uncountable cardinal κ and a set $A \subset L(\kappa)$ define a map $p_A : L(\kappa) \to L(\kappa)$ by the rule $p_A(x) = a$ if $x \notin A$ and $p_A(x) = x$ for all $x \in A$ (recall that $L(\kappa) = \kappa \cup \{a\}$ and a is the unique non-isolated point of $L(\kappa)$). Prove that

(i) p_A is a retraction on $L(\kappa)$ onto $A \cup \{a\}$ for any $A \subset L(\kappa)$;

(ii) if $B \subset L(\kappa)$ and F is a closed subset of $(L(\kappa))^\omega$ then there exists $A \subset L(\kappa)$ such that $B \subset A$, $|A| \le |B| \cdot \omega$ and $(p_A)^\omega(F) \subset F$. Here, as usual, the map $q_A = (p_A)^\omega : (L(\kappa))^\omega \to (L(\kappa))^\omega$ is the countable power of the map p_A defined by $q_A(x)(n) = p_A(x(n))$ for any $x \in (L(\kappa))^\omega$ and $n \in \omega$.

149. Prove that, for any primarily Lindelöf space X, the space $C_p(X)$ condenses linearly into $\Sigma(A)$ for some A.

150. Prove that the following conditions are equivalent for any compact space X:

(i) X is Corson compact;

(ii) $C_p(X)$ is primarily Lindelöf;

(iii) there is a primarily Lindelöf $P \subset C_p(X)$ which separates the points of X;

(iv) X embeds in $C_p(Y)$ for some primarily Lindelöf space Y.

151. Prove that a continuous image of a Corson compact space is Corson compact.

152. Observe that $\Sigma_*(A)$ and $\sigma(A)$ are invariant subsets of $\Sigma(A)$; prove that, for any infinite cardinal κ and any closed $F \subset \Sigma(A)$ we have

(i) if $B_\alpha \subset A$, $r_{B_\alpha}(F) \subset F$ for any $\alpha < \kappa$ and $\alpha < \beta < \kappa$ implies $B_\alpha \subset B_\beta$ then $r_B(F) \subset F$ where $B = \bigcup_{\alpha < \kappa} B_\alpha$;

(ii) for any non-empty $D \subset A$ with $|D| \le \kappa$ there is a set $E \subset A$ such that $|E| \le \kappa$, $D \subset E$ and $r_E(F) \subset F$.

In particular, F is invariant in $\Sigma(A)$.

153. Prove that the following properties are equivalent for any X:

(i) X is a Sokolov space;

(ii) if, for any $n \in \mathbb{N}$, a set $B_n \subset X^n$ is chosen then there exists a continuous map $f : X \to X$ such that $nw(f(X)) \leq \omega$ and $f^n(B_n) \subset \overline{B}_n$ for each $n \in \mathbb{N}$;

(iii) if F_{nm} is a closed subset of X^n for all $n, m \in \mathbb{N}$, then there exists a continuous map $f : X \to X$ such that $nw(f(X)) \leq \omega$ and $f^n(F_{nm}) \subset F_{nm}$ for all $n, m \in \mathbb{N}$.

154. Prove that if X is a Sokolov space then $X \times \omega$ is a Sokolov space and every closed $F \subset X$ is also a Sokolov space.

155. Given a Sokolov space X and a second countable space E, prove that $C_p(X, E)$ is also a Sokolov space.

156. Prove that X is a Sokolov space if and only if $C_p(X)$ is a Sokolov space.

157. Let X be a Sokolov space with $t^*(X) \leq \omega$. Prove that $C_p(X, E)$ is Lindelöf for any second countable space E.

158. Prove that

(i) any $\mathbb{R}$-quotient image of a Sokolov space is a Sokolov space;

(ii) if X is a Sokolov space then X^ω is also a Sokolov space;

(iii) a space with a unique non-isolated point is Sokolov if and only if it is Lindelöf.

159. Let X be a space with a unique non-isolated point. Prove that the following properties are equivalent:

(i) $l(X) \leq \omega$ and $t^*(X) \leq \omega$;

(ii) X is a Sokolov space and $t^*(X) \leq \omega$;

(iii) $C_{p,n}(X)$ is Lindelöf for all $n \in \mathbb{N}$;

(iv) $C_p(X)$ is Lindelöf.

160. Let X be an invariant subspace of $\Sigma(A)$. Prove that X is a Sokolov space. Deduce from this fact that every Corson compact space is Sokolov.

161. Prove that every Sokolov space is collectionwise normal and has countable extent. Deduce from this fact that $ext(C_{p,n}(X)) \leq \omega$ for any Sokolov space X and $n \in \mathbb{N}$.

162. Let X be a Sokolov space. Prove that

(i) if $t^*(X) \leq \omega$ then $C_{p,2n+1}(X)$ is Lindelöf for any $n \in \omega$.

(ii) if $l^*(X) \leq \omega$ then $C_{p,2n}(X)$ is Lindelöf for any $n \in \mathbb{N}$;

(iii) if $l^*(X) \cdot t^*(X) \leq \omega$ then $C_{p,n}(X)$ is Lindelöf for any $n \in \mathbb{N}$.

163. Prove that every Sokolov space is ω-stable and ω-monolithic. Deduce from this fact that every Sokolov compact space is Fréchet–Urysohn and has a dense set of points of countable character.

164. Prove that a metrizable space is Sokolov if and only if it is second countable.

165. Let X be a Sokolov space with $l^*(X) \cdot t^*(X) = \omega$. Prove that

(i) if X has a small diagonal then $nw(X) = \omega$;
(ii) if ω_1 is a caliber of X then $nw(X) = \omega$.

166. Prove that if X is a Sokolov space with a G_δ-diagonal then $nw(X) = \omega$.

167. Let X be a Lindelöf Σ-space. Prove that if X is Sokolov then $t(X) \leq \omega$ and $C_{p,n}(X)$ is Lindelöf for any $n \in \mathbb{N}$. In particular, if K is Sokolov compact (or Corson compact) then $C_{p,n}(K)$ is Lindelöf for any $n \in \mathbb{N}$.

168. Let T be an infinite set. Prove that, if $\mathcal{A}$ is an adequate family on T then $K_{\mathcal{A}}$ is a compact space.

169. Let T be an infinite set. Suppose that $\mathcal{A}$ is an adequate family on T. Prove that $K_{\mathcal{A}}$ is a Corson compact space if and only if all elements of $\mathcal{A}$ are countable.

170. Let T be an infinite set; suppose that $\mathcal{A}$ is an adequate family on T and u is the function on $K_{\mathcal{A}}$ which is identically zero. For any $t \in T$ let $e_t(f) = f(t)$ for any $f \in K_{\mathcal{A}}$. Observe that $Z = \{e_t : t \in T\} \cup \{u\} \subset C_p(K_{\mathcal{A}}, \mathbb{D})$; let $\varphi(\xi) = u$ and $\varphi(t) = e_t$ for any $t \in T$. Prove that $\varphi : T_{\mathcal{A}}^* \to Z$ is a homeomorphism and Z is closed in $C_p(K_{\mathcal{A}}, \mathbb{D})$. In particular, the space $T_{\mathcal{A}}^*$ is homeomorphic to a closed subspace of $C_p(K_{\mathcal{A}}, \mathbb{D})$.

171. Suppose that T is an infinite set and $\mathcal{A}$ is an adequate family on T. Prove that the spaces $C_p(K_{\mathcal{A}}, \mathbb{D})$ and $C_p(K_{\mathcal{A}})$ are both continuous images of the space $(T_{\mathcal{A}}^* \times \omega)^\omega$.

172. Let T be an infinite set. Suppose that $\mathcal{A}$ is an adequate family on T. Prove the space $C_p(K_{\mathcal{A}})$ is K-analytic if and only if $T_{\mathcal{A}}^*$ is K-analytic.

173. Let T be an infinite set. Suppose that $\mathcal{A}$ is an adequate family on T. Prove the space $C_p(K_{\mathcal{A}})$ is Lindelöf Σ if and only if $T_{\mathcal{A}}^*$ is Lindelöf Σ.

174. Observe that every adequate compact space is zero-dimensional. Give an example of a zero-dimensional Corson compact space which is not homeomorphic to any adequate compact space.

175. Let T be a subspace of $\mathbb{R}$ of cardinality ω_1. Consider some well-ordering $\prec$ on T and let $<$ be the order on T induced from the usual order on $\mathbb{R}$. Denote by $\mathcal{A}_1$ the family of all subsets of T on which the orders $<$ and $\prec$ coincide (i.e., $A \in \mathcal{A}_1$ if and only if, for any distinct $x, y \in A$, we have $x < y$ if and only if $x \prec y$). Let $\mathcal{A}_2$ be the family of all subsets of T on which the orders $<$ and $\prec$ are opposite (i.e., $A \in \mathcal{A}_2$ if and only if, for any distinct $x, y \in A$, we have $x < y$ if and only if $y \prec x$). Check that $\mathcal{A} = \mathcal{A}_1 \cup \mathcal{A}_2$ is an adequate family and that $X = K_{\mathcal{A}}$ is a Corson compact space for which $C_p(X)$ is not a continuous image of any Lindelöf k-space. In particular, $C_p(X)$ is not a Lindelöf Σ-space.

176. Give a ZFC example of a Corson compact space without a dense metrizable subspace.

177. Give an example of a compact space X for which $(C_p(X))^\omega$ is Lindelöf while X is not Corson compact.

178. Prove that any Corson compact space is a continuous image of a zero-dimensional Corson compact space.

179. Prove that every first countable space is a W-space and every W-space is Fréchet–Urysohn.

180. Suppose that $f : X \rightarrow Y$ is an open continuous onto map. Prove that if X is a W-space then so is Y.

181. Suppose that X is a separable space and a closed set $F \subset X$ has an outer base of closed neighbourhoods (i.e., for any $U \in \tau(F, X)$ there is $V \in \tau(F, X)$ such that $\overline{V} \subset U$). Prove that if F is a W-set in X then $\chi(F, X) \leq \omega$. In particular, if X is a separable W-space then $\chi(X) = \omega$.

182. Show that there exist W-spaces which are not first countable and Fréchet–Urysohn spaces which are not W-spaces.

183. Prove that any subspace of a W-space is a W-space and any countable product of W-spaces is a W-space.

184. Prove that any Σ-product of W-spaces is a W-space. Deduce from these facts that if X is a Corson compact space then every non-empty closed $F \subset X$ is a W-set; in particular, X is a W-space.

185. Prove that, if X is a compact space of countable tightness, then a non-empty closed $H \subset X$ is a W-set if and only if $X \backslash H$ is metalindelöf.

186. Let X be a compact scattered space. Prove that a non-empty closed $H \subset X$ is a W-set if and only if $X \backslash H$ is metacompact.

187. (Yakovlev's theorem) Prove that any Corson compact space is hereditarily metalindelöf.

188. Prove that the following are equivalent for any compact space X:

(i) X is Corson compact;
(ii) every closed subset of $X \times X$ is a W-set in $X \times X$;
(iii) the diagonal $\Delta = \{(x, x) : x \in X\}$ of the space X is a W-set in $X \times X$;
(iv) the space $(X \times X) \backslash \Delta$ is metalindelöf;
(v) the space $X \times X$ is hereditarily metalindelöf.

189. Give an example of a compact W-space X such that some continuous image of X is not a W-space.

190. Suppose that X is a compact space which embeds into a σ-product of second countable spaces. Prove that the space $X^2 \backslash \Delta$ is metacompact; here, as usual, the set $\Delta = \{(x, x) : x \in X\}$ is the diagonal of the space X.

191. Observe that any countably compact subspace of a Corson compact space is closed and hence compact. Deduce from this fact that there exists a countably compact space X which embeds into $\Sigma(A)$ for some A but is not embeddable into any Corson compact space.

192. Let M_α be a separable metrizable space for any $\alpha < \omega_1$. Prove that a dense subspace Y of the space $\prod \{M_\alpha : \alpha < \omega_1\}$ is normal if and only if Y is collectionwise normal.

193. Prove that if $2^{\omega_1} = \mathfrak{c}$ then there exists a dense hereditarily normal subspace Y in the space $\mathbb{D}^{\mathfrak{c}}$ such that $ext(Y) = \omega_1$. Deduce from this fact that it is independent of ZFC whether normality implies collectionwise normality in the class of dense subspaces of $\mathbb{D}^{\mathfrak{c}}$.

194. Let X be a monolithic compact space of countable tightness. Prove that any dense normal subspace of $C_p(X)$ is Lindelöf. In particular, if X is a Corson compact space and Y is a dense normal subspace of $C_p(X)$ then Y is Lindelöf.

195. Let X be a Corson compact space. Prove that there exists a σ-discrete set $Y \subset C_p(X)$ which separates the points of X.

196. Prove that, under CH, there exists a compact space X such that no σ-discrete $Y \subset C_p(X)$ separates the points of X.

197. Let X be a metrizable space. Prove that there is a discrete $Y \subset C_p(X)$ which separates the points of X.

198. Prove that, for each cardinal κ, there exists a discrete $Y \subset C_p(\mathbb{I}^\kappa)$ which separates the points of $\mathbb{I}^\kappa$.

199. Prove that $C_p(\beta\omega\backslash\omega)$ cannot be condensed into $\Sigma_*(A)$ for any A.

200. Prove that, for any Corson compact X and any $n \in \mathbb{N}$, the space $C_{p,n}(X)$ linearly condenses onto a subspace of $\Sigma(A)$ for some A.

1.3   More of Lindelöf Σ-Property. Gul'ko Compact Spaces

All spaces are assumed to Tychonoff. Given two families $\mathcal{A}$ and $\mathcal{B}$ of subsets of a space X, say that $\mathcal{A}$ *is a network with respect to* $\mathcal{B}$ if, for any $B \in \mathcal{B}$ and any open $U \supset B$, there is $A \in \mathcal{A}$ such that $B \subset A \subset U$. A space X is *a Lindelöf Σ-space,* if X has a cover $\mathcal{C}$ such that all elements of $\mathcal{C}$ are compact and there exists a countable family $\mathcal{F}$ which is a network with respect to $\mathcal{C}$. The class of Lindelöf Σ-spaces is denoted by $L(\Sigma)$.

Given a space X, denote by A the set $C(X, \mathbb{I})$ and, for each $f \in A$, let $\beta_x(f) = f(x)$. Then $\beta_x : A \to \mathbb{I}$ and the subspace $\tilde{X} = \{\beta_x : x \in X\} \subset \mathbb{I}^A$ is homeomorphic to X. Identifying the spaces X and $\tilde{X}$, we consider that $X \subset \mathbb{I}^A$. Denote by βX the closure of X in $\mathbb{I}^A$. The space βX is called the *Čech-Stone compactification of the space* X. Let $\upsilon X = \{y \in \beta X : H \cap X \neq \emptyset$ for any non-empty G_δ-set $H \subset \beta X$ such that $y \in H\}$. The space υX is called *the Hewitt realcompactification of the space* X. The space X is *realcompact* if $X = \upsilon X$. If $\varphi : X \to Y$ is a continuous mapping then its *dual map* $\varphi^* : C_p(Y) \to C_p(X)$ is defined by $\varphi^*(f) = f \circ \varphi$ for any $f \in C_p(Y)$.

Given a space X, we let $C_{p,0}(X) = X$ and $C_{p,n+1}(X) = C_p(C_{p,n}(X))$ for all $n \in \omega$. The spaces $C_{p,n}(X)$ are called *iterated function spaces of* X. A space X has *a small diagonal* if, for every uncountable $A \subset (X \times X) \backslash \Delta$, there is a neighbourhood U of the diagonal $\Delta = \{(x, x) : x \in X\}$ such that $A \backslash U$ is uncountable. A space Y is *Eberlein–Grothendieck* if it can be embedded into $C_p(K)$ for some compact space K. Say that a space X is $K_{\sigma\delta}$ if there exists a space Y such that $X \subset Y$ and $X = \bigcap\{Y_n : n \in \omega\}$, where each Y_n is a σ-compact subset of Y. A *K-analytic space* is a continuous image of a $K_{\sigma\delta}$-space.

A family $\mathcal{U} \subset \exp X$ is said to be *point-finite at* $x \in X$ if $\{U \in \mathcal{U} : x \in U\}$ is finite. The family $\mathcal{U}$ is *weakly σ-point-finite* if there exists a sequence $\{\mathcal{U}_n : n \in \omega\}$ of subfamilies of $\mathcal{U}$ such that, for every $x \in X$, we have $\mathcal{U} = \bigcup\{\mathcal{U}_n : n \in M_x\}$ where $M_x = \{n \in \omega : \mathcal{U}_n$ is point-finite at $x\}$. The family $\mathcal{U}$ is T_0-*separating* if, for any distinct $x, y \in X$, there exists $U \in \mathcal{U}$ such that $|U \cap \{x, y\}| = 1$. A set U is *a cozero set* in a space X if there is $f \in C_p(X)$ such that $U = X \backslash f^{-1}(0)$. The spaces which have a countable network are called *cosmic*.

Say that X is a *Gul'ko space* if $C_p(X)$ is a Lindelöf Σ-space. A compact Gul'ko space is called *Gul'ko compact.* The space X is *Talagrand* if $C_p(X)$ is K-analytic. A compact Talagrand space is called *Talagrand compact.*

A space X is called *d-separable,* if it has a dense σ-discrete subspace. If X is a space and $\mathcal{U}, \mathcal{V} \subset \tau^*(X)$, we say that $\mathcal{V}$ is *a π-base for* $\mathcal{U}$ if, for every $U \in \mathcal{U}$, there is $V \in \mathcal{V}$ such that $V \subset U$. The *point-finite cellularity* $p(X)$ of a space X is the supremum of cardinalities of point-finite families of non-empty open subsets of X. A space X is κ-*stable* if, for any continuous onto map $f : X \to Y$, we have $nw(Y) \leq \kappa$ whenever $iw(Y) \leq \kappa$. Given a cardinal κ, a space X is κ-*monolithic* if $A \subset X$ and $|A| \leq \kappa$ implies $nw(\overline{A}) \leq \kappa$. Now, X is a *Preiss–Simon space* if, for any closed $F \subset X$ and any non-isolated $x \in F$, there exists a sequence $\{U_n : n \in \omega\}$ of open non-empty subsets of F such that $U_n \to x$, i.e., any neighbourhood of x contains all but finitely many of U_n's.

An uncountable regular cardinal κ is *a caliber* of a space X if, for any family $\mathcal{U} \subset \tau^*(X)$ of cardinality κ, there exists $\mathcal{U}' \subset \mathcal{U}$ such that $|\mathcal{U}'| = \kappa$ and $\bigcap \mathcal{U}' \neq \emptyset$. An uncountable regular cardinal κ is called *a precaliber* of a space X if, for any family $\mathcal{U} \subset \tau^*(X)$ of cardinality κ, there exists $\mathcal{U}' \subset \mathcal{U}$ such that $|\mathcal{U}'| = \kappa$ and $\mathcal{U}'$ is centered ($\equiv$has the finite intersection property, i.e., $\bigcap \mathcal{V} \neq \emptyset$ for any finite $\mathcal{V} \subset \mathcal{U}'$).

The symbol $\mathbb{P}$ stands for the space of the irrationals which is identified with ω^ω; in particular, if $p, q \in \mathbb{P}$, we say that $p \leq q$ if $p(n) \leq q(n)$ for any $n \in \omega$. Let $\omega^0 = \{\emptyset\}$ and $\omega^{<\omega} = \bigcup\{\omega^n : n \in \omega\}$. Usually, when considering $\omega^{<\omega}$, we identify $n = 0$ with the empty set and every $n \in \mathbb{N}$ with the set $\{0, \dots, n-1\}$. If $s \in \omega^{<\omega}$ and $n \in \omega$ then $t = s^\frown n \in \omega^{<\omega}$ is defined as follows: there is a unique $k \in \omega$ with $s \in \omega^k$; let $t|k = s$ and $t(k) = n$.

Say that a space X is $\mathbb{P}$-*dominated* if there is a compact cover $\{K_p : p \in \mathbb{P}\}$ of the space X such that $p, q \in \mathbb{P}$ and $p \leq q$ imply $K_p \subset K_q$. In other words, the space X is $\mathbb{P}$-dominated if it has a $\mathbb{P}$-directed compact cover. A space X is said to be *strongly* $\mathbb{P}$-*dominated* if it has a $\mathbb{P}$-directed compact cover $\mathcal{C}$ such that, for any compact $K \subset X$ there is $C \in \mathcal{C}$ such that $K \subset C$, i.e., $\mathcal{C}$ "swallows" all compact subsets of X.

Given a space Z, the family of all compact subsets of Z is denoted by $\mathcal{K}(Z)$. A space X *is dominated by a space* Y if there is a compact cover $\{F_K : K \in \mathcal{K}(Y)\}$ of the space X such that $K, L \in \mathcal{K}(Y)$ and $K \subset L$ imply $F_K \subset F_L$.

Given a set A and a point $x \in \mathbb{R}^A$, let $\text{supp}(x) = \{a \in A : x(a) \neq 0\}$. If we have a family $s = \{A_n : n \in \omega\} \subset \exp A$ and $\bigcup s = A$ then $N_x = \{n \in \omega : A_n \cap \text{supp}(x)$ is finite$\}$ for any $x \in \mathbb{R}^A$. Let $\Sigma_s(A) = \{x \in \Sigma(A) : A = \bigcup\{A_n : n \in N_x\}\}$; here, as usual, $\Sigma(A) = \{x \in \mathbb{R}^A : |\text{supp}(x)| \leq \omega\}$ and $\Sigma_*(A) = \{x \in \mathbb{R}^A :$ for any $\varepsilon > 0$, the set $\{a \in A : |x(a)| > \varepsilon\}$ is finite$\}$. The spaces $\Sigma(A)$ (or $\Sigma_*(A)$) will be called Σ-products (Σ_*-products) of real lines. If X is a space and $M \subset X$, let μ_M be the topology generated by $\tau(X) \cup \{\{x\} : x \in X \backslash M\}$. The space (X, μ_M) is usually denoted by X_M.

The statement CH (called *Continuum Hypothesis*) says that the first uncountable ordinal is equal to the continuum, i.e., $\omega_1 = \mathfrak{c}$. The statement "$\kappa^+ = 2^\kappa$ for any infinite cardinal κ" is called *Generalized Continuum Hypothesis (GCH)*.

201. Suppose that $X = \upsilon Y$ and Z is a subspace of $\mathbb{R}^X$ such that $C_p(X) \subset Z$. Prove that there exists $Z' \subset \mathbb{R}^Y$ such that $C_p(Y) \subset Z'$ and Z' is a continuous image of Z.

202. Suppose that X is σ-compact. Prove that there exists a $K_{\sigma\delta}$-space Z such that $C_p(X) \subset Z \subset \mathbb{R}^X$.

203. Suppose that υX is σ-compact. Prove that there exists a K-analytic space Z such that $C_p(X) \subset Z \subset \mathbb{R}^X$.

204. Prove that X is pseudocompact if and only if there exists a σ-compact space Z such that $C_p(X) \subset Z \subset \mathbb{R}^X$.

205. Give an example of a Lindelöf space X for which there exists no Lindelöf space Z such that $C_p(X) \subset Z \subset \mathbb{R}^X$.

206. Prove that υX is a Lindelöf Σ-space if and only if $C_p(X) \subset Z \subset \mathbb{R}^X$ for some Lindelöf Σ-space Z. In particular,

    (i) if $C_p(X)$ is a Lindelöf Σ-space, then υX is a Lindelöf Σ-space;

    (ii) (Uspenskij's theorem) if X is a Lindelöf Σ-space then there exists a Lindelöf Σ-space Z such that $C_p(X) \subset Z \subset \mathbb{R}^X$;

    (iii) if $\upsilon(C_p(X))$ is a Lindelöf Σ-space then υX is Lindelöf Σ.

207. Given a natural $n \geq 1$, suppose that there exists a Lindelöf Σ-space Z such that $C_{p,n}(X) \subset Z \subset \mathbb{R}^{C_{p,n-1}(X)}$. Prove that there exists a Lindelöf Σ-space Y such that $C_p(X) \subset Y \subset \mathbb{R}^X$.

208. Suppose that $C_p(X)$ is a Lindelöf Σ-space. Prove that $C_{p,n}(X)$ is ω-stable and ω-monolithic for any natural n.

209. Prove that a space X is dominated by a space homeomorphic to the irrationals if and only if X is $\mathbb{P}$-dominated.

210. Suppose that X is dominated by a second countable space. Prove that there is a countable family $\mathcal{F}$ of subsets of X which is a network with respect to a cover of X with countably compact subspaces of X.

211. Suppose that a space X has a countable family $\mathcal{F}$ which is a network with respect to a cover of X with countably compact subspaces of X. Prove that υX is a Lindelöf Σ-space.

212. Prove that the property of being dominated by a second countable space is preserved by countable unions, products and intersections as well as by closed subspaces and continuous images.

213. Show that every Lindelöf Σ-space is dominated by a second countable space. Prove that X is a Lindelöf Σ-space if and only if X is Dieudonné complete and dominated by a second countable space.

214. Prove that, for any space X, the space $C_p(X)$ is dominated by a second countable space if and only if $C_p(X)$ is Lindelöf Σ.

215. Prove that, for any space X, the space $C_p(X)$ is $\mathbb{P}$-dominated if and only if $C_p(X)$ is K-analytic.

216. Prove that, for any space X, the space $C_p(X)$ is strongly $\mathbb{P}$-dominated if and only if X is countable and discrete.

217. Observe that there exist spaces X for which $C_p(X, \mathbb{I})$ is Lindelöf Σ while $C_p(X)$ is not Lindelöf. Supposing that υX and $C_p(X, \mathbb{I})$ are Lindelöf Σ-spaces prove that $C_p(X)$ is a Lindelöf Σ-space. In particular, if X is Lindelöf Σ then the space $C_p(X)$ is Lindelöf Σ if and only if $C_p(X, \mathbb{I})$ is a Lindelöf Σ-space.

218. (Okunev's theorem). Suppose that X and Y are Lindelöf Σ-spaces such that $Y \subset C_p(X)$. Prove that $C_p(Y)$ is a Lindelöf Σ-space.

219. Let X and $C_p(X)$ be Lindelöf Σ-spaces. Prove that, for every natural n, the space $C_{p,n}(X)$ is a Lindelöf Σ-space. In particular, if X is compact and $C_p(X)$ is Lindelöf Σ then all iterated function spaces of X are Lindelöf Σ-spaces.

220. For an arbitrary Lindelöf Σ-space X, prove that every countably compact subspace $Y \subset C_p(X)$ is Gul'ko compact.

221. Suppose that $C_p(X)$ is a Lindelöf Σ-space. Prove that every countably compact $Y \subset C_p(X)$ is Gul'ko compact.

222. (Reznichenko's compactum) Prove that there exists a compact space M with the following properties:

(i) $C_p(M)$ is a K-analytic space, i.e., M is Talagrand compact;
(ii) there is $x \in M$ such that $M \backslash \{x\}$ is pseudocompact and M is the Stone–Čech extension of $M \backslash \{x\}$.

As a consequence, there is an example of a K-analytic space X such that some closed pseudocompact subspace of $C_p(X)$ is not countably compact.

223. Suppose that, for a countably compact space X, there exists a condensation $f : X \to Z \subset C_p(Y)$, where $C_p(Y)$ is a Lindelöf Σ-space. Prove that f is a homeomorphism and X is Gul'ko compact.

224. Give an example of a pseudocompact non-countably compact space X which can be condensed onto a compact $K \subset C_p(Y)$, where $C_p(Y)$ is Lindelöf Σ.

225. Give an example of a space X such that $C_p(X)$ is Lindelöf Σ and some pseudocompact subspace of $C_p(X)$ is not countably compact.

226. Observe that if there exist spaces X such that $C_p(X)$ is a Lindelöf Σ-space while $t(C_p(X)) > \omega$. Prove that, if $C_p(X)$ is Lindelöf Σ and $Y \subset C_p(X)$ is pseudocompact then Y is Fréchet–Urysohn.

227. Show that there exists a space X such that $C_p(X)$ is a Lindelöf Σ-space and $t(Y) > \omega$ for some σ-compact subspace $Y \subset C_p(X)$.

228. Let X be an arbitrary space. Denote by $\pi : C_p(\upsilon X) \to C_p(X)$ the restriction map. Prove that, for any countably compact $Y \subset C_p(X)$, the space $\pi^{-1}(Y) \subset C_p(\upsilon X)$ is countably compact.

229. Give an example of a space X such that $\pi^{-1}(Y)$ is not pseudocompact for some pseudocompact $Y \subset C_p(X)$. Here $\pi : C_p(\upsilon X) \to C_p(X)$ is the restriction map.

230. Assume that υX is a Lindelöf Σ-space and $\pi : C_p(\upsilon X) \to C_p(X)$ is the restriction map. Prove that, for any compact $Y \subset C_p(X)$, the space $\pi^{-1}(Y) \subset C_p(\upsilon X)$ is also compact.

231. Assume that υX is a Lindelöf Σ-space and $\pi : C_p(\upsilon X) \to C_p(X)$ is the restriction map. Prove that, for any Lindelöf Σ-space Y contained in $C_p(X)$, the space $\pi^{-1}(Y) \subset C_p(\upsilon X)$ is Lindelöf Σ.

232. Let X be a pseudocompact space and denote by $\pi : C_p(\beta X) \to C_p(X)$ the restriction map. Prove that, for any Lindelöf Σ-space (compact space) $Y \subset C_p(X)$, the space $\pi^{-1}(Y) \subset C_p(\beta X)$ is Lindelöf Σ (or compact, respectively).

233. Give an example of a pseudocompact X such that $\pi^{-1}(Y) \subset C_p(\beta X)$ is not Lindelöf for some Lindelöf $Y \subset C_p(X)$. Here $\pi : C_p(\beta X) \to C_p(X)$ is the restriction map.

234. Observe that $C_p(X)$ is a Lindelöf Σ-space if and only if $C_p(\upsilon X)$ is Lindelöf Σ; prove that, for any X, the space $C_p(X)$ is K-analytic if and only if $C_p(\upsilon X)$ is K-analytic. In other words, X is a Talagrand space if and only if υX is Talagrand.

235. Suppose that $C_p(X)$ is a Lindelöf Σ-space. Prove that $C_{p,n}(\upsilon X)$ is a Lindelöf Σ-space for every $n \in \mathbb{N}$.

236. Given an arbitrary space X let $\pi : C_p(\upsilon X) \to C_p(X)$ be the restriction mapping. Let $\pi^*(\varphi) = \varphi \circ \pi$ for any function $\varphi \in \mathbb{R}^{C_p(X)}$ and observe that the map $\pi^* : \mathbb{R}^{C_p(X)} \to \mathbb{R}^{C_p(\upsilon X)}$ is an embedding. Identifying the space $\upsilon(C_p(C_p(X)))$ with the subspace $\{\varphi \in \mathbb{R}^{C_p(X)} : \varphi$ is strictly ω-continuous on $C_p(X)\}$ of the space $\mathbb{R}^{C_p(X)}$ (see TFS-438) prove that

(i) $\pi^*(C_p(C_p(X))) \subset \pi^*(\upsilon(C_p(C_p(X)))) \subset C_p(C_p(\upsilon X))$;
(ii) if $C_p(X)$ is normal then $\pi^*(\upsilon(C_p(C_p(X)))) = C_p(C_p(\upsilon X))$ and hence the spaces $\upsilon(C_p(C_p(X)))$ and $C_p(C_p(\upsilon X))$ are homeomorphic.

237. Suppose that $C_p(X)$ is a Lindelöf Σ-space. Prove that $C_{p,2n}(\upsilon X)$ is homeomorphic to $\upsilon(C_{p,2n}(X))$ for every $n \in \mathbb{N}$.

238. Suppose that $C_p(X)$ is a Lindelöf Σ-space. Prove that $C_{p,2n+1}(\upsilon X)$ can be condensed onto $C_{p,2n+1}(X)$ for every $n \in \omega$.

239. Suppose that $C_{p,2k+1}(X)$ is a Lindelöf Σ-space for some $k \in \omega$. Prove that $C_{p,2n+1}(X)$ is a Lindelöf Σ-space every $n \in \omega$.

240. Suppose that $C_{p,2k}(X)$ is a Lindelöf Σ-space for some $k \in \mathbb{N}$. Prove that $C_{p,2n}(X)$ is a Lindelöf Σ-space every $n \in \mathbb{N}$.

241. Give an example of a space X such that $C_p(X)$ is not Lindelöf while $C_{p,2n}(X)$ is a Lindelöf Σ-space for every $n \in \mathbb{N}$.

242. Give an example of a space X such that $C_pC_p(X)$ is not Lindelöf while $C_{p,2n+1}(X)$ is a Lindelöf Σ-space for every $n \in \omega$.

243. Prove that, for any space X, only the following distributions of the Lindelöf Σ-property in iterated function spaces are possible:

(i) $C_{p,n}(X)$ is not a Lindelöf Σ-space for any $n \in \mathbb{N}$;
(ii) $C_{p,n}(X)$ is a Lindelöf Σ-space for any $n \in \mathbb{N}$;
(iii) $C_{p,2n+1}(X)$ is a Lindelöf Σ-space and $C_{p,2n+2}(X)$ is not Lindelöf for any $n \in \omega$;
(iv) $C_{p,2n+2}(X)$ is a Lindelöf Σ-space and $C_{p,2n+1}(X)$ is not Lindelöf for any $n \in \omega$.

244. Suppose that $C_{p,2k+1}(X)$ is a Lindelöf Σ-space for some $k \in \omega$. Prove that, if $C_{p,2l+2}(X)$ is normal for some $l \in \omega$, then $C_{p,n}(X)$ is a Lindelöf Σ-space for any $n \in \mathbb{N}$.

245. Suppose that $C_{p,2k+2}(X)$ is a Lindelöf Σ-space for some $k \in \omega$. Prove that, if $C_{p,2l+1}(X)$ is normal for some $l \in \omega$, then $C_{p,n}(X)$ is a Lindelöf Σ-space for any $n \in \mathbb{N}$.

246. Prove that, if $C_p(X)$ is a Lindelöf Σ-space, then $\upsilon(C_pC_p(X))$ is a Lindelöf Σ-space.

247. Prove that, if X is normal and $\upsilon(C_p(X))$ is a Lindelöf Σ-space, then $\upsilon(C_pC_p(X))$ is a Lindelöf Σ-space.

248. Prove that, if X is realcompact and $\upsilon(C_p(X))$ is a Lindelöf Σ-space, then $\upsilon(C_pC_p(X))$ is a Lindelöf Σ-space.

249. Let ω_1 be a caliber of a space X. Prove that $C_p(X)$ is a Lindelöf Σ-space if and only if X has a countable network.

250. Prove that there exists a space X such that ω_1 is a precaliber of X, the space $C_{p,n}(X)$ is a Lindelöf Σ-space for all $n \in \omega$, while X does not have a countable network.

251. Let X be a Lindelöf Σ-space with ω_1 a caliber of X. Prove that any Lindelöf Σ-subspace of $C_p(X)$ has a countable network.

252. Prove that a Lindelöf Σ-space Y has a small diagonal if and only if it embeds into $C_p(X)$ for some X with ω_1 a caliber of X.

253. Prove that, if $C_p(X)$ is a Lindelöf Σ-space and has a small diagonal then X has a countable network.

254. Suppose that a space X has a dense subspace which is a continuous image of a product of separable spaces. Prove that any Lindelöf Σ-subspace of $C_p(X)$ has a countable network.

255. Prove that any first countable space is a Preiss–Simon space.

256. Prove that any Preiss–Simon space is Fréchet–Urysohn.

257. Give an example of a compact Fréchet–Urysohn space which does not have the Preiss–Simon property.

258. Let X be a space which has the Preiss–Simon property. Prove that each pseudocompact subspace of X is closed in X.

259. Suppose that X is a Preiss–Simon compact space. Prove that, for any proper dense $Y \subset X$, the space X is not the Čech-Stone extension of Y.

260. Prove that the following properties are equivalent for any countably compact space X:

   (i) X is a Preiss–Simon space;
   (ii) each pseudocompact subspace of X is closed in X;
   (iii) for each closed $F \subset X$ and any non-isolated $x \in F$, the space $F \setminus \{x\}$ is not pseudocompact.

261. Let X be a Lindelöf Σ-space. Suppose that $Y \subset C_p(X)$ and the set of non-isolated points of Y is Lindelöf Σ. Prove that $C_p(Y, \mathbb{I})$ is Lindelöf Σ.

262. Let X be an Eberlein–Grothendieck space. Suppose that the set of non-isolated points of X is σ-compact. Prove that $C_p(X, \mathbb{I})$ is $K_{\sigma\delta}$.

263. Let X be a second countable space. Prove that, for any $M \subset X$, the space $C_p(X_M, \mathbb{I})$ is Lindelöf Σ.

264. Let X be a σ-compact Eberlein–Grothendieck space. Prove that $C_p(X)$ is a $K_{\sigma\delta}$-space.

265. Give an example of a Lindelöf space X such that $C_p(X, \mathbb{I})$ is Lindelöf Σ and $X \times X$ is not Lindelöf.

266. Suppose that X is a space such that $\upsilon(C_p(X))$ is Lindelöf Σ and we have the equality $s(C_p(X)) = \omega$. Prove that $nw(X) = \omega$.

267. Suppose that $C_p(X)$ is hereditarily stable and υX is a Lindelöf Σ-space. Prove that $nw(X) = \omega$.

268. Show that if $C_p(X)$ is hereditarily stable then $nw(Y) = \omega$ for any Lindelöf Σ-subspace $Y \subset X$.

269. Suppose that $\upsilon(C_p(X))$ is a Lindelöf Σ-space and ω_1 is a caliber of $C_p(X)$. Prove that $nw(Y) = \omega$ for any Lindelöf Σ-subspace $Y \subset X$.

270. Give an example of a space X which has a weakly σ-point-finite family $\mathcal{U} \subset \tau^*(X)$ such that $\mathcal{U}$ is not σ-point-finite.

271. Let X be an arbitrary space with $s(X) \leq \kappa$. Prove that any weakly σ-point-finite family of non-empty open subsets of X has cardinality $\leq \kappa$.

272. Give an example of a non-cosmic Lindelöf Σ-space X such that any closed uncountable subspace of X has more than one (and hence infinitely many) non-isolated points.

273. Suppose that $C_p(X)$ is a Lindelöf Σ-space. Prove that, if all closed uncountable subspaces of $C_p(X)$ have more than one non-isolated points, then $C_p(X)$ has a countable network.

274. Let X be a Lindelöf Σ-space with a unique non-isolated point. Prove that any subspace of $C_p(X)$ has a weakly σ-point-finite T_0-separating family of cozero sets.

275. Let X be a space of countable spread. Prove that $C_p(X)$ is a Lindelöf Σ-space if and only if X has a countable network.

276. Show that, under CH, there exists a space X of countable spread for which there is a Lindelöf Σ-space $Y \subset C_p(X)$ with $nw(Y) > \omega$.

277. Let X be a space with a unique non-isolated point: $X = \{a\} \cup Y$, where all points of Y are isolated and $a \notin Y$. Prove that, for every infinite cardinal κ, the following conditions are equivalent:

   (i) $p(C_p(X)) \leq \kappa$;
   (ii) if $\{A_\alpha : \alpha < \kappa^+\}$ is a disjoint family of finite subsets of Y then there is an infinite $S \subset \kappa^+$ such that $a \notin \overline{\bigcup\{A_\alpha : \alpha \in S\}}$;
   (iii) if $\{A_\alpha : \alpha < \kappa^+\}$ is a family of finite subsets of Y then there is an infinite $S \subset \kappa^+$ such that $a \notin \overline{\bigcup\{A_\alpha : \alpha \in S\}}$.

278. Let X be a space with a unique non-isolated point. Prove that, if X has no non-trivial convergent sequences, then the point-finite cellularity of $C_p(X)$ is countable.

279. Call a family γ of finite subsets of a space X *concentrated* if there is no infinite $\mu \subset \gamma$ such that $\bigcup \mu$ is discrete and C^*-embedded in X. Prove that, if every concentrated family of finite subsets of X has cardinality $\leq \kappa$, then $p(C_p(X)) \leq \kappa$.

280. Prove that there exists a Lindelöf Σ-space X with a unique non-isolated point such that $C_p(X)$ is a Lindelöf Σ-space, $p(C_p(X)) = \omega$, all compact subsets of X are countable and $nw(X) = \mathfrak{c}$.

281. Prove that there exists a space X such that $C_p(X)$ is Lindelöf Σ-space, $nw(X) = \mathfrak{c}$ and $p(X) = \omega$.

282. Prove that any continuous image and any closed subspace of a Gul'ko compact space is a Gul'ko compact space.

283. Prove that any countable product of Gul'ko compact spaces is a Gul'ko compact space.

284. Let X be a Gul'ko compact space. Prove that for every second countable M, the space $C_p(X, M)$ is Lindelöf Σ.

285. Prove that if $C_p(X)$ is a Lindelöf Σ-space then X can be condensed into a Σ-product of real lines. Deduce from this fact that every Gul'ko compact space is Corson compact.

286. Prove that if X is Corson compact then the space $C_p(X)$ condenses linearly into a Σ_*-product of real lines. As a consequence, for any Gul'ko compact X the space $C_p(X)$ condenses linearly into a Σ_*-product of real lines.

287. Let X be a Corson compact space. Prove that, if $p(C_p(X)) = \omega$ then X is metrizable. Therefore if X is a Gul'ko compact space and $p(C_p(X)) = \omega$ then X is metrizable.

288. Suppose that X and $C_p(X)$ are Lindelöf Σ-spaces and $p(C_p(X)) = \omega$. Prove that $|X| \leq \mathfrak{c}$.

289. Prove that a compact space X is Gul'ko compact if and only if X has a weakly σ-point-finite T_0-separating family of cozero sets.

290. Prove that a compact X is Gul'ko compact if and only if there exists a set A such that X embeds into $\Sigma_s(A)$ for some family $s = \{A_n : n \in \omega\}$ of subsets of A with $\bigcup s = A$.

291. Suppose that X is a space, $n \in \mathbb{N}$ and a non-empty family $\mathcal{U} \subset \tau^*(X)$ has order $\leq n$, i.e., every $x \in X$ belongs to at most n elements of $\mathcal{U}$. Prove that there exist disjoint families $\mathcal{V}_1, \ldots, \mathcal{V}_n$ of non-empty open subsets of X such that $\mathcal{V} = \bigcup \{\mathcal{V}_i : i \leq n\}$ is a π-base for $\mathcal{U}$.

292. Suppose that a space X has the Baire property and $\mathcal{U}$ is a weakly σ-point-finite family of non-empty open subsets of X. Prove that there exists a σ-disjoint family $\mathcal{V} \subset \tau^*(X)$ which is a π-base for $\mathcal{U}$.

293. Prove that every Gul'ko compact space has a dense metrizable subspace.

294. Let X be a Gul'ko compact space. Prove that $w(X) = d(X) = c(X)$. In particular, each Gul'ko compact space with the Souslin property is metrizable.

295. Let X be a pseudocompact space with the Souslin property. Prove that any Lindelöf Σ-subspace of $C_p(X)$ has a countable network.

296. Let X be a Lindelöf Σ-space. Suppose that Y is a pseudocompact subspace of $C_p(X)$. Prove that Y is compact and metrizable if and only if $c(Y) = \omega$.

297. Prove that every Gul'ko compact space is hereditarily d-separable.

298. Let X be a compact space. Prove that $C_p(X)$ is a K-analytic space if and only if X has a T_0-separating family $\mathcal{U}$ of open F_σ-subsets of X and subfamilies $\{\mathcal{U}_s : s \in \omega^{<\omega}\}$ of the family $\mathcal{U}$ with the following properties:

(a) $\mathcal{U}_\emptyset = \mathcal{U}$ and $\mathcal{U}_s = \bigcup\{\mathcal{U}_{s\frown k} : k \in \omega\}$ for any $s \in \omega^{<\omega}$;

(b) for every $x \in X$ and every $f \in \omega^\omega$, there exists $m \in \omega$ such that the family $\mathcal{U}_{f|n}$ is point-finite at x for all $n \geq m$.

299. Let X be a compact space. Prove that $C_p(X)$ is a K-analytic space if and only if X can be embedded into some $\Sigma(A)$ in such a way that, for some family $\{A_s : s \in \omega^{<\omega}\}$ of subsets of A, the following conditions are fulfilled:

(a) $A_\emptyset = A$ and $A_s = \bigcup\{A_{s\frown k} : k \in \omega\}$ for any $s \in \omega^{<\omega}$;

(b) for any point $x \in X$ and any $f \in \omega^\omega$, there exists $m \in \omega$ such that the set $A_{f|n} \cap \operatorname{supp}(x)$ is finite for all $n \geq m$.

300. (Talagrand's example) Show that there exists a Gul'ko compact space X such that $C_p(X)$ is not K-analytic. In other words, not every Gul'ko compact space is Talagrand compact.

1.4 Eberlein Compact Spaces

All spaces are assumed to be Tychonoff. Given an arbitrary set A, we will need the spaces $\Sigma(A) = \{x \in \mathbb{R}^A : |\{a \in A : x(a) \neq 0\}| \leq \omega\}$, $\Sigma_*(A) = \{x \in \mathbb{R}^A :$ for any $\varepsilon > 0$, the set $\{a \in A : |x(a)| \geq \varepsilon\}$ is finite$\}$ and $\sigma(A) = \{x \in \mathbb{R}^A :$ the set $\{a \in A : x(a) \neq 0\}$ is finite$\}$. Suppose that we have a product $X = \prod\{X_t : t \in T\}$ and $x \in X$. Let $\Sigma(X, x) = \{y \in X : |\{t \in T : y(t) \neq x(t)\}| \leq \omega\}$. The space $\Sigma(X, x)$ is called *the Σ-product of* $\{X_t : t \in T\}$ *centered at the point* x and the point x is *the center* of the relevant Σ-product. If we have a product $X = \prod\{X_t : t \in T\}$ and $x \in X$, then $\sigma(X, x) = \{y \in X : |\{t \in T : y(t) \neq x(t)\}| < \omega\}$. The set $\sigma(X, x)$ is called *the σ-product of* $\{X_t : t \in T\}$ *centered at the point* x and the point x is *the center* of the relevant σ-product. If no center of a Σ-product or a σ-product is specified, then the formulated statements are valid (or must be proved) for any center of the mentioned Σ-product or σ-product. If X is a set, then $\Delta = \Delta_X = \{(x, x) : x \in X\}$ is the diagonal of X.

The space $\mathbb{D}$ is the two-point set $\{0, 1\}$ with the discrete topology. A space X is called *functionally perfect* if there is a compact $K \subset C_p(X)$ which separates the points of X. A functionally perfect compact space is called *Eberlein compact*. A compact space K is called *a uniform Eberlein compact* if, for some set A, the space K embeds into $\Sigma_*(A)$ in such a way that there exists a function $N : \mathbb{R}^+ \to \mathbb{N}$ such that $|\{a \in A : |x(a)| \geq \varepsilon\}| \leq N(\varepsilon)$ for all $x \in K$ and $\varepsilon > 0$. Here $\mathbb{R}^+$ is the set of all positive real numbers. Call X *strong Eberlein compact* if it is homeomorphic to a compact subspace of $\sigma_0(A) = \{x \in \mathbb{D}^A : |x^{-1}(1)| < \omega\}$ for some A. A space is called *k-separable* if it has a dense σ-compact subspace. The expression $X \simeq Y$ says that X and Y are homeomorphic. Given a locally compact non-compact X, let $\alpha(X) = X \cup \{a_0\}$, where $a_0 \notin X$. Let $\mu = \tau(X) \cup \{\{a_0\} \cup U : X \backslash U$ is compact$\}$. The space $(\alpha(X), \mu)$ is called *the one-point compactification of the space* X.

If $\mathcal{P}$ is a class of spaces, then $\mathcal{P}_\sigma$ consists of spaces representable as a countable union of elements of $\mathcal{P}$. The class $\mathcal{P}_\delta$ contains the spaces which are countable intersections of elements of $\mathcal{P}$ in some larger space. More formally, $X \in \mathcal{P}_\sigma$ if $X = \bigcup\{X_n : n \in \omega\}$ where $X_n \in \mathcal{P}$ for any $n \in \omega$. Analogously, $X \in \mathcal{P}_\delta$ if there exists a space Y and $Y_n \subset Y$ such that $Y_n \in \mathcal{P}$ for all $n \in \omega$ and $\bigcap\{Y_n : n \in \omega\} \simeq X$. Then $\mathcal{P}_{\sigma\delta} = (\mathcal{P}_\sigma)_\delta$. Say that X is a $\mathcal{K}_{\sigma\delta}$-space if $X \in \mathcal{K}_{\sigma\delta}$ where $\mathcal{K}$ is the class of compact spaces.

Let T be an infinite set. An arbitrary family $\mathcal{A} \subset \exp T$ is called *adequate* if $\bigcup \mathcal{A} = T$, $\exp A \subset \mathcal{A}$ for any $A \in \mathcal{A}$, and $A \in \mathcal{A}$ whenever all finite subsets of A belong to $\mathcal{A}$. Given $A \subset T$, let $\chi_A(t) = 1$ if $t \in A$ and $\chi_A(t) = 0$ if $t \notin A$. The map $\chi_A : T \to \{0, 1\}$ is called *the characteristic function of A in the set* T. If we have a set T and an adequate family $\mathcal{A}$ on T, let $K_\mathcal{A} = \{\chi_A \in \mathbb{D}^T : A \in \mathcal{A}\}$. Another object associated with $\mathcal{A}$, is the space $T_\mathcal{A}^*$ whose underlying set is $T \cup \{\xi\}$, where $\xi \notin T$, all points of T are isolated in $T_\mathcal{A}^*$ and the basic neighbourhoods of ξ are the complements of finite unions of elements of $\mathcal{A}$. A subspace $X \subset \mathbb{D}^T$ is called *adequate* if $X = K_\mathcal{A}$ for some adequate family $\mathcal{A}$ on T.

A family γ of subsets of X is T_0-*separating* if, for any distinct $x, y \in X$, there is $A \in \gamma$ such that $|A \cap \{x, y\}| = 1$. Now, γ is T_1-*separating* if, for any distinct $x, y \in X$, there are $A, B \in \gamma$ such that $A \cap \{x, y\} = \{x\}$ and $B \cap \{x, y\} = \{y\}$. A set $U \subset X$ is *cozero* if there is $f \in C_p(X)$ such that $U = f^{-1}(\mathbb{R} \backslash \{0\})$. Say that X is *a Preiss–Simon space* if, for any closed $F \subset X$ and for any non-isolated $x \in F$, there exists a sequence $\{U_n : n \in \omega\}$ of open non-empty subsets of F such that $U_n \to x$, i.e., any neighbourhood of x contains all but finitely many of U_n's. A space X is called *homogeneous* if, for every $x, y \in X$, there exists a homeomorphism $f : X \to X$ such that $f(x) = y$. A space X is σ-metacompact if any open cover of X has a σ-point-finite open refinement, i.e., a refinement which is a countable union of point-finite families.

All linear spaces in this book are considered over the space $\mathbb{R}$ of the reals. Let L be a linear topological space. A set $M \subset L$ is called *a linear subspace of L* if $\alpha x + \beta y \in M$ whenever $x, y \in M$ and $\alpha, \beta \in \mathbb{R}$. A linear space L, equipped with a topology τ, is called *linear topological space* if (L, τ) is a T_1-space and the linear operations $(x, y) \to x + y$ and $(t, x) \to tx$ are continuous with respect to τ. A subset A of a linear space L is called *convex* if $x, y \in A$ implies $tx + (1 - t)y \in A$ for any number $t \in [0, 1]$. A *convex hull* conv(A) of a set $A \subset L$ is the set $\{t_1 x_1 + \ldots + t_n x_n : n \in \mathbb{N}, x_1, \ldots, x_n \in A, t_1, \ldots, t_n \in [0, 1], t_1 + \ldots + t_n = 1\}$. A linear topological space L is called *locally convex* if it has a base which consists of convex sets. If L is a linear space and $A \subset L$ then *the linear span of A in L* is the intersection of all linear subspaces of L which contain A. A real-valued function $x \to ||x||$, defined on a linear space L, is *a norm* if it has the following properties:

(N1) $||x|| \geq 0$ for any $x \in L$; besides, $||x|| = 0$ if and only if $x = 0$;
(N2) $||\alpha x|| = |\alpha| \cdot ||x||$ for any $x \in L$ and $\alpha \in \mathbb{R}$;
(N3) $||x + y|| \leq ||x|| + ||y||$ for any $x, y \in L$.

If $||\cdot||$ is a norm on a linear space L, the pair $(L, ||\cdot||)$ is called *a normed space*. If the norm is clear, we write L instead of $(L, ||\cdot||)$. If L is a normed space, then the function $d_L(x, y) = ||x - y||$ is a metric on L. Every normed space L is considered to carry the topology generated by the metric d_L. If (L, d_L) is a complete metric space, L is called *a Banach space*. Given a linear space L, a function $f : L \to \mathbb{R}$ is called *a linear functional* if $f(\alpha x + \beta y) = \alpha f(x) + \beta f(y)$ for any $x, y \in L$ and $\alpha, \beta \in \mathbb{R}$; the functional f is called *trivial* if $f(x) = 0$ for any $x \in L$. The set of all continuous linear functionals on L is denoted by L^*; clearly, $L^* \subset C(L)$. Given a set X and $\mathcal{F} \subset \mathbb{R}^X$, let $\tau_{\mathcal{F}}$ be the topology generated by the family $\{f^{-1}(U) : f \in \mathcal{F}, U \in \tau(\mathbb{R})\}$ as a subbase. We call $\tau_{\mathcal{F}}$ *the topology generated by $\mathcal{F}$*. If L is a linear topological space, denote by L_w the space (L, τ_{L^*}). We call τ_{L^*} *the weak topology of L*. Given a topological property $\mathcal{P}$, a set $K \subset L$ is called *weakly $\mathcal{P}$* if the topology, induced in K from L_w, has the property $\mathcal{P}$. For example, a set $K \subset L$ is *weakly compact* if the topology, induced in K from L_w, is compact.

301. Prove that the following conditions are equivalent for any space X:

    (i) X is functionally perfect;
    (ii) X condenses onto a subspace of $C_p(Y)$ for some compact Y;
    (iii) X condenses onto a subspace of $C_p(Y)$ for some σ-compact Y;
    (iv) there exists a σ-compact $H \subset C_p(X)$ which separates the points of X;
    (v) the space $C_p(X)$ is k-separable.

302. Show that neither $\Sigma(A)$ nor $\mathbb{R}^A$ is functionally perfect whenever the set A is uncountable.

303. Prove that the spaces $\sigma(A)$ and $\Sigma_*(A)$ are functionally perfect for any set A.

304. Prove that $C_p(\omega_1)$ is functionally perfect.

305. Show that, if a space condenses onto a functionally perfect space, then it is functionally perfect.

306. Prove that any subspace of a functionally perfect space is functionally perfect.

307. Prove that a countable product of functionally perfect spaces is a functionally perfect space. In particular, a countable product of Eberlein compact spaces is Eberlein compact.

308. Prove that any σ-product of functionally perfect spaces is a functionally perfect space.

309. Prove that any product of k-separable spaces is k-separable.

310. Prove that a space X is hereditarily k-separable (i.e., every $Y \subset X$ is k-separable) if and only if X is hereditarily separable.

311. Suppose that $f : X \to Y$ is an irreducible perfect map. Show that the space X is k-separable if and only if so is Y.

312. Prove that, for any k-separable X, the space $C_p(X)$ is functionally perfect. In particular, the space $C_p(X)$ is functionally perfect for any compact X.

313. Give an example of a non-k-separable space X for which $C_p(X)$ is functionally perfect.

314. Prove that, for an arbitrary space X, the space $C_p(X)$ is a continuous image of $C_p(C_p(C_p(X)))$.

315. Prove that $C_p(X)$ is k-separable if and only if $C_pC_p(X)$ is functionally perfect. As a consequence, X is functionally perfect if and only if $C_p(C_p(X))$ is functionally perfect.

316. Prove that any metrizable space is functionally perfect. In particular, any second countable space is functionally perfect and hence any metrizable compact space is Eberlein compact.

317. Let X be a metrizable space. Prove that $C_p(X)$ is functionally perfect if and only if X is second countable.

318. Prove that any paracompact space with a G_δ-diagonal can be condensed onto a metrizable space. Deduce from this fact that any paracompact space with a G_δ-diagonal is functionally perfect.

319. Observe that any Eberlein–Grothendieck space is functionally perfect. Give an example of a functionally perfect space which is not Eberlein–Grothendieck.

320. Prove that every metrizable space embeds into an Eberlein compact space.

321. Prove that $C_p(X)$ is a $K_{\sigma\delta}$-space for any Eberlein compact X. In particular, each Eberlein compact space is Gul'ko compact and hence Corson compact.

322. Prove that a compact space X is Eberlein if and only if it embeds into $\Sigma_*(A)$ for some A.

323. Prove that a compact space is metrizable if and only if it has a T_1-separating σ-point-finite family of cozero sets.

324. Prove that a compact space X is Eberlein compact if and only if X has a T_0-separating σ-point-finite family of cozero sets.

325. Give an example of a scattered compact space which fails to be Corson compact and has a T_0-separating σ-point-finite family of open sets.

326. Suppose that a compact X has a T_0-separating point-finite family of open sets. Prove that X is Eberlein compact.

327. Prove that a non-empty compact X is Eberlein if and only if there is a compact $F \subset C_p(X)$ which separates the points of X and is homeomorphic to $A(\kappa)$ for some cardinal κ.

328. Say that a (not necessarily continuous) function $x : \mathbb{I} \to \mathbb{I}$ is increasing (decreasing) if $x(s) \leq x(t)$ (or $x(s) \geq x(t)$ respectively) whenever $s, t \in \mathbb{I}$ and $s \leq t$. A function $x : \mathbb{I} \to \mathbb{I}$ is called monotone if it is either increasing or decreasing. Prove that the *Helly space* $X = \{x \in \mathbb{I}^{\mathbb{I}} : x$ is a monotone function$\}$ is closed in $\mathbb{I}^{\mathbb{I}}$ and hence compact. Is it an Eberlein compact space?

329. Prove that a compact space X is Eberlein compact if and only if $C_p(X)$ is a continuous image of $(A(\kappa))^{\omega} \times \omega^{\omega}$ for some infinite cardinal κ.

330. Prove that a compact space X is Eberlein compact if and only if there is a compact space K and a separable space M such that $C_p(X)$ is a continuous image of $K \times M$.

331. Prove that any infinite Eberlein compact space X is a continuous image of a closed subspace of $(A(\kappa))^{\omega} \times M$, where $\kappa = w(X)$ and M is a second countable space.

332. Prove that each Eberlein compact space is a Preiss–Simon space.

333. Prove that, if a pseudocompact space X condenses onto a subspace of $C_p(K)$ for some compact K, then this condensation is a homeomorphism and X is Eberlein compact. In particular, any functionally perfect pseudocompact space is Eberlein compact.

334. Prove that a zero-dimensional compact space X is Eberlein compact if and only if X has a T_0-separating σ-point-finite family of clopen sets.

335. Let X be a zero-dimensional compact space. Prove that X is Eberlein if and only if $C_p(X, \mathbb{D})$ is σ-compact.

336. Prove that any Eberlein compact space is a continuous image of a zero-dimensional Eberlein compact space.

337. Prove that any continuous image of an Eberlein compact space is Eberlein compact.

338. Prove that there exists a compact X such that X is a union of countably many Eberlein compact spaces while $C_p(X)$ is not Lindelöf. In particular, a compact countable union of Eberlein compact spaces need not be Corson compact.

339. Prove that it is consistent with ZFC that there exists a Corson compact space which does not map irreducibly onto an Eberlein compact space.

340. Suppose that $\mathcal{P}$ is a class of topological spaces such that, for any $X \in \mathcal{P}$, all continuous images and all closed subspaces of X belong to $\mathcal{P}$. Prove that it is impossible that a compact space X be Eberlein compact if and only if $C_p(X) \in \mathcal{P}$, i.e., either there exists an Eberlein compact space X such that $C_p(X) \notin \mathcal{P}$ or there is a compact Y which is not Eberlein and $C_p(Y) \in \mathcal{P}$.

341. Let X be a Gul'ko compact space. Prove that there exists a countable family $\mathcal{F}$ of closed subsets of X such that $\bigcup \mathcal{F} = X$ and $K_x = \bigcap \{A : x \in A \in \mathcal{F}\}$ is Eberlein compact for any $x \in X$.

342. Let X be an Eberlein compact space with $|X| \leq \mathfrak{c}$. Prove that there exists a countable family $\mathcal{F}$ of closed subsets of X such that $\bigcup \mathcal{F} = X$ and the subspace $K_x = \bigcap \{A : x \in A \in \mathcal{F}\}$ is metrizable for any $x \in X$.

343. Observe that $c(X) = w(X)$ for any Eberlein compact space X. Prove that, for any infinite compact X, we have $c(X) = \sup\{w(K) : K \subset C_p(X) \text{ and } K \text{ is compact}\}$.

344. Given a pseudocompact space X and functions $f, g \in C(X)$, let $d(f, g) = \sup\{|f(x) - g(x)| : x \in X\}$. Prove that d is a complete metric on the set $C(X)$ and the topology of $C_u(X)$ is generated by d.

345. Prove that, for any pseudocompact X, the space $C_u(X)$ is separable if and only if X is compact and metrizable.

346. Suppose that X is a compact space and let $\|f\| = \sup\{|f(x)| : x \in X\}$ for any $f \in C(X)$. Assume additionally that $h \in C(X)$, $r > 0$ and $H = \{h_n : n \in \omega\} \subset C(X)$ is a sequence such that $\|h_n\| \leq r$ for all $n \in \omega$ and $h_n(x) \to h(x)$ for any $x \in X$ (i.e., the sequence H converges to h in the space $C_p(X)$). Prove that h belongs to the closure of the convex hull $\mathrm{conv}(H)$ of the set H in the space $C_u(X)$.

347. Suppose that X is a Čech-complete space and we are given a continuous map $\varphi : X \to C_p(K)$ for some compact space K. Prove that there exists a dense G_δ-set $P \subset X$ such that $\varphi : X \to C_u(K)$ is continuous at every point of P.

348. Prove that any Eberlein–Grothendieck Čech-complete space has a dense G_δ-subspace which is metrizable.

349. Prove that if X is an Eberlein–Grothendieck Čech-complete space then $c(X) = w(X)$.

350. Let X be a compact space. Assume that $X = X_1 \cup \ldots \cup X_n$, where every X_i is a metrizable (not necessarily closed) subspace of X. Prove that $\overline{X}_1 \cap \ldots \cap \overline{X}_n$ is metrizable. In particular, if all X_i's are dense in X then X is metrizable.

351. Suppose that X is a compact space which is a union of two metrizable subspaces. Prove that X is Eberlein compact which is not necessarily metrizable.

352. Observe that there exists a compact space K which is not Eberlein while being a union of three metrizable subspaces. Suppose that X is a compact space such that $X \times X$ is a union of its three metrizable subspaces. Prove that X is Eberlein compact.

353. Prove that, if X is compact and X^ω is a union of countably many of its Eberlein compact subspaces then X is Eberlein compact.

354. Give an example of an Eberlein compact space which cannot be represented as a countable union of its metrizable subspaces.

355. Let X be a Corson compact space such that X is a countable union of Eberlein compact spaces. Prove that $C_p(X)$ is K-analytic and hence X is Gul'ko compact.

356. Let X be a σ-product of an arbitrary family of Eberlein compact spaces. Prove that $C_p(X)$ is a $K_{\sigma\delta}$-space.

357. Prove that the one-point compactification of an infinite discrete union of non-empty Eberlein compact spaces is an Eberlein compact space.

358. Prove that the Alexandroff double of an Eberlein compact space is an Eberlein compact space.

359. Recall that a space X is homogeneous if, for any $x, y \in X$, there is a homeomorphism $h : X \to X$ such that $h(x) = y$. Construct an example of a homogeneous non-metrizable Eberlein compact space.

360. Give an example of a hereditarily normal but not perfectly normal Eberlein compact space.

361. Let X be an Eberlein compact space such that $X \times X$ is hereditarily normal. Prove that X is metrizable.

362. Prove that there exists an Eberlein compact space X such that $X^2 \backslash \Delta$ is not metacompact.

363. Prove that any Eberlein compact space is hereditarily σ-metacompact.

364. Prove that, for any compact X, the subspace $X^2 \backslash \Delta \subset X^2$ is σ-metacompact if and only if X is Eberlein compact.

365. Prove that a compact space X is Eberlein compact if and only if $X \times X$ is hereditarily σ-metacompact.

366. Prove that a compact space X has a closure-preserving cover by compact metrizable subspaces if and only if it embeds into a σ-product of compact metrizable spaces. In particular, if X has a closure-preserving cover by compact metrizable subspaces then it is an Eberlein compact space.

367. Construct an Eberlein compact space which does not have a closure-preserving cover by compact metrizable subspaces.

368. Observe that every strong Eberlein compact is Eberlein compact. Prove that a metrizable compact space is strong Eberlein compact if and only if it is countable.

369. Prove that a compact X is strong Eberlein compact if and only if it has a point-finite T_0-separating cover by clopen sets.

370. Prove that every σ-discrete compact space is scattered. Give an example of a scattered compact non-σ-discrete space.

371. Prove that every strong Eberlein compact space is σ-discrete and hence scattered.

372. Prove that a hereditarily metacompact scattered compact space is strong Eberlein compact.

373. Prove that the following conditions are equivalent for any compact X:

   (i) X is σ-discrete and Corson compact;
   (ii) X is scattered and Corson compact;
   (iii) X is strong Eberlein compact.

374. Prove that any continuous image of a strong Eberlein compact space is a strong Eberlein compact space.

375. Prove that any Eberlein compact space is a continuous image of a closed subset of a countable product of strong Eberlein compact spaces.

376. Let X be a strong Eberlein compact space. Prove that the Alexandroff double of X is also strong Eberlein compact.

377. Suppose that X_t is strong Eberlein compact for each $t \in T$. Prove that the Alexandroff one-point compactification of the space $\bigoplus\{X_t : t \in T\}$ is also strong Eberlein compact.

378. Observe that any uniform Eberlein compact is Eberlein compact. Prove that any metrizable compact space is uniform Eberlein compact.

379. Observe that any closed subspace of a uniform Eberlein compact space is uniform Eberlein compact. Prove that any countable product of uniform Eberlein compact spaces is uniform Eberlein compact.

380. Prove that if X is a uniform Eberlein compact space then it is a continuous image of a closed subspace of $(A(\kappa))^\omega$ for some infinite cardinal κ.

381. Prove that any continuous image of a uniform Eberlein compact space is uniform Eberlein compact. Deduce from this fact that a space X is uniform Eberlein compact if and only if it is a continuous image of a closed subspace of $(A(\kappa))^\omega$ for some infinite cardinal κ.

382. Given an infinite set T suppose that a space $X_t \neq \emptyset$ is uniform Eberlein compact for each $t \in T$. Prove that the Alexandroff compactification of the space $\bigoplus\{X_t : t \in T\}$ is also uniform Eberlein compact.

383. Let T be an infinite set. Suppose that $\mathcal{A}$ is an adequate family on T. Prove that the space $K_\mathcal{A}$ is Eberlein compact if and only if $T_\mathcal{A}^*$ is σ-compact.

384. Let T be an infinite set. Suppose that $\mathcal{A}$ is an adequate family on T. Prove that the space $K_\mathcal{A}$ is Eberlein compact if and only if there exists a disjoint family $\{T_i : i \in \omega\}$ such that $T = \bigcup\{T_i : i \in \omega\}$ and $x^{-1}(1) \cap T_i$ is finite for every $x \in K_\mathcal{A}$ and $i \in \omega$.

385. Let T be an infinite set and $\mathcal{A}$ an adequate family on T. Prove that the adequate compact $K_\mathcal{A}$ is uniform Eberlein compact if and only if there exists a disjoint family $\{T_i : i \in \omega\}$ and a function $N : \omega \to \omega$ such that $T = \bigcup\{T_i :\in \omega\}$ and $|x^{-1}(1) \cap T_i| \leq N(i)$ for any $x \in K_\mathcal{A}$ and $i \in \omega$.

386. For the set $T = \omega_1 \times \omega_1$ let us introduce an order $<$ on T declaring that $(\alpha_1, \beta_1) < (\alpha_2, \beta_2)$ if and only if $\alpha_1 < \alpha_2$ and $\beta_1 > \beta_2$. Denote by $\mathcal{A}$ the family of all subsets of T which are linearly ordered by $<$ (the empty set and the one-point sets are considered to be linearly ordered). Prove that $\mathcal{A}$ is an adequate family and $X = K_\mathcal{A}$ is a strong Eberlein compact space which is not uniform Eberlein compact.

387. (Talagrand's example) For any distinct $s, t \in \omega^\omega$, consider the number $\delta(s, t) = \min\{k \in \omega : s(k) \neq t(k)\}$. For each $n \in \omega$, let $\mathcal{A}_n = \{A \subset \omega^\omega :$ for any distinct $s, t \in A$ we have $\delta(s, t) = n\}$. Prove that $\mathcal{A} = \bigcup\{\mathcal{A}_n : n \in \omega\}$ is an adequate family and $X = K_\mathcal{A}$ is a Talagrand compact space (i.e., $C_p(X)$ is K-analytic and hence X is Gul'ko compact) while X is not Eberlein compact.

388. Given a compact space X let $||f|| = \sup\{|f(x)| : x \in X\}$ for any $f \in C(X)$. Prove that $(C(X), ||\cdot||)$ is a Banach space.

389. (Hahn–Banach Theorem) Assume that M is a linear subspace of a normed space $(L, ||\cdot||)$ and suppose that $f : M \to \mathbb{R}$ is a linear functional such that $|f(x)| \leq ||x||$ for any $x \in M$. Prove that there exists a linear functional $F : L \to \mathbb{R}$ such that $F|M = f$ and $|F(x)| \leq ||x||$ for all $x \in L$.

390. Given a normed space $(L, ||\cdot||)$ let $S = \{x \in L : ||x|| \leq 1\}$ be the unit ball of L. Prove that a linear functional $f : L \to \mathbb{R}$ is continuous if and only if there exists $k \in \mathbb{N}$ such that $|f(x)| \leq k$ for any $x \in S$.

391. Given a normed space $(L, ||\cdot||)$, let $S = \{x \in L : ||x|| \leq 1\}$ and consider the set $S^* = \{f \in L^* : f(S) \subset [-1, 1]\}$. Prove that S^* separates the points of L.

392. Let L be a linear space without any topology. Suppose that $\mathcal{F}$ is a family of linear functionals on L which separates the points of L. Prove that the topology on L generated by $\mathcal{F}$, is Tychonoff and makes L a locally convex linear topological space.

393. Let L be a linear space. Denote by L' the set of all linear functionals on L. Considering L' a subspace of $\mathbb{R}^L$, prove that L' is closed in $\mathbb{R}^L$.

394. Given a normed space $(L, ||\cdot||)$, consider the sets $S = \{x \in L : ||x|| \leq 1\}$ and $S^* = \{f \in L^* : f(S) \subset [-1, 1]\}$. Prove that, for any point $x \in L$, the set $S^*(x) = \{f(x) : f \in S^*\}$ is bounded in $\mathbb{R}$.

395. Given a normed space $(L, ||\cdot||)$, let $S = \{x \in L : ||x|| \leq 1\}$ and consider the set $S^* = \{f \in L^* : f(S) \subset [-1, 1]\}$. Denote by L_w the set L with the topology generated by L^*. Observe that $S^* \subset C(L_w)$ and give S^* the topology τ induced from $C_p(L_w)$. Prove that (S^*, τ) is a compact space.

396. Prove that L_w is functionally perfect for any normed space $(L, ||\cdot||)$. As a consequence, any compact subspace of L_w is Eberlein compact.

397. Let L be a linear topological space. Given a sequence $\{x_n : n \in \omega\} \subset L$, prove that $x_n \to x$ in the weak topology on L if and only if $f(x_n) \to f(x)$ for any $f \in L^*$.

398. For an arbitrary compact space K let $||f|| = \sup\{|f(x)| : x \in K\}$ for any $f \in C(K)$. Denote by $C_w(K)$ the space $C(K)$ endowed with the weak topology of the normed space $(C(K), ||\cdot||)$. Prove that $\tau(C_w(K)) \supset \tau(C_p(K))$ and show that, in the case of $K = \mathbb{I}$, this inclusion is strict, i.e., $\tau(C_w(\mathbb{I})) \neq \tau(C_p(\mathbb{I}))$.

399. Suppose that K is a compact space and let $||f|| = \sup\{|f(x)| : x \in K\}$ for any $f \in C(K)$. Denote by $C_w(K)$ the space $C(K)$ endowed with the weak topology of the normed space $(C(K), ||\cdot||)$. Prove that, for any $X \subset C(K, \mathbb{I})$, if X is compact as a subspace of $C_p(K)$ then the topologies, induced on X from the spaces $C_p(K)$ and $C_w(K)$, coincide.

400. (The original definition of an Eberlein compact space) Prove that X is an Eberlein compact space if and only if it is homeomorphic to a weakly compact subset of a Banach space.

1.5   Special Embeddings and Extension Operators

A space X is called *splittable* if, for every $f \in \mathbb{R}^X$, there exists a countable set $A \subset C_p(X)$ such that $f \in \overline{A}$ (the closure is taken in $\mathbb{R}^X$). Splittable spaces are also called *cleavable*, but we won't use the last term in this book. A space X is *weakly splittable* if, for every function $f \in \mathbb{R}^X$, there exists a σ-compact $A \subset C_p(X)$ such that $f \in \overline{A}$ (the closure is taken in $\mathbb{R}^X$). A compact weakly splittable space is called *weak Eberlein compact*.

A space X is *strongly splittable* if, for every $f \in \mathbb{R}^X$, there exists a sequence $S = \{f_n : n \in \omega\} \subset C_p(X)$ such that $f_n \to f$. A space X is *functionally perfect* if there exists a compact $K \subset C_p(X)$ which separates the points of X. Given a space X let $L_p(X)$ be the set of all continuous linear functionals on $C_p(X)$ with the topology induced from $C_p(C_p(X))$.

A space X has *a small diagonal* if, for any uncountable set $A \subset (X \times X) \backslash \Delta$, there exists an uncountable $B \subset A$ such that $\overline{B} \cap \Delta = \emptyset$. Here, as usual, the set $\Delta = \{(x, x) : x \in X\} \subset X \times X$ is the diagonal of the space X. A subspace $Y \subset X$ is *strongly discrete* if there exists a discrete family $\{U_y : y \in Y\} \subset \tau(X)$ such that $y \in U_y$ for each $y \in Y$. A space is *strongly σ-discrete* if it is a union of countably many of its closed discrete subspaces. The space $\mathbb{D}$ is the two-point set $\{0, 1\}$ endowed with the discrete topology.

An uncountable regular cardinal κ is *a caliber* of a space X if, for any family $\mathcal{U} \subset \tau^*(X)$ of cardinality κ, there exists $\mathcal{U}' \subset \mathcal{U}$ such that $|\mathcal{U}'| = \kappa$ and $\bigcap \mathcal{U}' \neq \emptyset$. An uncountable regular cardinal κ is called *a precaliber* of a space X if, for any family $\mathcal{U} \subset \tau^*(X)$ of cardinality κ, there exists $\mathcal{U}' \subset \mathcal{U}$ such that $|\mathcal{U}'| = \kappa$ and $\mathcal{U}'$ has finite intersection property, i.e., $\bigcap \mathcal{V} \neq \emptyset$ for any finite $\mathcal{V} \subset \mathcal{U}'$. Now, $p(X) = \sup\{|\mathcal{U}| : \mathcal{U}$ is a point-finite family of non-empty open subsets of $X\}$. The cardinal $p(X)$ is called *the point-finite cellularity* of X. Given a space X and $x \in X$, call a family $\mathcal{B} \subset \tau^*(X)$ a *π-base* of X at x if, for any $U \in \tau(x, X)$, there is $V \in \mathcal{B}$ such that $V \subset U$. Note that the elements of a π-base at x need not contain the point x. Now, $\pi\chi(x, X) = \min\{|\mathcal{B}| : \mathcal{B}$ is a π-base of X at $x\}$ and $\pi\chi(X) = \sup\{\pi\chi(x, X) : x \in X\}$. Let $iw(X) = \min\{|\kappa| :$ there is a condensation of X onto a space of weight $\leq \kappa\}$. The cardinal $iw(X)$ is called *the i-weight* of X.

A subspace Y of a space X is *C-embedded in X* if, for any function $f \in C(Y)$, there is $g \in C(X)$ such that $g|Y = f$. If any bounded continuous function on Y can be extended to a continuous function on X, we say that *Y is C^*-embedded in X*.

If X is a space, $Y \subset X$ and $E \subset C_p(Y)$, say that a map $\varphi : E \to C_p(X)$ is *an extender* if $\varphi(f)|Y = f$ for any $f \in E$. A set $Y \subset X$ is *t-embedded in X*, if there exists a continuous extender $\varphi : C_p(Y) \to C_p(X)$. Note that any t-embedded subspace of X is C-embedded in X. It is said that *Y is l-embedded in X* if there exists a linear continuous extender $\varphi : C_p(Y) \to C_p(X)$. Note that any l-embedded subspace of X is t-embedded in X.

A space X is called *extendial or l-extendial* if every closed subspace of X is l-embedded in X. If every closed subspace of X is t-embedded in X, then X is *t-extendial.* A space X is *extral or l-extral* if, for any space Y which contains X as a closed subspace, X is l-embedded in Y. Analogously, X is *t-extral* if, for any space Y which contains X as a closed subspace, X is t-embedded in Y. Given a space X and a subspace $F \subset X$, we say that F *is a retract of* X if there exists a continuous map $f : X \to F$ such that $f(x) = x$ for any $x \in F$. The map f is called *retraction of* X *onto* F. A compact space is called *dyadic* if it is a continuous image of the Cantor cube $\mathbb{D}^\kappa$ for some κ. A compact space X is *l-dyadic* if it is l-embedded in some dyadic space. If X can be t-embedded in some dyadic space, it is called *t-dyadic*.

Recall that a space X is *radial* if, for every $A \subset X$ and every $x \in \overline{A}$, there exists a transfinite sequence $s = \{x_\alpha : \alpha < \kappa\} \subset A$ such that $s \to x$. A space is *k-separable* if it has a dense σ-compact subspace. The symbol $\mathfrak{c}$ denotes the cardinality of the continuum. A space X is *Baire* if any countable intersection of open dense subsets of X is dense in X. The space X is Čech-complete if X is a G_δ-set in βX. A sequence $\{\mathcal{U}_n : n \in \omega\}$ of covers of X is called *complete* if, for any filter $\mathcal{F}$ on the set X such that $\mathcal{F} \cap \mathcal{U}_n \neq \emptyset$ for all $n \in \omega$, we have $\bigcap\{\overline{F} : F \in \mathcal{F}\} \neq \emptyset$.

A family $\mathcal{U}$ of subsets of a space X is *weakly σ-point-finite* if there exists a sequence $\{\mathcal{U}_n : n \in \omega\}$ of subfamilies of $\mathcal{U}$ such that, for every $x \in X$, we have $\mathcal{U} = \bigcup\{\mathcal{U}_n : n \in M_x\}$ where $M_x = \{n \in \omega : \mathcal{U}_n$ is point-finite at $x\}$.

401. Prove that every subspace of a splittable space is splittable.

402. Prove that every second countable space is splittable.

403. Prove that $\psi(X) \leq \omega$ for every splittable space X.

404. Prove that, if X condenses onto a splittable space, then X is splittable. In particular, any space of countable i-weight is splittable.

405. Give an example of a splittable space which does not condense onto a second countable space.

406. Give an example of a metrizable space which is not splittable.

407. Give an example of a splittable space whose square is not splittable.

408. Prove that a space X with a unique non-isolated point is splittable if and only if $\psi(X) \leq \omega$.

409. Let X be a non-discrete space. Prove that, for any $f \in \mathbb{R}^X$, there exists a countable $A \subset \mathbb{R}^X \backslash C_p(X)$ such that $f \in \overline{A}$ (the closure is taken in $\mathbb{R}^X$).

410. Let X be a splittable space. Prove that every regular uncountable cardinal is a caliber of $C_p(X)$.

411. Prove that every splittable space has a small diagonal.

412. Prove that $C_p(X)$ is splittable if and only if it condenses onto a second countable space.

413. Show that an open continuous image of a splittable space can fail to be splittable.

414. Let X be a space of cardinality $\leq \mathfrak{c}$. Prove that X is splittable if and only if the i-weight of X is countable.

415. Prove that a space X is splittable if and only if, for every $f \in \mathbb{D}^X$, there is a countable $A \subset C_p(X)$ such that $f \in \overline{A}$ (the closure is taken in $\mathbb{R}^X$).

416. Prove that a space X is splittable if and only if, for any $A \subset X$, there exists a continuous map $\varphi : X \to \mathbb{R}^\omega$ such that $A = \varphi^{-1}\varphi(A)$.

417. Prove that any pseudocompact splittable space must be compact and metrizable.

418. Prove that a Lindelöf space X is splittable if and only if $iw(X) \leq \omega$.

419. Prove that a Lindelöf Σ-space is splittable if and only if it has a countable network.

420. Prove that a Lindelöf p-space is splittable if and only if it is second countable.

421. Prove that any Čech-complete splittable paracompact space is metrizable.

422. Let X be a complete metrizable dense-in-itself space. Prove that X is splittable if and only if $|X| \leq \mathfrak{c}$.

423. Suppose that $X = \bigcup\{X_n : n \in \omega\}$, where $X_n \subset X_{n+1}$ for each $n \in \omega$, the subspace X_n is splittable and C^*-embedded in X for every n. Prove that X is splittable.

424. Prove that any normal strongly σ-discrete space is strongly splittable.

425. Give an example of a strongly σ-discrete space which is not splittable.

426. Show that, for any cardinal κ, there exists a normal strongly σ-discrete (and hence splittable) space X with $c(X) = \omega$ and $|X| \geq \kappa$.

427. Show that there exists a splittable space which cannot be condensed onto a first countable space.

428. Assuming the Generalized Continuum Hypothesis prove that, if X is a splittable space and $A \subset X$, $|A| \leq \mathfrak{c}$ then $|\overline{A}| \leq \mathfrak{c}$.

429. Prove that any Čech-complete splittable space has a dense metrizable subspace.

430. Prove that every subspace of a weakly splittable space must be weakly splittable.

431. Prove that, if X condenses onto a weakly splittable space then X is weakly splittable.

432. Give an example of a weakly splittable non-splittable space.

433. Prove that a separable weakly splittable space is splittable.

434. Let X be a space with ω_1 caliber of X. Prove that, under CH, X is splittable if and only if it is weakly splittable.

435. Show that every functionally perfect space is weakly splittable. In particular, every Eberlein compact space is weakly splittable.

436. Prove that every metrizable space is weakly splittable.

437. Let X be a weakly splittable space of cardinality $\leq \mathfrak{c}$. Prove that $C_p(X)$ is k-separable.

438. Prove that if X is a weak Eberlein compact space and $|X| \leq \mathfrak{c}$ then X is Eberlein compact.

439. Let X be a weak Eberlein compact space. Prove that $w(X) = c(X)$. In particular, a weak Eberlein compact space is metrizable whenever it has the Souslin property.

440. Prove that any weak Eberlein compact space X is ω-monolithic, Fréchet–Urysohn and $C_p(X)$ is Lindelöf.

441. Give an example of a Gul'ko compact space which fails to be weakly splittable.

442. Prove that any subspace of a strongly splittable space must be strongly splittable.

443. Prove that, if X condenses onto a strongly splittable space then X is strongly splittable.

444. Prove that, under MA+¬CH, there is a strongly splittable space which is not σ-discrete.

445. Prove that every subset of a strongly splittable space is a G_δ-set.

446. Show that there exists a space X in which every subset is a G_δ-set while X is not splittable.

447. Let X be a normal space in which every subset is G_δ. Prove that X is strongly splittable.

448. Suppose that $Y \subset X$ and $\varphi : C_p(Y) \to C_p(X)$ is a continuous (linear) extender. For $I = \{f \in C_p(X) : f(Y) \subset \{0\}\}$, define a map $\xi : C_p(Y) \times I \to C_p(X)$ by $\xi(f, g) = \varphi(f) + g$ for any $(f, g) \in C_p(Y) \times I$. Prove that ξ is a (linear) embedding and hence $C_p(Y)$ embeds in $C_p(X)$ as a closed (linear) subspace.

449. Given a space X define a map $e : X \to C_p(C_p(X))$ by $e(x)(f) = f(x)$ for any $x \in X$ and $f \in C_p(X)$. If X is a subspace of a space Y prove that X is t-embedded in Y if and only if there exists a continuous map $\varphi :$

$Y \rightarrow C_p(C_p(X))$ such that $\varphi|X = e$. Deduce from this fact that $e(X)$ is t-embedded in $C_p(C_p(X))$ and hence X is homeomorphic to a t-embedded subspace of $C_p(C_p(X))$.

450. Prove that, for any space X every t-embedded subspace of X is closed in X.

451. Prove that $\beta\omega\backslash\omega$ is not t-embedded in $\beta\omega$.

452. Suppose that Y is t-embedded in a space X. Prove that $p(Y) \leq p(X)$ and $d(Y) \leq d(X)$.

453. Suppose that Y is t-embedded in X and a regular cardinal κ is a caliber of X. Prove that κ is a caliber of Y.

454. Prove that any closed subspace of a t-extendial space is t-extendial.

455. Let X be a t-extendial space. Prove that $s(X) = p(X)$.

456. Let X be a t-extendial Baire space. Prove that $s(X) = c(X)$. In particular, if X is a pseudocompact t-extendial space, then $c(X) = s(X)$.

457. Prove that, for any compact t-extendial space X, we have $t(X) \leq c(X)$.

458. Prove that, under MA+¬CH, if $X \times X$ is a t-extendial compact space and $c(X) \leq \omega$ then X is metrizable.

459. Suppose that X is a t-extendial Čech-complete space such that ω_1 is a caliber of X. Prove that X is hereditary separable.

460. Assuming MA+¬CH prove that any t-extendial Čech-complete space with the Souslin property is hereditarily separable.

461. Prove that a t-extendial compact space cannot be mapped onto $\mathbb{I}^{\omega_1}$.

462. Prove that the set $\{x \in X : \pi\chi(x, X) \leq \omega\}$ is dense in any t-extendial compact space X.

463. Give an example of a countable space which is not t-extendial.

464. Prove that any strongly discrete subspace $A \subset X$ is l-embedded in X.

465. Prove that, if Y is a retract of X, then Y is l-embedded in X.

466. Given a space X define a map $e : X \rightarrow C_p(C_p(X))$ by $e(x)(f) = f(x)$ for any $x \in X$ and $f \in C_p(X)$. Observe that $e(X) \subset L_p(X)$; prove that $e(X)$ is l-embedded in $L_p(X)$ and hence any space X is homeomorphic to an l-embedded subspace of $L_p(X)$.

467. Given a space X define a map $e : X \rightarrow C_p(C_p(X))$ by $e(x)(f) = f(x)$ for any $x \in X$ and $f \in C_p(X)$. Observe that $e(X) \subset L_p(X)$ so we can consider that $e : X \rightarrow L_p(X)$. If X is a subspace of a space Y prove that X is l-embedded in Y if and only if there exists a continuous map $\varphi : Y \rightarrow L_p(X)$ such that $\varphi|X = e$.

468. Prove that any closed subspace of an extendial space is extendial.

469. Prove that every metrizable space is extendial.

470. Prove that, for any zero-dimensional linearly ordered compact space X any closed $F \subset X$ is a retract of X; in particular, the space X is extendial.

471. Give an example of a perfectly normal, hereditarily separable non-metrizable extendial compact space.

472. Give an example of an extendial compact space X such that $X \times X$ is not t-extendial.

473. Show that there exist extendial compact spaces of uncountable tightness.

474. Give an example of a non-linearly orderable extendial compact space.

475. Prove that every t-extral (and hence every extral) space X is compact and every uncountable regular cardinal is a caliber of X.

476. Prove that X is t-extral if and only if it can be t-embedded in $\mathbb{I}^{\kappa}$ for some cardinal κ.

477. Prove that any retract of a t-extral space is a t-extral space.

478. Let X be a t-extral space such that $w(X) \leq \mathfrak{c}$. Prove that X is separable.

479. Suppose that an ω-monolithic space X is t-extral and has countable tightness. Prove that X is metrizable.

480. Prove that a t-extral space X is metrizable whenever $C_p(X)$ is Lindelöf.

481. Give an example of a t-extral space which is not extral.

482. Prove that every metrizable compact space is extral.

483. Prove that X is extral if and only if it can be l-embedded in $\mathbb{I}^{\kappa}$ for some cardinal κ.

484. Prove that any retract of an extral space is an extral space.

485. Given an extral space X and an infinite cardinal κ prove that $w(X) > \kappa$ implies that $\mathbb{D}^{\kappa^+}$ embeds in X.

486. Let X be an extral space. Prove that $w(X) = t(X) = \chi(X)$.

487. Assuming that $\mathfrak{c} < 2^{\omega_1}$ prove that any extral space X with $|X| \leq \mathfrak{c}$ is metrizable.

488. Prove that every extral t-extendial space is metrizable.

489. Suppose that every closed subspace of X is extral. Prove that X is metrizable.

490. Prove that any extral linearly orderable space is metrizable.

491. Give an example of an extral space which is not dyadic.

492. Give an example of an extral space X such that some continuous image of X is not extral.

493. Prove that any zero-dimensional extral space is metrizable.

494. Prove that any continuous image of a t-dyadic space is a t-dyadic space.

495. Prove that any continuous image of an l-dyadic space is an l-dyadic space.

496. Given an l-dyadic space X prove that, for any infinite cardinal κ such that $\kappa < w(X)$, the space $\mathbb{D}^{\kappa^+}$ embeds in X.

497. Prove that, if βX is an l-dyadic space then X is pseudocompact.

498. Prove that any hereditarily normal l-dyadic space is metrizable.

499. Prove that any radial l-dyadic space is metrizable.

500. Give an example of an l-dyadic space which is not extral.

Bibliographic Notes to Chapter 1

The material of Chapter 1 consists of problems of the following types:

(1) textbook statements which give a gradual development of some topic;
(2) folkloric statements that might not be published but are known by specialists;
(3) famous theorems cited in textbooks and well-known surveys;
(4) comparatively recent results which have practically no presence in textbooks.

We will almost never cite the original papers for statements of the first three types. We are going to cite them for a very small sample of results of the fourth type. The selection of theorems to cite is made according to the preferences of the author and *does not mean that all statements of the fourth type are mentioned.* I bring my apologies to readers who might think that I did not cite something more important than what is cited. The point is that a selection like that has to be biased because it is impossible to mention all contributors. As a consequence, *there are quite a few statements of the main text, published as results in papers, which are never mentioned in our bibliographic notes.* A number of problems of the main text cite published or unpublished results of the author. However, those are treated exactly like the results of others: some are mentioned and some aren't. On the other hand, the Bibliography contains (to the best knowledge of the author) the papers and books of *all contributors to the material of this book.*

Section 1.1 introduces the technique for representing the whole $C_p(X)$ by means of subspaces which separate the points of a compact space X. A considerable part of the material of this section is covered in Arhangel'skii's book [1992a]. The theorem on equivalence of countability of fan tightness of $C_p(X)$ and Hurewicz property of all finite powers of X (Problem 057) was proved in Arhangel'skii (1986). The result on ω-monolithity of a compact X when $C_p(X)$ is normal (Problem 080) was proved under MA+¬CH by Reznichenko (see Arhangel'skii (1992a)). The fact that PFA implies countable tightness of all compact subspaces of $C_p(X)$, for a Lindelöf X (Problem 089), was established in Arhangel'skii (1992a). A very non-trivial example of a "big" separable σ-compact space X with $(C_p(X))^\omega$ Lindelöf (Problem 094) was constructed by Okunev and Tamano (1996).

Section 1.2 brings in the most important results on Corson compact spaces. Again, the book of Arhangel'skii (1992a) covers a significant part of this section. The invariance of the class of Corson compact spaces under continuous mappings (Problem 151) was proved in Michael and Rudin (1977) and Gul'ko (1977). The Lindelöf property in all iterated C_p's of any Corson compact space (Problem 167) was proved in Sokolov (1986). The example of a Corson compact space which is not Gul'ko compact (Problem 175) was constructed in Leiderman and Sokolov (1984). A very deep result stating that all iterated function spaces of a Corson compact space can be condensed into a Σ-product of real lines (Problem 200) was established in Gul'ko (1981).

Section 1.3 shows that the class of Lindelöf Σ-spaces is of great importance in C_p-theory. A good coverage of the topic is done in the book of Arhangel'skii (1992a) and his survey [1992b]. Okunev's theorem on Lindelöf Σ-property in iterated function spaces (Problems 218 and 219) was a crucial breakthrough. It was published in Okunev (1993a). A complete classification of the distributions of the Lindelöf Σ-property in iterated function spaces (Problem 243) was given in Tkachuk (2000). A very subtle example of a Gul'ko compact space which is not Preiss–Simon (Problem 222) was constructed by Reznichenko]. It is a famous theorem of Gul'ko (1979) that every Gul'ko compact space is Corson compact (Problem 285). Finally, it was proved in Leiderman (1985) that every Gul'ko compact space has a dense metrizable subspace (Problem 293).

Section 1.4 presents the present-day state of knowledge about Eberlein compact spaces. They were originally defined as weakly compact subsets of Banach spaces. Amir and Lindenstrauss (1968) proved that a compact X is an Eberlein compact iff it embeds into $\Sigma_*(A)$ (Problem 322). Rosenthal gave in [1974] a topological criterion in terms of T_0-separating families (Problem 324). The theorem on invariance of the class of Eberlein compact spaces under continuous maps (Problem 337) was proved in Benyamini et al. (1977) and Gul'ko [1977]. The characterization of Eberlein compacta in terms of the topology of X^2 (Problem 364) was established in Gruenhage (1984a). Talagrand constructed in [1979a] an example of a Gul'ko compact space which is not Eberlein compact (Problem 387). A theorem of Namioka (1974) on joint continuity and separate continuity has a great number of applications in functional analysis, topology and topological algebra; an important consequence of this theorem is formulated in Problem 347. The equivalence of the original definition of Eberlein compact spaces to the topological one (Problem 400) was proved by Grothendieck in [1952].

In Section 1.5 we study approximations of arbitrary functions by continuous ones and the classes of spaces with nice properties for extensions of continuous functions. Arhangel'skii and Shakhmatov proved in [1988] that any pseudocompact splittable space is compact and metrizable (Problem 417). Arhangel'skii and Choban introduced and studied in [1988] the classes of extral spaces and extendial spaces which could be considered far-reaching generalizations of absolute retracts and metrizable spaces, respectively.

Chapter 2
Solutions of Problems 001–500

This volume's background includes 1000 solutions of problems of the main text as well as more than 500 statements proved as auxiliary facts; some of these facts are quite famous and highly non-trivial theorems. As in the previous volume, the treatment of topology and C_p-theory is professional. When you read a solution of a problem of the main text, it has more or less the same level of exposition as a published paper on a similar topic.

The author hopes, however, that reading our solutions is more helpful than ploughing through the proofs in published papers; the reason is that we are not so constrained by the amount of the available space as a journal contributor; so we take much more care about all details of the proof. It is also easier to work with the references in our solutions than with those in research papers because in a paper the author does not need to bother about whether the reference is accessible for the reader whereas we only refer to what we have proved in this book apart from some very simple facts of calculus and set theory.

This volume has the same policy about the references as the second one; we use the textbook facts from general topology without giving a reference to them. This book is self-contained; so all the necessary results are proved in the previous volumes but the references to standard things have to stop sometime. This makes it difficult for a beginner to read this volume's results without some knowledge of the previous material. However, a reader who mastered the first four chapters of R. Engelking's book [1977] will have no problem with this.

We also omit references to some standard facts of C_p-theory. The reader can easily find the respective proofs using the index. Our reference omission rule can be expressed as follows: we omit references to textbook results from topology and C_p-theory *proved in the previous volumes*. There are quite a few phrases like "it is easy to see" or "it is an easy exercise"; the reader should trust the author's word and experience that the statements like that are *really* easy to prove as soon as one has the necessary background. On the other hand, the highest percentage of errors comes exactly from omissions of all kinds; so my recommendation is that, even though you

© Springer International Publishing Switzerland 2015
V.V. Tkachuk, *A Cp-Theory Problem Book*, Problem Books
in Mathematics, DOI 10.1007/978-3-319-16092-4_2

should trust the author's claim that the statement is easy to prove or disprove, you shouldn't take just his word for the truthfulness of *any* statement. Verify it yourself and if you find any errors communicate them to me to correct the respective parts.

U.001. *Prove that, if X is a normal space and $\dim X = 0$ then $\dim \beta X = 0$ and hence βX is zero-dimensional.*

Solution. Take any distinct $a, b \in \beta X$. There is a function $\varphi \in C_p(\beta X, [0, 1])$ such that $\varphi(a) = 0$ and $\varphi(b) = 1$. Therefore $U_a = \varphi^{-1}([0, \frac{1}{3})) \in \tau(a, \beta X)$ and $U_b = \varphi^{-1}((\frac{2}{3}, 1]) \in \tau(b, \beta X)$. The space X being dense in βX, the set $V_a = U_a \cap X$ is dense in U_a and $V_b = U_b \cap X$ is dense in U_b. Observe also that $\overline{U}_a \subset F_a = \varphi^{-1}([0, \frac{1}{3}])$ and $\overline{U}_b \subset F_b = \varphi^{-1}([\frac{2}{3}, 1])$ (the bar denotes the closure in βX) which shows that $\overline{U}_a \cap \overline{U}_b \subset F_a \cap F_b = \emptyset$.

As a consequence, $\mathrm{cl}_X(V_a) \cap \mathrm{cl}_X(V_b) \subset \overline{U}_a \cap \overline{U}_b = \emptyset$, i.e., $P_a = \mathrm{cl}_X(V_a)$ and $P_b = \mathrm{cl}_X(V_b)$ are non-empty disjoint closed subsets of X. We have $\mathrm{Ind}(X) = 0$ by SFFS-308 so there is a clopen set O in the space X such that $P_a \subset O \subset X \backslash P_b$. Let $g(x) = 0$ if $x \in O$ and $g(x) = 1$ for every $x \in X \backslash O$. It is immediate that $g : X \to \mathbb{D}$ is continuous; so there is a continuous $h : \beta X \to \mathbb{D}$ such that $h|X = g$. Since $a \in \overline{U}_a = \overline{V}_a$ and $V_a \subset P_a$, we have $a \in \overline{P}_a$; the function h is continuous and $h(P_a) = g(P_a) = \{0\}$ so $h(a) = 0$. Analogously, $b \in \overline{P}_b$ and $h(P_b) = g(P_b) = \{1\}$ so $h(b) = 1$.

Thus the set $A = C_p(\beta X, \mathbb{D})$ separates the points of βX. For any $x \in \beta X$ let $e_x(f) = f(x)$ for any $f \in A$; this gives a continuous map $e : \beta X \to \mathbb{D}^A$ defined by $e(x) = e_x$ for each $x \in X$ (see TFS-166). Furthermore, e is injective because A separates the points of X (see Fact 2 of S.351) so e is an embedding of βX in $\mathbb{D}^A$. Now apply SFFS-303 to conclude that βX is zero-dimensional; since it is compact, it is also strongly zero-dimensional by SFFS-306.

U.002. *Let X be a zero-dimensional compact space. Suppose that Y is second countable and $f : X \to Y$ is a continuous onto map. Prove that there exists a compact metrizable zero-dimensional space Z and continuous onto maps $g : X \to Z$ and $h : Z \to Y$ such that $f = h \circ g$.*

Solution. The space X being zero-dimensional, we can assume that $X \subset \mathbb{D}^A$ for some A (see SFFS-303); let $p_S : \mathbb{D}^A \to \mathbb{D}^S$ be the natural projection for any $S \subset A$. By TFS-298 there is a countable $B \subset A$ and a continuous map $h : p_B(X) \to Y$ such that $h \circ p_B = f$. The space $Z = p_B(X)$ is second countable because $Z \subset \mathbb{D}^B$ and $|B| = \omega$; besides, $Z \subset \mathbb{D}^B$ implies that Z is zero-dimensional (see SFFS-303). Consequently, the space Z and the maps h and $g = p_B|X$ are as promised.

U.003. *Prove that there exists a continuous map $k : \mathbb{K} \to \mathbb{I}$ such that, for any compact zero-dimensional space X and any continuous map $f : X \to \mathbb{I}$, there exists a continuous map $g_f : X \to \mathbb{K}$ such that $f = k \circ g_f$.*

Solution. Given a set T and $S \subset T$ define a function $\chi_S^T : T \to \mathbb{D}$ by $\chi_S^T(t) = 1$ if $t \in S$ and $\chi_S^T(t) = 0$ for any $t \in T \backslash S$. If Z is a space then $\mathcal{C}(Z)$ is the family of all clopen subsets of Z.

Fact 1. If $\dim Z_t = 0$ for any $t \in T$ and $Z = \bigoplus \{Z_t : t \in T\}$ then $\dim Z = 0$.

Proof. We consider that every Z_t is a clopen subspace of Z; it suffices to show that $\mathrm{Ind} Z = 0$ (see SFFS-308). To do this, take any closed $P \subset Z$ and $U \in \tau(P, Z)$. Letting $P_t = P \cap Z_t$ and $U_t = U \cap Z_t$ we have $U_t \in \tau(P_t, Z_t)$ for every $t \in T$.

It follows from SFFS-308 that $\mathrm{Ind}Z_t = 0$; so there is $W_t \in \mathcal{C}(Z_t)$ such that $P_t \subset W_t \subset U_t$ for each $t \in T$. It is immediate that $W = \bigcup\{W_t : t \in T\} \in \mathcal{C}(Z)$ and $P \subset W \subset U$. Thus $\mathrm{Ind}Z = 0$ and hence $\dim Z = 0$ by SFFS-308; so Fact 1 is proved.

Fact 2. Given an infinite cardinal κ a space Z is zero-dimensional and $w(Z) \leq \kappa$ if and only if Z is homeomorphic to a subspace of $\mathbb{D}^\kappa$.

Proof. If $Z \subset \mathbb{D}^\kappa$ then $w(Z) \leq w(\mathbb{D}^\kappa) = \kappa$ and Z is zero-dimensional by SFFS-303; so sufficiency is clear.

Now assume that $w(Z) \leq \kappa$ and Z is zero-dimensional; fix a base $\mathcal{B}'$ of the space Z such that every $B \in \mathcal{B}'$ is a clopen set. There is $\mathcal{B} \subset \mathcal{B}'$ for which $|\mathcal{B}| \leq \kappa$ and $\mathcal{B}$ is still a base in Z (see Fact 1 of T.102).

For every $B \in \mathcal{B}$ the map $\chi_B^Z : Z \to \mathbb{D}$ is continuous because $\{B, Z\backslash B\}$ is an open cover of Z on the elements of which χ_B^Z is constant (see Fact 1 of S.472).

Consequently, $\varphi = \Delta\{\chi_B^Z : B \in \mathcal{B}\} : Z \to \mathbb{D}^\mathcal{B}$ is a continuous map; let $Y = \varphi(Z)$. We claim that $\varphi : Z \to Y$ is a homeomorphism. In the first place observe that for any distinct $x, y \in Z$ there is $B \in \mathcal{B}$ such that $x \in B$ and $y \notin B$. Thus $\chi_B^Z(x) = 1$ while $\chi_B^Z(y) = 0$ which shows that $\varphi(x) \neq \varphi(y)$ and therefore φ is a bijection. For any $B \in \mathcal{B}$ let $p_B : \mathbb{D}^\mathcal{B} \to \mathbb{D}$ be the natural projection of $\mathbb{D}^\mathcal{B}$ onto its B-th factor.

To see that φ^{-1} is continuous take any $y \in Y$ and $U \in \tau(\varphi^{-1}(y), Z)$; let $x = \varphi^{-1}(y)$. Since $\mathcal{B}$ is a base in Z, there is $B \in \mathcal{B}$ such that $x \in B \subset U$. The set $W = \{z \in Y : p_B(z) = 1\} = p_B^{-1}(1) \cap Y$ is an open subset of Y because $p_B^{-1}(1)$ is open in $\mathbb{D}^\mathcal{B}$. Now, if $t \in W$ and $s = \varphi^{-1}(t)$ then, by definition of the diagonal product, $1 = p_B(t) = \chi_B^Z(s)$ and hence $s \in B$ which implies that $s = \varphi^{-1}(t) \in U$ for any $t \in W$. Thus $\varphi^{-1}(W) \subset U$, i.e., W witnesses continuity at the point y. The point y has been chosen arbitrarily, so the map φ^{-1} is continuous and hence φ is an embedding of Z in $\mathbb{D}^\mathcal{B}$. Since $|\mathcal{B}| \leq \kappa$, the space $\mathbb{D}^\mathcal{B}$ embeds in $\mathbb{D}^\kappa$ and hence Fact 2 is proved.

Returning to our solution denote by $\mathcal{F}$ the family of all closed non-empty subsets of $\mathbb{K}$; it is an easy exercise that $|\mathcal{F}| = \mathfrak{c}$. For every $F \in \mathcal{F}$ we have $|C_p(F, \mathbb{I})| = |C_p(F)| = \mathfrak{c}$ because $nw(C_p(F)) = \omega$. An immediate consequence is that $\mathcal{P} = \{(F, u) : F \in \mathcal{F}, u \in C_p(F, \mathbb{I})\}$ also has cardinality $\mathfrak{c}$; take an enumeration $\{(F_\alpha, u_\alpha) : \alpha < \mathfrak{c}\}$ of the family $\mathcal{P}$. Since every zero-dimensional second countable compact space M is homeomorphic to an element of $\mathcal{F}$ by Fact 2, we conclude that

(1) for any second countable zero-dimensional compact M and any $q \in C_p(M, \mathbb{I})$ there is $\alpha < \mathfrak{c}$ and a homeomorphism $j : M \to F_\alpha$ such that $q = u_\alpha \circ j$.

Observe next that the space $\Phi = \bigoplus\{F_\alpha : \alpha < \mathfrak{c}\}$ is metrizable because every F_α is metrizable (see Fact 1 of S.234). Furthermore, F_α is zero-dimensional and hence $\dim F_\alpha = 0$ for any $\alpha < \mathfrak{c}$ (see SFFS-306). Consequently, $\dim \Phi = 0$ by Fact 1; the space Φ is normal being metrizable, so we can apply Problem 001 to conclude that $\dim(\beta\Phi) = 0$ as well.

Let $u = \bigcup\{u_\alpha : \alpha < \mathfrak{c}\}$; then $u : \Phi \to \mathbb{I}$ is a continuous map such that $u|F_\alpha = u_\alpha$ for any $\alpha < \mathfrak{c}$. There is a continuous $v : \beta\Phi \to \mathbb{I}$ for which $v|\Phi = u$. Now apply Problem 002 to find a second countable compact zero-dimensional space C and continuous functions $c : C \to \mathbb{I}$ and $d : \beta\Phi \to C$ for which $c \circ d = v$. By Fact 2, we can consider that C is a closed subspace of $\mathbb{K}$; this implies that there is a continuous retraction $r : \mathbb{K} \to C$ (see SFFS-316). We claim that the map $k = c \circ r$ is as promised.

Indeed, take a zero-dimensional compact X and a continuous $f : X \to \mathbb{I}$. By Problem 002 there is a zero-dimensional second countable compact M and continuous functions $h : M \to \mathbb{I}$ and $g : X \to M$ for which $h \circ g = f$. By (1) there is $\alpha < \mathfrak{c}$ and a homeomorphism $j : M \to F_\alpha$ such that $h = u_\alpha \circ j$. Then $g_f = d \circ j \circ g : X \to C$; since $C \subset \mathbb{K}$, we can consider that $g_f : X \to \mathbb{K}$. Now, if $x \in X$ then $g_f(x) \in C$ and hence $r(g_f(x)) = g_f(x)$; besides, $j(g(x)) \in F_\alpha$ which implies $v(j(g(x))) = u(j(g(x))) = (u_\alpha \circ j)(g(x)) = h(g(x)) = f(x)$ so we finally obtain

$$k(g_f(x)) = c(r(g_f(x))) = c(g_f(x)) = c(d(j(g(x)))) = v(j(g(x))) = h(g(x)) = f(x).$$

Since $x \in X$ was chosen arbitrarily, this shows that $k \circ g_f = f$ and makes our solution complete.

U.004. *Prove that, for any zero-dimensional compact X, the space $C_p(X, \mathbb{I})$ is a continuous image of $C_p(X, \mathbb{D}^\omega)$.*

Solution. Take a map $k : \mathbb{D}^\omega \to \mathbb{I}$ such that for any function $f \in C_p(X, \mathbb{I})$ there is a continuous map $g_f : X \to \mathbb{D}^\omega$ for which $k \circ g_f = f$ (see Problem 003). For any function $h \in C_p(X, \mathbb{D}^\omega)$ let $\varphi(h) = k \circ h$. Then $\varphi : C_p(X, \mathbb{D}^\omega) \to C_p(X, \mathbb{I})$ is a continuous map by TFS-091 and $\varphi(g_f) = f$ for every $f \in C_p(X, \mathbb{I})$. Therefore $\varphi(C_p(X, \mathbb{D}^\omega)) = C_p(X, \mathbb{I})$ and hence $C_p(X, \mathbb{I})$ is a continuous image of $C_p(X, \mathbb{D}^\omega)$.

U.005. *Given a countably infinite space X prove that the following conditions are equivalent:*

(i) $C_p(X, \mathbb{D})$ is countable;
(ii) $C_p(X, \mathbb{D}) \simeq \mathbb{Q}$;
(iii) X is compact.

Solution. Observe first that X has to be zero-dimensional by SFFS-305; so the family $\mathcal{C}(X)$ of all clopen subsets of X forms a base in X.

The implication (ii)$\Longrightarrow$(i) is evident. If X is a countable compact space then $w(X) = \omega$ by Fact 4 of S.307. There exists a countable $\mathcal{B} \subset \mathcal{C}(X)$ which is a base in X (see Fact 1 of T.102).

Let $\mathcal{B}_0$ be the family of all finite unions of the elements of $\mathcal{B}$; then $\mathcal{B}_0 \subset \mathcal{C}(X)$ and $\mathcal{B}_0$ is countable. Given any $U \in \mathcal{C}(X)$ we can choose $O_x \in \mathcal{B}$ such that $x \in O_x \subset U$ for any $x \in U$. The family $\{O_x : x \in U\}$ is an open cover of a compact space U; so there is a finite $A \subset U$ for which $U = U_0 = \bigcup\{O_x : x \in A\}$. It is evident that $U_0 \in \mathcal{B}_0$; since $U = U_0$, we proved that $\mathcal{C}(X) = \mathcal{B}_0$ is countable. Since $C_p(X, \mathbb{D}) = \{\chi_U^X : U \in \mathcal{C}(X)\}$, the set $C_p(X, \mathbb{D})$ is also countable and we proved that (iii)$\Longrightarrow$(i).

Now if $C_p(X, \mathbb{D})$ is countable then $w(C_p(X, \mathbb{D})) \le w(C_p(X)) = \omega$. Furthermore, the space $C_p(X, \mathbb{D})$ is dense in $\mathbb{D}^X$ by Fact 1 of S.390; so it has no isolated points and hence $C_p(X, \mathbb{D}) \simeq \mathbb{Q}$ by SFFS-349. This settles (i)$\Longrightarrow$(ii) and hence (i) $\Longleftrightarrow$ (ii).

To see that (i)$\Longrightarrow$(iii) assume that X is not compact and $|C_p(X, \mathbb{D})| \le \omega$; it follows from $|X| = \omega$ that X is not pseudocompact and hence there is a faithfully indexed discrete family $\{U_n : n \in \omega\} \subset \tau^*(X)$. The family $\mathcal{C}(X)$ is a base in X so we can choose a non-empty $V_n \in \mathcal{C}(X)$ such that $V_n \subset U_n$ for all $n \in \omega$. The family $\{V_n : n \in \omega\}$ is also discrete; so the set $W_A = \bigcup\{V_n : n \in A\}$ is clopen in X for any $A \subset \omega$. The family $\exp \omega$ is uncountable, so we have $|\mathcal{C}(X)| > \omega$; since $f \to f^{-1}(0)$ is a surjective map from $C_p(X, \mathbb{D})$ onto $\mathcal{C}(X)$, we have $|C_p(X, \mathbb{D})| \ge |\mathcal{C}(X)| > \omega$; this contradiction shows that (i)$\Longrightarrow$(iii) and makes our solution complete.

U.006. *For an arbitrary space X prove that*

(i) *for any $P \subset C_p(X)$ there is an algebra $A(P) \subset C_p(X)$ such that $P \subset A(P)$ and $A(P)$ is minimal in the sense that, for any algebra $A \subset C_p(X)$, if $P \subset A$ then $A(P) \subset A$;*

(ii) *$A(P)$ is a countable union of continuous images of spaces which belong to $\mathcal{H}(P) = \{P^m \times K$ for some $m \in \mathbb{N}$ and metrizable compact $K\}$.*

(iii) *if Q is a weakly k-directed property and $P \vdash Q$ then $A(P) \vdash Q_\sigma$, i.e., $A(P)$ is a countable union of spaces with the property Q;*

Solution. Consider the sets $U(P) = \{f_1 \cdot \ldots \cdot f_n : n \in \mathbb{N}, f_i \in P$ for all $i \le n\}$ and $W(P) = \{\lambda_0 + \lambda_1 \cdot g_1 + \ldots + \lambda_m \cdot g_m : m \in \mathbb{N}, \lambda_i \in \mathbb{R}$ and $g_i \in U(P)$ for all $i \le m\}$. It is straightforward that $A(P) = W(P)$ is an algebra in $C_p(X)$ and $P \subset A(P)$. To see that $A(P)$ is minimal, take any algebra $A \supset P$. Then $U(P) \subset A$ because A has to contain all finite products of its elements. Since A also contains all finite sums of its elements, we have $A(P) = W(P) \subset A$ so $A(P)$ is minimal and (i) is proved.

The property (ii) was established in Fact 1 of S.312; it was shown in Fact 2 of S.312 (in other terms) that if Q is a weakly k-directed property and P has Q then $W(P)$ is a countable union of spaces with the property Q. Since $A(P) = W(P)$, we have $A(P) \vdash Q_\sigma$; so (iii) is also proved.

U.007. *Given a compact space X suppose that $A \subset C_p(X)$ is an algebra. Prove that both $\overline{A}$ and $cl_u(A)$ are algebras in $C_p(X)$.*

Solution. The set $B = cl_u(A)$ contains all constant functions on X because A is an algebra and $A \subset B$. Now if $f, g \in B$, fix sequences $\{f_n\}, \{g_n\} \subset A$ such that $f_n \rightrightarrows f$ and $g_n \rightrightarrows g$. Then $\{f_n + g_n\} \subset A$ because the set A is an algebra and $f_n + g_n \rightrightarrows f + g$ by TFS-035. This shows that $f + g \in B$. The sequence $\{f_n \cdot g_n\}$ also lies in A because A is an algebra. We will prove that $f_n \cdot g_n \rightrightarrows f \cdot g$.

Let us show first that there exists $K \in \mathbb{R}$ such that $|f(x)| \leq K$, $|f_n(x)| \leq K$ and $|g_n(x)| \leq K$ for all $n \in \omega$ and $x \in X$. Since f is continuous and X is compact, the functions f and g are bounded on X, i.e., there exists $M \in \mathbb{R}$ such that $|f(x)| \leq M$ and $|g(x)| \leq M$ for all $x \in X$. Applying the relevant uniform convergences we can find $m \in \omega$ such that $|f_n(x) - f(x)| < 1$ and $|g_n(x) - g(x)| < 1$ for all $n \geq m$ and $x \in X$. The functions $f_1, \ldots, f_m$ and $g_1, \ldots, g_m$ are bounded on X which implies that there is $N \in \mathbb{R}$ such that $|f_i(x)| + |g_i(x)| \leq N$ for all $i \leq m$ and $x \in X$. It is easy to verify that the number $K = M + N + 1$ is as promised.

Given an arbitrary $\varepsilon > 0$, we can find $l \in \omega$ such that $|f_n(x) - f(x)| < \frac{\varepsilon}{2K}$ and $|g_n(x) - g(x)| < \frac{\varepsilon}{2K}$ for all $n \geq l$ and $x \in X$. Then

$$|f_n(x)g_n(x) - f(x)g(x)| = |g_n(x)(f_n(x) - f(x)) + f(x)(g_n(x) - g(x))|$$
$$\leq |g_n(x)||f_n(x) - f(x)| + |f(x)||g_n(x) - g(x)| < K \cdot \frac{\varepsilon}{2K} + K \cdot \frac{\varepsilon}{2K} = \varepsilon$$

for all $n \geq l$ and $x \in X$ which proves that $f_n \cdot g_n \rightrightarrows f \cdot g$ and hence $f \cdot g \in B$. Thus $B = \mathrm{cl}_u(A)$ is an algebra.

To prove that the set $C = \overline{A}$ is an algebra, observe that the addition map $s : C_p(X) \times C_p(X) \to C_p(X)$ defined by $s(f, g) = f + g$ for all $f, g \in C_p(X)$, is continuous by TFS-115. We also have $s(A \times A) \subset A$ because A is an algebra. Furthermore, $\overline{A \times A} = C \times C$ and hence $s(C \times C) \subset \overline{s(A \times A)} \subset \overline{A} = C$ which shows that $s(C \times C) \subset C$ or, in other words, $f + g \in C$ for any $f, g \in C$.

Analogously, the multiplication map $m : C_p(X) \times C_p(X) \to C_p(X)$ defined by $s(f, g) = f \cdot g$ for all $f, g \in C_p(X)$, is continuous by TFS-116. We also have $m(A \times A) \subset A$ because A is an algebra. Furthermore, $\overline{A \times A} = C \times C$ and hence $m(C \times C) \subset \overline{m(A \times A)} \subset \overline{A} = C$ which shows that $m(C \times C) \subset C$ or, in other words, $f \cdot g \in C$ for any $f, g \in C$. Finally, observe that C contains all constant function on X because so does $A \subset C$. Therefore $C = \overline{A}$ is also an algebra in $C_p(X)$.

U.008. *Let X be a compact space. Suppose that $A \subset C_p(X)$ separates the points of X, contains the constant functions and has the following property: for each $f, g \in A$ and $a, b \in \mathbb{R}$ we have $af + bg \in A$, $\max(f, g) \in A$, $\min(f, g) \in A$. Prove that every $f \in C_p(X)$ is a uniform limit of some sequence from A.*

Solution. Given any functions $f_0, \ldots, f_n \in C_p(X)$, let

$$\max(f_0, \ldots, f_n)(x) = \max\{f_0(x), \ldots, f_n(x)\}$$

for every $x \in X$. This defines a function $\max(f_0, \ldots, f_n) \in C_p(X)$. Analogously, the expression $\min(f_0, \ldots, f_n)(x) = \min\{f_0(x), \ldots, f_n(x)\}$ for any point $x \in X$, defines the function $\min(f_0, \ldots, f_n) \in C_p(X)$. It is clear that if $f_0, \ldots, f_n \in A$ then $\max(f_0, \ldots, f_n) \in A$ and $\min(f_0, \ldots, f_n) \in A$. We will show first that

(1) for any function $f \in C_p(X)$ and $\varepsilon > 0$, there exists a function $f_\varepsilon \in A$ such that $|f_\varepsilon(x) - f(x)| < \varepsilon$ for any $x \in X$.

To prove (1) observe that, for every pair of distinct points $a, b \in X$ we can find a function $h \in A$ such that $h(a) \neq h(b)$. Since all linear combinations of elements of

A still belong to A, the function g defined by $g(x) = (h(x) - h(a))(h(b) - h(a))^{-1}$ for every $x \in X$, belongs to A. It is immediate that $g(a) = 0$ and $g(b) = 1$. Now let $f_{a,b}(x) = (f(b) - f(a))g(x) + f(a)$ for each $x \in X$. Of course, $f_{a,b} \in A$ and $f_{a,b}(a) = f(a)$ and $f_{a,b}(b) = f(b)$.

The sets

$$U_{a,b} = \{x \in X : f_{a,b}(x) < f(x) + \varepsilon\} \quad \text{and} \quad V_{a,b} = \{x \in X : f_{a,b}(x) > f(x) - \varepsilon\}$$

are open neighbourhoods of the points a and b respectively. Fix any $b \in X$; by compactness of X we can extract a finite subcover $\{U_{a_i,b} : i \in \{0, \ldots, n\}\}$ from the open cover $\{U_{a,b} : a \in X\}$. By our hypothesis about the set A, the function $f_b = \min(f_{a_0,b}, \ldots, f_{a_n,b})$ belongs to A. It is easy to see that $f_b(x) < f(x) + \varepsilon$ for all $x \in X$ and $f_b(x) > f(x) - \varepsilon$ for any $x \in V_b = \bigcap\{V_{a_i,b} : i \leq n\}$. Since X is compact, we can choose a finite subcover $\{V_{b_i} : 0 \leq i \leq k\}$ of the open cover $\{V_b : b \in X\}$ of the space X. The function $f_\varepsilon = \max(f_{b_0}, \ldots, f_{b_k})$ belongs to A and we have $|f_\varepsilon(x) - f(x)| < \varepsilon$ for all $x \in X$ so (1) is proved.

Finally, take an arbitrary $f \in C_p(X)$ and find, for any $n \in \mathbb{N}$, a function $f_n \in A$ such that $|f_n(x) - f(x)| < \frac{1}{n}$ for all $x \in X$. The existence of such f_n is guaranteed by (1). It is obvious that $f_n \rightrightarrows f$; so our solution is complete.

U.009. *Let X be a compact space and suppose that $Y \subset C_p(X)$ separates the points of X. Prove that*

(i) for any algebra $A \subset C_p(X)$ with $Y \subset A$, we have $cl_u(A) = C_p(X)$;
(ii) if Y contains a non-zero constant function then $cl_u(\bigcup S(Y)) = C_p(X)$.

Solution. For (i), observe that $B = cl_u(A)$ is an algebra in $C_p(X)$ by Problem 007; since $Y \subset A \subset B$, the algebra B also separates the points of X and hence $B = C_p(X)$ by TFS-191.

As to (ii), let $S = \bigcup S(Y)$; we have $af + bg \in S$ for any $a, b \in \mathbb{R}$ and $f, g \in S$. If $w \in Y$ is a non-zero constant function then $w(x) = c \neq 0$ for all $x \in X$. Therefore, for any $a \in \mathbb{R}$, the function $w_a = \frac{a}{c}w$ belongs to S and $w_a(x) = a$ for any $x \in X$. Thus S contains all constant functions on X. It is immediate from the definition of S that $\max(f, g) \in S$ and $\min(f, g) \in S$ for any $f, g \in S$. Furthermore, S separates the points of X because so does $Y \subset S$. Therefore we can apply Problem 008 to conclude that $cl_u(S) = C_p(X)$ finishing the proof of (ii).

U.010. *For a space X, suppose that $Y \subset C_p(X)$ and $cl_u(Y) = C_p(X)$. Prove that $C_p(X) \in (\mathcal{E}(Y))_\delta$.*

Solution. It was proved in Fact 1 of T.459 that, under our assumptions, we have $C_p(X) = \bigcap\{C_n : n \in \mathbb{N}\}$ where $C_n \subset \mathbb{R}^X$ is a continuous image of $Y \times [-\frac{1}{n}, \frac{1}{n}]^X$ for every $n \in \mathbb{N}$. Therefore $C_n \in \mathcal{E}(Y)$ for all $n \in \mathbb{N}$ and hence $C_p(X) \in (\mathcal{E}(Y))_\delta$.

U.011. *Prove that every k-directed non-empty class is weakly k-directed. Give an example of a weakly k-directed class which is not k-directed.*

Solution. Let Q be a non-empty k-directed class. If $Q \in \mathcal{Q}$ then, for every compact K, the relevant projection maps $Q \times K$ continuously onto K. Therefore all (metrizable) compact spaces belong to $\mathcal{Q}$. Since every k-directed class is preserved by finite products and continuous images, this proves that every k-directed class is weakly k-directed.

Now, if $\mathcal{Q}$ is the class of all metrizable compact spaces then it is evident that $\mathcal{Q}$ is weakly k-directed. However, the class $\mathcal{Q}$ is not k-directed because $\mathbb{I} \in \mathcal{Q}$ and $\mathbb{I}^{\omega_1} \simeq \mathbb{I} \times \mathbb{I}^{\omega_1} \in \mathcal{E}(\mathbb{I})$ while $\mathbb{I}^{\omega_1} \notin \mathcal{Q}$.

U.012. *Prove that if $\mathcal{K} \in \{$compact spaces, σ-compact spaces, k-separable spaces$\}$ then the class $\mathcal{K}$ is k-directed. How about the class of countably compact spaces?*

Solution. Observe first that the class $\mathcal{K}$ is finitely productive. This is evident for compact and σ-compact spaces; if we have k-separable spaces X and Y and X', Y' are their respective dense σ-compact subspaces then $X' \times Y'$ is a dense σ-compact subspace of $X \times Y$. Furthermore, $\mathcal{K}$ contains the class of compact spaces and every continuous image of an element from $\mathcal{K}$ also belongs to $\mathcal{K}$. Again, this clear for compact and σ-compact spaces; if X is k-separable and $f : X \to Y$ is a continuous onto map then take a dense σ-compact $X' \subset X$ and note that the set $f(X')$ is a dense σ-compact subspace of Y, i.e., Y is k-separable. This implies that any $\mathcal{K} \in \{$compact spaces, σ-compact spaces, k-separable spaces$\}$ is k-directed.

Finally observe that the class of countably compact spaces is not k-directed because it is not finitely productive (see Fact 2 of S.483).

U.013. *Let $\mathcal{P}$ be a weakly k-directed class. Prove that, for any $Y \subset C_p(X)$ such that $Y \in \mathcal{P}$, we have $S(Y) \subset \mathcal{P}$.*

Solution. Given a function $f \in C_p(X)$ let $|f|(x) = |f(x)|$ for any $x \in X$. It follows from TFS-091 that the map $\varphi_0 : C_p(X) \to C_p(X)$ defined by $\varphi_0(f) = |f|$ for any $f \in C_p(X)$ is continuous. Besides, it was proved in TFS-115 that the addition map $s : C_p(X) \times C_p(X) \to C_p(X)$ defined by $s(f, g) = f + g$ for all $f, g \in C_p(X)$ is also continuous. The multiplication $m : C_p(X) \times C_p(X) \to C_p(X)$ defined by $m(f, g) = f \cdot g$ for all $f, g \in C_p(X)$ is continuous as well: this was proved in TFS-116.

An immediate consequence is that the map $\varphi_1 : C_p(X) \times C_p(X) \to C_p(X)$ defined by $\varphi_1(f, g) = \max(f, g)$ is continuous because $\max(f, g) = \frac{1}{2}(f + g + |f - g|)$ for all $f, g \in C_p(X)$. Analogously, if $\varphi_2(f, g) = \min(f, g)$ for all $f, g \in C_p(X)$ then the map $\varphi_2 : C_p(X) \times C_p(X) \to C_p(X)$ is continuous because we have the equality $\varphi_2(f, g) = \frac{1}{2}(f + g - |f - g|)$ for any $f, g \in C_p(X)$.

Now, observe that $S_1(Y) = \{Y\} \subset \mathcal{P}$ and assume that we proved for some $n \in \mathbb{N}$ that $S_n(Y) \subset \mathcal{P}$. If $A, B \in S_n(Y)$ then both MIN(A, B) and MAX(A, B) are continuous images of $A \times B$ by our above remarks. Analogously, $G_k(A)$ is a continuous image of $A \times A \times [-k, k] \times [-k, k]$ for any $A \in S_n(Y)$ and $k \in \mathbb{N}$. Since the class $\mathcal{P}$ is weakly k-directed, this shows that MIN$(A, B) \in \mathcal{P}$, MAX$(A, B) \in \mathcal{P}$ and $G_k(A) \in \mathcal{P}$ for any $A, B \in S_n(Y)$ and $k \in \mathbb{N}$. Therefore $S_{n+1}(Y) \subset \mathcal{P}$ and hence our inductive procedure proves that $S(Y) = \bigcup\{S_n(Y) : n \in \mathbb{N}\} \subset \mathcal{P}$.

U.014. *Given a k-directed class Q and a compact space X assume that $Y \subset C_p(X)$ separates the points of X and $Y \in Q$. Prove that $C_p(X) \in Q_{\sigma\delta}$, i.e., there is a space Z such that $C_p(X) \subset Z$ and $C_p(X) = \bigcap \{C_n : n \in \omega\}$ where every $C_n \subset Z$ is a countable union of spaces with the property Q.*

Solution. It is not evident from the definition that weakly k-directed properties and classes are finitely additive; so let us formulate it for further references.

Fact 1. If $\mathcal{R}$ is a weakly k-directed property (or class) and $Z = Z_1 \cup \ldots \cup Z_n$ where $Z_i \vdash \mathcal{R}$ (or $Z_i \in \mathcal{R}$ respectively) for all $i \leq n$ then Z also has $\mathcal{R}$ (or $Z \in \mathcal{R}$ respectively). In particular, if $Z \backslash \{z\}$ has $\mathcal{R}$ for some $z \in Z$ then $Z \vdash \mathcal{R}$.

Proof. It suffices to show it for $n = 2$, so assume that $Z = Z_1 \cup Z_2$ and $Z_i \vdash \mathcal{R}$ for $i = 1, 2$. Since $\mathcal{R}$ is a weakly k-directed property, the space $T = Z_1 \times Z_2$ has $\mathcal{R}$ as well as $T' = T \times \mathbb{D}$. For any $i \in \{1, 2\}$ the relevant projection maps T continuously onto Z_i so $T \times \{0\}$ maps continuously onto Z_1 and $T \times \{1\}$ maps continuously onto Z_2. Since $T' \simeq (T \times \{0\}) \oplus (T \times \{1\})$, the space T' can be mapped continuously onto $Z_1 \cup Z_2 = Z$. The property $\mathcal{R}$ is preserved by continuous images; so the space Z has $\mathcal{R}$. Finally, if $Z \backslash \{z\} \vdash \mathcal{R}$ then $Z = (Z \backslash \{z\}) \cup \{z\}$ and both summands in this union have $\mathcal{R}$ (the first one by hypothesis and the second one because all metrizable compact spaces have $\mathcal{R}$) so $Z \vdash \mathcal{R}$ and Fact 1 is proved.

Returning to our solution observe that $Y \cup \{f\}$ has Q for any $f \in C_p(X)$ (see Fact 1). This shows that we can assume, without loss of generality, that $u \in Y$ where $u(x) = 1$ for all $x \in X$. It follows from Problems 013 and 011 that $S(Y) \subset Q$; besides, $\bigcup S(Y)$ is uniformly dense in $C_p(X)$ by Problem 009. The family $S(Y)$ being countable, the set $Z = \bigcup S(Y)$ has Q_σ. Now apply Problem 010 to conclude that $C_p(X) \in (\mathcal{E}(Z))_\delta$. It is easy to see that $\mathcal{E}(Z)$ also has Q_σ; so $C_p(X)$ has $Q_{\sigma\delta}$.

U.015. *For a compact space X suppose that $Y \subset C_p(X)$ separates the points of X. Prove that there exists a compact space K and a closed subspace $F \subset o_\omega(Y) \times K$ such that $C_p(X)$ is a continuous image of F.*

Solution. Observe that $Z = o_\omega(Y) = (Y \times \omega)^\omega \simeq Y^\omega \times \omega^\omega$. It is easy to see that $\omega^\omega = \omega \times \omega^{\omega \backslash \{0\}} \simeq \omega \times \omega^\omega$; as a consequence,

(1) $Z \times \omega \simeq Y^\omega \times \omega^\omega \times \omega \simeq Y^\omega \times \omega^\omega \simeq Z$.
Furthermore,

(2) $Z^\omega = ((Y \times \omega)^\omega)^\omega \simeq (Y \times \omega)^{\omega \times \omega} \simeq (Y \times \omega)^\omega = Z$.
Say that a space T has Q if T is a continuous image of a closed subspace of $Z \times P$ for some compact P. Suppose that we are given a family of spaces $\{E_i : i \in \omega\}$ where every E_i has Q. There is a family $\{P_i : i \in \omega\}$ of compact spaces and a collection $\{F_i : i \in \omega\}$ such that F_i is closed in $Z \times P_i$ and E_i is a continuous image of F_i for every $i \in \omega$. If $P = \prod \{P_i : i \in \omega\}$ then the set $F = \prod \{F_i : i \in \omega\}$ is closed in $\prod \{Z \times P_i : i \in \omega\} \simeq Z^\omega \times P \simeq Z \times P$ by (2); it immediate that $E = \prod_{i \in \omega} E_i$ is a continuous image of F so $E \in Q$, which proves that

(3) Q is countably productive.

It is evident that

(4) Q is preserved by continuous images and closed subspaces; so Q is k-directed. Now, assume that $T = \bigcup\{T_i : i \in \omega\}$ and $T_i \vdash Q$ for every $i \in \omega$; fix a compact P_i and a closed $F_i \subset Z \times P_i$ which maps continuously onto T_i. Let $P = \prod\{P_i : i \in \omega\}$. It is evident that $Z \times P_i$ embeds in $Z \times P$ as a closed subset; therefore every F_i can be considered to be a closed subset of $Z \times P$. Thus $F = \bigoplus\{F_i : i \in \omega\}$ is homeomorphic to a closed subset of $(Z \times P) \times \omega \simeq (Z \times \omega) \times P \simeq Z \times P$ (here we used (1)). It is clear that F maps continuously onto T so T has Q and hence

(5) Q is countably additive.

Finally, observe that Y has Q and apply Problem 014 to see that there exists a space C and a family $\{C_n : n \in \omega\}$ of subspaces of C such that $C_p(X) = \bigcap\{C_n : n \in \omega\}$ and $C_n \vdash Q_\sigma$ for every $n \in \omega$. The property (5) implies that $C_n \vdash Q$ for every $n \in \omega$; it follows from (3) and (4) that any countable intersection of spaces with the property Q also has Q (see Fact 7 of S.271); so $C_p(X) \vdash Q$, i.e., there is a compact space K such that $C_p(X)$ is a continuous image of a closed subset of $Z \times K = o_\omega(Y) \times K$.

U.016. *Prove that, for any compact space X, there exists a compact space K and a closed subspace $F \subset (C_p(X))^\omega \times K$ such that $C_p(X^\omega)$ is a continuous image of F.*

Solution. Denote by X_i a homeomorphic copy of the space X for any $i \in \omega$; then $X^\omega = \prod_{i \in \omega} X_i$. Let $p_i : X^\omega \to X_i$ be the natural projection for every $i \in \omega$. Then the dual map $p_i^* : C_p(X_i) \to C_p(X^\omega)$ embeds $C_p(X_i)$ is $C_p(X^\omega)$ (see TFS-163). Since the space $Z_i = p_i^*(C_p(X_i))$ is homeomorphic to $C_p(X)$ for all $i \in \omega$, the space $C_p(X) \times \omega$ maps continuously onto $Z = \bigcup_{n \in \omega} Z_n$.

To prove that Z separates the points of X^ω take any $x, y \in X^\omega$ such that $x \neq y$. There is $n \in \omega$ for which $p_n(x) \neq p_n(y)$. Now, take $f \in C_p(X_i)$ such that $f(p_n(x)) \neq f(p_n(y))$; then $h = p_n^*(f) \in Z$ and $h(x) = f(p_n(x)) \neq f(p_n(y)) = h(y)$.

The space $C_p(X)$ is never countably compact; so there is a countably infinite closed discrete $D \subset C_p(X)$. Since $D \simeq \omega$, the countable discrete space ω embeds in $C_p(X)$ as a closed subspace. The space $C_p(X) \times \omega$ maps continuously onto Z which implies that the space $(C_p(X) \times \omega) \times \omega \simeq C_p(X) \times \omega$ maps continuously onto $Z \times \omega$ so $o_\omega(C_p(X))$ maps continuously onto $o_\omega(Z)$. By Problem 015 there is a compact space K and a closed $G \subset o_\omega(Z) \times K$ which maps continuously onto $C_p(X^\omega)$. Therefore there is a closed $F \subset o_\omega(C_p(X)) \times K$ which maps continuously onto G and hence onto $C_p(X^\omega)$.

Since ω embeds in $C_p(X)$ as a closed subspace, the space ω^ω embeds in $(C_p(X))^\omega$ as a closed subspace; so $o_\omega(C_p(X)) = (C_p(X))^\omega \times \omega^\omega$ embeds as a closed subspace in the space $(C_p(X))^\omega \times (C_p(X))^\omega \simeq (C_p(X))^\omega$. Consequently, F can be considered to be a closed subspace of $(C_p(X))^\omega \times K$ which maps continuously onto $C_p(X^\omega)$.

U.017. *Let X be a compact space such that $(C_p(X))^\omega$ is Lindelöf. Show that $C_p(X^\omega)$ is Lindelöf. As a consequence, $C_p(X^n)$ is Lindelöf for each $n \in \mathbb{N}$.*

Solution. It was proved in Problem 016 that there is a compact space K and a closed $F \subset (C_p(X))^\omega \times K$ which maps continuously onto $C_p(X^\omega)$. Since $(C_p(X))^\omega$ is Lindelöf, the space $(C_p(X))^\omega \times K$ is also Lindelöf so F is Lindelöf as well. Hence $C_p(X^\omega)$ is Lindelöf being a continuous image of a Lindelöf space F.

Now, if $n \in \mathbb{N}$ then there is a continuous onto map $\varphi : X^\omega \to X^n$; the dual map $\varphi^* : C_p(X^n) \to C_p(X^\omega)$ embeds $C_p(X^n)$ in $C_p(X^\omega)$ as a closed subspace because φ is a closed map (see TFS-163). Therefore $C_p(X^n)$ is Lindelöf being homeomorphic to a closed subspace of the Lindelöf space $C_p(X^\omega)$.

U.018. *Assume that a space X is compact and $\mathcal{P}$ is an ω-perfect class. Prove that, if $C_p(X) \in \mathcal{P}$ then $C_p(X^\omega) \in \mathcal{P}$.*

Solution. We have $o_\omega(C_p(X)) \in \mathcal{P}$ because $\mathcal{P}$ is ω-perfect; since $(C_p(X))^\omega$ is a continuous image of $o_\omega(C_p(X))$, we also have $(C_p(X))^\omega \in \mathcal{P}$. By Problem 016, there is a compact space K and a closed $F \subset (C_p(X))^\omega \times K$ which maps continuously onto $C_p(X^\omega)$; since products with compact spaces, continuous images and closed subspaces of spaces from $\mathcal{P}$ are also in $\mathcal{P}$, we have $F \in \mathcal{P}$ and hence $C_p(X^\omega) \in \mathcal{P}$.

U.019. *Let $\mathcal{P}$ be an ω-perfect class of spaces. Prove that the following properties are equivalent for any compact X:*

(i)   *the space $C_p(X)$ belongs to $\mathcal{P}$;*
(ii) *there exists $Y \subset C_p(X)$ such that Y is dense in $C_p(X)$ and $Y \in \mathcal{P}$;*
(iii) *there exists $Y \subset C_p(X)$ which separates the points of X and belongs to $\mathcal{P}$;*
(iv) *the space X embeds into $C_p(Z)$ for some $Z \in \mathcal{P}$.*

Solution. The implication (i)$\Longrightarrow$(ii) is trivial; we have (ii)$\Longrightarrow$(iii) because every dense subset of $C_p(X)$ separates the points of X. Now assume that some $Y \subset C_p(X)$ separates the points of X and $Y \in \mathcal{P}$. By Problem 015, there is a compact space K and a closed $F \subset o_\omega(Y) \times K$ such that $C_p(X)$ is a continuous image of F. The class $\mathcal{P}$ being ω-perfect, we have $o_\omega(Y) \times K \in \mathcal{P}$ and hence $F \in \mathcal{P}$. Every ω-perfect class is invariant under continuous images, so $C_p(X) \in \mathcal{P}$; this settles (iii)$\Longrightarrow$(i) and shows that we have (i) $\Longleftrightarrow$ (ii) $\Longleftrightarrow$ (iii).

Since the space X embeds in $C_p(C_p(X))$ (see TFS-167), if $C_p(X) \in \mathcal{P}$ then, letting $Z = C_p(X)$, we obtain (i)$\Longrightarrow$(iv). Finally, assume that $X \subset C_p(Z)$ for some $Z \in \mathcal{P}$ and let $e_z(f) = f(z)$ for any $z \in Z$ and $f \in X$. Then $e_z \in C_p(X)$ for any $z \in Z$ and, letting $e(z) = e_z$ for every $z \in Z$ we obtain a continuous map $e : Z \to C_p(X)$ (see TFS-166). The property $\mathcal{P}$ being ω-perfect, we have $Y = e(Z) \in \mathcal{P}$. Given distinct $f, g \in X$ there is $z \in Z$ such that $f(z) \neq g(z)$; this, evidently, implies $e_z(f) \neq e_z(g)$ and hence Y separates the points of X. We proved, therefore, that (iv)$\Longrightarrow$(iii) and hence all the properties (i)–(iv) are equivalent.

U.020. *Prove that the class $L(\Sigma)$ of Lindelöf Σ-spaces is ω-perfect. As a consequence, for any compact X, the following properties are equivalent:*

(i) *the space $C_p(X)$ is Lindelöf Σ;*
(ii) *there exists $Y \subset C_p(X)$ such that Y is dense in $C_p(X)$ and $Y \in L(\Sigma)$;*

(iii) there exists $Y \subset C_p(X)$ which separates the points of X and belongs to $L(\Sigma)$;
(iv) the space X embeds into $C_p(Y)$ for some Lindelöf Σ-space Y.

Solution. Any compact space is Lindelöf Σ; besides, the class $L(\Sigma)$ is invariant under countable unions, continuous images, closed subspaces and countable products (see SFFS-256, SFFS-257, SFFS-243 and SFFS-224). This proves that $L(\Sigma)$ is an ω-perfect class; so we can apply Problem 019 to conclude that (i) $\Longleftrightarrow$ (ii) $\Longleftrightarrow$ (iii) $\Longleftrightarrow$ (iv).

U.021. *Let X be a compact space such that $C_p(X)$ is Lindelöf Σ. Show that $C_p(X^\omega)$ is a Lindelöf Σ-space and so is $C_p(X^n)$ for each $n \in \mathbb{N}$.*

Solution. The class $L(\Sigma)$ of Lindelöf Σ-spaces is ω-perfect by Problem 020; so we can apply Problem 018 to conclude that if $C_p(X) \in L(\Sigma)$ then $C_p(X^\omega) \in L(\Sigma)$.

Now, if $n \in \mathbb{N}$ then there is a continuous onto map $\varphi : X^\omega \to X^n$; the dual map $\varphi^* : C_p(X^n) \to C_p(X^\omega)$ embeds $C_p(X^n)$ in $C_p(X^\omega)$ as a closed subspace because φ is a closed map (see TFS-163). Therefore $C_p(X^n)$ is Lindelöf Σ being homeomorphic to a closed subspace of the Lindelöf Σ-space $C_p(X^\omega)$.

U.022. *Prove that the class $K(\mathcal{A})$ of K-analytic spaces is ω-perfect. Thus, for any compact X, the following properties are equivalent:*

(i) the space $C_p(X)$ is K-analytic;
(ii) there exists $Y \subset C_p(X)$ such that Y is dense in $C_p(X)$ and $Y \in K(\mathcal{A})$;
(iii) there exists $Y \subset C_p(X)$ which separates the points of X and belongs to $K(\mathcal{A})$;
(iv) the space X embeds into $C_p(Y)$ for some K-analytic space Y.

Solution. Any compact space is K-analytic; besides, the class $K(\mathcal{A})$ is invariant under countable unions, continuous images, closed subspaces and countable products (see SFFS-343 and SFFS-345). This proves that $K(\mathcal{A})$ is an ω-perfect class; so we can apply Problem 019 to conclude that (i) $\Longleftrightarrow$ (ii) $\Longleftrightarrow$ (iii) $\Longleftrightarrow$ (iv).

U.023. *Let X be a compact space such that $C_p(X)$ is K-analytic. Show that $C_p(X^\omega)$ is a K-analytic space and so is $C_p(X^n)$ for each $n \in \mathbb{N}$.*

Solution. The class $K(\mathcal{A})$ of K-analytic spaces is ω-perfect by Problem 022; so we can apply Problem 018 to conclude that if $C_p(X) \in K(\mathcal{A})$ then $C_p(X^\omega) \in K(\mathcal{A})$.

Now, if $n \in \mathbb{N}$ then there is a continuous onto map $\varphi : X^\omega \to X^n$; the dual map $\varphi^* : C_p(X^n) \to C_p(X^\omega)$ embeds $C_p(X^n)$ in $C_p(X^\omega)$ as a closed subspace because φ is a closed map (see TFS-163). Therefore $C_p(X^n)$ is K-analytic being homeomorphic to a closed subspace of the K-analytic space $C_p(X^\omega)$.

U.024. *Observe that any K-analytic space is Lindelöf Σ. Give an example of a space X such that $C_p(X)$ is Lindelöf Σ but not K-analytic.*

Solution. Any σ-compact space is Lindelöf Σ by SFFS-226; therefore, it follows from SFFS-258 that any $K_{\sigma\delta}$-space is Lindelöf Σ. Since every continuous image of a Lindelöf Σ-space is Lindelöf Σ (see SFFS-243), any K-analytic space is Lindelöf Σ.

Now, take any $\xi \in \beta\omega\backslash\omega$ and let $X = \omega \cup \{\xi\}$. Then X is a countable space, so $C_p(X)$ is Lindelöf Σ being second countable (see SFFS-228). However, the space $C_p(X)$ cannot be K-analytic because otherwise it would be analytic (see SFFS-346) which contradicts SFFS-371.

U.025. *Give an example of X such that $C_p(X)$ is K-analytic but not $K_{\sigma\delta}$.*

Solution. It was proved in SFFS-372 that there is a countable space X such that $C_p(X)$ is a Borel subspace of $\mathbb{R}^X$ and $C_p(X) \notin \Sigma_3^0(\mathbb{R}^X)$. It is evident that every $K_{\sigma\delta}$-subspace of $\mathbb{R}^X$ belongs to $\Pi_2^0(\mathbb{R}^X) \subset \Sigma_3^0(\mathbb{R}^X)$ (see SFFS-320). Furthermore, $C_p(X)$ is analytic (and hence K-analytic, see SFFS-346) because all Borel sets are analytic (see SFFS-334). Consequently, $C_p(X)$ is a K-analytic space which fails to be $K_{\sigma\delta}$.

U.026. *Let X be a Lindelöf Σ-space. Prove that $C_p(X)$ is normal if and only if $C_p(X)$ is Lindelöf. In particular, if X is compact then $C_p(X)$ is normal if and only if it is Lindelöf.*

Solution. If $C_p(X)$ is normal then $ext(C_p(X)) = \omega$ by Reznichenko's theorem (TFS-296) and hence $C_p(X)$ is Lindelöf by Baturov's theorem (SFFS-269). Since every Lindelöf space is normal, this proves that normality of $C_p(X)$ is equivalent to $l(C_p(X)) = \omega$ for any Lindelöf Σ-space X (and hence for any compact space X).

U.027. *Suppose that X is a Lindelöf Σ-space such that $C_p(X)\backslash\{f\}$ is normal for some $f \in C_p(X)$. Prove that X is separable. In particular, if X is ω-monolithic and $C_p(X)\backslash\{f\}$ is normal for some $f \in C_p(X)$ then X has a countable network.*

Solution. Since the space $C_p(X)$ is homogeneous, the space $C_p(X)\backslash\{g\}$ is normal for any $g \in C_p(X)$; in particular, if $u(x) = 1$ for all $x \in X$ then $C_p(X)\backslash\{u\}$ is normal. Observe that for $Z = C_p(X, \mathbb{I})$ we have $Z' = Z\backslash\{u\} = Z \cap (C_p(X)\backslash\{u\})$ so Z' is normal being a closed subspace of the normal space $C_p(X)\backslash\{u\}$.

Another important observation is that Z' is a convex subset of $\mathbb{I}^X$; indeed, if we are given $h, g \in Z'$ and $t \in [0, 1]$ then, for the function $v = tg + (1 - t)h$, we have $|v(x)| \leq t|g(x)| + (1 - t)|h(x)| \leq t + (1 - t) = 1$ so $v \in Z$. Assume that $v = u$; then $t \in (0, 1)$ because $\{g, h\} \subset Z\backslash\{u\}$. Besides, there is $x \in X$ for which $g(x) < 1$ and hence $v(x) = tg(x) + (1 - t)h(x) < t + (1 - t) = 1$ which is a contradiction. Therefore $v \in Z'$ and hence Z' is a dense convex subspace of $\mathbb{I}^X$. This makes it possible to apply Reznichenko's theorem (TFS-294) to conclude that $ext(Z') = \omega$ and hence $l(Z') = \omega$ by Baturov's theorem (SFFS-269).

Fact 1. For any space T and a closed $F \subset T$ we have $\psi(F, T) \leq l(T\backslash F)$. In particular, $\psi(t, T) \leq l(T\backslash\{t\})$ for any $t \in T$.

Proof. If $\kappa = l(T\backslash F)$ then choose, for any $z \in T\backslash F$ a set $U_z \in \tau(z, T)$ such that $\overline{U}_z \subset T\backslash F$. The family $\{U_z : z \in T\backslash F\}$ is an open cover of the space $T\backslash F$; so there is a set $A \subset T\backslash F$ for which $|A| \leq \kappa$ and $\bigcup\{U_z : z \in A\} = T\backslash F$. We also have the equality $\bigcup\{U_z : z \in A\} = T\backslash F$ and therefore $F = \bigcap\{T\backslash\overline{U}_z : z \in A\}$ which shows that $\psi(F, T) \leq |A| \leq \kappa = l(T\backslash F)$ and finishes the proof of Fact 1.

Returning to our solution apply Fact 1 to conclude that $\psi(u, Z) = \omega$. Consequently, $\psi(C_p(X)) = \psi(u, Z) = \omega$ (see Fact 3 of S.398 and Fact 1 of T.448) and hence X is separable by TFS-173.

U.028. *Let X and $C_p(X)$ be Lindelöf Σ-spaces and suppose that $C_p(X)\backslash\{f\}$ is normal for some $f \in C_p(X)$. Prove that X has a countable network.*

Solution. Observe that $C_p(X)$ is stable being a Lindelöf Σ-space (see SFFS-266); so X is ω-monolithic by SFFS-152. Now apply Problem 027 to conclude that X is separable and hence $nw(X) = \omega$.

U.029. *Let M_t be a separable metrizable space for all $t \in T$. Suppose that Y is dense in $M = \prod\{M_t : t \in T\}$ and Z is a continuous image of Y. Prove that, if $Z \times Z$ is normal then $ext(Z) = \omega$ and hence Z is collectionwise normal.*

Solution. For every $S \subset T$ the map $p_S : M \to M_S = \prod_{t \in S} M_t$ is the natural projection and $Y_S = p_S(Y)$; we will also need the map $q_S : M \times M \to M_S \times M_S$ defined by $q_S(a, b) = (p_S(a), p_S(b))$ for any $a, b \in M$.

Fix a continuous onto map $\varphi : Y \to Z$ and assume that $ext(Z) > \omega$ and hence there is a faithfully indexed closed discrete set $D = \{z_\alpha : \alpha < \omega_1\} \subset Z$. The set $D \times D$ is closed and discrete in $Z \times Z$; so $A = \{(z_\alpha, z_\alpha) : \alpha < \omega_1\}$ and $B = \{(z_\alpha, z_\beta) : \alpha, \beta < \omega_1$ and $\alpha \neq \beta\}$ are disjoint closed subsets of $Z \times Z$. By normality of Z there is a continuous function $u : Z \to \mathbb{R}$ such that $u(A) = \{0\}$ and $u(B) = \{1\}$.

Choose a point $y_\alpha \in \varphi^{-1}(z_\alpha)$ for any ordinal $\alpha < \omega_1$ and consider the sets $A' = \{(y_\alpha, y_\alpha) : \alpha < \omega_1\}$ and $B' = \{(y_\alpha, y_\beta) : \alpha, \beta < \omega_1$ and $\alpha \neq \beta\}$. For any element $(a, b) \in Y \times Y$ let $\eta(a, b) = (\varphi(a), \varphi(b)) \in Z \times Z$. Then $\eta : Y \times Y \to Z \times Z$ is a continuous onto map for which $\eta(A') = A$ and $\eta(B') = B$. Consequently, $v = u \circ \eta : Y \times Y \to \mathbb{R}$ is a continuous function such that $v(A') = \{0\}$ and $v(B') = \{1\}$.

The space $Y \times Y$ is dense in $M \times M$ which is still a product of second countable spaces; so there is a countable $S \subset T$ for which there exists a continuous map $h : Y_S \times Y_S \to \mathbb{R}$ such that $h \circ (q_S|(Y \times Y)) = v$ (see TFS-299).

We have $h(q_S(A')) = \{0\}$ and $h(q_S(B')) = \{1\}$; so the sets $A_1 = q_S(A')$ and $B_1 = q_S(B')$ are separated in $M_S \times M_S$, i.e., $\overline{A}_1 \cap B_1 = \emptyset$ and $\overline{B}_1 \cap A_1 = \emptyset$ (the bar denotes the closure in $M_S \times M_S$). For any $\alpha < \omega_1$ let $x_\alpha = p_S(y_\alpha)$; since $(x_\alpha, x_\alpha) \notin \overline{B}_1$, there is $W_\alpha \in \tau(x_\alpha, M_S)$ such that $(W_\alpha \times W_\alpha) \cap B_1 = \emptyset$.

The space M_S is second countable; so the set $P = \{x_\alpha : \alpha < \omega_1\} \subset M_S$ cannot be discrete. Therefore there is $\gamma < \omega_1$ such that x_γ is an accumulation point for P. In particular, there are distinct ordinals $\alpha, \beta \in \omega_1\backslash\{\gamma\}$ such that $x_\alpha, x_\beta \in W_\gamma$. As a consequence, $b = (y_\alpha, y_\beta) \in B'$ and $q_S(b) = (x_\alpha, x_\beta) \in B_1 \cap (W_\gamma \times W_\gamma)$ which is a contradiction with the choice of W_γ. This proves that $ext(Z) = \omega$ and hence Z is collectionwise normal by Fact 3 of S.294.

U.030. *Prove that, for any infinite zero-dimensional compact space X, there exists a closed $F \subset C_p(X, \mathbb{D}^\omega) \subset C_p(X)$ which maps continuously onto $(C_p(X))^\omega$.*

Solution. Since X is compact, we have $C_p(X) = \bigcup\{C_p(X, [-n, n]) : n \in \mathbb{N}\}$. Every $C_p(X, [-n, n])$ is homeomorphic to $C_p(X, \mathbb{I})$; so $C_p(X)$ is a continuous image of $C_p(X, \mathbb{I}) \times \omega$. The space $C_p(X, \mathbb{I})$ is, in turn, a continuous image of $C_p(X, \mathbb{D}^\omega)$ by Problem 004 which shows that $C_p(X)$ is a continuous image of $C_p(X, \mathbb{D}^\omega) \times \omega$. The space X being infinite it cannot be a P-space (see Fact 2 of T.090) so $C_p(X, \mathbb{I})$ is not countably compact by TFS-397. As a consequence, the space $C_p(X, \mathbb{D}^\omega)$ is not countably compact either; so we can find a countably infinite closed discrete $D \subset C_p(X, \mathbb{D}^\omega)$.

Observe that $C_p(X, \mathbb{D}^\omega) \simeq C_p(X, (\mathbb{D}^\omega)^\omega) \simeq (C_p(X, \mathbb{D}^\omega))^\omega$ (see TFS-112); since ω is homeomorphic to a closed subspace of $C_p(X, \mathbb{D}^\omega)$, the space $C_p(X, \mathbb{D}^\omega) \times \omega$ is homeomorphic to a closed subspace of $C_p(X, \mathbb{D}^\omega) \times C_p(X, \mathbb{D}^\omega) \simeq C_p(X, \mathbb{D}^\omega)$. Thus some closed $G \subset C_p(X, \mathbb{D}^\omega)$ maps continuously onto $C_p(X)$. Then $F = G^\omega$ is a closed subspace of $(C_p(X, \mathbb{D}^\omega))^\omega \simeq C_p(X, \mathbb{D}^\omega) \subset C_p(X)$ which maps continuously onto $(C_p(X))^\omega$.

U.031. *Prove that, for any infinite zero-dimensional compact space X, there exists a closed $F \subset C_p(X, \mathbb{D}^\omega) \subset C_p(X)$ which maps continuously onto $C_p(X^\omega)$.*

Solution. If $\mathcal{A}$ is a family of sets then $\bigvee \mathcal{A}$ is the family of all finite unions of elements of $\mathcal{A}$; analogously, $\bigwedge \mathcal{A}$ is the family of all finite intersections of elements of $\mathcal{A}$. If Z is a space then $\mathcal{C}(Z)$ is the family of all clopen subsets of Z.

Observe that $C_p(X^\omega) = \bigcup\{C_p(X^\omega, [-n, n]) : n \in \mathbb{N}\}$ because X^ω is compact. For every $n \in \mathbb{N}$, the space $C_p(X^\omega, [-n, n])$ is homeomorphic to $C_p(X^\omega, \mathbb{I})$; so $C_p(X^\omega)$ is a continuous image of the space $C_p(X^\omega, \mathbb{I}) \times \omega$. Since X^ω is zero-dimensional (see SFFS-302), the space $C_p(X^\omega, \mathbb{I})$ is a continuous image of $C_p(X^\omega, \mathbb{D}^\omega)$ by Problem 004; thus

(0) $C_p(X^\omega)$ is a continuous image of $C_p(X^\omega, \mathbb{D}^\omega) \times \omega$.

For any $i \in \omega$ let $p_i : X^\omega \to X$ be the natural projection of X^ω onto its i-th factor. Then p_i is continuous and onto so its dual map $p_i^* : C_p(X) \to C_p(X^\omega)$ is an embedding by TFS-163; let $Z_i = p_i^*(C_p(X, \mathbb{D}))$ for all $i \in \omega$. It turns out that

(1) the set $Z = \bigcup\{Z_i : i \in \omega\} \subset C_p(X^\omega, \mathbb{D})$ *strongly separates the points in X^ω*, i.e., for any distinct $x, y \in X^\omega$ and $i, j \in \mathbb{D}$ there is $g \in Z$ with $g(x) = i$ and $g(y) = j$.

Since constant functions belong to Z, there is nothing to prove if $i = j$ so assume that $i \neq j$ and hence $i + j = 1$. There is $n \in \omega$ such that $x(n) \neq y(n)$ and therefore we can find a clopen set U in the space X such that $x(n) \in U$ and $y(n) \notin U$. If $i = 1$ then let $f = \chi_U$ where χ_U is the characteristic function of U; if $i = 0$ then let $f = 1 - \chi_U$. Then $f \in C_p(X, \mathbb{D})$ and $f(x(n)) = i$ while $f(y(n)) = 1 - i = j$. Thus $p_n^*(f)(x) = f(x(n)) = i$ and $f(y(n)) = p_i^*(f)(y) = j$ which shows that $g = p_i^*(f) \in Z$ is as promised, i.e., (1) is proved.

Our next step is to show that

(2) $\bigvee(\bigwedge\{f^{-1}(1) : f \in Z\}) = \mathcal{C}(X^\omega)$.

To prove (2) let $\mathcal{C} = \{f^{-1}(1) : f \in Z\}$ and take any non-empty $U \in \mathcal{C}(X^\omega)$. For any $x \in U$ and $y \in X^\omega \backslash U$ apply (1) to find a function $f_{x,y} \in Z$ such that $f_{x,y}(x) = 1$ and $f_{x,y}(y) = 0$; then $O_{x,y} = f_{x,y}^{-1}(1) \in \mathcal{C}$. Furthermore, $\bigcap \{O_{x,y} : y \in X^\omega \backslash U\} \subset U$ so we can apply Fact 1 of S.326 to see that there is a finite $A \subset X^\omega \backslash U$ for which $W_x = \bigcap \{O_{x,y} : y \in A\} \subset U$. It is clear that $x \in W_x \in \bigwedge \mathcal{C}$ for any $x \in U$. The set U being compact there is a finite set $B \subset U$ such that $\bigcup \{W_x : \in B\} = U$ which shows that U is a finite union of elements of $\bigwedge \mathcal{C}$, i.e., $U \in \bigvee(\bigwedge \mathcal{C})$ and hence (2) is proved.

Let $M(Z) = \{f_0 \cdot \ldots \cdot f_n : n \in \omega, \ f_i \in Z, \ i \leq n\}$. It is easy to see that $\{f^{-1}(1) : f \in M(Z)\} = \bigwedge \mathcal{C}$. If $f, g \in C_p(X^\omega, \mathbb{D})$ then let $(f * g)(x) = f(x) + g(x) - f(x)g(x)$ for any $x \in X^\omega$. We have $f * g = g * f$ for any $f, g \in C_p(X^\omega, \mathbb{D})$ so, for any $n \in \omega$, we can define the operation $f_0 * \ldots * f_n$ inductively as $(f_0 * \ldots * f_{n-1}) * f_n$. It is straightforward that it is associative and $f_0 * \ldots * f_n \in C_p(X^\omega, \mathbb{D})$; besides, $(f_0 * \ldots * f_n)^{-1}(1) = f_0^{-1}(1) \cup \ldots \cup f_n^{-1}(1)$ for any $f_0, \ldots, f_n \in C_p(X^\omega, \mathbb{D})$. Thus, for the set $S(Z) = \{f_0 * \ldots * f_n : n \in \omega, \ f_i \in M(Z) \text{ for all } i \leq n\}$, we have $\{f^{-1}(1) : f \in S(Z)\} = \bigvee(\bigwedge \mathcal{C}) = \mathcal{C}(X^\omega)$ by (2). An immediate consequence is that $S(Z) = C_p(X^\omega, \mathbb{D})$.

For any $n \in \omega$ we will need the maps

$$m_n : (C_p(X^\omega, \mathbb{D}))^{n+1} \to C_p(X^\omega, \mathbb{D}) \quad \text{and} \quad a_n : (C_p(X^\omega, \mathbb{D}))^{n+1} \to C_p(X^\omega, \mathbb{D})$$

defined by $m_n(f_0, \ldots, f_n) = f_0 \cdot \ldots \cdot f_n$ and $a_n(f_0, \ldots, f_n) = f_0 * \ldots * f_n$ for any $(f_0, \ldots, f_n) \in (C_p(X^\omega, \mathbb{D}))^{n+1}$. All maps m_n ad a_n are continuous being arithmetical combinations of sums and products. Recalling the definitions of $M(Z)$ and $S(Z)$ we can see that $M(Z) = \bigcup \{m_n(Z^{n+1}) : n \in \omega\}$ and $S(Z) = \bigcup \{a_n((M(Z))^{n+1}) : n \in \omega\}$.

Given spaces P and Q we will use the expression $P \gg Q$ to abbreviate the phrase "some closed subset of P maps continuously onto Q". In particular, if P maps continuously onto Q, or Q is homeomorphic to a closed subset of P then $P \gg Q$.

Since every Z_i is a continuous image of $C_p(X, \mathbb{D})$, we have $C_p(X, \mathbb{D}) \times \omega \gg Z$. Therefore, $(C_p(X, \mathbb{D}) \times \omega)^\omega \simeq C_p(X, \mathbb{D}^\omega) \times \omega^\omega \gg Z^\omega \gg Z^{n+1}$ for any $n \in \omega$. Since X is compact and infinite, the space $C_p(X, \mathbb{I})$ is not countably compact by TFS-397; furthermore, $C_p(X, \mathbb{D}^\omega)$ maps continuously onto $C_p(X, \mathbb{I})$ by Problem 004 so the space $C_p(X, \mathbb{D}^\omega)$ is not countably compact either. As a consequence, ω can be considered to be a closed subspace of $C_p(X, \mathbb{D}^\omega)$ and hence ω^ω is a closed subspace of the space $(C_p(X, \mathbb{D}^\omega))^\omega \simeq C_p(X, \mathbb{D}^\omega)$; so $C_p(X, \mathbb{D}^\omega) \times \omega^\omega$ is homeomorphic to a closed subset of $C_p(X, \mathbb{D}^\omega) \times C_p(X, \mathbb{D}^\omega) \simeq C_p(X, \mathbb{D}^\omega)$ and hence we obtain

(3) $C_p(X.\mathbb{D}^\omega) \gg (C_p(X.\mathbb{D}^\omega))^\omega$ and $C_p(X.\mathbb{D}^\omega) \gg C_p(X, \mathbb{D}^\omega) \times \omega^\omega \gg C_p(X, \mathbb{D}^\omega) \times \omega$.

This shows that $C_p(X, \mathbb{D}^\omega) \gg Z^{n+1} \gg m_n(Z^{n+1})$ for any $n \in \omega$ and therefore $C_p(X, \mathbb{D}^\omega) \times \omega \gg M(Z)$. By (3), we have $C_p(X, \mathbb{D}^\omega) \gg C_p(X, \mathbb{D}^\omega) \times \omega$; so $C_p(X, \mathbb{D}^\omega) \gg M(Z)$.

Analogously, $(C_p(X, \mathbb{D}^\omega))^{n+1} \gg (M(Z))^{n+1} \gg a_n((M(Z))^{n+1})$ which, together with $(C_p(X, \mathbb{D}^\omega))^{n+1} \simeq C_p(X, \mathbb{D}^\omega)$ implies $C_p(X, \mathbb{D}^\omega) \gg a_n((M(Z))^{n+1})$ for any $n \in \omega$ and therefore $C_p(X, \mathbb{D}^\omega) \gg C_p(X, \mathbb{D}^\omega) \times \omega \gg S(Z) = C_p(X^\omega, \mathbb{D})$. Consequently, $C_p(X, \mathbb{D}^\omega) \simeq (C_p(X, \mathbb{D}^\omega))^\omega \gg (C_p(X^\omega, \mathbb{D}))^\omega \simeq C_p(X^\omega, \mathbb{D}^\omega)$. Finally, apply (0) to conclude that $C_p(X, \mathbb{D}^\omega) \gg C_p(X, \mathbb{D}^\omega) \times \omega \gg C_p(X^\omega, \mathbb{D}^\omega) \times \omega \gg C_p(X^\omega)$; this is the same as saying that there is a closed subset of $C_p(X, \mathbb{D}^\omega)$ which maps continuously onto $C_p(X^\omega)$.

U.032. *Prove that the following conditions are equivalent for an arbitrary zero-dimensional compact X:*

(i) $C_p(X, \mathbb{D}^\omega)$ *is normal;*
(ii) $C_p(X, \mathbb{I})$ *is normal;*
(iii) $C_p(X)$ *is normal;*
(iv) $C_p(X)$ *is Lindelöf;*
(v) $(C_p(X))^\omega$ *is Lindelöf;*
(vi) $C_p(X^\omega)$ *is Lindelöf.*

Solution. There is nothing to prove if X is finite; so assume that $|X| \geq \omega$. The implications (v)$\Longrightarrow$(iv)$\Longrightarrow$(iii) are evident. Since $C_p(X, \mathbb{I})$ is closed in $C_p(X)$ and $C_p(X, \mathbb{D}^\omega)$ is closed in $C_p(X, \mathbb{I})$, the implications (iii)$\Longrightarrow$(ii) and (ii)$\Longrightarrow$(i) are also clear.

Since $C_p(X, \mathbb{D}^\omega) \simeq (C_p(X, \mathbb{D}))^\omega$ and $C_p(X, \mathbb{D})$ is dense in $\mathbb{D}^X$ (see Fact 1 of S.390), the space $C_p(X, \mathbb{D}^\omega)$ can be considered to be a dense subspace of $(\mathbb{D}^X)^\omega$ so we can apply Problem 029 to conclude that it follows from normality of the space $C_p(X, \mathbb{D}^\omega) \simeq C_p(X, \mathbb{D}^\omega) \times C_p(X, \mathbb{D}^\omega)$ that $ext(C_p(X, \mathbb{D}^\omega)) = \omega$ and hence $C_p(X, \mathbb{D}^\omega)$ is Lindelöf by Baturov's theorem (SFFS-269).

By Problem 030, there is a closed $F \subset C_p(X, \mathbb{D}^\omega)$ which maps continuously onto $(C_p(X))^\omega$; the space F is Lindelöf being closed in $C_p(X, \mathbb{D}^\omega)$ so $(C_p(X))^\omega$ has also to be Lindelöf and hence we settled (i)$\Longrightarrow$(v). Therefore all properties (i)–(v) are equivalent.

The space X is a continuous image of X^ω; the respective dual map embeds $C_p(X)$ in $C_p(X^\omega)$ as a closed subspace (see TFS-163) so if $C_p(X^\omega)$ is Lindelöf then $C_p(X)$ is Lindelöf as well. This proves that (vi)$\Longrightarrow$(iv). Finally, if $(C_p(X))^\omega$ is Lindelöf then $C_p(X^\omega)$ is also Lindelöf by Problem 017 and hence (v)$\Longrightarrow$(vi). Therefore all properties (i)–(vi) are equivalent.

U.033. *Observe that $C_p(X)$ is monolithic for any compact X. Using this fact prove that, for any compact space X, each compact subspace $Y \subset C_p(X)$ is a Fréchet–Urysohn space.*

Solution. The space $C_p(X)$ is monolithic because X is stable (see SFFS-154). That every compact $Y \subset C_p(X)$ is a Fréchet–Urysohn space was proved in Fact 10 of S.351.

U.034. *Prove that, for any metrizable space M, there is a compact space K such that M embeds in $C_p(K)$.*

Solution. Fix a base $\mathcal{B} \subset \tau^*(M)$ in the space M such that $\mathcal{B} = \bigcup_{n \in \omega} \mathcal{B}_n$ and $\mathcal{B}_n$ is a discrete family for every $n \in \omega$ (this is possible because any metrizable space has a σ-discrete base by TFS-221). For any $U \in \mathcal{B}_n$ there is a continuous function $p_U^n : M \to [0, \frac{1}{n+1}]$ such that $(p_U^n)^{-1}(0) = M \backslash U$ (see Fact 2 of T.080).

Define a function $p : M \to \mathbb{R}$ by $p(x) = 0$ for all $x \in M$ and consider the space $K = \{p\} \cup \{p_U^n : n \in \omega, \ U \in \mathcal{B}_n\} \subset C_p(M)$. Take any $O \in \tau(p, C_p(M))$; there is $\varepsilon > 0$ and a finite set $A \subset M$ such that $W = \{f \in C_p(M) : f(A) \subset (-\varepsilon, \varepsilon)\} \subset O$. Take $m \in \omega$ for which $\frac{1}{m} < \varepsilon$; then $p_U^n(x) \leq \frac{1}{n+1} < \frac{1}{m} < \varepsilon$ for any $n \geq m$, $x \in M$ and $U \in \mathcal{B}_n$. In particular, $K_n = \{p_U^n : U \in \mathcal{B}_n\} \subset W$ for all $n \geq m$.

Now, if $n < m$ then only finitely many elements of $\mathcal{B}_n$ meet A because $\mathcal{B}_n$ is discrete. If $U \in \mathcal{B}_n$ and $A \cap U = \emptyset$ then $p_U^n(A) = \{0\}$ and hence $p_U^n \in W$. Therefore $K_n \backslash W$ is finite for any $n < m$ which shows that $K \backslash W \supset K \backslash O$ is finite and hence

(1) $K \backslash O$ is finite for any $O \in \tau(p, C_p(M))$.

An immediate consequence of (1) is that the space K is compact. Given $x \in M$ let $e_x(f) = f(x)$ for any $f \in K$; then $e_x \in C_p(K)$ and, letting $e(x) = e_x$ for any $x \in M$, we obtain a continuous map $e : M \to C_p(K)$ (see TFS-166).

Suppose that $x \in M$ and $G \subset M$ is a closed set such that $x \notin G$. Since $\mathcal{B}$ is a base in M, there is $n \in \omega$ and $U \in \mathcal{B}_n$ such that $x \in U \subset M \backslash G$. For the function $f = p_U^n \in K$ we have $f(x) > 0$ and $f(G) = \{0\}$; so $f(x) \notin \overline{f(G)}$. This proves that K separates the points and closed subsets of M so e is an embedding (see TFS-166). Thus K is a compact space such that M embeds in $C_p(K)$.

U.035. *Prove that the following conditions are equivalent for any compact X:*

  (i) there is a compact $K \subset C_p(X)$ which separates the points of X;
 (ii) there is a σ-compact $Y \subset C_p(X)$ which separates the points of X;
(iii) there is a σ-compact $Z \subset C_p(X)$ which is dense in $C_p(X)$;
(iv) X embeds into $C_p(K)$ for some compact K;
 (v) X embeds into $C_p(Y)$ for some σ-compact Y.

Solution. The implication (i)$\Longrightarrow$(ii) is evident. Now if the statement (ii) is true then let $Z = A(Y)$ where $A(Y)$ is the minimal algebra in $C_p(X)$ such that $Y \subset A(Y)$ (see Problem 006). It is clear that the class of σ-compact spaces is k-directed; so Z is a countable union of σ-compact spaces, i.e., Z is also σ-compact. The set Z is dense in $C_p(X)$ by TFS-192; so we proved that (ii)$\Longrightarrow$(iii). It was established in Fact 3 of S.312 that (iii) $\Longleftrightarrow$ (i) and, in particular, (iii)$\Longrightarrow$(i) so the properties (i)–(iii) are equivalent. Besides, it was proved in Fact 12 of S.351 that (i) $\Longleftrightarrow$ (iv) so (i) $\Longleftrightarrow$ (ii) $\Longleftrightarrow$ (iii) $\Longleftrightarrow$ (iv).

It is immediate that (iv)$\Longrightarrow$(v); now if $X \subset C_p(Y)$ for some σ-compact space Y then let $e_y(f) = f(y)$ for any $y \in Y$ and $f \in X$. Then $e_y \in C_p(X)$ and, letting $e(y) = e_y$ for every $y \in Y$, we obtain a continuous map $e : Y \to C_p(X)$ (see TFS-166). The set $T = e(Y)$ is σ-compact and it is an easy exercise that T separates the points of X. Thus some σ-compact $T \subset C_p(X)$ separates the points of X and hence we proved that (v)$\Longrightarrow$(ii).

U.036. *Suppose that X is compact and embeds into $C_p(Y)$ for some compact Y.*
Prove that it is possible to embed X into $C_p(Z)$ for some Fréchet–Urysohn compact
space Z.

Solution. We consider that $X \subset C_p(Y)$; for any point $y \in Y$ and $f \in X$ let
$e_y(f) = f(y)$. Then $e_y \in C_p(X)$ and the correspondence $y \to e_y$ defines a
continuous map $e : Y \to C_p(X)$ (see TFS-166). The space $Z = e(Y) \subset C_p(X)$
is compact and it is an easy exercise that Z separates the points of X. Besides, Z
is Fréchet–Urysohn by Problem 033; for any $f \in X$ and $z \in Z$ let $u_f(z) = z(f)$.
Then $u_f \in C_p(Z)$ and letting $u(f) = u_f$ for any $f \in X$ we obtain a continuous
map $u : X \to C_p(Z)$. Since Z separates the points of X, the map u is injective by
Fact 2 of S.351; the space X being compact, u is an embedding of X into $C_p(Z)$
and we already saw that Z is a compact Fréchet–Urysohn space.

U.037. *Give an example of a compact space X which embeds into $C_p(Y)$ for some*
compact Y but cannot be embedded into $C_p(Z)$ for any compact first countable
space Z.

Solution. If κ is a cardinal then $A(\kappa)$ is the Alexandroff one-point compactification
of a discrete space of cardinality κ. Fix any cardinal $\kappa > 2^c$; the space $Y = A(\kappa)$
is compact and it follows from $2^c < \kappa = c(A(\kappa)) = a(C_p(A(\kappa)))$ (see TFS-181),
that the compact space $X = A(2^c)$ embeds in $C_p(Y)$.

   Now, if the space X embeds in $C_p(Z)$ for some compact first countable space Z
then $|Z| \leq c$ by TFS-329 and hence $2^c = w(X) \leq w(C_p(Z)) = |Z| \leq c$ which is
a contradiction.

U.038. *Suppose that X embeds into $C_p(Y)$ for some compact Y. Prove that it is*
possible to embed X into $C_p(Z)$ for some zero-dimensional compact space Z.

Solution. We can consider that $Y \subset \mathbb{I}^\kappa$ for some infinite cardinal κ (see TFS-127).
There exists a continuous onto map $\varphi : \mathbb{D}^\kappa \to \mathbb{I}^\kappa$ (see Fact 2 of T.298). The space
$Z = \varphi^{-1}(Y)$ is zero-dimensional and compact (see SFFS-303) and $\eta = \varphi|Z$ maps
Z continuously onto Y. The dual map $\eta^* : C_p(Y) \to C_p(Z)$ embeds $C_p(Y)$ in
$C_p(Z)$ (see TFS-163) so X can also be embedded in $C_p(Z)$.

U.039. *Suppose that X embeds into $C_p(Y)$ for some countably compact Y. Prove*
that it is possible to embed X into $C_p(Z)$ for some zero-dimensional countably
compact space Z.

Solution. We can consider that $Y \subset \mathbb{I}^\kappa$ for some infinite cardinal κ (see TFS-127).
There exists a continuous onto map $\varphi : \mathbb{D}^\kappa \to \mathbb{I}^\kappa$ (see Fact 2 of T.298). The space
$Z = \varphi^{-1}(Y)$ is zero-dimensional by SFFS-303. Furthermore, $\eta = \varphi|Z$ maps Z
continuously onto Y and the map η is perfect by Fact 2 of S.261.

Fact 1. Any perfect preimage of a countably compact space is countably compact.

Proof. Let Q be a countably compact space and assume that $u : P \to Q$ is a
perfect map. If D is an infinite closed discrete subset of P then $D \cap u^{-1}(y)$ is finite
because $u^{-1}(y)$ is compact for any $y \in Q$. This implies that the set $E = u(D) \subset Q$

is infinite. For any $E' \subset E$ we have $E' = u(D')$ for some $D' \subset D$. The set D is closed and discrete and the map u is closed so D' is closed in P and hence E' has to be closed in Q. It turns out that $E \subset Q$ is infinite and every $E' \subset E$ is closed in Q. As a consequence, E' is an infinite closed discrete subspace of Q which contradicts countable compactness of Q; so Fact 1 is proved.

Returning to our solution observe that Z is countably compact by Fact 1. The dual map $\eta^* : C_p(Y) \rightarrow C_p(Z)$ embeds $C_p(Y)$ in $C_p(Z)$ (see TFS-163); so X can also be embedded in $C_p(Z)$; the space Z being countably compact and zero-dimensional, our solution is complete.

U.040. *Give an example of a space Y which embeds in $C_p(X)$ for a pseudocompact space X but does not embed in $C_p(Z)$ for any countably compact Z.*

Solution. It was proved in TFS-400 that there exists a pseudocompact non-compact space X such that every countable $A \subset X$ is closed and C^*-embedded in X. It follows from TFS-398 that the space $Y = C_p(X, \mathbb{I}) \subset C_p(X)$ is pseudocompact.

Assume that Y embeds in $C_p(Z)$ for some countably compact space Z. We can consider that $Y \subset C_p(Z)$ and hence $K = \overline{Y}$ is also a pseudocompact subspace of $C_p(Z)$ (see Fact 18 of S.351). Since K is closed in $C_p(Z)$, we can apply Fact 2 of S.307 to conclude that K is compact.

For any $z \in Z$ and $f \in K$ let $e_z(f) = f(z)$; then $e_z \in C_p(K)$ and the correspondence $z \rightarrow e_z$ gives a continuous map $e : Z \rightarrow C_p(K)$ (see TFS-166). The space $Z' = e(Z) \subset C_p(K)$ is countably compact; so we can use Fact 18 of S.351 and Fact 2 of S.307 again to see that $M = \overline{Z'}$ is compact. It is straightforward that Z' separates the points of K and hence so does M. Thus we have a compact space $M \subset C_p(K)$ which separates the points of K. Consequently, K embeds in $C_p(M')$ for some compact M' by Problem 035 and therefore K is Fréchet–Urysohn by Problem 033.

Now, $Y \subset K$ implies that the space Y is also Fréchet–Urysohn; since $C_p(X)$ embeds in $Y = C_p(X, \mathbb{I})$, it has to be Fréchet–Urysohn as well and therefore $t(C_p(X)) = \omega$. Thus X is Lindelöf which, together with pseudocompactness of X implies that X is compact. This contradiction shows that Y is a pseudocompact space which embeds in $C_p(X)$ for a pseudocompact X but does not embed in $C_p(Z)$ for any countably compact Z.

U.041. *Prove that a countable space Y embeds into $C_p(X)$ for some pseudocompact space X if and only if Y embeds into $C_p(Z)$ for some compact metrizable space Z.*

Solution. Sufficiency is evident; so assume that $Y \subset C_p(X)$ for some pseudocompact X. For any $x \in X$ let $e_x(f) = f(x)$ for any $f \in Y$; then $e_x \in C_p(Y)$ and the correspondence $x \rightarrow e_x$ represents a continuous map $e : X \rightarrow C_p(Y)$ (see TFS-166); let $Z = e(X)$. The space Z is pseudocompact because the map $e : X \rightarrow Z$ is continuous; besides, $w(Z) \leq w(C_p(Y)) = |Y| \leq \omega$ so Z is metrizable and hence compact. The dual map $e^* : C_p(Z) \rightarrow C_p(X)$ embeds $C_p(Z)$ in $C_p(X)$ (see TFS-163) and it is easy to check that $e^*(C_p(Z)) \supset Y$. Thus Y embeds in $C_p(Z)$ for a compact metrizable space Z.

U.042. *Give an example of a space Y which embeds into $C_p(X)$ for a countably compact space X but does not embed into $C_p(Z)$ for a compact space Z.*

Solution. Let X be the ordinal ω_1 with the interval topology. Then X is a countably compact non-compact space and $Y = C_p(X)$ embeds in $C_p(X)$ (even coincides with it). However, Y is not embeddable in $C_p(Z)$ for a compact space Z because otherwise $t(Y) = \omega$ which implies that $l(X) = \omega$ and hence X is compact, a contradiction.

U.043. *Let $\xi \in \beta\omega\setminus\omega$. Prove that the countable space $\omega_\xi = \omega \cup \{\xi\}$, considered with the topology inherited from $\beta\omega$, does not embed into $C_p(X)$ for a pseudocompact X.*

Solution. If ω_ξ embeds in $C_p(X)$ for some pseudocompact X then there is a metrizable compact K such that ω_ξ embeds in $C_p(K)$ (see Problem 041). There is a continuous onto map $\varphi : \mathbb{P} \to K$ (see e.g., SFFS-328) and hence $C_p(K)$ embeds in $C_p(\mathbb{P})$ (see TFS-163). Therefore ω_ξ embeds in $C_p(\mathbb{P})$ which is a contradiction with SFFS-371.

U.044. *(Grothendieck's theorem). Suppose that X is a countably compact space and $B \subset C_p(X)$ is a bounded subset of $C_p(X)$. Prove that $\overline{B}$ is compact. In particular, the closure of any pseudocompact subspace of $C_p(X)$ is compact.*

Solution. The set $F = \overline{B}$ is also bounded in $C_p(X)$ by Fact 2 of S.398. For any $x \in X$ the set $e_x(F) = \{f(x) : f \in F\} \subset \mathbb{R}$ is a continuous image of F (TFS-167); therefore $e_x(F)$ is bounded in $\mathbb{R}$ by Fact 1 of S.399 and hence there is $K_x > 0$ such that $e_x(F) \subset [-K_x, K_x]$. It is easy to see that $F \subset Q = \prod\{[-K_x, K_x] : x \in X\}$; since Q is compact, it suffices to show that F is closed in $\mathbb{R}^X$.

Suppose not, and fix any $f \in [F]\setminus F$ (the brackets denote the closure in $\mathbb{R}^X$). Since F is closed in $C_p(X)$, the function f must be discontinuous; so take any point $a \in X$ and $A \subset X$ such that $a \in \mathrm{cl}_X(A)$ while $f(a) \notin \overline{f(A)}$. Take $O, G \in \tau(\mathbb{R})$ such that $f(a) \in O$, $f(A) \subset G$ and $\overline{O} \cap \overline{G} = \emptyset$. We will construct sequences $\{f_n : n \in \omega\} \subset F$, $\{U_n : n \in \omega\} \subset \tau(a, X)$ and $\{a_n : n \in \omega\} \subset A$ with the following properties:

(1) $\mathrm{cl}_X(U_{n+1}) \subset U_n$ and $a_n \in U_n$ for all $n \in \omega$;
(2) $f_n(U_n) \subset O$ for all $n \in \omega$;
(3) $f_{n+1}(a_i) \in G$ for all $n \in \omega$ and $i \leq n$.

Since $f \in [F]$, there is $f_0 \in F$ such that $f_0(a) \in O$; the function f_0 being continuous there exists $U_0 \in \tau(a, X)$ such that $f_0(U_0) \subset O$. The point a belongs to the closure of A so there exists $a_0 \in A \cap U_0$. It is evident that (1)-(2) are fulfilled for a_0, f_0 and U_0. The property (3) is fulfilled vacuously.

Assume that we have a_i, f_i and U_i with the properties (1)–(3) for all $i \leq n$. Since $A_n = \{a_0, \ldots, a_n\} \subset A$ we have $f(A_n) \subset G$; it follows from $f \in [F]$ that there exists $f_{n+1} \in F$ such that $f_{n+1}(a) \in O$ and $f_{n+1}(A_n) \subset G$. The function f_{n+1} being continuous there exists $U_{n+1} \in \tau(a, X)$ such that $\mathrm{cl}_X(U_{n+1}) \subset U_n$ and

$f_{n+1}(U_{n+1}) \subset O$. Take any point $a_{n+1} \in U_{n+1} \cap A$ and observe that (1)-(3) are fulfilled for the sequence $\{a_i, f_i, U_i : i \leq (n+1)\}$; so our inductive construction can be carried out for all $n \in \omega$.

Once we have the sequences $\{f_n : n \in \omega\} \subset F$, $\{U_n : n \in \omega\} \subset \tau(a, X)$ and $S = \{a_n : n \in \omega\} \subset A$ with the properties (1)–(3) take an accumulation point b of the sequence $\{a_n : n \in \omega\}$ (which exists because X is countably compact). Note that, for any $x \in X\backslash(\bigcap\{U_n : n \in \omega\})$ we have $x \in V = X\backslash\mathrm{cl}_X(U_n)$ for some $n \in \omega$ and hence V is a neighbourhood of x which intersects only finitely many points of the sequence S so x cannot be an accumulation point of S.

This shows that $b \in P = \bigcap\{U_n : n \in \omega\} = \bigcap\{\mathrm{cl}_X(U_n) : n \in \omega\}$. If $Y = \{b\} \cup S$ then Y is countable and $\pi_Y(F) \subset C_p(Y)$ is a bounded subspace of $C_p(Y)$ (here, as usual, $\pi_Y : C_p(X) \to C_p(Y)$ is the restriction map). Since $w(C_p(Y)) = \omega$, we have $\upsilon(C_p(Y)) = C_p(Y)$ and hence the closure of $\pi_Y(F)$ if $C_p(Y)$ is compact by TFS-415. Therefore there exists an accumulation point $g \in C_p(Y)$ of the sequence $\{g_n = \pi_Y(f_n) : n \in \omega\} \subset \pi_Y(F)$. Thus $g(b)$ has to be in the closure of the set $\{g_n(b) : n \in \omega\} = \{f_n(b) : n \in \omega\}$. But $f_n(b) \in f_n(P) \subset f_n(U_n) \subset O$ for every $n \in \omega$; so $\overline{\{f_n(b) : n \in \omega\}} \subset \overline{O}$ and hence $g(b) \in \overline{O}$.

On the other hand, it immediately follows from continuity of the function g that $g(b) \in \overline{\{g(a_n) : n \in \omega\}}$; since $f_k(a_n) \in G$ for all $k > n$, we have $g(a_n) \in \overline{G}$ for each $n \in \omega$. An immediate consequence is that $g(b) \in \overline{G}$, i.e., $g(b) \in \overline{O} \cap \overline{G} = \emptyset$ which is a contradiction. We proved that F is closed in $\mathbb{R}^X$; so F is compact.

Finally observe that if $E \subset C_p(X)$ is pseudocompact then E is bounded in $C_p(X)$ and hence $\overline{E}$ is compact.

U.045. *Prove that there exists a pseudocompact space X for which there is a closed pseudocompact $Y \subset C_p(X)$ which is not countably compact.*

Solution. It was proved in TFS-400 that there exists a pseudocompact non-compact space X such that every countable $A \subset X$ is closed and C^*-embedded in X. The set $Y = C_p(X, \mathbb{I}) \subset C_p(X)$ is closed in $C_p(X)$ and it follows from TFS-398 that Y is pseudocompact.

Now, if Y is countably compact then X is a P-space by TFS-397; since X is not compact, it has to be infinite, so take a countably infinite $A \subset X$. It follows from Fact 1 of S.479 that A is C-embedded in X. If $\{a_n : n \in \omega\}$ is a faithful enumeration of A then let $f(a_n) = n$ for any $n \in \omega$. The function $f : A \to \mathbb{R}$ is continuous because A is discrete. If $g \in C_p(X)$ and $g|A = f$ (such a function g has to exist because A is C-embedded in X) then g is a continuous unbounded function on X which is a contradiction with pseudocompactness of X. Thus Y is a closed pseudocompact subspace of $C_p(X)$ which fails to be countably compact.

U.046. *Let X be a σ-compact space. Prove that any countably compact subspace of $C_p(X)$ is compact.*

Solution. If $Y \subset C_p(X)$ is countably compact then Y embeds in $C_p(K)$ for some compact space K (see Problem 035). The space $F = \overline{Y} \subset C_p(K)$ is compact by Problem 044; so we can apply Fact 19 of S.351 to conclude that Y is compact.

U.047. *Let X be a space and suppose that there is a point $x_0 \in X$ such that $\psi(x_0, X) = \omega$ and $x_0 \notin \overline{A}$ for any countable $A \subset X$. Prove that there is an infinite closed discrete $B \subset C_p(X)$ such that B is bounded in $C_p(X)$.*

Solution. Fix a sequence $\{U_n : n \in \omega\} \subset \tau(x_0, X)$ such that $\overline{U}_{n+1} \subset U_n$ for any $n \in \omega$ and $\bigcap\{U_n : n \in \omega\} = \{x_0\}$. There exists a function $f_n \in C_p(X)$ such that $f_n(x_0) = 1$ and $f_n(X \backslash U_n) = \{0\}$ for all $n \in \omega$. Let $f(x_0) = 1$ and $f(x) = 0$ for all $x \in X \backslash \{x_0\}$. Then $f \in \mathbb{R}^X \backslash C_p(X)$ and the sequence $B = \{f_n : n \in \omega\}$ converges to f; an easy consequence is that B is a discrete subspace of $C_p(X)$. Now take any countable $A \subset X$; if $P = A \backslash \{x_0\}$ then $x_0 \notin \overline{P}$ so there exists $g \in C_p(X)$ such that $g(x_0) = 1$ and $g|P \equiv 0$. It is immediate that $g|A = f$ and therefore f is strictly ω-continuous.

Denote by S_X the set of all strictly ω-continuous real-valued functions on X. Then S_X can be identified with $\upsilon(C_p(X))$ by TFS-438. Since $K = B \cup \{f\} \subset S_X$ is compact being a convergent sequence, the closure of B in S_X is compact and hence B is bounded in $C_p(X)$ by TFS-415. The equality $B = K \cap C_p(X)$ implies that B is closed in $C_p(X)$ so B is an infinite closed discrete subspace of $C_p(X)$ which is bounded in $C_p(X)$.

U.048. *Prove that there exists a σ-compact space X such that $C_p(X)$ contains an infinite closed discrete subspace which is bounded in $C_p(X)$.*

Solution. Consider the space $Y = \{x \in \mathbb{D}^{\omega_1} : |x^{-1}(1)| < \omega\} \subset \mathbb{D}^{\omega_1}$. We have $Y = \bigcup\{K_n : n \in \omega\}$ where $K_n = \{x \in \mathbb{D}^{\omega_1} : |x(1)| \leq n\}$ for every $n \in \omega$. It is easy to see that every K_n is compact being closed in $\mathbb{D}^{\omega_1}$ so the space Y is σ-compact. Let $x_0(\alpha) = 1$ for any $\alpha \in \omega_1$ and consider the space $X = Y \cup \{x_0\} \subset \mathbb{D}^{\omega_1}$. It is clear that X is σ-compact and $\psi(x_0, X) \leq \omega$.

Observe also that $Y \subset \Sigma = \{x \in \mathbb{D}^{\omega_1} : |x^{-1}(1)| \leq \omega\}$ so if A is a countable subset of Y then $\overline{A} \subset \Sigma$ (the bar denotes the closure in $\mathbb{D}^{\omega_1}$, see Fact 3 of S.307). Thus, for any countable $A \subset X \backslash \{x_0\} = Y$ we have $x_0 \notin \mathrm{cl}_X(A)$ and therefore we can apply Problem 047 to conclude that there is an infinite closed discrete $D \subset C_p(X)$ which is bounded in $C_p(X)$.

U.049. *Prove that there exists a σ-compact space X such that $C_p(X)$ does not embed as a closed subspace into $C_p(Y)$ for any countably compact space Y.*

Solution. It was proved in Problem 048 that there is a σ-compact space X such that some infinite closed discrete $D \subset C_p(X)$ is bounded in $C_p(X)$. Now, if $C_p(X)$ is a closed subspace of $C_p(Y)$ for some countably compact space Y then D is a closed discrete subspace of $C_p(Y)$. Besides, D is bounded in $C_p(Y)$ (it is evident that a bounded subset of a space is bounded in any larger space) so the infinite discrete space $\overline{D} = D$ must be compact by Problem 044; this contradiction shows that $C_p(X)$ does not embed as a closed subspace in $C_p(Y)$ for any countably compact Y.

U.050. *Given a metric space (M, ρ) call a family $\mathcal{U} \subset \exp M \backslash \{\emptyset\}$ ρ-vanishing if $\mathrm{diam}_\rho(U) < \infty$ for any $U \in \mathcal{U}$ and the diameters of the elements of $\mathcal{U}$ converge to* *zero, i.e., the family* $\{U \in \mathcal{U} : diam_\rho(U) \geq \varepsilon\}$ *is finite for any* $\varepsilon > 0$. *Prove that, for any separable metrizable* X, *the following conditions are equivalent:*

(i) X *is a Hurewicz space;*

(ii) *for any metric* ρ *which generates the topology of* X, *there is a* ρ-*vanishing family* $\mathcal{U} \subset \tau(X)$ *such that* $\bigcup \mathcal{U} = X$;

(iii) *for any metric* ρ *which generates the topology of* X, *there exists a* ρ-*vanishing base* $\mathcal{B}$ *of the space* X;

(iv) *there exists a metric* ρ *which generates the topology of* X, *such that, for any base* $\mathcal{B}$ *of the space* X, *there is a* ρ-*vanishing family* $\mathcal{U} \subset \mathcal{B}$ *for which* $\bigcup \mathcal{U} = X$;

(v) *for any metric* ρ *which generates the topology of* X *and any base* $\mathcal{B}$ *of the space* X *there is* ρ-*vanishing family* $\mathcal{B}' \subset \mathcal{B}$ *such that* $\mathcal{B}'$ *is also a base of* X;

(vi) *every base of* X *contains a family which is a locally finite cover of* X.

Solution. In this solution all spaces are assumed to be non-empty. For the sake of brevity we will say that ρ *is a metric on a space* Z if ρ generates the topology of Z. If (M, ρ) is a metric space and $\mathcal{A} \subset \exp M$, say that mesh$(\mathcal{A}) \leq r$ for some $r \geq 0$ if diam$_\rho(A) \leq r$ for any $A \in \mathcal{A}$.

Given a set Z a function $d : Z \times Z \to \mathbb{R}$ is called *a pseudometric on* Z if $d(x, x) = 0$, $d(x, y) = d(y, x) \geq 0$ and $d(x, y) \leq d(x, z) + d(z, y)$ for any $x, y, z \in Z$. In other words, a pseudometric d on a set Z is a function which has all properties of a metric except that $d(x, y) = 0$ need not imply that $x = y$. If Z is a space and d is a pseudometric on Z then we say that d is *a pseudometric on the space* Z if $d : Z \times Z \to \mathbb{R}$ is a continuous function. We can also define the notion of a diameter with respect to a pseudometric in the same way it is defined for a metric, i.e., if d is a pseudometric on Z and $A \subset Z$ then diam$_d(A) = \sup\{d(x, y) : x, y \in A\}$.

Fact 1. Given a metric space (M, ρ) a family $\mathcal{C} \subset \tau(M)$ is a base in M if and only if, for any $\varepsilon > 0$ there is a collection $\mathcal{C}' \subset \mathcal{C}$ such that $\bigcup \mathcal{C}' = M$ and mesh$(\mathcal{C}') \leq \varepsilon$.

Proof. The proof of necessity is straightforward and can be left to the reader. To prove sufficiency, observe that, for any $n \in \mathbb{N}$, there is $\mathcal{C}_n \subset \mathcal{C}$ such that $\bigcup \mathcal{C}_n = M$ and mesh$(\mathcal{C}_n) \leq \frac{1}{n}$. If $x \in U \in \tau(M)$ then there is $r > 0$ such that $B_\rho(x, r) \subset U$; choose $n \in \mathbb{N}$ for which $\frac{1}{n} < r$. There is $V \in \mathcal{C}_n$ such that $x \in V$; we have $\rho(x, y) \leq$ diam$_\rho(V) \leq \frac{1}{n} < r$ for any $y \in V$ and therefore $x \in V \subset B_\rho(x, r) \subset U$ which shows that $\bigcup\{\mathcal{C}_n : n \in \mathbb{N}\} \subset \mathcal{C}$ is a base in X; so Fact 1 is proved.

Fact 2. Suppose that Z is an arbitrary space. Then

(1) for any pseudometrics d_1 and d_2 on the space Z, the function $d = d_1 + d_2$ is a pseudometric on the space Z;

(2) if d is a pseudometric on the space Z then $a \cdot d$ is a pseudometric on Z for any $a > 0$;

(3) for any pseudometrics d_1 and d_2 on the space Z, the function $d = \max\{d_1, d_2\}$ is a pseudometric on the space Z;

(4) if d_1 is a pseudometric on the space Z and $a > 0$ then the function $d : Z \times Z \to$ $\mathbb{R}$ defined by $d(x, y) = \min\{d_1(x, y), a\}$ for all $x, y \in Z$ is a pseudometric on Z;

(5) if, for any $i \in \omega$, a function d_i is a pseudometric on the space Z and $d_i(x, y) \le$ 1 for any $x, y \in Z$ then $d = \sum_{i \in \omega} 2^{-i} \cdot d_i$ is a pseudometric on Z;

(6) if $f : Z \to \mathbb{R}$ is a continuous function then the function $d : Z \times Z \to \mathbb{R}$ defined by $d(x, y) = |f(x) - f(y)|$ for any $x, y \in Z$ is a pseudometric on the space Z;

(7) if d_1 is a metric and d_2 is a pseudometric on the space Z then $d = d_1 + d_2$ is a metric on the space Z.

Proof. In all items non-negativity of the respective function is evident; so we omit its proof.

(1) It is clear that d is continuous on $Z \times Z$. Given any $x, y, z \in Z$ we have $d(x, x) = d_1(x, x) + d_2(x, x) = 0$ and

$$d(x, y) = d_1(x, y) + d_2(x, y) = d_1(y, x) + d_2(y, x) = d(y, x);$$

besides,

$$d(x, y) = d_1(x, y) + d_2(x, y) \le d_1(x, z) + d_1(z, y) + d_2(x, z) + d_2(z, y)$$
$$= d(x, z) + d(z, y)$$

so (1) is proved.

The proof of the property (2) is even more straightforward than (1), so we omit it. As to the property (3), the function d is continuous by TFS-028. Furthermore, for any points $x, y \in Z$ we have the equalities $d(x, x) = \max\{d_1(x, x), d_2(x, x)\} = 0$ and $d(x, y) = \max\{d_1(x, y), d_2(x, y)\} = \max\{d_1(y, x), d_2(y, x)\} = d(y, x)$. To see that the triangle inequality is true take any points $x, y, z \in Z$. Note first that $d(x, y) = d_i(x, y) \le d_i(x, z) + d_i(z, y)$ for some $i \in \{1, 2\}$. Since $d_i(x, z) \le$ $d(x, z)$ and $d_i(z, y) \le d(z, y)$ by the definition of d, we have $d(x, y) \le d(x, z) +$ $d(z, y)$ so (3) is proved.

(4) Again, continuity of d follows from TFS-028. If $x, y \in Z$ then we have the equalities $d(x, x) = \min\{0, a\} = 0$ and

$$d(x, y) = \min\{d_1(x, y), a\} = \min\{d_1(y, x), a\} = d(y, x).$$

To prove the triangle inequality take any $x, y, z \in Z$. If $d(x, y) = a$ and one of the numbers $d_1(x, z), d_1(z, y)$ is greater than or equal to a then one of the numbers $d(x, z), d(z, y)$ is equal to a; so $d(x, y) = a \le d(x, z) + d(y, z)$. If $d_1(x, z) < a$ and $d_1(z, y) < a$ then $d(x, y) = a \le d_1(x, y) \le d_1(x, z) + d_1(z, y) = d(x, z) + d(z, y)$.

Now, if $d_1(x, y) < a$ and one of the numbers $d_1(x, z), d_1(z, y)$ is greater than or equal to a then one of the numbers $d(x, z), d(z, y)$ is equal to a and therefore $d(x, y) = d_1(x, y) < a \le d(x, z) + d(y, z)$. If $d_1(x, z) < a$ and $d_1(z, y) < a$ then $d(x, y) = d_1(x, y) \le d_1(x, z) + d_1(z, y) = d(x, z) + d(z, y)$ and (4) is proved.

(5) Continuity of the function d follows easily from TFS-029 and TFS-030. For any point $x \in Z$ we have $d_i(x, x) = 0$ for all $i \in \omega$ so $d(x, x) = 0$. If $x, y \in Z$ then $d_i(x, y) = d_i(y, x)$ for all $i \in \omega$ which implies $d(x, y) = d(y, x)$. Now, if $x, y, z \in Z$ then $2^{-i}d_i(x, y) \leq 2^{-i}d_i(x, z) + 2^{-i}d_i(z, y)$ for every $i \in \omega$. Passing to the sums of the respective series we obtain $d(x, y) \leq d(x, z) + d(z, y)$.

The proof of (6) is also straightforward and trivial; so we leave it to the reader. Finally, if d_1 is a metric and d_2 is a pseudometric on the space Z then $d_1 + d_2$ is a pseudometric on Z by (1). If $x, y \in Z$ and $x \neq y$ then $d_1(x, y) > 0$ because d_1 is a metric; so we have $d(x, y) \geq d_1(x, y) > 0$ which shows that d is also a metric; this settles (7) and finishes the proof of Fact 2.

Returning to our solution assume that X is a Hurewicz space and take any metric ρ on the space X. If $\mathcal{B}$ is a base of X then $\mathcal{B}_i = \{B \in \mathcal{B} : \operatorname{diam}_\rho(B) < 2^{-i}\}$ is also a base in X for any $i \in \omega$. Fix any $j \in \omega$; for any $i \in \omega$, we can choose a finite $\mathcal{U}_i^j \subset \mathcal{B}_{i+j}$ in such a way that $\mathcal{U}_j = \bigcup_{i \in \omega} \mathcal{U}_i^j$ is a cover of X. It is clear that $\operatorname{mesh}(\mathcal{U}_j) \leq 2^{-j}$ for every $j \in \omega$; so the family $\mathcal{B}' = \bigcup_{j \in \omega} \mathcal{U}_j \subset \mathcal{B}$ is a base of X by Fact 1.

Now take any $\varepsilon > 0$; there is $n \in \omega$ such that $2^{-j} < \varepsilon$ for any $j \geq n$. Thus the family $\mathcal{E} = \{B \in \mathcal{B}' : \operatorname{diam}_\rho(B) \geq \varepsilon\}$ is contained in $\bigcup_{j < n} \mathcal{U}_j$. Furthermore, $\mathcal{E} \cap \mathcal{U}_j \subset \bigcup\{\mathcal{U}_i^j : i < n\}$ and hence $\mathcal{E} \cap \mathcal{U}_j$ is finite for any $j < n$. Thus $\mathcal{E}$ is finite and therefore the family $\mathcal{B}'$ is ρ-vanishing. This proves that (i)$\Longrightarrow$(v).

Now assume that (v) is true and take any base $\mathcal{B}$ in X; fix a metric ρ on the space X. There exists a ρ-vanishing family $\mathcal{B}' \subset \mathcal{B}$ which is a base of X and hence $\bigcup \mathcal{B}' = X$. Since X is paracompact, for any $U \in \mathcal{B}'$ there is a closed $F_U \subset U$ such that $\bigcup\{F_U : U \in \mathcal{B}'\} = X$ and the family $\mathcal{F} = \{F_U : U \in \mathcal{B}'\}$ is locally finite (see Fact 2 of S.226).

The family $\mathcal{B}_1 = \{U \in \mathcal{B}' : F_U \neq \emptyset\} \subset \mathcal{B}$ is, clearly, a cover of X. Since $\mathcal{F}$ is locally finite, for any $x \in X$ there is $r > 0$ and a finite $\mathcal{C} \subset \mathcal{B}_1$ such that $B_\rho(x, r) \cap F_U = \emptyset$ whenever $U \in \mathcal{B}_1 \backslash \mathcal{C}$. The family $\mathcal{E} = \{U \in \mathcal{B}_1 : \operatorname{diam}_\rho(U) \geq \frac{r}{3}\}$ is also finite. For any $U \in \mathcal{B}_1 \backslash \mathcal{E}$ we have $\operatorname{diam}_\rho(U) < \frac{r}{3}$ so $U \cap B_\rho(x, \frac{r}{3}) \neq \emptyset$ implies that $U \subset B_\rho(x, r)$ and hence $F_U \subset B_\rho(x, r)$. As a consequence $B_\rho(x, \frac{r}{3}) \cap U = \emptyset$ for any $U \in \mathcal{B}_1 \backslash (\mathcal{C} \cup \mathcal{E})$. This shows that $\mathcal{B}_1$ is locally finite and proves (v)$\Longrightarrow$(vi).

Now assume that (vi) is true; we can consider that $X \subset \mathbb{I}^\omega$. Choose a metric d on the space $\mathbb{I}^\omega$ such that $d(x, y) \leq 1$ for any $x, y \in \mathbb{I}^\omega$. The metric $\rho = d|(X \times X)$ is totally bounded on X (see TFS-212 and Fact 1 of S.249); so we can fix a sequence $\{F_n : n \in \omega\}$ of finite subsets of X such that $\rho(x, F_n) = \min\{\rho(x, y) : y \in F_n\} \leq 2^{-n}$ for any $x \in X$ and $n \in \omega$.

Now take any base $\mathcal{B} \subset \tau^*(X)$ of the space X and consider, for every $i \in \omega$, the family $\mathcal{B}_i = \{U \in \mathcal{B} : 2^{-i-1} \leq \operatorname{diam}_\rho(U) \leq 2^{-i}\}$. For all $i \in \omega$ and $U \in \mathcal{B}_i$ choose $x \in U$ and $y \in F_i$ in such a way that $\rho(x, y) \leq 2^{-i}$ and let $O_U = U \cup B_\rho(y, 2^{-i})$; then $\operatorname{diam}_\rho(O_U) \leq 2^{-i+2}$.

Let D be the set of all isolated points of X; it is straightforward that the family $\mathcal{C} = \{\{x\} : x \in D\} \cup \{O_U : U \in \bigcup_{i \in \omega} \mathcal{B}_i\}$ is a base in the space X. By (vi), there exists a locally finite subfamily of $\mathcal{C}$ which covers X. In particular, for every $i \in \omega$, there is $\mathcal{B}'_i \subset \mathcal{B}_i$ such that the family $\mathcal{C}' = \bigcup\{\{O_U : U \in \mathcal{B}'_i\} : i \in \omega\}$ is locally finite and covers $X \backslash D$.

Suppose that $\mathcal{B}'_i$ is infinite for some $i \in \omega$ and let $\mathcal{C}'_i = \{O_U : U \in \mathcal{B}'_i\} \subset \mathcal{C}$. Since $O_U \cap F_i \neq \emptyset$ for any $U \in \mathcal{B}'_i$ and the set F_i is finite, there is $x \in F_i$ such that the family $\{U \in \mathcal{B}'_i : x \in O_U\}$ is infinite which is a contradiction with the fact that $\mathcal{C}'_i$ is locally finite. Thus $\mathcal{B}'_i$ is finite for every $i \in \omega$ and hence the family $\mathcal{C}'$ is ρ-vanishing. Then $\mathcal{C} \cup \{\{x\} : x \in D\}$ is also ρ-vanishing and covers X so we proved that (vi)$\Longrightarrow$(iv).

Let us assume that (iv) holds; denote by D the set of isolated points of the space X and let $\mathcal{D} = \{\{x\} : x \in D\}$. Take a sequence $\{\mathcal{U}_n : n \in \omega\}$ of open covers of X; say that a set $A \subset X$ is $\mathcal{U}_i$-small if there is $U \in \mathcal{U}_i$ such that $A \subset U$. For any $i \in \omega$ let $C_i = \{(x, y) \in X \times X : 2^{-i} \leq \rho(x, y) \leq 2^{-i+1}\}$. For every $i \in \omega$ and $z = (x, y) \in C_i$ choose $O_z \in \tau(x, X)$ and $V_z \in \tau(y, X)$ such that $\operatorname{diam}_\rho(O_z) \leq 2^{-i}, \operatorname{diam}_\rho(V_z) \leq 2^{-i}$ and both sets O_z, V_z are $\mathcal{U}_i$-small. Let $\mathcal{B}_i = \{O_z \cup V_z : z \in C_i\}$; it is an easy consequence of Fact 1 that the family $\mathcal{B} = \bigcup\{\mathcal{B}_i : i \in \omega\} \cup \mathcal{D}$ is a base of the space X. By (iv), there is a ρ-vanishing $\mathcal{B}' \subset \mathcal{B}$ such that $\bigcup \mathcal{B}' = X$. The family $\mathcal{B}'_i = \mathcal{B}' \cap \mathcal{B}_i$ has to be finite for every $i \in \omega$ because $\operatorname{diam}_\rho(W) \geq 2^{-i}$ for any $W \in \mathcal{B}'_i$.

Given $i \in \omega$ every element of $\mathcal{B}'_i$ is a union of two $\mathcal{U}_i$-small sets; so there is a finite $\mathcal{V}'_i \subset \mathcal{U}_i$ for which $\bigcup \mathcal{B}'_i \subset \bigcup \mathcal{V}_i$. The set D being countable we can choose a finite $\mathcal{V}''_i \subset \mathcal{U}_i$ for every $i \in \omega$ in such a way that $\bigcup\{\mathcal{V}''_i : i \in \omega\}$ covers D. Now, $\mathcal{V}_i = \mathcal{V}'_i \cup \mathcal{V}''_i$ is a finite subfamily of $\mathcal{U}_i$ for every $i \in \omega$ and $\bigcup\{\mathcal{V}_i : i \in \omega\}$ is a cover of X. This proves that X is a Hurewicz space, i.e., (iv)$\Longrightarrow$(i).

As a consequence, all properties (i),(iv),(v),(vi) are proved to be equivalent. The implications (v)$\Longrightarrow$(iii)$\Longrightarrow$(ii) are evident; so, to finish our solution, it suffices to establish that (ii)$\Longrightarrow$(i).

Suppose that (ii) holds and $\{\mathcal{U}_n : n \in \omega\}$ is a sequence of open covers of X. Since X is second countable, we can assume that $|\mathcal{U}_n| \leq \omega$ and $\mathcal{U}_n$ is locally finite for all $n \in \omega$. If some $\mathcal{U}_n$ is finite then there is nothing to prove; so there is no loss of generality to assume that, for every $n \in \omega$ we have a faithful enumeration $\{U^n_i : i \in \omega\}$ of the family $\mathcal{U}_n$. Apply Fact 2 of S.226 to find a family $\{F^n_i : i \in \omega\}$ of closed subsets of X such that $F^n_i \subset U^n_i$ for all $n, i \in \omega$ and $\mathcal{F}_n = \{F^n_i : i \in \omega\}$ is a cover of X for every $n \in \omega$. It follows from normality of the space X that we can construct a family $\{f^n_i : n, i \in \omega\} \subset C_p(X)$ with the following properties:

(a) $f^n_i|(X \backslash U^n_i) \equiv 0$ for any $n, i \in \omega$;
(b) $f^0_i|F^0_i \equiv i + 1$ for every $i \in \omega$;
(c) $f^n_i|F^n_i \equiv 1$ for any $n \in \mathbb{N}$ and $i \in \omega$.

Since every family $\mathcal{U}_n$ is locally finite, the function $g_n = \sum_{i \in \omega} f^n_i$ is continuous. For every $n \in \omega$ define $d_n : X \times X \to \mathbb{R}$ as follows: $d_0(x, y) = |g_0(x) - g_0(y)|$ and $d_n(x, y) = \min\{1, |g_n(x) - g_n(y)|\}$ for all $n \in \mathbb{N}$ and $x, y \in X$. It follows from Fact 2 that d_n is a continuous pseudometric on X for any $n \in \omega$ and hence $d = d_0 + \sum\{2^{-i} d_i : i \in \mathbb{N}\}$ is also a pseudometric on the space X. Take any metric σ on the space X and let $\rho = d + \sigma$; then ρ is a metric which generates the topology of X (see Fact 2).

By (ii) applied to the metric ρ, there is a ρ-vanishing family $\mathcal{O} \subset \tau(X)$ such that $\bigcup \mathcal{O} = X$. Consequently, $\mathcal{O}_0 = \{O \in \mathcal{O} : \operatorname{diam}_\rho(O) \geq \frac{1}{2}\}$ is finite as well as the family $\mathcal{O}_i = \{O \in \mathcal{O} : 2^{-i-1} \le \mathrm{diam}_\rho(O) < 2^{-i}\}$ for each $i \in \mathbb{N}$. We have the equality $\bigcup\{\mathcal{O}_i : i \in \omega\} = \mathcal{O}$; let $W_i = \bigcup \mathcal{O}_i$ for every $i \in \omega$. Since $\mathrm{diam}_\rho(O) < \infty$ for every $O \in \mathcal{O}$ and W_i is a finite union of elements of $\mathcal{O}$, we have $\mathrm{diam}_\rho(W_i) < \infty$ for every $i \in \omega$.

Assume first that infinitely many elements of $\mathcal{F}_0$ meet W_0. Then we can choose a strictly increasing sequence $\{n_k : n \in \omega\} \subset \omega$ and $z_k \in F_{n_k}^0 \cap W_0$ for any $k \in \omega$. The family $\mathcal{U}_0$ is locally finite; so z_0 belongs to finitely many elements of $\mathcal{U}_0$ which implies that $f_i^0(z_0) = 0$ for all but finitely many $i \in \omega$. Therefore $\rho(z_0, z_k) \ge d(z_0, z_k) \ge |f_{n_k}^0(z_0) - f_{n_k}^0(z_k)| = n_k + 1$ for all but finitely many k. Since $n_k \to \infty$, we have $\mathrm{diam}_\rho(W_0) = \infty$; this contradiction proves that

(d_0) the set $Q_0 = \{n \in \omega : F_n^0 \cap W_0 \ne \varnothing\}$ is finite.

Now assume that $i \in \mathbb{N}$ and W_i meets infinitely many elements of $\mathcal{F}_i$; since $W_i = \bigcup \mathcal{O}_i$ and $\mathcal{O}_i$ is finite, there is $O \in \mathcal{O}_i$ which intersects infinitely many elements of $\mathcal{F}_i$. We can choose a strictly increasing sequence $\{n_k : n \in \omega\} \subset \omega$ and $z_k \in F_{n_k}^i \cap O$ for any $k \in \omega$. The family $\mathcal{U}_i$ is locally finite; so z_0 belongs to finitely many elements of $\mathcal{U}_i$ which implies that $f_k^i(z_0) = 0$ for all but finitely many $k \in \omega$. Therefore $\rho(z_0, z_k) \ge d(z_0, z_k) \ge 2^{-i}|f_{n_k}^i(z_0) - f_{n_k}^i(z_k)| = 2^{-i}$ for all but finitely many k. Thus $\mathrm{diam}_\rho(O) \ge 2^{-i}$; this contradiction proves that

(d) the set $Q_i = \{n \in \omega : F_n^i \cap W_i \ne \varnothing\}$ is finite for any $i \in \mathbb{N}$.

It follows from (d_0) and (d) that the family $\mathcal{U}_i' = \{U_k^i : k \in Q_i\} \subset \mathcal{U}_i$ is finite for any $i \in \omega$. To see that $\mathcal{U}' = \bigcup\{\mathcal{U}_i' : i \in \omega\}$ is a cover of X take any $x \in X$. Since $\mathcal{O}$ is a cover of X, there is $i \in \omega$ for which $x \in O \in \mathcal{O}_i$ and hence $x \in W_i$. The family $\mathcal{F}_i$ being a cover of X there is $F_k^i \in \mathcal{F}_i$ for which $x \in F_k^i$. Then $F_k^i \cap W_i \ne \varnothing$ and hence $k \in Q_i$. Therefore $x \in F_k^i \subset U_k^i \in \mathcal{U}_i'$; so we proved that $\mathcal{U}'$ is a cover of X and hence X is a Hurewicz space. Thus (ii)$\Longrightarrow$(i) and our solution is complete.

U.051. *Prove that X^ω is a Hurewicz space if and only if X is compact.*

Solution. If X is compact then X^ω is also compact and hence Hurewicz. Now, assume, towards a contradiction, that X^ω is Hurewicz and non-compact. Since X is Lindelöf, it cannot be countably compact; so there is a countably infinite closed discrete $D \subset X$. Therefore D^ω is a closed subspace of X^ω homeomorphic to $\mathbb{P}$. Every closed subspace of a Hurewicz space is a Hurewicz space; so $\mathbb{P}$ has to be a Hurewicz space which it is not (see Fact 1 of T.132). This contradiction shows that X is countably compact and hence compact.

U.052. *Prove that any separable Luzin space is a Hurewicz space.*

Solution. Take a separable Luzin space X and let $D = \{d_n : n \in \omega\}$ be a countable dense subspace of X. If $\{\mathcal{U}_n : n \in \omega\}$ is a sequence of open covers of the space X then choose $U_n \in \mathcal{U}_n$ such that $d_n \in U_n$ for any $n \in \omega$. Then $D \subset U = \bigcup_{n\in\omega} U_n$ whence $F = X\backslash U$ is nowhere dense and hence countable; let $\{x_n : n \in \omega\}$ be an enumeration (with possible repetitions) of the set F. For every $n \in \omega$ choose $V_n \in \mathcal{U}_n$ such that $x_n \in V_n$; then $\mathcal{U}_n' = \{U_n, V_n\}$ is a finite subfamily of $\mathcal{U}_n$ and $\bigcup\{\bigcup \mathcal{U}_n' : n \in \omega\} = X$ so X is Hurewicz.

U.053. *Prove that any Hurewicz analytic space is σ-compact.*

Solution. Let X be a Hurewicz analytic space. If X is not σ-compact then there is a closed $Y \subset X$ such that $Y \simeq \mathbb{P}$ (see SFFS-352). Every closed subspace of a Hurewicz space is a Hurewicz space; so $\mathbb{P}$ has to be a Hurewicz space which is a contradiction (see Fact 1 of T.132). Therefore X has to be σ-compact.

U.054. *Give an example of a Hurewicz space which is not σ-compact.*

Solution. Let X be the Lindelöfication of the discrete space of cardinality ω_1, i.e., $X = \omega_1 \cup \{a\}$, where $a \notin \omega_1$ is the unique non-isolated point of X and $U \in \tau(a, X)$ if and only if $U = \{a\} \cup (\omega_1 \backslash B)$ where B is a countable subset of ω_1. Since every countable subset of X is closed and discrete, every compact subspace of X is finite and hence X is not σ-compact.

Now, assume that $\{\mathcal{U}_n : n \in \omega\}$ is a sequence of open covers of X. For every $n \in \omega$ choose $U_n \in \mathcal{U}_n$ such that $a \in U_n$; then $B_n = X \backslash U_n$ is countable and therefore so is the set $B = \bigcup\{B_n : n \in \omega\}$. Choose an enumeration $\{b_n : n \in \omega\}$ (with repetitions allowed) of the set B. For each $n \in \omega$ there is $V_n \in \mathcal{U}_n$ for which $b_n \in V_n$. The family $\mathcal{U}'_n = \{U_n, V_n\} \subset \mathcal{U}_n$ is finite for every $n \in \omega$ and $\bigcup\{\bigcup \mathcal{U}'_n : n \in \omega\} = X$ which shows that X is a Hurewicz non-σ-compact space.

U.055. *Prove that, under CH, there exists a Hurewicz space whose square is not normal.*

Solution. If Z is a set and $\mathcal{A} \subset \exp Z$ then $\mathcal{A}|Y = \{A \cap Y : A \in \mathcal{A}\}$ for any $Y \subset Z$. Under CH there is a dense Luzin subspace $H \subset \mathbb{R}$ (see Fact 1 of T.046); let $H_1 = \{-r : r \in H\}$ and $L = H \cup H_1$. Observe that, for any dense $Y \subset \mathbb{R}$, a set $N \subset Y$ is nowhere dense in Y if and only if it is nowhere dense in $\mathbb{R}$. Therefore, if $N \subset L$ is nowhere dense in L then N is nowhere dense in $\mathbb{R}$ and hence both sets $N \cap H$ and $N \cap H_1$ are nowhere dense in H and H_1 respectively. Since both H and H_1 are Luzin spaces, we have $|N \cap H| \leq \omega$ and $|N \cap H_1| \leq \omega$ whence $|N| \leq \omega$. This proves that

(1) there is a dense Luzin space $L \subset \mathbb{R}$ such that $-r \in L$ for any $r \in L$.

Let θ be the Sorgenfrey line topology on $\mathbb{R}$ (see TFS-165). Recall that the family $\mathcal{B} = \{[a, b) : a, b \in \mathbb{R}, a < b\}$ is a base for the space $S = (\mathbb{R}, \theta)$. We claim that $X = (L, \theta|L)$ is a Hurewicz space.

Indeed, let $\{\mathcal{U}_n : n \in \omega\}$ be a sequence of open covers of X; we can assume, without loss of generality, that $\mathcal{U}_n \subset \mathcal{B}|L$ for any $n \in \omega$. Since S is hereditarily separable, we can find a dense countable $A \subset X$; let $\{a_n : n \in \omega\}$ be an enumeration of the set A. For every $n \in \omega$ we can choose $U_n \in \mathcal{U}_n$ such that $a_n \in U_n$; we have $U_n = [a, b) \cap L$ for some $[a, b) \in \mathcal{B}$; let $U'_n = (a, b) \cap L$. It is evident that $\mathcal{U}' = \{U'_n : n \in \omega\} \subset \tau(\mathbb{R})|L$; besides, it follows from density of L in $\mathbb{R}$ that $a_n \in \overline{U'_n}$ for any $n \in \omega$ (the bar denotes the closure in $\mathbb{R}$).

As a consequence, $W = \bigcup \mathcal{U}'$ is an open dense subspace of L; since L is Luzin, the set $L \backslash W$ is countable. Let $\{w_n : n \in \omega\}$ be an enumeration of W. We can choose $V_n \in \mathcal{U}_n$ such that $w_n \in V_n$ for every $n \in \omega$. Then $\mathcal{U}'_n = \{U_n, V_n\}$ is a finite subfamily of $\mathcal{U}_n$ for each $n \in \omega$ and $\bigcup\{\bigcup \mathcal{U}'_n : n \in \omega\} = X$ which shows that X is a Hurewicz space.

To see that $X \times X$ is not normal consider the set $D = \{(r, -r) : r \in L\} \subset \mathbb{R} \times \mathbb{R}$; it follows from (1) that $D \subset X \times X$. Observe that the set D is closed in $L \times L$ because it is a graph of a continuous function $f : L \to L$ defined by $f(x) = -x$ for any $x \in L$ (see Fact 4 of S.390). The topology of X is stronger than that of L so D is also closed in $X \times X$. For any $r \in L$ the set $O_r = [r, r + 1) \times [-r, -r + 1)$ is open in $S \times S$ and $O_r \cap D = \{(r, -r)\}$ which shows that D is also discrete in $X \times X$. We have assumed that CH holds; so $|D| = \mathfrak{c}$, i.e., $X \times X$ is a separable space with $ext(X \times X) = \mathfrak{c}$. Consequently, $X \times X$ is not normal by TFS-164 and our solution is complete.

U.056. *Prove that X^n is a Hurewicz space for every $n \in \mathbb{N}$, if and only if, for any sequence $\{\gamma_k : k \in \omega\}$ of open ω-covers of the space X, we can choose, for each $k \in \omega$, a finite $\mu_k \subset \gamma_k$ such that the family $\bigcup\{\mu_k : k \in \omega\}$ is an ω-cover of X.*

Solution. This was proved in Fact 1 of T.188.

U.057. *Let X be any space. Prove that X^n is a Hurewicz space for all $n \in \mathbb{N}$ if and only if $vet(C_p(X)) \leq \omega$.*

Solution. This was proved in Fact 2 of T.188.

U.058. *Prove that if $C_p(X)$ is Fréchet–Urysohn then $vet(C_p(X)) \leq \omega$.*

Solution. If $C_p(X)$ is Fréchet–Urysohn then, for any sequence $\{\mathcal{U}_n : n \in \omega\}$ of open ω-covers of X, we can choose $U_n \in \mathcal{U}_n$ for every $n \in \omega$ in such a way that $\lim\{U_n : n \in \omega\} = X$ (see TFS-144). This implies that $\{U_n : n \in \omega\}$ is an ω-cover of X; so, for any $n \in \omega$, the family $\mathcal{U}'_n = \{U_n\} \subset \mathcal{U}_n$ is finite and $\bigcup\{\mathcal{U}'_n : n \in \omega\}$ is an ω-cover of X. Finally, apply Problem 056 to see that X^n is a Hurewicz space for all $n \in \mathbb{N}$ and hence $vet(C_p(X)) \leq \omega$ by Problem 057.

U.059. *Prove that, under MA+¬CH, there exists a second countable space X such that X^n is a Hurewicz space for each natural n, while X is not σ-compact.*

Solution. Let $X \subset \mathbb{R}$ be any subset with $|X| = \omega_1$. Then $|X| < \mathfrak{c}$; so every compact $K \subset X$ is countable by SFFS-353; this proves that X is not σ-compact. We have $t(C_p(X)) \leq nw(C_p(X)) = nw(X) = \omega$ and $\chi(C_p(X)) = |X| = \omega_1 < \mathfrak{c}$ so SFFS-054 can be applied to conclude that $C_p(X)$ is a Fréchet–Urysohn space. Thus $vet(C_p(X)) \leq \omega$ by Problem 058 and hence X^n is a Hurewicz space for all $n \in \mathbb{N}$ (see Problem 057).

U.060. *Say that a space is* subsequential *if it embeds in a sequential space. Prove that every sequential space has countable tightness and hence each subsequential space also has countable tightness.*

Solution. It was proved in Fact 3 of T.041 that every sequential space has countable tightness. It is immediate that countable tightness is a hereditary property; so every space which embeds in a sequential space has countable tightness.

U.061. *For any point $\xi \in \beta\omega \setminus \omega$ prove that the countable space $\omega \cup \{\xi\}$ is not subsequential.*

Solution. Given a space Z and a point $z \in Z$ call a family $\mathcal{A} \subset \exp Z \backslash \{\emptyset\}$ a π-network at z if, for any $U \in \tau(z, Z)$ there is $A \in \mathcal{A}$ such that $A \subset U$. For any $A \subset Z$ denote by $\text{Seq}(A)$ the set of all limits of sequences contained in A. It is clear that $A \subset \text{Seq}(A) \subset \overline{A}$. Let $S_0(A) = A$ and, if we have sets $\{S_\beta(A) : \beta < \alpha\}$ for some $\alpha < \omega_1$, let $S_\alpha(A) = \text{Seq}(\bigcup\{A_\beta : \beta < \alpha\})$. Observe that the family $\{S_\alpha(A) : \alpha < \omega_1\}$ has the following properties:

(1) $A \subset S_\alpha(A) \subset \overline{A}$ for all $\alpha < \omega_1$;
(2) $S_\alpha(A) \subset S_\beta(A)$ if $\alpha < \beta < \omega_1$

and therefore the set $S(A) = \bigcup\{S_\alpha(A) : \alpha < \omega_1\}$ also lies between A and the closure of A.

Fact 1. If Z is a sequential space, $A \subset Z$ and $z \in \overline{A} \backslash A$ then z has a countable π-network in Z which consists of infinite subsets of A.

Proof. It follows from sequentiality of Z that $S(A) = \overline{A}$ (see Fact 1 of T.041); so we can use the induction on $\gamma(z) = \min\{\alpha : z \in S_\alpha(A)\}$. If $\gamma(z) = 1$ then there is a sequence $T = \{a_n : n \in \omega\} \subset A$ such that z is the limit of T. Since $z \notin A$, the sequence T is infinite so we can consider that its enumeration is faithful, i.e., $a_n \neq a_m$ if $n \neq m$. It is evident that $\mathcal{E} = \{\{a_k : k \geq n\} : n \in \omega\} \subset \exp A$ is a countable π-network at z and all elements of $\mathcal{E}$ are infinite.

Now assume that $1 < \alpha < \omega_1$ and we proved our Fact for all points $z \in \overline{A} \backslash A$ with $\gamma(z) < \alpha$. If $\gamma(z) = \alpha$ then $z \in S_\alpha(A) \backslash (\bigcup\{S_\beta(A) : \beta < \alpha\})$ and there is a convergent sequence $T = \{z_n : n \in \omega\} \subset A' = \bigcup\{S_\beta(A) : \beta < \alpha\}$ such that z is a limit of T. The set $T \cap A$ has to be finite for otherwise there is a sequence in A which converges to z, a contradiction with $\gamma(z) > 1$.

Therefore we can apply the induction hypothesis to find a countable family $\mathcal{E}_n \subset \exp A$ such that $\mathcal{E}_n$ is a π-network at z_n and all elements of $\mathcal{E}_n$ are infinite for every $n \in \omega$. Then $\mathcal{E} = \bigcup\{\mathcal{E}_n : n \in \omega\} \subset \exp A$, all elements of $\mathcal{E}$ are infinite and it is straightforward that $\mathcal{E}$ is a countable π-network at z. Thus our inductive procedure can be carried out for all $\alpha < \omega_1$ which shows that every $z \in \overline{A} \backslash A$ has a countable π-network with infinite elements; so Fact 1 is proved.

Returning to our solution assume that $X = \omega \cup \{\xi\}$ can be embedded in some sequential space Y. Then there is a family $\mathcal{E} = \{A_n : n \in \omega\}$ of infinite subsets of ω which is a π-network at ξ in Y (see Fact 1). It is clear that the notion of π-network is hereditary; so $\mathcal{E}$ is a π-network at ξ in X.

Take any distinct points $a_0, b_0 \in A_0$; assume that $1 \leq n < \omega$ and we have chosen points $\{a_i, b_i : i < n\}$ with the following properties:

(3) $a_i, b_i \in A_i$ for all $i < n$;
(4) $a_i \neq b_j$ for all $i, j < n$;
(5) $i \neq j$ implies $a_i \neq a_j$ for all $i, j < n$.

Since the set $B = \{a_i, b_i : i < n\}$ is finite, we can choose distinct $a_n, b_n \in A_n \backslash B$. It is immediate that the properties (3)–(5) still hold for the set $\{a_i, b_i : i \leq n\}$; so our inductive construction can be carried out for all $n \in \omega$ giving us sets $P = \{a_i : i \in \omega\}$ and $Q = \{b_i : i \in \omega\}$ with the properties (3)–(5) fulfilled for all $i \in \omega$.

If $\xi \in \overline{P}$ then $P \cup \{\xi\}$ is an open neighbourhood of ξ in X (see Fact 1 of S.370) so there is $n \in \omega$ such that $A_n \subset P$ which is a contradiction with $b_n \in A_n \cap Q$. Therefore $\xi \notin \overline{P}$ and hence $(\omega \backslash P) \cup \{\xi\}$ is an open neighbourhood of ξ in X; so again there is $n \in \omega$ for which $A_n \subset (\omega \backslash P)$. However, this contradicts $a_n \in A_n \cap P$ and hence we proved that $X = \omega \cup \{\xi\}$ cannot be embedded in a sequential space.

U.062. *Prove that $C_p(\mathbb{I})$ is not subsequential.*

Solution. Recall that, for any spaces X and Y, a map $\varphi : C_p(X) \to C_p(Y)$ is called *linear* if $\varphi(\lambda f + \mu g) = \lambda \varphi(f) + \mu \varphi(g)$ for any $f, g \in C_p(X)$ and $\lambda, \mu \in \mathbb{R}$.

Fact 1. Let A be a non-empty closed subspace of a metrizable space M. Then there exists a continuous linear map $e : C_p(A) \to C_p(M)$ such that $e(f)|A = f$ for any $f \in C_p(A)$.

Proof. We can choose a metric d on M with $\tau(d) = \tau(M)$ and a Dugundji system $\{U_s, a_s : s \in S\}$ for $M \backslash A$ (see SFFS-103). Apply Fact 4 of T.104 to find a locally finite (in $M \backslash A$) family $\{F_s : s \in S\}$ such that F_s is closed in $M \backslash A$ and $F_s \subset U_s$ for each $s \in S$. By normality of the space $M \backslash A$ there exists a continuous function $c_s : M \backslash A \to [0, 1]$ such that $c_s(F_s) \subset \{1\}$ and $c_s((M \backslash A) \backslash U_s) \subset \{0\}$ for every $s \in S$.

Since the family $\{U_s : s \in S\}$ is locally finite in the space $M \backslash A$, the set $S(x) = \{s \in S : c_s(x) \neq 0\}$ is finite for every $x \in M \backslash A$ and hence the function $c = \sum_{s \in S} c_s$ is well defined. Given any $x \in M \backslash A$ there is $W \in \tau(x, M \backslash A)$ such that $S' = \{s \in S : W \cap U_s \neq \emptyset\}$ is finite; it is evident that $S(y) \subset S'$ for any $y \in W$ and hence $c|W = (\sum_{s \in S'} c_s)|W$ is continuous. Thus we can apply Fact 1 of S.472 to conclude that $c : M \backslash A \to \mathbb{R}$ is continuous.

Note that, for any $x \in M \backslash A$ there is $s \in S$ such that $x \in F_s$ and hence $c_s(x) = 1$; this shows that $c(x) \geq 1 > 0$ for any $x \in M \backslash A$ so the function $b_s = \frac{c_s}{c}$ is well defined. It is immediate that

(1) $(\sum_{s \in S} b_s)(x) = 1$ for any $x \in M \backslash A$.

For any function $f \in C_p(A)$ let $e(f)(x) = f(x)$ for every point $x \in A$ and $e(f)(x) = \sum_{s \in S} b_s(x) f(a_s)$ for each $x \in M \backslash A$. Observe that $e(f)(x)$ makes sense for any $x \in M \backslash A$ because $b_s(x) = 0$ for any $s \in S \backslash S(x)$. We have $e(f)|A = f$ by our definition of $e(f)$; let us prove that $e(f)$ is continuous. Given any point $x \in M \backslash A$ there is $W \in \tau(x, M \backslash A)$ such that the set $T(x) = \{s \in S : W \cap U_s \neq \emptyset\}$ is finite. As a consequence, $S(y) \subset T(x)$ for every $y \in W$ which shows that the function $e(f)|W = (\sum_{s \in T(x)} f(a_s) \cdot b_s)|W$ is continuous. Therefore $e(f)$ is continuous at the point x.

Now take any point $a \in A$ and any $\varepsilon > 0$. Since f is continuous on A, we can find $\delta > 0$ such that $a' \in A$ and $d(a, a') < \delta$ imply $|f(a') - f(a)| < \varepsilon$. Let $U = B(a, \frac{\delta}{3}) = \{a' \in M : d(a, a') < \frac{\delta}{3}\}$. If $x \in U \cap A$ then $|e(f)(x) - e(f)(a)| = |f(x) - f(a)| < \varepsilon$. Now, if $x \in U \backslash A$ then $d(x, A) \leq d(x, a) < \frac{\delta}{3}$. Besides, if $x \in U_s$ then $d(a_s, a) \leq d(a_s, x) + d(x, a) \leq 2d(x, A) + d(x, a) < \frac{2\delta}{3} + \frac{\delta}{3} = \delta$ (we used the fact that $\{U_s, a_s : s \in S\}$ is a Dugundji system and therefore

$d(x, a_s) \leq 2d(x, A)$ because $x \in U_s$). As a consequence, $|f(a_s) - f(a)| < \varepsilon$ for any $s \in S$ such that $x \in U_s$. Let $S' = \{s \in S : x \in U_s\}$; then $S(x) \subset S'$ and we have

$$|e(f)(x) - e(f)(a)| = |\sum_{s \in S'} b_s(x) f(a_s) - f(a)| =$$
$$= |\sum_{s \in S'} b_s(x)(f(a_s) - f(a))| < \varepsilon \cdot (\sum_{s \in S'} b_s(x)) = \varepsilon$$

(we used the equality $\sum_{s \in S'} b_s(x) = 1$). Thus we found $U \in \tau(a, M)$ such that $|e(f)(x) - e(f)(a)| < \varepsilon$ for any $x \in U$ so $e(f)$ is continuous at the point a. This proves that $e(f) \in C_p(M)$ for any $f \in C_p(A)$.

Now take any $f, g \in C_p(A)$ and $\lambda, \mu \in \mathbb{R}$; let $h = \lambda f + \mu g$. Then

$$e(h)(a) = h(a) = (\lambda f + \mu g)(a) = \lambda f(a) + \mu g(a) = \lambda e(f)(a) + \mu e(g)(a)$$

for any $a \in A$. If $x \in M \backslash A$ then

$$e(h)(x) = \sum_{s \in S} b_s(x)h(a_s) = \sum_{s \in S} b_s(x)(\lambda f + \mu g)(a_s) =$$
$$= \lambda(\sum_{s \in S} b_s(x)f(a_s)) + \mu(\sum_{s \in S} b_s(x)g(a_s)) = \lambda e(f)(x) + \mu e(g)(x),$$

which shows that $e(\lambda f + \mu g) = \lambda e(f) + \mu e(g)$, i.e., the map e is linear.

If $a \in A$ and $f \in C_p(A)$ then let $p_a(f) = f(a)$. Then $p_a : C_p(A) \to \mathbb{R}$ is continuous for any $a \in A$ (see TFS-166). For any $x \in M$ let $\pi_x(f) = f(x)$ for any $f \in C_p(M)$. To show that e is continuous it suffices to check that $\pi_x \circ e$ is continuous for any $x \in M$ (see TFS-102). If $x \in A$ then $\pi_x(e(f)) = e(f)(x) = f(x) = p_x(f)$ for any $f \in C_p(A)$; so $\pi_x \circ e = p_x$ is a continuous map.

If we are given a point $x \in M \backslash A$ then let $S' = \{s \in S : b_s(x) \neq 0\}$; then S' is a finite set and $\pi_x(e(f)) = e(f)(x) = \sum_{s \in S'} b_s(x) f(a_s)$ for any $f \in C_p(A)$ and therefore $\pi_x \circ e = \sum_{s \in S'} b_s(x) p_{a_s}$ is a continuous map being a linear combination of continuous maps $\{p_{a_s} : s \in S'\}$. We established that $\pi_x \circ e$ is continuous for any $x \in M$; so the map e is continuous and Fact 1 is proved.

Fact 2. Suppose that M is a metrizable space and $A \subset M$ is a non-empty closed subset of M; let $I_A = \{f \in C_p(M) : f|A \equiv 0\}$. Then there exists a linear homeomorphism between $C_p(M)$ and $C_p(A) \times I_A$ and, in particular, $C_p(A)$ embeds in $C_p(M)$ as a closed linear subspace.

Proof. There is a linear continuous map $e : C_p(A) \to C_p(M)$ such that $e(f)|A = f$ for any $f \in C_p(A)$ (see Fact 1). For an arbitrary function $f \in C_p(M)$ we can define $\varphi(f) = (f|A, f - e(f|A)) \in C_p(A) \times C_p(M)$. Now, $e(f|A)$ coincides with f on the set A so $(f - e(f|A))(x) = 0$ for any $x \in A$ which shows that $\varphi(f) \in C_p(A) \times I_A$ for any $f \in C_p(M)$. Therefore we have defined a map $\varphi : C_p(M) \to C_p(A) \times I_A$.

Take any $f \in C_p(A)$ and $g \in I_A$; it is immediate that $h = e(f) + g \in C_p(M)$ and $\varphi(h) = (f, g)$ so the map φ is surjective. If $f_1, f_2 \in C_p(M)$ and $\varphi(f_1) = \varphi(f_2)$ then $f = f_1|A = f_2|A$ and $g = f_1 - e(f) = f_2 - e(f)$ which shows that $f_1 = e(f) + g = f_2$ and hence φ is a bijection.

The restriction map $\pi : C_p(M) \to C_p(A)$ defined by $\pi(f) = f|A$ for any function $f \in C_p(M)$ is linear and continuous; this, together with linearity and continuity of e, implies that the map $\delta : C_p(M) \to C_p(M)$ defined by $\delta(f) = f - e(\pi(f))$ for all $f \in C_p(M)$ is also linear and continuous. Therefore φ is a linear continuous map being a diagonal product of linear continuous maps π and δ.

For any $(f, g) \in C_p(A) \times I_A$ let $\eta(f, g) = e(f) + g$; then $\eta : C_p(A) \times I_A \to C_p(M)$ is also a linear continuous map. Indeed, let $i : I_A \to I_A$ be the identity map; then $\alpha = e \times i : C_p(A) \times I_A \to C_p(M) \times C_p(M)$ is linear and continuous because any product of linear continuous maps is linear and continuous (see Fact 1 of S.271). Therefore η is linear and continuous because it is a composition of α and the addition map in $C_p(M)$ which is linear and continuous (see TFS-115). It is straightforward that η is the inverse of φ so φ is a linear homeomorphism. It is evident that any factor in a product of topological vector spaces embeds in that product as a closed linear subspace; so Fact 2 is proved.

Returning to our solution let $a = (0, 0) \in \mathbb{R} \times \mathbb{R}$; we will also need the set $E_k = \{0, \dots, k-1\} \times \{k\} \subset \mathbb{R} \times \mathbb{R}$ for any $k \in \mathbb{N}$. Let $E = \bigcup\{E_k : k \in \mathbb{N}\}$; to introduce a topology μ on the set $A = \{a\} \cup E$ we declare that all points of E are isolated and $a \in U \in \mu$ if and only if there is $m \in \omega$ such that $|E_k \backslash U| \leq m$ for any $k \in \mathbb{N}$. It is easy to see that this defines local bases of a topology μ at all points of A; let $X = (A, \mu)$. The T_1-property of X is clear; since X is also normal (see Claim 2 of S.018), it is a T_4-space.

Recall that a family $\mathcal{N} \subset \exp X \backslash \{\emptyset\}$ is a π-net at a if, for any $U \in \tau(a, X)$ there is $N \in \mathcal{N}$ such that $N \subset U$. We claim that

(1) no countable family $\mathcal{A}$ of infinite subsets of E can be a π-net at a.
Indeed, $\mathcal{A} = \{A_n : n \in \omega\} \subset \exp E$ where A_n is infinite for any $n \in \omega$. Consequently, the set $\{k \in \omega : A_n \cap E_k \neq \emptyset\}$ is infinite for any $n \in \omega$ and hence we can choose an increasing sequence $\{k_n : n \in \omega\} \subset \omega$ and $a_n \in A_n \cap E_{k_n}$ for any $n \in \omega$. Then $U = X \backslash \{a_n : n \in \omega\}$ is an open neighbourhood of a in X and $a_n \in A_n \backslash U$ for any $n \in \omega$, i.e., $\mathcal{A}$ is not a π-net at a. Now we can apply Fact 1 of U.061 to conclude that

(2) the space X is not embeddable in a sequential space.

Let $C = \{\chi_U : U \in \tau(a, X)\} \subset \mathbb{D}^X$; it is evident that every $f \in C$ is continuous on X, i.e., $C \subset C_p(X, \mathbb{D})$. For any $x \in X$ and $f \in C$ let $e_x(f) = f(x)$; then we have a map $e : X \to C_p(C)$ defined by $e(x) = e_x$ for every $x \in X$ (see TFS-166).

Suppose that F is a closed subset of X and $x \notin F$. If $x = a$ then for the set $U = X \backslash F$ the function $f = \chi_U \in C$ and $f(x) = 1$ while $f(F) \subset \{0\}$. If $x \neq a$ then let $U = X \backslash \{a\}$; again, $f = \chi_U \in C$ and $f(x) = 0$ while $f(F) \subset \{1\}$. Thus C separates the points and closed subsets of X and hence the map e is an embedding: this was also proved in TFS-166.

Given $n \in \omega$, let $C_n = \{f \in \mathbb{D}^X : f(a) = 1 \text{ and } |f^{-1}(0) \cap E_k| \leq n$ for all $k \in \omega\}$; it is immediate that $C = \bigcup\{C_n : n \in \omega\}$. Besides, every C_n is compact being closed in $\mathbb{D}^X$. As a consequence, C is σ-compact; the space X embeds in $C_p(C)$ and hence it can be embedded in $C_p(K)$ for some metrizable compact K (see Problems 035 and 041). Since the Cantor set $\mathbb{K}$ can be continuously mapped onto K (see TFS-128), the space $C_p(K)$ embeds in $C_p(\mathbb{K})$. Now, $\mathbb{K}$ is homeomorphic to a closed subset of $\mathbb{I}$ which, together with Fact 2, implies that $C_p(\mathbb{K})$ embeds in $C_p(\mathbb{I})$. Therefore X embeds in $C_p(\mathbb{I})$; so we can apply (2) to conclude that $C_p(\mathbb{I})$ cannot be embedded in a sequential space.

U.063. *Prove that the following are equivalent for any pseudocompact X:*

 (i) $C_p(X)$ *is a Fréchet–Urysohn space;*
 (ii) $C_p(X)$ *embeds in a sequential space;*
 (iii) X *is compact and scattered.*

Solution. It is evident that (i)$\Longrightarrow$(ii); the implication (iii)$\Longrightarrow$(i) was established in SFFS-134 so we only have to prove that (ii)$\Longrightarrow$(iii). Assume that $C_p(X)$ embeds in a sequential space S. Then $t(C_p(X)) \leq t(S) = \omega$ (see Problem 060) and hence X is Lindelöf (even in all finite powers) which, together with pseudocompactness of X shows that X is compact. If X is not scattered then it can be continuously mapped onto $\mathbb{I}$ by SFFS-133. This implies that $C_p(\mathbb{I})$ embeds in $C_p(X)$ and hence $C_p(\mathbb{I})$ can be embedded in our sequential space S. This contradiction with Problem 062 proves that X is scattered and hence (ii)$\Longrightarrow$(iii).

U.064. *Prove that radiality is a hereditary property; show that pseudoradiality is closed-hereditary. Give an example showing that pseudoradiality is not hereditary.*

Solution. Suppose that X is a radial space and $Y \subset X$. If $A \subset Y$ and $x \in \operatorname{cl}_Y(A)\backslash A$ then $x \in \operatorname{cl}_X(A)\backslash A$ and hence there exists a regular cardinal κ and a κ-sequence $S = \{a_\alpha : \alpha < \kappa\} \subset A$ with $S \to x$. It is evident that the sequence S witnesses radiality of Y; so we proved that any subspace of a radial space is radial.

Now assume that X is pseudoradial and Y is a closed subspace of X. If $A \subset Y$ is not closed in Y then it is not closed in X and hence there a regular cardinal κ and a κ-sequence $S = \{a_\alpha : \alpha < \kappa\} \subset A$ with $S \to x$ for some $x \in \operatorname{cl}_X(A)\backslash A$. The set Y being closed in X we have $\operatorname{cl}_X(A) \subset Y$ and hence $x \in \operatorname{cl}_Y(A)\backslash A$ which proves that pseudoradiality is closed-hereditary.

Fact 1. Let $\{S_m : m \in \omega\}$ be a disjoint family of copies of a convergent sequence, i.e., $S_m = \{a_n^m : n \in \omega\} \cup \{a_m\}$ where the enumeration of S_m is faithful and the sequence $\{a_n^m : n \in \omega\}$ converges to a_m for each $m \in \omega$. Given $k, m \in \omega$ denote by $S_m(k)$ the set $\{a_m\} \cup \{a_n^m : n \geq k\} \subset S_m$ and let $S = \bigcup_{m \in \omega} S_m$. Choose a point $a \notin S$ and let $\mathcal{B}_{mn} = \{\{a_n^m\}\}$ for all $m, n \in \omega$; to construct a topology on the set $\{a\} \cup S$ we will also need the family $\mathcal{B}_m = \{S_m(k) : k \in \omega\}$ which will be the future local base at a_m for every $m \in \omega$. For any $k \in \omega$ and $f \in \omega^{\omega\backslash k}$ let $O(f, k) = \{a\} \cup (\bigcup\{S_m(f(m)) : m \geq k\})$; the family $\mathcal{B}_a = \{O(f, k) : k \in \omega, \ f \in \omega^{\omega\backslash k}\}$ is the future local base at a. Then $\mathcal{B} = \bigcup\{\mathcal{B}_{mn} : m, n \in \omega\} \cup (\bigcup\{\mathcal{B}_m : m \in \omega\}) \cup \mathcal{B}_a$ generates a topology τ (as a collection of local bases at the relevant points) on the set $\{a\} \cup S$; let $T = (\{a\} \cup S, \tau)$. The space T is Tychonoff and sequential while no sequence from the set $A = \{a_n^m : m, n \in \omega\}$ converges to a and, in particular, T is a sequential space which is not Fréchet–Urysohn.

Proof. We omit the straightforward proof that the collection $\mathcal{B}$ satisfies the conditions (LB1)–(LB3) from TFS-007; so we can, indeed, generate a topology τ on the set $\{a\} \cup S$. It is evident that T is a T_0-space; since every point of A is isolated in T, all elements of $\mathcal{B}$ are clopen in T so T is zero-dimensional and hence Tychonoff.

To see that T is sequential suppose that $P \subset T$ and $P \neq \overline{P}$. If $a_m \in \overline{P} \setminus P$ for some $m \in \omega$ then $P' = P \cap S_m$ is infinite and $P' \subset P$ is a sequence that converges to a_m. This shows that we can assume, without loss of generality, that $a_m \in \overline{P}$ implies $a_m \in P$ for any $m \in \omega$. Now, if $a \in \overline{P} \setminus P$ then the set the $N = \{m \in \omega : P \cap S_m \text{ is infinite}\}$ is infinite. We have $a_m \in \overline{P}$ and hence $a_m \in P$ for any $m \in N$ so $P' = \{a_m : m \in N\} \subset P$ is a sequence that converges to a.

Finally observe that it follows from the definition of τ that $a \in \overline{A}$; however, if some sequence $P \subset A$ converges to a then $S \cap S_m$ is finite for any $m \in \omega$ (because otherwise $a_m \in \overline{P} \setminus P$) and hence there is a function $f \in \omega^\omega$ such that $P \cap S_m \subset \{0, \ldots, f(m) - 1\}$ for all $m \in \omega$; thus $U = O(f, 0)$ is an open neighbourhood of a with $U \cap P = \emptyset$. This contradiction shows that no sequence from A converges to a and hence Fact 1 is proved.

Returning to our solution consider the space T from Fact 1. It is sequential and hence pseudoradial. However, the subspace $Y = \{a\} \cup A$ of the space T is not pseudoradial. Indeed, assume, towards a contradiction, that Y is pseudoradial. The set A is not closed in Y and $a \in \overline{A} \setminus A$ is the unique point of Y which is in the closure of A and not in A. Therefore there is a regular cardinal κ and a κ-sequence $P = \{y_\alpha : \alpha < \kappa\}$ such that $P \to a$. The cardinal κ must be uncountable because otherwise P is a convergent sequence which contradicts Fact 1. Since A is countable, there is $y \in A$ such that the set $Q = \{\alpha < \kappa : y_\alpha = y\}$ has cardinality κ.

However, $Y \setminus \{y\}$ is an open neighbourhood of a in Y; so there is $\beta < \kappa$ such that $y_\alpha \neq y$ for all $\alpha \geq \beta$ and hence $Q \subset \{\alpha : \alpha < \beta\}$; this contradiction shows that P does not converge to a and hence Y is not pseudoradial. Consequently, T is a pseudoradial space such that $Y \subset T$ is not pseudoradial which shows that our solution is complete.

U.065. *Prove that any quotient (pseudo-open) image of a pseudoradial (radial) space is a pseudoradial (radial) space.*

Solution. Suppose that X is a pseudoradial space and $f : X \to Y$ is a quotient map. If $A \subset Y$ is not closed in Y then $B = f^{-1}(A)$ is not closed in X; so we can find a regular cardinal κ and a κ-sequence $S = \{b_\alpha : \alpha < \kappa\} \subset B$ such that $S \to x$ for some $x \in X \setminus B$. Then $y = f(x) \in Y \setminus A$ and the κ-sequence $T = \{f(b_\alpha) : \alpha < \kappa\} \subset A$ converges to y. This shows that Y is also pseudoradial.

Now, assume that X is radial and the mapping f is pseudo-open. Given a set $A \subset Y$ and a point $y \in \overline{A} \setminus A$ let $B = f^{-1}(A)$ and $P = f^{-1}(y)$. If $P \cap \overline{B} = \emptyset$ then $U = X \setminus \overline{B} \in \tau(P, X)$ and hence $y \in V = \text{Int}(f(U))$ so V is an open neighbourhood of y which does not meet A; this contradiction shows that we can choose $x \in \overline{B} \cap P$. By radiality of the space X there exists a regular cardinal κ and a κ-sequence $S = \{b_\alpha : \alpha < \kappa\} \subset B$ such that $S \to x$. It is evident that the κ-sequence $T = \{f(b_\alpha) : \alpha < \kappa\} \subset A$ converges to y so Y is radial as well.

U.066. *Prove that any radial space of countable tightness is Fréchet–Urysohn.*

Solution. Suppose that X is a radial space of countable tightness; take any $A \subset X$ and $x \in \overline{A}$. If $x \in A$ then $S = \{\{x\}\} \subset A$ is a (trivial) sequence which converges to x. If $x \in \overline{A}\backslash A$ then there is a countable $B \subset A$ such that $x \in \overline{B}$; of course, $x \in \overline{B}\backslash B$ so we can apply radiality of X to find a regular cardinal κ and a κ-sequence $S = \{x_\alpha : \alpha < \kappa\} \subset B$ such that $S \to x$.

If $\kappa = \omega$ then our proof is over; if not, then it follows from $|B| = \omega$ and regularity of κ that there is $b \in B$ such that the cardinality of the set $E = \{\alpha < \kappa : x_\alpha = b\}$ is equal to κ. However, $W = X\backslash\{b\}$ is an open neighbourhood of the point x; so there is $\beta < \kappa$ such that $\{x_\alpha : \alpha \geq \beta\} \subset W$ which implies that $E \subset \{\alpha : \alpha < \beta\}$ and hence $|E| < \kappa$. This contradiction shows that κ is countable and hence $S \subset A$ is a sequence which converges to x. Therefore X is a Fréchet–Urysohn space.

U.067. *Prove that a space is radial (pseudoradial) if and only if it is a pseudo-open (quotient) image of a linearly ordered space.*

Solution. If $(L, \preceq)$ is a linearly ordered set then, as usual, $\tau(\preceq)$ is the topology generated by the order $\preceq$; for any points $a, b \in L$ we let $a \prec b$ if and only if $a \preceq b$ and $a \neq b$. Observe that, to define the order $\preceq$ it suffices to indicate for which $a, b \in L$ we have $a \prec b$. We will need almost all types of intervals in $(L, \preceq)$ so we define

$$(\leftarrow, a)_{\preceq} = \{y \in L : y \prec a\}, \ (\leftarrow, a]_{\preceq} = \{y \in L : y \preceq a\}$$

and

$$(a, b)_{\preceq} = \{y \in L : a \prec y \prec b\}.$$

Analogously,

$$(a, \rightarrow)_{\preceq} = \{y \in L : a \prec y\}, \ [a, \rightarrow)_{\preceq} = \{y \in L : a \preceq y\}, \ [a, b)_{\preceq}$$
$$= \{y \in L : a \preceq y \prec b\}$$

and $(a, b]_{\preceq} = \{y \in L : a \prec y \preceq b\}$.

Fact 1. For an infinite cardinal κ let $\preceq$ be the lexicographic order on $D_\kappa = \kappa \times \mathbb{Z}$, i.e., for any $a, b \in D_\kappa$ such that $a = (\alpha, n)$, $b = (\beta, m)$ let $a \prec b$ if $\alpha < \beta$; if $\beta < \alpha$ then we let $b \prec a$. Now if $\alpha = \beta$ then $a \preceq b$ if $n \leq m$ and $b \preceq a$ if $m \leq n$. Then $\preceq$ is a linear order on D_κ and the space $(D_\kappa, \tau(\preceq))$ is discrete. Besides, if $\kappa > \omega$ then $|\{a \in D_\kappa : a \preceq b\}| < \kappa$ for any $b \in D_\kappa$. In particular, any discrete space X is linearly orderable.

Proof. It is easy to check that $\preceq$ is a linear order on D_κ; the space $(D_\kappa, \tau(\preceq))$ is discrete because for any $t = (\alpha, n) \in D_\kappa$ the points $a = (\alpha, n - 1)$ and

$b = (\alpha, n + 1)$ belong to D_κ and $(a, b)_{\preceq} = \{t\}$, i.e., every $t \in D_\kappa$ is isolated and hence the space $(D_\kappa, \tau(\preceq))$ is discrete. Now if $b = (\alpha, n) \in D_\kappa$ then

$$L_b = \{a \in D_\kappa : a \preceq b\} \subset (\alpha + 1) \times \mathbb{Z}$$

so $|L_b| \leq |(\alpha + 1) \times \mathbb{Z}| = \max\{|\alpha|, \omega\} < \kappa$.

Finally, assume that X is a discrete space. If X is finite then any well order on X generates its topology. If X is infinite then, for the cardinal $\kappa = |X|$ take a bijection $f : X \to D_\kappa$.

Using f, we transfer the order $\preceq$ to X, i.e., for any $x, y \in X$ we let $x \sqsubseteq y$ if and only if $f(x) \preceq f(y)$. The map f is an isomorphism between $(X, \sqsubseteq)$ and $(D_\kappa, \preceq)$; so it is a homeomorphism between $(X, \tau(\sqsubseteq))$ and $(D_\kappa, \tau(\preceq))$. Therefore $(X, \tau(\sqsubseteq))$ is also discrete and Fact 1 is proved.

Fact 2. Suppose that, for every $t \in T$, the topology of a space X_t is can be generated by a linear order $\preceq_t$ which has a maximal and a minimal element. Then the space $X = \bigoplus\{X_t : t \in T\}$ is linearly orderable.

Proof. We identify every X_t with the relevant clopen subspace of X. Apply Fact 1 to choose a linear order $\leq$ on T such that $(T, \tau(\leq))$ is discrete. For every $t \in T$ fix the minimal element a_t of $(X_t, \preceq_t)$ and its maximal element b_t (note that we do not discard the possibility of $a_t = b_t$).

Given distinct $x, y \in X$ there are $t, s \in T$ such that $x \in X_t$ and $y \in X_s$. If $s \neq t$ then let $x \prec y$ if $s < t$ and $y \prec x$ if $t < s$. Now, if $s = t$ then $x \prec y$ if and only if $x \prec_t y$. It is straightforward to verify that $\preceq$ is a linear order on X. For every $t \in T$ fix $t_l, t_r \in T$ for which $(t_l, t_r)_{\leq} = \{t\}$.

Take any $x, y \in X$ with $x \prec y$ and fix $t, s \in T$ for which $x \in X_t$ and $y \in X_s$. If $t = s$ then $(x, y)_{\preceq} = (x, y)_{\preceq_t}$; if $t \neq s$ then $t < s$ and

$$(x, y)_{\preceq} = (x, \rightarrow)_{\preceq_t} \cup (\leftarrow, y)_{\preceq_s} \cup \left(\bigcup\{X_u : t < u < s\} \right),$$

which shows that $(x, y)_{\preceq}$ is open in X for any $x, y \in X$ and hence $\tau(\preceq) \subset \tau(X)$.

To prove the converse inclusion observe that the family $\mathcal{B} = \bigcup\{\tau(X_t) : t \in T\}$ is a base in X. Take any $t \in T$, $x \in X_t$ and $U \in \tau(x, X_t)$. If $|X_t| = 1$ then $(b_{t_l}, a_{t_r})_{\preceq} = \{x\} \subset U$, i.e., we found $W \in \tau(\preceq)$ such that $x \in W \subset U$. If $|X_t| > 1$ then we have three cases:

(a) $a_t < x < b_t$; then there are $a, b \in X_t$ such that $x \in (a, b)_{\preceq_t} \subset U$. However, $W = (a, b)_{\preceq_t} = (a, b)_{\preceq}$ and hence we found $W \in \tau(\preceq)$ such that $x \in W \subset U$.

(b) $x = a_t$; then there is a point $b \in X_t$ such that $[a_t, b)_{\preceq_t} \subset U$. Observe that we have $W = [a_t, b)_{\preceq_t} = (b_{t_l}, b)_{\preceq} \in \tau(\preceq)$; so we found a set $W \in \tau(\preceq)$ such that $x \in W \subset U$.

(c) $x = b_t$; then there is $a \in X_t$ such that $(a, b_t]_{\preceq_t} \subset U$. Observe that we have $W = (a, b_t]_{\preceq_t} = (a, a_{t_r})_{\preceq} \in \tau(\preceq)$; so we found again a set $W \in \tau(\preceq)$ such that $x \in W \subset U$.

Thus we proved that for any $x \in X$ and $U \in \tau(x, X)$ there is $W \in \tau(\preceq)$ such that $x \in W \subset U$. This shows that $\tau(\preceq)$ is a base for $\tau(X)$ and hence $\tau(X) \subset \tau(\preceq)$. Therefore $\tau(X) = \tau(\preceq)$ and Fact 2 is proved.

Fact 3. Suppose that $X = \{x\} \cup \{x_\alpha : \alpha < \kappa\}$ where κ is an infinite regular cardinal, the enumeration of X is faithful and x is the unique non-isolated point of X. For every $\alpha < \kappa$ let $O_\alpha = \{x\} \cup \{x_\beta : \beta \geq \alpha\}$. If the family $\{O_\alpha : \alpha < \kappa\}$ is a local base at x in X then there is a linear order $\preceq$ on X such that $\tau(\preceq) = \tau(X)$, the point x is the maximal element of $(X, \preceq)$ and x_0 is its minimal element.

Proof. If $\kappa = \omega$ then let $x_n \preceq x$ for any $n \in \omega$ and $x_n \preceq x_m$ if and only if $n \leq m$. Then $\preceq$ is a well order on X and $(X, \preceq)$ is order-isomorphic to $\omega + 1$. It is evident that X is also homeomorphic to $\omega + 1$ and hence $\tau(\preceq) = \tau(X)$. It is evident that x the maximal element of $(X, \preceq)$ while x_0 is its minimal element.

Now, if $\kappa = |X| > \omega$ then let $Y = X \backslash (\{x\} \cup \{x_n : n < \omega\})$ and take any bijection $f : Y \to D_\kappa = \kappa \times \mathbb{Z}$.

Let $\leq$ be the lexicographic order on D_κ (see Fact 1) and transfer the order $\leq$ to Y using the bijection f, i.e., let $x \preceq y$ if and only if $f(x) \leq f(y)$. It follows from Fact 1 that all points of Y are isolated in the topology generated by $\preceq$ on Y. Let $x_\alpha \preceq x$ for any $\alpha < \kappa$. For any $m, n \in \omega$ we let $x_m \preceq x_n$ if and only if $m \leq n$. Furthermore, $x_m \prec x_\alpha$ for any $m < \omega$ and $\alpha \geq \omega$.

We omit a simple verification that $(X, \preceq)$ is a linearly ordered set in which all points of $X \backslash \{x\}$ are isolated in $\tau(\preceq)$ while x the maximal element of $(X, \preceq)$ and x_0 is its minimal element.

To see that $\tau(x, X) = \{W \in \tau(\preceq) : x \in U\}$ take any set $U \in \tau(x, X)$. By our hypothesis there is an ordinal $\alpha < \kappa$ such that $O_\alpha \subset U$. It follows from Fact 1 that the set $Q_\beta = \{y \in X : y \preceq x_\beta\}$ is countable for any $\beta \leq \alpha$ so we can choose a point $z \in (X \backslash \{x\}) \backslash (\bigcup \{Q_\beta : \beta \leq \alpha\})$. It is immediate that $x \in W = \{y \in X : z \prec y\} \subset U$ which shows that $\tau(\preceq)$ is a base for $\tau(X)$ and therefore $\tau(X) \subset \tau(\preceq)$.

Now observe that the family $\{(a, \to)_{\preceq} : a \in X \backslash \{x\}\}$ is a local base at the point x in $(X, \tau(\preceq))$ so if $W \in \tau(\preceq)$ and $x \in W$ then there is a point $a \in X \backslash \{x\}$ such that $(a, \to)_{\preceq} \subset W$. The set $B = (\leftarrow, a]_{\preceq}$ is countable by Fact 1 so we can choose $\alpha < \kappa$ such that $x_\beta \in B$ implies $\beta < \alpha$. It is immediate that $O_\alpha \subset W$ so $\tau(X)$ is also a base for $\tau(\preceq)$ and hence $\tau(\preceq) \subset \tau(X)$ whence $\tau(\preceq) = \tau(X)$. Thus X is linearly ordered by $\preceq$ and x, x_0 are the maximal and the minimal elements respectively of the set $(X, \preceq)$ which shows that Fact 3 is proved.

Returning to our solution let $(L, \preceq)$ be a linearly ordered set with the topology $\tau(\preceq)$. Given a set $A \subset L$ and a point $x \in \overline{A} \backslash A$ consider the sets $A_1 = A \cap (\leftarrow, x)_{\preceq}$ and $A_2 = A \cap (x, \to)_{\preceq}$. Since $A = A_1 \cup A_2$, we have either $x \in \overline{A_1} \backslash A_1$ or $x \in \overline{A_2} \backslash A_2$. The further reasonings for both cases are identical; so we consider that $y \prec x$ for any $y \in A$.

Choose $a_0 \in A$ arbitrarily and assume that β is an ordinal for which we have a set $A_\beta = \{a_\alpha : \alpha < \beta\} \subset A$ such that

(1) $\alpha < \gamma < \beta$ implies $a_\alpha \prec a_\gamma$.

If $x \in \overline{A}_\beta$ then our inductive construction stops; if not, then there is $a \in L$ such that $a < x$ and $(a, x]_{\preceq} \cap A_\beta = \emptyset$. It follows from $x \in \overline{A}$ and $A \subset (\leftarrow, x)_{\preceq}$ that there is $a_\beta \in (a, x)_{\preceq} \cap A$; it is clear that (1) still holds for the set $\{a_\alpha : \alpha \leq \beta\}$ so our inductive construction can be continued. Of course, it has to end before we arrive to $\beta = |L|^+$ so assume that $x \in \overline{A}_v$ for some ordinal v. Then $\kappa = \mathrm{cf}(v) = \min\{\alpha :$ there is a cofinal $M \subset v$ which is isomorphic (with the well order induced from v) to the ordinal $\alpha\}$ is an infinite regular cardinal. Choose a set $M = \{\mu_\alpha : \alpha < \kappa\} \subset v$ such that $\alpha < \gamma < \kappa$ implies $\mu_\alpha < \mu_\gamma$ and the set M is cofinal in v; let $y_\alpha = a_{\mu_\alpha}$ for any $\alpha < \kappa$. Then $\{y_\alpha : \alpha < \kappa\} \subset A$ is a κ-sequence which converges to x.

Indeed, take any set $U \in \tau(x, L)$; there is $a \in L$ such that $(a, x]_{\preceq} \subset U$. Since $x \in \overline{A}_v$, we have $A_v \cap (a, x]_{\preceq} \neq \emptyset$ and therefore $a_\gamma \in (a, x)_{\preceq}$ for some $\gamma < v$. The set M being cofinal in v there is $\beta < \kappa$ such that $\mu_\beta > \gamma$ and hence $a \prec a_\gamma \prec a_{\mu_\beta} = y_\beta$ (here we applied (1)). For any $\alpha \geq \beta$ we have $\mu_\beta \leq \mu_\alpha$ and therefore $a \prec y_\beta \preceq y_\alpha \prec x$ which shows that $\{y_\alpha : \alpha \geq \beta\} \subset (a, x) \subset U$. Thus we proved that $S \to x$ and hence

(2) any linearly ordered space is radial.

An immediate consequence of (2) and Problem 065 is that any pseudo-open (quotient) image of a linearly ordered space is radial (or pseudoradial respectively) so we established sufficiency for both cases.

The proofs of necessity for both radial and pseudoradial spaces are also parallel; so we will give them simultaneously. Assume that X is a radial (pseudoradial) space; if X is finite then it is linearly orderable by Fact 1 so there is nothing to prove. Thus we can assume that $|X| = \lambda \geq \omega$. Let $\mathcal{S}(X)$ be the family of all faithfully indexed κ-sequences $S = \{x_\alpha : \alpha < \kappa\} \subset X$ such that $\omega \leq \kappa = \mathrm{cf}(\kappa) \leq \lambda$ and the sequence S converges to some point $u_S \in X \setminus S$. Given $S = \{x_\alpha : \alpha < \kappa\} \in \mathcal{S}(X)$ we will also need a topology τ_S on the set $S \cup \{u_S\}$ (which might be distinct from the topology induced on S from X) namely, $\tau_S = \{A : A \subset S\} \cup \{B : u_S \in B$ and $|S \setminus B| < \kappa\}$. In other words, all points of S are isolated in τ_S and the sets $O_\alpha = \{u_S\} \cup \{x_\beta : \alpha \leq \beta\}$ form a local base at the point u_S in the space $P[S] = (\{u_S\} \cup S, \tau_S)$ when α runs over all elements of κ.

Denote by I the (discrete) subspace of all isolated points of the space X and let $L = I \oplus (\bigoplus\{P[S] : S \in \mathcal{S}(X)\})$. An immediate consequence of Facts 2 and 3 is that L is linearly orderable; so it suffices to show that there exists a pseudo-open (or a quotient respectively) map $f : L \to X$. As usual, we identify every summand of L with its respective clopen subspace.

Take a point $x \in L$; if $x \in I$ then let $f(x) = x$ recall that $I \subset X$ so $f(x) \in X$. If $x \in \bigoplus\{P[S] : S \in \mathcal{S}(X)\}$ then $x \in P[S]$ for some $S \in \mathcal{S}(X)$; again, let $f(x) = x$ and observe that $f(x)$ makes sense and belongs to X because $P[S] \subset X$. This gives us a map $f : L \to X$; let us check that f is pseudo-open (or quotient, respectively).

Suppose first that X is pseudoradial and take any $A \subset X$ such that $B = f^{-1}(A)$ is closed. If A is not closed in X then there is a regular cardinal $\kappa \leq \lambda$ and a faithfully indexed κ-sequence $S = \{x_\alpha : \alpha < \kappa\} \subset A$ such that $S \to x$ for some $x \in \overline{A} \setminus A$. Then $S \in \mathcal{S}(X)$ and $u_S = x$. By definition of f we have $f(u_S) = x$ and hence $u_S \notin B$ (here we consider that $u_S \in L$). It follows from $S \to x$ that

$u_S \in \mathrm{cl}_L(B) \backslash B$ which is a contradiction. Therefore A is closed in X and hence the map f is quotient. This proves necessity for a pseudoradial space X.

Finally, if the space X is radial and f is not pseudo-open then there is a point $x \in X$ and $O \in \tau(f^{-1}(x), L)$ such that $x \notin \mathrm{Int}(f(O))$. This implies that $x \in X \backslash f(O)$ and hence there is a regular cardinal $\kappa \leq \lambda$ and a faithfully indexed κ-sequence $S = \{x_\alpha : \alpha < \kappa\} \subset B = X \backslash f(O)$ such that $S \to x$ and hence $u_S \in f^{-1}(x)$. Observe that $S \subset B$ implies that $S \cap O = \emptyset$ (here we consider $S \subset P[S]$ to be a subset of L) while $u_S \in \mathrm{cl}_L(S) \cap O$ which is again a contradiction. Therefore f is a pseudo-open map if X is radial; this settles necessity for a radial X and makes our solution complete.

U.068. *Prove that any radial space of countable spread is Fréchet–Urysohn.*

Solution. Let X be a radial space with $s(X) = \omega$. If X is not Fréchet–Urysohn then there is $A \subset X$ and $x \in \overline{A} \backslash A$ such that no sequence from A converges to x. Since the space X is radial, there is a regular cardinal κ and a κ-sequence $S = \{x_\alpha : \alpha < \kappa\} \subset A$ such that $S \to x$. The cardinal κ has to be uncountable for otherwise $S \subset A$ is a sequence which converges to x.

Let $\mathcal{U} \subset \tau^*(A)$ be a maximal disjoint family such that $x \notin \overline{U}$ for any $U \in \mathcal{U}$. It is easy to see that $W = \bigcup \mathcal{U}$ is dense in A and hence $x \in \overline{W}$. It follows from $s(X) \leq \omega$ that $\mathcal{U}$ is countable; so $S' = S \cap U$ has cardinality κ for some $U \in \mathcal{U}$. However, $x \notin \overline{U}$ implies that there is $\beta < \kappa$ such that $x_\alpha \notin \overline{U}$ for all $\alpha \geq \beta$. Therefore $S' \subset \{x_\alpha : \alpha < \beta\}$ and hence $|S'| \leq |\beta| < \kappa$; this contradiction shows that X is a Fréchet–Urysohn space.

U.069. *Prove that any radial dyadic space is metrizable.*

Solution. Suppose that X is a radial dyadic space and fix a continuous onto map $f : \mathbb{D}^\kappa \to X$ for some cardinal κ. If $\kappa \leq \omega$ then $\mathbb{D}^\kappa$ is metrizable and hence so is X (see Fact 5 of S.307). Thus we can assume, without loss of generality, that $\kappa > \omega$. We will need the subspaces $\sigma = \{s \in \mathbb{D}^\kappa : |s^{-1}(1)| < \omega\}$ and $\Sigma = \{s \in \mathbb{D}^\kappa : |s^{-1}(1)| \leq \omega\}$ of the space $\mathbb{D}^\kappa$.

Observe that $\sigma = \bigcup_{n \in \omega} \sigma_n$ where $\sigma_n = \{s \in \mathbb{D}^\kappa : |s^{-1}(1)| \leq n\}$ is a compact subspace of $\mathbb{D}^\kappa$ for each $n \in \omega$. Besides, σ is dense in $\mathbb{D}^\kappa$ and hence $Y = f(\sigma)$ is dense in X. Fix an arbitrary $x \in X \backslash Y$; there is a regular cardinal λ and a λ-sequence $S = \{y_\alpha : \alpha < \lambda\} \subset Y$ with $S \to x$.

If $\lambda > \omega$ then it follows from $Y = \bigcup_{n \in \omega} f(\sigma_n)$ that there is $m \in \omega$ for which $T = S \cap f(\sigma_m)$ has cardinality λ. The space $K = f(\sigma_m)$ is compact; so $W = X \backslash K$ is an open neighbourhood of x. Therefore there is $\beta < \lambda$ such that $y_\alpha \in W$ and hence $y_\alpha \notin K$ for all $\alpha \geq \beta$. Consequently, $T \subset \{y_\alpha : \alpha < \beta\}$ which shows that $|T| \leq |\beta| < \lambda$; this contradiction proves that $\lambda = \omega$. Thus we can take a countable set $A \subset \sigma$ such that $f(A) \supset S$ and hence $x \in \overline{f(A)} = f(\overline{A})$ (the last equality is true because the map f is closed).

However, $\overline{A} \subset \Sigma$ by Fact 3 of S.307 and therefore $x \in f(\Sigma)$. Since the point $x \in X \backslash Y$ was chosen arbitrarily, we proved that $f(\Sigma) \supset X \backslash Y$; moreover $Y = f(\sigma) \subset f(\Sigma)$ as well so $f(\Sigma) = X$. Finally apply Fact 6 of S.307 to conclude that X is metrizable.

U.070. *Prove that $\beta\omega\backslash\omega$ is not pseudoradial.*

Solution. The space $\beta\omega\backslash\omega$ is infinite and compact (see Fact 1 of S.370 and Fact 1 of S.376) so we can find a countable non-closed set $A \subset \beta\omega\backslash\omega$ by Fact 2 of T.090. If the space $\beta\omega\backslash\omega$ is pseudoradial then there is a regular cardinal κ and a κ-sequence $S = \{x_\alpha : \alpha < \kappa\} \subset A$ such that $S \to x$ for some $x \in (\beta\omega\backslash\omega)\backslash A$. It can be left to the reader as an easy exercise to prove that κ cannot be uncountable; so S is a non-trivial convergent sequence in $\beta\omega$. However, $\beta\omega$ does not have non-trivial convergent sequences by Fact 2 of T.131; this contradiction proves that $\beta\omega\backslash\omega$ is not pseudoradial.

U.071. *Prove that $\mathbb{D}^{\omega_1}$ is not pseudoradial under CH and pseudoradial under MA+¬CH.*

Solution. If CH holds then $w(\beta\omega\backslash\omega) = \mathfrak{c} = \omega_1$ (see TFS-368 and TFS-371) so $\beta\omega\backslash\omega$ embeds in $\mathbb{I}^{\omega_1}$ and hence $\mathbb{I}^{\omega_1}$ is not pseudoradial by Problems 070 and 064. The space $\mathbb{D}^{\omega_1}$ can be continuously mapped onto $\mathbb{I}^{\omega_1}$ by Fact 2 of T.298. The relevant map has to be closed and hence quotient; so if $\mathbb{D}^{\omega_1}$ is pseudoradial then so is $\mathbb{I}^{\omega_1}$ (see Problem 065); this contradiction shows that $\mathbb{D}^{\omega_1}$ is not pseudoradial. To finish our solution it suffices to prove that

Fact 1. If MA+¬CH holds then any compact space X of weight at most ω_1 is pseudoradial. In particular, $\mathbb{D}^{\omega_1}$ is pseudoradial under MA+¬CH.

Proof. Suppose that $A \subset X$ is not closed in X. If there is a countable $B \subset A$ such that $\overline{B}\backslash A \neq \emptyset$ then take a point $x \in \overline{B}\backslash A$ and consider the space $Y = \{x\} \cup B$. We have $\chi(x, Y) \leq w(X) \leq \omega_1 < \mathfrak{c}$ and $|B| \leq \omega$ so SFFS-054 is applicable to conclude that there is a sequence $S = \{x_n : n \in \omega\} \subset B$ with $S \to x$. It is evident that the sequence S witnesses pseudoradiality of X.

Thus we can assume that $\overline{B} \subset A$ for any countable $B \subset A$ and hence A is countably compact (see Fact 1 of S.314); fix any point $x \in \overline{A}\backslash A$. It is impossible that $\chi(x, X) = \omega$ because then there is a sequence in A which converges to x; thus $\chi(x, X) = \omega_1$. Take a local base $\mathcal{B} = \{U_\alpha : \alpha < \omega_1\}$ at the point x and choose $V_\alpha \in \tau(x, X)$ such that $\overline{V}_\alpha \subset U_\alpha$ for any $\alpha < \omega_1$.

If $P \subset \omega_1$ is countable then take any enumeration $\{\alpha_n : n \in \omega\}$ of the set P and observe that $W_n = \bigcap_{i \leq n} V_{\alpha_i}$ is a neighbourhood of x; so $F_n = \overline{W}_n \cap A \neq \emptyset$ for any $n \in \omega$. By countable compactness of A we have $F = \bigcap_{n \in \omega} F_n \neq \emptyset$ and therefore $(\bigcap_{\alpha \in P} U_\alpha) \cap A \supset (\bigcap_{\alpha \in P} \overline{V}_\alpha) \cap A = F \neq \emptyset$ which proves that

(1) $(\bigcap_{\alpha \in P} U_\alpha) \cap A \neq \emptyset$ for any countable $P \subset \omega_1$.

Finally apply (1) to choose a point $x_\alpha \in (\bigcap\{U_\beta : \beta \leq \alpha\}) \cap A$ for every $\alpha < \omega_1$. It is immediate that the ω_1-sequence $S = \{x_\alpha : \alpha < \omega_1\} \subset A$ converges to x so X is pseudoradial and Fact 1 is proved.

U.072. *Prove that it is independent of ZFC whether every dyadic pseudoradial space is metrizable.*

Solution. Given a set T and $A \subset T$, the map $\pi_A^T : \mathbb{D}^T \to \mathbb{D}^A$ is the natural projection; if the set T is clear we write π_A instead of π_A^T. We say that a set $Z \subset \mathbb{D}^T$ *does not depend on* $A \subset T$ if $\pi_A^{-1}\pi_A(Z) = Z$. A set $U \in \tau(\mathbb{D}^T)$ is called *standard* if $U = \prod\{U_t : t \in T\}$ and the set $\mathrm{supp}(U) = \{t \in T : U_t \neq \mathbb{D}\}$ is finite.

Fact 1. Let λ be an infinite cardinal. If X is a dyadic space such that the set $C = \{x \in X : \pi\chi(x, X) \leq \lambda\}$ is dense in X then $w(X) \leq \lambda$. In particular, if X has a dense set of points of countable π-character then X is metrizable.

Proof. Fix a continuous onto map $f : \mathbb{D}^\kappa \to X$ for some infinite cardinal κ; if $\kappa \leq \lambda$ then there is nothing to prove, so we assume that $\kappa > \lambda$. We will show first that

(1) for any $x \in C$ there is a non-empty $A_x \subset \kappa$ such that $|A_x| \leq \lambda$ and there is a point $s_x \in \mathbb{D}^{A_x}$ for which $\pi_{A_x}^{-1}(s_x) \subset f^{-1}(x)$.

If $x \in C$ is an isolated point of X then $K = f^{-1}(x)$ is a clopen subset of $\mathbb{D}^\kappa$ so, being the closure of itself, it depends on countably many coordinates, i.e., there is a non-empty countable $A_x \subset \kappa$ such that $\pi_{A_x}^{-1}\pi_{A_x}(K) = K$ (see Fact 6 of T.298). Thus we can take any $s_x \in \pi_{A_x}(K)$ to satisfy (1).

Now, if $x \in C$ is not isolated in X fix a π-base $\mathcal{B}$ at x in X such that $|\mathcal{B}| \leq \lambda$; it is easy to see that we can choose $\mathcal{B}$ so that $x \notin \overline{U}$ for any $U \in \mathcal{B}$. It is an easy consequence of Fact 1 of S.226 that the family $\mathcal{V} = \{f^{-1}(U) : U \in \mathcal{B}\}$ is a π-base of the set $K = f^{-1}(x)$ in $\mathbb{D}^\kappa$, i.e., for any $O \in \tau(K, \mathbb{D}^\kappa)$ there is $V \in \mathcal{V}$ such that $V \subset O$. Since $\mathbb{D}^\kappa$ is a normal space, the family $\mathcal{V}_{cl} = \{\overline{V} : V \in \mathcal{V}\}$ is a π-net for the set K, i.e., for any $O \in \tau(K, \mathbb{D}^\kappa)$ there is $V \in \mathcal{V}$ such that $\overline{V} \subset O$.

For every $V \in \mathcal{V}$ the set $\overline{V}$ depends on countably many coordinates, i.e., there is a non-empty countable $S_V \subset \kappa$ such that $\overline{V} = \pi_{S_V}^{-1}\pi_{S_V}(\overline{V})$ (see Fact 6 of T.298). If $A_x = \bigcup\{S_V : V \in \mathcal{V}\}$ then $|A_x| \leq \lambda$; let us prove that A_x is as promised.

Observe first that, for every $V \in \mathcal{V}$, we have $\pi_{A_x}^{-1}\pi_{A_x}(\overline{V}) = \overline{V}$ because $S_V \subset A_x$. The set $F = \pi_{A_x}(K)$ is compact; assume that, for any point $s \in F$, there is a point $u_s \in \pi_{A_x}^{-1}(s) \setminus K$. Take $H_s \in \tau(u_s, \mathbb{D}^\kappa)$ for which $\overline{H}_s \cap K = \emptyset$; then $G_s = \pi_{A_x}(H_s)$ is an open neighbourhood of the point s for any $s \in F$. There is a finite $Q \subset F$ such that $G = \bigcup\{G_s : s \in Q\} \supset F$. The set $P = \bigcup\{\overline{H}_s : s \in Q\}$ is closed and disjoint from K; since $\mathcal{V}_{cl}$ is a π-net for K, there is $V \in \mathcal{V}$ such that $\overline{V} \subset (\mathbb{D}^\kappa \setminus P) \cap \pi_{A_x}^{-1}(G)$. Consequently, $\pi_{A_x}(\overline{V}) \subset G$ and hence $\pi_{A_x}(\overline{V}) \cap G_s \neq \emptyset$ for some $s \in Q$. The set $\overline{V}$ does not depend on A_x; so $\overline{V} \cap H_s \neq \emptyset$; this contradiction with the choice of V shows that (1) is proved.

For any $x \in C$ take a point $w_x \in \pi_{A_x}^{-1}(s_x)$ and let $C' = \{w_x : x \in C\}$; observe that $w_x \in f^{-1}(x)$ and hence $f(w_x) = x$ for any $x \in C$.

Take a point $d \in C$ arbitrarily and let $D_0 = \{d\}$, $B_0 = A_d$. Proceeding by induction assume that $m \in \mathbb{N}$ and we have constructed sets $D_0, \ldots, D_{m-1}$ and $B_0, \ldots, B_{m-1}$ with the following properties:

(2) $D_0 \subset D_1 \subset \ldots \subset D_{m-1} \subset C$ and $|D_i| \leq \lambda$ for any $i < m$;
(3) $B_0 \subset B_1 \subset \ldots \subset B_{m-1} \subset \kappa$ and $|B_i| \leq \lambda$ for any $i < m$;
(4) if $0 \leq i \leq m - 1$ then $A_x \subset B_i$ for any $x \in D_i$;
(5) if $0 \leq i < m - 1$ then $\pi_{B_i}(\{w_x : x \in D_{i+1}\})$ is dense in $\pi_{B_i}(C')$.

Observe that (2)–(5) are satisfied for $m = 1$ with (5) fulfilled vacuously. Since $\pi_{B_{m-1}}(C') \subset \mathbb{D}^{B_{m-1}}$, we have $w(\pi_{B_{m-1}}(C')) \leq \lambda$ so there is a set $\tilde{C} \subset C$ such that $|\tilde{C}| \leq \lambda$ and, for $C_0 = \{w_x : x \in \tilde{C}\}$, the set $\pi_{B_{m-1}}(C_0)$ is dense in $\pi_{B_{m-1}}(C')$; letting $D_m = D_{m-1} \cup \tilde{C}$ and $B_m = \bigcup\{A_x : x \in D_m\} \cup B_{m-1}$ we obtain sets D_m and B_m such that (2)–(5) are still fulfilled for the families $\{B_i : i \leq m\}$ and $\{D_i : i \leq m\}$. Therefore our inductive construction can be continued to obtain families $\{B_i : i \in \omega\}$ and $\{D_i : i \in \omega\}$ with the properties (2)–(5) satisfied for all $m \in \omega$.

Let $B = \bigcup_{n \in \omega} B_n$ and $D = \bigcup_{n \in \omega} D_n$. Consider also the set $F = \overline{C'}$ and $E = \{w_x : x \in D\}$; it is evident that $f(F) = X$. We claim that the map $g = f|F$ factorizes through the face $\mathbb{D}^B$, i.e., there exists a continuous map $h : \pi_B(F) \to X$ such that $g = h \circ (\pi_B|F)$. To prove this we will first establish that

(6) if $U, V \in \tau(\mathbb{D}^\kappa)$ and $\pi_B(U) \cap \pi_B(V) \cap \pi_B(E) \neq \emptyset$ then $f(U) \cap f(V) \neq \emptyset$.

To do it take a point $u \in E$ with $\pi_B(u) \in \pi_B(U) \cap \pi_B(V)$. By definition of E there is $x \in D$ such that $u = w_x$ and hence $\pi_B^{-1}(s_x) \subset \pi_{A_x}^{-1}(s_x) \subset f^{-1}(x)$. Thus it follows from $\pi_B^{-1}(s_x) \cap U \neq \emptyset$ and $\pi_B^{-1}(s_x) \cap V \neq \emptyset$ that $f^{-1}(x) \cap U \neq \emptyset$ and $f^{-1}(x) \cap V \neq \emptyset$ whence $x \in f(U) \cap f(V)$, i.e., (6) is proved.

Now assume that $s, t \in F$ and $\pi_B(s) = \pi_B(t)$. If $f(s) \neq f(t)$ then there are disjoint sets $U', V' \in \tau(X)$ such that $f(s) \in U'$ and $f(t) \in V'$. Therefore $U_1 = f^{-1}(U')$ and $V_1 = f^{-1}(V')$ are disjoint open neighbourhoods in $\mathbb{D}^\kappa$ of s and t respectively. It follows from $u = \pi_B(t) = \pi_B(s)$ that there are sets $U, V \in \tau(\mathbb{D}^\kappa)$ such that $s \in U \subset U_1$, $t \in V \subset V_1$ and $W = \pi_B(U) = \pi_B(V)$. There is a finite $B' \subset B$ and a standard open set G in the space $\mathbb{D}^B$ such that $u \in G \subset W$. The set $K = \text{supp}(G) \subset B$ being finite, there is $n \in \omega$ such that $K \subset B_n$. It follows from (5) that $\pi_{B_n}(E)$ is dense in $\pi_{B_n}(C')$; furthermore, $u \in \pi_B(F) = \overline{\pi_B(C')}$ which implies that $\pi_{B_n}^B(u) \in \overline{\pi_{B_n}(C')} \subset \overline{\pi_{B_n}(E)}$ and therefore $\pi_{B_n}(G) \cap \pi_{B_n}(E) \neq \emptyset$. We have $(\pi_{B_n}^B)^{-1}(G) = G$ so $\pi_B(E) \cap G \neq \emptyset$ and hence $\pi_B(E) \cap W \neq \emptyset$. This makes it possible to apply (6) to conclude that $f(U) \cap f(V) \neq \emptyset$ which is a contradiction with $f(U) \cap f(V) \subset f(U_1) \cap f(V_1) = U' \cap V' = \emptyset$. We finally proved that

(7) for any $s, t \in F$ if $\pi_B(s) = \pi_B(t)$ then $f(s) = f(t)$.

Thus, for any $u \in \pi_B(F)$ there is $x \in X$ for which $f(\pi_B^{-1}(u) \cap F) = \{x\}$; we let $h(u) = x$. Let $p = \pi|F : F \to \pi_B(F)$; it is immediate that $h \circ p = g$. The space F being compact the map p is closed and hence $\mathbb{R}$-quotient (see TFS-153 and TFS-154) so we can apply Fact 1 of T.268 to conclude that h is continuous. It turns out that $X = h(\pi_B(F))$ is a continuous image of the compact space $\pi_B(F)$ with $w(\pi_B(F)) \leq \lambda$ so $w(X) \leq \lambda$ (see Fact 1 of T.489) and Fact 1 is proved.

Returning to our solution observe that under MA+¬CH the space $\mathbb{D}^{\omega_1}$ is non-metrizable, pseudoradial and dyadic (see Problem 071); so it suffices to prove that

Fact 2. Under CH every pseudoradial dyadic space is metrizable.

Proof. Assume CH and let $f : \mathbb{D}^\kappa \to X$ be a continuous onto map for some infinite cardinal κ and a non-metrizable pseudoradial space X. Since CH holds, the space $\mathbb{D}^{\omega_1}$ is not pseudoradial (see Problem 071); this, together with Problem 064 implies that

(8) no closed subspace of X can be continuously mapped onto $\mathbb{D}^{\omega_1}$.

Suppose that $U \in \tau^*(X)$ and $\pi\chi(x, X) > \omega$ for any $x \in U$. Take $V \in \tau^*(X)$ such that $\overline{V} \subset U$. If $x \in \overline{V}$ and $\mathcal{U}$ is a countable π-base at the point x in $\overline{V}$ then the family $\{W \cap V : W \in \mathcal{U}\}$ is a countable π-base at x in X which contradicts the choice of U. Therefore $\pi\chi(x, \overline{V}) > \omega$ for any $x \in \overline{V}$ so we can apply Fact 5 of T.298 to see that some closed subspace of $\overline{V}$ maps continuously onto $\mathbb{D}^{\omega_1}$; this contradiction with (8) shows that the set $C = \{x \in X : \pi\chi(x, X) \leq \omega\}$ is dense in X and hence we can apply Fact 1 to conclude that X is metrizable obtaining the final contradiction to finish the proof of Fact 2 and our solution.

U.073. *Prove that, for any space X, the space $C_p(X)$ is radial if and only if it is Fréchet–Urysohn.*

Solution. It is evident that every Fréchet–Urysohn space is radial; so sufficiency is clear. To settle necessity assume that $C_p(X)$ is radial and fix an ω-cover $\mathcal{U}$ of the space X. Consider the set $A = \{f \in C_p(X) : f^{-1}(\mathbb{R}\backslash\{0\}) \subset U$ for some $U \in \mathcal{U}\}$. Observe first that

(1) the set A is dense in $C_p(X)$.

Indeed, take any function $f \in C_p(X)$ and $O \in \tau(f, C_p(X))$. There is a finite set $K \subset X$ and $\varepsilon > 0$ such that $W = \{g \in C_p(X) : |g(x) - f(x)| < \varepsilon$ for any $x \in K\} \subset O$. Since $\mathcal{U}$ is an ω-cover of X, there is $U \in \mathcal{U}$ such that $K \subset U$. It is easy to construct a function $g \in C_p(X)$ such that $g|K = f|K$ and $g(X\backslash U) \subset \{0\}$. It is immediate that $g \in A \cap W \subset A \cap O$ which proves that any neighbourhood of f meets A, i.e., $f \in \overline{A}$ and (1) is proved.

Let $p(x) = 1$ for all $x \in X$ and consider a maximal disjoint family $\mathcal{V} \subset \tau^*(A)$ such that $p \notin \overline{V}$ for any $V \in \mathcal{V}$. It is an easy exercise that $B = \bigcup \mathcal{V}$ is dense in A and hence $p \in \overline{B}$. Furthermore, $\mathcal{V}$ is countable because $c(A) = c(C_p(X)) = \omega$. By radiality of the space $C_p(X)$ there is a regular cardinal κ and a κ-sequence $S = \{f_\alpha : \alpha < \kappa\} \subset B$ such that $S \to p$.

If $\kappa > \omega$ then $|S \cap V| = \kappa$ for some $V \in \mathcal{V}$. However, $p \notin \overline{V}$ and hence there is $\beta < \kappa$ for which $f_\alpha \notin \overline{V}$ for any $\alpha \geq \beta$. Thus $S \cap V \subset \{f_\alpha : \alpha < \beta\}$ and therefore $|S \cap V| \leq |\beta| < \kappa$; this contradiction shows that $\kappa = \omega$, i.e., we have a sequence $S = \{f_n : n \in \omega\} \subset A$ which converges to p. By the choice of A, there is $U_n \in \mathcal{U}$ such that $f_n^{-1}(\mathbb{R}\backslash\{0\}) \subset U_n$ for every $n \in \omega$.

We have $p \in \overline{A}$ so, for any finite set $K \subset X$ there is $n \in \omega$ such that $f_n(x) > 0$ for every $x \in K$. This implies $K \subset f_n^{-1}(\mathbb{R}\backslash\{0\}) \subset U_n$ and therefore the family $\{U_n : n \in \omega\} \subset \mathcal{U}$ is an ω-cover of X. We established that every ω-cover of X has a countable ω-subcover; so $t(C_p(X)) = \omega$ by TFS-148 and TFS-149. Thus $C_p(X)$ is a radial space of countable tightness which shows that we can apply Problem 066 to conclude that $C_p(X)$ is Fréchet–Urysohn.

U.074. *An uncountable cardinal κ is called ω-inaccessible if $\lambda^\omega < \kappa$ for any cardinal $\lambda < \kappa$. Recall that, if ξ is an ordinal then $cf(\xi) = \min\{|A| : A$ is a cofinal subset of $\xi\}$. Prove that, for an infinite ordinal ξ, the space $C_p(\xi)$ is pseudoradial if and only if either $cf(\xi) \leq \omega$ or ξ is an ω-inaccessible regular cardinal (here,* *as usual, ξ is considered with its interval topology). Observe that ω-inaccessible regular cardinals exist in ZFC and hence there exist spaces X such that $C_p(X)$ is pseudoradial but not radial.*

Solution. If Z is a set then $\text{Fin}(Z)$ is the family of all non-empty finite subsets of Z. We will develop some methods of working with stationary sets in cardinals larger than ω_1; however, all proofs and concepts here are analogous to the ones given in SFFS-064 and SFFS-067. All cardinals and ordinals carry the interval topology of their natural well ordering. Any ordinal α is identified with the set $\{\beta : \beta < \alpha\}$. It will be always clear from the context what we mean except one situation in dealing with maps when confusion is possible.

To remedy this, we introduce the following notation: if $f : \alpha \to Z$ is a map and $\beta < \alpha$ then $f(\beta) \in Z$ is the respective image of β as an element of α and $f[\beta] = \{f(\gamma) : \gamma < \beta\} \subset Z$. If α and β are ordinals then $[\alpha, \beta] = \{\gamma : \alpha \leq \gamma \leq \beta\}$ and $(\alpha, \beta) = \{\gamma : \alpha < \gamma < \beta\}$; we will also need the intervals $(\alpha, \beta] = \{\gamma : \alpha < \gamma \leq \beta\}$ and $[\alpha, \beta) = \{\gamma : \alpha \leq \gamma < \beta\}$. As usual α is a *limit ordinal* if there exists no ordinal β such that $\alpha = \beta + 1$; observe that this definition says that $\alpha = 0$ is a limit ordinal.

Given a regular uncountable cardinal κ say that $C \subset \kappa$ is *(a) club* if C is closed and unbounded ($\equiv$cofinal) in κ. A set $A \subset \kappa$ is called *stationary* if $A \cap C \neq \emptyset$ for any club C. If Z is a space say that a set $A \subset Z$ is *radially closed in Z* if, for any regular cardinal λ and λ-sequence $S = \{z_\alpha : \alpha < \lambda\} \subset A$ if $S \to z$ then $z \in A$. If ξ is an ordinal, call a λ-sequence $S = \{\beta_\alpha : \alpha < \lambda\} \subset \xi$ *strictly increasing* if $\alpha < \alpha' < \lambda$ implies $\beta_\alpha < \beta_{\alpha'}$.

Fact 1. Given a regular uncountable cardinal κ suppose that $\lambda < \kappa$ and $C_\alpha \subset \kappa$ is a club for any $\alpha < \lambda$. Then $C = \bigcap\{C_\alpha : \alpha < \lambda\}$ is also a club.

Proof. The intersection of any number of closed sets is closed; so C is closed in κ. Thus we only have to prove that C is cofinal in κ. Observe first that,

(1) for any $\beta < \kappa$ there exists a strictly increasing λ-sequence $\{\mu_\alpha : \alpha < \lambda\} \subset \kappa$ such that $\beta < \mu_0$ and $\mu_\alpha \in C_\alpha$ for every $\alpha < \lambda$.

The proof of (1) can be done by making use of cofinality of each C_α by a trivial transfinite induction; so it can be left to the reader as an exercise. Fix an arbitrary ordinal $\gamma < \kappa$ and apply (1) to construct inductively a collection $\{S_n : n \in \omega\}$ of λ-sequences with the following properties:

(2) $S_n = \{\mu_\alpha^n : \alpha < \lambda\}$ and $\mu_\alpha^n \in C_\alpha$ for every $n \in \omega$ and $\alpha < \lambda$;
(3) $\gamma < \mu_0^0$ and $\mu_\alpha^n < \mu_\beta^n$ whenever $\alpha < \beta$ and $n \in \omega$;
(4) $\mu_\alpha^n < \mu_\beta^m$ for any $n < m < \omega$ and $\alpha, \beta < \lambda$.

In other words, all our λ-sequences are strictly increasing and lie in $\kappa \backslash \gamma$; besides, each one has elements larger the all elements of the preceding one. If $\mu_n = \sup S_n$ for all $n \in \omega$ and $\mu = \sup_{n \in \omega} \mu_n$ then $\mu = \lim_{n \to \infty} \mu_\alpha^n$; since C_α is closed, we have $\mu \in C_\alpha$ for every $\alpha < \lambda$, i.e., $\mu \in C$. Thus, for any $\gamma < \kappa$ there is $\mu \in C$ with $\mu > \gamma$. Therefore C is unbounded and Fact 1 is proved.

Fact 2. Let κ be an uncountable regular cardinal. Then

(i)   if $A \subset \kappa$ is stationary then $|A| = \kappa$;
(ii) if $A \subset B \subset \kappa$ and A is stationary then B is also stationary;
(iii) if $A \subset \kappa$ is stationary and $C \subset \kappa$ is a club then $A \cap C$ is stationary;
(iv) given a cardinal $\lambda < \kappa$ suppose that $A_\alpha \subset \kappa$ for all $\alpha < \lambda$ and $\bigcup\{A_\alpha : \alpha < \lambda\}$
      is stationary. Then A_α is stationary for some $\alpha < \lambda$.

Proof. If $|A| < \kappa$ then, by regularity of κ, the set A is not cofinal in κ and hence
there is $\beta < \kappa$ such that $\alpha < \beta$ for all $\alpha \in A$. Thus $C = \kappa \backslash \beta$ is a club with
$A \cap C = \emptyset$; this contradiction proves (i). The property (ii) is evident; as to (iii), let
$A' = A \cap C$. If $D \subset \kappa$ is a club then $C \cap D$ is also a club by Fact 1 so $A' \cap D =$
$A \cap (C \cap D) \neq \emptyset$ and hence A' is also stationary. Finally, if $A = \bigcup\{A_\alpha : \alpha < \lambda\}$
and every A_α is non-stationary then there is a club $C_\alpha \subset \kappa$ such that $A \cap C_\alpha = \emptyset$.
Then $C = \bigcap\{C_\alpha : \alpha < \lambda\}$ is a club by Fact 1 while $A \cap C = \emptyset$ which shows that
A is not stationary and proves (iv). Fact 2 is proved.

Fact 3. Suppose that κ is a regular uncountable cardinal and A is a stationary subset
of κ. Assume that $f : A \to \kappa$ and $f(\alpha) < \alpha$ for any $\alpha \in A$. Then there is $\beta < \kappa$
such that the set $\{\alpha \in A : f(\alpha) = \beta\}$ is stationary.

Proof. If the statement of this Fact is false then, for any $\beta < \kappa$ there is a club C_β
such that $C_\beta \cap f^{-1}(\beta) = \emptyset$. We claim that the set

(5) $C = \{\alpha < \kappa : \alpha \in C_\beta$ for any $\beta < \alpha\}$

is a club. To see that C is closed take any $\alpha \in \kappa \backslash C$; then there is $\beta < \alpha$ such that
$\alpha \notin C_\beta$. Since C_β is closed, there is $\alpha' < \alpha$ such that $\beta \leq \alpha'$ and $(\alpha', \alpha] \cap C_\beta = \emptyset$. It
is clear that, for every $\gamma \in (\alpha', \alpha]$ we have $\beta < \gamma$ and $\gamma \notin C_\beta$, i.e., $(\alpha', \alpha] \cap C = \emptyset$
which proves that C is closed in κ.

To see that C is cofinal fix any $\gamma < \kappa$ and let $\beta_{-1} = \gamma$. By Fact 1 the set
$D_0 = \bigcap\{C_\beta : \beta < \gamma\}$ is a club; so we can pick $\beta_0 \in D_0$ with $\gamma < \beta_0$. Assume
that $m \in \mathbb{N}$ and we have club sets $D_0, \ldots, D_{m-1}$ and ordinals $\beta_0, \ldots, \beta_{m-1}$ with
the following properties:

(6) $\gamma < \beta_0 < \ldots < \beta_{m-1}$ and $\beta_i \in D_i$ for any $i < m$;
(7) $D_{i+1} = \bigcap\{C_\beta : \beta < \beta_i\}$ for all $i < m - 1$.

The set $D_m = \bigcap\{C_\beta : \beta < \beta_{m-1}\}$ is a club by Fact 1; so we can choose
$\beta_m \in D_m$ such that $\beta_{m-1} < \beta_m$. It is clear that the conditions (6) and (7) are
still satisfied; so our inductive construction can be continued to obtain sequences
$\{D_n : n \in \omega\}$ and $\{\beta_n : n \in \omega\}$ for which (6) and (7) hold for all $m \in \omega$. It is
straightforward that $\beta = \sup_{n \in \omega} \beta_n \in C$ and $\beta > \gamma$ so C is, indeed, a club.

Finally observe that $C \cap A = \emptyset$ because, given any $\alpha \in A$, we have $\beta = f(\alpha) <$
α and it follows from $C_\beta \cap f^{-1}(\beta) = \emptyset$ that $\alpha \notin C_\beta$ and hence $\alpha \notin C$. This shows
that A is not stationary and gives a contradiction which finishes the proof of Fact 3.

Fact 4. For any ordinal ξ the space $\xi + 1$ is compact and scattered; so $C_p(\xi + 1)$ is
a Fréchet–Urysohn space.

Proof. The space $\xi + 1$ is well ordered and has a largest element; so it is compact by TFS-306. If $\emptyset \neq A \subset \xi + 1$ then the point $a = \min(A)$ is isolated in A which proves that every non-empty $A \subset \xi + 1$ has an isolated point and hence $\xi + 1$ is scattered. Thus $C_p(\xi)$ is Fréchet–Urysohn by SFFS-134 and Fact 4 is proved.

Fact 5. If ξ is any ordinal then any closed non-empty $F \subset \xi$ is a retract of ξ, i.e., there exists a continuous map $r : \xi \to F$ such that $r(\alpha) = \alpha$ for any $\alpha \in F$.

Proof. Let $\gamma_0 = \sup(F)$ (it is possible that $\gamma_0 = \xi$ and hence $\gamma_0 \notin \xi$). If $\alpha < \xi$ and $\alpha > \gamma_0$ then let $r(\alpha) = \gamma_0$. If $\alpha \leq \gamma_0$ then the ordinal $r(\alpha) = \min\{\beta \in F : \alpha \leq \beta\}$ is well defined; thus we have a map $r : \xi \to F$ and it is evident that $r(\alpha) = \alpha$ for any $\alpha \in F$.

To see that r is continuous observe first that $r|(\gamma_0, \xi)$ is constant on the open set (γ_0, ξ) so r is continuous at every $\alpha > \gamma_0$. Note next that $r(\alpha) \geq \alpha$ for any $\alpha \leq \gamma_0$. Given an arbitrary $\alpha \leq \gamma_0$ let $\beta = r(\alpha)$. If $\alpha \notin F$ then $(\alpha, \beta) \cap F = \emptyset$; since F is closed, there is $\alpha' < \alpha$ such that $(\alpha', \alpha] \cap F = \emptyset$. The set $U = (\alpha', \alpha]$ is a neighbourhood of α and $r(U) = \{\beta\}$ which shows that r is continuous at the point α. Now, if $\alpha \in F$ then $r(\alpha) = \alpha$; take any $U \in \tau(\alpha, \xi)$. There is $\alpha' < \alpha$ such that $(\alpha', \alpha] \subset U$; given any $\beta \in V = (\alpha', \alpha]$ we have $\alpha \geq r(\beta) \geq \beta > \alpha'$ which shows that $r(V) \subset V \subset U$, i.e., the neighbourhood V of the point α witnesses continuity of r at α. We showed that r is continuous at all point of ξ so $r : \xi \to F$ is, indeed, a retraction and hence Fact 5 is proved.

Fact 6. If ξ is an ordinal such that $\mathrm{cf}(\xi) > \omega$ then, for any second countable space M and a continuous map $f : \xi \to M$ there is $z \in M$ and $\eta < \xi$ such that $f(\alpha) = z$ for any $\alpha \in [\eta, \xi)$.

Proof. If $A \subset \xi$ is countable then there is $\beta < \xi$ such that $A \subset \beta + 1$ and hence $f(A) \subset f[\beta + 1]$. However, $\beta + 1$ is a scattered compact space (see Fact 4); so $f[\beta + 1]$ is countable by SFFS-129. Thus $\overline{f(A)} \subset f[\beta + 1]$ is countable for any countable $A \subset \xi$. Since $f[\xi]$ is separable, we conclude that $|f[\xi]| \leq \omega$. It follows from $\mathrm{cf}(\xi) > \omega$ that, for some $z \in M$ the set $K_z = f^{-1}(z)$ is cofinal in ξ. Suppose that $t \neq z$ and $K_t = f^{-1}(t)$ is also cofinal in ξ.

By an easy induction we can construct sequences $\{\alpha_n : n \in \omega\} \subset K_z$ and $\{\beta_n : n \in \omega\} \subset K_t$ such that $\alpha_n < \beta_n < \alpha_{n+1}$ for all $n \in \omega$. Since ξ is not ω-cofinal, the ordinal $\alpha = \sup_{n\in\omega} \alpha_n = \sup_{n\in\omega} \beta_n$ belongs to ξ and hence to $K_z \cap K_t$ because K_z and K_t are closed in ξ. However, $K_z \cap K_t = \emptyset$; this contradiction shows that, for every $t \in R = f[\xi]\setminus\{z\}$ there is $\alpha_t < \xi$ such that $f^{-1}(t) \subset \alpha_t$. Recalling once more that $\mathrm{cf}(\xi) > \omega$ we conclude that there is $\eta < \xi$ for which $\bigcup\{f^{-1}(t) : t \in R\} \subset \eta$ and therefore $f(\alpha) = z$ for all $\alpha \geq \eta$ as promised. Fact 6 is proved.

Fact 7. For any space Z we have $|C_p(Z)| \leq w(Z)^{l(Z)}$.

Proof. Fix a base $\mathcal{B}$ of the space Z such that $|\mathcal{B}| = \kappa = w(Z)$. Let $\lambda = l(Z)$ and consider the family $\mathcal{A} = \{\mathcal{B}' \subset \mathcal{B} : |\mathcal{B}'| \leq \lambda\}$. It is clear that $|\mathcal{A}| = \kappa^\lambda$ and hence $|\mathcal{A}^\omega| = (\kappa^\lambda)^\omega = \kappa^{\lambda\cdot\omega} = \kappa^\lambda$. Choose an enumeration $\{I_n : n \in \omega\}$ of all non-trivial rational intervals in $\mathbb{R}$. If $f \in C_p(Z)$ then $U_n = f^{-1}(I_n)$ is an F_σ-subset of Z; so

$l(U_n) \leq \lambda$ and hence we can choose a family $C_n(f) \subset \mathcal{B}$ such that $\bigcup C_n(f) = U_n$ and $|C_n(f)| \leq \lambda$ for every $n \in \omega$. Letting $\varphi(f)(n) = C_n(f)$ for all $f \in C_p(Z)$ and $n \in \omega$ we obtain a map $\varphi : C_p(Z) \to \mathcal{A}^\omega$.

Now, given distinct $f, g \in C_p(Z)$ there is $n \in \omega$ and $z \in Z$ such that $f(z) \in I_n$ and $g(z) \notin I_n$. There is $B \in C_n(f)$ such that $z \in B$; since $g(B) \ni g(z) \notin I_n$, we have $g(B) \not\subset I_n$ and hence $C_n(g) \neq C_n(f)$ which implies $\varphi(f) \neq \varphi(g)$. Thus $\varphi : C_p(Z) \to \mathcal{A}^\omega$ is an injection and therefore $|C_p(Z)| \leq |\mathcal{A}^\omega| = \kappa^\lambda$ so Fact 7 is proved.

Fact 8. Given an ordinal ξ assume that $\kappa = \mathrm{cf}(\xi) \geq \omega$ and $\mu \in \xi$. Then there exists a map $f : \kappa \to [\mu, \xi)$ such that $\alpha < \beta < \kappa$ implies $f(\alpha) < f(\beta)$, the set $F = f[\kappa]$ is closed in ξ and $f : \kappa \to F$ is a homeomorphism. In particular, κ embeds in $[\mu, \xi)$ as a closed subspace.

Proof. Fix a strictly increasing κ-sequence $\{\mu_\alpha : \alpha < \kappa\} \subset \xi$ such that A is cofinal in ξ and let $f(0) = \max\{\mu, \mu_0\}$. Proceeding inductively assume that $\gamma < \kappa$ and we defined $f(\alpha) \in [\mu, \xi)$ for any $\alpha < \gamma$ in such a way that

(8) the map $f : \gamma \to \xi$ is continuous;
(9) $\alpha < \beta < \gamma$ implies $f(\alpha) < f(\beta)$;
(10) $f(\alpha) \geq \mu_\alpha$ for any successor ordinal $\alpha < \gamma$.

If γ is a successor, i.e., $\gamma = \gamma' + 1$ then let $f(\gamma) = \max\{\mu_\gamma, f(\gamma')\} + 1$. It is immediate that (9) and (10) still hold for all $\alpha, \beta \leq \gamma$. Besides, the point γ is isolated in $\gamma + 1$; so $f : (\gamma + 1) \to [\mu, \xi)$ is continuous at the point γ and hence f is continuous on $\gamma + 1$. Thus the properties (8)–(10) are fulfilled for all $\alpha, \beta \leq \gamma$.

Now, if γ is a limit ordinal then let $f(\gamma) = \sup\{f(\alpha) : \alpha < \gamma\}$. Since the set $f[\gamma]$ has cardinality less than κ, the ordinal $f(\gamma)$ belongs to ξ. The property (10) does not require anything for limit ordinals, so it is still fulfilled for all $\alpha \leq \gamma$; it is obvious that (9) also holds for all $\alpha, \beta \leq \gamma$. As to (8), we only have to check continuity at the point γ; so take any $\beta < f(\gamma)$. There is $\nu < \gamma$ such that $f(\nu) \geq \beta$; so it follows from (9) that $f((\nu, \gamma]) \subset (\beta, f(\gamma)]$, i.e., the neighbourhood $(\nu, \gamma]$ witnesses continuity of f at the point γ.

Therefore our inductive construction can be continued to provide a mapping $f : \kappa \to [\mu, \xi)$ with the properties (8)–(10) satisfied for all $\gamma < \kappa$. In particular, f is continuous; let $F = f[\kappa]$. It follows from (9) that f is a bijection; the space $\gamma + 1$ is compact, so $f|(\gamma + 1)$ is a homeomorphism for any $\gamma < \kappa$. It follows from (10) that $f(\kappa)$ is cofinal in ξ. If $\alpha \in \xi \backslash F$ then there is $\gamma < \kappa$ such that $f(\gamma) > \alpha$ and hence $f(\nu) > \alpha$ for any $\nu \geq \gamma$. Since $f[\gamma + 1]$ is compact, there is $\beta < \alpha$ such that $(\beta, \alpha] \cap f[\gamma + 1] = \emptyset$; it is immediate that also $(\beta, \alpha] \cap F = \emptyset$ and hence F is closed in ξ.

Observe that a set $H \subset \xi$ is closed in ξ if and only if $H \cap (\nu + 1)$ is compact for any $\nu < \xi$. Given a closed $G \subset \kappa$ and $\nu < \xi$ there is $\gamma < \kappa$ such that $f(\gamma) > \nu$ and hence the set $f(G) \cap (\nu + 1) = f(G \cap (\gamma + 1)) \cap (\nu + 1)$ is compact because $f : (\gamma + 1) \to f[\gamma + 1]$ is a closed map. Thus $f(G)$ is a closed subset of ξ for any closed $G \subset \kappa$ and hence $f : \kappa \to F$ is a homeomorphism; so Fact 8 is proved.

Fact 9. For any ordinal α there exists a unique $n(\alpha) \in \omega$ and a unique limit ordinal $\mu(\alpha)$ such that $\alpha = \mu(\alpha) + n(\alpha)$.

Proof. Let $P = \{\beta \in \alpha + 1 : \text{for some } n \in \omega \text{ we have } \beta + n = \alpha\}$. The set P is non-empty because $0 \in P$. Let $\mu(\alpha)$ be the minimal element of P; by definition of $\mu(\alpha)$ we have $\alpha = \mu(\alpha) + n(\alpha)$ for some $n(\alpha) \in \omega$. The ordinal $\mu(\alpha)$ is limit for otherwise $\mu(\alpha) = \beta + 1$ and hence $\beta + (n(\alpha) + 1) = \alpha$, i.e., $\beta \in P$ and $\beta < \mu(\alpha)$ which is a contradiction. If $\alpha = \beta + n$ for some limit ordinal β and $n \in \omega$ we have $\beta \in P$ and hence $\mu(\alpha) \leq \beta$ by the choice of $\mu(\alpha)$. Now, if $\mu(\alpha) \neq \beta$ then $\mu(\alpha) < \beta \leq \mu(\alpha) + n(\alpha)$ which is a contradiction with the fact that β is a limit ordinal. Thus $\beta = \mu(\alpha)$ and $n = n(\alpha)$; so Fact 9 is proved.

Fact 10. For any space Z, if $C_p(Z)$ is Fréchet–Urysohn and $Y \neq \emptyset$ is an F_σ-subset of Z then $C_p(Y)$ is also Fréchet–Urysohn.

Proof. We have $Y = \bigcup_{n \in \omega} Y_n$ where $Y_n \subset Y_{n+1}$ and Y_n is closed in Z for every $n \in \omega$. Let $\mathcal{U}_n$ be an open ω-cover of Y for every $n \in \omega$. Given $n \in \omega$, for any $U \in \mathcal{U}_n$ take $U' \in \tau(Z)$ with $U' \cap Y = U$ and let $\mathcal{U}'_n = \{U' : U \in \mathcal{U}_n\}$. It is easy to see that $\mathcal{V}_n = \{V \cup (Z \backslash Y_n) : V \in \mathcal{U}'_n\}$ is an open ω-cover of Z for every $n \in \omega$.

Since $C_p(Z)$ is Fréchet–Urysohn, we can choose, for any $n \in \omega$, a set $W_n \in \mathcal{V}_n$ in such a way that $\mathcal{W} = \{W_n : n \in \omega\}$ converges to Z, i.e., any $z \in Z$ belongs to all except finitely many elements of $\mathcal{W}$ (see TFS-144). By definition of $\mathcal{V}_n$, there is $V_n \in \mathcal{U}'_n$ for which $W_n = V_n \cup (Z \backslash Y_n)$. Then $U_n = V_n \cap Y \in \mathcal{U}_n$ for every $n \in \omega$ and it is straightforward that the sequence $\{U_n : n \in \omega\}$ converges to Y. Applying TFS-144 again we conclude that $C_p(Y)$ is Fréchet–Urysohn and hence Fact 10 is proved.

Fact 11. Given spaces Y and Z assume that $f : Y \to Z$ is a continuous map such that there is $P \subset Y$ for which $f(P) = Z$ and $f|P : P \to Z$ is a quotient map. Then f is quotient. In particular, any retraction is a quotient map.

Proof. Consider the function $g = f|P$; if $U \subset Z$ and the set $f^{-1}(U)$ is open in Y then $W = f^{-1}(U) \cap P = g^{-1}(U)$ is an open subset of P. The map g being quotient, U is open in Z and hence f is also quotient. Finally, if Z is a space, $F \subset Z$ and $r : Z \to F$ is a retraction then the map $r|F : F \to F$ is quotient (even a homeomorphism); so r is quotient by what we established above. Fact 11 is proved.

Returning to our solution let us prove first that $C_p(\kappa)$ is pseudoradial for any regular ω-inaccessible cardinal κ. Let $u(\alpha) = 0$ for any $\alpha < \kappa$; given $d \in \mathbb{R}$ and $\alpha < \kappa$ let $C(\alpha, d) = \{f \in C_p(\kappa) : f(\beta) = d \text{ for any } \beta \geq \alpha\}$. If $K \in \text{Fin}(\kappa)$ and $\varepsilon > 0$ let $[K, \varepsilon] = \{f \in C_p(\kappa) : f(K) \subset (-\varepsilon, \varepsilon)\}$.

Take a radially closed set $A \subset C_p(\kappa)$; we must prove that A is closed in $C_p(\kappa)$, i.e., $f \in \overline{A}$ implies $f \in A$. By homogeneity of $C_p(\kappa)$ it suffices to show that $u \in \overline{A}$ implies $u \in A$; so assume that $u \in \overline{A}$.

Claim. For any finite $F \subset \kappa$ and $\varepsilon > 0$ there is a function $f = f_{F,\varepsilon} \in A$ and $\eta = \eta_{F,\varepsilon} < \kappa$ such that $|f(\alpha)| \leq \varepsilon$ for every $\alpha \in F$ and $f \in C(\eta, d)$ for some $d = d_{F,\varepsilon} \in [-\varepsilon, \varepsilon]$.

Proof of the Claim. It follows from $\omega^\omega \geq \omega_1$ that $\kappa > \omega_1$; denote by L the set $\{\alpha < \kappa : \mathrm{cf}(\alpha) = \omega_1\}$.

Given a club $C \subset \kappa$ we can choose, by an evident transfinite induction, a strictly increasing ω_1-sequence $\{\mu_\alpha : \alpha < \omega_1\} \subset C$. Since $\kappa = \mathrm{cf}(\kappa) > \omega_1$, the ordinal $\mu = \sup\{\mu_\alpha : \alpha < \omega_1\}$ belongs to κ and hence to C because C is closed in κ. Thus $\mu \in C \cap L$ which proves that L is a stationary subset of κ.

It is easy to see that the set $B' = \{f \in C_p(\kappa) : |f(\alpha)| \leq \varepsilon$ for any $\alpha \in F\}$ is closed in $C_p(\kappa)$, so $B = B' \cap A$ is radially closed; since B' is a neighbourhood of u, we still have $u \in \overline{B}$. For each $\alpha \in L$ there is $f_\alpha \in A$ such that $d_\alpha = f_\alpha(\alpha) \in (-\varepsilon, \varepsilon)$. By Fact 6 there exists $\eta_\alpha < \alpha$ such that $f_\alpha(\beta) = d_\alpha$ for any $\beta \in [\eta_\alpha, \alpha]$. Letting $\varphi(\alpha) = \eta_\alpha$ for all $\alpha \in L$ we obtain a function $\varphi : L \to \kappa$ such that $\varphi(\alpha) < \alpha$ for any $\alpha \in L$. By Fact 3 there is a cofinal $S \subset L$ and $\eta = \eta_{F,\varepsilon} < \kappa$ such that $\eta_\alpha = \eta$ for any $\alpha \in S$. We also have $|\mathbb{R}| = 2^\omega < \kappa$ which shows that there is $d \in (-\varepsilon, \varepsilon)$ such that $d_\alpha = d$ for κ-many $\alpha \in S$.

Furthermore, $|C_p(\eta + 1)| \leq |\eta|^\omega < \kappa$ (see Fact 7) which proves that there is $S' \subset S$ and $g \in C_p(\eta + 1)$ such that $|S'| = \kappa$, $d_\alpha = d = d_{F,\varepsilon}$ and $f_\alpha|(\eta + 1) = g$ for all $\alpha \in S'$. Let $f_{F,\varepsilon}(\alpha) = g(\alpha)$ for any $\alpha \leq \eta$ and $f_{F,\varepsilon}(\alpha) = d$ for all $\alpha > \eta$.

Choose by an evident transfinite induction a set $\{\beta_\alpha : \alpha < \kappa\} \subset S'$ such that $\alpha < \alpha' < \kappa$ implies $\beta_\alpha < \beta_{\alpha'}$. Letting $g_\alpha = f_{\beta_\alpha}$ for all $\alpha < \kappa$ we obtain a κ-sequence $S = \{g_\alpha : \alpha < \kappa\} \subset A$. It id clear that $g_\alpha|(\eta + 1) = g$ for all $\alpha < \kappa$. Besides, the set $\{\beta_\alpha : \alpha < \kappa\}$ is cofinal in κ; so for any $\nu \in \kappa \backslash (\eta + 1)$ there is $\gamma < \kappa$ such that $\beta_\alpha > \nu$ for any $\alpha \geq \gamma$. This implies $g_\alpha(\nu) = d$ for all $\alpha \geq \gamma$; so we proved that

(11) for any $\nu < \kappa$ there is $\gamma < \kappa$ for which $g_\alpha(\nu) = f_{F,\varepsilon}(\nu)$ for all $\alpha \geq \gamma$,

which shows that the κ-sequence S converges to $f_{F,\varepsilon}$ whence $f_{F,\varepsilon} \in A$ and our Claim is proved.

Let $\varepsilon_n = 2^{-n}$ for any $n \in \omega$; for arbitrary $n \in \omega$ and $F \in \mathrm{Fin}(\kappa)$ choose a function $g_{F,n} = f_{F,\varepsilon_n} \in A$ and an ordinal $\mu_{F,n} = \eta_{F,\varepsilon_n} < \kappa$ whose existence is granted by the Claim.

Let $\nu_0 = \omega$; proceeding inductively assume that $m \in \omega$ and we have ordinals $\nu_0 < \ldots < \nu_m < \kappa$ such that,

(12) for every $i < m$, we have $\mu_{F,n} < \nu_{i+1}$ for any $F \in \mathrm{Fin}(\nu_i)$ and $n \in \omega$.

The set $M = \{\mu_{F,n} : F \in \mathrm{Fin}(\nu_m)$ and $n \in \omega\}$ has cardinality $< \kappa$; so there is $\nu_{m+1} < \kappa$ such that $\alpha < \nu_{m+1}$ for all $\alpha \in M$. It is clear that (12) still holds for the ordinals $\{\nu_i : i \leq m + 1\}$; so our inductive construction can be continued giving us a strictly increasing sequence $\{\nu_i : i \in \omega\} \subset \kappa$ such that (12) holds for every $m \in \omega$. It is immediate that for the ordinal $\nu = \sup_{i \in \omega} \nu_i$ we have

(13) $\mu_{F,n} < \nu$ for any $F \in \mathrm{Fin}(\nu)$ and $n \in \omega$.

Consider the set $T = \{g_{F,n} : F \in \mathrm{Fin}(\nu)$ and $n \in \omega\} \subset A$. By definition of ν, for any $h \in T$ we have $h \in C(\nu, d)$ for some $d \in \mathbb{R}$, i.e., the function h is constant on $[\nu, \kappa)$. Now, take any finite $K \subset \kappa$ and $\varepsilon > 0$. Making K larger if necessary we can consider that $F = K \cap (\eta + 1) \neq \emptyset$. Take $n \in \omega$ for which $\varepsilon_n < \varepsilon$; then

$g = g_{F,n} \in T$ and $g(\alpha) \in [\varepsilon_n, \varepsilon_n] \subset (-\varepsilon, \varepsilon)$ for all $\alpha \in F$. By definition of g we have $g \in C(v, d)$ where $d \in [\varepsilon_n, \varepsilon_n] \subset (-\varepsilon, \varepsilon)$ and hence $g(\alpha) \in (-\varepsilon, \varepsilon)$ for all $\alpha \in K$. The family $\mathcal{E} = \{[K, \varepsilon] : K \in \text{Fin}(\kappa), \ \varepsilon > 0\}$ is a base at u in $C_p(\kappa)$; since $O \cap T \neq \emptyset$ for any $O \in \mathcal{E}$, we have $u \in \overline{T}$.

The space $C_p(v + 1)$ is Fréchet–Urysohn by Fact 4; so there is a sequence $E = \{h_n : n \in \omega\} \subset T$ such that $h_n|(\eta + 1)$ converges to $u|(\eta + 1)$. In particular, $h_n(\eta) \to 0$; since $h_n(\alpha) = h_n(\eta)$ for any $\alpha \geq \eta$ we have $h_n(\alpha) \to 0$ for any $\alpha \geq \eta$. Therefore $h_n(\alpha) \to 0$ for any $\alpha < \kappa$ which shows that $h_n \to u$ and hence $u \in A$. We finally proved that

(14) the space $C_p(\kappa)$ is pseudoradial for any ω-inaccessible regular cardinal κ.

Let us show next that

(15) for any uncountable regular cardinal κ the space $C_p(\kappa + \kappa)$ is not pseudoradial; here $\kappa + \kappa$ is the ordinal addition.

For any $\alpha \in (0, \kappa)$ let $n = n(\alpha)$ (see Fact 9) and define a function $f_\alpha \in C_p(\kappa + \kappa)$ as follows: $f_\alpha(\beta) = 0$ for any $\beta \in [0, \alpha] \cup [\kappa + 1, \kappa + \alpha]$; if $\beta \in (\alpha, \kappa]$ then $f_\alpha(\beta) = 2^{-n}$ and $f_\alpha(\beta) = 1$ for all $\beta \in (\kappa + \alpha, \kappa + \kappa)$. Let us show that the set $A = \{f_\alpha : \alpha < \kappa\}$ is radially closed.

Assume, towards a contradiction, that λ is a regular cardinal and we are given a λ-sequence $S = \{h_\alpha : \alpha < \lambda\} \subset A$ such that $S \to f \in C_p(\kappa + \kappa)\backslash A$. We have $h_\alpha = f_{\gamma(\alpha)}$ for any $\alpha < \lambda$. If $h \in A$ and $h_\alpha = h$ for λ-many α's then $S \to h \in A$; this contradiction shows that, passing to an appropriate subsequence of S if necessary, we can assume, without loss of generality, that the enumeration of S is faithful and $\alpha < \alpha' < \lambda$ implies $\gamma(\alpha) < \gamma(\alpha')$.

Consider first the case when $\lambda < \kappa$; then $\mu = \sup\{\gamma(\alpha) : \alpha < \lambda\} < \kappa$. Therefore $h_\alpha(\kappa + \mu) = 1$ for all $\alpha < \lambda$ and hence $f(\kappa + \mu) = 1$. However, for any $\beta \in [1, \mu)$ there is $\alpha' < \lambda$ such that $\gamma(\alpha) > \beta$ for any $\alpha \geq \alpha'$ and hence we have the equality $h_\alpha(\kappa + \beta) = f_{\gamma(\alpha)}(\kappa + \beta) = 0$ which shows that $f(\kappa + \beta) = 0$. It turns out that $f(\kappa + \beta + 1) = 0$ for any $\beta < \mu$ while $f(\kappa + \mu) = 1$, i.e., f is not continuous at the point $\kappa + \mu$ which is a contradiction.

The case of $\lambda < \kappa$ being settled, assume that $\lambda = \kappa$. Since $\kappa > \omega$, we can choose a κ-subsequence S' of our κ-sequence S again to assure that, for some $n \in \omega$ we have $n(\gamma(\alpha)) = n$ whenever $h_\alpha \in S'$ (see Fact 9); so we can assume that $n(\gamma(\alpha)) = n$ for any $\alpha < \kappa$. But then, for any $\alpha < \kappa$ we have $h_\alpha(\beta) = r = 2^{-n}$ for any $\beta \in (\gamma(\alpha), \kappa]$ and hence $h_\alpha(\kappa) = r$ whence $f(\kappa) = r$. Besides, for any $\beta < \kappa$ there is $\nu < \kappa$ such that $\gamma(\alpha) > \beta$ for all $\alpha \geq \nu$ and therefore $h_\alpha(\beta) = 0$ for all $\beta \geq \nu$. This implies $f(\beta) = 0$ for all $\beta < \kappa$ and hence $f[\kappa] = \{0\}$ while $f(\kappa) = r \neq 0$, i.e., f is discontinuous at the point κ. This contradiction concludes our proof that A is radially closed.

Finally, observe that the function u which is identically zero on $\kappa + \kappa$ is in the closure of A (this is an easy exercise which we leave to the reader). Thus $u \in \overline{A}\backslash A$ and hence A is a radially closed non-closed subset of $C_p(\kappa + \kappa)$ which shows that $C_p(\kappa + \kappa)$ is not pseudoradial and completes the proof of (15).

Our next step is to establish that

(16) if $\kappa > \omega$ is a regular not ω-inaccessible cardinal then $C_p(\kappa)$ is not pseudoradial.

Fix an infinite cardinal $\nu < \kappa$ such that $\nu^\omega \geq \kappa$. Observe first that,

(17) if $F \in \mathrm{Fin}(\nu + 1)$ and $I_F = \{f \in C_p(\nu + 1) : f|F \equiv 0\}$ then $|I_F| = \nu^\omega$.

Indeed, $I_F \subset C_p(\nu + 1)$ and hence $|I_F| \leq |C_p(\nu + 1)| \leq \nu^\omega$ (see Fact 7). Furthermore $C_p(\nu + 1)$ is homeomorphic to $I_F \times \mathbb{R}^F$ (see Fact 1 of S.494) which implies that $|C_p(\nu + 1)| = \max\{|I_F|, \mathfrak{c}\} = |I_F|$ because, obviously, $|I_F| \geq \mathfrak{c}$.

Now observe that the family $\{[0, \alpha] : \alpha < \nu\}$ consists of distinct clopen subsets of $\nu + 1$ and has cardinality ν; taking the respective characteristic functions we can see that $|C_p(\nu + 1, \mathbb{D})| \geq \nu$. Thus

$$\nu^\omega \leq |(C_p(\nu + 1, \mathbb{D}))^\omega| = |C_p(\nu + 1, \mathbb{D}^\omega)| \leq |C_p(\nu + 1)|,$$

i.e., $|C_p(\nu + 1)| = |I_F| = \nu^\omega$ as promised; so we proved (17).

Choose an enumeration $\{F_\alpha : \nu < \alpha < \kappa\}$ of the family $\mathrm{Fin}(\nu + 1)$ such that $|\{\alpha : F_\alpha = F\}| = \kappa$ for any $F \in \mathrm{Fin}(\nu + 1)$. Applying (17) and a trivial transfinite induction construct a family $\{f_\alpha : \nu < \alpha < \kappa\} \subset C_p(\nu + 1)$ such that $f_\alpha \neq f_\beta$ if $\alpha \neq \beta$ and $f_\alpha \in I_{F_\alpha}$ for any $\alpha \in (\nu, \kappa)$.

For any $\alpha \in (\nu, \kappa)$ define a function $g_\alpha \in C_p(\kappa)$ as follows: $g_\alpha|(\nu + 1) = f_\alpha$; if $\nu < \beta \leq \alpha$ then $g_\alpha(\beta) = 0$ and $g_\alpha(\beta) = 1$ for any $\beta \in (\alpha, \kappa)$. We claim that the set $A = \{g_\alpha : \nu < \alpha < \kappa\}$ is radially closed. Indeed, assume that λ is a regular cardinal such that some λ-sequence $S = \{h_\alpha : \alpha < \lambda\}$ converges to a function $f \in C_p(\kappa) \backslash A$. We have $h_\alpha = g_{\gamma(\alpha)}$ for every $\alpha < \lambda$. It is impossible that, for some $h \in A$ we have $h_\alpha = h$ for λ-many α's for otherwise $S \to h \in A$. Therefore we can substitute S by a relevant subsequence of S to guarantee that $\alpha < \alpha' < \lambda$ implies $\gamma(\alpha) < \gamma(\alpha')$.

If $\lambda = \kappa$ then it follows from convergence of S that $S' = \{h_\alpha|(\nu + 1) : \alpha < \kappa\}$ also converges to $h = f|(\nu + 1)$. Since $nw(\nu + 1) = nw(C_p(\nu + 1)) = \nu$, we have $\psi(h, C_p(\nu + 1)) \leq \nu$. Take a family $\mathcal{U} \subset \tau(C_p(\nu + 1))$ with $\bigcap \mathcal{U} = \{h\}$. For any $U \in \mathcal{U}$ there is $\alpha_U < \kappa$ such that $h_\alpha|(\nu + 1) \in U$ for all $\alpha \geq \alpha_U$. Since $\kappa > \nu$ is a regular cardinal, there is $\beta < \kappa$ such that $\alpha_U < \beta$ for all $U \in \mathcal{U}$ and hence $h_\alpha|(\nu + 1) \in \bigcap \mathcal{U} = \{h\}$, i.e., $h_\alpha|(\nu + 1) = h$ for all $\alpha \geq \beta$ which is a contradiction with the fact that $\alpha \neq \alpha'$ implies $\gamma(\alpha) \neq \gamma(\alpha')$ and hence

$$h_\alpha|(\nu + 1) = f_{\gamma(\alpha)} \neq f_{\gamma(\alpha')} = h_{\alpha'}|(\nu + 1).$$

Thus it is impossible that $\lambda = \kappa$.

Now, if $\lambda < \kappa$ then $\mu = \sup\{\gamma(\alpha) : \alpha < \nu\} < \kappa$ and hence $h_\alpha(\mu) = 1$ for all $\alpha < \lambda$; this implies $f(\mu) = 1$. If $\beta < \mu$ then there is $\alpha_0 < \lambda$ such that $\gamma(\alpha) > \beta$ for all $\alpha \geq \alpha_0$ and therefore $h_\alpha(\beta) = 0$ for all $\beta \geq \alpha_0$. Thus $f(\beta) = 0$ for any $\beta \in (\nu, \mu)$ which shows that f is discontinuous at the point μ. This contradiction proves that the case $\lambda < \kappa$ is also impossible and hence A is radially closed.

Finally, let $u(\alpha) = 0$ for all $\alpha < \kappa$. If F is a finite subset of κ then consider the set $F' = F \cap (\nu + 1)$; making F larger if necessary, we can consider that $F' \neq \emptyset$. Since there are κ-many α's for which $F_\alpha = F'$, there is $\alpha \in (\nu, \kappa)$ such that $F' = F_\alpha$ and $F \backslash F' \subset \alpha$. An immediate consequence is that $g_\alpha | F = u | F$; the finite set $F \subset \kappa$ was chosen arbitrarily so $u \in \overline{A} \backslash A$ and hence A is a radially closed non-closed subset of $C_p(\kappa)$ which proves that $C_p(\kappa)$ is not pseudoradial, i.e., (16) is settled.

Now take an arbitrary ordinal ξ. If ξ is a successor then the space $C_p(\xi)$ is Fréchet–Urysohn by Fact 4; so from now on we consider that ξ is a limit ordinal. If $cf(\xi) = \omega$ then $\xi = \bigcup\{(\xi_n + 1) : n \in \omega\}$ where $\xi_n < \xi$ for every $n \in \omega$. Since every $\xi_n + 1$ is compact, the set ξ is an F_σ-subset of $\xi + 1$. Since $C_p(\xi + 1)$ is Fréchet–Urysohn by Fact 4, the space $C_p(\xi)$ is also Fréchet–Urysohn by Fact 10.

The case of ω-cofinal ordinals being settled, assume that $\kappa = cf(\xi) > \omega$. If $\xi = \kappa$ then ξ is a regular cardinal; so the space $C_p(\xi)$ is pseudoradial if and only if ξ is ω-inaccessible by (14) and (16). Now, if $\kappa < \xi$ then we can apply Fact 8 to embed κ in $[\kappa + 1, \xi)$ as a closed subspace F; it is immediate that $\kappa \cup F$ is a closed subspace of ξ which is homeomorphic to $\kappa + \kappa$. By Fact 5 there is a retraction $r : \xi \to (\kappa \cup F)$; so there is quotient map of ξ onto $\kappa + \kappa$ (see Fact 11). Therefore $C_p(\kappa + \kappa)$ embeds in $C_p(\xi)$ as a closed subspace (see TFS-163). Since $C_p(\kappa + \kappa)$ is not pseudoradial by (15), the space $C_p(\xi)$ is not pseudoradial either by Problem 064. This settles all possible cases for ξ.

Finally observe that $\mathfrak{c}^+$ is a regular ω-inaccessible cardinal; so if $X = \mathfrak{c}^+$ then X is a countably compact space such that $C_p(X)$ is pseudoradial by (14). To see that $C_p(X)$ is not radial note that otherwise it is Fréchet–Urysohn by Problem 073; this implies $t(C_p(X)) = \omega$ whence X is Lindelöf and hence compact. However, $\mathfrak{c}^+$ does not have a largest element; so it is not compact by TFS-306. Thus we obtained a contradiction which shows that $C_p(X)$ is a pseudoradial non-radial space, i.e., our solution is complete.

U.075. *Let X be a compact space. Prove that, if $C_p(X)$ is pseudoradial then it is Fréchet–Urysohn (and hence X is scattered).*

Solution. Let us show first that $C_p(\mathbb{I})$ is not pseudoradial. It was proved in Problem 062 that $C_p(\mathbb{I})$ is not embeddable in a sequential space and, in particular, $C_p(\mathbb{I})$ is not sequential. Thus we can find a sequentially closed non-closed $A \subset C_p(\mathbb{I})$, i.e., for any sequence $S = \{f_n : n \in \omega\} \subset A$ if $S \to f \in C_p(\mathbb{I})$ then $f \in A$. We claim that A is also radially closed, i.e., for any regular cardinal κ and a κ-sequence $S = \{f_\alpha : \alpha < \kappa\} \subset A$ if $S \to f \in C_p(\mathbb{I})$ then $f \in A$.

This is, of course, clear for ω-sequences; so assume towards a contradiction that $\kappa > \omega$ and some κ-sequence $S = \{f_\alpha : \alpha < \kappa\} \subset A$ converges to some $f \in C_p(\mathbb{I}) \backslash A$. We have $\psi(C_p(\mathbb{I})) \leq nw(C_p(\mathbb{I})) = \omega$; so there is a countable family $\mathcal{U} \subset \tau(C_p(\mathbb{I}))$ such that $\bigcap \mathcal{U} = \{f\}$. For every $U \in \mathcal{U}$ there is $\alpha_U < \kappa$ such that $f_\alpha \in U$ for all $\alpha \geq \alpha_U$. Since κ is an uncountable regular cardinal, there is $\beta < \kappa$ such that $\alpha_U < \beta$ for all $U \in \mathcal{U}$. As a consequence, $f_\beta \in U$ for any $U \in \mathcal{U}$, i.e., $f_\beta = f$ which contradicts $f \notin A$. Therefore A is a radially closed non-closed subset of $C_p(\mathbb{I})$ which proves that $C_p(\mathbb{I})$ is not pseudoradial.

Finally, assume that X is a compact space such that $C_p(X)$ is pseudoradial. If X is not scattered then there exists a continuous onto map $\varphi : X \to \mathbb{I}$ (see SFFS-133). The dual map φ^* embeds $C_p(\mathbb{I})$ in $C_p(X)$ as a closed subspace (see TFS-163) and hence $C_p(\mathbb{I})$ has to be pseudoradial by Problem 064. This contradiction shows that X is scattered and hence $C_p(X)$ is Fréchet–Urysohn by SFFS-134.

U.076. *Let X be any space such that $C_p(X, \mathbb{D}) \times \omega^\omega$ is not Lindelöf. Prove that the space $C_p(X, \mathbb{D}^\omega)$ is not Lindelöf.*

Solution. Assume, towards a contradiction, that $C_p(X, \mathbb{D}^\omega)$ is Lindelöf. We have $C_p(X, \mathbb{D}^\omega) \simeq (C_p(X, \mathbb{D}))^\omega$; so $C_p(X, \mathbb{D})$ embeds in $C_p(X, \mathbb{D}^\omega)$ as a closed subspace. If $C_p(X, \mathbb{D}^\omega)$ is countably compact then, being Lindelöf, it is compact; so $C_p(X, \mathbb{D})$ is compact as well. But then $C_p(X, \mathbb{D}) \times \omega^\omega$ is Lindelöf being a perfect preimage of the Lindelöf space ω^ω (see Fact 5 of S.271). This contradiction proves that $C_p(X, \mathbb{D}^\omega)$ is not countably compact and hence we can choose a countably infinite closed discrete $D \subset C_p(X, \mathbb{D}^\omega)$. Then $D^\omega \simeq \omega^\omega$ is homeomorphic to a closed subspace of the space $(C_p(X, \mathbb{D}^\omega))^\omega \simeq C_p(X, \mathbb{D}^\omega)$. Consequently, $\omega^\omega \times C_p(X, \mathbb{D})$ is homeomorphic to a closed subspace of

$$C_p(X, \mathbb{D}^\omega) \times C_p(X, \mathbb{D}) \simeq (C_p(X, \mathbb{D}))^\omega \times C_p(X, \mathbb{D}) \simeq (C_p(X, \mathbb{D}))^\omega \simeq C_p(X, \mathbb{D}^\omega)$$

and therefore $\omega^\omega \times C_p(X, \mathbb{D})$ is Lindelöf which is again a contradiction. Thus $C_p(X, \mathbb{D}^\omega)$ is not Lindelöf.

U.077. *Suppose that X is a compact space such that a countable set $M \subset X$ is open and dense in X. Assume also that the set of isolated points of $Y = X \setminus M$ is uncountable and dense in Y. Prove that $ext(C_p(X, \mathbb{D}) \times \omega^\omega) > \omega$.*

Solution. Given a space Z denote by $\mathcal{C}(Z)$ the family of all clopen subsets of Z. The expression $Z \sqsubseteq T$ is an abbreviation of the phrase "the space Z can be embedded in the space T as a closed subspace".

Fact 1. If K is an infinite compact space then $|\mathcal{C}(K)| \leq w(K)$. In particular, for any metrizable compact K the family of all clopen subsets of K is countable.

Proof. Let $\mathcal{B}$ be a base of K such that $|\mathcal{B}| = \kappa = w(K)$. If $\mathcal{U}$ is the family of all finite unions of elements of $\mathcal{B}$ then $|\mathcal{U}| = \kappa$. Given any $W \in \mathcal{C}(K)$ there is $\mathcal{B}' \subset \mathcal{B}$ such that $W = \bigcup \mathcal{B}'$; since W is compact, there is a finite $\mathcal{B}'' \subset \mathcal{B}'$ such that $W = \bigcup \mathcal{B}''$ and hence $W \in \mathcal{U}$. We showed that every $W \in \mathcal{C}(K)$ belongs to $\mathcal{U}$ and hence $|\mathcal{C}(K)| \leq |\mathcal{U}| = \kappa$. Fact 1 is proved.

Fact 2. Given spaces Z, T and a continuous map $f : Z \to T$, for any $B \subset T$, the set $G(f, B) = \{(z, f(z)) : z \in f^{-1}(B)\} \subset Z \times B$ is closed in $Z \times B$.

Proof. The graph $G(f) = \{(z, f(z)) : z \in Z\}$ of the function f is closed in $Z \times T$ (see Fact 4 of S.390) and $G(f) \cap (Z \times B) = G(f, B)$; so $G(f, B)$ is closed in $Z \times B$ and Fact 2 is proved.

Returning to our solution denote by I the set of isolated points of Y and let $F = Y \setminus I$. For every $a \in I$ there is $U \in \tau(a, X)$ such that $\overline{U} \cap Y = \{a\}$; take $V \in \tau(a, X)$ for which $\overline{V} \subset U$. Since the set $\overline{U}$ is countable and compact, it is zero-dimensional; so there is a clopen subset O_a of the space $\overline{U}$ such that $a \in O_a \subset V$. The set O_a is open in $\overline{U}$; so it must be open in V and hence in X; besides, O_a is compact being closed in $\overline{U}$. As a consequence, we proved that

(1)   for every $a \in I$ there is $O_a \in \mathcal{C}(X)$ such that $O_a \cap Y = \{a\}$.

If $f_a = \chi_{O_a}$ is the characteristic function of the set O_a for every $a \in I$ then the set $D = \{f_a : a \in I\}$ is discrete and all of its accumulation points belong to the set $P = \{f \in C_p(X, \mathbb{D}) : f(Y) = \{0\}\}$. Indeed, all accumulation points of D belong to $C_p(X, \mathbb{D})$ because $\overline{D} \subset C_p(X, \mathbb{D})$; if g is an accumulation point of D then, for any $a \in I$ if $g(a) \neq 0$ then the set $U = \{f \in C_p(X) : f(a) \neq 0\}$ is an open neighbourhood of g with $U \cap D \subset \{f_a\}$ while every neighbourhood of g must contain infinitely many elements of D. This contradiction shows that $g(a) = 0$ for all $a \in I$ and hence $g(Y) = \{0\}$ because I is dense in Y.

Since the set P is closed in $C_p(X)$, we proved that $Q = P \cup D$ is a closed subset of $C_p(X)$ and all points of D are isolated in Q. Since M is countable, the set Y is G_δ in X and hence $\chi(Y, X) = \omega$ (see TFS-327). Choosing a countable decreasing outer base of Y in X and passing to the complements of its elements we obtain a sequence $\{K_n : n \in \omega\}$ of compact subsets of M such that $K_n \subset K_{n+1}$ for any $n \in \omega$ and, for any compact $K \subset M$ there is $n \in \omega$ for which $K \subset K_n$. Every K_n is metrizable; so $\mathcal{C}(K_n)$ is countable by Fact 1. Observe that every $f \in P$ is uniquely determined by the clopen compact set $K_f = f^{-1}(1) \subset M$; there is $n \in \omega$ such that $K_f \subset K_n$ and hence $K_f \in \mathcal{C}(K_n)$. It turns out that $f \to K_f$ is an injection of P in the countable family $\bigcup_{n \in \omega} \mathcal{C}(K_n)$ which shows that P is countable.

Let $\pi : Q \to \mathbb{D}^M$ be the restriction map. Since M is dense in X, the map π is injective. Letting $B = \mathbb{D}^M \setminus \pi(P)$ and noting that $D = \pi^{-1}(B)$ we can apply Fact 2 to conclude that the set $D' = \{(h, \pi(h)) : h \in D\} \subset Q \times B$ is a closed subspace of $Q \times B$. Since the projection of $Q \times B$ onto Q maps D' bijectively and continuously onto D, the set D' is discrete.

We proved that $D \sqsubseteq Q \times B$. Since $\pi(P)$ is countable, B is a G_δ-subspace of $\mathbb{D}^M$; let $B = \bigcap_{n \in \omega} O_n$ where $O_n \in \tau(\mathbb{D}^M)$ for every $n \in \omega$. It is evident that $\mathbb{D}^\omega \sqsubseteq \omega^\omega$ and hence $\mathbb{D}^M \sqsubseteq \omega^\omega$. The space $\mathbb{D}^M$ being compact, metrizable and zero-dimensional, every O_n is an F_σ-subset of $\mathbb{D}^M$; so we can apply Fact 5 of S.390 to see that $O_n \sqsubseteq \mathbb{D}^M \times \omega \sqsubseteq \omega^\omega \times \omega \sqsubseteq \omega^\omega$ for each $n \in \omega$.

Consequently, $B \sqsubseteq \prod_{n \in \omega} O_n \sqsubseteq (\omega^\omega)^\omega \simeq \omega^\omega$ (see Fact 7 of S.271); since Q is a closed subset of $C_p(X, \mathbb{D})$, we have $D \sqsubseteq Q \times B \sqsubseteq Q \times \omega^\omega \sqsubseteq C_p(X, \mathbb{D}) \times \omega^\omega$ whence $ext(C_p(X, \mathbb{D}) \times \omega^\omega) \geq |D| > \omega$, so our solution is complete.

U.078. *Suppose that X is a compact space such that a countable set $M \subset X$ is open and dense in X. Assume also that the set I of isolated points of $Y = X \setminus M$ is uncountable and dense in Y; let $F = Y \setminus I$. Prove that, under MA+¬CH, any uncountable subset of the set $E = \{f \in C_p(X, \mathbb{D}) : f(F) = \{0\}\}$ contains an uncountable set D which is closed and discrete in $C_p(X, \mathbb{D})$.*

Solution. If Z is a space then $\mathcal{C}(Z)$ is the family of all clopen subsets of Z.

Fact 1. Given a space Z and $f, g \in C_p(Z, \mathbb{D})$ define a function $h = f * g \in \mathbb{D}^Z$ as follows: for any $z \in Z$ let $h(z) = 1$ if $f(z) \neq g(z)$ and let $h(z) = 0$ whenever $f(z) = g(z)$. Then

(i) $f * g \in C_p(Z, \mathbb{D})$ for any $f, g \in C_p(Z, \mathbb{D})$;
(ii) if $q \in C_p(Z, \mathbb{D})$ and $s_q : C_p(Z, \mathbb{D}) \to C_p(Z, \mathbb{D})$ is defined by $s_q(f) = f * q$ for any $f \in C_p(Z, \mathbb{D})$, then s_q is a homeomorphism.

Proof. The set $O = (f^{-1}(0) \cap g^{-1}(1)) \cup (f^{-1}(1) \cap g^{-1}(0))$ is, evidently, clopen in Z; since $f * g$ is precisely the characteristic function of O, we have $f * g \in C_p(Z, \mathbb{D})$, i.e., (i) is proved.

As to (ii), to see that s_q is a continuous map observe that $s_q(f) = f + q - 2 \cdot f \cdot q$ for any $f \in C_p(Z, \mathbb{D})$; so continuity of s_q follows from TFS-115 and TFS-116. Another easy observation is that the map s_q is inverse to itself, i.e., $s_q(s_q(f)) = f$ for any $f \in C_p(Z, \mathbb{D})$; so s_q is a homeomorphism and Fact 1 is proved.

Returning to our solution let $S(f) = f^{-1}(1) \cap I$ for any $f \in C_p(X, \mathbb{D})$ and take an arbitrary uncountable set $P \subset E$; if $f \in P$ and $S(f)$ is infinite then, by compactness of Y, there is a point $z \in \overline{S(f)} \cap F$ and therefore $f(z) = 1$ which contradicts $f | F \equiv 0$. Thus $S(f)$ is finite for any $f \in P$. Apply the Δ-lemma (SFFS-038) to find an uncountable set $P_1 \subset P$ for which there exists a finite $A \subset I$ such that $S(f) \cap S(g) = A$ for any distinct $f, g \in P_1$.

Since M is countable, the set Y is G_δ in X and hence $\chi(Y, X) = \omega$ (see TFS-327). Choosing a countable decreasing outer base of Y in X and taking the complements of its elements we obtain a sequence $\{K_n : n \in \omega\}$ of compact subsets of M such that $K_n \subset K_{n+1}$ for any $n \in \omega$ and, for any compact $K \subset M$ there is $n \in \omega$ for which $K \subset K_n$. Every K_n is metrizable; so $C(K_n)$ is countable by Fact 1 of U.077. Let $E' = \{f \in C_p(X, \mathbb{D}) : f(Y) = \{0\}\}$; for any $f \in E'$ the set $K_f = f^{-1}(1)$ is compact, clopen and contained in M. There is $n \in \omega$ such that $K_f \subset K_n$ and hence $K_f \in C(K_n)$. Since f is uniquely determined by the set K_f, it turns out that $f \to K_f$ is an injection of E' into the countable family $\bigcup_{n \in \omega} C(K_n)$ which shows that E' is countable; thus $P_2 = P_1 \backslash E'$ is uncountable.

Fix any $q \in P_2$ and consider the set $Q = \{f * q : f \in P_2\}$. It is clear that $Q \subset E$ and $S(f) \cap S(g) = \emptyset$ for any distinct $f, g \in Q$. We have $Q = s_q(P_2)$ (see Fact 1); since s_q is a homeomorphism of $C_p(X, \mathbb{D})$ onto itself, it suffices to show that Q has an uncountable subset which is closed and discrete in $C_p(X, \mathbb{D})$.

Now, take an accumulation point g of the set Q. If $a \in I$ and $g(a) \neq 0$ then the set $W = \{f \in C_p(X, \mathbb{D}) : f(a) = 1\}$ is an open neighbourhood of g in $C_p(X, \mathbb{D})$ and $W \cap Q$ has at most one element because $\{S(f) : f \in Q\}$ is disjoint and hence $f(a) = 1$ for at most one function $f \in Q$. Since every neighbourhood of g must contain infinitely many elements of Q, we obtained a contradiction which shows that $g(a) = 0$ for any $a \in I$ and hence $g(Y) = \{0\}$ because I is dense in Y. Thus $g \in E'$.

An accumulation point g of the set Q was chosen arbitrarily so we proved that all accumulation points of Q belong to E'. Since E' is closed in $C_p(X, \mathbb{D})$, the set $H = Q \cup E'$ is also closed in $C_p(X, \mathbb{D})$ while all points of Q are isolated in H. Let $\pi : C_p(X, \mathbb{D}) \to \mathbb{D}^M$ be the restriction map. Then π is continuous and injective because M is dense in X. Therefore $\rho = \pi|H : H \to G = \rho(H)$ is a condensation. The space G is second countable and $B = \rho(Q)$ is an uncountable subset of G. Applying MA+¬CH we can find a disjoint family $\mathcal{F} = \{F_\alpha : \alpha < \omega_1\}$ of closed subsets of G such that $F_\alpha \cap B$ is uncountable for any $\alpha < \omega_1$ (see Fact 1 of T.063).

The set $\rho(E')$ is countable; so it is impossible that every element of the disjoint uncountable family $\mathcal{F}$ intersect $\rho(E')$. Take $\alpha < \omega_1$ for which $F_\alpha \cap \rho(E') = \emptyset$. Then $F_\alpha = F_\alpha \cap B \subset B$ is uncountable and closed in G. Therefore $D' = \rho^{-1}(F_\alpha)$ is uncountable and closed in H. Since $D' \subset Q$ and Q is discrete, the set D' is closed and discrete in H; the set H is closed in $C_p(X, \mathbb{D})$; so D' is closed and discrete in $C_p(X, \mathbb{D})$ as well. Consequently, $D = s_q(D')$ is an uncountable subset of P which is closed and discrete in $C_p(X, \mathbb{D})$; so our solution is complete.

U.079. *Let X be a compact space of weight ω_1 in which we have a countable dense set L and a nowhere dense closed non-empty set F. Assuming MA+¬CH prove that there exists $M \subset L$ such that $\overline{M} \backslash M = F$ and all points of M are isolated in the space $M \cup F$.*

Solution. Since F is nowhere dense in X, the set $L_0 = L \backslash F$ is still dense in the space X. Now, $\chi(X) \leq w(X) = \omega_1 < \mathfrak{c}$; so, for any point $x \in F$, we have $x \in \overline{L_0}$ and $\chi(x, \{x\} \cup L_0) \leq \chi(X) < \mathfrak{c}$. Thus we can apply SFFS-054 to see that there is a sequence $S_x \subset L_0$ that converges to x.

Observe also that $\chi(F, X) = \psi(F, X) \leq w(X) = \omega_1 < \mathfrak{c}$ (see TFS-327); so we can find an outer base $\mathcal{U}$ of the set F in the space X such that $|\mathcal{U}| < \mathfrak{c}$. Since $d(F) \leq w(F) \leq w(X) \leq \omega_1$, there exists $P \subset F$ for which $\overline{P} = F$ and $|P| = \omega_1$. Let $\mathcal{A} = \{S_x : x \in P\}$ and $\mathcal{B} = \{(X \backslash U) \cap L_0 : U \in \mathcal{U}\}$. This gives us families $\mathcal{A}$ and $\mathcal{B}$ on the countable set L_0 (which we can identify with ω) such that $|\mathcal{A}| < \mathfrak{c}$, $|\mathcal{B}| < \mathfrak{c}$ and $A \backslash \bigcup \mathcal{B}'$ is infinite for any $A \in \mathcal{A}$ and finite $\mathcal{B}' \subset \mathcal{B}$. Indeed, $\overline{B} \cap F = \emptyset$ for any $B \in \mathcal{B}$, so the complement of $\bigcup \mathcal{B}'$ is a neighbourhood of F; since A is a sequence which converges to a point of F, the set $A \cap (\bigcup \mathcal{B}')$ is finite and hence $A \backslash \bigcup \mathcal{B}'$ is infinite.

An immediate consequence of SFFS-051 is that there is a set $M \subset L_0$ such that $M \cap A$ is infinite for any $A \in \mathcal{A}$ and $M \cap B$ is finite for any $B \in \mathcal{B}$. If $x \in P$ then $A_x \in \mathcal{A}$ and hence $M \cap A_x$ is infinite which implies $x \in \overline{M}$. As a consequence, $F = \overline{P} \subset \overline{M}$ and hence $F \subset \overline{M} \backslash M$.

Now take an arbitrary accumulation point x of the set M; if $x \in X \backslash F$ then there is a set $U \in \mathcal{U}$ such that $x \notin \overline{U}$. Then $W = X \backslash \overline{U}$ is a neighbourhood of the point x for which $W \cap M \subset K = M \cap ((X \backslash U) \cap L_0)$ while the set K is finite because $B = (X \backslash U) \cap L_0 \in \mathcal{B}$. Since every neighbourhood of x must contain infinitely many points of M, we obtained a contradiction which shows that $\overline{M} \backslash M = F$. We also proved that no point of M is an accumulation point of M; so all points of M are isolated in the space $M \cup F$.

U.080. *Prove that, under* MA+¬CH, *if* X *is a compact space such that* $C_p(X)$ *is normal, then* X *is Fréchet–Urysohn,* ω-*monolithic and has a dense set of points of countable character.*

Solution. If Z is a set and $A \subset Z$ then the characteristic function χ_A of the set A in Z is defined by $\chi_A(z) = 1$ if $z \in A$ and $\chi_A(z) = 0$ otherwise.

Fact 1. If K is a compact ω-monolithic space of countable tightness then K is Fréchet–Urysohn and has a dense set of points of countable character. This is true in ZFC, i.e., no additional axioms are needed for the proof of this Fact.

Proof. If $A \subset K$ and $x \in \overline{A}$ then $w(\overline{A}) = nw(\overline{A}) = \omega$ because K is ω-monolithic (see Fact 4 of S.307). In particular, $\{x\} \cup A$ is second countable and hence metrizable. Thus there is a sequence $\{a_n : n \in \omega\} \subset A$ which converges to x. This proves that K is a Fréchet–Urysohn space.

Observe that a point $x \in K$ is of countable character in K if and only if $\{x\}$ is a G_δ-subset of K (see TFS-327); so we will call such x *a* G_δ-*point of* K. Thus $E = \{x \in K : \chi(x, K) \le \omega\}$ coincides with the set of all G_δ-points of K. To see that E is dense in K, take any $U \in \tau^*(K)$ and apply Fact 2 of S.328 to find a non-empty closed G_δ-set $F \subset U$.

Next, apply Fact 4 of T.041 to the compact space F to find a countable set $A \subset F$ such that there is a non-empty G_δ-set H in the space F such that $H \subset cl_F(A)$; pick any $x \in H$. It follows from Fact 2 of S.358 that H is also a G_δ-subset of K; besides, $cl_F(A) = \overline{A}$ because F is closed in K. By ω-monolithity of K the set $\overline{A}$ is second countable and hence so is H; thus x is a G_δ-point of H and hence it is also a G_δ-point of K (here we used Fact 2 of S.358 again). As a result, we found a G_δ-point $x \in U$, i.e., $E \cap U \ne \emptyset$ for any $U \in \tau^*(K)$. Thus E is dense in K and Fact 1 is proved.

Fact 2. Under MA+¬CH, if K is a separable compact space of weight $\le \omega_1$ such that $C_p(K)$ is Lindelöf then K is perfectly normal.

Proof. If K is not perfectly normal then $s(K) > \omega$ by SFFS-061; so take a discrete $D' \subset K$ with $|D'| = \omega_1$. Let L be a countable dense subset of K. The set I of all isolated points of K is contained in L; so the set $D = D'\backslash L$ has cardinality ω_1 and no point of D is isolated in K.

As a consequence, $Y = \overline{D}$ is a nowhere dense closed non-empty subset of K; so we can apply Problem 079 to find a set $M \subset L$ such that $\overline{M}\backslash M = Y$ and all points of M are isolated in $M \cup Y$. The set $M \cup Y$ is compact; so $C_p(M \cup Y)$ is Lindelöf being a continuous image of $C_p(K)$. Furthermore, the space $Y' = M \cup Y$ satisfies all assumptions of Problem 078; thus, for the set $F = \overline{D}\backslash D$ and any uncountable $A \subset P = \{f \in C_p(Y', \mathbb{D}) : f(F) = \{0\}\}$ we have an uncountable $B \subset A$ which is closed and discrete in $C_p(Y', \mathbb{D})$.

For every point $d \in D$ there is a set $U_d \in \tau(d, Y')$ such that $\overline{U}_d \cap Y = \{d\}$; an immediate consequence is that U_d is clopen in the space Y', so the characteristic function f_d of the set U_d belongs to P. It is evident that $d \ne d'$ implies $f_d \ne f_{d'}$; so $A = \{f_d : d \in D\}$ is an uncountable subset of P. Therefore we can apply

Problem 078 to find an uncountable set $B \subset A$ which is closed and discrete in the space $C_p(Y', \mathbb{D})$. As a consequence, $ext(C_p(Y')) \geq ext(C_p(Y', \mathbb{D})) \geq |B| \geq \omega_1$ while $C_p(Y')$ is Lindelöf being a continuous image of $C_p(K)$. This contradiction shows that Fact 2 is proved.

Fact 3. Under MA+¬CH, if K is a separable compact space such that $C_p(K)$ is Lindelöf then K is metrizable.

Proof. If K is not metrizable then it can be mapped continuously onto a compact space K' of weight ω_1 (see SFFS-094). It is immediate that $C_p(K')$ is also Lindelöf; besides, K' is separable so, to obtain a contradiction, we can assume that $K = K'$, i.e., K is a separable compact space of weight ω_1 such that $C_p(K)$ is Lindelöf.

If $K \times K$ is perfectly normal then the diagonal $\Delta = \{(z, z) : z \in K\} \subset K \times K$ of the space K is a G_δ-subset of $K \times K$; so K is metrizable by SFFS-091. This contradiction shows that $K \times K$ is not perfectly normal and hence we can apply SFFS-061 to find a discrete $D' \subset K \times K$ with $|D'| = \omega_1$. The space K is perfectly normal by Fact 2; so $D' \cap \Delta$ is countable because $\Delta \simeq K$. It follows from separability of K that the set I of isolated points of $K \times K$ is countable; so the set $D = D' \setminus (\Delta \cup I)$ is uncountable.

Fix a countable dense set L in the space $K \times K$; since $Y = \overline{D}$ is nowhere dense, we can apply Problem 079 again to obtain a set $M \subset L$ such that $Y = \overline{M} \setminus M$ and all points of M are isolated in $Z = M \cup Y$; let $F = \overline{D} \setminus D$.

Denote by u the function which is equal to zero at all points of K and let $\varphi_+(f) = \max(f, u)$, $\varphi_-(f) = \max(-f, u)$ for every $f \in C_p(K)$. The mappings $\varphi_+, \varphi_- : C_p(K) \to C_p(K)$ are continuous by TFS-082. Let $q_i : K \times K \to K$ be the natural projection of $K \times K$ onto its i-th factor for $i = 1, 2$. The dual maps $q_i^* : C_p(K) \to C_p(K \times K)$ are embeddings by TFS-163. Furthermore, the multiplication map $m : C_p(K \times K) \times C_p(K \times K) \to C_p(K \times K)$ is continuous by TFS-116. Now define $\varphi(f) = m(q_1^*(\varphi_+(f)), q_2^*(\varphi_-(f)))$ for any function $f \in C_p(K)$; then the map $\varphi : C_p(K) \to C_p(K \times K)$ is continuous being a composition of continuous maps. Note that, given a point $(x, y) \in K \times K$ we have $\varphi(f)(x, y) = \varphi_+(x) \cdot \varphi_-(y)$ for any $f \in C_p(K)$. If $\pi : C_p(K \times K) \to C_p(Z)$ is the restriction map then the map $\xi = \pi \circ \varphi : C_p(K) \to C_p(Z)$ is also continuous.

The set D is discrete; so for every $d = (x, y) \in D$ there is $W_d \in \tau(d, K \times K)$ such that $\overline{W}_d \cap Y = \{d\}$. Therefore $\overline{W}_d \cap Z$ is a compact space with the unique non-isolated point d. An immediate consequence is that

(1) if $W', W'' \in \tau(d, W_d)$ then $(W' \cap Z) \setminus (W'' \cap Z)$ is finite.

Since $D \cap \Delta = \emptyset$, we can take disjoint sets $U_d \in \tau(x, K)$ and $V_d \in \tau(y, K)$ such that $U_d \times V_d \subset W$. There exist $G \in \tau(x, K)$, $H \in \tau(y, K)$ for which $\overline{G} \subset U_d$ and $\overline{H} \subset V_d$. By (1), the set $T = ((U_d \times V_d) \cap Z) \setminus ((G \times H) \cap Z)$ is finite; so $q_1(T) \subset U_d$ is also finite as well as $q_2(T) \subset V_d$. It is easy to find $G', H' \in \tau(K)$ such that $q_1(T) \subset G' \subset \overline{G'} \subset U_d$ and $q_2(T) \subset H' \subset \overline{H'} \subset V_d$. It is immediate that for the sets $G_d = G \cup G'$ and $H_d = H \cup H'$ we have

(2) $\overline{G}_d \subset U_d$, $\overline{H}_d \subset V_d$ while $(U_d \times V_d) \cap Y = \{d\}$ and
   $(U_d \times V_d) \cap Z = (\overline{G}_d \times \overline{H}_d) \cap Z$.

The space K is compact and hence normal; so there are $f_d, g_d \in C_p(K)$ such that $f_d | G_d \equiv 1$, $f_d | (K \setminus U_d) \equiv 0$ and $g_d | H_d \equiv 1$, $g_d | (K \setminus V_d) \equiv 0$. Let $h_d = f_d - g_d$ for every $d \in D$ and consider the set $Q = \{h_d : d \in D\} \subset C_p(K)$.

For any $d \in H$ and $z = (x, y) \in \overline{G}_d \times \overline{H}_d$ we have $f_d(x) = 1$ and $g_d(y) = 1$ which implies $h_d(x) = 1$ and $h_d(y) = -1$; so $\varphi_+(x) = 1$ and $\varphi_-(y) = 1$ which shows that $\xi(h_d)(z) = \varphi(h_d)(z) = 1$. If, on the other hand, $z = (x, y) \in Z \setminus (U_d \times V_d)$ then either $x \notin U_d$ and hence $\varphi_+(h_d)(x) = 0$ or $y \notin V_d$ which implies $\varphi_-(h_d)(y) = 0$. In both cases we have $\xi(h_d)(z) = \varphi(h_d)(z) = \varphi_+(h_d)(x) \cdot \varphi_-(h_d)(y) = 0$. Thus it follows from (2) that $\xi(h_d)$ is the characteristic function of the set $B_d = (U_d \times V_d) \cap Z$ which is clopen in Z. Besides, for the function $\eta_d = \xi(h_d)$, we have $\eta_d^{-1}(1) \cap Y = \{d\}$ which proves that $d \neq d'$ implies $\eta_d \neq \eta_{d'}$ and $\eta_d \in I_F = \{f \in C_p(Z, \mathbb{D}) : f(F) = \{0\}\}$ for every $d \in D$.

Consequently, $E = \{\eta_d : d \in D\} \subset \xi(Q) \subset \xi(C_p(K))$; since E is an uncountable subset of I_F we can apply Problem 078 to conclude that there is an uncountable $E' \subset E$ which is closed and discrete in $C_p(Z, \mathbb{D})$ and hence in $C_p(Z)$. Thus E' is an uncountable closed and discrete subset of a Lindelöf space $\xi(C_p(K))$. This contradiction shows that K is metrizable; so Fact 3 is proved.

Returning to our solution observe that $C_p(X)$ is actually Lindelöf (see TFS-295 and SFFS-269) and therefore $t(X) = \omega$ (see TFS-189). Given a countable $A \subset X$ the compact space $\overline{A}$ is separable and $C_p(\overline{A})$ is Lindelöf being a continuous image of $C_p(X)$. Thus Fact 3 is applicable to conclude that $\overline{A}$ is metrizable, i.e., X is ω-monolithic. We proved that X is a compact ω-monolithic space of countable tightness; so X is Fréchet–Urysohn and has a dense set of points of countable character (see Fact 1), which means that our solution is complete.

U.081. *Assume MA+¬CH. Show that, if a compact space X has the Souslin property and $C_p(X)$ is normal then X is metrizable.*

Solution. Observe first that $C_p(X)$ is Lindelöf (see TFS-295 and SFFS-269) and therefore $t(X) = \omega$ (see TFS-189). Now apply SFFS-288 to see that ω_1 is a precaliber of X and hence ω_1 is a caliber of X by SFFS-279. It follows from TFS-332 that the space X has a point-countable π-base $\mathcal{B}$. Since ω_1 is a caliber of X, any point-countable family of non-empty open subsets of X is countable; so $|\mathcal{B}| \leq \omega$ and hence $d(X) \leq \pi w(X) \leq \omega$. The space X is ω-monolithic by Problem 080; so $w(X) = nw(X) = \omega$ (see Fact 4 of S.307) and hence X is metrizable.

U.082. *Prove that $w(X) = l(C_p(X))$ for any linearly orderable compact space X. In particular, if $C_p(X)$ is Lindelöf then X is metrizable.*

Solution. If Z is a set and $\mathcal{A} \subset \exp Z$ then $\mathrm{ord}(z, \mathcal{A}) = |\{A \in \mathcal{A} : z \in A\}|$ and $\mathrm{ord}(\mathcal{A}) = \sup\{\mathrm{ord}(z, \mathcal{A}) : z \in Z\}$. The cardinal $\mathrm{ord}(\mathcal{A})$ is called *the order of the family* $\mathcal{A}$.

Fact 1. If Z is a space and $l(Z) \leq \kappa$ for some infinite cardinal κ then any indexed set $Y = \{y_\alpha : \alpha < \kappa^+\} \subset Z$ has a complete accumulation point, i.e., there is $z \in Z$ such that $|\{\alpha < \kappa^+ : y_\alpha \in U\}| = \kappa^+$ for any $U \in \tau(z, Z)$.

Proof. If this is false then every $z \in Z$ has an open neighbourhood O_z such that the set $A_z = \{\alpha < \kappa^+ : y_\alpha \in O_z\}$ has cardinality at most κ. Since $l(Z) \leq \kappa$ and $\{O_z : z \in Z\}$ is an open cover of Z, we can choose a set $P \subset Z$ for which $|P| \leq \kappa$ and $\bigcup\{O_z : z \in P\} = Z$. As a consequence, $\bigcup\{A_z : z \in P\} = \kappa^+$ which is impossible because the cardinal κ^+ is regular, i.e., it cannot be represented as a union of $< \kappa^+$-many sets of cardinality $< \kappa^+$ each. Fact 1 is proved.

Now, let $\kappa = l(C_p(X))$ and take an order $\leq$ on the set X which generates $\tau(X)$. The case of a finite X is evident; so we assume that $|X| \geq \omega$. We will have the usual notation for the intervals in $(X, \leq)$, i.e., for any $a, b \in X$ we let

$$(a, b) = \{x \in X : a < x < b\}, \ [a, b) = \{x \in X : a \leq x < b\},$$
$$(a, b] = \{x \in X : a < x \leq b\} \text{ and } [a, b] = \{x \in X : a \leq x \leq b\}.$$

Furthermore,

$$[a, \to) = \{x \in X : a \leq x\} \text{ and } (a, \to) = \{x \in X : a < x\} \text{ while}$$
$$(\leftarrow, b] = \{x \in X : x \leq b\} \text{ and } (\leftarrow, b) = \{x \in X : x < b\} \text{ for any } a, b \in X.$$

The space X being compact, it has a minimal element a_* and a maximal element b_* (see TFS-305). It is clear that $\kappa \leq nw(C_p(X)) = nw(X) \leq w(X)$; so it suffices to prove that $w(X) \leq \kappa$. We will show first that

(1) the cardinal κ^+ is a caliber of X, i.e., $|\mathcal{U}| \leq \kappa$ for any family $\mathcal{U} \subset \tau^*(X)$ such that $\mathrm{ord}(\mathcal{U}) \leq \kappa$.

Assume, towards a contradiction, that (1) is false and hence there exists a family $\mathcal{U} = \{U_\alpha : \alpha < \kappa^+\} \subset \tau^*(X)$ such that the set $K_z = \{\alpha < \kappa^+ : z \in U_\alpha\}$ has cardinality $\leq \kappa$ for any $z \in X$. Making every U_α smaller we obtain a family with the same properties; so we can assume that each U_α is an element of the standard base of X, i.e., $U_\alpha = (a, b)$ for some $a, b \in L$ or there is $a \in L$ such that $U_\alpha = (a, \to)$ or $U_\alpha = (\leftarrow, a)$.

There are at most κ-many elements of $\mathcal{U}$ that intersect the set $\{a_*, b_*\}$; so we can throw them away and still have a family which witnesses that κ^+ is not a caliber of X. Thus we can assume, without loss of generality, that $U_\alpha \subset (a_*, b_*)$ for every $\alpha < \kappa^+$ and hence $U_\alpha = (a_\alpha, b_\alpha)$ for some $a_\alpha, b_\alpha \in X$.

For any $\alpha < \kappa^+$ the sets $P_\alpha = (\leftarrow, a_\alpha]$ and $Q_\alpha = [b_\alpha, \to)$ are closed and disjoint in X; so there is $f_\alpha \in C_p(X)$ such that $f_\alpha(P_\alpha) \subset \{0\}$ and $f_\alpha(Q_\alpha) \subset \{1\}$. For the indexed set $E = \{f_\alpha : \alpha < \kappa^+\} \subset C_p(X)$ we must have a complete accumulation point h (see Fact 1).

If there is $x \in X$ such that $h(x) \notin \mathbb{D}$ then $O = \{f \in C_p(X) : f(x) \notin \mathbb{D}\}$ is an open neighbourhood of h; if $f_\alpha \in O$ then $x \in U_\alpha$, so $\alpha \in K_x$ and hence $|\{\alpha : f_\alpha \in O\}| \leq |K_x| \leq \kappa$, i.e., h is not a complete accumulation point which is a contradiction. Thus $h \in C_p(X, \mathbb{D})$; since $f_\alpha(a_*) = 0$ and $f_\alpha(b_*) = 1$ for all $\alpha < \kappa^+$, we have $h(a_*) = 0$ and $h(b_*) = 1$.

The sets $H_0 = h^{-1}(0)$ and $H_1 = h^{-1}(1)$ are clopen, disjoint, non-empty and $X = H_0 \cup H_1$. Let $b = \min H_1$ (this minimal element exists by TFS-305). The set $G = [a_*, b)$ is non-empty because $a_* \in H_0$ and hence $a_* < b$; since $G \subset H_0$, we have $\overline{G} \subset H_0$ and hence $a \notin \overline{G}$. Therefore $\overline{G} \subset [a_*, b] \cap H_0 = [a_*, b) \cap H_0 \subset G$, i.e., the set G is closed in X. Apply TFS-305 once more conclude that G has a maximal element a. It is clear that $a < b$ and $(a, b) = \emptyset$. The set $W = \{f \in C_p(X) : f(a) < 1 \text{ and } f(b) > 0\}$ is an open neighbourhood of h in $C_p(X)$. If $f_\alpha \in W$ then $a_\alpha < b$ and $a < b_\alpha$.

If $b_\alpha \le b$ and $a \le a_\alpha$ then $(a_\alpha, b_\alpha) \subset (a, b)$ which is a contradiction because $(a, b) = \emptyset$ and $U_\alpha = (a_\alpha, b_\alpha) \ne \emptyset$. Thus either $a_\alpha < b < b_\alpha$ or $a_\alpha < a < b_\alpha$ and therefore $U_\alpha \cap \{a, b\} \ne \emptyset$. Consequently, $\{\alpha < \kappa^+ : f_\alpha \in W\} \subset K_a \cup K_b$ and hence $|W \cap E| \le |K_a \cup K_b| \le \kappa$. This proves that h is not a complete accumulation point for E and provides a contradiction which shows that κ^+ is a caliber of X, i.e., (1) is proved.

Call $x \in X$ *a jump point* if there exists $s_x > x$ such that $(x, s_x) = \emptyset$. It turns out that

(2) the set J of all jump points of X has cardinality at most κ.

To obtain a contradiction assume that $P = \{x_\alpha : \alpha < \kappa^+\} \subset J$ and the enumeration of P is faithful; it is evident that we can assume that $P \subset (a_*, b_*)$. Let $g_\alpha(x) = 1$ for all $x \ge s_{x_\alpha}$ and $g_\alpha(x) = 0$ whenever $x \le x_\alpha$. It is clear that $g_\alpha : X \to \mathbb{D}$ is continuous on X for any $\alpha < \kappa^+$. Let $G = \{g_\alpha : \alpha < \kappa^+\} \subset C_p(X, \mathbb{D})$; since $C_p(X, \mathbb{D})$ is closed in $C_p(X)$, we have $l(C_p(X, \mathbb{D})) \le \kappa$; so the set G has a complete accumulation point $g \in C_p(X, \mathbb{D})$ (see Fact 1). We have $g_\alpha(a_*) = 0$ and $g_\alpha(b_*) = 1$ for all $\alpha < \kappa^+$ so $g(a_*) = 0$ and $g(b_*) = 1$.

Let $b = \min g^{-1}(1)$; this point is well defined because $g^{-1}(1)$ is non-empty and closed in X (see TFS-305). The set $R = [a_*, b)$ is non-empty because $a_* < b$; besides, $R \subset g^{-1}(0)$ and hence $\overline{R} \subset g^{-1}(0)$ which implies that $\overline{R} \subset g^{-1}(0) \cap [a_*, b] = g^{-1}(0) \cap [a_*, b) \subset R$ and hence R is closed in X. Apply TFS-305 once more to take a point $a = \max[a_*, b)$; it is clear that $a \in g^{-1}(0)$, $b \in g^{-1}(1)$ and $(a, b) = \emptyset$.

The set $W = \{f \in C_p(X, \mathbb{D}) : f(a) = 0 \text{ and } f(b) = 1\}$ is an open neighbourhood of g in $C_p(X, \mathbb{D})$. Assume that $g_\alpha \in W$ for some $\alpha < \kappa^+$. Then $g_\alpha(a) = 0$ and hence $a \le x_\alpha$; furthermore, $g_\alpha(b) = 1$ implies that $s_{x_\alpha} \le b$ which shows that $(x_\alpha, s_{x_\alpha}) \subset (a, b)$. Since the interval (a, b) is empty, we have $x_\alpha = a$ and therefore at most one element of G belongs to W. Thus g is not a complete accumulation point of G; so we obtained a contradiction which completes the proof of (2).

Now observe that $t(X) \le \kappa$ by TFS-189; this implies that X has a π-base $\mathcal{B}$ with $\operatorname{ord}(\mathcal{B}) \le \kappa$ (see TFS-332). The cardinal κ^+ being a caliber of X we have $|\mathcal{B}| \le \kappa$ by (1); so $d(X) \le \pi w(X) \le |\mathcal{B}| \le \kappa$. Fix a dense $D \subset X$ such that $|D| \le \kappa$ and let $J' = \{s_x : x \in J\}$; then the set $E = D \cup J \cup J' \cup \{a_*, b_*\}$ also has cardinality at most κ. If $\mathcal{N} = \{[a, b) : a, b \in E\} \cup \{(a, b] : a, b \in E\}$ then $|\mathcal{N}| \le \kappa$. We claim that $\mathcal{N}$ is a network in X.

Indeed, take any point $x \in X$ and $U \in \tau(x, X)$. We have several cases to consider.

(a) $x \in J$; then $N = [x, s_x) = \{x\} \in \mathcal{N}$ and $x \in N \subset U$.

(b) $x \in J'$; then there is $y \in J$ such that $x = s_y$; so $N = (y, x] = \{x\} \in \mathcal{N}$ and $x \in N \subset U$.

(c) $x \in (E \backslash J) \cap (\leftarrow, b_*)$; then there is $b \in X$ such that $[x, b) \subset U$ and $(x, b) \neq \emptyset$. Since D is dense in X, we can find a point $d \in D \cap (x, b)$. It is immediate that $N = [x, d) \in \mathcal{N}$ and $x \in N \subset U$.

(d) $x \in (E \backslash J') \cap (a_*, \rightarrow)$; then there is $a \in X$ such that $(a, x] \subset U$ and $(a, x) \neq \emptyset$. Since D is dense in X, we can find a point $d \in D \cap (a, x)$. It is immediate that $N = (d, x] \in \mathcal{N}$ and $x \in N \subset U$.

(e) $x \in X \backslash E$; then there are $a, b \in X$ such that $x \in (a, b) \subset U$ while $(a, x) \neq \emptyset$ and $(x, b) \neq \emptyset$. Since the set D is dense in X, we can choose $d \in (a, x) \cap D$ and $d' \in (x, b) \cap D$. Then $N = [d, d'] \in \mathcal{N}$ and $x \in N \subset U$.

We proved that, in all possible cases, if $x \in X$ and $x \in U \in \tau(X)$ then there is $N \in \mathcal{N}$ such that $x \in N \subset U$ and hence $\mathcal{N}$ is a network in X. As a consequence, $w(X) = nw(X) \leq |\mathcal{N}| \leq \kappa$ (see Fact 4 of S.307); so $w(X) = l(C_p(X))$ and our solution is complete.

U.083. *Given an infinite compact space X prove that $|Y| \leq 2^{l(Y) \cdot c(X)}$ for any $Y \subset C_p(X)$.*

Solution. Let $\exp Z = \{A : A \subset Z\}$ and $\exp_\kappa(Z) = \{A \subset Z : |A| \leq \kappa\}$ for any set Z and any infinite cardinal κ.

Fact 1. Given a space Z and an infinite cardinal κ suppose that $\psi(Z) \leq 2^\kappa$ and $l(Z) \cdot t(Z) \leq \kappa$. Assume additionally that $|\overline{A}| \leq 2^\kappa$ for any $A \subset Z$ with $|A| \leq \kappa$. Then $|Z| \leq 2^\kappa$.

Proof. For every $z \in Z$ fix a family $\mathcal{B}_z \subset \tau(Z)$ such that $|\mathcal{B}_z| \leq 2^\kappa$ and $\bigcap \mathcal{B}_z = \{z\}$. Choose any point $z_0 \in Z$ and let $Z_0 = \{z_0\}$. To proceed inductively, assume that $\beta < \kappa^+$ and we have a family $\{Z_\alpha : \alpha < \beta\} \subset \exp Z$ with the following properties:

(1) Z_α is closed in Z and $|Z_\alpha| \leq 2^\kappa$ for any $\alpha < \beta$;
(2) $\alpha < \alpha' < \beta$ implies $Z_\alpha \subset Z_{\alpha'}$.
(3) Let $\mathcal{U}_\alpha = \bigcup \{\mathcal{B}_z : z \in Z_\alpha\}$ for any $\alpha < \beta$. Then $\alpha < \gamma < \beta$ implies that, for any $\mathcal{U} \subset \exp_\kappa(\mathcal{U}_\alpha)$ with $U = \bigcup \mathcal{U} \neq Z$, we have $Z_\gamma \cap (Z \backslash U) \neq \emptyset$.

It follows from (1) and (3) that the set $T = \bigcup \{Z_\alpha : \alpha < \beta\}$ and the family $\mathcal{B}_T = \bigcup \{\mathcal{B}_t : t \in T\}$ have cardinalities that do not exceed 2^κ. Therefore the collection $\mathcal{A} = \{\mathcal{V} \subset \exp_\kappa(\mathcal{B}_T) : \bigcup \mathcal{V} \neq Z\}$ also has cardinality at most 2^κ. For every $\mathcal{V} \in \mathcal{A}$ choose a point $a(\mathcal{V}) \in Z \backslash (\bigcup \mathcal{V})$ and let $Z_\beta = \overline{T \cup \{a(\mathcal{V}) : \mathcal{V} \in \mathcal{A}\}}$.

We have to verify that (1)–(3) still hold for all $\alpha \leq \beta$ and the only non-trivial thing to check is that $|Z_\beta| \leq 2^\kappa$. Let $T' = T \cup \{a(\mathcal{V}) : \mathcal{V} \in \mathcal{A}\}$. Since $t(Z) \leq \kappa$, we have $\overline{T'} = \bigcup \{\overline{C} : C \in \exp_\kappa(T')\}$. By our hypothesis, $|\overline{C}| \leq 2^\kappa$ for any $C \in \exp_\kappa(T')$; so $|Z_\beta| \leq \sum \{|\overline{C}| : C \in \exp_\kappa(T')\} \leq 2^\kappa \cdot 2^\kappa = 2^\kappa$ whence $|Z_\beta| \leq 2^\kappa$ and therefore (1)–(3) are fulfilled for all $\alpha \leq \beta$.

Thus our inductive procedure can be accomplished for all $\beta < \kappa^+$ giving us a family $\{Z_\alpha : \alpha < \kappa^+\}$ with the properties (1)–(3). It follows from (1) that $Z' = \bigcup\{Z_\alpha : \alpha < \kappa^+\}$ has cardinality at most 2^κ; so it suffices to prove that $Z' = Z$.

To obtain a contradiction, assume that there is $x \in Z \backslash Z'$. An easy consequence of (1) and $t(Z) \le \kappa$ is that Z' is closed in Z and hence $l(Z') \le \kappa$. For any $z \in Z'$ there is $W_z \in \mathcal{B}_z$ for which $x \notin W_z$. Since $\{W_z : z \in Z'\}$ is an open cover of Z', we can find $P \subset Z'$ such that $|P| \le \kappa$ and $Z' \subset \bigcup\{W_z : z \in P\}$. The condition (2) implies that there is $\alpha < \kappa^+$ such that $P \subset Z_\alpha$ and hence $\mathcal{V} = \{W_z : z \in P\} \in \exp_\kappa(\mathcal{U}_\alpha)$; besides, $x \notin V = \bigcup \mathcal{V}$; so we can apply (3) to conclude that $Z_{\alpha+1} \cap (Z \backslash V) \ne \emptyset$ which is a contradiction because $Z_{\alpha+1} \subset Z' \subset V$. Therefore $|Z| = |Z'| \le 2^\kappa$ and Fact 1 is proved.

Fact 2. Given a space Z let $\mathcal{R}(Z)$ be the family of all regular open subsets of Z, i.e., $\mathcal{R}(Z) = \{U \in \tau(Z) : U = \text{Int}(\overline{U})\}$. Then $|\mathcal{R}(Z)| \le \pi w(Z)^{c(Z)}$.

Proof. Fix a π-base $\mathcal{B}$ in Z with $|\mathcal{B}| \le \kappa = \pi w(Z)$. For any $U \in \mathcal{R}(Z)$ take a maximal disjoint $\mathcal{B}_U \subset \mathcal{B}$ such that $B \subset U$ for any $B \in \mathcal{B}_U$. Then $|\mathcal{B}_U| \le \lambda = c(Z)$ and hence we have a map $\varphi : \mathcal{R}(Z) \to \exp_\lambda(\mathcal{B})$ defined by $\varphi(U) = \mathcal{B}_U$ for every $U \in \mathcal{R}(Z)$. Since $\mathcal{B}$ is a π-base in Z, the set $\bigcup \mathcal{B}_U$ is dense in U for any $U \in \mathcal{R}(Z)$ (this is an easy exercise); so we have $U = \text{Int}(\overline{\bigcup \mathcal{B}_U})$ and therefore the map φ is an injection. As a consequence, $|\mathcal{R}(Z)| \le |\exp_\lambda(\mathcal{B})| \le \kappa^\lambda$ and Fact 2 is proved.

Fact 3. For any space Z we have $\pi w(Z) \le \pi \chi(Z) \cdot d(Z)$.

Proof. Let $\kappa = \pi \chi(Z)$ and $\lambda = d(Z)$. Fix a π-base $\mathcal{B}_z$ with $|\mathcal{B}_z| \le \kappa$ for any $z \in Z$ and take a dense $D \subset Z$ such that $|D| \le \lambda$. The family $\mathcal{B} = \bigcup\{\mathcal{B}_z : z \in D\}$ is a π-base in Z because, for any $U \in \tau^*(Z)$ there is $d \in U \cap D$ and hence we can find $B \in \mathcal{B}_d \subset \mathcal{B}$ for which $B \subset U$. Thus $\pi w(Z) \le |\mathcal{B}| \le \kappa \cdot \lambda$ and Fact 3 is proved.

Fact 4. For any space Z we have $w(Z) \le \pi \chi(Z)^{c(Z)}$.

Proof. Let $\pi \chi(Z) = \kappa$ and $c(Z) = \lambda$; we will us establish first that $d(Z) \le \kappa^\lambda$. For any $z \in Z$ fix a π-base $\mathcal{B}_z$ at the point z such that $|\mathcal{B}_z| \le \kappa$. Choose any point $z_0 \in Z$ and let $Z_0 = \{z_0\}$. To proceed inductively, assume that $\beta < \lambda^+$ and we have a family $\{Z_\alpha : \alpha < \beta\} \subset \exp Z$ with the following properties:

(4) $|Z_\alpha| \le \kappa^\lambda$ for any $\alpha < \beta$;
(5) $\alpha < \alpha' < \beta$ implies $Z_\alpha \subset Z_{\alpha'}$.
(6) Let $\mathcal{U}_\alpha = \bigcup\{\mathcal{B}_z : z \in Z_\alpha\}$ for any $\alpha < \beta$. Then $\alpha < \gamma < \beta$ implies that, for any $\mathcal{U} \subset \exp_\lambda(\mathcal{U}_\alpha)$ with $P = \overline{\bigcup \mathcal{U}} \ne Z$, we have $Z_\gamma \cap (Z \backslash P) \ne \emptyset$.

It follows from (4) and (6) that the set $T = \bigcup\{Z_\alpha : \alpha < \beta\}$ and the family $\mathcal{B}_T = \bigcup\{\mathcal{B}_t : t \in T\}$ have cardinalities that do not exceed κ^λ. Therefore the collection $\mathcal{A} = \{\mathcal{V} \subset \exp_\lambda(\mathcal{B}_T) : \overline{\bigcup \mathcal{V}} \ne Z\}$ also has cardinality at most κ^λ. For every $\mathcal{V} \in \mathcal{A}$ choose a point $a(\mathcal{V}) \in Z \backslash (\overline{\bigcup \mathcal{V}})$ and let $Z_\beta = T \cup \{a(\mathcal{V}) : \mathcal{V} \in \mathcal{A}\}$.

It is immediate that (4)–(6) still hold for all $\alpha \le \beta$; so our inductive procedure can be accomplished for all $\beta < \lambda^+$ giving us a family $\{Z_\alpha : \alpha < \kappa^+\}$ with the properties (4)–(6). It follows from (4) that $Z' = \bigcup\{Z_\alpha : \alpha < \kappa^+\}$ has cardinality at most κ^λ; so it suffices to prove that $\overline{Z'} = Z$; let $\mathcal{B} = \bigcup\{\mathcal{B}_z : z \in Z'\}$.

Assume towards a contradiction that $Z\backslash\overline{Z'} \neq \emptyset$ and choose $O \in \tau^*(Z)$ such that $\overline{O} \subset Z\backslash\overline{Z'}$. It is evident that the family $\mathcal{B}' = \{B \in \mathcal{B} : B \cap \overline{O} = \emptyset\}$ is still a π-base at any point $z \in Z'$. If $\mathcal{V}$ is a maximal disjoint family of the elements of $\mathcal{B}'$ and $V = \bigcup\mathcal{V}$ then $|\mathcal{V}| \leq \lambda$ and $Z' \subset \overline{V}$. Indeed, if $z \in Z'\backslash\overline{V}$ then there is $B \in \mathcal{B}_z$ for which $B \cap \overline{V} = \emptyset$; so $\mathcal{V} \cup \{B\}$ is a disjoint family of elements of $\mathcal{V}$ which is strictly larger than $\mathcal{V}$, a contradiction.

The condition (5) implies that there is an ordinal $\alpha < \kappa^+$ such that $\mathcal{V} \subset \mathcal{U}_\alpha$ and hence $\mathcal{V} \in \exp_\kappa(\mathcal{U}_\alpha)$; besides, $V \cap \overline{O} = \emptyset$, so we can apply (6) to conclude that $Z_{\alpha+1} \cap (Z\backslash\overline{V}) \neq \emptyset$ which is a contradiction because $Z_{\alpha+1} \subset Z' \subset \overline{V}$. Therefore Z' is dense in Z and hence $d(Z) \leq |Z'| \leq \kappa^\lambda$.

By Fact 3, we have $\pi w(Z) \leq d(Z) \cdot \pi\chi(Z) \leq \kappa^\lambda \cdot \kappa \leq \kappa^\lambda$; so we can apply Fact 2 to conclude that $|\mathcal{R}(Z)| \leq \pi w(Z)^{c(Z)} \leq (\kappa^\lambda)^\lambda = \kappa^\lambda$. By regularity of Z the family $\mathcal{R}(Z)$ is a base of Z; so $w(Z) \leq |\mathcal{R}(Z)| \leq \kappa^\lambda$ and Fact 4 is proved.

Fact 5. Suppose that κ is an uncountable regular cardinal, $a \notin \kappa$ and let $L = \kappa \cup \{a\}$. For any infinite cardinal $\lambda < \kappa$ let $\tau(\kappa, \lambda) = \exp\kappa \cup \{A : a \in A \subset L$ and $|\kappa\backslash A| \leq \lambda\}$. Then $\tau = \tau(\kappa, \lambda)$ is a Tychonoff topology on L and for the space $L(\kappa, \lambda) = (L, \tau)$ we have $l^*(L(\kappa, \lambda)) = \lambda$.

Proof. It is immediate that τ is a topology on L such that a is the unique non-isolated point of τ. Therefore $L(\kappa, \lambda)$ is a Tychonoff (and even normal) space by Claim 2 of S.018. Every subset of cardinality $\leq \lambda$ is closed and discrete in $L(\kappa, \lambda)$; so $l(L(\kappa, \lambda)) \geq \lambda$.

We will prove, by induction on $n \in \mathbb{N}$, that $l((L(\kappa, \lambda))^n) \leq \lambda$ for any $n \in \mathbb{N}$. If $n = 1$ and $\mathcal{U}$ is an open cover of $L(\kappa, \lambda)$ then take a set $U \in \mathcal{U}$ such that $a \in U$. Then $|L(\kappa, \lambda)\backslash U| \leq \lambda$; so there is $\mathcal{U}' \in \exp_\lambda(\mathcal{U})$ for which $L(\kappa, \lambda)\backslash U$ is covered by $\mathcal{U}'$. It is clear that $\mathcal{U}' \cup \{U\}$ is a subcover of $\mathcal{U}$ of cardinality $\leq \lambda$; so $l(L(\kappa, \lambda)) \leq \lambda$.

Now assume that $l((L(\kappa, \lambda))^k) \leq \lambda$ for some number $k \in \mathbb{N}$ and take an open cover $\mathcal{U}$ of the space $(L(\kappa, \lambda))^{k+1}$. Represent $(L(\kappa, \lambda))^{k+1}$ as $(L(\kappa, \lambda))^k \times L(\kappa, \lambda)$ and let $F = \{(x, a) : x \in (L(\kappa, \lambda))^k\} \subset (L(\kappa, \lambda))^{k+1}$. It is immediate that F is homeomorphic to $(L(\kappa, \lambda))^k$ so $l(F) \leq \lambda$ by the induction hypothesis. For any point $z = (x, a) \in F$ choose $U_z \in \tau(x, (L(\kappa, \lambda))^k)$ and $V_z \in \tau(a, L(\kappa, \lambda))$ such that $U_z \times V_z \subset W_z$ for some $W_z \in \mathcal{U}$. Since $l(F) \leq \lambda$, we can choose a set $P \subset (L(\kappa, \lambda))^k$ such that $|P| \leq \lambda$ and $F \subset \bigcup\{U_z \times V_z : z \in P\}$.

Observe that $Q_z = L(\kappa, \lambda)\backslash V_z$ has cardinality at most λ for any $z \in P$ and hence $Q = \bigcup\{Q_z : z \in P\}$ also has cardinality $\leq \lambda$. Let $\mathcal{U}' = \{W_z : z \in P\}$; if $y = (x, s) \notin \bigcup\mathcal{U}'$ then take a point $z \in P$ for which $x \in U_z$. Then $s \notin V_z$ which implies $z \in Q_z \subset Q$. We proved that $(L(\kappa, \lambda))^{k+1}\backslash(\bigcup\mathcal{U}')$ is contained in the set $G = (L(\kappa, \lambda))^k \times Q$ which is a union of $\leq \lambda$-many subspaces homeomorphic to $(L(\kappa, \lambda))^k$. This, together with the induction hypothesis, implies that $l(G) \leq \lambda$ and hence we can find $\mathcal{U}'' \subset \mathcal{U}$ such that $|\mathcal{U}''| \leq \lambda$ and $G \subset \bigcup\mathcal{U}''$. Thus $\mathcal{U}' \cup \mathcal{U}''$ is a subcover of $\mathcal{U}$ of cardinality $\leq \lambda$ which proves that $l((L(\kappa, \lambda))^{k+1}) \leq \lambda$.

The inductive step of our proof has been successfully accomplished; so we established that $l((L(\kappa,\lambda))^n) \leq \lambda$ for all $n \in \mathbb{N}$ and hence Fact 5 is proved.

Returning to our solution let $\lambda = l(Y) \cdot c(X)$. We will establish first that

(7) $|\overline{A}| \leq |A|^{\omega}$ for any $A \subset C_p(X)$.

Indeed, for any countable $B \subset C_p(X)$ we have $nw(\overline{B}) = \omega$ (see Problem 033) and hence $|\overline{B}| \leq \mathfrak{c}$. Besides, $t(C_p(X)) \leq \omega$; so

$$|\overline{A}| = |\bigcup\{\overline{B} : B \in \exp_{\omega}(A)\}| \leq \mathfrak{c} \cdot |\exp_{\omega}(A)| = \mathfrak{c} \cdot |A|^{\omega} = |A|^{\omega}$$

and hence (7) is proved. Let us show next that

(8) $\psi(Y) \leq 2^{\lambda}$.

Assume, for a contradiction, that we have $\psi(v, Y) \geq (2^{\lambda})^+$ for some $v \in Y$. Then $l(Y\backslash\{v\}) \geq (2^{\lambda})^+$ (see Fact 1 of U.027) and hence we can apply Baturov's theorem (SFFS-269) to conclude that there is a closed discrete subset D in the space $Y\backslash\{v\}$ such that $|D| = \kappa = (2^{\lambda})^+$. The set $E = D \cup \{v\}$ is closed in Y; so $l(E) \leq \lambda$. As a consequence $|E\backslash U| \leq \lambda$ for any $U \in \tau(v, E)$ and therefore the space E is a continuous image of the space $L(\kappa, \lambda)$ which implies $l^*(E) \leq \lambda$ (see Fact 5).

For any $x \in X$ let $\varphi(x)(f) = f(x)$ for any $f \in E$. Then $\varphi(x) \in C_p(E)$ for every $x \in X$ and the map $\varphi : X \to C_p(E)$ is continuous by TFS-166. The space $X' = \varphi(X)$ is compact; besides, $c(X') \leq c(X) \leq \lambda$ and $t(X') \leq t(C_p(E)) = l^*(E) \leq \lambda$. Furthermore, $\pi\chi(X') \leq t(X') \leq \lambda$ by TFS-331; so we can apply Fact 4 to see that $w(X') \leq \pi\chi(X)^{c(X')} \leq \lambda^{\lambda} = 2^{\lambda}$.

The dual map $\varphi^* : C_p(X') \to C_p(X)$ defined by $\varphi^*(h) = h \circ \varphi$ for any $h \in C_p(X')$ embeds $C_p(X')$ is $C_p(X)$ (see TFS-163) and it is easy to verify the inclusion $E \subset \varphi^*(C_p(X'))$. Thus $s(E) \leq nw(E) \leq nw(C_p(X')) = nw(X') \leq w(X') \leq 2^{\lambda}$ which is a contradiction because D is a discrete subspace of E and $|D| > 2^{\lambda}$. This shows that $\psi(Y) \leq 2^{\lambda}$; so (8) is proved.

Apply (7) to see that if $A \subset Y$ and $|A| \leq \lambda$ then $|\overline{A}| \leq |A|^{\omega} \leq \lambda^{\omega} \leq 2^{\lambda}$. The property (8) shows that we can apply Fact 1 to the space Y to conclude that $|Y| \leq 2^{\lambda}$. Finally apply (7) to see that $|\overline{Y}| \leq |Y|^{\omega} \leq (2^{\lambda})^{\omega} = 2^{\lambda}$ and hence our solution is complete.

U.084. *Suppose that X is a compact space with the Souslin property and $C_p(X)$ has a dense Lindelöf subspace. Prove that $w(X) \leq |C_p(X)| \leq 2^{\omega}$.*

Solution. We have $w(X) = nw(X) = nw(C_p(X)) \leq |C_p(X)|$ (see Fact 4 of S.307); so $w(X) \leq |C_p(X)|$. Now, if Z is a dense Lindelöf subspace of $C_p(X)$ then apply Problem 083 to see that $|C_p(X)| = |\overline{Z}| \leq 2^{l(Z) \cdot c(X)} \leq 2^{\omega \cdot \omega} = 2^{\omega}$.

U.085. *Prove that, for any uncountable regular cardinal κ, if $Z \subset C_p(\kappa + 1)$ separates the points of $\kappa + 1$ then $l(Z) \geq \kappa$.*

Solution. To obtain a contradiction, let us assume that $\lambda = l(Z) < \kappa$. Since the map $f \longmapsto (-f)$ is a homeomorphism of the space $C_p(\kappa + 1)$ onto itself, both sets $-Z = \{-f : f \in L\}$ and $Z \cup (-Z)$ have the Lindelöf number $\leq \lambda$. This shows that we can assume that $(-f) \in Z$ for any $f \in Z$.

For each $\alpha < \kappa$ fix rational numbers s_α, t_α and a function $f_\alpha \in Z$ such that $f_\alpha(\alpha) < s_\alpha < t_\alpha < f_\alpha(\kappa)$ or $f_\alpha(\alpha) > s_\alpha > t_\alpha > f_\alpha(\kappa)$. However, if we have the second inequality then, for the function $(-f_\alpha) \in Z$, we have the first one. Therefore we can assume that $f_\alpha(\alpha) < s_\alpha < t_\alpha < f_\alpha(\kappa)$ for all $\alpha < \kappa$. Since each f_α is continuous, there exists $\beta_\alpha < \alpha$ such that $f_\alpha(\gamma) < s_\alpha$ for each $\gamma \in (\beta_\alpha, \alpha]$.

The map $r : \kappa \to \kappa$ defined by $r(\alpha) = \beta_\alpha$ for all $\alpha < \kappa$ satisfies the hypothesis of Fact 3 of U.074; so there is $\beta < \kappa$ and a stationary set $R \subset \kappa$ such that $\beta_\alpha = \beta$ for all $\alpha \in R$. There is a set $R' \subset R$ with $|R'| = \kappa$ for which there are $s, t \in \mathbb{Q}$ such that $s_\alpha = s$ and $t_\alpha = t$ for all $\alpha \in R'$; let $E = \{f_\alpha : \alpha \in R'\}$.

For every $f \in Z$ let $O_f = Z \backslash \overline{E}$ if $f \notin \overline{E}$. Then O_f is an open neighbourhood of f in Z such that $O_f \cap E = \emptyset$ and hence $B_f = \{\alpha \in R' : f_\alpha \in O_f\} = \emptyset$. If $f \in \overline{E}$ then we have $f(\kappa) \geq t$ because $g(\kappa) > t$ for all $g \in E$. Choose any $s' \in (s, t)$ and observe that, by continuity of f, there is $\gamma > \beta$ such that $f(\gamma) > s' > s$. The set $O_f = \{g \in Z : g(\gamma) > s'\}$ is an open neighbourhood of f in Z. If $\alpha > \gamma$ and $\alpha \in R'$ then $\gamma \in (\beta, \alpha] = (\beta_\alpha, \alpha]$ which implies, by the choice of β_α, that $f_\alpha(\gamma) < s < s'$ whence $f_\alpha \notin O_f$. As a consequence, $O_f \cap E \subset \{f_\alpha : \alpha \leq \gamma\}$ and therefore, for the set $B_f = \{\alpha \in R' : f_\alpha \in O_f\}$, we have $|B_f| \leq |\gamma| < \kappa$.

Since $\mathcal{U} = \{O_f : f \in Z\}$ is an open cover of the space Z, there is a family $\mathcal{U}' \subset \mathcal{U}$ such that $Z \subset \bigcup \mathcal{U}'$ and $|\mathcal{U}'| \leq \lambda$. This implies $R' = \bigcup \{B_f : f \in \mathcal{U}'\}$, i.e., the set R' of cardinality κ is represented as a union of $\leq \lambda$-many subsets of cardinality $< \kappa$. This contradicts regularity of κ and proves that $l(Z) \geq \kappa$.

U.086. *Prove that, if X is a dyadic space and $Y \subset C_p(X)$ then $nw(Y) = l(Y)$. In particular, any Lindelöf subspace of $C_p(X)$ has a countable network.*

Solution. For any $n \in \mathbb{N}$ let $M_n = \{1, \ldots, n\}$; if A is a set then $\text{Fin}(A)$ is the family of all finite subsets of A and $\text{Fn}(A) = \bigcup \{\mathbb{D}^B : B \in \text{Fin}(A)\}$; observe that the unique element of $\mathbb{D}^{\emptyset}$ is the empty function. If f is a function then $\text{dom}(f)$ is its domain. If f and g are functions then $f \subset g$ says that $\text{dom}(f) \subset \text{dom}(g)$ and $g|\text{dom}(f) = f$. We consider that if g is a function and $f = \emptyset$ then $f \subset g$. Given two functions f and g such that $f(x) = g(x)$ for any point $x \in \text{dom}(f) \cap \text{dom}(g)$ the function $h = f \cup g$ is defined by letting $h(x) = f(x)$ for all $x \in \text{dom}(f)$ and $h(x) = g(x)$ for any $x \in \text{dom}(g) \backslash \text{dom}(f)$.

For any $h \in \text{Fn}(A)$ let $[h] = \{s \in \mathbb{D}^A : h \subset s\}$. It is easy to see that the family $\{[h] : h \in \text{Fn}(A)\}$ is a base in the space $\mathbb{D}^A$. Given a regular uncountable cardinal κ say that $C \subset \kappa$ is a *club* if C is closed and unbounded ($\equiv$cofinal) in κ. A set $A \subset \kappa$ is called *stationary* if $A \cap C \neq \emptyset$ for any club C.

Suppose that we have a family $\mathcal{F} = \{F_t^0, F_t^1 : t \in T\}$ of closed subsets of a space Z. Let $C_\emptyset = Z$ and $C_h = \bigcap \{F_t^{h(t)} : t \in \text{dom}(h)\}$ for any $h \in \text{Fn}(T) \backslash \{\emptyset\}$. We say that the family $\mathcal{F}$ is *dyadic* if $F_t^0 \cap F_t^1 = \emptyset$ for every $t \in T$ while $C_h \neq \emptyset$ for any $h \in \text{Fn}(T)$. In particular, $F_t^i \neq \emptyset$ for any $t \in T$ and $i \in \mathbb{D}$. A dyadic family

$\mathcal{F} = \{F_t^0, F_t^1 : t \in T\}$ will also be called κ-dyadic if $|T| = \kappa$. If $\mathcal{A}$ is a family of sets then $\bigwedge \mathcal{A}$ is the family of all non-empty finite intersections of the elements of $\mathcal{A}$.

Fact 1. Given an infinite cardinal κ if K is a compact space such that $\pi\chi(x, K) \geq \kappa$ for any $x \in K$ then there is a closed $P \subset K$ which maps continuously onto $\mathbb{D}^\kappa$ and hence K maps continuously onto $\mathbb{I}^\kappa$.

Proof. If $\kappa = \omega$ then K has no isolated points and hence there exists a continuous onto map $\varphi : K \to \mathbb{I}$ (see SFFS-133). Take a closed $F \subset \mathbb{I}$ homeomorphic to $\mathbb{D}^\omega$ (see TFS-128); then $P = \varphi^{-1}(F)$ is a closed subset of K which maps continuously onto $\mathbb{D}^\omega$; so K can be continuously mapped onto $\mathbb{I}^\omega$ (see Fact 4 of T.298).

From this moment on we consider that κ is an uncountable cardinal. Denote by $\mathcal{C}$ the family of all closed non-empty G_δ-subsets of K. Each $G \in \mathcal{C}$ has a countable outer base $\mathcal{B}_G$ in K by TFS-327. This shows that $G \subset U \in \tau(K)$ implies that there is $V \in \mathcal{B}_G$ with $V \subset U$.

Suppose that $\mathcal{C}' \subset \mathcal{C}$ and $|\mathcal{C}'| < \kappa$. Given a point $x \in K$, if every $U \in \tau(x, K)$ contains some $G \in \mathcal{C}'$ then it also contains some $V \in \mathcal{B}_G$ by the previous remark. This shows that $\mathcal{B} = \bigcup\{\mathcal{B}_G : G \in \mathcal{C}'\}$ is a local π-base at x with $|\mathcal{B}| < \kappa$ which is a contradiction. This proves that

(1) for any $x \in K$ and any $\mathcal{C}' \subset \mathcal{C}$ such that $|\mathcal{C}'| < \kappa$ there is $W \in \tau(x, K)$ such that $G \backslash W \neq \emptyset$ for all $G \in \mathcal{C}'$.

Our plan is to construct a κ-dyadic family in K; the second step on this way is to show that

(2) for any $\mathcal{C}' \subset \mathcal{C}$ with $|\mathcal{C}'| < \kappa$ there are $F, G \in \mathcal{C}$ such that $F \cap C \neq \emptyset$ and $G \cap C \neq \emptyset$ for any $C \in \mathcal{C}'$ but there exists $B \in \mathcal{C}'$ such that $F \cap G \cap B = \emptyset$.

To prove (2), apply (1) for every $x \in K$ to obtain a set $W_x \in \tau(x, K)$ such that $C \backslash W_x \neq \emptyset$ for every $C \in \mathcal{C}'$. Taking a smaller W_x if necessary we can assume that W_x is an F_σ-set and therefore $P_x = K \backslash W_x$ is a G_δ-set for all $x \in K$. Take a finite subcover $\{W_{x_1}, \ldots, W_{x_n}\}$ of the cover $\{W_x : x \in K\}$; then $P_{x_i} \cap C \neq \emptyset$ for any $C \in \mathcal{C}'$ and $i \in M_n$ while $P_{x_1} \cap \ldots \cap P_{x_n} = \emptyset$. Let $k \in M_n$ be the minimal number for which there exist $Q_1, \ldots, Q_k \in \{P_{x_1}, \ldots, P_{x_n}\}$ such that $Q_1 \cap \ldots \cap Q_k \cap B = \emptyset$ for some $B \in \mathcal{C}'$. Then $k \geq 2$ and the sets $F = Q_1$ and $G = Q_2 \cap \ldots \cap Q_k$ are as promised; so (2) is proved.

Our next step is to establish the following property.

(3) Let λ and μ be infinite cardinals with $\mu = \mathrm{cf}(\mu)$ and $\lambda < \mu \leq \kappa$. If $\mathcal{F} \subset \mathcal{C}$ is a λ-dyadic system then there is a μ-dyadic system $\mathcal{G} \subset \mathcal{C}$ such that $|\mathcal{F} \backslash \mathcal{G}| < \omega$.

Choose an arbitrary enumeration $\{F_\alpha^0, F_\alpha^1 : \alpha < \lambda\}$ of the family $\mathcal{F}$ which witnesses that $\mathcal{F}$ is λ-dyadic. Let $I_h = \bigcap\{F_\alpha^{h(\alpha)} : \alpha \in \mathrm{dom}(h)\}$ for any $h \in \mathrm{Fn}(\lambda)$; if $\mathrm{dom}(h) = \emptyset$ then $I_h = K$. We are going to construct by transfinite induction a family $\mathcal{K} = \{C_\alpha^0, C_\alpha^1, K_\alpha^0, K_\alpha^1 : \alpha < \mu\} \subset \mathcal{C}$. If $\beta < \mu$ and $g \in \mathrm{Fn}(\beta)$ then $J_g = \bigcap\{C_\alpha^{g(\alpha)} : \alpha \in \mathrm{dom}(g)\}$; if $\mathrm{dom}(g) = \emptyset$ then $J_g = K$.

The family $\mathcal{A}_0 = \mathcal{F}$ has cardinality strictly less than κ and hence $|\bigwedge \mathcal{A}_0| < \kappa$. Thus the property (2) is applicable to find $K_0^0, K_0^1 \in \mathcal{C}$ such that $K_0^i \cap H \neq \emptyset$ for any $H \in \bigwedge \mathcal{A}_0$ and $i \in \mathbb{D}$ while there is $F \in \bigwedge \mathcal{A}_0$ such that $K_0^0 \cap K_0^1 \cap F = \emptyset$. It is easy to see that there exists a function $h_0 \in \mathrm{Fn}(\lambda)$ such that $F = I_{h_0}$; let $g_0 = \emptyset$ and $C_0^i = F \cap K_0^i$ for every $i \in \mathbb{D}$.

Proceeding inductively assume that α is an ordinal with $0 < \alpha < \mu$ and we have a family $\mathcal{K}_\alpha = \{C_\beta^0, C_\beta^1, K_\beta^0, K_\beta^1 : \beta < \alpha\}$ with the following properties:

(4) if $\beta < \alpha$ and $\mathcal{A}_\beta = \mathcal{F} \cup \{C_\gamma^0, C_\gamma^1 : \gamma < \beta\}$ then $K_\beta^i \cap H \neq \emptyset$ for any $H \in \bigwedge \mathcal{A}_\beta$ and $i \in \mathbb{D}$;

(5) for every $\beta < \alpha$ we have functions $h_\beta \in \mathrm{Fn}(\lambda)$ and $g_\beta \in \mathrm{Fn}(\beta)$ such that $I_{h_\beta} \cap J_{g_\beta} \cap K_\beta^0 \cap K_\beta^1 = \emptyset$;

(6) if $\beta < \alpha$ then $C_\beta^0 \cap C_\beta^1 = \emptyset$ and $C_\beta^i = I_{h_\beta} \cap J_{g_\beta} \cap K_\beta^i$ for each $i \in \mathbb{D}$.

The family $\mathcal{A}_\alpha = \{C_\beta^0, C_\beta^1 : \beta < \alpha\} \cup \mathcal{F}$ has cardinality strictly less than κ and hence $|\bigwedge \mathcal{A}_\alpha| < \kappa$. Thus the property (2) is applicable to find $K_\alpha^0, K_\alpha^1 \in \mathcal{C}$ such that $K_\alpha^i \cap H \neq \emptyset$ for any $H \in \bigwedge \mathcal{A}_\alpha$ and $i \in \mathbb{D}$ while there is $F \in \bigwedge \mathcal{A}_\alpha$ such that $K_\alpha^0 \cap K_\alpha^1 \cap F = \emptyset$. It is easy to see that there exist $h_\alpha \in \mathrm{Fn}(\lambda)$ and $g_\alpha \in \mathrm{Fn}(\alpha)$ such that $F = I_{h_\alpha} \cap J_{g_\alpha}$; let $C_\alpha^i = F \cap K_\alpha^i$ for every $i \in \mathbb{D}$.

Evidently, the conditions (4)–(6) are now satisfied for all $\beta \leq \alpha$; so our inductive procedure can be continued to construct the family $\mathcal{K} = \{C_\beta^0, C_\beta^1, K_\beta^0, K_\beta^1 : \beta < \mu\}$ for which the properties (4)–(6) hold for all $\alpha < \mu$.

Our promised family $\mathcal{G}$ will be contained in the family $\mathcal{G}' = \{C_\alpha^0, C_\alpha^1 : \alpha < \mu\}$. To extract $\mathcal{G}$ from $\mathcal{G}'$ we will need the following property.

(7) If $h \in \mathrm{Fn}(\lambda)$, $g \in \mathrm{Fn}(\mu)$ and $I_h \cap J_g \neq \emptyset$ then there exists a finite set $A \subset \lambda$ such that $I_h \cap I_{h'} \cap J_g \neq \emptyset$ for any $h' \in \mathrm{Fn}(\lambda \backslash A)$.

We will prove (7) by induction on the maximal element δ_g of $\mathrm{dom}(g)$; if $g = \emptyset$ then let $\delta_g = -1$. Observe first that if $\delta_g = -1$ then we can take $A = \mathrm{dom}(h)$. If $h' \in \mathrm{Fn}(\lambda \backslash A)$ then $I_h \cap I_{h'} \cap J_g = I_h \cap I_{h'} = I_{h \cup h'} \neq \emptyset$ because $\mathcal{F}$ is λ-dyadic.

Now assume that $\alpha < \mu$ and (7) holds for any $g' \in \mathrm{Fn}(\mu)$ with $\delta_{g'} < \alpha$. Suppose that $I_h \cap J_g \neq \emptyset$ for some $h \in \mathrm{Fn}(\lambda)$ and $g \in \mathrm{Fn}(\mu)$ such that $\delta_g = \alpha$. Define a function $g' \in \mathrm{Fn}(\mu)$ be letting $\mathrm{dom}(g') = \mathrm{dom}(g) \backslash \{\alpha\}$ and $g'(\beta) = g(\beta)$ for any $\beta \in \mathrm{dom}(g')$. We have the equality $I_h \cap J_g = I_h \cap J_{g'} \cap C_\alpha^{g(\alpha)}$ which, together with (6) implies that $I_h \cap J_g = I_h \cap J_{g'} \cap I_{h_\alpha} \cap J_{g_\alpha} \cap K_\alpha^{g(\alpha)}$. Thus $I_h \cap I_{h_\alpha} \neq \emptyset$ which shows that $h(\beta) = h_\alpha(\beta)$ for any $\beta \in \mathrm{dom}(h) \cap \mathrm{dom}(h_\alpha)$; so the function $h \cup h_\alpha$ makes sense.

Analogously, the function $g' \cup g_\alpha$ is also consistently defined; the equalities $I_h \cap I_{h_\alpha} = I_{h \cup h_\alpha}$ and $J_{g'} \cap J_{g_\alpha} = J_{g' \cup g_\alpha}$ show that $I_{h \cup h_\alpha} \cap J_{g' \cup g_\alpha} = I_h \cap I_{h_\alpha} \cap J_{g'} \cap J_{g_\alpha}$ is a non-empty set. Besides, $\delta_{g' \cup g_\alpha} < \alpha$ and hence our induction hypothesis is applicable to conclude that there exists a finite set $A \subset \lambda$ such that, for any function $h' \in \mathrm{Fn}(\lambda \backslash A)$ the set $I_{h \cup h_\alpha} \cap I_{h'} \cap J_{g' \cup g_\alpha}$ is non-empty.

Let us show that the same set A is applicable to make our induction step. Indeed, if $h' \in \mathrm{Fn}(\lambda \backslash A)$ then the set $P = I_{h \cup h_\alpha} \cap I_{h'} \cap J_{g' \cup g_\alpha}$ is non-empty; since $P \in \bigwedge \mathcal{A}_\alpha$, we can apply (4) to see that $I_h \cap I_{h'} \cap J_g = P \cap K_\alpha^{g(\alpha)} \neq \emptyset$, so (7) is proved.

Let $p(\alpha) = \max \operatorname{dom}(g_\alpha)$; it is clear that $p(\alpha) < \alpha$ for any $\alpha < \mu$. It follows from Fact 3 of U.074 that there exists $\beta < \mu$ such that the set $\{\alpha < \mu : p(\alpha) = \beta\}$ has cardinality μ. It follows from $|\operatorname{Fn}(\lambda)| < \mu$ and $|\operatorname{Fn}(\beta)| < \mu$ that we can choose a set $E \subset \mu \backslash (\beta + 1)$ and functions $h \in \operatorname{Fn}(\lambda)$, $g \in \operatorname{Fn}(\beta)$ such that $|E| = \mu$ while $h_\alpha = h$ and $g_\alpha = g$ for all $\alpha \in E$. In particular, we have $I_h \cap J_g \cap K_\alpha^0 = C_\alpha^0 \neq \emptyset$ for any $\alpha \in E$; as an immediate consequence, $I_h \cap J_g \neq \emptyset$, so we can apply (7) to find a finite set $A \subset \lambda$ such that $I_h \cap I_{h'} \cap J_g \neq \emptyset$ for each $h' \in \operatorname{Fn}(\lambda \backslash A)$.

We claim that the set $\mathcal{G} = \{F_\alpha^0, F_\alpha^1 : \alpha \in \lambda \backslash A\} \cup \{C_\alpha^0, C_\alpha^1 : \alpha \in E\}$ is as promised in (3). It is trivial that $|\mathcal{G}| = \mu$ and $\mathcal{F} \backslash \mathcal{G}$ is finite; so we must only check that $\mathcal{G}$ is dyadic.

It is sufficient to show that, for any $h' \in \operatorname{Fn}(\lambda \backslash A)$ and $g' \in \operatorname{Fn}(E)$ we have $I_{h'} \cap J_{g'} \neq \emptyset$; so fix relevant h' and g'. If $g' = \emptyset$ then $I_{h'} \cap J_{g'} = I_{h'} \neq \emptyset$ because the family $\mathcal{F}$ is λ-dyadic. Thus we can consider that $g' \neq \emptyset$ and hence we can choose an enumeration $\{\alpha_1, \ldots, \alpha_k\}$ of the set $\operatorname{dom}(g')$ such that $\alpha_1 < \ldots < \alpha_k$; let $i_j = g'(\alpha_j)$ for all $j \leq k$.

Observe that $I_{h'} \cap J_{g'} = I_{h'} \cap C_{\alpha_0}^{i_0} \cap \ldots \cap C_{\alpha_k}^{i_k} = I_{h'} \cap I_h \cap J_g \cap K_{\alpha_0}^{i_0} \cap \ldots \cap K_{\alpha_k}^{i_k}$; since the condition (7) is fulfilled for the set A, the set $H = I_{h'} \cap I_h \cap J_g$ is non-empty. Furthermore, $H \in \bigwedge \mathcal{A}_{\alpha_0}$; so $H_0 = H \cap K_{\alpha_0}^{i_0} \neq \emptyset$ by the property (4).

Proceeding inductively assume that $j < k$ and $H_j = H \cap K_{\alpha_0}^{i_0} \cap \ldots \cap K_{\alpha_j}^{i_j} \neq \emptyset$. It is evident that $H_j \in \bigwedge \mathcal{A}_{\alpha_{j+1}}$; so it follows from (4) that $H_{j+1} = H_j \cap K_{\alpha_{j+1}}^{i_{j+1}} \neq \emptyset$. Thus our inductive procedure can be continued to establish that $I_{h'} \cap J_{g'} = H_k \neq \emptyset$; so (3) is proved.

Assume that $\kappa > \omega$ and κ is a regular cardinal; we saw at the beginning of our proof that there exists a continuous onto map $\varphi : K \to \mathbb{I}$. Take a closed $F \subset \mathbb{I}$ homeomorphic to $\mathbb{D}^\omega$; then $P = \varphi^{-1}(F)$ is a non-empty G_δ-subset of K which maps continuously onto $\mathbb{D}^\omega$. Let $\varphi' = \varphi | P$.

By Fact 4 of T.298 there is an ω-dyadic system $\mathcal{S} = \{G_n^0, G_n^1 : n \in \omega\}$ in the space F. It is clear that all elements of $\mathcal{S}$ are G_δ-subsets of F; so if we let $F_n^i = (\varphi')^{-1}(G_n^i)$ for all $i \in \mathbb{D}$ and $n \in \omega$ then $\mathcal{F} = \{F_n^0, F_n^1 : n \in \omega\}$ is an ω-dyadic system of G_δ-subsets of P. It follows from $P \in \mathcal{C}$ that $\mathcal{F} \subset \mathcal{C}$ and hence we can apply (3) to construct a κ-dyadic family $\mathcal{G}$ in the space K.

Finally assume that κ is an uncountable cardinal and $\lambda = \operatorname{cf}(\kappa) < \kappa$. Choose a λ-sequence $\{\kappa_\nu : \nu < \lambda\}$ of regular cardinals such that $\kappa_\nu > \lambda$ for all $\nu < \lambda$ and $\kappa = \sup\{\kappa_\nu : \nu < \lambda\}$ while $\kappa_\nu < \kappa_{\nu'}$ whenever $\nu < \nu' < \lambda$. It is an easy exercise that

(8) if μ is an infinite cardinal and $\{\mathcal{H}_\alpha : \alpha < \mu\}$ is a μ-sequence of dyadic families such that $\alpha < \beta < \mu$ implies $\mathcal{H}_\alpha \subset \mathcal{H}_\beta$ then $\mathcal{H} = \bigcup_{\alpha < \mu} \mathcal{H}_\alpha$ is a dyadic family.

From what was proved for regular cardinals, it follows that we can choose a κ_0-dyadic family $\mathcal{F}_0 \subset \mathcal{C}$. Suppose that $\mu < \lambda$ and we have a collection $\{\mathcal{F}_\nu : \nu < \mu\}$ of dyadic families such that

(9) $\mathcal{F}_\nu \subset \mathcal{C}$ and $|\mathcal{F}_\nu| = \kappa_\nu$ for any $\nu < \mu$ and $\nu' < \nu < \mu$ implies that $|\mathcal{F}_{\nu'} \backslash \mathcal{F}_\nu| < \lambda$.

Let $\mathcal{G}_\nu = \bigcap\{\mathcal{F}_{\nu'} : \nu \leq \nu' < \mu\}$ for any $\nu < \mu$. It follows from $\mathcal{G}_\nu \subset \mathcal{F}_\nu$ that $\mathcal{G}_\nu$ is a dyadic family for all $\nu < \mu$. Besides, it follows from (9) and regularity of λ that $|\mathcal{F}_\nu \backslash \mathcal{G}_\nu| < \lambda$ and hence $|\mathcal{G}_\nu| = \kappa_\nu < \kappa_\mu$ for each $\nu < \mu$. It is immediate that $\nu < \nu' < \mu$ implies $\mathcal{G}_\nu \subset \mathcal{G}_{\nu'}$; so the family $\mathcal{H} = \bigcup\{\mathcal{G}_\nu : \nu < \mu\}$ is dyadic by (8).

Observe also that $\kappa_\mu = \mathrm{cf}(\kappa_\mu) > \lambda > \mu$ so $|\mathcal{H}| < \kappa_\mu$ and hence we can apply (3) to the family $\mathcal{H} \subset \mathcal{C}$ and the cardinals $|\mathcal{H}|$ and κ_μ to find a κ_μ-dyadic family $\mathcal{F}_\mu$ such that $|\mathcal{H} \backslash \mathcal{F}_\mu| < \omega$. For any $\nu < \mu$ we have $|\mathcal{F}_\nu \backslash \mathcal{H}| \leq |\mathcal{F}_\nu \backslash \mathcal{G}_\nu| < \lambda$; as a consequence, $|\mathcal{F}_\nu \backslash \mathcal{F}_\mu| < \lambda$; so the property (9) holds for all $\nu \leq \mu$ and hence our inductive procedure can be continued to construct a collection $\{\mathcal{F}_\nu : \nu < \lambda\}$ such that the condition (9) is satisfied for all $\nu < \lambda$.

Let $\mathcal{F}'_\nu = \bigcap\{\mathcal{F}_\mu : \nu \leq \mu < \lambda\}$ for each $\nu < \lambda$. It follows easily from (9) that $|\mathcal{F}_\nu \backslash \mathcal{F}'_\nu| \leq \lambda$ and hence $|\mathcal{F}'_\nu| = \kappa_\nu$ for all $\nu < \lambda$. Furthermore, $\mathcal{F}'_\nu \subset \mathcal{F}'_{\nu'}$ whenever $\nu < \nu' < \lambda$. For the family $\mathcal{F} = \bigcup_{\nu < \mu} \mathcal{F}'_\nu$ we have $|\mathcal{F}| = \kappa$ and $\mathcal{F}$ is dyadic by (8). Therefore we found a κ-dyadic family in K in all possible cases. Applying Fact 4 of T.298 we conclude that there is a closed $P \subset K$ which maps continuously onto $\mathbb{D}^\kappa$ and hence K maps continuously onto $\mathbb{I}^\kappa$, i.e., Fact 1 is proved.

Fact 2. If κ is a regular uncountable cardinal and $\varphi : \mathbb{D}^\kappa \to \mathbb{I}^\kappa$ is a continuous onto map then there is a closed $F \subset \mathbb{D}^\kappa$ such that $F \simeq \mathbb{D}^\kappa$ and $\varphi|F$ is injective.

Proof. Let $\mathcal{O} = \{[h] : h \in \mathrm{Fn}(\kappa)\}$ be the standard base of $\mathbb{D}^\kappa$. For any $A \subset \kappa$ the map $\pi_A : \mathbb{I}^\kappa \to \mathbb{I}^A$ is the natural projection of $\mathbb{I}^\kappa$ onto its face $\mathbb{I}^A$. Analogously, $p_A : \mathbb{D}^\kappa \to \mathbb{D}^A$ is the natural projection of $\mathbb{D}^\kappa$ onto its face $\mathbb{D}^A$. If $A = \{\alpha\}$ for some $\alpha < \kappa$ we write π_α and p_α instead of $\pi_{\{\alpha\}}$ and $p_{\{\alpha\}}$ respectively. Say that a set $G \subset \mathbb{D}^\kappa$ depends on $A \subset \kappa$ if $G = p_A^{-1}(p_A(G))$. Denote by $\mathcal{C}$ the family of all non-empty clopen subsets of $\mathbb{D}^\kappa$. Every element of $\mathcal{C}$ is compact; so it is a finite union of elements of $\mathcal{O}$; since every $O \in \mathcal{O}$ depends on finitely many coordinates, we proved that

(10) every $U \in \mathcal{C}$ depends on finitely many coordinates, i.e., there is a finite set $S = S_U \subset \kappa$ such that $U = p_S^{-1}(p_S(U))$.

For any element $i \in \mathbb{D}$ and $\alpha < \kappa$ let $P_\alpha^i = \pi_\alpha^{-1}(i)$, $Q_\alpha^i = \varphi^{-1}(P_\alpha^i)$ and fix a set $O_\alpha^i \in \tau(P_\alpha^i, \mathbb{I}^\kappa)$ so that $O_\alpha^0 \cap O_\alpha^1 = \emptyset$. It follows from SFFS-303, SFFS-306 and SFFS-308 that we can find $U_\alpha^i \in \mathcal{C}$ such that $Q_\alpha^i \subset U_\alpha^i \subset \varphi^{-1}(O_\alpha^i)$ for every $\alpha < \kappa$ and $i \in \mathbb{D}$. Apply (10) to find a finite set $S_\alpha \subset \kappa$ such that $p_{S_\alpha}^{-1}(p_{S_\alpha}(U_\alpha^i)) = U_\alpha^i$ for any $i \in \mathbb{D}$ and $\alpha < \kappa$.

By the Δ-lemma (SFFS-038) we can find a set $T' \subset \kappa$ and finite $S \subset \kappa$ such that $|T'| = \kappa$ and $S_\alpha \cap S_\beta = S$ for any distinct $\alpha, \beta \in T'$. If $S = \emptyset$ let $T = T'$ and choose $h_\alpha^i \in p_{S_\alpha}(Q_\alpha^i)$ for any $i \in \mathbb{D}$ and $\alpha \in T$. If $S \neq \emptyset$ then the set $\mathbb{D}^S$ is finite and $\mathbb{D}^\kappa = \bigcup\{[s] : s \in \mathbb{D}^S\}$ so $\mathbb{I}^\kappa = \bigcup\{f([s]) : s \in \mathbb{D}^S\}$ which implies that there is $W \in \tau^*(\mathbb{I}^\kappa)$ with $W \subset f([s_0])$ for some $s_0 \in \mathbb{D}^S$.

Making W smaller if necessary, we can consider that W is a standard open subset of $\mathbb{I}^\kappa$ and hence there is a finite $B \subset \kappa$ such that $\pi_B^{-1}(\pi_B(W)) = W$. An immediate consequence is that $W \cap P_\alpha^i \neq \emptyset$ and hence $[s_0] \cap Q_\alpha^i \neq \emptyset$ for any $\alpha \in \kappa \backslash B$ and $i \in \mathbb{D}$. Let $T = T' \backslash B$ and choose $h_\alpha^i \in p_{S_\alpha}([s_0] \cap Q_\alpha^i)$ for any $i \in \mathbb{D}$ and $\alpha \in T$.

Observe that, in both cases, $|T| = \kappa$ and $h_\alpha^0 \neq h_\alpha^1$ for any element $\alpha \in T$. Let $F = \{s \in \mathbb{D}^\kappa : s(\kappa \backslash T) \subset \{0\}$ and $p_{S_\alpha}(s) \in H_\alpha = \{h_\alpha^0, h_\alpha^1\}$ for any $\alpha \in T\}$. To see that F is homeomorphic to $\mathbb{D}^\kappa$, it suffices to prove that there is a homeomorphism $g : F \to H = \prod\{H_\alpha : \alpha \in T\}$ because $H \simeq \mathbb{D}^\kappa$.

For any $s \in F$ and $\alpha \in T$ we have $s|S_\alpha = h_\alpha^i$ for some $i \in \mathbb{D}$; let $g(s)(\alpha) = h_\alpha^i$. This gives us a map $g : F \to H$. If $h \in H$ then for every $\alpha \in T$ we have $h(\alpha) = h_\alpha^{i(\alpha)}$ for some $i(\alpha) \in \mathbb{D}$. Let $s(\alpha) = 0$ for any $\alpha \in \kappa \backslash T$; if $\alpha \in S_\alpha \backslash S$ then $s(\alpha) = h_\alpha^{i(\alpha)}(\alpha)$ and $s(\alpha) = s_0(\alpha)$ for any $\alpha \in S$. This definition makes sense because $s_0 \subset h_\alpha^i$ for any $\alpha \in T$ and $i \in \mathbb{D}$. It is evident that $s \in F$ and $g(s) = h$; so the map g is onto.

Now, if $s, t \in F$ and $s \neq t$ then $s|T \neq t|T$ which shows that for some $\alpha \in T$ we have $s|S_\alpha \neq t|S_\alpha$, i.e., $g(s)(\alpha) \neq g(t)(\alpha)$ and hence $g(s) \neq g(t)$ so g is a bijection. To see that g is continuous, it suffices to prove that so is $g_\alpha = q_\alpha \circ g : F \to H_\alpha$ where $q_\alpha : H \to H_\alpha$ is the natural projection for all $\alpha \in T$. Since $g_\alpha^{-1}(h_\alpha^i) = [h_\alpha^i] \cap F$, the set $g_\alpha^{-1}(h_\alpha^i)$ is open in F for any $i \in \mathbb{D}$ which, evidently, implies continuity of g_α for every $\alpha \in T$. Therefore g is a condensation of F onto H. It is an easy exercise that F is closed in $\mathbb{D}^\kappa$; so g is a condensation between compact spaces F and H which implies that g is a homeomorphism and hence $F \simeq \mathbb{D}^\kappa$.

Finally, assume that $s, t \in F$ and $s \neq t$. There is $\alpha \in T$ such that $s|S_\alpha = h_\alpha^i$ and $t|S_\alpha = h_\alpha^{1-i}$. Therefore $s \in [h_\alpha^i] \subset U_\alpha^i$ and $t \in [h_\alpha^{1-i}] \subset U_\alpha^{1-i}$. Consequently, $\varphi(s) \in \varphi(U_\alpha^i)$ and $\varphi(t) \in \varphi(U_\alpha^{1-i})$. Since $f(U_\alpha^i) \cap f(U_\alpha^{1-i}) \subset O_\alpha^i \cap O_\alpha^{1-i} = \emptyset$, we have $\varphi(s) \neq \varphi(t)$ and hence $\varphi|F$ is injective; so Fact 2 is proved.

Fact 3. If K is a dyadic space and $w(K) > \kappa$ for some infinite cardinal κ then $\mathbb{D}^{\kappa^+}$ embeds in K.

Proof. If the set $C = \{x \in K : \pi\chi(x, K) \leq \kappa\}$ is dense in K then $w(K) \leq \kappa$ by Fact 1 of U.072. Thus there exists $U \in \tau^*(X)$ such that $\pi\chi(x, K) \geq \kappa^+$ for any $x \in U$. Take $V \in \tau^*(X)$ such that $\overline{V} \subset U$. If $x \in \overline{V}$ and $\mathcal{U}$ is a π-base at the point x in $\overline{V}$ with $|\mathcal{U}| \leq \kappa$ then the family $\mathcal{U}' = \{W \cap V : W \in \mathcal{U}\}$ is a π-base at x in K with $|\mathcal{U}'| \leq \kappa$ which contradicts the choice of U. Therefore $\pi\chi(x, \overline{V}) \geq \kappa^+$ for any $x \in \overline{V}$; so we can apply Fact 1 to see that some closed subspace of $\overline{V}$ maps continuously onto $\mathbb{D}^{\kappa^+}$ and hence there is a continuous onto map $f : K \to \mathbb{I}^{\kappa^+}$.

Since K is dyadic, there is an infinite cardinal λ and a continuous onto map $g : \mathbb{D}^\lambda \to K$. For any $A \subset \lambda$ let $p_A : \mathbb{D}^\lambda \to \mathbb{D}^A$ be the natural projection of $\mathbb{D}^\lambda$ onto its face $\mathbb{D}^A$. Since $w(\mathbb{I}^{\kappa^+}) = \kappa^+$, we can apply Fact 1 of T.109 to find $A \subset \lambda$ such that $|A| \leq \kappa^+$ and there is a continuous map $h : \mathbb{D}^A \to \mathbb{I}^{\kappa^+}$ for which $h \circ p_A = f \circ g$. Since $\mathbb{I}^{\kappa^+}$ is a perfect image of $\mathbb{D}^A$, we have $\kappa^+ = w(\mathbb{I}^{\kappa^+}) \leq w(\mathbb{D}^A) = |A|$ (see Fact 1 of T.489) and therefore $|A| = \kappa^+$.

Let $u(\alpha) = 0$ for all $\alpha < \lambda$; then the set $P = \mathbb{D}^A \times \{p_{\lambda \backslash A}(u)\} \subset \mathbb{D}^\lambda$ is closed in $\mathbb{D}^\lambda$ and homeomorphic to $\mathbb{D}^A \simeq \mathbb{D}^{\kappa^+}$. Given $y \in \mathbb{I}^{\kappa^+}$ there is $x \in \mathbb{D}^\lambda$ such that $(f \circ g)(x) = y$ and therefore $h(p_A(x)) = f(g(x)) = y$. If $x' = p_A(x) \cup p_{\lambda \backslash A}(u)$ then $x' \in P$ and $p_A(x') = p_A(x)$ which shows that $f(g(x')) = h(p_A(x')) = h(p_A(x)) = y$. Therefore

(11) there exists $P \subset \mathbb{D}^{\lambda}$ such that $P \simeq \mathbb{D}^{\kappa^+}$ and $\varphi = (f \circ g)|P : P \to \mathbb{I}^{\kappa^+}$ is an onto map.

By (11) and Fact 2 we can find a closed $F \subset P$ such that $F \simeq \mathbb{D}^{\kappa^+}$ and $\varphi_0 = \varphi|F$ is injective. The space F being compact, the map $\varphi_0 : F \to G = \varphi_0(F)$ is a homeomorphism. Let $K' = g(F)$; if $g_0 = g|F$ and $f_0 = f|K'$ then $f_0 \circ g_0 = \varphi_0$ and hence g_0 is a homeomorphism by Fact 2 of S.337. Thus $K' \subset K$ is homeomorphic to $F \simeq \mathbb{D}^{\kappa^+}$ and Fact 3 is proved.

Fact 4. For any space Z if $K \subset C_p(Z)$ is a dyadic space then $w(K) \leq l(Z)$.

Proof. Assume for a contradiction that $w(K) > \kappa = l(Z)$. By Fact 3 there is $K' \subset K$ with $K' \simeq \mathbb{D}^{\kappa^+}$. It is easy to see that $w(\kappa^+ + 1) = \kappa^+$; besides, $\kappa^+ + 1$ is a scattered compact space by Fact 4 of U.074; so it is zero-dimensional (see SFFS-129 and SFFS-305). Therefore $\kappa^+ + 1$ embeds in $\mathbb{D}^{\kappa^+}$; so there is a closed $F \subset K'$ such that $F \simeq \kappa^+ + 1$ (see SFFS-303).

For any $z \in Z$ let $\varphi(z)(f) = f(z)$ for every $f \in F$. Then $\varphi : Z \to C_p(F)$ is a continuous map; let $Z' = \varphi(Z)$. It is easy to see that Z' separates the points of F and we have $l(Z') \leq l(Z) = \kappa$. However, every subspace of $C_p(\kappa^+ + 1)$ that separates the points of $\kappa^+ + 1$ must have the Lindelöf number $\geq \kappa^+$ by Problem 085. This contradiction shows that $w(K) \leq \kappa$, i.e., Fact 4 is proved.

Fact 5. Given a space Z and $P \subset C_p(Z)$ let $\varphi(z)(f) = f(z)$ for any $z \in Z$ and $f \in P$. Then $\varphi(z) \in C_p(P)$ for any $z \in Z$ and $\varphi : Z \to C_p(P)$ is a continuous map; let $Z' = \varphi(Z)$. If $\varphi^* : C_p(Z') \to C_p(Z)$ is the dual map, i.e., $\varphi^*(f) = f \circ \varphi$ for any $f \in C_p(Z')$ then $P \subset \varphi^*(C_p(Z'))$.

Proof. It was proved in TFS-166 that $\varphi(z) \in C_p(P)$ for any $z \in Z$ and φ is a continuous map; so we only have to establish that $P \subset \Phi = \varphi^*(C_p(Z'))$. To do so, take any $g \in P$; if $z' \in Z'$ then $z' = \varphi(z)$ for some $z \in Z$; let $f(z') = g(z)$. We have to show that this definition is consistent, i.e., for any $y \in Z$ if $\varphi(y) = z'$ then $g(y) = g(z)$.

Indeed, if $\varphi(y) = z' = \varphi(z)$ then $\varphi(y)(h) = \varphi(z)(h)$ for any $h \in P$ which implies that $h(y) = h(z)$ for any $h \in P$ and, in particular, $g(y) = g(z)$; so our function $f : Z' \to \mathbb{R}$ is well defined. Next observe that $f \in C_p(Z')$ because $f(z') = z'(g)$ for any $z' \in Z'$ (recall that $Z' \subset C_p(P)$); so f is continuous on Z' by TFS-166.

Finally, note that $\varphi^*(f)(z) = (f \circ \varphi)(z) = f(\varphi(z)) = g(z)$ for any $z \in Z$ by the definition of f; so $g = \varphi^*(f) \in \Phi$. We proved that $g \in \Phi$ for an arbitrary $g \in P$, i.e., $P \subset \Phi$ and hence Fact 5 is proved.

Returning to our solution observe that $l(Z) \leq nw(Z)$ for any space Z; so we have $l(Y) \leq nw(Y)$. To convince ourselves that $nw(Y) \leq \kappa = l(Y)$ let $\varphi(x)(f) = f(x)$ for any $x \in X$ and $f \in Y$; then $\varphi : X \to C_p(Y)$ is a continuous map (see TFS-166). The space $X' = \varphi(X) \subset C_p(Y)$ is dyadic; so $w(X') \leq l(Y) = \kappa$ by Fact 4.

The dual map $\varphi^* : C_p(X') \to C_p(X)$ defined by $\varphi^*(f) = f \circ \varphi$ for any function $f \in C_p(X')$ embeds $C_p(X')$ in $C_p(X)$ (see TFS-163) and $Y \subset \varphi^*(C_p(X'))$ by Fact 5. Therefore

$$nw(Y) \leq nw(\varphi^*(C_p(X'))) = nw(C_p(X')) = nw(X') \leq w(X') \leq \kappa$$

so our solution is complete.

U.087. *Prove that, if X is a dyadic space and $C_p(X)$ has a dense Lindelöf subspace then X is metrizable.*

Solution. If $Y \subset C_p(X)$ is dense and $l(Y) \leq \omega$ then $d(Y) \leq nw(Y) = l(Y) \leq \omega$ by Problem 086. If D is a countable dense subset of Y then D is also dense in $C_p(X)$; so $d(C_p(X)) = \omega$ and hence X can be condensed onto a second countable space by TFS-174. Every condensation of a compact space is a homeomorphism; so $w(X) \leq \omega$ and hence X is metrizable.

U.088. *Given a space X suppose that $K \subset C_p(X)$ is a compact space of uncountable tightness. Show that there exists a closed $X_1 \subset X$ such that $C_p(X_1)$ contains a compact subspace of weight and tightness ω_1.*

Solution. Given a function $f \in C_p(X)$, a finite set $F \subset X$ and a number $\varepsilon > 0$ let $O(f, F, \varepsilon) = \{g \in C_p(X) : |f(x) - g(x)| < \varepsilon$ for all $x \in F\}$. It is clear that the family $\{O(f, F, \varepsilon) : F$ is a finite subset of X and $\varepsilon > 0\}$ is a local base at the point f in $C_p(X)$.

Since $t(K) > \omega$, there is a free sequence $S = \{f_\alpha : \alpha < \omega_1\} \subset K$ (see TFS-328). The space K being compact, we can find a complete accumulation point f for the set S (this means that $|U \cap S| = \omega_1$ for any $U \in \tau(f, K)$, see TFS-118). If $S_\alpha = \{f_\beta : \beta < \alpha\}$ then $\overline{S}_\alpha \cap \overline{S \backslash S_\alpha} = \emptyset$; so $f \notin \overline{S}_\alpha$ for any $\alpha < \omega_1$. Thus, for any $\alpha < \omega_1$ there is a finite set $F_\alpha \subset X$ and $\varepsilon_\alpha > 0$ such that $O(f, K_\alpha, \varepsilon_\alpha) \cap S_\alpha = \emptyset$.

The set $Y = \bigcup\{K_\alpha : \alpha < \omega_1\}$ has cardinality at most ω_1; so $X_1 = \overline{Y}$ has density $\leq \omega_1$ and therefore $C_p(X_1)$ condenses onto a space of weight $\leq \omega_1$ (see TFS-173). If $\pi : C_p(X) \to C_p(X_1)$ is the restriction map then $K_1 = \pi(K)$ is a compact subspace of $C_p(X_1)$ which implies that $w(K_1) = iw(K_1) \leq iw(C_p(X_1)) \leq \omega_1$.

Let $g_\alpha = \pi(f_\alpha)$ for every $\alpha < \omega_1$; if $T = \{g_\alpha : \alpha < \omega_1\} = \pi(S)$ then, by continuity of π, the function $g = \pi(f)$ belongs to the closure of T. If $G \subset T$ is countable then there is $\alpha < \omega_1$ such that $G \subset T_\alpha = \{g_\beta : \beta < \alpha\} = \pi(S_\alpha)$. The map π is open by TFS-152; so $U_\alpha = \pi(O(f, K_\alpha, \varepsilon_\alpha))$ is open in $C_p(X_1)$. Since $K_\alpha \subset X_1$, we have $\pi^{-1}(U_\alpha) = O_\alpha = O(f, K_\alpha, \varepsilon_\alpha)$ and it follows from $O_\alpha \cap S_\alpha = \emptyset$ that $U_\alpha \cap \pi(S_\alpha) = \emptyset$. Therefore U_α is an open neighbourhood of g which does not meet T_α, i.e., $g \notin \overline{T}_\alpha$ and hence $g \notin \overline{G}$.

Thus $g \in \overline{T} \subset K_1$ is a function which is not in the closure of any countable $G \subset T$, i.e., $t(K_1) \geq \omega_1$ and therefore $K_1 \subset C_p(X_1)$ is a compact space such that $t(K_1) = w(K_1) = \omega_1$.

U.089. *Prove that PFA implies that, for any Lindelöf space X and any compact $K \subset C_p(X)$, we have $t(K) \leq \omega$.*

Solution. Recall that our use of PFA is restricted to applying its topological consequences. This time the relevant consequence says: "for any compact space K with $t(K) = w(K) = \omega_1$ the space $\omega_1 + 1$ embeds in K".

Fact 1. If κ is an uncountable regular cardinal and Z is a space such that $l(Z) < \kappa$ then $\kappa + 1$ cannot be embedded in $C_p(Z)$. In particular, $\omega_1 + 1$ cannot be embedded in $C_p(Z)$ for a Lindelöf space Z.

Proof. Assume that $F \subset C_p(Z)$, where $l(Z) < \kappa$ and $F \simeq \kappa + 1$. For any $z \in Z$ and $f \in F$ let $\varphi(z)(f) = f(z)$. Then $\varphi(z) \in C_p(F)$ for any point $z \in Z$ and the map $\varphi : Z \to C_p(F)$ is continuous (see TFS-166); let $Z' = \varphi(Z)$. We have the inequalities $l(Z') \leq l(Z) < \kappa$; since Z' separates the points of F, we can apply Problem 085 to conclude that $l(Z') \geq \kappa$ thus obtaining a contradiction which proves Fact 1.

Returning to our solution assume that $t(K) > \omega$ for some compact $K \subset C_p(X)$. By Problem 088, there exists a closed $X_1 \subset X$ such that $t(K_1) = w(K_1) = \omega_1$ for some compact $K_1 \subset C_p(X_1)$. By PFA, the space $\omega_1 + 1$ embeds in K_1 and hence $\omega_1 + 1$ embeds in $C_p(X_1)$ for a Lindelöf space X_1. This contradiction with Fact 1 shows that $t(K) \leq \omega$ and finishes our solution.

U.090. *Given a space X and a set $A \subset X$ denote by τ_A the topology on X generated by the family $\tau(X) \cup \exp(X \setminus A)$ as a subbase; let $X[A] = (X, \tau_A)$. In other words, the space $X[A]$ is constructed by declaring isolated all points of $X \setminus A$ and keeping the same topology at the points of A. Prove that, for any uncountable Polish space M and $A \subset M$ the following conditions are equivalent:*

(i) the space $(M[A])^\omega$ is Lindelöf;

(ii) if $\mathcal{F}$ is a countable family of finite-to-one continuous maps from the Cantor set $\mathbb{K}$ to M then $\bigcap\{f^{-1}(A) : f \in \mathcal{F}\} \neq \emptyset$;

(iii) if $\mathcal{F}$ is a countable family of injective continuous maps from the Cantor set $\mathbb{K}$ to M then $\bigcap\{f^{-1}(A) : f \in \mathcal{F}\} \neq \emptyset$.

Deduce from this fact that, for any uncountable Polish space M there is a disjoint family $\{A_\alpha : \alpha < \mathfrak{c}\}$ of subsets of M such that $(M[A_\alpha])^\omega$ is Lindelöf for any $\alpha < \mathfrak{c}$.

Solution. As usual, we let $\mathbb{D}^0 = \{\emptyset\}$ and $\mathbb{D}^{\leq n} = \bigcup\{\mathbb{D}^i : i \leq n\}$ for any $n < \omega$; furthermore, $\mathbb{D}^{<\omega} = \bigcup\{\mathbb{D}^i : i < \omega\}$. For any element $s \in \mathbb{D}^{<\omega}$ we denote by $[s]$ the set $\{t \in \mathbb{D}^\omega : s \subset t\}$. Note first that $Z[A]$ is a Tychonoff space for any Tychonoff Z by Fact 1 of S.293. If we have two topologies τ and μ on a set Z then τ is *stronger than* μ if $\tau \supset \mu$. If $\tau \subset \mu$ then τ is said to be *weaker than* μ.

Fact 1. If Z is second countable and $A \subset Z$ then $Z[A]$ embeds in $C_p(S)$ for some σ-compact space S.

Proof. If $Z \setminus A$ is countable then it is easy to see that $w(Z[A]) = \omega$ so $Z[A]$ is an Eberlein–Grothendieck space by Problem 034. Now assume that $Z \setminus A$ is uncountable; for any $z \in Z \setminus A$ let $\chi_z(y) = 0$ if $y \in Z \setminus \{z\}$ and $\chi_z(z) = 1$. It is evident that $\chi_z \in C_p(Z[A])$ for any $z \in Z\backslash A$. Let $u(z) = 0$ for every $z \in Z[A]$; then the set $K = \{\chi_z : z \in Z\backslash A\} \cup \{u\}$ is compact and u is the unique non-isolated point of K.

Fix a countable base $\mathcal{B}$ in the space Z and let $\mathcal{C} = \{(U, V) \in \mathcal{B} \times \mathcal{B} : \overline{U} \subset V\}$. By normality of Z, for every $q = (U, V) \in \mathcal{C}$ there is $f_q \in C_p(Z)$ for which $f_q(U) \subset \{1\}$ and $f_q(Z\backslash V) \subset \{0\}$. It is straightforward that the set $S = \{f_q : q \in \mathcal{C}\} \cup K$ separates the points and the closed subsets of $Z[A]$; so $Z[A]$ embeds in $C_p(S)$ (see TFS-166). It is evident that S is σ-compact; so Fact 1 is proved.

Returning to our solution assume, towards a contradiction, that $M_A = (M[A])^\omega$ is Lindelöf and there is a set $\mathcal{F} = \{f_n : n \in \omega\} \subset C(\mathbb{K}, M)$ such that every f_n is a finite-to-one map and $\bigcap\{f_n^{-1}(A) : n \in \omega\} = \emptyset$. Let $B_n = f_n^{-1}(A)$ for every $n \in \omega$.

It is straightforward that the diagonal $\Delta = \{s \in \mathbb{K}^\omega : s(i) = s(j)$ for any $i, j \in \omega\}$ is closed in the space $\mathbb{K}^\omega$. Observe that $\Delta = \{d_s : s \in \mathbb{K}\}$ where, for every $s \in \mathbb{K}$, the point $d_s \in \mathbb{K}^\omega$ is defined by $d_s(n) = s$ for any $n \in \omega$. Therefore $|\Delta| = \mathfrak{c}$. Let $f = \prod \mathcal{F} : \mathbb{K}^\omega \to M^\omega$ be the product of the maps of $\mathcal{F}$ (recall that f is defined by $f(s)(n) = f_n(s(n))$ for any $s \in \mathbb{K}^\omega$ and $n \in \omega$). The map f is continuous (see Fact 1 of S.271); so the set $D = f(\Delta)$ is compact and hence closed in M^ω. The topology of M^ω is weaker than the topology of M_A; so D is also closed in M_A. Let $\pi_n : M^\omega \to M$ be the natural projection of M onto its n-th factor. For any $n \in \omega$ we have $\pi_n(D) = f_n(\mathbb{K})$; since f_n is finite-to-one, this implies $|\pi_n(D)| = |f_n(\mathbb{K})| = \mathfrak{c}$, so $|D| \geq \mathfrak{c}$ and, in particular, D is uncountable. From now on we consider that D carries the topology of the subspace of M_A.

Take an arbitrary point $d \in D$; there is $s \in \mathbb{K}$ such that $d = f(d_s)$; since $\bigcap_{n\in\omega} B_n = \emptyset$, there is $n \in \omega$ such that $s \notin B_n$ and therefore $x = f_n(s) = \pi_n(d) \notin A$. This shows that x is an isolated point of $M[A]$; so $W = \pi_n^{-1}(x) \cap D$ is an open subset of D. The set $E = f_n^{-1}(x)$ is finite and it is obvious that $W \subset \{f(d_t) : t \in E\}$. This proves that every $d \in D$ has a finite open neighbourhood in D, i.e., every $d \in D$ is isolated in D. Therefore D is an uncountable closed discrete subset of M_A; this contradiction with the Lindelöf property of M_A shows that (i)$\Longrightarrow$(ii).

The implication (ii)$\Longrightarrow$(iii) being evident let us establish that (iii)$\Longrightarrow$(i). Assume that (iii) holds and M_A is not Lindelöf. There is a σ-compact space S such that $M[A]$ embeds in $C_p(S)$ by Fact 1. Consequently, M_A embeds in $(C_p(S))^\omega$ which is homeomorphic to $C_p(S \times \omega)$ (see TFS-114). The space $S \times \omega$ is also σ-compact and hence Lindelöf Σ so we can apply Baturov's theorem (SFFS-269) to conclude that $l(M_A) = ext(M_A)$; this makes it possible to find an uncountable closed discrete set $D \subset M_A$.

Observe that every countable subspace of $M[A]$ is second countable; we have $D \subset P = \prod_{n\in\omega} \pi_n(D)$, so if $\pi_n(D)$ is countable for any $n \in \omega$ then D is an uncountable closed discrete subspace of the second countable space P which is a contradiction. Thus we can assume, without loss of generality, that $\pi_0(D)$ is uncountable. Choose a complete metric ρ on the space M such that $\rho(x, y) \leq 1$ for any $x, y \in M$. Call a set $Z \subset M$ *uniformly uncountable* if $U \in \tau(M)$ and

$U \cap Z \neq \emptyset$ implies $|U \cap Z| > \omega$. Every uncountable subset of M contains a uniformly uncountable subset by Fact 1 of S.343. Using this fact it is easy to prove by an evident induction that

(*) if $m \in \omega$, $\varepsilon > 0$ and $P_0, \ldots, P_m$ are uncountable subsets of M then there are uniformly uncountable sets $P'_0, \ldots, P'_m$ such that $P'_i \subset P_i$, $\text{diam}_\rho(P'_i) < \varepsilon$ for each $i \leq m$ and $\text{cl}_M(P'_i) \cap \text{cl}_M(P'_j) = \emptyset$ whenever $i \neq j$.

It is easy to find an uncountable set $E_0 \subset D$ such that $\pi_0|E_0$ is an injection. Apply (*) to find an uncountable $E_1 \subset E_0$ for which the set $G = \pi_0(E_1)$ is uniformly uncountable. Applying (*) again take uniformly uncountable $P', Q' \subset G$ such that $\text{cl}_M(P') \cap \text{cl}_M(Q') = \emptyset$. The space $\text{cl}_M(Q')$ is Polish and uncountable; so there is a set $K_0 \subset \text{cl}_M(Q')$ with $K_0 \simeq \mathbb{K}$ (see SFFS-353). Take an uncountable set $D_\emptyset \subset E_1 \cap \pi_0^{-1}(P')$ and let $\xi_0(\emptyset) = 1$. It follows from the choice of P' and Q' that $\text{cl}_M(\pi_0(D_\emptyset)) \cap K_0 = \emptyset$. Assume that $n < \omega$ and we have defined, for any $i \leq n$, a map $\xi_i : \mathbb{D}^i \to \mathbb{D}$, a set $K_i \subset M$ and a family $\{D_s : s \in \mathbb{D}^i\}$ with the following properties:

(1) D_s is an uncountable subset of D for any $s \in \mathbb{D}^{\leq n}$;
(2) if $s, t \in \mathbb{D}^{\leq n}$ and $s \subset t$ then $D_t \subset D_s$;
(3) $K_i \simeq \mathbb{K}$ and $K_i \cap \text{cl}_M(\pi_i(D_s)) = \emptyset$ for any $i \leq n$ and $s \in \mathbb{D}^i$;
(4) if $i \leq n$, $s \in \mathbb{D}^i$ and $\xi_i(s) = 0$ then $|\pi_i(D_s)| = 1$;
(5) if $i \leq n$, $s \in \mathbb{D}^i$ and $\xi_i(s) = 1$ then $\pi_i|D_s$ is injective;
(6) if $j \leq n$ and $s \in \mathbb{D}^j$ then $\text{diam}_\rho(\pi_i(D_s)) \leq 2^{-j}$ for any $i \leq j$;
(7) if $j \leq n$ and $s \in \mathbb{D}^j$ then $\text{cl}_M(\pi_i(D_{s^\frown 0})) \cap \text{cl}_M(\pi_i(D_{s^\frown 1})) = \emptyset$ for any $i \leq j$ such that $\xi_i(s|i) = 1$;
(8) if $i \leq n$ and $\xi_i^{-1}(1) \neq \emptyset$ then the family $\{\text{cl}_M(\pi_i(D_s)) : s \in \xi_i^{-1}(1)\}$ is disjoint.

Call a collection $\{D_s : s \in \mathbb{D}^{n+1}\}$ *suitable* if there is a map $\xi_{n+1} : \mathbb{D}^{n+1} \to \mathbb{D}$ such that (1)–(2) and (4)–(7) are fulfilled for the family $\{D_s : s \in \mathbb{D}^{\leq n+1}\}$ and the maps $\{\xi_i : i \leq n + 1\}$. It is evident that if a collection $\{D_s : s \in \mathbb{D}^{n+1}\}$ is suitable and we choose an uncountable $D'_s \subset D_s$ for any $s \in \mathbb{D}^{n+1}$ then we obtain a suitable collection $\{D'_s : s \in \mathbb{D}^{n+1}\}$.

Let us construct first a suitable collection $\{D_s : s \in \mathbb{D}^{n+1}\}$. Fix any $s \in \mathbb{D}^n$; if $\pi_{n+1}(D_s)$ is countable then choose an uncountable $E_0 \subset D_s$ such that $\pi_{n+1}(E_0)$ is a singleton and let $\xi_{n+1}(s^\frown j) = 0$ for every $j \in \mathbb{D}$.

If $\xi_i(s|i) = 0$ for all $i \leq n$ then apply the property (*) successively to choose an uncountable $E'_0 \subset E_0$ such that $\text{diam}_\rho(\pi_i(E'_0)) \leq 2^{-n-1}$ for all $i \leq n$ and let $D_{s^\frown 0} = D_{s^\frown 1} = E'_0$.

If the set $I = \{i \leq n : \xi_i(s|i) = 1\}$ is non-empty then apply (*) again to successively go over all $i \leq n$ and over all elements of I to obtain uncountable sets $E'_0, E'_1 \subset E_0$ such that $\text{diam}_\rho(\pi_i(E'_j)) \leq 2^{-n-1}$ for all $i \leq n$ and $j \in \mathbb{D}$ while $\text{cl}_M(\pi_i(E'_0)) \cap \text{cl}_M(\pi_i(E'_1)) = \emptyset$ for all $i \in I$. Let $D_{s^\frown j} = E'_j$ for each $j \in \mathbb{D}$.

Now, if the set $\pi_{n+1}(D_s)$ is uncountable then it is easy to choose uncountable $E_0, E_1 \subset D_s$ such that $\text{cl}_M(\pi_{n+1}(E_0)) \cap \text{cl}_M(\pi_{n+1}(E_1)) = \emptyset$ and the mapping $\pi_{n+1}|E_i$ is an injection for every $i \in \mathbb{D}$. Let $\xi_{n+1}(s^\frown j) = 1$ for every $j \in \mathbb{D}$.

If $\xi_i(s|i) = 0$ for all $i \leq n$ then apply $(*)$ successively to choose, for any $j \in \mathbb{D}$, an uncountable $E'_j \subset E_j$ such that $\mathrm{diam}_\rho(\pi_i(E'_j)) \leq 2^{-n-1}$ for all $i \leq n+1$ and let $D_{s^\frown j} = E'_j$.

If the set $I = \{i \leq n : \xi_i(s|i) = 1\}$ is non-empty then apply $(*)$ again to successively go over all $i \leq n+1$ and over all elements of I to obtain, for any $j \in \mathbb{D}$, an uncountable set $E'_j \subset E_j$ such that $\mathrm{diam}_\rho(\pi_i(E'_j)) \leq 2^{-n-1}$ for all $i \leq n+1$ while $\mathrm{cl}_M(\pi_i(E'_0)) \cap \mathrm{cl}_M(\pi_i(E'_1)) = \emptyset$ for all $i \in I$. Again, let $D_{s^\frown j} = E'_j$ for every $j \in \mathbb{D}$.

It is straightforward that, after we construct the sets $D_{s^\frown 0}$ and $D_{s^\frown 1}$ for all $s \in \mathbb{D}^n$ we obtain a suitable family $\{D_s : s \in \mathbb{D}^{n+1}\}$ and the relevant map ξ_{n+1}; let $Q = \pi_{n+1}(\bigcup\{D_s : s \in \mathbb{D}^{n+1}\})$.

If $\xi_{n+1}(s) = 0$ for every $s \in \mathbb{D}^{n+1}$ then it follows from (4) that the set Q is finite; so $M \backslash Q$ is an uncountable Polish space and therefore there is $K_{n+1} \subset M \backslash Q$ with $K_{n+1} \simeq \mathbb{K}$. It is clear that K_{n+1} witnesses the property (3).

Now, if the mapping ξ_{n+1} is not identically zero on $\mathbb{D}^{n+1}$ then let $\{s_1, \ldots, s_k\}$ be a faithful enumeration of the set $\xi_{n+1}^{-1}(1)$. Apply $(*)$ to find an uncountable $D'_{s_1} \subset D_{s_1}$ and $C_1 \subset \pi_{n+1}(D_{s_1})$ such that $\mathrm{cl}_M(\pi_{n+1}(D'_{s_1})) \cap \mathrm{cl}_M(C_1) = \emptyset$. Proceeding by an evident induction we can construct uncountable sets $C_1 \supset \ldots \supset C_k$ and uncountable $D'_{s_i} \subset D_{s_i}$ for all $i \leq k$ in such a way that $\mathrm{cl}_M(\pi_{n+1}(D'_{s_i})) \cap \mathrm{cl}_M(C_i) = \emptyset$ for all $i \leq k$. As a consequence, $\mathrm{cl}_M(\pi_{n+1}(D'_{s_i})) \cap \mathrm{cl}_M(C_k) = \emptyset$ for all $i \leq k$. Since $\pi_{n+1}(D_s)$ is a singleton for any $s \in \xi_{n+1}^{-1}(0)$, the set $Z = \pi_{n+1}(\bigcup\{D_s : s \in \xi_{n+1}^{-1}(0)\})$ is finite which shows that $\mathrm{cl}_M(C_k) \backslash Q = \mathrm{cl}_M(C_k) \backslash Z$ is an uncountable Polish space.

Apply SFFS-353 once more to find a set $K_{n+1} \subset \mathrm{cl}_M(C_k) \backslash Q$ such that $K_{n+1} \simeq \mathbb{K}$. It is immediate that, for the function ξ_{n+1}, the set K_{n+1} and the family $\mathcal{D}' = \{D'_{s_i} : i \leq k\} \cup \{D_s : s \in \xi_{n+1}^{-1}(0)\}$ all properties (1)–(7) are fulfilled. An evident application of $(*)$ allows to shrink the elements of $\mathcal{D}'$ to get the property (8) fulfilled as well.

Therefore our inductive procedure can be continued to construct sequences $\{\xi_n : n \in \omega\}$, $\{K_n : n \in \omega\}$ and a family $\{D_s : s \in \mathbb{D}^{<\omega}\}$ such that (1)–(8) are satisfied for all $n < \omega$.

For any $s \in \mathbb{D}^\omega$ let $Y_s(n) = \bigcap\{\mathrm{cl}_M(\pi_n(D_{s|i})) : i \in \omega\}$ for any $n \in \omega$; the properties (2) and (6), together with completeness of M imply that $Y_s(n) \neq \emptyset$ and $|Y_s(n)| = 1$, i.e., $Y_s(n)$ is a singleton. For any $s \in \mathbb{D}^\omega$ and $n \in \omega$ take $y_s(n) \in M$ such that $Y_s(n) = \{y_s(n)\}$. This defines a point $y_s \in M^\omega$ for every $s \in \mathbb{D}^\omega$.

Let $J = \{i \in \omega : \xi_i^{-1}(1) \neq \emptyset\}$ and observe that $J \neq \emptyset$ because $0 \in J$. Given $i \in J$ and $s \in W_i = \xi_i^{-1}(1)$ let $f_{is}(t) = y_t(i)$ for any $t \in \mathbb{D}^\omega$ with $t \supset s$. It follows from the property (7) that $f_{is} : [s] \to M$ is an injective continuous map; let $C_i = \bigcup\{[s] : s \in W_i\}$. To define a map $\tilde{f}_i : C_i \to M$ take any $t \in C_i$. There is a unique $s \in W_i$ such that $t \in [s]$; let $\tilde{f}_i(t) = f_{is}(t)$. The property (8) guarantees that $\tilde{f}_i$ is still injective; continuity of $\tilde{f}_i$ follows from continuity of f_{is} for any $s \in W_i$ and the fact that the family $\{[s] : s \in W_i\}$ is clopen and disjoint.

Since the set C_i is clopen in $\mathbb{K}$, the set $\mathbb{K} \backslash C_i$ is also clopen in $\mathbb{K}$ (possibly empty); let $g_i : \mathbb{K} \backslash C_i \to K_i$ be an embedding (which exists by (3)). The property (3) also shows that $\tilde{f}_i(C_i) \cap K_i = \emptyset$ and therefore the map $f_i = \tilde{f}_i \cup g_i : \mathbb{K} \to M$ is injective.

Apply (iii) to find a point $s \in \bigcap \{f_i^{-1}(A) : i \in J\}$; let $s_i = s|i$ for any $i \in \omega$. We claim that y_s is an accumulation point of D in the space M_A. To prove it take any $U \in \tau(y_s, M_A)$; we can find $n \in \omega$ and $O_0, \ldots, O_n \in \tau(M[A])$ such that $y_s(i) \in O_i$ for any $i \leq n$ and $O = \bigcap \{\pi_i^{-1}(O_i) : i \leq n\} \subset U$. We can consider that $O_i \in \tau(M)$ whenever $y_s(i) \in A$; so there is $k \in \omega$ such that, for every $i \leq n$, if $O_i \in \tau(y_s(i), M)$ then $B_i = \{x \in M : \rho(x, y_s(i)) < 2^{-k}\} \subset O_i$. Let $m = \max\{k, n\}$ and take any $i \leq n$. There are two possible cases:

a) $\xi_i(s_i) = 1$. Then $s_i \in W_i$ which shows that $s \in f_i^{-1}(A) \cap [s_i] = f_{is_i}^{-1}(A)$ and therefore $f_{is_i}(s) = y_s(i) \in A$. As a consequence, $O_i \in \tau(y_s(i), M)$ and hence $B_i \subset O_i$. Furthermore, $y_s(i) \in \mathrm{cl}_M(\pi_i(D_{s_m}))$ and $\mathrm{diam}(\pi_i(D_{s_m})) < 2^{-m} \leq 2^{-k}$ by (6). This implies $\pi_i(D_{s_m}) \subset B_i \subset O_i$.

b) $\xi_i(s_i) = 0$. Then $|\pi_i(D_{s_m})| = 1$ and hence $\pi_i(D_{s_m}) = \{y_s(i)\} \subset O_i$.

We proved that $\pi_i(D_{s_m}) \subset O_i$ for all $i \leq n$ and hence $D_{s_m} \subset O \subset U$. This shows that $|U \cap D| \geq |D_{s_m}| > \omega$, i.e., every neighbourhood U of the point y_s in M_A contains uncountably many points of D which is a contradiction with the set D being closed and discrete in M_A. Thus M_A is Lindelöf and we finally established that (i) $\Longleftrightarrow$ (ii) $\Longleftrightarrow$ (iii).

To construct the promised disjoint sets A_α observe that the collection $\mathcal{C}$ of all countable families of continuous injective maps from $\mathbb{K}$ to M has cardinality $\mathfrak{c}$; so we can choose an enumeration $\{\Psi_\beta : \beta < \mathfrak{c}\}$ of the family $\mathcal{C}$ such that every $\Psi \in \mathcal{C}$ occurs $\mathfrak{c}$-many times in this enumeration, i.e., $|\{\beta < \mathfrak{c} : \Psi = \Psi_\beta\}| = \mathfrak{c}$ for any $\Psi \in \mathcal{C}$.

Take an arbitrary point $x_0^0 \in \mathbb{K}$ and let $A_0^0 = \{\psi(x_0^0) : \psi \in \Psi_0^0\}$. Assume that $\beta < \mathfrak{c}$ and we have a set $K_\gamma = \{x_\alpha^\gamma : \alpha \leq \gamma\} \subset \mathbb{K}$ for any $\gamma < \beta$ with the following property

(9) if $A_\alpha^\gamma = \{\psi(x_\alpha^\gamma) : \psi \in \Psi_\gamma\}$ whenever $\alpha \leq \gamma < \beta$ and $\mathcal{A}_\gamma = \{A_\alpha^\gamma : \alpha \leq \gamma\}$ for every $\gamma < \beta$ then the family $\mathcal{A} = \bigcup\{\mathcal{A}_\gamma : \gamma < \beta\}$ is disjoint.

Since A_α^γ is a countable set for every $\gamma < \beta$ and $\alpha \leq \gamma$, we have $|\mathcal{A}| \leq |\beta| \cdot \omega < \mathfrak{c}$; so the set $P = \bigcup\{\psi^{-1}(A) : \psi \in \Psi_\beta, A \in \mathcal{A}\}$ has cardinality $< \mathfrak{c}$ because Ψ_β is countable and every $\psi \in \Psi_\beta$ is an injection. This makes it possible to pick a point $x_0^\beta \in \mathbb{K} \backslash P$. Now, suppose that $\gamma \leq \beta$ and we have a set $\{x_\alpha^\beta : \alpha < \gamma\}$ such that the family $\mathcal{B} = \mathcal{A} \cup \{\{\psi(x_\alpha^\beta) : \psi \in \Psi_\beta\} : \alpha < \gamma\}$ is disjoint. All elements of $\mathcal{B}$ are countable and $|\mathcal{B}| < \mathfrak{c}$; so the set $E = \bigcup\{\psi^{-1}(B) : B \in \mathcal{B}, \psi \in \Psi_\beta\}$ has cardinality $< \mathfrak{c}$ because every $\psi \in \Psi_\beta$ is an injection; take any $x_\gamma^\beta \in \mathbb{K} \backslash E$. It is evident that we can continue this inductive procedure to construct the set $K_\beta = \{x_\alpha^\beta : \alpha \leq \beta\}$ such that (9) still holds for all $\gamma \leq \beta$.

Therefore we can construct the set $K_\beta = \{x_\alpha^\beta : \alpha \leq \beta\}$ for any $\beta < \mathfrak{c}$ so that (9) is fulfilled for all $\gamma < \mathfrak{c}$; let $A_\alpha^\beta = \{\psi(x_\alpha^\beta) : \psi \in \Psi_\beta\}$ for any $\beta < \mathfrak{c}$ and $\alpha \leq \beta$. An immediate consequence of (9) is that the family $\{A_\alpha^\beta : \beta < \mathfrak{c}, \alpha \leq \beta\}$ is disjoint; so if $A_\alpha = \bigcup\{A_\alpha^\beta : \alpha \leq \beta < \mathfrak{c}\}$ for all $\alpha < \mathfrak{c}$ then the family $\{A_\alpha : \alpha < \mathfrak{c}\}$ is also disjoint.

To see that the collection $\{A_\alpha : \alpha < \mathfrak{c}\}$ is as promised fix any $\alpha < \mathfrak{c}$ and take a countable family Ψ of continuous injective maps from $\mathbb{K}$ to M. There is an ordinal $\beta \geq \alpha$ such that $\Psi = \Psi_\beta$. Therefore $\psi(x_\alpha^\beta) \in A_\alpha^\beta \subset A_\alpha$ for any $\psi \in \Psi$ and hence $\bigcap\{\psi^{-1}(A_\alpha) : \psi \in \Psi\} \supset \bigcap\{\psi^{-1}(A_\alpha^\beta) : \psi \in \Psi_\beta\} \ni x_\alpha^\beta$ which shows that $\bigcap\{\psi^{-1}(A_\alpha) : \psi \in \Psi\} \neq \emptyset$ for any $\Psi \in C$ and hence $(M[A_\alpha])^\omega$ is Lindelöf by (i) $\Longleftrightarrow$ (iii). Since $\alpha < \mathfrak{c}$ was chosen arbitrarily, we proved that $(M[A_\alpha])^\omega$ is Lindelöf for any $\alpha < \mathfrak{c}$ and hence our solution is complete.

U.091. *Given a space X and a set $A \subset X$ denote by τ_A the topology on X generated by the family $\tau(X) \cup \exp(X \setminus A)$ as a subbase; let $X[A] = (X, \tau_A)$. Prove that, if M is a Polish space, $A \subset M$ and $n \in \mathbb{N}$ then the following conditions are equivalent:*

(i) *the space $(M[A])^n$ is Lindelöf;*

(ii) *if $\mathcal{F}$ is a family of finite-to-one continuous maps from the Cantor set $\mathbb{K}$ to M and $|\mathcal{F}| \leq n$ then $\bigcap\{f^{-1}(A) : f \in \mathcal{F}\} \neq \emptyset$;*

(iii) *if $\mathcal{F}$ is a family of injective continuous maps from the Cantor set $\mathbb{K}$ to M and $|\mathcal{F}| \leq n$ then $\bigcap\{f^{-1}(A) : f \in \mathcal{F}\} \neq \emptyset$.*

Deduce from this fact that, for any uncountable Polish space M there is a disjoint family $\{A_\alpha : \alpha < \mathfrak{c}\}$ of subsets of M such that for every $\alpha < \mathfrak{c}$ the space $(M[A_\alpha])^k$ is Lindelöf for any $k \in \mathbb{N}$ while $(M[A_\alpha])^\omega$ is not Lindelöf.

Solution. As usual, we let $\mathbb{D}^0 = \{\emptyset\}$ and $\mathbb{D}^{\leq k} = \bigcup\{\mathbb{D}^i : i \leq k\}$ for any $k < \omega$; furthermore, $\mathbb{D}^{<\omega} = \bigcup\{\mathbb{D}^i : i < \omega\}$. Observe also that $Z[A]$ is a Tychonoff space for any Tychonoff Z by Fact 1 of S.293.

Assume, towards a contradiction, that $M_A = (M[A])^n$ is Lindelöf and there is a set $\mathcal{F} = \{f_0, \ldots, f_{n-1}\} \subset C(\mathbb{K}, M)$ such that every f_i is a finite-to-one map and $\bigcap\{f_i^{-1}(A) : i < n\} = \emptyset$. Let $B_i = f_i^{-1}(A)$ for every $i < n$. It is straightforward that the diagonal $\Delta = \{s \in \mathbb{K}^n : s(i) = s(j)$ for any $i, j < n\}$ is closed in $\mathbb{K}^n$. Observe that $\Delta = \{d_s : s \in \mathbb{K}\}$ where, for every $s \in \mathbb{K}$, the point $d_s \in \mathbb{K}^n$ is defined by $d_s(i) = s$ for any $i < n$. Therefore $|\Delta| = \mathfrak{c}$. Let $f = \prod \mathcal{F} : \mathbb{K}^n \to M^n$ be the product of the maps of $\mathcal{F}$ (recall that f is defined by $f(s)(i) = f_i(s(i))$ for any $s \in \mathbb{K}^n$ and $i < n$). The map f is continuous (see Fact 1 of S.271); so the set $D = f(\Delta)$ is compact and hence closed in M^n. The topology of M^n is weaker than the topology of M_A; so D is also closed in M_A. For any $i < n$ let $\pi_i : M^n \to M$ be the natural projection of M onto its i-th factor; then $\pi_i(D) = f_i(\mathbb{K})$; since f_i is finite-to-one, this implies $|\pi_i(D)| = |f_i(\mathbb{K})| = \mathfrak{c}$; so $|D| \geq \mathfrak{c}$ and, in particular, D is uncountable. From now on we consider that D carries the topology of the subspace of M_A.

Take an arbitrary point $d \in D$; there is $s \in \mathbb{K}$ such that $d = f(d_s)$; since $\bigcap_{i<n} B_i = \emptyset$, there is $i < n$ such that $s \notin B_i$ and therefore $x = f_i(s) = \pi_i(d) \notin A$. This shows that x is an isolated point of $M[A]$; so $W = \pi_i^{-1}(x) \cap D$ is an open subset of D. The set $E = f_i^{-1}(x)$ is finite and it is obvious that $W \subset \{f(d_t) : t \in E\}$. This proves that every $d \in D$ has a finite open neighbourhood in D, i.e., every $d \in D$ is isolated in D. Therefore D is an uncountable closed discrete subset of M_A; this contradiction with the Lindelöf property of M_A shows that (i)$\Longrightarrow$(ii).

The implication (ii)$\Longrightarrow$(iii) being evident let us establish that (iii)$\Longrightarrow$(i). Assume that (iii) holds and M_A is not Lindelöf. There is a σ-compact space S such that $M[A]$ embeds in $C_p(S)$ by Fact 1 of U.090. Consequently, M_A embeds in $(C_p(S))^n$ which is homeomorphic to $C_p(S \times n)$ (see TFS-114). The space $S \times n$ is also σ-compact and hence Lindelöf Σ so we can apply Baturov's theorem (SFFS-269) to conclude that $l(M_A) = ext(M_A)$; this makes it possible to find an uncountable closed discrete set $D \subset M_A$. Therefore it suffices to prove by induction on n that, if (iii) holds then $ext(M_A) \le \omega$.

If $n = 1$ and D is an uncountable closed discrete subset of $M_A = M[A]$ then $D \cap A$ is countable because $w(A) = \omega$; so we can consider that $D \subset M \setminus A$. Therefore $U = M \setminus D \in \tau(M[A])$ and $A \subset U$. This implies that there is $V \in \tau(M)$ such that $A \subset V \subset U$ and therefore $cl_M(D) \cap A = \emptyset$. The space $H = cl_M(D)$ is Polish and uncountable; so there is an injective map $f : \mathbb{K} \to M$ such that $f(\mathbb{K}) \subset H$ (see SFFS-353). Since $f^{-1}(A) = \emptyset$, we obtained a contradiction which proves (iii)$\Longrightarrow$(i) for $n = 1$.

Now assume that (iii)$\Longrightarrow$(i) is proved for any $n < m$, the property (iii) holds while $M_A = (M[A])^m$ is not Lindelöf and hence there is an uncountable closed discrete $D \subset M_A$. Note first that the property (iii) also holds for all $n < m$ and hence $(M[A])^n$ is Lindelöf for any $n < m$. This implies that

(1) for any uncountable $D' \subset D$ the set $\pi_i(D')$ is uncountable for any $i < m$,

for otherwise there is $x \in M$ and $i < m$ such that $\pi_i^{-1}(x) \cap D$ is uncountable; since $\pi_i^{-1}(x) \simeq (M[A])^{m-1}$, the extent of $(M[A])^{m-1}$ is uncountable which is a contradiction.

Now it is easy to choose, using (1), a sequence $D_0 \supset \ldots \supset D_{m-1}$ of uncountable subsets of D such that $\pi_i | D_i$ is injective for any $i < m$. It is clear that $\pi_i | D_{m-1}$ is injective for all $i < m$; so we can assume, without loss of generality, that $D = D_{m-1}$, i.e., $\pi_i | D$ is injective for any $i < m$.

Choose a complete metric ρ on the space M such that $\rho(x, y) \le 1$ for any points $x, y \in M$. Call a set $Z \subset M$ *uniformly uncountable* if $U \in \tau(M)$ and $U \cap Z \ne \emptyset$ implies $|U \cap Z| > \omega$. Every uncountable subset of M contains a uniformly uncountable subset by Fact 1 of S.343. Using this fact it is easy to prove by an evident induction that

(*) if $l \in \omega$, $\varepsilon > 0$ and $P_0, \ldots, P_l$ are uncountable subsets of M then there are uniformly uncountable sets $P'_0, \ldots, P'_l$ such that $P'_i \subset P_i$, $\text{diam}_\rho(P'_i) < \varepsilon$ for each $i \le l$ and $cl_M(P'_i) \cap cl_M(P'_j) = \emptyset$ whenever $i \ne j$.

Let $D_\emptyset = D$ and assume that $k < \omega$ and we have defined, for any $i \le k$, a family $\{D_s : s \in \mathbb{D}^i\}$ with the following properties:

(2) D_s is an uncountable subset of D for any $s \in \mathbb{D}^{\le k}$;
(3) if $s, t \in \mathbb{D}^{\le k}$ and $s \subset t$ then $D_t \subset D_s$;
(4) if $j \le k$ and $s \in \mathbb{D}^j$ then $\text{diam}_\rho(\pi_i(D_s)) \le 2^{-j}$ for any $i < m$;
(5) the family $\{cl_M(\pi_i(D_s)) : s \in \mathbb{D}^j\}$ is disjoint for any $i < m$ and $j \le k$.

Fix any $s \in \mathbb{D}^k$ and apply the property $(*)$ to successively go over all $i < m$ to obtain uncountable sets $E_0, E_1 \subset D_s$ such that $\mathrm{cl}_M(\pi_i(E_0)) \cap \mathrm{cl}_M(\pi_i(E_1)) = \emptyset$ while $\mathrm{diam}_\rho(\pi_i(E_j)) \le 2^{-k-1}$ for all $i < m$ and $j \in \mathbb{D}$. Let $D_{s\frown j} = E_j$ for each $j \in \mathbb{D}$.

After we construct the sets $D_{s\frown 0}$ and $D_{s\frown 1}$ for all $s \in \mathbb{D}^k$, we obtain a family $\{D_s : s \in \mathbb{D}^{k+1}\}$ with the properties (2)–(5) and hence our inductive procedure can be continued to construct a family $\{D_s : s \in \mathbb{D}^{<\omega}\}$ such that (2)–(5) are satisfied for all $k < \omega$.

For any $s \in \mathbb{D}^\omega$ let $Y_s(j) = \bigcap\{\mathrm{cl}_M(\pi_j(D_{s|i})) : i \in \omega\}$ for any $j < m$; the properties (3) and (4), together with completeness of M imply that $Y_s(j) \ne \emptyset$ and $|Y_s(j)| = 1$, i.e., $Y_s(j)$ is a singleton. For any $s \in \mathbb{D}^\omega$ and $j < m$ take $y_s(j) \in M$ such that $Y_s(j) = \{y_s(j)\}$. This defines a point $y_s \in M^m$ for every $s \in \mathbb{D}^\omega$. Let $f_i(s) = y_s(i)$ for any $s \in \mathbb{K}$ and $i < m$. It follows from (4) and (5) that $f_i : \mathbb{K} \to M$ is an injective continuous map for any $i < m$.

Apply (iii) to find a point $s \in \bigcap\{f_j^{-1}(A) : j < m\}$; let $s_i = s|i$ for any $i \in \omega$. We claim that y_s is an accumulation point of the set D in the space M_A. To prove it take any $U \in \tau(y_s, M_A)$; we can find $O_0, \ldots, O_{m-1} \in \tau(M[A])$ such that $y_s(j) \in O_j$ for any $j < m$ and $O = \prod\{O_i : i < m\} \subset U$. Besides, $y_s(j) = f_j(s) \in A$ for any $j < m$ and therefore the set O_j is a neighbourhood of the point $y_s(j)$ in the space M. Consequently, we can choose a number $k \in \omega$ such that, for every $i < m$ we have $B_i = \{x \in M : \rho(x, y_s(i)) < 2^{-k}\} \subset O_i$.

Fix any $i < m$; it follows from $y_s(i) \in \mathrm{cl}_M(\pi_i(D_{s_k}))$ and $\mathrm{diam}(\pi_i(D_{s_k})) \le 2^{-k}$ (see (4)) that $\pi_i(D_{s_k}) \subset B_i \subset O_i$ and therefore $\pi_i(D_{s_k}) \subset O_i$ for all $i < m$ which shows that $D_{s_k} \subset O \subset U$. Thus $|U \cap D| \ge |D_{s_k}| > \omega$, i.e., every neighbourhood U of the point y_s in M_A contains uncountably many points of D which is a contradiction with the set D being closed and discrete in M_A. Thus M_A is Lindelöf and we finally established that (i) $\Longleftrightarrow$ (ii) $\Longleftrightarrow$ (iii).

To construct the promised disjoint sets A_α observe that the collection $\mathcal{C}$ of all finite families of continuous injective maps from $\mathbb{K}$ to M has cardinality $\mathfrak{c}$; so we can choose an enumeration $\{\Psi_\beta : \beta < \mathfrak{c}\}$ of the family $\mathcal{C}$ such that every $\Psi \in \mathcal{C}$ occurs $\mathfrak{c}$-many times in this enumeration, i.e., $|\{\beta < \mathfrak{c} : \Psi = \Psi_\beta\}| = \mathfrak{c}$ for any $\Psi \in \mathcal{C}$.

Fix a sequence $F = \{f_n : n \in \omega\}$ of injective continuous maps of $\mathbb{K}$ to M such that the family $\{f_n(\mathbb{K}) : n \in \omega\}$ is disjoint (such a family exists because M contains a Cantor set $\mathbb{K}$ by SFFS-353 and $\mathbb{K} \simeq \mathbb{K} \times \mathbb{K}$). For any sets $\Phi \subset \mathcal{C}$, $P \subset \mathbb{K}$ and $Q \subset M$ let $\Phi(P) = \bigcup\{\varphi(P) : \varphi \in \Phi\}$ and $\Phi^{-1}(Q) = \bigcup\{\varphi^{-1}(Q) : \varphi \in \Phi\}$. Let $\prec$ be the lexicographic order on $\mathfrak{c} \times \mathfrak{c}$, i.e., $(\alpha, \beta) \prec (\alpha', \beta')$ if either $\alpha < \alpha'$ or $\alpha = \alpha'$ and $\beta < \beta'$.

Take an arbitrary point $x_0^0 \in \mathbb{K}$ and let $A_0^0 = \{\psi(x_0^0) : \psi \in \Psi_0^0\}$. Assume that $\beta < \mathfrak{c}$ and we have a set $K_\gamma = \{x_\alpha^\gamma : \alpha \le \gamma\} \subset \mathbb{K}$ for any $\gamma < \beta$ with the following properties

(6) if $A_\alpha^\gamma = \{\psi(x_\alpha^\gamma) : \psi \in \Psi_\gamma\}$ whenever $\alpha \le \gamma < \beta$ and $\mathcal{A}_\gamma = \{A_\alpha^\gamma : \alpha \le \gamma\}$ for every $\gamma < \beta$ then the family $\mathcal{A} = \bigcup\{\mathcal{A}_\gamma : \gamma < \beta\}$ is disjoint.

(7) if $\alpha \le \gamma < \beta$ then $x_\alpha^\gamma \notin \Psi_\gamma^{-1}(F(F^{-1}(A_{\alpha'}^{\gamma'})))$ for any $(\gamma', \alpha') \prec (\gamma, \alpha)$.

Since A_α^γ is finite for every $\gamma < \beta$ and $\alpha \leq \gamma$, we have $|A| \leq |\beta| \cdot \omega < \mathfrak{c}$; so both sets $P = \bigcup\{\psi^{-1}(A) : \psi \in \Psi_\beta, A \in \mathcal{A}\}$ and $P' = \bigcup\{\Psi_\beta^{-1}(F(F^{-1}(A_\alpha^\gamma))) : \alpha \leq \gamma < \beta\}$ have cardinality $< \mathfrak{c}$ because the set F is countable and Ψ_β is finite while every $\psi \in \Psi_\beta$ is an injection. This makes it possible to pick a point $x_0^\beta \in \mathbb{K}\backslash(P \cup P')$. Now, suppose that $\gamma \leq \beta$ and we have a set $\{x_\alpha^\beta : \alpha < \gamma\}$ such that the family $\mathcal{B} = \mathcal{A} \cup \{\{\psi(x_\alpha^\beta) : \psi \in \Psi_\beta\} : \alpha < \gamma\}$ is disjoint and $x_\alpha^\beta \notin \Psi_\beta^{-1}(F(F^{-1}(A_{\alpha'}^\beta))) \cup (P \cup P')$ for any $\alpha' < \alpha$.

All elements of the family $\mathcal{B}$ are finite and $|\mathcal{B}| < \mathfrak{c}$ so the set $E = \bigcup\{\psi^{-1}(B) : B \in \mathcal{B}, \psi \in \Psi_\beta\}$ has cardinality $< \mathfrak{c}$ because every $\psi \in \Psi_\beta$ is an injection. The same is true for the set $E' = \bigcup\{\Psi_\beta^{-1}(F(F^{-1}(A_{\alpha'}^\beta))) : \alpha' < \alpha\}$; so we can take a point $x_\gamma^\beta \in \mathbb{K}\backslash(E \cup E')$. It is evident that we can continue this inductive procedure to construct the set $K_\beta = \{x_\alpha^\beta : \alpha \leq \beta\}$ such that (6) and (7) still hold for all $\gamma \leq \beta$.

Therefore we can construct the set $K_\beta = \{x_\alpha^\beta : \alpha \leq \beta\}$ for any $\beta < \mathfrak{c}$; so that (6)–(7) are fulfilled for all $\gamma < \mathfrak{c}$; let $A_\alpha^\beta = \{\psi(x_\alpha^\beta) : \psi \in \Psi_\beta\}$ for any $\beta < \mathfrak{c}$ and $\alpha \leq \beta$. An immediate consequence of (6) is that the family $\{A_\alpha^\beta : \beta < \mathfrak{c}, \alpha \leq \beta\}$ is disjoint; so if $A_\alpha = \bigcup\{A_\alpha^\beta : \alpha \leq \beta < \mathfrak{c}\}$ for all $\alpha < \mathfrak{c}$ then the family $\{A_\alpha : \alpha < \mathfrak{c}\}$ is also disjoint.

To see that the collection $\{A_\alpha : \alpha < \mathfrak{c}\}$ is as promised fix $\alpha < \mathfrak{c}$ and take any $k \in \mathbb{N}$. Let Ψ be family of continuous injective maps from $\mathbb{K}$ to M with $|\Psi| \leq k$. There is an ordinal $\beta \geq \alpha$ such that $\Psi = \Psi_\beta$. Therefore $\psi(x_\alpha^\beta) \in A_\alpha^\beta \subset A_\alpha$ for any $\psi \in \Psi$ and hence $\bigcap\{\psi^{-1}(A_\alpha) : \psi \in \Psi\} \supset \bigcap\{\psi^{-1}(A_\alpha^\beta) : \psi \in \Psi_\beta\} \ni x_\alpha^\beta$ which shows that $\bigcap\{\psi^{-1}(A_\alpha) : \psi \in \Psi\} \neq \emptyset$ and hence $(M[A_\alpha])^k$ is Lindelöf by (i) $\Longleftrightarrow$ (iii). Since $\alpha < \mathfrak{c}$ and $k \in \omega$ were chosen arbitrarily, we proved that $(M[A_\alpha])^k$ is Lindelöf for any $\alpha < \mathfrak{c}$ and $k \in \omega$.

Finally take any $\alpha < \mathfrak{c}$; we claim that $H = \bigcap\{f_n^{-1}(A_\alpha) : n \in \omega\} = \emptyset$. Indeed, if $s \in H$ then $f_n(s) \neq f_m(s)$ for distinct m and n because the family $\{f_n(\mathbb{K}) : n \in \omega\}$ is disjoint. For any $n \in \omega$ there is $\beta_n \geq \alpha$ such that $f_n(s) \in A_\alpha^{\beta_n}$. The set $\{f_n(s) : n \in \omega\}$ is infinite and $A_\alpha^{\beta_n}$ is finite for any $n \in \omega$; so there are $m, n \in \omega$ such that $\beta_n < \beta_m$.

We have $f_n(s) \in A_\alpha^{\beta_n}$ and hence $s \in F^{-1}(A_\alpha^{\beta_n})$; so $f_m(s) \in F(F^{-1}(A_\alpha^{\beta_n}))$. On the other hand, $f_m(s) \in A_\alpha^{\beta_m}$ which implies that there is $\psi \in \Psi_{\beta_m}$ such that $\psi(x_\alpha^{\beta_m}) = f_m(s) \in F(F^{-1}(A_\alpha^{\beta_n}))$. Therefore $x_\alpha^{\beta_m} \in \Psi_{\beta_m}^{-1}(F(F^{-1}(A_\alpha^{\beta_n})))$ which contradicts (7) and shows that $H = \emptyset$; so $(M[A_\alpha])^\omega$ is not Lindelöf by Problem 090. The ordinal $\alpha < \mathfrak{c}$ was chosen arbitrarily; so we proved that, for any $\alpha < \mathfrak{c}$ we have $l^*(M[A_\alpha]) = \omega$ while $(M[A_\alpha])^\omega$ is not Lindelöf, i.e., our solution is complete.

U.092. *Suppose that $\mathcal{P}$ is an sk-directed class of spaces and $Y \in \mathcal{P}$. Prove that if $X \subset C_p(Y)$ and the set of non-isolated points of X is σ-compact then $C_p^*(X)$ belongs to the class $\mathcal{P}_{\sigma\delta}$.*

Solution. Given $\varepsilon > 0$, $n \in \mathbb{N}$, a point $y \in Y^n$ and $f, g \in C_p(Y)$ we say that $\rho_y(f, g) < \varepsilon$ if $|f(y(i)) - g(y(i))| < \varepsilon$ for any $i < n$. Let X' be the set of non-isolated points of the space X. We have $X' = \bigcup\{K_i : i \in \mathbb{N}\}$ where the set K_i

is compact for every $i \in \mathbb{N}$. For any $l, m, n \in \mathbb{N}$ consider the set $A(l, m, n) = \{\varphi \in [-n, n]^X : \text{there exists } y \in Y^n \text{ such that } |\varphi(f) - \varphi(g)| \leq \frac{1}{m}$ whenever $f \in K_l$, $g \in X$ and $\rho_y(f, g) < \frac{1}{n}\}$. Let $B(l, m) = \bigcup\{A(l, m, n) : n \in \mathbb{N}\}$ for any $l, m \in \mathbb{N}$; we claim that

(1) $C_p^*(X) = \bigcap\{B(l, m) : l, m \in \mathbb{N}\}$.

By definition, every element of the set $A(l, m, n)$ is a bounded function for any $l, m, n \in \mathbb{N}$. Take an arbitrary $\varphi \in \bigcap\{B(l, m) : l, m \in \mathbb{N}\}$; fix any $f \in X'$ and $\varepsilon > 0$. There are $l, m, n \in \mathbb{N}$ such that $f \in K_l$, $\frac{1}{m} < \varepsilon$ and $\varphi \in A(l, m, n)$, i.e., there is a point $y \in Y^n$ such that $\rho_y(f, g) < \frac{1}{n}$ implies $|\varphi(f) - \varphi(g)| \leq \frac{1}{m} < \varepsilon$ for any $g \in X$. The set $U = \{g \in X : \rho_y(f, g) < \frac{1}{n}\}$ is an open neighbourhood of f in X such that $\varphi(U) \subset (\varphi(f) - \varepsilon, \varphi(f) + \varepsilon)$, i.e., U witnesses continuity of φ at f. Thus φ is continuous at any point of X'; continuity of φ at the isolated points of X is clear; so $\varphi \in C_p^*(X)$ and we proved that $B = \bigcap\{B(l, m) : l, m \in \mathbb{N}\} \subset C_p^*(X)$.

To prove the converse inclusion take any $\varphi \in C_p^*(X)$ and fix $l, m \in \mathbb{N}$. The function φ being continuous, for every $h \in K_l$ there is $n(h) \in \mathbb{N}$ and $y(h) \in Y^{n(h)}$ such that $\rho_{y(h)}(g, h) < \frac{1}{n(h)}$ implies $|\varphi(h) - \varphi(g)| < \frac{1}{2m}$ for any $g \in X$. It is evident that the set $O_h = \{g \in X : \rho_{y(h)}(h, g) < \frac{1}{3n(h)}\}$ is an open neighbourhood of h in X; so there is a finite $E \subset X$ for which $K_l \subset \bigcup\{O_h : h \in E\}$. It is easy see that

(2) there exists a number $n \in \mathbb{N}$ and $y \in Y^n$ such that $\varphi(X) \subset [-n, n]$, $n > 3n(h)$ and $y(h)(i) \in \{y(1), \ldots, y(n-1)\}$ for any $h \in E$ and $i < n$.

We claim that $\varphi \in A(l, m, n)$. Indeed, take any $f \in K_l$ and $g \in X$ for which $\rho_y(f, g) < \frac{1}{n}$. There is a function $h \in E$ such that $f \in O_h$ and therefore we have $\rho_{y(h)}(f, h) < \frac{1}{3n(h)}$. It follows from the property (2) that $\rho_{y(h)}(f, g) < \frac{1}{n} < \frac{1}{3n(h)}$ and hence $\rho_{y(h)}(g, h) \leq \rho_{y(h)}(g, f) + \rho_{y(h)}(f, h) < \frac{2}{3} \cdot \frac{1}{n(h)} < \frac{1}{n(h)}$ which implies, by our choice of $n(h)$ and $y(h)$, that $|\varphi(g) - \varphi(h)| < \frac{1}{2m}$ and $|\varphi(f) - \varphi(h)| < \frac{1}{2m}$ whence $|\varphi(f) - \varphi(g)| < \frac{1}{m}$.

The functions $f \in K_l$ and $g \in X$ with $\rho_y(f, g) < \frac{1}{n}$ were chosen arbitrarily; so $y \in Y^n$ witnesses that $|\varphi(f) - \varphi(g)| < \frac{1}{m}$ for any $f \in K_l$ and $g \in X$ with $\rho_y(f, g) < \frac{1}{n}$, i.e., $\varphi \in A(l, m, n)$. Since the choice of $l, m \in \mathbb{N}$ was also arbitrary, we showed that $\varphi \in B$ and hence $C_p^*(X) \subset B$; so the equality (1) is proved.

For any numbers $l, m, n \in \mathbb{N}$ consider the set $Q(l, m, n) = \{(\varphi, y) \in [-n, n]^X \times Y^n : |\varphi(f) - \varphi(g)| \leq \frac{1}{m}$ whenever $f \in K_l$, $g \in X$ and $\rho_y(f, g) < \frac{1}{n}\}$. We will establish that the set $Q(l, m, n)$ is closed in the space $[-n, n]^X \times Y^n$. To do it, take an arbitrary point $w = (\varphi, y) \in ([-n, n]^X \times Y^n) \backslash Q(l, m, n)$. There are $f \in K_l$ and $g \in X$ such that $\rho_y(f, g) < \frac{1}{n}$ while $|\varphi(f) - \varphi(g)| > \frac{1}{m}$. Observe that the set $O_\varphi = \{\psi \in [-n, n]^X : |\psi(f) - \psi(g)| > \frac{1}{m}\}$ is an open neighbourhood of φ in $[-n, n]^X$ while $U_y = \{z \in Y^n : \rho_z(f, g) < \frac{1}{n}\}$ is an open neighbourhood of y in Y^n. It is immediate that $W = O_\varphi \times U_y$ is an open neighbourhood of w in $[-n, n]^X \times Y^n$ such that $W \cap Q(l, m, n) = \emptyset$. Thus we proved that

(3) the set $Q(l, m, n)$ is closed in $[-n, n]^X \times Y^n$ for any $l, m, n \in \mathbb{N}$.

Let $\pi : [-n, n]^X \times Y^n \to [-n, n]^X$ be the natural projection. It is straightforward that $\pi(Q(l, m, n)) = A(l, m, n)$ for any $l, m, n \in \mathbb{N}$. The class $\mathcal{P}$ being sk-directed, the set $Q(l, m, n)$ belongs to $\mathcal{P}$ and hence $A(l, m, n)$ also belongs to $\mathcal{P}$ for any $l, m, n \in \mathbb{N}$. Thus $B(l, m)$ belongs to $\mathcal{P}_\sigma$; so it follows from (1) that $C_p^*(X)$ belongs to $\mathcal{P}_{\sigma\delta}$ and hence our solution is complete.

U.093. *Prove that there exist separable, scattered σ-compact spaces X and Y such that both $(C_p(X))^\omega$ and $(C_p(Y))^\omega$ are Lindelöf while $C_p(X) \times C_p(Y)$ is not normal and contains a closed discrete set of cardinality* c.

Solution. Denote by J the set $(-1, 1)$ with the topology induced from the real line $\mathbb{R}$. If Z and T are arbitrary spaces, a multi-valued map $\varphi : Z \to \exp T$ is often denoted as $\varphi : Z \to T$; given sets $A \subset Z$ and $B \subset T$ let $\varphi(A) = \bigcup\{\varphi(z) : z \in A\}$ and $\varphi^{-1}(B) = \{z \in Z : \varphi(z) \subset B\}$.

If Z is a space and $A \subset Z$ then $Z[A]$ is the set Z with the topology generated by the family $\tau(Z) \cup \exp(Z \setminus A)$ as a subbase. In other words, the topology of $Z[A]$ is the same as in Z at all points of A while the points of $Z \setminus A$ are isolated in $Z[A]$.

Fact 1. If Z is a σ-compact space then there exists a $K_{\sigma\delta}$-space C such that $C_p(Z, J) \subset C \subset J^Z$.

Proof. We have $Z = \bigcup\{K_n : n \in \mathbb{N}\}$ where the set K_n is compact for any $n \in \mathbb{N}$. For technical purposes let $a_m = 1 - \frac{1}{m+1}$ for any $m \in \mathbb{N}$; given $m, n \in \mathbb{N}$ the set $K(m, n) = \{f \in \mathbb{I}^Z : f(K_n) \subset [-a_m, a_m]\}$ is compact being closed in $\mathbb{I}^Z$. The set $C(n) = \bigcup\{K(m, n) : m \in \mathbb{N}\}$ is σ-compact for any $n \in \mathbb{N}$; so $C = \bigcap\{C(n) : n \in \mathbb{N}\}$ is $K_{\sigma\delta}$.

Given any $f \in C_p(Z, J)$ and $n \in \mathbb{N}$ the set $f(K_n) \subset J$ is compact; so there is $m \in \mathbb{N}$ for which $f(K_n) \subset [-a_m, a_m]$, i.e., $f \in K(m, n)$. Therefore $f \in C(n)$ for any $n \in \mathbb{N}$, and hence $f \in \bigcap\{C(n) : n \in \mathbb{N}\} = C$; the function $f \in C_p(Z, J)$ was chosen arbitrarily, so we proved that $C_p(Z, J) \subset C$.

Now if $f \in C$ then take any $z \in Z$; there is $n \in \mathbb{N}$ such that $z \in K_n$. It follows from $f \in C(n)$ that $f(K_n) \subset [-a_m, a_m] \subset J$ for some $m \in \mathbb{N}$ and hence $f(z) \in J$. This shows that $f(z) \in J$ for any $z \in Z$ and hence $f \in J^Z$. Since $f \in C$ was chosen arbitrarily, we have $C_p(Z, J) \subset C \subset J^Z$; so Fact 1 is proved.

Fact 2. Suppose that $\varphi_t : E_t \to M_t$ is a compact-valued upper semicontinuous onto map for any $t \in T$. Let $E = \prod_{t \in T} E_t$, $M = \prod_{t \in T} M_t$ and define a multi-valued map $\varphi = \prod_{t \in T} \varphi_t : E \to M$ by $\varphi(x) = \prod_{t \in T} \varphi_t(x(t))$ for any $x \in E$. Then $\varphi : E \to M$ is a compact-valued upper semicontinuous onto map.

Proof. Let $\pi_t : E \to E_t$ be the natural projection for any $t \in T$. It is evident that $\varphi(x)$ is compact for any $x \in E$. If $y \in M$ then, for any $t \in T$ there is $x_t \in E_t$ such that $y(t) \in \varphi_t(x_t)$. Letting $x(t) = x_t$ for any $t \in T$ we obtain a point $x \in E$ such that $y \in \varphi(x)$; this shows that the map φ is onto.

To see that the map φ is upper semicontinuous fix $x \in E$ and $U \in \tau(\varphi(x), M)$. By Fact 3 of S.271 there exists a set $V = \prod_{t \in T} V_t \in \tau(M)$ such that the set $S = \text{supp}(V) = \{t \in T : V_t \neq M_t\}$ is finite and $\varphi(x) \subset V \subset U$. We have

$\varphi_t(x(t)) \subset V_t$ for any $t \in S$; so there is $O_t \in \tau(x(t), E_t)$ such that $\varphi_t(O_t) \subset V_t$. The set $O = \bigcap \{\pi_t^{-1}(O_t) : t \in S\}$ is open in E and it is straightforward that $\varphi(O) \subset V \subset U$ which proves that the map φ is upper semicontinuous at the point x. Consequently, φ is upper semicontinuous by Fact 1 of T.346 and hence Fact 2 is proved.

Fact 3. If Z^ω is Lindelöf then $Z^\omega \times T$ is also Lindelöf for any K-analytic space T.

Proof. Let us establish first that $Z^\omega \times \mathbb{P}$ is Lindelöf. If Z is countably compact then it is compact; so $Z^\omega \times \mathbb{P}$ Lindelöf by Fact 2 of T.490. If Z is not countably compact then ω embeds in Z as a closed subspace. Therefore $Z^\omega \times \omega^\omega \simeq (Z \times \omega)^\omega$ embeds as a closed subspace in $(Z \times Z)^\omega \simeq Z^\omega$ which shows that $Z^\omega \times \mathbb{P}$ embeds as a closed subspace in Z^ω; so $Z^\omega \times \mathbb{P}$ is Lindelöf.

There exists a compact-valued upper semicontinuous onto map $\varphi : \mathbb{P} \to T$ (see SFFS-388). Define a multi-valued map $\psi : Z^\omega \times \mathbb{P} \to Z^\omega \times T$ by $\psi(z, t) = \{z\} \times \varphi(t)$ for any $(z, t) \in Z^\omega \times T$. Since any continuous map is upper semicontinuous, the mapping ψ is a product of compact-valued upper semicontinuous onto maps; so ψ is also a compact-valued upper semicontinuous onto map by Fact 2. Therefore $l(Z^\omega \times T) \le l(Z^\omega \times \mathbb{P}) \le \omega$ (see SFFS-240). Fact 3 is proved.

Fact 4. If Z^ω is Lindelöf and $T \subset C_p(Z)$ is σ-compact then $(C_p(T))^\omega$ is also Lindelöf.

Proof. Given spaces G and H the expression $G \gg H$ abbreviates the phrase "G maps continuously onto H". Denote by $\mathcal{P}$ the class of spaces E such that there is a compact space K_E and a closed subset F_E of the space $Z^\omega \times K_E \times \mathbb{P}$ such that $F_E \gg E$. If $E \in \mathcal{P}$ and $E \gg E'$ then $F_E \gg E'$ and hence $E' \in \mathcal{P}$. It is evident that the irrationals and all compact spaces are in $\mathcal{P}$ and $E \in \mathcal{P}$ implies $F \in \mathcal{P}$ for any closed $F \subset E$.

Now, if $E_n \in \mathcal{P}$ for any number $n \in \omega$ then $F = \prod_{n \in \omega} F_{E_n}$ is a closed subspace of $\prod \{Z^\omega \times K_{E_n} \times \mathbb{P} : n \in \omega\}$ so, for the compact space $K = \prod \{K_{E_n} : n \in \omega\}$, the space F embeds in $Z^\omega \times K \times \mathbb{P}^\omega \simeq Z^\omega \times K \times \mathbb{P}$ as a closed subspace. It is clear that $F \gg \prod_{n \in \omega} E_n$ and therefore $E = \prod_{n \in \omega} E_n \in \mathcal{P}$. Since $E \gg E_n$ for every $n \in \omega$, we have $E \times \omega \gg \bigoplus_{n \in \omega} E_n$. Since $E \times \mathbb{P} \in \mathcal{P}$ and $E \times \mathbb{P} \gg E \times \omega$, we have $E \times \omega \in \mathcal{P}$ and therefore $\bigoplus_{n \in \omega} E_n \in \mathcal{P}$. An evident consequence is that $\mathcal{P}_\sigma = \mathcal{P}$. The class $\mathcal{P}$ being invariant under countable products and closed subspaces, we have $\mathcal{P}_\delta = \mathcal{P}$ (see Fact 7 of S.271).

Furthermore, $Z^\omega \times K \times \mathbb{P}$ is Lindelöf for any compact space K by Fact 3. An immediate consequence is that every $E \in \mathcal{P}$ is Lindelöf; since $E^\omega \in \mathcal{P}$ as well, the space E^ω is Lindelöf for any $E \in \mathcal{P}$.

We showed, in particular, the class $\mathcal{P}$ is sk-directed; so we can apply Problem 092 to conclude that $C_p^*(T)$ belongs to $\mathcal{P}_{\sigma\delta} = \mathcal{P}$. The set $C_p(T, \mathbb{I})$ is closed in $C_p^*(T)$; so $C_p(T, \mathbb{I})$ also belongs to $\mathcal{P}$. It follows from Fact 1 that there is a $K_{\sigma\delta}$-space C such that $C_p(T, J) \subset C \subset J^T$. It is evident that $C \cap C_p(T, \mathbb{I}) = C_p(T, J)$; so $C_p(T, J)$ is homeomorphic to a closed subspace of $C \times C_p(T, \mathbb{I})$ (we used again Fact 7 of S.271). Thus $(C_p(T, J))^\omega$ is homeomorphic to a closed subspace of

$C^\omega \times (C_p(T, \mathbb{I}))^\omega$; since C^ω is K-analytic, we can apply Fact 3 again to conclude that $C^\omega \times (C_p(T, \mathbb{I}))^\omega$ is Lindelöf and hence $(C_p(T))^\omega \simeq (C_p(T, J))^\omega$ is also Lindelöf. Fact 4 is proved.

Fact 5. Let M be an uncountable Polish space. Suppose that $A, B \subset M$ are disjoint sets such that $(M[A])^\omega$ and $(M[B])^\omega$ are Lindelöf. Then $M[A] \times M[B]$ is not normal and $ext(M[A] \times M[B]) = \mathfrak{c}$.

Proof. It follows from Problem 091 that $M[A]$ is Lindelöf if and only if, for every $K \subset M$ such that $K \simeq \mathbb{K}$, we have $K \cap A \neq \emptyset$. Since M is uncountable, we can find a set $K \subset M$ with $K \simeq \mathbb{K}$ and hence $K \simeq K \times K$ which shows that we can consider that $K \times K \subset M$ and hence $\mathcal{K} = \{K \times \{x\} : x \in K\}$ is a family of $\mathfrak{c}$-many disjoint copies of $\mathbb{K}$ in M. If $|A| < \mathfrak{c}$ or $|B| < \mathfrak{c}$ then A (or B respectively) cannot intersect all elements of $\mathcal{K}$. Thus $|A| = |B| = \mathfrak{c}$.

Let $\pi_A : \tilde{M} = M[A] \times M[B] \to M[A]$ and $\pi_B : \tilde{M} \to M[B]$ be the respective natural projections. The set $D = \{(x, x) : x \in B\}$ has cardinality $\mathfrak{c}$; besides, for any $z = (x, x) \in D$ the point $x = \pi_A(z)$ belongs to $M \backslash A$, i.e., x is isolated in $M[A]$. The map $\pi_A | D : D \to M[A]$ being continuous and injective, every $z \in D$ is isolated in D, i.e., D is a discrete subspace of $\tilde{M}$.

Observe next that the diagonal $\Delta = \{(x, x) : x \in M\}$ is closed in $M \times M$ and hence in $\tilde{M}$ because the topology of $\tilde{M}$ is stronger than $\tau(M \times M)$. The set B is closed in $M[B]$ and $D = \pi_B^{-1}(B) \cap \Delta$; so the set D is also closed in $\tilde{M}$. This proves that $ext(M[A] \times M[B]) = \mathfrak{c}$.

If $M[A] \times M[B]$ is normal then there is $O \in \tau(D, \tilde{M})$ such that $\overline{O} \cap \pi_A^{-1}(A) = \emptyset$ (the bar denotes the closure in $\tilde{M}$). Fix a countable base $\mathcal{B}$ in the space M. Since the family $\mathcal{B}$ contains local bases at all points of B in the space $M[B]$, for every $x \in B$ there is $U_x \in \mathcal{B}$ such that $x \in U_x$ and $\{x\} \times U_x \subset O$.

There is an uncountable set $B' \subset B$ and $U \in \mathcal{B}$ such that $U_x = U$ for any $x \in B'$. Fix a point $z \in U$ and observe that $P = B' \times \{z\} \subset O$. The set $B' \subset M \backslash A$ is discrete as a subspace of $M[A]$; so it cannot be closed in $M[A]$. Therefore there is a point $y \in A$ for which $y \in cl_{M[A]}(B')$. It is immediate that $t = (y, z) \in cl_{\tilde{M}}(P)$ which shows that $t \in \overline{O} \cap \pi_A^{-1}(A) = \emptyset$. The obtained contradiction shows that $M[A] \times M[B]$ is not normal; so Fact 5 is proved.

Returning to our solution take disjoint sets $A, B \subset I = [0, 1]$ such that both spaces $(I[A])^\omega$ and $(I[B])^\omega$ are Lindelöf (see Problem 090). Fix some countable base $\mathcal{B}$ in I such that $U \neq \emptyset$ for any $U \in \mathcal{B}$ and let $\mathcal{C} = \{(U, V) \in \mathcal{B} \times \mathcal{B} : \overline{U} \subset V\}$. By normality of I, for any $q = (U, V) \in \mathcal{C}$, we can choose a function $f_q \in C_p(I)$ such that $f_q(U) \subset \{1\}$ and $f_q(I \backslash V) \subset \{0\}$. It is evident that the set $\tilde{S} = \{f_q : q \in \mathcal{C}\}$ separates the points and closed subsets of I.

For any point $x \in I$ let $u_x(y) = 0$ for any $y \in I \backslash \{x\}$ and $u_x(x) = 1$; denote by u the function which is identically zero on the set I. It is easy to see that the set $K_A = \{u_x : x \in I \backslash A\} \cup \{u\}$ is compact and $K_A \subset C_p(I[A])$; besides, $K_A \cup \tilde{S}$ separates the points and closed subsets of $I[A]$ and $K_A \cap \tilde{S} = \emptyset$. Choose a countable $S'' \subset C_p(I) \backslash \{u\}$ which is dense in $C_p(I)$ and let $S = \tilde{S} \cup S''$; it is evident that $K_A \cap S = \emptyset$.

Analogously, $K_B = \{u_x : x \in I \setminus B\} \cup \{u\}$ is compact, $K_B \subset C_p(I[B])$ and $K_B \cup S$ separates the points and closed subsets of $I[B]$ while $K_B \cap S = \emptyset$. Observe also that u is the unique non-isolated point of both K_A and K_B. Our promised spaces are $X = (K_A \cup S)[K_A]$ and $Y = (K_B \cup S)[K_B]$. In other words, we declare the points of S isolated; if this is done in $K_A \cup S$ then we obtain X. Doing this in $K_B \cup S$ we obtain Y.

It is an easy exercise that the spaces X and Y are σ-compact and scattered. Furthermore, S is a countable dense set of isolated points of both X and Y while $K_A = X \setminus S$ and $K_B = Y \setminus S$ are uncountable compact spaces with a unique non-isolated point. Let us show next that

(1) $I[A]$ embeds in $C_p(X)$ as a closed subspace and $I[B]$ embeds as a closed subspace in $C_p(Y)$.

Of course, the proofs for $I[A]$ and $I[B]$ are analogous; so let us establish (1) for $I[A]$. For any $x \in I$ let $\varphi_x(f) = f(x)$ for any $f \in S$. Then $\varphi_x \in C_p(S)$ and the map $\varphi : I \to C_p(S)$ defined by $\varphi(x) = \varphi_x$ for any $x \in I$ is continuous (see TFS-166). Since S separates the points and closed subsets of I, the map φ is an embedding; let $I' = \varphi(I) \subset C_p(S)$.

Analogously, let $\psi_x(f) = f(x)$ for any $f \in K_A \cup S$ and $x \in I$. Define a map $\psi : I[A] \to C_p(K_A \cup S)$ by $\psi(x) = \psi_x$ for any $x \in I$; then ψ is an embedding because $K_A \cup S \subset C_p(I[A])$ separates the points and closed subsets of $I[A]$. Let $I_0 = \psi(I[A])$; observe that X has the underlying set $K_A \cup S$ but the topology of X is stronger; so $I_0 \subset C_p(K_A \cup S) \subset C_p(X)$. If S' is the set S with the discrete topology then S' is a subspace of X; since $\tau(S) \subset \tau(S')$, we have $I' \subset C_p(S) \subset C_p(S')$.

Let $\pi(f) = f|S'$ for any function $f \in C_p(X)$. This gives us the restriction map $\pi : C_p(X) \to Z = \pi(C_p(X)) \subset C_p(S')$ which is continuous and injective because S' is dense in X. It is easy to check that $\pi(\psi_x) = \varphi_x$ for any $x \in I$ and therefore $\pi(I_0) = I'$; the set I' is closed in Z being compact, so $I_0 = \pi^{-1}(I')$ is closed in $C_p(X)$. Since I_0 is homeomorphic to $I[A]$, the property (1) is proved.

For any $s \in S$ let $r_s(s) = 1$ and $r_s(x) = 0$ for all $x \in (K_A \cup S) \setminus \{s\}$. It is evident that $r_s \in C_p(X)$ for any $s \in S$; let $w(x) = 0$ for any $x \in K_A \cup S$. It is evident that the set $L = \{w\} \cup \{r_s : s \in S\}$ is compact. Besides, $I_0 \cup L$ generates the topology of X (see TFS-166); so X embeds in $C_p(I_0 \cup L)$. Since $I_0 \simeq I[A]$ and $|L| \leq \omega$, the space $I_0 \cup L$ is a continuous image of the space $I[A] \times \omega$; so $(I_0 \cup L)^\omega$ is a continuous image of $(I[A] \times \omega)^\omega \simeq (I[A])^\omega \times \omega^\omega$ which is Lindelöf by Fact 3. Therefore $(C_p(X))^\omega$ is also Lindelöf by Fact 4. Repeating the same reasoning for $K_B \cup S$ we conclude that $(C_p(Y))^\omega$ is Lindelöf as well. It follows from (1) that $I[A] \times I[B]$ embeds in $C_p(X) \times C_p(Y)$ as a closed subspace which implies that $ext(C_p(X) \times C_p(Y)) = \mathfrak{c}$ and $C_p(X) \times C_p(Y)$ is not normal (see Fact 5); so our solution is complete.

U.094. *Show that there is a separable scattered σ-compact space X and a countable space M such that $(C_p(X))^\omega$ is Lindelöf while $ext(C_p(X) \times C_p(M)) = \mathfrak{c}$ and the space $C_p(X) \times C_p(M)$ is not normal.*

Solution. If Z is a space and $A \subset Z$ then $Z[A]$ is the set Z with the topology generated by the family $\tau(Z) \cup \exp(Z \setminus A)$ as a subbase. In other words, the topology of $Z[A]$ is the same as in Z at all points of A while the points of $Z \setminus A$ are isolated in $Z[A]$.

Fact 1. For any second countable space Z there is a countable space T such that Z embeds in $C_p(T)$ as a closed subspace.

Proof. For any $B \subset C_p(Z)$ let $P(B) = \{f_1 \cdot \ldots \cdot f_n : n \in \mathbb{N}, \ f_i \in B \text{ for all } i \leq n\}$ and $A(B) = \{f_1 + \ldots + f_n : n \in \mathbb{N}, \ f_i \in P(B) \text{ for all } i \leq n\}$. Fix a countable base $\mathcal{B}$ in the space Z and let $\mathcal{C} = \{(U, V) \in \mathcal{B} \times \mathcal{B} : cl_Z(U) \subset V\}$. By normality of the space Z, for any $\mu = (U, V) \in \mathcal{C}$ there is a function $f_\mu \in C_p(Z)$ such that $f_\mu(U) \subset \{1\}$ and $f_\mu(Z \setminus V) \subset \{0\}$. For any point $q \in \mathbb{Q}$ let $u_q(z) = q$ for every $z \in Z$. The set $S = \{u_q : q \in \mathbb{Q}\} \cup \{f_\mu : \mu \in \mathcal{C}\} \subset C_p(Z)$ is countable and hence so is the set $M = A(S)$.

For any $z \in Z$ let $\varphi_z(f) = f(z)$ for any $f \in M$. The map $\varphi : Z \to C_p(M)$ defined by $\varphi(z) = \varphi_z$ for all $z \in Z$ is an embedding because M separates the points and closed subsets of Z (see TFS-166). Thus it suffices to show that $Z' = \varphi(Z)$ is closed in $C_p(M)$. We have $\xi(u_0) = 0$ for any $\xi \in Z'$; so $\overline{Z'} \subset F = \{\xi \in C_p(M) : \xi(u_0) = 0\}$ (the bar denotes the closure in $C_p(M)$). Since F is closed in $C_p(M)$, it suffices to show that $\xi \notin \overline{Z'}$ for any $\xi \in F \setminus Z'$

To do it, take an arbitrary $\xi \in F \setminus Z'$; since ξ is continuous at the point $u_0 \in M$ there are $z_1, \ldots, z_n \in Z$ and $\varepsilon > 0$ such that $\xi(f) \in (-\frac{1}{2}, \frac{1}{2})$ for any $f \in M$ such that $|f(z_i)| < \varepsilon$ for all $i \leq n$. In particular,

(1) if $f \in M$ and $f(z_i) = 0$ for all $i \leq n$ then $\xi(f) \in (-\frac{1}{2}, \frac{1}{2})$.

Since $\xi \notin \{z_1, \ldots, z_n\}$, it is easy to find $U_1, \ldots, U_n, V_1, \ldots, V_n \in \mathcal{B}$ such that $\xi \notin \bigcup\{\varphi(V_i) : i \leq n\}$, the family $\{V_i : i \leq n\}$ is disjoint and $z_i \in U_i \subset cl_Z(U_i) \subset V_i$ for all $i \leq n$. Therefore $\mu_i = (U_i, V_i) \in \mathcal{C}$ and hence $g_i = f_{\mu_i} \in S$ for all $i \leq n$.

Let $V = \bigcup\{V_i : i \leq n\}$; for the function $h = \prod_{i=1}^{n}(u_1 - g_i) \in M$ we have $h(z_i) = 0$ for all $i \leq n$ and $h(z) = 1$ for any $z \in Z \setminus V$. This shows that, for any $z \in Z \setminus V$ we have $\varphi(z)(h) = h(z) = 1$ whereas (1) implies that $\xi(h) \in (-\frac{1}{2}, \frac{1}{2})$ because $h(z_i) = 0$ for all $i \leq n$. Consequently, $\xi \notin \overline{\varphi(Z \setminus V)}$ which shows that $\xi \notin \overline{Z'}$. Thus Z' is closed in $C_p(M)$ and Fact 1 is proved.

Fact 2. Let Z be an uncountable Polish space. If $A \subset Z$, the set $Z \setminus A$ has cardinality $\mathfrak{c}$ and $Z[A]$ is Lindelöf then $Z[A] \times (Z \setminus A)$ is not normal and $ext(Z[A] \times (Z \setminus A)) = \mathfrak{c}$ (in this product the set $Z \setminus A$ is considered with the second countable topology induced from Z).

Proof. Let $\pi_A : \tilde{Z} = Z[A] \times (Z \setminus A) \to Z[A]$ be natural projection. Since $|Z \setminus A| = \mathfrak{c}$, the set $D = \{(x, x) : x \in Z \setminus A\} \subset Z[A] \times (Z \setminus A)$ has cardinality $\mathfrak{c}$; besides, for any $z = (x, x) \in D$ the point $x = \pi_A(z)$ belongs to $Z \setminus A$, i.e., x is isolated in $Z[A]$. The map $\pi_A|D : D \to Z[A]$ being continuous and injective, every $z \in D$ is isolated in D, i.e., D is a discrete subspace of $\tilde{Z}$.

Observe next that the diagonal $\Delta = \{(x, x) : x \in Z\}$ is closed in $Z \times Z$ and hence in $Z[A] \times Z$ because the topology of $Z[A] \times Z$ is stronger than $\tau(Z \times Z)$. Therefore $D = \Delta \cap \tilde{Z}$ is closed in $\tilde{Z}$. This proves that $ext(Z[A] \times (Z \backslash A)) = \mathfrak{c}$.

If $Z[A] \times (Z \backslash A)$ is normal then there is $O \in \tau(D, \tilde{Z})$ such that $\overline{O} \cap \pi_A^{-1}(A) = \emptyset$ (the bar denotes the closure in $\tilde{Z}$). Fix a countable base $\mathcal{B}$ in the space $Z \backslash A$. For every $x \in Z \backslash A$ there is $U_x \in \mathcal{B}$ such that $x \in U_x$ and $\{x\} \times U_x \subset O$.

Since $Z \backslash A$ is uncountable, there is an uncountable $B \subset Z \backslash A$ and $U \in \mathcal{B}$ such that $U_x = U$ for any $x \in B$. Fix a point $z \in U$ and observe that $P = B \times \{z\} \subset O$. The set $B \subset Z \backslash A$ is discrete as a subspace of $Z[A]$; so it cannot be closed in $Z[A]$ because $Z[A]$ is Lindelöf. Therefore there is $y \in A$ for which $y \in cl_{Z[A]}(B)$. It is clear that $t = (y, z) \in cl_{\tilde{Z}}(P)$ which shows that $t \in \overline{O} \cap \pi_A^{-1}(A) = \emptyset$. The obtained contradiction shows that $Z[A] \times (Z \backslash A)$ is not normal; so Fact 2 is proved.

Returning to our solution take disjoint sets $A, B \subset I = [0, 1]$ such that both spaces $(I[A])^\omega$ and $(I[B])^\omega$ are Lindelöf (see Problem 090). It follows from Problem 091 that $I[B]$ is Lindelöf if and only if, for every $K \subset I$ such that $K \simeq \mathbb{K}$, we have $K \cap B \neq \emptyset$. Since I is uncountable, we can find a set $K \subset I$ with $K \simeq \mathbb{K}$ and hence $K \simeq K \times K$ which shows that we can consider that $K \times K \subset I$ and hence $\mathcal{K} = \{K \times \{x\} : x \in K\}$ is a family of $\mathfrak{c}$-many disjoint copies of $\mathbb{K}$ in I. If $|B| < \mathfrak{c}$ then B cannot intersect all elements of $\mathcal{K}$. Thus $|B| = \mathfrak{c}$ and hence $|I \backslash A| = \mathfrak{c}$.

Fix some countable base $\mathcal{B}$ in I such that $U \neq \emptyset$ for any $U \in \mathcal{B}$ and let $\mathcal{C} = \{(U, V) \in \mathcal{B} \times \mathcal{B} : \overline{U} \subset V\}$. By normality of I, for any $\mu = (U, V) \in \mathcal{C}$, we can choose a function $f_\mu \in C_p(I)$ such that $f_\mu(U) \subset \{1\}$ and $f_\mu(I \backslash V) \subset \{0\}$. It is evident that the set $\tilde{S} = \{f_\mu : \mu \in \mathcal{C}\}$ separates the points and closed subsets of I.

For any point $x \in I$ let $u_x(y) = 0$ for any $y \in I \backslash \{x\}$ and $u_x(x) = 1$; denote by u the function which is identically zero on the set I. It is easy to see that the set $K_A = \{u_x : x \in I \backslash A\} \cup \{u\}$ is compact and $K_A \subset C_p(I[A])$; besides, $K_A \cup \tilde{S}$ separates the points and closed subsets of $I[A]$ and $K_A \cap \tilde{S} = \emptyset$. Choose a countable $S'' \subset C_p(I) \backslash \{u\}$ which is dense in $C_p(I)$ and let $S = \tilde{S} \cup S''$; it is evident that $K_A \cap S = \emptyset$.

Let $X = (K_A \cup S)[K_A]$; this means that we declare the points of S isolated and leave the same topology at all points of K_A. It is an easy exercise that the space X is σ-compact and scattered. Furthermore, S is a countable dense set of isolated points of X while $K_A = X \backslash S$ is an uncountable compact space with a unique non-isolated point. Let us show next that

(2) $I[A]$ embeds in $C_p(X)$ as a closed subspace.

For any $x \in I$ let $\varphi_x(f) = f(x)$ for any $f \in S$. Then $\varphi_x \in C_p(S)$ and the map $\varphi : I \to C_p(S)$ defined by $\varphi(x) = \varphi_x$ for any $x \in I$ is continuous (see TFS-166). Since S separates the points and closed subsets of I, the map φ is an embedding; let $I' = \varphi(I) \subset C_p(S)$.

Analogously, let $\psi_x(f) = f(x)$ for any $f \in K_A \cup S$ and $x \in I$. Define a map $\psi : I[A] \to C_p(K_A \cup S)$ by $\psi(x) = \psi_x$ for any $x \in I$; then ψ is an embedding because $K_A \cup S \subset C_p(I[A])$ separates the points and closed subsets of $I[A]$. Let $I_0 = \psi(I[A])$; observe that X has the underlying set $K_A \cup S$ but the topology of X is stronger; so $I_0 \subset C_p(K_A \cup S) \subset C_p(X)$. If S' is the set S with the discrete topology then S' is a subspace of X; since $\tau(S) \subset \tau(S')$, we have $I' \subset C_p(S) \subset C_p(S')$.

Let $\pi(f) = f|S'$ for any function $f \in C_p(X)$. This gives us the restriction map $\pi : C_p(X) \to Z = \pi(C_p(X)) \subset C_p(S')$ which is continuous and injective because S' is dense in X. It is easy to check that $\pi(\psi_x) = \varphi_x$ for any $x \in I$ and therefore $\pi(I_0) = I'$; the set I' is closed in Z being compact; so $I_0 = \pi^{-1}(I')$ is closed in $C_p(X)$. Since I_0 is homeomorphic to $I[A]$, the property (2) is proved.

For any $s \in S$ let $r_s(s) = 1$ and $r_s(x) = 0$ for all $x \in (K_A \cup S)\backslash\{s\}$. It is evident that $r_s \in C_p(X)$ for any $s \in S$; let $w(x) = 0$ for any $x \in K_A \cup S$. It is evident that the set $L = \{w\} \cup \{r_s : s \in S\}$ is compact. Besides, $I_0 \cup L$ generates the topology of X; so X embeds in $C_p(I_0 \cup L)$ (see TFS-166). Since $I_0 \simeq I[A]$ and $|L| \leq \omega$, the space $I_0 \cup L$ is a continuous image of the space $I[A] \times \omega$; so $(I_0 \cup L)^\omega$ is a continuous image of $(I[A] \times \omega)^\omega \simeq (I[A])^\omega \times \omega^\omega$ which is Lindelöf by Fact 3 of U.093. Therefore $(C_p(X))^\omega$ is also Lindelöf by Fact 4 of U.093.

Apply Fact 1 to find a countable space M such that $I\backslash A$ embeds in $C_p(M)$ as a closed subspace. It follows from (2) that $I[A] \times (I\backslash A)$ embeds in $C_p(X) \times C_p(M)$ as a closed subspace; this implies that $ext(C_p(X) \times C_p(M)) = \mathfrak{c}$ and $C_p(X) \times C_p(M)$ is not normal (see Fact 2), so our solution is complete.

U.095. *Prove that, under CH, there exists a separable scattered compact space X such that $(C_p(X, \mathbb{D}))^n$ is Lindelöf for any natural n, while $(C_p(X, \mathbb{D}))^\omega$ is not Lindelöf.*

Solution. Given a set Z and a family $\mathcal{A}$ of infinite subsets of Z we say that $\mathcal{A}$ is *almost disjoint* if $A \cap B$ is finite for any distinct $A, B \in \mathcal{A}$.

Fact 1. Assume that we have an uncountable space Z such that $w(Z) \leq \mathfrak{c}$ and there is a countable $Q \subset Z$ such that Z is concentrated around Q, i.e., $|Z\backslash U| \leq \omega$ for any $U \in \tau(Q, Z)$. Then the Continuum Hypothesis (CH) implies that there is an uncountable $T \subset Z$ such that $Q \subset T$ and T^n is Lindelöf for any $n \in \mathbb{N}$.

Proof. Since CH holds, for any $n \in \mathbb{N}$ there is a base $\mathcal{B}_n$ in the space Z^n such that $|\mathcal{B}_n| \leq \omega_1$ and $\bigcup \mathcal{U} \in \mathcal{B}_n$ for any countable $\mathcal{U} \subset \mathcal{B}_n$. It is clear that Z is Lindelöf; so $Z \in \mathcal{B}_1$ and hence we can choose an enumeration $\{U_\alpha : \alpha < \omega_1\}$ of the family $\bigcup\{\mathcal{B}_n : n \in \mathbb{N}\}$ in such a way that $U_0 = Z$. For every $\beta < \omega_1$ there is a unique $m_\beta \in \mathbb{N}$ such that $U_\beta \in \mathcal{B}_{m_\beta}$.

Our first step is to choose a point $z_0 \in Z\backslash Q$ arbitrarily. Next, assume that $\nu < \omega_1$ and we have a set $\{z_\alpha : \alpha < \nu\} \subset Z\backslash Q$ with the following properties:

(1) $\beta < \alpha < \nu$ implies $z_\beta \neq z_\alpha$;
(2) if $Z_\alpha = \{z_\gamma : \gamma < \alpha\}$ and $T_\alpha = Q \cup Z_\alpha$ for any $\alpha \leq \nu$, then the conditions $\beta \leq \alpha < \nu$, $m = m_\beta$ and $U_\beta \supset (T_\alpha)^m\backslash(Z_\alpha)^m$ imply $(z_{\alpha_1}, \ldots, z_{\alpha_m}) \in U_\beta$ whenever $\max\{\alpha_1, \ldots, \alpha_m\} = \alpha$.

For any $m \in \mathbb{N}$ the family $\Gamma_m = \{\beta \leq \nu : U_\beta \in \mathcal{B}_m$ and $U_\beta \supset (T_\nu)^m\backslash(Z_\nu)^m\}$ is countable. For each $m \in \mathbb{N}$ let $N_m = \{1, \ldots, m\}$ and denote by Φ_m the set of functions $\varphi : N_m \to \{\alpha : \alpha \leq \nu\}$ such that $\varphi(i) = \nu$ for at least one $i \in N_m$.

For every $\varphi \in \Phi_m$ define a map $g_\varphi : X \to X^m$ as follows: for any $x \in X$ we let $g_\varphi(x) = (y_1, \ldots, y_m)$ where $y_i = z_{\varphi(i)}$ if $\varphi(i) < \nu$ and $y_i = x$ whenever $\varphi(i) = \nu$. It is straightforward that g_φ is continuous for every $\varphi \in \Phi_m$. Furthermore, if $\alpha \in \Gamma_m$ then U_α contains all points of $(T_\nu)^m$ which have at least one coordinate from Q; consequently, $g_\varphi(Q) \subset U_\alpha$ for any $\alpha \in \Gamma_m$.

Therefore $W_m = \bigcap \{ g_\varphi^{-1}(U_\alpha) : \varphi \in \Phi_m, \ \alpha \in \Gamma_m \}$ is a G_δ-subset of Z such that $Q \subset W_m$ for every $m \in \mathbb{N}$. Thus $W = \bigcap \{ W_m : m \in \mathbb{N} \}$ is a G_δ-subset of Z which contains Q; so we can choose a point $z_\nu \in W \backslash Z_\nu$. It is evident that (1) is still true for all $\alpha \leq \nu$; to see that (2) is also fulfilled assume that $m \in \mathbb{N}, U_\beta \in \mathcal{B}_m$ and $U_\beta \supset (T_\nu)^m \backslash (Z_\nu)^m$ for some $\beta \leq \nu$. If $\max\{\alpha_1, \ldots, \alpha_m\} = \nu$ then define $\varphi \in \Phi_m$ by $\varphi(i) = \alpha_i$ for all $i \in N_m$. Then $g_\varphi(z_\nu) = (z_{\alpha_1}, \ldots, z_{\alpha_m})$ and it follows from $z_\nu \in W \subset W_m \subset g_\varphi^{-1}(U_\beta)$ that $g_\varphi(z_\nu) \in U_\beta$, i.e., $(z_{\alpha_1}, \ldots, z_{\alpha_m}) \in U_\beta$ as required.

This proves that our inductive procedure can be continued to construct a set $Y = \{ z_\alpha : \alpha < \omega_1 \}$ such that (1) and (2) are satisfied for all $\nu < \omega_1$; let $T = Q \cup Y$. Letting $\mathcal{C}_n = \mathcal{B}_n | T^n$ for any $n \in \mathbb{N}$ we obtain a sequence $\{ \mathcal{C}_n : n \in \mathbb{N} \}$ of bases in the respective finite powers of T such that

(3) for every $m \in \mathbb{N}$ the family $\mathcal{C}_m$ is closed under countable unions and $Y^m \backslash U$ is countable for any $U \in \mathcal{C}_m$ with $T^m \backslash Y^m \subset U$.

The first part of (3) being evident let us verify the second one. If $T^m \backslash Y^m \subset U$ for some $U \in \mathcal{C}_m$ then there is $U' \in \mathcal{B}_m$ such that $U = U' \cap T^m$. We have $U' = U_\beta$ for some $\beta < \omega_1$. Now, if $z = (z_{\alpha_1}, \ldots, z_{\alpha_m}) \in Y^m \backslash U$ then let $\alpha = \max\{\alpha_1, \ldots, \alpha_m\}$. If $\alpha \geq \beta$ then observe that $U_\beta \supset T^m \backslash Y^m \supset (T_\alpha)^m \backslash (Z_\alpha)^m$ and hence we can apply (2) to conclude that $z \in U_\beta \cap T^m = U$; this contradiction shows that $\alpha < \beta$ and hence $Y^m \backslash U \subset H = \{ (z_{\alpha_1}, \ldots, z_{\alpha_m}) : \max\{\alpha_1, \ldots, \alpha_m\} < \beta \}$. Since H is countable, so is $Y^m \backslash U$ and hence (3) is proved.

To finally see that T^n is Lindelöf for any $n \in \mathbb{N}$ observe that T is Lindelöf because it is concentrated around the set Q. Next, assume that $n \geq 2$ and T^m is Lindelöf for any $m < n$. To see that T^n is also Lindelöf take an open cover $\mathcal{U}$ of the space T^n. Since $\mathcal{C}_n$ is a base in T^n, we can assume, without loss of generality, that $\mathcal{U} \subset \mathcal{C}_n$. For every $i \in \{1, \ldots, n\}$ let $\pi_i : T^n \to T$ be the natural projection of T^n onto its i-th factor. It is immediate that $T^n \backslash Y^n = \bigcup \{ \pi_i^{-1}(q) : q \in Q, \ i \leq n \}$; since $\pi_i^{-1}(q)$ is homeomorphic to T^{n-1} for any $q \in Q$ and $i \leq n$, the space $T^n \backslash Y^n$ is a countable union of Lindelöf subspaces of T^n; so $l(T^n \backslash Y^n) = \omega$.

Thus we can choose a countable $\mathcal{U}'$ such that $U = \bigcup \mathcal{U}' \supset T^n \backslash Y^n$. Since $U \in \mathcal{C}_n$, the set $Y^n \backslash U$ is countable by (3) and hence there is a countable $\mathcal{U}'' \subset \mathcal{U}$ for which $Y^n \backslash U \subset \bigcup \mathcal{U}''$. It is immediate that $\mathcal{U}' \cup \mathcal{U}''$ is a countable subcover of $\mathcal{U}$; so T^n is Lindelöf. Thus our inductive procedure shows that T^n is Lindelöf for every $n \in \mathbb{N}$ and hence Fact 1 is proved.

Returning to our solution, let $\sigma(A) = \{ s \in \mathbb{D}^\omega : s^{-1}(1) \subset A \text{ and } |s^{-1}(1)| < \omega \}$ for any infinite $A \subset \omega$; it is easy to see that $\sigma(A)$ is a countable dense-in-itself set. An easy consequence of CH is that we can choose a family $\mathcal{W} = \{ W_\alpha : \alpha < \omega_1 \} \subset \tau(\mathbb{D}^\omega)$ such that $\sigma = \sigma(\omega) \subset \bigcap \mathcal{W}$ and $\mathcal{W}$ is an outer base of σ in $\mathbb{D}^\omega$, i.e., for any $U \in \tau(\sigma, \mathbb{D}^\omega)$ there is $\alpha < \omega_1$ such that $W_\alpha \subset U$. Letting $H_\alpha = \bigcap \{ W_\beta : \beta \leq \alpha \}$

for any $\alpha < \omega_1$ we obtain a decreasing ω_1-sequence $\{H_\alpha : \alpha < \omega_1\}$ of G_δ-subsets of $\mathbb{D}^\omega$ such that, for any $U \in \tau(\sigma, \mathbb{D}^\omega)$, there is $\beta < \omega_1$ for which $H_\alpha \subset U$ for all $\alpha \geq \beta$.

Take an almost disjoint family $\{A_\alpha : \alpha < \omega_1\}$ of infinite subsets of ω (such a family exists in ZFC, see TFS-141). The set $G_\alpha = \{s \in H_\alpha : s^{-1}(1) \subset A_\alpha\}$ is G_δ in $\mathbb{D}^\omega$ and it follows from $\sigma \subset H_\alpha$ that $\sigma(A_\alpha) \subset G_\alpha$ and $\sigma(A_\alpha)$ is dense in G_α. Therefore G_α is a completely metrizable space without isolated points. This implies that G_α is uncountable; so we can choose a point $s_\alpha \in G_\alpha \backslash \sigma$. It is evident that the set $S = \{s_\alpha : \alpha < \omega_1\}$ is concentrated around σ and $\mathcal{A} = \{s_\alpha^{-1}(1) : \alpha < \omega_1\}$ is an almost disjoint family because $s_\alpha^{-1}(1) \subset A_\alpha$ for any $\alpha < \omega_1$. Let $S_\alpha = s_\alpha^{-1}(1)$ for every $\alpha < \omega_1$.

To associate a point to every S_α take an injective map $r : \mathcal{A} \to \mathbb{P}$ and let $\xi_\alpha = r(S_\alpha)$; the family $\mathcal{B}_\alpha = \{\{\xi_\alpha\} \cup (S_\alpha \backslash F) : F$ is a finite subset of $\omega\}$ will be the respective local base at ξ_α for any $\alpha < \omega_1$. On the set $M(\mathcal{A}) = \{\xi_\alpha : \alpha < \omega_1\} \cup \omega$ we generate a topology τ by the family $\bigcup\{\mathcal{B}_\alpha : \alpha < \omega_1\} \cup \exp \omega$ as a base. It is easy to see that in the space $M = (M(\mathcal{A}), \tau)$ all points of ω are isolated while S_α is a sequence which converges to ξ_α for any $\alpha < \omega_1$. The space M is Tychonoff, locally countable and locally compact; let $\Omega = \{\xi_\alpha : \alpha < \omega_1\}$.

For every $\alpha < \omega_1$ let $f_\alpha(x) = 1$ for all $x \in S_\alpha \cup \{\xi_\alpha\}$ and $f_\alpha(x) = 0$ for all $x \in M \backslash (S_\alpha \cup \{\xi_\alpha\})$. It is immediate that $f_\alpha \in C_p(M, \mathbb{D})$ for every $\alpha < \omega_1$.

For each $s \in \sigma$ define a function $g_s : M \to \mathbb{D}$ by requiring that $g_s(x) = 0$ for any $x \in M \backslash \omega$ and $g_s|\omega = s$. We will prove that the set $F = \{f_\alpha : \alpha < \omega_1\}$ is concentrated around the set $Q = \{g_s : s \in \sigma\}$ (here both F and Q are considered as subspaces of $C_p(M, \mathbb{D})$).

Let U be an open subset of $C_p(M, \mathbb{D})$ such that $Q \subset U$. For every $s \in \sigma$ choose a finite $P_s \subset M$ such that $U_s = \{f \in C_p(M, \mathbb{D}) : f|P_s = g_s|P_s\} \subset U$; furthermore, let $V_s = \{f \in \mathbb{D}^\omega : f|(P_s \cap \omega) = s|(P_s \cap \omega)\}$. It is evident that $V = \bigcup_{s \in \sigma} V_s \in \tau(\sigma, \mathbb{D}^\omega)$ so there is $\beta < \omega_1$ such that $s_\alpha \in V$ for any $\alpha \geq \beta$. The set $P = (\bigcup\{P_s : s \in \sigma\}) \cup \omega$ is countable; so there is $\gamma < \omega_1$ such that $\gamma > \beta$ and $f_\alpha|(P_s \cap \Omega) \equiv 0$ for any $s \in \sigma$ and $\alpha \geq \gamma$. Now, if $\alpha \geq \gamma$ then there is $s \in \sigma$ for which $s_\alpha \in V_s$; consequently, $f_\alpha|(P_s \cap \omega) = s_\alpha|(P_s \cap \omega) = g_s|(P_s \cap \omega)$ while $f_\alpha|(P_s \cap \Omega) \equiv 0$ and $g_s|(P_s \cap \Omega) \equiv 0$. Therefore $f_\alpha|P_s = g_s|P_s$, i.e., $f_\alpha \in U_s \subset U$. It turns out that $f_\alpha \in U$ for all $\alpha \geq \gamma$; so the set F is, indeed, concentrated around Q.

It is clear that $w(Q \cup F) \leq \omega_1$; so we can apply Fact 1 to conclude that there is an uncountable $E \subset F$ such that the space $(Q \cup E)^n$ is Lindelöf for any $n \in \mathbb{N}$. The space $M' = \omega \cup \{\xi_\alpha : f_\alpha \in E\}$ is locally compact; so we can consider its one-point compactification X; let a be the unique point of the set $X \backslash M'$. To see that X is scattered, take any $A \subset X$. If $A \cap \omega \neq \emptyset$ then any point of $A \cap \omega$ is isolated in X and hence in A. If $A \subset X \backslash \omega$ then either A is a singleton (in which case there is nothing to prove) or $A \cap (M' \backslash \omega) \neq \emptyset$. It is evident that any point of $(M' \backslash \omega) \cap A$ is isolated in $(M' \backslash \omega) \cup \{a\}$ and hence in A. Thus X is a scattered compact space.

We will define next an addition operation which is natural for $C_p(X, \mathbb{D})$ but different from the usual addition operation in $C_p(X)$. Namely, if $f, g \in C_p(X, \mathbb{D})$

define $h = f \dotplus g \in C_p(X, \mathbb{D})$ by requiring, for every point $x \in X$, that $h(x) = 0$ if $f(x) = g(x)$ and $h(x) = 1$ otherwise. It is easy to see that $f \dotplus g = f + g - 2fg$; so this operation is continuous by TFS-115 and TFS-166.

If $\pi : C_p(M, \mathbb{D}) \to C_p(M', \mathbb{D})$ is the restriction map then $\pi|(Q \cup E)$ is a homeomorphism because $M \backslash M'$ is closed in M and $f|(M \backslash M') \equiv 0$ for any $f \in Q \cup E$. For every $f \in Q \cup E$ define a function $h_f \in C_p(X, \mathbb{D})$ by requiring that $h_f|M' = f|M'$ and $h_f(a) = 0$. It is easy to see that the set $H = \{h_f : f \in Q \cup E\} \subset C_p(X, \mathbb{D})$ is still homeomorphic to $Q \cup E$ and hence H^n is Lindelöf for any $n \in \mathbb{N}$.

The set $R_n = \{f_1 \dotplus \ldots \dotplus f_n : f_i \in H \text{ for any } i \leq n\}$ is a continuous image of H^n under an evident map for any $n \in \mathbb{N}$ (see TFS-115). This implies that $(R_n)^k$ is Lindelöf for any $k \in \mathbb{N}$. Let us prove that

(4) $\quad R = \bigcup\{R_n : n \in \mathbb{N}\} = I = \{f \in C_p(X, \mathbb{D}) : f(a) = 0\}.$

It is evident that $R \subset I$; to establish the opposite inclusion take any $f \in I$. Since $f(a) = 0$, the set $f^{-1}(1)$ is compact and contained in M'; as a consequence, there is a finite $K \subset \omega_1$ such that $f^{-1}(1) \subset \{\{\xi_\alpha\} \cup S_\alpha : \alpha \in K\}$. Since $\mathcal{A}$ is almost disjoint, there is a finite set $L \subset \omega$ such that the family $\{S_\alpha \backslash L : \alpha \in K\}$ is disjoint. Choose $s \in \sigma$ for which $s^{-1}(1) \subset L$ and $s|L = f|L$. For every $\alpha \in K$ let $t_\alpha(x) = 1$ for every $x \in L \cap S_\alpha$ and $t(x) = 0$ if $x \in X \backslash (L \cap S_\alpha)$.

It is immediate that $f = \sum\{h_{f_\alpha} \dotplus t_\alpha : \alpha \in K\} \dotplus h_s$; so $f \in R$ and (4) is proved. It is easy to see that $R_n \subset R_{n+1}$ for any $n \in \mathbb{N}$ which implies that $I^k = \bigcup\{(R_n)^k : n \in \mathbb{N}\}$ and therefore I^k is a Lindelöf space for any $k \in \mathbb{N}$. Now, $C_p(X, \mathbb{D})$ is homeomorphic to $I \times \mathbb{D}$ and hence $C_p(X, \mathbb{D})^k \simeq I^k \times \mathbb{D}^k$ is Lindelöf for any $k \in \mathbb{N}$.

Finally, assume that $(C_p(X, \mathbb{D}))^\omega$ is Lindelöf. Then $(C_p(X, \mathbb{D}))^\omega \times \omega^\omega$ is also Lindelöf by Fact 3 of U.093. However, X is a space which satisfies the hypothesis of Problem 077, so $C_p(X, \mathbb{D}) \times \omega^\omega$ is not Lindelöf; this contradiction shows that $(C_p(X, \mathbb{D}))^\omega$ is not Lindelöf and makes our solution complete.

U.096. *Prove that there is a scattered, separable, zero-dimensional σ-compact space X with $(C_p(X, \mathbb{D}))^n$ Lindelöf for each natural n, while $(C_p(X, \mathbb{D}))^\omega$ is not Lindelöf.*

Solution. There is a set $A \subset I = [0, 1]$ such that $(I[A])^\omega$ is not Lindelöf but $l^*(I[A]) = \omega$ and (see Problem 091). Fix some countable base $\mathcal{B}$ in I such that $U \neq \emptyset$ for any $U \in \mathcal{B}$ and let $\mathcal{C} = \{(U, V) \in \mathcal{B} \times \mathcal{B} : \overline{U} \subset V\}$. By normality of I, for any $\mu = (U, V) \in \mathcal{C}$, we can choose a function $f_\mu \in C_p(I)$ such that $f_\mu(U) \subset \{1\}$ and $f_\mu(I \backslash V) \subset \{0\}$. It is evident that the set $\tilde{S} = \{f_\mu : \mu \in \mathcal{C}\}$ separates the points and closed subsets of I.

For any point $x \in I$ let $u_x(y) = 0$ for any $y \in I \backslash \{x\}$ and $u_x(x) = 1$; denote by u the function which is identically zero on the set I. It is easy to see that the set $K_A = \{u_x : x \in I \backslash A\} \cup \{u\}$ is compact and $K_A \subset C_p(I[A])$; besides, $K_A \cup \tilde{S}$ separates the points and closed subsets of $I[A]$ and $K_A \cap \tilde{S} = \emptyset$. Choose a countable $S'' \subset C_p(I) \backslash \{u\}$ which is dense in $C_p(I)$ and let $S = \tilde{S} \cup S''$; it is evident that $K_A \cap S = \emptyset$.

Let $X = (K_A \cup S)[K_A]$; this means that we declare the points of S isolated and leave the same topology at all points of K_A. It is an easy exercise that the space X is σ-compact and scattered. Furthermore, S is a countable dense set of isolated points of X while $K_A = X \backslash S$ is an uncountable compact space with a unique non-isolated point. Let us show next that

(1) $I[A]$ embeds in $C_p(X)$ as a closed subspace.

For any $x \in I$ let $\varphi_x(f) = f(x)$ for any $f \in S$. Then $\varphi_x \in C_p(S)$ and the map $\varphi : I \to C_p(S)$ defined by $\varphi(x) = \varphi_x$ for any $x \in I$ is continuous (see TFS-166). Since S separates the points and closed subsets of I, the map φ is an embedding; let $I' = \varphi(I) \subset C_p(S)$.

Analogously, let $\psi_x(f) = f(x)$ for any $f \in K_A \cup S$ and $x \in I$. Define a map $\psi : I[A] \to C_p(K_A \cup S)$ by $\psi(x) = \psi_x$ for any $x \in I$; then ψ is an embedding because $K_A \cup S \subset C_p(I[A])$ separates the points and closed subsets of $I[A]$. Let $I_0 = \psi(I[A])$; observe that X has the underlying set $K_A \cup S$ but the topology of X is stronger; so $I_0 \subset C_p(K_A \cup S) \subset C_p(X)$. If S' is the set S with the discrete topology then S' is a subspace of X; since $\tau(S) \subset \tau(S')$, we have $I' \subset C_p(S) \subset C_p(S')$.

Let $\pi(f) = f|S'$ for any function $f \in C_p(X)$. This gives us the restriction map $\pi : C_p(X) \to Z = \pi(C_p(X)) \subset C_p(S')$ which is continuous and injective because S' is dense in X. It is easy to check that $\pi(\psi_x) = \varphi_x$ for any $x \in I$ and therefore $\pi(I_0) = I'$; the set I' is closed in Z being compact; so $I_0 = \pi^{-1}(I')$ is closed in $C_p(X)$. Since I_0 is homeomorphic to $I[A]$, the property (1) is proved.

An immediate consequence of (1) is that $(C_p(X))^\omega$ is not Lindelöf.

Fact 1. Suppose that T is a space and $Z \subset C_p(T)$. If the set Z' of non-isolated points of Z is compact then $C_p(Z, \mathbb{D})$ is a countable union of continuous images of closed subsets of products of finite powers of T with a compact space.

Proof. For any $m, n \in \mathbb{N}$ consider the set $M(m,n) = \{\varphi \in \mathbb{D}^Z :$ there exists a point $(t_1, \ldots, t_n) \in T^n$ such that, for any $f \in Z'$ and $g \in Z$, if $|f(t_i) - g(t_i)| < \frac{1}{m}$ for all $i \leq n$ then $\varphi(f) = \varphi(g)\}$. We claim that

(2) $C_p(Z, \mathbb{D}) = M = \bigcup\{M(m,n) : m, n \in \mathbb{N}\}$.

Assume first that $\varphi \in M$; to see that φ is continuous on Z it suffices to show that it is continuous at every point of Z'; so take any $f \in Z'$. There are $m, n \in \mathbb{N}$ and a point $(t_1, \ldots, t_n) \in T^n$ such that $|g(t_i) - f(t_i)| < \frac{1}{m}$ for all $i \leq n$ implies that $\varphi(f) = \varphi(g)$. The set $O = O(f, t_1, \ldots, t_n, m) = \{g \in Z : |g(t_i) - f(t_i)| < \frac{1}{m}$ for all $i \leq n\}$ is an open neighbourhood of f in Z and it is immediate that $\varphi(O) = \{\varphi(f)\}$ which proves that φ is continuous at the point f. Since $f \in Z'$ was chosen arbitrarily, we proved that φ is continuous on Z and therefore $M \subset C_p(Z, \mathbb{D})$.

To establish the opposite inclusion take any $\varphi \in C_p(Z, \mathbb{D})$. For any $f \in Z'$ there exist $k_f, n_f \in \mathbb{N}$ and $(t_1^f, \ldots, t_{n_f}^f) \in T^{n_f}$ such that $g \in Z$ and $|g(t_i^f) - f(t_i^f)| < \frac{1}{k_f}$ for all $i \leq n_f$ implies $\varphi(g) = \varphi(f)$.

Since $\{O(f, t_1^f, \ldots, t_{n_f}^f, 3k_f) : f \in Z'\}$ is an open cover of the compact space Z', there is a finite $A \subset Z'$ such that the family $\{O(f, t_1^f, \ldots, t_{n_f}^f, 3k_f) : f \in A\}$

covers Z'. Let $n = \sum\{n_f : f \in A\}$, $m = \max\{3k_f : f \in A\}$ and take any $t = (t_1, \ldots, t_n) \in T^n$ such that $t_i^f \in \{t_1, \ldots, t_n\}$ for any $f \in A$ and $i \leq n_f$.

Now, if $f \in Z'$, $g \in Z$ and $|f(t_i) - g(t_i)| < \frac{1}{m}$ for all $i \leq n$ then there is $h \in A$ for which $f \in O(h, t_1^h, \ldots, t_{n_h}^h, 3k_h)$ and hence $\varphi(f) = \varphi(h)$. Since $\{t_1^h, \ldots, t_{n_h}^h\} \subset \{t_1, \ldots, t_n\}$, we have

$$|g(t_i^h) - h(t_i^h)| \leq |g(t_i^h) - f(t_i^h)| + |f(t_i^h) - h(t_i^h)| < \frac{1}{m} + \frac{1}{3k_h} < \frac{1}{k_h}$$

for every $i \leq n_h$ which implies that $\varphi(g) = \varphi(h)$ and hence $\varphi(f) = \varphi(g)$. This shows that $\varphi \in M(m, n)$; so the equality (2) is proved.

Finally, consider the set $S(m, n) = \{(\varphi, t) \in \mathbb{D}^Z \times T^n : t = (t_1, \ldots, t_n)$ and, for any $f \in Z'$ and $g \in Z$, if $g \in O(f, t_1, \ldots, t_n, m)$ then $\varphi(f) = \varphi(g)\}$ for any numbers $m, n \in \mathbb{N}$. To prove that $S(n, m)$ is closed in $\mathbb{D}^Z \times T^n$ for any $m, n \in \mathbb{N}$, take a point $(\varphi, t) \in \mathbb{D}^Z \backslash S(m, n)$. Then $t = (t_1, \ldots, t_n)$ and there exist $f \in Z'$, $g \in Z$ for which $g \in O(f, t_1, \ldots, t_n, m)$ while $\varphi(g) \neq \varphi(f)$. Since the functions f and g are continuous on T, the set $V = \{(s_1, \ldots, s_n) \in T^n : g \in O(f, s_1, \ldots, s_n, m)\}$ is open in T^n; besides, the set $W = \{\psi \in \mathbb{D}^Z : \psi(g) \neq \psi(f)\}$ is open in $\mathbb{D}^Z$. It is immediate that $\varphi \in W$, $t \in V$ and $(W \times V) \cap S(m, n) = \emptyset$; so any point $(\varphi, t) \in \mathbb{D}^Z \backslash S(m, n)$ has a neighbourhood $W \times V$ disjoint from $S(m, n)$. Thus $\mathbb{D}^Z \backslash S(m, n)$ is open in $\mathbb{D}^Z \backslash S(m, n)$, i.e., $S(m, n)$ is closed in $\mathbb{D}^Z \times T^n$.

It is immediate that the set $M(m, n)$ is the image of $S(m, n)$ under the projection of $\mathbb{D}^Z \times T^n$ onto its first factor; so (2) implies that $C_p(Z, \mathbb{D})$ is a countable union of continuous images of the sets $S(m, n)$. Fact 1 is proved.

Returning to our solution, for any $s \in S$, let $r_s(s) = 1$ and $r_s(x) = 0$ for all points $x \in (K_A \cup S) \backslash \{s\}$. It is evident that $r_s \in C_p(X)$ for any $s \in S$; let $w(x) = 0$ for any $x \in K_A \cup S$. It is evident that the set $L = \{w\} \cup \{r_s : s \in S\}$ is compact. Besides, $I_0 \cup L$ generates the topology of X; so X embeds in $C_p(I_0 \cup L)$ (see TFS-166). If $T = I_0 \cup L$ then $l^*(T) = \omega$. By Fact 1, the space $C_p(X, \mathbb{D})$ is a countable union of continuous images of closed subsets of products of finite powers of T with a compact space. Observe that $l^*(T^n \times K) = \omega$ whenever $n \in \mathbb{N}$ and K is compact. As a consequence, if $F = \bigcup\{F_n : n \in \mathbb{N}\}$ and, for any $n \in \mathbb{N}$, the space F_n is a continuous image of a closed subset of $T^{m_n} \times K_n$ for some compact K_n and $m_n \in \mathbb{N}$ then $l^*(F) = \omega$. This proves that the space $(C_p(X, \mathbb{D}))^n$ is Lindelöf for any $n \in \mathbb{N}$; so our solution is complete.

U.097. *Assume MA+¬CH. Let X be a space with $l^*(X) = \omega$. Prove that any separable compact subspace of $C_p(X)$ is metrizable.*

Solution. If P is a set then $\text{Fin}(P)$ is the family of all finite subsets of P. Given a space Z and $Y \subset C_p(Z)$ let $\varphi_x(f) = f(x)$ for any $x \in Z$ and $f \in Y$. Then $\varphi_x \in C_p(Y)$ for any $x \in Z$ and we have a continuous map $\varphi : Z \to C_p(Y)$ defined by $\varphi(x) = \varphi_x$ for any $x \in Z$ (see TFS-166). We will call φ the Y-*reflection map*, *or the reflection map of Z with respect to Y.*

Say that a compact space K is a *Reznichenko space* if $w(K) = \omega_1$, the set I of isolated points of K is countable and dense in K and there is a discrete $D \subset K \backslash I$ such that $|D| = \omega_1$ and $\overline{D} = K \backslash I$.

Fact 1. Under MA+¬CH, if Z is a separable compact space with $w(Z) = \omega_1$ then there is a Reznichenko space $K \subset Z \times Z$.

Proof. Since Z is not metrizable, the diagonal of $Z \times Z$ is not a G_δ-subset of $Z \times Z$ (see SFFS-091) and hence $Z \times Z$ is not perfectly normal. By SFFS-061 there is a discrete $D' \subset Z \times Z$ such that $|D'| = \omega_1$.

The space $Z \times Z$ is separable; so the set S of isolated points of $Z \times Z$ is countable and hence $D = D' \backslash S$ has cardinality ω_1. Consequently, $M = \overline{D}$ is nowhere dense in $Z \times Z$. Now apply Problem 079 to find a countable $I \subset Z \times Z$ such that $\overline{I} \backslash I = M$ and all points of I are isolated in the compact space $K = I \cup M$. It is evident that $K \subset Z \times Z$ is a Reznichenko space; so Fact 1 is proved.

Fact 2. Under MA+¬CH, if K is a Reznichenko space and A is uniformly dense in $C_p(K)$ then A is not Lindelöf.

Proof. By definition, the set I of isolated points of K is countable and dense in K and there is a discrete $D \subset K \backslash I$ such that $|D| = \omega_1$ and $\overline{D} = K \backslash I$.

Let $\Phi = \{f \in C_p(K) : f(K \backslash I) \subset [-1, 1]\}$; it is evident that Φ is a closed subset of $C_p(K)$. We claim that

(1) $\Phi = \{f \in C_p(K) : \text{the set } \{x \in I : |f(x)| \geq 1 + \frac{1}{m}\} \text{ is finite for any } m \in \mathbb{N}\}$.

To prove (1) assume that $f \in C_p(K)$, $m \in \mathbb{N}$ and there is an infinite set $S \subset I$ such that $|f(x)| \geq 1 + \frac{1}{m}$ for any $x \in S$. Then there is an accumulation point $y \in K \backslash I$ for the set S and therefore $|f(y)| \geq 1 + \frac{1}{m} > 1$ which shows that $f \notin \Phi$.

Now, suppose that $f \in C_p(K)$ and, the set $\{x \in I : |f(x)| \geq 1 + \frac{1}{m}\}$ is finite for any $m \in \mathbb{N}$. If $f \notin \Phi$ then there is $y \in K \backslash I$ and $m \in \mathbb{N}$ with $|f(y)| > 1 + \frac{1}{m}$ which implies that the set $\{x \in I : |f(x)| > 1 + \frac{1}{m}\}$ is infinite because it contains y in its closure. This contradiction shows that $f \in \Phi$ and finishes the proof of (1).

Denote by $\pi : C_p(K) \to C = \pi(C_p(K)) \subset \mathbb{R}^I$ the restriction map given by $\pi(f) = f|I$ for any $f \in C_p(K)$. Then π is a condensation because I is dense in K (see TFS-152). The set $P(m, A) = \{f \in C : \{x \in I : |f(x)| \geq 1 + \frac{1}{m}\} \subset A\}$ is G_δ in C for any finite $A \subset I$ and $m \in \mathbb{N}$. To see it observe that, for any $x \in I$, the set $O_x^m = \{f \in C : |f(x)| < 1 + \frac{1}{m}\}$ is open in C and $P(m, A) = \bigcap\{O_x^m : x \in I \backslash A\}$. It is an immediate consequence of (1) that

(2) $\pi(\Phi) = \bigcap\{\bigcup\{P(m, A) : A \in \text{Fin}(I)\} : m \in \mathbb{N}\}$,

and hence $\pi(\Phi)$ is a Borel subset of C. Since all points of D are isolated in $K \backslash I$, for any $d \in D$, there is a function $f_d \in C_p(K, [0, 2])$ such that $f_d(d) = 2$ and $f_d(x) = 0$ for any $x \in (K \backslash I) \backslash \{d\}$. Since A is uniformly dense in $C_p(K)$, we can find $g_d \in A$ such that $|g_d(x) - f_d(x)| < \frac{1}{4}$ for any $x \in K$ and $d \in D$. The subspace $G = \{g_d : d \in D\}$ of the space $C_p(K)$ is discrete because the set $W_d = \{f \in C_p(K) : |f(d) - g_d(d)| < \frac{1}{4}\}$ is open in $C_p(K)$ and $W_d \cap G = \{g_d\}$ for any $d \in D$.

Next observe that $G \subset C_p(K) \backslash \Phi$ but all accumulation points of G are in Φ and hence $G \cup \Phi$ is closed in $C_p(K)$. Indeed, if $d \in D$ then $|g_e(d)| < \frac{1}{4}$ for any $e \in D \backslash \{d\}$ and hence $f \in \overline{D} \backslash D$ implies $|f(d)| \leq \frac{1}{4}$ for any $d \in D$. Since D is dense in $K \backslash I$, we have $|f(x)| \leq \frac{1}{4}$ for every $x \in K \backslash I$, and hence $f \in \Phi$.

It follows from (2) that $E = C \backslash \pi(\Phi)$ is a Borel subset of C and $\pi(G)$ is an uncountable subset of E. Let E' be a Borel subset of $\mathbb{R}^I$ such that $E' \cap C = E$ (see Fact 1 of T.319). Since $\mathbb{R}^I$ is completely metrizable, the set E' is analytic (see SFFS-334) and hence there is a family $\mathcal{C} = \{P_s : s \in \omega^\omega\}$ of compact subsets of $\mathbb{R}^I$ such that $E' = \bigcup \mathcal{C}$ and $s \leq t$ implies $P_s \subset P_t$ (see SFFS-391). In particular, $E \subset \bigcup \mathcal{C}$ and hence, for any $d \in D$ there is $s(d) \in \omega^\omega$ such that $\pi(g_d) \subset P_{s(d)}$. The set $T = \{s(d) : d \in D\} \subset \omega^\omega$ has cardinality $< \mathfrak{c}$; so we can apply MA+¬CH to conclude that there is $u \in \omega^\omega$ for which $s(d) \leq^* u$ for any $d \in D$ (see Fact 1 of T.395; recall that if $s, t \in \omega^\omega$ then $s \leq^* t$ means that there is $m \in \omega$ such that $s(n) \leq t(n)$ for any $n \geq m$).

As a consequence, the set $Q = \{s \in \omega^\omega : s(n) \neq u(n)$ for at most finitely many $n \in \omega\}$ is countable and

(3) for any $d \in D$ there is $s \in Q$ such that $s(d) \leq s$.

This implies that $P = \bigcup \{P_s : s \in Q\}$ is a σ-compact set with $\pi(G) \subset P$. Thus there is an uncountable set $G' \subset G$ for which $\pi(G') \subset P_s$ for some $s \in Q$; since $P_s \cap \pi(\Phi) = \emptyset$, we have $\mathrm{cl}_C(\pi(G')) \cap \pi(\Phi) = \emptyset$ which shows that the set $G' = \pi^{-1}(\mathrm{cl}_C(\pi(G'))) \cap (\Phi \cup G')$ is closed in $C_p(K)$. Since $G \supset G'$ is discrete, we found an uncountable closed discrete subset G' in the space A which shows A is not Lindelöf and hence Fact 2 is proved.

Fact 3. Suppose that Z is a compact space such that some $Y \subset C_p(Z)$ separates the points of Z and $l^*(Y) = \omega$. Then there is a uniformly dense $A \subset C_p(Z)$ such that $l^*(A) = \omega$ and, in particular, A is Lindelöf. No additional axioms are needed to prove this Fact.

Proof. Let $\mathcal{H} = \{T : T$ is a continuous image of $Y^m \times M$ for some $m \in \mathbb{N}$ and a metrizable compact space $M\}$. It is immediate that $l^*(H) = \omega$ for any $H \in \mathcal{H}$. If $A \subset C_p(K)$ is the minimal algebra that contains Y then $A = \bigcup_{n \in \omega} A_n$ where $A_n \in \mathcal{H}$ for any $n \in \omega$ (see Problem 006) and hence $l^*(A) = \omega$. Since A separates the points of K, it is uniformly dense in $C_p(K)$ by TFS-191; so Fact 3 is proved.

Fact 4. Under MA+¬CH, if K is a separable compact space and some $Y \subset C_p(K)$ with $l^*(Y) = \omega$ separates the points of K then K is metrizable.

Proof. The space K embeds in $C_p(Y)$ by TFS-166; so if Y is separable then we have $w(K) = iw(K) \leq iw(C_p(Y)) = d(Y) = \omega$ and hence K is metrizable. If Y is not separable then there is a left-separated $S \subset Y$ such that $|S| = \omega_1$ (see SFFS-004). The space $C_p(K)$ is monolithic (see SFFS-118 and SFFS-154); so $nw(\overline{S}) = \omega_1$.

For the set $T = \overline{S}$ let $e : K \to C_p(T)$ be the reflection map of K with respect to T; the space $K' = e(K)$ is separable, compact and non-metrizable because, for the dual map $e^* : C_p(K') \to C_p(K)$ defined by $e^*(f) = f \circ e$ for any $f \in C_p(K')$, we have $T \subset e^*(C_p(K'))$ (see Fact 5 of U.086) and hence T embeds in $C_p(K')$

(see TFS-163). Since e^* is an embedding, its inverse $j : e^*(C_p(K')) \to C_p(K')$ is also a homeomorphism and it is easy to see that $T' = j(T)$ separates the points of K'. Thus K' is a compact space of weight ω_1 such that $T' \subset C_p(K')$, $l^*(T') = \omega$ and T' separates the points of K'.

Let $\pi_i : K' \times K' \to K'$ be the natural projection of $K' \times K'$ onto its i-th factor; as before, we denote by $\pi_i^* : C_p(K') \to C_p(K' \times K')$ the dual map of π_i for each $i \in \{1, 2\}$. The space $Y' = \pi_1^*(T') \cup \pi_2^*(T') \subset C_p(K' \times K')$ separates the points of $K' \times K'$ and it is immediate that $l^*(Y') = \omega$. Since K' is not metrizable, we can apply Fact 1 to find a Reznichenko space $M \subset K' \times K'$. If $q : C_p(K' \times K') \to C_p(M)$ is the restriction map then $Y'' = q(Y')$ separates the points of M and $l^*(Y'') = \omega$. By Fact 3 there is a uniformly dense Lindelöf $A \subset C_p(M)$ which contradicts Fact 2 and shows that Fact 4 is proved.

Returning to our solution assume that K is a separable compact subspace of the space $C_p(X)$ and let $\varphi : X \to C_p(K)$ be the K-reflection map. The set $X' = \varphi(X) \subset C_p(K)$ separates the points of K (see TFS-166) and $l^*(X') = \omega$; so K is metrizable by Fact 4 and hence our solution is complete.

U.098. *Assume* MA$+\neg$CH. *Let X be a separable compact space. Prove that, for any $Y \subset C_p(X)$ with $l^*(Y) = \omega$, we have $nw(Y) = \omega$.*

Solution. For any $x \in X$ let $\varphi_x(f) = f(x)$ for each $f \in Y$. Then $\varphi_x \in C_p(Y)$ and the map $\varphi : X \to C_p(Y)$ defined by $\varphi(x) = \varphi_x$ for any $x \in X$ is continuous (see TFS-166); let $X' = \varphi(X)$. Since $l^*(Y) = \omega$ and the compact space $X' \subset C_p(Y)$ is separable, we can apply Problem 097 to conclude that X' is metrizable. Another consequence of TFS-166 is that Y embeds in the space $C_p(X')$ and hence $nw(Y) \le nw(C_p(X')) = nw(X') \le w(X') = \omega$.

U.099. *Prove that there exists a separable σ-compact space X such that $(C_p(X))^\omega$ is Lindelöf and $s(X) > \omega$.*

Solution. If Z is a space and $A \subset Z$ then $Z[A]$ is the set Z with the topology generated by the family $\tau(Z) \cup \exp(Z \backslash A)$ as a subbase. In other words, the topology of $Z[A]$ is the same as in Z at all points of A while the points of $Z \backslash A$ are isolated in $Z[A]$.

Take disjoint sets $A, B \subset I = [0, 1]$ such that both spaces $(I[A])^\omega$ and $(I[B])^\omega$ are Lindelöf (see Problem 090). It follows from Problem 091 that $I[B]$ is Lindelöf if and only if, for every $K \subset I$ such that $K \simeq \mathbb{K}$, we have $K \cap B \ne \emptyset$. Since I is uncountable, we can find a set $K \subset I$ with $K \simeq \mathbb{K}$ and hence $K \simeq K \times K$ which shows that we can consider that $K \times K \subset I$ and hence $\mathcal{K} = \{K \times \{x\} : x \in K\}$ is a family of $\mathfrak{c}$-many disjoint copies of $\mathbb{K}$ in I. If $|B| < \mathfrak{c}$ then B cannot intersect all elements of $\mathcal{K}$. Thus $|B| = \mathfrak{c}$ and hence $|I \backslash A| = \mathfrak{c}$.

For any point $x \in I$ let $u_x(y) = 0$ for any $y \in I \backslash \{x\}$ and $u_x(x) = 1$; denote by u the function which is identically zero on I. It is easy to see that the set $K_A = \{u_x : x \in I \backslash A\} \cup \{u\}$ is compact and $K_A \subset C_p(I[A])$. Choose a countable $S \subset C_p(I) \backslash \{u\}$ which is dense in $C_p(I)$; it is evident that $K_A \cap S = \emptyset$.

The space $X = K_A \cup S$ is separable because the countable set S is dense in X. Since $K_A \setminus \{u\}$ is a discrete subspace of X, we have $s(X) \geq |I \setminus A| = \mathfrak{c} > \omega$. The space $X \subset C_p(I[A])$ is σ-compact and $(I[A])^\omega$ is Lindelöf; so $(C_p(X))^\omega$ is also Lindelöf by Fact 4 of U.093.

U.100. *Assume* $MA + \neg CH$. *Prove that there is a separable* σ-*compact space* X *such that* $C_p(X)$ *does not embed into* $C_p(Y)$ *for a separable compact space* Y.

Solution. By Problem 099, there exists a separable σ-compact space X such that $(C_p(X))^\omega$ is Lindelöf and $s(X) > \omega$. If $C_p(X)$ can be embedded in $C_p(Y)$ for some separable compact space Y then it follows from Problem 098 that $nw(C_p(X)) = \omega$ and hence we have $s(X) \leq nw(X) = nw(C_p(X)) = \omega$ which is a contradiction.

U.101. *Let* M_t *be a metrizable space for each* $t \in T$. *For an arbitrary point* $a \in M = \prod \{M_t : t \in T\}$, *prove that* $\Sigma(M, a)$ *is a Fréchet–Urysohn space. In particular,* $\Sigma(A)$ *is a Fréchet–Urysohn space for any* A.

Solution. Given a point $x \in \Sigma = \Sigma(M, a)$ let $\mathrm{supp}(x) = \{t \in T : x(t) \neq a(t)\}$; if we have a set $A \subset \Sigma$ then $\mathrm{supp}(A) = \bigcup \{\mathrm{supp}(x) : x \in A\}$. For any $S \subset T$ define a map $r_S : \Sigma \to \Sigma_S = \{x \in \Sigma : \mathrm{supp}(x) \subset S\}$ as follows: $r_S(x)(t) = x(t)$ if $t \in S$ and $r_S(x)(t) = a(t)$ for all $t \in T \setminus S$. It is an easy exercise that r_S is a continuous retraction of Σ onto Σ_S. Furthermore, if S is countable then Σ_S is metrizable being homeomorphic to $\prod \{M_t : t \in S\}$ (see TFS-207).

Suppose that $A \subset \Sigma$ and $x \in \overline{A}$. The set $S_0 = \mathrm{supp}(x)$ is countable; so Σ_{S_0} is metrizable and hence there is a countable set $B_0 \subset A$ such that $r_{S_0}(x) \in \overline{r_{S_0}(B_0)}$. Assume that $m \in \omega$ and we have constructed countable sets $B_0, \ldots, B_m \subset A$ and $S_0, \ldots, S_m \subset T$ such that

(1) $S_0 = \mathrm{supp}(x)$, $B_i \subset B_{i+1}$, $S_i \subset S_{i+1}$ and $\mathrm{supp}(B_i) \subset S_{i+1}$ for all $i < m$;
(2) $r_{S_i}(x) \in \overline{r_{S_i}(B_i)}$ for any $i \leq m$.

Let $S_{m+1} = \mathrm{supp}(B_m)$; since $\Sigma_{S_{m+1}}$ is metrizable, there is a countable $B' \subset A$ such that $r_{S_{m+1}}(x) \in \overline{r_{S_{m+1}}(B')}$; it is straightforward that if we let $B_{m+1} = B_m \cup B'$ then (1) holds for all $i \leq m$ and (2) is satisfied for every $i \leq m + 1$. Therefore our inductive procedure can be continued to construct families $\{S_i : i \in \omega\}$ and $\{B_i : i \in \omega\}$ for which (1)–(2) are fulfilled for all $m \in \omega$. Let $S = \bigcup_{i \in \omega} S_i$ and $B = \bigcup_{i \in \omega} B_i$.

To see that $x \in \overline{B}$ take any $V \in \tau(x, \Sigma)$; we can choose $U_t \in \tau(x(t), M_t)$ such that $U = (\prod_{t \in T} U_t) \cap \Sigma \subset V$ and $Q = \{t \in T : U_t \neq M_t\}$ is finite. There is a number $m \in \omega$ such that $S \cap Q = S_m \cap Q$; it follows from (2) that there is $y \in B_m$ such that $y(t) \in U_t$ for any $t \in Q \cap S_m$. Besides, $y(t) = a(t)$ for any $t \in T \setminus S$ and $Q \setminus S_m \subset T \setminus S$; so $y(t) = a(t) = x(t)$ for any $t \in Q \setminus S_m$. Therefore $y(t) \in U_t$ for any $t \in Q$ and hence $y \in U \cap B \subset V \cap B$, which proves that $x \in \overline{B}$.

Finally observe that $B \subset \Sigma_S$; the set Σ_S is closed in Σ; so the set $\overline{B} \subset \Sigma_S$ is metrizable. This, together with $x \in \overline{B}$, implies that there is a sequence $C \subset B \subset A$ which converges to x. Therefore Σ is a Fréchet–Urysohn space.

U.102. *Suppose that M_t is a metrizable space for each $t \in T$. For an arbitrary point $a \in M = \prod\{M_t : t \in T\}$, prove that $\Sigma(M, a)$ is a collectionwise normal space. In particular, $\Sigma(A)$ is a collectionwise normal space for any A.*

Solution. In this solution all spaces are assumed to be non-empty. We will say that ρ *is a metric on a space* Z if ρ generates the topology of Z.

Given a set Z a function $d : Z \times Z \to \mathbb{R}$ is called *a pseudometric on* Z if $d(x, x) = 0$, $d(x, y) = d(y, x) \geq 0$ and $d(x, y) \leq d(x, z) + d(z, y)$ for any $x, y, z \in Z$. In other words, a pseudometric d on a set Z is a function which has all properties of a metric except that $d(x, y) = 0$ need not imply that $x = y$. If Z is a space and d is a pseudometric on Z then we say that d is *a pseudometric on the space* Z if $d : Z \times Z \to \mathbb{R}$ is a continuous function. We can also define the notion of a diameter with respect to a pseudometric in the same way it is defined for a metric, i.e., if d is a pseudometric on Z and $A \subset Z$ then $\operatorname{diam}_d(A) = \sup\{d(x, y) : x, y \in A\}$. If Z is a space and $\mathcal{A}, \mathcal{B} \subset \exp Z$ we say that $\mathcal{A}$ *is inscribed in* $\mathcal{B}$ if, for any $A \in \mathcal{A}$ there is $B \in \mathcal{B}$ such that $A \subset B$.

Fact 1. Given a space Z, any σ-locally finite open cover of Z has a locally finite refinement.

Proof. Take an open cover $\mathcal{U}$ of the space Z such that $\mathcal{U} = \bigcup\{\mathcal{U}_n : n \in \omega\}$ and $\mathcal{U}_n$ is locally finite for any $n \in \omega$. Let $U_n = \bigcup \mathcal{U}_n$ and $\mathcal{P}_n = \{U \setminus (\bigcup_{k < n} U_k) : U \in \mathcal{U}_n\}$ for any $n \in \omega$. If $z \in Z$ then let $m = \min\{n \in \omega : z \in U_n\}$ and take $U \in \mathcal{U}_m$ with $z \in U$. Then $P = U \setminus (\bigcup_{k < m} U_k) \in \mathcal{P}_m$ and $z \in P$. Thus $\mathcal{P} = \bigcup_{n \in \omega} \mathcal{P}_n$ is a cover of Z; since it is evident that $\mathcal{P}$ is inscribed in $\mathcal{U}$, it is a refinement of $\mathcal{U}$.

To see that $\mathcal{P}$ is locally finite take a point $z \in Z$; then $z \in U_n$ for some $n \in \omega$ and hence $P \cap U_n = \emptyset$ for any $P \in \bigcup\{\mathcal{P}_i : i > n\}$. It is obvious that $\bigcup\{\mathcal{P}_i : i \leq n\}$ is a locally finite family; so there is $W \in \tau(z, Z)$ which intersects only a finite number of elements of $\bigcup\{\mathcal{P}_i : i < n\}$. It is clear that $W \cap U_n$ is an open neighbourhood of z which intersects only finitely many elements of $\mathcal{P}$; so $\mathcal{P}$ is, indeed, locally finite and hence Fact 1 is proved.

Fact 2. Suppose that Z is a space and $\mathcal{F}$ is a discrete family of closed subsets of Z. If there exists a locally finite closed cover $\mathcal{C}$ of the space Z such that every $C \in \mathcal{C}$ meets at most one element of $\mathcal{F}$ then the family $\mathcal{F}$ is open-separated, i.e., for any $F \in \mathcal{F}$ we can choose $O_F \in \tau(F, Z)$ such that the family $\{O_F : F \in \mathcal{F}\}$ is disjoint.

Proof. For any $F \in \mathcal{F}$ let $O_F = Z \setminus (\bigcup\{C \in \mathcal{C} : C \cap F = \emptyset\})$. The family $\mathcal{C}$ is closure-preserving (see Fact 2 of S.221); so $O_F \in \tau(F, Z)$ for any $F \in \mathcal{F}$. If $z \in O_F \cap O_G$ for some distinct $F, G \in \mathcal{F}$ then pick $C \in \mathcal{C}$ with $z \in C$. Since C can meet at most one of the sets F, G, we have either $C \cap F = \emptyset$ or $C \cap G = \emptyset$. In the first case $C \cap O_F = \emptyset$ and hence $z \notin O_F$ while in the second case $C \cap O_G = \emptyset$ and hence $z \notin O_G$; so we get a contradiction in both cases. This shows that the family $\{O_F : F \in \mathcal{F}\}$ is disjoint and hence Fact 2 is proved.

Returning to our solution let $\operatorname{supp}(x) = \{t \in T : x(t) \neq a(t)\}$ for any point $x \in \Sigma = \Sigma(M, a)$; if $A \subset \Sigma$ then $\operatorname{supp}(A) = \bigcup\{\operatorname{supp}(x) : x \in A\}$. For any $S \subset T$ define a map $r_S : \Sigma \to \Sigma_S = \{x \in \Sigma : \operatorname{supp}(x) \subset S\}$ as follows:

$r_S(x)(t) = x(t)$ if $t \in S$ and $r_S(x)(t) = a(t)$ for all $t \in T \backslash S$. It is an easy exercise that r_S is a continuous retraction of Σ onto Σ_S. Furthermore, if S is countable then Σ_S is metrizable being homeomorphic to $\prod\{M_t : t \in S\}$ (see TFS-207); fix a metric σ_S on the space Σ_S and let $d_S(x, y) = \sigma_S(r_S(x), r_S(y))$ for any $x, y \in \Sigma$. It is easy to see that d_S is a pseudometric on the space Σ for any non-empty countable $S \subset T$. Given a non-empty set $S \subset T$, a set $U \subset \Sigma$ is called S-*saturated* if $U = r_S^{-1}(U)$.

Suppose that $\mathcal{F}$ is a discrete family of closed subsets of Σ. We start with the family $\mathcal{U}_0 = \{\Sigma\}$; besides, choose a non-empty countable $S_0 \subset T$ and let $A(\Sigma) = S_0$, $\rho_\Sigma = d_S$. Assume that $n \in \mathbb{N}$ and we have locally finite open covers $\mathcal{U}_0, \ldots, \mathcal{U}_{n-1}$ of the space Σ with the following properties:

(1) for every $i < n$ and $U \in \mathcal{U}_i$ there is a non-empty countable set $A(U) \subset T$ such that $U \cap U'$ is $A(U)$-saturated for any $U' \in \mathcal{U}_i$;
(2) for every $i < n$ and $U \in \mathcal{U}_i$ a pseudometric ρ_U is chosen on the space Σ in such a way that $\rho_U|(\Sigma_{A(U)} \times \Sigma_{A(U)})$ is a metric on the space $\Sigma_{A(U)}$ and $\rho_U(x, y) = \rho_U(r_{A(U)}(x), r_{A(U)}(y))$ for any $x, y \in \Sigma$;
(3) for any $i \in \{1, \ldots, n-1\}$ we have $\mathcal{U}_i = \bigcup\{\mathcal{V}_U : U \in \mathcal{U}_{i-1}\}$ where, for every $U \in \mathcal{U}_{i-1}$, the family $\mathcal{V}_U$ has the following properties:

    (3.1) $\mathcal{V}_U \subset \tau(\Sigma)$ is locally finite in Σ and $\bigcup \mathcal{V}_U = U$;
    (3.2) every $V \in \mathcal{V}_U$ is $A(U)$-saturated;
    (3.3) the set V meets at most finitely many elements of $\mathcal{U}_{i-1}$;
    (3.4) $\mathrm{diam}_{\rho_{A(U)}}(V) \leq 2^{-i}$ for any $V \in \mathcal{V}_U$;
    (3.5) for any $V \in \mathcal{V}_U$, if $W \cap V \neq \emptyset$ for some $W \in \mathcal{U}_{i-1}$ then $A(W) \subset A(V)$;
    (3.6) for any $V \in \mathcal{V}_U$, if there are distinct $F_0, F_1 \in \mathcal{F}$ such that $\overline{V} \cap F_0 \neq \emptyset$ and $\overline{V} \cap F_1 \neq \emptyset$ then a point $x_V^i \in \overline{V} \cap F_i$ is chosen for each $i = 0, 1$;
    (3.7) $\mathrm{supp}(x_V^i) \subset A(V)$ for $i \in \{0, 1\}$ if x_V^0, x_V^1 are defined for V.
    (3.8) $\rho_V = d_{A(V)} + \sum\{\rho_W : W \in \mathcal{U}_{i-1}$ and $W \cap V \neq \emptyset\}$ for any $V \in \mathcal{V}_U$.

Take any $U \in \mathcal{U}_{n-1}$ and consider the family $\mathcal{W}_U = \{W \cap U : W \in \mathcal{U}_{n-1}\}$. It follows from (1) that $r_{A(U)}(V) = V \cap \Sigma_{A(U)}$ for any $V \in \mathcal{W}_U$; so the family $\mathcal{W}' = \{V \cap \Sigma_{A(U)} : V \in \mathcal{W}_U\}$ is locally finite in Σ. By metrizability of $\Sigma_{A(U)}$ we can choose a locally finite cover $\mathcal{V} \subset \tau(\Sigma_{A(U)})$ of the space $\Sigma_{A(U)}$ such that, for every $G \in \mathcal{V}$ we have $\mathrm{diam}_{\rho_{A(U)}}(G) \leq 2^{-n}$ and G intersects only finitely many elements of $\mathcal{W}'$. It is straightforward that the family $\mathcal{V}_U = \{r_{A(U)}^{-1}(G \cap U) : G \in \mathcal{V}\}$ is locally finite in Σ and the properties (3.1)–(3.4) are fulfilled for $\mathcal{V}_U$ and $i = n$.

For any $V \in \mathcal{V}_U$, if there are distinct $F_0, F_1 \in \mathcal{F}$ such that $\overline{V} \cap F_0 \neq \emptyset$ and $\overline{V} \cap F_1 \neq \emptyset$ then choose a point $x_V^i \in \overline{V} \cap F_i$ for each $i = 0, 1$; it follows from (3.3) that, for any $V \in \mathcal{V}_U$, we can choose a countable $A(V) \subset T$ such that $A(W) \subset A(V)$ for any $W \in \mathcal{U}_{n-1}$ with $W \cap V \neq \emptyset$ and $\mathrm{supp}(x_V^i) \subset A(V)$ for every $i \in \{0, 1\}$ if the points x_V^0, x_V^1 are defined for V.

Letting $\rho_V = d_{A(V)} + \sum\{\rho_W : W \in \mathcal{U}_{n-1}$ and $W \cap V \neq \emptyset\}$ for any $V \in \mathcal{V}_U$ we complete the construction of $\mathcal{V}_U$; so the family $\mathcal{U}_n = \bigcup\{\mathcal{V}_U : U \in \mathcal{U}_{n-1}\}$ satisfies all conditions and subconditions in (3). It is also easy to check, using Fact 2 of U.050, that the condition (2) holds for $\mathcal{U}_n$ as well.

To see that (1) is fulfilled for $\{\mathcal{U}_0, \ldots, \mathcal{U}_n\}$ take any $V, V' \in \mathcal{U}_n$. If $V \cap V' = \emptyset$ then there is nothing to prove so assume that $V \cap V' \neq \emptyset$. There are $U, U' \in \mathcal{U}_{n-1}$ such that $V \in \mathcal{V}_U$ and $V' \in \mathcal{V}_{U'}$. The family $\mathcal{S}_V = \{W \in \mathcal{U}_{n-1} : W \cap V \neq \emptyset\}$ is finite; since $V \cap V' \subset V \cap U'$, we have $V \cap U' \neq \emptyset$, i.e., $U' \in \mathcal{S}_V$, so it follows from (3.5) that $A(U') \subset A(V)$. The set V' is $A(U')$-saturated by (3.2); so it is $A(V)$-saturated (it is an easy exercise that if a set is S-saturated for some $S \subset T$ then it is S'-saturated for any $S' \supset S$). The set V is $A(U)$-saturated by (3.2); so it is $A(V)$-saturated as well. It is straightforward that the intersection of two $A(V)$-saturated sets is an $A(V)$-saturated set so $V \cap V'$ is $A(V)$-saturated and hence (1) also holds for $\{\mathcal{U}_0, \ldots, \mathcal{U}_n\}$.

Therefore our inductive procedure can be continued to construct a sequence $\{\mathcal{U}_n : n \in \omega\}$ of locally finite open covers of Σ for which the properties (1)–(3) hold for all $n \in \omega$. Let $\mathcal{U}_n^+ = \{U \in \mathcal{U}_n : \overline{U} \text{ meets at most one element of } \mathcal{F}\}$ for each $n \in \omega$. It turns out that the family $\mathcal{U}^+ = \bigcup\{\mathcal{U}_n^+ : n \in \omega\}$ is a cover of Σ.

To see this, assume that $x \in \Sigma \setminus (\bigcup \mathcal{U}^+)$ and choose $U_n \in \mathcal{U}_n$ such that $x \in U_n$ for all $n \in \omega$. Since every U_n intersects at least two distinct elements of $\mathcal{F}$, the points $x_{U_n}^0$, $x_{U_n}^1$ are defined for every $n \in \omega$. Let $S = \bigcup\{A(U_n) : n \in \omega\}$; it follows from (3.5) that $A(U_n) \subset A(U_{n+1})$ for any $n \in \omega$. An immediate consequence of (3.8) and (3.5) is that $\rho_{U_n} \leq \rho_{U_{n+1}}$ for any $n \in \omega$. This, together with the property (3.4) implies that $\rho_{U_n}(x_{U_m}^i, x) \leq \rho_{U_m}(x_{U_m}^i, x) \leq 2^{-m}$ for any $m \geq n$. Now apply (3.7) to see that $r_S(x_{U_n}^i) = x_{U_n}^i$ for any $n \in \omega$ and $i \in \mathbb{D}$. Therefore

$$\rho_{U_n}(r_{A(U_n)}(x_{U_m}^i), r_{A(U_n)}(x)) \leq \rho_{U_m}(r_{A(U_n)}(x_{U_m}^i), r_{A(U_n)}(x)) \leq 2^{-m}$$

for any $m \geq n$ and $i \in \mathbb{D}$. Since ρ_{U_n} is a metric on the space $\Sigma_{A(U_n)}$ by (2), the sequence $\{r_{A(U_n)}(x_{U_m}^i) : m \in \omega\}$ converges to $r_{A(U_n)}(x)$ for any $n \in \omega$ and hence the sequence $E_i = \{x_{U_n}^i : n \in \omega\}$ converges to $r_S(x)$ for any $i \in \mathbb{D}$. The family $\mathcal{F}$ being discrete, there is a set $O \in \tau(r_S(x), \Sigma)$ which meets at most one element of $\mathcal{F}$. It follows from $E_0, E_1 \to r_S(x)$ that there is $n \in \omega$ such that $x_{U_n}^0, x_{U_n}^1 \in O$ which is a contradiction because $x_{U_n}^0, x_{U_n}^1$ belong to distinct elements of $\mathcal{F}$.

Thus the family $\mathcal{U}^+$ is a σ-locally finite cover of Σ; there is a locally finite refinement $\mathcal{A}$ of $\mathcal{U}^+$ (see Fact 1). Then $\mathcal{A}' = \{\overline{A} : A \in \mathcal{A}\}$ is a locally finite closed cover of Σ such that every $B \in \mathcal{A}'$ intersects at most one element of $\mathcal{F}$. By Fact 2 the family $\mathcal{F}$ is open-separated. Therefore every discrete family of closed sets in Σ is open-separated; so Σ is collectionwise normal (see Fact 1 of S.302) and hence our solution is complete.

U.103. Let M_t be a second countable space for each $t \in T$. For an arbitrary point $a \in M = \prod\{M_t : t \in T\}$, prove that $ext(\Sigma(M, a)) \leq \omega$. In particular, $ext(\Sigma(A)) = \omega$ for any set A.

Solution. For the space M we have $c(M) = \omega$ by TFS-109. Since $\Sigma = \Sigma(M, a)$ is dense in M, we have $c(\Sigma) = c(M) = \omega$. Now, if D is a closed uncountable subset of Σ then, by collectionwise normality of the space Σ (see Problem 102), there is a disjoint family $\mathcal{O} = \{O_d : d \in D\} \subset \tau(\Sigma)$ such that $d \in O_d$ for any $d \in D$. Thus $\mathcal{O} \subset \tau^*(\Sigma)$ is uncountable and disjoint; this contradiction with $c(\Sigma) = \omega$ shows that every closed discrete subset of Σ is countable, i.e., $ext(\Sigma) = \omega$.

U.104. *Assume that M_t is a second countable space for any $t \in T$ and take any point $a \in M = \prod\{M_t : t \in T\}$. Prove that, if a compact space X is a continuous image of a dense subspace of $\Sigma(M, a)$ then X is metrizable. In particular, if a compact X is a continuous image of $\sigma(M, a)$ or $\Sigma(M, a)$ then X is metrizable.*

Solution. Recall that a space is *cosmic* if has a countable network. For any $A \subset T$ let $p_A : M \to M_A = \prod\{M_t : t \in A\}$ be the natural projection of M onto its face M_A.

Fact 1. Suppose that K is a non-empty compact space with no points of countable character. Then K cannot be represented as a union of $\leq \omega_1$-many cosmic subspaces.

Proof. To get a contradiction assume that $K = \bigcup\{N_\alpha : \alpha < \omega_1\}$ where N_α is cosmic for each $\alpha < \omega_1$. Let $F_0 = K$; suppose that $0 < \alpha < \omega_1$ and we have a family $\{F_\beta : \beta < \alpha\}$ of non-empty closed G_δ-subsets of K with the following properties:

(1) $F_\gamma \subset F_\beta$ whenever $\beta < \gamma < \alpha$;
(2) if $\beta < \alpha$ then $F_\beta \cap N_\gamma = \emptyset$ for any $\gamma < \beta$.

It is evident that $F'_\alpha = \bigcap\{F_\beta : \beta < \alpha\}$ is a non-empty closed G_δ-subset of K and hence $\chi(x, F'_\alpha) > \omega$ for any $x \in F'_\alpha$ for otherwise $\{x\}$ is a G_δ-subset of K (see Fact 2 of S.358) and hence $\chi(x, K) \leq \omega$ (see TFS-327), which is a contradiction. In particular, F'_α is not cosmic and therefore we can pick a point $x \in F'_\alpha \backslash N_\alpha$. Since N_α is Lindelöf, we can apply Fact 3 of S.358 to see that there is a closed G_δ-set $G \subset K$ such that $x \in G \subset X \backslash N_\alpha$. It is clear that $F_\alpha = F'_\alpha \cap G$ is a non-empty closed G_δ-subset of K such that (1) and (2) are fulfilled for the family $\{F_\beta : \beta \leq \alpha\}$. Consequently, we can continue our inductive construction to obtain a family $\{F_\alpha : \alpha < \omega_1\}$ of closed non-empty G_δ-subsets of K with the properties (1)–(2) fulfilled for each $\alpha < \omega_1$. Since K is compact, the property (1) implies that $F = \bigcap\{F_\alpha : \alpha < \omega_1\} \neq \emptyset$. It follows from (2) that $x \notin \bigcup\{N_\alpha : \alpha < \omega_1\}$ for any $x \in F$, which is a contradiction. Fact 1 is proved.

Fact 2. Suppose that N_t is a cosmic space for each $t \in T$ and take any point $u \in N = \prod\{N_t : t \in T\}$. If $|T| \leq \omega_1$ then $\Sigma(N, u)$ is a union of $\leq \omega_1$-many cosmic spaces.

Proof. Given any point $x \in \Sigma(N, u)$ let $\text{supp}(x) = \{t \in T : x(t) \neq u(t)\}$. Choose an enumeration $\{t_\alpha : \alpha < \omega_1\}$ of the set T and let $T_\alpha = \{t_\beta : \beta < \alpha\}$ for every $\alpha < \omega_1$. If $E_\alpha = \{x \in \Sigma(N, u) : \text{supp}(x) \subset T_\alpha\}$ then $nw(E_\alpha) \leq \omega$ for each $\alpha < \omega_1$ because $E_\alpha = \prod\{N_t : t \in T_\alpha\} \times \{u|(T \backslash T_\alpha)\}$ is homeomorphic to the cosmic space $\prod\{N_t : t \in T_\alpha\}$. It is evident that $\Sigma(N, u) = \bigcup\{E_\alpha : \alpha < \omega_1\}$; so Fact 3 is proved.

Returning to our solution take a dense $S \subset M$ and fix a continuous onto map $\varphi : S \to X$. If X is not metrizable then there is a continuous onto map $\delta : X \to Y$ such that $w(Y) = \omega_1$ (see SFFS-094). Denote by C the set of points of countable character of Y; since Y is also a continuous image of S, we have $nw(C) = \omega$ by TFS-299. If $C = Y$ then $w(Y) = nw(Y) = \omega$ (see Fact 4 of S.307) which is a contradiction. Thus there is $x \in Y \backslash C$. Since C is Lindelöf, there is a closed G_δ-set F in the space Y such that $x \in F \subset Y \backslash C$ (see Fact 3 of S.358). No point $y \in F$ can be a G_δ-set in F because otherwise $\psi(y, Y) = \omega$ by Fact 2 of S.358 and hence $\chi(y, Y) = \psi(y, Y) = \omega$ (see TFS-327) which shows that $y \in C$, a contradiction.

Since $\eta = \delta \circ \varphi$ maps S continuously onto Y, we can apply Fact 1 of T.109 to find a set $A \subset T$ such that $|A| \le \omega_1$ and there is a continuous map $h : p_A(S) \to Y$ for which $\eta = h \circ (p_A|S)$. It is easy to see that $S_A = p_A(S)$ is contained in $\Sigma(M_A, p_A(a))$; so we can apply Fact 2 to conclude that $\Sigma(M_A, p_A(a))$ and hence S_A is a union of $\le \omega_1$-many cosmic subspaces. The class of cosmic spaces is invariant under continuous maps; so the space Y is a union of $\le \omega_1$-many cosmic subspaces.

Every subspace of a cosmic space is cosmic, so $F \subset Y$ is a union of $\le \omega_1$-many cosmic subspaces which contradicts Fact 1 and proves that X is metrizable. To finish our solution observe that $\sigma(M, a)$ is dense in $\Sigma(M, a)$; so if S is one of the spaces $\sigma(M, a)$, $\Sigma(M, a)$ then any compact continuous image of S is metrizable.

U.105. *Prove that, if $|A| = \kappa \ge \omega$ then the space $\Sigma_*(A)$ is homeomorphic to $C_p(A(\kappa))$.*

Solution. Suppose that we are given infinite sets D and E and a bijection $\varphi : D \to E$. If we consider D and E to be discrete topological spaces then φ is a homeomorphism and hence the dual map $\varphi^* : \mathbb{R}^E \to \mathbb{R}^D$ defined by $\varphi(f) = f \circ \varphi$ for any $f \in \mathbb{R}^E$, is a homeomorphism as well (see TFS-163). It is straightforward that $\varphi^*(\Sigma_*(E)) = \Sigma_*(D)$ which shows that

(1) if D and E are infinite sets of the same cardinality then $\Sigma_*(D)$ is homeomorphic to $\Sigma_*(E)$.

Now assume that D is an infinite set, take a point $x \notin D$ and let $E = D \cup \{x\}$. The restriction map $\pi : \mathbb{R}^E \to \mathbb{R}^D$ defined by $\pi(f) = f|D$ for any $f \in \mathbb{R}^E$ is continuous and $\pi(\Sigma_*(E)) = \Sigma_*(D)$. For any $f \in \Sigma_*(D)$ let $e(f)(x) = 0$ and $e(f)(d) = f(d)$ for any $d \in D$. It is evident that $e : \Sigma_*(D) \to \Sigma_*(E)$ is continuous (in fact, it is an embedding but we won't need that).

Given $r \in \mathbb{R}$ let $u_r(x) = r$ and $u_r(d) = 0$ for all $d \in D$; then $u_r \in \Sigma_*(E)$. For any $f \in \Sigma_*(E)$ let $\eta(f) = (\pi(f), f(x)) \in \Sigma_*(D) \times \mathbb{R}$. Then $\eta : \Sigma_*(E) \to \Sigma_*(D) \times \mathbb{R}$ is a continuous map being the diagonal product of two restrictions. Furthermore, if $(h, r) \in \Sigma_*(D) \times \mathbb{R}$ then let $\mu(h, r) = e(h) + u_r$. We omit a trivial verification that $\mu : \Sigma_*(D) \times \mathbb{R} \to \Sigma_*(E)$ is continuous. The maps η and μ are mutually inverse; so they are both homeomorphisms. Since D and E have the same cardinality, the property (1) implies that $\Sigma_*(D) \simeq \Sigma_*(E) \simeq \Sigma_*(D) \times \mathbb{R}$ and hence

(2) $\Sigma_*(D)$ is homeomorphic to $\Sigma_*(D) \times \mathbb{R}$ for any infinite set D.

We are now ready to deal with our set A. Choose a point $y \notin A$ and let $A' = A \cup \{y\}$. It is easy to see that the family $\tau = \exp A \cup \{U \subset A' : y \in U$ and $|A \backslash U| < \omega\}$ is a topology on A' and $K = (A', \tau)$ is homeomorphic to $A(\kappa)$.

Let $I = \{f \in C_p(K) : f(y) = 0\}$; if we have a function $f \in I$ and the set $P(f, \varepsilon) = \{a \in A : |f(a)| \ge \varepsilon\}$ is infinite for some $\varepsilon > 0$ then $y \in \overline{P}$ and hence $|f(y)| \ge \varepsilon$ by continuity of f. This contradiction with $f(y) = 0$ shows that $P(f, \varepsilon)$ is finite for any $f \in I$ and $\varepsilon > 0$. As a consequence, $f|A \in \Sigma_*(A)$ for any $f \in I$. On the other hand assume that $f \in C_p(K)$ and $f|A \in \Sigma_*(A)$. If $|f(y)| > \varepsilon > 0$ then there is $U \in \tau(y, K)$ for which $|f(a)| > \varepsilon$ for any $a \in U$. The set U has to be infinite; so $P(f, \varepsilon) \supset U$ is also infinite which is a contradiction. Therefore $f(y) = 0$ and we proved that $f \in I$ if and only if $f|A \in \Sigma_*(A)$. Let $\pi_A : C_p(K) \to C_p(A) = \mathbb{R}^A$ be the restriction map. It is immediate that $\pi_A|I : I \to \pi_A(I) = \Sigma_*(A)$ is a homeomorphism; so $I \simeq \Sigma_*(A)$. We have $C_p(K) \simeq I \times \mathbb{R}$ by Fact 1 of S.409 and therefore the property (2) implies that $C_p(K) \simeq \Sigma_*(A) \times \mathbb{R} \simeq \Sigma_*(A)$. Since K is homeomorphic to $A(\kappa)$, we proved that $C_p(A(\kappa)) \simeq C_p(K) \simeq \Sigma_*(A)$.

U.106. *Prove that, if* $|A| = \kappa > \omega$ *then the space* $\Sigma(A)$ *is homeomorphic to* $C_p(L(\kappa))$.

Solution. Suppose that we are given uncountable sets D and E and a bijection $\varphi : D \to E$. If we consider D and E to be discrete topological spaces then φ is a homeomorphism and hence the dual map $\varphi^* : \mathbb{R}^E \to \mathbb{R}^D$ defined by $\varphi(f) = f \circ \varphi$ for any $f \in \mathbb{R}^E$, is a homeomorphism as well (see TFS-163). It is straightforward that $\varphi^*(\Sigma(E)) = \Sigma(D)$ which shows that

(1) if D and E are uncountable sets of the same cardinality then $\Sigma(D)$ is homeomorphic to $\Sigma(E)$.

Now assume that D is an uncountable set, take a point $x \notin D$ and consider the set $E = D \cup \{x\}$. The restriction map $\pi : \mathbb{R}^E \to \mathbb{R}^D$ defined by $\pi(f) = f|D$ for any $f \in \mathbb{R}^E$ is continuous and $\pi(\Sigma(E)) = \Sigma(D)$. For any $f \in \Sigma(D)$ let $e(f)(x) = 0$ and $e(f)(d) = f(d)$ for any $d \in D$. It is evident that $e : \Sigma(D) \to \Sigma(E)$ is continuous (in fact, it is an embedding but we won't need it).

Given $r \in \mathbb{R}$ let $u_r(x) = r$ and $u_r(d) = 0$ for all $d \in D$; then $u_r \in \Sigma(E)$. For any $f \in \Sigma(E)$ let $\eta(f) = (\pi(f), f(x)) \in \Sigma(D) \times \mathbb{R}$. Then $\eta : \Sigma(E) \to \Sigma(D) \times \mathbb{R}$ is a continuous map being the diagonal product of two restrictions. Furthermore, if $(h, r) \in \Sigma(D) \times \mathbb{R}$ then let $\mu(h, r) = e(h) + u_r$. We omit a trivial verification that $\mu : \Sigma(D) \times \mathbb{R} \to \Sigma(E)$ is continuous. The maps η and μ are mutually inverse; so they are both homeomorphisms. Since D and E have the same cardinality, the property (1) implies that $\Sigma(D) \simeq \Sigma(E) \simeq \Sigma(D) \times \mathbb{R}$ and hence

(2) $\Sigma(D)$ is homeomorphic to $\Sigma(D) \times \mathbb{R}$ for any uncountable set D.

We are now ready to deal with our set A. Choose a point $y \notin A$ and let $A' = A \cup \{y\}$. It is easy to see that the family $\tau = \exp A \cup \{U \subset A' : y \in U$ and $|A \setminus U| \leq \omega\}$ is a topology on A' and $L = (A', \tau)$ is homeomorphic to $L(\kappa)$.

Let $I = \{f \in C_p(L) : f(y) = 0\}$; if $f \in I$ and the set $P(f) = \{a \in A : f(a) \neq 0\}$ is uncountable then there is $n \in \mathbb{N}$ such that $Q(f, n) = \{a \in A : |f(a)| \geq \frac{1}{n}\}$ is uncountable. It is clear that $y \in \overline{Q(f, n)}$ and hence $|f(y)| \geq \frac{1}{n}$ by continuity of f. This contradiction with $f(y) = 0$ shows that $P(f)$ is countable for any $f \in I$. As a consequence, $f|A \in \Sigma(A)$ for any $f \in I$. On the other hand assume that $f \in C_p(L)$ and $f|A \in \Sigma(A)$. Then the set $W = \{y\} \cup \{a \in A : f(a) = 0\}$ is an open neighbourhood of y

such that $f(W\backslash\{y\}) = \{0\}$. Therefore $f(y) = 0$ and we proved that $f \in I$ if and only if $f|A \in \Sigma(A)$. Let $\pi_A : C_p(L) \to C_p(A) = \mathbb{R}^A$ be the restriction map. It is immediate that $\pi_A|I : I \to \pi_A(I) = \Sigma(A)$ is a homeomorphism; so $I \simeq \Sigma(A)$. We have $C_p(L) \simeq I \times \mathbb{R}$ by Fact 1 of S.409; so the property (2) implies that $C_p(L) \simeq \Sigma(A) \times \mathbb{R} \simeq \Sigma(A)$. Since $L \simeq L(\kappa)$, we proved that $C_p(L(\kappa)) \simeq C_p(L) \simeq \Sigma(A)$.

U.107. *Prove that, for any κ, there is a compact subspace of $C_p(A(\kappa))$ which separates the points of $A(\kappa)$. As a consequence, $C_p(A(\kappa))$ and $\Sigma_*(\kappa)$ are $K_{\sigma\delta}$-spaces and hence Lindelöf Σ-spaces.*

Solution. As usual, we consider that $A(\kappa) = \kappa \cup \{a\}$ where a is the unique non-isolated point of the space $A(\kappa)$. For any $\alpha \in \kappa$ let $f_\alpha(\alpha) = 1$ and $f_\alpha(x) = 0$ for every $x \in A(\kappa)\backslash\{\alpha\}$. Denote by u the function which is identically zero on $A(\kappa)$. It is an easy exercise that $K = \{f_\alpha : \alpha < \kappa\} \cup \{u\}$ is compact (in fact, $K \simeq A(\kappa)$) and separates the points of $A(\kappa)$. The class $\mathcal{K}$ of compact spaces is k-directed; so we an apply Problem 014 to conclude that $C_p(A(\kappa))$ is a $K_{\sigma\delta}$-space. The space $\Sigma_*(\kappa)$ is homeomorphic to $C_p(A(\kappa))$ by Problem 105; so $\Sigma_*(\kappa)$ is a $K_{\sigma\delta}$-space as well. Since any $K_{\sigma\delta}$-space is Lindelöf Σ (see SFFS-261), both spaces $C_p(A(\kappa))$ and $\Sigma_*(\kappa)$ are Lindelöf Σ-spaces.

U.108. *Prove that $\sigma(A)$ is a σ-compact space (and hence a Lindelöf Σ-space) for any A.*

Solution. For any $x \in \mathbb{R}^A$ let $\text{supp}(x) = \{a \in A : x(a) \neq 0\}$; then $\text{supp}(x)$ is finite for any $x \in \sigma(A)$. For each $n \in \mathbb{N}$ let $K_n = \{x \in \sigma(A) : |\text{supp}(x)| \leq n$ and $x(a) \in [-n, n]$ for any $a \in A\}$; it is clear that $K_n \subset [-n, n]^A$. It turns out that every K_n is compact.

To see it take any point $z \in [-n, n]^A\backslash K_n$; then $|\text{supp}(z)| > n$ and hence we can choose distinct indices $a_1, \ldots, a_{n+1} \in A$ such that $z(a_i) \neq 0$ for all $i \leq n + 1$. The set $W = \{x \in [-n, n]^A : x(a_i) \neq 0$ for all $i \leq n + 1\}$ is open in the space $[-n, n]^A$ and $z \in W \subset [-n, n]^A\backslash K_n$. Thus $[-n, n]^A\backslash K_n$ is open in $[-n, n]^A$ and hence K_n is compact being closed in $[-n, n]^A$. Consequently, $\sigma(A) = \bigcup\{K_n : n \in \mathbb{N}\}$ is a σ-compact space.

U.109. *Prove that, for any uncountable set A, there is a closed countably compact non-compact subspace in $\Sigma(A)$ and hence $\Sigma(A)$ is not realcompact.*

Solution. Denote by u the function which is identically zero on the set A and let $C = \Sigma(A) \cap \mathbb{I}^A$; it is evident that C is closed in $\Sigma(A)$. If $M_a = \mathbb{I}$ for any $a \in A$ and $M = \prod\{M_a : a \in A\}$ then it is immediate that $C = \Sigma(M, u) \subset \mathbb{I}^A$. Since A is uncountable, the set C is dense in $\mathbb{I}^A$ and does not coincide with $\mathbb{I}^A$. However, $\overline{P} \subset C$ (the bar denotes the closure in $\mathbb{I}^A$) for any countable $P \subset C$ (see Fact 3 of S.307); this implies that C is countably compact (see Fact 1 of S.310). Thus C is a countably compact non-compact closed subspace of $\Sigma(A)$ and hence $\Sigma(A)$ is not realcompact.

U.110. *Prove that, for any infinite A, every pseudocompact subspace of $\Sigma_*(A)$ is compact.*

Solution. Let P be a pseudocompact subset of $\Sigma_*(A)$. Since $\Sigma_*(A) \simeq C_p(A(\kappa))$ where $\kappa = |A|$ (see Problem 105), we can apply Problem 044 to see that $K = \overline{P}$ is compact (the bar denotes the closure in $\Sigma_*(A)$). The space K embeds in $C_p(A(\kappa))$; so we can apply Fact 19 of S.351 to conclude that every pseudocompact subspace of K is compact and hence closed in K. Consequently, $P = \overline{P} = K$ and hence P is compact.

U.111. *Prove that any metrizable space M embeds in $\Sigma_*(A)$ for some A.*

Solution. Fix a base $\mathcal{B} \subset \tau^*(M)$ in the space M such that $\mathcal{B} = \bigcup_{n \in \omega} \mathcal{B}_n$ and $\mathcal{B}_n$ is a discrete family for every $n \in \omega$ (this is possible because any metrizable space has a σ-discrete base by TFS-221). For any $U \in \mathcal{B}_n$ there is a continuous function $p_U^n : M \to [0, \frac{1}{n+1}]$ such that $(p_U^n)^{-1}(0) = M \setminus U$ (see Fact 2 of T.080).

Define a function $p : M \to \mathbb{R}$ by $p(x) = 0$ for all $x \in M$ and consider the space $K = \{p\} \cup \{p_U^n : n \in \omega, \ U \in \mathcal{B}_n\} \subset C_p(M)$. Take any $O \in \tau(p, C_p(M))$; there is $\varepsilon > 0$ and a finite set $A \subset M$ such that $W = \{f \in C_p(M) : f(A) \subset (-\varepsilon, \varepsilon)\} \subset O$. Take $m \in \omega$ for which $\frac{1}{m} < \varepsilon$; then $p_U^n(x) \leq \frac{1}{n+1} < \frac{1}{m} < \varepsilon$ for any $n \geq m$, $x \in M$ and $U \in \mathcal{B}_n$. In particular, $K_n = \{p_U^n : U \in \mathcal{B}_n\} \subset W$ for all $n \geq m$.

Now, if $n < m$ then only finitely many elements of $\mathcal{B}_n$ meet A because $\mathcal{B}_n$ is discrete. If $U \in \mathcal{B}_n$ and $A \cap U = \emptyset$ then $p_U^n(A) = \{0\}$ and hence $p_U^n \in W$. Therefore $K_n \setminus W$ is finite for any $n < m$ which shows that $K \setminus W \supset K \setminus O$ is finite and hence

(1) $K \setminus O$ is finite for any $O \in \tau(p, C_p(M))$.

An immediate consequence of (1) is that the space K is compact. We claim that every $f \in K \setminus \{p\}$ is isolated in K. Indeed, there is $k \in \omega$ and $U \in \mathcal{B}_k$ such that $f = p_U^k$. Pick an point $x \in U$; then $\delta = f(x) > 0$ and therefore there is $m \in \omega$ such that $\frac{1}{m} < \frac{\delta}{2}$. Let $H = \{h \in C_p(K) : h(x) > \frac{\delta}{2}\}$; then H is an open neighbourhood of f in $C_p(K)$.

If $n \geq m$ and $g \in K_n$ then $g(x) \leq \frac{1}{n+1} \leq \frac{1}{m} < \frac{\delta}{2}$; so $g \notin H$. Now, if $n < m$ then at most one element of $\mathcal{B}_n$ contains x because $\mathcal{B}_n$ is discrete; besides, if $V \in \mathcal{B}_n$ and $x \notin V$ then $p_V^n(x) = 0$; so $p_V^n \notin H$. As a consequence, the set $H \cap K$ has at most m elements; so f is an isolated point of K. It is an easy exercise to see that any compact space with a unique non-isolated point is homeomorphic to the one-point compactification of a discrete space; so there is a cardinal κ such that $K \simeq A(\kappa)$.

Given $x \in M$ let $e_x(f) = f(x)$ for any $f \in K$; then $e_x \in C_p(K)$ and, letting $e(x) = e_x$ for any $x \in M$, we obtain a continuous map $e : M \to C_p(K)$ (see TFS-166).

Suppose that $x \in M$ and $G \subset M$ is a closed set such that $x \notin G$. Since $\mathcal{B}$ is a base in M, there is $n \in \omega$ and $U \in \mathcal{B}_n$ such that $x \in U \subset M \setminus G$. For the function $f = p_U^n \in K$ we have $f(x) > 0$ and $f(G) = \{0\}$; so $f(x) \notin f(G)$. This proves that K separates the points and closed subsets of M; so e is an embedding (see TFS-166). Thus K is a compact space such that M embeds in $C_p(K)$. We have $C_p(K) \simeq C_p(A(\kappa)) \simeq \Sigma_*(\kappa)$ (see Problem 105); so M embeds in $\Sigma_*(\kappa)$.

U.112. *Observe that any pseudocompact continuous image of* $\Sigma_*(A)$ *is compact and metrizable for any infinite A. Give an example of a countably compact non-compact space which is a continuous image of* $\Sigma(\omega_1)$.

Solution. The space $\Sigma_*(A)$ is Lindelöf by Problem 107; so any pseudocompact continuous image of $\Sigma_*(A)$ is compact and hence metrizable by Problem 104.

Now, let $L = L(\omega_1)$; then $\Sigma(\omega_1)$ is homeomorphic to $C_p(L)$ by Problem 106. Since L is a P-space, $K = C_p(L, \mathbb{I})$ is a countably compact non-compact space (see TFS-397). By TFS-092 $C_p(L)$ maps continuously onto K; so $\Sigma(\omega_1)$ maps continuously onto K as well.

U.113. *Prove that, for any uncountable A, the space* $\Sigma(A)$ *is not embeddable into* $\Sigma_*(B)$ *for any set B.*

Solution. By Problem 109, there is a countably compact non-compact subspace K in the space $\Sigma(A)$. Thus, if $\Sigma(A)$ embeds in some $\Sigma_*(B)$ then K embeds in $\Sigma_*(B)$ which contradicts Problem 110.

U.114. *Prove that, for any uncountable A, the space* $\Sigma_*(A)$ *is not embeddable into* $\sigma(B)$ *for any set B.*

Solution. Assume that A is uncountable and there is a set B such that $\Sigma_*(A)$ embeds in $\sigma(B)$; fix a set $S \subset \sigma(B)$ with $S \simeq \Sigma_*(A)$. Since $\sigma(B) \subset \Sigma(B)$, the space $\sigma(B)$ is Fréchet–Urysohn (see Problem 101). It was proved in Problem 108 that we can find a sequence $\{K_n : n \in \omega\}$ of compact subsets of $\sigma(B)$ such that $\sigma(B) = \bigcup_{n \in \omega} K_n$. If $F_n = K_n \cap S$ then F_n is a closed subset of S for every $n \in \omega$ and $S = \bigcup_{n \in \omega} F_n$. Now apply SFFS-432 to see that S and hence $\Sigma_*(A)$ embeds in $F_n \subset K_n$ for some $n \in \omega$.

Fix a set $T \subset F_n$ such that $T \simeq \Sigma_*(A)$; the space $K = \overline{T}$ is compact, Fréchet–Urysohn, and T is dense in K. Since $\Sigma_*(A) \simeq C_p(X)$ for an appropriate compact uncountable space X (see Problem 105), the cardinal ω_1 is a precaliber of T and hence of K (see SFFS-283 and SFFS-278). Since K is compact, ω_1 is a caliber of K by SFFS-279. Besides, K has a point-countable π-base by TFS-332; this π-base has to be countable because ω_1 is a caliber of K. Therefore K is separable; it follows from Problem 105 that there is a compact space Y for which $K \subset \sigma(B) \subset \Sigma_*(B) \simeq C_p(Y)$ and hence K is ω-monolithic (see SFFS-118 and SFFS-154). This shows that K is metrizable and hence so is $\Sigma_*(A) \simeq C_p(X)$ which is a contradiction with TFS-169.

U.115. *Prove that, for any uncountable A, neither of the spaces* $\Sigma(A)$ *and* $\Sigma_*(A)$ *maps continuously onto the other.*

Solution. Let $\kappa = |A|$; then $\Sigma_*(A) \simeq C_p(A(\kappa))$ by Problem 105. The space $\Sigma_*(A)$ cannot be mapped continuously onto $\Sigma(A)$ because the space $\Sigma_*(A)$ is Lindelöf (see Problem 107) and $\Sigma(A)$ is not (see Problem 109). Now assume that there exists a continuous onto map $\varphi : \Sigma(A) \to \Sigma_*(A)$. For any $B \subset A$ let $\pi_B : \mathbb{R}^A \to \mathbb{R}^B$ be the restriction map. Since we have the equality $\pi_B(\Sigma(A)) = \mathbb{R}^B$ for any countable $B \subset A$, the set $\Sigma(A)$ is C-embedded in $\mathbb{R}^A$ (see Fact 1 of T.455) and hence $\mathbb{R}^A$ can be identified with $\nu(\Sigma(A))$ (see Fact 1 of S.438). The space $\Sigma_*(A)$ being realcompact there exists a continuous map $\Phi : \mathbb{R}^A \to \Sigma_*(A)$ with $\Phi|\Sigma(A) = \varphi$ and hence $\Phi(\mathbb{R}^A) = \Sigma_*(A)$.

Let $K_n = [-n,n]^A$ for any $n \in \mathbb{N}$; then every K_n is a dyadic compact space and $P = \bigcup\{K_n : n \in \mathbb{N}\}$ is dense in $\mathbb{R}^A$. The space $\Sigma_*(A) \subset \Sigma(A)$ is Fréchet–Urysohn by Problem 101; so the dyadic compact space $L_n = \Phi(K_n)$ has countable tightness and hence $w(L_n) = \omega$ for each $n \in \mathbb{N}$ (see TFS-359). As a consequence, the space $L = \bigcup\{L_n : n \in \mathbb{N}\}$ is separable and hence so is $\Sigma_*(A)$ because $L = \Phi(P)$ is dense in $\Sigma_*(A)$. The space $\Sigma_*(A)$ is ω-monolithic (see Problem 105, SFFS-118 and SFFS-154); so $\omega = nw(\Sigma_*(A)) = nw(C_p(A(\kappa))) = nw(A(\kappa)) = \kappa > \omega$ which is a contradiction. Therefore $\Sigma(A)$ cannot be continuously mapped onto $\Sigma_*(A)$.

U.116. *Prove that, for any A, the space $\Sigma(A)$ embeds in a countably compact Fréchet–Urysohn space.*

Solution. If A is countable then the space $\Sigma(A) = \mathbb{R}^A$ is second countable; so it can be embedded in the metrizable compact space $\mathbb{I}^\omega$. Now, if $|A| = \kappa > \omega$ then $\Sigma(A) \simeq C_p(L)$ where $L = L(\kappa)$ (see Problem 106). Furthermore, it follows from Fact 1 of S.295 that $C_p(L) \simeq C_p(L, (-1, 1)) \subset C_p(L, \mathbb{I})$ and hence $\Sigma(A)$ embeds in $C_p(L, \mathbb{I})$. Since L is a P-space, $C_p(L, \mathbb{I})$ is countably compact by TFS-397. Besides, $C_p(L, \mathbb{I}) \subset C_p(L) \simeq \Sigma(A)$ is a Fréchet–Urysohn space (see Problem 101). Thus $\Sigma(A)$ embeds in the countably compact Fréchet–Urysohn space $C_p(L, \mathbb{I})$.

U.117. *Show that, if a set A is uncountable then $\Sigma_*(A)$ cannot be embedded in a σ-compact space of countable tightness. In particular, neither $\Sigma(A)$ nor $\Sigma_*(A)$ are embeddable in a compact space of countable tightness if $|A| > \omega$.*

Solution. Let $|A| = \kappa$; then $\Sigma_*(A) \simeq C_p(Z)$ for the space $Z = A(\kappa)$ by Problem 105. Suppose that Y is a space such that $t(Y) = \omega$ and $Y = \{K_n : n \in \omega\}$ where K_n is compact for any $n \in \omega$. If $\Sigma_*(A)$ embeds in Y then fix $S \subset Y$ with $S \simeq \Sigma_*(A)$. If $F_n = K_n \cap S$ then F_n is a closed subset of S for every $n \in \omega$ and $S = \bigcup_{n \in \omega} F_n$. Since $\Sigma_*(A) \simeq C_p(Z)$, we can apply SFFS-432 to see that S and hence $\Sigma_*(A)$ embeds in $F_n \subset K_n$ for some $n \in \omega$.

Fix a set $T \subset F_n$ such that $T \simeq \Sigma_*(A)$; the space $K = \overline{T}$ is compact, Fréchet–Urysohn, and T is dense in K. Since $T \simeq \Sigma_*(A) \simeq C_p(Z)$, the cardinal ω_1 is a precaliber of T and hence of K (see SFFS-283 and SFFS-278). Since K is compact, ω_1 is a caliber of K by SFFS-279. Besides, K has a point-countable π-base by TFS-332; this π-base has to be countable because ω_1 is a caliber of K. Therefore $\pi w(K) = \omega$ and hence $\pi w(T) = \omega$ because T is dense in K (see Fact 1 of T.187). As a consequence, $|Z| = w(C_p(Z)) = \pi w(C_p(Z)) = \pi w(T) = \omega$ (see Fact 2 of T.187) which is a contradiction with TFS-169 because $|Z| = \kappa > \omega$.

U.118. *Let X be a compact space. Prove that X is Corson compact if and only if X has a point-countable T_0-separating family of open F_σ-sets. Deduce from this fact that any metrizable compact space is Corson compact.*

Solution. Let X be Corson compact; so we can assume that $X \subset \Sigma(A)$ for some A. For any $a \in A$ let $p_a : \Sigma(A) \to \mathbb{R}$ be the natural projection of $\Sigma(A)$ onto the factor of $\mathbb{R}^A$ determined by a. Let $\mathcal{B} = \{(p, q) : p, q \in \mathbb{Q}, \ p < q \text{ and } p \cdot q > 0\}$. Observe that no element of $\mathcal{B}$ contains zero and $\mathcal{B}$ is a countable base in $\mathbb{R} \setminus \{0\}$. Let $\mathcal{U} = \{p_a^{-1}(B) \cap X : a \in A, \ B \in \mathcal{B}\}$. It is evident that every element of $\mathcal{U}$ is an open F_σ-subset of X. If $x \in X$ then the set $S = \{a \in A : x(a) \neq 0\}$ is countable and hence the family $\mathcal{U}_x = \{p_a^{-1}(B) : B \in \mathcal{B}, \ a \in S\}$ is countable as well. For any $U \in \mathcal{U} \setminus \mathcal{U}_x$ we have $U = p_a^{-1}(B)$ for some $B \in \mathcal{B}$ and $a \in A \setminus S$. But $p_a(x) = 0 \notin B$ and hence $x \notin U$. This shows that the family $\{U \in \mathcal{U} : x \in U\} \subset \mathcal{U}_x$ is countable; the point $x \in X$ was chosen arbitrarily; so we proved that $\mathcal{U}$ is point-countable.

Now take distinct $x, y \in X$. There is $a \in A$ such that $x(a) \neq y(a)$. One of the numbers $x(a), y(a)$, say $x(a)$, is not equal to zero and hence there is $B \in \mathcal{B}$ such that $x(a) \in B$ and $y(a) \notin B$. Then $U = p_a^{-1}(B) \cap X \in \mathcal{U}$ and $U \cap \{x, y\} = \{x\}$. This proves that $\mathcal{U}$ is a T_0-separating family in X; so we settled necessity.

To prove sufficiency suppose that there exists a T_0-separating family $\mathcal{U}$ of open F_σ-subsets of the space X. For every $U \in \mathcal{U}$ there exists a continuous function $f_U : X \to \mathbb{R}$ such that $f^{-1}(0) = X \setminus U$ (see Fact 1 of S.358). The diagonal product $f = \Delta\{f_U : U \in \mathcal{U}\} : X \to \mathbb{R}^{\mathcal{U}}$ of the family $\{f_U : U \in \mathcal{U}\}$ is a continuous map; since $\mathcal{U}$ is T_0-separating, the map f is injective; so $f : X \to X' = f(X)$ is a homeomorphism. Given any $x \in X$ the family $P(x) = \{U \in \mathcal{U} : x \in U\}$ is countable; so $f(x)(U) = f_U(x) = 0$ for any $U \in \mathcal{U} \setminus P(x)$. This shows that at most countably many coordinates of $f(x)$ are distinct from zero and therefore $X' \subset \Sigma(\mathcal{U})$. Thus X' is Corson compact and hence so is X being homeomorphic to X'.

Finally observe that if compact space X is metrizable then any countable base of X is T_0-separating, point-countable and consists of F_σ-subsets of X. Therefore every metrizable compact space is Corson compact.

U.119. *Let M_t be a second countable space for any $t \in T$. Prove that, for any point $a \in M = \prod\{M_t : t \in T\}$, any compact subset of $\Sigma(M, a)$ is Corson compact.*

Solution. Suppose that X is a compact subspace of $\Sigma(M, a)$. For any $t \in T$ let $p_t : M \to M_t$ be the natural projection. Choose a countable base $\mathcal{B}_t$ in the space $M_t \setminus \{a(t)\}$ for any $t \in T$. Let $\mathcal{U} = \{p_t^{-1}(B) \cap X : t \in T, \ B \in \mathcal{B}_t\}$. It is evident that every element of the family $\mathcal{U}$ is an open F_σ-subset of X. If $x \in X$ then the set $S = \{t \in A : x(t) \neq a(t)\}$ is countable and hence $\mathcal{U}_x = \{p_t^{-1}(B) : B \in \mathcal{B}_t, \ t \in S\}$ is countable as well. For any $U \in \mathcal{U} \setminus \mathcal{U}_x$ we have $U = p_t^{-1}(B)$ for some $B \in \mathcal{B}_t$ and $t \in T \setminus S$. But $p_t(x) = a(t) \notin B$ and hence $x \notin U$. This shows that the family $\{U \in \mathcal{U} : x \in U\} \subset \mathcal{U}_x$ is countable; the point $x \in X$ was chosen arbitrarily; so we proved that $\mathcal{U}$ is point-countable.

Now take distinct $x, y \in X$. There is $t \in T$ such that $x(t) \neq y(t)$. One of the points $x(t), y(t)$, say $x(t)$, is not equal to $a(t)$ and hence there is $B \in \mathcal{B}_t$ such that $x(t) \in B$ and $y(t) \notin B$. Then $U = p_t^{-1}(B) \cap X \in \mathcal{U}$ and $U \cap \{x, y\} = \{x\}$. This proves that $\mathcal{U}$ is a T_0-separating point-countable family of open F_σ-subsets of X; so we can apply Problem 118 to conclude that X is Corson compact.

U.120. *Prove that any Corson compact space is monolithic, Fréchet–Urysohn and has a dense set of points of countable character. As a consequence, $\omega_1 + 1$ is not Corson compact.*

Solution. Let X be a Corson compact space; we can assume that $X \subset \Sigma(A)$ for some A. The space $\Sigma(A)$ is Fréchet–Urysohn by Problem 101; so X is also Fréchet–Urysohn. Let $u \in \mathbb{R}^A$ be the function which is identically zero on A. For any $B \subset A$ let $p_B : \mathbb{R}^A \to \mathbb{R}^B$ be the natural projection.

To see that X is monolithic take any infinite cardinal κ and $Y \subset X$ with $|Y| = \kappa$. The set $\mathrm{supp}(y) = \{a \in A : y(a) \neq 0\}$ is countable for any $y \in Y$; so $S = \bigcup\{\mathrm{supp}(y) : y \in Y\}$ has cardinality at most κ. It is immediate that $Y \subset F = \mathbb{R}^S \times \{p_{A\backslash S}(u)\} \simeq \mathbb{R}^S$. As a consequence, $w(F) \leq |S| \leq \kappa$. Furthermore, $Y \subset F' = F \cap \Sigma(A)$; since F' is closed in $\Sigma(A)$, we have $\overline{Y} \subset F'$ (the bar denotes the closure in $\Sigma(A)$) and therefore $w(\overline{Y}) \leq w(F') \leq w(F) \leq \kappa$. This shows that X is monolithic. Finally, X has a dense set of points of countable character by Fact 1 of U.080. Finally, $\omega_1 + 1$ is not Corson compact because $t(\omega_1 + 1) = \omega_1$.

U.121. *Prove that $d(X) = w(X)$ for any Corson compact space. Thus, the two arrows space is not Corson compact.*

Solution. Let $\kappa = d(X)$ and fix a dense set $A \subset X$ with $|A| = \kappa$. The space X is monolithic by Problem 120; so $w(X) = nw(X) = nw(\overline{A}) \leq |A| = \kappa$ (see Fact 4 of S.307). Thus $w(X) \leq d(X)$; since always $d(X) \leq w(X)$, we have $w(X) = d(X)$. Finally observe that two arrows space K is separable and non-metrizable (see TFS-384) which implies that $\omega = d(K) < w(K)$ and hence K is not Corson compact.

U.122. *Let X be a Corson compact space such that $C_p(X)\backslash\{f\}$ is normal for some $f \in C_p(X)$. Prove that X is metrizable. In particular, if $C_p(X)$ is hereditarily normal, then X is metrizable.*

Solution. By Problem 027 the space X is separable and hence $w(X) = d(X) = \omega$ by Problem 121. Thus X is metrizable. In particular, if $C_p(X)$ is hereditarily normal then $C_p(X)\backslash\{f\}$ is normal for any $f \in C_p(X)$; so X is metrizable.

U.123. *Prove that any linearly ordered and any dyadic Corson compact space is metrizable.*

Solution. If X is a Corson dyadic compact space then X is Fréchet–Urysohn by Problem 120. Since every dyadic compact space of countable tightness is metrizable (see TFS-359), the space X is metrizable.

Now suppose that X is a linearly ordered Corson compact space. We can assume that $X \subset C_p(L(\kappa))$ for some cardinal κ (see Problem 106). Since $(L(\kappa))^\omega$ is Lindelöf (see TFS-354), we can apply Fact 4 of U.093 to see that $C_p(X)$ is Lindelöf. Finally, apply Problem 082 to conclude that X is metrizable.

U.124. *Let X be a Corson compact space. Prove that the Alexandroff double $AD(X)$ is also Corson compact. In particular, $AD(X)$ is Corson compact for any metrizable compact X.*

Solution. The space $AD(X)$ is compact by TFS-364. By Problem 118 the space X has a point-countable T_0-separating family $\mathcal{U}$ which consists of open F_σ-subsets of X. Recall that $X \times \{0, 1\}$ is the underlying set of the space $AD(X)$. Given $x \in X$ let $u_0(x) = (x, 0)$ and $u_1(x) = (x, 1)$. Then $AD(X) = u_0(X) \cup u_1(X)$. The points of $u_1(X)$ are isolated in $AD(X)$; if $z = (x, 0) \in AD(X)$ then the base at z is formed by the sets $u_0(V) \cup (u_1(V) \backslash \{u_1(x)\})$ where V runs over $\tau(x, X)$.

For any $P \subset X$ let $\pi(P) = u_0(P) \cup u_1(P)$. It is immediate that $\pi(U)$ is open in $AD(X)$ if $U \in \tau(X)$; besides $\pi(F)$ is closed in $AD(X)$ if F is closed in X. Therefore $\pi(U)$ is an open F_σ-subset of $AD(X)$ for any $U \in \mathcal{U}$ and hence the family $\mathcal{V} = \{\pi(U) : U \in \mathcal{U}\} \cup \{\{u_1(x)\} : x \in X\}$ consists of open F_σ-subsets of $AD(X)$. The family $\mathcal{V}_1 = \{\pi(U) : U \in \mathcal{U}\}$ is point-countable because so is $\mathcal{U}$. The family $\mathcal{V}_2 = \{\{u_1(x)\} : x \in X\}$ is disjoint so $\mathcal{V} = \mathcal{V}_1 \cup \mathcal{V}_2$ is point-countable as well.

To prove that $\mathcal{V}$ is T_0-separating take any distinct points $z = (x, i)$ and $t = (y, j)$ of the space $AD(X)$. If $x = y$ then $i \neq j$; letting $W = \{u_1(x)\}$ it is easy to see that $W \in \mathcal{V}$ and $W \cap \{z, t\}$ is a singleton. If $x \neq y$ then there is $U \in \mathcal{U}$ such that $U \cap \{x, y\}$ is a singleton. Then $W = \pi(U) \in \mathcal{V}$ and $W \cap \{z, t\}$ is a singleton as well. Thus $AD(X)$ has a point-countable T_0-separating family of open F_σ-sets and hence we can apply Problem 118 to conclude that $AD(X)$ is Corson compact.

U.125. *Let X_t be a Corson compact space for any $t \in T$. Prove that the one-point compactification of the space $\bigoplus \{X_t : t \in T\}$ is also Corson compact.*

Solution. We consider that X_t is a clopen subspace of $X = \bigoplus \{X_t : t \in T\}$ for every $t \in T$. Let $K = \{a\} \cup X$ be the one-point compactification of X; the space K is compact by Fact 1 of S.387. For every $t \in T$ there is a point-countable family $\mathcal{U}_t \subset \tau(X_t)$ such that every $U \in \mathcal{U}_t$ is σ-compact and $\mathcal{U}_t$ is T_0-separating in X_t. Since $\mathcal{V} = \{X_t : t \in T\}$ is disjoint, the family $\mathcal{U} = \bigcup \{\mathcal{U}_t : t \in T\} \cup \mathcal{V}$ is also point-countable. It is evident that $\mathcal{U}$ consists of open F_σ-subsets of K.

Let x and y be distinct points of K. If some of them, say x, coincides with a then $y \in X_t$ for some $t \in T$. Consequently, $W = X_t \in \mathcal{U}$ and $W \cap \{x, y\} = \{y\}$ is a singleton. If $a \notin \{x, y\}$ then there are two possibilities:

1) there is $t \in T$ such that $\{x, y\} \subset X_t$. Since $\mathcal{U}_t$ is T_0-separating in X_t, there is $W \in \mathcal{U}_t$ for which $W \cap \{x, y\}$ is a singleton. We also have $W \in \mathcal{U}$; so we found $W \in \mathcal{U}$ which T_0-separates x and y.

2) There are distinct $s, t \in T$ such that $x \in X_t$ and $y \in X_s$. Then $W = X_t \in \mathcal{U}$ and $W \cap \{x, y\} = \{x\}$ is a singleton.

This proves that $\mathcal{U}$ is a point-countable family of open F_σ-subsets of K which T_0-separates the points of K. Therefore Problem 118 can be applied to conclude that K is Corson compact.

U.126. *Prove that, under CH, there exists a compact space of countable spread which is not perfectly normal.*

Solution. It was proved in SFFS-099 that, under CH, there exists a compact space X such that $\chi(X) > \omega$ and $hl(C_p(X)) = \omega$. Then $s(X) \leq hd(X) \leq hl(C_p(X)) = \omega$ (see SFFS-017) and X is not perfectly normal because $\chi(X) = \omega$ for any perfectly normal compact X (see TFS-327).

U.127. *Let X be a Corson compact space such that $s(X) = \omega$. Prove that X is perfectly normal.*

Solution. We will need the following general statement.

Fact 1. If a space Z is κ-monolithic and $s(Z) \leq \kappa$ for some infinite cardinal κ then $hl(Z) \leq \kappa$.

Proof. Fix $Y \subset Z$ and take an open cover $\mathcal{U}$ of the space Y. Since $s(Y) \leq s(Z) \leq \kappa$, we can apply Fact 1 of T.007 to find a discrete $D \subset Y$ and $\mathcal{U}' \subset \mathcal{U}$ such that $|\mathcal{U}'| \leq \kappa$ and $Y = \operatorname{cl}_Y(D) \cup (\bigcup \mathcal{U}')$. By κ-monolithity of the space Z, we have $nw(\operatorname{cl}_Y(D)) \leq nw(\operatorname{cl}_X(D)) \leq \kappa$. Therefore $l(\operatorname{cl}_Y(D)) \leq nw(\operatorname{cl}_Y(D)) \leq \kappa$ and therefore there exists a family $\mathcal{U}'' \subset \mathcal{U}$ such that $\operatorname{cl}_Y(D) \subset \bigcup \mathcal{U}''$. It is immediate that $\mathcal{U}' \cup \mathcal{U}''$ is a subcover of $\mathcal{U}$ of cardinality $\leq \kappa$. This shows that $l(Y) \leq \kappa$ for any $Y \subset Z$, i.e., $hl(Z) \leq \kappa$ and hence Fact 1 is proved.

Returning to our solution observe that X is ω-monolithic by Problem 120; since also $s(X) = \omega$, we can apply Fact 1 to conclude that $hl(X) = \omega$ and hence X is perfectly normal by SFFS-001.

U.128. *Let X be an ω-monolithic compact space such that $s(C_p(X)) = \omega$. Prove that X is metrizable. In particular, a Corson compact space X is metrizable whenever $s(C_p(X)) = \omega$.*

Solution. We have $s(X \times X) \leq s(C_p(X)) \leq \omega$ by SFFS-016; besides, $X \times X$ is ω-monolithic by SFFS-114. Therefore we can apply Fact 1 of U.127 to conclude that $hl(X \times X) \leq \omega$ and hence $X \times X$ is perfectly normal by SFFS-001. Consequently, the diagonal of X is a G_δ-subset of $X \times X$; so X is metrizable by SFFS-091. Finally, if X is Corson compact and $s(C_p(X)) = \omega$ then X is metrizable because it is ω-monolithic (see Problem 120).

U.129. *Let X be a compact space of countable tightness. Prove that X maps irreducibly onto a Corson compact space.*

Solution. By TFS-367 any compact space X of countable tightness admits a continuous irreducible map onto a space $Y \subset \Sigma(\kappa)$ for some cardinal κ. The space Y is compact being a continuous image of X; so Y is Corson compact.

U.130. *Given spaces X and Y assume that there exists a closed continuous irreducible onto map $f : X \to Y$. Prove that $d(X) = d(Y)$ and $c(X) = c(Y)$.*

Solution. The equality $c(X) = c(Y)$ was proved in Fact 1 of S.228. Next observe that $d(Y) \leq d(X)$ (see TFS-157) and assume that $d(Y) \leq \kappa$. Pick a dense $D \subset Y$ with $|D| \leq \kappa$ and choose a point $q(y) \in f^{-1}(y)$ for every $y \in D$. If $E = \{q(y) : y \in D\}$ then $|E| \leq \kappa$. If E is not dense in X then $F = \overline{E} \neq X$; the map f being closed we have $f(F) = \overline{f(E)} = \overline{D} = Y$ which is a contradiction because the map f is irreducible. Therefore E is dense in X and hence $d(X) \leq |E| \leq \kappa$ which shows that $d(X) = d(Y)$.

U.131. *Prove that, under the Jensen's axiom* $(\diamondsuit)$, *there is a perfectly normal non-metrizable Corson compact space X. Therefore, under $\diamondsuit$, a Corson compact space X need not be metrizable if $c(X) = \omega$.*

Solution. It was proved in SFFS-073 that, under $\diamondsuit$, there exists a hereditarily Lindelöf non-separable compact space K. Since $t(K) = \omega$, there exists a Corson compact space X and a continuous irreducible onto map $f : K \to X$ (see Problem 129). We have $hl(X) \leq hl(K) \leq \omega$; so X is perfectly normal. However, X is not metrizable because $d(X) = d(K) > \omega$ (see Problem 130). It is evident that $c(X) \leq s(X) \leq hl(X) \leq \omega$ so X is a non-metrizable Corson compact space such that $c(X) = \omega$.

U.132. *Prove that any Corson compact space X, with ω_1 precaliber of X, is metrizable.*

Solution. We have $t(X) = \omega$ and hence there exists a point-countable π-base $\mathcal{B}$ in the space X (see TFS-332). Since ω_1 is a caliber of X (see SFFS-279), the family $\mathcal{B}$ has to be countable; so $d(X) \leq \pi w(X) \leq |\mathcal{B}| \leq \omega$. Therefore $w(X) = d(X) \leq \omega$ (see Problem 121) and hence X is metrizable.

U.133. *Assuming MA+¬CH, prove that any Corson compact space X, for which $c(X) = \omega$, is metrizable.*

Solution. Since $c(X) = \omega$ and we have MA+¬CH, the cardinal ω_1 is a precaliber of X (see SFFS-288); so X is metrizable by Problem 132.

U.134. *Prove that a compact space X can fail to be Corson compact being a countable union of Corson compact spaces.*

Solution. Let X be the one-point compactification of the Mrowka space M (see TFS-142). Then X has a countable dense set D of isolated points while the space $K = X \backslash D$ is homeomorphic to $A(\kappa)$ for an uncountable cardinal κ; let a be the unique non-isolated point of K. The space K is Corson compact because the family $\{\{x\} : x \in K \backslash \{a\}\}$ is disjoint, T_0-separating in K and consists of open compact subsets of K (see Problem 118). Since $\{x\}$ is also Corson compact for any $x \in D$, the space $X = \{\{x\} : x \in D\} \cup K$ is a countable union of its Corson compact subspaces. However, X is not Corson compact because it is separable and non-metrizable (see Problem 121).

U.135. *Prove that there exists a compact space X which is not Corson compact being a union of three metrizable subspaces.*

Solution. Let X be the one-point compactification of the Mrowka space M (see TFS-142). Then X has a countable dense set D of isolated points while the space $K = X \backslash D$ is homeomorphic to $A(\kappa)$ for an uncountable cardinal κ; let a be the unique non-isolated point of K. Since every discrete space is metrizable, the space $X = D \cup (K \setminus \{a\}) \cup \{a\}$ is the union of its three metrizable subspaces. However, X is not Corson compact because it is separable and non-metrizable (see Problem 121).

U.136. *Suppose that X is compact and X^ω is a countable union of Corson compact subspaces. Prove that X is Corson compact.*

Solution. Suppose that $X^\omega = \bigcup \{K_i : i \in \omega\}$ where the space K_i is Corson compact for any $i \in \omega$. Since the space X^ω has the Baire property, there is $i \in \omega$ such that $U = \mathrm{Int}(K_i) \neq \emptyset$ (the interior is taken in X^ω). By definition of the product topology there is $n \in \omega$ and $U_0, \ldots, U_{n-1} \in \tau(X)$ such that $V = \prod_{k<n} U_k \times \prod X^{\omega \setminus n} \subset U$. Since V is a product in which X is one of the factors, the space X embeds in $V \subset U$ and hence in K_i. It is evident that any subspace of a Corson compact space is Corson compact; so X is Corson compact.

U.137. *Prove that any countable product of Corson compact spaces is Corson compact. In particular, X^ω is Corson compact whenever X is Corson compact.*

Solution. Suppose that K_n is Corson compact and fix a point-countable family $\mathcal{U}_n$ of open F_σ-subsets of K_n which T_0-separates the points of K_n for any $n \in \omega$ (see Problem 118). For the space $K = \prod_{n \in \omega} K_n$ let $\pi_n : K \to K_n$ be the natural projection for each $n \in \omega$. The family $\mathcal{V}_n = \{\pi_n^{-1}(U) : U \in \mathcal{U}_n\}$ is point-countable for any $n \in \omega$; so $\mathcal{V} = \bigcup_{n \in \omega} \mathcal{V}_n$ is also point-countable; it is clear that $\mathcal{V}$ consists of open F_σ-subsets of K. If $x, y \in K$ and $x \neq y$ then $p_n(x) \neq p_n(y)$ for some $n \in \omega$ and therefore there is $U \in \mathcal{U}_n$ such that $U \cap \{p_n(x), p_n(y)\}$ is a singleton. Then $V = \pi_n^{-1}(U) \in \mathcal{V}$ and the set $V \cap \{x, y\}$ is a singleton. This proves that $\mathcal{V}$ is a point-countable family of open F_σ-subsets of K which is T_0-separating in K; hence K is Corson compact by Problem 118.

U.138. *Let X be a Corson compact space. Prove that X has a dense metrizable subspace if and only if it has a σ-disjoint π-base.*

Solution. If Z is a space and $\mathcal{A}$ is a family of subsets of Z say that a family $\mathcal{B} \subset \exp Z$ is *inscribed* in $\mathcal{A}$ if, for every $B \in \mathcal{B}$ there is $A \in \mathcal{A}$ such that $B \subset A$. If $\mathcal{A}, \mathcal{B} \subset \exp Z$ and $Y \subset Z$ then $\mathcal{A}|Y = \{A \cap Y : A \in \mathcal{A}\}$ and $\mathcal{A} \wedge \mathcal{B} = \{A \cap B : A \in \mathcal{A}, B \in \mathcal{B}\}$.

Fact 1. If a space Z has a dense metrizable subspace then it has a σ-disjoint π-base. For first countable spaces the converse is also true, i.e., a first countable space Z has a dense metrizable subspace if and only if it has a σ-disjoint π-base.

Proof. There is no loss of generality to assume that all spaces we consider are not empty. To prove the first part, suppose that M is a dense metrizable subspace of Z. By TFS-221, we can choose a base $\mathcal{B} = \bigcup_{n \in \omega} \mathcal{B}_n \subset \tau^*(M)$ of the space M such that $\mathcal{B}_n$ is discrete in M for any $n \in \omega$. For every $U \in \mathcal{B}$ fix $O_U \in \tau(Z)$ such that $O_U \cap M = U$. Since M is dense in Z, the family $\mathcal{V}_n = \{O_U : U \in \mathcal{B}_n\}$ is disjoint for any $n \in \omega$. We claim that the family $\mathcal{V} = \bigcup_{n \in \omega} \mathcal{V}_n$ is a π-base in Z. Indeed, take any $W \in \tau^*(Z)$; there is $W_1 \in \tau^*(Z)$ such that $\overline{W}_1 \subset W$. Choose any $U \in \mathcal{B}$ with $U \subset W_1 \cap M$. Then $O_U \in \mathcal{V}$ and $O_U \subset \overline{O}_U = \overline{U} \subset \overline{W}_1 \subset W$ and hence $\mathcal{V}$ is a σ-disjoint π-base in the space Z.

Now assume that Z is first countable and there is a π-base $\mathcal{B}$ in Z such that $\mathcal{B} = \bigcup_{n \in \omega} \mathcal{B}_n$ where $\mathcal{B}_n$ is a disjoint family for any $n \in \omega$. If we add a non-empty open set to a π-base, we will still have a π-base; if we take a non-empty open subset in every element of a π-base we will obtain a π-base. These observations make it possible to assume that $\bigcup \mathcal{B}_n$ is dense in Z and $\mathcal{B}_{n+1}$ is inscribed in $\mathcal{B}_n$ for any $n \in \omega$. For any $z \in Z$ let $\{O_n^z : n \in \omega\}$ be a local base at z with $O_{n+1}^z \subset O_n^z$ for each $n \in \omega$. For every $n \in \omega$ we will inductively construct a set $D_n \subset Z$ and a number $k(n, z) \in \omega$ for any $z \in D_n$ in such a way that

(1) $D_n \subset D_{n+1}$ for any $n \in \omega$;
(2) $k(n, z) \geq n$ and $k(n + 1, z) > k(n, z)$ for any $n \in \omega$ and $z \in D_n$;
(3) the family $\mathcal{U}_n = \{O_{k(n,z)}^z : z \in D_n\}$ is disjoint and $\bigcup \mathcal{U}_n$ is dense in Z for any $n \in \omega$;
(4) the family $\mathcal{U}_{n+1}$ is inscribed in $\mathcal{U}_n$ for any $n \in \omega$;
(5) the family $\{O_{k(n+1,z)}^{z} : z \in D_{n+1}\backslash D_n\}$ is inscribed in $\mathcal{B}_{n+1}$ for any $n \in \omega$.

To construct D_0, take any $z_0 \in Z$ and let $n(0, z_0) = 0$, $D_0[1] = \{O_{n(0,z_0)}^{z_0}\}$. Suppose that β is an ordinal and we have $\{z_\alpha : \alpha < \beta\}$ and $\{k(0, z_\alpha) : \alpha < \beta\}$ such that the family $D_0[\beta] = \{O_{k(0,z_\alpha)}^{z_\alpha} : \alpha < \beta\}$ is disjoint. If $\bigcup D_0[\beta]$ is dense in Z, then our construction stops. If not, then we can choose $z_\beta \in Z\backslash\bigcup D_0[\beta]$ and $k(0, z_\beta)$ so that $O_{k(0,z_\beta)}^{z_\beta} \subset Z\backslash\overline{\bigcup D_0[\beta]}$. It is clear that the family $D_0[\beta + 1] = \{O_{k(0,z_\alpha)}^{z_\alpha} : \alpha \leq \beta\}$ is still disjoint; so our construction can be continued until $\bigcup D_0[\beta]$ is dense in Z for some ordinal β (evidently, this will happen for some $\beta < |Z|^+$). It is obvious that the property (3) is fulfilled for the sets $D_0 = \{z_\alpha : \alpha < \beta\}$ and $\{k(0, z_\alpha) : \alpha < \beta\}$.

Now assume that $m \in \omega$ and we have sets $\{D_0, \ldots, D_m\}$ such that for each $i \leq m$ a number $k(i, z) \in \omega$ is chosen for every $z \in D_i$ in such a way that (3) is true for all $n \leq m$ and (1),(2),(4),(5) are satisfied for any $n < m$.

For any $z \in D_m$ let $k(m + 1, z) = k(m, z) + 1$; it follows from (3) that the family $D_{m+1}[0] = \{O_{k(m+1,z)}^z : z \in D_m\}$ is disjoint. If $\bigcup D_{m+1}[0]$ is dense in Z then we let $D_{m+1} = D_m$; it is clear that (3) still fulfilled for all $n \leq m + 1$ while (1),(2),(4),(5) are satisfied for any $n \leq m$. If $\bigcup D_{m+1}[0]$ is not dense in Z then it follows from (3) that we can find $B \in \mathcal{B}_{m+1}$ and $O \in \mathcal{U}_m$ such that $W = B \cap O \neq \emptyset$ and $W \cap (\bigcup D_{m+1}[0]) = \emptyset$. Pick $z_0 \in W$ and $k(m + 1, z_0) \geq m + 1$ such that $O_{k(m+1,z_0)}^{z_0} \subset W$ and let $D_{m+1}[1] = \{O_{k(m+1,z)}^z : z \in D_m\} \cup \{O_{k(m+1,z_0)}^{z_0}\}$.

Suppose that β is an ordinal and we have constructed $\{z_\alpha : \alpha < \beta\}$ and $\{k(m + 1, z_\alpha) : \alpha < \beta\}$ such that the family

$$D_{m+1}[\beta] = \{O_{k(m+1,z)}^z : z \in D_m\} \cup \{O_{k(m+1,z_\alpha)}^{z_\alpha} : \alpha < \beta\}$$

is disjoint and $\{O_{k(m+1,z_\alpha)}^{z_\alpha} : \alpha < \beta\}$ is inscribed in $\mathcal{U}_m \wedge \mathcal{B}_{m+1}$. If the set $\bigcup D_{m+1}[\beta]$ is dense in Z then our construction stops. If not, then we can find $B \in \mathcal{B}_{m+1}$ and $O \in \mathcal{U}_m$ such that $W = B \cap O \neq \emptyset$ and $W \cap (\bigcup D_{m+1}[\beta]) = \emptyset$.

Now choose a point $z_\beta \in W$ and a number $k(m + 1, z_\beta) \geq m + 1$ in such a way that $O^{z_\beta}_{k(m+1,z_\beta)} \subset W$. It is clear that the family $D_{m+1}[\beta + 1]$ is still disjoint; so our construction can be continued until $\bigcup D_{m+1}[\beta]$ is dense in Z for some ordinal β (evidently, this will happen for some $\beta < |Z|^+$). It is clear that, for the sets $D_0, \ldots, D_m, D_{m+1} = D_m \cup \{z_\alpha : \alpha < \beta\}$, the property (3) is true for all $n \leq m + 1$ and (1),(2),(4),(5) are satisfied for any $n \leq m$. Thus our inductive procedure can be continued to construct the sets $\{D_n : n \in \omega\}$ with the properties (1)–(5).

We claim that the set $D = \bigcup_{n \in \omega} D_n$ is dense in Z and metrizable. Indeed, it follows from (3) and (4) that $\mathcal{D}_n = \mathcal{U}_n | D$ is an open disjoint cover of D; so $\mathcal{D}_n$ is a discrete family in D for any $n \in \omega$. Besides, the property (2) implies that $\mathcal{D} = \bigcup_{n \in \omega} \mathcal{D}_n$ contains a local base at every $z \in D$; so $\mathcal{D}$ is a σ-discrete base of D and therefore D is metrizable.

To show that the set D is dense in the space Z it suffices to prove that $D \cap B \neq \emptyset$ for any $B \in \mathcal{B}$. There is $n \in \omega$ such that $B \in \mathcal{B}_n$; it follows from (3) that there is $z \in D_n$ such that $V = O^z_{k(n,z)} \cap B \neq \emptyset$. If $z \in B$ then there is nothing to prove so we assume that $z \notin B$.

Since $k(i, z) \to \infty$ when $i \to \infty$ and $\bigcap \{\mathrm{cl}_Z(O^z_{k(i,z)}) : i \geq n\} = \{z\}$, the number $m = \min\{i - 1 : i > n$ and V is not contained in $\mathrm{cl}_Z(O^z_{k(i,z)})\}$ is well defined. It is easy to see that the set $P = (O^z_{k(m,z)} \setminus \mathrm{cl}_Z(O^z_{k(m+1,z)})) \cap V$ is non-empty and hence there is $d \in D_{m+1}$ with $O^d_{k(m+1,d)} \cap P \neq \emptyset$. But $d \in D_{m+1} \setminus D_m$; so $O^d_{k(m+1,d)} \subset B'$ for some $B' \in \mathcal{B}_{m+1}$. It follows from $B' \cap B \neq \emptyset$ that $B' \subset B$ and hence $d \in B$. This proves that $D \cap B \neq \emptyset$ for any $B \in \mathcal{B}$; so D is dense in Z and Fact 1 is proved.

Returning to our solution observe that Fact 1 implies that if X has a dense metrizable subspace then it has a σ-disjoint π-base; so necessity is clear. Now if $\mathcal{B}$ is a σ-disjoint π-base in X then the set P of the points of countable character is dense in X by Problem 120. It is immediate that $\mathcal{B} | P$ is a σ-disjoint π-base in P; so P has a dense metrizable subspace D by Fact 1. It is clear that D is also dense in X; this settles sufficiency and makes our solution complete.

U.139. *Prove that $\mathcal{M}(\kappa)$ is an ω-perfect class for any κ.*

Solution. Given spaces Y and Z say that $Y \sqsubseteq Z$ if Y embeds in Z as a closed subspace; the fact that Y is a continuous image of Z will be denoted by $Z \gg Y$.

Suppose that $X \in \mathcal{M}(\kappa)$; there is a compact space K and $F \sqsubseteq L(\kappa)^\omega \times K$ such that $F \gg X$. Then $F \times \omega \gg X \times \omega$ and $(F \times \omega)^\omega \gg (X \times \omega)^\omega = o_\omega(X)$. Since any countable subset of κ is clopen in the space $L(\kappa)$, we have $L(\kappa) \gg \omega$ and therefore $L(\kappa) \times L(\kappa) \gg L(\kappa) \times \omega$ which shows that $L(\kappa)^\omega \gg (L(\kappa) \times \omega)^\omega \simeq L(\kappa)^\omega \times \omega^\omega$.

Furthermore, $(F \times \omega)^\omega \sqsubseteq (L(\kappa)^\omega \times \omega \times K)^\omega \simeq L(\kappa)^\omega \times \omega^\omega \times K^\omega$. Besides, $L(\kappa)^\omega \gg L(\kappa)^\omega \times \omega^\omega$, we have $L(\kappa)^\omega \times K^\omega \gg L(\kappa)^\omega \times \omega^\omega \times K^\omega$ and hence there exists $G \sqsubseteq L(\kappa)^\omega \times K^\omega$ such that $G \gg (F \times \omega)^\omega \gg o_\omega(X)$ whence $G \gg o_\omega(X)$. This proves that $o_\omega(X) \in \mathcal{M}(\kappa)$.

Now assume that $Y \in \mathcal{E}(X)$ and fix a compact space L such that $X \times L \gg Y$. We have $F \times L \gg X \times L$ and $F \times L \sqsubseteq L(\kappa)^\omega \times (K \times L)$. Since also $F \times L \gg Y$, we have $Y \in \mathcal{M}(\kappa)$ and hence $\mathcal{E}(X) \subset \mathcal{M}(\kappa)$. Finally, if H is closed in X then there is a closed $F' \subset F$ such that $F' \gg H$; we have $F' \sqsubseteq L(\kappa)^\omega \times K$ so $H \in \mathcal{M}(\kappa)$ and therefore $\mathcal{M}(\kappa)$ is a perfect class.

U.140. *Prove that, for any Corson compact space X the space $C_p(X)$ belongs to $\mathcal{M}(\kappa)$ for some uncountable κ.*

Solution. The space X embeds in $C_p(L(\kappa))$ for some uncountable κ by Problem 106. It is clear that $L(\kappa) \in \mathcal{M}(\kappa)$. The class $\mathcal{M}(\kappa)$ being ω-perfect by Problem 139, the space $C_p(X)$ belongs to $\mathcal{M}(\kappa)$ by Problem 019.

U.141. *Prove that if κ is an uncountable cardinal and $Y \in \mathcal{M}(\kappa)$ then Y^ω is Lindelöf. In particular, $(C_p(X))^\omega$ is Lindelöf for any Corson compact space X.*

Solution. It was proved in TFS-354 that the space $L(\kappa)^\omega$ is Lindelöf. As a consequence, $L(\kappa)^\omega \times K$ is also Lindelöf for any compact space K (see Fact 2 of T.490 and Fact 3 of S.288). Since any closed subspace and any continuous image of a Lindelöf space is Lindelöf, any element of $\mathcal{M}(\kappa)$ is Lindelöf. Now, it follows from $Y \in \mathcal{M}(\kappa)$ that $Y^\omega \in \mathcal{M}(\kappa)$ (see Problem 139) so Y^ω is also Lindelöf.

Finally, if X is a Corson compact space then $C_p(X) \in \mathcal{M}(\lambda)$ for some uncountable cardinal λ (see Problem 140) so, by what we proved above, the space $(C_p(X))^\omega$ is Lindelöf.

U.142. *Prove that any countable union of primarily Lindelöf spaces is a primarily Lindelöf space.*

Solution. Given spaces Y and Z say that $Y \sqsubseteq Z$ if Y embeds in Z as a closed subspace; the fact that Y is a continuous image of Z will be denoted by $Z \gg Y$.

Suppose that $X = \bigcup_{n \in \omega} X_n$ and X_n is a primarily Lindelöf space for all $n \in \omega$. Fix an uncountable cardinal κ_n and $F_n \sqsubseteq L(\kappa_n)^\omega$ for which $F_n \gg X_n$ for every $n \in \omega$. If $\kappa = \sup\{\kappa_n : n \in \omega\}$ then $F_n \sqsubseteq L(\kappa)^\omega$ for any $n \in \omega$. Any countable subset of κ is clopen in $L(\kappa)$ so $L(\kappa) \gg \omega$. Thus $L(\kappa) \times L(\kappa) \gg L(\kappa) \times \omega$ which shows that $L(\kappa)^\omega \gg (L(\kappa) \times \omega)^\omega \gg L(\kappa)^\omega \times \omega$. It is clear that $F = \bigoplus_{n \in \omega} F_n \sqsubseteq L(\kappa)^\omega \times \omega$; so there is $G \sqsubseteq L(\kappa)^\omega$ with $G \gg F \gg X$ and therefore $G \gg X$, i.e., X is primarily Lindelöf.

U.143. *Prove that any countable product of primarily Lindelöf spaces is a primarily Lindelöf space.*

Solution. Given spaces Y and Z say that $Y \sqsubseteq Z$ if Y embeds in Z as a closed subspace; the fact that Y is a continuous image of Z will be denoted by $Z \gg Y$.

Suppose that $X = \prod_{n \in \omega} X_n$ and X_n is a primarily Lindelöf space for all $n \in \omega$. Fix an uncountable cardinal κ_n and $F_n \sqsubseteq L(\kappa_n)^\omega$ for which $F_n \gg X_n$ for every $n \in \omega$. If $\kappa = \sup\{\kappa_n : n \in \omega\}$ then $F_n \sqsubseteq L(\kappa)^\omega$ for any $n \in \omega$. It is clear that $F = \prod_{n \in \omega} F_n \sqsubseteq (L(\kappa)^\omega)^\omega \simeq L(\kappa)^\omega$ and $F \gg X$ which shows that X is primarily Lindelöf.

U.144. *Prove that any continuous image as well as any closed subspace of a primarily Lindelöf space is a primarily Lindelöf space.*

Solution. Given spaces Y and Z say that $Y \sqsubseteq Z$ if Y embeds in Z as a closed subspace; the fact that Y is a continuous image of Z will be denoted by $Z \gg Y$.

Suppose that X is a primarily Lindelöf space and $X \gg Y$. Fix an uncountable cardinal κ and $F \sqsubseteq L(\kappa)^\omega$ for which $F \gg X$; then $F \gg Y$ and hence Y is primarily Lindelöf. Therefore every continuous image of a primarily Lindelöf spaces is primarily Lindelöf. Now, if H is closed in X then there is a closed $F' \subset F$ such that $F' \gg H$. Since also $F' \sqsubseteq L(\kappa)^\omega$, the space H is primarily Lindelöf.

U.145. *Prove that any countable intersection of primarily Lindelöf spaces is a primarily Lindelöf space.*

Solution. Suppose that Z is a space and $X_n \subset Z$ is primarily Lindelöf for any $n \in \omega$. Then $X = \bigcap_{n \in \omega} X_n$ embeds in $\prod_{n \in \omega} X_n$ as a closed subspace (see Fact 7 of S.271). Applying Problems 143 and 144 we can conclude that X is primarily Lindelöf.

U.146. *Prove that primarily Lindelöf spaces form a weakly k-directed class.*

Solution. Being primarily Lindelöf is preserved by finite products and continuous images by Problems 143 and 144. Any countable subset of ω_1 is a clopen subset of $L(\omega_1)$ so $L(\omega_1)$ maps continuously onto ω. As a consequence, $L(\omega_1)^\omega$ maps continuously onto ω^ω, i.e., ω^ω is primarily Lindelöf. Any metrizable compact space is a continuous image of ω^ω (see SFFS-328); so all metrizable compact spaces are primarily Lindelöf by Problem 144. This proves that primarily Lindelöf spaces form a weakly k-directed class.

U.147. *Given a space X assume that $r : X \to X$ is a retraction. For any $f \in C_p(X)$ let $r_1(f) = f \circ r$. Prove that $r_1 : C_p(X) \to C_p(X)$ is also a retraction.*

Solution. Let $F = r(X)$; then F is closed in X and $r(x) = x$ for any $x \in F$ (see Fact 1 of S.351). If $\pi_F : C_p(X) \to C_p(F)$ is the restriction map then π_F is continuous and $\pi_F(C_p(X)) = C_p(F)$ because F is C-embedded in X (see Fact 1 of S.398). Furthermore, we can also consider that $r : X \to F$ and therefore the dual map $r^* : C_p(F) \to C_p(X)$ defined by $r^*(f) = f \circ r$ for any $f \in C_p(F)$ is an embedding by TFS-163. It is evident that $r_1 = r^* \circ \pi_F$; so r_1 is a continuous map.

For any $x \in F$ and $f \in C_p(X)$ we have $r_1(f)(x) = (f|F)(r(x)) = f(x)$ which shows that $r_1(f)|F = f|F$, i.e., $\pi_F(r_1(f)) = \pi_F(f)$. As a consequence, for any function $f \in C_p(X)$ we obtain $r_1(r_1(f)) = r^*(\pi_F(r_1(f))) = r^*(\pi_F(f)) = r_1(f)$ which proves that $r_1 \circ r_1 = r_1$ and therefore r_1 is a retraction.

U.148. *Given an uncountable cardinal κ and a set $A \subset L(\kappa)$ define a map $p_A :$ $L(\kappa) \to L(\kappa)$ by the rule $p_A(x) = a$ if $x \notin A$ and $p_A(x) = x$ for all $x \in A$ (recall that $L(\kappa) = \kappa \cup \{a\}$ and a is the unique non-isolated point of $L(\kappa)$). Prove that*

(i) *p_A is a retraction on $L(\kappa)$ onto $A \cup \{a\}$ for any $A \subset L(\kappa)$;*

(ii) *if $B \subset L(\kappa)$ and F is a closed subset of $(L(\kappa))^\omega$ then there exists $A \subset L(\kappa)$ such that $B \subset A$, $|A| \leq |B| \cdot \omega$ and $(p_A)^\omega(F) \subset F$. Here, as usual, the map $q_A = (p_A)^\omega : (L(\kappa))^\omega \to (L(\kappa))^\omega$ is the countable power of the map p_A defined by $q_A(x)(n) = p_A(x(n))$ for any $x \in (L(\kappa))^\omega$ and $n \in \omega$.*

Solution. To see that $p_A : L(\kappa) \to A \cup \{a\}$ is continuous, it suffices to prove continuity at the point a. So, take any $U \in \tau(a, A \cup \{a\})$; there exists a countable $A' \subset A$ such that $(A \backslash A') \cup \{a\} \subset U$. It is evident that $V = (\kappa \backslash A') \cup \{a\}$ is an open neighbourhood of a in $L(\kappa)$. Since $p_A(V) = (A \backslash A') \cup \{a\} \subset U$, the set V witnesses continuity of p_A at the point a. Thus the map p_A is continuous.

Furthermore, $A \cup \{a\}$ is closed in $L(\kappa)$ and $p_A(x) = x$ for any $x \in A \cup \{a\}$. Therefore p_A is a retraction by Fact 1 of S.351, i.e., (i) is proved.

The proof of (ii) will require more effort. Let $\pi_n : (L(\kappa))^{\omega} \to (L(\kappa))^n$ be the natural projection for any $n \in \mathbb{N}$. Given $n \in \mathbb{N}$ and $z = (z_0, \ldots, z_{n-1}) \in (L(\kappa))^n$ call a set $U \subset (L(\kappa))^n$ a *canonic neighbourhood of* z if there is $V \in \tau(a, L(\kappa))$ such that $U = U_0 \times \ldots \times U_{n-1}$ where $U_i = \{z_i\}$ if $z_i \neq a$ and $U_i = V$ whenever $z_i = a$; such a set U will be denoted by $[z, V]$. It is easy to see that the family of canonic neighbourhoods of z is a local base at z in $(L(\kappa))^n$. Therefore the family $\{\pi_n^{-1}([z, V]) : n \in \mathbb{N}, z \in (L(\kappa))^n \text{ and } V \in \tau(a, L(\kappa))\}$ is a base in $(L(\kappa))^{\omega}$.

Given a number $n \in \mathbb{N}$, a point $z \in (L(\kappa))^n$ will be called *marked* if there exists a set $O \in \tau(z, (L(\kappa))^n)$ such that $\pi_n^{-1}(O) \cap F = \emptyset$; it is evident that z is marked if and only if $z \notin \overline{\pi_n(F)}$. For every $n \in \mathbb{N}$ and a marked $z \in (L(\kappa))^n$ fix a set $O_z \in \tau(a, L(\kappa))$ such that $\pi_n^{-1}([z, O_z]) \cap F = \emptyset$. Let $A_0 = B \cup \{a\}$; assume that $m \in \mathbb{N}$ and we have constructed sets $A_0 \subset \ldots \subset A_{m-1}$ such that

(1) $|A_n| \leq |B| \cdot \omega$ for any $n < m$;
(2) for any $n < m$ if $k < n$ and, for some points $z_0, \ldots, z_{j-1} \in A_k$, the element $z = (z_0, \ldots, z_{j-1}) \in (L(\kappa))^j$ is marked, then $L(\kappa) \backslash O_z \subset A_n$.

Let $P_j = \{z = (z_0, \ldots, z_{j-1}) \in (L(\kappa))^j : \{z_0, \ldots, z_{j-1}\} \subset A_{m-1}$ and z is marked$\}$ for any $j \in \mathbb{N}$. If $P = \bigcup\{P_j : j \in \mathbb{N}\}$ then $|P| \leq |A_{m-1}| \cdot \omega \leq |B| \cdot \omega$ by the property (1). An immediate consequence is that the properties (1) and (2) are still satisfied for the set $A_m = A_{m-1} \cup (\bigcup\{L(\kappa) \backslash O_z : z \in P\})$.

Therefore our inductive procedure can be continued to construct an increasing sequence $\{A_n : n \in \omega\}$ of subsets of $L(\kappa)$ such that $B \cup \{a\} \subset A_0$ and the properties (1) and (2) hold for every $n \in \omega$; let $A = \bigcup_{n \in \omega} A_n$. It is evident that $B \subset A$ and $|A| \leq |B| \cdot \omega$.

To see that we have $q_A(F) \subset F$ assume that there exists a point $x \in F$ such that $y = q_A(x) \notin F$. The set F being closed in $(L(\kappa))^{\omega}$ there is $j \in \mathbb{N}$ such that $z = \pi_j(y) \notin \pi_j(F)$ and hence the point z is marked; furthermore, we have $[z, O_z] \cap \pi_j(F) = \emptyset$. By the definition of the map $q_A = p_A^{\omega}$ all coordinates of the point $z = (z_0, \ldots, z_{j-1})$ belong to A and hence there is $n \in \omega$ for which $z_0, \ldots, z_{j-1} \in A_n$.

If we show that $\pi_j(x) \in [z, O_z]$ then the contradiction with $[z, O_z] \cap \pi_j(F) = \emptyset$ will finish the proof. Recall that $[z, O_z] = U_0 \times \ldots \times U_{j-1}$ where $U_k = \{z_k\}$ if $z_k \neq a$ and $U_k = O_z$ whenever $z_k = a$.

So, take any $k < j$; if $y(k) \neq a$ then $y(k) \in A$ and hence we have the equalities $p_A(x(k)) = y(k) = x(k)$. As a consequence, $z_k = y(k) = x(k)$ whence $x(k) \in U_k$. Now, if $y(k) = a$ then we have two possibilities: $x(k) \in A$ or $x(k) \notin A$.

If $x(k) \in A$ then $p_A(x(k)) = x(k) = y(k) = z_k$ and hence $x(k) \in U_k$. Finally, if $x(k) \notin A$ then $x(k) \notin A_{n+1}$. But $L(\kappa) \backslash O_z \subset A_{n+1}$ by the property (2); so we have $x(k) \in O_z = U_k$. We proved that, in all cases, $x(k) \in U_k$ for all $k < j$ and therefore $\pi_j(x) \in U_0 \times \ldots \times U_{j-1} = [z, O_z]$ which is a contradiction. Thus $q_A(F) \subset F$ and hence we proved (ii) which shows that our solution is complete.

U.149. *Prove that, for any primarily Lindelöf space X, the space $C_p(X)$ condenses linearly into $\Sigma(A)$ for some A.*

Solution. Recall that, for any cardinal λ, in the space $L(\lambda) = \lambda \cup \{a\}$ all points of λ are isolated, $a \notin \lambda$ and a set $U \ni a$ is open if and only if $L(\lambda) \backslash U$ is countable. This definition also makes sense for $\lambda = \omega$, in which case the resulting space is countable and discrete.

Suppose that a space Y is a continuous image of a closed $F \subset (L(\lambda))^\omega$ for some uncountable cardinal λ. The respective dual map embeds $C_p(Y)$ in $C_p(F)$ as a linear subspace; so if we want to prove that $C_p(Y)$ condenses linearly into some $\Sigma(A)$ it suffices to show the existence of the promised linear condensation for $C_p(F)$. Therefore we can assume, without loss of generality, that X is a closed subset of $(L(\lambda))^\omega$ for some uncountable cardinal λ.

Our proof will proceed by induction on λ. For the first step assume that $F \subset (L(\omega))^\omega$; then F is separable. If Q is a countable dense subset of F then the restriction map π_Q condenses $C_p(F)$ linearly in $\mathbb{R}^Q$. It is an easy exercise to see that $\mathbb{R}^Q$ condenses linearly into $\Sigma(\omega_1)$; so $C_p(F)$ also condenses linearly into $\Sigma(\omega_1)$. Now suppose that κ is an uncountable cardinal and we proved the existence of the relevant condensations for all cardinals smaller than κ. In other words,

(*) for any cardinal $\lambda < \kappa$, if F is a closed subset of $(L(\lambda))^\omega$ then there exists a linear condensation of $C_p(F)$ into $\Sigma(A)$ for some uncountable set A.

Take an arbitrary closed $X \subset (L(\kappa))^\omega$; for any $B \subset L(\kappa)$ we will need the retraction $p_B : L(\kappa) \to B \cup \{a\}$ defined as follows: $p_B(x) = x$ for any $x \in B$ and $p_B(x) = a$ for every $x \in L(\kappa) \backslash B$ (it was proved in Problem 148 that every p_B is, indeed, a retraction). The map $q_B = (p_B)^\omega : (L(\kappa))^\omega \to (L(\kappa))^\omega$ is defined by $q_B(x)(n) = p_B(x(n))$ for any $x \in (L(\kappa))^\omega$ and $n \in \omega$. It is immediate that q_B is a retraction as well.

Let us prove that for the cardinal $\mu = \mathrm{cf}(\kappa)$ there is a μ-sequence $\{B_\alpha : \alpha < \mu\}$ of subsets of $L(\kappa)$ such that

(1) $a \in B_0$, $B_\alpha \subset B_{\alpha+1}$, $B_{\alpha+1} \backslash B_\alpha \neq \emptyset$ and $|B_\alpha| < \kappa$ for all $\alpha < \mu$;
(2) $L(\kappa) = \bigcup\{B_\alpha : \alpha < \mu\}$;
(3) $B_\beta = \bigcup\{B_\alpha : \alpha < \beta\}$ whenever $\beta < \mu$ is a limit ordinal;
(4) $q_{B_\alpha}(X) \subset X$ for any $\alpha < \mu$.

First represent $L(\kappa)$ as $\bigcup\{M_\alpha : \alpha < \mu\}$ where $|M_\alpha| < \kappa$ for any $\alpha < \mu$. By Problem 148, there exists a set $B_0 \subset L(\kappa)$ such that $B_0 \supset M_0 \cup \{a\}$, $|B_0| < \kappa$ and $q_{B_0}(X) \subset X$. If $B_\alpha \subset L(\kappa)$ is chosen then apply Problem 148 to find $B_{\alpha+1} \subset L(\kappa)$ such that $M_{\alpha+1} \cup B_\alpha \subset B_{\alpha+1}$, $B_{\alpha+1} \backslash B_\alpha \neq \emptyset$, $|B_{\alpha+1}| < \kappa$ and $q_{B_{\alpha+1}}(X) \subset X$. If $\beta < \mu$ is a limit ordinal let $B_\beta = \bigcup\{B_\alpha : \alpha < \beta\}$.

Observe that (1)–(3) hold trivially for the μ-sequence $\{B_\alpha : \alpha < \mu\}$ and the property (4) is clear for successor ordinals. Now, if $\beta < \mu$ is a limit ordinal assume that $q_{B_\beta}(X)$ is not contained in X. Then there is $x \in X$ such that $y = q_{B_\beta}(x) \notin X$. Since X is closed in $(L(\kappa))^\omega$, there is $n \in \mathbb{N}$ such that $\pi_n(y) \notin \overline{\pi_n(X)}$ where $\pi_n : (L(\kappa))^\omega \to (L(\kappa))^n$ is the natural projection.

We have $y(i) \in B_\beta$ for all $i < n$; the property (3) implies that there is a successor ordinal $\alpha < \beta$ such that $y(i) \in B_\alpha$ for all $i < n$. Since the property (4) is satisfied for α, the point $t = q_{B_\alpha}(x)$ belongs to X. We claim that $\pi_n(t) = \pi_n(y)$. Indeed, if $i < n$ and $x(i) \in B_\beta$ then $x(i) = y(i) \in B_\alpha$ and therefore $t(i) = x(i)$ which implies $t(i) = y(i)$. If, on the other hand, $x(i) \notin B_\beta$ then $x(i) \notin B_\alpha$ and hence $t(i) = a$; since also $y(i) = a$, we have $t(i) = y(i)$ again. This proves that $t(i) = y(i)$ for all $i < n$ and hence $\pi_n(y) = \pi_n(t) \in \pi_n(X)$; this contradiction with $\pi_n(y) \notin \overline{\pi_n(X)}$ shows that the property (4) holds for all $\alpha < \mu$.

Let $X_\alpha = q_{B_\alpha}(X)$ for any $\alpha < \mu$; the map $r_\alpha = q_{B_\alpha}|X : X \to X_\alpha$ is a retraction and $X_\alpha = X \cap (B_\alpha)^\omega$. It is clear that B_α is homeomorphic to some $L(\lambda)$ so, for any $\alpha < \mu$, the space X_α is a closed subset of $(L(\lambda))^\omega$ for some $\lambda < \kappa$. The property (∗) shows that every $C_p(X_\alpha)$ can be linearly condensed onto a subspace of $\Sigma(A_\alpha)$ for some set A_α.

The dual map $r_\alpha^* : C_p(X_\alpha) \to C_p(X)$ is a linear embedding (see TFS-163); so the space $T_\alpha = r_\alpha^*(C_p(X_\alpha))$ is linearly homeomorphic to $C_p(X_\alpha)$. Therefore we can consider that there is a linear condensation $\varphi_\alpha : T_\alpha \to Z_\alpha$ where Z_α is a subspace of $\Sigma(A_\alpha)$ for any $\alpha < \mu$.

Given any ordinal $\alpha < \mu$, it is easy to see that $X_\alpha \subset X_{\alpha+1}$; so we have a retraction $\tilde{r}_\alpha = r_\alpha|X_{\alpha+1} : X_{\alpha+1} \to X_\alpha$. The map $(\tilde{r}_\alpha)^* : C_p(X_\alpha) \to C_p(X_{\alpha+1})$ is an embedding; let $T_\alpha' = (\tilde{r}_\alpha)^*(C_p(X_\alpha))$. It is straightforward that $r_{\alpha+1}^*(T_\alpha') = T_\alpha$ and hence $T_\alpha \subset T_{\alpha+1}$; let $\pi_\alpha : C_p(X) \to C_p(X_\alpha)$ be the restriction map for any ordinal $\alpha < \mu$. It is clear that the mappings $r_\alpha^* \circ \pi_\alpha : C_p(X) \to T_\alpha$ and $r_{\alpha+1}^* \circ \pi_{\alpha+1} : C_p(X) \to T_{\alpha+1}$ are continuous; this, together with the fact that $T_{\alpha+1} \supset T_\alpha$ is a linear subspace of $C_p(X)$ implies that $r_{\alpha+1}^* \circ \pi_{\alpha+1} - r_\alpha^* \circ \pi_\alpha$ maps $C_p(X)$ continuously into $T_{\alpha+1}$; so the mapping $u_\alpha = \varphi_{\alpha+1} \circ (r_{\alpha+1}^* \circ \pi_{\alpha+1} - r_\alpha^* \circ \pi_\alpha)$ is well defined, linear and continuous for all $\alpha < \mu$.

Consequently, the diagonal product $u = \Delta\{u_\alpha : \alpha < \mu\}$ is a linear continuous map from $C_p(X)$ to $\Sigma = \prod\{\Sigma(A_\alpha) : \alpha < \mu\}$. It is evident that, for the set $A = \bigcup\{A_\alpha : \alpha < \mu\}$, we can consider that Σ is a linear subspace of $\mathbb{R}^A$. We will prove that u is injective and $u(C_p(X)) \subset \Sigma(A)$; this will carry out our inductive step.

To see that u is injective take distinct $f, g \in C_p(X)$. Observe first that, for any limit ordinal $\beta \le \mu$ the space $Y_\beta = \bigcup\{X_\alpha : \alpha < \beta\}$ is dense in X_β (here $X_\beta = X$ and $B_\beta = L(\kappa)$ for $\beta = \mu$). Indeed, fix any point $y \in X_\beta$; we have $y = r_\beta(x)$ for some $x \in X$ and hence $y(i) \in B_\beta$ for any $i \in \omega$. Given $n \in \omega$, it follows from (3) that there exists $\alpha_n < \beta$ for which $\{y(0), \ldots, y(n-1)\} \subset B_{\alpha_n}$. Now, if $y_n = r_{\alpha_n}(y) \in X_{\alpha_n}$ for every $n \in \omega$ then $y_n(i) = y(i)$ for every $i < n$ and hence the sequence $S = \{y_n : n \in \omega\}$ converges to y. Since $S \subset \bigcup\{X_\alpha : \alpha < \beta\}$, we proved that $y \in \overline{Y}_\beta$ and therefore $\overline{Y}_\beta = X_\beta$.

As a consequence, we have $f|Y_\mu \ne g|Y_\mu$ (see Fact 0 of S.351); so the ordinal $\gamma = \min\{\alpha < \mu : f|X_\alpha \ne g|X_\alpha\}$ is well defined. The ordinal γ has to be a successor for otherwise we will have $f|Y_y \neq g|Y_y$ and hence there exists $\alpha < \gamma$ such that $f|X_\alpha \neq g|X_\alpha$ which is a contradiction. Therefore $\gamma = \alpha + 1$ and $f|X_\alpha = g|X_\alpha$ while $f|X_{\alpha+1} \neq g|X_{\alpha+1}$. This implies $r_\alpha^*(\pi_\alpha(f)) = (f|X_\alpha) \circ r_\alpha = (g|X_\alpha) \circ r_\alpha = r_\alpha^* \circ \pi_\alpha(g)$ while $r_{\alpha+1}^*(\pi_{\alpha+1}(f)) = (f|X_{\alpha+1}) \circ r_{\alpha+1} \neq (g|X_{\alpha+1}) \circ r_{\alpha+1} = r_{\alpha+1}^* \circ \pi_{\alpha+1}(g)$. Thus $(r_{\alpha+1}^* \circ \pi_{\alpha+1} - r_\alpha^* \circ \pi_\alpha)(f) \neq (r_{\alpha+1}^* \circ \pi_{\alpha+1} - r_\alpha^* \circ \pi_\alpha)(g)$; since the map $\varphi_{\alpha+1}$ is injective, we have $u_\alpha(f) \neq u_\alpha(g)$ whence $u(f) \neq u(g)$; so u is an injective map.

For every $\alpha < \mu$ let $\mathbf{0}_\alpha \in \Sigma(A_\alpha)$ be the function which is identically zero on A_α; we will denote by $\mathbf{0} \in C_p(X)$ the function which is identically zero on X. To prove that u maps $C_p(X)$ into $\Sigma(A)$ it suffices to show that, for any $f \in C_p(X)$, we have $u_\alpha(f) = \mathbf{0}_\alpha$ for all but countably many α. Since the map $\varphi_{\alpha+1}$ is linear, it is sufficient to show that $(r_{\alpha+1}^* \circ \pi_{\alpha+1} - r_\alpha^* \circ \pi_\alpha)(f) = \mathbf{0}$ or, equivalently, $f \circ r_{\alpha+1} = f \circ r_\alpha$ for all $\alpha < \mu$ except countably many of them.

If this is not true then there is an uncountable set $M' \subset \mu$ and a point $x_\alpha \in X$ such that $f(r_{\alpha+1}(x_\alpha)) \neq f(r_\alpha(x_\alpha))$ for any $\alpha \in M'$. There exist an uncountable $M \subset M'$ and $\varepsilon > 0$ such that $|f(r_{\alpha+1}(x_\alpha)) - f(r_\alpha(x_\alpha))| > \varepsilon$ for every $\alpha \in M$. The space X is Lindelöf being a closed subspace of the Lindelöf space $(L(\kappa))^\omega$ (see TFS-354) so, for the set $Y = \{r_\alpha(x_\alpha) : \alpha \in M\}$, there exists a point $z \in X$ such that $U \cap Y$ is uncountable for any $U \in \tau(z, X)$.

The function f being continuous on X there exists a set $W \in \tau(z, X)$ such that $\mathrm{diam}(f(W)) < \varepsilon$. Making the set W smaller if necessary, we can consider that there exist $V \in \tau(a, L(\kappa))$ and a number $n \in \omega$ such that $W = \pi_n^{-1}([V, n]) \cap X$; here $[V, n] = U_0 \times \ldots \times U_{n-1}$ where $U_i = \{z(i)\}$ if $z(i) \neq a$ and $U_i = V$ if $z(i) = a$ for all $i < n$. By the choice of z, the set $N = \{\alpha \in M : r_\alpha(x_\alpha) \in W\}$ is uncountable.

The set $L(\kappa)\backslash V$ is countable; since the family $\{B_{\alpha+1}\backslash B_\alpha : \alpha \in N\}$ is uncountable and consists of non-empty sets, we can choose $\alpha \in N$ such that $B_{\alpha+1}\backslash B_\alpha \subset V$. Let $t = r_\alpha(x_\alpha)$; we claim that $y = r_{\alpha+1}(x_\alpha) \in W$.

Indeed, take any $i < n$. If $x_\alpha(i) \notin B_{\alpha+1}$ then $y(i) = t(i) = a$. Since $t(i) \in U_i$, we have $y(i) \in U_i$ in this case. Now, if $x_\alpha(i) \in B_{\alpha+1}$ then, by the definition of $r_{\alpha+1}$, we have $y(i) = x_\alpha(i)$; if $x_\alpha(i) \in B_\alpha$ then $y(i) = x_\alpha(i) = t(i) \in U_i$. If, on the other hand, $x_\alpha(i) \in B_{\alpha+1}\backslash B_\alpha$ then $t(i) = a$ and it follows from $t \in W$ that $z(i) = a$. As a consequence, $y(i) \in B_{\alpha+1}\backslash B_\alpha \subset V = U_i$. This proves that $y(i) \in U_i$ for all $i < n$ and hence $y \in W$. Therefore $|f(y) - f(t)| \leq \mathrm{diam}(f(W)) < \varepsilon$ which contradicts $|f(y) - f(t)| \geq \varepsilon$. Therefore the map u is a linear condensation of X into $\Sigma(A)$ which finishes our inductive proof and shows that our solution is complete.

U.150. *Prove that the following conditions are equivalent for any compact space X:*

(i) X is Corson compact;
(ii) $C_p(X)$ is primarily Lindelöf;
(iii) there is a primarily Lindelöf $P \subset C_p(X)$ which separates the points of X;
(iv) X embeds in $C_p(Y)$ for some primarily Lindelöf space Y.

Solution. If $\mathcal{A}$ is a family of sets then $\bigvee \mathcal{A}$ is the family of all finite unions of elements of $\mathcal{A}$; analogously, $\bigwedge \mathcal{A}$ is the family of all finite intersections of elements of $\mathcal{A}$. If Z is a space then $\mathcal{C}(Z)$ is the family of all clopen subsets of Z.

Fact 1. Given a space Z and $f, g \in C_p(Z)$ let $(f * g)(z) = f(z) + g(z) - f(z)g(z)$ for any point $z \in Z$. Then $f * g \in C_p(Z, \mathbb{D})$ whenever $f, g \in C_p(Z, \mathbb{D})$, the operation $*$ is commutative and associative and the map $a_n : (C_p(Z, \mathbb{D}))^{n+1} \to C_p(Z, \mathbb{D})$ defined by $a_n(f_0, \ldots, f_n) = f_0 * \ldots * f_n$ for any $(f_0, \ldots, f_n) \in (C_p(Z, \mathbb{D}))^{n+1}$ is continuous for any $n \in \omega$.

Proof. Given $f, g \in C_p(Z, \mathbb{D})$ the sets $A = f^{-1}(1)$ and $B = g^{-1}(1)$ are clopen in Z. It is straightforward that $(f * g)(z) = 1$ if $z \in C = A \cup B$ and $(f * g)(z) = 0$ for all $z \in Z \backslash C$. Therefore $f * g$ is the characteristic function of the clopen set C, i.e., $f * g \in C_p(Z, \mathbb{D})$.

It is immediate from the definition that $f * g = g * f$ for any $f, g \in C_p(Z)$. Now, if $f, g, h \in C_p(Z)$ then

$$(f * g) * h = (f + g - fg) * h = f + g - fg + h - (f + g - fg)h$$
$$= f + g + h - fg - fh - gh + fgh;$$

analogously,

$$f * (g * h) = f + g * h - f(g * h) = f + g + h - gh - f(g + h - gh)$$
$$= f + g + h - fg - fh - gh + fgh = (f * g) * h,$$

which proves that $*$ is associative. Continuity of the map a_n for any $n \in \omega$ follows easily from TFS-115 and TFS-116; so Fact 1 is proved.

Fact 2. Suppose that K is a zero-dimensional compact space and $A \subset C_p(K, \mathbb{D})$ separates the points of the space K. Consider the sets $A_0 = A \cup \{1 - f : f \in A\}$ and $B_0 = \{f_0 \cdot \ldots \cdot f_n : n \in \omega, \ f_i \in A_0 \text{ for any } i \leq n\}$; it turns out that the set $D(A) = \{f_0 * \ldots * f_n : n \in \omega, \ f_i \in B_0 \text{ for all } i \leq n\}$ coincides with $C_p(K, \mathbb{D})$. In particular, $C_p(K, \mathbb{D}) = \bigcup_{n \in \omega} C_n$ where every C_n is a continuous image of a finite power of A.

Proof. Observe that

(1) for any distinct $x, y \in K$ there is $f \in A_0$ such that $f(x) = 1$ and $f(y) = 0$.

because there is $g \in A$ with $g(x) \neq g(y)$; if $g(x) = 1$ then $f = g$ works; if $g(x) = 0$ then $f = 1 - g$ is as promised.

Our next step is to show that

(2) $\bigvee(\bigwedge\{f^{-1}(1) : f \in A_0\}) = \mathcal{C}(K)$.

To prove (2) let $\mathcal{C} = \{f^{-1}(1) : f \in A_0\}$ and take any non-empty $U \in \mathcal{C}(K)$. For any $x \in U$ and $y \in K \backslash U$ apply (1) to find a function $f_{x,y} \in A_0$ such that $f_{x,y}(x) = 1$ and $f_{x,y}(y) = 0$; then $O_{x,y} = f_{x,y}^{-1}(1) \in \mathcal{C}$. Furthermore, $\bigcap\{O_{x,y} : y \in K \backslash U\} \subset U$; so we can apply Fact 1 of S.326 to see that there is a finite

$P \subset K\backslash U$ for which $W_x = \bigcap\{O_{x,y} : y \in P\} \subset U$. It is clear that $x \in W_x \in \bigwedge C$ for any $x \in U$. The set U being compact there is a finite set $Q \subset U$ such that $\bigcup\{W_x : x \in Q\} = U$ which shows that U is a finite union of elements of $\bigwedge C$, i.e., $U \in \bigvee(\bigwedge C)$ and hence (2) is proved.

Given any set $U \in \bigwedge C$ there are $n \in \omega$ and functions $f_0, \ldots, f_n \in A_0$ such that $U = f_0^{-1}(1) \cap \ldots \cap f_n^{-1}(1)$; it is easy to see that $U = f^{-1}(1)$ for $f = f_0 \cdot \ldots \cdot f_n$ which shows that

(3) $\bigwedge\{f^{-1}(1) : f \in A_0\} \subset \{f^{-1}(1) : f \in B_0\}$.

Now fix any $f \in C_p(K, \mathbb{D})$ and let $U = f^{-1}(1)$; by (2) there are $n \in \omega$ and $U_0, \ldots, U_n \in \bigwedge C$ such that $U = U_0 \cup \ldots \cup U_n$. It follows from (3) that there are $g_0, \ldots, g_n \in B_0$ for which $U_i = g^{-1}(1)$ for any $i \le n$. It is easy to check that, for $g = g_0 * \ldots * g_n$ we have $g^{-1}(1) = U_0 \cup \ldots \cup U_n = U$; thus $g \in D(A)$ and $g = f$ which shows that $D(A) = C_p(K, \mathbb{D})$.

Consider the class $\mathcal{A}$ of spaces representable as a countable union of continuous images of finite powers of A. It is evident that A_0 is a continuous image of $A \oplus A$; so $A_0 \in \mathcal{A}$. The class $\mathcal{A}$ is invariant under finite products and countable unions; so $A_0^{n+1} \in \mathcal{A}$ for any $n \in \omega$.

Let $m_n(f_0, \ldots, f_n) = f_0 \cdot \ldots \cdot f_n$ for any $(f_0, \ldots, f_n) \in C_p(K, \mathbb{D})$; then the map $m_n : (C_p(K, \mathbb{D}))^{n+1} \to C_p(K, \mathbb{D})$ is continuous for any $n \in \omega$ by TFS-116. We have $B_0 = \bigcup\{m_n(A_0^{n+1}) : n \in \omega\}$ so $B_0 \in \mathcal{A}$ and hence $B_0^{n+1} \in \mathcal{A}$ for any $n \in \omega$.

It follows from $C_p(K, \mathbb{D}) = D(A) = \bigcup\{a_n(B_0^{n+1}) : n \in \omega\}$ (see Fact 1) that $C_p(K, \mathbb{D}) \in \mathcal{A}$, i.e., $C_p(K, \mathbb{D})$ is representable as a countable union of continuous images of finite powers of A; so Fact 2 is proved.

Returning to our solution let X be a Corson compact space. By Problem 118, the space X has a point-countable T_0-separating family $\mathcal{U}$ of open F_σ-subsets of X. For any $U \in \mathcal{U}$ choose a continuous function $f_U : X \to I = [0, 1] \subset \mathbb{R}$ such that $X\backslash U = f_U^{-1}(0)$ (see Fact 1 of S.358 and Fact 1 of S.499). Then the mapping $f = \Delta\{f_U : U \in \mathcal{U}\} : X \to I^{\mathcal{U}}$ is an embedding. To see it take distinct $x, y \in X$. There is $U \in \mathcal{U}$ such that $U \cap \{x, y\}$ is a singleton. Then $f_U(x) \ne f_U(y)$ and therefore $f(x) \ne f(y)$ which proves that f is, indeed, an embedding being an injection; let $X' = f(X)$.

Observe that $X' \subset \{x \in I^{\mathcal{U}} :$ the set $\{U \in \mathcal{U} : x(U) \ne 0\}$ is countable$\}$ because if $x \in X'$ then $x = f(y)$ for some $y \in X$ and $x(U) = f_U(y)$ for any $U \in \mathcal{U}$ which implies that the family $\{U \in \mathcal{U} : x(U) \ne 0\} \subset \{U \in \mathcal{U} : y \in U\}$ is countable.

To simplify the notation we will reformulate the obtained result as follows:

(1) there exists a set T such that X embeds in $\Sigma = \{x \in I^T : |x^{-1}(I\backslash\{0\})| \le \omega\}$, so we can assume that $X \subset \Sigma$.

Denote by K the Cantor set $\mathbb{D}^\omega$; fix a point $a \in K$ and let $I_n = [\frac{1}{n+2}, \frac{1}{n+1}] \subset I$ for any $n \in \omega$. Since K is zero-dimensional, there is a local base $\mathcal{O} = \{O_n : n \in \omega\}$ at the point a in K such that the set O_n is clopen in K and $O_{n+1} \subset O_n$ for any $n \in \omega$. Making the relevant changes in $\mathcal{O}$ if necessary, we can assume that $O_0 = K$ and $K_n = O_n\backslash O_{n+1} \ne \emptyset$ for any $n \in \omega$.

Since no point of K is isolated, the same is true for any non-empty clopen subset of K and hence every non-empty clopen subset of K is homeomorphic to K (see SFFS-348). This shows that K_n is homeomorphic to K and hence there is a continuous onto map $\varphi_n : K_n \to I_n$ for any $n \in \omega$ (see TFS-128). Let $\varphi(a) = 0$; if $x \in K \backslash \{a\}$ then there is a unique $n \in \omega$ such that $x \in K_n$; let $\varphi(x) = \varphi_n(x)$. It is an easy exercise that $\varphi : K \to I$ is a continuous onto map such that $\varphi^{-1}(0) = \{a\}$.

Let $\Phi : K^T \to I^T$ be the product of T-many copies of φ, i.e., $\Phi(x)(t) = \varphi(x(t))$ for any $x \in K^T$ and $t \in T$. The map Φ is continuous by Fact 1 of S.271 and it is easy to see that $\Phi(K^T) = I^T$. The space $Y = \Phi^{-1}(X)$ is compact being closed in the compact space K^T. Take any point $y \in Y$ and let $x = \Phi(y)$. If $y(t) \neq a$ then $x(t) = \varphi(y(t)) \neq 0$ and hence

$$S_y = \{t \in T : y(t) \neq a\} \subset \operatorname{supp}(x) = \{t \in T : x(t) \neq 0\}.$$

It follows from $X \subset \Sigma$ that $\operatorname{supp}(x)$ is countable; so S_y is countable for any $y \in Y$ which shows that Y is a subset of a Σ-product of T-many copies of K. Therefore Y is a Corson compact space by Problem 119. It is clear that $\Phi | Y$ maps Y continuously onto X. Besides, Y is zero-dimensional because so is K^T (see SFFS-301 and SFFS-302). Therefore Y is a zero-dimensional Corson compact space which maps continuously onto X.

There exists a T_0-separating point-countable family $\mathcal{V}$ of open F_σ-subsets of Y (see Problem 118). For any $U \in \mathcal{V}$ we have $U = \bigcup_{n \in \omega} F_n^U$ where F_n^U is compact for any $n \in \omega$. The space Y being zero-dimensional and compact, we can apply SFFS-306 to find a clopen set C_n^U such that $F_n^U \subset C_n^U \subset U$ for any $U \in \mathcal{V}$ and $n \in \omega$. The family $\{C_n^U : U \in \mathcal{V}\}$ is point-countable for any $n \in \omega$ because so is $\mathcal{V}$. Consequently, the family $\mathcal{C} = \{C_n^U : n \in \omega, U \in \mathcal{V}\}$ is point-countable as well and it is straightforward that $\mathcal{C}$ is T_0-separating.

Denote by u the function which is identically zero on Y and consider the subspace $L = \{\chi_C : C \in \mathcal{C}\} \cup \{u\}$ of the space $C_p(Y, \mathbb{D})$. The family $\mathcal{C}$ being point-countable, the set $L \backslash U$ is countable for any $U \in \tau(u, C_p(Y, \mathbb{D}))$. Thus, for $\kappa = |L|$, there is an evident continuous map of $L(\kappa)$ onto L. Therefore the space L is primarily Lindelöf; since $\mathcal{C}$ is T_0-separating, the set L separates the points of Y. It follows from Fact 2 that $C_p(Y, \mathbb{D}) = \bigcup\{C_n : n \in \omega\}$ where every C_n is a continuous image of a finite power of L. Therefore $C_p(Y, \mathbb{D})$ is primarily Lindelöf by Problems 142, 143 and 144. Apply Problem 143 once more to see that $C_p(Y, \mathbb{D}^\omega) \simeq (C_p(Y, \mathbb{D}))^\omega$ is primarily Lindelöf as well.

The space $C_p(Y, \mathbb{I})$ is a continuous image of $C_p(Y, \mathbb{D}^\omega)$ by Problem 004; so the set $C_p(Y, \mathbb{I})$ is also primarily Lindelöf. The space Y being compact, we have the equality $C_p(Y) = \bigcup\{C_p(Y, [-n, n]) : n \in \mathbb{N}\}$. Since every $C_p(Y, [-n, n])$ is homeomorphic to $C_p(Y, \mathbb{I})$, the space $C_p(Y)$ is primarily Lindelöf.

Letting $q = \Phi | Y$ we obtain a continuous onto map $q : Y \to X$; the dual map q^* embeds $C_p(X)$ in $C_p(Y)$ as a closed subspace (see TFS-163); so $C_p(X)$ is primarily Lindelöf. This proves that (i)$\Longrightarrow$(ii).

The implication (ii)$\Longrightarrow$(iii) is obvious; so assume that X is compact and there is a primarily Lindelöf $P \subset C_p(X)$ which separates the points of X. For any $x \in X$

and $f \in P$ let $\varphi_x(f) = f(x)$. Then $\varphi_x \in C_p(P)$ for any $x \in X$ and the map $\varphi : X \to C_p(P)$ defined by $\varphi(x) = \varphi_x$ for any $x \in X$, is continuous and injective (see TFS-166). Since X is compact, the map φ embeds X in $C_p(P)$ so the space $Y = P$ is primarily Lindelöf and X embeds in $C_p(Y)$. This shows that (iii)$\Longrightarrow$(iv).

Finally, if $X \subset C_p(Y)$ for some primarily Lindelöf space Y then $C_p(Y)$ condenses onto a subspace of $\Sigma(A)$ for some uncountable set A (see Problem 149). Consequently, the space X also condenses onto a subset of $\Sigma(A)$. Since every condensation of a compact space is a homeomorphism, the space X embeds in $\Sigma(A)$, i.e., X is Corson compact. This settles (iv)$\Longrightarrow$(i) and finishes our solution.

U.151. *Prove that a continuous image of a Corson compact space is Corson compact.*

Solution. Suppose that X is a Corson compact space and $\varphi : X \to Y$ is a continuous onto map. The dual map $\varphi^* : C_p(Y) \to C_p(X)$ defined by $\varphi^*(f) = f \circ \varphi$ for any $f \in C_p(Y)$, is an embedding and $T = \varphi^*(C_p(Y))$ is closed in $C_p(X)$ (see TFS-163). The space $C_p(X)$ is primarily Lindelöf by TFS-150; so T is also primarily Lindelöf by Problem 144. Since $C_p(Y)$ is homeomorphic to T, it is also primarily Lindelöf which shows that Y is Corson compact by Problem 150.

U.152. *Observe that $\Sigma_*(A)$ and $\sigma(A)$ are invariant subsets of $\Sigma(A)$; prove that, for any infinite cardinal κ and any closed $F \subset \Sigma(A)$ we have*

(i) *if $B_\alpha \subset A, r_{B_\alpha}(F) \subset F$ for any $\alpha < \kappa$ and $\alpha < \beta < \kappa$ implies $B_\alpha \subset B_\beta$ then $r_B(F) \subset F$ where $B = \bigcup_{\alpha < \kappa} B_\alpha$;*
(ii) *for any non-empty set $D \subset A$ with $|D| \le \kappa$ there exists a set $E \subset A$ such that $|E| \le \kappa$, $D \subset E$ and $r_E(F) \subset F$.*

In particular, F is invariant in $\Sigma(A)$.

Solution. Given a point $x \in \Sigma(A)$ let $\operatorname{supp}(x) = \{a \in A : x(a) \ne 0\}$. It is evident that if $x \in \Sigma_*(A)$ (or $x \in \sigma(A)$) then $r_B(x) \in \Sigma_*(A)$ ($r_B(x) \in \sigma(A)$ respectively) for any $B \subset A$. This proves that $\Sigma_*(A)$ and $\sigma(A)$ are invariant subsets of $\Sigma(A)$. Now assume that F is a closed subset of $\Sigma(A)$ and let $\mathcal{F} = \{B \subset A : r_B(F) \subset F\}$.

Given points $a_1, \ldots, a_n \in A$ and sets $O_1, \ldots, O_n \in \tau(\mathbb{R})$ consider the set

$$[a_1, \ldots, a_n; O_1, \ldots, O_n] = \{x \in \Sigma(A) : x(a_i) \in O_i \text{ for all } i = 1, \ldots, n\};$$

it is evident that the sets $[a_1, \ldots, a_n; O_1, \ldots, O_n]$ form a base in $\Sigma(A)$. By our definition of $\mathcal{F}$ we have $B_\alpha \in \mathcal{F}$ for any $\alpha < \kappa$. If $B \notin \mathcal{F}$ then take $x \in F$ such that $y = r_B(x) \notin F$; since F is closed, there are $a_1, \ldots, a_n \in A$ and $O_1, \ldots, O_n \in \tau(\mathbb{R})$ such that $y \in [a_1, \ldots, a_n; O_1, \ldots, O_n] \subset \Sigma(A) \backslash F$.

The set $C = \{a_1, \ldots, a_n\}$ being finite, there is $\alpha < \kappa$ such that $B_\alpha \cap C = B \cap C$. The point $z = r_{B_\alpha}(x)$ belongs to F; we claim that $z(a_i) = y(a_i)$ for all $i \le n$. Indeed, if $a_i \in B_\alpha$ then $y(a_i) = x(a_i) = z(a_i)$; if $a_i \notin B_\alpha$ then $a_i \notin B$ and therefore $y(a_i) = 0 = z(a_i)$. Aa a consequence, $z \in [a_1, \ldots, a_n; O_1, \ldots, O_n] \cap F$ which is a contradiction with the choice of the set $[a_1, \ldots, a_n; O_1, \ldots, O_n]$. This proves (i).

To prove (ii) let $E_0 = D$; we have $w(r_{E_0}(F)) \leq \kappa$; so we can find a set $P_0 \subset F$ such that $|P_0| \leq \kappa$ and $r_{E_0}(P_0)$ is dense in $r_{E_0}(F)$. Assume that, for some $n \in \omega$, we have sets $E_0 \subset \ldots \subset E_n \subset A$ and $P_0 \subset \ldots \subset P_n \subset F$ with the following properties:

(1) $D \subset E_0$ and $r_{E_i}(P_i)$ is dense in $r_{E_i}(F)$ for any $i \leq n$;
(2) $\bigcup \{\text{supp}(x) : x \in P_i\} \subset E_{i+1}$ for any $i < n$.

Let $E_{n+1} = E_n \cup \{\text{supp}(x) : x \in P_n\}$; then the set E_{n+1} has cardinality at most κ, so there exists a set $P_{n+1} \subset F$ such that $|P_{n+1}| \leq \kappa$ and $P_n \subset P_{n+1}$ while $r_{E_{n+1}}(P_{n+1})$ is dense in $r_{E_{n+1}}(F)$. It is immediate that the properties (1) and (2) still hold for the sets $\{E_i : i \leq n+1\}$ and $\{P_i : i \leq n+1\}$; so our inductive procedure can be continued to construct sequences $\{E_n : n \in \omega\}$ and $\{P_n : n \in \omega\}$ for which the properties (1) and (2) are fulfilled for all $n \in \omega$. The set $E = \bigcup_{n \in \omega} E_n$ has cardinality at most κ and $D \subset E$. To see that $r_E(F) \subset F$ suppose not. Then there is $x \in F$ such that $y = r_E(x) \notin F$. Since F is closed, there exist $a_1, \ldots, a_n \in A$ and $O_1, \ldots, O_n \in \tau(\mathbb{R})$ such that $y \in U = [a_1, \ldots, a_n; O_1, \ldots, O_n] \subset \Sigma(A) \backslash F$.

The set $C = \{a_1, \ldots, a_n\}$ being finite, there is $k \in \omega$ such that $E_k \cap C = B \cap C$. The point $z = r_{E_k}(x)$ is in the closure of the set $r_{E_k}(P_k)$ by (1) and we have $z(a_i) = x(a_i)$ if $a_i \in E_k$. Thus there is $t \in P_k$ such that $t(a_i) \in O_i$ whenever $a_i \in E_k$. However, $a_i \notin E_k$ implies $a_i \notin E$ so $y(a_i) = 0$; besides, $a_i \notin E_{k+1}$ while $\text{supp}(t) \subset E_{k+1}$ by (2) so $t(a_i) = 0$ and hence $0 = t(a_i) = y(a_i) \in O_i$. We proved that $t(a_i) \in O_i$ for all $i \leq n$ and therefore $t \in U \cap F$. This contradiction shows that $r_E(F) \subset F$ and hence (ii) is proved. The properties (i) and (ii), evidently, imply that the set F is invariant in $\Sigma(A)$.

U.153. *Prove that the following properties are equivalent for any X:*

(i) *X is a Sokolov space;*
(ii) *if, for any $n \in \mathbb{N}$, a set $B_n \subset X^n$ is chosen then there exists a continuous map $f : X \to X$ such that $nw(f(X)) \leq \omega$ and $f^n(B_n) \subset \overline{B}_n$ for each $n \in \mathbb{N}$;*
(iii) *if F_{nm} is a closed subset of X^n for all $n, m \in \mathbb{N}$, then there exists a continuous map $f : X \to X$ such that $nw(f(X)) \leq \omega$ and $f^n(F_{nm}) \subset F_{nm}$ for all $n, m \in \mathbb{N}$.*

Solution. To deal with finite powers of the space X we will use the following convention: if we have points $x = (x_1, \ldots, x_n) \in X^n$ and $y = (y_1, \ldots, y_k) \in X^k$ then $\langle x, y \rangle = (x_1, \ldots, x_n, y_1, \ldots, y_k) \in X^{n+k}$; if $x \in X$ and $k \in \mathbb{N}$ then the point $x^k = (x, \ldots, x) \in X^k$ is the k-tuple with all its coordinates equal to x.

It is evident that (iii)$\Longrightarrow$(i). If X is a Sokolov space and $B_n \subset X^n$ then let $F_n = \overline{B}_n$ for any $n \in \mathbb{N}$. There exists a continuous map $f : X \to X$ for which $nw(f(X)) \leq \omega$ and $f^n(F_n) \subset F_n$ for any $n \in \mathbb{N}$; an immediate consequence is that $f^n(B_n) \subset f^n(F_n) \subset F_n = \overline{B}_n$ for any $n \in \mathbb{N}$ and hence (ii) holds. Therefore (i)$\Longrightarrow$(ii); it is also clear that (ii)$\Longrightarrow$(i).

Finally, assume that X is a Sokolov space and fix a closed subset F_{nm} of the space X^n for any $n, m \in \mathbb{N}$. It is easy to construct an injection $\varphi : \mathbb{N} \times \mathbb{N} \to \mathbb{N}$ such that $\varphi(n, m) > n + m$ for any $n, m \in \mathbb{N}$. Choose a point $a \in X$ and let $P_k = \emptyset$ if $k \notin \varphi(\mathbb{N} \times \mathbb{N})$; if $k \in \varphi(\mathbb{N} \times \mathbb{N})$ then there is a unique pair $(n, m) \in \mathbb{N} \times \mathbb{N}$ with

$\varphi(n,m) = k$; let $P_k = \{\langle x, a^{k-n}\rangle : x \in F_{nm}\}$. This gives a closed set $P_k \subset X^k$ for any $k \in \mathbb{N}$; so there exists a continuous map $f : X \to X$ for which $nw(f(X)) \leq \omega$ and $f^k(P_k) \subset P_k$ for every $k \in \mathbb{N}$.

Take an arbitrary pair $(n,m) \in \mathbb{N}$ and let $k = \varphi(n,m)$; since $f^k(P_k) \subset P_k$, for any point $x = (x_1, \ldots, x_n) \in F_{nm}$ the point $f^k(\langle x, a^{k-n}\rangle) = \langle f^n(x), f^{k-n}(a)\rangle$ belongs to P_k and therefore we have the equality $\langle f^n(x), f^{k-n}(a)\rangle = \langle y, a^{k-n}\rangle$ for some point $y = (y_1, \ldots, y_n) \in F_{nm}$. This implies $y_i = f(x_i)$ for every number $i \leq n$ and hence $y = f^n(x) \in F_{nm}$. Thus $f^n(F_{nm}) \subset F_{nm}$ for any $n, m \in \mathbb{N}$; so (iii) is fulfilled for X and hence all properties (i)–(iii) are equivalent.

U.154. *Prove that if X is a Sokolov space then $X \times \omega$ is a Sokolov space and every closed $F \subset X$ is also a Sokolov space.*

Solution. Suppose that we have a closed set F_n in the space $(X \times \omega)^n$ for each $n \in \mathbb{N}$. For every $n \in \mathbb{N}$ and $(k_1, \ldots, k_n) \in \omega^n$ consider the set

$$Y(k_1, \ldots, k_n) = \{((x_1, k_1), \ldots, (x_n, k_n)) : x_i \in X \text{ for all } i \leq n\};$$

it is straightforward that the family $\mathcal{Y}_n = \{Y(k_1, \ldots, k_n) : (k_1, \ldots, k_n) \in \omega^n\}$ is disjoint and $(X \times \omega)^n = \bigcup \mathcal{Y}_n$. For every $n \in \mathbb{N}$ and $(k_1, \ldots, k_n) \in \omega^n$ the map $((x_1, k_1), \ldots, (x_n, k_n)) \to (x_1, \ldots, x_n)$ is a homeomorphism between $Y(k_1, \ldots, k_n)$ and X^n; so there exists a closed set $G(k_1, \ldots, k_n) \subset X^n$ such that

$$F_n \cap Y(k_1, \ldots, k_n) = \{((x_1, k_1), \ldots, (x_n, k_n)) : (x_1, \ldots, x_n) \in G(k_1, \ldots, k_n)\}.$$

Since X is a Sokolov space, we can apply Problem 153 to conclude that there exists a continuous map $g : X \to X$ such that $nw(g(X)) \leq \omega$ and we have the inclusion $g^n(G(k_1, \ldots, k_n)) \subset G(k_1, \ldots, k_n)$ for any $n \in \mathbb{N}$ and $(k_1, \ldots, k_n) \in \omega^n$. For every $t = (x, k) \in X \times \omega$ let $f(t) = (g(x), k)$; then $f : X \times \omega \to X \times \omega$ and it is evident that $f(X \times \omega) \subset g(X) \times \omega$ which implies that $nw(f(X \times \omega)) \leq \omega$.

Now fix any $n \in \mathbb{N}$ and $y \in F_n$; there exists an n-tuple $(k_1, \ldots, k_n) \in \omega^n$ and a point $x = (x_1, \ldots, x_n) \in G(k_1, \ldots, k_n)$ for which $y = ((x_1, k_1), \ldots, (x_n, k_n))$. Consequently, the point $f^n(y) = ((g(x_1), k_1), \ldots, (g(x_n), k_n))$ belongs to F_n because $(g(x_1), \ldots, g(x_n)) = g^n(x) \in G(k_1, \ldots, k_n)$. Thus $f^n(F_n) \subset F_n$ for any $n \in \mathbb{N}$ which proves that $X \times \omega$ is a Sokolov space.

Finally, suppose that F is a closed subset of X. Assume that F_n is a closed subset of F^n for any $n \in \mathbb{N}$; then F_n is also a closed subset of X^n for any $n \in \mathbb{N}$. It follows from Problem 153 that there exists a continuous map $g : X \to X$ such that $nw(g(X)) \leq \omega$ while $g(F) \subset F$ and $g^n(F_n) \subset F_n$ for all $n \in \mathbb{N}$. If $f = g|F$ then $f : F \to F$ is a continuous map such that $nw(f(F)) \leq nw(g(X)) \leq \omega$ and $g^n(F_n) = f^n(F_n) \subset F_n$ for any $n \in \mathbb{N}$. Therefore F is a Sokolov space.

U.155. *Given a Sokolov space X and a second countable space E, prove that $C_p(X, E)$ is also a Sokolov space.*

Solution. Let $\mathcal{E}$ be a countable base in the space E. For any $x_1, \ldots, x_n \in X$ and $B_1, \ldots, B_n \in \mathcal{E}$ the set

$$[x_1, \ldots, x_n; B_1, \ldots, B_n] = \{f \in C_p(X, E) : f(x_i) \in B_i \text{ for all } i \leq n\}$$

is open in $C_p(X, E)$ and the family $\mathcal{B} = \{[x_1, \ldots, x_n; B_1, \ldots, B_n] : n \in \mathbb{N}, x_i \in X$ and $B_i \in \mathcal{E}$ for all $i \leq n\}$ is a base in $C_p(X, E)$.

For any $m \in \mathbb{N}$ assume that F_m is a closed subset of $(C_p(X, E))^m$; given m-tuples $l = (l_1, \ldots, l_m) \in \mathbb{N}^m$ and $D = (D_1, \ldots, D_m)$ such that $D_i = (B_1^i, \ldots, B_{l_i}^i) \in \mathcal{E}^{l_i}$ for every $i \in \{1, \ldots, m\}$ (such m-tuple D will be called $(l, \mathcal{E})$-admissible), take the numbers $l_0 = 0$, $n = l_1 + \ldots + l_m$ and consider the set

$$P(l, D, n) = \{x = (x_1, \ldots, x_n) \in X^n : (U_1^x \times \ldots \times U_m^x) \cap F_m = \emptyset\},$$

where, for every $i = 1, \ldots, m$, the set U_i^x is defined by calculating the number $p_i = l_0 + \ldots + l_{i-1}$ and letting $U_i^x = [x_{p_i+1}, \ldots, x_{p_i+l_i}; B_1^i, \ldots, B_{l_i}^i]$.

We will prove first that $P(l, D, n)$ is closed in X^n for any $l = (l_1, \ldots, l_m) \in \mathbb{N}^m$ and any $(l, \mathcal{E})$-admissible m-tuple $D = (D_1, \ldots, D_m)$ where $D_i = (B_1^i, \ldots, B_{l_i}^i) \in \mathcal{E}^{l_i}$ for every number $i \leq m$. Indeed, if $x = (x_1, \ldots, x_n) \in X^n \backslash P(l, D, n)$ and then there is $f \in F_m \cap (U_1^x \times \ldots \times U_m^x)$. We have $n = l_1 + \ldots + l_m$ so if $l_0 = 0$ and $p_i = l_0 + \ldots + l_{i-1}$ for all $i = 1, \ldots, m$ then, for every $j \in \{1, \ldots, n\}$ there are unique $i(j) \in \{1, \ldots, m\}$ and $k(j) \in \{1, \ldots, l_{i(j)}\}$ such that $j = p_{i(j)} + k(j)$ and therefore $f(x_j) \in B_{k(j)}^{i(j)}$. If $O_j = f^{-1}(B_{k(j)}^{i(j)})$ for every $j \in \{1, \ldots, n\}$ then $x \in O = O_1 \times \ldots \times O_n$ and $O \cap P(l, D, n) = \emptyset$ because $y = (y_1, \ldots, y_n) \in O$ implies $f \in F_m \cap (U_1^y \times \ldots \times U_n^y)$.

For each $n \in \mathbb{N}$ the family $\{P(l, D, n) :$ there exists a number $m \in \mathbb{N}$ for which $l = (l_1, \ldots, l_m) \in \mathbb{N}^m$, $n = l_1 + \ldots + l_m$ and D is $(l, \mathcal{E})$-admissible$\}$ is easily seen to be countable; so we can apply Problem 153 to find a continuous map $\varphi : X \to X$ such that $nw(\varphi(X)) \leq \omega$ and $\varphi^n(P(l, D, n)) \subset P(l, D, n)$ for any $l = (l_1, \ldots, l_m) \in \mathbb{N}^m$ with $l_1 + \ldots + l_m = n$ and any $(l, \mathcal{E})$-admissible m-tuple D; let $Y = \varphi(X)$.

Let $\varphi^*(f) = f \circ \varphi$ for any function $f \in C_p(X, E)$. To see that the map $\varphi^* : C_p(X, E) \to C_p(X, E)$ is continuous take any function $f \in C_p(X, E)$ and a set $U = [x_1, \ldots, x_n; B_1, \ldots, B_n] \in \mathcal{B}$ such that $\varphi^*(f) = f \circ \varphi \in U$. Then the set $V = [\varphi(x_1), \ldots, \varphi(x_n); B_1, \ldots, B_n]$ is an open neighbourhood of f in $C_p(X, E)$ such that $\varphi^*(V) \subset U$ which shows that φ^* is continuous at the point f. Furthermore, we can consider that $E \subset \mathbb{R}^\omega$ and hence $C_p(Y, E) \subset C_p(Y, \mathbb{R}^\omega) \simeq (C_p(Y))^\omega$ which shows that $nw(\varphi^*(C_p(X, E))) \leq nw(C_p(Y, E)) \leq nw((C_p(Y))^\omega) = \omega$ so, to finish our proof it suffices to establish that $(\varphi^*)^m(F_m) \subset F_m$ for any $m \in \mathbb{N}$.

Assume, towards a contradiction, that $m \in \mathbb{N}$, $f = (f_1, \ldots, f_m) \in F_m$ and $g = (\varphi^*(f_1), \ldots, \varphi^*(f_m))$ does not belong to F_m. Since the set F_m is closed in the space $(C_p(X, E))^m$, for every $i \in \{1, \ldots, m\}$, there exist $y_1^i, \ldots, y_{l_i}^i \in X$ and $B_1^i, \ldots, B_{l_i}^i \in \mathcal{E}$ such that $\varphi^*(f_i) \in D_i = [y_1^i, \ldots, y_{l_i}^i; B_1^i, \ldots, B_{l_i}^i]$ for every number $i \leq m$ while we have the equality $(D_1 \times \ldots \times D_m) \cap F_m = \emptyset$.

Let $l_0 = 0$, $l = (l_1, \ldots, l_m)$ and $p_i = l_0 + \ldots + l_{i-1}$ for every $i = 1, \ldots, m$. For $n = l_1 + \ldots + l_m$ let us construct a point $(x_1, \ldots, x_n) \in X^n$ by putting together all y_j^i in a row, i.e., for each $j \in \{1, \ldots, n\}$ we find the unique numbers $i(j) \in \{1, \ldots, m\}$ and $k(j) \in \{1, \ldots, l_{i(j)}\}$ such that $j = p_{i(j)} + k(j)$ and let $x_j = y_{k(j)}^{i(j)}$. If $x = (x_1, \ldots, x_n)$ and $D = (D_1, \ldots, D_m)$ then $D_i = U_i^x$ for any $i = 1, \ldots, m$ and $x \in P(l, D, n)$.

By our choice of the function φ, the point $z = \varphi^n(x) = (\varphi(x_1), \ldots, \varphi(x_n))$ still belongs to $P(l, D, n)$. It follows from $\varphi^*(f_i) \in D_i$ that $f_i(\varphi(y_k^i)) \in B_k^i$ for all $k \leq l_i$ and $i \leq m$. An immediate consequence is that $f_i \in [\varphi(y_1^i), \ldots, \varphi(y_{l_i}^i); B_1^i, \ldots, B_{l_i}^i]$ for any $i \leq m$; so $f \in (U_1^z \times \ldots \times U_m^z) \cap F_m$ which is a contradiction with the fact that $z \in P(l, D, n)$. Thus $(\varphi^*)^m(F_m) \subset F_m$ for every $m \in \mathbb{N}$; so $C_p(X, E)$ is a Sokolov space and hence our solution is complete.

U.156. *Prove that X is a Sokolov space if and only if $C_p(X)$ is a Sokolov space.*

Solution. If X is a Sokolov space then $C_p(X)$ is also a Sokolov space by Problem 155. If, on the other hand, the space $C_p(X)$ is Sokolov then $C_p(C_p(X))$ is also Sokolov by Problem 155 and hence X is Sokolov as well being homeomorphic to a closed subset of $C_p(C_p(X))$ (see TFS-167 and Problem 154).

U.157. *Let X be a Sokolov space with $t^*(X) \leq \omega$. Prove that $C_p(X, E)$ is Lindelöf for any second countable space E.*

Solution. Fix a countable base $\mathcal{E}$ in the space E. If $x = (x_1, \ldots, x_n) \in X^n$ and $B = (B_1, \ldots, B_n) \in \mathcal{E}^n$ then let $[x, B] = \{f \in C_p(X, E) : f(x_i) \in B_i$ for each $i \leq n\}$. It is evident that the family $\mathcal{B} = \{[x, B] :$ there is $n \in \mathbb{N}$ for which $x \in X^n$, $B \in \mathcal{E}^n\}$ is a base in $C_p(X, E)$.

Observe that, for any $n \in \mathbb{N}$ and $B = (B_1, \ldots, B_n) \in \mathcal{E}^n$,

(1) if $A \subset X^n$ and $x = (x_1, \ldots, x_n) \in \overline{A}$ then $[x, B] \subset \bigcup\{[a, B] : a \in A\}$.

Indeed, if $f \in [x, B]$ then let $O_i = f^{-1}(B_i)$ for all $i \leq n$. Since $f(x_i) \in B_i$ for every $i \leq n$, we have $x \in O = O_1 \times \ldots \times O_n$. It follows from $x \in \overline{A}$ that there is $a = (a_1, \ldots, a_n) \in A \cap O$. Then $f(a_i) \in B_i$ for all $i \leq n$ and hence $f \in [a, B]$ which proves (1).

To prove that $C_p(X, E)$ is Lindelöf take an open cover $\mathcal{U}$ of the space $C_p(X, E)$; we can assume, without loss of generality, that $\mathcal{U} \subset \mathcal{B}$. For any $n \in \mathbb{N}$ and $B \in \mathcal{E}^n$ let $F(n, B) = \{x \in X^n : [x, B] \in \mathcal{U}\}$. We can also assume that $F(n, B)$ is closed in X^n for any $n \in \mathbb{N}$ and $B \in \mathcal{E}^n$ because $x \in \overline{F(n, B)}$ and $t^*(X) = \omega$ imply that there is a countable $P \subset F(n, B)$ with $x \in \overline{P}$ and hence $[x, B] \subset \bigcup\{[y, B] : y \in P\}$ by (1). Now, if we are able to extract a countable subcover from the cover $\mathcal{U}' = \{[x, B] :$ there is $n \in \mathbb{N}$ and $B \in \mathcal{E}^n$ such that $x \in \overline{F(n, B)}\}$, then every $U \in \mathcal{U}' \setminus \mathcal{U}$ can be covered with countably many elements of $\mathcal{U}$; so $\mathcal{U}$ also has a countable subcover.

The family $\{F(n, B) : B \in \mathcal{E}^n\}$ is countable; so we can apply Problem 153 to find a continuous map $\varphi : X \to X$ for which $nw(\varphi(X)) \leq \omega$ and $\varphi^n(F(n, B)) \subset F(n, B)$ for any $n \in \mathbb{N}$ and $B \in \mathcal{E}^n$. The set $\varphi^n(F(n, B))$ has a countable network;

so we can take a countable $C(n, B) \subset \varphi^n(F(n, B))$ which is dense in $\varphi^n(F(n, B))$ for every $n \in \mathbb{N}$ and $B \in \mathcal{E}^n$. The family

$$\mathcal{V} = \{[x, B] : x \in C(n, B) \text{ for some } n \in \mathbb{N} \text{ and } B \in \mathcal{E}^n\} \subset \mathcal{U}$$

is countable.

Take an arbitrary $f \in C_p(X, E)$; then $f \circ \varphi \in [x, B] \in \mathcal{U}$ for some point $x = (x_1, \ldots, x_n) \in X^n$ and $B = (B_1, \ldots, B_n) \in \mathcal{E}^n$. Now, if $O_i = f^{-1}(B_i)$ for any $i \leq n$ then $\varphi^n(x) \in O_1 \times \ldots \times O_n$. Since $x \in F(n, B)$, we have $\varphi^n(x) \in \overline{C(n, B)}$; so there is $y = (y_1, \ldots, y_n) \in C(n, B) \cap O$. As a consequence, $f(y_i) \in B_i$ for all $i \leq n$, i.e., $f \in [y, B] \in \mathcal{V}$. The point $f \in C_p(X, E)$ was taken arbitrarily; so we proved that the cover $\mathcal{U}$ has a countable subcover $\mathcal{V}$. Therefore $C_p(X, E)$ is Lindelöf.

U.158. *Prove that*

(i) any $\mathbb{R}$-quotient image of a Sokolov space is a Sokolov space;
(ii) if X is a Sokolov space then X^ω is also a Sokolov space;
(iii) a space with a unique non-isolated point is Sokolov if and only if it is Lindelöf.

Solution. Suppose that X is a Sokolov space and $\varphi : X \to Y$ is an $\mathbb{R}$-quotient map. For any $f \in C_p(Y)$ let $\varphi^*(f) = f \circ \varphi$. Then $\varphi^* : C_p(Y) \to C_p(X)$ is an embedding and the set $T = \varphi^*(C_p(Y))$ is closed in $C_p(X)$ (see TFS-163). Since X is Sokolov, the space $C_p(X)$ is also Sokolov by Problem 156. Therefore T is Sokolov by Problem 154. The space $C_p(Y)$ is also Sokolov being homeomorphic to T; so we can apply Problem 156 again to conclude that Y is a Sokolov space. This proves (i).

Next, assume that X is Sokolov and observe that there is $T \subset C_p(X)$ such that $C_p(X) \simeq T \times \mathbb{R}$ (see Fact 1 of S.409); so we can apply TFS-177 to see that $C_p(C_p(X)) \simeq (C_p(C_p(X)))^\omega$; since X embeds in $C_p(C_p(X))$ as a closed subspace, this implies that X^ω also embeds in $C_p(C_p(X))$ as a closed subspace. Applying Problem 156 twice, we can conclude that $C_p(C_p(X))$ is a Sokolov space; so X^ω is Sokolov as well by Problem 154. This settles (ii).

To prove (iii) let X be a space such that $a \in X$ is the unique non-isolated point of X. Assume first that X is a Sokolov space. If X is not Lindelöf then $D = X \backslash U$ is uncountable for some $U \in \tau(a, X)$ and hence D is an uncountable closed discrete subspace of X; besides, D is Sokolov by Problem 154. Since $t^*(D) = \omega$, the space $C_p(D) = \mathbb{R}^D$ is Lindelöf by Problem 157. The space $\mathbb{R}^{\omega_1}$ embeds in $\mathbb{R}^D$ as a closed subspace; so $\mathbb{R}^{\omega_1}$ is also Lindelöf which contradicts Fact 2 of S.215. Therefore, Sokolov property of X implies that X is Lindelöf.

To establish the converse, suppose that the space X is Lindelöf and fix a closed set $F_n \subset X^n$ for any $n \in \mathbb{N}$. Observe that the set $X \backslash U$ is countable for any $U \in \tau(a, X)$ for otherwise the space X would contain an uncountable closed discrete subspace. Given $n \in \mathbb{N}$, $x = (x_1, \ldots, x_n) \in X^n$ and $U \in \tau(a, X)$ consider the set $O(x, U) = O_1 \times \ldots \times O_n$ where, for every $i \in \{1, \ldots, n\}$, we let $O_i = \{x_i\}$ if $x_i \neq a$ and $O_i = U$ if $x_i = a$. It is immediate that the family $\{O(x, U) : U \in \tau(a, X)\}$ is a local base at the point x in the space X^n. Therefore we can fix, for any $n \in \mathbb{N}$ and any $x \in X^n \backslash F_n$, a set $U_x \in \tau(a, X)$ such that $O(x, U_x) \cap F_n = \emptyset$.

For any $A \subset X$ let $r_A(x) = x$ if $x \in A$ and $r_A(x) = a$ if $x \in X \backslash A$. We omit a simple verification of the fact that $r_A : X \to A \cup \{a\}$ is a continuous map for any $A \subset X$. Let $A_0 = \{a\}$ and assume that we have, for some $k \in \omega$, a collection $A_0, \ldots, A_k$ of countable subsets of X with the following properties:

(1) $\{a\} = A_0 \subset \ldots \subset A_k$;
(2) if $i < k$ and $x \in A_i^n \backslash F_n$ for some $n \in \mathbb{N}$ then $X \backslash U_x \subset A_{i+1}$.

It follows from $|A_k| \leq \omega$ that the set $P = \bigcup\{A_k^n \backslash F_n : n \in \mathbb{N}\}$ is countable; so the set $A_{k+1} = A_k \cup (\bigcup\{X \backslash U_x : x \in P\})$ is countable as well. It is evident that (1) and (2) still hold for the sets $A_0, \ldots, A_{k+1}$; so our inductive procedure can be continued to construct a sequence $\{A_i : i \in \omega\}$ of countable subsets of X such that (1) and (2) are satisfied for all $k \in \omega$.

Let $A = \bigcup_{k \in \omega} A_k$; since $r_A(X)$ is countable, we have $nw(r_A(X)) \leq \omega$. Our purpose is to prove that $(r_A)^n(F_n) \subset F_n$ for any $n \in \mathbb{N}$. To this end assume, towards a contradiction, that there exists $n \in \mathbb{N}$ such that $y = (y_1, \ldots, y_n) = (r_A)^n(x) \notin F_n$ for some $x = (x_1, \ldots, x_n) \in F_n$. We have $O(y, U_y) \cap F_n = \emptyset$ by our choice of U_y; we claim, however, that $x \in O(y, U_y)$.

To see it observe first that $y_i \in A$ for all $i \leq n$ and therefore there is $m \in \omega$ such that $\{y_1, \ldots, y_n\} \subset A_m$. By definition, $O(y, U_y) = O_1 \times \ldots \times O_n$ where, for every $i \leq n$, we have $O_i = \{y_i\}$ if $y_i \neq a$ and $O_i = U_y$ if $y_i = a$. Consider first the case when $y_i \neq a$. Then $x_i \in A$ and hence $x_i = y_i \in O_i$. Now, if $y_i = a$ and $x_i \in A$ then again $x_i = y_i \in O_i$. If $x_i \notin A$ then $x_i \notin A_{m+1}$; besides, $y \in A_m^n \backslash F_n$; so $X \backslash U_y \subset A_{m+1}$ by (2). This implies that $x_i \in U_y = O_i$.

We proved that $x_i \in O_i$ for all $i \leq n$ and therefore $x \in O_1 \times \ldots \times O_n = O(y, U_y)$ whence $x \in O(y, U_y) \cap F_n$ which is a contradiction with the choice of U_y. Thus $(r_A)^n(F_n) \subset F_n$ for every $n \in \mathbb{N}$; so the Lindelöf property of X implies that X is a Sokolov space. This completes the proof of (iii) and finishes our solution.

U.159. *Let X be a space with a unique non-isolated point. Prove that the following properties are equivalent:*

(i) $l(X) \leq \omega$ and $t^*(X) \leq \omega$;
(ii) X is a Sokolov space and $t^*(X) \leq \omega$;
(iii) $C_{p,n}(X)$ is Lindelöf for all $n \in \mathbb{N}$;
(iv) $C_p(X)$ is Lindelöf.

Solution. If X is countable then all properties (i)–(iv) hold for X so there is nothing to prove; we assume, therefore, that X is uncountable. It follows from Problem 158 that (i)$\Longrightarrow$(ii).

If (ii) holds then X is Lindelöf (this was also proved in Problem 158). If b is the unique non-isolated point of X then $X \backslash U$ is countable for any $U \in \tau(b, X)$ for otherwise X has an uncountable closed discrete subspace. Let $\kappa = |X|$ and fix a surjective map $\varphi : L(\kappa) \to X$ such that $\varphi(a) = b$ and $\varphi(\kappa) = X \backslash \{b\}$. It is immediate that φ is continuous; so X^n is Lindelöf for any $n \in \mathbb{N}$ being a continuous image of $(L(\kappa))^n$ (see TFS-354). The space $C_p(X)$ is Sokolov by Problem 156; it follows from Problem 157 that $l^*(C_p(X)) = \omega$. Furthermore, $t^*(C_p(X)) = \omega$ because $l^*(X) = \omega$ (see TFS-149 and TFS-150).

Suppose that $n \in \mathbb{N}$ and we proved that the space $C_{p,n}(X)$ is Sokolov and $l^*(C_{p,n}(X)) = t^*(C_{p,n}(X)) = \omega$. Then $C_{p,n+1}(X)$ is Sokolov by Problem 156, it follows from $l^*(C_{p,n}(X)) = \omega$ that $t^*(C_{p,n+1}(X)) = \omega$. Applying Problem 157 again we conclude that $l^*(C_{p,n+1}(X)) = \omega$. Continuing this inductive procedure we convince ourselves that $l^*(C_{p,n}(X)) = t^*(C_{p,n}(X)) = \omega$ for any $n \in \mathbb{N}$ and hence (ii)$\Longrightarrow$(iii).

The implication (iii)$\Longrightarrow$(iv) being evident assume that $C_p(X)$ is Lindelöf. Then $t^*(X) = \omega$ by TFS-189. The space X is normal by Claim 2 of S.018; so if D is an uncountable closed discrete subset of X then the restriction π_D maps $C_p(X)$ onto $C_p(D) = \mathbb{R}^D$ whence $\mathbb{R}^D$ is Lindelöf, which is a contradiction (see Fact 2 of S.215). Therefore $X \backslash U$ is countable for any $U \in \tau(b, X)$; an easy consequence is that X is Lindelöf and hence we proved that (iv)$\Longrightarrow$(i).

U.160. *Let X be an invariant subspace of $\Sigma(A)$. Prove that X is a Sokolov space. Deduce from this fact that every Corson compact space is Sokolov.*

Solution. For any $x \in \Sigma = \Sigma(A)$ let supp$(x) = \{a \in A : x(a) \neq 0\}$. Call a set $U \subset \Sigma$ *standard* if there are $a_1, \ldots, a_n \in A$ and $O_1, \ldots, O_n \in \tau(\mathbb{R})$ such that $U = [a_1, \ldots, a_n; O_1, \ldots, O_n] = \{x \in \Sigma : x(a_i) \in O_i$ for all $i = 1, \ldots, n\}$. It is evident that standard sets form a base in Σ. If $U = [a_1, \ldots, a_n; O_1, \ldots, O_n]$ is a standard set then $E(U) = \{a_1, \ldots, a_n\}$.

Given a set $B \subset A$ define a map $r_B : \Sigma \rightarrow \Sigma$ as follows: for any $x \in \Sigma$ and $a \in A$ we let $r_B(x)(a) = 0$ if $a \notin B$ and $r_A(x)(a) = x(a)$ for any $a \in B$. The map $r_B : \Sigma \rightarrow \Sigma$ is continuous; to see it let $\pi_a : \mathbb{R}^A \rightarrow \mathbb{R}$ be the natural projection onto the factor determined by a. If $a \in B$ then $\pi_a \circ r_B(x) = r_B(x)(a) = x(a) = \pi_a(x)$ for any $x \in \Sigma$; thus $\pi_a \circ r_B$ is continuous being equal to $\pi_a | \Sigma$. If $a \notin B$ then $\pi_a \circ r_B(x) = 0$ for any $x \in \Sigma$; so the map $\pi_a \circ r_B$ is constant and hence continuous. Therefore $\pi_a \circ r_B$ is continuous for any $a \in A$ whence r_B is continuous by TFS-102. If $B \subset A$ is countable then $r_B(\Sigma) \simeq \mathbb{R}^B$; so $w(r_B(\Sigma)) \leq \omega$.

The set X being invariant, the family $\mathcal{U}_X = \{B \subset A : B$ is countable and $r_B(X) \subset X\}$ is ω-continuous and ω-cofinal in A. Take a closed set $F_n \subset X^n$ for any $n \in \mathbb{N}$. Choose a non-empty set $B_0 \in \mathcal{U}_X$; since $(r_{B_0})^n(F_n)$ is second countable, we can find a countable $H_n(0) \subset F_n$ such that $(r_{B_0})^n(H_n(0))$ is dense in $(r_{B_0})^n(F_n)$ for all $n \in \mathbb{N}$. Suppose that $k \in \omega$ and we have sets $B_0, \ldots, B_k \in \mathcal{U}_X$ and sequences $\{\{H_n(i) : n \in \mathbb{N}\} : i \leq k\}$ with the following properties:

(1) $B_0 \subset \ldots \subset B_k$ and $H_n(0) \subset \ldots \subset H_n(k)$ for any $n \in \mathbb{N}$;
(2) for any $i \leq k$ the set $H_n(i) \subset F_n$ is countable and $(r_{B_i})^n(H_n(i))$ is dense in $(r_{B_i})^n(F_n)$ for all $n \in \mathbb{N}$;
(3) given an arbitrary $j < k$, for each $n \in \mathbb{N}$, if $x = (x_1, \ldots, x_n) \in H_n(j)$ then $S[x] = \bigcup\{$supp$(x_i) : i \in \{1, \ldots, n\}\} \subset B_{j+1}$.

Let $C_{k+1} = \bigcup\{S[x] :$ there is $n \in \mathbb{N}$ such that $x \in H_n(k)\}$; it is evident that $C_{k+1} \subset A$ is countable, so there is $B_{k+1} \in \mathcal{U}_X$ such that $C_{k+1} \cup B_k \subset B_{k+1}$. The set $(r_{B_{k+1}})^n(F_n)$ is second countable; so there is a countable $H_n(k+1) \subset F_n$ for which $H_n(k) \subset H_n(k+1)$ and $(r_{B_{k+1}})^n(H_n(k+1))$ is dense in $(r_{B_{k+1}})^n(F_n)$ for any $n \in \mathbb{N}$.

Now, the properties (1)–(3) are satisfied for the collection $\{B_0, \ldots, B_k, B_{k+1}\}$ and the sequences $\{\{H_n(i) : n \in \mathbb{N}\} : i \leq k+1\}$; so our inductive procedure can be continued to construct a sequence $\{B_i : i \in \omega\} \subset \mathcal{U}_X$ as well as a collection $\{\{H_n(i) : n \in \mathbb{N}\} : i \in \omega\}$ such that (1)–(3) hold for all $k \in \omega$. Let $B = \bigcup_{i \in \omega} B_i$; then $B \in \mathcal{U}_X$ and hence $r_B(X) \subset X$. Therefore $\varphi = r_B|X$ is a continuous map from X in X and we have $w(\varphi(X)) \leq w(r_B(\Sigma)) \leq \omega$. Our purpose is to prove that $\varphi^n(F_n) = (r_B)^n(F_n) \subset F_n$ for any $n \in \mathbb{N}$.

So, fix $n \in \mathbb{N}$; the set $H_n = \bigcup_{i \in \omega} H_n(i) \subset F_n$ is countable; the property (3) implies that $S[x] \subset B$ and therefore $(r_B)^n(x) = x$ for any $x \in H_n$. Now take an arbitrary $y = (y_1, \ldots, y_n) \in F_n$; we claim that $z = (z_1, \ldots, z_n) = (r_B)^n(y) \in \overline{H}_n$ (the bar denotes the closure in X^n). To prove it take any $V \in \tau(z, X^n)$; there exist standard sets $U_1, \ldots, U_n$ such that $z \in U = (U_1 \times \ldots \times U_n) \cap X^n \subset V$. The set $D = \bigcup\{E(U_i) : i \leq n\}$ is finite; so there is $m \in \omega$ such that $B_m \cap D = B \cap D$.

We have $U_i = [a_1^i, \ldots, a_{k_i}^i ; O_1^i, \ldots, O_{k_i}^i]$ and there is no loss of generality to assume that there exists a number $p_i \leq k_i$ for which $\{a_1^i, \ldots, a_{p_i}^i\} \subset B_m$ and $\{a_{p_i+1}^i, \ldots, a_{k_i}^i\} \subset D \backslash B_m \subset A \backslash B$ for all $i \leq n$. For any $i \leq n$ we have $r_{B_m}(y_i)(a) = y_i(a) = z_i(a)$ whenever $a \in B_m$; so $r_{B_m}(y_i)(a_j^i) = z_i(a_j^i) \in O_j^i$ for any $j \leq p_i$. Furthermore, $z_i(a) = 0$ for any $a \in A \backslash B$; so $r_{B_m}(y_i)(a_j^i) = 0 = z_i(a_j^i) \in O_j^i$ for any $i \leq n$ and $j > p_i$.

As a consequence, U_i is an open neighbourhood of $r_{B_m}(y_i)$ for any $i \leq n$ and hence U is an open neighbourhood of $(r_{B_m})^n(y)$ in X^n. The set $(r_{B_m})^n(H_n)$ is dense in $(r_{B_m})^n(F_n)$ by (2), so there is $h = (h_1, \ldots, h_n) \in H_n$ such that $r_{B_m}(h_i) \in U_i$ for all $i \leq n$. Since $S[h] \subset B$, we have $r_B(h) = h$. Furthermore $r_B(h_i)(a) = 0$ for any $a \in A \backslash B$ and therefore $r_B(h_i)(a_j^i) = 0$ for any $j > p_i$ (recall that $a_j^i \in A \backslash B$ for any $j > p_i$) and $i \leq n$. As a consequence, $r_B(h_i)(a_j^i) = 0 = z_i(a_j^i) \in O_j^i$ for any $j > p_i$.

We also have $r_B(h_i)(a_j^i) = h_i(a_j^i) = r_{B_m}(h_i)(a_j^i) \in O_j^i$ for any $j \leq p_i$ which implies, together with the observations of the previous paragraph, that if $i \leq n$ then $r_B(h_i)(a_j^i) \in O_j^i$ for every $j \in \{1, \ldots, k_i\}$ and hence $h = (r_B)^n(h) \in U \cap F_n \subset V \cap F_n$. Therefore $(r_B)^n(y) \in \overline{H}_n \subset F_n$ (the last inclusion is true because F_n is closed in X^n). This proves that $(r_B)^n(y) \in F_n$ for any $y \in F_n$, i.e., $(r_B)^n(F_n) \subset F_n$ for any $n \in \mathbb{N}$. Thus we have a map $\varphi : X \to X$ such that $nw(\varphi(X)) \leq w(\varphi(X)) \leq \omega$ and $\varphi^n(F_n) \subset F_n$ for any $n \in \mathbb{N}$ which shows that X is a Sokolov space.

Finally observe that if X is a Corson compact space then we can consider that X is a closed subspace of Σ. By Problem 152, X is invariant in Σ; so it is Sokolov space.

U.161. *Prove that every Sokolov space is collectionwise normal and has countable extent. Deduce from this fact that, $ext(C_{p,n}(X)) \leq \omega$ for any Sokolov space X and $n \in \mathbb{N}$.*

Solution. Suppose that Y is a Sokolov space. If D is a closed discrete subspace of Y then D is Sokolov by Problem 154; since $t^*(D) = \omega$, we can apply Problem 157 to see that $C_p(D) = \mathbb{R}^D$ is Lindelöf. If D is uncountable then $\mathbb{R}^{\omega_1}$ embeds in $\mathbb{R}^D$ as a closed subset; so $\mathbb{R}^{\omega_1}$ has to be Lindelöf which is a contradiction with Fact 2 of S.215. This proves that $ext(Y) = \omega$ for any Sokolov space Y.

Now, if X is a Sokolov space then $C_{p,n}(X)$ is Sokolov by Problem 156 and therefore $ext(C_{p,n}(X)) = \omega$ for any $n \in \mathbb{N}$.

Furthermore, if Y is a Sokolov space and $F, G \subset Y$ are disjoint closed subsets of Y then we can let $F_n = \emptyset$ for any $n \geq 2$ and apply Problem 153 to find a continuous map $\varphi : Y \to Y$ such that $nw(\varphi(Y)) \leq \omega$ while $F' = \varphi(F) \subset F$ and $G' = \varphi(G) \subset G$; let $Z = \varphi(Y)$. We have $P = cl_Z(F') \subset \overline{F'} \subset F$ and $Q = cl_Z(G') \subset \overline{G'} \subset G$ which shows that $P \cap Q \subset F \cap G = \emptyset$. Thus P and Q are disjoint closed subsets of Z; the space Z is normal because $nw(Z) \leq \omega$; so there is a continuous function $f : Z \to [0,1]$ such that $f(P) \subset \{0\}$ and $f(Q) \subset \{1\}$. It is clear that $g = f \circ \varphi : Y \to [0,1]$ is continuous while $g(F) \subset \{0\}$ and $g(G) \subset \{1\}$. Therefore Y is normal.

Finally observe that any normal space of countable extent is collectionwise normal by Fact 3 of S.294; so every Sokolov space is collectionwise normal and has countable extent.

U.162. *Let X be a Sokolov space. Prove that*

(i) if $t^*(X) \leq \omega$ then $C_{p,2n+1}(X)$ is Lindelöf for any $n \in \omega$.
(ii) if $l^*(X) \leq \omega$ then $C_{p,2n}(X)$ is Lindelöf for any $n \in \mathbb{N}$;
(iii) if $l^*(X) \cdot t^*(X) \leq \omega$ then $C_{p,n}(X)$ is Lindelöf for any $n \in \mathbb{N}$.

Solution. Suppose that X is a Sokolov space. It takes a trivial induction using Problem 156 to see that $C_{p,n}(X)$ is Sokolov for any $n \in \mathbb{N}$. If $t^*(X) = \omega$ then the space $(C_p(X))^\omega \simeq C_p(X, \mathbb{R}^\omega)$ is Lindelöf by Problem 157. Assume that $k \in \omega$ and we proved that $(C_{p,2k+1}(X))^\omega$ is Lindelöf. Then the space $C_{p,2k+2}(X)$ is Sokolov and $t^*(C_{p,2k+2}(X)) = \omega$; so we can apply Problem 157 to conclude that $((C_{p,2k+3}(X))^\omega$ is Lindelöf. This inductive procedure shows that $((C_{p,2n+1}(X))^\omega$ is Lindelöf for any $n \in \omega$ and hence we proved (i).

Now, if $l^*(X) = \omega$ then the space $Y = C_p(X)$ is Sokolov and $t^*(Y) = \omega$. Therefore (i) can be applied to see that $C_{p,2n+2}(X) = C_{p,2n+1}(Y)$ is Lindelöf for any $n \in \omega$; this settles (ii). Finally, (iii) is a trivial consequence of (i) and (ii); so our solution is complete.

U.163. *Prove that every Sokolov space is ω-stable and ω-monolithic. Deduce from this fact that every Sokolov compact space is Fréchet–Urysohn and has a dense set of points of countable character.*

Solution. Let X be a Sokolov space. If $A \subset X$ is countable then the family $\{\{a\} : a \in A\}$ is countable and consists of closed subsets of the space X; thus we can apply Problem 153 to conclude that there is a continuous map $\varphi : X \to X$ such that $nw(\varphi(X)) \leq \omega$ and $\varphi(\{a\}) \subset \{a\}$, i.e., $\varphi(a) = a$ for any $a \in A$. An immediate consequence is that $\varphi(x) = x$ for any $x \in \overline{A}$ and hence $\overline{A} \subset \varphi(X)$ which implies that $nw(\overline{A}) \leq nw(\varphi(X)) \leq \omega$. This proves that every Sokolov space is ω-monolithic.

As a consequence, if X is a Sokolov space then $C_p(X)$ is also Sokolov by SFFS-156; so $C_p(X)$ is ω-monolithic and hence X is ω-stable by SFFS-154.

Finally assume that X is a compact Sokolov space. Then $ext(C_p(X)) = \omega$ by Problem 161 and hence $C_p(X)$ is Lindelöf by Baturov's theorem (SFFS-269). This implies $t(X) \leq \omega$ by TFS-189. Therefore X is Fréchet–Urysohn and a dense set of points of countable character by Fact 1 of U.080.

U.164. *Prove that a metrizable space is Sokolov if and only if it is second countable.*

Solution. If X is a metrizable Sokolov space then $ext(X) = \omega$ by Problem 161; so X is second countable by TFS-214. If, on the other hand, X is second countable (or has a countable network) and a closed set $F_n \subset X^n$ is taken for any $n \in \mathbb{N}$ then, for the identity map $\varphi : X \to X$, we have $nw(\varphi(X)) = nw(X) \leq \omega$ and $\varphi^n(F_n) = F_n$ for any $n \in \mathbb{N}$; thus X is Sokolov (evidently, metrizability is not needed to prove this implication).

U.165. *Let X be a Sokolov space with $l^*(X) \cdot t^*(X) = \omega$. Prove that*

(i) if X has a small diagonal then $nw(X) = \omega$;
(ii) if ω_1 is a caliber of X then $nw(X) = \omega$.

Solution. Suppose that X has a small diagonal. By Problem 163 it suffices to show that X is separable; so assume, towards a contradiction, that $d(X) > \omega$. Then there is a left-separated subspace $Y = \{x_\alpha : \alpha < \omega_1\} \subset X$ (see SFFS-004). Let $Y_\alpha = \{x_\beta : \beta < \alpha\}$ and $F_\alpha = \overline{Y}_\alpha$ for any $\alpha < \omega_1$. Applying Problem 163 once more we convince ourselves that $nw(F_\alpha) \leq \omega$ for any $\alpha < \omega_1$. It follows from $t(X) \leq \omega$ that $F = \bigcup\{F_\alpha : \alpha < \omega_1\}$ is closed in X and $nw(F) = \omega_1$.

The small diagonal is a hereditary property; so the space F has a small diagonal. It follows from $l^*(F) = \omega$ that ω_1 is a caliber of $C_p(F)$ (see SFFS-294). Furthermore, $d(C_p(F)) \leq nw(C_p(F)) = nw(F) = \omega_1$; so fix a set $D = \{f_\alpha : \alpha < \omega_1\} \subset C_p(F)$ which is dense in $C_p(F)$. If $D_\alpha = \{f_\beta : \beta < \alpha\}$ and $H_\alpha = \overline{D}_\alpha$ then H_α is closed in $C_p(F)$ for any $\alpha < \omega_1$. Every Sokolov space is ω-monolithic (see Problem 163) so $nw(H_\alpha) = \omega$; since $nw(C_p(F)) = \omega_1$, the set $U_\alpha = C_p(F) \backslash H_\alpha$ is non-empty for any $\alpha < \omega_1$.

It follows from $l^*(F) \leq l^*(X) \leq \omega$ that $t(C_p(F)) = \omega$ and therefore the set $H = \bigcup\{H_\alpha : \alpha < \omega_1\}$ is closed in $C_p(F)$; since the dense set D is contained in H, we have $H = C_p(F)$. Consequently, the family $\mathcal{U} = \{U_\alpha : \alpha < \omega_1\}$ consists of non-empty open subsets of $C_p(F)$; since $\mathcal{U}$ is decreasing and $\bigcap \mathcal{U} = \emptyset$, it is point-countable which is impossible because ω_1 is a caliber of $C_p(F)$. This contradiction shows that $d(X) = \omega$ and hence $nw(X) = \omega$; so (i) is proved.

As to (ii), if ω_1 is a caliber of X then $C_p(X)$ has a small diagonal (see SFFS-290). Since $C_p(X)$ is a Sokolov space by Problem 156, we can apply (i) to conclude that $nw(C_p(X)) = \omega$ and hence $nw(X) = nw(C_p(X)) = \omega$; so our solution is complete.

U.166. *Prove that if X is a Sokolov space with a G_δ-diagonal then $nw(X) = \omega$.*

Solution. Let $\Delta = \{(x, x) : x \in X\} \subset X \times X$ be the diagonal of the space X. Since Δ is a G_δ-subset of $X \times X$, there is a sequence $\{F_n : n \in \omega\}$ of closed subsets of $X \times X$ such that $F_n \subset F_{n+1}$ for every $n \in \omega$ and $(X \times X) \backslash \Delta = \bigcup_{n \in \omega} F_n$. It follows from Problem 153 that there is a continuous map $\varphi : X \to X$ such that $nw(\varphi(X)) \leq \omega$ and $\varphi^2(F_n) \subset F_n$ for any $n \in \omega$; let $Y = \varphi(X)$.

Given any distinct $x, y \in X$ the point $z = (x, y)$ does not belong to Δ; so there is $n \in \omega$ such that $z \in F_n$. Therefore $\varphi^2(z) = (\varphi(x), \varphi(y)) \in F_n$ which implies that $(\varphi(x), \varphi(y)) \notin \Delta$ and hence $\varphi(x) \neq \varphi(y)$. Thus the map φ is injective.

Let $\varphi^*(f) = f \circ \varphi$ for any $f \in C_p(Y)$; then the map $\varphi^* : C_p(Y) \to C_p(X)$ is an embedding and the set $T = \varphi^*(C_p(Y))$ is dense in the space $C_p(X)$ (see TFS-163). Consequently, $d(C_p(X)) \leq d(T) = d(C_p(Y)) \leq nw(C_p(Y)) = nw(Y) \leq \omega$. It turns out that $C_p(X)$ is a separable Sokolov space; so $nw(C_p(X)) = \omega$ by Problem 163. Therefore $nw(X) = nw(C_p(X)) = \omega$.

U.167. *Let X be a Lindelöf Σ-space. Prove that if X is Sokolov then $t(X) \leq \omega$ and $C_{p,n}(X)$ is Lindelöf for any $n \in \mathbb{N}$. In particular, if K is Sokolov compact (or Corson compact) then $C_{p,n}(K)$ is Lindelöf for any $n \in \mathbb{N}$.*

Solution. Recall that $C_p(X)$ has countable extent by Problem 161; by Baturov's theorem (SFFS-269) the space $C_p(X)$ is Lindelöf and therefore $t^*(X) = \omega$ by TFS-189. Since we also have $l^*(X) = \omega$ (see SFFS-256), we can apply Problem 162 to conclude that $C_{p,n}(X)$ is Lindelöf for any $n \in \mathbb{N}$.

U.168. *Let T be an infinite set. Prove that, if $\mathcal{A}$ is an adequate family on T then $K_{\mathcal{A}}$ is a compact space.*

Solution. It suffices to show that the set $K_{\mathcal{A}}$ is closed in $\mathbb{D}^T$. If $x \in \mathbb{D}^T \backslash K_{\mathcal{A}}$ then $B = x^{-1}(1) \notin \mathcal{A}$ and hence there is a finite $C \subset B$ such that $C \notin \mathcal{A}$. The set $O_x = \{y \in \mathbb{D}^T : y(t) = 1 \text{ for any } t \in C\}$ is open in $\mathbb{D}^T$ and $x \in O_x$. Furthermore, $O_x \cap K_{\mathcal{A}} = \emptyset$ because $y \in O_x \cap K_{\mathcal{A}}$ implies $D = y^{-1}(1) \in \mathcal{A}$ and hence $C \subset D$ also belongs to $\mathcal{A}$ which is a contradiction. This contradiction shows that $O_x \cap K_{\mathcal{A}} = \emptyset$ and hence every $x \in \mathbb{D}^T \backslash K_{\mathcal{A}}$ has an open neighbourhood $O_x \subset \mathbb{D}^T \backslash K_{\mathcal{A}}$. Thus $\mathbb{D}^T \backslash K_{\mathcal{A}}$ is open in $\mathbb{D}^T$; so $K_{\mathcal{A}}$ is compact being closed in $\mathbb{D}^T$.

U.169. *Let T be an infinite set. Suppose that $\mathcal{A}$ is an adequate family on T. Prove that $K_{\mathcal{A}}$ is a Corson compact space if and only if all elements of $\mathcal{A}$ are countable.*

Solution. The space $K_{\mathcal{A}}$ is compact by Problem 168; if $K_{\mathcal{A}}$ is Corson compact and some $A \in \mathcal{A}$ is uncountable then the set $X = \{\chi_B : B \subset A\}$ is contained in $K_{\mathcal{A}}$ and $X \simeq \mathbb{D}^A$. However, $t(X) > \omega$ because X is a non-metrizable dyadic compact space (see TFS-359); on the other hand, $t(X) \leq t(K_{\mathcal{A}}) = \omega$ (see Problem 120) which is a contradiction. This proves necessity.

Now, if $|A| \leq \omega$ for any $A \in \mathcal{A}$ then $K_{\mathcal{A}} \subset \Sigma = \{x \in \mathbb{D}^T : x^{-1}(1) \text{ is countable}\}$. The space Σ is a Σ-product of second countable spaces so $K_{\mathcal{A}} \subset \Sigma$ is Corson compact by Problem 119 and hence we have established sufficiency.

U.170. *Let T be an infinite set; suppose that A is an adequate family on T and u is the function on K_A which is identically zero. For any $t \in T$ let $e_t(f) = f(t)$ for any $f \in K_A$. Observe that $Z = \{e_t : t \in T\} \cup \{u\} \subset C_p(K_A, \mathbb{D})$; let $\varphi(\xi) = u$ and $\varphi(t) = e_t$ for any $t \in T$. Prove that $\varphi : T_A^* \to Z$ is a homeomorphism and Z is closed in $C_p(K_A, \mathbb{D})$. In particular, the space T_A^* is homeomorphic to a closed subspace of $C_p(K_A, \mathbb{D})$.*

Solution. To see that $Z \subset C_p(K_A, \mathbb{D})$ it suffices to prove that e_t is continuous on K_A for each $t \in T$. But this is an immediate consequence of the fact that the sets $e_t^{-1}(0) = \{f \in K_A : f(t) = 0\}$ and $e_t^{-1}(1) = \{f \in K_A : f(t) = 1\}$ are open in K_A by definition of the product topology on $\mathbb{D}^T$.

Every e_t is an isolated point of Z; indeed, it follows from $\bigcup A = T$ that $f = \chi_{\{t\}} \in K_A$; so $W = \{p \in C_p(K_A, \mathbb{D}) : p(f) = 1\}$ is an open neighbourhood of e_t in $C_p(K_A, \mathbb{D})$ and $W \cap Z = \{e_t\}$. Thus it suffices to show that φ and φ^{-1} are continuous at the points ξ and u respectively.

Take any set $U \in \tau(\xi, T_A^*)$; by definition of the topology of T_A^*, there are sets $A_1, \dots, A_n \in A$ such that $\{\xi\} \cup (T \setminus (\bigcup_{i \le n} A_i)) \subset U$; then $f_i = \chi_{A_i} \in K_A$ for any $i \le n$. If $W = \{p \in C_p(K_A, \mathbb{D}) : p(f_i) = 0$ for all $i \le n\}$ then W is an open neighbourhood of u in $C_p(K_A, \mathbb{D})$ and $\varphi^{-1}(W \cap Z) \subset \{\xi\} \cup (T \setminus (\bigcup_{i \le n} A_i))$ which shows that φ^{-1} is continuous at the point u.

To see that φ is continuous at ξ take any set $W \in \tau(u, Z)$; there exist functions $f_1, \dots, f_n \in K_A$ such that $V = \{p \in Z : p(f_i) = 0$ for all $i \le n\} \subset W$. If $A_i = f_i^{-1}(1)$ for all $i \le n$ then $U = \{\xi\} \cup (T \setminus (\bigcup_{i \le n} A_i))$ is an open neighbourhood of ξ in T_A^* and $\varphi(U) \subset V \subset W$ which shows that φ is continuous at ξ and hence $\varphi : T_A^* \to Z$ is a homeomorphism.

To finally prove that Z is closed in $C_p(K_A, \mathbb{D})$ take any point $p \in C_p(K_A, \mathbb{D}) \setminus Z$; the set $H = p^{-1}(1)$ is non-empty and open in K_A. Fix $A \in A$ such that $f = \chi_A \in H$ and let $P = \{g \in K_A : g^{-1}(1) \subset A$ and $g^{-1}(1)$ is finite$\}$. If $W \in \tau(f, K_A)$ then there is a finite $C \subset T$ such that $W_1 = \{g \in K_A : g|C = f|C\} \subset W$. If $B = C \cap A$ then $g = \chi_B \in P$ and $g \in W_1 \cap P \subset W \cap P$. This shows that $f \in \overline{P}$ and therefore there exists a finite $D \subset A$ such that $g = \chi_D \in H$.

The set $O = \{q \in C_p(K_A, \mathbb{D}) : q(g) = 1\}$ is open in $C_p(K_A, \mathbb{D})$; besides, $p \in O$ and $O \cap Z$ is finite because $e_t \in O$ if and only if $t \in D$. Thus every point $p \in C_p(K_A, \mathbb{D}) \setminus Z$ has a neighbourhood O such that $O \cap Z$ is finite. An evident consequence is that Z is closed in $C_p(K_A, \mathbb{D})$ and hence our solution is complete.

U.171. *Suppose that T is an infinite set and A is an adequate family on T. Prove that the spaces $C_p(K_A, \mathbb{D})$ and $C_p(K_A)$ are both continuous images of the space $(T_A^* \times \omega)^\omega$.*

Solution. If Y and Z are spaces then the expression $Y \gg Z$ says that Y maps continuously onto Z.

Denote by u the function on K_A which is identically zero. For any $t \in T$ let $e_t(f) = f(t)$ for any $f \in K_A$. Then $Z = \{e_t : t \in T\} \cup \{u\} \subset C_p(K_A, \mathbb{D})$ is closed in $C_p(K_A, \mathbb{D})$ and $T_A^* \simeq Z$ (see Problem 170). Therefore we can identify

Z and $T^*_{\mathcal{A}}$. It is easy to see that $T^*_{\mathcal{A}}$ separates the points of $K_{\mathcal{A}}$; so we can apply Fact 2 of U.150 to conclude that $C_p(K_{\mathcal{A}}, \mathbb{D}) = \bigcup\{C_n : n \in \omega\}$ where every C_n is a continuous image of a finite power of $T^*_{\mathcal{A}}$. In particular, we have $(T^*_{\mathcal{A}})^\omega \gg C_n$ for any $n \in \omega$ and therefore $(T^*_{\mathcal{A}})^\omega \times \omega \gg C_p(K_{\mathcal{A}}, \mathbb{D})$. Since $(T^*_{\mathcal{A}} \times \omega)^\omega \simeq (T^*_{\mathcal{A}})^\omega \times \omega^\omega \gg (T^*_{\mathcal{A}})^\omega \times \omega$ we conclude that the space $(T^*_{\mathcal{A}} \times \omega)^\omega$ maps continuously onto $C_p(K_{\mathcal{A}}, \mathbb{D})$.

Thus $(T^*_{\mathcal{A}} \times \omega)^\omega \simeq ((T^*_{\mathcal{A}} \times \omega)^\omega)^\omega \gg (C_p(K_{\mathcal{A}}, \mathbb{D}))^\omega \simeq C_p(K_{\mathcal{A}}, \mathbb{D}^\omega)$. Now, the space $K_{\mathcal{A}}$ is compact and zero-dimensional; so $C_p(K_{\mathcal{A}}, \mathbb{D}^\omega) \gg C_p(K_{\mathcal{A}}, \mathbb{I})$ (see Problem 004) and hence $(T^*_{\mathcal{A}} \times \omega)^\omega \gg C_p(K_{\mathcal{A}}, \mathbb{I})$. Another consequence of compactness of $K_{\mathcal{A}}$ is that $C_p(K_{\mathcal{A}}) = \bigcup\{C_p(K_{\mathcal{A}}, [-n,n]) : n \in \mathbb{N}\}$; every space $C_p(K_{\mathcal{A}}, [-n,n])$ is homeomorphic to $C_p(K_{\mathcal{A}}, \mathbb{I})$, so we have $C_p(K_{\mathcal{A}}, \mathbb{I}) \times \omega \gg C_p(K_{\mathcal{A}})$. Therefore $(T^*_{\mathcal{A}} \times \omega)^\omega \times \omega \gg C_p(K_{\mathcal{A}}, \mathbb{D}^\omega) \times \omega \gg C_p(K_{\mathcal{A}}, \mathbb{I}) \times \omega \gg C_p(K_{\mathcal{A}})$.

Finally, $(T^*_{\mathcal{A}} \times \omega)^\omega \times \omega \simeq (T^*_{\mathcal{A}})^\omega \times \omega^\omega \times \omega \simeq (T^*_{\mathcal{A}})^\omega \times \omega^\omega \simeq (T^*_{\mathcal{A}} \times \omega)^\omega$; so the space $(T^*_{\mathcal{A}} \times \omega)^\omega$ maps continuously onto $C_p(K_{\mathcal{A}})$ as well.

U.172. *Let T be an infinite set. Suppose that $\mathcal{A}$ is an adequate family on T. Prove the space $C_p(K_{\mathcal{A}})$ is K-analytic if and only if $T^*_{\mathcal{A}}$ is K-analytic.*

Solution. If $C_p(K_{\mathcal{A}})$ is K-analytic then $T^*_{\mathcal{A}}$ is K-analytic because $T^*_{\mathcal{A}}$ embeds in $C_p(K_{\mathcal{A}})$ as a closed subspace (see Problem 170). If, on the other hand, the set $T^*_{\mathcal{A}}$ is K-analytic then $Z = (T^*_{\mathcal{A}} \times \omega)^\omega$ is also K-analytic; since $C_p(K_{\mathcal{A}})$ is a continuous image of Z (see Problem 171), it is K-analytic as well (see SFFS-343).

U.173. *Let T be an infinite set. Suppose that $\mathcal{A}$ is an adequate family on T. Prove the space $C_p(K_{\mathcal{A}})$ is Lindelöf Σ if and only if $T^*_{\mathcal{A}}$ is Lindelöf Σ.*

Solution. If $C_p(K_{\mathcal{A}})$ is a Lindelöf Σ-space then $T^*_{\mathcal{A}}$ is Lindelöf Σ because $T^*_{\mathcal{A}}$ embeds in $C_p(K_{\mathcal{A}})$ as a closed subspace (see Problem 170). If, on the other hand, $T^*_{\mathcal{A}}$ is a Lindelöf Σ-space then $Z = (T^*_{\mathcal{A}} \times \omega)^\omega$ is also Lindelöf Σ; since $C_p(K_{\mathcal{A}})$ is a continuous image of Z (see Problem 171), it is a Lindelöf Σ-space as well (see SFFS-254, SFFS-256 and SFFS-243).

U.174. *Observe that every adequate compact space is zero-dimensional. Give an example of a zero-dimensional Corson compact space which is not homeomorphic to any adequate compact space.*

Solution. If T is an infinite set and $\mathcal{A}$ is an adequate family on T then the adequate compact space $K_{\mathcal{A}}$ is contained in $\mathbb{D}^T$; so it is zero-dimensional (see SFFS-303).

Fact 1. Suppose that X is a countably compact σ-discrete space, i.e., $X = \bigcup_{n \in \omega} X_n$ where each X_n is a discrete subspace of X. Then X is scattered.

Proof. If X is not scattered then there is $Y \subset X$ such that Y has no isolated points. It is clear that $K = \overline{Y}$ is also dense-in-itself. If $K_n = X_n \cap K$ then K_n is a discrete subspace of K for any $n \in \omega$. It is an easy exercise that, in a dense-in-itself space, the closure of any discrete subspace is nowhere dense; so $F_n = \overline{K_n}$ is a closed nowhere dense subspace of K for any $n \in \omega$. The space K is countably compact and $K = \bigcup_{n \in \omega} F_n$; this contradiction with the Baire property of K shows that Fact 1 is proved.

Fact 2. Given an infinite set A and $n \in \omega$ let $\sigma_n(A) = \{x \in \mathbb{D}^A : |x^{-1}(1)| \le n\}$. Then $\sigma_n(A)$ is a scattered compact space.

Proof. Take an arbitrary $x \in \mathbb{D}^A \backslash \sigma_n(A)$; there are distinct $a_1, \ldots, a_{n+1} \in A$ such that $x(a_i) = 1$ for any $i = 1, \ldots, n + 1$. The set $O = \{y \in \mathbb{D}^A : y(a_i) = 1$ for all $i \le n + 1\}$ is open in $\mathbb{D}^A$ and $x \in O \subset \mathbb{D}^A \backslash \sigma_n(A)$. This proves that the set $\sigma_n(A)$ is compact being closed in $\mathbb{D}^A$.

If $S_i = \{x \in \mathbb{D}^A : |x^{-1}(1)| = i\}$ then $S_i \subset \sigma_n(A)$ for any $i \le n$. It is clear that S_0 consists of the unique point which is identically zero on A. If $1 \le i \le n$ and $x \in S_i$ then there are distinct $b_1, \ldots, b_i \in A$ such that $\{b_1, \ldots, b_i\} = x^{-1}(1)$. The set $W = \{y \in \mathbb{D}^A : y(b_j) = 1$ for all $j \le i\}$ is open in $\mathbb{D}^A$ and $W \cap S_i = \{x\}$; this proves that S_i is discrete for any $i \le n$. Since $\sigma_n(A) = S_0 \cup \ldots \cup S_n$, the compact space $\sigma_n(A)$ is a finite union of discrete subspaces; so it is scattered by Fact 1 and therefore Fact 2 is proved.

Returning to our solution let K be the Alexandroff double $AD(Y)$ of the space $Y = \mathbb{D}^\omega$. Recall that $K = K_0 \cup K_1$ where $K_0 \cap K_1 = \emptyset$, the set K_0 is closed in K and all points of K_1 are isolated in K. Besides, the space K_0 is a homeomorphic copy of Y and a bijection $\varphi : K_0 \to K_1$ is chosen in such a way that the family $\mathcal{B}_x = \{U \cup (\varphi(U) \backslash \{\varphi(x)\}) : U \in \tau(x, K_0)\}$ is a local base at any $x \in K_0$.

The space K is Corson compact by Problem 124; since Y is zero-dimensional, the family $\mathcal{C}$ of all clopen subsets of Y is a base in Y. It is immediate that the family $\mathcal{C}_x = \{U \cup (\varphi(U) \backslash \{\varphi(x)\}) : x \in U \in \mathcal{C}\}$ is also a local base at x in K and every element of $\mathcal{C}_x$ is clopen in K. Thus $\bigcup \{\mathcal{C}_x : x \in K_0\} \cup \{\{x\} : x \in K_1\}$ is a clopen base in K, i.e., K is zero-dimensional. The family $\mathcal{C}$ is countable (see Fact 1 of U.077); so every $\mathcal{C}_x$ is a countable local base at any $x \in K_0$. Therefore K is first countable.

Now assume that there is an infinite set T and an adequate family $\mathcal{A}$ of subsets of T such that $K_\mathcal{A}$ is homeomorphic to K. If $\sigma(T) = \{x \in \mathbb{D}^T : x^{-1}(1)$ is finite$\}$ then $N = \sigma(T) \cap K_\mathcal{A}$ is dense in the space $K_\mathcal{A}$. Indeed, if $x = \chi_A \in K_\mathcal{A}$ for some $A \in \mathcal{A}$ then take any $W \in \tau(x, K_\mathcal{A})$. There exists a finite set $C \subset T$ such that $W_1 = \{y \in K_\mathcal{A} : y|C = x|C\} \subset W$. If $B = C \cap A$ then $y = \chi_B \in W_1 \cap N \subset W \cap N$. Thus $W \cap N \ne \emptyset$ for any $W \in \tau(x, K_\mathcal{A})$ and hence $\overline{N} = K_\mathcal{A}$. Since $K_\mathcal{A} \simeq K$, the set D of isolated points of $K_\mathcal{A}$ is uncountable; the set N being dense in $K_\mathcal{A}$, we have $D \subset N$.

Furthermore, $N = \bigcup_{m \in \omega} N_m$ where $N_m = \{z \in N : |z^{-1}(1)| \le m\}$ for every $m \in \omega$. If $\sigma_m(T) = \{x \in \mathbb{D}^T : |x^{-1}(1)| \le m\}$ then $N_m = N \cap \sigma_m(T)$ for any $m \in \omega$. Thus N_m is uncountable for some $m \in \omega$ and hence $F = \overline{N}_m \subset \sigma_m(T)$ because $\sigma_m(T)$ is compact by Fact 2. Therefore F is an uncountable scattered compact subspace of $K_\mathcal{A}$; the space $K_\mathcal{A} \simeq K$ being first countable, F is also first countable.

Finally, let F' be the set F with the topology generated by the family of all G_δ-subsets of F. It follows from $\chi(F) = \omega$ that F' is an uncountable discrete space. However, F' has to be Lindelöf by SFFS-128. Since no uncountable discrete space is Lindelöf, we obtained a contradiction which shows that K is not adequate and completes our solution.

U.175. *Let T be a subspace of $\mathbb{R}$ of cardinality ω_1. Consider some well-ordering $\prec$ on T and let $<$ be the order on T induced from the usual order on $\mathbb{R}$. Denote by $\mathcal{A}_1$ the family of all subsets of T on which the orders $<$ and $\prec$ coincide (i.e., $A \in \mathcal{A}_1$ if and only if, for any distinct $x, y \in A$, we have $x < y$ if and only if $x \prec y$). Let $\mathcal{A}_2$ be the family of all subsets of T on which the orders $<$ and $\prec$ are opposite (i.e., $A \in \mathcal{A}_2$ if and only if, for any distinct $x, y \in A$, we have $x < y$ if and only if $y \prec x$). Check that $\mathcal{A} = \mathcal{A}_1 \cup \mathcal{A}_2$ is an adequate family and that $X = K_{\mathcal{A}}$ is a Corson compact space for which $C_p(X)$ is not a continuous image of any Lindelöf k-space. In particular, $C_p(X)$ is not a Lindelöf Σ-space.*

Solution. Observe that, by definition, if $A \subset T$ and $|A| \leq 1$ then $A \in \mathcal{A}_1 \cap \mathcal{A}_2$ and hence $\bigcup \mathcal{A} = T$. If $A \in \mathcal{A}_1$ then the orders $<$ and $\prec$ coincide on A and hence on any subset of A; thus $\exp A \subset \mathcal{A}_1$. Analogously, if $A \in \mathcal{A}_2$ then the orders $<$ and $\prec$ are opposite on A and on any subset of A; so $\exp A \subset \mathcal{A}_2$. Now assume that $A \subset T$ is an infinite set such that any finite $B \subset A$ belongs to $\mathcal{A}$. If $A \notin \mathcal{A}$ then the orders $<$ and $\prec$ do not coincide on A and therefore there exist distinct $x, y \in A$ such that $x < y$ but $y \prec x$. The orders $<$ and $\prec$ cannot be opposite on A either; so there are distinct $z, t \in A$ for which $z < t$ and $z \prec t$. The set $B = \{x, y, z, t\} \subset A$ is finite; so $B \in \mathcal{A}$. If $B \in \mathcal{A}_1$ then the orders $<$ and $\prec$ have to coincide on B for which the pair $\{x, y\}$ gives a contradiction. If $B \in \mathcal{A}_2$ then the pair $\{z, t\}$ provides a contradiction again. Thus $A \in \mathcal{A}$ and hence $\mathcal{A}$ is, indeed, an adequate family.

Next, we prove that all elements of $\mathcal{A}$ are countable. Assume towards a contradiction, that $A \in \mathcal{A}_1$ and $|A| > \omega$. It follows from Fact 1 of S.151 that there is an uncountable $B \subset A$ such that B has no isolated points (considered as a subspace of $\mathbb{R}$). Since the orders $<$ and $\prec$ coincide on B, the set B is well ordered by the natural order of $\mathbb{R}$. In particular, there is $a \in B$ which is the $<$-minimal element of B. The set $B \backslash \{a\}$ has no isolated points; since it is well ordered, it has a unique minimal element b. We have $a < b$ and $(a, b) \cap B = \emptyset$ because the existence of a number $c \in (a, b) \cap B$ contradicts minimality of b in $B \backslash \{a\}$. As a consequence, for $c = \frac{a+b}{2}$ we have $\{a\} = (-\infty, c) \cap B$, i.e., a is isolated in B; this contradiction shows that $\mathcal{A}_1$ has no uncountable elements.

Now assume that $A \in \mathcal{A}_2$ and $|A| > \omega$. Apply Fact 1 of S.151 again to see that there is an uncountable $B \subset A$ such that B has no isolated points (considered as a subspace of $\mathbb{R}$). The set B is well ordered by $\prec$; so B has a $\prec$-minimal element a. Since the orders $<$ and $\prec$ are opposite on B, the point $a \in B$ is the $<$-maximal element of B. The set $B \backslash \{a\}$ has no isolated points; since it is well ordered, by $\prec$ it has a unique $<$-maximal element b. We have $b < a$ and $(b, a) \cap B = \emptyset$ because the existence of a number $c \in (b, a) \cap B$ contradicts maximality of b in $B \backslash \{a\}$. As a consequence, for $c = \frac{a+b}{2}$ we have $\{a\} = (c, +\infty) \cap B$, i.e., a is isolated in B; this contradiction shows that $\mathcal{A}_2$ cannot have uncountable elements either.

Therefore all elements of $\mathcal{A}$ are countable; so $K_{\mathcal{A}}$ is a Corson compact space by Problem 169. It turns out that

(1) if $A \subset T$ is an infinite set then there is an infinite $B \subset A$ such that $B \in \mathcal{A}$.

Call a sequence $S = \{r_n : n \in \omega\} \subset T$ monotone if either $r_n < r_{n+1}$ or $r_n > r_{n+1}$ for all $n \in \omega$; in the first case S is called increasing and in the second case S is decreasing. Observe first that it is possible to extract a monotone sequence $\{a_n : n \in \omega\} \subset A$. Indeed, if A has the property

(2) there is an infinite $A' \subset A$ such that for any $a \in A'$ the set $\{b \in A' : a < b\}$ is infinite,

then an increasing sequence $\{a_n : n \in \omega\} \subset A' \subset A$ can be constructed by an evident induction. If, the property (2) does not hold for A then

(3) any infinite $A' \subset A$ has a maximal element,

so we can let $a_0 = \max(A)$ and $a_{n+1} = \max(A \setminus \{a_0, \ldots, a_n\})$ for any $n \in \omega$ which gives us a decreasing sequence $\{a_n : n \in \omega\} \subset A$.

Thus we can assume, without loss of generality, that $A = \{a_n : n \in \omega\}$ is a monotone sequence. There is $n_0 \in \omega$ such that a_{n_0} is the $\prec$-minimal element of A; it is evident that $a_{n_0} \prec a_i$ for any $i > n_0$. Suppose that we have $n_0, \ldots, n_k \in \omega$ such that $n_0 < \ldots < n_k$ and $a_{n_0} \prec \ldots \prec a_{n_k}$ while $a_{n_k} \prec a_i$ for any $i > n_k$. Then there is $n_{k+1} > n_k$ such that $a_{n_{k+1}}$ is the $\prec$-minimal element of the set $\{a_i : i > n_k\}$ and our induction properties are fulfilled for the numbers $n_0, \ldots, n_k, n_{k+1}$.

Therefore there exists an increasing sequence $B = \{n_i : i \in \omega\} \subset \omega$ such that $a_{n_i} \prec a_{n_{i+1}}$ for any $i \in \omega$. Since our sequence A is $<$-monotone, the orders $<$ and $\prec$ either coincide or are opposite on B, i.e., $B \in \mathcal{A}$ and hence (1) is proved.

Observe that any $A \in \mathcal{A}$ is closed and discrete in $T_{\mathcal{A}}^*$; so it follows from (1) that every infinite subset of $T_{\mathcal{A}}^*$ has an infinite closed and discrete subset and, in particular, any compact subset of $T_{\mathcal{A}}^*$ is finite.

Fact 1. If Z is an uncountable space which is a continuous image of a Lindelöf k-space then there is an infinite compact $K \subset Z$.

Proof. Fix a Lindelöf k-space L such that there is a continuous onto map $f : L \to Z$. If $f^{-1}(z)$ is open in L for any $z \in Z$ then $\{f^{-1}(z) : z \in Z\}$ is a disjoint uncountable open cover of L which contradicts the Lindelöf property of L.

Thus there exists a point $z \in Z$ such that $f^{-1}(z)$ is not open in L and therefore there is a compact $P \subset L$ such that $P \cap (L \setminus f^{-1}(z)) = P \setminus f^{-1}(z)$ is not closed in the space L. If the compact set $K = f(P)$ is finite then $Q = K \setminus \{z\}$ is closed in Z and hence $P \setminus f^{-1}(z) = P \cap f^{-1}(Q)$ is closed in L which is a contradiction. Thus K is an infinite compact subset of Z and Fact 1 is proved.

Returning to our solution assume that a Lindelöf k-space maps continuously onto $C_p(K_{\mathcal{A}})$. Now, $T_{\mathcal{A}}^*$ is a closed subspace of $C_p(K_{\mathcal{A}})$, so some Lindelöf k-space maps continuously onto $T_{\mathcal{A}}^*$ as well; since $|T_{\mathcal{A}}^*| > \omega$ and all compact subsets of $T_{\mathcal{A}}^*$ are finite, this gives a contradiction with Fact 1.

Finally, observe that $C_p(K_{\mathcal{A}})$ cannot be a Lindelöf Σ-space because any Lindelöf Σ-space is a continuous image of a Lindelöf p-space (SFFS-253) and every p-space is a k-space (SFFS-230).

U.176. *Give a ZFC example of a Corson compact space without a dense metrizable subspace.*

Solution. Given a tree $(T, \le)$ and $p \in T$ let $\hat{p} = \{q \in T : q \le p\}$; say that a set $P \subset T$ a σ-*antichain* if $P = \bigcup_{n \in \omega} P_n$ and every P_n is an antichain. A set $Q \subset T$ is *dense* in T if, for any $p \in T$ there is $q \in Q$ such that $p \le q$. Let $\mathrm{Lim}(\omega_1)$ be the set of all limit ordinals of ω_1; if $F \subset \omega_1$ is compact then it has a unique maximal element which we will denote by $\max(F)$.

Fact 1. Suppose that $A \subset \omega_1$ is a stationary set such that $\omega_1 \backslash A$ is also stationary and let $T(A) = \{F \subset A : F \text{ is closed in } \omega_1\}$. Then all elements of $T(A)$ are countable and hence compact; given $F, G \in T(A)$ say that $F \le G$ if F is an initial segment of G, i.e., for the ordinal $\alpha = \max(F)$, we have $G \cap (\alpha + 1) = F$. Then $(T(A), \le)$ is a tree which has no uncountable chains and no dense σ-antichains.

Proof. It is straightforward that $\le$ is a partial order on $T = T(A)$. If $p \in T$ then $\hat{p}$ is well ordered because the correspondence $q \to \max(q)$ is an order-isomorphism of $\hat{p}$ onto a subset of ω_1 (with the well order inherited from ω_1). Thus T is, indeed, a tree; all elements of T are countable because an uncountable closed subset of A would miss a stationary set $\omega_1 \backslash A$ which is impossible.

Now, assume that $C \subset T$ is an uncountable chain. Since $\hat{p}$ is countable for any $p \in C$, it is easy to construct by transfinite induction a set $\{p_\alpha : \alpha < \omega_1\} \subset C$ such that $p_\alpha < p_\beta$ whenever $\alpha < \beta < \omega_1$. Then $p = \bigcup\{p_\alpha : \alpha < \omega_1\} \in T$ because the union of an increasing ω_1-sequence of closed sets is a closed set in a space of countable tightness (in our case the relevant space ω_1 is even first countable). Therefore p is an uncountable element of T which is a contradiction. Thus T has no uncountable chains.

Next, assume towards a contradiction, that U_n is an antichain in T for every $n \in \omega$ and $U = \bigcup_{n \in \omega} U_n$ is dense in T. Since every antichain of T is contained in a maximal antichain, we can assume, without loss of generality, that every U_n is a maximal antichain of T. As a consequence,

(1) for any $t \in T$ and $n \in \omega$ there exist $u \in U_n$ and $s \in T$ such that $\{u, t\} \subset \hat{s}$.
    We are going to construct by transfinite induction a family $\{E_\alpha : \alpha < \omega_1\}$ and an ω_1-sequence $\{\delta_\alpha : \alpha < \omega_1\} \subset \omega_1$ with the following properties:
(2) $\delta_0 = 0$ and $E_0 = \emptyset$;
(3) $E_\alpha \subset T$ is countable and $\max(p) \le \delta_\alpha$ for any $p \in E_\alpha$ and $\alpha < \omega_1$;
(4) if $\alpha \in \mathrm{Lim}(\omega_1)$ then $\delta_\alpha = \sup\{\delta_\beta : \beta < \alpha\}$ and $E_\alpha = \bigcup\{E_\beta : \beta < \alpha\}$;
(5) if $\alpha < \beta < \omega_1$ then $\alpha \le \delta_\alpha < \delta_\beta$, $E_\alpha \subset E_\beta$ and, for any $p \in E_\alpha$ and $n \in \omega$ there is $u \in U_n$ and $q \in E_\beta$ such that $\{p, u\} \subset \hat{q}$ and $\max(q) > \delta_\alpha$.

To satisfy the condition (2) we must start with $\delta_0 = 0$, $E_0 = \emptyset$; if $\alpha < \omega_1$ is a limit ordinal and we have the set $\{\delta_\beta : \beta < \alpha\}$ and the family $\{E_\beta : \beta < \alpha\}$ then let $\delta_\alpha = \sup\{\delta_\beta : \beta < \alpha\}$ and $E_\alpha = \bigcup\{E_\beta : \beta < \alpha\}$. This guarantees (4).

Now, suppose that, for some ordinal $\nu < \omega_1$, we have the set $\{\delta_\alpha : \alpha \le \nu\}$ and the family $\{E_\alpha : \alpha \le \nu\}$ with the properties (2)–(5) fulfilled for all $\alpha, \beta \le \nu$.

For every element $t \in E_\nu$ and $n \in \omega$ fix $u(t,n) \in U_n$ and $s(t,n) \in T$ such that $t \leq s(t,n)$, $u(t,n) \leq s(t,n)$ and $\max(s(t,n)) > \max\{\nu, \delta_\nu\}$ (this is possible by (1)).

Let $E'_{\nu+1} = E_\nu \cup \{s(t,n) : t \in E_\nu, n \in \omega\}$, $E_{\nu+1} = \bigcup\{\hat{p} : p \in E'_{\nu+1}\}$ and $\delta_{\nu+1} = \sup\{\max(p) : p \in E_{\nu+1}\}$. It is straightforward that the properties (2)–(5) hold for the set $\{\delta_\alpha : \alpha \leq \nu + 1\}$ and the family $\{E_\alpha : \alpha \leq \nu + 1\}$; so our inductive procedure can be continued to construct the promised ω_1-sequence $\{\delta_\alpha : \alpha < \omega_1\}$ and the family $\{E_\alpha : \alpha < \omega_1\}$ with the properties (2)–(5).

Observe that

(6) the set $H = \{\alpha < \omega_1 : \alpha = \delta_\alpha\}$ is closed and unbounded in ω_1.

Indeed, if $\{\alpha_n\}_{n \in \omega} \subset H$ is an increasing sequence and $\alpha_n \to \alpha$ then it follows from (4) that $\delta_\alpha = \sup\{\delta_{\alpha_n} : n \in \omega\} = \sup\{\alpha_n : n \in \omega\} = \alpha$ because $\delta_{\alpha_n} = \alpha_n$ for any $n \in \omega$. This proves that the set H is closed. Given any $\beta < \omega_1$, let $\alpha_0 = \beta$ and $\alpha_{n+1} = \delta_{\alpha_n} + 1$ for any $n \in \omega$. A consequence of (5) is that $\alpha_n < \alpha_{n+1}$ for any $n \in \omega$; if $\alpha = \sup_{n \in \omega} \alpha_n$ then $\alpha \in H$ and $\alpha > \beta$ which shows that H is cofinal in ω_1 and (6) is proved.

Our set A being stationary, it follows from (6) that there is $\alpha \in H \cap A$; fix an increasing sequence $\{\alpha_n : n \in \omega\}$ such that $\sup_{n \in \omega} \alpha_n = \alpha$. Applying the property (5) once more we conclude that $\sup\{\delta_{\alpha_n} : n \in \omega\} = \alpha$.

Take an element $p_0 \in E_{\alpha_0}$ arbitrarily. Suppose that $n \in \omega$ and we have sets $\{p_i : i \leq n\} \subset T$ and $\{u_i : i < n\} \subset T$ with the following properties:

(7) $p_i \in E_{\alpha_i}$ for all $i \leq n$;
(8) $p_i \leq p_{i+1}$ and $\max(p_{i+1}) > \alpha_i$ for any $i < n$;
(9) $u_i \in U_i$ and $u_i \leq p_{i+1}$ for every $i < n$.

The property (5) implies that we can choose $p_{n+1} \in E_{\alpha_{n+1}}$ and $u_n \in U_n$ such that $u_n \leq p_{n+1}$, $p_n \leq p_{n+1}$ and $\max(p_{n+1}) > \delta_{\alpha_n} \geq \alpha_n$. It is immediate that (7)–(9) are fulfilled for the sets $\{p_i : i \leq n + 1\}$ and $\{u_i : i < n + 1\}$; so our inductive procedure gives us sequences $\{p_n : n \in \omega\}$ and $\{u_n : n \in \omega\}$ with the properties (7)–(9).

It follows from (5) and (7) that $\alpha \notin p_n$ for any $n \in \omega$. It is easy to see that the set $p = (\bigcup_{n \in \omega} p_n) \cup \{\alpha\}$ belongs to T and $p_n < p$ for every $n \in \omega$. Since U is dense in T, there is $n \in \omega$ and $u \in U_n$ for which $p < u$ and hence $u_n \leq p_{n+1} < p < u$ (see (8) and (9)). This is a contradiction with the fact that U_n is an antichain; so Fact 1 is proved.

Fact 2. Given an infinite set T let $[P, Q] = \{x \in \mathbb{D}^T : x(P) \subset \{1\}$ and $x(Q) \subset \{0\}\}$ for any disjoint finite sets $P, Q \subset T$. Suppose additionally that we have a family $\mathcal{U} = \{[P_a, Q_a] : a \in A\}$ such that A is infinite and $\sup\{|P_a \cup Q_a| : a \in A\} < \omega$. Then $\mathcal{U}$ is not disjoint.

Proof. Observe first that the family $\mathcal{U}$ is not disjoint if and only if there are distinct $a, b \in A$ such that $(P_a \cup P_b) \cap (Q_a \cup Q_b) = \emptyset$. Our proof will be by induction on $n = \sup\{|P_a \cup Q_a| : a \in A\}$. If $n = 0$ then $[P_a, Q_a] = [\emptyset, \emptyset] = \mathbb{D}^T$ for any $a \in A$; so $\mathcal{U}$ is not disjoint.

Now suppose that $\sup\{|P_a \cup Q_a| : a \in A\} = k \in \mathbb{N}$ and we have proved that, for any infinite set B, a family $\mathcal{V} = \{[U_b, V_b] : b \in B\}$ is not disjoint whenever $\sup\{|U_b \cup V_b| : b \in B\} < k$.

Let us first consider the case when the family $\mathcal{P} = \{P_a : a \in A\} \cup \{Q_a : a \in A\}$ is point-finite. Fix $a_0 \in A$; then only finitely many elements of $\mathcal{P}$ can intersect either of the sets P_{a_0}, Q_{a_0}; so there is $a_1 \in A$ for which $(P_{a_0} \cup P_{a_1}) \cap (Q_{a_0} \cup Q_{a_1}) = \emptyset$ and hence $[P_{a_0}, Q_{a_0}] \cap [P_{a_1}, Q_{a_1}] \neq \emptyset$.

If the family $\mathcal{P}$ is not point-finite then one of the collections $\mathcal{P}_0 = \{P_a : a \in A\}$ or $\mathcal{P}_1 = \{Q_a : a \in A\}$ is not point-finite. Since both cases are analogous, we assume that $\mathcal{P}_0$ is not point-finite. Passing to a smaller infinite family if necessary, we can consider that $\bigcap \mathcal{P}_0 \neq \emptyset$; fix a point $a_0 \in \bigcap \mathcal{P}_0$. For any $a \in A$ let $P'_a = P_a \backslash \{a_0\}$; then $|P'_a| < |P_a|$. Besides, $a_0 \notin Q_a$ for any $a \in A$ because every Q_a is disjoint from $P_a \supset \bigcap \mathcal{P}_0$. As a consequence, $\sup\{|P'_a \cup Q_a| : a \in A\} < k$; so we can apply the induction hypothesis to conclude that the family $\{[P'_a, Q_a] : a \in A\}$ is not disjoint and therefore we can find distinct elements $a, b \in A$ such that $(P'_a \cup P'_b) \cap (Q_a \cup Q_b) = \emptyset$. But then $(P_a \cup P_b) \cap (Q_a \cup Q_b) = (P'_a \cup P'_b \cup \{a_0\}) \cap (Q_a \cup Q_b) = \emptyset$ and hence the family $\mathcal{U}$ is not disjoint.

Thus our inductive procedure can be continued to show that the family $\mathcal{U}$ is not disjoint for any $n = \sup\{|P_a \cup Q_a| : a \in A\}$; so Fact 2 is proved.

Returning to our solution take a stationary set $A \subset \omega_1$ such that $\omega_1 \backslash A$ is also stationary (such a set exists by SFFS-066) and construct the tree $T = T(A)$ as in Fact 1. The family $\mathcal{A} = \{C \subset T : C \text{ is a chain}\}$ is adequate if we consider that the empty set is also a chain. Indeed, it is clear that any subset of a chain is a chain; since the singletons are also chains, we have $\bigcup \mathcal{A} = T$. Now, if every finite subset of a set $C \subset T$ is a chain then C is a chain for otherwise there are incomparable $p, q \in C$ and the finite set $\{p, q\} \subset C$ gives a contradiction. Furthermore, all chains of T are countable by Fact 1; so the adequate compact space $K = K_A$ is Corson compact.

If K has a dense metrizable subspace then it has a σ-disjoint π-base by Problem 138; so we can choose, for any $i \in \omega$, a family $\mathcal{U}_i = \{[P_b, Q_b] : b \in B_i\} \subset \tau(\mathbb{D}^T)$ (for the definition of the sets $[P_b, Q_b]$ see Fact 2) such that $\mathcal{V}_i = \{U \cap K : U \in \mathcal{U}_i\}$ is disjoint and $\mathcal{V} = \bigcup\{\mathcal{V}_i : i \in \omega\}$ is a π-base in K. Splitting every $\mathcal{V}_i$ into countably many subfamilies if necessary, we can assume without loss of generality, that for any $i \in \omega$ there are $k(i), m(i) \in \omega$ such that $|P_b| = k(i)$ and $|Q_b| = m(i)$ for any $b \in B_i$.

For every $b \in B_i$ there is $x \in [P_b, Q_b] \cap K$, so the set $P_b \subset x^{-1}(1)$ has to be a chain; let z_b be the maximal element of P_b with respect to the tree order on T. The set $Z_i = \{z_b : b \in B_i\}$ has only finite chains for any $i \in \omega$. To see it take any $i \in \omega$ and suppose that $C \subset B_i$ is an infinite set such that $\{z_b : b \in C\}$ is a chain.

Fix any distinct elements $b_1, b_2 \in C$; the fact that z_{b_1} and z_{b_2} are comparable implies that $P = P_{b_1} \cup P_{b_2}$ is a chain and hence $x = \chi_P \in K$. If the sets $[P_{b_1}, Q_{b_1}]$ and $[P_{b_2}, Q_{b_2}]$ are not disjoint then $(P_{b_1} \cup P_{b_2}) \cap (Q_{b_1} \cup Q_{b_2}) = \emptyset$ and hence x belongs to the set $[P_{b_1}, Q_{b_1}] \cap [P_{b_2}, Q_{b_2}] \cap K$ which is a contradiction because the family $\mathcal{V}_i$ is disjoint. Therefore the family $\{[P_b, Q_b] : b \in C\}$ is disjoint in contradiction with Fact 2. Thus the set Z_i has no infinite chains for any $i \in \omega$.

Now let $Z_{ij} = \{z \in Z_i : |\hat{z} \cap Z_i| = j\}$ for any $i, j \in \omega$. Since all chains in Z_i are finite, we have $Z_i = \bigcup\{Z_{ij} : j \in \omega\}$ for every $i \in \omega$. Each Z_{ij} is an antichain for if $x, y \in Z_{ij}$ and $x < y$ then $\hat{y} \cap Z_i$ has more elements than $\hat{x} \cap Z_i$. Therefore every Z_i is a σ-antichain and hence so is $Z = \bigcup_{i \in \omega} Z_i$.

Given any $t \in T$ the set $H = [\{t\}, \emptyset] \cap K$ is non-empty and open; so there are $i \in \omega$ and $b \in B_i$ such that $[P_b, Q_b] \cap K \subset H$; fix a point $x \in [P_b, Q_b] \cap K$. Then $P_b \subset x^{-1}(1)$ is a chain and therefore $y = \chi_{P_b} \in K \cap [P_b, Q_b]$. Since also $y \in H$, we must have $y(t) = 1$ and hence $t \in P_b$ which implies $t \le z_b \in Z_i$. As a consequence, our σ-antichain Z is dense in T; this contradicts Fact 1 and shows that K is a Corson compact space without a dense metrizable subspace, i.e., our solution is complete.

U.177. *Give an example of a compact space X for which $(C_p(X))^\omega$ is Lindelöf while X is not Corson compact.*

Solution. Given ordinals $\alpha, \beta < \omega_1$ let $(\alpha, \beta] = \{\gamma < \omega_1 : \alpha < \gamma \le \beta\}$. Denote by L the set of all limit ordinals of ω_1 and let $I = \omega_1 \backslash L$. For any ordinal $\alpha \in L$ choose an increasing sequence $S'_\alpha = \{\mu_\alpha(n) : n \in \omega\} \subset I$ such that $S'_\alpha \to \alpha$ and consider the set $S_\alpha = S'_\alpha \cup \{\alpha\}$. Say that a space Z has *strong condensation property* if, for any uncountable $A \subset Z$ there is point $z_0 \in Z$ such that some uncountable $B \subset A$ is *concentrated around* z_0, i.e., $|B \backslash U| \le \omega$ for any $U \in \tau(z_0, Z)$. A function $f : Z \to \mathbb{R}$ *separates points* $x, y \in Z$ if $f(x) \ne f(y)$.

Fact 1. In a Lindelöf space Z every uncountable $A \subset Z$ has a condensation point, i.e., there is $z_0 \in Z$ for which $|A \cap U| > \omega$ for any $U \in \tau(z_0, Z)$. In addition, if Z is a space with $l(Z) \le \omega_1$ then Z is Lindelöf if and only if every uncountable $A \subset Z$ has a condensation point,

Proof. Suppose that Z is Lindelöf; if $A \subset Z$ is uncountable and there is no condensation point for A then choose, for any $z \in Z$ a set $U_z \in \tau(z, Z)$ such that $|U_z \cap A| \le \omega$. There is a countable $P \subset Z$ such that $\bigcup\{U_z : z \in P\} = Z$; so $A = \bigcup\{A \cap U_z : z \in P\}$ is countable which is a contradiction.

Now assume that $l(Z) \le \omega_1$ and every uncountable subset of Z has a condensation point. If Z is not Lindelöf then there is an open cover $\mathcal{U}$ of the space Z such that no countable subfamily of $\mathcal{U}$ covers Z. Since $l(Z) \le \omega_1$, we can assume that $|\mathcal{U}| = \omega_1$; choose an enumeration $\{U_\alpha : \alpha < \omega_1\}$ of the family $\mathcal{U}$.

The set $F_\alpha = Z \backslash (\bigcup\{U_\beta : \beta < \alpha\})$ is non-empty for any ordinal $\alpha < \omega_1$ and $\bigcap\{F_\alpha : \alpha < \omega_1\} = \emptyset$. Take a point $z_\alpha \in F_\alpha$ for any $\alpha < \omega_1$; then $A = \{z_\alpha : \alpha < \omega_1\}$ is uncountable for otherwise there is $\alpha < \omega_1$ such that $A \subset \bigcup\{U_\beta : \beta < \alpha\}$ which is a contradiction with $A \cap F_\alpha \ne \emptyset$.

For any $z \in Z$ there is $\alpha < \omega_1$ such that $z \in U_\alpha$ and therefore $W = X \backslash F_{\alpha+1}$ is an open neighbourhood of the point z such that $W \cap A \subset \{z_\beta : \beta \le \alpha\}$ is countable. Thus A has no condensation point in Z; this contradiction shows that the space Z is Lindelöf and hence Fact 1 is proved.

Fact 2. If Z is a space with strong condensation property and $l(Z^\omega) \leq \omega_1$ then Z^ω is Lindelöf.

Proof. Let $p_n : Z^\omega \to Z$ be the natural projection of Z^ω onto its n-th factor for any $n \in \omega$. Take an arbitrary uncountable set $A \subset Z^\omega$. If $p_0(A)$ is countable then there is $z_0 \in p_0(A)$ such that $A_0 = p_0^{-1}(z_0) \cap A$ is uncountable. If $p_0(A)$ is uncountable then there is $z_0 \in Z$ and an uncountable $A_0 \subset A$ such that $p_0|A_0$ is injective and $p_0(A_0)$ is concentrated around z_0.

Assume that $k \in \omega$ and we have uncountable sets $A_0 \supset \ldots \supset A_k$ and $z_0, \ldots, z_k \in Z$ such that $A \supset A_0$ and, for any $i \leq k$, either $p_i(A_i) = \{z_i\}$ or the map $p_i|A_i$ is injective and $p_i(A_i)$ is concentrated around z_i.

If $p_{k+1}(A_k)$ is countable then there is a point $z_{k+1} \in Z$ such that the set $A_{k+1} = p_{k+1}^{-1}(z_{k+1}) \cap A_k$ is uncountable; if $p_{k+1}(A_k)$ is uncountable then there is an uncountable $A_{k+1} \subset A_k$ and $z_{k+1} \in Z$ for which $p_{k+1}|A_{k+1}$ is injective and $p_{k+1}(A_{k+1})$ is concentrated around z_{k+1}. Thus our inductive procedure can be continued to obtain a decreasing family $\{A_i : i \in \omega\}$ of uncountable subsets of A and a sequence $\{z_i : i \in \omega\} \subset Z$ such that, for any $i \in \omega$, either $p_i(A_i) = \{z_i\}$ or $p_i|A_i$ is injective and $p_i(A_i)$ is concentrated around z_i.

Then $z = (z_i : i \in \omega) \in Z^\omega$ is a condensation point for the set A. Indeed, if we are given any set $U \in \tau(z, Z^\omega)$ then there exist a number $n \in \omega$ and $U_0, \ldots, U_n \in \tau(Z)$ such that $z \in W = U_0 \times \ldots \times U_n \times Z^{\omega \setminus (n+1)} \subset U$. It is easy to see that, for every $i \leq n$, the set $B_i = \{u \in A_{n+1} : p_i(u) \notin U_i\}$ is countable (possibly empty); so the set $A' = A_{n+1} \setminus (B_0 \cup \ldots \cup B_n)$ is uncountable and $A' \subset W \cap A \subset U \cap A$. Thus every uncountable $A \subset Z^\omega$ has a condensation point; so Z^ω is Lindelöf by Fact 1 and hence Fact 2 is proved.

Returning to our solution define a topology τ on ω_1 as follows: For every $\alpha \in I$ let $\mathcal{B}_\alpha = \{\{\alpha\}\}$; if $\alpha \in L$ then $\mathcal{B}_\alpha = \{\{\mu_\alpha(n) : n \geq k\} : k \in \omega\}$. Let τ be the topology generated by the family $\{\mathcal{B}_\alpha : \alpha < \omega_1\}$ as local bases. It is evident that the space $Y = (\omega_1, \tau)$ is locally compact and all points of I are isolated in Y. Since L is an uncountable closed discrete subspace of Y, the space Y is not compact. Let X be the one-point compactification of Y and denote by w the unique point of the set $X \setminus Y$.

If $K \subset Y$ is compact then $K \cap L$ is finite because K is closed and discrete in Y. The set $K' = K \setminus (\bigcup\{S_\alpha : \alpha \in K \cap L\})$ is also finite being closed and discrete in K; so $K \subset K' \cup (\bigcup\{S_\alpha : \alpha \in K \cap L\})$ is countable. This proves that

(1) every compact subspace of Y is countable.

Assume that X is Corson compact and hence there exists a point-countable family $\mathcal{U}$ of open F_σ-subsets of X which T_0-separates the points of X (see Problem 118). Given $x, y \in X$ say that a set $U \in \mathcal{U}$ *separates x and y* if $U \cap \{x, y\}$ is a singleton. Since every F_σ-subspace of X is σ-compact, the property (1) implies that

(2) if $U \in \mathcal{U}$ and $w \notin U$ then U is countable,

and hence $\sup(U) < \omega_1$ for any $U \in \mathcal{U}_0 = \mathcal{U} \setminus \tau(w, X)$. There are at most countably many elements of $\mathcal{U}$ which contain w. Since $X \setminus U \subset Y$ is compact and hence countable for any $U \in \tau(w, X)$, there exists an ordinal $\gamma < \omega_1$ such that

(3) for any countable ordinal $\alpha \geq \gamma$, if $U \in \mathcal{U}$ separates α and w then $U \in \mathcal{U}_0$ and hence $\alpha \in U$.

Take any $\gamma_0 \in L$ with $\gamma_0 \geq \gamma$ and choose $U_0 \in \mathcal{U}$ which separates γ_0 and w. The property (3) shows that $U_0 \in \mathcal{U}_0$ and $\gamma_0 \in U_0$. Suppose that $\beta < \omega_1$ and we have chosen a set $\{\gamma_\alpha : \alpha < \beta\} \subset L$ and a family $\{U_\alpha : \alpha < \beta\} \subset \mathcal{U}_0$ with the following properties:

(4) $\gamma_\alpha \in U_\alpha$ for any $\alpha < \beta$;
(5) $\alpha' < \alpha < \beta$ implies $\alpha' \leq \gamma_{\alpha'} < \gamma_\alpha$;
(6) if $\alpha' < \alpha < \beta$ then $\sup\{\sup(U_\delta) : \delta < \alpha'\} < \gamma_\alpha$;
(7) if $\alpha < \beta$ is a limit ordinal then $\gamma_\alpha = \sup\{\gamma_\delta : \delta < \alpha\}$.

If β is a limit ordinal then we have to let $\gamma_\beta = \sup\{\gamma_\alpha : \alpha < \beta\}$; since $\mathcal{U}$ is a T_0-separating family in X, there is $U_\beta \in \mathcal{U}$ which separates γ_β and w. It follows from (3) that $\gamma_\beta \in U_\beta \in \mathcal{U}_0$. It is easy to see that the properties (4)–(7) hold for all $\alpha \leq \beta$.

Now, if $\beta = \beta' + 1$ then $\gamma' = \sup\{\sup(U_\alpha) : \alpha \leq \beta'\} < \omega_1$ by (2); so if we take an ordinal $\gamma_\beta \in L$ with $\gamma_\beta > \max\{\gamma_{\beta'}, \gamma'\} + 1$ then some $U_\beta \in \mathcal{U}$ separates γ_β and w which implies, by (3), that $\gamma_\beta \in U_\beta \in \mathcal{U}_0$. It is clear that (4)–(7) still hold for any $\alpha \leq \beta$; so our inductive procedure can be continued to construct an ω_1-sequence $G = \{\gamma_\alpha : \alpha < \omega_1\} \subset L$ and a family $\{U_\alpha : \alpha < \omega_1\} \subset \mathcal{U}_0$ with the properties (4)–(7) fulfilled for any $\beta < \omega_1$.

An immediate consequence of (4) and (6) is that $U_\alpha \neq U_\beta$ if $\alpha \neq \beta$; besides, it follows from (5) and (7) that G is a closed unbounded subset of ω_1. For any $\alpha < \omega_1$, the set U_α is an open neighbourhood of γ_α; so there is $k_\alpha \in \omega$ such that $\mu_{\gamma_\alpha}(n) \in U_\alpha$ for any $n \geq k_\alpha$; let $f(\gamma_\alpha) = \mu_{\gamma_\alpha}(k_\alpha)$.

This gives us a function $f : G \to \omega_1$ such that $f(\beta) < \beta$ for any $\beta \in G$; so we can apply SFFS-067 to find an uncountable $H \subset G$ and $\beta < \omega_1$ such that $f(v) = \beta$ for any $v \in H$. In other words, there is an uncountable $E \subset \omega_1$ such that $f(\gamma_\alpha) = \beta$ for any $\alpha \in E$. By our choice of the function f we have $\beta \in U_\alpha$ for any $\alpha \in E$. Since $U_\alpha \neq U_{\alpha'}$ for distinct $\alpha, \alpha' \in E$, the point β belongs to uncountably many elements of $\mathcal{U}$; this contradiction shows that X is not Corson compact.

Let $D = \{f \in C_p(X, \mathbb{D}) : |f^{-1}(1) \cap L| \leq 1\}$. The set D separates the points of X. Indeed, if $\alpha \in I$ then $\chi_{\{\alpha\}} \in D$ separates the points w and α. If $\alpha \in L$ then χ_{S_α} also separates α and w.

If we are given distinct ordinals $\alpha, \beta < \omega_1$ such that $\{\alpha, \beta\} \cap I \neq \emptyset$ then one of the functions $\chi_{\{\alpha\}}, \chi_{\{\beta\}}$ belongs to D and separates α and β. Finally, if $\alpha, \beta \in L$ and $\alpha \neq \beta$ then $\chi_{S_\alpha} \in D$ separates α and β.

Let $D_0 = \{f \in C_p(X, \mathbb{D}) : f^{-1}(1) \cap L = \emptyset\}$ and suppose that E is an uncountable subset of D_0. It is easy to see that $f^{-1}(1)$ is finite for any $f \in E$; so there is an uncountable $E' \subset E$ and a finite $P \subset I$ such that $f^{-1}(1) \cap g^{-1}(1) = P$ for any distinct $f, g \in E'$. If $h = \chi_P$ then it is an easy exercise that the set E' is concentrated around h. This shows that

(8) the set $D_0 \subset D$ has strong condensation property.

Our purpose is to prove that D has strong condensation property so suppose that E is an uncountable subset of D. It follows from (8) that we do not lose generality assuming that $E \cap D_0$ is countable.

Consider first the case when there is an uncountable $G \subset E$ and $\alpha \in L$ such that $f(\alpha) = 1$ for all $f \in G$. Then $H = \{f - \chi_{S_\alpha} : f \in G\}$ is an uncountable subset of D_0; so we can apply (8) to find a function $g \in D_0$ and an uncountable set $H' \subset H$ which is concentrated around g. It is straightforward that the set $G' = \{f + \chi_{S_\alpha} : f \in H'\} \subset G$ is uncountable and concentrated around $g + \chi_{S_\alpha} \in D$.

Assume that $E_\alpha = \{f \in E : f(\alpha) = 1\}$ is countable for any $\alpha \in L$ and hence the set $\{\alpha \in L : \text{there exists } f \in E \text{ with } f(\alpha) = 1\}$ is uncountable. It takes an evident transfinite induction to construct sets $\{f_\alpha : \alpha < \omega_1\} \subset E$ and $\{\delta_\alpha : \alpha < \omega_1\} \subset L$ such that

(9) $f_\alpha(\delta_\alpha) = 1$ and $\delta_\alpha > \nu_\alpha = \sup\{\delta_\beta : \beta < \alpha\}$ for any $\alpha < \omega_1$.

The family $\{(\nu_\alpha, \delta_\alpha] : \alpha < \omega_1\}$ is disjoint; let $J_\alpha = f_\alpha^{-1}(1) \cap (\nu_\alpha, \delta_\alpha]$, $g_\alpha = \chi_{J_\alpha}$ and $h_\alpha = f_\alpha - g_\alpha$ for any $\alpha < \omega_1$. If $u \in D_0$ is identically zero on X then the set $\{g_\alpha : \alpha < \omega_1\}$ is concentrated around u because the family $\{g_\alpha^{-1}(1) : \alpha < \omega_1\}$ is disjoint. Furthermore, $h_\alpha \in D_0$ for any $\alpha < \omega_1$; so we can apply (8) again to find a function $h \in D_0$ and an uncountable $\Omega \subset \omega_1$ such that $\{h_\alpha : \alpha \in \Omega\}$ is concentrated around h.

It turns out that the set $P = \{f_\alpha : \alpha \in \Omega\}$ is also concentrated around h. Indeed, take any open neighbourhood O of the function $h = h + u$. Since the sum is continuous in $C_p(X) \times C_p(X)$ (see TFS-115), it is also continuous in $C_p(X, \mathbb{D}) \times C_p(X, \mathbb{D})$; so there exist $W_1 \in \tau(h, C_p(X, \mathbb{D}))$, $W_2 \in \tau(u, C_p(X, \mathbb{D}))$ such that $p + q \in O$ for any $p \in W_1$ and $q \in W_2$. There exists $\gamma < \omega_1$ such that $h_\alpha \in W_1$ and $g_\alpha \in W_2$ for any $\alpha \in \Omega$ with $\alpha \geq \gamma$. Then $f_\alpha = h_\alpha + g_\alpha \in O$ for any $\alpha \in \Omega$ with $\alpha \geq \gamma$ and hence P is, indeed, concentrated around h. Finally, observe that P is uncountable because $f_\alpha \neq f_\beta$ whenever $\alpha \neq \beta$. Therefore we proved that D has strong condensation property; this, together with $l(D) \leq w(D) \leq w(C_p(X)) \leq \omega_1$, implies that D^ω is Lindelöf (see Fact 2).

Since D separates the points of X, the space X condenses and hence embeds in $C_p(D)$ (see TFS-166). Finally, apply Fact 4 of U.093 to conclude that $(C_p(X))^\omega$ is Lindelöf and complete our solution.

U.178. *Prove that any Corson compact space is a continuous image of a zero-dimensional Corson compact space.*

Solution. Let X be a Corson compact space. By Problem 118, the space X has a point-countable T_0-separating family $\mathcal{U}$ of open F_σ-subsets of X. For any $U \in \mathcal{U}$ choose a continuous function $f_U : X \to I = [0, 1] \subset \mathbb{R}$ such that $X \setminus U = f_U^{-1}(0)$ (see Fact 1 of S.358 and Fact 1 of S.499). The map $f = \Delta\{f_U : U \in \mathcal{U}\} : X \to I^{\mathcal{U}}$ is an embedding. To see it take distinct $x, y \in X$. There is $U \in \mathcal{U}$ such that $U \cap \{x, y\}$ is a singleton. Then $f_U(x) \neq f_U(y)$ and therefore $f(x) \neq f(y)$ which proves that f is, indeed, an embedding being an injection; let $X' = f(X)$.

Observe that $X' \subset \{x \in I^{\mathcal{U}} :$ the set $\{U \in \mathcal{U} : x(U) \neq 0\}$ is countable$\}$ because if $x \in X'$ then $x = f(y)$ for some $y \in X$ and $x(U) = f_U(y)$ for any $U \in \mathcal{U}$ which implies that the family $\{U \in \mathcal{U} : x(U) \neq 0\} \subset \{U \in \mathcal{U} : y \in U\}$ is countable.

To simplify the notation we will reformulate the obtained result as follows:

(1) there exists a set T such that X embeds in $\Sigma = \{x \in I^T : |x^{-1}(I \setminus \{0\})| \leq \omega\}$, so we can assume that $X \subset \Sigma$.

Let $K = \mathbb{D}^\omega$ be the Cantor set; fix a point $a \in K$ and let $I_n = [\frac{1}{n+2}, \frac{1}{n+1}] \subset I$ for any $n \in \omega$. The space K being zero-dimensional, there exists a local base $\mathcal{O} = \{O_n : n \in \omega\}$ at the point a in K such that the set O_n is clopen in K and $O_{n+1} \subset O_n$ for any $n \in \omega$. Making the relevant changes in $\mathcal{O}$ if necessary, we can assume that $O_0 = K$ and $K_n = O_n \setminus O_{n+1} \neq \emptyset$ for any $n \in \omega$.

Since no point of K is isolated, the same is true for any non-empty clopen subset of K and hence every non-empty clopen subset of K is homeomorphic to K (see SFFS-348). This shows that K_n is homeomorphic to K and hence there is a continuous onto map $\varphi_n : K_n \to I_n$ for any $n \in \omega$ (see TFS-128). Let $\varphi(a) = 0$; if $x \in K \setminus \{a\}$ then there is a unique $n \in \omega$ such that $x \in K_n$; let $\varphi(x) = \varphi_n(x)$. It is an easy exercise that $\varphi : K \to I$ is a continuous onto map such that $\varphi^{-1}(0) = \{a\}$.

Let $\Phi : K^T \to I^T$ be the product of T-many copies of φ, i.e., $\Phi(x)(t) = \varphi(x(t))$ for any $x \in K^T$ and $t \in T$. The map Φ is continuous by Fact 1 of S.271 and it is easy to see that $\Phi(K^T) = I^T$. The space $Y = \Phi^{-1}(X)$ is compact being closed in the compact space K^T. Take any point $y \in Y$ and let $x = \Phi(y)$. If $y(t) \neq a$ then $x(t) = \varphi(y(t)) \neq 0$ and hence

$$S_y = \{t \in T : y(t) \neq a\} \subset \operatorname{supp}(x) = \{t \in T : x(t) \neq 0\}.$$

It follows from $X \subset \Sigma$ that $\operatorname{supp}(x)$ is countable; so S_y is countable for any $y \in Y$ which shows that Y is a subset of a Σ-product of T-many copies of K. Therefore Y is a Corson compact space by Problem 119. It is clear that $\Phi|Y$ maps Y continuously onto X. Besides, Y is zero-dimensional because so is K^T (see SFFS-301 and SFFS-302). Therefore Y is a zero-dimensional Corson compact space which maps continuously onto X.

U.179. *Prove that every first countable space is a W-space and every W-space is Fréchet–Urysohn.*

Solution. Suppose that X is a first countable space and fix a point $x \in X$. There is a countable local base $\mathcal{B} = \{O_n : n \in \mathbb{N}\}$ of X at x such that $O_{n+1} \subset O_n$ for any $n \in \mathbb{N}$. Now, let $\sigma(\emptyset) = O_1$ and, if moves $U_1, x_1, \ldots, U_n, x_n$ have been made, let $\sigma(U_1, x_1, \ldots, U_n, x_n) = O_{n+1}$. It is clear that σ is a strategy for the player OP.

If we have a play $\{(U_i, x_i) : i \in \mathbb{N}\}$ in which OP applied the strategy σ then $x_i \in U_i = O_i$ for any $i \in \mathbb{N}$ so it is evident that the sequence $\{x_i\}$ converges to x. Thus σ is a winning strategy for OP and therefore $\{x\}$ is a W-set; the point $x \in X$ was chosen arbitrarily; so X is a W-space, i.e., we proved that any first countable space is a W-space.

Now, assume that X is a W-space. Given a set $A \subset X$ and a point $x \in \overline{A}$ fix the respective winning strategy σ for the player OP at the point x and consider a play $\{(U_i, x_i) : i \in \mathbb{N}\}$ where $U_1 = \sigma(\emptyset)$, $O_{i+1} = \sigma(U_1, x_1, \ldots, U_i, x_i)$ and $x_i \in U_i \cap A$ for any $i \in \mathbb{N}$. To see that such a play is possible, observe that every move of OP is an open neighbourhood of x; so $U_i \cap A \neq \emptyset$ for any $i \in \mathbb{N}$ and hence PT can choose a point $x_i \in U_i \cap A$ at his i-th move. It is evident that in the play $\{(U_i, x_i) : i \in \mathbb{N}\}$ the player OP applied the strategy σ; so $\{x_i : i \in \mathbb{N}\} \subset A$ is a sequence which converges to x. Thus X is a Fréchet–Urysohn space.

U.180. *Suppose that $f : X \to Y$ is an open continuous onto map. Prove that if X is a W-space then so is Y.*

Solution. Take any point $y \in Y$ and fix $x \in X$ such that $f(x) = y$. The player OP has a winning strategy σ in X at the point x. We will define a strategy s for the Gruenhage game on Y at y in such a way that, for any play $P = \{(U_i, y_i) : i \in \mathbb{N}\}$ of this game where OP applies the strategy s, there is a play $P' = \{(V_i, x_i) : i \in \mathbb{N}\}$ of the same game on X at x such that the player OP applies σ in P' and we have $f(V_i) = U_i$, $f(x_i) = y_i$ for all $i \in \mathbb{N}$.

Let $s(\emptyset) = f(\sigma(\emptyset))$; if PT chooses $y_1 \in U_1 = s(\emptyset)$ there is $x_1 \in V_1 = \sigma(\emptyset)$ such that $f(x_1) = y_1$. Let $V_2 = \sigma(V_1, x_1)$, $U_2 = f(V_2)$ and $s(U_1, y_1) = U_2$. Suppose that $n \in \mathbb{N}$ and moves $U_1, y_1, \ldots, U_n, y_n, U_{n+1}$ and $V_1, x_1, \ldots, V_n, x_n, V_{n+1}$ have been made in the respective games on Y and X in such a way that the player OP applies the strategies s and σ in Y and X respectively while $f(V_i) = U_i$ for all $i \leq n + 1$ and $f(x_i) = y_i$ for all $i \leq n$.

If PT plays a point $y_{n+1} \in U_{n+1}$ then there exists a point $x_{n+1} \in V_{n+1}$ such that $f(x_{n+1}) = y_{n+1}$; let $V_{n+2} = \sigma(V_1, x_1, \ldots, V_{n+1}, x_{n+1})$, $U_{n+2} = f(V_{n+2})$ and $s(U_1, y_1, \ldots, U_{n+1}, y_{n+1}) = U_{n+2}$. It is evident that all mentioned properties of s now hold if we substitute n by $n + 1$. The induction step in defining the strategy s being carried out, our inductive procedure can be continued and hence we have the promised play $P' = \{(V_i, x_i) : i \in \mathbb{N}\}$ on X where OP applies σ for any play $P = \{(U_i, y_i) : i \in \mathbb{N}\}$ on Y where OP applies the strategy s. Since the strategy σ is winning, the sequence $S' = \{x_n : n \in \mathbb{N}\}$ converges to x and therefore the sequence $S = \{y_n : n \in \mathbb{N}\} = \{f(x_n) : n \in \mathbb{N}\}$ converges to $f(x) = y$.

Observe that we only defined the strategy s for the plays where s has to be applied. For the rest of the respective initial segments of our game the function s can be defined arbitrarily, e.g., we can let $s(U_1, y_1, \ldots, U_n, y_n) = Y$ for any sequence $\{U_1, y_1, \ldots, U_n, y_n\}$ where s was not defined yet. We already proved that if OP applies the strategy s in a play $\{(U_i, y_i) : i \in \mathbb{N}\}$ then $y_i \to y$; so s is a winning strategy and therefore Y is, indeed, a W-space.

U.181. *Suppose that X is a separable space and a closed set $F \subset X$ has an outer base of closed neighbourhoods (i.e., for any $U \in \tau(F, X)$ there is $V \in \tau(F, X)$ such that $\overline{V} \subset U$). Prove that if F is a W-set in X then $\chi(F, X) \leq \omega$. In particular, if X is a separable W-space then $\chi(X) = \omega$.*

Solution. Let A be a countable dense subset of the space X and fix a winning strategy σ for the player OP in the W-game at the set F in X. Consider the family $\mathcal{W} = \{W \in \tau(F, X) :$ there is $n \in \mathbb{N}$ and an initial segment $W_1, a_1, \ldots, W_{n-1}, a_{n-1}, W_n$ of a play in a W-game at F where OP applies the strategy σ, the point a_i belongs to A for any $i < n$ and $W_n = W\}$. Any $W \in \mathcal{W}$ is uniquely determined by a finite (maybe empty, in which case $W = \sigma(\emptyset)$) sequence of points from A; so $|\mathcal{W}| \leq \omega$ and hence it suffices to show that $\mathcal{W}$ is an outer base at F.

To obtain a contradiction assume that there exits a set $U' \in \tau(F, X)$ such that $W \backslash U' \neq \emptyset$ for any $W \in \mathcal{W}$. Fix a set $U \in \tau(F, X)$ for which $\overline{U} \subset U'$; then $W \backslash \overline{U} \neq \emptyset$ for any $W \in \mathcal{W}$. The set A being dense in X there is $a_1 \in A \cap (W_1 \backslash \overline{U})$ where $W_1 = \sigma(\emptyset) \in \mathcal{W}$. Suppose that $n \in \mathbb{N}$ and $\{W_1, a_1, \ldots, W_n, a_n, W_{n+1}\}$ is an initial segment of the W-game at F with the following properties:

(1) $W_{i+1} = \sigma(W_1, a_1, \ldots, W_i, a_i)$ for any $i \leq n$, i.e., the player OP applies the strategy σ;
(2) $a_i \in A \cap (W_i \backslash \overline{U})$ for any $i \leq n$.

As a consequence, $W_{n+1} \in \mathcal{W}$; so it follows from our choice of U that we can take a point $a_{n+1} \in A \cap (W_{n+1} \backslash \overline{U})$; let $W_{n+2} = \sigma(W_1, a_1, \ldots, W_{n+1}, a_{n+1})$. It is evident that the conditions (1)–(2) are still satisfied for all $i \leq n+1$; so our inductive procedure gives a play $\{W_i, a_i : i \in \mathbb{N}\}$ for which (1) and (2) hold. In particular, the player OP applies the strategy σ and hence the sequence $S = \{a_n : n \in \mathbb{N}\}$ has to converge to F. On the other hand, $S \subset X \backslash U$ by (2) which is a contradiction.

Thus $\mathcal{W}$ is, indeed, an outer base of F in X and therefore $\chi(F, X) \leq |\mathcal{W}| \leq \omega$. Finally observe that if F is a singleton then it has a base of closed neighbourhoods by regularity of X. Consequently, $\chi(x, X) \leq \omega$ for any $x \in X$ whenever X is a separable W-space.

U.182. *Show that there exist W-spaces which are not first countable and Fréchet–Urysohn spaces which are not W-spaces.*

Solution. Let $X = A(\omega_1) = \omega_1 \cup \{a\}$ be the one-point compactification of the discrete space of cardinality ω_1. Recall that $a \in X$ is the unique non-isolated point of X and $\tau(a, X) = \{X \backslash K : K \subset X \backslash \{a\}$ is finite$\}$.

If $x \in X \backslash \{a\}$ then it is evident that the n-th move $U_n = \{x\}$ for any $n \in \mathbb{N}$ gives a winning strategy for OP in the Gruenhage game on X at x. Now, for the case $x = a$ let $U_1 = X$ and $\sigma(\emptyset) = U_1$; if $n \in \mathbb{N}$ and moves $U_1, x_1, \ldots, U_n, x_n$ are made then let $\sigma(U_1, x_1, \ldots, U_n, x_n) = X \backslash \{x_1, \ldots, x_n\}$. This defines a strategy σ on the space X at the point a.

If we have a play $\{(U_n, x_n) : n \in \mathbb{N}\}$ where σ is applied then $x_n \neq x_m$ for distinct $m, n \in \mathbb{N}$; so the sequence $\{x_n : n \in \mathbb{N}\}$ converges to a. Thus σ is a winning strategy, so X is a W-space; since $\psi(a, X) = \chi(a, X) > \omega$ we have $\chi(X) > \omega$.

Now let M be a countable second countable space for which there is a closed map $f : M \to Y$ such that Y is not metrizable (see TFS-227). A countable space is metrizable if and only if it is first countable; so $\chi(Y) > \omega$. It follows from TFS-225 that Y is a Fréchet–Urysohn space. If Y is a W-space then it is first countable by

Problem 181; this contradiction shows that Y is an example of a Fréchet–Urysohn space which does not have the W-property.

U.183. *Prove that any subspace of a W-space is a W-space and any countable product of W-spaces is a W-space.*

Solution. Suppose that X is a W-space and $Y \subset X$. Given a point $y \in Y$ there is a winning strategy s in the W-game on X at y. Let $\sigma(\emptyset) = s(\emptyset) \cap Y$; then $W_1 = \sigma(\emptyset) \in \tau(y, Y)$. Now suppose that $n \in \mathbb{N}$ and $\{W_1, y_1, \ldots, W_{n-1}, y_{n-1}, W_n\}$ is an initial segment of the play in the W-game on Y at y where our future strategy σ is defined and applied, i.e., $\sigma(W_1, y_1, \ldots, W_i, y_i) = W_{i+1}$ for any $i < n$ and, besides, we have an initial segment $\{U_1, y_1, \ldots, U_{n-1}, y_{n-1}, U_n\}$ of a play in the W-game on X at y where OP applies the strategy s and $W_i = U_i \cap Y$ for any $i \leq n$. If PT plays with a point $y_n \in W_n$ then $y_n \in U_n$, so the strategy s is applicable; let $\sigma(W_1, y_1, \ldots, W_n, y_n) = s(U_1, y_1, \ldots, U_n, y_n) \cap Y$; it is clear that $W_{n+1} = \sigma(W_1, y_1, \ldots, W_n, y_n)$ is an open neighbourhood of y in Y, so our inductive procedure defines a strategy σ in the W-game on Y at the point y.

To see that σ is winning observe that, for every play $\mathcal{P} = \{(W_n, y_n) : n \in \mathbb{N}\}$ in the W-game on Y at y where σ is applied we have a play $\mathcal{P}' = \{(U_n, y_n) : n \in \mathbb{N}\}$ in the W-game on X at the point y where the player OP applies the strategy s. Since s is a winning strategy, the sequence $\{y_n\}$ converges to y and hence σ is also a winning strategy, i.e., Y is a W-space. This proves that any subspace of a W-space is a W-space.

Now assume that X_n is a W-space for any $n \in \mathbb{N}$; we must prove that the space $X = \prod\{X_n : n \in \mathbb{N}\}$ is a W-space. If $x \in X$ then there is a winning strategy σ_n in the W-game on X_n at the point $x(n)$ for any $n \in \mathbb{N}$. Denote by $p_n : X \to X_n$ the respective natural projection for any $n \in \mathbb{N}$ and let $\sigma(\emptyset) = p_1^{-1}(\sigma_1(\emptyset))$. It is clear that the set $U_1 = \sigma(\emptyset)$ is an open neighbourhood of x in X; let $O_1^1 = \sigma_1(\emptyset)$.

Suppose that $n \in \mathbb{N}$ and $\{U_1, x_1, \ldots, U_{n-1}, x_{n-1}, U_n\}$ is an initial segment of a play in the W-game on X at x where our future strategy σ of the player OP is defined and applied, i.e., $\sigma(U_1, x_1, \ldots, U_i, x_i) = U_{i+1}$ for any $i < n$ and, besides, for any $j \leq n$ an initial segment $\mathcal{S}_j = \{O_1^j, x_j(j), \ldots, O_{n-j}^j, x_{n-1}(j), O_{n-j+1}^j\}$ of a play in the W-game on X_j at the point $x(j)$ is chosen in such a way that the strategy σ_j is applied in every $\mathcal{S}_j$ and

(1) $O_1^i = \sigma_i(\emptyset)$ and $U_i = \bigcap\{p_j^{-1}(O_{i-j+1}^j) : j \leq i\}$ for any $i \leq n$.

If the player PT makes a move $x_n \in U_n$ then it follows from the property (1) that $x_n(j) \in O_{n-j+1}^j$ and therefore σ_j is applicable; let $O_1^{n+1} = \sigma_{n+1}(\emptyset)$ and $O_{n-j+2}^j = \sigma_j(O_1^j, x_j(j), \ldots, O_{n-j}^j, x_{n-1}(j), O_{n-j+1}^j, x_n(j))$ for any $j \leq n$; since O_{n-j+2}^j is an open neighbourhood of the point $x(j)$ for any $j \in \{1, \ldots, n + 1\}$, the set $U_{n+1} = \bigcap\{p_j^{-1}(O_{n-j+2}^j) : j \leq n + 1\}$ is an open neighbourhood of the point x ao we can let $\sigma(U_1, x_1, \ldots, U_{n-1} x_{n-1}, U_n, x_n) = U_{n+1}$. The property (1) is still fulfilled when we substitute n by $n + 1$ so our inductive procedure defines a strategy σ on X for which the property (1) holds for all $n \in \mathbb{N}$.

To see that σ is winning suppose that $\{(U_n, x_n) : n \in \mathbb{N}\}$ is a play where OP applies the strategy σ. By our choice of σ there is a play $\{(O^i_j, x_{i+j-1}(i)) : j \in \mathbb{N}\}$ in the W-game on the space X_i at the point $x(i)$ where OP applies the strategy σ_i for any $i \in \mathbb{N}$.

As a consequence, the sequence $\{x_{i+j-1}(i) : j \in \mathbb{N}\}$ converges to $x(i)$ for any $i \in \mathbb{N}$; so it is an easy exercise to see that $x_n \to x$ and hence σ is a winning strategy. This shows that X is a W-space and therefore any countable product of W-spaces is a W-space.

U.184. *Prove that any Σ-product of W-spaces is a W-space. Deduce from these facts that if X is a Corson compact space then every non-empty closed $F \subset X$ is a W-set; in particular, X is a W-space.*

Solution. Suppose that X_t is a W-space and a point $a_t \in X_t$ is fixed for any $t \in T$. Let $a(t) = a_t$ for any $t \in T$; then $a \in X = \prod_{t \in T} X_t$. We must prove that $\Sigma(X, a) = \{x \in X : \text{the set supp}(x) = \{t \in T : x(t) \neq a_t\} \text{ is countable}\}$ is a W-space.

For any $t \in T$ denote by $p_t : \Sigma(X, a) \to X_t$ the restriction to $\Sigma(X, a)$ of the natural projection of X onto the factor X_t. Fix a point $y \in \Sigma = \Sigma(X, a)$ and a winning strategy σ_t of the player OP in the W-game on the space X_t at the point $y_t = y(t)$ for any $t \in T$. For any $x \in \Sigma$ choose a countable set $T^x = \{t^x_n : n \in \mathbb{N}\} \subset T$ such that $\text{supp}(x) \subset T^x$ and let $T^x_n = \{t^x_i : i \leq n\}$ for any $n \in \mathbb{N}$.

Let $\sigma(\emptyset) = p^{-1}_{t^y_1}(\sigma_{t^y_1}(\emptyset))$; it is clear that the set $U_1 = \sigma(\emptyset)$ is an open neighbourhood of the point y; let $S_1 = \{t^y_1\}$. To construct a strategy σ by induction assume that $n \in \mathbb{N}$ and $\{U_1, x_1, \ldots, U_{n-1}, x_{n-1}, U_n\}$ is an initial segment of a play in the W-game on Σ at y where our future strategy σ of the player OP is defined and applied, i.e., $\sigma(U_1, x_1, \ldots, U_i, x_i) = U_{i+1}$ for any $i < n$ and, besides, we have finite sets $S_1 \subset \ldots \subset S_n \subset T$ with the following properties:

(1) $T^y_i \cup T^{x_1}_i \cup \ldots \cup T^{x_{i-1}}_i \subset S_i$ for any $i \leq n$;
(2) for any $t \in S_n$ if $k_t = \min\{i : t \in S_i\}$ then a play $\mathcal{P}_t$ in the W-game on X_t at the point y_t starts at the step k_t and OP applies the strategy σ_t in $\mathcal{P}_t$;
(3) for any $i < n$ and $t \in S_n$ if $k_t \leq i$ then at the step i the player OP makes his/her strategy move number $(i - k_t + 1)$ (starting with $\sigma_t(\emptyset)$ if $i = k_t$) following the move number $(i - k_t)$ of the player PT (in case when $k_t < i$) who plays with $x_{i-1}(t)$;
(4) for any $i \leq n$ we have $U_i = \bigcap\{p^{-1}_t(O_t) : t \in S_i, \ k_t \leq i \text{ and } O_t \text{ is the move number } (i - k_t + 1) \text{ of the player OP}\}$.

Now, suppose that the player PT makes a move $x_n \in U_n$ and consider the respective initial segment $\mathcal{S}_t = \{O^t_1, z^t_1, \ldots, O^t_{m_t-1}, z^t_{m_t-1}, O^t_{m_t}\}$ of the play $\mathcal{P}_t$ for any $t \in S_n$. It follows from the property (4) that $x_n(t) \in O^t_{m_t}$ for any $t \in S_n$ so the strategy σ_t is applicable; let $O_t = \sigma_t(O^t_1, z^t_1, \ldots, O^t_{m_t-1}, z^t_{m_t-1}, O^t_{m_t}, x_n(t))$ for any $t \in S_n$. Take a finite set $S_{n+1} \subset T$ such that $S_n \cup T^y_{n+1} \cup T^{x_1}_{n+1} \cup \ldots \cup T^{x_n}_{n+1} \subset S_{n+1}$ and let $O_t = \sigma_t(\emptyset)$ for any $t \in S_{n+1} \setminus S_n$.

It is immediate that $U_{n+1} = \bigcap\{p_t^{-1}(O_t) : t \in S_{n+1}\}$ is an open neighbourhood of y in Σ; let $\sigma(U_1, x_1, \ldots, U_n, x_n) = U_{n+1}$. It is evident that the conditions (1)–(4) are still satisfied if we substitute $n + 1$ in place of n. Thus our strategy σ can be inductively constructed so that the properties (1)-(4) hold for any $n \in \mathbb{N}$.

To see that σ is winning, assume that $\{(U_n, x_n) : n \in \mathbb{N}\}$ is a play where OP applies the strategy σ. We also have a sequence $\{S_n : n \in \mathbb{N}\}$ with the properties (1)–(4); let $S = \bigcup\{S_n : n \in \mathbb{N}\}$. It follows from (1) that $\text{supp}(y) \cup \text{supp}(x_i) \subset S$ for any $i \in \mathbb{N}$.

For any $t \in S$ we have a play $\{(O_n^t, z_n^t) : n \in \mathbb{N}\}$ in the W-game on X_t at $y(t)$ in which OP applies the strategy σ_t and $\{z_n^t : n \in \mathbb{N}\} = \{x_{k_t+n-1}(t) : n \in \mathbb{N}\}$. The strategy σ_t is winning so the sequence $\{z_n^t\}$ converges to $y(t)$ for any $t \in S$; thus $x_n(t) \to y(t)$ for any $t \in S$. Furthermore, it follows from (1) that $x_n(t) = a_t = y(t)$ for any $t \in T \backslash S$ so $x_n(t) \to y(t)$ for any $t \in T$ and hence the sequence $\{x_n\}$ converges to y. Thus the strategy σ is winning; so we proved that Σ is a W-space.

Therefore every Σ-product of the real lines is a W-space; applying Problem 183 we can conclude that

(5) every Corson compactum is a W-space.

The following useful fact will help us to show that every closed subset of a Corson compact space is a W-set.

Fact 1. If $f : Y \to Z$ is a perfect map and $z \in Z$ then $\{z\}$ is a W-set in Z if and only if $f^{-1}(z)$ is a W-set in Y.

Proof. Suppose that $\{z\}$ is a W-set in Z and fix a winning strategy s for the player OP in the W-game in Z at the point z. Let $V_1 = s(\emptyset)$ and $U_1 = f^{-1}(V_0)$; then U_1 is an open neighbourhood of the set $F = f^{-1}(z)$. Suppose that $n \in \mathbb{N}$ and $\{U_1, y_1, \ldots, U_{n-1}, y_{n-1}, U_n\}$ is an initial segment of a play in the W-game on Y at F where our future strategy σ of the player OP is defined and applied, i.e., $\sigma(U_1, y_1, \ldots, U_i, y_i) = U_{i+1}$ for any $i < n$ and, besides, we have an initial segment $\{V_1, z_1, \ldots, V_{n-1}, z_{n-1}, V_n\}$ of a play of the W-game on Y at y in which OP applies the strategy s while $U_i = f^{-1}(V_i)$ and $z_i = f(y_i)$ for any $i < n$.

If the player PT makes a move $y_n \in U_n$ then $z_n = f(y_n) \in V_n$ and hence the strategy s can be applied; let $V_{n+1} = s(V_1, z_1, \ldots, V_n, z_n)$ and $U_{n+1} = f^{-1}(V_{n+1})$. Then the formula $\sigma(U_1, y_1, \ldots, U_n, y_n) = U_{n+1}$ completes the inductive definition of the strategy σ.

To see that σ is winning suppose that $\{(U_i, y_i) : i \in \mathbb{N}\}$ is a play where OP applies the strategy σ. Then there is a play $\{(V_i, z_i) : i \in \mathbb{N}\}$ in the W-game on Z at the point z in which OP applies the strategy s while $f(y_i) = z_i$ and $U_i = f^{-1}(V_i)$ for any $i \in \mathbb{N}$. Since the strategy s is winning, the sequence $\{z_i\}$ converges to z. Now, if $O \in \tau(F, Y)$ then $W = Z \backslash f(Y \backslash O)$ is an open neighbourhood of z such that $f^{-1}(W) \subset O$. There exists $m \in \mathbb{N}$ such that $z_i \in W$ for all $i \geq m$; it is immediate that $y_i \in f^{-1}(W) \subset O$ for all $i \geq m$ and hence the sequence $\{y_i\}$ converges to F. This proves that F is a W-set and therefore we established sufficiency.

To prove necessity, assume that F is a W-set. Given an open set $U \subset Y$ the set $V = Z \backslash f(Y \backslash U)$ is an open subset of Z (maybe empty) such that $f^{-1}(V) \subset U$; let $f^{\#}(U) = V$.

Fix a winning strategy σ for the player OP in the W-game on the space Y at the set F and let $U_1 = \sigma(\emptyset)$; then $V_1 = f^{\#}(U_1) \in \tau(z, Z)$ so we can let $s(\emptyset) = V_1$. Suppose that $n \in \mathbb{N}$ and $\{V_1, z_1, \ldots, V_{n-1}, z_{n-1}, V_n\}$ is an initial segment of a play in the W-game on Z at z where our future strategy s of the player OP is defined and applied, i.e., $s(V_1, z_1, \ldots, V_i, z_i) = V_{i+1}$ for any $i < n$ and, besides, we have an initial segment $\{U_1, y_1, \ldots, U_{n-1}, y_{n-1}, U_n\}$ in a play of the W-game on Y at F in which OP applies the strategy σ while $V_i = f^{\#}(U_i)$ and $z_i = f(y_i)$ for any $i < n$.

If the player PT makes a move $z_n \in V_n$ then take any point $y_n \in f^{-1}(z_n)$; since $f^{-1}(z_n) \subset f^{-1}(V_n) \subset U_n$, we have $y_n \in U_n$ and therefore the strategy σ is applicable; let $U_{n+1} = \sigma(U_1, y_1, \ldots, U_n, y_n)$ and $V_{n+1} = f^{\#}(U_{n+1})$. Then V_{n+1} is an open neighbourhood of z in Z; so the formula $\sigma(V_1, z_1, \ldots, V_n, z_n) = V_{n+1}$ completes our inductive definition of the strategy s.

To see that s is winning suppose that $\{(V_i, z_i) : i \in \mathbb{N}\}$ is a play where OP applies the strategy s. Then there is a play $\{(U_i, y_i) : i \in \mathbb{N}\}$ in the W-game on Y at the set F in which OP applies the strategy σ while $f(y_i) = z_i$ for any $i \in \mathbb{N}$. Since the strategy σ is winning, the sequence $\{y_i\}$ converges to F. An immediate consequence of continuity of f is that the sequence $\{z_i\} = \{f(y_i)\}$ converges to z. This proves that $\{z\}$ is a W-set and therefore we established necessity. Fact 1 is proved.

Finally, take any Corson compact space K and fix a non-empty closed set $F \subset K$. Let us consider the map $f : K \to N$ which is obtained by collapsing F to a point. Recall that $N = \{x_F\} \cup (K \backslash F)$, the base of the topology of N at a point $x \in K \backslash F$ is given by the family $\tau(x, K \backslash F)$ and the local base at x_F is the family $\{\{x_F\} \cup (U \backslash F) : U \in \tau(F, K)\}$. The map f is defined by $f(x) = x_F$ for any $x \in F$ and $f(x) = x$ whenever $x \in K \backslash F$.

Then N is a compact Hausforff space and the map f is surjective and continuous (see Fact 2 of T.245). The space N is Corson compact by Problem 151 and therefore $\{x_F\}$ is a W-set in N by (5). Since $F = f^{-1}(x_F)$ and f is perfect, we can apply Fact 1 to conclude that F is a W-set in K and hence our solution is complete.

U.185. *Prove that, if X is a compact space of countable tightness, then a non-empty closed $H \subset X$ is a W-set if and only if $X \backslash H$ is metalindelöf.*

Solution. We will deduce sufficiency from a more general statement.

Fact 1. Suppose that Z is a compact space, F is non-empty and closed in Z and, additionally, there is a point-countable open cover $\mathcal{U}$ of the set $Z \backslash F$ such that $\overline{U} \subset Z \backslash F$ for any $U \in \mathcal{U}$. Then F is a W-set in Z.

Proof. Given any $x \in Z \backslash F$ let $\mathcal{U}_x = \{U_n^x : n \in \mathbb{N}\}$ be a subfamily of $\mathcal{U}$ such that $x \in U \in \mathcal{U}$ implies $U \in \mathcal{U}_x$; we will also need the family $\mathcal{U}_x^n = \{U_k^x : k \leq n\}$ for any $n \in \mathbb{N}$.

To define inductively a winning strategy σ for the player OP let $\sigma(\emptyset) = Z$ and assume that $n \in \mathbb{N}$ and we have an initial segment $\{U_1, x_1, \ldots, U_{n-1} x_{n-1}, U_n\}$ of the W-game on Z at the set F in which our future strategy σ is defined and applied, i.e., $U_1 = \sigma(\emptyset)$ and $U_{i+1} = \sigma(U_1, x_1, \ldots, U_i, x_i)$ for any $i < n$ and, besides

(1) for any $i < n$ we have $U_{i+1} \cap \overline{\bigcup \mathcal{U}_{x_j}^i} = \emptyset$ for any $j \leq i$.

If the player PT makes a move $x_n \in U_n$ then the closed set $P_i = \overline{\bigcup \mathcal{U}_{x_i}^n}$ does meet F for any $i \leq n$ so the set $U_{n+1} = U_n \backslash (P_1 \cup \ldots \cup P_n)$ is an open neighbourhood of the set F; let $\sigma(U_1, x_1, \ldots, U_n, x_n) = U_{n+1}$. It is evident that (1) still holds if we substitute $n + 1$ in place of n; so our inductive procedure defines a strategy of the player OP in the W-game on Z at F with the property (1).

To see that σ is winning suppose that $\{(U_i, x_i) : i \in \mathbb{N}\}$ is a play in which OP applies the strategy σ. Given any point $y \in Z \backslash F$ fix $V \in \mathcal{U}$ such that $y \in V$. If there is $k \in \mathbb{N}$ such that $x_k \in V$ then $V = U_m^{x_k}$ and hence $V \in \mathcal{U}_{x_k}^m$ for some $m \in \mathbb{N}$. We have $x_n \in U_n$ for every $n \in \mathbb{N}$; it follows from the property (1) that $U_n \cap V = \emptyset$ for any $n \geq m+k+1$ and therefore $V \cap \{x_i : i \in \mathbb{N}\} \subset \{x_1, \ldots, x_{m+k}\}$. Thus, for any $y \in Z \backslash F$ there is $V \in \tau(y, X)$ such that $V \cap \{x_i : i \in \mathbb{N}\}$ is finite, and hence $D = \{x_i : i \in \mathbb{N}\}$ is closed and discrete in $Z \backslash F$. Now, if $O \in \tau(F, Z)$ then $K = Z \backslash O$ is a compact subspace of $Z \backslash F$; so $D \cap K = D \backslash O$ is finite which shows that D converges to F and hence the strategy σ is winning, i.e., F is a W-set in Z. Fact 1 is proved.

Fact 2. Suppose that Z is a set and $\mathcal{A}$ is a family of subsets of Z. Assume also that we have a map $\varphi : \exp \mathcal{A} \to \exp \mathcal{A}$ such that $|\varphi(\mathcal{B})| \leq \max\{\omega, |\mathcal{B}|\}$ for any $\mathcal{B} \subset \mathcal{A}$. A family $\mathcal{B} \subset \mathcal{A}$ will be called φ-closed if $\varphi(\mathcal{B}) \subset \mathcal{B}$. Suppose, additionally, that we have a property $\mathcal{P}$ such that

(a) any countable φ-closed family $\mathcal{B} \subset \mathcal{A}$ has $\mathcal{P}$;
(b) for any ordinal κ, if a family $\mathcal{B}_\alpha \subset \mathcal{A}$ is φ-closed and has $\mathcal{P}$ for any $\alpha < \kappa$ and, besides, $\alpha < \beta < \kappa$ implies $\mathcal{B}_\alpha \subset \mathcal{B}_\beta$, then the family $\mathcal{B} = \bigcup\{\mathcal{B}_\alpha : \alpha < \kappa\}$ has $\mathcal{P}$.

Then the family $\mathcal{A}$ has the property $\mathcal{P}$.

Proof. We will prove this by induction on $\lambda = |\mathcal{A}|$. If $\lambda = \omega$ then $\mathcal{A}$ has $\mathcal{P}$ by (a). Now assume that $|\mathcal{A}| = \kappa$ and our Fact is proved for all families of cardinality less than κ. Write $\mathcal{A}$ as $\{A_\beta : \beta < \kappa\}$; we will construct an increasing κ-sequence $\{\mathcal{B}_\beta : \beta < \kappa\}$ of subfamilies of $\mathcal{A}$ with the following properties:

(2) the family $\mathcal{B}_\beta$ is φ-closed and $|\mathcal{B}_\beta| < \kappa$ for any $\beta < \kappa$;
(3) $\{A_\gamma : \gamma < \beta\} \subset \mathcal{B}_\beta$ for every $\beta < \kappa$.

We can start with $\mathcal{B}_0 = \emptyset$. If α is a limit ordinal and we have $\mathcal{B}_\beta$ for any $\beta < \alpha$ then let $\mathcal{B}_\alpha = \bigcup\{\mathcal{B}_\beta : \beta < \alpha\}$. It is clear that the properties (2) and (3) are fulfilled for every $\beta \leq \alpha$.

Now, if $\alpha = \alpha' + 1$ then let $C_0 = \mathcal{B}_{\alpha'} \cup \{A_{\alpha'}\}$ and $C_{n+1} = C_n \cup \varphi(C_n)$ for any $n \in \omega$. It is immediate that $\mathcal{B}_\alpha = \bigcup_{n \in \omega} C_n$ is φ-closed, has cardinality $< \kappa$ and $\{A_\beta : \beta < \alpha\} \subset \mathcal{B}_\alpha$. Therefore our inductive procedure gives us an increasing κ-sequence $\mathcal{S} = \{\mathcal{B}_\alpha : \alpha < \kappa\}$ with the properties (2) and (3). Since $A_\alpha \in \mathcal{B}_{\alpha+1}$ for any $\alpha < \kappa$ by (3), we have $\bigcup \mathcal{S} = \mathcal{A}$.

Furthermore, the induction hypothesis is applicable to every $\mathcal{B}_\alpha$ and the map $\varphi|\exp(\mathcal{B}_\alpha)$ because $\mathcal{B}_\alpha$ is φ-closed and $|\mathcal{B}_\alpha| < \kappa$ by (2). Thus every $\mathcal{B}_\alpha$ has $\mathcal{P}$; the collection $\mathcal{S}$ being increasing, the family $\mathcal{A} = \bigcup \mathcal{S}$ has $\mathcal{P}$ by (b); so Fact 2 is proved.

Returning to our solution suppose that $X \backslash H$ is metalindelöf. Every $x \in X \backslash H$ has an open neighbourhood O_x such that $\overline{O}_x \subset X \backslash H$; there is a point-countable open refinement $\mathcal{U}$ of the open cover $\{O_x : x \in X \backslash H\}$ of the space $X \backslash H$. It is clear that $\overline{U} \subset X \backslash H$ for any $U \in \mathcal{U}$; so H is a W-set by Fact 1. This proves sufficiency.

Now assume that H is a W-set and fix a winning strategy σ for the player OP in the W-game on X at the set H. The following observations will be useful to simplify our notation.

First note that the values of the strategy σ only matter in the plays where it is applied. Thus, when we calculate σ for an initial segment $\{U_1, x_1, \ldots, U_n, x_n\}$, the set $\sigma(U_1, x_1, \ldots, U_n, x_n)$ actually depends only on the points $\{x_1, \ldots, x_n\}$ because $U_1 = \sigma(\emptyset)$, $U_2 = \sigma(\sigma(\emptyset), x_1)$, $U_3 = \sigma(\sigma(\emptyset), x_1, \sigma(\sigma(\emptyset), x_1), x_2)$ and so on by trivial induction.

So our first simplification will consist in writing $\sigma(x_1, \ldots, x_n)$ instead of the expression $\sigma(U_1, x_1, \ldots, U_n, x_n)$. Formally, however, the set $\sigma(x_1, \ldots, x_n)$ is defined only for those n-tuples $(x_1, \ldots, x_n)$ for which $x_1 \in \sigma(\emptyset)$ and $x_{i+1} \in \sigma(x_1, \ldots, x_i)$ for every $i < n$. But it is easy to see that we can define $\sigma(x_1, \ldots, x_n)$ arbitrarily, say $\sigma(x_1, \ldots, x_n) = X$ for all other n-tuples $(x_1, \ldots, x_n)$; the resulting strategy will still be winning.

The next step is to define a map σ' for any finite subset F of the space X. Let $\sigma'(\emptyset) = \sigma(\emptyset)$; if $|F| = n > 0$ then define the value of σ' at F by the formula $\sigma'(F) = \bigcap \{\sigma(x_1, \ldots, x_n) : x_i \in F \text{ for any } i \leq n\}$. It is an easy exercise to see that the function σ' is still a winning strategy in the sense that for any sequence $S = \{x_i : i \in \mathbb{N}\}$ such that $x_1 \in \sigma(\emptyset)$ and $x_{n+1} \in \sigma(x_1, \ldots, x_n)$ for any $n \in \mathbb{N}$ the sequence S converges to H. Our next simplification is to consider that $\sigma = \sigma'$, i.e., $\sigma(F)$ is defined for all finite sets $F \subset X$ and $\sigma(F)$ does not depend on enumeration of F. Call $S = \{x_i : i \in \mathbb{N}\}$ a σ-sequence if $x_1 \in \sigma(\emptyset)$ and $x_{n+1} \in \sigma(x_1, \ldots, x_n)$ for any $n \in \mathbb{N}$. We have observed already that any σ-sequence converges to H.

To prove that $X \backslash H$ is metalindelöf take an open cover $\mathcal{U}$ of the space $X \backslash H$. We can assume, without loss of generality, that $\overline{U} \cap H = \emptyset$ for any $U \in \mathcal{U}$. For any finite family $\mathcal{U}' \subset \mathcal{U}$ fix a finite $\mathcal{O}(\mathcal{U}') \subset \mathcal{U}$ such that $\bigcup\{\overline{U} : U \in \mathcal{U}'\} \subset \bigcup \mathcal{O}(\mathcal{U}')$; if $\mathcal{V} \subset \mathcal{U}$ then let $\mathcal{B}(\mathcal{V}) = \bigcup\{\mathcal{O}(\mathcal{U}') : \mathcal{U}' \text{ is a finite subfamily of } \mathcal{U}\}$.

For an arbitrary $\mathcal{V} \subset \mathcal{U}$ let $\mathcal{V}^*$ be the minimal family of subsets of X such that $\mathcal{V} \subset \mathcal{V}^*$ and $A, B \in \mathcal{V}^*$ implies $A \cup B \in \mathcal{V}^*$, $A \cap B \in \mathcal{V}^*$ and $A \backslash B \in \mathcal{V}^*$. It is an easy exercise that $|\mathcal{V}^*| \leq \max\{\omega, |\mathcal{V}|\}$ for any $\mathcal{V} \subset \mathcal{U}$; observe also that the elements of $\mathcal{V}^*$ are not necessarily open and $\mathcal{V}_0 \subset \mathcal{V}_1 \subset \mathcal{U}$ implies $\mathcal{V}_0^* \subset \mathcal{V}_1^*$. For any non-empty $A \in \mathcal{U}^*$ choose a point $y(A) \in A$ and let $Y(\mathcal{V}) = \{y(A) : A \in \mathcal{V}^* \backslash \{\emptyset\}\}$ for any $\mathcal{V} \subset \mathcal{U}$.

Given $\mathcal{V} \subset \mathcal{U}$ and a finite set $F \subset Y(\mathcal{V})$ the set $X \backslash \sigma(F)$ is compact; since it is contained in $X \backslash H$, we can choose a finite family $\mathcal{Q}(F) \subset \mathcal{U}$ which covers $X \backslash \sigma(F)$. Let $\mathcal{A}(\mathcal{V}) = \bigcup\{\mathcal{Q}(F) : F \text{ is a finite subset of } Y(\mathcal{V})\}$.

For an arbitrary family $\mathcal{V} \subset \mathcal{U}$ let $\varphi(\mathcal{V}) = \mathcal{A}(\mathcal{V}) \cup \mathcal{B}(\mathcal{V})$; this gives us a map $\varphi : \exp \mathcal{U} \to \exp \mathcal{U}$. It is straightforward that $|\varphi(\mathcal{V})| \leq \max\{\omega, |\mathcal{V}|\}$ for any $\mathcal{V} \subset \mathcal{U}$. Say that a family $\mathcal{V} \subset \mathcal{U}$ has the property $\mathcal{P}$ if the cover $\mathcal{V}$ of the space $\bigcup \mathcal{V}$ has a point-countable open refinement (from now on we will abbreviate this phrase saying simply that $\mathcal{V}$ has a point-countable open refinement). It is evident that any countable $\mathcal{V} \subset \mathcal{U}$ has $\mathcal{P}$ so it suffices, by Fact 2, to prove that if κ is an ordinal and $\{\mathcal{V}_\alpha : \alpha < \kappa\}$ is an increasing κ-sequence of subfamilies of $\mathcal{U}$ with the property $\mathcal{P}$ then the family $\mathcal{V} = \bigcup\{\mathcal{V}_\alpha : \alpha < \kappa\}$ also has $\mathcal{P}$. Let $\mathcal{W}_\alpha$ be a point-countable refinement of the family $\mathcal{V}_\alpha$ for any $\alpha < \kappa$.

If the ordinal κ is countably cofinal then $\mathcal{V}$ is a countable union of families with the property $\mathcal{P}$; so it is an easy exercise that $\mathcal{V}$ also has $\mathcal{P}$. Thus we can assume, without loss of generality, that the cofinality of κ is uncountable. The following property is crucial for our proof:

(4) if $\mathcal{E} \subset \mathcal{U}$ is a φ-closed family then $\overline{Y(\mathcal{E})} \subset \bigcup \mathcal{E} \cup H$ and $\overline{\bigcup \mathcal{E}'} \subset \bigcup \mathcal{E}$ for any finite $\mathcal{E}' \subset \mathcal{E}$.

Assume first that $\mathcal{E}' \subset \mathcal{E}$ is finite. Since $\mathcal{E}$ is φ-closed, we have $\mathcal{O}(\mathcal{E}') \subset \mathcal{E}$ and therefore $\overline{\bigcup \mathcal{E}'} \subset \bigcup \mathcal{O}(\mathcal{E}') \subset \bigcup \mathcal{E}$.

To prove the first part of (4) assume towards a contradiction that, for the set $P = Y(\mathcal{E})$, there is a point $z \in \overline{P}\backslash(\bigcup \mathcal{E} \cup H)$; fix $W \in \tau(z, X)$ such that $\overline{W} \cap H = \emptyset$. If $U_1 = \sigma(\emptyset)$ then $X\backslash U_1 \subset \bigcup \mathcal{Q}(\emptyset)$ which, together with $\mathcal{Q}(\emptyset) \subset \mathcal{E}$ implies that $X\backslash U_1 \subset \bigcup \mathcal{E}$ and therefore $z \in U_1$. It follows from $z \in \overline{P}$ that we can choose $x_1 \in P \cap U_1 \cap W$. Suppose that we have an initial segment $\{U_1, x_1, \ldots, U_n, x_n\}$ of a play in the W-game on X at H in which OP applies the strategy σ and $x_i \in P \cap W$ for any $i \leq n$.

Since the family $\mathcal{E}$ is φ-closed, for the set $F = \{x_1, \ldots, x_n\} \subset P$ we have $X\backslash \sigma(F) \subset \bigcup \mathcal{Q}(F)$ and $\mathcal{Q}(F) \subset \mathcal{E}$ which implies that for the set $U_{n+1} = \sigma(F)$ we have $X\backslash U_{n+1} \subset \bigcup \mathcal{E}$ and therefore $z \in U_{n+1}$; so we can find $x_{n+1} \in P \cap U_{n+1} \cap W$. Thus our inductive procedure gives us a σ-sequence $S = \{x_n : n \in \mathbb{N}\} \subset W$; however, $X\backslash \overline{W}$ is an open neighbourhood of H which contains no elements of S. This contradiction with the fact that σ is winning strategy, shows that the property (4) is true.

For any $\alpha < \kappa$ let $\mathcal{N}_\alpha = \bigcup\{\mathcal{V}_\beta : \beta < \alpha\}$, $\mathcal{V}'_\alpha = \mathcal{V}_\alpha \backslash \mathcal{N}_\alpha$ and $Z_\alpha = Y(\mathcal{N}_\alpha)$. Fix $\alpha < \kappa$, an element $V \in \mathcal{V}'_\alpha$ and observe that $\mathcal{N}_\alpha$ is a φ-closed family (it is an easy exercise that any union of φ-closed families is φ-closed). The property (4) implies that $\overline{Z_\alpha} \subset \bigcup \mathcal{N}_\alpha \cup H$; so $K = \overline{V} \cap \overline{Z_\alpha}$ is a compact subset of $O_\alpha = \bigcup \mathcal{N}_\alpha$. The family $\mathcal{N}_\alpha$ being φ-closed, we can find a finite $\mathcal{N}' \subset \mathcal{N}_\alpha$ such that $K \subset \bigcup \mathcal{N}'$; let $O_V = V\backslash(\bigcup \mathcal{N}')$. It is clear that $O_V \in (\mathcal{V}'_\alpha)^*$ and $\overline{O}_V \cap \overline{Z}_\alpha = \emptyset$ for any $V \in \mathcal{V}'_\alpha$. Let $\mathcal{M}_\alpha = \{O_V : V \in \mathcal{V}'_\alpha\}$; if $\beta < \alpha$ then $Z_\beta \subset Z_\alpha$; so we have

(5) $\overline{A} \cap \overline{Z}_\beta = \emptyset$ for any $A \in \mathcal{M}_\alpha$ and $\beta < \alpha$.

Furthermore, $O_V \supset V\backslash O_\alpha$ for any $V \in \mathcal{V}'_\alpha$ and hence $\bigcup\{\mathcal{M}_\alpha : \alpha < \kappa\}$ is a cover of $X\backslash H$. It turns out that

(6) the family $\{\bigcup \mathcal{M}_\alpha : \alpha < \kappa\}$ is point-countable.

If (6) is not true then there exists an increasing ω_1-sequence $\{\beta(\alpha) : \alpha < \omega_1\}$ of ordinals such that, for any $\alpha < \omega_1$, we can choose $A_\alpha \in \mathcal{M}_{\beta(\alpha)}$ in such a way that $A = \bigcap\{A_\alpha : \alpha < \omega_1\} \neq \emptyset$; take a point $z \in A$ and let $\gamma = \sup\{\beta(\alpha) : \alpha < \omega_1\}$.

The family $\mathcal{C} = \{A_\alpha \cap Z_\gamma : \alpha < \omega_1\}$ is centered. Indeed, if $\alpha_1 < \ldots < \alpha_n < \omega_1$ then $Q = A_{\alpha_1} \cap \ldots \cap A_{\alpha_n}$ is a non-empty element of $(\mathcal{V}_{\beta(\alpha_n)})^*$ which shows that $y(Q) \in Z_{\beta(\alpha_n)} \subset Z_\gamma$ and hence $y(Q) \in Q \cap Z_\gamma$.

Thus $\mathcal{C}$ is centered and hence $R = \bigcap\{\overline{A_\alpha \cap Z_\gamma} : \alpha < \omega_1\} \neq \emptyset$; fix a point $y \in R$. Since $t(X) \leq \omega$ and $Z_\gamma = \bigcup\{Z_{\beta(\alpha)} : \alpha < \omega_1\}$, we have $\overline{Z}_\gamma = \bigcup\{\overline{Z}_{\beta(\alpha)} : \alpha < \omega_1\}$ which implies that there is $\alpha_0 < \omega_1$ for which $y \in \overline{Z}_{\beta(\alpha_0)}$. However, for $\alpha = \alpha_0 + 1$ we have $y \in \overline{A}_\alpha$ and hence $\overline{A}_\alpha \cap \overline{Z}_{\beta(\alpha_0)} \neq \emptyset$ which is a contradiction with (5); so the property (6) is proved.

Finally, recall that every $\mathcal{V}_\alpha$ has a point-countable open refinement $\mathcal{W}_\alpha$. Let $\mathcal{W}'_\alpha = \{W \cap \text{Int}(\bigcup \mathcal{M}_\alpha) : W \in \mathcal{W}_\alpha\}$ for any $\alpha < \kappa$. It is clear that $\mathcal{W}'_\alpha$ is a family of open subsets of X. We claim that $\mathcal{W} = \bigcup\{\mathcal{W}'_\alpha : \alpha < \kappa\}$ is a point-countable refinement of $\mathcal{V}$.

Since every $\mathcal{W}_\alpha$ is a refinement of $\mathcal{V}_\alpha$, the family $\mathcal{W}$ is inscribed in $\mathcal{V}$, i.e., for any $W \in \mathcal{W}'_\alpha$ there is $V \in \mathcal{V}$ such that $W \subset V$. To see that $\mathcal{W}$ covers $\bigcup \mathcal{V}$ take any $x \in \bigcup \mathcal{V}$ and let $\alpha = \min\{\beta < \kappa : x \in \bigcup \mathcal{V}_\beta\}$. Then there is $V \in \mathcal{V}'_\alpha$ with $x \in V$. Observe that $O_V = V \backslash (\bigcup \mathcal{F})$ for some finite $\mathcal{F} \subset \bigcup\{\mathcal{V}_\beta : \beta < \alpha\}$ and therefore $x \notin \overline{\bigcup \mathcal{F}}$ by (4). Consequently, $x \in V \backslash \overline{\bigcup \mathcal{F}} \subset \text{Int}(O_V) \subset \text{Int}(\bigcup \mathcal{M}_\alpha)$. Now if $W \in \mathcal{W}_\alpha$ and $x \in W$ then $x \in W \cap \text{Int}(\bigcup \mathcal{M}_\alpha) \in \mathcal{W}'_\alpha$ and hence $\mathcal{W}$ covers the set $\bigcup \mathcal{V}$, i.e., $\mathcal{W}$ is a refinement of $\mathcal{V}$.

To finally see that $\mathcal{W}$ is point-countable take any point $x \in \bigcup \mathcal{W}$. The set $G = \{\alpha < \kappa : x \in \bigcup \mathcal{M}_\alpha\}$ is countable by (6) and it is immediate that $x \notin W$ whenever $W \in \mathcal{W}'_\alpha$ and $\alpha \notin G$. Therefore x can only belong to the elements of the family $\mathcal{W}' = \bigcup\{\mathcal{W}'_\alpha : \alpha \in G\}$ which is point-countable being a countable union of point-countable families. Thus x can belong to at most countably many elements of $\mathcal{W}$ i.e., $\mathcal{W}$ is point-countable.

We proved that our property $\mathcal{P}$ satisfies all premises of Fact 2; so the family $\mathcal{U}$ has the property $\mathcal{P}$, i.e., there exists a point-countable open refinement of $\mathcal{U}$. Therefore $X \backslash H$ is metalindelöf; this settles necessity and makes our solution complete.

U.186. *Let X be a compact scattered space. Prove that a non-empty closed $H \subset X$ is a W-set if and only if $X \backslash H$ is metacompact.*

Solution. If K is a scattered compact space then we denote by $I(K)$ the set of isolated points of K. Given a scattered compact space Z let $F_0(Z) = Z$; if α is a limit ordinal and we have constructed a decreasing α-sequence $\{F_\beta(Z) : \beta < \alpha\}$ then let $F_\alpha(Z) = \bigcap\{F_\beta(Z) : \beta < \alpha\}$. If $\alpha = \beta + 1$ and we have the set $F_\beta(Z)$ then let $F_\alpha(Z) = F_\beta(Z) \backslash I(F_\beta(Z))$. Observe that $F_\alpha(Z) \neq \emptyset$ implies that $F_{\alpha+1}(Z)$ is strictly smaller than $F_\alpha(Z)$; so if $\lambda = |Z|^+$ then $F_\lambda(Z) = \emptyset$.

If α is a limit ordinal and $F_\beta(Z) \neq \emptyset$ for any $\beta < \alpha$ then $F_\alpha(Z) \neq \emptyset$ because Z is compact and every $F_\beta(Z)$ is closed in Z. Therefore the *dispersion index* $di(Z) = \min\{\alpha < \lambda : F_{\alpha+1}(Z) = \emptyset\}$ is well defined. If $\alpha = di(Z)$ then $F_\alpha(Z)$ is finite being a compact discrete space; we will call the set $F_\alpha(Z)$ *the top level* of Z.

Suppose first that $X \setminus H$ is metacompact. Every $x \in X \setminus H$ has an open neighbourhood O_x such that $\overline{O}_x \subset X \setminus H$; there is a point-finite open refinement $\mathcal{U}$ of the open cover $\{O_x : x \in X \setminus H\}$ of the space $X \setminus H$. It is clear that $\overline{U} \subset X \setminus H$ for any $U \in \mathcal{U}$; so H is a W-set by Fact 1 of U.185. This proves sufficiency.

Now assume that H is a W-set and fix a winning strategy σ for the player OP in the W-game on X at the set H. The following observations will be useful to simplify our notation.

First note that the values of the strategy σ only matter in the plays where it is applied. Thus, when we calculate σ for an initial segment $\{U_1, x_1, \ldots, U_n, x_n\}$, the set $\sigma(U_1, x_1, \ldots, U_n, x_n)$ actually depends only on the points $\{x_1, \ldots, x_n\}$ because $U_1 = \sigma(\emptyset)$, $U_2 = \sigma(\sigma(\emptyset), x_1)$, $U_3 = \sigma(\sigma(\emptyset), x_1, \sigma(\sigma(\emptyset), x_1), x_2)$ and so on by trivial induction.

So our first simplification will consist in writing $\sigma(x_1, \ldots, x_n)$ instead of the expression $\sigma(U_1, x_1, \ldots, U_n, x_n)$. Formally, however, the set $\sigma(x_1, \ldots, x_n)$ is defined only for those n-tuples $(x_1, \ldots, x_n)$ for which $x_1 \in \sigma(\emptyset)$ and $x_{i+1} \in \sigma(x_1, \ldots, x_i)$ for every $i < n$. But it is easy to see that we can define $\sigma(x_1, \ldots, x_n)$ arbitrarily, say $\sigma(x_1, \ldots, x_n) = X$ for all other n-tuples $(x_1, \ldots, x_n)$; the resulting strategy will still be winning.

The next step is to define a map σ' for any finite subset F of the space X. Let $\sigma'(\emptyset) = \sigma(\emptyset)$; if $|F| = n > 0$ then define the value of σ' at F by the formula $\sigma'(F) = \bigcap\{\sigma(x_1, \ldots, x_n) : x_i \in F$ for any $i \leq n\}$. It is an easy exercise to see that the function σ' is still a winning strategy in the sense that for any sequence $S = \{x_i : i \in \mathbb{N}\}$ such that $x_1 \in \sigma(\emptyset)$ and $x_{n+1} \in \sigma(x_1, \ldots, x_n)$ for any $n \in \mathbb{N}$ the sequence S converges to H. Our next simplification is to consider that $\sigma = \sigma'$, i.e., $\sigma(F)$ is defined for all finite sets $F \subset X$ and $\sigma(F)$ does not depend on enumeration of F. Call $S = \{x_i : i \in \mathbb{N}\}$ a σ-sequence if $x_1 \in \sigma(\emptyset)$ and $x_{n+1} \in \sigma(x_1, \ldots, x_n)$ for any $n \in \mathbb{N}$. We have observed already that any σ-sequence converges to H.

To prove that $X \setminus H$ is metacompact take an open cover $\mathcal{U}$ of the space $X \setminus H$. The space X is zero-dimensional; so we can assume, without loss of generality, that every $U \in \mathcal{U}$ is a clopen subset of X.

For an arbitrary $\mathcal{V} \subset \mathcal{U}$ let $\mathcal{V}^*$ be the minimal family of subsets of X such that $\mathcal{V} \subset \mathcal{V}^*$ and $A, B \in \mathcal{V}^*$ implies $A \cup B \in \mathcal{V}^*$, $A \cap B \in \mathcal{V}^*$ and $A \setminus B \in \mathcal{V}^*$. It is an easy exercise that $|\mathcal{V}^*| \leq \max\{\omega, |\mathcal{V}|\}$ for any $\mathcal{V} \subset \mathcal{U}$; observe also that the elements of $\mathcal{V}^*$ are clopen in X and $\mathcal{V}_0 \subset \mathcal{V}_1 \subset \mathcal{U}$ implies $\mathcal{V}_0^* \subset \mathcal{V}_1^*$. For any non-empty $A \in \mathcal{U}^*$ the set $L(A)$ is the top level of A; let $Y(\mathcal{V}) = \bigcup\{L(A) : A \in \mathcal{V}^* \setminus \{\emptyset\}\}$ for any $\mathcal{V} \subset \mathcal{U}$.

Given $\mathcal{V} \subset \mathcal{U}$ and a finite set $F \subset Y(\mathcal{V})$ the set $X \setminus \sigma(F)$ is compact; since it is contained in the set $X \setminus H$, we can choose a finite family $\mathcal{Q}(F) \subset \mathcal{U}$ which covers $X \setminus \sigma(F)$. Let $\varphi(\mathcal{V}) = \bigcup\{\mathcal{Q}(F) : F$ is a finite subset of $Y(\mathcal{V})\}$; this gives us a map $\varphi : \exp\mathcal{U} \to \exp\mathcal{U}$. It is clear that $|\varphi(\mathcal{V})| \leq \max\{\omega, |\mathcal{V}|\}$ for any $\mathcal{V} \subset \mathcal{U}$. Say that a family $\mathcal{V} \subset \mathcal{U}$ has the property $\mathcal{P}$ if the cover $\mathcal{V}$ of the space $\bigcup \mathcal{V}$ has a point-finite open refinement (from now on we will abbreviate this phrase saying simply that $\mathcal{V}$ has a point-finite open refinement).

If a family $\mathcal{V} = \{V_n : n \in \omega\} \subset \mathcal{U}$ is countable then letting $V'_0 = V_0$ and $V'_n = V_n \backslash (V_0 \cup \ldots \cup V_{n-1})$ for any $n \in \mathbb{N}$ we obtain a disjoint refinement $\{V'_n : n \in \omega\}$ of the family $\mathcal{V}$. Thus any countable $\mathcal{V} \subset \mathcal{U}$ has $\mathcal{P}$ so it suffices, by Fact 2 of U.185, to prove that if κ is an ordinal and $\{\mathcal{V}_\alpha : \alpha < \kappa\}$ is an increasing κ-sequence of subfamilies of $\mathcal{U}$ with the property $\mathcal{P}$ then the family $\mathcal{V} = \bigcup\{\mathcal{V}_\alpha : \alpha < \kappa\}$ also has $\mathcal{P}$. Let $\mathcal{W}_\alpha$ be a point-finite refinement of the family $\mathcal{V}_\alpha$ for any $\alpha < \kappa$.

If the ordinal κ is a successor then there is an ordinal β with $\kappa = \beta + 1$ and hence $\mathcal{V} = \mathcal{V}_\beta$ has a point-finite refinement. Thus we can assume, without loss of generality, that κ is a limit ordinal. The following property is crucial for our proof:

(1) if $\mathcal{E} \subset \mathcal{U}$ is a φ-closed family then $\overline{Y(\mathcal{E})} \subset \bigcup \mathcal{E} \cup H$.

To prove (1) assume towards a contradiction that, for the set $P = Y(\mathcal{E})$, there is a point $z \in \overline{P} \backslash (\bigcup \mathcal{E} \cup H)$; fix $W \in \tau(z, X)$ such that $\overline{W} \cap H = \emptyset$. If $U_1 = \sigma(\emptyset)$ then $X \backslash U_1 \subset \bigcup \mathcal{Q}(\emptyset)$ which, together with $\mathcal{Q}(\emptyset) \subset \mathcal{E}$ implies that $X \backslash U_1 \subset \bigcup \mathcal{E}$ and therefore $z \in U_1$. It follows from $z \in \overline{P}$ that we can choose $x_1 \in P \cap U_1 \cap W$. Suppose that we have an initial segment $\{U_1, x_1, \ldots, U_n, x_n\}$ of a play in the W-game on X at H in which OP applies the strategy σ and $x_i \in P \cap W$ for any $i \leq n$.

Since the family $\mathcal{E}$ is φ-closed, for the set $F = \{x_1, \ldots, x_n\} \subset P$ we have $X \backslash \sigma(F) \subset \bigcup \mathcal{Q}(F)$ and $\mathcal{Q}(F) \subset \mathcal{E}$ which implies that for the set $U_{n+1} = \sigma(F)$ we have $X \backslash U_{n+1} \subset \bigcup \mathcal{E}$ and therefore $z \in U_{n+1}$ so we can find $x_{n+1} \in P \cap U_{n+1} \cap W$. Thus our inductive procedure gives us a σ-sequence $S = \{x_n : n \in \mathbb{N}\} \subset W$; however, $X \backslash \overline{W}$ is an open neighbourhood of H which contains no elements of S. This contradiction with the fact that σ is winning strategy, shows that the property (1) is true.

For any $\alpha < \kappa$ let $\mathcal{N}_\alpha = \bigcup\{\mathcal{V}_\beta : \beta < \alpha\}$, $\mathcal{V}'_\alpha = \mathcal{V}_\alpha \backslash \mathcal{N}_\alpha$ and $Z_\alpha = Y(\mathcal{N}_\alpha)$. Fix $\alpha < \kappa$, an element $V \in \mathcal{V}'_\alpha$ and observe that $\mathcal{N}_\alpha$ is a φ-closed family (it is an easy exercise that any union of φ-closed families is φ-closed). The property (1) implies that $\overline{Z_\alpha} \subset \bigcup \mathcal{N}_\alpha \cup H$; so $K = \overline{V} \cap \overline{Z_\alpha}$ is a compact subset of $O_\alpha = \bigcup \mathcal{N}_\alpha$. The family $\mathcal{N}_\alpha$ being φ-closed, we can find a finite $\mathcal{N}' \subset \mathcal{N}_\alpha$ such that $K \subset \bigcup \mathcal{N}'$; let $O_V = V \backslash (\bigcup \mathcal{N}')$. It is clear that $O_V \in (\mathcal{V}'_\alpha)^*$ and $\overline{O}_V \cap \overline{Z_\alpha} = \emptyset$ for any $V \in \mathcal{V}'_\alpha$. Let $\mathcal{M}_\alpha = \{O_V : V \in \mathcal{V}'_\alpha\}$; if $\beta < \alpha$ then $Z_\beta \subset Z_\alpha$; so we have

(2) $\overline{A} \cap \overline{Z}_\beta = \emptyset$ for any $A \in \mathcal{M}_\alpha$ and $\beta < \alpha$.

Furthermore, $O_V \supset V \backslash O_\alpha$ for any $V \in \mathcal{V}'_\alpha$ and hence $\bigcup\{\mathcal{M}_\alpha : \alpha < \kappa\}$ is a cover of $X \backslash H$. It turns out that

(3) the family $\{\bigcup \mathcal{M}_\alpha : \alpha < \kappa\}$ is point-finite.

If (3) is not true then there exists an increasing sequence $\{\beta(n) : n \in \omega\} \subset \kappa$ of ordinals such that, for any $n < \omega$, we can choose $A_n \in \mathcal{M}_{\beta(n)}$ in such a way that $A = \bigcap\{A_n : n < \omega\} \neq \emptyset$; let $\gamma = \sup\{\beta(n) : n < \omega\}$.

If $B_n = \bigcap \{A_i : i \leq n\}$ for any $n \in \omega$ then every B_n is a scattered compact space; so the ordinal $\mu_n = di(B_n)$ is well defined. Let D_n be the top level of B_n for each $n \in \omega$. Since the sequence $\{B_n : n \in \omega\}$ is decreasing, we have $\mu_0 \geq \mu_1 \geq \dots$ which shows that there is $k \in \omega$ such that $\mu_i = \mu_k$ for any $i \geq k$.

Observe that $D_k \subset B_k \subset B_n$ for any $n \leq k$. If $n > k$ then $B_n \subset B_k$ and therefore $F_\alpha(B_k) \supset F_\alpha(B_n)$ for any ordinal α. In particular, $D_k = F_{\mu_k} \supset F_{\mu_k}(B_n) = D_n$. This proves that

(4) $D_k \cap B_n \neq \emptyset$ for any $n \in \omega$.

However, $B_k \in \mathcal{V}^*_{\beta(k)}$ and hence $D_k = L(B_k) \subset Z_{\beta(k)+1} \subset Z_{\beta(k+1)}$ which shows that $B_n \cap D_k \subset A_n \cap D_k \subset A_n \cap Z_{\beta(k+1)} = \emptyset$ for any $n > k + 1$ by (2). This contradiction with the property (4) completes the proof of (3).

Recall that every family $\mathcal{V}_\alpha$ has a point-finite open refinement $\mathcal{W}_\alpha$ and let $\mathcal{W}'_\alpha = \{W \cap (\bigcup \mathcal{M}_\alpha) : W \in \mathcal{W}_\alpha\}$ for any $\alpha < \kappa$. It is clear that $\mathcal{W}'_\alpha$ is a family of open subsets of X. We claim that $\mathcal{W} = \bigcup \{\mathcal{W}'_\alpha : \alpha < \kappa\}$ is a point-finite refinement of $\mathcal{V}$.

Since every $\mathcal{W}_\alpha$ is a refinement of $\mathcal{V}_\alpha$, the family $\mathcal{W}$ is inscribed in $\mathcal{V}$, i.e., for any $W \in \mathcal{W}'_\alpha$ there is $V \in \mathcal{V}$ such that $W \subset V$. To see that $\mathcal{W}$ covers $\bigcup \mathcal{V}$ take any $x \in \bigcup \mathcal{V}$ and let $\alpha = \min\{\beta < \kappa : x \in \bigcup \mathcal{V}_\beta\}$. Then there is $V \in \mathcal{V}'_\alpha$ with $x \in V$. Observe that $O_V = V \setminus (\bigcup \mathcal{F})$ for some finite $\mathcal{F} \subset \bigcup \{\mathcal{V}_\beta : \beta < \alpha\}$ and therefore $x \in O_V \subset \bigcup \mathcal{M}_\alpha$. Now if $W \in \mathcal{W}_\alpha$ and $x \in W$ then $x \in W \cap (\bigcup \mathcal{M}_\alpha) \in \mathcal{W}'_\alpha$ and hence $\mathcal{W}$ covers the set $\bigcup \mathcal{V}$, i.e., $\mathcal{W}$ is a refinement of $\mathcal{V}$.

To finally see that the family $\mathcal{W}$ is point-finite take any point $x \in \bigcup \mathcal{W}$. The set $G = \{\alpha < \kappa : x \in \bigcup \mathcal{M}_\alpha\}$ is finite by (3) and it is immediate that $x \notin W$ whenever $W \in \mathcal{W}'_\alpha$ and $\alpha \notin G$. Therefore x can only belong to the elements of the family $\mathcal{W}' = \bigcup \{\mathcal{W}'_\alpha : \alpha \in G\}$ which is point-finite being a finite union of point-finite families. Thus x can belong to at most finitely many elements of $\mathcal{W}$, i.e., $\mathcal{W}$ is point-finite.

We proved that our property $\mathcal{P}$ satisfies all premises of Fact 2 of U.185; so the family $\mathcal{U}$ has the property $\mathcal{P}$, i.e., there exists a point-finite open refinement of $\mathcal{U}$. Therefore $X \setminus H$ is metacompact; this settles necessity and makes our solution complete.

U.187. *(Yakovlev's theorem) Prove that any Corson compact space is hereditarily metalindelöf.*

Solution. Given a Corson compact space K take any $X \subset K$; to prove that X is metalindelöf take an arbitrary open cover $\mathcal{U}$ of the space X. For any $U \in \mathcal{U}$ fix a set $O_U \in \tau(K)$ such that $O_U \cap X = U$. The family $\mathcal{V} = \{O_U : U \in \mathcal{U}\}$ is an open cover of the set $G = \bigcup \mathcal{V}$; let $F = K \setminus G$. If $F = \emptyset$ then G is compact; if $F \neq \emptyset$ then it follows from Problem 184 that F is a W-set in K and hence $G = K \setminus F$ is metalindelöf by Problem 185 (recall that every Corson compact space has countable tightness by Problem 120). Therefore G is metalindelöf in all possible cases; let $\mathcal{V}'$ be a point-countable open refinement of $\mathcal{V}$. It is straightforward that, in the space X, the family $\{V \cap X : V \in \mathcal{V}'\}$ is a point-countable open refinement of $\mathcal{U}$ and hence X is metalindelöf.

U.188. *Prove that the following are equivalent for any compact space X:*

(i) *X is Corson compact;*
(ii) *every closed subset of $X \times X$ is a W-set in $X \times X$;*
(iii) *the diagonal $\Delta = \{(x,x) : x \in X\}$ of the space X is a W-set in $X \times X$;*
(iv) *the space $(X \times X) \backslash \Delta$ is metalindelöf;*
(v) *the space $X \times X$ is hereditarily metalindelöf.*

Solution. If X is Corson compact then so is $X \times X$ (see Problem 137) and hence every closed $F \subset X \times X$ is a W-set by Problem 184. Besides, $X \times X$ is hereditarily metalindelöf by Problem 187. Therefore (i)$\Longrightarrow$(ii) and (i)$\Longrightarrow$(v). The implications (ii)$\Longrightarrow$(iii) and (v)$\Longrightarrow$(iv) are trivial; so let us prove that (iii)$\Longrightarrow$(iv). Assuming that (iii) holds fix a point $x \in X$ and a winning strategy σ for the player OP in the W-game on $X \times X$ at the set Δ. If $U_1 = \sigma(\emptyset)$ then there is $V_1 \in \tau(x, X)$ such that $V_1 \times V_1 \subset U_1$; let $s(\emptyset) = V_1$.

Suppose that $n \in \mathbb{N}$ and we have an initial segment $\{V_1, x_1, \ldots, V_{n-1}, x_{n-1}, V_n\}$ of a play in the W-game on X at the point x in which our future strategy s is defined and applied, i.e., $V_{i+1} = s(V_1, x_1, \ldots, V_i, x_i)$ for any $i < n$ and there is an initial segment $\{U_1, z_1, \ldots, U_{n-1}, z_{n-1}, U_n\}$ of a play in the W-game on $X \times X$ at the set Δ where OP applies the strategy σ while

(1) $z_i = (x_i, x)$ for any $i < n$ and $V_i \times V_i \subset U_i$ for any $i \leq n$.

If the player PT makes a move $x_n \in V_n$ then $z_n = (x_n, x) \in V_n \times V_n \subset U_n$, so the strategy σ is applicable; let $U_{n+1} = \sigma(U_1, z_1, \ldots, U_n, z_n)$. Then we can choose $V_{n+1} \in \tau(x, X)$ such that $V_{n+1} \times V_{n+1} \subset U_{n+1}$; let $s(V_1, x_1, \ldots, V_n, x_n) = V_{n+1}$. It is immediate that (1) is fulfilled if n is replaced by $n + 1$; so our inductive procedure defines a strategy s on the space X at the point x.

To see that s is winning take a play $\{(V_n, x_n) : n \in \mathbb{N}\}$ in which s is applied. We have a play $\{(U_n, z_n) : n \in \mathbb{N}\}$ in which OP applies σ and (1) is satisfied for all $n \in \mathbb{N}$. Since σ is a winning strategy, $z_n \to \Delta$. Let $S = \{x_n : n \in \mathbb{N}\}$; given any $W \in \tau(x, X)$ assume that $S \backslash W$ is infinite. Then $D = \{(x_n, x) : x_n \notin W\}$ is an infinite subset of $S' = \{z_n : n \in \mathbb{N}\}$. It is clear that $\overline{D} \subset P = (X \backslash W) \times \{x\}$. However, P is closed in $X \times X$ and $P \cap \Delta = \emptyset$. Thus $(X \times X) \backslash P$ is an open neighbourhood of Δ outside of which we have an infinite subset D of the set S'. This contradiction with $S' \to \Delta$ shows that $S \backslash W$ is finite and hence $S \to x$. Therefore s is a winning strategy; the point $x \in X$ was chosen arbitrarily so we established that X is a W-space. Therefore $X \times X$ is a W-space by Problem 183; in particular, $t(X \times X) = \omega$ (see Problem 179) and hence Problem 185 can be applied to conclude that $(X \times X) \backslash \Delta$ is metalindelöf. This settles (iii)$\Longrightarrow$(iv); so to finish our proof it suffices to show that (iv)$\Longrightarrow$(i).

Fact 1. Given a space Z suppose that $\mathcal{U}$ is an open cover of Z such that $\overline{U}$ is Lindelöf for any $U \in \mathcal{U}$. Then $\mathcal{U}$ can be shrunk, i.e., for any $U \in \mathcal{U}$ there is a closed set $F_U \subset U$ such that $\{F_U : U \in \mathcal{U}\}$ is a cover of Z.

Proof. Call a subcollection $\mathcal{U}' \subset \mathcal{U}$ enveloping if $\overline{U} \subset \bigcup \mathcal{U}'$ for any $U \in \mathcal{U}$. Observe first that there is a set T such that

(2) $\mathcal{U} = \bigcup\{\mathcal{U}_t : t \in T\}$ where $\mathcal{U}_t \subset \mathcal{U}$ is countable and enveloping for any $t \in T$.

To prove (2), it suffices to show that every $U \in \mathcal{U}$ can be included in a countable enveloping subcollection of $\mathcal{U}$. So, let $\mathcal{U}_0 = \{U\}$; if a we have a countable $\mathcal{U}_n \subset \mathcal{U}$ then $L = \bigcup\{\overline{V} : V \in \mathcal{U}_n\}$ is Lindelöf so there is a countable $\mathcal{U}_{n+1} \subset \mathcal{U}_n$ such that $\mathcal{U}_n \subset \mathcal{U}_{n+1}$ and $L \subset \bigcup\mathcal{U}_{n+1}$. This gives us a sequence $\{\mathcal{U}_n : n \in \omega\}$ of subfamilies of $\mathcal{U}$; it is evident that $U \in \mathcal{U}' = \bigcup\{\mathcal{U}_n : n \in \omega\}$ and $\mathcal{U}'$ is enveloping; so (2) is proved.

Now, if a we are given the decomposition (2) of the family $\mathcal{U}$ then the set $L_t = \bigcup\mathcal{U}_t = \bigcup\{\overline{U} : U \in \mathcal{U}_t\}$ is Lindelöf and hence paracompact for any $t \in T$. Therefore the cover $\mathcal{U}_t$ can be shrunk in L_t by Fact 2 of S.226. For any $U \in \mathcal{U}_t$ there is $F_U^t \subset U$ such that the set F_U^t is closed in L_t and $\bigcup\{F_U^t : U \in \mathcal{U}_t\} = L_t$. Consequently, $F_U^t \subset \overline{U}$ is closed in $\overline{U}$ because $\overline{U} \subset L_t$; this proves that every F_U^t is closed in Z.

Choose a well order $<$ on the set T and let $s(U) = \min\{t \in T : U \in \mathcal{U}_t\}$ for any $U \in \mathcal{U}$; then $F_U = F_U^{s(U)}$ is closed in Z and contained in U. We claim that $\mathcal{F} = \{F_U : U \in \mathcal{U}\}$ is a shrinking of the cover $\mathcal{U}$. To see this take any $z \in Z$ and let $t_0 = \min\{t \in T : z \in L_t\}$. Since $\{F_U^{t_0} : U \in \mathcal{U}_{t_0}\}$ is a cover of L_{t_0}, there is $U \in \mathcal{U}_{t_0}$ such that $z \in F_U^{t_0}$; observe that $s(U) = t_0$ and hence $z \in F_U^{s(U)} = F_U$ which shows that $\mathcal{F}$ is a cover of Z and hence Fact 1 is proved.

Returning to our solution assume that $Y = (X \times X)\backslash\Delta$ is metalindelöf. Then there is a point-countable open cover $\mathcal{Q}$ of the set Y such that $\overline{U} \subset Y$ for any $U \in \mathcal{Q}$. This implies that $\overline{U} = \mathrm{cl}_Y(U)$ is a compact set; so Fact 1 is applicable to find a shrinking $\{F_U : U \in \mathcal{Q}\}$ of the family $\mathcal{Q}$. It is easy to see that every F_U is compact; so there is a finite family $\mathcal{O}_U$ of open subsets of $X \times X$ such that $F_U \subset \bigcup\mathcal{O}_U \subset \bigcup\overline{\mathcal{O}_U} \subset U$ and every $O \in \mathcal{O}_U$ is *standard*, i.e., $O = O_1 \times O_2$ for some σ-compact sets $O_1, O_2 \in \tau(X)$ such that $\overline{O}_1 \cap \overline{O}_2 = \emptyset$. It is an easy exercise that the family $\mathcal{O} = \{\overline{O} : O \in \mathcal{O}_U, U \in \mathcal{Q}\}$ is point-countable.

The map $\varphi : X \times X \to X \times X$ defined by the formula $\varphi(x, y) = (y, x)$ for any $x, y \in X$ is a homeomorphism such that $\varphi(Y) = Y$; so the family $\mathcal{O} \cup \{\varphi(O) : O \in \mathcal{O}\}$ is also point-countable. Choosing an indexation $\{U_t \times V_t : t \in T\}$ of the family $\mathcal{O}' = \{O : O \in \mathcal{O}_U, U \in \mathcal{Q}\} \cup \{\varphi(O) : O \in \mathcal{O}_U, U \in \mathcal{Q}\}$ we will have the following properties:

(3) U_t, V_t are open σ-compact subsets of X for any $t \in T$;
(4) $\overline{U}_t \cap \overline{V}_t = \emptyset$ for any $t \in T$;
(5) the family $\{\overline{U}_t \times \overline{V}_t : t \in T\}$ is point-countable;
(6) $\bigcup\mathcal{O}' = Y$ and $U \times V \in \mathcal{O}'$ implies $V \times U \in \mathcal{O}'$.

Let $\kappa = d(X)$ and choose a dense set $D = \{p_\alpha : \alpha < \kappa\}$ in the space X; let $X_\alpha = \overline{\{p_\beta : \beta < \alpha\}}$ for any $\alpha < \kappa$; call a family $\mathcal{F} \subset \exp X$ a *minimal cover of X_α* if $X_\alpha \subset \bigcup\mathcal{F}$ and $X_\alpha \not\subset \bigcup(\mathcal{F}\backslash\{F\})$ for any $F \in \mathcal{F}$. Given a finite $F \subset T$ we will often use the set $U_F = \bigcap\{U_t : t \in F\}$; for any $\alpha < \kappa$ let $\mathcal{U}_\alpha = \{U_F : F \subset T$ is finite and $\{\overline{V}_t : t \in F\}$ is a minimal cover of $X_\alpha\}$. It is easy to see that $\mathcal{U}_\alpha$ consists of σ-compact open subsets of X for any $\alpha < \kappa$. It turns out that

(7) the family $\mathcal{U} = \bigcup\{\mathcal{U}_\alpha : \alpha < \kappa\}$ is point-countable.

If we assume the contrary then there exists a family $\{F_\alpha : \alpha < \omega_1\}$ of finite subsets of T such that

(8) there is $n \in \mathbb{N}$ for which $|F_\alpha| = n$ for any $\alpha < \omega_1$;
(9) for every $\alpha < \omega_1$ there is $\beta(\alpha) < \kappa$ such that $U_{F_\alpha} \in \mathcal{U}_{\beta(\alpha)}$ and the ω_1-sequence $\{\beta(\alpha) : \alpha < \omega_1\}$ is non-decreasing;
(10) there a set $F \subset T$ such that $F_\alpha \cap F_\beta = F$ for any distinct $\alpha, \beta < \omega_1$;
(11) there is a point $x \in X$ with $x \in \bigcap\{U_{F_\alpha} : \alpha < \omega_1\}$.

The properties (8) and (10) can be guaranteed taking an uncountable family with the property (11) and passing to an appropriate uncountable subfamily applying the Δ-lemma (SFFS-038). Once we have (8) and (10) choose a function $\varphi : \omega_1 \to \kappa$ such that $U_{F_\alpha} \in \mathcal{U}_{\varphi(\alpha)}$ for any $\alpha < \omega_1$. Suppose first that there is an ordinal $\alpha_0 < \kappa$ for which the set $S = \varphi^{-1}(\alpha_0)$ is uncountable. Passing to $\{F_\alpha : \alpha \in S\}$ gives a set for which $\beta(\alpha)$ is the same ordinal for all $\alpha < \omega_1$; so the condition (9) is satisfied.

Now if $\varphi^{-1}(\alpha)$ is countable for any $\alpha < \kappa$ then use a trivial transfinite induction to get the relevant subfamily with the property (9).

Since we have $U_{F_\alpha} \neq U_{F_\beta}$ for distinct ordinals $\alpha, \beta < \omega_1$ we have $F_\alpha \neq F$ for all $\alpha < \omega_1$. The family $\mathcal{V}_0 = \{V_t : t \in F_0\}$ is a minimal cover of $X_{\beta(0)}$; so there is a point $y \in X_{\beta(0)} \backslash (\bigcup\{\overline{V}_t : t \in F\})$.

We have $\beta(0) \leq \beta(\alpha)$ for any $\alpha < \omega_1$; so $y \in \bigcup\{\overline{V}_t : t \in F_\alpha\}$ and hence there is $t_\alpha \in F_\alpha \backslash F$ for which $y \in \overline{V}_{t_\alpha}$. It follows from (10) that $t_\alpha \neq t_\beta$ if $\alpha < \beta$; since $(x, y) \in \bigcap\{U_{t_\alpha} \times \overline{V}_{t_\alpha} : 0 < \alpha < \omega_1\}$ we obtain a contradiction with the property (5). Thus the family $\mathcal{U}$ is point-countable and (7) is proved.

Our final step is to show that

(12) the family $\mathcal{U}$ is T_0-separating in X.

Take distinct points $x_1, x_2 \in X$ and assume first that there is $\alpha < \kappa$ such that $x_i \in X_\alpha$ while $z = x_{2-i} \notin X_\alpha$. For any $y \in X_\alpha$ there is $t_y \in T$ such that $(z, y) \in U_{t_y} \times V_{t_y}$. There exists a finite $P \subset X_\alpha$ for which $X_\alpha \subset \bigcup\{V_{t_y} : y \in P\}$; so the set $F' = \{t_y : y \in P\}$ is finite and $X_\alpha \subset \bigcup\{V_t : t \in F'\}$. Take $F \subset F'$ such that $\{\overline{V}_t : t \in F\}$ is a minimal cover of X_α; then $U_F \in \mathcal{U}$ while $z \in U_F$ and $U_F \cap X_\alpha = \emptyset$ which shows that U_F separates the points x_1 and x_2.

Therefore we can assume, without loss of generality, that there is $\alpha < \kappa$ such that $x_1, x_2 \in X_\alpha$ and $\{x_1, x_2\} \cap X_\beta = \emptyset$ for any $\beta < \alpha$. Let $\mathcal{V}_1 = \{V_t : x_1 \in U_t$ and $x_2 \notin U_t\}$ and $\mathcal{V}_2 = \{V_t : x_2 \in U_t$ and $x_1 \notin U_t\}$. Observe that $x_2 \in \bigcup \mathcal{V}_1$ and $x_1 \in \bigcup \mathcal{V}_2$.

The space $K = X \backslash \bigcup(\mathcal{V}_1 \cup \mathcal{V}_2)$ is compact; for any point $z \in K$ there is $t(z) \in T$ such that $x_1 \in U_{t(z)}$ and $z \in V_{t(z)}$. Since $V_{t(z)} \notin \mathcal{V}_1 \cup \mathcal{V}_2$, we must have $x_2 \in U_{t(z)}$. As a consequence, there is a finite $F \subset T$ such that $K \subset \bigcup\{V_t : t \in F\}$ and $\{x_1, x_2\} \subset U_F$. It follows from $\{x_1, x_2\} \subset X_\alpha$ that $U_F \cap \{p_\beta : \beta < \alpha\} \neq \emptyset$ and therefore the set $H = \bigcup\{\overline{V}_t : t \in F\}$ does not contain $\{p_\beta : \beta < \alpha\}$; let $\delta = \min\{\beta : p_\beta \notin H\}$.

The point p_δ has to belong to $X \backslash K = \bigcup (V_1 \cup V_2)$; so there is $s \in T$ for which $V_s \in V_1 \cup V_2$ and $p_\delta \in V_s$. Then the set U_s separates the points x_1 and x_2. Since $\{\overline{V}_s\} \cup \{\overline{V}_t : t \in F\}$ covers $X_{\delta+1}$, there is $F' \subset F$ such that the family $\{\overline{V}_s\} \cup \{\overline{V}_t : t \in F'\}$ is a minimal cover of $X_{\delta+1}$. As a consequence, $U = U_s \cap U_{F'} \in \mathcal{U}$ and U separates the points x_1 and x_2 because so does U_s while $\{x_1, x_2\} \subset U_t$ for any $t \in F'$. Thus the property (12) is proved.

It follows from (7) and (12) that $\mathcal{U}$ is a point-countable T_0-separating family of open F_σ-subsets of X; so X is Corson compact by Problem 118. This proves the implication (iv)$\Longrightarrow$(i) and makes our solution complete.

U.189. *Give an example of a compact W-space X such that some continuous image of X is not a W-space.*

Solution. Take a separable first countable non-metrizable compact space K, e.g., the two arrows space (see TFS-384). Then $X = K \times K$ is first countable; so it is a W-space (see Problem 179). Let $\Delta = \{(x, x) : x \in K\} \subset K \times K$ be the diagonal of the space K. Define a map $p : X \to Y$ by collapsing the set Δ to a point, i.e., $Y = \{a_\Delta\} \cup (X \backslash \Delta)$ and $\tau(Y) = \tau(X \backslash \Delta) \cup \{\{a_\Delta\} \cup (U \backslash \Delta) : U \in \tau(\Delta, X)\}$ while the map p is defined by $p(x) = a_\Delta$ for any $x \in \Delta$ and $p(x) = x$ whenever $x \in X \backslash \Delta$.

It was proved in Fact 2 of T.245 that the space Y is Tychonoff and the map p is continuous. By compactness of X, the space Y is also compact and the map p is perfect; besides, $\Delta = p^{-1}(a_\Delta)$. If $\{a_\Delta\}$ is a W-set in the space Y then Δ is a W-set in X by Fact 1 of U.184. Apply Problem 188 to conclude that the space K must be Corson compact and hence metrizable because any separable Corson compact space is metrizable by Problem 121. This contradiction shows that Y is not a W-space; so X is a compact W-space whose continuous image Y fails to be a W-space.

U.190. *Suppose that X is a compact space which embeds into a σ-product of second countable spaces. Prove that the space $X^2 \backslash \Delta$ is metacompact; here, as usual, $\Delta = \{(x, x) : x \in X\}$ is the diagonal of the space X.*

Solution. If Z is a space and $\mathcal{A} \subset \exp Z$ then a family $\mathcal{B} \subset \exp Z$ is *inscribed* in $\mathcal{A}$ if for any $B \in \mathcal{B}$ there is $A \in \mathcal{A}$ such that $B \subset A$. The family $\mathcal{B}$ is a shrinking of $\mathcal{A}$ if $\mathcal{B} = \{B_A : A \in \mathcal{A}\}$ and $B_A \subset A$ for any $A \in \mathcal{A}$. Suppose that M_t is a second countable space for any $t \in T$ and we are given a point $a \in M = \prod_{t \in T} M_t$. If the space Z embeds in $\sigma(M, a) = \{x \in M : |\{t \in T : x(t) \neq a(t)\}| < \omega\}$ then Z^2 embeds in $\sigma(M, a) \times \sigma(M, a) \subset M \times M$. Let $T_i = T \times \{i\}$ for any $i \in \{0, 1\}$; we want T_i be a copy of T; so let $t' = (t, 0)$ and $t'' = (t, 1)$ for any $t \in T$. Let $M_{t'} = M_{t''} = M_t$ for any $t \in T$ and define a map

$$\varphi : M \times M \to N = \prod \{M_s : s \in T_0 \cup T_1\}$$

by requiring that $\varphi(x, y)(s) = x(t)$ if $s = t' \in T_0$ and $\varphi(x, y)(s) = y(t)$ in the case when $s = t'' \in T_1$. It is straightforward that the map φ is a homeomorphism such that $\varphi(\sigma(M,a) \times \sigma(M,a)) = \sigma(N,b)$ where $b = \varphi(a,a)$. It is evident that $\sigma(N,b)$ is also a σ-product of second countable spaces; so we proved that

(1) if a space Z embeds in a σ-product of second countable spaces then Z^2 also embeds in a σ-product of second countable spaces.

We will deduce metacompactness of $X^2 \backslash \Delta$ from a more general statement.

Fact 1. Any subspace of a σ-product of second countable spaces is metacompact.

Proof. Suppose that M_t is a second countable space for any $t \in T$ and we are given a point $a \in M = \prod\{M_t : t \in T\}$; let $\sigma = \sigma(M,a)$. For any $S \subset T$ the map $p_S : \sigma \to M_S = \prod_{t \in S} M_t$ defined by the formula $p_S(x) = x|S$ for any $x \in \sigma$, is the restriction of the natural projection of M onto the face M_S. Denote by $\text{Fin}(T)$ the family of all finite subsets of T. If $F \in \text{Fin}(T)$ then let $O_F = \{x \in \sigma : x(t) \neq a(t)$ for any $t \in F\}$. It is convenient to agree that $O_\emptyset = \sigma$; we will also need the set $Q_F = \{x \in O_F : x(t) = a(t)$ for any $t \in T \backslash F\}$ for any finite $F \subset T$. Observe that $Q_\emptyset = \{a\}$. It is straightforward that

(2) $\bigcup\{Q_F : F \in \text{Fin}(T)\} = \sigma$,

so our next step is to show that

(3) the family $\mathcal{O} = \{O_F : F \in \text{Fin}(T)\}$ is point-finite.

Indeed, if $\mathcal{A}$ is an infinite subfamily of $\text{Fin}(T)$ then $S = \bigcup \mathcal{A}$ has to be infinite; if $x \in \bigcap\{O_F : F \in \mathcal{A}\}$ then $x(t) \neq a(t)$ for any $t \in S$ which is a contradiction with $x \in \sigma$. Thus the intersection of any infinite subfamily of $\mathcal{O}$ is empty, i.e., $\mathcal{O}$ is point-finite.

Given a family $\mathcal{U}$ of open subsets of a space Z say that a collection $\mathcal{V} \subset \tau(Z)$ is a refinement of $\mathcal{U}$ (in the space Z) if $\mathcal{V}$ is inscribed in $\mathcal{U}$ and $\bigcup \mathcal{V} = \bigcup \mathcal{U}$. It turns out that

(4) any family $\mathcal{U} \subset \tau(\sigma)$ has a point-finite refinement in σ.

To prove the property (4) let $\mathcal{U}_F = \{U \cap Q_F : U \in \mathcal{U}\}$ for any $F \in \text{Fin}(T)$. The space Q_F is homeomorphic to $N_F = \prod\{M_t \backslash \{a(t)\} : t \in F\}$; so it is paracompact being second countable. Therefore there exists a point-finite refinement $\mathcal{V}_F$ of the family $\mathcal{U}_F$ in the space Q_F. The map $p_F|Q_F : Q_F \to N_F$ is a homeomorphism; so the family $\{p_F(V) : V \in \mathcal{V}_F\}$ is point-finite and consists of open subsets of N_F. For any $V \in \mathcal{V}_F$ choose a set $G(V) \in \tau(\sigma)$ which is contained in an element of $\mathcal{U}$ and $G(V) \cap Q_F = V$. Let $\tilde{V} = p_F^{-1}(p_F(V)) \cap G(V)$ for any $V \in \mathcal{V}_F$. The set $\tilde{V}$ is open in σ, contained in an element of $\mathcal{U}$ and $\tilde{V} \cap Q_F = V$ for any $V \in \mathcal{V}_F$.

Observe that every family $\mathcal{W}_F = \{\tilde{V} : V \in \mathcal{V}_F\}$ is point-finite because it is a shrinking of the point-finite family $\{p_F^{-1}(p_F(V)) : V \in \mathcal{V}_F\}$. Furthermore, $\mathcal{W}_F$ is inscribed in $\mathcal{U}$; so $\mathcal{W} = \bigcup\{\mathcal{W}_F : F \in \text{Fin}(T)\}$ is also inscribed in $\mathcal{U}$.

To see that $\mathcal{W}$ is point-finite take any point $x \in \sigma$. It follows from (3) that the family $\mathcal{E} = \{F \in \text{Fin}(T) : x \in O_F\}$ is finite. Since $\bigcup \mathcal{W}_F \subset p_F^{-1}(Q_F) = O_F$ for any finite $F \subset T$, the point x does not belong to any $W \in \mathcal{W} \backslash (\bigcup\{\mathcal{W}_F : F \in \mathcal{E}\})$. Therefore x can only belong to the elements of the family $\bigcup\{\mathcal{W}_F : F \in \mathcal{E}\}$ which is point-finite being a finite union of point-finite families. This proves that W is a point-finite family inscribed in $\mathcal{U}$. Finally, it follows easily from (2) that $\bigcup W = \bigcup \mathcal{U}$; so W is a point-finite refinement of $\mathcal{U}$ and (4) is proved.

Now, take any $Y \subset \sigma$; if $\mathcal{U}'$ is an open cover of Y then choose $W_U \in \tau(\sigma)$ such that $W_U \cap Y = U$ for any $U \in \mathcal{U}'$. The family $\mathcal{U} = \{W_U : U \in \mathcal{U}'\}$ is a collection of open subsets of σ; so we can apply (4) to find a point-finite open refinement $\mathcal{V}$ of the family $\mathcal{U}$. It is evident that $\mathcal{V}' = \{V \cap Y : V \in \mathcal{V}\}$ is a point-finite open refinement of the cover $\mathcal{U}'$; so Y is metacompact and Fact 1 is proved.

Returning to our solution observe that the property (1) implies that X^2 embeds in a σ-product of second countable spaces; so $X^2 \backslash \Delta$ also embeds in a σ-product of second countable spaces. Finally, apply Fact 1 to conclude that $X^2 \backslash \Delta$ is metacompact and finish our solution.

U.191. *Observe that any countably compact subspace of a Corson compact space is closed and hence compact. Deduce from this fact that there exists a countably compact space X which embeds into $\Sigma(A)$ for some A but is not embeddable into any Corson compact space.*

Solution. Suppose that K is a Corson compact space and $P \subset K$ is countably compact. If P is not closed in K then there is a sequence $S = \{a_n : n \in \omega\} \subset P$ which converges to some $a \in K \backslash P$ (see Problem 120). It is an easy exercise to see that S is an infinite closed discrete subspace of P which is a contradiction with countable compactness of P. Thus any countably compact subspace of a Corson compact space is compact.

Now let $X = C_p(L(\omega_1), \mathbb{I})$ (here, as usual, the space $L(\omega_1)$ is the one-point Lindelöfication of the discrete space of cardinality ω_1). Since $L(\omega_1)$ is a P-space, our space X is countably compact by TFS-397. Furthermore, X is not compact because $L(\omega_1)$ is not discrete (see TFS-396). Observe also that the space X embeds in $C_p(L(\omega_1))$ which is a Σ-product of real lines (see Problem 106); by the conclusion of the previous paragraph, X is not embeddable in a Corson compact space because it is countably compact and non-compact.

U.192. *Let M_α be a separable metrizable space for any $\alpha < \omega_1$. Prove that a dense subspace Y of the space $\prod\{M_\alpha : \alpha < \omega_1\}$ is normal if and only if Y is collectionwise normal.*

Solution. Given a space Z and $A, B \subset Z$ say that A and B are *separated in Z* if $\overline{A} \cap B = \emptyset = \overline{B} \cap A$. The sets A and B are *open-separated in Z* if there are disjoint open sets $U, V \subset Z$ for which $A \subset U$ and $B \subset V$.

Fact 1. Suppose that Z is a set and κ is an infinite regular cardinal. Given a non-empty family $\mathcal{A} \subset \exp Z$ such that $|\mathcal{A}| \leq \kappa$ and $|A| \geq \kappa$ for any $A \in \mathcal{A}$ there is a disjoint family $\{Z_\alpha : \alpha < \kappa\}$ such that $Z = \bigcup\{Z_\alpha : \alpha < \kappa\}$ and $Z_\alpha \cap A \neq \emptyset$ for any $A \in \mathcal{A}$ and $\alpha < \kappa$.

Proof. We have $\mathcal{A} = \{A_\alpha : \alpha < \kappa\}$ (repetitions are allowed in this enumeration); choose a point $z_0^0 \in A_0$. Assume that $\alpha < \kappa$ and we have a set $P = \{z_\beta^\gamma : \gamma, \beta < \alpha\}$ with the following properties:

(1) if $\gamma, \beta, \gamma', \beta' < \alpha$ and $(\beta, \gamma) \neq (\beta', \gamma')$ then $z_\beta^\gamma \neq z_{\beta'}^{\gamma'}$;
(2) $z_\beta^\gamma \in A_\gamma$ for any $\beta, \gamma < \alpha$.

Since $|P| \leq |\alpha \times \alpha| = |\alpha| < \kappa$, we have the equality $|A_\beta \backslash P| = \kappa$ for any $\beta \leq \alpha$ which shows that we can choose points $z_\beta^\alpha \in A_\alpha \backslash P$ for any $\beta \leq \alpha$ and $z_\alpha^\beta \in A_\beta \backslash P$ for every $\beta < \alpha$ in such a way that $z_\beta^\gamma \neq z_{\beta'}^{\gamma'}$ whenever $(\beta, \gamma) \neq (\beta', \gamma')$ and $\max\{\beta, \gamma\} = \max\{\beta', \gamma'\} = \alpha$.

It is immediate that the set $\{z_\beta^\gamma : \gamma, \beta \leq \alpha\}$ still has the properties (1) and (2); so our inductive procedure gives us a set $\{z_\beta^\gamma : \gamma, \beta < \kappa\}$ for which (1) and (2) are satisfied for any $\alpha < \kappa$.

If $1 \leq \alpha < \kappa$ then let $Z_\alpha = \{z_\alpha^\beta : \beta < \kappa\}$; if $Z_0 = Z \backslash (\bigcup\{Z_\alpha : 1 \leq \alpha < \kappa\})$ then the family $\mathcal{D} = \{Z_\alpha : \alpha < \kappa\}$ is as promised. Indeed, it is clear that $\mathcal{D}$ is disjoint and $\bigcup \mathcal{D} = Z$. If $A \in \mathcal{A}$ and $\alpha < \kappa$ then $A = A_\beta$ for some $\beta < \kappa$ and hence $z_\alpha^\beta \in Z_\alpha \cap A_\beta$, i.e., $Z_\alpha \cap A \neq \emptyset$ for any $A \in \mathcal{A}$; so Fact 1 is proved.

Returning to our solution observe first that collectionwise normality implies normality for any space; so we only have to prove that if a subspace Y of the space $M = \prod\{M_\alpha : \alpha < \omega_1\}$ is normal then it is collectionwise normal. We will do it by showing that $ext(Y) = \omega$ so assume, towards a contradiction, that $D \subset Y$ is closed, discrete and $|D| = \omega_1$.

Choose a countable base $\mathcal{H}_\alpha$ in the space M_α such that $M_\alpha \in \mathcal{H}_\alpha$ for every $\alpha < \omega_1$. Given $\alpha_1, \ldots, \alpha_n < \omega_1$ and $O_i \in \mathcal{H}_{\alpha_i}$ for any $i \leq n$, consider the set

$$[\alpha_1, \ldots, \alpha_n; O_1, \ldots, O_n] = \{x \in M : x(\alpha_i) \in O_i \text{ for any } i \leq n\}.$$

It is clear that the family $\mathcal{B} = \{[\alpha_1, \ldots, \alpha_n; O_1, \ldots, O_n] : n \in \mathbb{N}, \alpha_1 < \ldots < \alpha_n < \omega_1 \text{ and } O_i \in \mathcal{H}_{\alpha_i} \text{ for any } i \leq n\}$ is a base of the space M. For any $S \subset \omega_1$ let $p_S : M \to M_S = \prod\{M_\alpha : \alpha \in S\}$ be the projection of M onto its face M_S.

Consider the family $\mathcal{A} = \{B \in \mathcal{B} : B \cap D \text{ is uncountable}\}$; then $|\mathcal{A}| \leq \omega_1$. Besides, $\mathcal{A} \neq \emptyset$ because $M = [\alpha; M_\alpha] \in \mathcal{A}$ for any $\alpha < \omega_1$. Applying Fact 1 we can find disjoint sets $E, G' \subset D$ such that $E \cap B \neq \emptyset \neq G' \cap B$ for any $B \in \mathcal{A}$; let $G = D \backslash E$. One of the sets E, G is uncountable; so we can assume without loss of generality that $|E| > \omega$.

The space Y being normal, the sets E and G are open-separated in M and hence there is a countable $S \subset \omega_1$ such that the sets $p_S(E)$ and $p_S(G)$ are separated in M_S (see Fact 3 of S.291). Let $\mathcal{B}' = \{[\alpha_1, \ldots, \alpha_n : O_1, \ldots, O_n] \in \mathcal{B} : \alpha_1, \ldots, \alpha_n \in S\}$. We have two cases to consider.

Case 1. The set $p_S(E)$ is countable. Then there exists a point $z \in E$ such that $p_S^{-1}(p_S(z)) \cap E$ is uncountable. If $z \in B = [\alpha_1, \ldots, \alpha_n; O_1, \ldots, O_n] \in \mathcal{B}'$ then $p_S^{-1}(p_S(z)) \cap E \subset B$ and hence $B \in \mathcal{A}$.

Case 2. If the second countable space $p_S(E)$ is uncountable then it must have a complete accumulation point t; fix a point $z \in E$ with $p_S(z) = t$. If we assume that $z \in B = [\alpha_1, \ldots, \alpha_n; O_1, \ldots, O_n] \in \mathcal{B}'$ then $p_S(B) \in \tau(t, M_S)$ and therefore the set $E' = \{x \in E : p_S(x) \in p_S(B)\}$ is uncountable. Since $p_S^{-1}(p_S(B)) = B$, we have $E' \subset B$ and therefore $B \in \mathcal{A}$. This shows that

(*) there is $z \in E$ such that $z \in B = [\alpha_1, \ldots, \alpha_n; O_1, \ldots, O_n] \in \mathcal{B}'$ implies $B \in \mathcal{A}$.

It follows from (*) that $G \cap B \neq \emptyset$ for any $B \in \mathcal{B}'$ with $z \in B$. It is easy to check that the family $\{p_S(B) : B \in \mathcal{B}'\}$ is a local base in M_S at the point $t = p_S(z)$. It follows from $G \cap B \neq \emptyset$ that $p_S(B) \cap p_S(G) \neq \emptyset$ for any $B \in \mathcal{B}'$ and therefore $t \in cl_{M_S}(p_S(G)) \cap p_S(E)$ which is a contradiction with the fact that $p_S(E)$ and $p_S(G)$ are separated in M_S. This contradiction shows that $ext(Y) = \omega$ and hence Y is collectionwise normal by Fact 3 of S.294. We proved that any normal $Y \subset M$ is collectionwise normal; so our solution is complete.

U.193. *Prove that if $2^{\omega_1} = \mathfrak{c}$ then there exists a dense hereditarily normal subspace Y in the space $\mathbb{D}^\mathfrak{c}$ such that $ext(Y) = \omega_1$. Deduce from this fact that it is independent of ZFC whether normality implies collectionwise normality in the class of dense subspaces of $\mathbb{D}^\mathfrak{c}$.*

Solution. If we are given a space Z say that sets $A, B \subset Z$ are *separated* in Z if $\overline{A} \cap B = \emptyset = \overline{B} \cap A$. The sets A and B are *open-separated* in Z if there are disjoint $U, V \in \tau(Z)$ such that $A \subset U$ and $B \subset V$.

We denote by K the Cantor set $\mathbb{D}^\omega$. As usual, any $n \in \mathbb{N}$ is identified with the set $\{0, \ldots, n-1\}$. The set of all *non-empty* finite subsets of a set A is denoted by $Fin(A)$. For every $S \in Fin(\omega)$ and $x \in K^S$ let $b_S(x) = \{x(i) : i \in S\} \subset K$. If $|S| \geq 2$ then $\Delta_S = \{x \in K^S : $ there are distinct $m, n \in S$ for which $x(m) = x(n)\}$; if $|S| = 1$ then $\Delta_S = \emptyset$. Given a set $S \in Fin(\omega)$ say that $F \subset K^S \backslash \Delta_S$ is *proper* in K^S if there is $H \subset F$ with $|H| = \mathfrak{c}$ such that the family $\mathcal{F}_H = \{b_S(x) : x \in H\}$ is disjoint. Given distinct points $x_1, \ldots, x_n \in K$ and (not necessarily distinct) $i_1, \ldots, i_n \in \mathbb{D}$ let $[x_1, \ldots, x_n; i_1, \ldots, i_n] = \{f \in \mathbb{D}^K : f(x_j) = i_j$ for all $j \leq n\}$. It is evident that the family $\mathcal{B} = \{[x_1, \ldots, x_n; i_1, \ldots, i_n] : n \in \mathbb{N}, x_j \in K$ and $i_j \in \mathbb{D}$ for any $j \leq n\}$ is a base of $\mathbb{D}^K$; the elements of $\mathcal{B}$ are called *the standard open subsets of* $\mathbb{D}^K$. If $U = [x_1, \ldots, x_n; i_1, \ldots, i_n]$ is a standard open subset of the space $\mathbb{D}^K$ then $supp(U) = \{x_1, \ldots, x_n\}$.

If $T, T' \in Fin(\omega)$ and $T' \subset T$ then $p_{T'}^T : K^T \to K^{T'}$ is the projection of K^T onto its face $K^{T'}$. A set $F \subset K^T$ is *almost proper* in K^T if F is either a singleton or proper in K^T or else there is a non-empty $S \subset T$, $S \neq T$ such that $p_S^T(F)$ is a proper subset of K^S and $p_{T \backslash S}^T(F)$ is a singleton.

Fact 1. Suppose that $S \in Fin(\omega)$ and a set $F \subset K^S \backslash \Delta_S$ is compact and non-empty. Then $F = \bigcup\{F_i : i \in \omega\}$ where F_i is compact and almost proper in K^S for any $i \in \omega$.

Proof. We will carry out induction along $n = |S|$. If $|S| = 1$ and $F \subset K^S$ is countable then we can take as F_i the respective singletons whose union is F. If F is uncountable then $|F| = \mathfrak{c}$ (see e.g., SFFS-353); so it is immediate that F is proper in K^S.

Assume that $n \in \mathbb{N}$ and our Fact has been proved for all $S \in \text{Fin}(\omega)$ with $|S| \leq n$. Take any $S = \{l_0, \ldots, l_n\} \subset \omega$ such that $l_i \neq l_j$ whenever $i \neq j$ and assume that F is a compact subset of $K^S \setminus \Delta_S$. For the sake of brevity denote the map $p^S_{\{l_i\}}$ by q_i for any $i \leq n$. Suppose first that

(1) there is a countable set $A \subset K$ such that $F \subset \bigcup\{q_i^{-1}(A) : i \leq n\}$ (we identify K and $K^{\{l_i\}}$ for every $i \leq n$).

If $A = \{x_m : m \in \omega\}$ then let $G(m, i) = q_i^{-1}(x_m) \cap F$ for any numbers $m \in \omega$ and $i \leq n$. If $G(m, i) \neq \emptyset$ then $H(m, i) = p^S_{S \setminus \{l_i\}}(G(m, i)) \subset K^{S \setminus \{l_i\}} \setminus \Delta_{S \setminus \{l_i\}}$; so the induction hypothesis is applicable to $H(m, i)$, i.e., $H(m, i) = \bigcup\{G_k : k \in \omega\}$ where every G_k is almost proper in $K^{S \setminus \{l_i\}}$.

Given $k \in \omega$, for every $x \in G_k$ let $\varphi(x)|(S \setminus \{l_i\}) = x$ and $\varphi(x)(l_i) = x_m$. Then $\varphi : G_k \to K^S$ and it is immediate that $G'_k = \varphi(G_k)$ is almost proper in K^S. It is evident that $G(m, i) = \bigcup\{G'_k : k \in \omega\}$; so we represented every set $G(m, i)$ as a countable union of almost proper subsets of K^S. It follows from the equality $F = \bigcup\{G(m, i) : m \in \omega, \ i \leq n \text{ and } G(m, i) \neq \emptyset\}$ that F is also representable as a countable union of almost proper subsets of K^S; so our proof is complete in this case.

If the property (1) does not hold then

(2) $F \not\subset \bigcup\{q_i^{-1}(A) : i \leq n\}$ for any $A \subset K$ with $|A| < \mathfrak{c}$.

To see that (2) is true, for any $x \in K^S$ let $\xi(x)(i) = x(l_i)$ for any $i \leq n$. Then $\xi(x) \in K^{n+1}$ and hence we have a map $\xi : K^S \to K^{n+1}$. It is easy to see that ξ is a homeomorphism which preserves the properties (1) and (2), i.e., a set $G \subset K^S$ has either (1) or (2) if and only if $\xi(G)$ has the respective property in K^{n+1}. However, the properties (1) and (2) for $S = \{0, \ldots, n\}$ have already been introduced in S.151 where (1) stood for $|F|_{n+1} \leq \omega$ and (2) was denoted by $|F|_{n+1} \geq \mathfrak{c}$. Now we can apply Fact 4 of S.151 to see that $|\xi(F)|_{n+1} > \omega$ implies $|\xi(F)|_{n+1} = \mathfrak{c}$, i.e., (2) holds for $\xi(F)$ in K^{n+1} and therefore (2) is fulfilled for F.

Now we can prove that F is itself proper. Take a point $x_0 \in F$ arbitrarily. Suppose that $\alpha < \mathfrak{c}$ and we have constructed a set $\{x_\beta : \beta < \alpha\}$ such that the family $\mathcal{A} = \{b_S(x_\beta) : \beta < \alpha\}$ is disjoint. Then $A = \bigcup \mathcal{A}$ has cardinality strictly less than $\mathfrak{c}$; so (2) implies that there is a point $x_\alpha \in F \setminus (\bigcup\{q_i^{-1}(A) : i \leq n\})$. It is immediate that the family $\{b_S(x_\beta) : \beta \leq \alpha\}$ is still disjoint; so our inductive procedure gives us a set $B = \{x_\beta : \beta < \mathfrak{c}\} \subset F$ such that $\mathcal{F}_B$ is disjoint. This shows that the set F is proper; so we can take $F_i = F$ for all $i \in \omega$; this finishes the proof of Fact 1.

Fact 2. There exist disjoint sets $P_0, P_1 \subset K$ such that $Q = K \setminus (P_0 \cup P_1)$ has cardinality $\mathfrak{c}$ and, for any $S \in \text{Fin}(\omega)$, $u \in \mathbb{D}^S$ and a proper set $F \subset K^S$ there is $x \in F$ such that $x(m) \in P_{u(m)}$ for any $m \in S$.

Proof. Let $\{F_\alpha : \alpha < \mathfrak{c}\}$ be an enumeration of all compact proper subsets of K^S for all $S \in \text{Fin}(\omega)$. For every $\alpha < \mathfrak{c}$ fix the set $S_\alpha \in \text{Fin}(\omega)$ such that F_α is proper in K^{S_α} and a set $H_\alpha \subset F_\alpha$ such that $|H_\alpha| = \mathfrak{c}$ and $\mathcal{F}_{H_\alpha}$ is disjoint.

Since H_0 is infinite, for any $u \in \mathbb{D}^{S_0}$ we can pick a point $x_u \in H_0$ so that $u \neq v$ implies $x_u \neq x_v$ and consider the set $P_i^u(0) = \{x_u(m) : m \in S_0 \text{ and } u(m) = i\}$; let $P_i(0) = \bigcup\{P_i^u(0) : u \in \mathbb{D}^{S_0}\}$ for each $i \in \mathbb{D}$ and choose a point $z_0 \in K\backslash(P_0(0) \cup P_1(0))$.

Suppose that $\alpha < \mathfrak{c}$ and we have a set $Z_\alpha = \{z_\beta : \beta < \alpha\} \subset K$ and a collection $\{P_i(\beta) : \beta < \alpha\}$ of finite subsets of K for every $i \in \mathbb{D}$ with the following properties:

(3) the sets Z_α, $T_\alpha^0 = \bigcup\{P_0(\beta) : \beta < \alpha\}$ and $T_\alpha^1 = \bigcup\{P_1(\beta) : \beta < \alpha\}$ are disjoint;

(4) for any $\beta < \alpha$ and $u \in \mathbb{D}^{S_\beta}$ there is a point $x \in F_\beta$ such that $x(m) \in P_{u(m)}(\beta)$ for any $m \in S_\beta$.

Since the set $Z = Z_\alpha \cup T_\alpha^0 \cup T_\alpha^1$ has cardinality strictly less than $\mathfrak{c}$, the set $H = \{x \in H_\alpha : b_{S_\alpha}(x) \cap Z = \emptyset\}$ has cardinality $\mathfrak{c}$. For any $u \in \mathbb{D}^{S_\alpha}$ pick an element $x_u \in H$ such that $u \neq v$ implies $x_u \neq x_v$ and consider the set $P_i^u(\alpha) = \{x_u(m) : m \in S_\alpha \text{ and } u(m) = i\}$; let $P_i(\alpha) = \bigcup\{P_i^u(\alpha) : u \in \mathbb{D}^{S_\alpha}\}$ for every $i \in \mathbb{D}$. Since the set $T = Z \cup P_0(\alpha) \cup P_1(\alpha)$ has cardinality strictly less than $\mathfrak{c}$, we can take a point $z_\alpha \in K\backslash T$.

It is evident that $\{z_\beta : \beta \leq \alpha\} \subset K$ and the families $\{P_i(\beta) : \beta \leq \alpha\}$, $i \in \mathbb{D}$ still satisfy (3) and (4); so our inductive procedure can be continued to construct a set $\{z_\beta : \beta < \mathfrak{c}\}$ and the collection $\{P_i(\beta) : \beta < \mathfrak{c}\}$ for every $i \in \mathbb{D}$ in such a way that the conditions (3) and (4) are fulfilled for all $\alpha < \mathfrak{c}$.

Let $P_i = \bigcup\{P_i(\beta) : \beta < \mathfrak{c}\}$ for every element $i \in \mathbb{D}$. It follows from the inclusion $\{z_\alpha : \alpha < \mathfrak{c}\} \subset Q = K\backslash(P_0 \cup P_1)$ that $|Q| = \mathfrak{c}$. The condition (3) shows that we have $P_0 \cap P_1 = \emptyset$. Now, if $S \in \text{Fin}(\omega)$ and $F \subset K^S$ is a compact proper subset of K^S then there is $\beta < \mathfrak{c}$ such that $F = F_\beta$ and $S = S_\beta$. The condition (4) implies that, for any $u \in \mathbb{D}^S = \mathbb{D}^{S_\beta}$ there is $x \in F_\beta = F$ such that $x(m) \in P_{u(m)}(\beta) \subset P_{u(m)}$ for all $m \in S$. Thus P_0 and P_1 have all promised properties; so Fact 2 is proved.

Fact 3. A space Z is hereditarily normal if and only if any pair of separated subsets of Z are open-separated.

Proof. If Z is hereditarily normal and $A, B \subset Z$ are separated in Z then they are open-separated in Z: this was proved in Fact 1 of S.291, so we have necessity.

Now assume that if $A, B \subset Z$ are separated then they are open-separated and fix any subspace $Y \subset Z$. If F and G are disjoint closed subsets of the space Y then $\text{cl}_Z(F) \cap G = \text{cl}_Y(F) \cap G = F \cap G = \emptyset$ and, analogously, $\text{cl}_Z(G) \cap F = \emptyset$ which shows that F and G are separated in Z. Thus there are disjoint $U', V' \in \tau(Z)$ such that $F \subset U'$ and $G \subset V'$. The sets $U = U' \cap Y$ and $V = V' \cap Y$ are open in Y, disjoint and $F \subset U$, $G \subset V$. Thus any disjoint closed subsets of Y are open-separated, i.e., Y is normal. Fact 3 is proved.

Returning to our solution observe that $C = C_p(K, \mathbb{D})$ is a countable dense subspace of $\mathbb{D}^K$ (see Fact 1 of U.077 and Fact 1 of S.390) and let $P_0, P_1, Q \subset K$ be the sets provided by Fact 2. Since $2^{\omega_1} = \mathfrak{c}$ and $|Q| = \mathfrak{c}$, we can choose a surjective map $\varphi : Q \to \exp(\omega_1)$. For any $\alpha < \omega_1$ and $i \in \mathbb{D}$ let $h_\alpha(x) = i$ for every $x \in P_i$; if $x \in Q$ then $h_\alpha(x) = 0$ if $\alpha \notin \varphi(x)$ and $h_\alpha(x) = 1$ whenever $\alpha \in \varphi(x)$. This gives us a set $D = \{h_\alpha : \alpha < \omega_1\} \subset \mathbb{D}^K$; we will prove that the space $Y = C \cup D$ is hereditarily normal and $ext(Y) = \omega_1$.

Observe that any open subset of $K = K^1$ is proper in K; so $P_i \cap U \neq \emptyset$ for any $U \in \tau^*(K)$ and $i \in \mathbb{D}$. Therefore every P_i is dense in K which shows that every $h \in D$ is discontinuous on K, i.e., $D \cap C = \emptyset$. Given a function $f \in C$ either $f^{-1}(0)$ or $f^{-1}(1)$ is infinite; if $U = f^{-1}(i)$ is infinite then we can take $x_0 \in P_0 \cap U$ and $x_1 \in P_1 \cap U$. It is immediate that $V = [x_0, x_1; i, i] \in \tau(f, \mathbb{D}^K)$ and $V \cap D = \emptyset$. This proves that D is closed in Y.

If $E \subset D$ then there is $A \subset \omega_1$ such that $E = \{h_\alpha : \alpha \in A\}$ and $x \in Q$ for which $\varphi(x) = A$. It is straightforward that $h_\alpha(x) = 1$ for all $\alpha \in A$ and $h_\alpha(x) = 0$ whenever $\alpha \in \omega_1 \backslash A$. Therefore $[x; 1]$ and $[x; 0]$ are disjoint open neighbourhoods of E and $D \backslash E$ respectively. Thus any $E \subset D$ is open in D, i.e., D is a closed discrete subspace of Y. Besides, if we take $E = \{f_\alpha\}$ then we proved that E and $D \backslash E$ are contained in disjoint open subsets of $\mathbb{D}^K$; therefore $f_\alpha \neq f_\beta$ whenever $\alpha \neq \beta$; so $ext(Y) = |D| = \omega_1$.

As a consequence, the space Y cannot be collectionwise normal for otherwise we would have ω_1-many disjoint non-empty open subsets which separate the points of D; however, this is impossible by $c(Y) = \omega$ (recall that Y is dense in $\mathbb{D}^K$).

Let us finally prove that Y is hereditarily normal; by Fact 3 it suffices to show that if F and G are separated in Y then they are open-separated in Y.

For any $A \subset K$ let $\pi_A : \mathbb{D}^K \to \mathbb{D}^A$ be the natural projection of $\mathbb{D}^K$ onto its face $\mathbb{D}^A$. It is sufficient to establish that there is a countable $E \subset K$ such that $\pi_E(F)$ and $\pi_E(G)$ are separated in $\mathbb{D}^E$ (see Fact 3 of S.291). To simplify the terminology we will say that sets $R, T \subset \mathbb{D}^K$ are A-separated if $\pi_A(R)$ and $\pi_A(T)$ are separated in $\mathbb{D}^A$.

We will use several times the following trivial observations.

(5) If $A \subset K$ and $R, T \subset \mathbb{D}^K$ are A-separated then R and T are B-separated for any $B \subset K$ with $A \subset B$.

(6) Suppose that $R, T \subset \mathbb{D}^K$, $R = R_1 \cup R_2$, $T = T_1 \cup T_2$ and there exist sets $A_{11}, A_{12}, A_{21}, A_{22} \subset K$ such that the sets R_i and T_j are A_{ij}-separated for any $i, j \in \{1, 2\}$. Then R and T are A-separated for the set $A = A_{11} \cup A_{12} \cup A_{21} \cup A_{22}$.

(7) If $R, T \subset \mathbb{D}^K$ and there exist families $\mathcal{U}, \mathcal{V}$ of standard open subsets of $\mathbb{D}^K$ such that $U = \bigcup \mathcal{U} \supset R$, $T \cap U = \emptyset$ and $V = \bigcup \mathcal{V} \supset T$ while $V \cap R = \emptyset$ then R and T are A-separated where $A = \bigcup \{supp(W) : W \in \mathcal{U} \cup \mathcal{V}\}$.

For any $n \in \mathbb{N}$ let $M_n = \{1, \ldots, n\}$. Let $F_1 = F \cap C$ and $F_2 = F \cap D$; analogously, define $G_1 = G \cap C$ and $G_2 = G \cap D$. The sets F_1 and G_1 are countable and separated in C; so there exits countable families $\mathcal{U}_1$ and $\mathcal{V}_1$ of standard open subsets of $\mathbb{D}^K$ such that $F_1 \subset \bigcup \mathcal{U}_1 \subset \mathbb{D}^K \backslash G_1$ and $G_1 \subset \bigcup \mathcal{V}_1 \subset \mathbb{D}^K \backslash F_1$. Thus the property (7) guarantees existence of a countable $A_{11} \subset K$ such that F_1 and G_1 are A_{11}-separated.

The set F_1 being countable and separated from G_2 there is a countable $\mathcal{U}_2 \subset \mathcal{B}$ such that $F_1 \subset \bigcup \mathcal{U}_2 \subset \mathbb{D}^K \backslash G_2$. Let $\Omega = \{\alpha < \omega_1 : h_\alpha \in G_2\}$; for any $\alpha \in \Omega$

there is a standard $O_\alpha = [x_1^\alpha, \ldots, x_{k_\alpha}^\alpha; i_1^\alpha, \ldots, i_{k_\alpha}^\alpha]$ such that $h_\alpha \in O_\alpha \subset \mathbb{D}^K \backslash F_1$. It is easy to see that we can choose a countable family $\{\Omega_n : n \in \omega\}$ of subsets of ω_1 such that $\bigcup\{\Omega_n : n \in \omega\} = \Omega$ while $k_\alpha = k(n)$ for any $\alpha \in \Omega_n$ and there is $u_n \in \mathbb{D}^{M_{k(n)}}$ such that $i_j^\alpha = u_n(j)$ for any $j \in M_{k(n)}$.

Let $x_\alpha = (x_1^\alpha, \ldots, x_{k_\alpha}^\alpha) \in K^{M_{k_\alpha}}$ and $e_\alpha = (i_1^\alpha, \ldots, i_{k_\alpha}^\alpha) \in \mathbb{D}^{M_{k_\alpha}}$ for any $\alpha < \omega_1$; for any $n \in \omega$ there is $e(n) \in \mathbb{D}^{M_{k(n)}}$ such that $e_\alpha = e(n)$ for any $\alpha \in \Omega_n$. If $y = (y_1, \ldots, y_m) \in K^{M_m}$ and $e = (i_1, \ldots, i_m) \in \mathbb{D}^{M_m}$ then we will also denote the set $[y_1, \ldots, y_m; i_1, \ldots, i_m]$ by $O(y, e)$; then $O_\alpha = O(x_\alpha, e_\alpha)$ for any $\alpha < \omega_1$.

Let $P_n = \{x_\alpha : \alpha \in \Omega_n\} \subset K^{M_{k(n)}} \backslash \Delta_{M_{k(n)}}$ for any number $n \in \omega$. The set $\Delta_{M_{k(n)}}$ is closed in the space $K^{M_{k(n)}}$; so $P_n = \bigcup\{P_n^i : i \in \omega\}$ where $\mathrm{cl}(P_n^i) \cap \Delta_{M_{k(n)}} = \emptyset$ for any $i \in \omega$. Observe that $O(x, e(n)) \cap F_1 = \emptyset$ for any $x \in \mathrm{cl}(P_n^i)$ and $i \in \omega$. Indeed, if $f \in F_1 \cap O(x, e(n))$ for some $x = (x_1, \ldots, x_n) \in Q_n^i = \mathrm{cl}(P_n^i)$ then $f(x_j) = u_n(j)$ for all $j \leq k(n)$. Since the function f is continuous, there exists $x_\alpha = (x_1^\alpha, \ldots, x_{k(n)}^\alpha) \in P_n^i$ such that $f(x_j^\alpha) = u_n(j) = i_j^\alpha$ for any $j \leq k(n)$. Therefore $f \in O(x_\alpha, e_\alpha) = O_\alpha$ which is a contradiction with the choice of O_α.

Fix $i, n \in \omega$; by Fact 1 we have a representation $Q_n^i = \bigcup\{F_j : j \in \omega\}$ where the set F_j is compact and almost proper in $K^{M_{k(n)}}$ for every $j \in \omega$. We claim that,

(8) for every $j \in \omega$, the set $G' = G_2 \cap (\bigcup\{O(x, e(n)) : x \in F_j\})$ can be covered by just one element of the family $\{O(x, e(x)) : x \in F_j\}$.

This is evident if F_j is a singleton. If not then either F_j is proper or there is $L \subset M_{k(n)}$, $L \neq M_{k(n)}$ and a proper compact set P in K^L such that $\pi_L^{M_{k(n)}}(F_j) = P$ and $\pi_{M_{k(n)}\backslash L}^{M_{k(n)}}(F_j)$ is a singleton. Recalling the main property of the sets P_0 and P_1 from Fact 2 we see that there is a point $z \in P$ such that $z(m) \in P_{u_n(m)}$ for any $m \in L$. By the definition of the set D we have $h(z(m)) = u_n(m)$ for any $h \in D$ which shows that all elements of D take the same value $u_n(m)$ at $z(m)$ for any $m \in L$. But outside of L the coordinates of all points of F_j are constant; so if a we take a point $x = (x_1, \ldots, x_{k(n)}) \in F_j$ such that $m \in L$ implies $x_m = z(m)$ then $G' \subset O(x, e(n))$. Indeed, if $m \notin L$ then $h(x_m) = u_n(m)$ for all $h \in G'$ because G' is covered by the sets $\{O(y, e(n)) : y \in F_j\}$ while $y_m = x_m$ for any $y \in F_j$ and $m \in M_{k(n)}\backslash L$.

On the other hand, if $m \in L$ then all points of G_2 (and even of D) have the same values at x_m; so again $h(x_m) = u_n(m)$ for any $m \in L$ and $h \in G'$; so (8) is proved.

An immediate consequence of (8) is that $G_2 \cap (\bigcup\{O(x, e(n)) : x \in Q_n^i\})$ can be covered by countably many elements of the family $\{O(x, e(n)) : x \in Q_n^i\}$ and hence $G_2 \cap (\bigcup\{O(x, e(n)) : x \in P_n\})$ can also be covered by countably many elements of the family $\{O(x, e(n)) : x \in Q_n^i, i \in \omega\}$ for any $n \in \omega$. This implies that there is a countable family $\mathcal{V}_2 \subset \mathcal{B}$ such that $G_2 \subset \bigcup\mathcal{V}_2 \subset \mathbb{D}^K \backslash F_1$. The set $A_{12} = \bigcup\{\mathrm{supp}(U) : U \in \mathcal{U}_2 \cup \mathcal{V}_2\}$ is countable; applying (7) we conclude that the sets F_1 and G_2 are A_{12}-separated. An analogous proof shows that there is a countable $A_{21} \subset K$ such that F_2 and G_1 are A_{21}-separated.

Finally, choose $x \in Q$ such that $\varphi(x) = \Omega$ (recall that $G_2 = \{h_\alpha : \alpha \in \Omega\}$). Then $h(x) = 1$ for any $h \in G_2$ while $h(x) = 0$ if $h \in D\backslash G_2$. In particular, $h(x) = 0$ for any $h \in F_2$. Thus F_2 and G_2 are A_{22}-separated if $A_{22} = \{x\}$. Finally, apply (6) to conclude that the set $A = A_{11} \cup A_{12} \cup A_{21} \cup A_{22}$ is countable while F and G are A-separated. This proves that Y is a hereditarily normal dense subspace of $\mathbb{D}^K \simeq \mathbb{D}^{\mathfrak{c}}$ with $ext(Y) = \omega_1$.

We have already noted that Y cannot be collectionwise normal; so under $2^{\omega_1} = \mathfrak{c}$ there is a dense normal subspace of $\mathbb{D}^{\mathfrak{c}}$ which is not collectionwise normal. However, under CH, we have $\mathfrak{c} = \omega_1$; so every dense normal subspace of $\mathbb{D}^{\mathfrak{c}} = \mathbb{D}^{\omega_1}$ is collectionwise normal by Problem 192. Therefore it is independent of ZFC whether normality of a dense subspace of $\mathbb{D}^{\mathfrak{c}}$ implies its collectionwise normality and hence our solution is complete.

U.194. *Let X be a monolithic compact space of countable tightness. Prove that any dense normal subspace of $C_p(X)$ is Lindelöf. In particular, if X is a Corson compact space and Y is a dense normal subspace of $C_p(X)$ then Y is Lindelöf.*

Solution. If Z is a space then $A, B \subset Z$ are *separated in Z* if $\overline{A} \cap B = \overline{B} \cap A = \emptyset$. The sets A and B are *open-separated in Z* if there are disjoint $U, V \in \tau(Z)$ such that $A \subset U$ and $B \subset V$. Let $\mathcal{O}$ be the family of all non-empty intervals in $\mathbb{R}$ with rational endpoints; it is clear that the family $\mathcal{O}$ is countable. If we have points $z_1, \ldots, z_n \in Z$ and $O_1, \ldots, O_n \in \mathcal{O}$ then $W(z_1, \ldots, z_n; O_1, \ldots, O_n) = \{f \in C_p(Z) : f(z_i) \in O_i$ for all $i \leq n\}$. It is clear that the family $\{W(z_1, \ldots, z_n; O_1, \ldots, O_n) : n \in \mathbb{N}, z_i \in Z$ and $O_i \in \mathcal{O}$ for all $i \leq n\}$ is a base in the space $C_p(Z)$.

Fact 1. Suppose that Z is a space and $\mathcal{N} \subset \exp Z$ is a network in Z. Given $M_1, \ldots, M_n \in \mathcal{N}$ and $O_1, \ldots, O_n \in \mathcal{O}$ let $[M_1, \ldots, M_n; O_1, \ldots, O_n] = \{f \in C_p(K) : f(M_i) \subset O_i$ for any $i \leq n\}$. Then the family

$$\mathcal{M} = \{[M_1, \ldots, M_n; O_1, \ldots, O_n] : n \in \mathbb{N}, M_i \in \mathcal{N}, O_i \in \mathcal{O} \quad \text{for every} \quad i \leq n\}$$

is a network in $C_p(Z)$.

Proof. If $f \in C_p(Z)$ and $U \in \tau(f, C_p(Z))$ then there exist points $z_1, \ldots, z_n \in Z$ and $O_1, \ldots, O_n \in \mathcal{O}$ for which $f \in G = W(z_1, \ldots, z_n; O_1, \ldots, O_n) \subset U$. By continuity of f there are $V_i \in \tau(z_i, Z)$ such that $f(V_i) \subset O_i$ for any $i \leq n$. The family $\mathcal{N}$ being a network of Z there are $M_1, \ldots, M_n \in \mathcal{N}$ for which $z_i \in M_i \subset V_i$ for every $i \leq n$. Then $H = [M_1, \ldots, M_n; O_1, \ldots, O_n] \in \mathcal{M}$ and $f \in H \subset G \subset U$. Thus $\mathcal{M}$ is a network in $C_p(Z)$ and Fact 1 is proved.

Returning to our solution suppose that N is normal and dense in $C_p(X)$. If N is not Lindelöf then $ext(N) > \omega$ by Baturov's theorem (SFFS-269); so we can fix a closed discrete subspace D of the space N such that $D = \omega_1$. If A and B are disjoint subsets of D then they are open-separated in N by normality of N; take disjoint $U', V' \in \tau(N)$ such that $A \subset U'$ and $B \subset V'$. There exist $U, V \in \tau(C_p(X))$ such that $U \cap N = U'$ and $V \cap N = V'$. An immediate consequence of density of N in $C_p(X)$ is that $U \cap V = \emptyset$. This shows that

(1) any disjoint $A, B \subset D$ are open-separated in $C_p(X)$.

For any $x \in X$ let $\varphi(x)(f) = f(x)$ for any $f \in D$; then $\varphi(x) \in C_p(D)$ for any $x \in X$ and the map $\varphi : X \to C_p(D)$ must be continuous (see TFS-166). Then $K = \varphi(X)$ is an ω-monolithic compact space with $w(K) \le \omega_1$. The dual map $\varphi^* : C_p(K) \to C_p(X)$ is defined by $\varphi^*(f) = f \circ \varphi$ for any $f \in C_p(K)$; it is an embedding such that $D \subset \varphi^*(C_p(K))$ (see Fact 5 of U.086).

It is evident that D is a discrete subspace of $C = \varphi^*(C_p(K))$ and it follows from (1) that any disjoint $A, B \subset D$ are open-separated in C. Since C is homeomorphic to $C_p(K)$,

(2) there exists a discrete subspace $E \subset C_p(K)$ such that $|E| = \omega_1$ and any disjoint $A, B \subset E$ are open-separated in $C_p(K)$.

If $S = \{s_\alpha : \alpha < \omega_1\}$ is a dense subspace of K then let $K_\alpha = \overline{\{s_\beta : \beta < \alpha\}}$ for any $\alpha < \omega_1$. Every K_α is second countable by ω-monolithity of K and it follows from $t(K) = \omega$ that $K = \bigcup\{K_\alpha : \alpha < \omega_1\}$. If $L_\alpha = K_\alpha \backslash (\bigcup\{K_\beta : \beta < \alpha\})$ for any $\alpha < \omega_1$ then the family $\{L_\alpha : \alpha < \omega_1\}$ of second countable spaces is disjoint and $\bigcup\{L_\alpha : \alpha < \omega_1\} = K$. If we fix a countable network $\mathcal{N}_\alpha$ in the space L_α for all $\alpha < \omega_1$ then the family $\mathcal{N} = \bigcup\{\mathcal{N}_\alpha : \alpha < \omega_1\}$ is a point-countable network in K such that $|\mathcal{N}| \le \omega_1$.

The family $\mathcal{M} = \{[M_1, \ldots, M_n; O_1, \ldots, O_n] : n \in \mathbb{N}, M_i \in \mathcal{N}, O_i \in \mathcal{O}$ for every $i \le n\}$ is a network in $C_p(K)$ by Fact 1. Let $\mathcal{M}' = \{M \in \mathcal{M} : M \cap E$ is uncountable$\}$. Applying Fact 1 of U.192 to the family $\{M \cap E : M \in \mathcal{M}'\}$ we can find disjoint sets $A, B' \subset E$ such that $A \cap M \ne \emptyset \ne B' \cap M$ for any $M \in \mathcal{M}'$. If $B = E \backslash A$ then $A \cap B = \emptyset$, $A \cup B = E$ and $A \cap M \ne \emptyset \ne B \cap M$ for any $M \in \mathcal{M}'$. One of the sets A, B is uncountable; so we can assume without loss of generality that $|A| > \omega$.

The sets A and B are open-separated in $C_p(K)$ by (2); so we can apply Fact 3 of S.291 to see that there exists a countable $T \subset K$ such that $\pi_T(A)$ and $\pi_T(B)$ are separated in $\mathbb{R}^T$ (here $\pi_T : C_p(K) \to C_p(T) \subset \mathbb{R}^T$ is the restriction map). The set A is uncountable while $w(\mathbb{R}^T) \le \omega$; so there is $f_0 \in A$ such that, for the function $g_0 = \pi_T(f_0)$, the set $\{f \in A : \pi_T(f) \in U\}$ is uncountable for any $U \in \tau(g_0, \mathbb{R}^T)$.

Since $\pi_T(A)$ and $\pi_T(B)$ are separated in $\mathbb{R}^T$, there exists a set $U \in \tau(g_0, \mathbb{R}^T)$ such that $U \cap \pi_T(B) = \emptyset$. There are $t_1, \ldots, t_n \in T$ and $O_1, \ldots, O_n \in \mathcal{O}$ such that $g_0 \in V = \{g \in C_p(T) : g(t_i) \in O_i$ for all $i \le n\} \subset U$; we already observed that the set $H = \{f \in A : \pi_T(f) \in V\}$ has to be uncountable. It is immediate that any $f \in H$ belongs to the set $W(t_1, \ldots, t_n; O_1, \ldots, O_n)$; so continuity of f implies that there are $M_1^f, \ldots, M_n^f \in \mathcal{N}$ such that $t_i \in M_i^f$ and $f(M_i^f) \subset O_i$ for any $i \le n$.

Since the set T is countable and $\mathcal{N}$ is point-countable, only countably many elements of the family $\mathcal{N}$ meet the set T. As a consequence, there is an uncountable set $H' \subset H$ and $M_1, \ldots, M_n \in \mathcal{N}$ such that $M_i^f = M_i$ for any $f \in H'$ and $i \le n$. In particular, $f \in M = [M_1, \ldots, M_n; O_1, \ldots, O_n]$ for any $f \in H'$ and therefore $M \in \mathcal{M}'$. Thus $B \cap M \ne \emptyset$ by our choice of B; if $f \in B \cap M$ then $f(M_i) \subset O_i$ and therefore $f(t_i) \in O_i$ for any $i \le n$. This shows that $\pi_T(f) \in V \cap \pi_T(B) = \emptyset$ which is a contradiction.

Recall that our contradiction was obtained assuming that some dense normal $N \subset C_p(X)$ is not Lindelöf; so we proved that any normal dense subspace of $C_p(X)$ is Lindelöf. Finally, if X is a Corson compact space then it is ω-monolithic and has countable tightness (see Problem 120); so any normal dense $Y \subset C_p(X)$ is Lindelöf and hence our solution is complete.

U.195. *Let X be a Corson compact space. Prove that there exists a σ-discrete set $Y \subset C_p(X)$ which separates the points of X.*

Solution. Fix a point-countable T_0-separating family $\mathcal{U}$ of open F_σ-subsets of the space X (such a family exists by Problem 118); there is no loss of generality to assume that $U \neq X$ for any $U \in \mathcal{U}$. For any $U \in \mathcal{U}$ there is a function $f_U \in C_p(X)$ such that $X \setminus U = f^{-1}(0)$ (see Fact 1 of S.358). The family $\mathcal{U}$ being T_0-separating, the set $F = \{f_U : U \in \mathcal{U}\}$ separates the points of X.

It turns out that

(1) the space F is locally countable, i.e., for any function $f_0 \in F$ there is a countable $H \in \tau(f_0, F)$.

To see that the property (1) holds, let u be the function which is identically zero on X. It follows from $f_0 \neq u$ that there are $x_1, \ldots, x_n \in X$ and $O_1, \ldots, O_n \in \tau^*(\mathbb{R})$ such that $f_0 \in W = \{f \in C_p(X) : f(x_i) \in O_i$ for any $i \leq n\}$ and $u \notin W$. Thus there is $k \leq n$ such that $0 \notin O_k$. The family $\mathcal{U}' = \{U \in \mathcal{U} : x_k \in U\}$ is countable; since $f_U(x_k) = 0$, we have $f_U \notin W$ for any $U \in \mathcal{U} \setminus \mathcal{U}'$. Therefore the set $H = W \cap F \in \tau(f_0, F)$ is countable and (1) is proved.

Now let $\mathcal{V}$ be a maximal disjoint family of non-empty countable open subsets of F. It is an easy exercise to prove, using (1), that $Y = \bigcup \mathcal{V}$ is dense in F. The set Y also separates the points of X (see Fact 2 of S.351). Choose an enumeration $\{f_V^i : i \in \omega\}$ of every set $V \in \mathcal{V}$. It is evident that the set $Y_i = \{f_V^i : V \in \mathcal{V}\}$ is discrete for any $i \in \omega$. Therefore $Y = \bigcup \{Y_i : i \in \omega\}$ is a σ-discrete subspace of $C_p(X)$ which separates the points of X.

U.196. *Prove that, under Continuum Hypothesis, there exists a compact space X such that no σ-discrete $Y \subset C_p(X)$ separates the points of X.*

Solution. If CH holds then there is a compact non-metrizable space X such that $hd^*(X) = \omega$ (see SFFS-099 and SFFS-027). Now we can apply SFFS-025 to see that $s^*(C_p(X)) = s^*(X) \leq hd^*(X) = \omega$. Thus any σ-discrete $Y \subset C_p(X)$ is countable; so if Y separates the points of X then X has to be metrizable (see TFS-166) which is a contradiction. Therefore no σ-discrete subspace $Y \subset C_p(X)$ separates the points of X.

U.197. *Let X be a metrizable space. Prove that there is a discrete $Y \subset C_p(X)$ which separates the points of X.*

Solution. If X is finite then there is a finite subspace of $C_p(X)$ which separates the points of X; so assume that X is infinite. Fix a base $\mathcal{B} \subset \tau^*(X)$ in the space X such that $B \neq X$ for any $B \in \mathcal{B}$ and $\mathcal{B} = \bigcup_{n \in \omega} \mathcal{B}_n$ while every family $\mathcal{B}_n$ is discrete (see TFS-221). Pick a point $x_B \in B$ for any $B \in \mathcal{B}$. Every open subset of

X is a cozero-set (see Fact 1 of S.358) so, for any $n \in \omega$ and $B \in \mathcal{B}_n$ there exists a function $f_B : X \to [0, 2^{-n}]$ such that $f(x_B) = 2^{-n}$ and $X \setminus B = f_B^{-1}(0)$.

It is easy to see that the set $Y = \{f_B : B \in \mathcal{B}\} \subset C_p(X)$ separates the points of X. To prove that Y is discrete take any $h \in Y$; there are $m \in \omega$ and $B \in \mathcal{B}_m$ such that $h = f_B$. Let $\varepsilon = 2^{-m-1}$; the set $O = \{f \in C_p(X) : |f(x_B) - f_B(x_B)| < \varepsilon\}$ is an open neighbourhood of f_B in $C_p(X)$. If $n \geq m + 1$ and $U \in \mathcal{B}_n$ then $f_U(x_B) \leq 2^{-n}$ and hence $|f_U(x_B) - f_B(x_B)| \geq 2^{-m} - 2^{-n} \geq 2^{-m} - 2^{-m-1} = 2^{-m-1} = \varepsilon$ which shows that $f_U \notin O$. Since every family $\mathcal{B}_n$ is discrete, only finitely many elements of $\mathcal{B}' = \bigcup\{\mathcal{B}_n : n \leq m\}$ contain x_B. If $U \in \mathcal{B}'$ and $x_B \notin U$ then $f_U(x_B) = 0$ and hence $|f_U(x_B) - f_B(x_B)| = 2^{-m} > \varepsilon$, i.e., $f_U \notin O$. This proves that the set $O \in \tau(h, C_p(X))$ contains only finitely many elements of Y. The function $h \in Y$ was chosen arbitrarily; so every $h \in Y$ has an open neighbourhood O in $C_p(X)$ such that $O \cap Y$ is finite. Therefore Y is a discrete subspace of $C_p(X)$ which separates the points of X.

U.198. *Prove that, for each cardinal κ, there exists a discrete $Y \subset C_p(\mathbb{I}^\kappa)$ which separates the points of $\mathbb{I}^\kappa$.*

Solution. Let us consider that κ carries a discrete topology, i.e., κ is a discrete space; then $\mathbb{R}^\kappa = C_p(\kappa)$ and $\mathbb{I}^\kappa = C_p(\kappa, \mathbb{I})$. For any $\alpha \in \kappa$ let $\varphi_\alpha(f) = f(\alpha)$ for any $f \in \mathbb{I}^\kappa$; then $\varphi_\alpha \in C_p(\mathbb{I}^\kappa)$ for any $\alpha \in \kappa$ and the map $\varphi : \kappa \to C_p(\mathbb{I}^\kappa)$ defined by $\varphi(\alpha) = \varphi_\alpha$ for any $\alpha \in \kappa$, is continuous (see TFS-166). In any Tychonoff space Z the set $C_p(Z, \mathbb{I})$ separates the points and the closed subsets of Z so the same is true for $Z = \kappa$ and hence φ is an embedding by TFS-166. Therefore the space $Y = \varphi(\kappa) \subset C_p(\mathbb{I}^\kappa)$ is discrete. Finally, observe that Y separates the points of $C_p(\mathbb{I}^\kappa)$ (this was also proved in TFS-166); so Y is a discrete subspace of $C_p(\mathbb{I}^\kappa)$ which separates the points of the space $\mathbb{I}^\kappa$.

U.199. *Prove that $C_p(\beta\omega \setminus \omega)$ cannot be condensed into $\Sigma_*(A)$ for any A.*

Solution. Assume towards a contradiction that A is a set such that there exists a condensation $\varphi : C_p(\beta\omega \setminus \omega) \to Y$ for some $Y \subset \Sigma_*(A)$. Observe that A cannot be countable for otherwise $iw(C_p(\beta\omega \setminus \omega)) = \omega$ and hence $\beta\omega \setminus \omega$ is separable which is a contradiction with TFS-371. Thus A is uncountable and hence we can consider that there is an uncountable cardinal κ such that $Y \subset C_p(A(\kappa))$ (see Problem 105).

Let $\pi : C_p(C_p(A(\kappa))) \to C_p(Y)$ be the restriction map; there exists a subspace $F \subset C_p(C_p(A(\kappa)))$ such that $F \simeq A(\kappa)$ and F separates the points of $C_p(A(\kappa))$ (see TFS-166 and TFS-167). Consequently, the set $G = \pi(F) \subset C_p(Y)$ separates the points of the space Y. It is easy to see that any continuous image of $A(\kappa)$ is either finite or homeomorphic to the one-point compactification of a discrete space; so $G \simeq A(\lambda)$ for some cardinal λ.

If G is countable then $iw(Y) \leq \omega$ (see TFS-166); since $C_p(\beta\omega \setminus \omega)$ condenses onto Y, we have $iw(C_p(\beta\omega \setminus \omega)) = \omega$ which we already saw to be impossible. Thus $\lambda > \omega$; since the dual map φ^* embeds $C_p(Y)$ in $C_p(C_p(\beta\omega \setminus \omega))$ (see TFS-163), the space $A(\lambda)$ embeds in $C_p(C_p(\beta\omega \setminus \omega))$ which shows that $p(C_p(\beta\omega \setminus \omega)) \geq \lambda > \omega$ (see TFS-178). However, $C_p(\beta\omega \setminus \omega)$ is a continuous image of $C_p(\beta\omega)$ (TFS-380) which, together with $p(C_p(\beta\omega)) = \omega$ (TFS-382) implies that $p(C_p(\beta\omega\backslash\omega)) \leq \omega$; this contradiction shows that $C_p(\beta\omega\backslash\omega)$ cannot be condensed into $\Sigma_*(A)$.

U.200. *Prove that, for any Corson compact X and any $n \in \mathbb{N}$, the space $C_{p,n}(X)$ linearly condenses onto a subspace of $\Sigma(A)$ for some A.*

Solution. The symbol $\mathbb{P}$ stands for the space of the irrationals which we identify with ω^ω. A space Z *condenses into* a space Y if there is a condensation of Z onto a subspace of Y. Such a condensation is called *a condensation of Z into Y*.

Fact 1. If Z is a primarily Lindelöf space then $C_{p,2n}(Z)$ contains a dense primarily Lindelöf subspace for any $n \in \mathbb{N}$.

Proof. If $n = 1$ then let $e_z(f) = f(z)$ for any point $z \in Z$ and function $f \in C_p(Z)$. Then $e_z \in C_p(C_p(Z))$ for any $z \in Z$ and the subspace $Z' = \{e_z : z \in Z\}$ of the space $C_p(C_p(Z))$ is homeomorphic to Z by TFS-167. Furthermore, the minimal subalgebra $A(Z')$ of $C_p(C_p(Z))$ which contains Z' is dense in $C_p(C_p(Z))$ because Z' and hence $A(Z')$ separates the points of $C_p(Z)$ (see TFS-192).

Since primarily Lindelöf spaces form a weakly k-directed class by Problem 146, the set $A(Z')$ is a countable union of primarily Lindelöf spaces by Problem 006; so $A(Z')$ is primarily Lindelöf by Problem 142. Thus $A(Z')$ is a dense primarily Lindelöf subspace of $C_p(C_p(Z))$ which proves our Fact for $n = 1$.

Now assume that $k \in \mathbb{N}$ and we have proved that $Y = C_{p,2k}(Z)$ has a dense primarily Lindelöf subspace L. The space $C_{p,2(k+1)}(Z)$ is homeomorphic to $C_p(C_p(Y))$. Again, let $e_y(f) = f(y)$ for any $y \in Y$ and $f \in C_p(Y)$; then $e_y \in C_p(C_p(Y))$ for any $y \in Y$ and the map $e : Y \to C_p(C_p(Y))$ defined by $e(y) = e_y$ for any $y \in Y$ is an embedding by TFS-167. Thus the subspace $Y' = \{e_y : y \in Y\}$ of the space $C_p(C_p(Y))$ is homeomorphic to Y and $L' = \{e_y : y \in L\}$ is a dense primarily Lindelöf subspace of the set Y'. The set Y' separates the points of $C_p(Y)$ and hence so does L' being dense in Y'. Therefore the minimal subalgebra $A(L')$ of the space $C_p(C_p(Y))$ which contains L' is dense in $C_p(C_p(Y))$ by TFS-192. As before, we can observe, applying Problems 146, 006 and 142 that $A(L')$ is primarily Lindelöf. Thus, $A(L')$ is a dense primarily Lindelöf subspace of $C_p(C_p(Y)) \simeq C_{p,2(k+1)}(Z)$; so our inductive procedure shows that $C_{p,2n}(Z)$ contains a dense primarily Lindelöf subspace for any $n \in \mathbb{N}$, i.e., Fact 1 is proved.

Fact 2. If F is a non-empty closed subspace of a Σ-product of real lines then $C_p(F)$ condenses linearly into $\Sigma(B)$ for some B.

Proof. Suppose that F is a closed subspace of $\Sigma(T)$ for some T; we will proceed by induction on the cardinal $w(F) = d(F)$; if $d(F) = \omega$ take a dense countable $D \subset F$ and observe that the restriction map condenses $C_p(F)$ linearly into $\mathbb{R}^D$ which embeds linearly in $\mathbb{R}^\omega$. Since $\mathbb{R}^\omega$ embeds linearly in $\Sigma(\omega_1)$, we proved our Fact for all separable sets F.

Now assume that $\kappa = d(F)$ is an uncountable cardinal and we proved, for any cardinal $\lambda < \kappa$ that, for any closed $G \subset \Sigma(T)$ with $d(G) \leq \lambda$, the space $C_p(G)$ condenses linearly into a Σ-product of real lines. Fix a dense set $D = \{x_\alpha : \alpha < \kappa\}$

in the space F. For any $x \in \Sigma(T)$ let $\mathrm{supp}(x) = \{t \in T : x(t) \neq 0\}$; then $\mathrm{supp}(x)$ is a countable set for any $x \in \Sigma(T)$. Given a set $S \subset T$ and $x \in \Sigma(T)$ let $r_S(x)(t) = x(t)$ if $t \in S$ and $r_S(x)(t) = 0$ for all $t \in T \backslash S$. Then $r_S(x) \in \Sigma(T)$ for any $x \in \Sigma(T)$ and the map $r_S : \Sigma(T) \to \Sigma(T)$ is a retraction for any $S \subset T$. We will also need the family $\mathcal{F} = \{S \subset T : r_S(F) \subset F\}$.

It follows from Problem 152 that there exists a non-empty countable set $E_0 \in \mathcal{F}$. Assume that $\alpha < \kappa$ and we have a family $\{E_\beta : \beta < \alpha\} \subset \mathcal{F}$ with the following properties:

(1) $\gamma < \beta < \alpha$ implies $E_\gamma \subset E_\beta$ and $\mathrm{supp}(x_\gamma) \subset E_\beta$;
(2) $|E_\beta| \leq |\beta| \cdot \omega$ for any $\beta < \alpha$;
(3) if $\beta < \alpha$ is a limit ordinal then $E_\beta = \bigcup\{E_\gamma : \gamma < \beta\}$.

If α is a limit ordinal then let $E_\alpha = \bigcup_{\beta < \alpha} E_\beta$; then $E_\alpha \in \mathcal{F}$ by Problem 152 and it is evident that the properties (1)–(3) still hold for all $\beta \leq \alpha$. If $\alpha = \alpha_0 + 1$ then it is a consequence of Problem 152 that there exists a set $E_\alpha \in \mathcal{F}$ such that $E_{\alpha_0} \cup \mathrm{supp}(x_{\alpha_0}) \subset E_\alpha$ and $|E_\alpha| \leq |E_{\alpha_0} \cup \mathrm{supp}(x_{\alpha_0})| \cdot \omega \leq |\alpha| \cdot \omega$. It is also clear that (1)–(3) hold for all $\beta \leq \alpha$; so our inductive procedure can be continued to construct a family $\{E_\alpha : \alpha < \kappa\} \subset \mathcal{F}$ with (1),(2) and (3) fulfilled for any $\beta < \kappa$.

It is evident that the map $r_\alpha = r_{E_\alpha}|F$ is a retraction on F. We claim that

(4) the set $\Omega(f) = \{\alpha < \kappa : f \circ r_\alpha \neq f \circ r_{\alpha+1}\}$ is countable for any $f \in C_p(F)$.

To see that the property (4) is true suppose not; then there is $f \in C_p(F)$ such that $\Omega(f)$ is uncountable and hence we can choose a point $z_\alpha \in F$ such that $f(r_\alpha(z_\alpha)) \neq f(r_{\alpha+1}(z_\alpha))$ for any $\alpha \in \Omega(f)$. There is an uncountable set $\Omega \subset \Omega(f)$ and $\varepsilon > 0$ such that $|f(r_\alpha(z_\alpha)) - f(r_{\alpha+1}(z_\alpha))| > \varepsilon$ for any $\alpha \in \Omega$. Since $ext(F) \leq \omega$ (see Problems 152, 160 and 161), the set $\{r_\alpha(z_\alpha) : \alpha \in \Omega\}$ has an accumulation point $z \in F$.

The function f being continuous at z, there is a set $S = \{t_1, \ldots, t_n\} \subset T$ and $O_1, \ldots, O_n \in \tau(\mathbb{R})$ such that $\mathrm{diam}(f(U)) < \varepsilon$ where $z \in U = \{x \in F : x(t_i) \in O_i \text{ for all } i \leq n\}$. Therefore the set $M = \{\alpha \in \Omega : r_\alpha(z_\alpha) \in U\}$ is infinite; the family $\{E_{\alpha+1} \backslash E_\alpha : \alpha \in M\}$ is disjoint; so there is $\alpha \in M$ such that $(E_{\alpha+1} \backslash E_\alpha) \cap S = \emptyset$.

Now observe that $r_{\alpha+1}(z_\alpha) = r_{E_{\alpha+1}}(z_\alpha)$ can have coordinates distinct from the coordinates of the point $r_\alpha(z_\alpha) = r_{E_\alpha}(z_\alpha)$ only on the set $E_{\alpha+1} \backslash E_\alpha$. Thus $r_{\alpha+1}(z_\alpha)(t_i) = r_\alpha(z_\alpha)(t_i) \in O_i$ for every $i \leq n$ which implies that $r_{\alpha+1}(z_\alpha) \in U$. However, then $\varepsilon < |f(r_\alpha(z_\alpha)) - f(r_{\alpha+1}(z_\alpha))| \leq \mathrm{diam}(f(U)) < \varepsilon$; this contradiction shows that (4) is proved.

For any $\alpha < \kappa$ let $F_\alpha = r_\alpha(F)$; it is easy to see that the family $\{F_\alpha : \alpha < \kappa\}$ is non-decreasing and $d(F_\alpha) = w(F_\alpha) \leq |E_\alpha| < \kappa$ for every $\alpha < \kappa$ (see (2)). We will need the following property of the family $\{F_\alpha : \alpha < \kappa\}$:

(5) if $\alpha < \kappa$ is a non-zero limit ordinal then $H_\alpha = \bigcup\{F_\beta : \beta < \alpha\}$ is dense in F_α.

If (5) is not true then we can choose a point $y \in F_\alpha \backslash \overline{H}_\alpha$. There is a set $S' = \{s_1, \ldots, s_m\} \subset T$ and $W_1, \ldots, W_m \in \tau(\mathbb{R})$ such that $y \in W = \{x \in F : x(s_i) \in W_i \text{ for any } i \leq m\} \subset F \backslash H_\alpha$. It follows from (3) that there is $\beta < \alpha$ such that $E_\beta \cap S' = E_\alpha \cap S'$. We have $r_\beta(y) \in F_\beta$ while the distinct coordinates of y and

$r_\beta(y)$ lie in $E_\alpha \backslash E_\beta$; so it follows from our choice of β that $(E_\alpha \backslash E_\beta) \cap S' = \emptyset$. As a consequence, $r_\beta(y)(s_i) = y(s_i) \in W_i$ for any $i \le m$, i.e., $r_\beta(y) \in W \cap F_\beta$ which is a contradiction with $F_\beta \subset H_\alpha$ and $W \cap H_\alpha = \emptyset$. This proves the property (5).

Denote by π_α the restriction map from $C_p(F)$ onto $C_p(F_\alpha)$. Define the dual map $r_\alpha^* : C_p(F_\alpha) \to C_p(F)$ by $r_\alpha^*(f) = f \circ r_\alpha$ for any $f \in C_p(F_\alpha)$; then r_α^* is a linear embedding for any $\alpha < \kappa$. Furthermore, $s_\alpha = r_\alpha^* \circ \pi_\alpha : C_p(F) \to C_\alpha = r_\alpha^*(C_p(F_\alpha))$ is a linear retraction by Problem 147; we leave to the reader the straightforward verification that $\alpha < \beta < \kappa$ implies $C_\alpha \subset C_\beta$. Since F_α is a closed subspace of $\Sigma(T)$ with $d(F_\alpha) < \kappa$ and $C_\alpha \simeq C_p(F_\alpha)$, our induction hypothesis shows that, for any $\alpha < \kappa$, there exists a linear injective map $\delta_\alpha : C_\alpha \to \Sigma(B_\alpha)$ for some set B_α; we can assume, without loss of generality, that the family $\{B_\alpha : \alpha < \kappa\}$ is disjoint. Given $\alpha < \kappa$ let $u_\alpha \in \Sigma(B_\alpha)$ be defined by $u_\alpha(b) = 0$ for all $b \in B_\alpha$.

Now let $\mu_0(f) = \delta_0(s_0(f))$ and $\mu_{\alpha+1}(f) = \delta_{\alpha+1}(s_{\alpha+1}(f) - s_\alpha(f))$ for any $\alpha < \kappa$ and $f \in C_p(F)$. The maps μ_0 and $\mu_{\alpha+1} : C_p(F) \to \Sigma(B_{\alpha+1})$ are continuous and linear for any $\alpha < \kappa$; observe also that every $\mu_{\alpha+1}$ is well defined because $C_\alpha \subset C_{\alpha+1}$ and hence $s_{\alpha+1}(f) - s_\alpha(f) \in C_{\alpha+1}$ for any $f \in C_p(F)$ and $\alpha < \kappa$. The map δ_α being linear, an immediate consequence of (4) is that

(6) $\mu_{\alpha+1}(f) = u_{\alpha+1}$ for any $\alpha \notin \Omega(f)$.

We are finally ready to construct the promised linear condensation of the space $C_p(F)$; let $B = B_0 \cup \bigcup\{B_{\alpha+1} : \alpha < \kappa\}$ and $\mu = \mu_0 \Delta(\Delta\{\mu_{\alpha+1} : \alpha < \kappa\})$; it is immediate that $\mu : C_p(F) \to \prod\{\Sigma(B_{\alpha+1}) : \alpha < \kappa\} \times \Sigma(B_0) \subset \mathbb{R}^B$. The diagonal product of linear continuous maps is, evidently, a linear continuous map; so μ is linear and continuous.

To see that μ is injective take distinct functions $f, g \in C_p(F)$. Since $W = \{x \in F : f(x) \ne g(x)\} \in \tau^*(F)$ and D is dense in F, there is $\alpha < \kappa$ such that $f(x_\alpha) \ne g(x_\alpha)$. We have $\text{supp}(x_\alpha) \subset E_{\alpha+1}$ which implies $r_{\alpha+1}(x_\alpha) = x_\alpha \in F_{\alpha+1}$ and therefore $\pi_{\alpha+1}(f) \ne \pi_{\alpha+1}(g)$. As a consequence, the set $J = \{\alpha < \kappa : \pi_\alpha(f) \ne \pi_\alpha(g)\}$ is non-empty; let $\alpha = \min J$. If $\alpha = 0$ then $\pi_0(f) \ne \pi_0(g)$ which implies $s_0(f) \ne s_0(g)$ and hence $\mu_0(f) \ne \mu_0(g)$ because δ_0 is an injection.

Now assume that $\alpha > 0$ is a limit ordinal. Then we have $f|F_\alpha \ne g|F_\alpha$ and hence $f|H_\alpha \ne g|H_\alpha$ by the property (5). Therefore there is $\beta < \alpha$ for which $f(q) \ne g(q)$ for some $q \in F_\beta$, i.e., $f|F_\beta \ne g|F_\beta$ which is a contradiction with the choice of α. Therefore $\alpha > 0$ cannot be a limit and hence $\alpha = \beta + 1$. By the choice of α we have $\pi_\beta(f) = \pi_\beta(g)$ while $\pi_{\beta+1}(f) \ne \pi_{\beta+1}(g)$ and therefore $s_\beta(f) = s_\beta(g)$ while $s_{\beta+1}(f) \ne s_{\beta+1}(g)$. This implies that $s_{\beta+1}(f) - s_\beta(f) \ne s_{\beta+1}(g) - s_\beta(g)$ whence $\mu_{\beta+1}(f) \ne \mu_{\beta+1}(g)$ because $\delta_{\beta+1}$ is an injection. Thus $\mu(f) \ne \mu(g)$ and hence we proved that μ is injective.

The last thing we have to show is that $\mu(C_p(F)) \subset \Sigma(B)$ so take any function $f \in C_p(F)$. The property (6) implies that $\mu(f)(b) = \mu_{\alpha+1}(f)(b) = u_{\alpha+1}(b) = 0$ whenever $\alpha \notin \Omega(f)$. The set $\Omega(f)$ must be countable by (4); if $\alpha \in \Omega(f)$ then $\mu_{\alpha+1}(f) \in \Sigma(B_{\alpha+1})$ by the definition of $\delta_{\alpha+1}$. Thus the set $\{b \in B : \mu(f)(b) \ne 0\}$ is contained in the set $\bigcup\{\{b \in B_{\alpha+1} : \mu_{\alpha+1}(f)(b) \ne 0\} : \alpha \in \Omega(f)\} \cup \{b \in B_0 : \mu_0(f)(b) \ne 0\}$ which is countable so $\mu(f) \in \Sigma(B)$ for any $f \in C_p(F)$ and therefore μ is a linear continuous injective map from $C_p(F)$ to $\Sigma(B)$. Fact 2 is proved.

Fact 3. For any space Z and $n \in \omega$ the space $C_{p,2n+1}(Z)$ condenses linearly into $C_p(Z^\omega \times \mathbb{P})$.

Proof. Let $\mathcal{P}$ be the class of spaces representable as a continuous image of $Z^\omega \times \mathbb{P}$. Any metrizable compact space K is a continuous image of $\mathbb{P}$ (see e.g., SFFS-328); so K is a continuous image of $Z^\omega \times \mathbb{P}$, i.e., $K \in \mathcal{P}$. Since $\mathbb{P}^\omega \simeq \mathbb{P}$, it is an easy exercise to see that $P \in \mathcal{P}$ implies $P^\omega \in \mathcal{P}$. It is evident that a continuous image of a space from $\mathcal{P}$ is in $\mathcal{P}$; so $\mathcal{P}$ is a weakly k-directed class. Besides, a countable union of spaces from $\mathcal{P}$ is also in $\mathcal{P}$; this easily follows from $\mathbb{P} \times \omega \simeq \mathbb{P}$.

Let us show by induction that, for any $n \in \mathbb{N}$, the space $C_{p,2n}(Z)$ has a dense subspace which belongs to $\mathcal{P}$. Let $e_z(f) = f(z)$ for any $z \in Z$ and $f \in C_p(Z)$. Then $e_z \in C_p(C_p(Z))$ for any $z \in Z$ and the subspace $Z' = \{e_z : z \in Z\}$ of the space $C_p(C_p(Z))$ is homeomorphic to Z by TFS-167. Furthermore, the minimal subalgebra $A(Z')$ of $C_p(C_p(Z))$ which contains Z' is dense in $C_p(C_p(Z))$ because Z' and hence $A(Z')$ separates the points of $C_p(Z)$ (see TFS-192).

Since $\mathcal{P}$ is a weakly k-directed class, the set $A(Z')$ is a countable union of elements of $\mathcal{P}$ (see Problem 006); so $A(Z') \in \mathcal{P}$, i.e., $A(Z')$ is a dense subspace of $C_p(C_p(Z))$ which belongs to $\mathcal{P}$. This proves our Fact for $n = 1$.

Now assume that $k \in \mathbb{N}$ and we have proved that $Y = C_{p,2k}(Z)$ has a dense subspace $L \in \mathcal{P}$. The space $C_{p,2(k+1)}(Z)$ is homeomorphic to $C_p(C_p(Y))$. Again, let $e_y(f) = f(y)$ for any $y \in Y$ and $f \in C_p(Y)$; then $e_y \in C_p(C_p(Y))$ for any $y \in Y$ and the map $e : Y \to C_p(C_p(Y))$ defined by $e(y) = e_y$ for any $y \in Y$ is an embedding by TFS-167. Thus the subspace $Y' = \{e_y : y \in Y\}$ of the space $C_p(C_p(Y))$ is homeomorphic to Y and $L' = \{e_y : y \in L\} \in \mathcal{P}$ is a dense subspace of the set Y'. The set Y' separates the points of $C_p(Y)$ and hence so does L' being dense in Y'. Therefore the minimal subalgebra $A(L')$ of the space $C_p(C_p(Y))$ which contains L' is dense in $C_p(C_p(Y))$ by TFS-192. As before, we can observe, that $A(L') \in \mathcal{P}$. Thus, $A(L') \in \mathcal{P}$ is a dense subspace of $C_p(C_p(Y)) \simeq C_{p,2(k+1)}(Z)$; so our inductive procedure shows that $C_{p,2n}(Z)$ contains a dense subspace $L_n \in \mathcal{P}$ for any $n \in \mathbb{N}$.

Thus the restriction map π condenses $C_{p,2n+1}(Z) = C_p(C_{p,2n}(Z))$ linearly into $C_p(L_n)$. There is a continuous onto map $\varphi : Z^\omega \times \mathbb{P} \to L_n$; so the dual map φ^* linearly embeds $C_p(L_n)$ in $C_p(Z^\omega \times \mathbb{P})$. It is immediate that $\varphi^* \circ \pi$ linearly condenses the space $C_{p,2n+1}(Z)$ into $C_p(Z^\omega \times \mathbb{P})$; so Fact 3 is proved.

Returning to our solution take any $n \in \mathbb{N}$ and let $Y = C_{p,2n-1}(X)$. Since X is Corson compact, the space $Z = C_p(X)$ is primarily Lindelöf by Problem 150; therefore $C_{p,2n-1}(X)$ is primarily Lindelöf for $n = 1$. If $n > 1$ then $C_{p,2n-1}(X) = C_{p,2n-2}(Z)$; so Y has a dense primarily Lindelöf subspace L by Fact 1. Therefore the restriction map $\pi : C_p(Y) \to C_p(L)$ is linear and injective. By Problem 149, there exists a linear injective map $\varphi : C_p(L) \to \Sigma(A)$ for some set A. Consequently, the map $\varphi \circ \pi$ linearly condenses the space $C_p(Y) \simeq C_{p,2n}(X)$ into $\Sigma(A)$. Therefore $C_{p,2n}(X)$ linearly condenses into a Σ-product of real lines for any $n \in \mathbb{N}$.

Finally, take any number $n \in \omega$; there exists a linear injective continuous mapping $\varphi : C_{p,2n+1}(X) \to C_p(X^\omega \times \mathbb{P})$ by Fact 3. The space X^ω is Corson compact by Problem 137; so it can be embedded (as a closed subspace) in $\Sigma(B)$ for some set B. Besides, $\mathbb{P}$ can be embedded in $\mathbb{R}^\omega$ as a closed subspace by TFS-273. Therefore $F = X^\omega \times \mathbb{P}$ is embeddable in a Σ-product of real lines as a closed subspace. Therefore there exists a linear injective continuous map $\delta : C_p(F) \to \Sigma(A)$ for some set A (see Fact 2). It is clear that $\delta \circ \varphi : C_{p,2n+1}(X) \to \Sigma(A)$ is an injective linear continuous map; so $C_{p,2n+1}(X)$ can be linearly condensed into a Σ-product of real lines for any $n \in \omega$ and hence our solution is complete.

U.201. *Suppose that $X = \upsilon Y$ and Z is a subspace of $\mathbb{R}^X$ such that $C_p(X) \subset Z$. Prove that there exists $Z' \subset \mathbb{R}^Y$ such that $C_p(Y) \subset Z'$ and Z' is a continuous image of Z.*

Solution. Let $\pi : \mathbb{R}^X \to \mathbb{R}^Y$ be the restriction map; since every $f \in C_p(Y)$ extends to a continuous map on X (see TFS-412), we have $\pi(C_p(X)) = C_p(Y)$. Therefore $C_p(Y) = \pi(C_p(X)) \subset \pi(Z) \subset \mathbb{R}^Y$ and $Z' = \pi(Z)$ is a continuous image of Z.

U.202. *Suppose that X is σ-compact. Prove that there exists a $K_{\sigma\delta}$-space Z such that $C_p(X) \subset Z \subset \mathbb{R}^X$.*

Solution. We have $X = \bigcup\{X_i : i \in \omega\}$ where every X_i is compact; let $Y_0 = X_0$ and $Y_{i+1} = X_{i+1} \backslash (\bigcup\{X_k : k \leq i\})$ for any $i \in \omega$. Then $Y_i \subset X_i$ for every $i \in \omega$, the family $\mathcal{Y} = \{Y_i : i \in \omega\}$ is disjoint and $X = \bigcup\{Y_i : i \in \omega\}$; throwing away the empty elements of the family $\mathcal{Y}$ if necessary we can assume, without loss of generality, that $Y_i \neq \emptyset$ for any $i \in \omega$.

For every $i \in \omega$ let $\pi_i : \mathbb{R}^X \to \mathbb{R}^{Y_i}$ be the natural projection (which coincides with the restriction map) onto the face $\mathbb{R}^{Y_i}$. The set $P(n, i) = \{f \in \mathbb{R}^{Y_i} : |f(x)| \leq n$ for all $x \in Y_i\}$ is compact for all $n, i \in \omega$ being homeomorphic to the space $[-n, n]^{Y_i}$. The set Y_i is contained in a compact subspace $X_i \subset X$; so every $f \in C_p(X)$ is bounded on Y_i which shows that $\pi_i(C_p(X)) \subset Q(i) = \bigcup\{P(n, i) : n \in \omega\}$ for every $i \in \omega$.

Therefore $C_p(X) \subset Z = \prod_{i<\omega} Q(i) \subset \mathbb{R}^X = \prod_{i<\omega} \mathbb{R}^{Y_i}$ and, to finish the proof, it suffices to observe that Z is a $K_{\sigma\delta}$-space being a countable product of σ-compact spaces (see TFS-338).

U.203. *Suppose that υX is σ-compact. Prove that there exists a K-analytic space Z such that $C_p(X) \subset Z \subset \mathbb{R}^X$.*

Solution. The space υX being σ-compact, there exists a $K_{\sigma\delta}$-space T such that $C_p(\upsilon X) \subset T \subset \mathbb{R}^{\upsilon X}$ by Problem 202. Apply Problem 201 to conclude that there is a space $Z \subset \mathbb{R}^X$ such that Z is a continuous image of T and $C_p(X) \subset Z$. Thus $C_p(X) \subset Z \subset \mathbb{R}^X$ and the space Z is K-analytic because it is a continuous image of the $K_{\sigma\delta}$-space T.

U.204. *Prove that X is pseudocompact if and only if there exists a σ-compact space Z such that $C_p(X) \subset Z \subset \mathbb{R}^X$.*

Solution. If X is pseudocompact then every function $f \in C_p(X)$ is bounded on X; so $C_p(X) \subset Z = \bigcup\{[-n,n]^X : n \in \omega\} \subset \mathbb{R}^X$ and it is evident that Z is σ-compact. This proves necessity.

Now, assume that X is not pseudocompact while there is a σ-compact Z such that $C_p(X) \subset Z \subset \mathbb{R}^X$. There is a countably infinite closed discrete $D \subset X$ which is C-embedded in X (see Fact 1 of S.350); let $\pi : \mathbb{R}^X \to \mathbb{R}^D$ be the restriction map. We have $\mathbb{R}^D = \pi(C_p(X)) \subset \pi(Z)$ and therefore $\mathbb{R}^D = \pi(Z)$ which implies that $\mathbb{R}^\omega \simeq \mathbb{R}^D$ is σ-compact being a continuous image of a σ-compact space Z. This contradiction with Fact 2 of S.399 shows that there is no σ-compact space Z with $C_p(X) \subset Z \subset \mathbb{R}^X$, i.e., we established sufficiency.

U.205. *Give an example of a Lindelöf space X for which there exists no Lindelöf space Z such that $C_p(X) \subset Z \subset \mathbb{R}^X$.*

Solution. Let $X = L(\omega_1)$ be the Lindelöfication of the discrete space of cardinality ω_1. The space X is Lindelöf; so assume, towards a contradiction, that there is a Lindelöf Z such that $C_p(X) \subset Z \subset \mathbb{R}^X$. Since X is a Lindelöf P-space, every countable subset of X is closed and C-embedded in X. This implies that the space $\mathbb{R}^X$ is canonically homeomorphic to $\upsilon(C_p(X))$ (see TFS-485) and hence $C_p(X)$ has to be C-embedded in $\mathbb{R}^X$ (see TFS-413). But then $C_p(X)$ is also C-embedded in its Lindelöf (and hence realcompact) extension Z. Applying Fact 1 of S.438 we conclude that $Z \simeq \upsilon(C_p(X)) \simeq \mathbb{R}^X \simeq \mathbb{R}^{\omega_1}$ which is a contradiction (see e.g., Fact 3 of S.215). As a consequence, there exists no Lindelöf space Z for which $C_p(X) \subset Z \subset \mathbb{R}^X$.

U.206. *Prove that υX is a Lindelöf Σ-space if and only if $C_p(X) \subset Z \subset \mathbb{R}^X$ for some Lindelöf Σ-space Z. In particular,*

  (i) *if $C_p(X)$ is a Lindelöf Σ-space, then υX is a Lindelöf Σ-space;*
  (ii) *(Uspenskij's theorem) if X is a Lindelöf Σ-space then there exists a Lindelöf Σ-space Z such that $C_p(X) \subset Z \subset \mathbb{R}^X$;*
  (iii) *if $\upsilon(C_p(X))$ is a Lindelöf Σ-space then υX is Lindelöf Σ.*

Solution. If υX is a Lindelöf Σ-space then there exists a Lindelöf Σ-space Y such that $C_p(\upsilon X) \subset Y \subset \mathbb{R}^{\upsilon X}$ by Fact 1 of T.399. Apply Problem 201 to see that there exists a set $Z \subset \mathbb{R}^X$ such that $C_p(X) \subset Z$ and Z is a continuous image of Y. Therefore Z is a Lindelöf Σ-space such that $C_p(X) \subset Z \subset \mathbb{R}^X$, i.e., we proved necessity.

Now suppose that there is a Lindelöf Σ-space Z such that $C_p(X) \subset Z \subset \mathbb{R}^X$. For every $x \in X$ let $e_x(f) = f(x)$ for any $f \in \mathbb{R}^X$. Then $e_x : \mathbb{R}^X \to \mathbb{R}$ is a continuous function because it coincides with the natural projection of $\mathbb{R}^X$ onto the factor of $\mathbb{R}^X$ determined by x. Let $u_x = e_x | C_p(X)$ for any $x \in X$; then the space $X' = \{u_x : x \in X\} \subset C_p(C_p(X)) \subset \mathbb{R}^{C_p(X)}$ is homeomorphic to X by TFS-167. Besides, X' is C-embedded in $\mathbb{R}^{C_p(X)}$ by TFS-168.

Let $\pi : \mathbb{R}^Z \to \mathbb{R}^{C_p(X)}$ be the restriction map; observe first that $X' \subset \pi(C_p(Z))$ because every $u_x \in X'$ extends to $e_x | Z$, i.e., $u_x = \pi(e_x | Z)$ for any $x \in X$. Apply Fact 1 of T.399 once more to find a Lindelöf Σ-space L such that $C_p(Z) \subset L \subset \mathbb{R}^Z$. The space $M = \pi(L) \subset \mathbb{R}^{C_p(X)}$ is Lindelöf Σ and $X' \subset \pi(C_p(Z)) \subset \pi(L) = M$. Therefore $Y = \mathrm{cl}_M(X')$ is a Lindelöf Σ-space which is an extension of the space X'. The space X' being C-embedded in a larger space $\mathbb{R}^{C_p(X)}$, it is also C-embedded in Y. The space Y is Lindelöf Σ and hence realcompact; so $Y \simeq \upsilon(X')$ by Fact 1 of S.438.

Therefore $\upsilon X \simeq \upsilon(X') \simeq Y$ is a Lindelöf Σ-space and hence we proved the main statement of our Problem. The assertions (i) and (ii) follow trivially. To see that (iii) also holds observe that there is a set $S_X \subset \mathbb{R}^X$ such that $C_p(X) \subset S_X$ and S_X is canonically homeomorphic to $\upsilon(C_p(X))$ (see TFS-438). By our assumption, the space $\upsilon(C_p(X))$ and hence S_X is Lindelöf Σ so we can use the main statement to conclude that υX is Lindelöf Σ and make our solution complete.

U.207. *Given a natural $n \geq 1$, suppose that there exists a Lindelöf Σ-space Z such that $C_{p,n}(X) \subset Z \subset \mathbb{R}^{C_{p,n-1}(X)}$. Prove that there exists a Lindelöf Σ-space Y such that $C_p(X) \subset Y \subset \mathbb{R}^X$.*

Solution. Let us first prove by induction that

(*) if $\upsilon(C_{p,k}(X))$ is a Lindelöf Σ-space for some number $k \in \omega$ then υX also has the Lindelöf Σ-property.

Since (*) evidently holds for $k = 0$ assume that it is fulfilled for all $k \leq m$ and $\upsilon(C_{p,m+1}(X))$ is a Lindelöf Σ-space. An immediate consequence of Problem 206 is that $\upsilon(C_{p,m}(X))$ is Lindelöf Σ so the induction hypothesis shows that υX is Lindelöf Σ and hence (*) is proved.

Returning to our solution observe that if $n = 1$ then we can take $Y = Z$. If $n > 1$ and there is a Lindelöf Σ-space Z with $C_{p,n}(X) \subset Z \subset \mathbb{R}^{C_{p,n-1}(X)}$ then let $T = C_{p,n-1}(X)$. We have $C_p(T) \subset Z \subset \mathbb{R}^T$; so Problem 206 is applicable again to conclude that $\upsilon T = \upsilon(C_{p,n-1}(X))$ is a Lindelöf Σ-space. Thus we can apply (*) to see that υX is a Lindelöf Σ-space. Applying Problem 206 once more we conclude that there exists a Lindelöf Σ-space Y for which $C_p(X) \subset Y \subset \mathbb{R}^X$.

U.208. *Suppose that $C_p(X)$ is a Lindelöf Σ-space. Prove that $C_{p,n}(X)$ is ω-stable and ω-monolithic for any natural n.*

Solution. Any Lindelöf Σ-space is ω-stable by SFFS-266; so $C_p(X)$ is ω-stable and hence X is ω-monolithic by SFFS-152. Furthermore, it follows from Problem 206 that υX is a Lindelöf Σ-space; so we can apply SFFS-267 to conclude that X is also ω-stable. Apply SFFS-152 and SFFS-154 to see that a space Z is both ω-stable and ω-monolithic if and only if so is $C_p(Z)$. Now, it takes a trivial induction to conclude that $C_{p,n}(X)$ is both ω-stable and ω-monolithic for any $n \in \omega$.

U.209. *Prove that a space X is dominated by a space homeomorphic to the irrationals if and only if X is $\mathbb{P}$-dominated.*

Solution. Given a space Z we denote by $\mathcal{K}(Z)$ the family of all compact subsets of Z. Suppose that X is $\mathbb{P}$-dominated and hence there is a $\mathbb{P}$-ordered compact cover $\{F_p : p \in \mathbb{P}\}$ of the space X.

For any $n \in \omega$ let $\pi_n : \omega^\omega \to \omega$ be the natural projection of ω^ω onto its n-th factor. If K is a compact subset of ω^ω then $\pi_n(K)$ is a compact and hence finite subset of ω. Consequently, there is $p \in \omega^\omega$ such that $\pi_n(K) \subset \{0, \ldots, p(n)\}$ for any $n \in \omega$. Therefore the number $u_n(K) = \min\{m \in \omega : \pi_n(K) \subset \{0, \ldots, m\}\}$ is well defined for any $n \in \omega$. Letting $p_K(n) = u_n(K)$ for any $n \in \omega$ we obtain an element $p_K \in \omega^\omega$ for any $K \in \mathcal{K}(\omega^\omega)$.

It is immediate that if $K, L \in \mathcal{K}(\omega^\omega)$ and $K \subset L$ then $p_K \leq p_L$ and therefore $F_{p_K} \subset F_{p_L}$. This shows that letting $Q_K = F_{p_K}$ for any $K \in \mathcal{K}(\omega^\omega)$ we get a family $\mathcal{F} = \{Q_K : K \in \mathcal{K}(\omega^\omega)\}$ of compact subsets of X such that $K \subset L$ implies $Q_K \subset Q_L$. Furthermore, if $x \in X$ then there is $p \in \omega^\omega$ for which $x \in F_p$; it is straightforward that, for the compact set $K = \prod\{\{0, \ldots, p(n)\} : n \in \omega\} \subset \omega^\omega$ we have $p_K = p$ and hence $x \in Q_K$. Thus $\mathcal{F}$ is also a cover of X and hence X is dominated by ω^ω, i.e., we proved sufficiency.

Now assume that X is dominated by ω^ω, i.e., there exists a compact cover $\{F_K : K \in \mathcal{K}(\omega^\omega)\}$ of the space X such that $K \subset L$ implies $F_K \subset F_L$. For any $p \in \omega^\omega$ the set $K_p = \prod\{\{0, \ldots, p(n)\} : n \in \omega\} \subset \omega^\omega$ is compact and it is easy to see that $p \leq q$ implies $K_p \subset K_q$. Let $G_p = F_{K_p}$ for any $p \in \omega^\omega$. Then $\mathcal{G} = \{G_p : p \in \mathbb{P}\}$ is a family of compact subsets of X such that $p \leq q$ implies $G_p \subset G_q$, i.e., $\mathcal{G}$ is $\mathbb{P}$-ordered.

Given an arbitrary point $x \in X$ there is $K \in \mathcal{K}(\omega^\omega)$ such that $x \in F_K$. For any $n \in \omega$ the set $\pi_n(K) \subset \omega$ is compact and hence finite; so there is $p \in \omega^\omega$ such that $\pi_n(K) \subset \{0, \ldots, p(n)\}$ for any $n \in \omega$ which shows that $K \subset K_p$. Consequently, $F_K \subset F_{K_p} = G_p$ whence $x \in G_p$ and therefore $\mathcal{G}$ is a cover of X. Thus X is $\mathbb{P}$-dominated; so we proved necessity and hence our solution is complete.

U.210. *Suppose that X is dominated by a second countable space. Prove that there is a countable family $\mathcal{F}$ of subsets of X which is a network with respect to a cover of X with countably compact subspaces of X.*

Solution. As usual, given a space Z, the symbol $\mathcal{K}(Z)$ stands for the family of all compact subsets of Z. Suppose that X is dominated by a second countable space M; let $\{F_K : K \in \mathcal{K}(M)\}$ be the respective compact cover of X. Fix a countable base $\mathcal{B}$ in M such that $\mathcal{B}$ is closed under finite unions and intersections. It is easy to see that $\mathcal{B}$ is a network for all compact subsets of M, i.e., if $K \subset M$ is compact and $O \in \tau(K, M)$ then there is $B \in \mathcal{B}$ for which $K \subset B \subset O$.

For any set $B \in \mathcal{B}$ let $Q(B) = \bigcup\{F_K : K \in \mathcal{K}(M)$ and $K \subset B\}$; then the family $\mathcal{F} = \{Q(B) : B \in \mathcal{B}\} \subset \exp X$ is countable. Since the family $\mathcal{B}$ is closed under finite intersections, for every $K \in \mathcal{K}(M)$, we can choose an outer base $\mathcal{B}_K = \{B_n^K : n \in \omega\} \subset \mathcal{B}$ of the set K in M in such a way that $\operatorname{cl}(B_{n+1}^K) \subset B_n^K$ for any $n \in \omega$. Let $C_K = \bigcap\{Q(B_n^K) : n \in \omega\}$ for any $K \in \mathcal{K}(M)$. It is immediate that $K \subset C_K$; so $\mathcal{C} = \{C_K : K \in \mathcal{K}(M)\}$ is a cover of the space X. It turns out that all elements of $\mathcal{C}$ are countably compact and $\mathcal{F}$ is a network with respect to $\mathcal{C}$. To prove it we will need the following property.

(*) If $x_n \in Q(B_n^K)$ for any $n \in \omega$ then the sequence $S = \{x_n : n \in \omega\}$ has an accumulation point which belongs to the set C_K.

For any $n \in \omega$ take $K_n \in \mathcal{K}(M)$ such that $K_n \subset B_n^K$ and $x_n \in F_{K_n}$. It is easy to see that, for any $m \in \omega$ the set $H_m = \bigcup\{K_n : n \geq m\} \cup K \subset B_m^K$ is compact; so we have $\{x_n : n \geq m\} \subset F_{H_m}$ for every $m \in \omega$. In particular, $S \subset F_{H_0}$; the set F_{H_0} being compact, the sequence S has accumulation points in F_{H_0}. On the other hand, the set $\{n \in \omega : x_n \notin F_{H_m}\}$ is finite for any $m \in \omega$; so all accumulation points of S have to belong to the set $\bigcap\{F_{H_n} : n \in \omega\} \subset C_K$ which shows that $(*)$ is proved.

Finally, assume that $K \in \mathcal{K}(M)$ and $C_K \subset O \in \tau(X)$. We have $C_K \subset Q(B_n^K)$ for any $n \in \omega$. If there is $x_n \in Q(B_n^K)\backslash O$ for any $n \in \omega$ then $S = \{x_n : n \in \omega\}$ has an accumulation point in C_K by $(*)$. However, $S \subset X\backslash O$; so all accumulation points of S have to belong to $X\backslash O$ which does not meet C_K. This contradiction shows that $C_K \subset Q(B_n^K) \subset O$ for some $n \in \omega$ and therefore $\mathcal{F}$ is a countable network with respect to the cover $\mathcal{C}$. Finally, if $S = \{x_n : n \in \omega\} \subset C_K$ then $x_n \in Q(B_n^K)$ for any $n \in \omega$ so S has an accumulation point in C_K by $(*)$. Therefore C_K is countably compact for any $K \in \mathcal{K}(M)$, i.e., all elements of the cover $\mathcal{C}$ are countably compact.

U.211. *Suppose that a space X has a countable family $\mathcal{F}$ which is a network with respect to a cover of X with countably compact subspaces of X. Prove that υX is a Lindelöf Σ-space.*

Solution. Let $\mathcal{C}$ be a cover of X such that every $C \in \mathcal{C}$ is countably compact and $\mathcal{F}$ is a countable network with respect to $\mathcal{C}$. Let $Y = \bigcup\{\overline{C} : C \in \mathcal{C}\}$ (the bar denotes the closure in υX). The set $\overline{C}$ is compact for any $C \in \mathcal{C}$ (see TFS-415); so $\mathcal{D} = \{\overline{C} : C \in \mathcal{C}\}$ is a compact cover of Y; besides, $X \subset Y \subset \upsilon X$.

The family $\mathcal{G} = \{\overline{F} \cap Y : F \in \mathcal{F}\}$ is countable; suppose that $C \in \mathcal{C}$ and $D = \overline{C} \subset O \in \tau(Y)$. The set D being compact there exists $V \in \tau(D, Y)$ such that $\text{cl}_Y(V) \subset O$. Since $W = V \cap X \in \tau(C, X)$, there exists $F \in \mathcal{F}$ for which $C \subset F \subset W$. Consequently, $D = \overline{C} \subset \overline{F} \subset \overline{W}$; so $G = \overline{F} \cap Y \in \mathcal{G}$ and $D \subset G$. Furthermore, $G \subset \overline{W} \cap Y \subset \overline{V} \cap Y = \text{cl}_Y(V) \subset O$. This proves that $D \subset G \subset O$ and therefore $\mathcal{G}$ is a countable network with respect to a compact cover $\mathcal{D}$ of the space Y. Thus Y is a Lindelöf Σ-space; by TFS-414 and TFS-406 we have $\upsilon Y \simeq Y \simeq \upsilon X$; so υX is a Lindelöf Σ-space.

U.212. *Prove that the property of being dominated by a second countable space is preserved by countable unions, products and intersections as well as by closed subspaces and continuous images.*

Solution. Given a space Z the symbol $\mathcal{K}(Z)$ stands for the family of all compact subsets of Z. Let X be dominated by a second countable space M; fix the respective compact cover $\{F_K : K \in \mathcal{K}(M)\}$ of the space X. If $f : X \to Y$ is a continuous onto map then $\{f(F_K) : K \in \mathcal{K}(M)\}$ is a compact cover of Y which witnesses the domination of Y by M. Therefore

(1) if X is dominated by a second countable space then any continuous image of X is dominated by the same second countable space.

Now, if A is a closed subset of X then the compact cover $\{F_K \cap A : K \in \mathcal{K}(M)\}$ of the set A witnesses the domination of A by M. Thus

(2) if X is dominated by a second countable space then any closed subspace of X is dominated by the same second countable space.

Suppose that a space X_n is dominated by a second countable space M_n and let $\{F_K : K \in \mathcal{K}(M_n)\}$ be the respective compact cover of X_n for any $n \in \omega$. Let $X = \prod_{n<\omega} X_n$ and $M = \prod_{n<\omega} M_n$; the map $\pi_n : M \to M_n$ is the natural projection for any $n \in \omega$.

For any $K \in \mathcal{K}(M)$ the set $G_K = \prod_{n<\omega} F_{\pi_n(K)} \subset X$ is compact and it is immediate that $\{G_K : K \in \mathcal{K}(M)\}$ is a cover of X which witnesses the domination of X by M. This shows that

(3) if X_n is dominated by a second countable space M_n for every $n \in \omega$ then $X = \prod_{n<\omega} X_n$ is dominated by the second countable space $M = \prod_{n<\omega} M_n$.

The properties (2) and (3) together with Fact 7 of S.271 imply that

(4) if Y is a space and $X_n \subset Y$ is dominated by a second countable space then $X = \bigcap_{n<\omega} X_n$ is also dominated by a second countable space.

Finally assume that $X = \bigcup_{n<\omega} X_n$ and every X_n is dominated by a second countable space M_n; let $\{F_K : K \in \mathcal{K}(M_n)\}$ be the respective compact cover of X_n. We can assume, without loss of generality, that the family $\{M_n : n \in \omega\}$ is disjoint. The space $M = \bigoplus_{n<\omega} M_n$ is second countable; we can consider that every M_n is a clopen subspace of M.

For any $K \in \mathcal{K}(M)$ let $m_K = \min\{n : K \subset M_0 \cup \ldots \cup M_n\}$; it is clear that $m_K \in \omega$ is well defined and $K \subset L$ implies $m_K \leq m_L$. Now, if $K \in \mathcal{K}(M)$ let $K_n = K \cap M_n$ for every $n \in \omega$. Then $G_K = F_{K_0} \cup \ldots \cup F_{K_{m_K}}$ is a compact subset of X and it is easy to check that $K \subset L$ implies $G_K \subset G_L$. Since $\mathcal{K}(M_n) \subset \mathcal{K}(M)$ for any $n \in \omega$, the family $\{G_K : K \in \mathcal{K}(M)\}$ covers X and hence X is dominated by M. As a consequence,

(5) if $X = \bigcup_{n<\omega} X_n$ and every X_n is dominated by a second countable space then X is also dominated by a second countable space.

The properties (1)–(5) show that our solution is complete.

U.213. *Show that every Lindelöf Σ-space is dominated by a second countable space. Prove that X is a Lindelöf Σ-space if and only if X is Dieudonné complete and dominated by a second countable space.*

Solution. Given a space Z the symbol $\mathcal{K}(Z)$ stands for the family of all compact subsets of Z. If X is a Lindelöf Σ-space then there exists a second countable space M and a compact-valued upper semicontinuous onto map $\varphi : M \to X$ (see SFFS-249). If $K \subset M$ is compact then $\varphi(K) = \bigcup\{\varphi(x) : x \in K\}$ is also compact by SFFS-241. It is evident that $K \subset L$ implies $\varphi(K) \subset \varphi(L)$; so $\{\varphi(K) : K \in \mathcal{K}(M)\}$ is a family which witnesses domination of X by M. Thus every Lindelöf Σ-space is Dieudonné complete (see TFS-454 and TFS-406) and dominated by a second countable space, i.e., we proved necessity.

Now assume that X is Dieudonné complete and dominated by a second countable space. By Problem 210, there exists a countable family $\mathcal{F} \subset \exp X$ and a cover $\mathcal{C}$

of the space X such that every $C \in \mathcal{C}$ is countably compact and $\mathcal{F}$ is a network with respect to $\mathcal{C}$. The set $\overline{C}$ is compact for any $C \in \mathcal{C}$ (see TFS-455); so the family $\mathcal{D} = \{\overline{C} : C \in \mathcal{C}\}$ is a compact cover of X. The family $\mathcal{G} = \{\overline{F} : F \in \mathcal{F}\}$ is countable. Given $D \in \mathcal{D}$ there is $C \in \mathcal{C}$ such that $D = \overline{C}$. If $U \in \tau(D, X)$ there exists $V \in \tau(D, X)$ for which $\overline{V} \subset U$.

The family $\mathcal{F}$ being a network with respect to $\mathcal{C}$, there is $F \in \mathcal{F}$ such that $C \subset F \subset V$. Then $G = \overline{F} \in \mathcal{G}$ and $D = \overline{C} \subset G \subset \overline{V} \subset U$ which shows that $\mathcal{G}$ is a countable network with respect to the compact cover $\mathcal{D}$ of the space X. Therefore X is Lindelöf Σ and hence we have established sufficiency.

U.214. *Prove that, for any space X, the space $C_p(X)$ is dominated by a second countable space if and only $C_p(X)$ is Lindelöf Σ.*

Solution. Given a space Z the symbol $\mathcal{K}(Z)$ stands for the family of all compact subsets of Z. If $C_p(X)$ is Lindelöf Σ then it is dominated by a second countable space (see Problem 213); so sufficiency is clear.

To prove necessity assume that $C_p(X)$ is dominated by a second countable space M and fix the respective compact cover $\{F_K : K \in \mathcal{K}(M)\}$ of the space $C_p(X)$. Apply Problem 210 and Problem 211 to see that $\upsilon(C_p(X))$ is a Lindelöf Σ-space and hence υX is also Lindelöf Σ (see Problem 206). Let $\pi : C_p(\upsilon X) \to C_p(X)$ be sthe restriction map. Since $\pi|A : A \to \pi(A)$ is a homeomorphism for any countable $A \subset C_p(\upsilon X)$ (see TFS-437), the set $G_K = \pi^{-1}(F_K)$ is countably compact for any $K \in \mathcal{K}(M)$. Indeed, G_K is closed in $C_p(\upsilon X)$; so if it is not countably compact then there is a countably infinite $D \subset G_K$ which is closed and discrete in G_K and hence in $C_p(\upsilon X)$. The set $E = \pi(D) \subset F_K$ cannot be closed and discrete in F_K; so it has an accumulation point $f \in F_K$. If $g = \pi^{-1}(f)$ then g is an accumulation point of D because $\pi|(D \cup \{g\})$ is a homeomorphism between $D \cup \{g\}$ and $E \cup \{f\}$.

This contradiction shows that every G_K is, indeed, countably compact and hence $ext(G_K) = \omega$; applying Baturov's theorem (SFFS-269) we conclude that G_K is Lindelöf and hence compact. Therefore $\{G_K : K \in \mathcal{K}(M)\}$ is a compact cover of $C_p(\upsilon X)$; it is evident that $K \subset L$ implies $G_K \subset G_L$; so the space $C_p(\upsilon X)$ is dominated by our second countable space M.

Observe also that $ext(C_p(\upsilon X)) = \omega$ for otherwise there is an uncountable closed discrete $D \subset C_p(\upsilon X)$ and hence D is dominated by a second countable space by Problem 212. However, D is metrizable and hence Dieudonné complete; so it has to be Lindelöf by Problem 213. Since any discrete Lindelöf space is countable, we have a contradiction which shows that $ext(C_p(\upsilon X)) = \omega$ and hence $C_p(\upsilon X)$ is Lindelöf by Baturov's theorem (SFFS-269). Any Lindelöf space is Dieudonné complete (see TFS-454 and TFS-406); so we can apply Problem 213 again to see that $C_p(\upsilon X)$ is Lindelöf Σ. Therefore $C_p(X)$ is also Lindelöf Σ being a continuous image of $C_p(\upsilon X)$. This settles necessity and makes our solution complete.

U.215. *Prove that, for any space X, the space $C_p(X)$ is $\mathbb{P}$-dominated if and only if $C_p(X)$ is K-analytic.*

Solution. Any K-analytic space is $\mathbb{P}$-dominated by SFFS-391; so sufficiency is clear. To prove necessity, assume that $C_p(X)$ is $\mathbb{P}$-dominated. Then $C_p(X)$ is dominated by the second countable space ω^ω (see Problem 209); applying Problem 214 we conclude that $C_p(X)$ is a Lindelöf Σ-space. Any Lindelöf space is realcompact (TFS-406); so we can apply SFFS-391 again to see that $C_p(X)$ is K-analytic.

U.216. *Prove that, for any space X, the space $C_p(X)$ is strongly $\mathbb{P}$-dominated if and only if X is countable and discrete.*

Solution. If X is countable and discrete then $C_p(X) = \mathbb{R}^X$ is Polish and hence strongly $\mathbb{P}$-dominated by SFFS-365; so sufficiency is clear. To prove necessity assume that $C_p(X)$ is strongly $\mathbb{P}$-dominated and let $\{K_p : p \in \mathbb{P}\}$ be the respective compact cover of $C_p(X)$.

Let $\pi : C_p(\upsilon X) \to C_p(X)$ be the restriction map. Since $\pi|A : A \to \pi(A)$ is a homeomorphism for any countable $A \subset C_p(\upsilon X)$ (see TFS-437), the set $G_p = \pi^{-1}(K_p)$ is countably compact for any $p \in \mathbb{P}$. Indeed, G_p is closed in $C_p(\upsilon X)$ so, if it is not countably compact then there is a countably infinite $D \subset G_p$ which is closed and discrete in G_p and hence in $C_p(\upsilon X)$. The set $E = \pi(D) \subset K_p$ cannot be closed and discrete in K_p; so it has an accumulation point $f \in K_p$. If $g = \pi^{-1}(f)$ then g is an accumulation point of D because $\pi|(D \cup \{g\})$ is a homeomorphism between $D \cup \{g\}$ and $E \cup \{f\}$.

This contradiction shows that every G_p is, indeed, countably compact and hence $ext(G_p) = \omega$; applying Baturov's theorem (SFFS-269) we conclude that G_p is Lindelöf and hence compact. Therefore $\mathcal{G} = \{G_p : p \in \mathbb{P}\}$ is a compact cover of $C_p(\upsilon X)$; it is evident that $\mathcal{G}$ swallows compact subsets of $C_p(\upsilon X)$ and $p \le q$ implies $G_p \subset G_q$; so the space $C_p(\upsilon X)$ is also strongly $\mathbb{P}$-dominated. It is easy to see that strong $\mathbb{P}$-domination is closed-hereditary so every closed $F \subset C_p(\upsilon X)$ is strongly $\mathbb{P}$-dominated.

Fact 1. If Z is a space and $K \subset Z$ is a non-empty metrizable compact subspace of Z then there exists a linear continuous map $e : C_p(K) \to C_p(Z)$ such that $e(f)|K = f$ for any $f \in C_p(K)$.

Proof. There exists a countable set $A \subset C_p(K)$ which separates the points of K. The set K is C-embedded in Z (see Fact 1 of T.218) so, for any $f \in A$ there is $u(f) \in C_p(Z)$ such that $u(f)|K = f$. Let $\varphi = \Delta\{u(f) : f \in A\} : Z \to \mathbb{R}^A$; the spaces $Y = \varphi(Z)$ and $L = \varphi(K)$ are second countable and $L \subset Y$. Besides, L is closed in Y being compact. Since the family $\{u(f) : f \in A\}$ separates the points of K, the map $\varphi|K : K \to L$ is a homeomorphism; let $\upsilon : L \to K$ be its inverse.

Apply Fact 1 of U.062 to find a linear continuous map $\delta : C_p(L) \to C_p(Y)$ such that $\delta(f)|L = f$ for any $f \in C_p(L)$. The dual map $\varphi^* : C_p(Y) \to C_p(Z)$ defined by $\varphi^*(f) = f \circ \varphi$ for any $f \in C_p(Y)$ is a linear embedding by TFS-163. Therefore the map $\mu = \varphi^* \circ \delta : C_p(L) \to C_p(Z)$ is also linear and continuous. The dual map $\upsilon^* : C_p(K) \to C_p(L)$ of the map υ is a linear homeomorphism; so the map $e = \mu \circ \upsilon^* : C_p(K) \to C_p(Z)$ is linear and continuous. Given any function $f \in C_p(K)$ we have $e(f) = \mu(\upsilon^*(f)) = \mu(f \circ \upsilon) = \varphi^*(\delta(f \circ \upsilon)) = \delta(f \circ \upsilon) \circ \varphi$.

For an arbitrary point $x \in K$ we have $e(f)(x) = \delta(f \circ v)(\varphi(x))$. Since $\varphi(x) \in L$, we have $\delta(f \circ v)(\varphi(x)) = (f \circ v)(\varphi(x))$ by the choice of δ. Therefore $e(f)(x) = f(v(\varphi(x))) = f(x)$ because $v(\varphi(y)) = y$ for any $y \in K$ by the choice of v.

Thus we established that $e(f)(x) = f(x)$ for any $x \in K$ which shows that $e(f)|K = f$ for any $f \in C_p(K)$; so Fact 1 is proved.

Fact 2. If Z is a space and K is a non-empty metrizable compact subspace of Z then $C_p(Z)$ is linearly homeomorphic to $C_p(K) \times I$ where $I = \{ f \in C_p(Z) : f(K) = \{0\} \}$. In particular, $C_p(K)$ embeds in $C_p(Z)$ as a closed linear subspace.

Proof. By Fact 1, there exists a linear continuous map $e : C_p(K) \to C_p(Z)$ such that $e(f)|K = f$ for any $f \in C_p(K)$. Let $\pi_K : C_p(Z) \to C_p(K)$ be the restriction map. Given any $f \in C_p(Z)$ let $\delta(f) = f - e(\pi_K(f))$; it is evident that the mapping $\delta : C_p(Z) \to C_p(Z)$ is linear, continuous and we have the inclusion $\delta(C_p(Z)) \subset I$. Consequently, letting $\varphi(f) = (\pi_K(f), \delta(f))$ for any function $f \in C_p(Z)$, we obtain a map $\varphi : C_p(Z) \to C_p(K) \times I$ which is linear and continuous being the diagonal product of linear continuous maps.

For any $(f, g) \in C_p(K) \times I$, let $\mu(f, g) = e(f) + g$; it is straightforward that $\mu : C_p(K) \times I \to C_p(Z)$ is linear, continuous and inverse to φ. Thus the spaces $C_p(Z)$ and $C_p(K) \times I$ are linearly homeomorphic; it is evident that any factor of a product of topological vector spaces embeds in that product as a linear closed subspace; so $C_p(K)$ embeds in $C_p(Z)$ as a closed linear subspace and hence Fact 2 is proved.

Returning to our solution observe that vX contains no non-trivial convergent sequences; indeed, if $S \subset vX$ is a non-trivial convergent sequence then it is metrizable and compact so $C_p(S)$ embeds in $C_p(vX)$ as a closed subspace by Fact 2. Thus $C_p(S)$ is strongly $\mathbb{P}$-dominated and hence Polish by SFFS-365 which contradicts TFS-265.

Now take an arbitrary compact subset K of vX; since $C_p(vX)$ is Lindelöf Σ (see Problem 215), the space vX and hence K is ω-monolithic (see Problem 208). It is evident that any infinite ω-monolithic compact space has non-trivial convergent sequences; so if K is infinite then there are non-trivial convergent sequences in vX which is a contradiction. Therefore every compact subspace of vX is finite; the space vX being Lindelöf Σ (see Problem 206) we can apply Fact 2 of T.227 to conclude that vX is countable and hence $C_p(vX)$ is second countable. Apply SFFS-365 once more to see that $C_p(vX)$ is Polish and hence vX is discrete. Thus X is also countable and discrete; this settles necessity and makes our solution complete.

U.217. *Observe that there exist spaces X for which $C_p(X, \mathbb{I})$ is Lindelöf Σ while $C_p(X)$ is not Lindelöf. Supposing that vX and $C_p(X, \mathbb{I})$ are Lindelöf Σ-spaces prove that $C_p(X)$ is a Lindelöf Σ-space. In particular, if X is Lindelöf Σ then the space $C_p(X)$ is Lindelöf Σ if and only if $C_p(X, \mathbb{I})$ is a Lindelöf Σ-space.*

Solution. If X is a discrete space of cardinality ω_1 then $C_p(X, \mathbb{I}) = \mathbb{I}^X$ is even compact while $C_p(X) = \mathbb{R}^X$ is not normal (see Fact 2 of S.215). Now assume that vX and $C_p(X, \mathbb{I})$ are Lindelöf Σ-spaces. By Problem 206, there exists a Lindelöf Σ-space L such that $C_p(X) \subset L \subset \mathbb{R}^X$.

If $J = (-1, 1) \subset \mathbb{I}$ then let $v : \mathbb{R} \to J$ be a homeomorphism. If $v_*(f) = v \circ f$ for any $f \in \mathbb{R}^X$ then the map $v_* : \mathbb{R}^X \to J^X$ is a homeomorphism such that $v_*(C_p(X)) = C_p(X, J)$ (see TFS-091). Thus $M = v_*(L)$ is a Lindelöf Σ-space for which $C_p(X, J) \subset M \subset J^X \subset \mathbb{I}^X$. Therefore $C_p(X, J) = M \cap C_p(X, \mathbb{I})$ is a Lindelöf Σ-space by SFFS-258. Consequently, $C_p(X) \simeq C_p(X, J)$ is a Lindelöf Σ-space.

U.218. *(Okunev's theorem). Suppose that X and Y are Lindelöf Σ-spaces such that $Y \subset C_p(X)$. Prove that $C_p(Y)$ is a Lindelöf Σ-space.*

Solution. Since Y is Lindelöf Σ, there exists a countable family $\mathcal{F} \subset \exp Y$ which is a network with respect to a compact cover $\mathcal{C}$ of the space Y. We denote by Q_0 the set $\mathbb{Q} \cap (0, 1)$.

Given numbers $n \in \mathbb{N}$, $\delta > 0$, a point $x = (x_1, \ldots, x_n) \in X^n$ and $f \in Y$ let $O(f, x, \delta) = \{g \in Y : |g(x_i) - f(x_i)| < \delta$ for every $i \leq n\}$. It is evident that the family $\{O(f, x, \delta) : \delta > 0$ and there is $n \in \mathbb{N}$ such that $x = (x_1, \ldots, x_n) \in X^n\}$ is a local base in Y at the point f.

For arbitrary numbers $\varepsilon > 0, \delta > 0$, $n \in \mathbb{N}$ and a set $P \subset Y$ consider the set $M(\varepsilon, \delta, n, P) = \{(\varphi, x) \in \mathbb{I}^Y \times X^n : \varphi \in \mathbb{I}^Y, x = (x_1, \ldots, x_n) \in X^n$ and $|\varphi(f) - \varphi(g)| \leq \varepsilon$ for any $f \in Y$ and $g \in P$ such that $f \in O(g, x, \delta)\}$. We claim that

(1) $M(\varepsilon, \delta, n, P)$ is closed in $\mathbb{I}^Y \times X^n$ for any $\varepsilon > 0$, $\delta > 0$, $n \in \mathbb{N}$ and $P \subset Y$.

To see that the property (1) holds take a point $(\varphi, x) \in (\mathbb{I}^Y \times X^n) \backslash M(\varepsilon, \delta, n, P)$ where $x = (x_1, \ldots, x_n) \in X^n$. By the definition of the set $M(\varepsilon, \delta, n, P)$ there exist $f \in Y$ and $g \in P$ such that $|\varphi(f) - \varphi(g)| > \varepsilon$ and $f \in O(g, x, \delta)$. Observe that the set $W = \{\eta \in \mathbb{I}^Y : |\eta(f) - \eta(g)| > \varepsilon\}$ is open in $\mathbb{I}^Y$ and $\varphi \in W$. Furthermore, the functions f and g are continuous on X; so the set $V = \{y \in X : |f(y) - g(y)| < \delta\}$ is open in X and $x_i \in V$ for any $i \leq n$. Thus V^n is open in X^n and $x \in V^n$. It is straightforward that $(W \times V^n) \cap M(\varepsilon, \delta, n, P) = \emptyset$; since $(\varphi, x) \in W \times V^n$, we showed that any point $(\varphi, x) \in (\mathbb{I}^Y \times X^n) \backslash M(\varepsilon, \delta, n, P)$ has a neighbourhood $W \times V^n$ contained in the complement of $M(\varepsilon, \delta, n, P)$. Therefore $(\mathbb{I}^Y \times X^n) \backslash M(\varepsilon, \delta, n, P)$ is open in $\mathbb{I}^Y \times X^n$, i.e., $M(\varepsilon, \delta, n, P)$ is closed in $\mathbb{I}^Y \times X^n$; so (1) is proved.

For any $n \in \mathbb{N}$ let $\pi : \mathbb{I}^Y \times X^n \to \mathbb{I}^Y$ be the natural projection; given any numbers $\varepsilon > 0, \delta > 0$, and $P \subset Y$ let $L(\varepsilon, \delta, n, P) = \pi(M(\varepsilon, \delta, n, P))$. Observe that $\mathbb{I}^Y \times X^n$ is a Lindelöf Σ-space because so is X; it follows from (1) that $M(\varepsilon, \delta, n, P)$ is also Lindelöf Σ and hence $L(\varepsilon, \delta, n, P)$ is Lindelöf Σ as well being a continuous image of the Lindelöf Σ-space $M(\varepsilon, \delta, n, P)$. Thus

(2) $L(\varepsilon, \delta, n, P) \subset \mathbb{I}^Y$ is a Lindelöf Σ-space for any $\varepsilon > 0$, $\delta > 0$, $n \in \mathbb{N}$ and $P \subset Y$.

The family $\mathcal{L} = \{L(\varepsilon, \delta, n, P) : n \in \mathbb{N}, \varepsilon, \delta \in Q_0$ and $P \in \mathcal{F}\}$ is countable and consists of Lindelöf Σ-subspaces of $\mathbb{I}^Y$. We will establish next that

(3) the family $\mathcal{L}$ separates the set $C_p(Y, \mathbb{I})$ from $\mathbb{I}^Y \backslash C_p(Y, \mathbb{I})$ in the sense that, for any $\varphi \in C_p(Y, \mathbb{I})$ and $\eta \in \mathbb{I}^Y \backslash C_p(Y, \mathbb{I})$ there is $L \in \mathcal{L}$ such that $\varphi \in L$ while $\eta \notin L$.

To prove (3) take any $\varphi \in C_p(Y, \mathbb{I})$ and $\eta \in \mathbb{I}^Y \backslash C_p(Y, \mathbb{I})$; there is some point $g \in Y$ such that η is discontinuous at g and hence there is $\varepsilon \in Q_0$ such that,

(4) for any $n \in \mathbb{N}$, $y = (y_1, \ldots, y_n) \in X^n$ and $\delta > 0$ there is $f \in O(g, y, \delta)$ for which $|\eta(f) - \eta(g)| > \varepsilon$.

The family $\mathcal{C}$ being a cover of the space Y we can choose a set $C \in \mathcal{C}$ such that $g \in C$. For any element $h \in C$ the map φ is continuous at the point h; so there exist $n_h \in \mathbb{N}$, $x_h = (x_1^h, \ldots, x_{n_h}^h) \in X^{n_h}$ and $\delta_h \in Q_0$ such that for any $u \in O(h, x_h, 3\delta_h)$ we have $|\varphi(u) - \varphi(h)| < \frac{\varepsilon}{2}$. The open cover $\{O(h, x_h, \delta_h) : h \in C\}$ of the compact set C has a finite subcover and therefore there exists a finite set $E \subset C$ such that $C \subset G = \bigcup\{O(h, x_h, \delta_h) : h \in E\}$; then $\delta = \min\{\delta_h : h \in E\} \in Q_0$. Take $m \in \mathbb{N}$ and $x = (x_1, \ldots, x_m) \in X^m$ such that $x_i^h \in \{x_1, \ldots, x_m\}$ for any $h \in E$ and $i \leq n_h$. There exists $P \in \mathcal{F}$ such that $C \subset P \subset G$; it is easy to check that if $v \in P$ and $u \in O(v, x, \delta)$ then $|\varphi(u) - \varphi(v)| < \varepsilon$.

Recalling the definition of $M(\varepsilon, \delta, m, P)$ we conclude that $(\varphi, x) \in M(\varepsilon, \delta, m, P)$ and therefore $\varphi \in L = L(\varepsilon, \delta, m, P)$. On the other hand, $\eta \notin L$ because $g \in P$; so the property (4) says exactly that it is not possible to find $n \in \mathbb{N}$ and $y \in X^n$ such that $(\eta, y) \in M(\varepsilon, \delta, n, P)$. Thus the property (3) is proved.

The property (3) shows that $\mathcal{L}$ is a countable family of Lindelöf Σ-subspaces of $\mathbb{I}^Y$ which separates $C_p(Y, \mathbb{I})$ from $\mathbb{I}^Y \backslash C_p(Y, \mathbb{I})$. Since $\mathbb{I}^Y$ is a compact extension of $C_p(Y, \mathbb{I})$, it follows from SFFS-233 that $C_p(Y, \mathbb{I})$ is a Lindelöf Σ-space. Finally, apply Problem 217 to conclude that $C_p(Y)$ is a Lindelöf Σ-space and complete our solution.

U.219. Let X and $C_p(X)$ be Lindelöf Σ-spaces. Prove that, for every natural n, the space $C_{p,n}(X)$ is a Lindelöf Σ-space. In particular, if X is compact and $C_p(X)$ is Lindelöf Σ then all iterated function spaces of X are Lindelöf Σ-spaces.

Solution. This is easily done by induction on $n \in \omega$. By our assumptions, $C_{p,n}(X)$ is a Lindelöf Σ-space for $n \in \{0, 1\}$. Now assume that we are given a natural number $k > 1$ and the space $C_{p,n}(X)$ is Lindelöf Σ for any $n < k$. In particular, $Y = C_{p,k-2}(X)$ and $Z = C_{p,k-1}(X) = C_p(Y)$ are Lindelöf Σ-spaces. Applying Problem 218 we conclude that $C_{p,k}(X) = C_p(Z)$ is a Lindelöf Σ-space. Thus our inductive procedure shows that $C_{p,n}(X)$ is a Lindelöf Σ-space for any $n \in \omega$.

U.220. For an arbitrary Lindelöf Σ-space X, prove that any countably compact subspace $Y \subset C_p(X)$ is Gul'ko compact.

Solution. By countable compactness of Y we have $ext(Y) = \omega$; so Baturov's theorem (SFFS-269) implies that Y is Lindelöf and hence compact. Finally apply Problem 020 to conclude that $C_p(Y)$ is a Lindelöf Σ-space, i.e., Y is Gul'ko compact.

U.221. *Suppose that $C_p(X)$ is a Lindelöf Σ-space. Prove that any countably compact $Y \subset C_p(X)$ is Gul'ko compact.*

Solution. The space υX is Lindelöf Σ by Problem 206; let $\pi : C_p(\upsilon X) \to C_p(X)$ be the restriction map. We claim that the set $Z = \pi^{-1}(Y) \subset C_p(\upsilon X)$ is also countably compact. Indeed, if D is a countably infinite closed discrete subspace of Z then $E = \pi(D)$ has to have an accumulation point $f \in Y$. Then $g = \pi^{-1}(f) \in Z$ and g is an accumulation point of D because $\pi | (D \cup \{g\})$ is a homeomorphism between $D \cup \{g\}$ and $E \cup \{f\}$ (see TFS-437); this contradiction shows that Z is countably compact.

Recall that $Z \subset C_p(\upsilon X)$ and υX is a Lindelöf Σ-space; besides, countable compactness of Z implies $ext(Z) = \omega$; so Baturov's theorem (SFFS-269) shows that Z is Lindelöf and hence compact. Apply Problem 020 to conclude that $C_p(Z)$ is also a Lindelöf Σ-space, i.e., Z is Gul'ko compact. The map $\pi | Z : Z \to Y$ is a condensation and hence homeomorphism; so Y is also Gul'ko compact.

U.222. *(Reznichenko's compactum) Prove that there exists a compact space M with the following properties:*

(i) $C_p(M)$ is a K-analytic space, i.e., M is Talagrand compact;
(ii) there is $x \in M$ such that $M \setminus \{x\}$ is pseudocompact and M is the Stone–Čech extension of $M \setminus \{x\}$.

As a consequence, there is an example of a K-analytic space X such that some closed pseudocompact subspace of $C_p(X)$ is not countably compact.

Solution. As usual, $A(\kappa)$ is the one-point compactification of the discrete space of cardinality κ. We let $\omega^0 = \{\emptyset\}$ and $\omega^{<\omega} = \bigcup \{\omega^n : n \in \omega\}$; for any $n \in \mathbb{N}$, we identify ω^n with the set of all maps from $n = \{0, \ldots, n-1\}$ to ω. If $n \in \omega$ and $s \in \omega^n$ then $t = s^\frown k \in \omega^{n+1}$ is defined by $t | n = s$ and $t(n) = k$ for any $k \in \omega$. If $f \in \omega^\omega$ and $n \in \omega$ then $f | 0 = \emptyset$ and $f | n = f | \{0, \ldots, n-1\}$ for any $n \in \mathbb{N}$.

The symbol $\mathbb{P}$ denotes the space of the irrationals which we identify with ω^ω. If Z is a space then $z \in Z$ is called *a π-point* if there exists a finite family $\mathcal{U} \subset \tau(Z)$ such that $\{z\} = \bigcap \{\overline{U} : U \in \mathcal{U}\}$. A family $\{U_n : n \in \omega\}$ of subsets of a space Z *converges to a point* $z \in Z$ if, for any $U \in \tau(z, Z)$ there is $m \in \omega$ such that $U_n \subset U$ for all $n \geq m$. Given a set A let $\sigma(\mathbb{D}^A) = \{x \in \mathbb{D}^A : |x^{-1}(1)| < \omega\}$ and $\Sigma(\mathbb{D}^A) = \{x \in \mathbb{D}^A : |x^{-1}(1)| \leq \omega\}$.

Fact 1. Suppose that K is a compact space and some $x \in K$ is not a π-point. Then $K \setminus \{x\}$ is pseudocompact and K is canonically homeomorphic to $\beta(K \setminus \{x\})$, i.e., there exists a homeomorphism $\varphi : \beta(K \setminus \{x\}) \to K$ such that $\varphi(y) = y$ for any point $y \in K \setminus \{x\}$.

Proof. Observe first that x cannot be an isolated point because otherwise $\{x\} = \overline{U}$ where $U = \{x\} \in \tau(K)$. If the space $K \setminus \{x\}$ is not pseudocompact then there is a discrete family $\mathcal{U} = \{U_n : n \in \omega\} \subset \tau^*(K \setminus \{x\})$. Given any $U \in \tau(x, K)$ if $U_n \setminus U \neq \emptyset$ for infinitely many $n \in \omega$ then take $V \in \tau(x, K)$ with $\overline{V} \subset U$ and observe that the family $\{U_n\backslash\overline{V} : n \in \omega$ and $U_n\backslash U \neq \emptyset\} \subset \tau^*(K\backslash V)$ is infinite and discrete in the compact space $K\backslash V$; this contradiction shows that $U_n\backslash U = \emptyset$ for all but finitely many n, i.e., the family $\mathcal{U}$ converges to the point x.

It is now easy to check that for the open sets $G = \bigcup\{U_{2n} : n \in \omega\}$ and $H = \bigcup\{U_{2n+1} : n \in \omega\}$ we have $\{x\} = \overline{G} \cap \overline{H}$, i.e., x is a π-point which is again a contradiction. Therefore $K\backslash\{x\}$ is pseudocompact.

Let $L = \beta(K\backslash\{x\})$; since the space K is a compact extension of $K\backslash\{x\}$, there exists a continuous map $\varphi : L \to K$ such that $\varphi|(K\backslash\{x\}) \to K\backslash\{x\}$ is the identity mapping. If $R = L\backslash(K\backslash\{x\})$ is a singleton then φ is a canonical homeomorphism; so assume that there are distinct points $y_1, y_2 \in R$. Take $U_1, U_2 \in \tau(L)$ such that $\mathrm{cl}_L(U_1) \cap \mathrm{cl}_L(U_2) = \emptyset$.

The set $V_i = U_i \cap (K\backslash\{x\})$ is open in the space K for any $i \in \{1,2\}$. Since $\varphi(R) = \{x\}$ (see Fact 1 of S.259), we have $\varphi(y_i) = x$; furthermore, $y_i \in \mathrm{cl}_L(V_i)$ and hence $x = \varphi(y_i) \in \mathrm{cl}_K(\varphi(V_i)) = \mathrm{cl}_K(V_i)$ for every $i \in \{1,2\}$. As a consequence, $\{x\} \subset \mathrm{cl}_K(V_1) \cap \mathrm{cl}_K(V_2)$. If, on the other hand, $y \in K\backslash\{x\}$ and $y \in \mathrm{cl}_K(V_1) \cap \mathrm{cl}_K(V_2)$ then $y \in \mathrm{cl}_L(V_1) \cap \mathrm{cl}_L(V_2) = \emptyset$ which gives us a contradiction. This proves that $\{x\} = \mathrm{cl}_K(V_1) \cap \mathrm{cl}_K(V_2)$, i.e., x is a π-point which is a contradiction again. Thus K is canonically homeomorphic to $\beta(K\backslash\{x\})$; so Fact 1 is proved.

Fact 2. Given a set A let $u \in \mathbb{D}^A$ be defined by $u(a) = 0$ for any $a \in A$. Suppose that $S_n \subset \sigma(\mathbb{D}^A)\backslash\{u\}$ is a sequence which converges to u for any $n \in \omega$. Then there exists a sequence $S \subset \bigcup\{S_n : n \in \omega\}$ such that $S \to u$, the set $S \cap S_n$ is infinite for any $n \in \omega$ and the family $\{x^{-1}(1) : x \in S\}$ is disjoint.

Proof. Let $\{m_k : k \in \omega\}$ be an enumeration of ω where every $n \in \omega$ occurs infinitely many times. Choose $x_0 \in S_{m_0}$ arbitrarily; assume that $n \in \omega$ and we have chosen $x_i \in S_{m_i}$ for each $i \leq n$ in such a way that the family $\{x_i^{-1}(1) : i \leq n\}$ is disjoint. Since $S_{m_{n+1}}$ converges to u, the family $\mathcal{S} = \{x^{-1}(1) : x \in S_{m_{n+1}}\}$ is point-finite; so at most finitely many elements of $\mathcal{S}$ meet the finite set $A = \bigcup\{x_i^{-1}(1) : i \leq n\}$. Therefore we can choose $x_{n+1} \in S_{m_{n+1}}$ in such a way that $x_{n+1}^{-1}(1) \cap A = \emptyset$; it is evident that the family $\{x_i^{-1}(1) : i \leq n + 1\}$ is still disjoint; so our inductive procedure gives us a set $S = \{x_i : i \in \omega\}$ such that $x_i \in S_{m_i}$ for any $i \in \omega$ and the family $\{x_i^{-1}(1) : i \in \omega\}$ is disjoint. It is straightforward that S is as promised; so Fact 2 is proved.

If Z is a space and $z \in Z$ say that a sequence $S \subset Z$ *converges flexibly* to z if S converges to z and for any family $\{G_x : x \in S\}$ of G_δ-subsets of Z with $x \in G_x$ for any $x \in S$, we can choose a point $y(x) \in G_x$ for any $x \in S$ so that the sequence $\{y(x) : x \in S\}$ converges to a point $y \in Z\backslash\{z\}$.

Fact 3. Suppose that K is a Fréchet–Urysohn compact space and $Y \subset K$ is dense in K. Suppose additionally that a point $u \in K\backslash Y$ has the following property:

(*) if $S_n \subset Y$ is a sequence which converges to u for any $n \in \omega$ then there is a sequence $S \subset K$ such that $S \cap S_n$ is infinite for any $n \in \omega$ and S flexibly converges to u.

Then, for any countable family $\mathcal{U} \subset \tau(K)$ we have $\{u\} \neq \bigcap\{\overline{U} : U \in \mathcal{U}\}$ and, in particular, u is not a π-point in K.

Proof. Suppose that $U_n \in \tau(K)$ for any $n \in \omega$ and $\{u\} = \bigcap\{\overline{U}_n : n \in \omega\}$; there is no loss of generality to assume that $U_0 = K$. Then $u \in \overline{U_n \cap Y}$ and hence we can pick a sequence $S_n \subset U_n \cap Y$ such that $S_n \to u$ for any $n \in \omega$. Use the property $(*)$ to find a sequence $S \subset K$ such that S flexibly converges to u and $S \cap S_n$ is infinite for any $n \in \omega$. Then $G_x = \bigcap\{U_n : x \in U_n\}$ is a G_δ-set and $x \in G_x$ for any $x \in S$.

By our choice of S we can take $y(x) \in G_x$ for any $x \in S$ such that the sequence $\{y(x) : x \in S\}$ converges to a point $y \in K \backslash \{u\}$. Given any number $n \in \omega$ the set $\{x \in S : x \in U_n\}$ is infinite; so $S' = \{x \in S : y(x) \in U_n\}$ is infinite as well which shows that $S' \subset U_n$ converges to y and hence $y \in \overline{U}_n$ for any $n \in \omega$. Thus $y \in \bigcap\{\overline{U}_n : n \in \omega\}$ whence $\{u\} \neq \bigcap\{\overline{U}_n : n \in \omega\}$; this contradiction shows that Fact 3 is proved.

Fact 4. Suppose that A is a set and $K \subset \Sigma(A)$ is a compact subspace for which there exists a family $\{A_s : s \in \omega^{<\omega}\}$ of subsets of A with the following properties:

(i) $A_\emptyset = A$ and $A_s = \bigcup\{A_{s^\frown n} : n \in \omega\}$ for any $s \in \omega^{<\omega}$;
(ii) for any $x \in K$ and $f \in \omega^\omega$ there exists $m \in \omega$ such that $A_{f|n} \cap x^{-1}(\mathbb{R}\backslash\{0\})$ is finite for all $n \geq m$.

Then $C_p(K)$ is a K-analytic space.

Proof. Let $\chi_a : A \to \mathbb{D}$ be the characteristic function of the set $\{a\}$ for any $a \in A$, i.e., $\chi_a(a) = 1$ while $\chi_a(b) = 0$ for any $b \in A\backslash\{a\}$; let $z_0(a) = 0$ for any $a \in A$. The set $L_0 = \{z_0\} \cup \{\chi_a : a \in A\}$ is compact; so the set $K' = K \cup L_0 \supset K$ is compact as well and therefore it suffices to prove that $C_p(K')$ is K-analytic. Since the property (ii) still holds for K', we can assume, without loss of generality, that $K' = K$, i.e., $L_0 \subset K$.

For any point $x \in \Sigma(A)$ let $\mathrm{supp}(x) = x^{-1}(\mathbb{R}\backslash\{0\})$. Denote by u the function on K which is identically zero. For any $a \in A$ let $e_a(x) = x(a)$ for any $x \in K$; then $e_a \in C_p(K)$ because e_a coincides with the restriction to K of the natural projection of $\mathbb{R}^A$ onto the factor determined by a. The set $T = \{u\} \cup \{e_a : a \in A\}$ separates the points of K; so it suffices to establish that T is K-analytic (see Problem 022). Observe that the set $W_a = \{v \in C_p(K) : v(\chi_a) > 0\}$ is an open neighbourhood of e_a in $C_p(K)$ such that $W_a \cap T = \{e_a\}$; therefore every e_a is an isolated point of T.

For every $s \in \omega^{<\omega}$ let $Q_s = \{u\} \cup \{e_a : a \in A_s\}$; then every Q_s is closed in T and hence so is the set $P_f = \bigcap\{Q_{f|n} : n \in \omega\}$ for any $f \in \omega^\omega$. Letting $\varphi(f) = P_f$ for any $f \in \mathbb{P}$ we obtain a multi-valued map $\varphi : \mathbb{P} \to T$. It turns out that P_f is compact for every $f \in \mathbb{P}$, i.e., φ is compact-valued.

To prove this take any $U \in \tau(u, C_p(K))$; there is a finite set $F \subset K$ and $\varepsilon > 0$ such that $W = \{v \in C_p(K) : |v(x)| < \varepsilon$ for all $x \in F\} \subset U$. Consider the set $C = \bigcup\{\mathrm{supp}(x) : x \in F\}$. It follows from (ii) that there is $m \in \omega$ such that $C \cap A_{f|m}$ is finite and therefore $e(x) = 0$ whenever $x \in F$ for all $e \in Q_{f|m}$ except for finitely many of them. Since $P_f \subset Q_{f|m}$, it turns out that $e(x) = 0$ whenever $x \in F$ for all but finitely many $e \in P_f$. Thus $P_f\backslash U \subset P_f\backslash W$ is finite, i.e., we found a point $u \in P_f$ such that $P_f \setminus U$ is finite for any open neighbourhood of u in $C_p(K)$. Therefore P_f is a compact space with at most one non-isolated point and hence the map φ is, indeed, compact-valued.

Given any $a \in A$, it follows from the property (i) that there is a function $f \in \mathbb{P}$ with $a \in \bigcap \{A_{f|n} : n \in \omega\}$ which shows that $e_a \in P_f$ and therefore $\{P_f : f \in \omega^\omega\}$ is a cover of T.

Now assume that a sequence $S = \{f_n : n \in \omega\} \subset \omega^\omega$ converges to $f \in \omega^\omega$ and $t_n \in P_{f_n}$ for any $n \in \omega$. Passing to a subsequence of S if necessary, we can assume, without loss of generality, that $f_n|n = f|n$ and hence $t_n \in Q_{f|n}$ for any $n \in \omega$.

If there exists a point $t \in T$ such that $t = t_n \in Q_{f|n}$ for infinitely many $n \in \omega$ then $t \in P_f$ is an accumulation point of S. If not, then, passing to a subsequence of S if necessary, we can assume that $t_n \neq t_m$ if $n \neq m$. To see that $S \to u$ take any $x \in K$; there is $m \in \omega$ for which the set $D = \text{supp}(x) \cap A_{f|m}$ is finite. We have $S' = \{t_n : n \geq m\} \subset Q_{f|m}$ and $t_n = e_{a_n}$ where $a_n \in A_{f|m}$ for any $n \geq m$. Therefore $t_n(x) = x(a_n) = 0$ for all $n \geq m$ such that $a_n \notin D$. This proves that $t_n(x) = 0$ for all but finitely many n for any $x \in K$ and therefore the sequence $\{t_n : n \in \omega\}$ converges to $t = u \in P_f$.

We checked that, in all cases, there is an accumulation point $t \in P_f$ for the sequence $\{t_n\}$; so we can apply SFFS-389 to conclude that T is K-analytic and hence $C_p(K)$ is K-analytic as well by Problem 022. Fact 4 is proved.

Fact 5. There exists a compact space K with the following properties:

(i) $C_p(K)$ is K-analytic, i.e., K is Talagrand compact;
(ii) there is a point $u \in K$ such that for any family $\{U_n : n \in \omega\} \subset \tau(K)$ we have $\{u\} \neq \bigcap \{\overline{U}_n : n \in \omega\}$ and, in particular, u is not a π-point;
(iii) we have $K = \bigcup \{K_n : n \in \omega\}$ where the family $\{K_n : n \in \omega\}$ is disjoint, $K_0 \simeq A(\mathfrak{c})$ and u is the unique non-isolated point of K_0; besides, K_n is clopen in K and homeomorphic to a closed subspace of $(A(\mathfrak{c}))^\omega$ for any $n \in \mathbb{N}$.

Proof. For any ordinals $\alpha, \beta \leq \omega_1$ the ordinal intervals are defined in the usual way, i.e., $[\alpha, \beta] = \{\gamma : \gamma \in \omega_1 \text{ and } \alpha \leq \gamma \leq \beta\}$, $[\alpha, \beta) = \{\gamma : \gamma \in \omega_1 \text{ and } \alpha \leq \gamma < \beta\}$; analogously, $(\alpha, \beta) = \{\gamma : \gamma \in \omega_1 \text{ and } \alpha < \gamma < \beta\}$. Let $I = [0, 1] \subset \mathbb{R}$ and denote by T the set $(\{0\} \times \{\frac{1}{n} : n \in \mathbb{N}\}) \cup ((0, \omega_1) \times I)$. Thus $T \subset [0, \omega_1) \times I$; let $\pi : T \to [0, \omega_1)$ be the projection, i.e., if $\alpha \in [0, \omega_1)$, $r \in I$ and $t = (\alpha, r) \in T$ then $\pi(t) = \alpha$. It is clear that $a_n = (0, \frac{1}{n}) \in T$ for any $n \in \mathbb{N}$.

Call a set $G \subset T$ *thin* if $\pi|G$ is injective and denote by $\mathcal{F}$ the family of all non-empty finite thin subsets of T. Let $\mathcal{A}_0 = \{\{a_n\} : n \in \mathbb{N}\}$; it is straightforward that $\mathcal{A}_0 \subset \mathcal{F}$. Assume that $\alpha < \omega_1$ and we have families $\{\mathcal{A}_\beta : \beta < \alpha\}$ with the following properties:

(1) $\mathcal{A}_\beta \subset \mathcal{F}$ and $|\mathcal{A}_\beta| \leq \mathfrak{c}$ for any $\beta < \alpha$;
(2) if $\beta < \alpha$ and $A \in \mathcal{A}_\beta$ then $\{0, \beta\} \subset \pi(A)$ and $A \subset \pi^{-1}([0, \beta])$;
(3) for any $\beta \in (0, \alpha)$ and $r \in I$ the family $\mathcal{A}_\beta^r = \{A \in \mathcal{A}_\beta : (\beta, r) \in A\}$ is countably infinite, while the collection $\mathcal{B}_\beta^r = \{A \setminus \{(\beta, r)\} : A \in \mathcal{A}_\beta^r\}$ is disjoint and contained in $\bigcup \{\mathcal{A}_\gamma : \gamma < \beta\}$;

(4) if $\beta \in (0, \alpha)$ and $C \subset \bigcup \{A_\gamma : \gamma < \beta\}$ is a countably infinite disjoint collection then there are $\mathfrak{c}$-many $r \in I$ such that $\mathcal{A}_\beta^r = \{C \cup \{(\beta, r)\} : C \in \mathcal{C}\}$.

Let $\mathbb{M} = \{C : C \subset \bigcup \{A_\beta : \beta < \alpha\}$ is a countably infinite disjoint collection$\}$; it is clear that $|\mathbb{M}| \leq \mathfrak{c}$; so there is a map $g : I \to \mathbb{M}$ such that $|g^{-1}(C)| = \mathfrak{c}$ for any $C \in \mathbb{M}$. The family $\mathcal{A}_\alpha^r = \{C \cup \{(\alpha, r)\} : C \in g(r)\}$ is countable for any $r \in I$; so the collection $\mathcal{A}_\alpha = \bigcup \{\mathcal{A}_\alpha^r : r \in I\}$ has cardinality $\mathfrak{c}$.

It is immediate that the conditions (1)–(4) are still satisfied for all $\beta \leq \alpha$; so our inductive procedure gives us a collection $\{\mathcal{A}_\alpha : \alpha < \omega_1\}$ for which the properties (1)–(4) hold for any $\beta < \omega_1$; let $\mathcal{A} = \bigcup \{\mathcal{A}_\alpha : \alpha < \omega_1\}$. It follows from (2) that, for any $A \in \mathcal{A}$ there are unique $\alpha < \omega_1$ and $r \in I$ such that $A \in \mathcal{A}_\alpha^r$; call the point $t = (\alpha, r) \in A$ the *maximal element of A* and denote it by $\max(A)$.

Given a set $A \subset T$ the characteristic function $\chi_A \in \mathbb{D}^T$ of the set A is defined as usual: $\chi_A(t) = 1$ if $t \in A$ and $\chi_A(t) = 0$ for all $t \in T \backslash A$. Let $u = \chi_\emptyset$ and $Y = \{\chi_A : A \in \mathcal{A}\}$; we are going to prove that the compact space $K = \overline{Y}$ (the bar denotes the closure in $\mathbb{D}^T$) is as promised.

Observe that $\{a_n\} \in \mathcal{A}$ and hence $\chi_{\{a_n\}} \in Y$ for any $n \in \mathbb{N}$; if $\alpha > 0$ and $r \in I$ then it follows from (3) that the sequence $\{\chi_A : A \in \mathcal{A}_\alpha^r\}$ converges to $\chi_{\{(\alpha, r)\}}$. Therefore

(5) $\chi_{\{t\}} \in K$ for any $t \in T$.

It is straightforward that u belongs to the closure of the set $\{\chi_{\{t\}} : t \in T\}$ and hence $u \in K$.

It is important to note that

(6) the set $\text{supp}(x) = x^{-1}(1)$ is thin for any $x \in K$.

Indeed, if $r_1 \neq r_2$, $t_1 = (\alpha, r_1)$, $t_2 = (\alpha, r_2)$ and $\{t_1, t_2\} \subset \text{supp}(x)$ then it follows from $x \in \overline{Y}$ that there exists $A \in \mathcal{A}$ for which $\chi_A(t_i) = 1$ for every $i \in \{1, 2\}$, i.e., $\{t_1, t_2\} \subset A$ which is a contradiction with $A \in \mathcal{F}$. This proves (6).

We will show next that $K \subset \Sigma(\mathbb{D}^T)$ and hence K is Corson compact. Take an arbitrary $A \in \mathcal{A}$; then $A \in \mathcal{A}_\alpha$ for some $\alpha < \omega_1$ and hence $(\alpha, r) \in A$ for some $r \in I$. If $\alpha > 0$ then it follows from (3) that $A \backslash \{(\alpha, r)\} \in \mathcal{A}$; proceed in the same way inductively (i.e., at every step throw away the maximal element of the current set) and observe that successively taking away some number of maximal elements is the same as considering the set $\pi^{-1}([0, \beta]) \cap A$ for some $\beta < \omega_1$. Thus

(7) for any $A \in \mathcal{A}$ and $\beta < \omega_1$ the set $\pi^{-1}([0, \beta]) \cap A$ also belongs to $\mathcal{A}$.

The following property of the family $\mathcal{A}$ is crucial.

(8) if $A, B \in \mathcal{A}$ and $\{t_1, t_2\} \subset A \cap B$ for some $t_1, t_2 \in T$ with $\beta = \pi(t_1) < \alpha = \pi(t_2)$ then $\pi^{-1}([0, \alpha]) \cap A = \pi^{-1}([0, \alpha]) \cap B$.

To see that the property (8) is true observe that $A' = \pi^{-1}([0, \alpha]) \cap A \in \mathcal{A}$ and $B' = \pi^{-1}([0, \alpha]) \cap B \in \mathcal{A}$ by (7). There is $r \in I$ with $t_2 = (\alpha, r) \in A' \cap B'$; so it follows from (2) that $A', B' \in \mathcal{A}_\alpha^r$; the family $\mathcal{B}_\alpha^r$ is disjoint by (3) and both sets $A' \backslash \{t_2\}$ and $B' \backslash \{t_2\}$ belong to $\mathcal{B}_\alpha^r$. Now it follows from $t_1 \in (B' \backslash \{t_2\}) \cap (A' \backslash \{t_2\})$ that $A' = B'$; so (8) is proved.

Now assume that $A = \text{supp}(x)$ is uncountable for some $x \in K$. Then there is $\alpha \in \pi(A)$ such that $\pi(A) \cap [0, \alpha]$ is infinite; pick $t = (\alpha, r) \in A$ and $s = (\beta, q) \in A$ with $\beta < \alpha$. The set $O = \{y \in \mathbb{D}^T : y(s) = \overline{y(t)} = 1\}$ is an open neighbourhood of the point x in $\mathbb{D}^T$; so it follows from $x \in \overline{Y}$ that $x \in \overline{Y \cap O}$. Let $\mathcal{U} = \{B \in \mathcal{A} : \{t, s\} \subset B\}$; it is evident that $Y_0 = Y \cap O = \{\chi_B : B \in \mathcal{U}\}$. It follows from (8) that there exists a finite set $H \subset \pi^{-1}([0, \alpha])$ such that $B \cap \pi^{-1}([0, \alpha]) = H$ for any $B \in \mathcal{U}$. Since $\pi(A) \cap [0, \alpha]$ is infinite, we can find $\gamma \in (\pi(A) \cap [0, \alpha]) \backslash H$. If $t' \in A$ and $\pi(t') = \gamma$ then there exists $y = \chi_B \in Y_0$ for which $y(t') = 1$ which shows that $B \in \mathcal{U}$ while $t' \in (B \cap \pi^{-1}([0, \alpha])) \backslash H$ which is a contradiction. Thus $A = x^{-1}(1)$ is countable for any $x \in K$ and hence $K \subset \Sigma(\mathbb{D}^T)$ is Corson compact.

To see that the space $C_p(K)$ is K-analytic consider, for any $m, n \in \omega$, the set $S_n^m = \{t = (\alpha, r) \in T : \text{either } a_{n+1} \notin A \text{ for any } A \in \mathcal{A}_\alpha \text{ with } t \in A \text{ or there is } A \in \mathcal{A}_\alpha \text{ such that } t \in A, \ a_{n+1} \in A \text{ and } |A| = m + 1\}$. Let $T_\emptyset = T$; if $s \in \omega^{<\omega}$ and $\text{dom}(s) = \{0, \ldots, m - 1\}$ for some $m \in \mathbb{N}$ then let $T_s = S_0^{s(0)} \cap \ldots \cap S_{m-1}^{s(m-1)}$. We will check that the family $\mathcal{T} = \{T_s : s \in \omega^{<\omega}\}$ satisfies the conditions (i) and (ii) of Fact 4.

Fix $s \in \omega^{<\omega}$ and $t = (\alpha, r) \in T_s$; if $s \in \omega^n$ and $a_{n+1} \notin A$ for any $A \in \mathcal{A}$ with $\max(A) = t$ then $t \in S_n^0 \cap T_s = T_{s^\frown 0}$. If there exists $A \in \mathcal{A}$ such that $\max(A) = t$ and $a_{n+1} \in A$ then $t \in S_n^m \cap T_s = T_{s^\frown m}$ for $m = |A| - 1$. This shows that we have $T_s = \bigcup\{T_{s^\frown k} : k \in \omega\}$ for any $s \in \omega^{<\omega}$, i.e., the family $\mathcal{T}$ satisfies (i) of Fact 4.

To prove that the condition (ii) of Fact 4 is also satisfied take $x \in K$ and $f \in \omega^\omega$; if $|\text{supp}(x)| < \omega$ then there is nothing to prove; so we can apply (6) to fix $t, s \in \text{supp}(x)$ such that $0 < \pi(t) < \pi(s)$. Since $x \in \overline{Y}$, there is $A \in \mathcal{A}$ with $\chi_A(t) = \chi_A(s) = x(s) = x(t) = 1$ and therefore $\{t, s\} \subset A$. By (2) and (6), there is a unique $m \in \omega$ with $a_{m+1} \in A$. We claim that

(9) $|S_m^k \cap \text{supp}(x)| \leq 1$ for any $k \in \omega$.

Assume, towards a contradiction that (9) is false; since $\text{supp}(x)$ is thin by (6), there are $s', t' \in \text{supp}(x) \cap S_m^k$ such that $\alpha = \pi(s') < \beta = \pi(t')$. It follows from $x \in \overline{Y}$ that there is $B \in \mathcal{A}$ such that $\{t, s, t', s'\} \subset B$. The property (8) implies that $a_{m+1} \in B$; furthermore, $D = \pi^{-1}([0, \beta]) \cap B \in \mathcal{A}$ by (7) and $t' = \max(D)$. If $D' \in \mathcal{A}$, $a_{m+1} \in D'$ and $t' = \max(D')$ then $D' = D$ (see (8)); so it follows from $t' \in S_m^k$ that $|D| = k + 1$.

Analogously, $E = \pi^{-1}([0, \alpha]) \cap B \in \mathcal{A}$ while $a_{m+1} \in E$ and $\max(E) = s'$. Again, $s' \in S_m^k$ implies that $|E| = k + 1$; however, this is a contradiction because $E \subset D$ and $E \neq D$; so it is impossible that $|D| = |E| = k + 1$. Thus we established the property (9) and therefore $|\text{supp}(x) \cap T_{f|(m+1)}| \leq |\text{supp}(x) \cap S_m^{f(m)}| \leq 1$; since $T_{f|n} \subset T_{f|(m+1)}$ for any $n \geq m + 1$, we proved that $|\text{supp}(x) \cap T_{f|n}| < \omega$ for any $n \geq m + 1$, i.e., the condition (ii) of Fact 4 is also satisfied for $\mathcal{T}$. Thus $C_p(K)$ is K-analytic by Fact 4.

To prove that the point $u \in K$ satisfies the condition (ii) of our Fact, it suffices to show that the set $Y \subset K$ has the property $(*)$ from Fact 3. So, assume that $S_n \subset Y$ is a sequence which converges to u for any $n \in \omega$. It follows from Fact 2 that there is a sequence $S \subset Y$ for which $S \cap S_n$ is infinite for any $n \in \omega$ and the family $\{\text{supp}(x) : x \in S\}$ is disjoint. To see that S flexibly converges to u let

$B_x = \text{supp}(x)$ and suppose that G_x is a G_δ-subset of K with $x \in G_x$ for any $x \in S$. Making every G_x smaller if necessary we can assume that there exists a countable $A_x \subset T \backslash B_x$ such that $G_x = \{y \in K : B_x \subset \text{supp}(y) \subset T \backslash A_x\}$.

The set $A = \bigcup\{A_x : x \in S\}$ is countable and it follows from $u \in \overline{S} \backslash S$ that the family $\mathcal{C} = \{B_x : x \in S\} \subset \mathcal{A}$ is countably infinite; there exists $\alpha < \omega_1$ such that $\mathcal{C} \subset \bigcup\{\mathcal{A}_\beta : \beta < \alpha\}$ and $A \subset \pi^{-1}([0, \alpha))$; so it follows from (4) that $\mathcal{A}_\alpha^r = \{C \cup \{(\alpha, r)\} : C \in \mathcal{C}\}$ for some $r \in I$.

We have $B_x \subset E_x = B_x \cup \{(\alpha, r)\} \subset T \backslash A_x$ and therefore $y(x) = \chi_{E_x} \subset G_x$ for any $x \in S$. It is clear that the sequence $\{y(x) : x \in S\} = \{\chi_B : B \in \mathcal{A}_\alpha^r\}$ converges to $y = \chi_{\{(\alpha, r)\}} \neq u$ and hence S flexibly converges to u which, together with Fact 3, shows that (ii) is proved.

To finally prove that the property (iii) holds for K observe that the subspace $K_0 = \{u\} \cup \{\chi_{\{t\}} : t \in (0, \omega_1) \times I\} \subset K$ is homeomorphic to the space $A(\mathfrak{c})$. Let $\mathcal{A}_n = \{A \in \mathcal{A} : a_n \in A\}$ for any $n \in \mathbb{N}$; then $\mathcal{A} = \bigcup\{\mathcal{A}_n : n \in \mathbb{N}\}$ and $\mathcal{A}_n \cap \mathcal{A}_m = \emptyset$ if $n \neq m$.

If $Y_n = \{\chi_A : A \in \mathcal{A}_n\}$ and $K_n = \overline{Y}_n$ for any $n \in \mathbb{N}$ then $K_n \subset K$ is compact. The set $U_n = \{x \in K : x(a_n) = 1\}$ is clopen in K and it is immediate that $K_n = U_n$ so K_n is clopen in K for every $n \in \mathbb{N}$. It is straightforward that the family $\{K_n : n \in \omega\}$ is disjoint.

To see that $K = \bigcup_{n \in \omega} K_n$ take any $x \in K \backslash K_0$; If $x = \chi_{\{a_n\}}$ for some $n \in \mathbb{N}$ then $x \in K_n$. If not then there are $t, s \in \text{supp}(x)$ such that $\pi(t) < \pi(s)$. Since $x \in \overline{Y}$, there is $B \in \mathcal{A}$ such that $\{t, s\} \subset B$; there is a unique $n \in \mathbb{N}$ with $a_n \in B$. It follows from (8) that $A \in \mathcal{A}_n$ for any $A \in \mathcal{A}$ with $\{t, s\} \subset A$. The set $O = \{y \in K : y(t) = y(s) = 1\}$ is open in K and $x \in O$; so $x \in \overline{Y \cap O}$. But $Y \cap O = \{\chi_A : A \in \mathcal{A}$ and $\{t, s\} \subset A\} \subset Y_n$ and hence $x \in \overline{Y}_n = K_n$; so we proved that $K = \bigcup_{n \in \omega} K_n$.

Now fix $m \in \mathbb{N}$; to prove that K_m is homeomorphic to a closed subspace of the space $(A(\mathfrak{c}))^\omega$ consider the set $R = \bigcup\{\text{supp}(x) : x \in K_m\} = \bigcup\{\text{supp}(x) : x \in Y_m\}$. For every $t = (\alpha, r) \in R$ there is $A \in \mathcal{A}$ such that $\{a_m, t\} \subset A$; so it follows from (8) that the set $A' \cap \pi^{-1}([0, \alpha])$ is the same for any $A' \in \mathcal{A}_m$ with $t \in A'$. Thus the number $\mu(t) = |A \cap \pi^{-1}([0, \alpha])|$ is well defined and depends only on t. Let $R_n = \{t \in R : \mu(t) = n\}$ for any $n \in \mathbb{N}$. Then $R = \bigcup\{R_n : n \in \mathbb{N}\}$ and the family $\{R_n : n \in \mathbb{N}\}$ is disjoint. Besides,

(10) $|\text{supp}(x) \cap R_n| \leq 1$ for any $x \in K_m$ and $n \in \mathbb{N}$.

Since Y_m is dense in K_m it suffices to prove (10) for every $x \in Y_m$. Suppose that $x \in Y_m$, $t, s \in \text{supp}(x)$ while $\alpha = \pi(t) < \beta = \pi(s)$ and $\{t, s\} \subset R_n$. The set $A = \text{supp}(x)$ belongs to $\mathcal{A}_m$; let $B = A \cap \pi^{-1}([0, \beta])$ and $C = A \cap \pi^{-1}([0, \alpha])$. It follows from $s \in A \cap R_n$ that $|B| = n$; since also $t \in R_n$, we have $|C| = n$ which is impossible because B and C are distinct finite sets with $C \subset B$. This contradiction shows that (10) is true.

Let $\pi_R : \mathbb{D}^T \to \mathbb{D}^R$ be the projection of $\mathbb{D}^T$ onto its face $\mathbb{D}^R$; analogously π_{R_n} is the projection of $\mathbb{D}^T$ onto its face $\mathbb{D}^{R_n}$ for any $n \in \mathbb{N}$. It is not difficult to see that $\pi_R | K_m : K_m \to \pi_R(K_m)$ is a homeomorphism. Besides, $\pi_R(K_m)$ is homeomorphic to a closed subset of the product $\prod\{\pi_{R_n}(K_m) : n \in \mathbb{N}\}$. Given $t \in R_n$ let $v_t \in \mathbb{D}^{R_n}$ be the characteristic function of $\{t\}$ and denote by w_n the function on R_n which is identically zero on R_n. It follows from the property (10) that, for any $n \in \mathbb{N}$, we have $\pi_{R_n}(K_m) \subset L_n = \{w_n\} \cup \{v_t : t \in R_n\}$. Since every L_n can be embedded in $A(\mathfrak{c})$ as a closed subspace, the space K_m embeds in $(A(\mathfrak{c}))^\omega$ for any $m \in \omega$. We finally checked (iii) and hence Fact 5 is proved.

Returning to our solution let M be the compact space K whose existence is stated by Fact 5. The property (i) of Fact 5 says that $C_p(M)$ is K-analytic. The property (ii) of Fact 5 together with Fact 1 imply that there is a point $x \in M$ such that $M \backslash \{x\}$ is pseudocompact and M is the Stone–Čech extension of $M \backslash \{x\}$. Observe that M embeds in $C_p(C_p(M))$; since $C_p(M)$ is Lindelöf Σ, the space $C_p(C_p(M))$ and hence M is ω-monolithic and has countable tightness. Consequently, M is Fréchet–Urysohn (see Fact 1 of U.080); so there exists a sequence $S \subset Y = M \backslash \{x\}$ which converges to x. It is evident that S is an infinite closed and discrete subset of Y; so Y is not countably compact.

Finally, let $X = C_p(Y)$; if $\pi_Y : C_p(M) \to C_p(Y)$ is the restriction map then it follows from pseudocompactness of Y and $\beta Y = M$ that $C_p(Y) = \pi_Y(C_p(M))$ and hence $X = C_p(Y)$ is also K-analytic. Furthermore, Y embeds in $C_p(X)$ as a closed subspace by TFS-167. Thus X is a K-analytic space such that the subspace $Y \subset C_p(X)$ is closed, pseudocompact and not countably compact.

U.223. *Suppose that, for a countably compact space X, there exists a condensation $f : X \to Z \subset C_p(Y)$, where $C_p(Y)$ is a Lindelöf Σ-space. Prove that f is a homeomorphism and X is Gul'ko compact.*

Solution. If F is a closed subset of X then it is countably compact and hence so is $f(F)$ which implies, together with Problem 221, that $f(F)$ is Gul'ko compact and hence closed in Z; in particular, Z is Gul'ko compact. Thus our condensation f is a closed map; this shows that f is a homeomorphism and X is Gul'ko compact being homeomorphic to a Gul'ko compact space Z.

U.224. *Give an example of a pseudocompact non-countably compact space X which can be condensed onto a compact $K \subset C_p(Y)$, where $C_p(Y)$ is Lindelöf Σ.*

Solution. Let M be Reznichenko's compactum (see Problem 222). Then there is a point $x \in M$ such that $X = M \backslash \{x\}$ is pseudocompact while M is canonically homeomorphic to βX. Observe that M embeds in $C_p(C_p(M))$; since $C_p(M)$ is Lindelöf Σ, the space $C_p(C_p(M))$ and hence M is ω-monolithic and has countable tightness. Consequently, M is Fréchet–Urysohn (see Fact 1 of U.080); so there exists a sequence $S \subset X$ which converges to x. It is evident that S is an infinite closed and discrete subset of X; so X is not countably compact.

The space X being locally compact, there is a condensation $\varphi : X \to K$ of X onto some compact space K (see Fact 3 of T.357). There is a continuous mapping $\Phi : M \to K$ such that $\Phi|X = \varphi$ (see TFS-257). The dual map $\Phi^* : C_p(K) \to C_p(M)$ embeds $C_p(K)$ in $C_p(M)$ as a closed subspace (see TFS-163). As a consequence, the space $Y = C_p(K)$ is Lindelöf Σ being a closed subspace of a Lindelöf Σ-space $C_p(M)$. By TFS-167 we can consider that $K \subset C_p(Y)$; it follows from Problem 219 that $C_p(Y)$ is also a Lindelöf Σ-space; so our condensation φ is what we looked for.

U.225. *Give an example of a space X such that $C_p(X)$ is Lindelöf Σ and some pseudocompact subspace of $C_p(X)$ is not countably compact.*

Solution. Let M be Reznichenko's compactum (see Problem 222). Then there is a point $x \in M$ such that $Y = M\backslash\{x\}$ is pseudocompact while M is canonically homeomorphic to βY. Observe that M embeds in $C_p(C_p(M))$; since $C_p(M)$ is Lindelöf Σ, the space $C_p(C_p(M))$ and hence M is ω-monolithic and has countable tightness. Consequently, M is Fréchet–Urysohn (see Fact 1 of U.080); so there exists a sequence $S \subset Y$ which converges to x. It is evident that S is an infinite closed and discrete subset of Y; so Y is not countably compact.

Since $X = C_p(M)$ is a Lindelöf Σ-space, the space $C_p(X)$ is also Lindelöf Σ by Problem 219. The restriction map $\pi : C_p(M) \to C_p(Y)$ is onto because Y is pseudocompact and $M = \beta Y$; thus the dual map $\pi^* : C_p(C_p(Y)) \to C_p(C_p(M)) = C_p(X)$ is an embedding (see TFS-163). Since Y embeds in $C_p(C_p(Y))$, it also embeds in $C_p(X)$. Therefore $C_p(X)$ is a Lindelöf Σ-space which contains a pseudocompact non-countably compact subspace Y.

U.226. *Observe that there exist Gul'ko spaces X such that $t(C_p(X)) > \omega$. Prove that, if $C_p(X)$ is Lindelöf Σ and $Y \subset C_p(X)$ is pseudocompact then Y must be Fréchet–Urysohn.*

Solution. Let M be Reznichenko's compactum (see Problem 222). Then there exists a point $x \in M$ such that the space $X = M\backslash\{x\}$ is pseudocompact while M is canonically homeomorphic to βX; an immediate consequence is that the restriction map $\pi : C_p(M) \to C_p(X)$ is onto so $C_p(X)$ is a Lindelöf Σ-space being a continuous image of a Lindelöf Σ-space $C_p(M)$. The space X is not compact and hence not Lindelöf because it is a pseudocompact proper dense subspace of M. Therefore $t(C_p(X)) > \omega$ (see TFS-149); so $C_p(X)$ is a Lindelöf Σ-space of uncountable tightness.

Now assume that X is a space such that $C_p(X)$ is Lindelöf Σ and $Y \subset C_p(X)$ is pseudocompact. The space $Z = \overline{Y}$ is also pseudocompact (see Fact 18 of S.351); since Z is closed in the Lindelöf space $C_p(X)$, it has to be compact. Therefore Z is Gul'ko compact (see Problem 221); so $t(Z) = \omega$ by TFS-189; besides, $C_p(X)$ is ω-monolithic (see Problem 208) and hence so is Z which shows that Z is a Fréchet–Urysohn space (see Fact 1 of U.080). Finally, Y has to be a Fréchet–Urysohn space being a subspace of a Fréchet–Urysohn space Z.

U.227. *Show that there exists a space X such that $C_p(X)$ is a Lindelöf Σ-space and $t(Y) > \omega$ for some σ-compact subspace $Y \subset C_p(X)$.*

Solution. Let $S = \{x \in \mathbb{D}^{\omega_1} : x^{-1}(1)$ is finite$\}$; since $S = \mathbb{D}^{\omega_1} \cap \sigma(\omega_1)$, the space S is homeomorphic to a closed subspace of $\sigma(\omega_1)$. The space $\sigma(\omega_1)$ is σ-compact by Problem 108; so S is also σ-compact. Furthermore, $S \subset \Sigma_*(\omega_1) \simeq C_p(A(\omega_1))$ (see Problem 105). The class $L(\Sigma)$ of Lindelöf Σ-spaces is sk-directed (see SFFS-254) and $A(\omega_1)$ is also Lindelöf Σ being compact; so we can apply Problem 092 to see that $C_p^*(S)$ belongs to $(L(\Sigma))_{\sigma\delta} = L(\Sigma)$ and hence $C_p(S, \mathbb{I})$ is a Lindelöf Σ-space. Now apply Problem 217 to conclude that $C_p(S)$ is a Lindelöf Σ-space.

Let $z(\alpha) = 1$ for any $\alpha < \omega_1$; then $z \in \mathbb{D}^{\omega_1} \setminus \Sigma(\omega_1)$. Since $S \subset \Sigma(\omega_1)$, for any countable $B \subset S$ we have $z \notin \overline{B}$ (the bar denotes the closure in $\mathbb{D}^{\omega_1}$). Therefore $Z = S \cup \{z\}$ is also σ-compact and $t(Z) > \omega$.

Let us show that $C_p(C_p(Z))$ is a Lindelöf Σ-space. Since the restriction map $\pi : C_p(\upsilon(C_p(Z))) \to C_p(C_p(Z))$ is continuous and onto, it suffices to show that $C_p(\upsilon(C_p(Z)))$ is Lindelöf Σ. The space $\upsilon(C_p(Z))$ is canonically homeomorphic to the set $H = \{f \in \mathbb{R}^Z : f$ is strictly ω-continuous on $Z\}$ (see TFS-438). Observe that $S \subset Z$ is a Fréchet–Urysohn space so, for every $f \in H$, the function $f|S$ is continuous. On the other hand, if $f \in \mathbb{R}^Z$ and $f|S$ is continuous then f is strictly ω-continuous on Z; this is an easy consequence of the fact that $z \notin \overline{B}$ for any countable $B \subset S$.

Since $\upsilon(C_p(Z)) \simeq H$, we established that the space $\upsilon(C_p(Z))$ is homeomorphic to the set $\{f \in \mathbb{R}^Z : f|S \in C_p(S)\} = C_p(S) \times \mathbb{R}$ where $\mathbb{R}$ is the factor of $\mathbb{R}^Z$ determined by the point z. If $T = S \oplus \{z\}$ then $\upsilon(C_p(Z)) \simeq C_p(S) \times \mathbb{R} \simeq C_p(T)$.

The space T is σ-compact and hence Lindelöf Σ; besides, $C_p(T) = C_p(S) \times \mathbb{R}$ is also Lindelöf Σ because so is $C_p(S)$. This makes it possible to apply Okunev's theorem (see Problem 218) to conclude that $C_p(C_p(T)) = C_p(\upsilon(C_p(Z)))$ is a Lindelöf Σ-space. Thus $C_p(C_p(Z))$ is a Lindelöf Σ-space as well being a continuous image (under π) of the space $C_p(C_p(T)) = C_p(\upsilon(C_p(Z)))$.

Finally observe that, for $X = C_p(Z)$, the space Z embeds in $C_p(X)$ by TFS-167 and hence there is $Y \subset C_p(X)$ with $Y \simeq Z$ and hence $t(Y) > \omega$. Therefore $C_p(X)$ is a Lindelöf Σ-space such that $t(Y) > \omega$ for a σ-compact subspace $Y \subset C_p(X)$.

U.228. *Let X be a space and denote by $\pi : C_p(\upsilon X) \to C_p(X)$ the restriction map. Prove that, for any countably compact $Y \subset C_p(X)$, the space $\pi^{-1}(Y) \subset C_p(\upsilon X)$ is countably compact.*

Solution. To see that the set $Z = \pi^{-1}(Y) \subset C_p(\upsilon X)$ is also countably compact suppose that D is a countably infinite closed discrete subspace of Z. Then $E = \pi(D)$ has to have an accumulation point $f \in Y$. Then $g = \pi^{-1}(f) \in Z$ and g is an accumulation point of D because $\pi|(D \cup \{g\})$ is a homeomorphism between $D \cup \{g\}$ and $E \cup \{f\}$ (see TFS-437); this contradiction shows that Z is countably compact.

U.229. *Give an example of a space X such that $\pi^{-1}(Y)$ is not pseudocompact for some pseudocompact $Y \subset C_p(X)$. Here $\pi : C_p(\upsilon X) \to C_p(X)$ is the restriction map.*

Solution. There exists an infinite pseudocompact space X such that the space $Y = C_p(X, \mathbb{I})$ is also pseudocompact (see TFS-400 and TFS-398). Since every countable subset of X is closed and discrete in X by TFS-398, the space X is not countably compact; besides, $\upsilon X = \beta X$ (see TFS-415 and TFS-417).

If the set $Z = \pi^{-1}(Y) \subset C_p(\upsilon X) = C_p(\beta X)$ is pseudocompact then it is bounded in $C_p(\beta X)$ and hence compact by Grothendieck's theorem (Problem 044). Therefore $Y = C_p(X, \mathbb{I})$ is also compact being a continuous image of Z. Thus X is discrete by TFS-396; this contradiction proves that Y is a pseudocompact subspace of $C_p(X)$ such that $\pi^{-1}(Y)$ is not pseudocompact.

U.230. *Assume that υX is a Lindelöf Σ-space and $\pi : C_p(\upsilon X) \to C_p(X)$ is the restriction map. Prove that, for any compact subspace $Y \subset C_p(X)$, the space $\pi^{-1}(Y) \subset C_p(\upsilon X)$ is also compact.*

Solution. The space $Z = \pi^{-1}(Y) \subset C_p(\upsilon X)$ is countably compact by Problem 228; therefore $ext(Z) = \omega$ and hence we can apply Baturov's theorem SFFS-269 to see that Z is Lindelöf and hence compact.

U.231. *Assume that υX is a Lindelöf Σ-space and $\pi : C_p(\upsilon X) \to C_p(X)$ is the restriction map. Prove that, for any Lindelöf Σ-space $Y \subset C_p(X)$, the space $\pi^{-1}(Y) \subset C_p(\upsilon X)$ is Lindelöf Σ.*

Solution. Given a space Z we denote by $\mathcal{K}(Z)$ the family of all compact subsets of Z. There exists a second countable space M for which there is a compact cover $\{F_K : K \in \mathcal{K}(M)\}$ of the space Y such that $K \subset L$ implies $F_K \subset F_L$ (see Problem 213). If $T = \pi^{-1}(Y)$ then $G_K = \pi^{-1}(F_K)$ is compact for any $K \in \mathcal{K}(M)$ (see Problem 230). Therefore $\{G_K : K \in \mathcal{K}(M)\}$ is a compact cover of T such that $K \subset L$ implies $G_K \subset G_L$.

If F is a closed discrete subspace of T then let $H_K = G_K \cap F$ for any $K \in \mathcal{K}(M)$; it is evident that $\{H_K : K \in \mathcal{K}(M)\}$ is a compact cover of F such that $K \subset L$ implies $H_K \subset H_L$. Thus F is dominated by a second countable space; since F is metrizable, it is Dieudonné complete; so we can apply Problem 213 again to conclude that F is Lindelöf Σ and hence countable (it is evident that every Lindelöf discrete space is countable). This proves that $ext(T) = \omega$ and hence we can apply Baturov's theorem (SFFS-269) to conclude that T is Lindelöf. Any Lindelöf space is Dieudonné complete (see TFS-462); so we can apply Problem 213 once more to conclude that $T = \pi^{-1}(Y)$ is a Lindelöf Σ-space.

U.232. *Let X be a pseudocompact space and denote by $\pi : C_p(\beta X) \to C_p(X)$ the restriction map. Prove that, for an arbitrary Lindelöf Σ-space (compact space) $Y \subset C_p(X)$, the space $\pi^{-1}(Y) \subset C_p(\beta X)$ is Lindelöf Σ (or compact, respectively).*

Solution. Observe that $\beta X = \upsilon X$ because X is pseudocompact (see TFS-415 and TFS-417). Now if $Y \subset C_p(X)$ is Lindelöf Σ then $\pi^{-1}(Y)$ is a Lindelöf Σ-space by Problem 231. If Y is compact then $\pi^{-1}(Y)$ is also compact by Problem 230.

U.233. *Give an example of a pseudocompact X such that $\pi^{-1}(Y) \subset C_p(\beta X)$ is not Lindelöf for some Lindelöf $Y \subset C_p(X)$. Here $\pi : C_p(\beta X) \to C_p(X)$ is the restriction map.*

Solution. Let X be the ordinal ω_1 with its usual order topology. Then X is countably compact and $\beta X = \omega_1 + 1$ (see TFS-314). The space $Y = C_p(X)$ is Lindelöf by TFS-316 while $\pi^{-1}(Y) = C_p(\omega_1 + 1)$ is not Lindelöf (see TFS-320).

U.234. *Observe that $C_p(X)$ is a Lindelöf Σ-space if and only if $C_p(\upsilon X)$ is Lindelöf Σ; prove that, for any X, the space $C_p(X)$ is K-analytic if and only if $C_p(\upsilon X)$ is K-analytic. In other words, X is a Talagrand space if and only if υX is Talagrand.*

Solution. We identify the space $\mathbb{P}$ of the irrationals with ω^ω; if $p, q \in \mathbb{P}$ then $p \leq q$ says that $p(n) \leq q(n)$ for any $n \in \omega$.

The restriction map $\pi : C_p(\upsilon X) \rightarrow C_p(X)$ is a condensation of $C_p(\upsilon X)$ onto $C_p(X)$; thus the space $C_p(X)$ is a continuous image of $C_p(\upsilon X)$ and hence the Lindelöf Σ-property of $C_p(\upsilon X)$ implies that $C_p(X)$ is also Lindelöf Σ. Now, if $C_p(X)$ is Lindelöf Σ then so is υX by Problem 206 which makes it possible to apply Problem 231 to see that $C_p(\upsilon X) = \pi^{-1}(C_p(X))$ is also a Lindelöf Σ-space. This shows that $C_p(X)$ is a Lindelöf Σ-space if and only if so is $C_p(\upsilon X)$.

If $C_p(\upsilon X)$ is K-analytic then $C_p(X)$ is also K-analytic being a continuous image of $C_p(\upsilon X)$. Now, if $C_p(X)$ is K-analytic then there is a compact cover $\{K_p : p \in \mathbb{P}\}$ of the space $C_p(X)$ such that $p \leq q$ implies $K_p \subset K_q$ (see SFFS-391). The space υX is Lindelöf Σ by Problem 206; so we can apply Problem 230 to see that $F_p = \pi^{-1}(K_p)$ is compact for any $p \in \mathbb{P}$. Therefore $\{F_p : p \in \mathbb{P}\}$ is a compact cover of $C_p(\upsilon X)$ such that $p \leq q$ implies $F_p \subset F_q$, i.e., the space $C_p(\upsilon X)$ is $\mathbb{P}$-dominated. Finally, apply Problem 215 to conclude that $C_p(\upsilon X)$ is K-analytic. Thus $C_p(X)$ is K-analytic if and only if $C_p(\upsilon X)$ is K-analytic.

U.235. *Suppose that $C_p(X)$ is a Lindelöf Σ-space. Prove that $C_{p,n}(\upsilon X)$ is a Lindelöf Σ-space for every $n \in \mathbb{N}$.*

Solution. The space υX is Lindelöf Σ by Problem 206 and $C_p(\upsilon X)$ is Lindelöf Σ by Problem 234. Thus Problem 219 is applicable to conclude that $C_{p,n}(\upsilon X)$ is Lindelöf Σ for any $n \in \mathbb{N}$.

U.236. *Given an arbitrary space X let $\pi : C_p(\upsilon X) \rightarrow C_p(X)$ be the restriction mapping. Let $\pi^*(\varphi) = \varphi \circ \pi$ for any function $\varphi \in \mathbb{R}^{C_p(X)}$ and observe that the map $\pi^* : \mathbb{R}^{C_p(X)} \rightarrow \mathbb{R}^{C_p(\upsilon X)}$ is an embedding. Identifying the space $\upsilon(C_p(C_p(X)))$ with the subspace $\{\varphi \in \mathbb{R}^{C_p(X)} : \varphi$ is strictly ω-continuous on $C_p(X)\}$ of the space $\mathbb{R}^{C_p(X)}$ (see TFS-438) prove that*

(i) $\pi^*(C_p(C_p(X))) \subset \pi^*(\upsilon(C_p(C_p(X)))) \subset C_p(C_p(\upsilon X))$;
(ii) if $C_p(X)$ is normal then $\pi^*(\upsilon(C_p(C_p(X)))) = C_p(C_p(\upsilon X))$ and hence the spaces $\upsilon(C_p(C_p(X)))$ and $C_p(C_p(\upsilon X))$ are homeomorphic.

Solution. If we consider that the spaces $C_p(X)$ and $C_p(\upsilon X)$ are discrete then $\pi^* : \mathbb{R}^{C_p(X)} \rightarrow \mathbb{R}^{C_p(\upsilon X)}$ is the usual dual map between their spaces of continuous functions. Thus TFS-163 is applicable to conclude that π^* is an embedding.

If $\varphi \in \upsilon(C_p(C_p(X)))$ then φ is strictly ω-continuous on $C_p(X)$ and hence $\pi^*(\varphi) = \varphi \circ \pi$ is strictly ω-continuous on $C_p(\upsilon X)$. The space υX being realcompact, we have $t_m(C_p(\upsilon X)) = \omega$ (see TFS-434); so $\pi^*(\varphi)$ is continuous on $C_p(\upsilon X)$ which shows that $\pi^*(\upsilon(C_p(C_p(X)))) \subset C_p(C_p(\upsilon X))$. This proves (i).

Furthermore, if the space $C_p(X)$ is normal and $\varphi : C_p(\upsilon X) \rightarrow \mathbb{R}$ is a continuous function then $\varphi \circ \pi^{-1} : C_p(X) \rightarrow \mathbb{R}$ is ω-continuous because the map π is a homeomorphism if restricted to a countable subset of $C_p(\upsilon X)$ (see TFS-437). The space $C_p(X)$ being normal, the map $\xi = \varphi \circ \pi^{-1}$ has to be strictly ω-continuous (see TFS-421) and $\pi^*(\xi) = \xi \circ \pi = \varphi \circ \pi^{-1} \circ \pi = \varphi$.

This shows that we have the inclusion $C_p(C_p(\upsilon X)) \subset \pi^*(\upsilon(C_p(C_p(X))))$, i.e., $C_p(C_p(\upsilon X)) = \pi^*(\upsilon(C_p(C_p(X))))$; since π^* is a homeomorphism, we have $C_p(C_p(\upsilon X)) \simeq \upsilon(C_p(C_p(X)))$; so (ii) is proved.

U.237. *Suppose that $C_p(X)$ is a Lindelöf Σ-space. Prove that $C_{p,2n}(\upsilon X)$ is homeomorphic to $\upsilon(C_{p,2n}(X))$ for every $n \in \mathbb{N}$.*

Solution. Since $C_p(X)$ is normal, the space $C_p(C_p(\upsilon X))$ is homeomorphic to $\upsilon(C_p(C_p(X)))$ by Problem 236; so the statement of this Problem is true for $n = 1$. Proceeding by induction assume that $\upsilon(C_{p,2k}(X)) \simeq C_{p,2k}(\upsilon X)$ for some $k \geq 1$. If $Y = C_{p,2k}(\upsilon X)$ then $C_p(Y) = C_{p,2k+1}(\upsilon X)$ is a Lindelöf Σ-space by Problem 235. Thus we can apply Problem 236 to see that $\upsilon(C_{p,2k+2}(X)) = \upsilon(C_p(C_p(Y))) \simeq C_p(C_p(\upsilon Y))$. Recalling that $\upsilon Y = Y$ because Y is a Lindelöf Σ-space (see Problem 235), we conclude that $\upsilon(C_{p,2k+2}(X)) \simeq C_p(C_p(Y)) = C_{p,2k+2}(\upsilon X)$. This concludes the induction step and proves that $\upsilon(C_{p,2n}(X))$ is homeomorphic to $C_{p,2n}(\upsilon X)$ for any $n \in \mathbb{N}$.

U.238. *Suppose that $C_p(X)$ is a Lindelöf Σ-space. Prove that $C_{p,2n+1}(\upsilon X)$ can be condensed onto $C_{p,2n+1}(X)$ for every $n \in \omega$.*

Solution. The space X is dense and C-embedded in υX and hence the restriction map $r : C_p(\upsilon X) \to C_p(X)$ is a condensation. Therefore the statement of our Problem is true for $n = 0$. Now, if $n > 0$ then we can apply Problem 237 to see that $Y = C_{p,2n}(\upsilon X)$ is homeomorphic to $\upsilon(C_{p,2n}(X))$; let $Z = C_{p,2n}(X)$. The restriction map $\pi : C_p(\upsilon Z) \to C_p(Z) = C_{p,2n+1}(X)$ is a condensation; since $Y \simeq \upsilon Z$, there is a homeomorphism ξ between the spaces $C_p(Y) = C_{p,2n+1}(\upsilon X)$ and $C_p(\upsilon Z)$. It is evident that $\pi \circ \xi$ is a condensation of $C_{p,2n+1}(\upsilon X)$ onto $C_{p,2n+1}(X)$.

U.239. *Suppose that $C_{p,2k+1}(X)$ is a Lindelöf Σ-space for some $k \in \omega$. Prove that $C_{p,2n+1}(X)$ is a Lindelöf Σ-space every $n \in \omega$.*

Solution. Given spaces Y and Z, the expression $Y \sqsubseteq Z$ says that Y embeds in Z as a closed subspace. Observe that $Z \sqsubseteq C_p(C_p(Z))$ for any space Z (see TFS-167). Therefore $C_p(X) \sqsubseteq C_{p,3}(X) \sqsubseteq C_{p,5}(X) \sqsubseteq \ldots$, i.e., proceeding by a trivial induction we can establish that $C_p(X) \sqsubseteq C_{p,2n+1}(X)$ for any $n \in \omega$. In particular, $C_p(X) \sqsubseteq C_{p,2k+1}(X)$ and hence $C_p(X)$ is a Lindelöf Σ-space.

Given any $n \in \omega$ the space $C_{p,2n+1}(\upsilon X)$ is Lindelöf Σ by Problem 235; besides, the space $C_{p,2n+1}(X)$ is a continuous image of $C_{p,2n+1}(\upsilon X)$ by Problem 238. Therefore $C_{p,2n+1}(X)$ is a Lindelöf Σ-space for any $n \in \omega$.

U.240. *Suppose that $C_{p,2k}(X)$ is a Lindelöf Σ-space for some $k \in \mathbb{N}$. Prove that $C_{p,2n}(X)$ is a Lindelöf Σ-space every $n \in \mathbb{N}$.*

Solution. Given spaces Y and Z, the expression $Y \sqsubseteq Z$ says that Y embeds in Z as a closed subspace. Observe that $Z \sqsubseteq C_p(C_p(Z))$ for any space Z (see TFS-167). Therefore $C_p(C_p(X)) \sqsubseteq C_{p,4}(X) \sqsubseteq C_{p,6}(X) \sqsubseteq \ldots$, i.e., proceeding by a trivial induction we can establish that $C_p(C_p(X)) \sqsubseteq C_{p,2n}(X)$ for any $n \in \mathbb{N}$. In particular, $C_p(C_p(X)) \sqsubseteq C_{p,2k}(X)$ and hence $C_p(C_p(X))$ is a Lindelöf Σ-space; let $Y = C_p(X)$. Finally, if $n \in \mathbb{N}$ then $C_{p,2n}(X) = C_{p,2n-1}(Y)$ is Lindelöf Σ by Problem 239; so $C_{p,2n}(X)$ is a Lindelöf Σ-space for any $n \in \mathbb{N}$.

U.241. *Give an example of a space* X *such that* $C_p(X)$ *is not Lindelöf while* $C_{p,2n}(X)$ *is a Lindelöf* Σ*-space for every* $n \in \mathbb{N}$.

Solution. Let $S = \{x \in \mathbb{D}^{\omega_1} : x^{-1}(1) \text{ is finite}\}$; since $S = \mathbb{D}^{\omega_1} \cap \sigma(\omega_1)$, the space S is homeomorphic to a closed subspace of $\sigma(\omega_1)$. The space $\sigma(\omega_1)$ is σ-compact by Problem 108; so S is also σ-compact. Furthermore, $S \subset \Sigma_*(\omega_1) \simeq C_p(A(\omega_1))$ (see Problem 105). The class $L(\Sigma)$ of Lindelöf Σ-spaces is sk-directed (see SFFS-254) and $A(\omega_1)$ is also Lindelöf Σ being compact; so we can apply Problem 092 to see that $C_p^*(S)$ belongs to $(L(\Sigma))_{\sigma\delta} = L(\Sigma)$ and hence $C_p(S, \mathbb{I})$ is a Lindelöf Σ-space. Now apply Problem 217 to conclude that $C_p(S)$ is a Lindelöf Σ-space.

Let $z(\alpha) = 1$ for any $\alpha < \omega_1$; then $z \in \mathbb{D}^{\omega_1} \setminus \Sigma(\omega_1)$. Since $S \subset \Sigma(\omega_1)$, for any countable $B \subset S$ we have $z \notin \overline{B}$ (the bar denotes the closure in $\mathbb{D}^{\omega_1}$). Therefore $X = S \cup \{z\}$ is also σ-compact and $t(X) > \omega$.

Let us show that $C_p(C_p(X))$ is a Lindelöf Σ-space. Since the restriction map $\pi : C_p(\upsilon(C_p(X))) \to C_p(C_p(X))$ is continuous and onto, it suffices to show that $C_p(\upsilon(C_p(X)))$ is Lindelöf Σ. The space $\upsilon(C_p(X))$ is canonically homeomorphic to the set $H = \{f \in \mathbb{R}^X : f \text{ is strictly } \omega\text{-continuous on } X\}$ (see TFS-438). Observe that $S \subset X$ is a Fréchet–Urysohn space so, for every $f \in H$, the function $f|S$ is continuous. On the other hand, if $f \in \mathbb{R}^X$ and $f|S$ is continuous then f is strictly ω-continuous on X; this is an easy consequence of the fact that $z \notin \overline{B}$ for any countable $B \subset S$.

Since $\upsilon(C_p(X)) \simeq H$, we proved that the space $\upsilon(C_p(X))$ is homeomorphic to the set $\{f \in \mathbb{R}^X : f|S \in C_p(S)\} = C_p(S) \times \mathbb{R}$ where $\mathbb{R}$ is the factor of $\mathbb{R}^X$ determined by the point z. If $T = S \oplus \{z\}$ then $\upsilon(C_p(X)) \simeq C_p(S) \times \mathbb{R} \simeq C_p(T)$.

The space T is σ-compact and hence Lindelöf Σ; besides, $C_p(T) = C_p(S) \times \mathbb{R}$ is also Lindelöf Σ because so is $C_p(S)$. This makes it possible to apply Okunev's theorem (Problem 218) to conclude that $C_p(C_p(T)) = C_p(\upsilon(C_p(X)))$ is a Lindelöf Σ-space. Thus $C_p(C_p(X))$ is a Lindelöf Σ-space as well being a continuous image (under π) of the space $C_p(C_p(T)) = C_p(\upsilon(C_p(X)))$.

However, the space $C_p(X)$ is not Lindelöf because $t(X) > \omega$ (see TFS-189); applying Problem 240 we conclude that $C_{p,2n}(X)$ is a Lindelöf Σ-space for all $n \in \mathbb{N}$, i.e., X is the promised example.

U.242. *Give an example of a space* X *such that* $C_p C_p(X)$ *is not Lindelöf while* $C_{p,2n+1}(X)$ *is a Lindelöf* Σ*-space for every* $n \in \omega$.

Solution. It was proved in Problem 222 that there exists a compact space M such that $C_p(M)$ is K-analytic while there is a point $x \in M$ such that $X = M \setminus \{x\}$ is pseudocompact and M is the Stone–Čech extension of X. Observe that M embeds in $C_p(C_p(M))$; since $C_p(M)$ is Lindelöf Σ, the space $C_p(C_p(M))$ and hence M is ω-monolithic and has countable tightness. Consequently, M is Fréchet–Urysohn (see Fact 1 of U.080); so there exists a sequence $S \subset X = M \setminus \{x\}$ which converges to x. It is evident that S is an infinite closed and discrete subset of X; so X is not countably compact and hence not Lindelöf.

If $\pi : C_p(M) \to C_p(X)$ is the restriction map then it follows from pseudo-compactness of X and $\beta X = M$ that $C_p(X) = \pi(C_p(M))$ and hence $C_p(X)$ is also K-analytic. Furthermore, X embeds in $C_p(C_p(X))$ as a closed subspace (see TFS-167) which shows that $C_p(C_p(X))$ is not Lindelöf. Finally observe that $C_p(X)$ is Lindelöf Σ so we can apply Problem 239 to conclude that $C_{p,2n+1}(X)$ is a Lindelöf Σ-space for any $n \in \omega$.

U.243. *Prove that, for any space X, only the following distributions of the Lindelöf Σ-property in iterated function spaces are possible:*

(i) $C_{p,n}(X)$ *is not a Lindelöf Σ-space for any $n \in \mathbb{N}$;*
(ii) $C_{p,n}(X)$ *is a Lindelöf Σ-space for any $n \in \mathbb{N}$;*
(iii) $C_{p,2n+1}(X)$ *is a Lindelöf Σ-space and $C_{p,2n+2}(X)$ is not Lindelöf for any $n \in \omega$;*
(iv) $C_{p,2n+2}(X)$ *is a Lindelöf Σ-space and $C_{p,2n+1}(X)$ is not Lindelöf for any $n \in \omega$.*

Solution. If X is a discrete space of cardinality ω_1 then $C_p(X) = \mathbb{R}^{\omega_1}$ is not Lindelöf by Fact 2 of S.215; furthermore, X embeds in $C_p(C_p(X))$ as a closed subspace; so $C_p(C_p(X))$ is not Lindelöf either. Since $C_p(X)$ is not Lindelöf Σ, it follows from Problem 239 that $C_{2n+1}(X)$ is not Lindelöf Σ for any $n \in \omega$. The space $C_p(C_p(X))$ is not Lindelöf Σ so Problem 240 implies that $C_{2n}(X)$ is not Lindelöf Σ for any $n \in \mathbb{N}$. Thus, (i) holds for the space X.

To prove that there exists a space X for which (ii) is true it suffices to consider $X = \mathbb{R}$. Then $C_{p,n}(X)$ is a Lindelöf Σ-space because it has a countable network for any $n \in \mathbb{N}$. Next, observe that the space X from Problem 241 has the property (iv) while the space X from Problem 242 has the property (iii).

To see that no other cases can occur, observe that if a space X does not have the property (i) then $C_{p,k}(X)$ is Lindelöf Σ for some $k \in \mathbb{N}$. Assume first that there are $m, l \in \mathbb{N}$ with m even and l odd such that both spaces $C_{p,m}(X)$ and $C_{p,l}(X)$ are Lindelöf Σ. It is an immediate consequence of Problem 239 and Problem 240 that $C_{p,n}(X)$ is Lindelöf Σ both for even and odd $n \in \mathbb{N}$, i.e., $C_{p,n}(X)$ is a Lindelöf Σ-space for all $n \in \mathbb{N}$. Thus we have the case (ii).

Now, if $C_{p,n}(X)$ is Lindelöf Σ for some even $n \in \mathbb{N}$ and there exists no odd $m \in \omega$ with $C_{p,m}(X)$ Lindelöf Σ, then it follows from Problem 241 that we have case (iv). Finally, if $C_{p,n}(X)$ is Lindelöf Σ for some odd n and there exists no even $m \in \omega$ with $C_{p,m}(X)$ Lindelöf Σ, then it follows from Problem 242 that we have the case (iii).

U.244. *Suppose that $C_{p,2k+1}(X)$ is a Lindelöf Σ-space for some $k \in \omega$. Prove that, if $C_{p,2l+2}(X)$ is normal for some $l \in \omega$, then $C_{p,n}(X)$ is a Lindelöf Σ-space for any $n \in \mathbb{N}$.*

Solution. Given spaces Y and Z, the expression $Y \sqsubseteq Z$ says that Y embeds in Z as a closed subspace. Observe that $Z \sqsubseteq C_p(C_p(Z))$ for any space Z (see TFS-167). Therefore $C_p(C_p(X)) \sqsubseteq C_{p,4}(X) \sqsubseteq C_{p,6}(X) \sqsubseteq \ldots$, i.e., proceeding by a trivial induction we can establish that $C_p(C_p(X)) \sqsubseteq C_{p,2n+2}(X)$ for any $n \in \omega$. In particular, $C_p(C_p(X)) \sqsubseteq C_{p,2l+2}(X)$ and hence $C_p(C_p(X))$ is a normal space. It follows from TFS-295 that $ext(C_p(C_p(X))) = \omega$.

Apply Problem 239 to see that $C_p(X)$ is a Lindelöf Σ-space; thus Baturov's theorem (SFFS-269) is applicable to see that $l(C_p(C_p(X))) = ext(C_p(C_p(X))) = \omega$, i.e., the space $C_p(C_p(X))$ is Lindelöf and hence realcompact which shows that we have the equality $\upsilon(C_p(C_p(X))) = C_p(C_p(X))$. The space $C_{p,3}(X)$ is also Lindelöf Σ by Problem 239; so $\upsilon(C_p(C_p(X))) = C_p(C_p(X))$ has to be a Lindelöf Σ-space by Problem 206. Finally, apply Problem 240 to see that $C_{p,n}(X)$ is a Lindelöf Σ-space for all even $n \in \mathbb{N}$; by Problem 239 the space $C_{p,n}(X)$ is Lindelöf Σ for all odd $n \in \mathbb{N}$; so $C_{p,n}(X)$ is Lindelöf Σ for all $n \in \mathbb{N}$.

U.245. *Suppose that $C_{p,2k+2}(X)$ is a Lindelöf Σ-space for some $k \in \omega$. Prove that, if $C_{p,2l+1}(X)$ is normal for some $l \in \omega$, then $C_{p,n}(X)$ is a Lindelöf Σ-space for any $n \in \mathbb{N}$.*

Solution. Given spaces Y and Z, the expression $Y \sqsubseteq Z$ says that Y embeds in Z as a closed subspace. Observe that $Z \sqsubseteq C_p(C_p(Z))$ for any space Z (see TFS-167). Therefore $C_p(X) \sqsubseteq C_{p,3}(X) \sqsubseteq C_{p,5}(X) \sqsubseteq \ldots$, i.e., proceeding by a trivial induction we can establish that $C_p(X) \sqsubseteq C_{p,2n+1}(X)$ for any $n \in \omega$. In particular, $C_p(X) \sqsubseteq C_{p,2l+1}(X)$ and hence $C_p(X)$ is a normal space. It follows from TFS-295 that $ext(C_p(X)) = \omega$.

Apply Problem 240 to see that $C_p(C_p(X))$ is a Lindelöf Σ-space; since X embeds in $C_p(C_p(X))$ as a closed subspace, it also has to be a Lindelöf Σ-space. Thus Baturov's theorem (SFFS-269) is applicable to see that $l(C_p(X)) = ext(C_p(X)) = \omega$, i.e., $C_p(X)$ is Lindelöf and hence realcompact which shows that $\upsilon(C_p(X)) = C_p(X)$. The space $C_p(C_p(X))$ being Lindelöf Σ we can apply Problem 206 to conclude that $\upsilon(C_p(X)) = C_p(X)$ is a Lindelöf Σ-space. Finally, apply Problem 240 to see that $C_{p,n}(X)$ is a Lindelöf Σ-space for all even $n \in \mathbb{N}$; by Problem 239 the space $C_{p,n}(X)$ is Lindelöf Σ for all odd $n \in \mathbb{N}$; so $C_{p,n}(X)$ is Lindelöf Σ for all $n \in \mathbb{N}$.

U.246. *Prove that, if $C_p(X)$ is a Lindelöf Σ-space, then $\upsilon(C_pC_p(X))$ is a Lindelöf Σ-space.*

Solution. It follows from Problem 239 that $C_p(C_p(C_p(X)))$ is a Lindelöf Σ-space; so $\upsilon(C_p(C_p(X)))$ is also Lindelöf Σ by Problem 206.

U.247. *Prove that, if X is normal and $\upsilon(C_p(X))$ is a Lindelöf Σ-space, then $\upsilon(C_pC_p(X))$ is a Lindelöf Σ-space.*

Solution. There exists a space Y such that $C_p(Y) \simeq \upsilon(C_p(X))$ (see TFS-439). Thus $C_p(Y)$ is a Lindelöf Σ-space and therefore $\upsilon(C_p(C_p(Y)))$ is also a Lindelöf Σ-space by Problem 246. The restriction map $\pi : C_p(\upsilon(C_p(X))) \to C_p(C_p(X))$ can be extended to a continuous map $\tilde{\pi} : \upsilon(C_p(\upsilon(C_p(X)))) \to \upsilon(C_p(C_p(X)))$ (see TFS-413). For the set $H = \tilde{\pi}(\upsilon(C_p(\upsilon(C_p(X)))))$ we have

(1) $C_p(C_p(X)) \subset H \subset \upsilon(C_p(C_p(X)))$.

The space H is Lindelöf Σ (and hence realcompact) because H is a continuous image of the space $\upsilon(C_p(\upsilon(C_p(X))))$ which is homeomorphic to a Lindelöf

Σ-space $\upsilon(C_p(C_p(Y)))$. It follows from (1) that $C_p(C_p(X))$ is C-embedded in H; so we can apply Fact 1 of S.438 to conclude that $\upsilon(C_p(C_p(X))) = H$ is a Lindelöf Σ-space.

U.248. *Prove that, if X is realcompact and $\upsilon(C_p(X))$ is a Lindelöf Σ-space, then $\upsilon(C_p C_p(X))$ is a Lindelöf Σ-space.*

Solution. It follows from Problem 206 that $\upsilon X = X$ is a Lindelöf Σ-space. Thus X is normal and hence we can apply Problem 247 to see that $\upsilon(C_p(C_p(X)))$ is a Lindelöf Σ-space.

U.249. *Let ω_1 be a caliber of a space X. Prove that $C_p(X)$ is a Lindelöf Σ-space if and only if X has a countable network.*

Solution. If network weight of X is countable then $nw(C_p(X)) = \omega$, so $C_p(X)$ is a Lindelöf Σ-space; this proves sufficiency. To deal with necessity, say that a space Y is *a counterexample* if ω_1 is a caliber of Y while the space $C_p(Y)$ is Lindelöf Σ and $nw(Y) > \omega$. Our aim is to show that there are no counterexamples.

Assume towards a contradiction that a space X is a counterexample. Then $nw(\upsilon X) \geq nw(X) > \omega$ and ω_1 is a caliber of υX because X is dense in υX (see SFFS-278). Furthermore, $C_p(\upsilon X)$ is a Lindelöf Σ-space by Problem 234; so υX is also a counterexample. Thus we can assume, without loss of generality, that $X = \upsilon X$, i.e., X is realcompact and hence $X = \upsilon X$ is Lindelöf Σ (see Problem 206).

The space $C_p(X)$ being ω-monolithic by Problem 208, if $d(C_p(X)) = \omega$ then $nw(C_p(X)) = nw(X) = \omega$, i.e., X is not a counterexample. Therefore $C_p(X)$ is not separable and hence there exists a left-separated $A \subset C_p(X)$ such that $|A| = \omega_1$ (see SFFS-004).

Let $e_x(f) = f(x)$ for any point $x \in X$ and function $f \in A$. Then $e_x \in C_p(A)$ and the map $e : X \to C_p(A)$ defined by $e(x) = e_x$ for any $x \in X$ is continuous by TFS-166; let $Y = e(X)$. It is clear that $w(Y) \leq w(C_p(A)) \leq |A| = \omega_1$. There exists a space Z and continuous onto maps $\varphi : Z \to Y$ and $\xi : X \to Z$ such that φ is a condensation, ξ is $\mathbb{R}$-quotient and $\varphi \circ \xi = e$ (see Fact 2 of T.139). Any Lindelöf Σ-space is stable by SFFS-266; so $nw(Z) \leq \omega_1$.

The dual map $\xi^* : C_p(Z) \to C_p(X)$ is a closed embedding of $C_p(Z)$ in $C_p(X)$ (see TFS-163); so $C_p(Z)$ is also a Lindelöf Σ-space. Since Z is a continuous image of X, the cardinal ω_1 is a caliber of Z. We also have $A \subset e^*(C_p(Y)) \subset \xi^*(C_p(Z))$ (see Fact 5 of U.086) which shows that $nw(Z) = nw(C_p(Z)) \geq nw(A) > \omega$; so Z is still a counterexample.

Since $nw(Z) = \omega_1$, we can take a dense set D in the space Z with $|D| \leq \omega_1$; let $\{z_\alpha : \alpha < \omega_1\}$ be an enumeration of the set D. If $Z_\alpha = \overline{\{z_\beta : \beta < \alpha\}}$ for each $\alpha < \omega_1$ then $\bigcup\{Z_\alpha : \alpha < \omega_1\} = Z$ because $t(Z) \leq l(C_p(Z)) = \omega$. If $Z_\alpha = Z$ for some $\alpha < \omega_1$ then Z is separable and hence $nw(Z) = \omega$ because any counterexample is ω-monolithic.

This contradiction shows that $U_\alpha = Z \setminus Z_\alpha \in \tau^*(Z)$ for any $\alpha < \omega_1$. Besides, $\alpha < \beta < \omega_1$ implies $U_\beta \subset U_\alpha$ and $\bigcap\{U_\alpha : \alpha < \omega_1\} = \emptyset$. Observe that the family $\mathcal{U} = \{U_\alpha : \alpha < \omega_1\} \subset \tau^*(Z)$ is uncountable because otherwise there is $\alpha_0 < \omega_1$ such that $U_\alpha = U_{\alpha_0}$ for any $\alpha \geq \alpha_0$ and hence $\bigcap \mathcal{U} = U_{\alpha_0} \neq \emptyset$ which is a contradiction.

Now, if $x \in Z$ then there is $\beta < \omega_1$ such that $x \notin U_\beta$. This implies that $x \notin U_\alpha$ for any $\alpha \geq \beta$ which proves that $\mathcal{U}$ is point-countable. As a consequence, the uncountable family $\mathcal{U} \subset \tau^*(Z)$ is point-countable which contradicts the fact that ω_1 is a caliber of Z. This final contradiction shows that counterexamples do not exist; therefore we proved necessity and made our solution complete.

U.250. *Prove that there exists a space X such that ω_1 is a precaliber of X, the space $C_{p,n}(X)$ is a Lindelöf Σ-space for all $n \in \omega$, while X does not have a countable network.*

Solution. If $Y = A(\omega_1)$ then the space Y is compact and hence Lindelöf Σ. The space $X = C_p(Y)$ is also Lindelöf Σ (see Problem 107); so $C_{p,n}(X) = C_{p,n-1}(Y)$ is a Lindelöf Σ-space for any $n \in \mathbb{N}$ by Problem 219. The cardinal ω_1 is a precaliber of X by SFFS-283 while $nw(X) = nw(Y) = \omega_1$; so our space X has all the promised properties.

U.251. *Let X be a Lindelöf Σ-space with ω_1 a caliber of X. Prove that any Lindelöf Σ-subspace of $C_p(X)$ has a countable network.*

Solution. Recall that a space is called *cosmic* if it has a countable network. Suppose that $Y \subset C_p(X)$ is Lindelöf Σ and not cosmic; since Y is monolithic (see SFFS-266), it is not separable; so there is a left-separated $A \subset Y$ such that $|A| = \omega_1$ (see SFFS-004). Clearly, the set $B = \overline{A} \cap Y$ is also Lindelöf Σ and not cosmic.

Let $e_x(f) = f(x)$ for any $x \in X$ and $f \in A$. Then $e_x \in C_p(A)$ for any $x \in X$ and the map $e : X \to C_p(A)$ defined by $e(x) = e_x$ for any $x \in X$ is continuous by TFS-166; let $X' = e(X)$. Then $w(X') \leq w(C_p(A)) = \omega_1$. There exists a space Z and continuous onto maps $\varphi : Z \to X'$ and $\xi : X \to Z$ such that ξ is $\mathbb{R}$-quotient, φ is a condensation and $e = \varphi \circ \xi$ (see Fact 2 of T.139). If $\xi^* : C_p(Z) \to C_p(X)$ is the dual map of ξ then it embeds $C_p(Z)$ in $C_p(X)$ as a closed subset by TFS-163. Besides, $A \subset e^*(C_p(X')) \subset \xi^*(C_p(Z))$ (see Fact 5 of U.086) and hence $B \subset \xi^*(C_p(Z))$; so Z is a Lindelöf Σ-space with ω_1 caliber of Z while a non-cosmic Lindelöf Σ-space B embeds in $C_p(Z)$.

The space X being stable by SFFS-266, we have $nw(Z) \leq \omega_1$ so we can choose a set $D = \{z_\alpha : \alpha < \omega_1\} \subset Z$ which is dense in Z. If $U_\alpha = Z \backslash \overline{\{z_\beta : \beta < \alpha\}} \neq \emptyset$ for every $\alpha < \omega_1$ then $\{U_\alpha : \alpha < \omega_1\} \subset \tau^*(Z)$ is an uncountable point-countable family which contradicts the fact that ω_1 is a caliber of Z. Thus $\overline{\{z_\beta : \beta < \alpha\}} = Z$ for some $\alpha < \omega_1$, i.e., Z is separable. Therefore $iw(B) \leq iw(C_p(Z)) = d(Z) = \omega$; the space B is stable being Lindelöf Σ so $nw(B) = \omega$ which is a contradiction with $A \subset B$ and $nw(A) > \omega$. This contradiction shows that every Lindelöf Σ-space $Y \subset C_p(X)$ is cosmic.

U.252. *Prove that a Lindelöf Σ-space Y has a small diagonal if and only if it embeds into $C_p(X)$ for some X with ω_1 a caliber of X.*

Solution. If Y has a small diagonal then ω_1 is a caliber of $X = C_p(Y)$ by SFFS-294. The space Y embeds in $C_p(X) = C_p(C_p(Y))$ by TFS-167; so we proved necessity. Now, if X is a space and ω_1 is a caliber of X then $C_p(X)$ has a small diagonal by SFFS-290; so if $Y \subset C_p(X)$ then Y also has a small diagonal (it is an easy exercise that having a small diagonal is a hereditary property).

U.253. *Prove that, if $C_p(X)$ is a Lindelöf Σ-space and has a small diagonal then X has a countable network.*

Solution. The cardinal ω_1 is a caliber of the space X by SFFS-290; so X has a countable network by Problem 249.

U.254. *Suppose that a space X has a dense subspace which is a continuous image of a product of separable spaces. Prove that any Lindelöf Σ-subspace of $C_p(X)$ has a countable network.*

Solution. Fix a Lindelöf Σ-space $L \subset C_p(X)$ and let Y be a dense subspace of X such that some product of separable spaces maps continuously onto Y. The restriction map $\pi : C_p(X) \to C_p(Y)$ is injective; so $\varphi = \pi|L : L \to M = \pi(L)$ is a condensation.

The cardinal ω_1 is a caliber of the space Y (see SFFS-282 and SFFS-277); so the diagonal of $C_p(Y)$ is small by SFFS-290. Therefore M has a small diagonal as well. Furthermore, every compact subspace of $C_p(Y)$ is metrizable by TFS-307; so all compact subspace of M are also metrizable. Thus we can apply Fact 1 of T.300 to see that $nw(M) = \omega$ and hence $iw(M) = \omega$ (see TFS-156). Since L condenses onto M, we have $iw(L) = \omega$ and hence $nw(L) = \omega$ because every Lindelöf Σ-space is stable by SFFS-266. Thus every Lindelöf Σ-space $L \subset C_p(X)$ has a countable network.

U.255. *Prove that any first countable space is a Preiss–Simon space.*

Solution. Suppose that $\chi(X) = \omega$ and take a closed $F \subset X$. For any $x \in F$ we have $\chi(x, F) \leq \omega$; so there is a local base $\{U_n : n \in \omega\}$ at the point x in F such that $U_{n+1} \subset U_n$ for any $n \in \omega$. If $U \in \tau(x, X)$ then $U \cap F \in \tau(x, F)$; so there is $m \in \omega$ with $U_m \subset U \cap F$. Consequently, $U_n \subset U_m \subset U$ for any $n \geq m$ which shows that the sequence $\{U_n : n \in \omega\}$ converges to x. This proves that X is a Preiss–Simon space.

U.256. *Prove that any Preiss–Simon space is Fréchet–Urysohn.*

Solution. Suppose that X is a Preiss–Simon space; given $A \subset X$ and $x \in \overline{A}$, the set $F = \overline{A}$ is closed in X and $x \in F$; so there exists a sequence $\{U_n : n \in \omega\} \subset \tau^*(F)$ which converges to x. Since A is dense in F, we have $A \cap U_n \neq \emptyset$; choose a point $x_n \in A \cap U_n$ for any $n \in \omega$. It is immediate that the sequence $S = \{x_n : n \in \omega\} \subset A$ converges to x; so X is a Fréchet–Urysohn space.

U.257. *Give an example of a compact Fréchet–Urysohn space which does not have the Preiss–Simon property.*

Solution. It was proved in Problem 222 that there exists a compact space X such that $C_p(X)$ is a Lindelöf Σ-space while $X \setminus \{x\}$ is pseudocompact for some non-isolated point $x \in X$. The space X is ω-monolithic because $C_p(X)$ is stable (see SFFS-266 and SFFS-152); besides, $t(X) \leq l(C_p(X)) = \omega$ (see TFS-189); so X is a Fréchet–Urysohn space by Fact 1 of U.080.

Now, suppose that $\mathcal{U} = \{U_n : n \in \omega\} \subset \tau^*(X)$ is a sequence that converges to the point x and let $W_n = U_n \backslash \{x\}$ for any $n \in \omega$. It is easy to check that the family $\mathcal{W} = \{W_n : n \in \omega\} \subset \tau^*(X \backslash \{x\})$ is locally finite in the space $X \backslash \{x\}$. If $\mathcal{W}$ is finite, say $\mathcal{W} = \{V_1, \ldots, V_k\}$ then pick a point $y_i \in V_i$ for every $i \leq k$. Since $\mathcal{U} \to x$, there is $U \in \mathcal{U}$ such that $\{y_1, \ldots, y_k\} \cap U = \emptyset$ which implies that $\mathcal{W} \ni U \backslash \{x\} \notin \{V_1, \ldots, V_k\}$. Therefore $\mathcal{W}$ is an infinite locally finite family of non-empty open subsets of a pseudocompact space $X \backslash \{x\}$; this contradiction shows that X is not a Preiss–Simon space.

U.258. *Let X be a space which has the Preiss–Simon property. Prove that each pseudocompact subspace of X is closed in X.*

Solution. Suppose that P is a pseudocompact subspace of X. If there exists a point $x \in \overline{P} \backslash P$ then we can apply the Preiss–Simon property to x and the closed set $F = \overline{P}$ to obtain a sequence $\mathcal{S} = \{U_n : n \in \omega\} \subset \tau^*(F)$ such that $\mathcal{S} \to x$. The point x is not isolated in F; so $U_n \neq \{x\}$ and hence $V_n = U_n \backslash \{x\} \neq \emptyset$ for any $n \in \omega$. It is easy to see that the family $\mathcal{V} = \{V_n : n \in \omega\} \subset \tau^*(F)$ is locally finite in $F \backslash \{x\}$. If $\mathcal{V}$ is finite, say $\mathcal{V} = \{W_1, \ldots, W_k\}$ then choose a point $z_i \in W_i$ for any $i \leq k$. Since $x \notin \{z_1, \ldots, z_k\}$, it follows from $\mathcal{S} \to x$ that there is $U \in \mathcal{S}$ with $U \cap \{z_1, \ldots, z_k\} = \emptyset$. It is immediate that $U \backslash \{x\} \in \mathcal{V}$ while $U \backslash \{x\} \notin \{W_1, \ldots, W_k\}$ which is a contradiction.

Therefore $\mathcal{V}$ is an infinite locally finite family of non-empty open subsets of $F \backslash \{x\}$ while $F \backslash \{x\}$ is pseudocompact because P is dense in $F \backslash \{x\}$ (see Fact 18 of S.351). This final contradiction shows that there are no points $x \in \overline{P} \backslash P$, i.e., P is closed in X.

U.259. *Suppose that X is a Preiss–Simon compact space. Prove that, for any proper dense $Y \subset X$, the space X is not the Čech–Stone extension of Y.*

Solution. Assume, towards a contradiction, that $X = \beta Y$ for some proper dense $Y \subset X$ and fix a point $x \in X \backslash Y$. By the Preiss–Simon property of X there is a sequence $\{U_n : n \in \omega\} \subset \tau^*(X)$ which converges to x. The set Y is dense in X; so we can pick a point $y_0 \in U_0 \cap Y$ and a set $V_0 \in \tau(y_0, X)$ such that $\overline{V}_0 \subset U_0$ and $x \notin \overline{V}_0$; let $m(0) = 0$.

Assume that $k \in \omega$ and we have chosen points $y_0, \ldots, y_k$, open sets $V_0, \ldots, V_k$ and natural numbers $m(0), \ldots, m(k)$ with the following properties:

(1) $m(i) < m(i+1)$ for any $i < k$;
(2) $y_i \in V_i \subset \overline{V}_i \subset U_{m(i)}$ for any $i \leq k$;
(3) $x \notin \overline{V}_i$ for any $i \leq k$ and the family $\{\overline{V}_i : i \leq k\}$ is disjoint.

Since $H = X \backslash (\bigcup \{\overline{V}_i : i \leq k\}) \in \tau(x, X)$, there exists a number $m(k+1) \in \omega$ such that $m(k+1) > m(k)$ and $U_{m(k+1)} \subset H$; the set Y being dense in X we can find a point $y_{k+1} \in U_{m(k+1)} \cap Y$ and a set $V_{k+1} \in \tau(y_{k+1}, X)$ for which $x \notin \overline{V}_{k+1}$ and $\overline{V}_{k+1} \subset U_{m(k+1)}$. It is evident that (1)–(3) hold for all $i \leq k+1$;

so our inductive procedure can be continued to construct sequences $\{y_i : i \in \omega\} \subset X\backslash\{x\}$, $\{m(i) : i \in \omega\} \subset \omega$ and $\{V_i : i \in \omega\} \subset \tau(X)$ such that (1)–(3) are fulfilled for all $i < \omega$.

An evident consequence of the properties (1) and (2) is that the sequence $\{V_n : n \in \omega\}$ converges to x; the property (3) implies that the family $\mathcal{V} = \{V_i : i \in \omega\}$ is discrete in $X\backslash\{x\}$. It follows from Fact 5 of T.132 that there is a continuous function $f : X\backslash\{x\} \to [0, 1]$ such that $f(y_{2i}) = 0$ and $f(y_{2i+1}) = 1$ for all $i \in \omega$. The function $g = f|Y$ is continuous on Y. Since $X = \beta Y$, there is $h \in C_p(X, [0, 1])$ such that $h|Y = g$ and, in particular, $h(y_{2i}) = 0$ and $h(y_{2i+1}) = 1$ for all $i \in \omega$. However, $y_{2i} \to x$ and $y_{2i+1} \to x$ when $i \to \infty$; so it follows from continuity of h that $h(x) = 0$ and $h(x) = 1$ at the same time; this contradiction shows that our solution is complete.

U.260. *Prove that the following properties are equivalent for any countably compact space X:*

(i) *X is a Preiss–Simon space;*
(ii) *each pseudocompact subspace of X is closed in X;*
(iii) *for each closed $F \subset X$ and any non-isolated $x \in F$, the space $F\backslash\{x\}$ is not pseudocompact.*

Solution. The implication (i)$\Longrightarrow$(ii) was proved in Problem 258. Suppose that (ii) holds and a set F is closed in X; if $x \in F$ is not isolated in F then $F\backslash\{x\}$ is not closed in X; so it is not pseudocompact. This proves (ii)$\Longrightarrow$(iii).

Finally, assume that (iii) takes place. Given a closed $F \subset X$ and a non-isolated point $x \in F$, the set $G = F\backslash\{x\}$ is not pseudocompact; so there is a family $\mathcal{U} = \{U_n : n \in \omega\} \subset \tau^*(G)$ which is discrete in G. To see that $\mathcal{U} \to x$ take any $U \in \tau(x, X)$. If the set $A = \{n \in \omega : U_n\backslash U \neq \emptyset\}$ is infinite then take $x_n \in U_n\backslash U$ for any $n \in A$. Since $D = \{x_n : n \in A\} \subset X\backslash U$, the set U is an open neighbourhood of x which does not meet D.

Assume that $y \neq x$; if $y \in X\backslash F$ then $X\backslash F$ is a neighbourhood of y which does not meet D. If $y \in F$ then $y \in F\backslash\{x\}$; the family $\mathcal{U}$ being discrete in $F\backslash\{x\}$ there exists a set $W \in \tau(y, F\backslash\{x\})$ such that W meets at most one element of $\mathcal{U}$ and hence $|W \cap D| \leq 1$. The set $V = W \cup (X\backslash F)$ is an open neighbourhood of y in X and $|V \cap D| \leq 1$. Thus every $y \in X$ has a neighbourhood which meets at most one element of D. This shows that D is an infinite closed discrete subspace of X which contradicts countable compactness of X.

Consequently, $U_n\backslash U = \emptyset$ for all but finitely many $n \in \omega$ which shows that $U_n \subset U$ eventually and hence $\mathcal{U} \to x$. Thus X is a Preiss–Simon space; this settles (iii)$\Longrightarrow$(i) and completes our solution.

U.261. *Let X be a Lindelöf Σ-space. Suppose that $Y \subset C_p(X)$ and the set of non-isolated points of Y is Lindelöf Σ. Prove that $C_p(Y, \mathbb{I})$ is Lindelöf Σ.*

Solution. Let E be the set of non-isolated points of Y; since E is Lindelöf Σ, there exists a countable family $\mathcal{F} \subset \exp E$ which is a network with respect to a compact cover $\mathcal{C}$ of the space E. We denote by Q_0 the set $\mathbb{Q} \cap (0, 1)$.

Given numbers $n \in \mathbb{N}$, $\delta > 0$, a point $x = (x_1, \ldots, x_n) \in X^n$ and $f \in Y$ let $O(f, x, \delta) = \{g \in Y : |g(x_i) - f(x_i)| < \delta$ for every $i \le n\}$. It is evident that the family $\{O(f, x, \delta) : \delta > 0$ and there is $n \in \mathbb{N}$ such that $x = (x_1, \ldots, x_n) \in X^n\}$ is a local base in Y at the point f.

For arbitrary numbers $\varepsilon > 0, \delta > 0$, $n \in \mathbb{N}$ and a set $P \subset E$ consider the set $M(\varepsilon, \delta, n, P) = \{(\varphi, x) \in \mathbb{I}^Y \times X^n : \varphi \in \mathbb{I}^Y, \ x = (x_1, \ldots, x_n) \in X^n$ and $|\varphi(f) - \varphi(g)| \le \varepsilon$ for any $f \in Y$ and $g \in P$ such that $f \in O(g, x, \delta)\}$. We claim that

(1) $M(\varepsilon, \delta, n, P)$ is closed in $\mathbb{I}^Y \times X^n$ for any $\varepsilon > 0$, $\delta > 0$, $n \in \mathbb{N}$ and $P \subset E$.

To see that (1) is true take any point $(\varphi, x) \in (\mathbb{I}^Y \times X^n) \backslash M(\varepsilon, \delta, n, P)$ where $x = (x_1, \ldots, x_n) \in X^n$. By the definition of the set $M(\varepsilon, \delta, n, P)$ there exist $f \in Y$ and $g \in P$ such that $|\varphi(f) - \varphi(g)| > \varepsilon$ and $f \in O(g, x, \delta)$. The set $W = \{\eta \in \mathbb{I}^Y : |\eta(f) - \eta(g)| > \varepsilon\}$ is open in $\mathbb{I}^Y$ and $\varphi \in W$. Furthermore, the functions f and g are continuous on X; so the set $V = \{y \in X : |f(y) - g(y)| < \delta\}$ is open in X and $x_i \in V$ for any $i \le n$. Thus V^n is open in X^n and $x \in V^n$. It is straightforward that $(W \times V^n) \cap M(\varepsilon, \delta, n, P) = \emptyset$; since $(\varphi, x) \in W \times V^n$, we showed that any point $(\varphi, x) \in (\mathbb{I}^Y \times X^n) \backslash M(\varepsilon, \delta, n, P)$ has a neighbourhood $W \times V^n$ contained in the complement of $M(\varepsilon, \delta, n, P)$. Therefore $(\mathbb{I}^Y \times X^n) \backslash M(\varepsilon, \delta, n, P)$ is open in $\mathbb{I}^Y \times X^n$, i.e., $M(\varepsilon, \delta, n, P)$ is closed in $\mathbb{I}^Y \times X^n$; so (1) is proved.

For any number $n \in \mathbb{N}$ let $\pi : \mathbb{I}^Y \times X^n \to \mathbb{I}^Y$ be the natural projection; for arbitrary $\varepsilon > 0, \delta > 0$, and $P \subset Y$ let $L(\varepsilon, \delta, n, P) = \pi(M(\varepsilon, \delta, n, P))$. Observe that $\mathbb{I}^Y \times X^n$ is a Lindelöf Σ-space because so is X; it follows from (1) that $M(\varepsilon, \delta, n, P)$ is also Lindelöf Σ and hence $L(\varepsilon, \delta, n, P)$ is Lindelöf Σ as well being a continuous image of the Lindelöf Σ-space $M(\varepsilon, \delta, n, P)$. Thus

(2) $L(\varepsilon, \delta, n, P) \subset \mathbb{I}^Y$ is a Lindelöf Σ-space for any $\varepsilon > 0$, $\delta > 0$, $n \in \mathbb{N}$ and $P \subset E$.

The family $\mathcal{L} = \{L(\varepsilon, \delta, n, P) : n \in \mathbb{N}, \varepsilon, \delta \in Q_0$ and $P \in \mathcal{F}\}$ is countable and consists of Lindelöf Σ-subspaces of $\mathbb{I}^Y$. We will establish next that

(3) the family $\mathcal{L}$ separates the set $C_p(Y, \mathbb{I})$ from $\mathbb{I}^Y \backslash C_p(Y, \mathbb{I})$ in the sense that, for any $\varphi \in C_p(Y, \mathbb{I})$ and $\eta \in \mathbb{I}^Y \backslash C_p(Y, \mathbb{I})$ there is $L \in \mathcal{L}$ such that $\varphi \in L$ while $\eta \notin L$.

To prove (3) take any $\varphi \in C_p(Y, \mathbb{I})$ and $\eta \in \mathbb{I}^Y \backslash C_p(Y, \mathbb{I})$; since all points of $Y \backslash E$ are isolated, there is some point $g \in E$ such that η is discontinuous at g and hence there is $\varepsilon \in Q_0$ such that,

(4) for any $n \in \mathbb{N}$, $y = (y_1, \ldots, y_n) \in X^n$ and $\delta > 0$ there is $f \in O(g, y, \delta)$ for which $|\eta(f) - \eta(g)| > \varepsilon$.

The family $\mathcal{C}$ being a cover of the set E we can choose a set $C \in \mathcal{C}$ such that $g \in C$. For any element $h \in C$ the map φ is continuous at the point h; so there are $n_h \in \mathbb{N}$, $x_h = (x_1^h, \ldots, x_{n_h}^h) \in X^{n_h}$ and $\delta_h \in Q_0$ such that for any $u \in O(h, x_h, 3\delta_h)$ we have $|\varphi(u) - \varphi(h)| < \frac{\varepsilon}{2}$. The open cover $\{O(h, x_h, \delta_h) : h \in C\}$ of the compact set C must have a finite subcover and therefore there exists a finite $D \subset C$ such that $C \subset G = \bigcup\{O(h, x_h, \delta_h) : h \in D\}$; then $\delta = \min\{\delta_h : h \in D\} \in Q_0$.

Take $m \in \mathbb{N}$ and $x = (x_1, \ldots, x_m) \in X^m$ such that $x_i^h \in \{x_1, \ldots, x_m\}$ for any $h \in D$ and $i \leq n_h$. There exists $P \in \mathcal{F}$ such that $C \subset P \subset G$; it is easy to check that if $v \in P$ and $u \in O(v, x, \delta)$ then $|\varphi(u) - \varphi(v)| < \varepsilon$.

Recalling the definition of $M(\varepsilon, \delta, m, P)$ we conclude that $(\varphi, x) \in M(\varepsilon, \delta, m, P)$ and therefore $\varphi \in L = L(\varepsilon, \delta, m, P)$. On the other hand, we have $\eta \notin L$ because $g \in P$; so the property (4) says exactly that there is no $n \in \mathbb{N}$ and $y \in X^n$ such that $(\eta, y) \in M(\varepsilon, \delta, n, P)$. Thus the property (3) is proved.

The property (3) shows that $\mathcal{L}$ is a countable family of Lindelöf Σ-subspaces of $\mathbb{I}^Y$ which separates $C_p(Y, \mathbb{I})$ from $\mathbb{I}^Y \backslash C_p(Y, \mathbb{I})$. Since $\mathbb{I}^Y$ is a compact extension of $C_p(Y, \mathbb{I})$, it follows from SFFS-233 that $C_p(Y, \mathbb{I})$ is a Lindelöf Σ-space.

U.262. *Let X be an Eberlein–Grothendieck space. Suppose that the set of non-isolated points of X is σ-compact. Prove that $C_p(X, \mathbb{I})$ is $K_{\sigma\delta}$.*

Solution. There is a compact space K such that $X \subset C_p(K)$; let E be the set of non-isolated points of X. There exists a countable family $\mathcal{C}$ of compact subspaces of E such that $E = \bigcup \mathcal{C}$. We denote by Q_0 the set $\mathbb{Q} \cap (0, 1)$.

Given numbers $n \in \mathbb{N}$, $\delta > 0$, a point $x = (x_1, \ldots, x_n) \in K^n$ and $f \in X$ let $O(f, x, \delta) = \{g \in X : |g(x_i) - f(x_i)| < \delta$ for every $i \leq n\}$. It is evident that the family $\{O(f, x, \delta) : \delta > 0$ and there is $n \in \mathbb{N}$ such that $x = (x_1, \ldots, x_n) \in K^n\}$ is a local base in X at the point f.

For arbitrary numbers $\varepsilon > 0$, $\delta > 0$, $n \in \mathbb{N}$ and a set $P \subset E$ consider the set $M(\varepsilon, \delta, n, P) = \{(\varphi, x) \in \mathbb{I}^X \times K^n : \varphi \in \mathbb{I}^X$, $x = (x_1, \ldots, x_n) \in K^n$ and $|\varphi(f) - \varphi(g)| \leq \varepsilon$ for any $f \in X$ and $g \in P$ such that $f \in O(g, x, \delta)\}$. We claim that

(1) $M(\varepsilon, \delta, n, P)$ is closed in $\mathbb{I}^X \times K^n$ for any $\varepsilon > 0$, $\delta > 0$, $n \in \mathbb{N}$ and $P \subset E$.

To see that (1) is true take an arbitrary point $(\varphi, x) \in (\mathbb{I}^X \times K^n) \backslash M(\varepsilon, \delta, n, P)$ where $x = (x_1, \ldots, x_n) \in K^n$. By the definition of the set $M(\varepsilon, \delta, n, P)$ there exist $f \in X$ and $g \in P$ such that $|\varphi(f) - \varphi(g)| > \varepsilon$ and $f \in O(g, x, \delta)$. The set $W = \{\eta \in \mathbb{I}^X : |\eta(f) - \eta(g)| > \varepsilon\}$ is open in $\mathbb{I}^X$ and $\varphi \in W$. Furthermore, the functions f and g are continuous on K; so the set $V = \{y \in K : |f(y) - g(y)| < \delta\}$ is open in K and $x_i \in V$ for any $i \leq n$. Thus V^n is open in K^n and $x \in V^n$. It is straightforward that $(W \times V^n) \cap M(\varepsilon, \delta, n, P) = \emptyset$; since $(\varphi, x) \in W \times V^n$, we showed that any point $(\varphi, x) \in (\mathbb{I}^X \times K^n) \backslash M(\varepsilon, \delta, n, P)$ has a neighbourhood $W \times V^n$ contained in the complement of $M(\varepsilon, \delta, n, P)$. Therefore $(\mathbb{I}^X \times K^n) \backslash M(\varepsilon, \delta, n, P)$ is open in $\mathbb{I}^X \times K^n$, i.e., $M(\varepsilon, \delta, n, P)$ is closed in $\mathbb{I}^X \times K^n$; so (1) is proved.

For any number $n \in \mathbb{N}$ let $\pi : \mathbb{I}^X \times K^n \to \mathbb{I}^X$ be the natural projection; for arbitrary $\varepsilon > 0$, $\delta > 0$, and $P \subset X$ let $L(\varepsilon, \delta, n, P) = \pi(M(\varepsilon, \delta, n, P))$. Observe that $\mathbb{I}^X \times K^n$ is a compact space because so is K; it follows from (1) that $M(\varepsilon, \delta, n, P)$ is also compact and hence $L(\varepsilon, \delta, n, P)$ is compact as well being a continuous image of the compact space $M(\varepsilon, \delta, n, P)$. Thus

(2) $L(\varepsilon, \delta, n, P) \subset \mathbb{I}^X$ is compact for any $\varepsilon > 0$, $\delta > 0$, $n \in \mathbb{N}$ and $P \subset E$.

The family $\mathcal{L} = \{L(\varepsilon, \delta, n, C) : n \in \mathbb{N},\ \varepsilon, \delta \in Q_0 \text{ and } C \in \mathcal{C}\}$ is easily seen to be countable; besides, it consists of compact subspaces of the space $\mathbb{I}^X$. Therefore the set $H[C, \varepsilon] = \bigcup\{L(\varepsilon, \delta, n, C) : n \in \mathbb{N} \text{ and } \delta \in Q_0\}$ is σ-compact for any $C \in C$ and $\varepsilon \in Q_0$. We will establish next that

(3) $C_p(X, \mathbb{I}) = \bigcap\{H[C, \varepsilon] : \varepsilon \in Q_0,\ C \in \mathcal{C}\}$.

To prove (3) take $\eta \in \mathbb{I}^X \backslash C_p(X, \mathbb{I})$; since all points of $X \backslash E$ are isolated, there is some point $g \in E$ such that η is discontinuous at g and hence there is $\varepsilon \in Q_0$ such that,

(4) for any $n \in \mathbb{N}$, $y = (y_1, \ldots, y_n) \in K^n$ and $\delta > 0$ there is $f \in O(g, y, \delta)$ for which $|\eta(f) - \eta(g)| > \varepsilon$.

The family $\mathcal{C}$ being a cover of the set E we can choose $C \in \mathcal{C}$ such that $g \in C$. Observe that $\eta \notin H[C, \varepsilon]$ because $g \in C$; so the property (4) says exactly that there are no $n \in \mathbb{N}$, $\delta \in Q_0$ and $y \in K^n$ such that $(\eta, y) \in M(\varepsilon, \delta, n, C)$. Thus $\eta \notin H = \bigcap\{H[C, \varepsilon] : C \in \mathcal{C},\ \varepsilon \in Q_0\}$ for any $\eta \in \mathbb{I}^X \backslash C_p(X, \mathbb{I})$, i.e., $H \subset C_p(X, \mathbb{I})$.

To prove the opposite inclusion take any $\varphi \in C_p(X, \mathbb{I})$ and fix $C \in \mathcal{C}$ and $\varepsilon \in Q_0$. For any element $h \in C$ the map φ is continuous at the point h and hence there exist $n_h \in \mathbb{N}$, $x_h = (x_1^h, \ldots, x_{n_h}^h) \in K^{n_h}$ and $\delta_h \in Q_0$ such that for any $u \in O(h, x_h, 3\delta_h)$ we have $|\varphi(u) - \varphi(h)| < \frac{\varepsilon}{2}$. The open cover $\{O(h, x_h, \delta_h) : h \in C\}$ of the compact set C has a finite subcover; so there is a finite $D \subset C$ such that $C \subset G = \bigcup\{O(h, x_h, \delta_h) : h \in D\}$; then $\delta = \min\{\delta_h : h \in D\} \in Q_0$. Take $m \in \mathbb{N}$ and $x = (x_1, \ldots, x_m) \in K^m$ such that $x_i^h \in \{x_1, \ldots, x_m\}$ for any $h \in D$ and $i \leq n_h$. It is easy to check that if $v \in C$ and $u \in O(v, x, \delta)$ then $|\varphi(u) - \varphi(v)| < \varepsilon$. Recalling the definition of $M(\varepsilon, \delta, m, C)$ we conclude that $(\varphi, x) \in M(\varepsilon, \delta, m, C)$ and therefore $\varphi \in H[C, \varepsilon]$. The set $C \in \mathcal{C}$ and $\varepsilon \in Q_0$ were chosen arbitrarily; so $\varphi \in H$ and hence $H = C_p(X, \mathbb{I})$, i.e., the property (3) is proved.

Finally, apply (3) to conclude that the space $C_p(X, \mathbb{I})$ is a countable intersection of σ-compact subspaces of $\mathbb{I}^X$, i.e., $C_p(X, \mathbb{I})$ is a $K_{\sigma\delta}$-space.

U.263. *Let X be a second countable space. Prove that, for any $M \subset X$, the space $C_p(X_M, \mathbb{I})$ is Lindelöf Σ.*

Solution. Observe that the space X_M is Tychonoff and all points of $X \backslash M$ are isolated in X_M; besides, the topologies induced on M from X_M and X coincide which shows that M is also second countable if considered as a subspace of X_M (see Fact 1 of S.293).

Fix a countable base $\mathcal{B}$ in the space X; since the topology of X_M is stronger than the topology of X, every $B \in \mathcal{B}$ is a cozero-set in X_M; so we can take a function $f_B \in C_p(X_M, \mathbb{I})$ such that $X_M \backslash U = f_B^{-1}(0)$ for any $B \in \mathcal{B}$.

Let $\chi_x(x) = 1$ and $\chi_x(y) = 0$ for each $y \in X_M \backslash \{x\}$; then $\chi_x \in C_p(X_M)$ for any $x \in X \backslash M$. If u is the function which is identically zero on X_M then the set $F = \{\chi_x : x \in X_M \backslash M\} \cup \{u\}$ is compact because $F \backslash U$ is finite for any $U \in \tau(u, C_p(X_M))$. It is straightforward that the set $Y = F \cup \{f_B : B \in \mathcal{B}\}$ separates the points and the closed subsets of X_M.

Let $e_x(f) = f(x)$ for any point $x \in X_M$ and function $f \in Y$. Then $e_x \in C_p(Y)$ and the map $e : X \to C_p(Y)$ defined by $e(x) = e_x$ for any $x \in X$ is an embedding by TFS-166. Therefore X can be embedded in $C_p(Y)$ for a σ-compact (and hence

Lindelöf Σ-) space Y. We have noticed already that the set E of non-isolated points of X_M is contained in M; so $w(E) \leq \omega$ and, in particular, E is a Lindelöf Σ-space. Therefore Problem 261 can be applied to conclude that $C_p(X_M, \mathbb{I})$ is a Lindelöf Σ-space.

U.264. *Let X be a σ-compact Eberlein–Grothendieck space. Prove that $C_p(X)$ is a $K_{\sigma\delta}$-space.*

Solution. Let $J = (-1, 1) \subset \mathbb{I}$ and fix a homeomorphism $\xi : \mathbb{R} \to J$. Then the map $\eta : \mathbb{R}^X \to J^X$ defined by $\eta(f) = \xi \circ f$ for any $f \in \mathbb{R}^X$ is a homeomorphism and $\eta(C_p(X)) = C_p(X, J)$ (see TFS-091).

There exists a $K_{\sigma\delta}$-space Z such that $C_p(X) \subset Z \subset \mathbb{R}^X$ (see Problem 202). Then $T = \eta(Z)$ is also a $K_{\sigma\delta}$-space and $C_p(X, J) \subset T \subset J^X$. The space $C_p(X, \mathbb{I})$ is also $K_{\sigma\delta}$ by Problem 262; so $H = C_p(X, \mathbb{I}) \cap T$ is a $K_{\sigma\delta}$-space as well (see TFS-338 and Fact 7 of S.271). It is clear that $C_p(X, J) \subset H$. Given $f \in H$, it follows from $f \in C_p(X, \mathbb{I})$ that f is continuous on X; besides, $f \in J^X$ and therefore $f \in C_p(X, J)$. This proves that $C_p(X, J) = H$ is a $K_{\sigma\delta}$-space and hence so is $C_p(X)$ being homeomorphic to $C_p(X, J)$.

U.265. *Give an example of a Lindelöf space X such that $C_p(X, \mathbb{I})$ is Lindelöf Σ and $X \times X$ is not Lindelöf.*

Solution. Given a space Z and a set $A \subset Z$ we will denote the space Z_A by $Z[A]$; recall that the underlying set of $Z[A]$ is Z and its topology is generated by the family $\tau(Z) \cup \{\{z\} : z \in Z \setminus A\}$. It follows from Problem 090 that there exist disjoint sets $A, B \subset \mathbb{I}$ such that both spaces $(\mathbb{I}[A])^\omega$ and $(\mathbb{I}[B])^\omega$ are Lindelöf. Therefore the space $X = \mathbb{I}[A] \oplus \mathbb{I}[B]$ is also Lindelöf. Both spaces $C_p(\mathbb{I}[A], \mathbb{I})$ and $C_p(\mathbb{I}[B], \mathbb{I})$ are Lindelöf Σ by Problem 263 which shows that $C_p(X, \mathbb{I}) \simeq C_p(\mathbb{I}[A], \mathbb{I}) \times C_p(\mathbb{I}[B], \mathbb{I})$ (see TFS-114) is also Lindelöf Σ. However, the space $\mathbb{I}[A] \times \mathbb{I}[B]$ is not normal (see Fact 5 of U.093) and embeds in $X \times X$ as a closed subspace; so $X \times X$ is not normal and hence not Lindelöf.

U.266. *Suppose that $\upsilon(C_p(X))$ is a Lindelöf Σ-space and $s(C_p(X)) = \omega$. Prove that $nw(X) = \omega$.*

Solution. The space $C_p(X)$ is ω-stable by SFFS-267; so the space X and hence $X \times X$ is ω-monolithic (see SFFS-152 and SFFS-114). Furthermore, $s(X \times X) \leq s(C_p(X)) = \omega$ (see SFFS-016) which, together with Fact 1 of U.127 shows that $hl(X \times X) = \omega$. Therefore $\Delta(X) = \omega$ and the space X is realcompact being Lindelöf. Thus $\upsilon X = X$ is a Lindelöf Σ-space by Problem 206; so it follows from $\Delta(X) = \omega$ that $nw(X) = \omega$ (see SFFS-300).

U.267. *Suppose that $C_p(X)$ is hereditarily stable and υX is a Lindelöf Σ-space. Prove that $nw(X) = \omega$.*

Solution. A metrizable space is ω-stable if and only if it is separable (see SFFS-106); in particular, a discrete space is ω-stable if and only if it is countable. Thus, hereditary stability of $C_p(X)$ implies $s(C_p(X)) = \omega$. Therefore $s(X \times X) \leq s(C_p(X)) = \omega$ (see SFFS-016); since $C_p(X)$ is stable, the space X and hence $X \times X$ is monolithic which, together with Fact 1 of U.127 shows that $hl(X \times X) \leq \omega$.

Therefore $\Delta(X) = \omega$ and the space X is realcompact being Lindelöf. Thus $X = \upsilon X$ is a Lindelöf Σ-space; so we can apply SFFS-300 to conclude that X has a countable network.

U.268. *Show that if $C_p(X)$ is hereditarily stable then $nw(Y) = \omega$ for any Lindelöf Σ-subspace $Y \subset X$.*

Solution. A metrizable space is ω-stable if and only if it is separable (see SFFS-106); in particular, a discrete space is ω-stable if and only if it is countable. Thus, hereditary stability of $C_p(X)$ implies $s(C_p(X)) = \omega$. Therefore $s(X \times X) \le s(C_p(X)) = \omega$ (see SFFS-016); since $C_p(X)$ is stable, the space X and hence $X \times X$ is monolithic which, together with Fact 1 of U.127 shows that $hl(X \times X) \le \omega$ and therefore $\Delta(X) = \omega$, i.e., X has a G_δ-diagonal. Finally, if $Y \subset X$ is a Lindelöf Σ-space then $\Delta(Y) \le \Delta(X) = \omega$; so we can apply SFFS-300 to conclude that Y has a countable network.

U.269. *Suppose that $\upsilon(C_p(X))$ is a Lindelöf Σ-space and ω_1 is a caliber of $C_p(X)$. Prove that $nw(Y) = \omega$ for any Lindelöf Σ-subspace $Y \subset X$.*

Solution. If K is a compact subspace of the space X then the restriction map $\pi : C_p(X) \to C_p(K)$ is continuous and onto (see Fact 1 of T.218). Therefore ω_1 is a caliber of the space $C_p(K)$. There is a continuous map $\xi : \upsilon(C_p(X)) \to \upsilon(C_p(K))$ such that $\xi | C_p(X) = \pi$ (see TFS-413); let $C = \xi(\upsilon(C_p(X)))$. The space C is Lindelöf Σ and $C_p(K) \subset C \subset \upsilon(C_p(K))$ which shows that $C = \upsilon C \simeq \upsilon(C_p(K))$ (see TFS-414); so $\upsilon(C_p(K))$ is a Lindelöf Σ-space.

Now let us look at the restriction map $r : C_p(\upsilon(C_p(K))) \to C_p(C_p(K))$; we can consider that $K \subset C_p(C_p(K))$ (see TFS-167). It follows from Problem 230 that $K' = r^{-1}(K)$ is compact; so the condensation $s = r|K' : K' \to K$ is a homeomorphism. We have $t(K') \le t(C_p(\upsilon(C_p(X)))) \le \omega$; so $t(K) = \omega$ and hence $C_p(K)$ is realcompact by TFS-429. Thus $C_p(K) = \upsilon(C_p(K))$ is a Lindelöf Σ-space. By Okunev's theorem (Problem 218) the space $C_p(C_p(K))$ is also Lindelöf Σ which makes it possible to apply Problem 249 to conclude that $nw(C_p(K)) = \omega$ and hence $w(K) = nw(K) = \omega$. As a consequence

(1) any compact subspace of X is metrizable.

Finally, if $Y \subset X$ is a Lindelöf Σ-subspace of X then any compact subspace of Y is metrizable by (1); the space X has a small diagonal by SFFS-293 and hence the diagonal of Y is small as well. Thus we can apply Fact 1 of T.300 to conclude that $nw(Y) = \omega$.

U.270. *Show that there exists an example of a space X that has a weakly σ-point-finite family $\mathcal{U} \subset \tau^*(X)$ which is not σ-point-finite.*

Solution. If Z is a space and $\mathcal{A} \subset \exp Z$ say that $\mathcal{A}$ is *point-finite at a point* $z \in Z$ if the family $\{A \in \mathcal{A} : z \in A\}$ is finite. The space of the irrationals is denoted by $\mathbb{P}$; it is identified with the space ω^ω. Given $p, q \in \mathbb{P}$ we let $p \le q$ if $p(n) \le q(n)$ for any $n \in \omega$. Take a point $a \notin \mathbb{P}$ and introduce a topology μ on the set $T = \{a\} \cup \mathbb{P}$ declaring all the points of $\mathbb{P}$ isolated and taking as a base at a the family of all complements of closed discrete subspaces of $\mathbb{P}$. It is easy to see that the space $Y = (T, \mu)$ is Tychonoff and zero-dimensional. Let $X = \{f \in C_p(Y, \mathbb{D}) : f(a) = 0\}$. As usual, if $A \subset T$ then $\chi_A \in \mathbb{D}^T$ is the characteristic function of A, i.e., $\chi_A(t) = 1$ for all $t \in A$ and $\chi_A(t) = 0$ whenever $t \in T \backslash A$.

The space $\mathbb{P}$ being second countable it is easy to find a countable family $\mathcal{B}$ of open subsets of $\mathbb{P}$ such that $\mathcal{B}$ is an outer base at any compact $K \subset \mathbb{P}$, i.e., for any $U \in \tau(K, \mathbb{P})$ there is $B \in \mathcal{B}$ such that $K \subset B \subset U$. For any point $p \in \mathbb{P}$ the set $O_p = \{f \in X : f(p) = 1\}$ is open in X. We claim that $\mathcal{U} = \{O_p : p \in \mathbb{P}\}$ is the promised family. Observe first that, for any $p \in \mathbb{P}$ we have $\chi_{\{p\}} \in O_p$, i.e., $O_p \neq \emptyset$.

Assume that $\mathcal{V}_n \subset \tau^*(X)$ is point-finite for any $n \in \omega$ and $\mathcal{U} = \bigcup\{\mathcal{V}_n : n \in \omega\}$. Let $P_n = \{p \in \mathbb{P} : O_p \in \mathcal{V}_n\}$ for any $n \in \omega$. The space $\mathbb{P}$ is not σ-compact; so it follows from $\mathbb{P} = \bigcup\{P_n : n \in \omega\}$ that there is $m \in \omega$ such that $\overline{P}_m$ is not compact; it is an easy exercise to see that there is an infinite $D \subset P_m$ which is closed and discrete in $\mathbb{P}$. Therefore $f = \chi_D \in X$ and the point f belongs to every element of the infinite subfamily $\{O_p : p \in D\}$ of the family $\mathcal{V}_m$, i.e., $\mathcal{V}_m$ is not point-finite at f. This contradiction proves that $\mathcal{U}$ is not σ-point-finite.

To see that $\mathcal{U}$ is weakly σ-point-finite let $\mathcal{U}_B = \{O_p : p \in B\}$ for any $B \in \mathcal{B}$. Then $\mathcal{C} = \{\mathcal{U}_B : B \in \mathcal{B}\}$ is a countable collection of subfamilies of $\mathcal{U}$. To prove that $\mathcal{C}$ witnesses that $\mathcal{U}$ is weakly σ-point-finite take any $f \in X$ and $p \in \mathbb{P}$. The set $K = \{q \in \mathbb{P} : q \leq p\}$ is compact and $p \in K$. The set $E = f^{-1}(1)$ being closed and discrete in $\mathbb{P}$, the intersection $F = E \cap K$ has to be finite. The set $W = \mathbb{P}\backslash(E\backslash F)$ is an open neighbourhood of K in $\mathbb{P}$ so there is $B \in \mathcal{B}$ for which $K \subset B \subset W$. In particular, $B \cap E = F$ is a finite set and therefore $\mathcal{U}_B$ is point-finite at f. Since also $O_p \in \mathcal{U}_B$, we proved that $\mathcal{U} = \bigcup\{\mathcal{U}_B : \mathcal{U}_B$ is point-finite at $f\}$, i.e., $\mathcal{U}$ is weakly σ-point-finite.

U.271. *Suppose that X is a space and $s(X) \leq \kappa$. Prove that any weakly σ-point-finite family of non-empty open subsets of X has cardinality $\leq \kappa$.*

Solution. Given a space Z and a family $\mathcal{A} \subset \exp Z$ let $\mathcal{A}(z) = \{A \in \mathcal{A} : z \in A\}$ and $\mathrm{ord}(z, \mathcal{A}) = |\mathcal{A}(z)|$ for any $z \in Z$; the family $\mathcal{A}$ is *point-finite* at a point $z \in Z$ if $\mathrm{ord}(z, \mathcal{A}) < \omega$. Say that a set $D \subset Z$ is *dense* in $\mathcal{A}$ if $D \cap A \neq \emptyset$ for any $A \in \mathcal{A}$.

Fact 1. Given a space Z and a family $\mathcal{U} \subset \tau^*(Z)$ there is a discrete $D \subset Z$ such that $\bigcup\{U \in \mathcal{U} : D \cap U \neq \emptyset\} = \bigcup\mathcal{U}$.

Proof. If $\mathcal{U} = \emptyset$ then there is nothing to prove. If not, take any $U \in \mathcal{U}$ and $z_0 \in U$. Suppose that $\beta < |Z|^+$ is an ordinal and we have a set $\{z_\alpha : \alpha < \beta\} \subset Z$ with the following properties:

(1) $W_\alpha = \bigcup\{\bigcup\mathcal{U}(z_\gamma) : \gamma < \alpha\} \neq \bigcup\mathcal{U}$ for any $\alpha < \beta$;
(2) $z_\alpha \in (\bigcup\mathcal{U})\backslash W_\alpha$ for any $\alpha < \beta$.

If $W_\beta = \bigcup\{\bigcup\mathcal{U}(z_\alpha) : \alpha < \beta\} = \bigcup\mathcal{U}$ then our inductive construction stops. If $W_\beta \neq \bigcup\mathcal{U}$ then we can choose $V \in \mathcal{U}$ and $z_\beta \in V\backslash W_\beta$. It is evident that the properties (1) and (2) are still fulfilled for all $\alpha \leq \beta$; so our inductive construction can be continued.

Observe that it follows from the property (2) that $\gamma \neq \alpha$ implies $z_\gamma \neq z_\alpha$. Since we cannot have a faithfully indexed set $\{z_\alpha : \alpha < |Z|^+\} \subset Z$, our inductive procedure has to stop for some $\beta < |Z|^+$ and hence $\bigcup\{\bigcup \mathcal{U}(z_\alpha) : \alpha < \beta\} = \bigcup \mathcal{U}$; this shows that, for the set $D = \{z_\alpha : \alpha < \beta\}$ we have $\bigcup\{U \in \mathcal{U} : D \cap U \neq \emptyset\} = \bigcup \mathcal{U}$. Finally, the set D is discrete because $(\bigcup \mathcal{U}(z_\alpha)) \cap D = \{z_\alpha\}$ for any $\alpha < \beta$; so Fact 1 is proved.

Fact 2. Suppose that λ is an infinite cardinal, Z is a space and $\mathcal{B} \subset \tau^*(Z)$ is a family with $\mathrm{ord}(z, \mathcal{B}) < \lambda$ for any $z \in Z$. Then there exists a family $\{D_\alpha : \alpha < \lambda\}$ of discrete subspaces of Z such that, for the set $D = \bigcup\{D_\alpha : \alpha < \lambda\}$, we have $D \cap B \neq \emptyset$ for any $B \in \mathcal{B}$. In particular, if $\mathcal{B}$ is point-finite then there is a σ-discrete subset of Z which is dense in $\mathcal{B}$.

Proof. Given a set $A \subset Z$ and $\mathcal{C} \subset \mathcal{B}$ let $\mathcal{C}(A) = \{B \in \mathcal{C} : B \cap A \neq \emptyset\}$ and $\mathcal{C}[A] = \mathcal{C}\backslash\mathcal{C}(A)$; assume towards a contradiction that $\mathcal{B}[A] \neq \emptyset$ for any set A which can be represented as a union of $\leq \lambda$-many discrete subspaces of Z.

Apply Fact 1 to find a discrete subspace $D_0 \subset Z$ such that $\bigcup \mathcal{B}(D_0) = \bigcup \mathcal{B}$ and let $\mathcal{D}_0 = \mathcal{B}(D_0)$. Assume that, for some $\beta < \lambda$, we have a family $\{D_\alpha : \alpha < \beta\}$ of discrete subsets of Z and a collection $\{\mathcal{D}_\alpha : \alpha < \beta\}$ of subfamilies of $\mathcal{B}$ such that

(3) $\mathcal{D}_\alpha \subset \mathcal{B}(D_\alpha)$ for any $\alpha < \beta$;
(4) if $E_\alpha = \bigcup\{D_\gamma : \gamma < \alpha\}$, $\mathcal{C}_\alpha = \mathcal{B}[E_\alpha]$ then $\mathcal{D}_\alpha = \mathcal{C}_\alpha(D_\alpha)$ and $\bigcup \mathcal{D}_\alpha = \bigcup \mathcal{C}_\alpha$ for any $\alpha < \beta$;

Let $E_\beta = \bigcup\{D_\alpha : \alpha < \beta\}$; our assumption about $\mathcal{B}$ shows that $\mathcal{C}_\beta = \mathcal{B}[E_\beta] \neq \emptyset$; so we can apply Fact 1 to find a discrete set $D_\beta \subset Z$ such that $\bigcup \mathcal{C}_\beta(D_\beta) = \bigcup \mathcal{C}_\beta$. If we let $\mathcal{D}_\beta = \mathcal{C}_\beta(D_\beta)$ then it is evident that the properties (3) and (4) are still fulfilled for all $\alpha \leq \beta$; so our inductive procedure can be continued to construct a family $\{D_\alpha : \alpha < \lambda\}$ of discrete subsets of Z and a collection $\{\mathcal{D}_\alpha : \alpha < \lambda\}$ of subfamilies of $\mathcal{B}$ for which (3) and (4) hold for all $\beta < \lambda$.

Let $D = \bigcup\{D_\alpha : \alpha < \lambda\}$ and use again our assumption about $\mathcal{B}$ to find a set $U \in \mathcal{B}[D]$; fix a point $x \in U$. It follows from (4) that $U \in \mathcal{C}_\alpha$ and hence there is $B_\alpha \in \mathcal{D}_\alpha$ such that $x \in B_\alpha$ for any $\alpha < \lambda$. Another consequence of (3) and (4) is that the collection $\{\mathcal{D}_\alpha : \alpha < \lambda\}$ is disjoint; so $B_\alpha \neq B_\beta$ whenever $\alpha \neq \beta$. Since $x \in \bigcap\{B_\alpha : \alpha < \lambda\}$, we have $\mathrm{ord}(x, \mathcal{B}) \geq \lambda$ which is a contradiction. Fact 2 is proved.

Returning to our solution suppose that $\mathcal{U} \subset \tau^*(X)$ is weakly σ-point-finite and fix a collection $\{\mathcal{U}_n : n \in \omega\}$ of subfamilies of $\mathcal{U}$ which witnesses this. If $z \in X$ then let $A = \{n \in \omega : \text{the family } \mathcal{U}_n \text{ is point-finite at } z\}$; we have $\mathcal{U} = \bigcup\{\mathcal{U}_n : n \in A\}$ which shows that $\mathcal{U}(z) \subset \bigcup\{\mathcal{U}_n(z) : n \in A\}$ and hence

(5) the family $\mathcal{U}(z)$ is countable for any $z \in Z$.

Consider the set $X_n = \{x \in X : \mathcal{U}_n \text{ is point-finite at } x\}$ for any $n \in \omega$. Then $\bigcup\{X_n : n \in \omega\} = X$ because, for any $x \in X$ there is $n \in \omega$ such that the family $\mathcal{U}_n$ is point-finite at x. By Fact 2, there exists a σ-discrete $D_n \subset X_n$ such that D_n is dense in $\mathcal{V}_n = \{U \cap X_n : U \in \mathcal{U}_n \text{ and } U \cap X_n \neq \emptyset\}$.

The set $D = \bigcup\{D_n : n \in \omega\}$ is also σ-discrete; we claim that D is dense in $\mathcal{U}$. Indeed, given $U \in \mathcal{U}$ pick a point $x \in U$. By the choice of the collection $\{\mathcal{U}_n : n \in \omega\}$, we have $\mathcal{U} = \bigcup\{\mathcal{U}_n :$ the family $\mathcal{U}_n$ is point-finite at $x\}$. Therefore there is $n \in \omega$ with $\text{ord}(x, \mathcal{U}_n) < \omega$ and $U \in \mathcal{U}_n$. As a consequence, $x \in X_n$ and therefore $U \cap X_n \in V_n$ which shows that $D_n \cap (U \cap X_n) \neq \emptyset$ and hence $D \cap U \neq \emptyset$.

The set D being σ-discrete, it follows from $s(X) \leq \kappa$ that $|D| \leq \kappa$. Therefore $\mathcal{U} = \bigcup\{\mathcal{U}(z) : z \in D\}$ which, together with (5), implies $|\mathcal{U}| \leq \kappa$ and makes our solution complete.

U.272. *Give an example of a non-cosmic Lindelöf Σ-space X such that any closed uncountable subspace of X has more than one (and hence infinitely many) non-isolated points.*

Solution. Such a space can even be compact. Indeed, let $X = \beta\omega$ where ω is taken with the discrete topology and take any infinite closed $F \subset X$. There is an infinite discrete subspace $D \subset F$ (see Fact 4 of S.382); so $|\overline{D}| = 2^{\mathfrak{c}}$ by Fact 1 of S.483. Every point of $\overline{D} \setminus D$ is not isolated in F; so we have $2^{\mathfrak{c}}$ non-isolated points in the subspace F. Finally, X is not cosmic because $|X| = 2^{\mathfrak{c}}$ (see TFS-368) while $|Z| \leq \mathfrak{c}$ for any cosmic space Z (see TFS-156, TFS-159 and SFFS-015).

U.273. *Suppose that $C_p(X)$ is a Lindelöf Σ-space. Prove that, if all closed uncountable subspaces of $C_p(X)$ have more than one non-isolated points, then $C_p(X)$ has a countable network.*

Solution. If $Y = \upsilon X$ then both spaces Y and $C_p(Y)$ are Lindelöf Σ (see Problems 235 and 206). Let $u(y) = 0$ for all $y \in Y$; the spaces Y and $C_p(Y)$ are ω-monolithic by Problem 208; so if $nw(X) > \omega$ then $nw(Y) > \omega$ and hence $\psi(C_p(Y)) = d(Y) > \omega$ which implies $l(C_p(Y) \setminus \{u\}) > \omega$ (see Fact 1 of U.027). Apply Baturov's theorem (SFFS-269) to find a closed discrete $D \subset C_p(Y) \setminus \{u\}$ with $|D| = \omega_1$. It is evident that the set $H = D \cup \{u\}$ is closed in $C_p(Y)$ and has a unique non-isolated point u.

If $\pi : C_p(Y) \to C_p(X)$ is the restriction map then the set $E = \pi(H)$ is uncountable because π is a condensation; let $v = \pi(u)$. The set H is concentrated around the point u in the sense that $H \setminus U$ is countable for any $U \in \tau(u, C_p(Y))$. This property is, evidently, preserved by continuous maps; so the set E is concentrated around the point v.

To see that the set E is closed in $C_p(X)$ take any function $f \in C_p(X) \setminus E$; then $f \neq v$; so we can choose $V \in \tau(f, C_p(X))$ such that $v \notin \overline{V}$. Since $W = C_p(X) \setminus \overline{V}$ is an open neighbourhood of v, the set $E_0 = E \setminus W$ is countable; let $H_0 = \pi^{-1}(E_0)$. It is clear that $f \notin \overline{(W \cap E)}$; besides, if $f \in \overline{E_0}$ then $g = \pi^{-1}(f) \in \overline{H_0}$ because the map $\pi|(H_0 \cup \{g\}) \to E_0 \cup \{f\}$ is a homeomorphism (see TFS-436). However, $\overline{H_0} \subset H$ because H is closed in $C_p(Y)$ so $g \in H$ which is a contradiction. This proves that $f \notin \overline{E}$ for any $f \in C_p(X) \setminus E$, i.e., E is closed in $C_p(X)$.

Finally observe that v is the unique non-isolated point of the subspace E; indeed, if $f \in E \setminus \{v\}$ is not isolated in E then take $V \in \tau(f, C_p(X))$ such that $v \notin \overline{V}$. Since E is concentrated around v, the set $V \cap E$ is countable and hence there is a countable

$M \subset E$ such that $f \in \overline{M} \backslash M$; let $N = \pi^{-1}(M)$ and $g = \pi^{-1}(f)$. Apply TFS-436 once more to see that g is an accumulation point of the set N because $\pi | (N \cup \{g\}) : N \cup \{g\} \to M \cup \{f\}$ is a homeomorphism. As a consequence, $g \neq u$ is a non-isolated point of H which is a contradiction. Thus E is a closed uncountable subspace of $C_p(X)$ with a unique non-isolated point. This final contradiction with our assumption about $C_p(X)$ shows that $nw(C_p(X)) = nw(X) = \omega$.

U.274. *Let X be a Lindelöf Σ-space with a unique non-isolated point. Prove that any subspace of $C_p(X)$ has a weakly σ-point-finite T_0-separating family of cozero sets.*

Solution. It is evident that having a weakly σ-point-finite T_0-separating family of cozero sets is a hereditary property so it suffices to construct such a family in $C_p(X)$. Denote by a the unique non-isolated point of X and let $D = X \backslash \{a\}$. The space X being Lindelöf Σ, there is a countable family $\mathcal{T}$ of closed subsets of X which is a network with respect to a compact cover $\mathcal{C}$ of the space X. Let $Q_+ = \mathbb{Q} \cap (0, +\infty)$ and $Q_- = \mathbb{Q} \cap (-\infty, 0)$. If $q \in Q_+$ then let $O_q = (q, +\infty)$; if $q \in Q_-$ then $O_q = (-\infty, q)$.

Fact 1. If Z is a space with a unique non-isolated point then $Z \oplus \{t\} \simeq Z$ for any $t \notin Z$.

Proof. We can consider that both Z and $\{t\}$ are clopen subspaces of $Z \oplus \{t\}$. Let w be the non-isolated point of Z; there are two cases to consider.

a) There is an infinite $Y \subset Z \backslash \{w\}$ such that $w \notin \overline{Y}$. Then Y is a clopen discrete subspace of Z; so $Y \oplus \{t\}$ is also a discrete space of the same cardinality as Y. Thus $Y \oplus \{t\} \simeq Y$ and therefore

$$Z \simeq (Z \backslash Y) \oplus Y \simeq (Z \backslash Y) \oplus (Y \oplus \{t\}) \simeq ((Z \backslash Y) \oplus Y) \oplus \{t\} \simeq Z \oplus \{t\}.$$

b) $w \in \overline{Y}$ for any infinite $Y \subset Z \backslash \{w\}$. Then $Z \backslash U$ is finite for any $U \in \tau(w, Z)$. Let $\xi : (Z \backslash \{w\}) \to (Z \backslash \{w\}) \cup \{t\}$ be a bijection (which exists because $Z \backslash \{w\}$ has to be infinite). Now construct a map $f : Z \to Z \oplus \{t\}$ letting $f(a) = a$ and $f(z) = \xi(z)$ for any $z \in Z$. Then f is a bijection and we have only to check continuity of f at the point a.

Given $U \in \tau(a, Z \oplus \{t\})$ the set $(Z \oplus \{t\}) \backslash U$ is finite; so $V = f^{-1}(U)$ contains a and $Z \backslash V$ is also finite which shows that V is open in Z. Thus f is a homeomorphism being a condensation of a compact space Z onto $Z \oplus \{t\}$. Fact 1 is proved.

Returning to our solution let $C = \{f \in C_p(X) : f(a) = 0\}$; then $C_p(X) \simeq C \times \mathbb{R}$ (see Fact 1 of S.409). Take a point $t \notin X$; the space $Y = X \oplus \{t\}$ is homeomorphic to X by Fact 1. If $\varphi : X \to Y$ is a homeomorphism then $\varphi(a) = a$ and the dual map $\varphi^* : C_p(Y) \to C_p(X)$ is a homeomorphism as well (see TFS-163).

If $I = \{f \in C_p(Y) : f(a) = 0\}$ then it is routine to check that $\varphi^*(I) = C$ and therefore $C \simeq I$. On the other hand, if $f \in C$ and $r \in \mathbb{R}$ then define a function

$g = \eta(f, r) \in I$ by the equalities $g|X = f$ and $g(t) = r$. It is straightforward that $\eta : C \times \mathbb{R} \to I$ is a homeomorphism; so $C_p(X) \simeq C \times \mathbb{R} \simeq I \simeq C$ which shows that $C_p(X)$ is homeomorphic to C and hence it suffices to find a weakly σ-point-finite T_0-separating family of cozero sets in C.

For any $x \in D$ and $q \in \mathbb{Q}\backslash\{0\}$ let $W(x, q) = \{f \in C : f(x) \in O_q\}$. Let $e_x(f) = f(x)$ for any $x \in D$ and $f \in C$. Then $e_x : C \to \mathbb{R}$ is a continuous map by TFS-166; since $W(x, q) = e_x^{-1}(O_q)$, the set $W(x, q)$ is cozero for any $q \in \mathbb{Q}\backslash\{0\}$ and $x \in D$ (see Fact 1 of T.252).

Take distinct $f, g \in C$; there is $x \in D$ such that $f(x) \neq g(x)$. We can assume, without loss of generality, that $f(x) < g(x)$ and hence there is $q \in \mathbb{Q}\backslash\{0\}$ with $f(x) < q < g(x)$. It is immediate that $W(x, q) \cap \{f, g\}$ is a singleton; so the family $\mathcal{W} = \{W(x, q) : x \in D, q \in \mathbb{Q}\backslash\{0\}\}$ is T_0-separating.

For any $q \in \mathbb{Q}\backslash\{0\}$ let $\mathcal{W}_q = \{W(x, q) : x \in D\}$; then $\mathcal{W} = \bigcup\{\mathcal{W}_q : q \in \mathbb{Q}\backslash\{0\}\}$. We leave it to the reader as an easy exercise to prove that a countable union of weakly σ-point-finite families is a weakly σ-point-finite family; so it suffices to show that $\mathcal{W}_q$ is weakly σ-point-finite for any $q \in \mathbb{Q}\backslash\{0\}$.

Let $\mathcal{W}_q^T = \{W(x, q) : x \in T \cap D\}$ for any $T \in \mathcal{T}$; then $\mathcal{S} = \{\mathcal{W}_q^T : T \in \mathcal{T}\}$ is a countable collection of subfamilies of $\mathcal{W}_q$. To see that $\mathcal{S}$ witnesses that $\mathcal{W}_q$ is weakly σ-point-finite take any $f \in C$ and $x \in D$. There is $K \in \mathcal{C}$ such that $x \in K$. Observe that the set $P = f^{-1}(O_q) \cap K$ is finite for otherwise a is a limit point of P and therefore $|f(a)| \geq |q|$ which is a contradiction with $f(a) = 0$.

Furthermore, a is not in the closure of the open set $G = f^{-1}(O_q)$; so G is closed in X whence $G\backslash P$ is a clopen subset of X as well. The family $\mathcal{T}$ is a network with respect to $\mathcal{C}$; so there is $T \in \mathcal{T}$ such that $K \subset T \subset X\backslash(G\backslash P)$. It is immediate that $T \cap G = K \cap G$ is finite; so only finitely many elements of $\mathcal{W}_q^T$ contain f, i.e., the family $\mathcal{W}_q^T$ is point-finite at f. It follows from $x \in K \subset T$ that $W(x, q) \in \mathcal{W}_q^T$; so $\mathcal{W}_q = \bigcup\{\mathcal{W}_q^T :$ the family $\mathcal{W}_q^T$ is point-finite at $f\}$, i.e., $\mathcal{W}_q$ is weakly σ-point-finite.

Thus $\mathcal{W}$ is a weakly σ-point-finite T_0-separating family of cozero subsets of C; since $C_p(X) \simeq C$, such a family also exists in $C_p(X)$ and hence our solution is complete.

U.275. *Let X be a space of countable spread. Prove that $C_p(X)$ is a Lindelöf Σ-space if and only if X has a countable network.*

Solution. If X has a countable network then $nw(C_p(X)) = nw(X) = \omega$ and hence $C_p(X)$ is a Lindelöf Σ-space; so sufficiency is clear. Now assume that $s(X) = \omega$ and $C_p(X)$ is a Lindelöf Σ-space; let $u(x) = 0$ for any $x \in X$. If X does not have a countable network then there is an uncountable set $D \subset C_p(X)$ such that $E = D \cup \{u\}$ is closed in $C_p(X)$ and u is the unique non-isolated point of E (see Problem 273). The space E is Lindelöf Σ being closed in $C_p(X)$; for any $x \in X$ and $f \in E$ let $e_x(f) = f(x)$. Then $e_x \in C_p(E)$ for any $x \in X$ and the map $e : X \to C_p(E)$ defined by $e(x) = e_x$ for any $x \in X$ is continuous by TFS-166; let $Y = e(X)$.

Since E is a Lindelöf Σ-space with a unique non-isolated point, we can apply Problem 274 to see that Y has a weakly σ-point-finite T_0-separating family $\mathcal{U} \subset \tau^*(Y)$ of cozero-sets. Since $s(Y) \leq s(X) = \omega$, the family $\mathcal{U}$ is countable by Problem 271. For any $U \in \mathcal{U}$ fix $f_U \in C_p(Y)$ such that $X \backslash U = f_U^{-1}(0)$. Since $\mathcal{U}$ is T_0-separating, the set $F = \{f_U : U \in \mathcal{U}\}$ separates the points of Y and hence Y condenses into $\mathbb{R}^F$ (see TFS-166); we have $w(\mathbb{R}^F) \leq |F| \cdot \omega = \omega$; so Y can be condensed onto a second countable space, i.e., $iw(Y) = \omega$. Therefore $d(C_p(Y)) = \omega$ (see TFS-174); the space $C_p(Y)$ embeds in $C_p(X)$ (see TFS-163), so $C_p(Y)$ is ω-monolithic because so is $C_p(X)$ (see Problem 208). This implies $nw(C_p(Y)) = \omega$.

On the other hand the set E embeds in $C_p(Y)$ by TFS-166 and hence $nw(E) = \omega$ which is impossible because $D \subset E$ is uncountable and all points of D are isolated in E. The obtained contradiction shows that $nw(X) = \omega$; this settles necessity and completes our solution.

U.276. *Show that, under CH, there exists a space X of countable spread for which there is a Lindelöf Σ-space $Y \subset C_p(X)$ with $nw(Y) > \omega$.*

Solution. The promised subspace Y can even be compact. Indeed, under CH, there exists a compact non-metrizable space K such that $hd^*(K) = \omega$ (see SFFS-099 and SFFS-027). If $X = C_p(K)$ then $s(X) \leq s^*(X) = s^*(K) \leq hd^*(K) = \omega$ (see SFFS-025) while K embeds in $C_p(X) = C_p(C_p(K))$ (see TFS-167) and hence there is $Y \subset C_p(X)$ with $Y \simeq K$. It is clear that Y is Lindelöf Σ and $nw(Y) = w(Y) > \omega$.

U.277. *Let X be a space with a unique non-isolated point: $X = \{a\} \cup Y$, where all points of Y are isolated and $a \notin Y$. Prove that, for every infinite cardinal κ, the following conditions are equivalent:*

(i) $p(C_p(X)) \leq \kappa$;
(ii) *if $\{A_\alpha : \alpha < \kappa^+\}$ is a disjoint family of finite subsets of Y then there is an infinite $S \subset \kappa^+$ such that $a \notin \bigcup\{A_\alpha : \alpha \in S\}$;*
(iii) *if $\{A_\alpha : \alpha < \kappa^+\}$ is a family of finite subsets of Y then there is an infinite $S \subset \kappa^+$ such that $a \notin \overline{\bigcup\{A_\alpha : \alpha \in S\}}$.*

Solution. To see that (i)$\Longrightarrow$(ii) suppose that $\mathcal{A} = \{A_\alpha : \alpha < \kappa^+\}$ is a disjoint family of finite subsets of Y; there is nothing to prove if infinitely many elements of $\mathcal{A}$ are empty; so we can assume, throwing away the empty elements of $\mathcal{A}$ if necessary, that $A_\alpha \neq \emptyset$ for all $\alpha < \kappa^+$. Since κ^+ is uncountable, there exists $H \subset \kappa^+$ such that $|H| = \kappa^+$ and there is $m \in \mathbb{N}$ for which $|A_\alpha| = m$; let $A_\alpha = \{y_\alpha^1, \ldots, y_\alpha^m\}$ for any $\alpha \in H$. The set $O_\alpha = \{f \in C_p(X) : f(a) \in (-1, 1)$ and $f(y_\alpha^i) \in (i, i+1)$ for any $i \leq m\}$ is non-empty and open in $C_p(X)$ for any $\alpha \in H$. We have $p(C_p(X)) \leq \kappa$, so the family $\{O_\alpha : \alpha \in H\}$ cannot be point-finite; let $S \subset H$ be an infinite set such that $W = \bigcap\{O_\alpha : \alpha \in S\} \neq \emptyset$ and hence we can pick a function $f \in W$.

Assume that we have $a \in \overline{\bigcup\{A_\alpha : \alpha \in S\}}$; recalling that $f(y_\alpha^i) \in (i, i+1)$ and hence $f(y_\alpha^i) \geq 1$ for any $\alpha \in S$ and $i \leq m$ we conclude that $f(y) \geq 1$ for any point $y \in A = \bigcup \{A_\alpha : \alpha \in S\}$ and hence $f(a) \geq 1$ by continuity of f. Since this contradicts $f(a) \in (-1, 1)$, we proved that $a \notin \overline{A}$ and settled the implication (i)$\Longrightarrow$(ii).

To see that (ii)$\Longrightarrow$(iii) is true take a family $\mathcal{A} = \{A_\alpha : \alpha < \kappa^+\}$ of finite subsets of Y. Apply the Δ-lemma (SFFS-038) to find a set $E \subset \kappa$ such that $|E| = \kappa^+$ and there is a set $R \subset \kappa$ such that $A_\alpha \cap A_\beta = R$ for any distinct $\alpha, \beta \in E$. If $B_\alpha = A_\alpha \backslash R$ for any $\alpha \in E$ then the family $\{B_\alpha : \alpha \in E\}$ is disjoint; so we can apply (ii) to find an infinite $F \subset E$ for which $a \notin \bigcup \{B_\alpha : \alpha \in F\}$. The set $R \subset Y$ is finite; so $a \notin \bigcup \{B_\alpha : \alpha \in F\} \cup R = \bigcup \{A_\alpha : \alpha \in F\}$. This proves that (ii)$\Longrightarrow$(iii).

Finally, assume that the condition (iii) is satisfied and there is a point-finite family $\mathcal{U} \subset \tau^*(C_p(X))$ with $|\mathcal{U}| = \kappa^+$. It is easy to see that every $U \in \tau^*(C_p(X))$ contains a standard open set $[a, x_1, \ldots, x_n; O, O_1, \ldots, O_n] = \{f \in C_p(X) : f(a) \in O$ and $f(x_i) \in O_i$ for any $i \leq n\}$ where O and every O_i is a nonempty interval with rational endpoints and the family $\{O, O_1, \ldots, O_n\}$ is disjoint. Choosing such a standard open subset in every element of $\mathcal{U}$ we will still have a point-finite family of cardinality κ^+; so we can assume, without loss of generality, that all elements of $\mathcal{U}$ are the standard open sets described above.

Since there are only countably many finite families of rational intervals, we can pass, if necessary, to an appropriate subfamily of $\mathcal{U}$ of cardinality κ^+ to assume that there is $n \in \mathbb{N}$ and rational non-empty intervals $O, O_1, \ldots, O_n \subset \mathbb{R}$ such that every element of the family $\mathcal{U}$ is of the form $[a, x_1, \ldots, x_n; O, O_1, \ldots, O_n]$ where $\{x_1, \ldots, x_n\} \subset Y$. Applying the Δ-lemma (SFFS-038) once more and passing to a relevant subfamily of $\mathcal{U}$ of cardinality κ^+, we can consider that there exists a set $R = \{y_1, \ldots, y_m\} \subset Y$ and non-empty disjoint intervals $W_1, \ldots, W_m$ with rational endpoints such that

$$\mathcal{U} = \{[a, y_1, \ldots, y_m, x_\alpha^1, \ldots, x_\alpha^n; O, W_1, \ldots, W_m, O_1, \ldots, O_n] : \alpha < \kappa^+\}$$

where the set $A_\alpha = \{x_\alpha^1, \ldots, x_\alpha^n\}$ is contained in Y for any ordinal $\alpha < \kappa^+$ and the family $\mathcal{A} = \{A_\alpha : \alpha < \kappa^+\} \cup \{R\}$ is disjoint. Pick $r \in O$, $s_i \in W_i$ for any $i \leq m$ and $r_i \in O_i$ for any $i \leq n$.

By (iii), there is an infinite set $S \subset \kappa^+$ such that $a \notin \bigcup\{A_\alpha : \alpha \in S\}$. Since the family $\mathcal{A}$ is disjoint, there exists a function $f : (\{a\} \cup (\bigcup\{A_\alpha : \alpha \in S\})) \to \mathbb{R}$ such that $f(a) = r$, $f(x_\alpha^i) = r_i$ for all $i \leq n$ and $f(y_i) = s_i$ for all $i \leq m$. The set $A = \{a\} \cup (\bigcup\{A_\alpha : \alpha \in S\})$ is closed and discrete in X while the space X is normal (see Claim 2 of S.018); so there exists a function $g \in C_p(X)$ with $g|A = f$. It is immediate that $g \in [a, y_1, \ldots, y_m, x_\alpha^1, \ldots, x_\alpha^n; O, W_1, \ldots, W_m, O_1, \ldots, O_n]$ for any $\alpha \in S$ which shows that g belongs to infinitely many elements of the point-finite family $\mathcal{U}$; this contradiction settles (iii)$\Longrightarrow$(i) and completes our solution.

U.278. *Let X be a space with a unique non-isolated point. Prove that, if X has no non-trivial convergent sequences, then the point-finite cellularity of $C_p(X)$ is countable.*

Solution. Let a be the unique non-isolated point of the space X and denote by Y the set $X \backslash \{a\}$. Suppose that $\mathcal{A}$ is a disjoint family of finite subsets of Y such that

$|\mathcal{A}| = \omega_1$. Passing, if necessary, to an appropriate subfamily of $\mathcal{A}$ of cardinality ω_1, we can assume that there is $n \in \mathbb{N}$ such that $|A| = n$ for any $A \in \mathcal{A}$. Thus $\mathcal{A} = \{A_\alpha : \alpha < \omega_1\}$ where $A_\alpha = \{x_\alpha^1, \ldots, x_\alpha^n\} \subset Y$ for any $\alpha < \omega_1$.

The set $\{x_\alpha^1 : \alpha < \omega_1\}$ does not have a sequence which converges to a; so there is an infinite set $S_1 \subset \omega_1$ such that a is not in the closure of the set $\{x_\alpha^1 : \alpha \in S_1\}$. Suppose that we have infinite subsets $S_1 \supset \ldots \supset S_k$ of ω_1 such that a is not in the closure of the set $\{x_\alpha^i : \alpha \in S_i\}$ for any $i \leq k$. Since there is no sequence in the set $\{x_\alpha^k : \alpha \in S_k\}$ which converges to a, we can find an infinite $S_{k+1} \subset S_k$ such that a is not in the closure of the set $\{x_\alpha^{k+1} : \alpha \in S_{k+1}\}$.

This inductive procedure shows that we can continue our construction to obtain infinite subsets $S_1 \supset \ldots \supset S_n$ of ω_1 such that a is not in the closure of the set $\{x_\alpha^i : \alpha \in S_i\}$ for any $i \leq n$. An immediate consequence is that a is not in the closure of the set $\bigcup\{A_\alpha : \alpha \in S_n\}$ which shows that we can apply Problem 277 to conclude that $p(C_p(X)) \leq \omega$.

U.279. *Call a family γ of finite subsets of a space X concentrated if there is no infinite $\mu \subset \gamma$ such that $\bigcup \mu$ is discrete and C^*-embedded in X. Prove that, if every concentrated family of finite subsets of X has cardinality $\leq \kappa$, then $p(C_p(X)) \leq \kappa$.*

Solution. Assume that there exists a point-finite family $\mathcal{U} \subset \tau^*(C_p(X))$ with $|\mathcal{U}| = \kappa^+$. It is easy to see that every $U \in \tau^*(C_p(X))$ contains a standard open set $[x_1, \ldots, x_n; O_1, \ldots, O_n] = \{f \in C_p(X) : f(x_i) \in O_i \text{ for any } i \leq n\}$ where every $O_i \subset \mathbb{R}$ is a non-empty interval with rational endpoints and the family $\{O_1, \ldots, O_n\}$ is disjoint. Choosing such a standard open subset in every element of $\mathcal{U}$ we will still have a point-finite family of cardinality κ^+; so we can assume, without loss of generality, that all elements of $\mathcal{U}$ are the standard open sets described above.

Since there are only countably many finite families of rational intervals, we can pass, if necessary, to an appropriate subfamily of $\mathcal{U}$ of cardinality κ^+ to assume that there is $n \in \mathbb{N}$ and rational non-empty intervals $O_1, \ldots, O_n \subset \mathbb{R}$ such that every element of the family $\mathcal{U}$ is of the form $[x_1, \ldots, x_n; O_1, \ldots, O_n]$. Applying the Δ-lemma (SFFS-038) and passing to a relevant subfamily of $\mathcal{U}$ of cardinality κ^+, we can consider that there exists a set $R = \{y_1, \ldots, y_m\} \subset X$ and non-empty disjoint intervals $W_1, \ldots, W_m \subset \mathbb{R}$ with rational endpoints such that $\mathcal{U} = \{[y_1, \ldots, y_m, x_\alpha^1, \ldots, x_\alpha^n; W_1, \ldots, W_m, O_1, \ldots, O_n] : \alpha < \kappa^+\}$ and the family $\mathcal{A} = \{\{x_\alpha^1, \ldots, x_\alpha^n\} : \alpha < \kappa^+\} \cup \{R\}$ is disjoint; let $A_\alpha = \{x_\alpha^1, \ldots, x_\alpha^n\}$ for any $\alpha < \kappa^+$. Pick $s_i \in W_i$ for any $i \leq m$ and $r_i \in O_i$ for any $i \leq n$.

Since the family $\{A_\alpha \cup R : \alpha < \kappa^+\}$ cannot be concentrated, there is an infinite set $S \subset \kappa^+$ such that $\bigcup\{A_\alpha \cup R : \alpha \in S\}$ is discrete and C^*-embedded in X. The family $\mathcal{A}$ being disjoint, there exists a function $f : \bigcup\{A_\alpha \cup R : \alpha \in S\} \to \mathbb{R}$ such that $f(x_\alpha^i) = r_i$ for all $i \leq n$ and $f(y_i) = s_i$ for all $i \leq m$. The function f is bounded and continuous on $A = \bigcup\{A_\alpha \cup R : \alpha \in S\}$ because A is discrete. The set A is also C^*-embedded; so there is $g \in C_p(X)$ such that $g|A = f$. It is immediate that $g \in \bigcap\{[y_1, \ldots, y_m, x_\alpha^1, \ldots, x_\alpha^n; W_1, \ldots, W_m, O_1, \ldots, O_n] : \alpha \in S\}$, i.e., g belongs to infinitely many elements of the point-finite family $\mathcal{U}$. This contradiction shows that, in $C_p(X)$, there are no point-finite families of non-empty open sets of cardinality κ^+, i.e., $p(C_p(X)) \leq \kappa$.

U.280. *Prove that there exists a Lindelöf Σ-space X with a unique non-isolated point such that $C_p(X)$ is a Lindelöf Σ-space, $p(C_p(X)) = \omega$, all compact subsets of X are countable and $nw(X) = \mathfrak{c}$.*

Solution. Given a set A let $\mathrm{Fin}(A)$ be the family of all finite subsets of A. Consider the interval $I = [0, 1] \subset \mathbb{R}$ with the topology inherited from $\mathbb{R}$; recall that the Alexandroff double $AD(I)$ of the space I is the set $I \times \mathbb{D}$ with the topology generated by the family

$$\exp(I \times \{1\}) \cup \{(U \times \mathbb{D}) \backslash F : U \in \tau(I) \text{ and } F \text{ is a finite subset of } I \times \{1\}\},$$

as a base. Letting $I_0 = \{(t, 0) : t \in I\}$ and $I_1 = \{(t, 1) : t \in I\}$ we have the equality $AD(I) = I_0 \cup I_1$. If $x = (t, i) \in AD(I)$ then $\pi(x) = t$; the map $\pi : AD(I) \to I$ is called the projection. The space $AD(I)$ is compact and the projection π is a continuous and hence perfect map (see TFS-364). It is easy to see that all points of the set I_1 are isolated in $AD(I)$ and the local base at any $x = (t, 0) \in I_0$ is given by the family $\{((t - \frac{1}{n}, t + \frac{1}{n}) \times \mathbb{D}) \backslash \{(t, 1)\} : n \in \mathbb{N}\}$ which shows that $\chi(AD(I)) = \omega$.

For any $y \in I_1$ let $\chi_y(y) = 1$ and $\chi_y(x) = 0$ for any $x \in AD(I) \backslash \{y\}$. Since every $y \in I_1$ is isolated in Y, the function χ_y is continuous on $AD(I)$. If $u(x) = 0$ for all $x \in AD(I)$ then the set $K = \{u\} \cup \{\chi_y : y \in I_1\}$ is compact; therefore the set $K_1 = K \cup \{\pi\} \subset C_p(AD(I))$ is also compact. It is easy to see that K_1 separates the points of $AD(I)$; so there is a continuous injective map of $AD(I)$ into $C_p(K_1)$ (see TFS-166); the space $AD(I)$ being compact, this injective map is a homeomorphism and hence $AD(I)$ embeds in $C_p(K_1)$, i.e., we proved that

(1) any subspace of $AD(I)$ is an Eberlein–Grothendieck space.

Given a set $A \subset I$ we will also need the space I_A which has the underlying set I and the topology generated by the family $\exp(I \backslash A) \cup \tau(I)$ as a subbase. In other words, we declare all points of $I \backslash A$ isolated while the topology is the same at all points of A. It was proved in Problem 090 that there exist disjoint dense sets $A, B \subset I$ such that $|A| = |B| = \mathfrak{c}$, $A \cup B = I$ and the space $(I_A)^\omega$ is Lindelöf. Observe that

(2) the space $Y = B \times \mathbb{D} \subset AD(I)$ is Lindelöf Σ,

because $\pi | Y : Y \to B$ is a perfect map of Y onto the second countable space B (see SFFS-243 and Fact 2 of S.261).

The space $Y \subset AD(I)$ is Eberlein–Grothendieck by the property (1); since the set $B_0 = B \times \{0\}$ of non-isolated points of Y is homeomorphic to B with the topology induced from $\mathbb{R}$, the subspace B_0 is Lindelöf Σ being second countable. Thus we can apply Problem 261 to see that $C_p(Y, \mathbb{I})$ is Lindelöf Σ which implies, together with (2), that $C_p(Y)$ is also Lindelöf Σ (see Problem 217).

Now, considering B as a subspace of $\mathbb{R}$ we will establish that

(3) for any uncountable disjoint family $\mathcal{F} \subset \mathrm{Fin}(B)$ there is an infinite $\mathcal{G} \subset \mathcal{F}$ such that $\bigcup \{F : F \in \mathcal{G}\}$ is closed and discrete in B.

Passing to a subfamily of $\mathcal{F}$ of cardinality ω_1 if necessary, we can assume that there is $n \in \mathbb{N}$ and a faithfully indexed set $F_\alpha = \{x_\alpha^1, \ldots, x_\alpha^n\} \subset B$ for any $\alpha < \omega_1$ such that $\mathcal{F} = \{F_\alpha : \alpha < \omega_1\}$. Let $z_\alpha = (x_\alpha^1, \ldots, x_\alpha^n) \in B^n$ for any $\alpha < \omega_1$. Then $Z = \{z_\alpha : \alpha < \omega_1\} \subset B^n$; if we consider Z to be a subspace of $(I_A)^n$ then it must have an accumulation point $z = (z_1, \ldots, z_n) \in (I_A)^n$ because $(I_A)^n$ is Lindelöf.

Let $p_i : (I_A)^n \to I_A$ be the natural projection of $(I_A)^n$ onto its i-th factor for any $i \leq n$. The family $\mathcal{F}$ being disjoint, the map $p_i | Z$ is injective for any $i \in \{1, \ldots, n\}$. If $z_i \in B$ for some $i \leq n$ then $p_i^{-1}(z_i)$ is an open neighbourhood of z in $(I_A)^n$ (recall that all point of B are isolated in I_A) which intersects at most one element of Z. This contradiction shows that $z_i \in A$ for all $i \leq n$.

The space $(I_A)^n$ being first countable, we can find a faithfully indexed set $\{\alpha_n : n \in \omega\} \subset \omega_1$ such that the sequence $\{z_{\alpha_n} : n \in \omega\}$ converges to z. Therefore the sequence $D_i = \{x_{\alpha_n}^i : n \in \omega\}$ converges to $z_i \in A$ and hence D_i is a closed discrete subset of B for any $i \leq n$; this implies that $D = D_1 \cup \ldots \cup D_n$ is also closed and discrete in B. Now, if $\mathcal{G} = \{F_{\alpha_n} : n \in \omega\}$ then $\mathcal{G}$ is and infinite subfamily of $\mathcal{F}$ such that $\bigcup\{F : F \in \mathcal{G}\} = D$ is closed and discrete in B; so (3) is proved.

Next let X be the space obtained from Y by collapsing the closed set B_0 to a point. Recall that $X = \{x_0\} \cup (Y \backslash B_0)$ where all points of $B_1 = Y \backslash B_0$ are isolated in X and $\tau(x_0, X) = \{\{x_0\} \cup (B_1 \backslash F) : F \subset B_1 \text{ and } F \text{ is closed in } Y\}$. We have a natural quotient map $h : Y \to X$ defined by $h(x) = x$ for any $x \in B_1$ and $h(x) = x_0$ for any $x \in B_0$ (see Fact 2 of T.245). Therefore X is a Lindelöf Σ-space with a unique non-isolated point x_0.

Since h is a quotient map of X onto Y, the dual map $h^* : C_p(X) \to C_p(Y)$ embeds $C_p(X)$ in a Lindelöf Σ-space $C_p(Y)$ as a closed subspace (see TFS-163); so $C_p(X)$ is also a Lindelöf Σ-space.

To compute the cardinal $p(C_p(X))$ assume that $\mathcal{H}$ is an uncountable disjoint family of finite subsets of B_1. For any point $x \in B_1$ there is a unique $\xi(x) \in B$ such that $x = (\xi(x), 1)$; it is evident that $\xi : B_1 \to B$ is a bijection. Therefore the family $\mathcal{F} = \{\xi(P) : P \in \mathcal{H}\} \subset \text{Fin}(B)$ is disjoint and uncountable; so we can apply (3) to extract an infinite $\mathcal{G} \subset \mathcal{H}$ such that the set $D = \bigcup\{\xi(F) : F \in \mathcal{G}\}$ is closed and discrete in B.

Given a point $x = (t, 0) \in B_0$ we can find $W \in \tau(t, I)$ such that $|W \cap D| \leq 1$; then $W \times \mathbb{D} \in \tau(x, Y)$ also contains at most one point of $E = \bigcup\{F : F \in \mathcal{G}\}$. This proves that $E \subset B_1$ is closed in Y and hence $\{x_0\} \cup (B_1 \backslash E)$ is an open neighbourhood of x_0 in X which shows that $x_0 \notin \text{cl}_X(E)$. We proved that, for any uncountable disjoint family $\mathcal{H}$ of finite subsets of B_1 there exists an infinite family $\mathcal{G} \subset \mathcal{H}$ such that $x_0 \notin \text{cl}_X(\bigcup\{F : F \in \mathcal{G}\})$; this makes it possible to apply Problem 277 to conclude that $p(C_p(X)) = \omega$.

To see that $nw(X) = \mathfrak{c}$ observe that $B_1 \subset X$ is a discrete subspace of X with $|B_1| = |X| = \mathfrak{c}$. Now, if $K \subset X$ is compact and uncountable then it is possible to apply (3) to the family $\{\{\xi(x)\} : x \in K \backslash \{x_0\}\}$ to extract an infinite $T \subset K \backslash \{x_0\}$ such that $\xi(T)$ is closed and discrete in B. As before, we can prove that T is closed in Y and hence in X. It turns out that T is an infinite closed discrete subspace of a compact space K; this contradiction shows that any compact subspace of X is countable; so we have proved all promised properties of the space X.

U.281. *Prove that there exists a space* X *such that* $C_p(X)$ *is Lindelöf* Σ-*space,* $nw(X) = \mathfrak{c}$ *and* $p(X) = \omega$.

Solution. It was proved in Problem 280 that there exists a Lindelöf Σ-space Y such that $nw(Y) = \mathfrak{c}$, the space $X = C_p(Y)$ is Lindelöf Σ and $p(X) = \omega$. Applying Problem 218 we conclude that $C_p(X)$ is also a Lindelöf Σ-space; finally, $nw(X) = nw(Y) = \mathfrak{c}$; so the space X is as required.

U.282. *Prove that any continuous image and any closed subspace of a Gul'ko compact space is a Gul'ko compact space.*

Solution. Suppose that K is Gul'ko compact and $f : K \to L$ is a continuous onto map. Then f is closed and hence quotient; so the dual map $f^* : C_p(L) \to C_p(K)$ embeds $C_p(L)$ in $C_p(K)$ as a closed subspace (see TFS-163). Any closed subspace of a Lindelöf Σ-space is Lindelöf Σ so $C_p(L)$ is a Lindelöf Σ-space, i.e., L is Gul'ko compact.

Now, if $F \subset K$ is closed then the restriction map $\pi : C_p(K) \to C_p(F)$ is continuous and onto because K is normal. Therefore $C_p(F)$ is a Lindelöf Σ-space being a continuous image of a Lindelöf Σ-space $C_p(X)$ (see SFFS-243). Thus F is Gul'ko compact as well.

U.283. *Prove that any countable product of Gul'ko compact spaces is a Gul'ko compact space.*

Solution. Suppose that a space K_n is Gul'ko compact for any $n \in \omega$ and let $K = \prod_{n \in \omega} K_n$; we will need the natural projection $p_n : K \to K_n$ for any $n \in \omega$. The dual map $p_n^* : C_p(K_n) \to C_p(K)$ is an embedding (see TFS-163); let $C_n = p_n^*(C_p(K_n))$ for any $n \in \omega$. It is an easy exercise that the set $C = \bigcup_{n \in \omega} C_n$ separates the points of K. Since C is Lindelöf Σ by SFFS-257, we can apply Problem 020 to see that $C_p(K)$ is also Lindelöf Σ and hence K is Gul'ko compact.

U.284. *Let* X *be a Gul'ko compact space. Prove that for every second countable* M, *the space* $C_p(X, M)$ *is Lindelöf* Σ.

Solution. Given a space Z and $Y \subset Z$ say that a family $\mathcal{A} \subset \exp Z$ *separates* Y *from* $Z \backslash Y$ if, for any $y \in Y$ and $z \in Z \backslash Y$ there is $A \in \mathcal{A}$ such that $y \in A$ and $z \notin A$.

Fact 1. Suppose that Z is a Lindelöf Σ-space, $Y \subset Z$ and there is a countable family $\mathcal{A}$ of Lindelöf Σ-subspaces of Z that separates Y from $Z \backslash Y$. Then Y is a Lindelöf Σ-space.

Proof. The set $P = \overline{Y}$ is Lindelöf Σ being closed in Z; by SFFS-233, there exists a countable family $\mathcal{F}$ of compact subsets of βZ that separates Z from $\beta Z \backslash Z$. The space $K = \mathrm{cl}_{\beta Z}(Y)$ is a compactification of the space Y. It is evident that the family $\mathcal{G} = \{F \cap K : F \in \mathcal{F}\}$ consists of compact subspaces of K while all elements of the family $\mathcal{B} = \{A \cap P : A \in \mathcal{A}\}$ are Lindelöf Σ-spaces.

Thus all elements of the family $\mathcal{H} = \mathcal{G} \cup \mathcal{B}$ are Lindelöf Σ-subspaces of the space K. Take any $y \in Y$ and $z \in K \backslash Y$. If $z \notin Z$ then there is $F \in \mathcal{F}$ with $y \in F$

and $z \notin F$. The set $H = F \cap K$ belongs to $\mathcal{H}$ while we have $y \in H$ and $z \notin H$. Now, if $z \in Z$ then $z \in Z \backslash Y$ and hence there is $A \in \mathcal{A}$ for which $y \in A$ and $z \notin A$; it is clear that $B = A \cap P$ belongs to $\mathcal{H}$ while $y \in B$ and $z \notin B$. This proves that the family $\mathcal{H}$ separates Y from $K \backslash Y$; so we can apply SFFS-233 to conclude that Y is a Lindelöf Σ-space. Fact 1 is proved.

Returning to our solution observe that we can consider that M is a subspace of $\mathbb{I}^\omega$ and therefore $C_p(X, M) \subset C_p(X, \mathbb{I}^\omega)$. Observe that $C_p(X, \mathbb{I}^\omega) \simeq (C_p(X, \mathbb{I}))^\omega$ by TFS-112 and hence $C_p(X, \mathbb{I}^\omega)$ is a Lindelöf Σ-space (see SFFS-256). If $K \subset \mathbb{I}^\omega$ is a closed set then $C_p(X, K)$ is a closed subspace of $C_p(X, \mathbb{I}^\omega)$; so we have

(1) $C_p(X, K)$ is a Lindelöf Σ-space for any closed $K \subset \mathbb{I}^\omega$.

The space $\mathbb{I}^\omega$ is second countable so, given a non-empty open subspace U of the space $\mathbb{I}^\omega$ it is easy to find a sequence $\{U_n : n \in \omega\} \subset \tau^*(\mathbb{I}^\omega)$ such that $U_n \subset \overline{U}_n \subset U_{n+1}$ for any $n \in \omega$ and $\bigcup \{\overline{U}_n : n \in \omega\} = U$. It takes a moment's reflection to understand that, for any compact $K \subset U$, there is $n \in \omega$ with $K \subset U_n$. The set $K = f(X)$ being compact for any $f \in C_p(X, U)$, we obtain the equality $C_p(X, U) = \bigcup \{C_p(X, \overline{U}_n) : n \in \omega\}$ which, together with (1), shows that

(2) the space $C_p(X, U)$ is Lindelöf Σ for any open $U \subset \mathbb{I}^\omega$.

Fix a countable base $\mathcal{B}$ in the space $\mathbb{I}^\omega$ which is closed under finite unions and finite intersections. It is easy to see that $\mathcal{B}$ is a network with respect to any compact $K \subset \mathbb{I}^\omega$, i.e., for any $U \in \tau(K, \mathbb{I}^\omega)$ there is $B \in \mathcal{B}$ with $K \subset B \subset U$. The space $C_B = C_p(X, B)$ is Lindelöf Σ for any $B \in \mathcal{B}$ by (2). Let us show that

(3) The family $\mathcal{C} = \{C_B : B \in \mathcal{B}\}$ separates $C_p(X, M)$ from $C_p(X, \mathbb{I}^\omega) \backslash C_p(X, M)$.

To see that (3) holds take any $f \in C_p(X, M)$ and $g \in C_p(X, \mathbb{I}^\omega) \backslash C_p(X, M)$; fix a point $x \in X$ such that $g(x) \notin M$. The set $K = f(X)$ being compact there is $B \in \mathcal{B}$ such that $K \subset B \subset \mathbb{I}^\omega \backslash \{g(x)\}$. It is evident that $f \in C_B$ and $g \notin C_B$; so (3) is proved.

Finally, apply (2) and (3) together with Fact 1 to conclude that $C_p(X, M)$ is a Lindelöf Σ-space.

U.285. *Prove that if $C_p(X)$ is a Lindelöf Σ-space then X can be condensed into a Σ-product of real lines. Deduce from this fact that every Gul'ko compact space is Corson compact.*

Solution. Given a space X and $A \subset X$ let $\pi_A^X : C_p(X) \to C_p(A)$ be the restriction map. For any $B \subset C_p(X)$ let $e_B^X(x)(f) = f(x)$ for any $f \in B$; then $e_B^X(x) \in C_p(B)$ for any $x \in X$ and the map $e_B^X : X \to C_p(B)$ is continuous (see TFS-166). If X is clear then we will write π_A and e_B instead of π_A^X and e_B^X respectively. For a map $\varphi : P \to Q$ its *dual map* $\varphi^* : C_p(Q) \to C_p(P)$ is defined by $\varphi^*(f) = f \circ \varphi$ for any $f \in C_p(Q)$. If X is a space, $\mathcal{B} \subset \exp X$ and $Y \subset X$ then $\mathcal{B}|Y = \{U \cap Y : U \in \mathcal{B}\}$. A family $\mathcal{F} \subset \exp X$ is called a *Σ-family in X* if $\mathcal{F}$ is a network with respect to a compact cover of X. In this terminology, a space is Lindelöf Σ if and only if it has a countable Σ-family.

In most of our proofs we will have a fixed pair (X, Y) of spaces such that $Y \subset C_p(X)$. If, additionally, $P \subset X$ and $Q \subset Y$ we will say that sets $M \subset P$ and $L \subset Q$ are (P, Q)-*conjugate* if $\pi_M(L) = \pi_M(Q)$ and $e_L(M) = e_L(P)$; the sets M and L are (P, Q)-*preconjugate* if $\pi_M(L)$ is dense in $\pi_M(Q)$ and $e_L(M)$ is dense in $e_L(P)$. If no confusion can be made then (X, Y)-conjugate sets will be called *conjugate* and (X, Y)-preconjugate sets will be simply called *preconjugate*.

Fact 1. Given a space X suppose that $Y \subset C_p(X)$ generates the topology of X. Then $\pi_M(Y)$ generates the topology of M for any $M \subset X$.

Proof. It is evident that, for any function $f \in C_p(X)$ and any set $U \subset \mathbb{R}$ we have $f^{-1}(U) \cap M = (f|M)^{-1}(U) = (\pi_M(f))^{-1}(U)$. By our assumption on the set Y the family $\mathcal{B} = \{f^{-1}(U) : f \in Y, \ U \in \tau(\mathbb{R})\}$ is a subbase in X; so the family $\mathcal{B}|M = \{(\pi_M(f))^{-1}(U) : f \in Y \ U \in \tau(\mathbb{R})\}$ is a subbase in M which shows that $\pi_M(Y)$ generates the topology of M. Fact 1 is proved.

Fact 2. Suppose that X is a space and a set $Y \subset C_p(X)$ generates the topology of X. Assume also that $M \subset X$ and $L \subset Y$ are conjugate sets. Then the maps $u = e_L|M : M \to e_L(M)$ and $v = \pi_M|L : L \to \pi_M(L)$ are homeomorphisms; besides, the maps $r = u^{-1} \circ e_L : X \to M$ and $q = v^{-1} \circ \pi_M : Y \to L$ are continuous retractions such that $q = r^*|Y$. The maps r and q are called *the pair of retractions corresponding to the conjugate pair* (M, L).

Proof. The set $L_0 = \pi_M(L) = \pi_M(Y)$ generates the topology of M by Fact 1; so the evaluation map $e^M_{L_0} : M \to C_p(L_0)$ is an embedding by TFS-166. Since the map $v : L \to L_0$ is continuous and onto, the dual map $v^* : C_p(L_0) \to C_p(L)$ is also an embedding (see TFS-163); so the map $v^* \circ e^M_{L_0}$ embeds M in $C_p(L)$. It is straightforward that $u = v^* \circ e^M_{L_0}$; so the map u is a homeomorphism.

To convince ourselves that the map $v : L \to L_0$ is a homeomorphism observe that the set $M_0 = e_L(M) = e_L(X)$ generates the topology of L by TFS-166. Therefore the map $e^L_{M_0} : L \to C_p(M_0)$ is an embedding. The map $u : M \to M_0$ is continuous and onto so $u^* : C_p(M_0) \to C_p(M)$ is an embedding as well and it is easy to check that $u^* \circ e^L_{M_0} = v$; so v is, indeed, a homeomorphism.

An immediate consequence is that the maps r and q are continuous being compositions of continuous maps. If $x \in M$ then $e_L(x) = u(x)$ and therefore $r(x) = u^{-1}(e_L(x)) = u^{-1}(u(x)) = x$ which shows that r is a retraction (see Fact 1 of S.351). Analogously, if $f \in L$ then $\pi_M(f) = v(f)$ and hence we have the equality $q(f) = v^{-1}(\pi_M(f)) = v^{-1}(v(f)) = f$; so q is a retraction as well.

Finally, take any $f \in Y$; to prove that $q(f) = r^*(f)$ pick any $x \in X$. By the definition of r we have $r(x) = u^{-1}(e_L(x))$ so $e_L(r(x)) = u(r(x)) = e_L(x)$. The equality $e_L(x) = e_L(r(x))$ shows that $g(x) = g(r(x))$ for any $g \in L$ and, in particular, $q(f)(x) = q(f)(r(x))$. Recalling that $q(f)|M = f|M$ we can see that $q(f)(x) = q(f)(r(x)) = f(r(x))$ and therefore $q(f)(x) = f(r(x))$ for any $x \in X$ which shows that $q(f) = f \circ r = r^*(f)$. Thus $q(f) = r^*(f)$ for any $f \in Y$ and hence $q = r^*|Y$; so Fact 2 is proved.

Fact 3. Suppose that X is a space and $Y \subset C_p(X)$ generates the topology of X. Assume that we are given sets $M_0, M_1 \subset X$ and $L_0, L_1 \subset Y$ for which both (M_0, L_0) and (M_1, L_1) are conjugate pairs such that $M_0 \subset M_1$ and $L_0 \subset L_1$. If (r_i, q_i) is the pair of retractions corresponding to the conjugate pair (M_i, L_i) for every $i \in \mathbb{D}$ then $r_0 \circ r_1 = r_1 \circ r_0 = r_0$ and $q_0 \circ q_1 = q_1 \circ q_0 = q_0$.

Proof. If $u_i = e_{L_i}|M_i$ then $r_i = u_i^{-1} \circ e_{L_i}$ for any $i \in \mathbb{D}$. Fix any $x \in X$. We have $r_1(x) = u_1^{-1}(e_{L_1}(x))$; so $e_{L_1}(r_1(x)) = u_1(r_1(x)) = e_{L_1}(x)$ which shows that $f(r_1(x)) = f(x)$ for any $f \in L_1$. Since $L_0 \subset L_1$, we have $f(r_1(x)) = f(x)$ for any $f \in L_0$ and hence $e_{L_0}(r_1(x)) = e_{L_0}(x)$ which implies that

$$r_0(r_1(x)) = u_0^{-1}(e_{L_0}(r_1(x))) = u_0^{-1}(e_{L_0}(x)) = r_0(x).$$

This proves that $r_0 \circ r_1 = r_0$.

Now, $r_0(x) \in M_0 \subset M_1$ which shows that $e_{L_1}(r_0(x)) = u_1(r_0(x))$ and hence $r_1(r_0(x)) = u_1^{-1}(e_{L_1}(r_0(x))) = r_0(x)$; as a consequence, $r_1 \circ r_0 = r_0$ which implies that $r_0 = r_1 \circ r_0 = r_0 \circ r_1$ and therefore we proved the first statement of our Fact.

To finish the proof observe that $q_i = r_i^*$ for any $i \in \mathbb{D}$ by Fact 2 and therefore $q_0 \circ q_1(f) = q_0(q_1(f)) = r_0^*(r_1^*(f)) = r_0^*(f \circ r_1) = f \circ r_1 \circ r_0 = f \circ r_0 = r_0^*(f) = q_0(f)$ for any $f \in Y$. Thus $q_0 \circ q_1 = q_0$. Analogously, $q_1(q_0(f)) = f \circ r_0 \circ r_1 = f \circ r_0 = q_0(f)$ for any $f \in Y$ and therefore $q_0 \circ q_1 = q_1 \circ q_0 = q_0$ which shows that Fact 3 is proved.

Fact 4. Suppose that X is a space and $Y \subset C_p(X)$ generates the topology of X; assume also that we have sets $P \subset X$ and $Q \subset Y$. Given a limit ordinal β assume that we have a (P, Q)-preconjugate pair (M_α, L_α) for every $\alpha < \beta$ such that $\alpha < \gamma < \beta$ implies $M_\alpha \subset M_\gamma$ and $L_\alpha \subset L_\gamma$. If $M = \bigcup\{M_\alpha : \alpha < \beta\}$ and $L = \bigcup\{L_\alpha : \alpha < \beta\}$ then (M, L) is a (P, Q)-preconjugate pair.

Proof. To see that the set $M' = e_L(M)$ is dense in $P' = e_L(P)$ take any $x \in P$ and $U \in \tau(e_L(x), P')$; let $x' = e_L(x)$. There are $f_1, \dots, f_n \in L$ and $\varepsilon > 0$ such that $V = \{y' \in P' : |y'(f_i) - x'(f_i)| < \varepsilon$ for any $i \leq n\} \subset U$. There exists an ordinal $\alpha < \beta$ such that $\{f_1, \dots, f_n\} \subset L_\alpha$. Since $e_{L_\alpha}(M_\alpha)$ is dense in $e_{L_\alpha}(P)$, there is a point $y \in M_\alpha$ such that $|e_{L_\alpha}(y)(f_i) - e_{L_\alpha}(x)(f_i)| < \varepsilon$ for any $i \leq n$. This is equivalent to saying that $|f_i(y) - f_i(x)| < \varepsilon$ for every $i \leq n$ which, in turn, is equivalent to saying that $|e_L(y)(f_i) - e_L(x)(f_i)| < \varepsilon$ for any $i \leq n$. Thus $y' = e_L(y) \in V \cap M' \subset U \cap M'$; so the point x' is in the closure of M'. Thus $M' = e_L(M)$ is dense in $P' = e_L(P)$.

To show that the set $\pi_M(L)$ is dense in $\pi_M(Q)$ take any function $f \in Q$ and a set $U \in \tau(\pi_M(f), \pi_M(Q))$. There exist points $x_1, \dots, x_k \in M$ and $\varepsilon > 0$ such that $V = \{g \in \pi_M(Q) : |g(x_i) - f(x_i)| < \varepsilon$ for all $i \leq k\} \subset U$. Take $\alpha < \beta$ for which $\{x_1, \dots, x_k\} \subset M_\alpha$. The set $\pi_{M_\alpha}(L_\alpha)$ being dense in $\pi_{M_\alpha}(Q)$ there is $g \in L_\alpha$ such that $|\pi_{M_\alpha}(g)(x_i) - \pi_{M_\alpha}(f)(x_i)| < \varepsilon$ for all $i \leq k$ which is equivalent to the inequality $|g(x_i) - f(x_i)| < \varepsilon$ for all $i \leq k$.

An immediate consequence is that $|\pi_M(g)(x_i) - \pi_M(f)(x_i)| < \varepsilon$ for all $i \leq k$ and therefore $\pi_M(g) \in V \cap \pi_M(L) \subset U \cap \pi_M(L)$. Thus any $f \in \pi_M(Q)$ is in the closure of $\pi_M(L)$, i.e., $\pi_M(L)$ is dense in $\pi_M(Q)$ and hence (M, L) is a (P, Q)-preconjugate pair. Fact 4 is proved.

Fact 5. Suppose that X is a Lindelöf Σ-space and $\mathcal{F}$ is a fixed countable network with respect to a compact cover $\mathcal{C}$ of the space X. Assume additionally that $\mathcal{F}$ is closed under finite intersections and $f : X \to Y$ is a continuous onto map. If $A \subset X$ is a set such that $f(A \cap F)$ is dense in $f(F)$ for any $F \in \mathcal{F}$ then $f(\overline{A}) = Y$.

Proof. Assume, towards a contradiction that there is a point $y \in Y \backslash f(\overline{A})$ and fix a set $C \in \mathcal{C}$ such that $C \cap f^{-1}(y) \neq \emptyset$. The set $K = \overline{A} \cap C$ is compact; so $U = Y \backslash f(K)$ is an open neighbourhood of y; take $V \in \tau(y, Y)$ such that $\overline{V} \subset U$. Since $\mathcal{F}$ is closed under finite intersections, we can choose a sequence $\mathcal{S} = \{F_n : n \in \omega\} \subset \mathcal{F}$ such that $C \subset F_n$ and $F_n \supset F_{n+1}$ for any $n \in \omega$ while $\mathcal{S}$ is a network at C, i.e., for any $O \in \tau(C, X)$ there is $n \in \omega$ with $F_n \subset O$. We have $y \in f(F_n)$; the set $f(A \cap F_n)$ being dense in $f(F_n)$, we can pick a point $a_n \in A \cap F_n$ such that $f(a_n) \in V$ for every $n \in \omega$.

Next, observe that the sequence $\mathcal{S} = \{a_n : n \in \omega\}$ has an accumulation point in C. Indeed, if every $z \in C$ has an open neighbourhood O_z such that the set $N_z = \{n \in \omega : a_n \in O_z\}$ is finite then, by compactness of C, there is a finite $D \subset C$ with $C \subset O = \bigcup\{O_z : z \in D\}$. Then there are only finitely many $n \in \omega$ such that $a_n \in O$ while there exists $m \in \omega$ with $F_m \subset O$ and therefore $a_n \in O$ for all $n \geq m$. This contradiction shows that there is an accumulation point $a \in C$ for the sequence $\mathcal{S}$. Then $a \in \overline{A} \cap C = K$; since $f(a_n) \in V$ for all $n \in \omega$, we have $f(a) \in \overline{V}$ by continuity of f. Thus $f(a) \in U \cap f(K)$ which is a contradiction. Fact 5 is proved.

Fact 6. Suppose that κ is an infinite cardinal, X is a space and a set $Y \subset C_p(X)$ generates the topology of X. Assume also that $\mathcal{M} \subset \exp X$ and $\mathcal{L} \subset \exp Y$ are countable families and we have sets $A \subset X$, $B \subset Y$ with $|A| \leq \kappa$ and $|B| \leq \kappa$. Then there exist sets $M \subset X$ and $L \subset Y$ such that $A \subset M$, $B \subset L$, $|M| \leq \kappa$, $|L| \leq \kappa$ and the pair $(M \cap P, L \cap Q)$ is (P, Q)-preconjugate for any $P \in \mathcal{M}$ and $Q \in \mathcal{L}$.

Proof. Let $M_0 = A$, $L_0 = B$ and choose an enumeration $\{(P_n, Q_n) : n \in \omega\}$ of the set $\mathcal{M} \times \mathcal{L}$ such that every pair $(P, Q) \in \mathcal{M} \times \mathcal{L}$ occurs infinitely many times in this enumeration.

Assume that $n \in \omega$ and we have sets $M_0, \ldots, M_n \subset X$ and $L_0, \ldots, L_n \subset Y$ with the following properties:

(1) $A = M_0 \subset \ldots \subset M_n$ and $B = L_0 \subset \ldots \subset L_n$;
(2) $|M_i| \leq \kappa$ and $|L_i| \leq \kappa$ for all $i \leq n$;
(3) the set $e_{L_i}(M_{i+1} \cap P_i)$ is dense in $e_{L_i}(P_i)$ for any $i < n$;
(4) the set $\pi_{M_i}(L_{i+1} \cap Q_i)$ is dense in $\pi_{M_i}(Q_i)$ for any $i \leq n$.

The sets $e_{L_n}(P_n) \subset C_p(L_n)$ and $\pi_{M_n}(Q_n) \subset C_p(M_n)$ have weight $\leq \kappa$ by the property (2); so there exist $M_{n+1} \subset X$ and $L_{n+1} \subset Y$ such that $M_n \subset M_{n+1}$, $L_n \subset L_{n+1}$ and $|M_{n+1}| \leq \kappa$, $|L_{n+1}| \leq \kappa$ while $\pi_{M_n}(L_{n+1} \cap Q_n)$ is dense in $\pi_{M_n}(Q_n)$ and $e_{L_n}(M_{n+1} \cap P_n)$ is dense in $e_{L_n}(P_n)$.

It is immediate that the properties (1)–(4) still hold for all $i \leq n$; so our inductive procedure can be continued to construct sequences $\{M_i : i \in \omega\}$ and $\{L_i : i \in \omega\}$ for which (1)–(4) are fulfilled for any $n \in \omega$.

To see that the sets $M = \bigcup_{n \in \omega} M_n$ and $L = \bigcup_{n \in \omega} L_n$ are as promised take any pair $(P, Q) \in \mathcal{M} \times \mathcal{L}$ and let $M' = M \cap P$, $L' = L \cap Q$; we must prove that the pair (M', L') is (P, Q)-preconjugate. Take any $f \in Q$, and $U \in \tau(\pi_{M'}(f), \pi_{M'}(Q))$. By the definition of the topology of pointwise convergence there are $x_1, \ldots, x_k \in M'$ and $\varepsilon > 0$ such that $V = \{h \in \pi_{M'}(Q) : |h(x_i) - f(x_i)| < \varepsilon$ for all $i \leq k\} \subset U$.

It follows from (1) and the choice of our enumeration of $\mathcal{M} \times \mathcal{L}$ that there is $n \in \omega$ such that $(P_n, Q_n) = (P, Q)$ and $\{x_1, \ldots, x_k\} \subset M_n$. The property (4) shows that there is $g \in L_{n+1} \cap Q$ such that $|g(x_i) - f(x_i)| < \varepsilon$ for all $i \leq k$. It is immediate that $\pi_{M'}(g) \in V \cap \pi_{M'}(L') \subset U \cap \pi_{M'}(L')$. Therefore $\pi_{M'}(L')$ is dense in $\pi_{M'}(Q)$.

To show that the set $e_{L'}(M')$ is dense in the space $e_{L'}(P)$ fix any $x \in P$ and $U \in \tau(e_{L'}(x), e_{L'}(P))$. There are functions $f_1, \ldots, f_k \in L'$ and $\varepsilon > 0$ such that $V = \{z \in e_{L'}(P) : |z(f_i) - e_{L'}(x)(f_i)| < \varepsilon$ for all $i \leq k\} \subset U$. It follows from (1) and the choice of our enumeration of $\mathcal{M} \times \mathcal{L}$ that there is $n \in \omega$ such that $(P_n, Q_n) = (P, Q)$ and $\{f_1, \ldots, f_k\} \subset L_n$. The property (3) shows that there is $y \in M_{n+1} \cap P$ such that $|e_{L_n}(y)(f_i) - e_{L_n}(x)(f_i)| < \varepsilon$ for all $i \leq k$. It is immediate that $e_{L'}(y) \in V \cap e_{L'}(M') \subset U \cap e_{L'}(M')$. Therefore $e_{L'}(M')$ is dense in $e_{L'}(P)$ and Fact 6 is proved.

Fact 7. Suppose that X is a Lindelöf Σ-space and a Lindelöf Σ-space $Y \subset C_p(X)$ generates the topology of X. Assume additionally that some countable families $\mathcal{P} \subset \exp X$, $\mathcal{Q} \subset \exp Y$ are closed under finite intersections and finite unions and there exist compact covers $\mathcal{K}$ and $\mathcal{C}$ of the spaces X and Y respectively such that $\mathcal{P}$ is a network with respect to $\mathcal{K}$ and $\mathcal{Q}$ is a network with respect to $\mathcal{C}$. Assume that we are given sets $M \subset X$ and $L \subset Y$ such that the pair $(M \cap P, L \cap Q)$ is (P, Q)-preconjugate for any $(P, Q) \in \mathcal{P} \times \mathcal{Q}$. Then the pair $(\overline{M}, \mathrm{cl}_Y(L))$ is conjugate.

Proof. Let $F = \overline{M}$ and $G = \mathrm{cl}_Y(L)$. If $x \in F$, $y \in X$ and $e_L(x) = e_L(y)$ then $e_G(x) = e_G(y)$ because the functions $e_G(x)$ and $e_G(y)$ are continuous on G while $e_L(x)$ and $e_L(y)$ are their restrictions to the dense subspace L of the space G (see Fact 0 of S.351). Thus it is sufficient to show that $e_L(F) = e_L(X)$.

If $X' = e_L(X)$ then $e_L : X \to X'$ is continuous and onto; fix any $P \in \mathcal{P}$ and let $L_Q = L \cap Q$ for any $Q \in \mathcal{Q}$. If $M' = M \cap P$ then the pair (M', L_Q) is (P, Q)-conjugate for any $Q \in \mathcal{Q}$.

Take any point $x \in P$ and $U \in \tau(e_L(x), e_L(P))$. There are functions $f_1, \ldots, f_n \in L$ and $\varepsilon > 0$ such that $V = \{z \in e_L(P) : |z(f_i) - e_L(x)(f_i)| < \varepsilon$ for all $i \leq n\} \subset U$. Since $\mathcal{Q}$ is closed under finite unions, there is $Q \in \mathcal{Q}$ such that $A = \{f_1, \ldots, f_n\} \subset Q$ and hence $A \subset L_Q$. The pair (M', L_Q) being (P, Q)-preconjugate there is a point $y \in M'$ such that $|e_{L_Q}(y)(f_i) - e_{L_Q}(x)(f_i)| < \varepsilon$ for all $i \leq n$. This is the same as saying that $|f_i(y) - f_i(x)| < \varepsilon$ and hence $|e_L(y)(f_i) - e_L(x)(f_i)| < \varepsilon$ for any $i \leq n$. This shows that $e_L(y) \in V \cap e_L(M') \subset U \cap e_L(M')$ and therefore $e_L(M')$ is dense in $e_L(P)$.

The set P was chosen arbitrarily; so $e_L(M \cap P)$ is dense in $e_L(P)$ for any $P \in \mathcal{P}$. This makes it possible to apply Fact 5 to see that $e_L(F) = X'$ and hence $e_G(F) = e_G(X)$.

Analogously, if $\pi_M(G) = \pi_M(Y)$ then $\pi_F(G) = \pi_F(Y)$; so it suffices to show that $\pi_M(G) = \pi_M(Y)$. Let $Y' = \pi_M(Y)$; then $\pi_M : Y \to Y'$ is a continuous onto map. Fix an element $Q \in \mathcal{Q}$; let $Q' = Q \cap L$ and $M_P = M \cap P$ for any $P \in \mathcal{P}$. Take any $f \in Q$ and $U \in \tau(\pi_M(f), \pi_M(Q))$. There are points $x_1, \ldots, x_n \in M$ and a number $\varepsilon > 0$ such that $V = \{g \in \pi_M(Q) : |g(x_i) - f(x_i)| < \varepsilon$ for all $i \leq n\} \subset U$. The family $\mathcal{P}$ being closed under finite unions, there is $P \in \mathcal{P}$ such that $A = \{x_1, \ldots, x_n\} \subset P$ and therefore $A \subset M_P$. The pair (M_P, Q') is (P, Q)-preconjugate; so there is $g \in Q'$ such that $|g(x_i) - f(x_i)| < \varepsilon$ for all $i \leq n$. It is clear that $|\pi_M(g)(x_i) - \pi_M(f)(x_i)| < \varepsilon$ for all $i \leq n$ and hence $\pi_M(g) \in V \cap \pi_M(Q') \subset U \cap \pi_M(Q')$. This proves that $\pi_M(L \cap Q)$ is dense in $\pi_M(Q)$ for any $Q \in \mathcal{Q}$; so we can apply Fact 5 again to conclude that $\pi_M(G) = Y' = \pi_M(Y)$. Thus the pair $(\overline{M}, \mathrm{cl}_Y(L))$ is conjugate and Fact 7 is proved.

In the following Fact and its proof a pair of sets $M \subset X$ and $L \subset C_p(X)$ is called conjugate if they are $(X, C_p(X))$-conjugate.

Fact 8. Suppose that κ is an infinite cardinal and X is a Lindelöf Σ-space such that $C_p(X)$ is also Lindelöf Σ and $d(X) = \kappa$. Assume additionally that a set $D = \{x_\alpha : \alpha < \kappa\}$ is dense in X and $E = \{f_\alpha : \alpha < \kappa\}$ is dense in $C_p(X)$; observe that this assumption cannot be contradictory because both X and $C_p(X)$ are monolithic (see SFFS-266) and hence $d(X) = nw(X) = nw(C_p(X)) = d(C_p(X))$. Then there exists families $\{M_\alpha : \alpha < \kappa\} \subset \exp X$ and $\{L_\alpha : \alpha < \kappa\} \subset \exp(C_p(X))$ with the following properties:

(a) M_α is closed in X and L_α is closed in $C_p(X)$ while the pair (M_α, L_α) is conjugate for any $\alpha < \kappa$;
(b) if $\alpha < \kappa$ is a limit ordinal then $M_\alpha = \overline{\bigcup\{M_\beta : \beta < \alpha\}}$ and $L_\alpha = \overline{\bigcup\{L_\beta : \beta < \alpha\}}$;
(c) if $\beta < \alpha < \kappa$ then $M_\beta \subset M_\alpha$ and $L_\beta \subset L_\alpha$;
(d) $x_\alpha \in M_\alpha$ and $f_\alpha \in L_\alpha$ for any $\alpha < \kappa$;
(e) $nw(M_\alpha) \leq |\alpha| \cdot \omega$ and $nw(L_\alpha) \leq |\alpha| \cdot \omega$ for any $\alpha < \kappa$.

Proof. Let $Y = C_p(X)$ and fix countable Σ-families $\mathcal{P}$ and $\mathcal{Q}$ in the spaces X and Y respectively such that both $\mathcal{P}$ and $\mathcal{Q}$ are closed under finite unions and finite intersections.

Next, apply Fact 6 to the sets $A = \{x_0\}$ and $B = \{f_0\}$ to find countable $A_0 \subset X$ and $B_0 \subset Y$ for which the pair $(A_0 \cap P, B_0 \cap Q)$ is (P, Q)-preconjugate for any pair $(P, Q) \in \mathcal{P} \times \mathcal{Q}$. By Fact 7, the pair $(\overline{A_0}, \mathrm{cl}_Y(B_0))$ is conjugate; let $M_0 = \overline{A_0}$ and $L_0 = \mathrm{cl}_Y(B_0)$. It is clear that the conditions (a)–(e) are satisfied for $\alpha = 0$.

Now assume that $\xi < \kappa$ and we have constructed families $\{M_\alpha : \alpha < \xi\}$ and $\{A_\alpha : \alpha < \xi\}$ of subsets of X as well as families and $\{L_\alpha : \alpha < \xi\}$ and $\{B_\alpha : \alpha < \xi\}$ of subsets of Y such that the properties (a)–(e) hold for any $\alpha < \xi$ and we also have the following properties:

(4) the pair $(A_\alpha \cap P, B_\alpha \cap Q)$ is (P, Q)-preconjugate for any ordinal $\alpha < \xi$ and
    $(P, Q) \in \mathcal{P} \times \mathcal{Q}$.
(5) $|A_\alpha| \leq |\alpha| \cdot \omega$ and $|B_\alpha| \leq |\alpha| \cdot \omega$ for any $\alpha < \xi$;
(6) $M_\alpha = \overline{A}_\alpha$ and $L_\alpha = \mathrm{cl}_Y(B_\alpha)$ for any $\alpha < \xi$;
(7) if $\beta < \alpha < \xi$ then $A_\beta \subset A_\alpha$ and $B_\beta \subset B_\alpha$.

If $\xi = \eta + 1$ then we can apply Fact 6 to find sets $A_\xi \subset X$ and $B_\xi \subset Y$ such that
$A_\eta \cup \{x_\xi\} \subset A_\xi$, $B_\eta \cup \{f_\xi\} \subset B_\xi$, the pair $(A_\xi \cap P, B_\xi \cap Q)$ is (P, Q)-preconjugate
for any pair $(P, Q) \in \mathcal{P} \times \mathcal{Q}$ and $|A_\xi| \leq |A_\eta| \cdot \omega$, $|B_\xi| \leq |B_\eta| \cdot \omega$. It follows from
Fact 7 that the sets $M_\xi = \overline{A}_\xi$ and $L_\xi = \mathrm{cl}_Y(B_\xi)$ form a conjugate pair. It is evident
that the conditions (a)–(e) and (4)–(7) are still fulfilled for all $\alpha \leq \xi$ (we have to use
monolithity of X and Y to see that the condition (e) is satisfied).

Now, if ξ is a limit ordinal then consider the sets $A_\xi = \bigcup\{A_\alpha : \alpha < \xi\}$ and
$B_\xi = \bigcup\{B_\alpha : \alpha < \xi\}$. The pair $(A_\xi \cap P, B_\xi \cap Q)$ has to be (P, Q)-preconjugate
for any $(P, Q) \in \mathcal{P} \times \mathcal{Q}$ (see Fact 4); so the sets $M_\xi = \overline{A}_\xi$ and $L_\xi = \mathrm{cl}_Y(B_\xi)$ are
conjugate by Fact 7. The property (e) holds by monolithity of X and Y, properties
(a)–(d) are evident as well as (4)–(7) for all $\alpha \leq \xi$. This shows that our inductive
procedure can be continued to construct the promised families $\{M_\alpha : \alpha < \kappa\} \subset$
$\exp X$ and $\{L_\alpha : \alpha < \kappa\} \subset \exp(C_p(X))$. Fact 8 is proved.

Fact 9. If X and $C_p(X)$ are Lindelöf Σ-spaces then $C_p(X)$ can be linearly
condensed into a Σ-product of real lines.

Proof. Observe that both spaces X and $C_p(X)$ are monolithic (see SFFS-266); so
we have $d(X) = nw(X) = nw(C_p(X)) = d(C_p(X))$. Our proof will be by
induction on $\kappa = d(X)$. If $\kappa = \omega$ then fix a countable dense set $D \subset X$ and
observe that the restriction map $\pi_D : C_p(X) \to \mathbb{R}^D$ condenses $C_p(X)$ linearly in
$\mathbb{R}^D$ which, turn, is linearly homeomorphic to a subspace of $\Sigma(\omega_1)$.

Now assume that κ is an uncountable cardinal and we have proved that, for any
cardinal $\lambda < \kappa$, if Z and $C_p(Z)$ are Lindelöf Σ-spaces with $d(Z) < \lambda$ then $C_p(Z)$
condenses linearly into a Σ-product of real lines.

Assume that $d(X) = \kappa$ and fix a dense set $D = \{x_\alpha : \alpha < \kappa\}$ in the space
X. There exists a dense set $E = \{f_\alpha : \alpha < \kappa\}$ in the space $C_p(X)$. We can
take families $\{M_\alpha : \alpha < \kappa\} \subset \exp X$ and $\{L_\alpha : \alpha < \kappa\} \subset \exp(C_p(X))$ whose
existence is guaranteed by Fact 8. Let (r_α, q_α) be the pair of the retractions which
correspond to the conjugate pair (M_α, L_α) for any $\alpha < \kappa$. We will first consider the
case when $\mathrm{cf}(\kappa) = \omega$ and hence there is an increasing sequence $\{\lambda_n : n \in \omega\}$ of
infinite cardinals such that $\kappa = \sup\{\lambda_n : n \in \omega\}$.

The property (e) of Fact 8 shows that $d(M_{\lambda_n}) \leq \lambda_n < \kappa$ for any $n \in \omega$; so there
exists a linear condensation Φ_n of the space $C_p(M_{\lambda_n})$ into $\Sigma(A_n)$ for some set A_n;
there is no loss of generality to assume that the family $\{A_n : n \in \omega\}$ is disjoint.
Let $\pi_n : C_p(X) \to C_p(M_{\lambda_n})$ be the restriction map. Then the diagonal product
$\Phi = \Delta\{\Phi_n \circ \pi_n : n \in \omega\}$ maps $C_p(X)$ linearly into $\prod\{\Sigma(A_n) : n \in \omega\} \subset \Sigma(A)$
where $A = \bigcup\{A_n : n \in \omega\}$. Therefore $\Phi : C_p(X) \to \Sigma(A)$ is a linear map. To see
that Φ is injective take distinct $f, g \in C_p(X)$.

The properties (c) and (d) imply that $D \subset M = \bigcup\{M_{\lambda_n} : n \in \omega\}$; so the set M is dense in X. Therefore $f|M \neq g|M$ and hence there is $n \in \omega$ and $x \in M_{\lambda_n}$ such that $f(x) \neq g(x)$ and therefore $\pi_n(f) \neq \pi_n(g)$. The mapping Φ_n being an injection, we have $\Phi_n(\pi_n(f)) \neq \Phi_n(\pi_n(g))$ which implies that $\Phi(f) \neq \Phi(g)$; so Φ is injective. Thus Φ condenses $C_p(X)$ linearly into $\Sigma(A)$.

Now assume that the cofinality of the cardinal κ is uncountable. It follows from $t(X) = t(C_p(X)) = \omega$ that $\bigcup\{M_\alpha : \alpha < \kappa\} = X$ and $\bigcup\{L_\alpha : \alpha < \kappa\} = C_p(X)$. The following property is crucial.

(8) for any $f \in C_p(X)$ the set $E_f = \{\alpha < \kappa : f \circ r_\alpha \neq f \circ r_{\alpha+1}\}$ is countable.

To prove (8) fix a function $f \in C_p(X)$; observe first that $f \circ r_\alpha = r_\alpha^*(f) = q_\alpha(f)$ for any $\alpha < \kappa$ (see Fact 2). There is $\eta < \kappa$ such that $f \in L_\eta$; this means that $q_\alpha(f) = f$ for any $\alpha \geq \eta$; so $E_f \subset \eta$. If E_f is uncountable then we can choose an increasing ω_1-sequence $S = \{\xi_\alpha : \alpha < \omega_1\} \subset E_f$; let $\xi = \sup S$. It follows from $E_f \subset \eta$ that $\xi \leq \eta < \kappa$.

The function $g = q_\xi(f)$ belongs to L_ξ; it follows from (b) of Fact 8 and $t(C_p(X)) = \omega$ that $g \in L_\gamma$ for some $\gamma < \xi$. There exists $\alpha < \omega_1$ such that $\gamma < \xi_\alpha$ and therefore $q_{\xi_\alpha}(g) = q_{\xi_\alpha+1}(g) = g$. Now apply Fact 3 to conclude that $q_{\xi_\alpha}(g) = q_{\xi_\alpha}(q_\xi(f)) = q_{\xi_\alpha}(f)$. Analogously, $q_{\xi_\alpha+1}(g) = q_{\xi_\alpha+1}(q_\xi(f)) = q_{\xi_\alpha+1}(f)$. As a consequence, $q_{\xi_\alpha}(f) = q_{\xi_\alpha}(g) = g$ and $q_{\xi_\alpha+1}(f) = q_{\xi_\alpha+1}(g) = g$ which shows that $q_{\xi_\alpha+1}(f) = q_{\xi_\alpha}(f)$; this contradiction with $\xi_\alpha \in E_f$ completes the proof of (8).

Now it is time to apply the induction hypothesis and take linear condensations $\Phi : C_p(M_0) \to \Sigma(B)$ and $\Phi_\alpha : C_p(M_{\alpha+1}) \to \Sigma(B_\alpha)$ for any $\alpha < \kappa$. We can assume, without loss of generality, that the family $\{B\} \cup \{B_\alpha : \alpha < \kappa\}$ is disjoint; let $A = B \cup (\bigcup\{B_\alpha : \alpha < \kappa\})$.

For any function $f \in C_p(X)$ consider the set $\Omega(f) = \Phi \circ \pi_{M_0}(f)$ and, for any ordinal $\alpha < \kappa$, let $\Omega_\alpha(f) = \Phi_\alpha(\pi_{M_{\alpha+1}}(q_{\alpha+1}(f) - q_\alpha(f)))$. Then $\Omega : C_p(X) \to \Sigma(B)$ and $\Omega_\alpha : C_p(X) \to \Sigma(B_\alpha)$ for any $\alpha < \kappa$. Observe that $q_\alpha : C_p(X) \to L_\alpha$ is a linear map; so L_α is a linear subspace of the space $C_p(X)$ for any $\alpha < \kappa$. Therefore $q_{\alpha+1}(f) - q_\alpha(f) \in L_{\alpha+1}$ for any $\alpha < \kappa$ and $f \in C_p(X)$.

It is evident that Ω and Ω_α are continuous linear maps for any $\alpha < \kappa$; so the diagonal product $\varphi = \Omega \Delta (\Delta\{\Omega_\alpha : \alpha < \kappa\})$ maps $C_p(X)$ linearly and continuously into the space $\Sigma(B) \times (\prod\{\Sigma(B_\alpha) : \alpha < \kappa\}) \subset \mathbb{R}^A$. We claim that

(9) the map φ is an injection.

Indeed, take distinct functions $f, g \in C_p(X)$ and let $U = \{x \in X : f(x) \neq g(x)\}$. If $f|M_0 \neq g|M_0$ then $\Omega(f) \neq \Omega(g)$ because Φ is a condensation; thus $\varphi(f) \neq \varphi(g)$. If $\pi_{M_0}(f) = \pi_{M_0}(g)$ then $\xi = \min\{\alpha : U \cap M_\alpha \neq \emptyset\} > 0$. Observe that ξ has to be a successor ordinal for otherwise the set $V = U \cap M_\xi$ is non-empty and open in M_ξ; so it follows from (b) of Fact 8 that $V \cap M_\alpha \neq \emptyset$ for some $\alpha < \xi$, a contradiction.

Thus $\xi = \eta + 1$ which shows that $f|M_\eta = g|M_\eta$ while $f|M_{\eta+1} \neq g|M_{\eta+1}$. Therefore $q_\eta(f) = f \circ r_\eta = \pi_{M_\eta}(f) \circ r_\eta = \pi_{M_\eta}(g) \circ r_\eta = g \circ r_\eta = q_\eta(g)$. Fix a point $x \in U \cap M_{\eta+1}$. We have $f(r_{\eta+1}(x)) = f(x) \neq g(x) = g(r_{\eta+1}(x))$

which shows that $f \circ r_{\eta+1} \neq g \circ r_{\eta+1}$, i.e., $q_{\eta+1}(f) \neq q_{\eta+1}(g)$. As an immediate consequence, $q_{\eta+1}(f) - q_\eta(f) \neq q_{\eta+1}(g) - q_\eta(g)$ and hence $\Omega_\eta(f) \neq \Omega_\eta(g)$ because the maps $\pi_{M_{\eta+1}}|L_{\eta+1}$ and Φ_η are injective. This implies $\varphi(f) \neq \varphi(g)$ and concludes the proof of (9).

We will finally show that φ actually maps $C_p(X)$ in $\Sigma(A)$, i.e., $\varphi(f) \in \Sigma(A)$ for any $f \in C_p(X)$. Let $u_\alpha \in \mathbb{R}^{B_\alpha}$ be the function which is identically zero on B_α for any $\alpha < \kappa$. The property (8) shows that, for any $\alpha \in \kappa \setminus E_f$ we have $q_{\alpha+1}(f) = q_\alpha(f)$ and hence $\Omega_\alpha(f) = u_\alpha$ because $\Phi_\alpha \circ \pi_{M_{\alpha+1}}$ is a linear map. An immediate consequence is that the set $H_f = \{a \in A : \varphi(f)(a) \neq 0\}$ is contained in $B \cup (\bigcup\{B_\alpha : \alpha \in E_f\})$. Recalling that $\Omega(C_p(X)) \subset \Sigma(B)$ and $\Omega_\alpha(C_p(X)) \subset \Sigma(B_\alpha)$ for any ordinal $\alpha \in E_f$ we conclude that H_f is countable and hence φ is a linear condensation of $C_p(X)$ into $\Sigma(A)$. Fact 9 is proved.

Returning to our solution, suppose that X is an arbitrary space such that $C_p(X)$ is Lindelöf Σ. If $Y = \upsilon X$ then Y is Lindelöf Σ (see Problem 206) and $C_p(Y)$ is a Lindelöf Σ-space as well (see Problem 234). Therefore $C_p(C_p(Y))$ is a Lindelöf Σ-space by Okunev's theorem (see Problem 218); so we can apply Fact 9 to the space $Z = C_p(Y)$ to conclude that the space $C_p(Z) = C_p(C_p(Y))$ condenses into a Σ-product of real lines. Since Y embeds in $C_p(C_p(Y))$ by TFS-167, the space Y can be condensed into $\Sigma(A)$ for some A. The space X being a subspace of Y we conclude that X also condenses into a Σ-product of real lines. Finally, if X is a Gul'ko compact space then it condenses into some $\Sigma(A)$; this condensation has to be a homeomorphism; so X is Corson compact and hence our solution is complete.

U.286. *Prove that if X is Corson compact then the space $C_p(X)$ condenses linearly into a Σ_*-product of real lines. As a consequence, for any Gul'ko compact X the space $C_p(X)$ condenses linearly into a Σ_*-product of real lines.*

Solution. If Z is a compact space then let $\|f\| = \sup\{|f(z)| : z \in Z\}$ for any function $f \in C_p(Z)$. For any set A we can also define a norm in the space $\Sigma_*(A)$ letting $\|x\| = \sup\{|x(a)| : a \in A\}$ for any $x \in \Sigma_*(A)$. If a confusion can happen with norms in different spaces we will use the norm symbol with indices to make it clear in which space the norm is taken.

Fix a set T such that $X \subset \Sigma(T)$; we will prove by induction on the cardinal $d(X)$ that there exists a set B and a linear condensation $\varphi : C_p(X) \to \Sigma_*(B)$ such that $\|\varphi(f)\| \leq \|f\|$ for any $f \in C_p(X)$ (we will say that such a condensation *does not increase the norm*). If $d(X) = \omega$ then such a condensation of $C_p(X)$ into $\Sigma_*(\mathbb{N})$ exists by Fact 14 of S.351.

Now assume that $\kappa = d(X)$ is an uncountable cardinal and we proved, for any cardinal $\lambda < \kappa$ that, for any compact $G \subset \Sigma(T)$ with $d(G) \leq \lambda$, there is a set A and a linear condensation $\varphi : C_p(G) \to \Sigma_*(A)$ such that $\|\varphi(f)\| \leq \|f\|$ for any $f \in C_p(G)$. Fix a dense set $D = \{x_\alpha : \alpha < \kappa\}$ in the space X. For any $x \in \Sigma(T)$ let $\text{supp}(x) = \{t \in T : x(t) \neq 0\}$; then $\text{supp}(x)$ is a countable set for any $x \in \Sigma(T)$. Given a set $S \subset T$ and $x \in \Sigma(T)$ let $r_S(x)(t) = x(t)$ if $t \in S$ and $r_S(x)(t) = 0$ for all $t \in T \setminus S$. Then $r_S(x) \in \Sigma(T)$ for any $x \in \Sigma(T)$ and the map $r_S : \Sigma(T) \to \Sigma(T)$ is a retraction for any $S \subset T$. We will also need the family $\mathcal{F} = \{S \subset T : r_S(X) \subset X\}$.

It follows from Problem 152 that there exists a countably infinite set $E_0 \in \mathcal{F}$. Assume that $\alpha < \kappa$ and we have a family $\{E_\beta : \beta < \alpha\} \subset \mathcal{F}$ with the following properties:

(1) $\gamma < \beta < \alpha$ implies $E_\gamma \subset E_\beta$ and $\text{supp}(x_\gamma) \subset E_\beta$;
(2) $|E_\beta| \leq |\beta| \cdot \omega$ for any $\beta < \alpha$;
(3) if $\beta < \alpha$ is a limit ordinal then $E_\beta = \bigcup\{E_\gamma : \gamma < \beta\}$.

If α is a limit ordinal then let $E_\alpha = \bigcup_{\beta < \alpha} E_\beta$; then $E_\alpha \in \mathcal{F}$ by Problem 152 and it is evident that the properties (1)–(3) still hold for all $\beta \leq \alpha$. If $\alpha = \alpha_0 + 1$ then it is a consequence of Problem 152 that there exists a set $E_\alpha \in \mathcal{F}$ such that $E_{\alpha_0} \cup \text{supp}(x_{\alpha_0}) \subset E_\alpha$ and $|E_\alpha| \leq |E_{\alpha_0} \cup \text{supp}(x_{\alpha_0})| \cdot \omega \leq |\alpha| \cdot \omega$. It is also clear that (1)–(3) hold for all $\beta \leq \alpha$; so our inductive procedure can be continued to construct a family $\{E_\alpha : \alpha < \kappa\} \subset \mathcal{F}$ with (1),(2) and (3) fulfilled for any $\beta < \kappa$.

It is evident that every map $r_\alpha = r_{E_\alpha}|X$ is a retraction on X. We claim that

(4) the set $\Omega(f, \varepsilon) = \{\alpha < \kappa : \text{there is } x \in X \text{ such that } |f(r_\alpha(x)) - f(r_{\alpha+1}(x))| \geq \varepsilon\}$ is finite for any $\varepsilon > 0$ and $f \in C_p(X)$.

To see that (4) is true suppose not; then there exists a function $f \in C_p(X)$ and a number $\varepsilon > 0$ such that the set $\Omega(f, \varepsilon)$ is infinite. Choose a point $z_\alpha \in X$ such that $|f(r_\alpha(z_\alpha)) - f(r_{\alpha+1}(z_\alpha))| \geq \varepsilon$ for any $\alpha \in \Omega(f, \varepsilon)$. By compactness of X, the set $\{r_\alpha(z_\alpha) : \alpha \in \Omega(f, \varepsilon)\}$ has an accumulation point $z \in X$.

The function f being continuous at z, there is a set $S = \{t_1, \ldots, t_n\} \subset T$ and $O_1, \ldots, O_n \in \tau(\mathbb{R})$ such that $\text{diam}(f(U)) < \varepsilon$ where $z \in U = \{x \in X : x(t_i) \in O_i \text{ for all } i \leq n\}$. Therefore the set $M = \{\alpha \in \Omega(f, \varepsilon) : r_\alpha(z_\alpha) \in U\}$ is infinite; the family $\{E_{\alpha+1} \setminus E_\alpha : \alpha \in M\}$ is disjoint; so there is $\alpha \in M$ such that $(E_{\alpha+1} \setminus E_\alpha) \cap S = \emptyset$.

Now observe that $r_{\alpha+1}(z_\alpha) = r_{E_{\alpha+1}}(z_\alpha)$ can have coordinates distinct from the coordinates of the point $r_\alpha(z_\alpha) = r_{E_\alpha}(z_\alpha)$ only on the set $E_{\alpha+1} \setminus E_\alpha$. Thus $r_{\alpha+1}(z_\alpha)(t_i) = r_\alpha(z_\alpha)(t_i) \in O_i$ for every $i \leq n$ which implies that $r_{\alpha+1}(z_\alpha) \in U$. However, then $\varepsilon \leq |f(r_\alpha(z_\alpha)) - f(r_{\alpha+1}(z_\alpha))| \leq \text{diam}(f(U)) < \varepsilon$; this contradiction shows that (4) is proved.

For any $\alpha < \kappa$ let $X_\alpha = r_\alpha(X)$; it is easy to see that the family $\{X_\alpha : \alpha < \kappa\}$ is non-decreasing and $d(X_\alpha) = w(X_\alpha) \leq |E_\alpha| < \kappa$ for every $\alpha < \kappa$ (see (2)). We will need the following property of the family $\{X_\alpha : \alpha < \kappa\}$:

(5) if $\alpha < \kappa$ is a non-zero limit ordinal then $H_\alpha = \bigcup\{X_\beta : \beta < \alpha\}$ is dense in X_α.

If (5) is not true then we can choose a point $y \in X_\alpha \setminus \overline{H}_\alpha$. There is a set $S' = \{s_1, \ldots, s_m\} \subset T$ and $W_1, \ldots, W_m \in \tau(\mathbb{R})$ such that

$$y \in W = \{x \in X : x(s_i) \in W_i \text{ for any } i \leq m\} \subset X \setminus H_\alpha.$$

It follows from the property (3) that there is $\beta < \alpha$ such that $E_\beta \cap S' = E_\alpha \cap S'$. We have $r_\beta(y) \in X_\beta$ while the distinct coordinates of y and $r_\beta(y)$ lie in $E_\alpha \setminus E_\beta$; by our choice of β, we have $(E_\alpha \setminus E_\beta) \cap S' = \emptyset$ and therefore $r_\beta(y)(s_i) = y(s_i) \in W_i$

for any $i \leq m$, i.e., $r_\beta(y) \in W \cap X_\beta$ which is a contradiction with $X_\beta \subset H_\alpha$ and $W \cap H_\alpha = \emptyset$. This proves the property (5).

Denote by π_α the restriction map from $C_p(X)$ onto $C_p(X_\alpha)$; observe that we have the inequality $||\pi_\alpha(f)|| \leq ||f||$ for any $f \in C_p(X)$ and $\alpha < \kappa$. Our induction hypothesis shows that, for any $\alpha < \kappa$, there exists a set B_α and a linear injective map $\delta_\alpha : C_p(X_\alpha) \to \Sigma(B_\alpha)$ such that $||\delta(f)||_\alpha \leq ||f||$ for any $f \in C_p(X_\alpha)$ (here $|| \cdot ||_\alpha$ is the norm in $\Sigma_*(B_\alpha)$).

We can assume, without loss of generality, that the family $\{B_\alpha : \alpha < \kappa\}$ is disjoint. Let $\mu_0(f) = \delta_0(\pi_0(f))$ and $\mu_{\alpha+1}(f) = \frac{1}{2}\delta_{\alpha+1}(\pi_{\alpha+1}(f \circ r_{\alpha+1} - f \circ r_\alpha))$ for any ordinal $\alpha < \kappa$ and $f \in C_p(X)$. The maps μ_0 and $\mu_{\alpha+1} : C_p(X) \to \Sigma(B_{\alpha+1})$ are continuous and linear for any $\alpha < \kappa$. It is evident that $||\mu_0(f))||_0 \leq ||f||$ for any $f \in C_p(X)$; given $\alpha < \kappa$ and $x \in X$ we have

$$|f(r_{\alpha+1}(x)) - f(r_\alpha(x))| \leq |f(r_{\alpha+1}(x))| + |f(r_\alpha(x))| \leq 2||f||$$

so $||f \circ r_{\alpha+1} - f \circ r_\alpha|| \leq 2||f||$. Since $\pi_{\alpha+1}$ and $\delta_{\alpha+1}$ do not increase the norm, we have $||\mu_{\alpha+1}(f)||_{\alpha+1} \leq ||f||$ for any $f \in C_p(X)$. This, together with the property (4) implies that

(6) for any $\varepsilon > 0$ if $\alpha \notin \Omega(f, \varepsilon)$ then $|\mu_{\alpha+1}(f)(b)| < \varepsilon$ for any $b \in B_{\alpha+1}$.

We are finally ready to construct the promised linear condensation of the space $C_p(X)$; let $B = B_0 \cup \bigcup\{B_{\alpha+1} : \alpha < \kappa\}$ and $\mu = \mu_0\Delta(\Delta\{\mu_{\alpha+1} : \alpha < \kappa\})$; it is immediate that $\mu : C_p(X) \to \prod\{\Sigma(B_{\alpha+1}) : \alpha < \kappa\} \times \Sigma(B_0) \subset \mathbb{R}^B$. The diagonal product of linear continuous maps is, evidently, a linear continuous map; so μ is linear and continuous. Furthermore, if $b \in B$ then there is $\alpha < \kappa$ such that $b \in B_\alpha$ and hence $|\mu(f)(b)| \leq ||\mu_\alpha(f)||_\alpha \leq ||f||$ which shows that

(7) $\sup\{|\mu(f)(b)| : b \in B\} \leq ||f||$ for any $f \in C_p(X)$.

To see that the map μ is injective take distinct functions $f, g \in C_p(X)$. Since $W = \{x \in X : f(x) \neq g(x)\} \in \tau^*(X)$ and D is dense in X, there is $\alpha < \kappa$ such that $f(x_\alpha) \neq g(x_\alpha)$. We have $\text{supp}(x_\alpha) \subset E_{\alpha+1}$ which implies $r_{\alpha+1}(x_\alpha) = x_\alpha \in X_{\alpha+1}$ and therefore $\pi_{\alpha+1}(f) \neq \pi_{\alpha+1}(g)$. Therefore the set $J = \{\alpha < \kappa : \pi_\alpha(f) \neq \pi_\alpha(g)\}$ is non-empty; let $\alpha = \min J$. If $\alpha = 0$ then $\pi_0(f) \neq \pi_0(g)$ which implies $\mu_0(f) \neq \mu_0(g)$ because δ_0 is an injection.

Now assume that $\alpha > 0$ is a limit ordinal. Then we have $f|X_\alpha \neq g|X_\alpha$ and hence $f|H_\alpha \neq g|H_\alpha$ by (5). Therefore there is $\beta < \alpha$ for which $f(q) \neq g(q)$ for some $q \in X_\beta$, i.e., $f|X_\beta \neq g|X_\beta$ which is a contradiction with the choice of α. Therefore $\alpha > 0$ cannot be a limit and hence $\alpha = \beta + 1$ for some $\beta < \kappa$. By the choice of α we have $\pi_\beta(f) = \pi_\beta(g)$ while $\pi_{\beta+1}(f) \neq \pi_{\beta+1}(g)$ and therefore $f \circ r_\beta = g \circ r_\beta$ while $f \circ r_{\beta+1} \neq g \circ r_{\beta+1}$. There exists $z \in X$ such that $f(r_{\beta+1}(z)) \neq g(r_{\beta+1}(z))$ so, for the point $y = r_{\beta+1}(z) \in X_{\beta+1}$ we have $f(r_{\beta+1}(y)) = f(y) \neq g(y) = g(r_{\beta+1}(y))$. This shows that $\pi_{\beta+1}(f \circ r_{\beta+1}) \neq \pi_{\beta+1}(g \circ r_{\beta+1})$; an immediate consequence is that $\pi_{\beta+1}(f \circ r_{\beta+1} - f \circ r_\beta) \neq \pi_{\beta+1}(g \circ r_{\beta+1} - g \circ r_\beta)$ whence $\mu_{\beta+1}(f) \neq \mu_{\beta+1}(g)$ because $\delta_{\beta+1}$ is injective. Thus $\mu(f) \neq \mu(g)$ and hence we proved that μ is injective.

The last thing we have to show is that $\mu(C_p(X)) \subset \Sigma_*(B)$ so take any function $f \in C_p(X)$ and $\varepsilon > 0$. The property (6) implies that $|\mu(f)(b)| = |\mu_{\alpha+1}(f)(b)| < \varepsilon$ whenever $\alpha \notin \Omega(f,\varepsilon)$ and $b \in B_{\alpha+1}$. The set $\Omega(f,\varepsilon)$ is finite by (4); if $\alpha \in \Omega(f,\varepsilon)$ then $\mu_{\alpha+1}(f) \in \Sigma_*(B_{\alpha+1})$ by the definition of the mapping $\delta_{\alpha+1}$. Therefore the set $\{b \in B : |\mu(f)(b)| \geq \varepsilon\}$ is contained in the set

$$\bigcup\{\{b \in B_{\alpha+1} : |\mu_{\alpha+1}(f)(b)| \geq \varepsilon\} : \alpha \in \Omega(f,\varepsilon)\} \cup \{b \in B_0 : |\mu_0(f)(b)| \geq \varepsilon\}$$

which is finite; so $\mu(f) \in \Sigma_*(B)$ for any $f \in C_p(X)$ and therefore μ is a linear continuous injective map from $C_p(X)$ to $\Sigma_*(B)$.

The property (7) implies that the condensation μ does not increase the norm; so our induction step is accomplished and hence we proved that, for any Corson compact X the space $C_p(X)$ condenses into a Σ_*-product of real lines. Finally, observe that if X is Gul'ko compact then it is Corson compact by Problem 285; so $C_p(X)$ can still be condensed into a Σ_*-product of real lines.

U.287. *Let X be a Corson compact space. Prove that, if $p(C_p(X)) = \omega$ then X is metrizable. Therefore if X is a Gul'ko compact space and $p(C_p(X)) = \omega$ then X is metrizable.*

Solution. By Problem 286 we can find a set A such that there exists a condensation $\varphi : C_p(X) \to Y$ for some $Y \subset \Sigma_*(A)$. If A is countable then $w(Y) \leq w(\Sigma_*(A)) = \omega$ which shows that $d(X) = iw(C_p(X)) \leq \omega$; so X is metrizable by Problem 121.

If A is uncountable then we can consider that there is an uncountable cardinal κ such that $Y \subset C_p(A(\kappa))$ (see Problem 105). Let $\pi : C_p(C_p(A(\kappa))) \to C_p(Y)$ be the restriction map; there exists a subspace $F \subset C_p(C_p(A(\kappa)))$ such that $F \simeq A(\kappa)$ and F separates the points of $C_p(A(\kappa))$ (see TFS-166 and TFS-167). Consequently, the set $G = \pi(F) \subset C_p(Y)$ separates the points of the space Y. If G is countable then $iw(Y) \leq \omega$ (see TFS-166); thus $d(X) = iw(C_p(X)) \leq \omega$; so X is metrizable by Problem 121.

It is easy to see that any continuous image of $A(\kappa)$ is either finite or homeomorphic to the one-point compactification of a discrete space so, if $|G| > \omega$ then $G \simeq A(\lambda)$ for some uncountable cardinal λ. The dual map $\varphi^* : C_p(Y) \to C_p(C_p(X))$ embeds $C_p(Y)$ in $C_p(C_p(X))$ (see TFS-163) and therefore G embeds in $C_p(C_p(X))$. Since $G \simeq A(\lambda)$, we have $p(C_p(X)) \geq \lambda > \omega$ (see TFS-178) which is a contradiction. Therefore G cannot be uncountable and hence X is metrizable.

Finally, if X is a Gul'ko compact space with $p(C_p(X)) = \omega$ then X is Corson compact by Problem 285; so X is metrizable.

U.288. *Suppose that X and $C_p(X)$ are Lindelöf Σ-spaces and $p(C_p(X)) = \omega$. Prove that $|X| \leq \mathfrak{c}$.*

Solution. Fix a compact cover $\mathcal{C}$ of the space X such that there is a countable family $\mathcal{F} \subset \exp X$ which is a network with respect to $\mathcal{C}$. For any $C \in \mathcal{C}$ there is $\mathcal{F}_C \subset \mathcal{F}$ which is a network at C, i.e., for any $U \in \tau(C, X)$ there is $F \in \mathcal{F}_C$

such that $C \subset F \subset U$. An immediate consequence is that $\bigcap \mathcal{F}_C = C$ which proves that the correspondence $C \to \mathcal{F}_C$ is an injection from $\mathcal{C}$ in $\exp \mathcal{F}$. Therefore $|\mathcal{C}| \leq |\exp \mathcal{F}| \leq \mathfrak{c}$.

For any set $C \in \mathcal{C}$ the restriction map $\pi_C : C_p(X) \to C_p(C)$ is surjective; so $p(C_p(C)) = \omega$ (it is an easy exercise that the point-finite cellularity is not raised by continuous maps). Since C is Gul'ko compact by Problem 220, we can apply Problem 287 to see that C is metrizable. Thus $|C| \leq \mathfrak{c}$ for any $C \in \mathcal{C}$ and hence $|X| \leq |\mathcal{C}| \cdot \mathfrak{c} = \mathfrak{c}$.

U.289. *Prove that a compact space X is Gul'ko compact if and only if X has a weakly σ-point-finite T_0-separating family of cozero sets.*

Solution. Given a space Z say that $\mathcal{U} \subset \tau(Z)$ is *a Gul'ko family* if $\mathcal{U}$ is weakly σ-point-finite, T_0-separating and consists of cozero subsets of Z. In this terminology we have to prove that a compact space Z is Gul'ko compact if and only if there exists a Gul'ko family in Z.

Call a space Z *concentrated around a point* $z \in Z$ if $Z \setminus U$ is countable for any $U \in \tau(z, Z)$. If $\mathcal{U}$ is an open cover of a space Z say that a set $Y \subset Z$ is $\mathcal{U}$-compact if there is a finite $\mathcal{U}' \subset \mathcal{U}$ such that $Y \subset \bigcup \mathcal{U}'$. If Z is a space and $\mathcal{A} \subset \exp Z$ then $\mathcal{A}$ is said to be *point-finite at a point* $z \in Z$ if $|\{A \in \mathcal{A} : z \in A\}| < \omega$.

Fact 1. Suppose that a set $Z \subset \Sigma(A)$ is concentrated around a point $t \in Z$. Then there exists a sequence $\{Z_n : n \in \omega\} \subset \exp Z$ such that $Z \setminus \{t\} = \bigcup \{Z_n : n \in \omega\}$ while, for every $n \in \omega$, the set $K_n = \{t\} \cup Z_n$ is compact and all points of Z_n are isolated in K_n.

Proof. If the set Z is countable then $Z \setminus \{t\} = \{Z_n : n \in \omega\}$ where $|Z_n| \leq 1$ for any $n \in \omega$; thus every K_n is finite so the statement of our Fact is true.

Now assume that $|Z| > \omega$; there is no loss of generality to consider that $t(a) = 0$ for any $a \in A$. Since Z is concentrated around t, the set $Z \setminus H$ is countable whenever H is a G_δ-subset of Z with $t \in H$.

For any $x \in \Sigma(A)$ let $\mathrm{supp}(x) = \{a \in A : x(a) \neq 0\}$. Given a point $a \in A$ the set $H(a) = \{z \in Z \setminus \{t\} : z(a) \neq 0\}$ has to be countable because $Z \setminus H(a)$ is a G_δ-subset of the space Z such that $t \in Z \setminus H(a)$. For any $a \in A$ let $P_0(a) = H(a)$ and $P_{n+1}(a) = \bigcup \{H(a) : a \in \bigcup \{\mathrm{supp}(z) : z \in P_n(a)\}\}$ for any $n \in \omega$. It is evident that every $P_n(a)$ is countable so the set $P(a) = \bigcup \{P_n(a) : n \in \omega\}$ is countable for every $a \in A$.

Given any point $z \in Z \setminus \{t\}$, there exists an index $a \in A$ with $z(a) \neq 0$ and therefore $z \in H(a) \subset P(a)$. This shows that $\bigcup \{P(a) : a \in A\} = Z \setminus \{t\}$. Since $P(a)$ is countable, the set $Q(a) = \bigcup \{\mathrm{supp}(z) : z \in P(a)\}$ is also countable for any $a \in A$. Given $a, b \in A$ say that $a \sim b$ if there exists a finite sequence $\{z_0, \ldots, z_n\} \subset Z$ such that $a \in \mathrm{supp}(z_0)$, $b \in \mathrm{supp}(z_n)$ and $\mathrm{supp}(z_i) \cap \mathrm{supp}(z_{i+1}) \neq \emptyset$ for any $i < n$. It is straightforward that $a \sim a$ for any $a \in A$; besides, $a \sim b$ is equivalent to $b \sim a$ and $a \sim b \sim c$ implies $a \sim c$. Therefore $\sim$ is an equivalence relation; it is easy to see that $Q(a) = \{b \in A : b \sim a\}$; so we have

(1) if $a, b \in A$ and $Q(a) \cap Q(b) \neq \emptyset$ then $Q(a) = Q(b)$.

As a consequence, we can find a set $B \subset A$ such that the family $\{Q(b) : b \in B\}$ is disjoint and $\bigcup\{P(b) : b \in B\} = Z\backslash\{t\}$. Since every $P(b)$ is countable, there is a sequence $\{Z_n : n \in \omega\} \subset \exp(Z\backslash\{t\})$ such that $\bigcup\{Z_n : n \in \omega\} = Z\backslash\{t\}$ and we have $|Z_n \cap P(b)| \leq 1$ for any $b \in B$ and $n \in \omega$. Thus the family $\{\text{supp}(z) : z \in Z_n\}$ is disjoint; so every neighbourhood of t contains all but finitely many points of Z_n for any $n \in \omega$. Therefore every $K_n = Z_n \cup \{t\}$ is compact and the points of Z_n are isolated in K_n; so Fact 1 is proved.

Fact 2. A space Z has the Lindelöf Σ-property if and only if there exists a countable family $\mathcal{F} \subset \exp Z$ such that, for any open cover $\mathcal{U}$ of the space Z the collection $\mathcal{F}' = \{F \in \mathcal{F} : F \text{ is } \mathcal{U}\text{-compact}\}$ covers Z.

Proof. If Z is Lindelöf Σ then fix a compact cover $\mathcal{C}$ of the space Z such that there exists a countable network $\mathcal{F}$ with respect to $\mathcal{C}$. Given any open cover $\mathcal{U}$ of the space Z, for any $z \in Z$ there is $C \in \mathcal{C}$ with $z \in C$. Pick a finite $\mathcal{U}' \subset \mathcal{U}$ such that $C \subset \bigcup\mathcal{U}'$. Since $\mathcal{F}$ is a network at C, there is $F \in \mathcal{F}$ such that $C \subset F \subset \bigcup\mathcal{U}'$. Thus F is $\mathcal{U}$-compact and $x \in F$. This shows that the family of all $\mathcal{U}$-compact elements of $\mathcal{F}$ covers Z and hence we proved necessity.

To establish sufficiency, fix the respective countable family $\mathcal{F} \subset \exp Z$ and let $\mathcal{P} = \{\text{cl}_{\beta Z}(F) : F \in \mathcal{F}\}$. Then $\mathcal{P}$ is a countable family of closed subsets of βZ. Given any points $z \in Z$ and $t \in \beta Z\backslash Z$ observe that $\mathcal{U} = \{Z\backslash\text{cl}_{\beta Z}(U) : U \in \tau(t, \beta Z)\}$ is an open cover of the space Z. Since $\mathcal{U}$-compact elements of $\mathcal{F}$ cover Z, there is $F \in \mathcal{F}$ such that F is $\mathcal{U}$-compact and $z \in F$. Take $U_1, \ldots, U_n \in \tau(t, \beta Z)$ such that $F \subset (Z\backslash\text{cl}_{\beta Z}(U_1)) \cup \ldots \cup (Z\backslash\text{cl}_{\beta Z}(U_n))$ and hence $F \cap U = \emptyset$ where $U = U_1 \cap \ldots \cap U_n$ which shows that $z \in G = \text{cl}_{\beta Z}(F)$ and $t \notin G$.

This proves that the family $\mathcal{P}$ separates Z from $\beta Z\backslash Z$; so we can apply SFFS-233 to conclude that Z is a Lindelöf Σ-space. Fact 2 is proved.

Fact 3. If there exists a Gul'ko family in a space Z_n for any $n \in \omega$ then there is a Gul'ko family in the space $Z = \prod\{Z_n : n \in \omega\}$.

Proof. Fix a Gul'ko family $\mathcal{V}_n$ on the space Z_n and let $\{\mathcal{V}_m^n : m \in \omega\}$ be the collection of subfamilies of $\mathcal{V}_n$ which witnesses the Gul'ko property of $\mathcal{V}_n$ for any $n \in \omega$. We will need the natural projection $p_n : Z \to Z_n$; let $\mathcal{U}_n = \{p_n^{-1}(V) : V \in \mathcal{V}_n\}$ for any $n \in \omega$.

The family $\mathcal{U} = \bigcup\{\mathcal{U}_n : n \in \omega\}$ consists of cozero subsets of Z (see Fact 1 of T.252). If y, z are distinct elements of Z then there is $n \in \omega$ with $p_n(y) \neq p_n(z)$. The family $\mathcal{V}_n$ is T_0-separating in Z_n; so there is $V \in \mathcal{V}_n$ such that $V \cap \{p_n(y), p_n(z)\}$ is a singleton. It is immediate that $U = p_n^{-1}(V) \in \mathcal{U}$ and $U \cap \{y, z\}$ is a singleton; so $\mathcal{U}$ is a T_0-separating family in Z.

To see that $\mathcal{U}$ is a Gul'ko family, we must prove that it is weakly σ-point-finite. To do so consider the family $\mathcal{U}_m^n = \{p_n^{-1}(V) : V \in \mathcal{V}_m^n\}$ for any $m, n \in \omega$. Given a point $z \in Z$ and $n \in \omega$ let $A_n = \{m \in \omega : \mathcal{V}_m^n \text{ is point-finite at } z(n)\}$. It is evident that the family $\mathcal{U}_m^n$ is point-finite at z for any $n \in \omega$ and $m \in A_n$. Since every $\mathcal{V}_n$ is a Gul'ko family on Z_n, we have $\bigcup\{\mathcal{V}_m^n : m \in A_n\} = \mathcal{V}_n$ for any $n \in \omega$ and therefore

$\bigcup\{\mathcal{U}_m^n : \mathcal{U}_m^n$ is point-finite at $z\} \supset \bigcup\{\mathcal{U}_m^n : n \in \omega,\ m \in A_n\} = \bigcup\{\mathcal{U}_n : n \in \omega\} = \mathcal{U}$. Thus the countable collection $\{\mathcal{U}_m^n : m, n \in \omega\}$ of subfamilies of $\mathcal{U}$ guarantees that $\mathcal{U}$ is weakly σ-point-finite. Consequently, $\mathcal{U}$ is a Gul'ko family on Z and hence Fact 3 is proved.

Returning to our solution assume that X is Gul'ko compact. We can consider that $X \subset \Sigma(A)$ for some set A by Problem 285. As usual, $\mathrm{supp}(x) = \{a \in A : x(a) \neq 0\}$ for any $x \in \Sigma(A)$. Let u be the function on X with $u(x) = 0$ for all $x \in X$. For any $a \in A$ let $p_a : \Sigma(A) \to \mathbb{R}$ be the natural projection onto the factor determined by a; then $u_a = p_a|X \in C_p(X)$. It is evident that the set $T = \{u\} \cup \{u_a : a \in A\} \subset C_p(X)$ separates the points of X.

Observe that T is concentrated around the point u. Indeed, if $U \in \tau(u, T)$ then there are $x_1, \ldots, x_n \in X$ and $\varepsilon > 0$ such that $\{t \in T : |t(x_i)| < \varepsilon$ for all $i \leq n\} \subset U$. Since every $\mathrm{supp}(x_i)$ is countable, the set $B = \mathrm{supp}(x_1) \cup \ldots \cup \mathrm{supp}(x_n)$ is countable as well and it is immediate that $u_a \in U$ for any $a \in A \backslash B$. Thus $T \backslash U \subset \{u_a : a \in B\}$ is a countable set.

Apply Problem 286 to find a condensation $\varphi : C_p(X) \to Y$ where $Y \subset \Sigma(S)$ for some set S. The set $\varphi(T)$ is concentrated around the point $\varphi(u)$; so we can apply Fact 1 to conclude that there exists a sequence $\{T_n : n \in \omega\} \subset \exp T \backslash \{u\}$ such that $\bigcup\{T_n : n \in \omega\} = T \backslash \{u\}$ and $K_n = \varphi(T_n) \cup \{\varphi(u)\}$ is a compact subspace of $\Sigma(S)$ such that all points of $\varphi(T_n)$ are isolated in K_n. Therefore $T_n \cup \{u\} = \varphi^{-1}(K_n)$ is closed in $C_p(X)$ and all points of T_n are isolated in $H_n = T_n \cup \{u\}$. The space $H = \bigoplus_{n \in \omega} H_n$ maps continuously onto $T = \bigcup_{n \in \omega} H_n$; so $C_p(T)$ embeds in $C_p(H)$. Since T separates the points of X, the space X condenses and hence embeds in $C_p(T)$ (see TFS-166). Consequently, X embeds in $C_p(H) = \prod_{n \in \omega} C_p(H_n)$ as well.

If H_n is countable then $C_p(H_n)$ has a countable base which is, evidently, a Gul'ko family in $C_p(H_n)$. If H_n is uncountable then u is the unique non-isolated point of H_n. The space H_n is Lindelöf Σ being closed in $C_p(X)$; so we can apply Problem 274 to conclude that there exists a Gul'ko family in $C_p(H_n)$. This proves that every $C_p(H_n)$ has a Gul'ko family and hence so has $C_p(H)$ by Fact 3. Having a Gul'ko family is a hereditary property; so X has a Gul'ko family and hence we established necessity.

Now assume that X is compact and $\mathcal{U} \subset \tau^*(X)$ is a Gul'ko family in X; choose a sequence $\{\mathcal{U}_n : n \in \omega\}$ of subfamilies of $\mathcal{U}$ which witnesses the Gul'ko property of $\mathcal{U}$. Note first that $\mathcal{U}$ is point-countable because, for any $x \in X$ the family $\{U \in \mathcal{U} : x \in U\}$ is contained in the collection $\bigcup\{\{U \in \mathcal{U}_n : x \in U\} : n \in \omega$ and $\mathcal{U}_n$ is point-finite at $x\}$ which is countable. Thus X is Corson compact by Problem 118.

For any $U \in \mathcal{U}$ fix a function $f_U \in C_p(X, [0, 1])$ such that $X \backslash U = f^{-1}(0)$. Let $u(x) = 0$ for all $x \in X$; it is clear that the space $T = \{u\} \cup \{f_U : U \in \mathcal{U}\}$ separates the points of X. Observe first that the set T is concentrated around the point u. Indeed, if $O \in \tau(u, T)$ then there are points $x_1, \ldots, x_n \in X$ and a number $\varepsilon > 0$ such that $O' = \{t \in T : |t(x_i)| < \varepsilon\} \subset O$. The family $\mathcal{U}$ being point-countable there is a countable $\mathcal{V} \subset \mathcal{U}$ such that $U \cap \{x_1, \ldots, x_n\} = \emptyset$ for any $U \in \mathcal{U} \backslash \mathcal{V}$. It is immediate that $f_U \in O' \subset O$ for any $U \in \mathcal{U} \backslash \mathcal{V}$ and hence $T \backslash O \subset \{f_U : U \in \mathcal{V}\}$ is countable.

Apply Problem 286 again to find a condensation $\varphi : C_p(X) \to Y$ where $Y \subset \Sigma(S)$ for some set S. The set $\varphi(T)$ is concentrated around the point $\varphi(u)$; so we can apply Fact 1 to conclude that there exists a sequence $\{T_n : n \in \omega\} \subset \exp T \backslash \{u\}$ such that $\bigcup \{T_n : n \in \omega\} = T \backslash \{u\}$ and $K_n = \varphi(T_n) \cup \{\varphi(u)\}$ is a compact subspace of $\Sigma(S)$ such that all points of $\varphi(T_n)$ are isolated in K_n. Therefore $T_n \cup \{u\} = \varphi^{-1}(K_n)$ is closed in $C_p(X)$ and all points of T_n are isolated in $H_n = T_n \cup \{u\}$.

We claim that every H_n is a Lindelöf Σ-space; this is evident if H_n is countable; so we can assume, without loss of generality, that $|T_n| > \omega$. Take a family $\mathcal{U}_n \subset \mathcal{U}$ such that $T_n = \{f_U : U \in \mathcal{U}_n\}$. It is clear that $\mathcal{U}_n$ is weakly σ-point-finite; so fix a collection $\{\mathcal{U}_n^m : m \in \omega\}$ witnessing this. If $P_m = \{f_U : U \in \mathcal{U}_n^m\} \cup \{u\}$ for any $m \in \omega$ then the family $\mathcal{P} = \{P_m : m \in \omega\} \subset \exp(H_n)$ is countable; let $\mathcal{Q}$ be the family of all finite intersections of the family $\mathcal{P}$.

Take any open cover $\mathcal{E}$ of the space H_n and fix $O \in \mathcal{E}$ with $u \in O$; there are $x_1, \ldots, x_k \in X$ and $\varepsilon > 0$ such that $O' = \{t \in H_n : |t(x_i)| < \varepsilon \text{ for all } i \leq k\} \subset O$. Fix any $t \in T_n$ and $U \in \mathcal{U}_n$ with $t = f_U$. The family $\mathcal{U}_n$ being weakly σ-point-finite, for any $i \leq k$ there is $m_i \in \omega$ such that the family $\mathcal{U}_n^{m_i}$ is point-finite at x_i and $U \in \mathcal{U}_n^{m_i}$. If $\mathcal{V} = \bigcap \{\mathcal{U}_n^{m_i} : i \leq k\}$ then $U \in \mathcal{V}$ and only finitely many elements of $\mathcal{V}$ meet the set $\{x_1, \ldots, x_k\}$.

Therefore the set $Q = \{f_V : V \in \mathcal{V}\} \cup \{u\} = P_n^{m_1} \cap \ldots \cap P_n^{m_k}$ belongs to $\mathcal{Q}$ and $t \in Q$. Furthermore, $Q \backslash O \subset Q \backslash O'$ is finite which shows that Q is $\mathcal{E}$-compact. We also have $\{u, t\} \subset Q$ and hence for any open cover $\mathcal{E}$ of the space H_n, the family of $\mathcal{E}$-compact elements of $\mathcal{Q}$ covers H_n. Thus H_n is a Lindelöf Σ-space for any $n \in \omega$ by Fact 2. As a consequence, $T = \bigcup_{n \in \omega} H_n$ is also Lindelöf Σ (see SFFS-257). Therefore we found a Lindelöf Σ-space $T \subset C_p(X)$ that separates the points of X. This implies that $C_p(X)$ is a Lindelöf Σ-space by Problem 020 and finishes the proof of sufficiency making our solution complete.

U.290. *Prove that a compact X is Gul'ko compact if and only if there exists a set A such that X embeds into $\Sigma_s(A)$ for some family $s = \{A_n : n \in \omega\}$ of subsets of A with $\bigcup s = A$.*

Solution. Given an arbitrary space Z say that $\mathcal{U} \subset \tau(Z)$ *a Gul'ko family* in Z if $\mathcal{U}$ is weakly σ-point-finite, T_0-separating and consists of cozero subsets of Z. A family $\mathcal{A} \subset \exp Z$ is *point-finite at* $z \in Z$ if the family $\{A \in \mathcal{A} : z \in A\}$ is finite.

Fact 1. Suppose that A is a set and we have a family $s = \{A_n : n \in \omega\} \subset \exp A$ such that $\bigcup s = A$. Then there exists a Gul'ko family in the space $\Sigma_s(A)$.

Proof. Given a rational number $r \neq 0$ let $O_r = (r, +\infty)$ if $r > 0$ and $O_r = (-\infty, r)$ if $r < 0$. Let $U_r^a = \{x \in \Sigma_s(A) : x(a) \in O_r\}$ for any $r \in \mathbb{Q} \backslash \{0\}$ and $a \in A$. Every U_r^a is a cozero set in $\Sigma_s(A)$ being the inverse image of O_r under the projection on the factor determined by a (see Fact 1 of T.252); so the family $\mathcal{U}_r = \{U_r^a : a \in A\}$ consists of cozero sets for any $r \in \mathbb{Q} \backslash \{0\}$. If x and y are distinct points of $\Sigma_s(A)$ then there is $a \in A$ such that $x(a) \neq y(a)$. There is no loss of generality to assume that $x(a) < y(a)$. Pick a number $r \in \mathbb{Q} \backslash \{0\}$ with $x(a) < r < y(a)$; it is immediate that $O_r \cap \{x(a), y(a)\}$ is a singleton and therefore $U_r^a \cap \{x, y\}$ is a singleton as well.

Thus the family $\mathcal{U} = \bigcup\{\mathcal{U}_r : r \in \mathbb{Q}\backslash\{0\}\}$ is T_0-separating in $\Sigma_s(A)$; so it suffices to show that $\mathcal{U}$ is weakly σ-point-finite. It is an easy exercise that a countable union of weakly σ-point-finite families is a weakly σ-point-finite family; so it is sufficient to prove that $\mathcal{U}_r$ is weakly σ-point-finite for any $r \in \mathbb{Q}\backslash\{0\}$.

To do so let $\mathcal{V}_n = \{U_r^a : a \in A_n\} \subset \mathcal{U}_r$ for any $n \in \omega$. Given any $x \in \Sigma_s(A)$ let $N_x = \{n \in \omega : \text{supp}(x) \cap A_n \text{ is finite}\}$. Then $A = \bigcup\{A_n : n \in N_x\}$ and hence $\mathcal{U}_r = \bigcup\{\mathcal{V}_n : n \in N_x\}$. Furthermore, if $n \in N_x$ and $x \in U_r^a$ for some $a \in A_n$ then $x(a) \in O_r$ and hence $a \in \text{supp}(x) \cap A_n$ which shows that

$$\{V \in \mathcal{V}_n : x \in V\} \subset \{U_r^a : a \in \text{supp}(x) \cap A_n\}$$

and therefore $\mathcal{V}_n$ is point-finite at x for every $n \in N_x$. As a consequence, we have $\mathcal{U}_r = \bigcup\{\mathcal{V}_n : n \in N_x\} \subset \bigcup\{\mathcal{V}_n : \mathcal{V}_n \text{ is point-finite at } x\}$ which proves that every $\mathcal{U}_r$ is weakly σ-point-finite. Thus $\mathcal{U}$ is a Gul'ko family in $\Sigma_s(A)$ and hence Fact 1 is proved.

Fact 2. If Z is a space and $\mathcal{U}$ is a weakly σ-point-finite family of subsets of Z then $\mathcal{U}$ is point-countable.

Proof. Fix a sequence $\{\mathcal{U}_n : n \in \omega\}$ of subfamilies of $\mathcal{U}$ which witnesses that $\mathcal{U}$ is weakly σ-point-finite. For any $z \in Z$ let $N_z = \{n \in \omega : \mathcal{U}_n \text{ is point-finite at } z\}$. We have $\mathcal{U} = \bigcup\{\mathcal{U}_n : n \in N_z\}$; so the family $\{U \in \mathcal{U} : z \in U\}$ is contained in the countable family $\bigcup\{\{U \in \mathcal{U}_n : z \in U\} : n \in N_z\}$ and therefore every $z \in Z$ belongs to at most countably many elements of $\mathcal{U}$, i.e., $\mathcal{U}$ is point-countable. Fact 2 is proved.

Returning to our solution assume that X is Gul'ko compact. Then there exists a Gul'ko family $\mathcal{U}$ in X by Problem 289. For any $U \in \mathcal{U}$ fix a function $f_U \in C_p(X, [0, 1])$ such that $X\backslash U = f^{-1}(0)$. Then $A = \{f_U : U \in \mathcal{U}\} \subset C_p(X)$ separates the points of X. Let $e_x(f) = f(x)$ for any $x \in X$ and $f \in A$. Then $e_x \in C_p(A)$ and the map $e : X \to C_p(A)$ defined by $e(x) = e_x$ for any $x \in X$ is continuous; since A separates the points of X, the map e is an embedding (see TFS-166); so the space $Y = \{e_x : x \in X\} \subset \mathbb{R}^A$ is homeomorphic to X.

There is a sequence $\{\mathcal{U}_n : n \in \omega\}$ of subfamilies of $\mathcal{U}$ which witnesses the Gul'ko property of $\mathcal{U}$; let $A_n = \{f_U : U \in \mathcal{U}_n\}$ for any $n \in \omega$. Then $A = \bigcup_{n \in \omega} A_n$; for the family $s = \{A_n : n \in \omega\}$ we will prove that $Y \subset \Sigma_s(A)$.

Observe first that $e_x(f_U) = f_U(x) \neq 0$ if and only if $x \in U$; besides, the family $\mathcal{U}(x) = \{U \in \mathcal{U} : x \in U\}$ is countable by Fact 2 and $e_x(f_U) = 0$ for any $U \in \mathcal{U}\backslash\mathcal{U}(x)$. This shows that $e_x(f) \neq 0$ for at most countably many $f \in A$; so $Y \subset \Sigma(A)$.

Now take any $x \in X$ and let $N_x = \{n \in \omega : \mathcal{U}_n \text{ is point-finite at } x\}$. If $n \in N_x$ then $\text{supp}(e_x) = \{f \in A : e_x(f) \neq 0\} = \{f_U : x \in U\}$; since $\mathcal{U}_n$ is point-finite at x, the set $\text{supp}(e_x) \cap A_n = \{f_U : U \in \mathcal{U}_n \text{ and } x \in U\}$ is finite for any $n \in N_x$. Thus $\bigcup\{A_n : \text{supp}(e_x) \cap A_n \text{ is finite}\} \supset \bigcup\{A_n : n \in N_x\} = A$ for any $x \in X$ and therefore $Y \subset \Sigma_s(A)$. Thus every Gul'ko compact X is homeomorphic to a space $Y \subset \Sigma_s(A)$ for some set A and a sequence $s = \{A_n : n \in \omega\}$ of subsets of A, i.e., we proved necessity.

Finally, if X is a compact space and $X \subset \Sigma_s(A)$ for some set A and a sequence $s = \{A_n : n \in \omega\}$ of subsets of A then X has a Gul'ko family because so does $\Sigma_s(A)$ by Fact 1 (it is an easy exercise that having a Gul'ko family is a hereditary property). Therefore X is a Gul'ko compact by Problem 289; so we checked sufficiency and hence our solution is complete.

U.291. *Suppose that X is a space, $n \in \mathbb{N}$ and a non-empty family $\mathcal{U} \subset \tau^*(X)$ has order $\leq n$, i.e., every $x \in X$ belongs to at most n elements of the family $\mathcal{U}$. Prove that there exist disjoint families $\mathcal{V}_1, \ldots, \mathcal{V}_n$ of non-empty open subsets of X such that $\mathcal{V} = \bigcup\{\mathcal{V}_i : i \leq n\}$ is a π-base for $\mathcal{U}$.*

Solution. If we have a family $\mathcal{V} \subset \exp X$ then $\mathcal{V}(x) = \{V \in \mathcal{V} : x \in V\}$ and $\mathrm{ord}(x, \mathcal{V}) = |\mathcal{V}(x)|$ for any $x \in X$; let $\mathrm{ord}(\mathcal{V}) = \sup\{\mathrm{ord}(x, \mathcal{V}) : x \in X\}$.

Our solution will be done by induction on $n = \mathrm{ord}(\mathcal{U})$; if $n = 1$ then $\mathcal{U}$ is disjoint; so we can take $\mathcal{V}_1 = \mathcal{U}$. Now assume that we proved our statement for all $n \leq k$ and take a non-empty $\mathcal{U} \subset \tau^*(X)$ with $\mathrm{ord}(\mathcal{U}) \leq k + 1$. For any point x of the set $O = \{x \in X : \mathrm{ord}(x, \mathcal{U}) = k + 1\}$ let $\{U_1^x, \ldots, U_{k+1}^x\}$ be an enumeration of the family $\mathcal{U}(x)$; then it has to be faithful and hence $U_x = U_1^x \cap \ldots \cap U_{k+1}^x \subset O$ which shows that O is open in X.

Now assume that $x, y \in O$ and $z \in U_x \cap U_y$; then $z \in (\bigcap \mathcal{U}(x)) \cap (\bigcap \mathcal{U}(y))$. It follows from $\mathrm{ord}(z, \mathcal{U}) \leq k + 1$ and $|\mathcal{U}(x)| = |\mathcal{U}(y)| = k + 1$ that $\mathcal{U}(x) = \mathcal{U}(y)$ and therefore $U_x = U_y$. Thus the family $\mathcal{V}_1 = \{U_x : x \in O\}$ is disjoint and it is immediate that $\mathcal{V}_1$ is a π-base for the family $\mathcal{U}_1 = \{U \in \mathcal{U} : U \cap O \neq \emptyset\}$.

It is evident that $\mathrm{ord}(\mathcal{U} \backslash \mathcal{U}_1) \leq k$; if the family $\mathcal{U} \backslash \mathcal{U}_1$ is empty then we can let $\mathcal{V}_i = \mathcal{V}_1$ for any $i \in \{2, \ldots, k + 1\}$. If $\mathcal{U} \backslash \mathcal{U}_1 \neq \emptyset$ then can apply the induction hypothesis to find disjoint families $\mathcal{V}_2, \ldots, \mathcal{V}_{k+1}$ of non-empty open subsets of X such that $\mathcal{V}_2 \cup \ldots \cup \mathcal{V}_{k+1}$ is a π-base for $\mathcal{U} \backslash \mathcal{U}_1$. It is straightforward that the families $\mathcal{V}_1, \ldots, \mathcal{V}_{k+1}$ are as required; so our induction procedure shows that for any $n \in \mathbb{N}$ and any non-empty family $\mathcal{U} \subset \tau^*(X)$ if $\mathrm{ord}(\mathcal{U}) \leq n$ there are disjoint families $\mathcal{V}_1, \ldots, \mathcal{V}_n \subset \tau^*(X)$ such that $\mathcal{V}_1 \cup \ldots \cup \mathcal{V}_n$ is a π-base for $\mathcal{U}$.

U.292. *Suppose that a we are given a space X with the Baire property and $\mathcal{U}$ is a weakly σ-point-finite family of non-empty open subsets of X. Prove that there exists a σ-disjoint family $\mathcal{V} \subset \tau^*(X)$ which is a π-base for $\mathcal{U}$.*

Solution. If we have a family $\mathcal{W} \subset \exp X$ then $\mathcal{W}(x) = \{W \in \mathcal{W} : x \in W\}$ and $\mathrm{ord}(x, \mathcal{W}) = |\mathcal{W}(x)|$ for any $x \in X$. The family $\mathcal{W}$ is called *point-finite* at $x \in X$ if $\mathrm{ord}(x, \mathcal{W}) < \omega$; if $A \subset X$ then $\mathcal{W}[A] = \{W \in \mathcal{W} : W \cap A \neq \emptyset\}$.

Fix a sequence $\{\mathcal{U}_n : n \in \omega\}$ which witnesses that $\mathcal{U}$ is weakly σ-point-finite; the set $X_{nm} = \{x \in X : \mathrm{ord}(\mathcal{U}_n) \leq m\}$ is closed in X for any $n, m \in \omega$. Given $U \in \mathcal{U}$ let $N_U = \{n \in \omega : U \in \mathcal{U}_n\}$. It turns out that

(1) $\bigcup\{X_{nm} : n \in N_U, m \in \omega\} = X$,

because, for any $x \in X$ there is $n \in \omega$ such that $U \in \mathcal{U}_n$ (i.e., $n \in N_U$) and $\mathcal{U}_n$ is point-finite at x. Let $U_{nm} = \mathrm{Int}(X_{nm})$; it follows from Problem 291 that there exists a σ-disjoint family $\mathcal{V}_{nm} \subset \tau^*(X)$ (which might be empty) such that $\mathcal{V}_{nm}$ is a π-base for $\mathcal{U}_n[U_{nm}]$ for any $n, m \in \omega$. We claim that $\mathcal{V} = \bigcup\{\mathcal{V}_{nm} : n, m \in \omega\}$ is a π-base for $\mathcal{U}$. Indeed, take any $U \in \mathcal{U}$. It follows from (1) and the Baire property of X that $\bigcup \{U_{nm} : n \in N_U, \, m \in \omega\}$ is dense in X; so there is $n \in N_U$ and $m \in \omega$ such that $U \cap U_{nm} \neq \emptyset$. Thus $U \in \mathcal{U}_n[U_{nm}]$ and therefore there is $V \in \mathcal{V}_{nm}$ for which $V \subset U$. Therefore $\mathcal{V}$ is a σ-disjoint π-base for the family $\mathcal{U}$.

U.293. *Prove that every Gul'ko compact space has a dense metrizable subspace.*

Solution. If X is a set and $\mathcal{W} \subset \exp X$ then $\mathcal{W}(x) = \{W \in \mathcal{W} : x \in W\}$ and $\mathrm{ord}(x, \mathcal{W}) = |\mathcal{W}(x)|$ for any $x \in X$. The family $\mathcal{W}$ is called *point-finite* at $x \in X$ if $\mathrm{ord}(x, \mathcal{W}) < \omega$; if $A \subset X$ then $\mathcal{W}|A = \{W \cap A : W \in \mathcal{W}\}$.

Fact 1. For an arbitrary space Z,

(a) a countable union of weakly σ-point-finite families of subsets of Z is a weakly σ-point-finite family;

(b) if $\mathcal{U} \subset \exp Z$ is weakly σ-point-finite and $\mathcal{V}_U \subset \exp U$ is disjoint for any $U \in \mathcal{U}$ then the family $\mathcal{V} = \bigcup \{\mathcal{V}_U : U \in \mathcal{U}\}$ is weakly σ-point-finite.

Proof. Suppose that, for any $n \in \omega$, a family $\mathcal{W}_n \subset \exp Z$ is weakly σ-point-finite and fix a sequence $\{\mathcal{W}_{nm} : m \in \omega\}$ of subfamilies of $\mathcal{W}_n$ which witnesses this. To see that $\mathcal{W} = \bigcup \{\mathcal{W}_n : n \in \omega\}$ is weakly σ-point-finite take any $x \in Z$ and $W \in \mathcal{W}$. There is $n \in \omega$ with $W \in \mathcal{W}_n$; so there exists $m \in \omega$ such that $W \in \mathcal{W}_{nm}$ and the family $\mathcal{W}_{nm}$ is point-finite at x. This shows that the countable collection $\{\mathcal{W}_{nm} : n, m \in \omega\}$ witnesses that $\mathcal{W}$ is weakly σ-point-finite; so (a) is proved.

As to (b), fix a sequence $\{\mathcal{U}_n : n \in \omega\}$ which witnesses that the family $\mathcal{U}$ is weakly σ-point-finite and let $\mathcal{G}_n = \bigcup \{\mathcal{V}_U : U \in \mathcal{U}_n\}$ for any $n \in \omega$. If $x \in Z$ and $V \in \mathcal{V}$ then there is $U \in \mathcal{U}$ such that $V \in \mathcal{V}_U$. Take $n \in \omega$ such that the family $\mathcal{U}_n$ is point-finite at x and $U \in \mathcal{U}_n$. Then $V \in \mathcal{G}_n$ and it is easy to check that $\mathcal{G}_n$ is point-finite at x. Thus the sequence $\{\mathcal{G}_n : n \in \omega\} \subset \exp \mathcal{V}$ witnesses that $\mathcal{V}$ is weakly σ-point-finite. This settles (b); so Fact 1 is proved.

Returning to our solution let us establish that

(1) given a set A and a sequence $s = \{A_n : n \in \omega\} \subset \exp A$ such that $A = \bigcup s$, any compact subset of $\Sigma_s(A)$ has a σ-disjoint π-base.

To prove (1) fix an arbitrary set A with a sequence $s = \{A_n : n \in \omega\} \subset \exp A$ such that $A = \bigcup s$. Given $B, C \subset A$ with $B \subset C$ let $\pi_B^C : \mathbb{R}^C \to \mathbb{R}^B$ be the natural projection of $\mathbb{R}^C$ onto its face $\mathbb{R}^B$. For any $B \subset A$, the family $s_B = \{A_n \cap B : n \in \omega\}$ defines the set $\Sigma_{s_B}(B)$ and it is straightforward that $\pi_B^A(\Sigma_s(A)) \subset \Sigma_{s_B}(B)$. Let us call the set $\Sigma_{s_B}(B)$ the B-face of $\Sigma_s(A)$. A set $U \subset \mathbb{R}^B$ is called *a standard open subset of* $\mathbb{R}^B$ if $U = \prod \{U_b : b \in B\}$ where $U_b \in \tau(\mathbb{R})$ for any $b \in B$ and the set $\mathrm{supp}(U) = \{b \in B : U_b \neq \mathbb{R}\}$ is finite. It is evident that standard open sets of $\mathbb{R}^B$ constitute a base of $\mathbb{R}^B$ for any $B \subset A$.

Given any $B \subset A$ and $Y \subset \Sigma_{s_B}(B)$ denote by λ_Y the minimal cardinal λ such that Y can be embedded in a C-face of $\Sigma_s(A)$ for some $C \subset A$ with $|C| = \lambda$; we will call λ_Y *the embedding index of* Y. Call a compact space $Y \subset \Sigma_{s_B}(B)$ *solid* if $\lambda_{\overline{U}} = \lambda_Y$ for any $U \in \tau^*(Y)$. Let us show first that

(2) if, for any $B \subset A$, every solid compact $Y \subset \Sigma_{s_B}(B)$ has a σ-disjoint π-base then, for any $B \subset A$, every compact subspace of $\Sigma_{s_B}(B)$ has a σ-disjoint π-base.

To prove (2) assume that, for any $B \subset A$, every solid compact subspace of $\Sigma_{s_B}(B)$ has a σ-disjoint π-base and take an arbitrary compact $Y \subset \Sigma_{s_B}(B)$ for some $B \subset A$.

The family $\mathcal{G} = \{G \in \tau^*(Y) : \overline{G}$ is solid$\}$ is a π-base in Y. Indeed, for any $U \in \tau^*(Y)$ we can choose a set $G \in \tau^*(U)$ such that the embedding index of $\overline{G}$ is minimal; then $\lambda_{\overline{G}} = \lambda_{\overline{V}}$ for any $V \in \tau^*(\overline{G})$, i.e., $\overline{G}$ is solid. Thus $G \in \mathcal{G}$ and $G \subset U$.

Take a maximal disjoint family $\mathcal{U} \subset \tau^*(Y)$ of non-empty solid open sets. Since the non-empty solid open sets form a π-base in Y, the set $\bigcup \mathcal{U}$ is dense in Y. By our assumption about solid sets we can take a σ-disjoint π-base $\mathcal{G}_U$ in the space $\overline{U}$ for any $U \in \mathcal{U}$. It is an easy exercise that $\bigcup \{\mathcal{G}_U | U : U \in \mathcal{U}\}$ is a σ-disjoint π-base in Y; so (2) is proved.

We will establish next that, for any $E \subset A$, every compact solid subspace $Y \subset \Sigma_{s_E}(E)$ has a σ-disjoint π-base. If $\lambda_Y \leq \omega$ then the space Y has a countable base; so there is nothing to prove. Assume, towards a contradiction, that some compact solid space $Y \subset \Sigma_{s_B}(B)$ does not have a σ-disjoint π-base and let κ be the minimal cardinal for which there exists $B \subset A$ and a solid compact subspace $X \subset \Sigma_{s_B}(B)$ such that $\lambda_X = \kappa$ and X has no σ-disjoint π-base. Then $\kappa > \omega$ and we can consider that $|B| = \kappa$. Fix a faithful enumeration $\{b_\alpha : \alpha < \kappa\}$ of the set B and consider the set $C_\alpha = \{b_\beta : \beta \leq \alpha\}$ for any $\alpha < \kappa$.

The set $O_\alpha = \{x \in X : x(b_\alpha) \neq 0\}$ is open in X (possibly empty) and the set $H_\alpha = \pi_{C_\alpha}^B(\overline{O}_\alpha)$ is compact for any $\alpha < \kappa$. It is clear that $\lambda_{H_\alpha} \leq |\alpha| \cdot \omega < \kappa$; so we can find a σ-disjoint π-base $\mathcal{U}_\alpha$ in the space H_α (if $H_\alpha = \emptyset$ then $\mathcal{U}_\alpha = \emptyset$). The map $p_\alpha = \pi_{C_\alpha}^B | \overline{O}_\alpha : \overline{O}_\alpha \to H_\alpha$ is continuous, so $p_\alpha^{-1}(U) \cap O_\alpha \in \tau^*(X)$ for any $\alpha < \kappa$ and $U \in \mathcal{U}_\alpha$; fix a sequence $\{\mathcal{U}_{\alpha,n} : n \in \omega\}$ of disjoint subfamilies of $\mathcal{U}_\alpha$ such that $\mathcal{U}_\alpha = \bigcup \{\mathcal{U}_{\alpha,n} : n \in \omega\}$.

Let $V_\alpha = \{p_\alpha^{-1}(U) \cap O_\alpha : U \in \mathcal{U}_\alpha\}$ for any $\alpha < \kappa$; then $V = \bigcup \{V_\alpha : \alpha < \kappa\}$ is a π-base in X. To see it take any $O \in \tau^*(X)$. There is a standard set U in the space $\mathbb{R}^B$ such that $\emptyset \neq V = U \cap X \subset O$. The space X being solid there is $\alpha < \kappa$ such that supp$(U) \subset \bigcup \{b_\beta : \beta < \alpha\}$ and $O_\alpha \cap V \neq \emptyset$. It follows from Fact 1 of S.298 that $W = p_\alpha(\overline{O}_\alpha \cap V)$ is a non-empty open subset of H_α such that $p_\alpha^{-1}(W) = \overline{O}_\alpha \cap V$. The family $\mathcal{U}_\alpha$ being a π-base in H_α, there is $G \in \mathcal{U}_\alpha$ with $G \subset W$. Then $G' = p_\alpha^{-1}(G) \cap O_\alpha \in V$ and $G' \subset V$; so V is, indeed, a π-base in X. It turns out that

(3) the family $\mathcal{O} = \{O_\alpha : \alpha < \kappa\}$ is weakly σ-point-finite in X.

Consider the family $\mathcal{O}_n = \{O_\alpha : b_\alpha \in A_n \cap B\}$ for any $n \in \omega$. Given any $x \in X$ and $\alpha < \kappa$, it follows from $x \in \Sigma_{s_B}(A)$ that there is $n \in \omega$ such that $b_\alpha \in A_n \cap B$ and supp$(x) \cap A_n \cap B$ is finite. It is immediate that $O_\alpha \in \mathcal{O}_n$ and only finitely many elements of $\mathcal{O}_n$ can contain the point x; this proves that the sequence $\{\mathcal{O}_n : n \in \omega\} \subset \exp \mathcal{O}$ witnesses that $\mathcal{O}$ is weakly σ-point-finite, i.e., (3) is proved.

Our next step is to show that the family $\mathcal{V}$ is also weakly σ-point-finite. The family $\mathcal{V}_{\alpha,n} = \{p_\alpha^{-1}(U) \cap O_\alpha : U \in \mathcal{U}_{\alpha,n}\}$ is disjoint for any $\alpha < \kappa$ and $n \in \omega$. Therefore the family $\mathcal{G}_n = \bigcup\{\mathcal{V}_{\alpha,n} : \alpha < \kappa\}$ is weakly σ-point-finite for any $n \in \omega$ (see Fact 1). Since $\mathcal{V} = \bigcup\{\mathcal{G}_n : n \in \omega\}$, we can apply Fact 1 again to see that $\mathcal{V}$ is weakly σ-point-finite.

Thus $\mathcal{V}$ is a weakly σ-point-finite π-base in X. Now apply Problem 292 to conclude that there is a σ-disjoint family $\mathcal{B} \subset \tau^*(X)$ which is a π-base for $\mathcal{V}$. An immediate consequence is that $\mathcal{B}$ is a σ-disjoint π-base in X. This contradiction with the choice of X shows that, for any $B \subset A$, every solid compact subset of $\Sigma_{s_B}(B)$ has a σ-disjoint π-base. Now we can apply (2) to see that for any $B \subset A$, every compact subset of $\Sigma_{s_B}(B)$ has a σ-disjoint π-base; so (1) is proved.

Finally, take a Gul'ko compact space X. By Problem 290, there is a set A and a sequence $s = \{A_n : n \in \omega\}$ of subsets of A such that $\bigcup s = A$ and X embeds in $\Sigma_s(A)$. By (1), the space X has a σ-disjoint π-base; so it follows from Problems 285 and 138 that X has a dense metrizable subspace.

U.294. *Let X be a Gul'ko compact space. Prove that $w(X) = d(X) = c(X)$. In particular, each Gul'ko compact space with the Souslin property is metrizable.*

Solution. The inequalities $c(X) \le d(X) \le w(X)$ are evident. Now, assume that κ is a cardinal and $c(X) \le \kappa$; take a dense metrizable $M \subset X$ (this is possible by Problem 293) and observe that $c(M) \le c(X) \le \kappa$, so $d(M) = c(M) \le \kappa$ (see TFS-214) and therefore $d(X) \le \kappa$. Finally, apply Problem 285 and Problem 121 to see that $w(X) = d(X) \le \kappa$. This proves that $w(X) \le c(X)$ and hence $c(X) = d(X) = w(X)$.

U.295. *Let X be a pseudocompact space with the Souslin property. Prove that any Lindelöf Σ-subspace of $C_p(X)$ has a countable network.*

Solution. Take a Lindelöf Σ-subspace $Y \subset C_p(X)$ and let $\pi : C_p(\beta X) \to C_p(X)$ be the restriction map. Then π is a condensation and the space $Z = \pi^{-1}(Y)$ is Lindelöf Σ (see Problem 232). For any $x \in \beta X$ and $f \in Z$ let $e_x(f) = f(x)$; then $e_x \in C_p(Z)$ and the map $e : \beta X \to C_p(Z)$ defined by $e(x) = e_x$ for any $x \in \beta X$ is continuous by TFS-166. The space $K = e(\beta X)$ is Gul'ko compact (see Problem 220) and $c(K) \le c(\beta X) = c(X) = \omega$. Thus $w(K) = \omega$ by Problem 294; the space Z embeds in $C_p(K)$ by TFS-166, so $nw(Z) \le nw(C_p(K)) = nw(K) \le w(K) = \omega$ which implies that Z has a countable network. This shows that $Y = \pi(Z)$ also has a countable network being a continuous image of Z.

U.296. *Let X be a Lindelöf Σ-space. Suppose that Y is a pseudocompact subspace of $C_p(X)$. Prove that Y is compact and metrizable if and only if $c(Y) = \omega$.*

Solution. If Y is compact and metrizable then $c(Y) \le w(Y) = \omega$ (see TFS-212); so necessity is clear. Now, assume that $c(Y) = \omega$. For any $x \in X$ and $f \in Y$ let $e_x(f) = f(x)$; then $e_x \in C_p(Y)$ and the map $e : X \to C_p(Y)$ defined by $e(x) = e_x$ for any $x \in X$ is continuous by TFS-166. The space $Z = e(X) \subset C_p(Y)$ is Lindelöf Σ so $nw(Z) = \omega$ by Problem 295. Furthermore, Y embeds in $C_p(Z)$ by TFS-166; so $nw(Y) \le nw(C_p(Z)) = nw(Z) = \omega$.

As a consequence, $iw(Y) \leq nw(Y) = \omega$ (see TFS-156) and hence $w(Y) = \omega$ by TFS-140. Thus Y is metrizable and hence compact by TFS-212. This settles sufficiency and makes our solution complete.

U.297. *Prove that every Gul'ko compact space is hereditarily d-separable.*

Solution. Let K be a Gul'ko compact space. Given any $Y \subset K$ the set $F = \overline{Y}$ is also Gul'ko compact by Problem 282. Therefore F has a dense metrizable subspace by Problem 293 which, together with Problems 285 and 138, implies that the space F has a σ-disjoint π-base $\mathcal{B}$. It is evident that the family $\mathcal{C} = \{B \cap Y : B \in \mathcal{B}\}$ is a σ-disjoint π-base in Y.

If we pick a point $y_C \in C$ for any $C \in \mathcal{C}$ then the set $D = \{y_C : C \in \mathcal{C}\}$ is dense in Y. We have $\mathcal{C} = \bigcup\{\mathcal{C}_n : n \in \omega\}$ where $\mathcal{C}_n$ is disjoint for any $n \in \omega$. It is straightforward that the subspace $D_n = \{y_C : C \in \mathcal{C}_n\}$ is discrete for each $n \in \omega$; so $D = \bigcup_{n \in \omega} D_n$ is a dense σ-discrete subspace of Y which shows that Y is d-separable.

U.298. *Let X be a compact space. Prove that $C_p(X)$ is a K-analytic space if and only if X has a T_0-separating family $\mathcal{U}$ of open F_σ-subsets of X and subfamilies $\{\mathcal{U}_s : s \in \omega^{<\omega}\}$ of the family $\mathcal{U}$ with the following properties:*

(a) $\mathcal{U}_\emptyset = \mathcal{U}$ and $\mathcal{U}_s = \bigcup\{\mathcal{U}_{s\frown k} : k \in \omega\}$ for any $s \in \omega^{<\omega}$;
(b) for every $x \in X$ and every $f \in \omega^\omega$, there exists $m \in \omega$ such that the family $\mathcal{U}_{f|n}$ is point-finite at x for all $n \geq m$.

Solution. Given a space Z let $u_Z \in C_p(Z)$ be the function which is identically zero on Z; say that a set $A \subset Z$ is *concentrated around a point* $z \in Z$ if $A \backslash U$ is countable for any $U \in \tau(z, Z)$. The space $\mathbb{P}$ of the irrationals is identified with ω^ω; given $p, q \in \mathbb{P}$, the expression $p \leq q$ says that $p(n) \leq q(n)$ for any $n \in \omega$. For $s \in \omega^{<\omega}$ and $p \in \omega^\omega$ we write $s \subset p$ if $p|\mathrm{dom}(s) = s$.

A family $\mathcal{U} \subset \exp Z$ is said to be $\mathbb{P}$-*point-finite* if there is exists a collection $\{\mathcal{U}_s : s \in \omega^{<\omega}\}$ of subfamilies of $\mathcal{U}$ such that $\mathcal{U}_\emptyset = \mathcal{U}$, $\mathcal{U}_s = \bigcup\{\mathcal{U}_{s\frown k} : k \in \omega\}$ for any $s \in \omega^{<\omega}$ and, for any $x \in Z$ and $f \in \omega^\omega$ there is $m \in \omega$ such that $\mathcal{U}_{f|m}$ is point-finite at x (observe that in this case $\mathcal{U}_{f|n}$ is automatically point-finite at x for any $n \geq m$).

Fact 1. If K is Corson compact then there is a set $A \subset C_p(K)$ such that $A \cup \{u_K\}$ is closed in $C_p(K)$, all points of A are isolated in $A \cup \{u_K\}$ and A separates the points of K.

Proof. There exists a point-countable T_0-separating family $\mathcal{U}$ of non-empty cozero subsets of K (see Problem 118). For any $U \in \mathcal{U}$ take a function $f_U \in C_p(K, [0, 1])$ such that $K \backslash U = f^{-1}(0)$. The family $\mathcal{U}$ being point-countable the set $H = \{f_U : U \in \mathcal{U}\}$ is concentrated around the point u_K.

Apply Problem 286 to convince ourselves that there exists a linear condensation $\varphi : C_p(K) \rightarrow \Sigma(B)$ for some set B. It is evident that the set $E = \varphi(H)$ is concentrated around $w = \varphi(u_K)$; so there is a sequence $\{E_n : n \in \omega\} \subset \exp E$ such that $E \backslash \{w\} = \bigcup\{E_n : n \in \omega\}$ while, for every $n \in \omega$, the set $E_n \cup \{w\}$

is compact and all points of E_n are isolated in $E_n \cup \{w\}$ (see Fact 1 of U.289). If $H_n = \varphi^{-1}(E_n)$ then $H_n \cup \{u_K\} = \varphi^{-1}(E_n \cup \{w\})$ is closed in $C_p(K)$ and all points of H_n are isolated in $H_n \cup \{u_K\}$ for any $n \in \omega$.

Let $A_0 = H_0$ and $A_n = \frac{1}{n} \cdot H_n = \{\frac{1}{n} \cdot f : f \in H_n\}$ for any $n \in \mathbb{N}$. Since the multiplication by $\frac{1}{n}$ is a homeomorphism of $C_p(K)$ onto itself, the set $A_n \cup \{u_K\}$ is closed in $C_p(K)$ for any $n \in \omega$; let $A = \bigcup\{A_n : n \in \omega\}$.

Take a function $f \in C_p(K)\backslash(A \cup \{u_K\})$. Then $f(x) \neq 0$ for some $x \in K$; so there is $m \in \mathbb{N}$ such that $|f(x)| > \frac{1}{m}$. The set $W = \{g \in C_p(K) : |g(x)| > \frac{1}{m}\}$ is open in $C_p(K)$ and $A_n \cap W = \emptyset$ for any $n \geq m$. Since $A_n \cup \{u_K\}$ is closed in $C_p(K)$ for any $n < m$, the set $W' = W\backslash(\bigcup\{A_n \cup \{u_K\} : n < m\})$ is an open neighbourhood of f which does not meet $A \cup \{u_K\}$. This proves that

(1) the set $A \cup \{u_K\}$ is closed in $C_p(K)$.

It is evident that the set H separates the points of K; since all functions of A were obtained from the elements of H by multiplication by a non-zero constant, we conclude that

(2) the set A separates the points of K.

Now, if $f \in A$ then $f \in A_k$ for some $k \in \omega$ and $f \neq u_K$; so there is $x \in K$ for which $f(x) > 0$ and hence there is $m \in \omega$ such that $f(x) > \frac{1}{m}$. The set $W = \{g \in C_p(K) : |g(x)| > \frac{1}{m}\}$ is open in $C_p(K)$ and $A_n \cap W = \emptyset$ for any $n \geq m$. Furthermore, the set $A_n \cup \{u_K\}$ is closed in the space $C_p(K)$ for any number $n \in \omega$; so the set $W' = W\backslash(\bigcup\{A_n \cup \{u_K\} : n < m, n \neq k\})$ is an open neighbourhood of f in $C_p(K)$ such that $W' \cap A \subset A_k$. Since f is an isolated point of $A_k \cup \{u_K\}$, there is $U \in \tau(f, C_p(K))$ such that $U \cap (A_k \cup \{u_K\}) = \{f\}$. It is immediate that $U' = U \cap W'$ is an open neighbourhood of f in $C_p(K)$ and $U \cap (A \cup \{u_K\}) = \{f\}$, i.e., f is isolated in $A \cup \{u_K\}$. This, together with the properties (1) and (2), shows that Fact 1 is proved.

Fact 2. If Z is a space and $\mathcal{U}_n \subset \exp Z$ is $\mathbb{P}$-point-finite for any $n \in \omega$ then $\mathcal{U} = \bigcup\{\mathcal{U}_n : n \in \omega\}$ is also $\mathbb{P}$-point-finite.

Proof. For any $s \in \omega^{<\omega}\backslash\{\emptyset\}$ let $[s] \in \omega^{<\omega}$ be the finite sequence obtained from s by "cutting off" its first element, i.e., $\mathrm{dom}([s]) = \mathrm{dom}(s) - 1$ and $[s](m) = s(m+1)$ for any $m \in \mathrm{dom}([s])$. Analogously, if $f \in \omega^\omega$ then $g = [f] \in \omega^\omega$ is defined by $g(n) = f(n+1)$ for any $n \in \omega$.

Fix a collection $\{\mathcal{U}_s^n : s \in \omega^{<\omega}\}$ of subfamilies of $\mathcal{U}_n$ which witnesses that $\mathcal{U}_n$ is $\mathbb{P}$-point-finite for any $n \in \omega$. Let $\mathcal{U}_\emptyset = \mathcal{U}$ and $\mathcal{U}_s = \mathcal{U}_{[s]}^{s(0)}$ for any $s \in \omega^{<\omega}\backslash\{\emptyset\}$. Observe first that, for any $s \in \omega^1$, we have $[s] = \emptyset$ and therefore $\mathcal{U}_{\emptyset \frown k} = \mathcal{U}_\emptyset^k = \mathcal{U}$ for any $k \in \omega$ so $\mathcal{U} = \mathcal{U}_\emptyset = \bigcup\{\mathcal{U}_{\emptyset \frown k} : k \in \omega\}$.

Now, if we are given an element $s \in \omega^{<\omega}\backslash\{\emptyset\}$ then, for $t = [s]$ and $n = s(0)$, we have $\mathcal{U}_t^n = \bigcup\{\mathcal{U}_{t \frown k}^n : k \in \omega\}$ which is the same as saying that $\mathcal{U}_s = \bigcup\{\mathcal{U}_{s \frown k} : k \in \omega\}$.

Now if $x \in Z$ and $f \in \omega^\omega$ then, for $k = f(0)$ and $g = [f]$, there is $m \in \omega$ such that $\mathcal{U}_{g|n}^k$ is point-finite at x for all $n \geq m$ and this is equivalent to saying that

$\mathcal{U}_{f|n}$ is point-finite at x for all $n \geq m+1$. This verifies all required properties of the family $\mathcal{U}$; so $\mathcal{U}$ is $\mathbb{P}$-point-finite and Fact 2 is proved.

Fact 3. If Z is a space and a family $\mathcal{U} \subset \exp Z$ is $\mathbb{P}$-point-finite then $\mathcal{U}$ is weakly σ-point-finite and hence point-countable.

Proof. Take a collection $\mathcal{C} = \{\mathcal{U}_s : s \in \omega^{<\omega}\}$ which witnesses that the family $\mathcal{U}$ is $\mathbb{P}$-point-finite. Let us check that $\mathcal{C}$ also witnesses that $\mathcal{U}$ is weakly σ-point-finite. Fix a point $x \in Z$ and $U \in \mathcal{U}$; it follows from the definition of a $\mathbb{P}$-point-finite family that there is $f \in \omega^\omega$ such that $U \in \mathcal{U}_{f|n}$ for any $n \in \omega$. Besides, there exists $m \in \omega$ for which $\mathcal{U}_{f|m}$ is point-finite at x, i.e., for $s = f|m$ we have $U \in \mathcal{U}_s$ and $\mathcal{U}_s$ is point-finite at x. This shows that $\bigcup\{\mathcal{U}_s : U \in \mathcal{U}_s$ and $\mathcal{U}_s$ is point-finite at $x\} = \mathcal{U}$; so $\mathcal{U}$ is, indeed, weakly σ-point-finite. Finally, apply Fact 2 of U.290 to conclude that $\mathcal{U}$ is point-countable. Fact 3 is proved.

Returning to our solution suppose that $\mathcal{U}$ is a $\mathbb{P}$-point-finite T_0-separating family of cozero subsets of X and fix $f_U \in C_p(X, [0,1])$ such that $X\backslash U = f_U^{-1}(0)$ for any $U \in \mathcal{U}$. Let a collection $\{\mathcal{U}_s : s \in \omega^{<\omega}\}$ witness that $\mathcal{U}$ is $\mathbb{P}$-point-finite. The set $A = \{f_U : U \in \mathcal{U}\}$ separates the points of X. For any $f \in A$ and $x \in X$ let $e_x(f) = f(x)$. Then $e_x \in \mathbb{R}^A$ and the map $e : X \to \mathbb{R}^A$ defined by $e(x) = e_x$ for any $x \in X$ is continuous and injective by TFS-166; so e embeds X in $\mathbb{R}^A$; let $Y = e(X)$.

The family $\mathcal{U}$ is point-countable by Fact 3; an immediate consequence is that $Y \subset \Sigma(A)$. Let $A_s = \{f_U : U \in \mathcal{U}_s\}$ for any $s \in \omega^{<\omega}$. It is evident that $A_\emptyset = A$ and $A_s = \bigcup\{A_{s\frown k} : k \in \omega\}$ for any $s \in \omega^{<\omega}$. Now if $y \in Y$ then $y = e_x$ for some $x \in X$; for any $f \in \omega^\omega$ there exists $m \in \omega$ such that the family $\mathcal{U}_{f|n}$ is point-finite at x for any $n \geq m$. This implies that $\operatorname{supp}(y) \cap A_{f|n}$ is finite for any $n \geq m$. Therefore the set $Y \subset \Sigma(A)$ satisfies the premises of Fact 4 of U.222 which implies that $C_p(Y)$ is K-analytic. Since $X \simeq Y$, the space $C_p(X)$ is also K-analytic and hence we proved sufficiency.

To establish necessity assume that $C_p(X)$ is a K-analytic space and let $u = u_X$. The space X is Corson compact by Problem 285 and hence there is a set $A \subset C_p(X)\backslash\{u\}$ such that the set $Y = A \cup \{u\}$ is closed in $C_p(X)$, all points of A are isolated in $A \cup \{u\}$ and A separates the points of X (see Fact 1). If A is countable then $w(X) = \omega$; so it follows from Fact 2 that any countable base of X is a $\mathbb{P}$-point-finite, T_0-separating family of cozero open subsets of X which proves necessity in this case. Therefore we can only consider the case when A is uncountable.

The space Y is K-analytic; so there is a compact cover $\{K_p : p \in \mathbb{P}\}$ of the space Y such that $p \leq q$ implies $K_p \subset K_q$ (see SFFS-391). For any $s \in \omega^{<\omega}$ let $H_s = \bigcup\{K_p : s \subset p\}$. Then $H_\emptyset = Y$ and $H_s = \bigcup\{H_{s\frown k} : k \in \omega\}$ for any $s \in \omega^{<\omega}$.

For any $q \in \mathbb{Q}$ with $q > 0$ let $O_q = (q, +\infty)$; if $q \in \mathbb{Q}$ and $q < 0$ then let $O_q = (-\infty, q)$. We claim that

(3) the family $\mathcal{U}^q = \{f^{-1}(O_q) : f \in A\}$ is $\mathbb{P}$-point-finite for any $q \in \mathbb{Q}\backslash\{0\}$.

To prove (3) let $\mathcal{U}_s^q = \{f^{-1}(O_q) : f \in A \cap H_s\}$ for any $s \in \omega^{<\omega}$. It is immediate that $\mathcal{U}_\emptyset^q = \mathcal{U}^q$ and $\mathcal{U}_s^q = \bigcup\{\mathcal{U}_{s^\frown k}^q : k \in \omega\}$ for any $s \in \omega^{<\omega}$. Given a point $x \in X$ and $p \in \omega^\omega$ the set $F = \bigcap\{H_{p|n} : n \in \omega\}$ is countably compact by Fact 1 of S.391; since Y has a unique non-isolated point, the set F is compact.

The family $\mathcal{H} = \{H_{p|n} : n \in \omega\}$ is a network at F in the sense that, for any $U \in \tau(F, Y)$ there is $n \in \omega$ such that $H_{p|n} \subset U$. Indeed, if it were not so then there is $U \in \tau(F, Y)$ such that we can choose a point $f_n \in H_{p|n} \backslash U$ for any $n \in \omega$. The sequence $S = \{f_n : n \in \omega\}$ has an accumulation point $f \in F$ (this was also proved in Fact 1 of S.391); however, $S \subset Y \backslash U$ while $f \in U$ which is a contradiction.

If the set $P = \{f \in F \backslash \{u\} : x \in f^{-1}(O_q)\}$ is infinite then P must have an accumulation point in F and this accumulation point has to be u. However, the set $W = \{f \in Y : |f(x)| < |q|\}$ is an open neighbourhood of u in Y while $|f(x)| > |q|$ for any $f \in P$. Thus $W \cap P = \emptyset$; this contradiction shows that P is finite.

The closure of the set $G = \{f \in Y : f(x) \in O_q\}$ in Y does not contain u because $W \cap G = \emptyset$. Therefore G is clopen in Y which shows that $Y \backslash (G \backslash P)$ is an open neighbourhood of F in Y. By our observation about $\mathcal{H}$ there is $m \in \omega$ such that $H_{p|m} \subset Y \backslash (G \backslash P)$. Since $G \cap H_{p|m} = \{f \in H_{p|m} : x \in f^{-1}(O_q)\} \subset P$ is finite, the family $\mathcal{U}_{p|m}^q$ is point-finite at x. If $n \geq m$ then $\mathcal{U}_{p|n}^q \subset \mathcal{U}_{p|m}^q$; so $\mathcal{U}_{p|n}^q$ is point-finite at x for any $n \geq m$. Therefore the family $\mathcal{U}^q$ is $\mathbb{P}$-point-finite for any $q \in \mathbb{Q} \backslash \{0\}$, i.e., (3) is proved.

By Fact 2, the family $\mathcal{U} = \bigcup\{\mathcal{U}^q : q \in \mathbb{Q} \backslash \{0\}\}$ is $\mathbb{P}$-point-finite as well. All elements of $\mathcal{U}$ are cozero sets in X by Fact 1 of T.252. Besides, $\mathcal{U}$ is T_0-separating; indeed, if x and y are distinct points of X then there is $f \in Y$ with $f(x) \neq f(y)$. We can assume, without loss of generality, that $f(x) < f(y)$. There is $q \in \mathbb{Q} \backslash \{0\}$ such that $f(x) < q < f(y)$. It is immediate that $U = f^{-1}(O_q) \in \mathcal{U}$ and $U \cap \{x, y\}$ is a singleton. This shows that the family $\mathcal{U}$ has all required properties, i.e., we settled necessity and hence our solution is complete.

U.299. *Let X be a compact space. Prove that $C_p(X)$ is a K-analytic space if and only if X can be embedded into some $\Sigma(A)$ in such a way that, for some family $\{A_s : s \in \omega^{<\omega}\}$ of subsets of A, the following conditions are fulfilled:*

(a) $A_\emptyset = A$ and $A_s = \bigcup\{A_{s^\frown k} : k \in \omega\}$ for any $s \in \omega^{<\omega}$;
(b) for any point $x \in X$ and any $f \in \omega^\omega$, there exists $m \in \omega$ such that the set $A_{f|n} \cap supp(x)$ is finite for all $n \geq m$.

Solution. A family $\mathcal{U}$ of subsets of Z is $\mathbb{P}$-*point-finite* if there is exists a collection $\{\mathcal{U}_s : s \in \omega^{<\omega}\}$ of subfamilies of $\mathcal{U}$ such that $\mathcal{U}_\emptyset = \mathcal{U}$, $\mathcal{U}_s = \bigcup\{\mathcal{U}_{s^\frown k} : k \in \omega\}$ for any $s \in \omega^{<\omega}$ and, for any $x \in Z$ and $f \in \omega^\omega$ there is $m \in \omega$ such that $\mathcal{U}_{f|m}$ is point-finite at x (observe that in this case $\mathcal{U}_{f|n}$ is automatically point-finite at x for any $n \geq m$).

Observe that sufficiency was established in Fact 4 of U.222. Now, if $C_p(X)$ is a K-analytic space then there exists a $\mathbb{P}$-point-finite T_0-separating family $\mathcal{U}$ of cozero subsets of X (see Problem 298). Fix $f_U \in C_p(X, [0, 1])$ such that $X \backslash U = f_U^{-1}(0)$ for any $U \in \mathcal{U}$. Let a collection $\{\mathcal{U}_s : s \in \omega^{<\omega}\}$ witness that $\mathcal{U}$ is $\mathbb{P}$-point-finite.

The set $A = \{f_U : U \in \mathcal{U}\}$ separates the points of X. For any $f \in A$ and $x \in X$ let $e_x(f) = f(x)$. Then $e_x \in \mathbb{R}^A$ and the map $e : X \to \mathbb{R}^A$ defined by $e(x) = e_x$ for any $x \in X$ is continuous and injective by TFS-166; so e embeds X in $\mathbb{R}^A$; let $Y = e(X)$.

The family $\mathcal{U}$ is point-countable by Fact 3 of U.298; an immediate consequence is that $Y \subset \Sigma(A)$. Let $A_s = \{f_U : U \in \mathcal{U}_s\}$ for any $s \in \omega^{<\omega}$. It is evident that $A_\emptyset = A$ and $A_s = \bigcup\{A_{s^\frown k} : k \in \omega\}$ for any $s \in \omega^{<\omega}$. Now if $y \in Y$ then $y = e_x$ for some $x \in X$; for any $f \in \omega^\omega$ there exists $m \in \omega$ such that the family $\mathcal{U}_{f|n}$ is point-finite at x for any $n \geq m$. This implies that $\mathrm{supp}(y) \cap A_{f|n}$ is finite for any $n \geq m$. Therefore the set $Y \subset \Sigma(A)$ and the family $\{A_s : s \in \omega^{<\omega}\}$ show that there is an embedding of X in $\Sigma(A)$ with all required properties. This proves necessity and finishes our solution.

U.300. *(Talagrand's example) Show that there exists a Gul'ko compact space X such that $C_p(X)$ is not K-analytic. In other words, not every Gul'ko compact space is Talagrand compact.*

Solution. As usual, we identify any ordinal with the set of its predecessors; in particular, $0 = \emptyset$ and $n = \{0, \ldots, n-1\}$ for any $n \in \mathbb{N}$. Given $s, t \in \omega^{<\omega}$ such that $n = \mathrm{dom}(s)$ and $m = \mathrm{dom}(t)$ we define $u = s^\frown t \in \omega^{<\omega}$ with $\mathrm{dom}(u) = n + m$ by concatenating s and t, i.e., we let $u|n = s$ and $u(n + i) = t(i)$ for any $i \in \mathrm{dom}(t)$. If $s \in \omega^{<\omega}$, $\mathrm{dom}(s) = k$ and $n \in \omega$ then $t = s^\frown n \in \omega^{<\omega}$ is defined by letting $\mathrm{dom}(t) = k + 1$, $t|k = s$ and $t(k) = n$. For any functions s and t the expression $t \subset s$ stands for $\mathrm{dom}(t) \subset \mathrm{dom}(s)$ and $s|\mathrm{dom}(t) = t$. The space $\mathbb{P}$ of the irrationals is identified with ω^ω.

Let $\Omega = \{s \in \omega^{<\omega} : i, j \in \mathrm{dom}(s)$ and $i < j$ imply $s(i) < s(j)\}$. In other words, Ω consists of strictly increasing elements of $\omega^{<\omega}$. For any $n \in \omega$ consider the set $\Omega_n = \{s \in \Omega : s(i) \leq n$ for any $i \in \mathrm{dom}(s)\}$. It is clear that every Ω_n is a finite set with $\Omega_n \subset \Omega_{n+1}$ and $\bigcup\{\Omega_n : n \in \omega\} = \Omega$.

A set $A \subset \Omega$ is *a tree* if $s \in A$ implies $t \in A$ for any $t \in \Omega$ with $t \subset s$. If $A \subset \Omega$ is a tree then a set $B \subset A$ is called *a branch* of A if, for any $s, t \in B$ either $s \subset t$ or $t \subset s$. Let $\mathcal{E} = \{A \subset \Omega : A$ is a tree without infinite branches$\}$. Observe first that

(1) the set $M = \{x \in \mathbb{D}^\Omega : x^{-1}(1)$ is a tree$\} \subset \mathbb{D}^\Omega$ is closed in $\mathbb{D}^\Omega$ and hence compact.

Indeed, if $x \in \mathbb{D}^\Omega \backslash M$ then there are $s, t \in \Omega$ such that $s \in x^{-1}(1)$, $t \subset s$ and $t \notin x^{-1}(1)$. The set $U = \{y \in \mathbb{D}^\Omega : y(s) = 1$ and $y(t) = 0\}$ is an open neighbourhood of x in $\mathbb{D}^\Omega$ and $U \cap M = \emptyset$. Thus $\mathbb{D}^\Omega \backslash M$ is open in $\mathbb{D}^\Omega$, i.e., M is closed; so (1) is proved.

The set we are after is $T = \{x \in M : x^{-1}(1) \neq \emptyset$ and $x^{-1}(1) \in \mathcal{E}\}$. For any $x \in M$ and $n \in \omega$ let $O_n(x) = \{y \in M : y|\Omega_n = x|\Omega_n\}$. It is clear that $O_n(x) \in \tau(x, M)$ for any $x \in M$ and $n \in \omega$; besides, the family $\{O_n(x) : n \in \omega\}$ is a local base at x in the space M.

Let $\mathcal{A}_0 = \{A \subset M :$ there is $x \in M$, $s \in x^{-1}(1)$ and an increasing sequence $i_0, \ldots, i_n \in \mathrm{dom}(s)$ such that $A = \{x_0, \ldots, x_n\}$ and $x_k \in O_{s(i_k)}(x)$ for any $k \leq n\}$. It is clear from the definition that all elements of $\mathcal{A}_0$ are finite. We are going to work with the family $\mathcal{A} = \{A \subset T : \text{there is a sequence } \{A_n : n \in \omega\} \subset \mathcal{A}_0 \text{ such that } A_n \subset A_{n+1} \text{ for any } n \in \omega \text{ and } A \subset \bigcup\{A_n : n \in \omega\}\}$.

Observe first that

(2) $\{x\} \in \mathcal{A}$ for any $x \in T$ and hence $\bigcup \mathcal{A} = T$,

because $x \in O_{s(0)}(x)$ for any $s \in x^{-1}(1)$. It is immediate from the definition that

(3) if $A \in \mathcal{A}$ and $B \subset A$ then $B \in \mathcal{A}$.

Our purpose is to prove that $\mathcal{A}$ is an adequate family on T. The following property of $\mathcal{A}_0$ is crucial.

(4) Suppose that $B \subset T$ and every finite $C \subset B$ belongs to $\mathcal{A}_0$. Then any accumulation point of B in the space M belongs to $M \setminus T$.

To prove (4) assume the contrary and fix a point $z \in T$ and a faithfully indexed sequence $S = \{z_n : n \in \omega\} \subset B \setminus \{z\}$ which converges to z. For every $n \in \omega$ there is $p(n) \in \omega$ such that $z_n \in O_{p(n)}(z) \setminus O_{p(n)+1}(z)$. Passing to a subsequence of S if necessary we can assume, without loss of generality, that $p(n) < p(n+1)$ for every $n \in \omega$; let $l = 2p(0) + 2$.

Fix any $n \in \omega$, $n \geq l$; then $r(n) = \max\{i \in \omega : p(i) \leq \frac{n}{2} - 1\}$ is well defined because $p(0) \leq \frac{n}{2} - 1$. Since $\{z_0, \ldots, z_n\}$ is a finite subset of B, it belongs to $\mathcal{A}_0$ and therefore there exist $y \in M$ and $s_n \in y^{-1}(1)$ such that for some (not necessarily increasing) sequence $i_0, \ldots, i_n \in \text{dom}(s_n)$ we have $z_k \in O_{s_n(i_k)}(y)$ for any $k \leq n$. According to the definition we could have taken an increasing sequence of elements of $\text{dom}(s_n)$ but then we would have to reorder the set $\{z_0, \ldots, z_n\}$. However, what is left after reordering the respective elements of $\text{dom}(s_n)$ is their faithful enumeration; therefore $i_k \neq i_m$ whenever $k \neq m$.

The sequence $\{p(k) : k \leq n\}$ being increasing, we have $p(k) \geq \frac{n}{2}$ for any $k \geq \frac{n}{2}$. The numbers which belong to the set $\{s_n(i_k) : k \geq \frac{n}{2}\}$ are distinct and there are at least $\frac{n}{2}$-many of them which shows that $s_n(i_k) \geq \frac{n}{2}$ for some $k \geq \frac{n}{2}$. For the number $m = \min\{p(k), s_n(i_k)\} \geq \frac{n}{2}$ we have $z_k \in O_m(z)$ and $z_k \in O_m(y)$ which implies $O_m(z) = O_m(y)$.

For any $j \leq r(n)$ we have the inequality $p(j) + 1 \leq \frac{n}{2} \leq m$; an immediate consequence is that $z_j \notin O_{p(j)+1}(z) = O_{p(j)+1}(y)$. This, together with $z_j \in O_{s_n(i_j)}(y)$ implies that $s_n(i_j) \leq p(j)$. If $t_n = s_n|(r(n) + 1)$ then $t_n \in y^{-1}(1)$ (recall that $y^{-1}(1)$ is a tree). Besides, there is $j \leq \frac{n}{2}$ such that $t_n \in \Omega_j$; it follows from $O_m(z) = O_m(y)$ that $O_j(z) = O_j(y)$ and therefore $t_n \in z^{-1}(1)$.

We have established that, for any $n \geq l$, there exists $t_n \in z^{-1}(1)$ such that $\text{dom}(t_n) = \{0, \ldots, r(n)\}$ and $t_n(j) \leq p(j)$ for any $j \leq r(n)$. Since every $t_n(j)$ takes finitely many values, there is an increasing sequence $\{n_j : j \in \omega\} \subset \omega$ and a sequence $\{m_k : k \in \omega\} \subset \omega$ such that $t_{n_j}(k) = m_k$ for any $j \in \omega$ and $k \leq r(n_j)$. Since $r(n_j) \to +\infty$, the sequence $\{t_{n_j} : j \in \omega\}$ is an infinite branch in $z^{-1}(1)$; this contradiction with $z \in T$ shows that (4) is proved.

Now it is easy to prove that $\mathcal{A}$ is an adequate family. To do it take a set $A \subset T$ such that, for every finite $B \subset A$, we have $B \in \mathcal{A}$ and hence there is a sequence $\{A_n : n \in \omega\} \subset \mathcal{A}_0$ such that $A_n \subset A_{n+1}$ for any $n \in \omega$ and $B \subset \bigcup_{n \in \omega} A_n$.

There is $n \in \omega$ such that $B \subset A_n$ and therefore $B \in \mathcal{A}_0$. Thus every finite subset of A belongs to $\mathcal{A}_0$. If A is uncountable then it follows from $w(A) = \omega$ (we consider the topology induced in A from M) that some $a \in A$ is an accumulation point of A. This contradiction with (4) shows that A is countable; let $\{a_n : n \in \omega\}$ be an enumeration of A.

We saw that $B_n = \{a_0, \ldots, a_n\} \in \mathcal{A}_0$ for any $n \in \omega$ so it follows from the definition of $\mathcal{A}$ that $A = \bigcup\{B_n : n \in \omega\} \in \mathcal{A}$. Thus $\mathcal{A}$ is an adequate family on T. If $A \subset T$ then $\chi_A \in \mathbb{D}^T$ is the characteristic function of A on T, i.e., $\chi_A(x) = 1$ for any $x \in A$ and $\chi_A(x) = 0$ whenever $x \in T \backslash A$.

Take a point $\xi \notin T$ and define a topology μ on $T \cup \{\xi\}$ by declaring all points of T isolated while the local base at ξ is given by the complements of all finite unions of elements of $\mathcal{A}$. Denote the space $(T \cup \{\xi\}, \mu)$ by $T^*_{\mathcal{A}}$ and let $X = \{\chi_A : A \in \mathcal{A}\} \subset \mathbb{D}^T$. Then X is a compact space by Problem 168. Besides, all elements of $\mathcal{A}$ are countable; so X is Corson compact by Problem 169.

Another consequence of (4) is that no $A \in \mathcal{A}$ has an accumulation point in T and therefore every $A \in \mathcal{A}$ is closed and discrete in the topology on T induced from M. For any $x \in T$ let $\varphi(x) = \{\xi, x\}$; then $\varphi : T \to T^*_{\mathcal{A}}$ is a compact-valued map. It is evident that $\varphi(T) = \bigcup\{\varphi(x) : x \in T\} = T^*_{\mathcal{A}}$. If $x \in T$ and $U \in \tau(\varphi(x), T^*_{\mathcal{A}})$ then $\xi \in U$ and hence $F = T \backslash U \subset A_1 \cup \ldots \cup A_n$ for some $A_1, \ldots, A_n \in \mathcal{A}$. Since every A_i is closed and discrete in T, the set $A = \bigcup\{A_i : i \leq n\}$ is closed and discrete in T as well so $W = T \backslash A$ is an open neighbourhood of x in T. It is immediate that $\varphi(W) = \bigcup\{\varphi(y) : y \in W\} \subset U$ which proves that φ is upper semicontinuous. Since T is second countable, we conclude that $T^*_{\mathcal{A}}$ is a Lindelöf Σ-space (see SFFS-249). Consequently, $C_p(X)$ is also a Lindelöf Σ-space (see Problem 173), i.e., X is a Gul'ko compact.

For any $s, t \in \Omega$ let $s < t$ if $s \subset t$ and $s \neq t$. If $H \subset \Omega$ then an element $s \in H$ is called *maximal in H* if there is no $t \in H$ with $s < t$. Let us prove that

(5) if $H \in \mathcal{E}$ then, for any $s \in H$ there is a maximal element $t \in H$ with $s \subset t$.

Indeed, let $s_0 = s$ and assume that $n \in \omega$ and we have a sequence $s_0, \ldots, s_n \in H$ such that $s_0 < \ldots < s_n$. If s_n is a maximal element of H then we are done. If not, then there is $s_{n+1} \in H$ with $s_n < s_{n+1}$. If we have not obtained a maximal element s_n in H for any $n \in \omega$ then $\{s_n : n \in \omega\}$ is an infinite branch in H; this contradiction proves (5).

For any non-empty $H \in \mathcal{E}$ let $D(H) = \{s \in H : \text{there is } t \in H \text{ with } s < t\}$. In other words, to obtain $D(H)$, we throw away all maximal elements of H.

Now we can define the order of any $H \in \mathcal{E}$. Let $A_0(H) = H$. If we have $A_\alpha(H)$ then $A_{\alpha+1}(H) = D(A_\alpha(H))$; if β is a limit ordinal and we have the family $\{A_\alpha(H) : \alpha < \beta\}$ then $A_\beta(H) = \bigcap\{A_\alpha(H) : \alpha < \beta\}$.

It is easy to see that every $A_\alpha(H)$ is a tree; it follows from $A_\alpha(H) \subset H$ that $A_\alpha(H)$ has no infinite branches and therefore (5) implies that $A_{\alpha+1}(H) \neq A_\alpha(H)$. Since every $A_\alpha(H)$ is countable, there is a countable ordinal α such that $A_\alpha(H) = \emptyset$. Therefore the ordinal $\eta(H) = \min\{\alpha < \omega_1 : A_\alpha(H) = \emptyset\}$ is well defined. If $x \in T$ then $x^{-1}(1) \in \mathcal{E}$; so we can define $\text{ord}(x) = \eta(x^{-1}(1))$.

If $Z \subset T$ then let $\mathrm{ord}(Z) = \sup\{\mathrm{ord}(z) : z \in Z\}$. Call a set $Z \subset T$ *unbounded* if $\mathrm{ord}(Z) = \omega_1$; otherwise Z will be called *bounded*.

It is evident that

(6) if $Z_n \subset T$ is bounded for any $n \in \omega$ then $\bigcup\{Z_n : n \in \omega\}$ is bounded.

We claim that

(7) the set T is unbounded.

To prove (7) it suffices to show that, for any $\alpha < \omega_1$, there is $H_\alpha \in \mathcal{E}$ such that $\eta(H_\alpha) \geq \alpha$. If we take any $s \in \Omega$ then, for the set $H_0 = \{s\}$ we have $\eta(H_0) = 1 \geq 0$.

Assume that $\alpha < \omega_1$ is a limit ordinal and, for any $\beta < \alpha$, we have a set $H_\beta \subset \Omega$ such that $\eta(H_\beta) \geq \beta$. Take a surjective map $\varphi : \omega \to \alpha$ and define $s_n \in \omega^1$ by $s_n(0) = n$ for any $n \in \omega$. For any $n \in \omega$ and $s \in H_{\varphi(n)}$ let $t_s(i) = s(i) + n + 1$ for any $i \in \mathrm{dom}(s)$. Then $w_s = s_n ^\frown t_s \in \Omega$ for any $s \in H_{\varphi(n)}$; let $G_n = \{w_s : s \in H_{\varphi(n)}\}$. It is easy to see that the family $\{G_n : n \in \omega\}$ is disjoint and $\eta(G_n) \geq \eta(H_{\varphi(n)}) \geq \varphi(n)$ for any $n \in \omega$; let $G = \bigcup\{G_n : n \in \omega\}$.

It is straightforward that $A_\beta(G) = \bigcup\{A_\beta(G_n) : n \in \omega\}$ for any $\beta < \omega_1$. An immediate consequence is that $\eta(G) \geq \sup\{\eta(G_n) : n \in \omega\} \geq \sup\{\beta : \beta < \alpha\} = \alpha$. Therefore we can take $H_\alpha = G$.

Now assume that $\alpha = \beta + 1$ and we have a set H_β such that $\eta(H_\beta) \geq \beta$. For any $s \in H_\beta$ let $s'(n) = s(n) + 2$ for any $n \in \mathrm{dom}(s)$. It is clear that, for the set $H' = \{s' : s \in H_\beta\}$ we have $\eta(H') \geq \beta$. Then $s_0 ^\frown s' \in \Omega$ for any $s \in H_\beta$; so $H_\alpha = \{s_0\} \cup \{s_0 ^\frown s' : s \in H_\beta\} \in \mathcal{E}$. Since s_0 is the minimal element of H_α, we have $s_0 \in A_\gamma(H_\alpha)$ for any $\gamma \leq \beta$. Consequently, $\eta(H_\alpha) \geq \beta + 1 = \alpha$; so our inductive procedure gives us a set $H_\alpha \in \mathcal{E}$ such that $\eta(H_\alpha) \geq \alpha$ for any $\alpha < \omega_1$. If $x_\alpha = \chi_{H_\alpha}$ then $x_\alpha \in T$ and $\mathrm{ord}(x_\alpha) \geq \alpha$ for any $\alpha < \omega_1$, i.e., $\mathrm{ord}(T) = \omega_1$ and (7) is proved.

Assume towards a contradiction that $T_{\mathcal{A}}^*$ is K-analytic. By SFFS-388, there exists a compact-valued upper semicontinuous onto map $\varphi : \mathbb{P} \to T_{\mathcal{A}}^*$. For any $Z \subset \mathbb{P}$ let $\varphi(Z) = \bigcup\{\varphi(p) : p \in Z\}$. Given any $s \in \omega^{<\omega}$ let $A_s = \{p \in \mathbb{P} : s \subset p\}$ and $B_s = \varphi(A_s) \cap T$. For any $n \in \omega$ define $u_n \in \omega^1$ by $u_n(0) = n$. Given $s \in \Omega$ and $x \in T$ let $Y(x, s) = \{t \in \Omega : s ^\frown t \in x^{-1}(1)\}$ and $x[s] = \chi_{Y(x,s)}$.

It follows from $\mathbb{P} = \bigcup\{A_{u_n} : n \in \omega\}$ that $T = \bigcup\{B_{u_n} : n \in \omega\}$; so the properties (6) and (7) imply that there is $n_0 \in \omega$ for which $B_{u_{n_0}}$ is unbounded; let $t_0 = u_{n_0}$.

It is easy to see that $\mathrm{ord}(x) \leq \sup\{\mathrm{ord}(x[u_n]) + 1 : n \in \omega \text{ and } u_n \in x^{-1}(1)\}$ for any $x \in T$. Therefore there is $m_0 \in \omega$ such that the set $Z_0 = \{x[u_{m_0}] : x \in B_{t_0} \text{ and } u_{m_0} \in x^{-1}(1)\}$ is unbounded; let $s_0 = u_{m_0}$.

The set $\{z|\Omega_{m_0} : z \in Z_0\}$ being finite, we can apply the property (6) again to see that there exists $x_0 \in T$ such that the set $Z_1 = \{x[s_0] : x \in O_{m_0}(x_0) \cap B_{t_0} \text{ and } s_0 \in x^{-1}(1)\}$ is unbounded. We have $s_0 \in \Omega_{m_0}$; if $s_0 \notin x_0^{-1}(1)$ then $s_0 \notin x^{-1}(1)$ for any point $x \in O_{m_0}(x_0)$. The set Z_1 being unbounded, there exists $x \in O_{m_0}(x_0)$ such that $x[s_0] \in T$; however $Y(x, s_0) = \emptyset$ and hence $x[s_0] \notin T$; this contradiction shows that $s_0 \in x_0^{-1}(1)$.

Assume that $k \in \omega$ and we have constructed

$$n_0, \ldots, n_k, m_0, \ldots, m_k \in \omega, t_0, \ldots, t_k \in \omega^{<\omega}, s_0, \ldots, s_k \in \Omega \text{ and } x_0, \ldots, x_k \in T$$

with the following properties:

(8) $m_i < m_{i+1}$ for any $i < k$;
(9) $t_0 = u_{n_0}$ and $t_{i+1} = t_i \frown n_{i+1}$ for any $i < k$;
(10) $s_0 = u_{m_0}$ and $s_{i+1} = s_i \frown m_{i+1}$ for any $i < k$;
(11) $s_i \in x_i^{-1}(1)$ for all $i \le k$;
(12) $\{x[s_i] : x \in O_{m_i}(x_i) \cap B_{t_i} \text{ and } s_i \in x^{-1}(1)\}$ is unbounded for any $i \le k$;
(13) $O_{m_i}(x_i) = O_{m_i}(x_k)$ for any $i \le k$.

We have $B_{t_k} = \bigcup \{B_{t_k \frown n} : n \in \omega\}$; so it follows from (6) that there exists $n_{k+1} \in \omega$ such that, for $t_{k+1} = t_k \frown n_{k+1}$, the set $\{x[s_k] : x \in O_{m_k}(x_k) \cap B_{t_{k+1}}$ and $s_k \in x^{-1}(1)\}$ is unbounded.

For any $n > n_k$ the point $w_n = s_k \frown n$ belongs to Ω; it is easy to see that $\mathrm{ord}(x[s_k]) \le \sup\{\mathrm{ord}(x[w_n]) + 1 : n \in \omega$ and $w_n \in x^{-1}(1)\}$ for any $x \in T$ with $s_k \in x^{-1}(1)$. As a consequence, there is $m_{k+1} \in \omega$, $m_{k+1} > m_k$ such that, for the element $s_{k+1} = s_k \frown m_{k+1} \in \Omega$, the set $Z_k = \{x[s_{k+1}] : x \in O_{m_k}(x_k) \cap B_{t_{k+1}}$ and $s_{k+1} \in x^{-1}(1)\}$ is unbounded.

Since the set $\{z|\Omega_{m_{k+1}} : z \in Z_k\}$ is finite, we can apply the property (6) once more to see that there is $x_{k+1} \in T$ such that $O_{m_{k+1}}(x_{k+1}) \subset O_{m_k}(x_k)$ and the set $Z_{k+1} = \{x[s_{k+1}] : x \in O_{m_{k+1}}(x_{k+1}) \cap B_{t_{k+1}}$ and $s_{k+1} \in x^{-1}(1)\}$ is unbounded. We have $s_{k+1} \in \Omega_{m_{k+1}}$; if $s_{k+1} \notin x_{k+1}^{-1}(1)$ then $s_{k+1} \notin x^{-1}(1)$ for any $x \in O_{m_{k+1}}(x_{k+1})$. The set Z_{k+1} being unbounded, there exists $x \in O_{m_{k+1}}(x_{k+1})$ such that $x[s_{k+1}] \in T$; however $Y(x, s_{k+1}) = \emptyset$ and hence $x[s_{k+1}] \notin T$; this contradiction shows that $s_{k+1} \in x_{k+1}^{-1}(1)$.

Now it is clear that the properties (8)–(10) still hold for all $i \le k$ and (11)–(13) are fulfilled for all $i \le k + 1$. Thus our inductive procedure makes it possible to construct sequences $\{n_i : i \in \omega\} \subset \omega, \{m_i : i \in \omega\} \subset \omega, \{t_i : i \in \omega\} \subset \omega^{<\omega}$ as well as sequences $\{s_i : i \in \omega\} \subset \Omega$ and $\{x_i : i \in \omega\} \subset T$ for which the conditions (8)–(13) are satisfied for all $i < \omega$.

It follows from (13) that there exists $y \in M$ such that $O_{m_n}(y) = O_{m_n}(x_n)$ and hence $s_n \in y^{-1}(1)$ for any $n \in \omega$.

For any $n \in \omega$ choose a point $y_n \in B_{t_n} \cap O_{m_n}(y)$; then $\{y_1, \ldots, y_k\} \in \mathcal{A}_0$ for any $k \in \omega$; so $B = \{y_i : i \in \omega\} \in \mathcal{A}$. Therefore the set B is closed and discrete in $T_{\mathcal{A}}^*$. The set B is contained in T while $y \in M \backslash T$ because an infinite branch $\{s_n : n \in \omega\}$ is a subset of $y^{-1}(1)$. This implies that B is an infinite set. Let $p(k) = n_k$ for any $k \in \omega$; then $p \in \mathbb{P}$. Since $\varphi(p)$ is compact, the set $\varphi(p) \cap B$ is finite; so the set $C = B \backslash \varphi(p)$ is infinite.

The set C is closed and disjoint from $\varphi(p)$. The family $\{A_{t_k} : k \in \omega\}$ is a local base at p in $\mathbb{P}$; the map φ being upper semicontinuous, there is $k \in \omega$ such that $B_{t_k} \cap C = \emptyset$. Observe that $y_n \in B_{t_n} \subset B_{t_k}$ for all $n \ge k$ which shows that $B \backslash B_{t_k}$ is finite which is a contradiction with $C \subset B \backslash B_{t_k}$.

This contradiction shows that the space T_A^* is not K-analytic and therefore the space $C_p(X)$ is not K-analytic either (see Problem 172). Thus X is a compact space such that $C_p(X)$ is Lindelöf Σ (i.e., X is Gul'ko compact) but not K-analytic.

U.301. *Prove that the following conditions are equivalent for any space X:*

(i) *X is functionally perfect;*
(ii) *X condenses onto a subspace of $C_p(Y)$ for some compact Y;*
(iii) *X condenses onto a subspace of $C_p(Y)$ for some σ-compact Y;*
(iv) *there exists a σ-compact $H \subset C_p(X)$ which separates the points of X;*
(v) *the space $C_p(X)$ is k-separable.*

Solution. Suppose that X is functionally perfect and fix a compact set $Y \subset C_p(X)$ which separates the points of X. For any $x \in X$ let $e_x(f) = f(x)$ for any $f \in Y$. Then $e_x \in C_p(Y)$ and the map $e : X \to C_p(Y)$ defined by $e(x) = e_x$ for any $x \in X$, is continuous (see TFS-166); let $Z = e(X)$. The map $e : X \to Z$ is a condensation (this was also proved in TFS-166); so X condenses onto the subspace Z of the space $C_p(Y)$. This proves (i)$\Longrightarrow$(ii).

The implication (ii)$\Longrightarrow$(iii) being trivial assume that there is a σ-compact space Y for which there exists a condensation $\varphi : X \to Z$ for some $Z \subset C_p(Y)$. For any $y \in Y$ let $q_y(f) = f(y)$ for any $f \in Z$. Then $q_y \in C_p(Z)$ and the map $q : Y \to C_p(Z)$ defined by $q(y) = q_y$ for any $y \in Y$, is continuous. This shows that the space $Y' = q(Y)$ is σ-compact and separates the points of Z (see TFS-166).

Given $f \in C_p(Z)$ let $\varphi^*(f) = f \circ \varphi$. Then the map $\varphi^* : C_p(Z) \to C_p(X)$ is continuous by TFS-163; so the space $H = \varphi^*(Y') \subset C_p(X)$ is σ-compact. If x and y are distinct points of X then $\varphi(x) \neq \varphi(y)$; since Y' separates the points of Z, there is $f \in Y'$ with $f(\varphi(x)) \neq f(\varphi(y))$ or, in other words, $\varphi^*(f)(x) \neq \varphi^*(f)(y)$. Thus $\varphi^*(f)$ is a function from H which separates x and y. This proves that H is a σ-compact subspace of $C_p(X)$ which separates the points of X, i.e., we settled (iii)$\Longrightarrow$(iv).

Now assume that there exists a σ-compact $H \subset C_p(X)$ which separates the points of X and consider the minimal subalgebra Y of $C_p(X)$ which contains H; the σ-compactness $\mathcal{P}$ being a k-directed property we can apply Problem 006 to see that Y belongs to the class $\mathcal{P}_\sigma = \mathcal{P}$, i.e., Y is also σ-compact. The set $Y \supset H$ separates the points of X because so does H; therefore TFS-192 is applicable to conclude that Y is dense in $C_p(X)$. This shows that $C_p(X)$ is k-separable and hence (iv)$\Longrightarrow$(v) is proved.

Finally suppose that the space $C_p(X)$ has a dense σ-compact subspace; since $C_p(X, (-1, 1)) \simeq C_p(X)$, we can fix a set $Y \subset C_p(X, (-1, 1))$ such that Y is dense in $C_p(X, (-1, 1))$ and $Y = \bigcup_{n \in \omega} Y_n$ where every Y_n is compact. It is evident that Y separates the points of X.

For any $n \in \omega$ the set $K_n = \{\frac{1}{n+1}f : f \in Y_n\}$ is compact because the multiplication by $\frac{1}{n+1}$ maps Y_n continuously onto K_n (see TFS-116). It is an easy exercise that $\bigcup_{n \in \omega} K_n$ still separates the points of X. Let u be the function which is identically zero on X and consider the set $K = \{u\} \cup (\bigcup\{K_n : n \in \omega\})$.

Fix any $U \in \tau(u, C_p(X))$; there is $\varepsilon > 0$ and a finite set $F \subset X$ such that $V = \{f \in C_p(X) : |f(x)| < \varepsilon$ for any $x \in F\} \subset U$. Choose $m \in \omega$ such that $\frac{1}{m+1} < \varepsilon$; then $|f(x)| < \frac{1}{n+1} \leq \frac{1}{m+1} < \varepsilon$ for any $f \in K_n$, $x \in X$ and $n \geq m$ which implies that $K_n \subset V \subset U$ for all $n \geq m$ and therefore $K \backslash U \subset K_0 \cup \ldots \cup K_m$ is a compact set.

We have shown that $K \backslash U$ is compact for any open $U \ni u$; an evident consequence is that K is a compact subspace of $C_p(X)$ which separates the points of X, i.e., X is functionally perfect. This settles the implication (v)$\Longrightarrow$(i) and makes our solution complete.

U.302. *Show that neither $\Sigma(A)$ nor $\mathbb{R}^A$ is functionally perfect whenever the set A is uncountable.*

Solution. Let $\kappa = |A|$ and observe that $\mathbb{I}^\kappa \subset \mathbb{R}^\kappa$ while $\mathbb{R}^\kappa \simeq (-1, 1)^\kappa \subset \mathbb{I}^\kappa$. As a consequence, $nw(\mathbb{I}^\kappa) \leq nw(\mathbb{R}^\kappa) \leq nw(\mathbb{I}^\kappa)$ which shows that we have the equalities $nw(\mathbb{R}^\kappa) = nw(\mathbb{I}^\kappa) = w(\mathbb{I}^\kappa) = \kappa$ (see Fact 3 of S.368 and Fact 4 of S.307).

Assume that a compact $K \subset C_p(\mathbb{R}^\kappa)$ separates the points of $\mathbb{R}^\kappa$. Then K is metrizable by TFS-307 and the evaluation map $e : \mathbb{R}^\kappa \to C_p(K)$ is injective (see TFS-166); let $H = e(\mathbb{R}^\kappa)$. We have $iw(H) \leq nw(H) \leq nw(C_p(K)) = nw(K) = \omega$ (see TFS-156). Since $\mathbb{R}^\kappa$ condenses onto H, we have the inequality $iw(\mathbb{R}^\kappa) \leq iw(H) \leq \omega$. The space $\mathbb{R}^\kappa$ being stable by SFFS-268, we must have $nw(\mathbb{R}^\kappa) = \omega$; this contradiction shows that $\mathbb{R}^\kappa$ is not functionally perfect because $nw(\mathbb{R}^\kappa) = \kappa = |A|$ is uncountable. Since $\mathbb{R}^A \simeq \mathbb{R}^\kappa$, the space $\mathbb{R}^A$ is not functionally perfect either.

Next observe that $\Sigma(A) \simeq C_p(L(\kappa))$ by Problem 106; it is easy to see that $L(\kappa)$ is a P-space; so every countable $A \subset L(\kappa)$ is closed and C-embedded in $L(\kappa)$ (see Fact 1 of S.479). This implies that the space $C_p(L(\kappa))$ is pseudocomplete and hence $\upsilon(C_p(L(\kappa))) \simeq \mathbb{R}^{L(\kappa)} \simeq \mathbb{R}^\kappa$ (see TFS-485). This proves that we can identify $\mathbb{R}^A$ with $\upsilon(\Sigma(A))$; let $\pi : C_p(\mathbb{R}^A) \to C_p(\Sigma(A))$ be the respective restriction map.

Given a compact subspace $K \subset C_p(\Sigma(A))$ the space $K' = \pi^{-1}(K)$ is countably compact by Problem 228. Therefore we can apply TFS-307 once more to see that K' is compact and metrizable; consequently, K is also metrizable being a continuous image of K'. If K separates the points of the space $\Sigma(A)$ then the evaluation map $q : \Sigma(A) \to C_p(K)$ is injective by TFS-166; let $G = q(\Sigma(A))$. Since q condenses $\Sigma(A)$ onto G, we have $iw(\Sigma(A)) \leq iw(G) \leq nw(G) \leq nw(C_p(K)) = nw(K) = \omega$; so it is possible to apply ω-stability of the space $\Sigma(A)$ to conclude that $nw(\Sigma(A)) = \omega$ (see SFFS-268).

For any $a \in A$ let $f_a(a) = 1$ and $f_a(b) = 0$ for all $b \in A \backslash \{a\}$. It is straightforward that the set $D = \{f_a : a \in A\} \subset \Sigma(A)$ is a discrete subspace of $\Sigma(A)$; since $|D| = |A| = \kappa > \omega$, we conclude that $nw(\Sigma(A)) \geq |D| > \omega$; this final contradiction shows that $\Sigma(A)$ is not functionally perfect as well.

U.303. *Prove that the spaces $\sigma(A)$ and $\Sigma_*(A)$ are functionally perfect for any A.*

Solution. If A is finite then the spaces $\sigma(A)$ and $\Sigma_*(A)$ coincide with the second countable space $\mathbb{R}^A$. Therefore the space $C_p(\sigma(A)) = C_p(\Sigma_*(A))$ is separable and hence k-separable. Applying Problem 301 we can see that $\sigma(A) = \Sigma_*(A)$ is functionally perfect.

Now, if $|A| = \kappa \geq \omega$ then $\Sigma_*(A) \simeq C_p(A(\kappa))$ (see Problem 105). The space $A(\kappa)$ being compact, we can apply Problem 301 to see that $C_p(A(\kappa))$ and hence $\Sigma_*(A)$ is functionally perfect. Finally, $\sigma(A)$ is a subspace of $\Sigma_*(A)$; so it embeds in $C_p(A(\kappa))$ which, together with Problem 301, implies that $\sigma(A)$ is functionally perfect.

U.304. *Prove that $C_p(\omega_1)$ is functionally perfect.*

Solution. Given an ordinal $\alpha < \omega_1$ let $e_\alpha(f) = f(\alpha)$ for any function $f \in C_p(\omega_1)$. Then $e_\alpha \in C_p(C_p(\omega_1))$ for every $\alpha < \omega_1$ and the set $E = \{e_\alpha : \alpha < \omega_1\}$ is homeomorphic to ω_1 (see TFS-167). Let u be the function which is identically zero on $C_p(\omega_1)$ and consider the space

$$K = \{u\} \cup \{e_0\} \cup \{e_\alpha - e_{\alpha+1} : \alpha < \omega_1\} \subset C_p(C_p(\omega_1)).$$

Let us prove first that

(1) the set K separates the points of $C_p(\omega_1)$.

To see that (1) is true take distinct functions $f, g \in C_p(\omega_1)$ and consider the ordinal $\beta = \min\{\alpha < \omega_1 : f(\alpha) \neq g(\alpha)\}$. If $\beta = 0$ then $e_0(f) = f(0) \neq g(0) = e_0(g)$; so $e_0 \in K$ separates f and g.

Now, if $\beta > 0$ is a limit ordinal then $A = \{\alpha : \alpha < \beta\}$ is dense in $A \cup \{\beta\}$; so it follows from $f|A = g|A$ that $f(\beta) = g(\beta)$ (see Fact 0 of S.351). This contradiction shows that β is a successor ordinal, i.e., $\beta = \alpha + 1$ for some $\alpha < \omega_1$. By our definition of the ordinal β, we have $f(\alpha+1) \neq g(\alpha+1)$ while $f(\alpha) = g(\alpha)$. As a consequence, $e_\alpha(f) = e_\alpha(g)$ and $e_{\alpha+1}(f) \neq e_{\alpha+1}(g)$ which shows that $(e_\alpha - e_{\alpha+1})(f) \neq (e_\alpha - e_{\alpha+1})(g)$, i.e., (1) is proved.

Let us show next that

(2) for any $U \in \tau(u, C_p(C_p(\omega_1)))$, the set $K \backslash U$ is finite.

Indeed, given any $U \in \tau(u, C_p(C_p(\omega_1)))$, there is a finite set $F \subset C_p(\omega_1)$ and $\varepsilon > 0$ such that $V = \{\varphi \in C_p(C_p(\omega_1)) : |\varphi(f)| < \varepsilon \text{ for all } f \in F\} \subset U$. Therefore it suffices to show that $K \backslash V$ is finite. If $V_f = \{\varphi \in C_p(C_p(\omega_1)) : |\varphi(f)| < \varepsilon\}$ for any $f \in F$ then $V = \bigcap_{f \in F} V_f$; so it suffices to establish that $K \backslash V_f$ is finite for any $f \in F$.

Suppose for a contradiction that there is a function $f \in F$ for which the set $K \backslash V_f$ is infinite. Then there is a strictly increasing sequence $\{\alpha_n : n \in \omega\} \subset \omega_1$ such that $|(e_{\alpha_n} - e_{\alpha_n+1})(f)| \geq \varepsilon$ for any $n \in \omega$. If $\alpha = \sup\{\alpha_n : n \in \omega\}$ then the sequence $S = \{\alpha_n : n \in \omega\}$ converges to α so, by continuity of f, there is $\gamma < \alpha$ such that $|f(\beta) - f(\alpha)| < \frac{\varepsilon}{2}$ whenever $\gamma < \beta < \alpha$. The sequence S being convergent to α, there exists a number $n \in \omega$ for which $\gamma < \alpha_n < \alpha$; an immediate consequence is that $|f(\alpha_n) - f(\alpha)| < \frac{\varepsilon}{2}$ and $|f(\alpha_n + 1) - f(\alpha)| < \frac{\varepsilon}{2}$. Thus

$$|(e_{\alpha_n} - e_{\alpha_n+1})(f)| = |f(\alpha_n) - f(\alpha_n + 1)|$$
$$\leq |f(\alpha_n) - f(\alpha)| + |f(\alpha_n + 1) - f(\alpha)| < \frac{\varepsilon}{2} + \frac{\varepsilon}{2} = \varepsilon;$$

this contradiction shows that (2) is proved.

An evident consequence of (2) is that K is compact; the property (1) shows that K separates the points of $C_p(\omega_1)$ so $C_p(\omega_1)$ is functionally perfect.

U.305. *Show that, if a space condenses onto a functionally perfect space, then it is functionally perfect.*

Solution. Suppose that Y is functionally perfect and $\varphi : X \to Y$ is a condensation. By Problem 301, there is a condensation $\mu : Y \to Z$ where $Z \subset C_p(K)$ for some compact space K. Evidently, $\mu \circ \varphi$ condenses X onto Z; so we can apply Problem 301 again to see that X is also functionally perfect.

U.306. *Prove that any subspace of a functionally perfect space is functionally perfect.*

Solution. Suppose that X is functionally perfect and $Y \subset X$. By Problem 301, there is a condensation $\varphi : X \to Z$ where $Z \subset C_p(K)$ for some compact space K. If $Y' = \varphi(Y)$ then $\varphi_0 = \varphi|Y$ condenses Y onto $Y' \subset C_p(K)$. Applying Problem 301 again we conclude that Y is also functionally perfect.

U.307. *Prove that a countable product of functionally perfect spaces is a functionally perfect space. In particular, a countable product of Eberlein compact spaces is Eberlein compact.*

Solution. Suppose that a space X_n is functionally perfect and apply Problem 301 to fix a condensation $\varphi_n : X_n \to Y_n$ where $Y_n \subset C_p(K_n)$ for some compact space K_n for any $n \in \omega$. Let $X = \prod_{n\in\omega} X_n$ and $Y = \prod_{n\in\omega} Y_n$; it is immediate that the map $\varphi = \prod_{n\in\omega} \varphi_n : X \to Y$ is a condensation. Besides, $Y \subset \prod_{n\in\omega} C_p(K_n) \simeq C_p(K)$ where $K = \bigoplus\{K_n : n \in \omega\}$ (see TFS-114). It turns out that X condenses onto a subspace of $C_p(K)$ while the space K is σ-compact; this makes it possible to apply Problem 301 again to conclude that $X = \prod_{n\in\omega} X_n$ is functionally perfect.

U.308. *Prove that any σ-product of functionally perfect spaces is a functionally perfect space.*

Solution. Suppose that X_t is functionally perfect and a point $a_t \in X_t$ is fixed for any $t \in T$. Let $X = \prod_{t\in T} X_t$ and define a point $a \in X$ by $a(t) = a_t$ for any $t \in T$. We must prove that the space $\sigma(X, a) = \{x \in X : |\{t \in T : x(t) \neq a_t\}| < \omega\}$ is functionally perfect. For any $x \in \sigma(X, a)$ let $\mathrm{supp}(x) = \{t \in T : x(t) \neq a_t\}$.

Fix a compact set $F_t \subset C_p(X_t)$ which separates the points of X_t and consider, for any $t \in T$, the map $\varphi_t : C_p(X_t) \to C_p(X_t)$ defined by the formula $\varphi_t(f) = f - f(a_t)$ for each $f \in C_p(X_t)$. It is easy to see that every φ_t is continuous and the compact set $K_t = \varphi_t(F_t)$ still separates the points of X_t. Furthermore,

(1) $f(a_t) = 0$ for any $f \in K_t$ and $t \in T$,

because $\varphi_t(f)(a_t) = f(a_t) - f(a_t) = 0$ for every $f \in C_p(X_t)$.

Given $t \in T$, we will need the natural projection $\pi_t : \sigma(X, a) \to X_t$; its dual map $\pi_t^* : C_p(X_t) \to C_p(\sigma(X, a))$ is continuous (see TFS-163); so the set $G_t = \pi_t^*(K_t)$ is a compact subspace of $C_p(\sigma(X, a))$. Define a function $u \in C_p(\sigma(X, a))$ to be identically zero on $\sigma(X, a)$ and consider the set $G = \{u\} \cup (\bigcup\{G_t : t \in T\})$. We claim that

(2) for any $U \in \tau(u, C_p(\sigma(X, a)))$ there is a finite $A \subset T$ such that $G_t \subset U$ for any $t \in T \backslash A$.

Take any $U \in \tau(u, C_p(\sigma(X, a)))$; there is a finite set $E \subset \sigma(X, a)$ and $\varepsilon > 0$ such that $V = \{f \in C_p(\sigma(X, a)) : |f(x)| < \varepsilon$ for all $x \in E\} \subset U$. The set $A = \bigcup\{\text{supp}(x) : x \in E\}$ is finite. Fix any $t \in T \backslash A$ and $x \in E$; then $\pi_t(x) = a_t$ and therefore $f(\pi_t(x)) = 0$ for each $f \in K_t$. For any $g \in G_t$ there is $f \in K_t$ such that $g = \pi_t^*(f) = f \circ \pi_t$. Thus $g(x) = f(\pi_t(x)) = 0$ which shows that $g(x) = 0$ for any $g \in G_t$ and $x \in E$. This implies $G_t \subset V \subset U$ for all $t \in T \backslash A$, i.e., (2) is proved.

It follows from (2) that, for any $U \in \tau(u, C_p(\sigma(X, a)))$ there is a finite $A \subset T$ such that the set $G \backslash U$ is contained in a compact set $P = \bigcup\{G_t : t \in A\}$. Since $G \backslash U$ is closed in G, it also closed in P; the space P being compact, the set $G \backslash U$ is compact for any $U \in \tau(u, C_p(\sigma(X, a)))$. This, evidently, implies that G is compact.

Finally, take any distinct $x, y \in \sigma(X, a)$. There exists $t \in T$ with $\pi_t(x) \neq \pi_t(y)$. Since the set K_t separates the points of X_t, there is $f \in K_t$ with $f(\pi_t(x)) \neq f(\pi_t(y))$. The function $g = \pi_t^*(f)$ belongs to G and we have $g(x) = f(\pi_t(x)) \neq f(\pi_t(y)) = g(y)$, i.e., g separates x and y. This proves that G is a compact subset of $\sigma(X, a)$ which separates the points of $\sigma(X, a)$, i.e., $\sigma(X, a)$ is functionally perfect.

U.309. *Prove that any product of k-separable spaces is k-separable.*

Solution. Suppose that X_t is k-separable and fix a σ-compact $Y_t \subset X_t$ for any $t \in T$; we must prove that the space $X = \prod_{t \in T} X_t$ is k-separable. The space $Y = \prod_{t \in T} Y_t$ is dense in X; so it suffices to show that Y is k-separable. We can choose a compact subset $Y_t^n \subset Y_t$ of the space Y_t such that $Y_t^n \subset Y_t^{n+1}$ for any $n \in \omega$ and $Y_t = \bigcup_{n \in \omega} Y_t^n$ for any $t \in T$. We can assume, without loss of generality, that $Y_t^0 \neq \emptyset$ and hence we can fix a point $a_t \in Y_t^0$ for any $t \in T$.

The product space $K_n = \prod_{t \in T} Y_t^n \subset Y$ is compact for any number $n \in \omega$ and hence the set $K = \bigcup_{n \in \omega} K_n$ is σ-compact. Given any point $y \in Y$ and $U \in \tau(y, Y)$ there exists a finite set $S \subset T$ and a set $O_s \in \tau(Y_s)$ for each index $s \in S$ such that $y \in V = \prod_{s \in S} O_s \times \prod_{t \in T \backslash S} Y_t \subset U$. There is $m \in \omega$ such that $y(s) \in Y_s^m$ for any $s \in S$. Let $z(s) = y(s)$ for all $s \in S$ and $z(t) = a_t$ for any $t \in T \backslash S$. It is immediate that $z \in K_m \cap V \subset K \cap U$; so $K \cap U \neq \emptyset$ for any $U \in \tau(y, Y)$, i.e., $y \in \overline{K}$. Since the point $y \in Y$ was chosen arbitrarily, we have proved that $\overline{K} = Y$, i.e., the σ-compact set K is dense in Y. Thus Y is k-separable; so X is k-separable as well.

U.310. *Prove that a space X is hereditarily k-separable (i.e., every subspace $Y \subset X$ is k-separable) if and only if X is hereditarily separable.*

Solution. It is evident that any hereditarily separable space has to be hereditarily k-separable; so assume that the space X is hereditarily k-separable. Observe that a discrete k-separable space must be countable; so every discrete subspace of X is countable, i.e., $s(X) = \omega$. If X is not hereditarily separable then there exists a set $Y = \{y_\alpha : \alpha < \omega_1\} \subset X$ such that $y_\alpha \notin \overline{\{y_\beta : \beta < \alpha\}}$ for any $\alpha < \omega_1$ (see SFFS-004). It is easy to see that

(1) the closure in Y of any countable subset of Y is countable and hence the space Y is not separable.

However, it is k-separable by our assumption about X; so we can take a family $\{K_n : n \in \omega\}$ of compact subspaces of Y such that $K = \bigcup\{K_n : n \in \omega\}$ is dense in Y. It is easy to see that any free sequence in a space Z is a discrete subspace of Z; so any free sequence in K_n is countable; this implies that $t(K_n) \leq \omega$ for any $n \in \omega$ (see TFS-328). The property (1) implies that every K_n is ω-monolithic; so it is Fréchet–Urysohn by Fact 1 of U.080.

If $n \in \omega$ and $F \subset K_n$ is a closed dense-in-itself closed subspace of K_n then we can apply Fact 1 of T.045 to see that there is a separable closed dense-in-itself set $G \subset F$. It follows from (1) that G has to be countable; so G is a countable dense-in-itself compact space which is a contradiction with the Baire property of G. Therefore every K_n is scattered and hence the set D_n of isolated points of K_n is dense in K_n. Since D_n is discrete, it has to be countable so the space K_n is also countable by (1). As a consequence, K is a dense countable subset of Y; this last contradiction shows that X is hereditarily separable.

U.311. *Let $f : X \to Y$ be an irreducible perfect map. Show that X is k-separable if and only if so is Y.*

Solution. If X is k-separable then it has a dense σ-compact subspace A; it is evident that $f(A)$ is a dense σ-compact subspace of Y; so Y is k-separable as well. Here we only used continuity of the map f.

Now assume that f is perfect and irreducible while Y is k-separable. Fix a family $\{K_n : n \in \omega\}$ of compact subspaces of Y such that $K = \bigcup\{K_n : n \in \omega\}$ is dense in Y. The set $F_n = f^{-1}(K_n)$ is compact for any $n \in \omega$ (see Fact 2 of S.259). If the set $F = \bigcup\{F_n : n \in \omega\}$ is not dense in X then $E = \overline{F}$ is a proper closed subset of X; by irreducibility of f, the set $G = f(E)$ is a proper closed subset of Y and therefore $U = Y \backslash G \neq \emptyset$. The set K being dense in the space Y, there exists a number $n \in \omega$ such that $K_n \cap U \neq \emptyset$. But $K_n = f(F_n) \subset f(F) \subset f(E)$ which is a contradiction with $K_n \cap (Y \backslash f(E)) = K_n \cap U \neq \emptyset$. Thus F is a dense σ-compact subspace of X, i.e., X is k-separable.

U.312. *Prove that, for any k-separable X, the space $C_p(X)$ is functionally perfect. In particular, the space $C_p(X)$ is functionally perfect for any compact X.*

Solution. Fix a family $\{K_n : n \in \omega\}$ of compact subspaces of the space X such that $K = \bigcup\{K_n : n \in \omega\}$ is dense in X. The restriction map $\pi : C_p(X) \to C_p(K)$ is an injection (see TFS-152); let $Z = \pi(C_p(X))$. It turns out that $C_p(X)$ condenses onto the subspace Z of $C_p(K)$ where the space K is σ-compact. This makes it possible to apply Problem 301 to conclude that $C_p(X)$ is functionally perfect.

U.313. *Give an example of a non-k-separable space X for which $C_p(X)$ is functionally perfect.*

Solution. The space $C_p(\omega_1)$ is functionally perfect by Problem 304. If $K \subset \omega_1$ is an uncountable set then the family $\{K \cap \{\beta : \beta < \alpha\} : \alpha < \omega_1\}$ is an open cover of

K which has no countable subcover; therefore K is not Lindelöf. This shows that every Lindelöf subspace of ω_1 is countable. In particular, every σ-compact subspace of ω_1 is countable; since the closure of every countable subset of ω_1 is countable, ω_1 is not separable and hence not k-separable. Thus $X = \omega_1$ is an example of a non-k-separable space such that $C_p(X)$ is functionally perfect.

U.314. *Prove that, for an arbitrary space X, the space $C_p(X)$ is a continuous image of $C_p(C_p(C_p(X)))$.*

Solution. Let $e_x(f) = f(x)$ for any $f \in C_p(X)$; then $e_x \in C_p(C_p(X))$ and the map $e : X \to C_p(C_p(X))$ defined by $e(x) = e_x$ for any $x \in X$, is an embedding (see TFS-167). Therefore the space $E = \{e_x : x \in X\}$ is homeomorphic to X; besides, the set E is C-embedded in the space $C_p(C_p(X))$ by TFS-168. As an immediate consequence, the restriction $\pi : C_p(C_p(C_p(X))) \to C_p(E)$ maps $C_p(C_p(C_p(X)))$ continuously onto $C_p(E) \simeq C_p(X)$.

U.315. *Prove that $C_p(X)$ is k-separable if and only if $C_pC_p(X)$ is functionally perfect. As a consequence, X is functionally perfect if and only if $C_p(C_p(X))$ is functionally perfect.*

Solution. If $C_p(X)$ is k-separable then $C_p(C_p(X))$ is functionally perfect by Problem 312. Now, if $C_p(C_p(X))$ is functionally perfect then $C_p(C_p(C_p(X)))$ is k-separable by Problem 301. Being a continuous image of $C_p(C_p(C_p(X)))$ by Problem 314, the space $C_p(X)$ is also k-separable. Finally, apply Problem 301 once more to see that a space X is functionally perfect $\iff$ $C_p(X)$ is k-separable $\iff$ $C_p(C_p(X))$ is functionally perfect.

U.316. *Prove that any metrizable space is functionally perfect. In particular, any second countable space is functionally perfect and hence any metrizable compact space is Eberlein compact.*

Solution. If M is a metrizable space then there exists a compact K such that M embeds in $C_p(K)$ (see Problem 034). Therefore Problem 301 is applicable to conclude that M is functionally perfect.

U.317. *Let X be a metrizable space. Prove that $C_p(X)$ is functionally perfect if and only if X is second countable.*

Solution. If X is second countable then it is separable (and hence k-separable); so $C_p(X)$ is functionally perfect by Problem 312. Now, if $C_p(X)$ is functionally perfect and $w(X) > \omega$ then $ext(X) > \omega$ and hence we can apply Fact 1 of S.215 to conclude that $\mathbb{R}^{\omega_1}$ embeds in $C_p(X)$. Therefore $\mathbb{R}^{\omega_1}$ is also functionally perfect by Problem 306; this contradiction with Problem 302 shows that X has to be second countable.

U.318. *Prove that any paracompact space with a G_δ-diagonal can be condensed onto a metrizable space. Deduce from this fact that any paracompact space with a G_δ-diagonal is functionally perfect.*

Solution. Given a space Z and families $\mathcal{A}, \mathcal{B}$ of subsets of Z we will need the family $\mathcal{A} \wedge \mathcal{B} = \{A \cap B : A \in \mathcal{A}, B \in \mathcal{B}\}$. Recall that *a G_δ-diagonal sequence of a space Z is a family* $\{\mathcal{D}_n : n \in \omega\}$ *of open covers of Z such that* $\{z\} = \bigcap\{\text{St}(z, \mathcal{D}_n) : n \in \omega\}$ for each $z \in Z$. It was proved in Fact 1 of T.235 that a space has a G_δ-diagonal if and only if it has a G_δ-diagonal sequence.

Fact 1. A space Z can be condensed onto a metrizable space if and only if it has a G_δ-diagonal sequence $\{\mathcal{U}_n : n \in \omega\}$ such that $\mathcal{U}_{n+1}$ is a star refinement of $\mathcal{U}_n$ for any $n \in \omega$.

Proof. Suppose that $f : Z \to M$ is a condensation of Z onto a metric space (M, d). If $\mathcal{D}_n = \{B_d(y, \frac{1}{n+1}) : y \in M\}$ for any $n \in \omega$ then it is straightforward that $\{\mathcal{D}_n : n \in \omega\}$ is a G_δ-diagonal sequence in M. Any metrizable space is paracompact by TFS-218; so we can construct inductively, using TFS-230, a sequence $\{\mathcal{V}_n : n \in \omega\}$ of open covers of M such that $\mathcal{V}_0 = \mathcal{D}_0$ and $\mathcal{V}_{n+1}$ is a star refinement of $\mathcal{V}_n \wedge \mathcal{D}_n$ for any $n \in \omega$. It is immediate that $\{\mathcal{V}_n : n \in \omega\}$ is still a G_δ-diagonal sequence in M such that $\mathcal{V}_{n+1}$ is a star refinement of $\mathcal{V}_n$ for all $n \in \omega$.

If $\mathcal{U}_n = \{f^{-1}(V) : V \in \mathcal{V}_n\}$ for every $n \in \omega$ then $\{\mathcal{U}_n : n \in \omega\}$ is the promised G_δ-diagonal sequence in Z; this settles necessity.

Now assume that we have a G_δ-diagonal sequence $\{\mathcal{U}_n : n \in \omega\}$ in the space Z such that $\mathcal{U}_{n+1}$ is a star refinement of $\mathcal{U}_n$ for any $n \in \omega$. Given any points $x, y \in Z$ let $\varphi(x, y) = 0$ if $x = y$; if $x \neq y$ then let $\varphi(x, y) = 2^{-n(x,y)}$ where $n(x, y) = \min\{n \in \omega : x \notin \text{St}(y, \mathcal{U}_n)\}$. It follows from $\{y\} = \bigcap\{\text{St}(y, \mathcal{U}_n) : n \in \omega\}$ that the numbers $n(x, y)$ and $\varphi(x, y)$ are well defined. Observe first that

(1) $n(x, y) = n(y, x)$ and hence $\varphi(x, y) = \varphi(y, x)$ for any $x, y \in Z$.

Fix $x, y \in Z$; it suffices to show that $n(x, y) = n(y, x)$ which is evident if $x = y$; so assume that $x \neq y$. The point x belongs to $\text{St}(y, \mathcal{U}_n)$ if and only if there is $U \in \mathcal{U}_n$ with $x, y \in U$ which holds if and only if $y \in \text{St}(x, \mathcal{U}_n)$. Therefore $x \in \text{St}(y, \mathcal{U}_n) \iff y \in \text{St}(x, \mathcal{U}_n)$ for any $n \in \omega$ which shows that $n(x, y) = n(y, x)$; so (1) is proved. An immediate consequence of the definition of φ is that

(2) $\varphi(x, y) = 0$ if $x = y$ and $\varphi(x, y) > 0$ whenever $x \neq y$.

Call *a chain* any indexed sequence $\{x_0, \ldots, x_n\}$ of the points of Z; say that a chain $C = \{x_0, \ldots, x_n\}$ *connects points* $x, y \in Z$ if $x_0 = x$ and $x_n = y$. For any chain $C = \{x_0, \ldots, x_n\}$ let $l(C) = 0$ if $n = 0$ and $l(C) = \varphi(x_0, x_1) + \ldots + \varphi(x_{n-1}, x_n)$ if $n > 0$. If $x, y \in Z$ then let $d(x, y) = \inf\{l(C) : C$ is a chain which connects the points x and $y\}$. Let us show that

(3) the function d is a metric on Z.

It is immediate that $d(x, y) \geq 0$ for any $x, y \in Z$. If $x = y$ then the chain $C = \{x\}$ connects the points x and y; so $d(x, y) \leq l(C) = 0$, i.e., $d(x, y) = 0$.

To see that d is symmetric observe that if a chain $C_0 = \{x_0, \ldots, x_n\}$ connects x and y then the chain $C_1 = \{x_n, \ldots, x_0\}$ connects y and x; since $l(C_0) = l(C_1)$, we have $d(y, x) \leq l(C)$ for any chain C which connects the points x and y.

As a consequence, $d(y, x) \leq d(x, y)$; changing the roles of x and y we obtain the equality $d(x, y) = d(y, x)$ for any $x, y \in Z$, i.e., d is, indeed, symmetric. The following property of d is crucial.

(4) $d(x, y) \leq \varphi(x, y) \leq 2d(x, y)$ for any $x, y \in Z$.

The first inequality follows from the fact that $\{x, y\}$ is a chain which connects the points x and y. If we have a chain $C = \{x_0, \ldots, x_n\}$, call n *the number of links of* C.

To verify the second inequality, it suffices to show that, for any chain C which connects the points x and y, we have $l(C) \geq \frac{1}{2}\varphi(x, y)$. We will do that by induction on the number n of links of the chain C. If $n = 1$ then $C = \{x, y\}$ and hence $l(C) = \varphi(x, y) \geq \frac{1}{2}\varphi(x, y)$.

Suppose that the inequality $l(C) \geq \frac{1}{2}\varphi(x, y)$ has been established for all points $x, y \in Z$ and chains C with at most k links which connect x and y. Take any chain $C = \{x_0, \ldots, x_k, x_{k+1}\}$ which connects some points x and y; we have $\varphi(x, y) = 2^{-m}$ for some $m \in \omega$ which implies that $y \notin \mathrm{St}(x, \mathcal{U}_m)$. If $\varphi(x_i, x_{i+1}) \geq \frac{1}{2}\varphi(x, y)$ for some $i \leq k$ then there is nothing to prove; so assume that $\varphi(x_i, x_{i+1}) < \frac{1}{2}\varphi(x, y) = 2^{-m-1}$ and therefore $x_{i+1} \in \mathrm{St}(x_i, \mathcal{U}_{m+1})$ for all $i \leq k$.

If $x_k \in \mathrm{St}(x, \mathcal{U}_{m+1})$ then there exist sets $U, V \in \mathcal{U}_{m+1}$ such that $\{x, x_k\} \subset U$ and $\{x_k, y\} \in V$ which shows that $\{x, y\} \subset \mathrm{St}(x_k, \mathcal{U}_{m+1})$. The family $\mathcal{U}_{m+1}$ being a star refinement of $\mathcal{U}_m$, there is a set $W \in \mathcal{U}_m$ such that $\mathrm{St}(x_k, \mathcal{U}_{m+1}) \subset W$ and therefore $\{x, y\} \subset W$ whence $y \in \mathrm{St}(x, \mathcal{U}_m)$ which is a contradiction. Thus $x_k \notin \mathrm{St}(x, \mathcal{U}_{m+1})$ and hence the number $p = \min\{l \leq k : x_l \notin \mathrm{St}(x, \mathcal{U}_{m+1})\}$ is well defined. Furthermore, it follows from $x_1 \in \mathrm{St}(x, \mathcal{U}_{m+1})$ that $p > 1$.

If we have $y \in \mathrm{St}(x_p, \mathcal{U}_{m+1})$ then it follows from $x_p \in \mathrm{St}(x_{p-1}, \mathcal{U}_{m+1})$ and $x_{p-1} \in \mathrm{St}(x, \mathcal{U}_{m+1})$ that there exist elements U, V, W of the family $\mathcal{U}_{m+1}$ such that $\{x, x_{p-1}\} \subset U$, $\{x_{p-1}, x_p\} \subset V$ and $\{x_p, y\} \subset W$. An immediate consequence is that $\{x, y\} \subset \mathrm{St}(V, \mathcal{U}_{m+1})$; using again the fact that $\mathcal{U}_{m+1}$ is a star refinement of $\mathcal{U}_m$ we conclude that there is $G \in \mathcal{U}_m$ with $\{x, y\} \subset \mathrm{St}(V, \mathcal{U}_{m+1}) \subset G$; so $\{x, y\} \subset G$, i.e., $y \in \mathrm{St}(x, \mathcal{U}_m)$ which is a contradiction.

Thus, $y \notin \mathrm{St}(x_p, \mathcal{U}_{m+1})$ which shows that

$$\varphi(x, x_p) \geq 2^{-m-1} = \frac{1}{2}\varphi(x, y) \text{ and } \varphi(x_p, y) \geq 2^{-m-1} = \frac{1}{2}\varphi(x, y).$$

Furthermore, both chains $C_0 = \{x_0, \ldots, x_p\}$ and $C_1 = \{x_p, \ldots, x_{k+1}\}$ have at most k-many links; so by the induction hypothesis we have

$$l(C_0) \geq \frac{1}{2}\varphi(x, x_p) \geq \frac{1}{4}\varphi(x, y) \text{ and } l(C_1) \geq \frac{1}{2}\varphi(x_p, y) \geq \frac{1}{4}\varphi(x, y)$$

which implies that $l(C) = l(C_0) + l(C_1) \geq \frac{1}{2}\varphi(x, y)$. This completes our induction step showing that $l(C) \geq \frac{1}{2}\varphi(x, y)$ for any chain C which connects the points x and y. Therefore $d(x, y) \geq \frac{1}{2}\varphi(x, y)$ and (4) is proved.

Now it follows from (4) that $x \neq y$ implies $d(x, y) \geq \frac{1}{2}\varphi(x, y) > 0$; so $d(x, y) = 0$ if and only if $x' = y$. To finally check the triangle inequality take any $x, y, z \in Z$. Given any $\varepsilon > 0$ there exist chains $C_0 = \{x_0, \ldots, x_n\}$ which connects x and y and a chain $C_1 = \{y_0, \ldots, y_k\}$ which connects y and z such that $l(C_0) < d(x, y) + \frac{\varepsilon}{2}$ and $l(C_1) < d(y, z) + \frac{\varepsilon}{2}$. It is clear that $C_2 = \{x_0, \ldots, x_{n-1}, y_0, \ldots, y_k\}$ is a chain which connects x and z; so $d(x, z) \leq l(C_2) = l(C_0) + l(C_1) < d(x, y) + d(y, z) + \varepsilon$. Since this inequality holds for any $\varepsilon > 0$, we have $d(x, z) \leq d(x, y) + d(y, z)$, i.e., d is, indeed, a metric on Z and hence (3) is proved.

We claim that the topology τ generated by the metric d is contained in $\tau(Z)$. Indeed, fix any $U \in \tau$; for any $x \in U$ there is $\varepsilon > 0$ such that $B_d(x, \varepsilon) \subset U$. Choose $m \in \omega$ for which $2^{-m} < \varepsilon$. If $y \in \mathrm{St}(x, \mathcal{U}_{m+1})$ then $\varphi(x, y) < 2^{-m-1}$; so $d(x, y) \leq 2\varphi(x, y) < 2^{-m} < \varepsilon$ (here we used the property (4)). This proves that $O_x = \mathrm{St}(x, \mathcal{U}_{m+1}) \subset B_d(x, \varepsilon) \subset U$ for any point $x \in U$ and therefore the set $U = \bigcup\{O_x : x \in U\}$ is open in Z. We have taken a set $U \in \tau$ arbitrarily; so every $U \in \tau$ is open in Z and therefore $\tau \subset \tau(Z)$.

Finally, it follows from $\tau \subset \tau(Z)$ that the identity map $i : Z \to (Z, \tau)$ is a condensation of Z onto a metric space (Z, τ); this shows that we established sufficiency, i.e., Fact 1 is proved.

Returning to our solution assume that X is a paracompact space which has a G_δ-diagonal. By Fact 1 of T.235 there exists a G_δ-diagonal sequence $\{\mathcal{D}_n : n \in \omega\}$ in the space X. Using paracompactness of X and TFS-230 we can construct inductively a sequence $\{\mathcal{U}_n : n \in \omega\}$ of open covers of X such that $\mathcal{U}_0 = \mathcal{D}_0$ and $\mathcal{U}_{n+1}$ is a star refinement of $\mathcal{U}_n \wedge \mathcal{D}_n$ for each $n \in \omega$. It is evident that the collection $\{\mathcal{U}_n : n \in \omega\}$ is still a G_δ-diagonal sequence which satisfied the premises of Fact 1. Therefore X can be condensed onto a metrizable space. Finally apply Problems 316 and 305 to conclude that X is functionally perfect and complete our solution.

U.319. *Observe that any Eberlein–Grothendieck space is functionally perfect. Give an example of a functionally perfect space which is not Eberlein–Grothendieck.*

Solution. It is an immediate consequence of Problem 301 that every Eberlein–Grothendieck space is functionally perfect. Now let X be the Sorgenfrey line, i.e., the underlying set of X is $\mathbb{R}$ and the family $\mathcal{B} = \{[a, b) : a, b \in \mathbb{R}, a < b\}$ is a base of the topology of X. It is evident that the identity mapping condenses X onto $\mathbb{R}$; so X is functionally perfect by Problems 305 and 316. However, X is not Eberlein–Grothendieck because $d(X) = \omega$ and $nw(X) > \omega$ (see TFS-165), while every Eberlein–Grothendieck space is monolithic (see SFFS-118 and SFFS-154).

U.320. *Prove that every metrizable space embeds into an Eberlein compact space.*

Solution. Given an infinite cardinal κ, denote by $J(\kappa)$ the Kowalsky hedgehog with κ-many spines. Any metrizable space embeds in $(J(\kappa))^\omega$ for some κ (see TFS-222) so it suffices to show that $(J(\kappa))^\omega$ embeds in an Eberlein compact space for any infinite cardinal κ. It follows from Problem 307 that it suffices to embed every $J(\kappa)$ in an Eberlein compact space; so fix an infinite cardinal κ.

Recall that we have $J(\kappa) = \{0\} \cup (\bigcup\{J_\alpha : \alpha < \kappa\})$ where $J_\alpha = (0, 1] \times \{\alpha\}$ for any $\alpha < \kappa$. The topology of $J(\kappa)$ is generated by a metric d defined as follows: $d(x, x) = 0$ for any $x \in J(\kappa)$. If $x = 0$ and $y = (t, \alpha) \in J_\alpha$ then $d(x, y) = t$; if $x = (t, \alpha)$ and $y = (s, \alpha)$ then $d(x, y) = |t - s|$. Finally, if $x = (t, \alpha)$, $y = (s, \beta)$ and $\alpha \neq \beta$ then $d(x, y) = s + t$. It is easy to see that, for every $\alpha < \kappa$, the map $(t, \alpha) \to t$ is a homeomorphism between J_α and $J = (0, 1] \subset \mathbb{R}$.

Let $I = [0, 1] \subset \mathbb{R}$ and consider the space $I \times A(\kappa)$; here, as usual, $A(\kappa)$ is the one-point compactification of a discrete space of cardinality κ. We consider that $A(\kappa) = \kappa \cup \{a\}$ where $a \notin \kappa$ is the unique non-isolated point of $A(\kappa)$. The set $P = \{0\} \times A(\kappa)$ is closed in the compact space $I \times A(\kappa)$; so we can collapse it to a point to obtain the space $K = (I \times A(\kappa))/P$.

Recall that we have $K = ((I \times A(\kappa))\backslash P) \cup \{x_P\}$ where the topology on the subset $(I \times A(\kappa))\backslash P = J \times A(\kappa)$ is induced from $I \times A(\kappa)$ and a set $U \subset K$ with $x_P \in U$ is open in K if and only if $(U \cap (J \times A(\kappa))) \cup P \in \tau(P, I \times A(\kappa))$. The space K is compact being a continuous image of $I \times A(\kappa)$ (see Fact 2 of T.245). Since we have not proved yet that a continuous image of an Eberlein compact space is Eberlein compact, let us establish directly that K is Eberlein compact.

Define a function $g \in C_p(K)$ as follows: for any $x = (t, \alpha) \in J \times A(\kappa)$ let $g(x) = t$; if $x = x_P$ then let $g(x) = 0$. Furthermore, for any $x = (t, \alpha) \in J_\alpha$ let $f_\alpha(x) = t$; if $x = x_P$ or $x = (t, \beta)$ for some $\beta \neq \alpha$ let $f_\alpha(x) = 0$. Then $f_\alpha : K \to \mathbb{R}$ is a continuous function on K such that $f_\alpha|(J_\alpha \cup \{x_P\})$ is a homeomorphism between $J_\alpha \cup \{x_P\}$ and I for any $\alpha < \kappa$. Let u be the function which is identically zero on K; then the set $F = \{f_\alpha : \alpha < \kappa\} \cup \{u\} \cup \{g\}$ separates the points of K. Besides, the set F is compact because $F\backslash U$ is finite for any $U \in \tau(u, C_p(K))$. This proves that K is an Eberlein compact space.

To see that $J(\kappa)$ embeds in K let $\varphi(x) = x$ for any $x \in J \times \kappa$ and $f(0) = x_P$; then $\varphi : J(\kappa) \to Y = (J \times \kappa) \cup \{x_P\} \subset K$ is a bijection. It is evident that $\varphi|J_\alpha : J_\alpha \to J_\alpha$ is a homeomorphism for any $\alpha < \kappa$. Since every J_α is open both in $J(\kappa)$ and in Y, the maps φ and φ^{-1} are continuous at all points of $J \times \kappa$.

Take any set $U \in \tau(x_P, Y)$; by the definition of the topology at x_P, there exists $V \in \tau(P, I \times A(\kappa))$ such that $V \cap (J \times \kappa) \subset U$. By Fact 3 of S.271 there is $\varepsilon > 0$ such that $[0, \varepsilon) \times A(\kappa) \subset V$. It is immediate that the set $W = ((0, \varepsilon) \times \kappa) \cup \{0\}$ is an open neighbourhood of the point 0 in the space $J(\kappa)$ and $\varphi(W) = ((0, \varepsilon) \times \kappa) \cup \{x_P\} \subset U$. This shows that the map φ is continuous at the point 0.

To see that φ^{-1} is continuous at the point x_P take any $U \in \tau(0, J(\kappa))$. There is $\varepsilon > 0$ such that $B_d(0, \varepsilon) \subset U$. It is straightforward that $B_d(0, \varepsilon) = ((0, \varepsilon) \times \kappa) \cup \{0\}$; the set $W = (([0, \varepsilon) \times A(\kappa)) \cap (J \times \kappa)) \cup \{x_P\}$ is an open neighbourhood of x_P in Y and it is easy to see that $\varphi^{-1}(W) = ((0, \varepsilon) \times \kappa) \cup \{0\} = B_d(0, \varepsilon) \subset U$. This shows that φ^{-1} is continuous at the point x_P so φ is an embedding of $J(\kappa)$ into an Eberlein compact space K.

U.321. *Prove that $C_p(X)$ is a $K_{\sigma\delta}$-space for any Eberlein compact X. In particular, each Eberlein compact space is Gul'ko compact and hence Corson compact.*

Solution. If X is Eberlein compact then there is a compact $K \subset C_p(X)$ which separates the points of X. The class $\mathcal{K}$ of compact spaces is k-directed; since $K \in \mathcal{K}$, we can apply Problem 014 to conclude that $C_p(X)$ belongs to $\mathcal{K}_{\sigma\delta}$, i.e., $C_p(X)$ is a $\mathcal{K}_{\sigma\delta}$-space.

Since every $\mathcal{K}_{\sigma\delta}$-space is Lindelöf Σ (see SFFS-257 and SFFS-258), any Eberlein compact space is Gul'ko compact and hence Corson compact by Problem 285.

U.322. *Prove that a compact space X is Eberlein if and only if it embeds into $\Sigma_*(A)$ for some A.*

Solution. This was proved in Fact 16 of S.351.

U.323. *Prove that a compact space X is metrizable if and only if the space X has a T_1-separating σ-point-finite family of cozero sets.*

Solution. If X is metrizable then it is second countable; so it has a countable base $\mathcal{B}$. It is evident that $\mathcal{B}$ is a T_1-separating σ-point-finite family of cozero open sets in X; this proves necessity.

As to sufficiency, if $\mathcal{U}$ is a T_1-separating σ-point-finite family of open subsets of X (we don't even need them to be cozero sets) then $\mathcal{U}$ is point-countable so we can apply Fact 1 of T.203 to see that X is metrizable.

U.324. *Prove that a compact space X is Eberlein compact if and only if X has a T_0-separating σ-point-finite family of cozero sets.*

Solution. If X is Eberlein compact then we can consider that $X \subset \Sigma_*(A)$ for some set A by Problem 322. Observe that the property of having a T_0-separating σ-point-finite family of cozero subsets in hereditary; so it suffices to show that $\Sigma_*(A)$ has such a family.

Given $a \in A$, for any rational number $q > 0$ let $U_q(a) = \{x \in \Sigma_*(A) : x(a) > q\}$; if $q \in \mathbb{Q} \cap (-\infty, 0)$ then let $U_q(a) = \{x \in \Sigma_*(A) : x(a) < q\}$. It is evident that $U_q(a)$ is a cozero set in $\Sigma_*(A)$ for any $q \in \mathbb{Q} \backslash \{0\}$ and $a \in A$. For any $q \in \mathbb{Q} \backslash \{0\}$ the family $\mathcal{U}_q = \{U_q(a) : a \in A\}$ is point-finite for otherwise there is a point $x \in \Sigma_*(A)$ such that $|x(a)| > |q|$ for infinitely many $a \in A$.

Therefore the family $\mathcal{U} = \bigcup \{\mathcal{U}_q : q \in \mathbb{Q} \backslash \{0\}\}$ is σ-point-finite and consists of cozero subsets of $\Sigma_*(A)$. To see that $\mathcal{U}$ is T_0-separating take distinct $x, y \in \Sigma_*(A)$; there is $a \in A$ for which $x(a) \neq y(a)$. It is evident that there exists $q \in \mathbb{Q} \backslash \{0\}$ which lies between the points $x(a)$ and $y(a)$. Then $U_q(a) \in \mathcal{U}$ and $U_q(a) \cap \{x, y\}$ is a singleton. Thus $\mathcal{U}$ is a T_0-separating σ-point-finite family of cozero subsets of $\Sigma_*(A)$. We already noted that this implies existence of such a family in X; so we proved necessity.

To establish sufficiency assume that $\mathcal{U} = \bigcup \{\mathcal{U}_n : n \in \omega\}$ is a T_0-separating family of cozero subsets of X such that $\mathcal{U}_n$ is point-finite for any $n \in \omega$. Denote by u the function of X which is identically zero. Given any $U \in \mathcal{U}_n$ fix a function $f_U^n \in C_p(X, [0, 2^{-n}])$ such that $X \backslash U = (f_U^n)^{-1}(0)$.

We claim that the set $K = \{f_U^n : n \in \omega, U \in \mathcal{U}_n\} \cup \{u\}$ is compact. Indeed, given any $G \in \tau(u, C_p(X))$ there is a finite set $F \subset X$ and $\varepsilon > 0$ such that $H = \{f \in C_p(X) : |f(x)| < \varepsilon$ for any $x \in F\} \subset G$. Pick $m \in \omega$ such that

$2^{-m} < \varepsilon$; then $f_U^n \in H$ for all $n \geq m$ and $U \in \mathcal{U}_n$ because $|f_U^n(x)| = f_U^n(x) \leq 2^{-n} \leq 2^{-m} < \varepsilon$ for any $x \in X$. The family $\mathcal{V} = \bigcup\{\mathcal{U}_n : n < m\}$ being point-finite there is a finite $\mathcal{V}' \subset \mathcal{V}$ such that $x \notin U$ and hence $f_U^n(x) = 0$ for any $x \in F$, $n < m$ and $U \in \mathcal{U}_n \backslash \mathcal{V}'$. As a consequence, $f \in H$ for all except finitely many $f \in K$ which shows that $K\backslash G \subset K\backslash H$ is a finite set. Since $K\backslash G$ is finite for any $G \in \tau(u, C_p(X))$, the set K is compact.

Finally, if $x, y \in X$ and $x \neq y$ then there are $n \in \omega$ and $U \in \mathcal{U}_n$ such that $U \cap \{x, y\}$ is a singleton. It is clear that f_U^n is equal to zero at exactly one of the points x, y; so $f_U^n(x) \neq f_U^n(y)$. We proved that the compact set $K \subset C_p(X)$ separates the points of our compact space X; so X is Eberlein compact. This settles sufficiency and makes our solution complete.

U.325. *Give an example of a scattered compact space which fails to be Corson compact and has a T_0-separating σ-point-finite family of open sets.*

Solution. Let L be the set of all limit ordinals of ω_1. For every $\alpha \in L$ choose a strictly increasing sequence $S_\alpha = \{\mu_\alpha(n) : n \in \omega\} \subset \alpha\backslash L$ which converges to α and let $\mathcal{B}_\alpha = \{\{\alpha\} \cup (S_\alpha\backslash K) : K$ is a finite subset of $S_\alpha\}$. Let τ be the topology on ω_1 generated by the family $\mathcal{B} = \{\{\alpha\} : \alpha \in \omega_1\backslash L\} \cup (\bigcup\{\mathcal{B}_\alpha : \alpha \in L\})$ as a base. It is easy to see that $Y = (\omega_1, \tau)$ is a locally compact space in which all points of $\omega_1\backslash L$ are isolated and $\mathcal{B}_\alpha$ is a countable base at the point α for any $\alpha \in L$. Furthermore, L is a closed discrete subspace of Y; so Y is not compact.

If $K \subset Y$ is compact then $K \cap L$ is finite because K is closed and discrete in Y. The set $K' = K\backslash(\bigcup\{S_\alpha : \alpha \in K \cap L\})$ is also finite being closed and discrete in K; so $K \subset K' \cup (\bigcup\{S_\alpha : \alpha \in K \cap L\})$ is countable. This proves that

(1) every compact subspace of Y is countable.

Denote by X the one-point compactification of Y; it is straightforward that X is compact and scattered. Let $w \in X$ be the point which compactifies Y, i.e., $\{w\} = X\backslash Y$. Assume that X is Corson compact and hence there exists a point-countable family $\mathcal{U}$ of open F_σ-subsets of X which T_0-separates the points of X (see Problem 118). Given $x, y \in X$ say that a set $U \in \mathcal{U}$ *separates x and y* if $U \cap \{x, y\}$ is a singleton. Since every F_σ-subspace of X is σ-compact, the property (1) implies that

(2) if $U \in \mathcal{U}$ and $w \notin U$ then U is countable,

and hence $\sup(U) < \omega_1$ for any $U \in \mathcal{U}_0 = \mathcal{U}\backslash\tau(w, X)$. There are at most countably many elements of $\mathcal{U}$ which contain w. Since $X\backslash U \subset Y$ is compact and hence countable for any $U \in \tau(w, X)$, there exists an ordinal $\gamma < \omega_1$ such that

(3) for any countable ordinal $\alpha \geq \gamma$, if $U \in \mathcal{U}$ separates α and w then $U \in \mathcal{U}_0$ and hence $\alpha \in U$.

Take any $\gamma_0 \in L$ with $\gamma_0 \geq \gamma$ and choose $U_0 \in \mathcal{U}$ which separates γ_0 and w. The property (3) shows that $U_0 \in \mathcal{U}_0$ and $\gamma_0 \in U_0$. Suppose that $\beta < \omega_1$ and we have a set $\{\gamma_\alpha : \alpha < \beta\} \subset L$ and a family $\{U_\alpha : \alpha < \beta\} \subset \mathcal{U}_0$ with the following properties:

(4) $\gamma_\alpha \in U_\alpha$ for any $\alpha < \beta$;
(5) $\alpha' < \alpha < \beta$ implies $\alpha' \leq \gamma_{\alpha'} < \gamma_\alpha$;
(6) if $\alpha' < \alpha < \beta$ then $\sup\{\sup(U_\delta) : \delta < \alpha'\} < \gamma_\alpha$;
(7) if $\alpha < \beta$ is a limit ordinal then $\gamma_\alpha = \sup\{\gamma_\delta : \delta < \alpha\}$.

If β is a limit ordinal then we have to let $\gamma_\beta = \sup\{\gamma_\alpha : \alpha < \beta\}$; since $\mathcal{U}$ is a T_0-separating family in X, there is $U_\beta \in \mathcal{U}$ which separates γ_β and w. It follows from (3) that $\gamma_\beta \in U_\beta \in \mathcal{U}_0$. It is easy to see that the properties (4)–(7) hold for all $\alpha \leq \beta$.

Now, if $\beta = \beta' + 1$ then $\gamma' = \sup\{\sup(U_\alpha) : \alpha \leq \beta'\} < \omega_1$ by (2); so if we take $\gamma_\beta \in L$ with $\gamma_\beta > \max\{\gamma_{\beta'}, \gamma'\} + 1$ then some $U_\beta \in \mathcal{U}$ separates γ_β and w which implies, by (3), that $\gamma_\beta \in U_\beta \in \mathcal{U}_0$. It is clear that (4)–(7) still hold for any $\alpha \leq \beta$; so our inductive procedure can be continued to construct an ω_1-sequence $G = \{\gamma_\alpha : \alpha < \omega_1\} \subset L$ and a family $\{U_\alpha : \alpha < \omega_1\} \subset \mathcal{U}_0$ with the properties (4)–(7) fulfilled for any $\beta < \omega_1$.

An immediate consequence of (4) and (6) is that $U_\alpha \neq U_\beta$ if $\alpha \neq \beta$; besides, it follows from (5) and (7) that G is a closed unbounded subset of ω_1. For any $\alpha < \omega_1$, the set U_α is an open neighbourhood of γ_α; so there is $k_\alpha \in \omega$ such that $\mu_{\gamma_\alpha}(n) \in U_\alpha$ for any $n \geq k_\alpha$; let $f(\gamma_\alpha) = \mu_{\gamma_\alpha}(k_\alpha)$.

This gives us a function $f : G \to \omega_1$ such that $f(\beta) < \beta$ for any $\beta \in G$; so we can apply SFFS-067 to find an uncountable $H \subset G$ and $\beta < \omega_1$ such that $f(\nu) = \beta$ for any $\nu \in H$. In other words, there is an uncountable $E \subset \omega_1$ such that $f(\gamma_\alpha) = \beta$ for any $\alpha \in E$. By our choice of the function f we have $\beta \in U_\alpha$ for any $\alpha \in E$. Since $U_\alpha \neq U_{\alpha'}$ for distinct $\alpha, \alpha' \in E$, the point β belongs to uncountably many elements of $\mathcal{U}$; this contradiction shows that X is not Corson compact.

To finally see that the space X has a σ-point-finite T_0-separating family of open sets fix a countable base $\mathcal{O}$ in $\mathbb{R}$ and an injective map $\xi : L \to \mathbb{R}$; then the family $\mathcal{H} = \{\xi^{-1}(O) : O \in \mathcal{O}\} \subset \exp X$ is also countable. For any $H \in \mathcal{H}$ the set $F_H = (L \backslash H) \cup \{w\}$ is closed in X being either finite or homeomorphic to the one-point compactification of the discrete space $L \backslash H$. Therefore the set $W_H = (\omega_1 \backslash L) \cup H = X \backslash F_H$ is open in X for any $H \in \mathcal{H}$. Thus the family $\mathcal{W} = \{\{\alpha\} : \alpha \in \omega_1 \backslash L\} \cup \{W_H : H \in \mathcal{H}\}$ is σ-disjoint (and hence σ-point-finite) and consists of open subsets of X.

To see that $\mathcal{W}$ is T_0-separating, take distinct points $x, y \in X$. If $x = \alpha \in \omega_1 \backslash L$ or $y = \alpha \in \omega_1 \backslash L$ then the set $W = \{\alpha\}$ belongs to $\mathcal{W}$ and separates x and y. If $\{x, y\} \subset X \backslash (\omega_1 \backslash L)$ and one of the points x, y, say, x coincides with w then there is $O \in \mathcal{O}$ with $\xi(y) \in O$; then $H = (\omega_1 \backslash L) \cup \xi^{-1}(O) \in \mathcal{W}$ and H separates the points x and y. Finally, if $x, y \in L$ then $\xi(x) \neq \xi(y)$; so there is $O \in \mathcal{O}$ with $\xi(x) \in O$ and $\xi(y) \notin O$. The set $H = (\omega_1 \backslash L) \cup \xi^{-1}(O)$ belongs to $\mathcal{W}$ and $x \in H$ while $y \notin H$, i.e., H also separates the points x and y. Thus, X is the promised scattered compact space which has a T_0-separating σ-point-finite family of open sets and fails to be Corson.

U.326. *Suppose that a compact X has a T_0-separating point-finite family of open sets. Prove that X is Eberlein compact.*

Solution. Let $\mathcal{U}$ be a T_0-separating point-finite family of open subsets of X. Say that an open set U *separates some points* $x, y \in X$ if $U \cap \{x, y\}$ is a singleton. For any $x \in X$ let $\mathcal{U}_x = \{U \in \mathcal{U} : x \in U\}$ and $O_x = \bigcap \mathcal{U}_x$ (if $\mathcal{U}_x = \emptyset$ then $O_x = X$). Observe that

(1) for any distinct $x, y \in X$ we have $\mathcal{U}_x \neq \mathcal{U}_y$,

because otherwise no $U \in \mathcal{U}$ separates the points x and y. Our next observation is that

(2) the family $\mathcal{O} = \{O_x : x \in X\}$ is point-finite.

To see that (2) is true assume that there is $z \in X$ and an infinite set $A \subset X$ such that $z \in \bigcap\{O_x : x \in A\}$. For any $x \in A$ it follows from $z \in O_x$ that $\mathcal{U}_x \subset \mathcal{U}_z$ so the family $\{\mathcal{U}_x : x \in \cdot A\}$ consists of finite collections contained in $\mathcal{U}_z$. The family $\mathcal{U}_z$ being finite, there are distinct $x, y \in A$ such that $\mathcal{U}_x = \mathcal{U}_y$; this contradiction with (1) shows that no point $z \in X$ can belong to infinitely many elements of $\mathcal{O}$, i.e., (2) is proved.

For any $x \in X$ fix a cozero set W_x such that $x \in W_x \subset O_x$; it follows from (2) that the family $\mathcal{W} = \{W_x : x \in X\}$ is point-finite. If x and y are distinct points of X then there is $U \in \mathcal{U}$ which separates x and y, say, $x \in U$ and $y \notin U$. Then $x \in W_x \subset O_x \subset U \subset X \backslash \{y\}$ and hence W_x separates the points x and y (if $y \in U$ then, analogously, the set W_y separates the points x and y). This shows that $\mathcal{W}$ is a point-finite T_0-separating family of cozero subsets of X and hence X is Eberlein compact by Problem 324.

U.327. *Prove that a non-empty compact X is Eberlein if and only if there is a compact $F \subset C_p(X)$ which separates the points of X and is homeomorphic to $A(\kappa)$ for some cardinal κ.*

Solution. Sufficiency is clear; so assume that X is an Eberlein compact space. If some finite $A \subset C_p(X)$ can separate the points of X then, for any $n \in \omega$, let $f_n(x) = 2^{-n}$ for any $x \in X$ and denote by u the function of X which is identically zero. It is clear that the set $K = A \cup \{f_n : n \in \omega\} \cup \{u\}$ separates the points of X and $K \simeq A(\omega)$.

Now assume that no finite subset of $C_p(X)$ separates the points of X. By Problem 324, there is a T_0-separating family $\mathcal{U} = \bigcup\{\mathcal{U}_n : n \in \omega\}$ of cozero subsets of X such that $\mathcal{U}_n$ is point-finite for any $n \in \omega$. Given any set $U \in \mathcal{U}_n$ fix a function $f_U^n \in C_p(X, [0, 2^{-n}])$ such that $X \backslash U = (f_U^n)^{-1}(0)$.

Let $K = \{f_U^n : n \in \omega, U \in \mathcal{U}_n\} \cup \{u\}$; given $x, y \in X$ with $x \neq y$ there are $n \in \omega$ and $U \in \mathcal{U}_n$ such that $U \cap \{x, y\}$ is a singleton. It is clear that f_U^n is equal to zero at exactly one of the points x, y; so $f_U^n(x) \neq f_U^n(y)$. This proves that the set $K \subset C_p(X)$ separates the points of the space X and hence K is infinite.

Given any $G \in \tau(u, C_p(X))$ there is a finite set $F \subset X$ and $\varepsilon > 0$ such that $H = \{f \in C_p(X) : |f(x)| < \varepsilon$ for any $x \in F\} \subset G$. Pick a number $m \in \omega$ such that $2^{-m} < \varepsilon$; then $f_U^n \in H$ for all $n \geq m$ and $U \in \mathcal{U}_n$ because $|f_U^n(x)| = f_U^n(x) \leq 2^{-n} \leq 2^{-m} < \varepsilon$ for any $x \in X$. The family $\mathcal{V} = \bigcup\{\mathcal{U}_n : n < m\}$ being point-finite there is a finite $\mathcal{V}' \subset \mathcal{V}$ such that $x \notin U$ and hence $f_U^n(x) = 0$

for any $x \in F$, $n < m$ and $U \in \mathcal{U}_n \backslash \mathcal{V}'$. As a consequence, $f \in H$ for all except finitely many $f \in K$ which shows that $K \backslash G \subset K \backslash H$ is a finite set. Since the space K is infinite and the set $K \backslash G$ is finite for any $G \in \tau(u, C_p(X))$, the set K is homeomorphic to $A(\kappa)$ for some κ and hence our solution is complete.

U.328. *Say that a function $x : \mathbb{I} \to \mathbb{I}$ is increasing (decreasing) if $x(s) \leq x(t)$ (or $x(s) \geq x(t)$ respectively) whenever $s, t \in \mathbb{I}$ and $s \leq t$. A function $x : \mathbb{I} \to \mathbb{I}$ is called monotone if it is either increasing or decreasing. Prove that the* Helly space $X = \{x \in \mathbb{I}^{\mathbb{I}} : x$ *is a monotone function$\}$ is closed in $\mathbb{I}^{\mathbb{I}}$ and hence compact. Is it an Eberlein compact space?*

Solution. We have the equality $X = X_0 \cup X_1$ where $X_0 = \{x \in \mathbb{I}^{\mathbb{I}} : x$ is increasing$\}$ and $X_1 = \{x \in \mathbb{I}^{\mathbb{I}} : x$ is decreasing$\}$. If $z \in \mathbb{I}^{\mathbb{I}} \backslash X_0$ then there are $s, t \in \mathbb{I}$ such that $s < t$ and $z(s) > z(t)$. For $\varepsilon = z(s) - z(t) > 0$ the set $U = \{x \in \mathbb{I}^{\mathbb{I}} : x(s) > z(s) - \frac{\varepsilon}{2}$ and $x(t) < z(t) + \frac{\varepsilon}{2}\}$ is open in $\mathbb{I}^{\mathbb{I}}$ and $z \in U \subset \mathbb{I}^{\mathbb{I}} \backslash X_0$; so X_0 is closed in $\mathbb{I}^{\mathbb{I}}$ and hence compact. It is now evident how to make the relevant changes in this proof to establish compactness of X_1; so the space X is compact being the union of two compact sets.

For any $a \in (-1, 1)$ define a function $x_a \in \mathbb{I}^{\mathbb{I}}$ as follows: $x_a(t) = -1$ if $t < a$, $x_a(a) = 0$ and $x_a(t) = 1$ for any $t > a$. It is clear that every x_a is increasing; so $D = \{x_a : a \in (-1, 1)\} \subset X_0$. The set $U_a = \{x \in \mathbb{I}^{\mathbb{I}} : |x(a)| < 1\}$ is open in $\mathbb{I}^{\mathbb{I}}$ and $U_a \cap D = \{x_a\}$ for any $a \in (-1, 1)$. This proves that

(1) the set D is a discrete subspace of X_0.

Let $Q = \mathbb{Q} \cap (-1, 1)$; for any $p, q \in Q$ with $p < q$ consider the function $\varphi_{p,q} \in \mathbb{I}^{\mathbb{I}}$ defined by $\varphi_{p,q}(t) = -1$ if $t \leq p$, $\varphi_{p,q}(t) = 1$ for all $t \geq q$ and $\varphi_{p,q}(t) = \frac{2}{q-p} t - \frac{q+p}{q-p}$ for any $t \in (p, q)$.
The set $A = \{\varphi_{p,q} : p, q \in Q$ and $p < q\} \subset X_0$ is countable. We claim that

(2) the set D is contained in the closure of A.

To prove (2), fix any $a \in (-1, 1)$ and take an arbitrary $U \in \tau(x_a, X)$. There is $\varepsilon > 0$ and a finite set $K \subset \mathbb{I}$ such that $a \in K$ and $V = \{x \in X : |x(t) - x_a(t)| < \varepsilon$ for any $t \in K\} \subset U$. It is easy to find $\delta > 0$ such that $\delta < \min\{1, \varepsilon, \frac{1}{2}(1+a), \frac{1}{2}(1 - a)\}$ and $K \backslash \{a\} \subset [-1, a - \delta] \cup [a + \delta, 1]$.

Observe that $(a - \delta, a + \delta) \subset (-1, 1)$; besides, $\delta < 1$ and hence $\delta^2 < \delta$. Choose a point $r \in Q \cap (a - \frac{\delta^2}{4}, a)$ and pick $s \in Q$ for which $\frac{\delta}{4} < s < \frac{\delta}{2}$. The numbers $p = r - s$ and $q = r + s$ belong to Q and $p > a - \frac{\delta^2}{4} - \frac{\delta}{2} > a - \frac{3\delta}{4} > a - \delta$. Furthermore, $q > a - \frac{\delta^2}{4} + \frac{\delta}{4} > a$ and $q < a + \frac{\delta}{2} < a + \delta$. As a consequence, $\varphi_{p,q}(t) = x_a(t)$ for any $t \in K \backslash \{a\}$.

Now, $|\varphi_{p,q}(a) - x_a(a)| = |\varphi_{p,q}(a)| = |\frac{2a-p-q}{q-p}| = \frac{|a-r|}{s} < \frac{\delta^2/4}{\delta/4} = \delta < \varepsilon$. Thus $|\varphi_{p,q}(t) - x_a(t)| < \varepsilon$ for any $t \in K$ which shows that $\varphi_{p,q} \in V \cap A \subset U \cap A$; the open neighbourhood U of the point x_a was chosen arbitrarily; so $x_a \in \overline{A}$ for any $a \in (-1, 1)$ and therefore $D \subset \overline{A}$, i.e., (2) is proved.

In a discrete space network is equal to the cardinality of the space; so it follows from the property (1) that $nw(D) = |D| > \omega$. The property (2) implies that

$nw(\overline{A}) \geq nw(D) = |D| > \omega$. Therefore the set $A \subset X_0 \subset X$ witnesses that the space X is not ω-monolithic; since every Corson compact space is ω-monolithic (see Problem 120), our space X is not even Corson compact.

U.329. *Prove that a compact space X is Eberlein compact if and only if $C_p(X)$ is a continuous image of $(A(\kappa))^\omega \times \omega^\omega$ for some infinite cardinal κ.*

Solution. Given $r > 0$ let $I_r = [-r, r] \subset \mathbb{R}$.

Fact 1. If T is a space, $r > 0$ and $\varphi_n : T \to I_r$ is a continuous function for any $n \in \omega$ then, letting $\varphi(t) = \sum_{n \in \omega} 2^{-n-1} \varphi_n(t)$ for any $t \in T$, we obtain a continuous function $\varphi : T \to I_r$.

Proof. Since $|\varphi_n(t)| \leq r$ for any $n \in \omega$, the series which defines $\varphi(t)$, converges for any $t \in T$; so the function φ is well defined. Besides, $|\varphi(t)| \leq \sum_{n \in \omega} (2^{-n-1} \cdot r) = r$ for any $t \in T$; so $\varphi : T \to I_r$. Since $|2^{-n-1} \varphi_n(t)| \leq r \cdot 2^{-n-1}$ for any $t \in T$ and $n \in \omega$ while the series $\sum_{n \in \omega} (2^{-n-1} \cdot r)$ converges, we can apply TFS-030 to see that the series $\sum_{n \in \omega} 2^{-n-1} \varphi_n$ converges uniformly and hence $\varphi = \sum_{n \in \omega} 2^{-n-1} \varphi_n$ is continuous by TFS-029. Fact 1 is proved.

Fact 2. For an arbitrary space Z and $r > 0$ the map $\delta : (C_p(Z, I_r))^\omega \to C_p(Z, I_r)$, defined by the formula $\delta(f)(z) = \sum \{2^{-n-1} \cdot f_n(z) : n \in \omega\}$ for every $z \in Z$ and $f = (f_n : n \in \omega) \in (C_p(Z, I_r))^\omega$, is continuous.

Proof. It follows from Fact 1 that, indeed, δ maps $(C_p(Z, I_r))^\omega$ to the space $C_p(Z, I_r)$. For any $m \in \omega$ and $f = (f_n : n \in \omega) \in (C_p(Z, I_r))^\omega$, let $p_m(f) = f_m$; then $p_m : (C_p(Z, I_r))^\omega \to C_p(Z, I_r)$ is the natural projection of $(C_p(Z, I_r))^\omega$ onto its m-th factor.

For any $z \in Z$ let $\pi_z(f) = f(z)$ for any $f \in C_p(Z, I_r)$. Then the map π_z is the restriction to $C_p(X, I_r)$ of the natural projection of $(I_r)^Z$ onto its factor determined by z. We have $C_p(Z, I_r) \subset (I_z)^Z$ so, to check that the map δ is continuous, it suffices to show that $\pi_z \circ \delta$ is continuous for any $z \in Z$ (see TFS-102).

To do so, fix a point $z \in Z$; the map $\varphi_n = \pi_z \circ p_n : (C_p(Z, I_r))^\omega \to I_r$ is continuous for any $n \in \omega$ and it is immediate that $\pi_z \circ \delta = \sum_{n \in \omega} 2^{-n-1} \varphi_n$ so we can apply Fact 1 again to conclude that $\pi_z \circ \delta$ is continuous for any $z \in Z$ and hence the map δ is continuous. Fact 2 is proved.

Fact 3. If K is a compact space and $Y \subset C_p(K)$ separates the points of K then $C_p(K)$ is a continuous image of $(Y \times \omega)^\omega$.

Proof. Denote by $\mathcal{A}$ the family of all continuous images of the space $(Y \times \omega)^\omega$. We will use the expression $P \gg Q$ to abbreviate the phrase "the space P can be continuously mapped onto the space Q". It is evident that

(1) if $Z \in \mathcal{A}$ then any continuous image of Z also belongs to $\mathcal{A}$.

Furthermore, if $Z \in \mathcal{A}$ then $(Y \times \omega)^\omega \simeq Y^\omega \times \omega^\omega \gg Z$ and therefore we have $Y^\omega \times \omega^\omega \times \omega^\omega \gg Z \times \omega^\omega$; it follows from $\omega^\omega \times \omega^\omega \simeq \omega^\omega$ that the space $Y^\omega \times \omega^\omega \simeq Y \times \omega^\omega \times \omega^\omega$ maps continuously onto $Z \times \omega^\omega$, i.e., $Z \in \mathcal{A}$. This proves that

(2) if $Z \in \mathcal{A}$ then $Z \times \omega^\omega \in \mathcal{A}$ and hence $Z \times \omega \in \mathcal{A}$.

Now, if $Z \in \mathcal{A}$ and M is a metrizable compact space then ω^ω maps continuously onto M (see SFFS-328); so $Z \times \omega^\omega \gg Z \times M$ which, together with (1) and (2), shows that $Z \times M \in \mathcal{A}$. This implies that

(3) if $Z \in \mathcal{A}$ and M is a metrizable compact space then any continuous image of $Z \times M$ belongs to $\mathcal{A}$.

Now let A be the minimal algebra which contains Y; then $A = \bigcup \{A_n : n \in \omega\}$ where every A_n is a continuous image of $Y^{k_n} \times M_n$ for some $k_n \in \mathbb{N}$ and metrizable compact M_n (see Problem 006). Since $Y^\omega \gg Y^{k_n}$ and $\omega^\omega \gg M_n$ for any $n \in \omega$, we have $Y^\omega \times \omega^\omega \gg Y^{k_n} \times M_n \gg A_n$ which shows that $Y^\omega \times \omega^\omega \gg A_n$ for any $n \in \omega$. An immediate consequence is that $Y^\omega \times \omega^\omega \times \omega \gg A$ and hence the properties (1) and (2) imply that $A \in \mathcal{A}$.

For every $r \in (0, 1]$ define a function $a_r : \mathbb{R} \to I_r$ as follows: $a(t) = -r$ for all $t \le -r$ and $a(t) = r$ for every $t \ge r$; if $t \in (-r, r)$ then let $a(t) = t$. It is clear that a_r is continuous; let $a = a_1$. For any $f \in C_p(K)$ the function $\eta(f) = a \circ f$ is continuous and the map $\eta : C_p(K) \to C_p(K, \mathbb{I})$ is also continuous by TFS-091. Therefore the set $B = \eta(A)$ belongs to the class $\mathcal{A}$ by the property (1). For any function $f \in C_p(K)$ let $||f|| = \sup\{|f(x)| : x \in K\}$. The algebra A separates the points of K; so we can apply TFS-191 to see that

(4) A is uniformly dense in $C_p(K)$, i.e., for any $f \in C_p(K)$ there is a sequence $\{f_n : n \in \omega\} \subset A$ such that $f_n \rightrightarrows f$.

Let us establish next that

(5) for any $r \in (0, 1]$ and $f \in C_p(K)$ such that $||f|| \le r$, there is a function $g \in A$ such that $||g|| \le \frac{r}{2}$ and $||g - f|| \le \frac{r}{2}$.

It follows from the property (4) that there is $h \in A$ such that $||h - f|| \le \frac{r}{2}$; let $g = a_{r/2} \circ h$. Now, if $x \in K$ and $|h(x)| \le \frac{r}{2}$ then $g(x) = h(x)$ and hence $|g(x) - f(x)| = |h(x) - f(x)| \le \frac{r}{2}$. If $h(x) > \frac{r}{2}$ then it follows from $||h - f|| \le \frac{r}{2}$ and $||f|| \le r$ that $f(x) \in [0, r]$; so $|g(x) - f(x)| = |\frac{r}{2} - f(x)| \le \frac{r}{2}$. Analogously, if $g(x) < -\frac{r}{2}$ then $f(x) \in [-r, 0]$; so $|g(x) - f(x)| = |\frac{r}{2} + f(x)| \le \frac{r}{2}$. This shows that $|g(x) - f(x)| \le \frac{r}{2}$ for any $x \in K$, i.e., $||g - f|| \le \frac{r}{2}$; it is evident that $||g|| \le \frac{r}{2}$; so (5) is proved.

Now take any function $f \in C_p(K, \mathbb{I})$. Then $||f|| \le 1$; so there is $g_0 \in A$ such that $||g_0|| \le \frac{1}{2}$ and $||f - g_0|| \le \frac{1}{2}$. Proceeding by induction suppose that $n \in \omega$ and we have functions $g_0, \ldots, g_n \in A$ with the following properties:

(6) $||g_i|| \le 2^{-i-1}$ for all $i \le n$;
(7) $||f - (g_0 + \ldots + g_i)|| \le 2^{-i-1}$ for any $i \le n$.

Applying (5) to the function $h = f - (g_0 + \ldots + g_n)$ and the number $r = 2^{-n-1}$ we can find $g_{n+1} \in A$ such that $||g_{n+1}|| \le 2^{-n-2}$ and

$$||f - (g_0 + \ldots + g_{n+1})|| = ||g_{n+1} - h|| \le 2^{-n-2}.$$

Since the properties (6) and (7) now hold for all $i \leq n + 1$, our inductive procedure shows that we can construct a sequence $\{g_n : n \in \omega\} \subset A$ such that (6) and (7) are fulfilled for all $i \in \omega$. The property (6) implies that $f_n = 2^{n+1} g_n \in A \cap C_p(K, \mathbb{I}) \subset B$ for any $n \in \omega$. The property (7) says that the series $\sum_{n \in \omega} g_n = \sum_{n \in \omega} 2^{-n-1} f_n$ converges uniformly to f. Thus

(8) for any function $f \in C_p(K, \mathbb{I})$ there is a sequence $\{f_n\}_{n \in \omega} \subset B$ such that
$f = \sum \{2^{-n-1} f_n : n \in \omega\}$.

For any $h = (h_n : n \in \omega) \in B^\omega$ let $\delta(h) = \sum_{n \in \omega} 2^{-n-1} h_n$. Then $\delta : B^\omega \to C_p(K, \mathbb{I})$ is a continuous map by Fact 2. The property (8) shows that $\delta(B^\omega) = C_p(K, \mathbb{I})$ and hence $B^\omega \gg C_p(K, \mathbb{I})$. Since $(Y \times \omega)^\omega \gg B$, we have $(Y \times \omega)^\omega \simeq ((Y \times \omega)^\omega)^\omega \gg B^\omega$ and hence $(Y \times \omega)^\omega \gg C_p(K, \mathbb{I})$, i.e., $C_p(K, \mathbb{I}) \in \mathcal{A}$. The space K being compact we have $C_p(K) = \bigcup\{C_p(K, [-n, n]) : n \in \mathbb{N}\}$; since $C_p(K, [-n, n]) \simeq C_p(K, \mathbb{I})$, we have $C_p(K, \mathbb{I}) \gg C_p(K, [-n, n])$ for any $n \in \mathbb{N}$, and hence $C_p(K, \mathbb{I}) \times \omega \gg C_p(K)$. Now it follows from (1) and (2) that $C_p(K) \in \mathcal{A}$, i.e., $C_p(K)$ is a continuous image of $(Y \times \omega)^\omega$; so Fact 3 is proved.

Returning to our solution assume that X is an Eberlein compact space. Then there is $Y \subset C_p(X)$ such that $Y \simeq A(\kappa)$ for some infinite cardinal κ and Y separates the points of X (see Problem 327). By Fact 3, the space $C_p(X)$ is a continuous image of $(Y \times \omega)^\omega \simeq Y^\omega \times \omega^\omega \simeq (A(\kappa))^\omega \times \omega^\omega$; so we proved necessity.

Finally, assume that a space X is compact and there exists a continuous onto map $\varphi : (A(\kappa))^\omega \times \omega^\omega \to C_p(X)$. If E is a countable dense subspace of ω^ω then the space $H = (A(\kappa))^\omega \times E$ is σ-compact and dense in $(A(\kappa))^\omega \times \omega^\omega$. Therefore $\varphi(H)$ is a σ-compact dense subspace of $C_p(X)$ which implies, by Problem 035, that X is Eberlein compact. This settles sufficiency and makes our solution complete.

U.330. *Prove that a compact space X is Eberlein compact if and only if there is an compact space K and a separable space M such that $C_p(X)$ is a continuous image of $K \times M$.*

Solution. If X is Eberlein compact then $C_p(X)$ is a continuous image of the space $(A(\kappa))^\omega \times \omega^\omega$ (see Problem 329); so letting $K = (A(\kappa))^\omega$ and $M = \omega^\omega$ we settle necessity.

Now, if there exists a compact space K and a separable space M such that there is a continuous onto map $\varphi : K \times M \to C_p(X)$ then take a countable dense set $E \subset M$. Then $H = K \times E$ is a σ-compact dense subspace of $K \times M$. Therefore $\varphi(H)$ is a σ-compact dense subspace of $C_p(X)$ which implies, by Problem 035, that X is Eberlein compact.

U.331. *Prove that any infinite Eberlein compact space X is a continuous image of a closed subspace of $(A(\kappa))^\omega \times M$, where $\kappa = w(X)$ and M is a second countable space.*

Solution. There exists an infinite cardinal κ such that some $Y \subset C_p(X)$ separates the points of X and $Y \simeq A(\kappa)$ (see Problem 327). By compactness of X and $A(\kappa)$, we have $\kappa = w(A(\kappa)) \leq nw(C_p(X)) = nw(X) = w(X)$.

Let $e_x(f) = f(x)$ for any $x \in X$ and $f \in Y$. Then $e_x \in C_p(Y)$ for any $x \in X$ and the map $e : X \rightarrow C_p(Y)$ defined by $e(x) = e_x$ for any $x \in X$, is an embedding by TFS-166. Therefore $w(X) = nw(X) \leq nw(C_p(Y)) = nw(Y) = w(Y) = \kappa$ and hence $\kappa = w(X)$.

Since $e : X \rightarrow C_p(Y)$ is an embedding, we can assume, without loss of generality, that $X \subset C_p(Y)$. The space $Y \simeq A(\kappa)$ is Eberlein compact and it is easy to see that there is a space $Z \subset C_p(Y)$ such that Z separates the points of Y and $Z \simeq A(\kappa)$. Thus there exists a continuous onto map $\varphi : (A(\kappa))^\omega \times \omega^\omega \rightarrow C_p(Y)$ by Problem 329. The set $F = \varphi^{-1}(X)$ is closed in $(A(\kappa))^\omega \times \omega^\omega$ and $\delta = \varphi|F$ maps F continuously onto X. Therefore we can take $M = \omega^\omega$ to conclude that there is a closed subset F of the space $(A(\kappa))^\omega \times M$ which maps continuously onto X.

U.332. *Prove that each Eberlein compact space is a Preiss–Simon space.*

Solution. Suppose that X is Eberlein compact and F is a closed subset of X. Then F is also an Eberlein compact space so, for any point $x \in F$, there is a sequence $\{U_n : n \in \omega\} \subset \tau^*(F)$ which converges to x (this was proved in Fact 17 of S.351). Therefore X is a Preiss–Simon space.

U.333. *Prove that, if a pseudocompact space X condenses onto a subspace of $C_p(K)$ for some compact K, then this condensation is a homeomorphism and X is Eberlein compact. In particular, any functionally perfect pseudocompact space is Eberlein compact.*

Solution. Suppose that K is a compact space and $\varphi : X \rightarrow Y$ is a condensation for some $Y \subset C_p(K)$.

Suppose that H is a pseudocompact subspace of X. Then $G = \varphi(H)$ is pseudocompact; so Problem 044 is applicable to see that $P = \overline{G}$ is a compact subspace of $C_p(K)$. The space P is Eberlein compact by Problem 035; so we can apply Fact 19 of S.351 to conclude that G is closed in P and hence $G = P$, i.e., G is Eberlein compact. Thus,

(1) if H is a pseudocompact subspace of X then $\varphi(H)$ is Eberlein compact.

In particular, the space $Y = \varphi(X)$ is Eberlein compact. Fix a closed set $F \subset X$; for any $U \in \tau(F, X)$ the set $\overline{U}$ is pseudocompact (see Observation 2 of S.140). Therefore $\varphi(\overline{U})$ is a compact subspace of Y by (1). An easy consequence of the Hausdorffness of X is the equality $F = \bigcap\{\overline{U} : U \in \tau(F, X)\}$. The map φ being a bijection, we have $\varphi(F) = \bigcap\{\varphi(\overline{U}) : U \in \tau(F, X)\}$. Since every $\varphi(\overline{U})$ is compact by (1), the space $\varphi(F)$ is also compact and hence closed in Y. We proved that

(2) $\varphi(F)$ is closed in Y for any closed $F \subset X$, i.e., the map φ is closed.

Since every closed condensation is a homeomorphism (see TFS-153, TFS-154 and TFS-155), the map φ is a homeomorphism of X onto an Eberlein compact space Y. Therefore X is also Eberlein compact.

Finally, if X is a functionally perfect pseudocompact space then X condenses onto a subspace of $C_p(K)$ for some compact K (see Problem 301); so X is Eberlein compact.

U.334. *Prove that a zero-dimensional compact space X is Eberlein compact if and only if X has a T_0-separating σ-point-finite family of clopen sets.*

Solution. Suppose that X is compact and $\mathcal{U}$ is a T_0-separating σ-point-finite family of clopen subsets of X. Since every clopen set is a cozero set, we can apply Problem 324 to conclude that X is Eberlein compact. This settles sufficiency.

To prove necessity, assume that X is a zero-dimensional Eberlein compact space and apply Problem 324 again to find a T_0-separating σ-point-finite family $\mathcal{U}$ of cozero subsets of X. Since every $U \in \mathcal{U}$ is an F_σ-subset of X, we can take a family $\{F_n^U : n \in \omega\}$ of compact subsets of X such that $U = \{F_n^U : n \in \omega\}$.

The space X being strongly zero-dimensional (see SFFS-306) there exists a clopen set C_n^U such that $F_n^U \subset C_n^U \subset U$ for all $U \in \mathcal{U}$ and $n \in \omega$. For any $n \in \omega$, the family $C_n = \{C_n^U : U \in \mathcal{U}\}$ is σ-point-finite because so is $\mathcal{U}$. Therefore the family $C = \bigcup\{C_n : n \in \omega\}$ is also σ-point-finite and consists of clopen subsets of X.

Given distinct $x, y \in X$ there is $U \in \mathcal{U}$ such that $U \cap \{x, y\}$ is a singleton, say, $U \cap \{x, y\} = \{x\}$. Therefore there is a number $n \in \omega$ such that $x \in F_n^U \subset C_n^U$ and hence $C_n^U \cap \{x, y\} = \{x\}$; in the case when $U \cap \{x, y\} = \{y\}$ an analogous reasoning shows that there is $n \in \omega$ such that $C_n^U \cap \{x, y\} = \{y\}$; so we have established that the family C is T_0-separating. Thus we have constructed a T_0-separating σ-point-finite family C of clopen subsets of X.

U.335. *Let X be a zero-dimensional compact space. Prove that X is Eberlein if and only if $C_p(X, \mathbb{D})$ is σ-compact.*

Solution. Take distinct points $x, y \in X$. Since X is zero-dimensional, there is a clopen set $U \subset X$ such that $x \in U \subset X \setminus \{y\}$. Then $f = \chi_U \in C_p(X, \mathbb{D})$ and $f(x) \neq f(y)$; this shows that the set $C_p(X, \mathbb{D})$ separates the points of X. If $Y = C_p(X, \mathbb{D})$ is σ-compact then Y is a σ-compact subspace of $C_p(X)$ which separates the points of X; so X is Eberlein compact by Problem 035. This gives sufficiency.

Now, if the space X is Eberlein compact then there exists a T_0-separating family $\mathcal{U}$ of clopen subsets of X such that $\mathcal{U} = \bigcup\{\mathcal{U}_n : n \in \omega\}$ and every $\mathcal{U}_n$ is point-finite (see Problem 334); denote by u the function which is identically zero on X. Let $K_n = \{u\} \cup \{\chi_U : U \in \mathcal{U}_n\}$; it is an easy exercise to see that $K_n \setminus O$ is finite for any $O \in \tau(u, C_p(X))$; so the space K_n is compact for any $n \in \omega$.

Consequently, $K = \bigcup\{K_n : n \in \omega\} \subset C_p(X, \mathbb{D})$ is a σ-compact subspace of $C_p(X, \mathbb{D})$ and it is straightforward that K separates the points of X. Apply Fact 2 of U.150 to conclude that $C_p(X, \mathbb{D}) = \bigcup\{C_n : n \in \omega\}$ where every C_n is a continuous image of K^{m_n} for some $m_n \in \mathbb{N}$. The space K^{m_n} is σ-compact for each $n \in \omega$; so every C_n is also σ-compact and therefore $C_p(X, \mathbb{D}) = \bigcup\{C_n : n \in \omega\}$ is σ-compact as well. This settles necessity.

U.336. *Prove that any Eberlein compact space is a continuous image of a zero-dimensional Eberlein compact space.*

Solution. Let X be Eberlein compact; by Problem 324, the space X has a T_0-separating family $\mathcal{U}$ of open F_σ-subsets of X such that $\mathcal{U} = \bigcup\{\mathcal{U}_n : n \in \omega\}$ and every $\mathcal{U}_n$ is point-finite. There is no loss of generality to assume that $\mathcal{U}_m \cap \mathcal{U}_n = \emptyset$ whenever $m \neq n$. For every $U \in \mathcal{U}$ choose a continuous function $f_U : X \to I = [0, 1] \subset \mathbb{R}$ such that $X \setminus U = f_U^{-1}(0)$ (see Fact 1 of S.358 and Fact 1 of S.499).

For any $n \in \omega$ and $U \in \mathcal{U}_n$ let $g_U = \frac{1}{n+1} \cdot f_U$; then $\varphi_n = \Delta\{g_U : U \in \mathcal{U}_n\}$ maps X continuously into $I^{\mathcal{U}_n}$ in such a way that $\varphi_n(X) \subset [0, \frac{1}{n+1}]^{\mathcal{U}_n}$. The map

$$\varphi = \Delta\{\varphi_n : n \in \omega\} : X \to \prod\{I^{\mathcal{U}_n} : n \in \omega\} = I^{\mathcal{U}}$$

is an embedding. To see it take distinct $x, y \in X$. There is $n \in \omega$ and $U \in \mathcal{U}_n$ such that $U \cap \{x, y\}$ is a singleton. Then $f_U(x) \neq f_U(y)$ and hence $g_U(x) \neq g_U(y)$ which shows that $\varphi_n(x) \neq \varphi_n(y)$ and therefore $\varphi(x) \neq \varphi(y)$. This proves that φ is, indeed, an embedding being an injection; let $X' = \varphi(X)$.

Observe that $X' \subset \{x \in I^{\mathcal{U}} :$ the set $\{U \in \mathcal{U} : |x(U)| \geq \varepsilon\}$ is finite for any $\varepsilon > 0\}$. Indeed, if $x \in X'$ and $\varepsilon > 0$ then $x = \varphi(y)$ for some $y \in X$ and there is $n \in \omega$ such that $\frac{1}{n+1} < \varepsilon$. We have $|x(U)| = g_U(y) \leq \frac{1}{k+1} \leq \frac{1}{n+1} < \varepsilon$ for any $k \geq n$ and $U \in \mathcal{U}_k$. The family $\mathcal{V} = \bigcup\{\mathcal{U}_k : k < n\}$ is point-finite; so $\{U \in \mathcal{U} : |x(U)| \geq \varepsilon\} \subset \{U \in \mathcal{V} : y \in U\}$ is finite.

To simplify the notation we will reformulate the obtained result as follows:

(1) there exists a set T such that X embeds in $\Sigma = \{x \in I^T :$ the set $\{t : x(t) \geq \varepsilon\}$ is finite for any $\varepsilon > 0\}$,; so we can assume that $X \subset \Sigma$.

Let $K = \mathbb{D}^\omega$ be the Cantor set; fix a point $a \in K$ and let $I_n = [\frac{1}{n+2}, \frac{1}{n+1}] \subset I$ for any $n \in \omega$. The space K being zero-dimensional, we can find a local base $\mathcal{O} = \{O_n : n \in \omega\}$ at the point a in K such that the set O_n is clopen in K and $O_{n+1} \subset O_n$ for any $n \in \omega$. Making the relevant changes in $\mathcal{O}$ if necessary, we can assume that $O_0 = K$ and $K_n = O_n \setminus O_{n+1} \neq \emptyset$ for any $n \in \omega$.

Since no point of K is isolated, the same is true for any non-empty clopen subset of K and hence every non-empty clopen subset of K is homeomorphic to K (see SFFS-348). This shows that K_n is homeomorphic to K and hence there is a continuous onto map $\xi_n : K_n \to I_n$ for any $n \in \omega$ (see TFS-128). Let $\xi(a) = 0$; if $x \in K \setminus \{a\}$ then there is a unique $n \in \omega$ such that $x \in K_n$; let $\xi(x) = \xi_n(x)$. It is an easy exercise that $\xi : K \to I$ is a continuous onto map such that $\xi^{-1}(0) = \{a\}$. We proved that

(2) for any $a \in \mathbb{D}^\omega$ there is a continuous onto map $\xi : \mathbb{D}^\omega \to I$ with $\xi^{-1}(0) = \{a\}$.

From now on we will consider that $K \subset I$ and $0 \in K$ (see TFS-128); apply (2) to fix a continuous onto map $\xi : K \to I$ such that $\xi^{-1}(0) = \{0\}$. Given any $\varepsilon > 0$, the set $L = K \setminus [0, \varepsilon)$ is compact, so $\xi(L)$ is compact in I; since $0 \notin \xi(L)$, there is $r(\varepsilon) > 0$ such that $[0, r(\varepsilon)) \cap \xi(L) = \emptyset$. This proves that

(3) for any $\varepsilon > 0$ there is $r(\varepsilon) > 0$ such that $a \in K$ and $a \geq \varepsilon$ implies $\xi(a) \geq r(\varepsilon)$.

Let $\Phi : K^T \to I^T$ be the product of T-many copies of ξ, i.e., $\Phi(x)(t) = \xi(x(t))$ for any $x \in K^T$ and $t \in T$. The map Φ is continuous by Fact 1 of S.271 and it is easy to see that $\Phi(K^T) = I^T$. The space $Y = \Phi^{-1}(X)$ is compact being closed in the compact space $K^T \subset I^T \subset \mathbb{R}^T$.

Take any $x \in \Sigma$ and $y \in \Phi^{-1}(x)$. For any $t \in T$, if $\varepsilon > 0$ and $y(t) \geq \varepsilon$ then $x(t) = \xi(y(t)) \geq r(\varepsilon)$ by (3); so the set $S(y, \varepsilon) = \{t \in T : y(t) \geq \varepsilon\}$ is contained in $\mathrm{supp}(x) = \{t \in T : x(t) \geq r(\varepsilon)\}$. Since $X \subset \Sigma$, the property (1) shows that $\mathrm{supp}(x)$ is finite; so $S(y, \varepsilon)$ is finite for any $\varepsilon > 0$. This proves that $\Phi^{-1}(\Sigma) \subset \Sigma$ and hence $Y \subset \Sigma \subset \Sigma_*(T) = \{x \in \mathbb{R}^T : \text{the set } \{t \in T : |x(t)| \geq \varepsilon\} \text{ is finite for any } \varepsilon > 0\}$.

It follows from $Y \subset \Sigma_*(T)$ that Y is an Eberlein compact space (see Problem 322). It is clear that $\Phi|Y$ maps Y continuously onto X. Besides, Y is zero-dimensional because so is $K^T \supset Y$ (see SFFS-301 and SFFS-302). Therefore Y is a zero-dimensional Eberlein compact space which maps continuously onto X.

U.337. *Prove that any continuous image of an Eberlein compact space is Eberlein compact.*

Solution. Suppose that X is an Eberlein compact space and $f : X \to Y$ is a continuous onto map. If X is metrizable then Y is also metrizable and hence Eberlein (see Problem 316 and Fact 5 of S.307). Thus we can assume that X is not metrizable and hence $|X| > \omega$. For any set $A \subset X$ let $f^\#(A) = Y \backslash f(X \backslash A)$. It is easy to see that if U is an open subset of X and $V = f^\#(U)$ then V is open in Y and $f^{-1}(V) \subset U$. It was proved in Problem 336 that there exists a zero-dimensional Eberlein compact space X' such that X is a continuous image of X'. Then Y is a continuous image of X'; so we can assume, without loss of generality, that $X' = X$, i.e., X is a zero-dimensional Eberlein compact space.

Our first observation is that $C_p(Y)$ embeds in $C_p(X)$ as a closed subspace (see TFS-163); thus $C_p(Y)$ is a Lindelöf Σ-space because so is $C_p(X)$ (see Problem 020). Therefore Y is Gul'ko compact; so it is ω-monolithic and Fréchet–Urysohn (see Problem 208, and Fact 1 of U.080).

Apply Problem 334 to find a T_0-separating family $\mathcal{U}$ of clopen subsets of X such that $\mathcal{U} = \bigcup\{\mathcal{U}_n : n \in \omega\}$ and every $\mathcal{U}_n$ is point-finite. Given a family $\mathcal{A}$ of subsets of X we consider that $\bigcap \mathcal{A} = X$ if $\mathcal{A}$ is empty.

Fact 1. Let K be an infinite compact space with $|K| = \kappa$. Suppose that U is an open subset of K and F is a closed subset of the subspace U. Then there exist families $\{S(\alpha, n) : \alpha < \kappa, n \in \omega\}$ and $\{V(\alpha, n) : \alpha < \kappa, n \in \omega\}$ with the following properties:

(i) the set $S(\alpha, n)$ is compact and $S(\alpha, n) \subset F$ for any $\alpha < \kappa$ and $n \in \omega$;
(ii) $V(\alpha, n)$ is open in K and $S(\alpha, n) \subset V(\alpha, n) \subset U$ for any $\alpha < \kappa$ and $n \in \omega$;
(iii) $\bigcup\{S(\alpha, n) : \alpha < \kappa, n \in \omega\} = F$;
(iv) if $\alpha < \beta < \kappa$ then $S(\beta, n) \cap V(\alpha, m) = \emptyset$ for any $m, n \in \omega$.

Proof. Since $|K| = \kappa$, we can choose an enumeration $\{x_\alpha : \alpha < \kappa\}$ of the space K. For any $\alpha < \kappa$, Let $P(\alpha, 0) = \{x_\alpha\}$ if $x_\alpha \in U$ and $P(\alpha, 0) = \emptyset$ if $x_\alpha \notin U$.

Proceeding by induction suppose that $n \in \omega$ and we have constructed a family $\{P(\alpha, n) : \alpha < \kappa\}$ such that $P(\alpha, n) \subset U$ for any $\alpha < \kappa$. The space K is normal; so we can find an open set $V(\alpha, n)$ such that $P(\alpha, n) \subset V(\alpha, n) \subset \overline{V(\alpha, n)} \subset U$; let $P(\alpha, n + 1) = \overline{V(\alpha, n)}$ for any $\alpha < \kappa$. This shows that we can construct a compact set $P(\alpha, n)$ and an open set $V(\alpha, n)$ such that $P(\alpha, n) \subset V(\alpha, n) \subset \overline{V(\alpha, n)} \subset U$ for any $\alpha < \kappa$ and $n \in \omega$. It is clear from our construction that $\bigcup \{P(\alpha, n) : \alpha < \kappa\} = U$ for all $n \in \omega$.

Every set $P(\alpha) = \bigcup_{n \in \omega} P(\alpha, n) = \bigcup_{n \in \omega} V(\alpha, n)$ is open in K, so the set $Q(\alpha, n) = P(\alpha, n) \backslash (\bigcup \{P(\beta) : \beta < \alpha\})$ is compact; let $S(\alpha, n) = Q(\alpha, n) \cap F$ for all $\alpha < \kappa$ and $n \in \omega$. It is immediate that the properties (i),(ii) and (iv) are satisfied for our families $\{S(\alpha, n) : \alpha < \kappa, n \in \omega\}$ and $\{V(\alpha, n) : \alpha < \kappa, n \in \omega\}$. To see that (iii) is also holds, take any $x \in F$ and let $\alpha = \min\{\beta < \kappa : x \in P(\beta, n)$ for some $n \in \omega\}$. Then $x \notin \bigcup \{P(\beta) : \beta < \alpha\}$ and $x \in P(\alpha, n)$ for some $n \in \omega$. This implies $x \in Q(\alpha, n)$ so $x \in S(\alpha, n)$. Thus (i)–(iv) are satisfied for our families $\{S(\alpha, n) : \alpha < \kappa, n \in \omega\}$ and $\{V(\alpha, n) : \alpha < \kappa, n \in \omega\}$; so Fact 1 is proved.

Fact 2. If $n \in \omega$ and T is an infinite set then for any family $\mathcal{N} = \{N_t : t \in T\}$ such that $|N_t| \leq n$ for any $t \in T$, there is a set D and an infinite $T' \subset T$ such that $N_s \cap N_t = D$ for any distinct $s, t \in T'$.

Proof. If $n = 0$ then $D = \emptyset$ and $T' = T$ work. Proceeding inductively, assume that $m \in \omega$ and we have proved the statement of our Fact for all $n \leq m$. Now, if $|N_t| \leq m + 1$ for all $t \in T$ then consider the following two cases.

(a) There is an infinite $S \subset T$ such that the family $\mathcal{M} = \{N_t : t \in S\}$ is point-finite. Then every finite set intersects only finitely many elements of $\mathcal{M}$; so we can construct, by a trivial induction, an infinite $T' \subset S$ such that the family $\{N_t : t \in T'\}$ is disjoint. Then T' and $D = \emptyset$ are as promised.
(b) There is an infinite $S \subset T$ such that $M = \bigcap \{N_t : t \in S\} \neq \emptyset$; choose a point $x \in M$. Then every element of the family $\mathcal{M} = \{N_t \backslash \{x\} : t \in S\}$ has at most m elements; so our inductive assumption is applicable to find an infinite $T' \subset S$ and a set D' such that $(N_t \backslash \{x\}) \cap (N_s \backslash \{x\}) = D'$ for any distinct $s, t \in T'$. It is straightforward that, letting $D = D' \cup \{x\}$ we will have the equality $N_t \cap N_s = D$ for any distinct $t, s \in T'$. Thus the case of $n = m + 1$ is settled; so our induction procedure shows that Fact 2 is proved.

Returning to our solution denote by $\mathbb{F}$ the collection of all finite subfamilies of the family $\mathcal{U}$. We will also need the families $\mathcal{U}[x] = \{U \in \mathcal{U} : x \in U\}$ and $\mathcal{U}[x, n] = \{U \in \bigcup \{\mathcal{U}_k : k \leq n\} : x \in U\}$ for any $x \in X$ and $n \in \omega$. For any $\mathcal{A} \in \mathbb{F}$ the set $U(\mathcal{A}) = f^{\#}(\bigcup \mathcal{A})$ is open in Y; so $F(\mathcal{A}) = U(\mathcal{A}) \backslash (\bigcup \{U(\mathcal{A}') : \mathcal{A}' \subset \mathcal{A}$ and $\mathcal{A}' \neq \mathcal{A}\})$ is closed in $U(\mathcal{A})$. We claim that the collection $\{F(\mathcal{A}), U(\mathcal{A}) : \mathcal{A} \in \mathbb{F}\}$ separates the points of Y in the sense that

(1) for any distinct points $x, y \in Y$ there exists a family $\mathcal{A} \in \mathbb{F}$ such that either $x \in F(\mathcal{A}) \subset U(\mathcal{A}) \subset Y \backslash \{y\}$ or $y \in F(\mathcal{A}) \subset U(\mathcal{A}) \subset Y \backslash \{x\}$.

To prove (1) we will first establish that

(2) either there exists a point $a \in f^{-1}(x)$ such that $\mathcal{U}[b] \not\subset \mathcal{U}[a]$ for any $b \in f^{-1}(y)$ or there is $a \in f^{-1}(y)$ such that $\mathcal{U}[b] \not\subset \mathcal{U}[a]$ for any $b \in f^{-1}(x)$.

Indeed, if (2) is not true then we can find sequences $A = \{a_n : n \in \omega\} \subset f^{-1}(x)$
and $B = \{b_n : n \in \omega\} \subset f^{-1}(y)$ such that $\mathcal{U}[a_n] \supset \mathcal{U}[b_n] \supset \mathcal{U}[a_{n+1}]$ for any
$n \in \omega$. This property will still hold if a we pass to any subsequences $\{a_{k_n} : n \in \omega\}$
and $\{b_{k_n} : n \in \omega\}$ of our sequences A and B. Since Y is a Fréchet–Urysohn space,
we can assume, without loss of generality, that there are points $a \in f^{-1}(x)$ and
$b \in f^{-1}(y)$ such that $A \to a$ and $B \to b$.

The family $\mathcal{U}$ being T_0-separating we can pick a set $U \in \mathcal{U}$ such that $a \in U$
and $b \notin U$ or vice versa. If we have the first case then the sequence A is eventually
in U and B is eventually in $X \backslash U$ (here we use the fact that U is a clopen set).
Therefore there is $n \in \omega$ for which $a_{n+1} \in U$ and $b_n \notin U$. This contradiction
with $\mathcal{U}[a_{n+1}] \subset \mathcal{U}[b_n]$ shows that the first case is impossible. In the second case the
contradiction is obtained in an analogous way; so (2) is proved.

Now apply (2) to find a point $a \in f^{-1}(x)$ such that $\mathcal{U}[b] \not\subset \mathcal{U}[a]$ for any
$b \in f^{-1}(y)$ (or vice versa). Thus we can fix a set $U_b \in \mathcal{U}[b]$ such that $a \notin U_b$;
the set $f^{-1}(y)$ being compact, there is a finite $B' \subset B$ for which $\bigcup \{U_b : b \in$
$B'\} \supset f^{-1}(y)$. Since the family $\mathcal{U}' = \{U_b : b \in B'\}$ is finite, we can choose a
minimal $\mathcal{A} \subset \mathcal{U}'$ for which $f^{-1}(y) \subset \bigcup \mathcal{A}$. Thus $y \in f^\#(\bigcup \mathcal{A}) = U(\mathcal{A})$ and it
follows from minimality of $\mathcal{A}$ that $f^{-1}(y) \not\subset \bigcup \mathcal{A}'$ for any proper $\mathcal{A}' \subset \mathcal{A}$. As a
consequence, $y \in F(\mathcal{A})$; it follows from $a \in f^{-1}(x) \backslash (\bigcup \mathcal{A})$ that $x \notin U(\mathcal{A})$ so the
pair $(F(\mathcal{A}), U(\mathcal{A}))$ separates the points x and y. The case when $a \in f^{-1}(y)$ and
$\mathcal{U}[b] \not\subset \mathcal{U}[a]$ for any $b \in f^{-1}(x)$ is considered analogously; so (1) is proved.

Let us show that the family $\mathcal{U}$ has another useful property.

(3) If F and G are disjoint closed subsets of the space X, then there is $k \in \omega$ such
that $\mathcal{U}[x, k] \neq \mathcal{U}[y, k]$ for any $x \in F$ and $y \in G$.

Indeed, if (3) is not true then there exist sequences $\{x_n : n \in \omega\} \subset F$ and
$\{y_n : n \in \omega\} \subset G$ for which $\mathcal{U}[x_n, n] = \mathcal{U}[y_n, n]$ for any $n \in \omega$. It is easy to
see that we can assume, without loss of generality, that there are points $x \in F$ and
$y \in G$ such that $x_n \to x$ and $y_n \to y$. Fix $U \in \mathcal{U}$ such that $x \in U$ and $y \notin U$ (or
vice versa). There is $k \in \omega$ with $U \in \mathcal{U}_k$; the sequence $\{x_n : n \in \omega\}$ is eventually
in U while $\{y_n : n \in \omega\}$ is eventually outside U (here we used again the fact that U
is a clopen set). Therefore there exists $n > k$ such that $x_n \in U$ and $y_n \notin U$. This,
together with $U \in \mathcal{U}[x_n, n]$ implies that $\mathcal{U}[x_n, n] \neq \mathcal{U}[y_n, n]$ which is contradiction.
In the symmetric case the contradiction is obtained analogously; so (3) is proved.

It follows from Fact 1 that, for the cardinal $\kappa = |Y|$, we can construct, for any
$\mathcal{A} \in \mathbb{F}$, a family $\{S(\alpha, n, \mathcal{A}) : \alpha < \kappa, n \in \omega\}$ of compact subsets of Y and a family
$\{V(\alpha, n, \mathcal{A}), \alpha < \kappa, n \in \omega\}$ of open subsets of Y such that

(4) $\bigcup \{S(\alpha, n, \mathcal{A}) : \alpha < \kappa, n \in \omega\} = F(\mathcal{A})$ for any $\mathcal{A} \in \mathbb{F}$;
(5) $S(\alpha, n, \mathcal{A}) \subset V(\alpha, n, \mathcal{A}) \subset U(\mathcal{A})$ for any $\mathcal{A} \in \mathbb{F}$;
(6) $S(\alpha, n, \mathcal{A}) \cap V(\beta, m, \mathcal{A}) = \emptyset$ for any $\beta < \alpha$, $m, n \in \omega$ and $\mathcal{A} \in \mathbb{F}$.

Given an ordinal $\alpha < \kappa$ a family $\mathcal{A} \in \mathbb{F}$ and $n \in \omega$, the sets $f^{-1}(S(\alpha, n, \mathcal{A}))$
and $f^{-1}(Y \backslash V(\alpha, n, \mathcal{A}))$ are closed and disjoint in the space X; so it follows
from the property (3) that there is $k = k(\alpha, n, \mathcal{A})$ such that the conditions
$x \in f^{-1}(S(\alpha, n, \mathcal{A}))$ and $y \in f^{-1}(Y \backslash V(\alpha, n, \mathcal{A}))$ imply $\mathcal{U}[x, k] \neq \mathcal{U}[y, k]$.
An immediate consequence is that

(7) given any $A \in \mathbb{F}$, $\alpha < \kappa$ and $n \in \omega$, if $k = k(\alpha, n, A)$ and $x \in f^{-1}(S(\alpha, n, A))$ then $\mathcal{U}[y, k] = \mathcal{U}[x, k]$ implies $f(y) \in V(\alpha, n, A)$.

Now fix $\alpha < \kappa$, $n \in \omega$ and $A \in \mathbb{F}$; if $S(\alpha, n, A) = \emptyset$ then let $G(\alpha, n, A) = \emptyset$. If the set $S(\alpha, n, A)$ is non-empty then let $k = k(\alpha, n, A)$. It is evident that the set $O(\alpha, n, A) = \bigcup\{\bigcap \mathcal{U}[x, k] : x \in f^{-1}(S(\alpha, n, A))\}$ is open in X and contains $f^{-1}(S(\alpha, n, A))$; so $W(\alpha, n, A) = f^{\#}(O(\alpha, n, A))$ is an open neighbourhood of the set $S(\alpha, n, A)$; by normality of Y there exists an F_σ-set $G(\alpha, n, A) \in \tau(Y)$ such that $S(\alpha, n, A) \subset G(\alpha, n, A) \subset W(\alpha, n, A) \cap V(\alpha, n, A)$. We claim that

(8) the family $\mathcal{G} = \{G(\alpha, n, A) : \alpha < \kappa, \ n \in \omega \text{ and } A \in \mathbb{F}\}$ is T_0-separating.

Indeed, given distinct $x, y \in Y$, we can apply (1) to find $A \in \mathbb{F}$ such that either $x \in F(A)$ and $y \notin U(A)$ or $y \in F(A)$ and $x \notin U(A)$. In the first case apply (4) to see that there are $\alpha < \kappa$ and $n \in \omega$ for which $x \in S(\alpha, n, A)$. Then $x \in G(\alpha, n, A)$ and $y \notin G(\alpha, n, A)$ because $G(\alpha, n, A) \subset V(\alpha, n, A) \subset U(A)$ while $y \notin U(A)$. The second case can be considered analogously; so (8) is proved.

The next property of $\mathcal{G}$ is crucial.

(9) the family $\mathcal{G}$ is σ-point-finite.

If $\mathcal{H}(m) = \{G(\alpha, n, A) : \alpha < \kappa, \ n \le m, \ k(\alpha, n, A) \le m, \ |A| \le m$ and we have the inclusion $A \subset \mathcal{U}_0 \cup \ldots \cup \mathcal{U}_m\}$ for any $m \in \omega$ then $\bigcup\{\mathcal{H}(m) : m \in \omega\} = \mathcal{G}$; so it suffices to show that every $\mathcal{H}(m)$ is point-finite.

Assume, towards a contradiction, that $m \in \omega$, $y \in Y$ and there is a countably infinite $\mathcal{H} \subset \mathcal{H}(m)$ such that $y \in \bigcap \mathcal{H}$. Passing, if necessary, to an infinite subfamily of $\mathcal{H}$ we can assume that there are $k, p, r \in \omega$ such that $k(\alpha, n, A) = k$, $|A| = r$ and $n = p$ whenever $G(\alpha, n, A) \in \mathcal{H}$.

It follows from Fact 2 that, passing once more to an infinite subfamily of $\mathcal{H}$ if necessary, we can assume that there is $\mathcal{D} \subset \mathcal{U}$ such that $A \cap A' = \mathcal{D}$ for any distinct $G(\alpha, p, A) \in \mathcal{H}$ and $G(\beta, p, A') \in \mathcal{H}$; let $\{G(\alpha_i, p, A_i) : i \in \omega\}$ be a faithful enumeration of the family $\mathcal{H}$. Pick a point $x \in f^{-1}(y)$; then, for any $i \in \omega$, we have $x \in f^{-1}(W(\alpha_i, n, A_i)) \subset O(\alpha_i, n, A_i)$; so there is $x_i \in f^{-1}(S(\alpha_i, p, A_i))$ such that $x \in \bigcap \mathcal{U}[x_i, k]$ (here we used the fact that $k(\alpha_i, p, A_i) = k$ for any $i \in \omega$).

As a consequence, $\mathcal{U}[x_i, k] \subset \mathcal{U}[x, k]$ for any $i \in \omega$; the family $\mathcal{U}[x, k]$ being finite, we can pass again to an infinite subfamily of $\mathcal{H}$ to guarantee the equality $\mathcal{U}[x_i, k] = \mathcal{U}[x_j, k]$ for any $i, j \in \omega$. To finally obtain the promised contradiction consider the following cases.

(a) There are distinct numbers $i, j \in \omega$ such that $A_i = A_j = \mathcal{D}$. Then $\alpha_i \neq \alpha_j$, say, $\alpha_i < \alpha_j$. It follows from (6) that $S(\alpha_j, p, A_j) \cap V(\alpha_i, p, A_i) = \emptyset$ and hence $f(x_j) \notin V(\alpha_i, p, A_i)$. On the other hand, it follows from (7) and $\mathcal{U}[x_i, k] = \mathcal{U}[x_j, k]$ that $f(x_j) \in V(\alpha_i, p, A_i)$ which is a contradiction.

(b) We have $A_i \neq \mathcal{D}$ for any $i \in \omega$ (to see that this case covers the complement of (a) observe that $|A_i| = |A_j| = r$ for any $i, j \in \omega$). Since $f(x_i) \in F(A_i)$, there is $y_i \in f^{-1}(f(x_i)) \setminus (\bigcup \mathcal{D})$ for any $i \in \omega$.

The family $\mathcal{U}[y_0, m]$ is finite while the collection $\{A_i \setminus \mathcal{D} : i \in \omega \setminus \{0\}\}$ is infinite and disjoint. Therefore there is $i > 0$ for which $\mathcal{U}[y_0, m] \cap (A_i \setminus \mathcal{D}) = \emptyset$. Since $A_i \setminus \mathcal{D} \subset \mathcal{U}_0 \cup \ldots \cup \mathcal{U}_m$ and $\mathcal{D} \cap \mathcal{U}[y_0, m] = \emptyset$, we have $A_i \cap \mathcal{U}[y_0] = \emptyset$ and therefore $f^{-1}(f(y_0)) \not\subset \bigcup A_i$. This shows that $f(x_0) = f(y_0) \notin U(A_i)$; on the other hand, the property (7) and $\mathcal{U}[x_0, k] = \mathcal{U}[x_i, k]$ imply that $f(x_0) \in V(\alpha_i, p, A_i) \subset U(A_i)$. This final contradiction shows that every family $\mathcal{H}(m)$ is, indeed, point-finite; so (9) is proved.

Finally, observe that the properties (8) and (9) show that $\mathcal{G}$ is a σ-point-finite T_0-separating family of open F_σ-subsets of Y; so Problem 324 implies that Y is Eberlein compact and completes our solution.

U.338. *Prove that there exists a compact X such that X is a union of countably many Eberlein compact spaces while $C_p(X)$ is not Lindelöf. In particular, a compact countable union of Eberlein compact spaces need not be Corson compact.*

Solution. Let X be the one-point compactification of the Mrowka space M (see TFS-142). Then X has a countable dense set D of isolated points while the space $K = X \setminus D$ is homeomorphic to $A(\kappa)$ for an uncountable cardinal κ; let a be the unique non-isolated point of K. Observe that the space K is Eberlein compact because the family $\{\{x\} : x \in K \setminus \{a\}\}$ is disjoint, T_0-separating in K and consists of open compact subsets of K (see Problem 324). Since $\{x\}$ is also Eberlein compact for any $x \in D$, the space $X = \{\{x\} : x \in D\} \cup K$ is a countable union of its Eberlein compact subspaces. However, X is not Corson compact because it is separable and non-metrizable (see Problem 121).

To see that $C_p(X)$ is not Lindelöf observe that D is a countable open dense subset of X while the set of isolated points $X \setminus D$ is dense in $X \setminus D$. This makes it possible to apply Problem 077 to see that $C_p(X, \mathbb{D}) \times \omega^\omega$ is not Lindelöf. This implies that $C_p(X, \mathbb{D}^\omega)$ is not Lindelöf either (see Problem 076). Since $C_p(X, \mathbb{D}^\omega)$ is a closed subspace of $C_p(X)$, the space $C_p(X)$ is not Lindelöf.

U.339. *Prove that it is consistent with ZFC that there exists a Corson compact space which does not map irreducibly onto an Eberlein compact space.*

Solution. It was proved in Problem 131 that, under Jensen's axiom, there exists a Corson non-metrizable compact space X such that $c(X) = \omega$. If X maps irreducibly onto an Eberlein compact space Y then $w(Y) = c(Y) \leq c(X) = \omega$ because every Eberlein compact space is Gul'ko compact (see Problems 020 and 294). Therefore $\pi w(Y) = \omega$ and hence $\pi w(X) = \omega$ by Fact 1 of S.228. This shows that $d(X) \leq \pi w(X) = \omega$ and hence X is metrizable by Problem 121. This contradiction shows that X is a Corson compact space which cannot be irreducibly mapped onto an Eberlein compact space.

U.340. *Suppose that $\mathcal{P}$ is a class of topological spaces such that, for any $X \in \mathcal{P}$, all continuous images and all closed subspaces of X belong to $\mathcal{P}$. Prove that it is impossible that a compact space X be Eberlein compact if and only if $C_p(X) \in \mathcal{P}$, i.e., either there exists an Eberlein compact space with $C_p(X) \notin \mathcal{P}$ or there is a compact Y which is not Eberlein and $C_p(Y) \in \mathcal{P}$.*

Solution. Recall that a point x of a space Z is called *a π-point* if there exists a finite family $\mathcal{U} \subset \tau(Z)$ such that $\{x\} = \bigcap\{\overline{U} : U \in \mathcal{U}\}$. Say that a set $Y \subset Z$ is *concentrated around a point* $x \in Z$ if $Y \setminus U$ is countable for any $U \in \tau(x, Z)$.

Fact 1. The space $A(\kappa)$ is Eberlein compact for any infinite cardinal κ.

Proof. It was proved in TFS-129 that every $A(\kappa)$ is compact. Now, $A(\kappa) = \kappa \cup \{a\}$ where $a \notin \kappa$ is the unique non-isolated point of $A(\kappa)$. It is straightforward that $\{\{x\} : x \in \kappa\}$ is a disjoint (and hence point-finite) T_0-separating family of cozero subsets of $A(\kappa)$; so $A(\kappa)$ is Eberlein compact by Problem 324. Fact 1 is proved.

Fact 2. Suppose that K_n is an Eberlein compact space for any $n \in \omega$. Then the space $\prod\{C_p(K_n, \mathbb{I}) : n \in \omega\}$ embeds as a closed subspace in $C_p(K)$ for some Eberlein compact space K.

Proof. Let K be the one-point compactification of the space $L = \bigoplus\{K_n : n \in \omega\}$; denote by a the unique element of the set $K \setminus L$. We will identify every K_n with the relevant clopen subspace of L.

Since every K_n is Eberlein compact, we can find a σ-point-finite family $\mathcal{U}_n$ of cozero subsets of K_n which T_0-separates the points of K_n and $K_n \in \mathcal{U}_n$ (see Problem 324). It is evident that the family $\mathcal{U} = \bigcup\{\mathcal{U}_n : n \in \omega\}$ is σ-point-finite, consists of cozero subsets of K and T_0-separates the points of K. Thus K is Eberlein compact by Problem 324.

The set $Z = \{f \in C_p(K) : f | K_n \in C_p(K_n, [0, \frac{1}{n+1}])$ for any $n \in \omega\}$ is closed in the space $C_p(K)$; it is straightforward that $f(a) = 0$ for any $f \in Z$. For every $n \in \omega$ the restriction map $\pi_n : Z \to C_p(K_n)$ is continuous and hence the diagonal product $\pi = \Delta_{n \in \omega} \pi_n : Z \to C = \prod\{C_p(K_n, [0, \frac{1}{n+1}]) : n \in \omega\}$ is continuous as well.

If $f, g \in Z$ and $f \neq g$ then there are $n \in \omega$ and $x \in K_n$ such that $f(x) \neq g(x)$. Then $\pi_n(f) \neq \pi_n(g)$ and hence $\pi(f) \neq \pi(g)$. Therefore π is injective. If we have a function $f \in \mathbb{R}^K$ such that $f(a) = 0$ and $f | K_n \in C_p(K_n, [0, \frac{1}{n+1}])$ for any $n \in \omega$ then f is continuous, i.e., $f \in Z$. An immediate consequence is that $\pi(Z) = C$, i.e., π is surjective.

Let $\delta = \pi^{-1} : C \to Z$; to see that δ is continuous take any $x \in K$ and denote by $p_x : \mathbb{R}^K \to \mathbb{R}$ the projection onto the factor determined by x, i.e., $p_x(f) = f(x)$ for any $f \in \mathbb{R}^K$. we will also need the natural projection $q_n : C \to C_p(K_n, [0, \frac{1}{n+1}])$ for any $n \in \omega$.

If $x = a$ then $(p_x \circ \delta)(f) = 0$ for any $f \in C$; so $p_x \circ \delta$ is continuous. If $x \in K_n$ for some $n \in \omega$ then $p_x \circ \delta$ coincides with the map $r_x \circ q_n$ where $r_x(f) = f(x)$ for any $f \in C_p(K_n, [0, \frac{1}{n+1}])$. The maps r_x and q_n being continuous for each $n \in \omega$, we proved that $p_x \circ \delta$ is continuous for any $x \in K$ and therefore $\delta = \pi^{-1}$ is continuous by TFS-102. Thus δ is a homeomorphism of the space C onto a closed subspace Z of the space $C_p(K)$; it follows from $C \simeq \prod\{C_p(K_n, \mathbb{I}) : n \in \omega\}$ that Fact 2 is proved.

Fact 3. Suppose that Z is a space such that $C_p(Z, \mathbb{I})$ is not pseudocompact. Then $C_p(Z, \mathbb{I}) \times \omega$ embeds in $C_p(Z, \mathbb{I})$ as a closed subspace.

Proof. For any $f \in C_p(Z)$ and $\varepsilon > 0$ let $I(f, \varepsilon) = \{g \in C_p(Z) : |g(x) - f(x)| \leq \varepsilon$ for any $x \in Z\}$. There exists a discrete family $\{U_n : n \in \omega\}$ of non-empty open subsets of $C_p(Z, \mathbb{I})$. Fix any $n \in \omega$; there is a set $\{z_1, \ldots, z_k\} \subset Z$ and a family $\{O_1, \ldots, O_k\}$ of non-empty open subsets of the interval $\mathbb{I}$ such that $V = \{f \in C_p(Z, \mathbb{I}) : f(z_i) \in O_i$ for all $i \leq k\} \subset U_n$. Making every O_i smaller if necessary, we can consider that $O_i \subset (-1, 1)$; pick a point $r_i \in O_i$.

There exists $\varepsilon > 0$ such that $(r_i - 2\varepsilon, r_i + 2\varepsilon) \subset O_i$ and therefore $|r_i| + 2\varepsilon < 1$ for all $i \leq k$. It is easy to find a function $h_n \in C_p(Z, \mathbb{I})$ such that $h_n(z_i) = r_i$ for all $i \leq k$ (use TFS-034 to choose $g_n \in C_p(Z)$ with this property and then consider the function $h_n = \xi \circ g_n$ where $\xi(t) = \frac{1}{2}(|t + 1| - |t - 1|)$ for all $t \in \mathbb{R}$). Let $f_n = (1 - \varepsilon)h_n$; if $f \in I(f_n, \varepsilon)$ then $|f(x)| \leq |f_n(x)| + \varepsilon \leq 1$ for any $x \in Z$. Besides, $|f(z_i) - r_i| \leq |f(z_i) - f_n(z_i)| + |f_n(z_i) - r_i| \leq \varepsilon + |h_n(z_i) - \varepsilon h(z_i) - r_i| = \varepsilon(1 + |r_i|) < 2\varepsilon$ which shows that $f(z_i) \in O_i$ for any $i \leq k$. This implies that $f \in V \subset U_n$; since $|f(x)| \leq 1$ for any $x \in Z$, we have established that

(1) for any $n \in \omega$ there are $f_n \in U_n$ and $\varepsilon_n > 0$ such that $I(f_n, \varepsilon_n) \subset C_p(Z, \mathbb{I})$
    $\cap U_n$.

An immediate consequence of (1) is that the family $\{I(f_n, \varepsilon_n) : n \in \omega\}$ is discrete. The set $I(f_n, \varepsilon_n)$ is closed in $C_p(Z, \mathbb{I})$ and homeomorphic to $C_p(Z, \mathbb{I})$ for any $n \in \omega$ (see Fact 3 of S.398) which shows that $F = \bigcup\{I(f_n, \varepsilon_n) : n \in \omega\}$ is homeomorphic to $C_p(Z, \mathbb{I}) \times \omega$. It is evident that F is closed in $C_p(Z, \mathbb{I})$; so Fact 3 is proved.

Fact 4. If K is an infinite compact space then any F_σ-subset of $C_p(K)$ is a continuous image of a closed subset of $C_p(K, \mathbb{I})$.

Proof. Suppose that F_n is closed in $C_p(K)$ for every $n \in \omega$ and $F = \bigcup_{n \in \omega} F_n$. The set $F_{nm} = F_n \cap C_p(K, [-m, m])$ is closed in $C_p(K)$ for any $m \in \mathbb{N}$ and we have the equality $F_n = \bigcup_{m \in \mathbb{N}} F_{nm}$ for each $n \in \omega$. Since every $C_p(K, [-m, m])$ is homeomorphic to $C_p(K, \mathbb{I})$, we can consider that $F = \bigcup_{n \in \omega} G_n$ where every G_n is homeomorphic to a closed subspace of $C_p(K, \mathbb{I})$.

The set $G = \bigoplus_{n \in \omega} G_n$ is homeomorphic to a closed subspace of $C_p(K, \mathbb{I}) \times \omega$. Since K is compact, the set $C_p(K, \mathbb{I})$ is not pseudocompact (see TFS-398 and Fact 2 of T.090); so we can apply Fact 3 to see that $C_p(K, \mathbb{I}) \times \omega$ embeds in $C_p(K, \mathbb{I})$ as a closed subspace. Therefore G also embeds in $C_p(K, \mathbb{I})$ as a closed subspace. Since G maps continuously onto F, Fact 4 is proved.

Returning to our solution assume, towards a contradiction, that the class $\mathcal{P}$ is *dually Eberlein*, i.e., a compact space X belongs to $\mathcal{P}$ if and only if $C_p(X) \in \mathcal{P}$. It was proved in Fact 5 of U.222 that there exists a Talagrand compact space K such that some $p \in K$ is not a π-point and $K = \bigcup\{K_n : n \in \omega\}$ where every K_n is homeomorphic to a closed subset of $(A(\mathfrak{c}))^\omega$ and, in particular, K_n is Eberlein compact by Fact 1 and Problem 307. However, $K \setminus \{p\}$ is pseudocompact and dense in K by Fact 1 of U.222. This, together with Fact 19 of S.351 shows that

(2) the space K is not Eberlein compact.

The space K is Corson compact by Problem 285; so we assume from now on that $K \subset \Sigma(T)$ for some uncountable set T. Let $u(t) = 0$ for all $t \in T$. We will also need the characteristic function u_t of the singleton $\{t\}$ for any $t \in T$; recall that $u_t(t) = 1$ and $u_t(s) = 0$ for any $s \in T \backslash \{t\}$.

It is straightforward that the space $E = \{u\} \cup \{u_t : t \in T\} \subset \Sigma(T)$ is homeomorphic to $A(\kappa)$ for $\kappa = |T|$. The space $K' = K \cup E$ is also Corson compact; by Fact 1 the space K' is still a countable union of Eberlein compacta and $E \subset K'$. Thus we can enumerate the relevant Eberlein compact subspaces of K' as $\{L_n : n \in \omega\}$ in such a way that $L_0 = E$, $L_n \subset L_{n+1}$ for any $n \in \omega$ and $K' = \bigcup\{L_n : n \in \omega\}$.

Let $h(x) = 0$ for all $x \in K'$; given $t \in T$ let $h_t(x) = x(t)$ for all $x \in K'$. Then the function $h_t : K' \to \mathbb{R}$ is continuous because it coincides with the restriction of the natural projection of $\mathbb{R}^T$ onto its factor determined by t. It is immediate that the space $H = \{h\} \cup \{h_t : t \in T\}$ separates the points of K' and H is concentrated around h. There exists a condensation $\eta : C_p(K') \to S$ of the space $C_p(K')$ into a Σ-product S of real lines by Problem 200. The set $\eta(H)$ is concentrated around the point $\eta(h)$ so we can apply Fact 1 of U.289 to see that $\eta(H)$ is σ-compact and therefore H is an F_σ-subset of $C_p(K')$.

Let $\pi_n : C_p(K') \to C_p(L_n)$ be the restriction map for any $n \in \omega$.

We leave it to the reader to verify that

(3) $\pi_0 | H$ is injective, $\pi_0(h)$ is the unique non-isolated point of $\pi_0(H)$ and $\pi_0(H)$ is homeomorphic to $A(\kappa)$.

If $\pi_0^n : C_p(L_n) \to C_p(L_0)$ is the restriction map then $\pi_0^n(\pi_n(H)) = \pi_0(H)$; so (3) implies that $\pi_n(h)$ is the unique non-isolated point of $G_n = \pi_n(H)$ for all $n \in \omega$.

For any $n \in \omega$ there exists a condensation $\eta_n : C_p(L_n) \to S_n$ of the space $C_p(L_n)$ into a Σ-product S_n of real lines by Problem 200. The set $\eta_n(G_n)$ is concentrated around the point $\eta_n(\pi_n(h))$; so we can apply Fact 1 of U.289 to see that $\eta_n(G_n)$ is σ-compact and therefore G_n is an F_σ-subset of $C_p(L_n)$.

By Fact 4, every set G_n is a continuous image of a closed subset Q_n of $C_p(L_n, \mathbb{I})$; by Fact 2, the space $(\prod\{C_p(L_n, \mathbb{I}) : n \in \omega\})^\omega$ embeds as a closed subspace in $C_p(R)$ for some Eberlein compact space R. Therefore $Q = (\prod_{n\in\omega} Q_n)^\omega$ also embeds in $C_p(R)$ as a closed subspace. This implies $Q \in \mathcal{P}$; the space $(\prod_{n\in\omega} G_n)^\omega$ being a continuous image of Q, we proved that

(4) the space $(\prod_{n\in\omega} G_n)^\omega$ belongs to the class $\mathcal{P}$.

We will show next that

(5) the space H embeds in $G = \prod_{n\in\omega} G_n$ as a closed subspace.

To prove (5) consider the diagonal product $\pi = \Delta_{n\in\omega}\pi_n$ which maps $C_p(K')$ to $\prod_{n\in\omega} C_p(L_n)$. It is clear that $\pi(H) \subset G$. Let $j_n : L_n \to K'$ be the identity map for any $n \in \omega$. Then we can define a map $j : L = \bigoplus_{n\in\omega} L_n \to K'$ by letting $j(x) = j_n(x)$ whenever $x \in L_n$ (we identify every L_n with the respective clopen subset of L). The map j is continuous, onto and it is clear that the dual map $j^* : C_p(K') \to C_p(L)$ coincides with π. Thus π is an embedding by TFS-163.

To see that $\pi(H)$ is closed in the space G take any $g = (g_n)_{n\in\omega} \in G\backslash\pi(H)$ and fix $f_n \in H$ such that $g_n = \pi_n(f_n)$ for any $n \in \omega$. If there is $f \in H$ for which $g_n = \pi_n(f)$ for all $n \in \omega$ then $g = \pi(f) \in \pi(H)$ which is a contradiction. Therefore there are distinct $v, w \in H$ such that $g_k = \pi_k(v)$ and $g_m = \pi_m(w)$ for some distinct $k, m \in \omega$.

One of the points v, w, say, v is isolated in H. Therefore g_k is isolated in G_k by (3) and hence $W = \{a = (a_n)_{n\in\omega} \in G : a_k = g_k\}$ is an open neighbourhood of g such that $W \cap \pi(H)$ cannot contain any point distinct from $\pi(v)$. Thus $W\backslash\{\pi(v)\}$ is an open neighbourhood of g in G disjoint from $\pi(H)$. This shows that $\pi(H)$ is closed in G, i.e., (5) is proved.

Assume that the space H is countably compact. The space $C_p(K')$ is Lindelöf by Problem 150; since H is an F_σ-subset of $C_p(K')$, it is Lindelöf as well. This shows that H is compact; recalling that H separates the points of K' we conclude that K' is Eberlein compact which is false because $K' \supset K$ and K is not Eberlein compact by the property (2).

This contradiction shows that H is not countably compact and hence ω embeds in H as a closed subspace. Therefore ω^ω embeds in H^ω as a closed subspace and hence the space $H^\omega \times \omega^\omega$ embeds as a closed subspace in $H^\omega \times H^\omega \simeq H^\omega$. It follows from (5) that H^ω embeds in G^ω as a closed subspace. This, together with the property (4), implies that the space H^ω belongs to $\mathcal{P}$ and hence $H^\omega \times \omega^\omega$ also belongs to $\mathcal{P}$ being homeomorphic to a closed subspace of H^ω. The space $C_p(K')$ is a continuous image of $H^\omega \times \omega^\omega$ by Fact 3 of U.329; so $C_p(K') \in \mathcal{P}$; since the class $\mathcal{P}$ is dually Eberlein, the space K' has to be Eberlein compact. Therefore the space $K \subset K'$ is also Eberlein compact; this contradiction with (2) shows that $\mathcal{P}$ is not dually Eberlein and completes our solution.

U.341. *Let X be a Gul'ko compact space. Prove that there exists a countable family $\mathcal{F}$ of closed subsets of X such that $\bigcup\mathcal{F} = X$ and $K_x = \bigcap\{A : x \in A \in \mathcal{F}\}$ is Eberlein compact for any $x \in X$.*

Solution. If A is a set then $\sigma(A) = \{x \in \mathbb{R}^A : x^{-1}(\mathbb{R}\backslash\{0\})$ is a finite set$\}$. There exists an uncountable T and a family $\{T_n : n \in \omega\} \subset \exp T$ such that $X \subset \mathbb{R}^T$ and, given any point $x \in X$, if $N_x = \{n \in \omega : \operatorname{supp}(x) \cap T_n$ is finite$\}$ then we have the equality $T = \bigcup\{T_n : n \in N_x\}$ (see Problem 290). Here, as usual, $\operatorname{supp}(x) = x^{-1}(\mathbb{R}\backslash\{0\})$.

The space $\sigma(T_n)$ is σ-compact by Problem 108; so choose a countable family $\mathcal{G}_n$ of compact subsets of $\sigma(T_n)$ such that $\sigma(T_n) = \bigcup\mathcal{G}_n$ for any $n \in \omega$. We will make use of the natural projection $\pi_n : \mathbb{R}^T \to \mathbb{R}^{T_n}$ for every $n \in \omega$. If $p_n = \pi_n|X$ then the family $\mathcal{F}_n = \{p_n^{-1}(F) : F \in \mathcal{G}_n\}$ is countable and consists of compact subsets of X for all $n \in \omega$. We claim that the family $\mathcal{F} = \bigcup\{\mathcal{F}_n : n \in \omega\}$ is as required.

To see it, take any point $x \in X$; if $n \in N_x$ then $\operatorname{supp}(x) \cap T_n$ is finite and hence $p_n(x) \in \sigma(T_n)$; thus $p_n(x) \in \bigcup\mathcal{G}_n$ and therefore $x \in \bigcup\mathcal{F}_n$. This proves that $\bigcup\mathcal{F} = X$.

Fix a point $x \in X$; it is clear that K_x is compact. To see that K_x is Eberlein compact consider the map $\varphi = \Delta\{p_n : n \in N_x\} : X \to \prod\{\mathbb{R}^{T_n} : n \in N_x\}$. If y and z are distinct points of X then there is $t \in T$ for which $y(t) \neq z(t)$.

Since $T = \bigcup\{T_n : n \in N_x\}$, we have $t \in T_n$ for some $n \in N_x$ and therefore $p_n(y) \neq p_n(z)$ whence $\varphi(y) \neq \varphi(z)$. This shows that φ is an injection; so φ embeds X in $\prod\{\mathbb{R}^{T_n} : n \in N_x\}$.

Given $y \in K_x$ and $n \in N_x$ it follows from $p_n(x) \in \sigma(T_n)$ that there is $G \in \mathcal{G}_n$ with $p_n(x) \in G$. Therefore $x \in F = p_n^{-1}(G) \in \mathcal{F}$; an immediate consequence is that $y \in F$ and hence $p_n(y) \in G \subset \sigma(T_n)$. This proves that $p_n(y) \in \sigma(T_n)$ for any $y \in K_x$ and hence $L_n = p_n(K_x) \subset \sigma(T_n)$ for each $n \in N_x$. The compact space L_n is Eberlein because $\sigma(T_n) \subset \Sigma_*(T_n)$ (see Problem 322). Since the map φ embeds K_x in $\prod_{n\in\omega} L_n$, the space K_x is Eberlein compact by Problems 306 and 307. Thus $\mathcal{F}$ is a family of closed subsets of X such that $\bigcup \mathcal{F} = X$ and K_x is Eberlein compact for any $x \in X$.

U.342. *Let X be an Eberlein compact space with $|X| \leq \mathfrak{c}$. Prove that there exists a countable family $\mathcal{F}$ of closed subsets of X such that $\bigcup \mathcal{F} = X$ and the subspace $K_x = \bigcap\{A : x \in A \in \mathcal{F}\}$ is metrizable for any $x \in X$.*

Solution. Given a set A let $\sigma_0(A) = \{x \in \mathbb{D}^A : |x^{-1}(1)| < \omega\}$. Call a compact space K *weakly metrizably fibered* if there is closed countable cover $\mathcal{C}$ of the space K such that $\bigcap\{C : x \in C \in \mathcal{C}\}$ is metrizable for any $x \in K$; we will say that the family $\mathcal{C}$ *weakly metrizably fibers* K. In this terminology we must prove that any Eberlein compact space of cardinality at most $\mathfrak{c}$ is weakly metrizably fibered.

Fact 1. Suppose that K is compact and $K = \bigcup\{K_n : n \in \omega\}$ where K_n is weakly metrizably fibered and closed in K for any $n \in \omega$. Then K is also weakly metrizably fibered.

Proof. Take a countable family $\mathcal{F}_n$ which weakly metrizably fibers K_n for any $n \in \omega$. The family $\mathcal{F} = \{K_n : n \in \omega\} \cup (\bigcup\{\mathcal{F}_n : n \in \omega\})$ is countable and consists of closed subsets of K. Besides, $\bigcup \mathcal{F} \supset \bigcup_{n\in\omega} K_n = K$; so $\mathcal{F}$ is a cover of K. Given any $x \in K$ fix $n \in \omega$ with $x \in K_n$; the set $P_x = \bigcap\{F : x \in F \in \mathcal{F}\}$ is contained in the metrizable space $Q_x = \bigcap\{F : x \in F \in \mathcal{F}_n\}$. Thus P_x is metrizable for any $x \in K$ and hence the family $\mathcal{F}$ weakly metrizably fibers K. Fact 1 is proved.

Fact 2. If K_n is a compact weakly metrizably fibered space for any $n \in \omega$ then $K = \prod_{n\in\omega} K_n$ is also weakly metrizably fibered.

Proof. Let $\pi_n : K \to K_n$ be the natural projection for any $n \in \omega$. Fix a family $\mathcal{F}_n$ of compact subsets of K_n which weakly metrizably fibers K_n for any $n \in \omega$. The family $\mathcal{F} = \{\pi_n^{-1}(F) : n \in \omega, F \in \mathcal{F}_n\}$ is countable and consists of closed subsets of K. For any $n \in \omega$ we have $\bigcup \mathcal{F} \supset \bigcup \mathcal{F}_n = \pi_n^{-1}(K_n) = K$; so the family $\mathcal{F}$ covers K. Given $x = (x_n)_{n\in\omega} \in K$ the set $C_n^x = \bigcap\{F : x_n \in F \in \mathcal{F}_n\}$ is metrizable for any $n \in \omega$. It is straightforward that $C_x = \bigcap\{F : x \in F \in \mathcal{F}\} \subset \prod_{n\in\omega} C_n^x$; so C_x is metrizable for any $x \in K$, i.e., the family $\mathcal{F}$ weakly metrizably fibers K. Fact 2 is proved.

Fact 3. Any closed subspace and any continuous image of a compact weakly metrizably fibered space is weakly metrizably fibered.

Proof. Suppose that K is a compact weakly metrizably fibered space and A is closed in K. If $\mathcal{F}$ is a countable family of closed subsets of K which weakly metrizably fibers K then the family $\mathcal{G} = \{F \cap A : F \in \mathcal{F}\}$ is countable and consists of closed subsets of A. It is evident that $P_x = \bigcap\{F : x \in F \in \mathcal{G}\} \subset Q_x = \bigcap\{F : x \in F \in \mathcal{F}\}$. The space Q_x being metrizable, the subspace P_x is also metrizable for any $x \in A$ and hence the family $\mathcal{G}$ weakly metrizably fibers A.

Now suppose that $f : K \to L$ is a continuous onto map. Let $\mathcal{H}$ be the family of all finite intersections of the elements of $\mathcal{F}$. It is clear that $\mathcal{H}$ still weakly metrizably fibers K. The family $\mathcal{E} = \{f(H) : H \in \mathcal{H}\}$ is countable and consists of closed subsets of L. Besides, $\bigcup\mathcal{E} = f(\bigcup\mathcal{H}) = f(K) = L$, i.e., $\mathcal{E}$ is a cover of L. Take any point $y \in L$ and pick $x \in f^{-1}(y)$. The space $P_x = \bigcap\{H : x \in H \in \mathcal{H}\}$ is compact and metrizable so $Q_x = f(P_x)$ is metrizable as well by Fact 5 of S.307.

Assume that the set $E_y = \bigcap\{E : y \in E \in \mathcal{E}\}$ is not contained in Q_x; then there is a point $z \in E_y$ such that $f^{-1}(z) \cap P_x = \emptyset$. We can apply Fact 1 of S.326 to the set P_x and $U = K \backslash f^{-1}(z) \in \tau(P_x, K)$ to see that there exists a finite family $\mathcal{H}' \subset \{H \in \mathcal{H} : x \in H\}$ such that $H' = \bigcap\mathcal{H}' \subset U$. The family $\mathcal{H}$ is closed under finite intersections; so $H' \in \mathcal{H}$; since $x \in H'$, we have $y \in f(H')$ and hence $z \in E_y \subset f(H')$ which is a contradiction with $H' \cap f^{-1}(z) = \emptyset$. Therefore $E_y \subset Q_x$ is metrizable for any $y \in L$, i.e., the family $\mathcal{E}$ weakly metrizably fibers L. Fact 3 is proved.

Fact 4. Given an infinite cardinal κ let $S_n = \{x \in \mathbb{D}^\kappa : |x^{-1}(1)| \leq n\}$ for any $n \in \omega$. Then S_n is a continuous image of $(A(\kappa))^n$ for any $n \in \omega$.

Proof. Recall that $(A(\kappa))^0 = \{\emptyset\}$; it is clear that $S_0 = \{u\}$ where $u(\alpha) = 0$ for any $\alpha < \kappa$, so S_0 is a continuous image of $(A(\kappa))^0$. Thus our statement is true for $n = 0$. Let $u_\alpha \in \mathbb{D}^\kappa$ be the characteristic function of the singleton $\{\alpha\}$ for any $\alpha < \kappa$. Recall that $u_\alpha(\alpha) = 1$ and $u_\alpha(\beta) = 0$ for any $\beta \neq \alpha$. It is straightforward that the space $F = \{u\} \cup \{u_\alpha : \alpha < \kappa\} = S_1$ is homeomorphic to $A(\kappa)$; so our statement is also true for $n = 1$.

For any $x, y \in \mathbb{D}^\kappa$ let $(x * y)(\alpha) = x(\alpha) + y(\alpha) - x(\alpha)y(\alpha)$ for any $\alpha < \kappa$. Then $x * y \in \mathbb{D}^\kappa$ for any $x, y \in \mathbb{D}^\kappa$ and the operation $\varphi_n : (\mathbb{D}^\kappa)^n \to \mathbb{D}^\kappa$ defined by $\varphi_n(x_0, \ldots, x_{n-1}) = x_0 * \ldots * x_{n-1}$ is continuous for any $n \in \mathbb{N}$ (to see this apply Fact 1 of U.150 to the space $Z = D(\kappa)$ for which we have $C_p(Z, \mathbb{D}) = \mathbb{D}^\kappa$).

It is evident that $\varphi_n(F^n) \subset S_n$ for any $n \in \mathbb{N}$. Given any $x \in S_n \backslash \{u\}$ there exist distinct $\alpha_0, \ldots, \alpha_{m-1} \in \kappa$ such that $m \leq n$ and $x^{-1}(1) = \{\alpha_0, \ldots, \alpha_{m-1}\}$. Let $y = (u_{\alpha_0}, \ldots, u_{\alpha_{m-1}}, y_m, \ldots, y_{n-1})$ where $y_i = u$ for all $i \geq m$. It is clear that $y \in F^n$ and $\varphi_n(y) = x$. This proves that $\varphi_n(F^n) = S_n$; so S_n is a continuous image of the space $F^n \simeq (A(\kappa))^n$ for any $n \in \mathbb{N}$. Fact 4 is proved.

Fact 5. If $\kappa \leq \mathfrak{c}$ is an infinite cardinal then $A(\kappa)$ is weakly metrizably fibered. As a consequence, if A is a set such that $|A| \leq \mathfrak{c}$ then any compact subspace of $\sigma_0(A)$ is weakly metrizably fibered.

Proof. Recall that $A(\kappa) = \{a\} \cup \kappa$ where $a \notin \kappa$ is the unique non-isolated point of $A(\kappa)$. Take an injection $\xi : A(\kappa) \to \mathbb{R}$ and fix some countable base $\mathcal{B}$ of the space $\mathbb{R}$.

The set $F_B = \xi^{-1}(B) \cup \{a\}$ is closed in $A(\kappa)$ for any $B \in \mathcal{B}$; let $\mathcal{F} = \{F_B : B \in \mathcal{B}\}$. It follows from $\bigcup \mathcal{B} = \mathbb{R}$ that $\bigcup \mathcal{F} = A(\kappa)$.

Given a point $x \in A(\kappa)$ consider the set $P_x = \bigcap \{F : x \in F \in \mathcal{F}\}$. If $y \notin \{x, a\}$ then $\xi(y) \neq \xi(x)$ and hence there is $B \in \mathcal{B}$ such that $\xi(x) \in B$ and $\xi(y) \notin B$. It is immediate that $y \notin F_B$ while $x \in F_B$; this proves that $P_x \subset \{x, a\}$ and hence P_x is metrizable for any $x \in A(\kappa)$. Therefore $\mathcal{F}$ weakly metrizably fibers $A(\kappa)$.

Finally, assume that $|A| = \kappa \leq \mathfrak{c}$ and K is a compact subspace of $\sigma_0(A)$. If κ is finite then there is nothing to prove; so we assume that $\kappa \geq \omega$. We have $\sigma_0(A) = \bigcup \{S_n : n \in \omega\}$ where $S_n = \{x \in \mathbb{D}^A : |x^{-1}(1)| \leq n\}$; let $K_n = S_n \cap K$ for any $n \in \omega$.

The space $A(\kappa)$ is weakly metrizably fibered because $\kappa \leq \mathfrak{c}$. Therefore $(A(\kappa))^n$ is weakly metrizably fibered for any $n \in \omega$ (see Fact 2) and hence every S_n is also weakly metrizably fibered by Fact 4 and Fact 3. Applying Fact 3 once more we can see that every K_n is weakly metrizably fibered; since $K = \bigcup_{n \in \omega} K_n$, the space K is weakly metrizably fibered (see Fact 1); so Fact 5 is proved.

Returning to our solution apply Problem 336 to find an Eberlein compact zero-dimensional space Y for which there is a continuous onto map $f : Y \to X$. By TFS-366, there is a closed $F \subset Y$ such that $f(F) = X$ and $f_0 = f|F$ is irreducible. We have $\pi w(X) \leq w(X) = nw(X) \leq |X| \leq \mathfrak{c}$ (see Fact 4 of S.307); so we can take a π-base $\mathcal{B}$ in the space X such that $|\mathcal{B}| \leq \mathfrak{c}$. The family $\mathcal{U} = \{f_0^{-1}(U) : U \in \mathcal{B}\}$ is a π-base in F (see Fact 2 of S.373) and $|\mathcal{U}| \leq \mathfrak{c}$. Choosing a point in every element of $\mathcal{U}$ we obtain a dense $A \subset F$ with $|A| \leq \mathfrak{c}$. Since F is monolithic (see Problem 120), we have $nw(F) = w(F) \leq \mathfrak{c}$.

The space F being Eberlein compact and zero-dimensional, apply Problem 334 to see that there exists a T_0-separating family $\mathcal{V} = \bigcup_{n \in \omega} \mathcal{V}_n$ of clopen subsets of F such that $\mathcal{V}_n$ is point-finite for any $n \in \omega$. Apply Fact 1 of U.077 to conclude that $|\mathcal{V}| \leq \mathfrak{c}$ and hence $|\mathcal{V}_n| \leq \mathfrak{c}$ for any $n \in \omega$. Let $\chi_V : F \to \mathbb{D}$ be the characteristic function of V for every $V \in \mathcal{V}$. Recall that $\chi_V(x) = 1$ if $x \in V$ and $\chi_V(x) = 0$ for all $x \in F \backslash V$.

If $\pi_n = \Delta\{\chi_V : V \in \mathcal{V}_n\}$ then $\pi_n : F \to \mathbb{D}^{\mathcal{V}_n}$ is a continuous map. The family $\mathcal{V}_n$ is point-finite; so $\pi_n(F) \subset \sigma_0(\mathcal{V}_n)$; let $F_n = \pi_n(F)$ for any $n \in \omega$. If $\pi = \Delta_{n \in \omega} \pi_n$ then $\pi : F \to \prod\{F_n : n \in \omega\}$ is a continuous map. Given distinct points $x, y \in F$ there is $n \in \omega$ and $V \in \mathcal{V}_n$ such that $V \cap \{x, y\}$ is a singleton. Then $\chi_V(x) \neq \chi_V(y)$ which shows that $\pi_n(x) \neq \pi_n(y)$ and hence $\pi(x) \neq \pi(y)$.

Therefore the map π is an embedding; so F is homeomorphic to a subspace of $\prod_{n \in \omega} F_n$. By Fact 5, every F_n is weakly metrizably fibered. Apply Fact 2 and Fact 3 to see that F is also weakly metrizably fibered. Since X is a continuous image of F, we can apply Fact 3 again to conclude that X is weakly metrizably fibered and finish our solution.

U.343. *Observe that $c(X) = w(X)$ for any Eberlein compact space X. Prove that, for any infinite compact X, we have $c(X) = \sup\{w(K) : K \subset C_p(X)$ and K is compact$\}$.*

Solution. The equality $c(X) = w(X)$ is true even for any Gul'ko compact space X (see Problem 294). Now assume that X is an infinite compact space and consider the cardinal $\lambda = \sup\{w(K) : K \subset C_p(X)$ and K is compact$\}$. Observe first that we have the equality $c(X) = p(X) = \sup\{\kappa : A(\kappa)$ embeds in $C_p(X)\}$ (see TFS-178 and TFS-282). An immediate consequence is that $c(X) \leq \lambda$.

To see that $\lambda \leq c(X)$ take any compact $K \subset C_p(X)$. For any $x \in X$ let $e_x(f) = f(x)$ for any $f \in K$. Then $e_x \in C_p(K)$ and the map $e : X \to C_p(K)$ defined by $e(x) = e_x$ for every $x \in X$ is continuous by TFS-166; let $Y = e(X)$. The space Y is Eberlein compact (see Problem 035) and K embeds in $C_p(Y)$ by TFS-166. Observe that $w(Y) = c(Y) \leq c(X)$; so

$$w(K) = nw(K) \leq nw(C_p(Y)) = nw(Y) \leq w(Y) \leq c(X)$$

(see Fact 4 of S.307). Thus $w(K) \leq c(X)$ for any compact $K \subset C_p(X)$; this implies $\lambda \leq c(X)$ and hence $\lambda = \sup\{w(K) : K \subset C_p(X)$ and K is compact$\} = c(X)$.

U.344. *Suppose that X is a pseudocompact space and we have functions $f, g \in C(X)$. Let $d(f, g) = \sup\{|f(x) - g(x)| : x \in X\}$. Prove that d is a complete metric on the set $C(X)$ and the topology of $C_u(X)$ is generated by d.*

Solution. The fact that d is a complete metric on $C^*(X)$ was proved in Fact 2 of S.237. The metric d generates the topology of $C_u^*(X)$: this was established in Fact 1 of T.357. Since X is pseudocompact, we have $C(X) = C^*(X)$; so $(C(X), d)$ is a complete metric space whose topology coincides with the topology of $C_u(X)$.

U.345. *Prove that, for any pseudocompact X, the space $C_u(X)$ is separable if and only if X is compact and metrizable.*

Solution. If $C_u(X)$ is separable then $C_p(X)$ is also separable being a continuous image of the space $C_u(X)$ (see TFS-086). Therefore X can be condensed onto a second countable space by TFS-174. This condensation has to be a homeomorphism by TFS-140; so X is compact and metrizable (see TFS-209 and TFS-212).

For the converse, apply Fact 2 of T.357 to see that if X is compact and metrizable then $C_u(X)$ is second countable and hence separable.

U.346. *Suppose that X is a compact space and we area given a function $f \in C(X)$; let $\|f\| = \sup\{|f(x)| : x \in X\}$. Assume additionally that $h \in C(X)$, $r > 0$ and we have a sequence $H = \{h_n : n \in \omega\} \subset C(X)$ such that $\|h_n\| \leq r$ for all $n \in \omega$ and $h_n(x) \to h(x)$ for any $x \in X$ (i.e., the sequence H converges to h in the space $C_p(X)$). Prove that h belongs to the closure of the convex hull $\mathrm{conv}(H)$ of the set H in the space $C_u(X)$.*

Solution. We must first develop an elementary technique of dealing with simple measures to avoid using deep facts of measure theory. Given a set Z say that μ is *a measure on Z* if $\mu : Z \to [0, 1]$, the set $\mathrm{supp}(\mu) = \{z \in Z : \mu(z) \neq 0\}$ is finite and $\sum\{\mu(z) : z \in \mathrm{supp}(\mu)\} = 1$. If μ is a measure on Z and $A \subset Z$ let $\mu[A] = \sum\{\mu(z) : z \in A \cap \mathrm{supp}(\mu)\}$.

If Z is a set, $T \subset Z$, $\mathcal{E} \subset \exp Z$ and $\varepsilon > 0$ then let $M(T, \mathcal{E}, \varepsilon) = \{\mu : \mu$ is a measure on Z such that $\text{supp}(\mu) \subset T$ and $\mu[E] < \varepsilon$ for any $E \in \mathcal{E}\}$. For any $A \subset Z$ let $\mathcal{E}(A) = \{E \in \mathcal{E} : E \cap A \neq \emptyset\}$.

Fact 1. Suppose that Z is a set and T is a non-empty subset of Z. Assume also that $\mathcal{E} \subset \exp Z$, $\varepsilon > 0$ and $0 < \delta < \varepsilon$. If $M(T, \mathcal{E}, \varepsilon) = \emptyset$ then there is a non-empty finite set $F \subset T$ such that $M(T, \mathcal{E}(F), \delta) = \emptyset$.

Proof. Assume that $M(T, \mathcal{E}(F), \delta) \neq \emptyset$ for any non-empty finite $F \subset T$ and choose $m \in \mathbb{N}$ such that $\delta + \frac{1}{m} < \varepsilon$. Pick a point $z_0 \in T$ and let $A_0 = \{z_0\}$. By our assumption, there exists $\mu_0 \in M(T, \mathcal{E}(A_0), \delta)$. Proceeding by induction, assume that we have constructed finite subsets $A_0, \ldots, A_n$ of the set T and measures $\mu_0, \ldots, \mu_n$ on Z such that

(1) $A_{i+1} = A_i \cup \text{supp}(\mu_i)$ for all $i < n$ and $\mu_i \in M(T, \mathcal{E}(A_i), \delta)$ for every $i \leq n$.

Let $A_{n+1} = A_n \cup \text{supp}(\mu_n)$; the set A_{n+1} being non-empty and finite, there exists a measure $\mu_{n+1} \in M(T, \mathcal{E}(A_{n+1}), \delta)$; so our inductive procedure can be continued to obtain finite sets $A_0, \ldots, A_{m-1}$ and measures $\mu_0, \ldots, \mu_{m-1}$ which have the property (1).

It is evident that $\mu = \frac{1}{m}(\mu_0 + \ldots + \mu_{m-1})$ is a measure on Z and $\text{supp}(\mu) \subset T$. Fix any $E \in \mathcal{E}$ and consider the sequence $\mu_0[E], \ldots, \mu_{m-1}[E]$. If $k \leq m - 1$ is the first number such that $\mu_k[E] \neq 0$ then $\mu_l[E] < \delta$ for any $l > k$. Indeed, $\mu_k[E] \neq 0$ implies that $\text{supp}(\mu_k) \cap E \neq \emptyset$; since $\text{supp}(\mu_k) \subset A_{k+1} \subset A_l$, we have $E \cap A_l \neq \emptyset$, i.e., $E \in \mathcal{E}(A_l)$. Now $\mu_l \in M(Z, \mathcal{E}(A_l), \delta)$ implies $\mu_l[E] < \delta$. An immediate consequence is that $\mu_0[E] + \ldots + \mu_{m-1}[E] \leq 1 + (m - 1)\delta$ and hence $\mu[E] \leq \frac{1}{m}(1 + (m - 1)\delta) < \delta + \frac{1}{m} < \varepsilon$.

We established that $\mu[E] < \varepsilon$ for any $E \in \mathcal{E}$ which, together with $\text{supp}(\mu) \subset T$, implies that $\mu \in M(T, \mathcal{E}, \varepsilon)$. This contradiction shows that Fact 1 is proved.

Fact 2. Suppose that Z is a set and $\mathcal{E} \subset \exp Z$. If there exists an infinite set $S \subset Z$ such that $M(S, \mathcal{E}, \varepsilon) = \emptyset$ for some $\varepsilon > 0$ then there is a faithfully indexed sequence $\{z_n : n \in \omega\} \subset Z$ and a family $\{E_n : n \in \omega\} \subset \mathcal{E}$ such that $\{z_0, \ldots, z_n\} \subset E_n$ for any $n \in \omega$.

Proof. Let $\varepsilon_n = 2^{-n-1}\varepsilon$ for all $n \in \omega$. Then $\varepsilon_0 < \varepsilon$ and $\varepsilon_{n+1} < \varepsilon_n$ for every $n \in \omega$. Apply Fact 1 to the non-empty set $S_0 = S$ to find a non-empty finite $K_0 \subset S_0$ such that $M(S_0, \mathcal{E}(K_0), \varepsilon_0) = \emptyset$. Observe that $\mathcal{E}(K_0) \neq \emptyset$ because otherwise the characteristic function of any point of K_0 would belong to $M(S, \mathcal{E}, \varepsilon)$. Proceeding inductively, assume that $n \in \omega$ and we have a sequence $K_0, \ldots, K_n$ of finite subsets of S with the following properties:

(2) $K_{i+1} \subset S_{i+1} = S \setminus (K_0 \cup \ldots \cup K_i)$ for any $i < n$;
(3) $\mathcal{E}(K_0) \cap \ldots \cap \mathcal{E}(K_i) \neq \emptyset$ and $M(S_i, \mathcal{E}(K_0) \cap \ldots \mathcal{E}(K_i), \varepsilon_i) = \emptyset$ for all $i \leq n$.

Let $\mathcal{E}' = \mathcal{E}(K_0) \cap \ldots \cap \mathcal{E}(K_n)$ and $S_{n+1} = S \setminus (K_0 \cup \ldots \cup K_n)$. It follows from $M(S_n, \mathcal{E}', \varepsilon_n) = \emptyset$ that $M(S_{n+1}, \mathcal{E}', \varepsilon_n) = \emptyset$; the set S_{n+1} is infinite and hence non-empty; so we can apply Fact 1 find a finite non-empty set $K_{n+1} \subset S_{n+1}$ such that $M(S_{n+1}, \mathcal{E}'(K_{n+1}), \varepsilon_{n+1}) = \emptyset$. Since $\mathcal{E}'(K_{n+1}) = \mathcal{E}(K_0) \cap \ldots \cap \mathcal{E}(K_n) \cap \mathcal{E}(K_{n+1})$, the condition (2) and the second part of (3) are satisfied if we replace n by $n + 1$. To see that the first part of (3) also holds for $i = n + 1$ observe that if $\mathcal{E}'(K_{n+1}) = \emptyset$ then the characteristic function of any point of K_{n+1} would belong to the empty set $M(S_n, \mathcal{E}', \varepsilon_n)$.

This proves that our induction procedure can be continued to construct a family $\mathcal{K} = \{K_n : n \in \omega\}$ of non-empty finite subsets of S for which the properties (2) and (3) hold for any $n \in \omega$. It follows from (2) that $\mathcal{K}$ is a disjoint family. The property (3) implies that there is a family $\mathcal{D} = \{D_n : n \in \omega\} \subset \mathcal{E}$ such that

(4) $D_n \cap K_i \neq \emptyset$ for any $i \leq n$.

The set K_0 is finite and $D_n \cap K_0 \neq \emptyset$ for any $n \in \omega$ so there exists a point $z_0 \in K_0$ and an infinite $\mathcal{D}_0 \subset \mathcal{D}$ such that $z_0 \in \bigcap \mathcal{D}_0$. Given $n \in \omega$ suppose that we have chosen points $z_0, \ldots, z_n$ and infinite families $\mathcal{D}_0, \ldots, \mathcal{D}_n$ such that

(5) $\mathcal{D} \supset \mathcal{D}_0 \supset \ldots \supset \mathcal{D}_n$ and $z_i \in K_i \cap (\bigcap \mathcal{D}_n)$ for all $i \leq n$.

The family $\mathcal{D}_n$ being infinite, it follows from (4) that infinitely many elements of $\mathcal{D}_n$ intersect K_{n+1}. Therefore there is $z_{n+1} \in K_{n+1}$ and an infinite $\mathcal{D}_{n+1} \subset \mathcal{D}_n$ such that $z_{n+1} \in \bigcap \mathcal{D}_{n+1}$ and hence $\{z_0, \ldots, z_{n+1}\} \subset \bigcap \mathcal{D}_{n+1}$. Thus our inductive procedure shows that we can construct sequences $\{z_n : n \in \omega\}$ and $\{\mathcal{D}_n : n \in \omega\}$ such that the condition (5) is satisfied for all $n \in \omega$. The family $\{K_i : i \in \omega\}$ being disjoint, the property (5) implies that $z_n \neq z_m$ if $n \neq m$. If we take any $E_n \in \mathcal{D}_n$ then it follows from (5) that $\{z_0, \ldots, z_n\} \subset E_n$ for all $n \in \omega$, i.e., Fact 2 is proved.

Returning to our solution let $u_n = h_n - h$ for any $n \in \omega$. If u is identically zero on X then $u_n(x) \to u(x)$ for any $x \in X$. It is easy to see that $||h|| \leq r$ and therefore $||u_n|| \leq 2r$ for any $n \in \omega$. Given $A \subset C(X)$ we will denote by $\mathrm{cl}_u(A)$ the closure of A in $C_u(X)$. The following fact is evident.

(6) if $f \in C(X)$ and $A \subset C(X)$ then $f \in \mathrm{cl}_u(A)$ if and only if for any $\varepsilon > 0$ there is $g \in A$ such that $||f - g|| < \varepsilon$.

To prove that $h \in \mathrm{cl}_u(\mathrm{conv}(H))$ fix an arbitrary $\varepsilon > 0$ and let $\delta = \min\{\frac{\varepsilon}{4r}, \frac{\varepsilon}{2}\}$. For any $x \in X$ let $W(x) = \{n \in \omega : |u_n(x)| \geq \delta\}$. We claim that

(7) If $\mathcal{E} = \{W(x) : x \in X\} \subset \exp \omega$ then $M(\omega, \mathcal{E}, \delta) \neq \emptyset$.

Indeed, if the assertion (7) is false then we can apply Fact 2 to see that

(8) there exist sequences $\{m_i : i \in \omega\} \subset \omega$ and $\{x_i : i \in \omega\} \subset X$ such that $i \neq j$ implies $m_i \neq m_j$ and $\{m_0, \ldots, m_n\} \subset W(x_n)$ for any $n \in \omega$.

The set $F_i = \{x \in X : |u_{m_i}(x)| \geq \delta\}$ is closed in X for any $i \in \omega$ and it follows from (8) that $x_n \in F_0 \cap \ldots \cap F_n$ for any $n \in \omega$. Therefore the family $\{F_n : n \in \omega\}$ is centered; the space X being compact, there exists a point $x \in \bigcap\{F_n : n \in \omega\}$. We have $|u_{m_i}(x)| \geq \delta$ for every $i \in \omega$; this contradiction with the fact that $u_{m_i}(x) \to 0$ shows that (7) is proved.

By (7), we can fix a measure $\mu \in M(\omega, \mathcal{E}, \delta)$; let $N = \mathrm{supp}(\mu)$. It is immediate that the function $w = \sum\{\mu(n)h_n : n \in N\}$ belongs to $\mathrm{conv}(H)$. Given any $x \in X$ we have $|w(x) - h(x)| = |\sum_{n \in N}(\mu(n)h_n(x) - \mu(n)h(x))| \leq \sum_{n \in N} \mu(n)|u_n(x)|$.

Let $N_0 = \{n \in N : |u_n(x)| < \delta\}$; it is clear that $N_1 = N \setminus N_0 = W(x)$. We have $\mu[W(x)] < \delta$; so $\sum_{n \in N_1} \mu(n)|u_n(x)| < \delta \cdot 2r \leq \frac{\varepsilon}{2}$.

Furthermore, $\sum_{n \in N_0} \mu(n)|u_n(x)| \leq \delta \cdot (\sum_{n \in N_0} \mu(n)) \leq \delta \leq \frac{\varepsilon}{2}$ which shows that $\sum_{n \in N} \mu(n)|u_n(x)| = \sum_{n \in N_0} \mu(n)|u_n(x)| + \sum_{n \in N_1} \mu(n)|u_n(x)| < \varepsilon$. We proved that $|w(x) - h(x)| < \varepsilon$ for any $x \in X$; so it follows from compactness of X that $||w - h|| < \varepsilon$. Thus, for any $\varepsilon > 0$ there is $w \in \text{conv}(H)$ such that $||w - h|| < \varepsilon$. Finally apply (6) to conclude that $h \in \text{cl}_u(\text{conv}(H))$ and finish our solution.

U.347. *Suppose that X is a Čech-complete space and we are given a continuous map $\varphi : X \to C_p(K)$ for some compact space K. Prove that there exists a dense G_δ-set $P \subset X$ such that $\varphi : X \to C_u(K)$ is continuous at every point of P.*

Solution. Let (M, ρ) be a metric space; as usual, if $x \in M$ and $\varepsilon > 0$ then $B_\rho(x, \varepsilon) = \{y \in M : \rho(x, y) < \varepsilon\}$ is the ε-ball centered at the point x. If $A \subset Z$ then $\text{diam}_\rho(A) = \sup\{\rho(x, y) : x, y \in A\}$ is the ρ-diameter of the set A. Given a space Z, a map $f : Z \to M$, and a point $z \in Z$, *the oscillation $\text{osc}(f, z)$ of the function f at the point z* is defined by $\text{osc}(f, z) = \inf\{\text{diam}_\rho(f(U)) : U \in \tau(z, Z)\}$.

Fact 1. Suppose that Z is a space, (M, ρ) is a metric space and we have a map $f : Z \to M$. Then

(a) for any $\varepsilon > 0$ the set $O_\varepsilon = \{z \in Z : \text{osc}(f, z) < \varepsilon\}$ is open in Z;
(b) the map f is continuous at a point $z \in Z$ if and only if $\text{osc}(f, z) = 0$.

Proof. (a) If $z \in O_\varepsilon$ then there is $U \in \tau(z, Z)$ such that $\text{diam}_\rho(f(U)) < \varepsilon$. The set U witnesses that $\text{osc}(f, y) < \varepsilon$ for any $y \in U$; so $U \subset O_\varepsilon$. This proves that O_ε is open in Z.
(b) If f is continuous at z then, for any $\varepsilon > 0$ there is $U \in \tau(z, Z)$ such that $f(U) \subset B_\rho(f(z), \frac{\varepsilon}{3})$. Therefore $\text{diam}_\rho(f(U)) \leq \text{diam}_\rho(B_\rho(f(z), \frac{\varepsilon}{3})) < \varepsilon$. Since $\varepsilon > 0$ was taken arbitrarily, this proves that $\text{osc}(f, z) = 0$, i.e., we established necessity.

Now, if $\text{osc}(f, z) = 0$ for some $z \in Z$, then, for any $\varepsilon > 0$ there is $U \in \tau(z, Z)$ such that $\text{diam}_\rho(f(U)) < \varepsilon$. Then, for any point $y \in U$, we have the inequality $\rho(f(y), f(z)) \leq \text{diam}_\rho(f(U)) < \varepsilon$ which shows that $f(U) \subset B_\rho(f(z), \varepsilon)$, i.e., f is continuous at the point z. This settles sufficiency in (b) and shows that Fact 1 is proved.

Returning to our solution let $||f|| = \sup\{|f(x)| : x \in K\}$ for any $f \in C_p(K)$. The topology on $C_u(K)$ is generated by the metric d defined by $d(f, g) = ||f - g||$ for any $f, g \in C_p(K)$ (see Problem 344); so we identify $C_u(K)$ with the metric space $(C(K), d)$. For any $f \in C(K)$ and $r > 0$ let $B(f, r) = \{g \in C(K) : ||g - f|| < r\}$. It is clear that the set $I(f, r) = \{g \in C(K) : ||g - f|| \leq r\}$ is closed both in $C_u(K)$ and $C_p(K)$ for any $f \in C(K)$ and $r > 0$. From now on we will also consider that $\varphi : X \to C_u(K)$ indicating each time whether we use the topology of $C_p(K)$ or the topology of $C_u(K)$. In particular, we will sometimes say that φ is τ_p-continuous (at a point $x \in X$) instead of saying that $\varphi : X \to C_p(K)$ is continuous (at a point $x \in X$). Analogously, the phrase "φ is τ_u-continuous at the point $x \in X$" says that $\varphi : X \to C_u(K)$ is continuous at the point $x \in X$.

Considering that $\varphi : X \to C_u(K)$ we can define the oscillation $\mathrm{osc}(\varphi, z)$ of φ at any point $z \in Z$. By Fact 1, the set $O_\varepsilon = \{x \in X : \mathrm{osc}(\varphi, x) < \varepsilon\}$ is open in X for any $\varepsilon > 0$; so $P = \bigcap_{n \in \omega} O_{2^{-n}}$ is a G_δ-set. It is evident that P coincides with the set of points at which the oscillation of φ is equal to zero; so φ is τ_u-continuous at any point of P by Fact 1. If every O_ε is dense in X then P is also dense in X because X is Čech-complete and hence has the Baire property (see TFS-274). Therefore it suffices to show that

(1) for any $\varepsilon > 0$ the set O_ε is dense in X.

Assume towards a contradiction that (1) is false. Then the closed set $F = X \backslash O_\varepsilon$ has non-empty interior U. It is clear that $\mathrm{diam}_d (\varphi(V)) \geq \varepsilon$ for any non-empty open $V \subset U$. Consider the set $E_r = \{f \in C_p(K) : \|f\| \leq r\}$ for any $r \in \omega$. Then $C_p(K) = \bigcup_{r \in \omega} E_r$; every E_r is closed in $C_p(K)$; so $D_r = \varphi^{-1}(C_r)$ is closed in X. The equality $X = \bigcup_{r \in \omega} D_r$ implies that $U \subset \bigcup_{r \in \omega} D_r$; so we can apply the Baire property of X again to see that there is a non-empty open $U' \subset U$ such that $U' \subset D_r$ for some $r \in \omega$. The set U' is also Čech-complete so, to obtain a contradiction, we can replace X by U'. Therefore we can assume, without loss of generality, that $X = U'$, i.e.,

(2) $\mathrm{diam}_d (\varphi(V)) \geq \varepsilon$ for any $V \in \tau^*(X)$ and there is $r > 0$ such that $\|\varphi(x)\| \leq r$ for any $x \in X$.

Recall that, for any $A \subset C(K)$, the set $\mathrm{conv}(A)$ is the convex hull of A. We will first prove that

(3) If $A = \{f_0, \ldots, f_n\} \subset C(K)$ then $\mathrm{conv}(A)$ is a compact subset of $C_u(K)$.

To prove (3) consider the set $C = \{t = (t_0, \ldots, t_n) \in [0, 1]^{n+1} : \sum_{i=0}^n t_i = 1\}$. It is an easy exercise that the set C is closed in $[0, 1]^{n+1}$ and hence compact. For any point $t = (t_0, \ldots, t_n) \in C$ let $\xi(t) = \sum_{i=0}^n t_i \cdot f_i$; then $\xi : C \to C_u(K)$ and it is straightforward that $\xi(C) = \mathrm{conv}(A)$. It follows from Fact 1 of T.105 that $C_u(K)$ is a linear topological space; so ξ is a continuous map. Therefore $\mathrm{conv}(A)$ is compact being a continuous image of the compact space C. This settles (3).

Apply Čech-completeness of X to fix a family $\{G_n : n \in \omega\} \subset \tau(\beta X)$ such that $G_{n+1} \subset G_n$ for any $n \in \omega$ and $\bigcap_{n \in \omega} G_n = X$. Choose a point $x_0 \in X$ arbitrarily and take a set $U_0 \in \tau(x_0, \beta X)$ such that $\mathrm{cl}_{\beta X}(U_0) \subset G_0$. Proceeding inductively suppose that we have constructed, for some $n \in \omega$, a set $\{x_0, \ldots, x_n\} \subset X$ and a family $\{U_0, \ldots, U_n\} \subset \tau(\beta X)$ with the following properties:

(4) $x_i \in U_i$ for all $i \leq n$;
(5) $\mathrm{cl}_{\beta X}(U_0) \subset G_0$ and $\mathrm{cl}_{\beta X}(U_{i+1}) \subset U_i \cap G_{i+1}$ for all $i < n$;
(6) $\|\varphi(x) - f\| \geq \frac{\varepsilon}{3}$ whenever $x \in U_{i+1} \cap X$ and $f \in \mathrm{conv}(\{\varphi(x_0), \ldots, \varphi(x_i)\})$ for any $i < n$;

The set $\Phi_n = \mathrm{conv}(\{\varphi(x_0), \ldots, \varphi(x_n)\})$ is compact by (3); so we can choose a finite set $A_n \subset \Phi_n$ such that $\Phi_n \subset \bigcup\{B(f, \frac{\varepsilon 12}{}) : f \in A_n\}$. As a consequence, $\Phi_n \subset R_n = \bigcup\{I(f, \frac{5\varepsilon}{12}) : f \in A_n\}$. Assume that $\|g - f\| < \frac{\varepsilon}{3}$ for some $g \in C(K)$

and $f \in \Phi_n$. There is $u \in A_n$ such that $||f - u|| < \frac{\varepsilon}{12}$; so $||g - u|| < \frac{\varepsilon}{3} + \frac{\varepsilon}{12} = \frac{5\varepsilon}{12}$ and therefore $g \in B(u, \frac{5\varepsilon}{12})$ which implies $g \in R_n$. This shows that

(7) if $g \notin R_n$ then $||g - f|| \geq \frac{\varepsilon}{3}$ for any $f \in \Phi_n$.

If $u \in A_n$ and $f, g \in I(u, \frac{5\varepsilon}{12})$ then

$$||f - g|| \leq ||f - u|| + ||u - g|| \leq \frac{5\varepsilon}{12} + \frac{5\varepsilon}{12} \leq \frac{5\varepsilon}{6}$$

which shows that $\text{diam}_d (I(u, \frac{5\varepsilon}{12})) \leq \frac{5\varepsilon}{6} < \varepsilon$. The set $T(u) = \varphi^{-1}(I(u, \frac{5\varepsilon}{12}))$ is closed in X because $\varphi : X \to C_p(K)$ is continuous and $I(u, \frac{5\varepsilon}{12})$ is closed in $C_p(K)$. If $V \in \tau^*(X)$ and $V \subset T(u)$ then $\text{diam}_d (\varphi(V)) \leq \text{diam}_d (I(u, \frac{5\varepsilon}{12})) < \varepsilon$ which is a contradiction with (2). Therefore $T(u)$ is nowhere dense in X for any $u \in A_n$. This implies that $T = \bigcup \{T(u) : u \in A_n\}$ is nowhere dense in X and hence in βX.

Thus we can choose a set $U_{n+1} \in \tau(\beta X)$ such that $\text{cl}_{\beta X}(U_{n+1}) \subset U_n \cap G_{n+1}$ and $U_{n+1} \cap T = \emptyset$. Since X is dense in βX, we can pick a point $x_{n+1} \in U_{n+1} \cap X$. It is immediate that the properties (4) and (5) still hold if we replace n by $n + 1$.

To see that (6) is also true for $i = n$ take any $x \in U_{n+1} \cap X$. Since $x \notin T$, we have $\varphi(x) \notin R_n$; so $||\varphi(x) - f|| \geq \frac{\varepsilon}{3}$ for any $f \in \Phi_n$ by (7). Thus our inductive procedure can be continued to construct a set $Y = \{x_i : i \in \omega\} \subset X$ and a family $\{U_i : i \in \omega\} \subset \tau(\beta X)$ for which the properties (4)–(6) are true for all $n \in \omega$.

The set $F = \bigcap_{n \in \omega} U_n = \bigcap_{n \in \omega} \text{cl}_{\beta X}(U_n)$ compact and it follows from (5) that $F \subset X$. For any $z \in \beta X \backslash F$ there is $n \in \omega$ with $z \notin \text{cl}_{\beta X}(U_n)$ and hence the set $V = \beta X \backslash \text{cl}_{\beta X}(U_n)$ is an open neighbourhood of z whose intersection with Y has at most $n + 1$ points. This shows that $Y \cup F \subset X$ is closed in βX and all accumulation points of Y belong to F.

Thus $\varphi(Y \cup F) \subset C_p(K)$ is a Fréchet–Urysohn space being Eberlein compact. Let $y \in F$ be an accumulation point of Y; then $\varphi(y)$ is in the closure of $\varphi(Y)$ in $C_p(K)$. Since $\varphi(Y \cup F)$ is a Fréchet–Urysohn space, there is a sequence $H \subset Y$ such that $\varphi(H)$ converges to $\varphi(y)$ (in the space $C_p(K)$). It follows from the property (2) that $||f|| \leq r$ for any function $f \in \varphi(H)$; so we can apply Problem 346 to see that $\varphi(y) \in \text{cl}_u(\text{conv}(\varphi(H)))$ and therefore $\varphi(y) \in \text{cl}_u(\text{conv}(\varphi(Y)))$.

It is clear that $\text{conv}(\varphi(Y)) = \bigcup_{n \in \omega} \Phi_n$. However, for any number $n \in \omega$, it follows from $y \in U_n$ that $||\varphi(y) - f|| \geq \frac{\varepsilon}{3}$ for any $f \in \Phi_n$. Thus $||\varphi(y) - f|| \geq \frac{\varepsilon}{3}$ for any $f \in \text{conv}(\varphi(Y))$ which implies that $\varphi(y) \notin \text{cl}_u(\text{conv}(\varphi(Y)))$, a contradiction.

This contradiction shows that (1) is true, i.e., every O_ε is a dense subset of X. Thus P is a dense G_δ-subset of X such that $\varphi : X \to C_u(K)$ is continuous at every point of P, i.e., our solution is complete.

U.348. *Prove that any Eberlein–Grothendieck Čech-complete space has a dense G_δ-subspace which is metrizable.*

Solution. Suppose that X is a Čech-complete space and $X \subset C_p(K)$ for some compact space K. If $i(x) = x$ for any $x \in X$ then $i : X \to C_p(K)$ is a continuous map; so we can apply Problem 347 to find a dense G_δ-subset P of the space X such that $i : X \to C_u(K)$ is continuous at every point of P and therefore the mapping $i_0 = i|P : P \to C_u(K)$ is continuous.

The topology of $C_u(K)$ is metrizable (see Problem 344); so the topology μ induced on P from $C_u(K)$ is metrizable as well. Let τ be the topology on P induced from X; of course, τ can also be considered as the topology induced from $C_p(K)$. Since $\tau(C_p(K)) \subset \tau(C_u(K))$, we have $\tau \subset \mu$. The map i_0 being continuous, the set $U = i_0^{-1}(U)$ has to be open in (P, τ) for any $U \in \mu$. Thus, $U \in \tau$ for any $U \in \mu$ which shows that $\tau = \mu$ and hence $P = (P, \tau)$ is metrizable.

U.349. *Prove that if X is an Eberlein–Grothendieck Čech-complete space then $c(X) = w(X)$.*

Solution. Suppose that X is a Čech-complete Eberlein–Grothendieck space. The inequality $c(X) \leq w(X)$ is true for any space X so we are only going to establish that $w(X) \leq c(X)$. By Problem 348, there is a dense metrizable $P \subset X$. If $\kappa = c(X)$ then $c(P) \leq \kappa$ and hence $d(P) \leq \kappa$ by TFS-214. This implies that $d(X) \leq \kappa$; the space X being monolithic (see SFFS-118 and SFFS-154), we have $nw(X) \leq \kappa$. Finally, apply TFS-270 to conclude that $w(X) = nw(X) \leq \kappa = c(X)$ and hence $w(X) = c(X)$.

U.350. *Let X be a compact space. Assume that $X = X_1 \cup \ldots \cup X_n$, where every X_i is a metrizable (not necessarily closed) subspace of X. Prove that $\overline{X}_1 \cap \ldots \cap \overline{X}_n$ is metrizable. In particular, if all X_i's are dense in X then X is metrizable.*

Solution. Every X_i has a σ-discrete base; since any σ-discrete base is σ-disjoint, we can apply Fact 1 of T.412 to see that $F = \overline{X}_1 \cap \ldots \cap \overline{X}_n$ has a σ-disjoint base. Any σ-disjoint base is a point-countable T_1-separating family; so we can apply Fact 1 of T.203 to conclude that F is metrizable. If every X_i is dense in X then X is metrizable because it coincides with F.

U.351. *Suppose that X is a compact space which is a union of two metrizable subspaces. Prove that X is Eberlein compact which is not necessarily metrizable.*

Solution. If $X = A(\omega_1)$ then X is a non-metrizable Eberlein compact space representable as the union of two discrete (and hence metrizable) subspaces.

Now assume that $X = M_0 \cup M_1$ where both spaces M_0 and M_1 are metrizable. The space $P = \overline{M}_0 \cap \overline{M}_1$ is metrizable by Problem 350. If $L_i = M_i \backslash P$ for $i \in \{0, 1\}$ then $\overline{L}_0 \cap L_1 = \emptyset$ and $L_0 \cap \overline{L}_1 = \emptyset$ which shows that every L_i is clopen in $L_0 \cup L_1$, i.e., $L_0 \cup L_1$ is homeomorphic to $L_0 \oplus L_1$.

Since every L_i is metrizable, the space $L_0 \oplus L_1$ is also metrizable and hence $U = X \backslash P = L_0 \cup L_1 \simeq L_0 \oplus L_1$ is metrizable as well. Let $\mathcal{B}$ be a σ-discrete base in the space U. It is easy to see that the family $\mathcal{B}' = \{B \in \mathcal{B} : \overline{B} \subset U\}$ is also a base of U.

Metrizability of U also implies that every $V \in \tau(U)$ is an F_σ-subset of U. Therefore we can fix, for any $B \in \mathcal{B}'$ a family $\{F_B^n : n \in \omega\}$ of closed subsets of U such that $\bigcup_{n \in \omega} F_B^n = B$. By our choice of $\mathcal{B}'$, we have $\overline{B} \subset U$ and hence every F_B^n is closed in the compact space $\overline{B}$. Thus, every F_B^n is closed in X which shows that

(1) the family $\mathcal{B}'$ is a σ-disjoint base in U which consists of open F_σ-subsets of X.

Given $x, y \in X$ say that a set $V \subset X$ *separates the points x and y* if $V \cap \{x, y\}$ is a singleton. Denote by $\mathcal{Q}$ the family of open intervals in $\mathbb{R}$ with rational endpoints. The space P being second countable, there exists a countable $A \subset C(P)$ which separates the points of P. By normality of X, for any $f \in A$ there is $u_f \in C(X)$ such that $u_f | P = f$. The family $\mathcal{U} = \{u_f^{-1}(I) : f \in A, I \in \mathcal{Q}\}$ is countable and consists of open F_σ-subsets of X. This, together with (1), implies that the family $\mathcal{V} = \mathcal{U} \cup \mathcal{B}'$ is σ-disjoint (and hence σ-point-finite) and consists of open F_σ-subsets (which are, therefore, cozero subsets) of X. Given distinct points $x, y \in X$, consider the following cases.

  (i) $x, y \in U$; since $\mathcal{B}'$ is a base in U, there is $B \in \mathcal{B}'$ with $x \in B$ and $y \notin B$; so $B \in \mathcal{V}$ separates x and y.
 (ii) if $x \in U$ and $y \in F$ (or $x \in F$ and $y \in U$) then there is $B \in \mathcal{B}'$ with $x \in B$ (or $y \in B$ respectively). Since $B \subset X \backslash F$, we have $y \notin B$ (or $x \notin B$ respectively) i.e., the set $B \in \mathcal{V}$ separates the points x and y.
(iii) $x, y \in F$; then there is $f \in A$ such that $f(x) \neq f(y)$ and hence there is $I \in \mathcal{Q}$ such that $f(x) \in I$ and $f(y) \notin I$. The set $V = u_f^{-1}(I)$ belongs to $\mathcal{V}$; since $u_f(x) = f(x) \in I$ and $u_f(y) = f(y) \notin I$, we have $x \in V$ while $y \notin V$, i.e., the set $V \in \mathcal{V}$ again separates the points x and y.

This proves that $\mathcal{V}$ is a σ-point-finite T_0-separating family of cozero subsets of X; so we can apply Problem 324 to conclude that X is Eberlein compact.

U.352. *Observe that there exists a compact space K which is not Eberlein while being a union of three metrizable subspaces. Suppose that X is a compact space such that $X \times X$ is a union of its three metrizable subspaces. Prove that X is Eberlein compact.*

Solution. Let K be the one-point compactification of the Mrowka space M (see TFS-142). Then K is a separable non-metrizable compact space so K is not Eberlein compact by Problem 120. If a is the unique point of the set $K \backslash M$ then it follows from $K = \{a\} \cup \omega \cup (M \backslash \omega)$ that K is the union of its three discrete (and hence metrizable) subspaces (see TFS-142). Thus K witnesses that a compact space need not be Eberlein if it is representable as the union of three metrizable spaces.

Fact 1. Any finite union of Eberlein compact spaces is an Eberlein compact space.

Proof. Suppose that $K = E_0 \cup \ldots \cup E_n$ and every E_i is Eberlein compact. The space $E = E_0 \oplus \ldots \oplus E_n$ is compact and maps continuously onto K; so it suffices to prove that E is Eberlein compact (see Problem 337).

Apply Problem 324 to fix a T_0-separating σ-point-finite family $\mathcal{U}_i$ of cozero subsets of E_i for any $i \leq n$. The family $\mathcal{U} = \mathcal{U}_0 \cup \ldots \cup \mathcal{U}_n \cup \{E_0, \ldots, E_n\}$ is easily seen to be σ-point-finite; it consists of cozero subsets of E and T_0-separates the points of E. Thus E is Eberlein compact; so K is also Eberlein compact, i.e., Fact 1 is proved.

Returning to our solution assume that X is compact and $X \times X = Y_0 \cup Y_1 \cup Y_2$ where every Y_i is metrizable. By Problem 350 the set $F = \overline{Y}_0 \cap \overline{Y}_1 \cap \overline{Y}_2$ is metrizable. Let $\pi : X \times X \to X$ be the natural projection of $X \times X$ onto its first factor. If $\pi(F) = X$ then X is metrizable and hence Eberlein (see Fact 5 of S.307 and Problem 316). If there is some $a \in X \backslash \pi(F)$ then the set $X_a = \pi^{-1}(a)$ is homeomorphic to X and $X_a \cap F = \emptyset$.

This implies that, for any $z \in X_a$, there is $i \in \{0, 1, 2\}$ such that $z \notin \overline{Y}_i$ and hence we can choose $O_z \in \tau(z, X \times X)$ such that $\overline{O}_z \cap Y_i = \emptyset$. Therefore $\overline{O}_z$ is covered by the family $\{Y_0, Y_1, Y_2\} \backslash \{Y_i\}$ of two metrizable spaces. Apply Problem 351 to see that $\overline{O}_z$ is Eberlein compact for every $z \in X_a$. By compactness of X_a, there is a finite set $A \subset X_a$ such that $X_a \subset K = \bigcup \{\overline{O}_z : z \in A\}$. The space K is Eberlein compact by Fact 1; so $X \simeq X_a$ is also Eberlein compact by Problem 306.

U.353. *Prove that, if X is compact and X^ω is a union of countably many of its Eberlein compact subspaces then X is Eberlein compact.*

Solution. Suppose that $X^\omega = \bigcup_{n \in \omega} E_n$ and every subspace E_n is Eberlein compact. For any $n \in \omega$ let $\pi_n : X^\omega \to X$ be the natural projection of X^ω onto its n-th factor. If $\pi_n(E_n) \neq X$ for any $n \in \omega$, then choose $x_n \in X \backslash \pi_n(E_n)$ for each n and observe that the point $x = (x_n)_{n \in \omega} \in X^\omega$ does not belong to any E_n. This contradiction shows that $X = \pi_n(E_n)$ for some $n \in \omega$ and hence X is Eberlein compact by Problem 337.

U.354. *Give an example of an Eberlein compact space which cannot be represented as a countable union of its metrizable subspaces.*

Solution. The space $A(\omega_1)$ is Eberlein compact and non-metrizable. Therefore the space $K = (A(\omega_1))^\omega$ is Eberlein compact (see Problem 307) which cannot be represented as a countable union of its metrizable subspaces (see SFFS-417).

U.355. *Let X be a Corson compact space such that X is a countable union of Eberlein compact spaces. Prove that $C_p(X)$ is K-analytic and hence X is Gul'ko compact.*

Solution. Given a space Z say that a set $A \subset Z$ is *concentrated around a point* $z \in Z$ if $A \backslash U$ is countable for any $U \in \tau(z, Z)$. Suppose that $X = \bigcup \{K_n : n \in \omega\}$ where every K_n is Eberlein compact. Since X is Corson compact, we can assume from now on that $X \subset \Sigma(T)$ for some uncountable set T. Let $u(t) = 0$ for all $t \in T$. We will also need the characteristic function u_t of the singleton $\{t\}$ for any $t \in T$; recall that $u_t(t) = 1$ and $u_t(s) = 0$ for any $s \in T \backslash \{t\}$.

It is straightforward that the space $E = \{u\} \cup \{u_t : t \in T\} \subset \Sigma(T)$ is homeomorphic to $A(\kappa)$ for $\kappa = |T|$. The space $K = X \cup E$ is also Corson compact; by Fact 1 of U.340, the space K is still a countable union of Eberlein compacta and $E \subset K$. Thus we can enumerate the relevant Eberlein compact subspaces of K as $\{L_n : n \in \omega\}$ in such a way that $L_0 = E$, $L_n \subset L_{n+1}$ for any $n \in \omega$ and $K = \bigcup \{L_n : n \in \omega\}$.

Let $h(x) = 0$ for all $x \in K$; given $t \in T$ let $h_t(x) = x(t)$ for all $x \in K$. Then the function $h_t : K \to \mathbb{R}$ is continuous because it coincides with the restriction of the natural projection of $\mathbb{R}^T$ onto its factor determined by t. It is immediate that the space $H = \{h\} \cup \{h_t : t \in T\}$ separates the points of K and H is concentrated around h.

Let $\pi_n : C_p(K) \to C_p(L_n)$ be the restriction map for any $n \in \omega$.

We leave it to the reader to verify that

(1) $\pi_0|H$ is injective, $\pi_0(h)$ is the unique non-isolated point of $\pi_0(H)$ and $\pi_0(H)$ is homeomorphic to $A(\kappa)$.

If $\pi_0^n : C_p(L_n) \to C_p(L_0)$ is the restriction map then $\pi_0^n(\pi_n(H)) = \pi_0(H)$; so (1) implies that $\pi_n(h)$ is the unique non-isolated point of $G_n = \pi_n(H)$ for all $n \in \omega$.

For any $n \in \omega$ there exists a condensation $\eta_n : C_p(L_n) \to S_n$ of the space $C_p(L_n)$ into a Σ-product S_n of real lines by Problem 200. The set $\eta_n(G_n)$ is concentrated around the point $\eta_n(\pi_n(h))$; so we can apply Fact 1 of U.289 to see that $\eta_n(G_n)$ is σ-compact and therefore G_n is an F_σ-subset of $C_p(L_n)$. The space $C_p(L_n)$ is K-analytic by Problem 321; so G_n is K-analytic as well for any $n \in \omega$. We will show next that

(2) the space H embeds in $G = \prod_{n\in\omega} G_n$ as a closed subspace.

To prove (2) consider the diagonal product $\pi = \Delta_{n\in\omega}\pi_n$ which maps $C_p(K)$ to $\prod_{n\in\omega} C_p(L_n)$. It is clear that $\pi(H) \subset G$. Let $j_n : L_n \to K$ be the identity map for any $n \in \omega$. Then we can define a map $j : L = \bigoplus_{n\in\omega} L_n \to K$ by letting $j(x) = j_n(x)$ whenever $x \in L_n$ (we identify every L_n with the respective clopen subset of L). The map j is continuous, onto and it is clear that the dual map $j^* : C_p(K) \to C_p(L)$ coincides with π. Thus π is an embedding by TFS-163.

To see that $\pi(H)$ is closed in the space G take any $g = (g_n)_{n\in\omega} \in G\backslash\pi(H)$ and fix $f_n \in H$ such that $g_n = \pi_n(f_n)$ for any $n \in \omega$. If there is $f \in H$ for which $g_n = \pi_n(f)$ for all $n \in \omega$ then $g = \pi(f) \in \pi(H)$ which is a contradiction. Therefore there are distinct $v, w \in H$ such that $g_k = \pi_k(v)$ and $g_m = \pi_m(w)$ for some distinct $k, m \in \omega$.

One of the points v, w, say, v is isolated in H. Therefore g_k is isolated in G_k by (1) and hence $W = \{a = (a_n)_{n\in\omega} \in G : a_k = g_k\}$ is an open neighbourhood of g such that $W \cap \pi(H)$ cannot contain any point distinct from $\pi(v)$. Thus $W\backslash\{\pi(v)\}$ is an open neighbourhood of g in G disjoint from $\pi(H)$. This shows that $\pi(H)$ is closed in G, i.e., (2) is proved.

Any countable product of K-analytic spaces is a K-analytic space and any closed subspace of a K-analytic space is K-analytic; so it follows from (2) that H is K-analytic. Since H separates the points of K, we can apply Problem 022 to conclude that $C_p(K)$ is K-analytic. We have $X \subset K$; the space K being normal, the restriction map $\pi_X : C_p(K) \to C_p(X)$ is surjective. This proves that $C_p(X) = \pi_X(C_p(K))$ is also K-analytic and completes our solution.

U.356. *Let X be a σ-product of an arbitrary family of Eberlein compact spaces. Prove that $C_p(X)$ is a $K_{\sigma\delta}$-space.*

Solution. Suppose that, for any $t \in T$, we are given an Eberlein compact space K_t and fix a point $q_t \in K_t$. In the product space $K = \prod_{t \in T} K_t$ consider the subspace $X = \{x \in K : \text{the set } \text{supp}_T(x) = \{t \in T : x(t) \neq q_t\} \text{ is finite}\}$; we must show that $C_p(X)$ is a $K_{\sigma\delta}$-space. By Problem 322, we can assume that there is an infinite set E_t such that $K_t \subset \Sigma_*(E_t)$ for any $t \in T$. There is no loss of generality to assume that the family $\{E_t : t \in T\}$ is disjoint; by homogeneity of $\Sigma_*(E_t) \simeq C_p(A(|E_t|))$ (see Problem 105) we can also assume that q_t coincides with the zero function of $\Sigma_*(E_t)$, i.e., $q_t(a) = 0$ for all $a \in E_t$.

Let $E = \bigcup\{E_t : t \in T\}$; in the product space $\Sigma = \prod\{\Sigma_*(E_t) : t \in T\}$ consider the σ-product $\sigma = \{x \in \Sigma : |\{t \in T : x(t) \neq q_t\}| < \omega\}$. It is straightforward that $X \subset \sigma$; therefore every $x \in X$, apart from being a function on T (because $X \subset K$), is also a function on E (because $X \subset K \subset \prod_{t \in T} \mathbb{R}^{E_t} \simeq \mathbb{R}^E$). Since the family $\{E_t : t \in T\}$ is disjoint, we will cause no confusion writing $x(a)$ instead of $(x(t))(a)$ for any $t \in T$ and $a \in E_t$.

If $x \in \sigma$ then there is a finite set $Q \subset T$ such that $x(a) = 0$ whenever $a \in E_t$ for some $t \in T \backslash Q$. Given any $\varepsilon > 0$, the set $S_\varepsilon^t = \{a \in E_t : |x(a)| \geq \varepsilon\}$ is finite for any $t \in T$ and hence so is the set $S_\varepsilon = \bigcup\{S_\varepsilon^t : t \in Q\}$. Thus the set $\{a \in E : |x(a)| \geq \varepsilon\} \subset S_\varepsilon$ is finite and hence $x \in \Sigma_*(E)$. This proves that $\sigma \subset \Sigma_*(E)$ and therefore

(1) $X \subset \sigma \subset \Sigma_*(E)$ is an Eberlein–Grothendieck space.

Consider the set $X_n = \{x \in X : |\text{supp}_T(x)| \leq n\} \subset K$ for any $n \in \omega$; it is evident that $X = \bigcup_{n \in \omega} X_n$. If $n \in \omega$ and $x \in K \backslash X_n$ then there are distinct $t_0, \ldots, t_n \in T$ such that $x(t_i) \neq q_{t_i}$ for all $i \leq n$. The set $U = \{y \in K : y(t_i) \in X_{t_i} \backslash \{q_{t_i}\} \text{ for all } i \leq n\}$ is open in K and $x \in U \subset K \backslash X_n$; this proves that X_n is closed in K and hence compact for any $n \in \omega$. Now, it follows from (1) that X is a countable union of Eberlein compact spaces. By Problem 105, the space X embeds in $C_p(A(\kappa))$ for $\kappa = |E|$; the space $A(\kappa)$ is compact and the class of compact spaces is sk-directed; so we can apply Problem 092 to see that $C_p(X, \mathbb{I})$ is a $K_{\sigma\delta}$-space.

The space X being σ-compact, we can apply Problem 202 to see that there exists a $K_{\sigma\delta}$-space H such that $C_p(X) \subset H \subset \mathbb{R}^X$. If $\xi(t) = \frac{2}{\pi} \arctan(t)$ for any $t \in \mathbb{R}$ then $\xi : \mathbb{R} \to J = (-1, 1)$ is a homeomorphism. Let $\varphi(f) = \xi \circ f$ for any $f \in \mathbb{R}^X$. Then $\varphi : \mathbb{R}^X \to J^X$ is a homeomorphism such that $\varphi(C_p(X)) = C_p(X, J)$ (see TFS-091). If $H' = \varphi(H)$ then H' is a $K_{\sigma\delta}$-space such that $C_p(X, J) \subset H' \subset J^X$. It is easy to see that $H' \cap C_p(X, \mathbb{I}) = C_p(X, J)$; so the space $C_p(X, J)$ is a $K_{\sigma\delta}$-space being the intersection of two $K_{\sigma\delta}$-spaces (see TFS-338 and Fact 7 of S.271). Therefore $C_p(X) \simeq C_p(X, J)$ is also a $K_{\sigma\delta}$-space.

U.357. *Prove that the one-point compactification of an infinite discrete union of non-empty Eberlein compact spaces is an Eberlein compact space.*

Solution. Given an infinite set T suppose that K_t is a non-empty Eberlein compact space for any $t \in T$; let $L = \bigoplus_{t \in T} K_t$. To see that the one-point compactification K of the space L is Eberlein compact denote by a the unique point of the set $K \backslash L$. We will identify every K_t with the respective clopen subspace of L; therefore K_t is also a clopen subspace of K for any $t \in T$. that By Problem 324, we can take a σ-point-finite family $\mathcal{U}_t$ of cozero subsets of K_t which T_0-separates the points of K_t. It is easy to check that the family $\mathcal{U} = \{K_t : t \in T\} \cup (\bigcup_{t \in T} \mathcal{U}_t)$ consists of cozero subsets of K and T_0-separates the points of K.

For any $t \in T$ we have $\mathcal{U}_t = \bigcup\{\mathcal{U}_t^n : n \in \omega\}$ where the family $\mathcal{U}_t^n$ is point-finite for any $n \in \omega$. It is immediate that $\mathcal{U}_n = \{K_t : t \in T\} \cup (\bigcup_{t \in T} \mathcal{U}_t^n)$ is point-finite for any $n \in \omega$. Since $\mathcal{U} = \bigcup_{n \in \omega} \mathcal{U}_n$, the family $\mathcal{U}$ is σ-point-finite, consists of cozero subsets of K and T_0-separates the points of K; therefore K is Eberlein compact by Problem 324.

U.358. *Prove that the Alexandroff double of an Eberlein compact space is an Eberlein compact space.*

Solution. Let K be an Eberlein compact space; recall that the Alexandroff double $AD(K)$ of the space K has $K \times \mathbb{D}$ as the underlying set. All points of $K \times \{1\}$ are isolated in $AD(K)$ and the local base $\mathcal{O}_z$ at any point $z = (x, 0)$ is defined by $\mathcal{O}_z = \{(U \times \mathbb{D}) \setminus \{(x, 1)\} : U \in \tau(x, K)\}$. It was proved in TFS-364 that $AD(K)$ is compact.

The space K being Eberlein compact, we can fix a σ-point-finite family $\mathcal{V}$ of cozero subsets of K which T_0-separates the points of K (see Problem 324). It is easy to see that the family $\mathcal{U} = \{V \times \mathbb{D} : V \in \mathcal{V}\} \cup \{\{z\} : z \in K \times \{1\}\}$ is σ-point-finite, consists of cozero subsets of $AD(K)$ and T_0-separates the points of K. Therefore K is Eberlein compact by Problem 324.

U.359. *Recall that a space X is homogeneous if, for any $x, y \in X$, there is a homeomorphism $h : X \to X$ such that $h(x) = y$. Construct an example of a homogeneous non-metrizable Eberlein compact space.*

Solution. If X is a space then $\mathcal{C}(X)$ is the family of all clopen subspaces of X. Denote by C the Cantor set $\mathbb{D}^\omega$ with its usual topology and fix a point $a \in C$; we will denote by $\mathcal{A}$ the family of all clopen neighbourhoods of a in C. For any set B the map $\mathrm{id}_B : B \to B$ is the *identity*, i.e., $\mathrm{id}_B(b) = b$ for any $b \in B$.

Given a space X, a point $t \in X$, a set $U \in \tau(t, X)$ and $A \in \mathcal{A}$ we will denote by $O(t, U, A)$ the subset $((U \setminus \{t\}) \times C) \cup (\{t\} \times A)$ of $X \times C$. If $x = (t, c) \in X \times C$ for some $c \neq a$ then let $\mathcal{U}_x = \{\{t\} \times V : c \in V \in \mathcal{C}(C) \text{ and } a \notin V\}$; if $x = (t, a)$ then $\mathcal{U}_x = \{O(t, U, A) : U \in \tau(t, X) \text{ and } A \in \mathcal{A}\}$.

Let $X[C]$ be the space whose underlying set is $X \times C$ and the topology is generated by the collection $\{\mathcal{U}_x : x \in X[C]\}$ as local bases.

Fact 1. Suppose that X is a compact space. Then

  (i) the space $X[C]$ is compact and Hausdorff;
 (ii) if X is Eberlein compact then $X[C]$ is also Eberlein compact;
(iii) the map defined by $t \to (t, a)$ for any $t \in X$ is an embedding of X in $X[C]$;
(iv) the natural projection $\pi : X \times C \to X$ is also continuous considered as a map from $X[C]$ onto X; we will call π *the projection of $X[C]$ onto X*.

Proof. We omit the straightforward proof that the families $\{\mathcal{U}_x : x \in X[C]\}$ indeed generate a topology as local bases. Suppose that $x = (t, c)$ and $y = (z, d)$ are distinct points of $X[C]$. If $t \neq z$ then fix disjoint $U, U' \in \tau(X)$ such that $t \in U$ and $z \in U'$. Then the sets $U \times C$ and $U' \times C$ are open (in $X[C]$) disjoint neighbourhoods of the points x and y respectively. Now, if $x = (t, c)$ and $y = (t, d)$ then one of the points c, d, say d, is distinct from a. Take any $V \in \mathcal{C}(C)$ with $d \in V$ such that $\{c, a\} \subset A = C \backslash V$. It is clear that $O(t, X, A)$ and $\{t\} \times V$ are open disjoint neighbourhoods of the points c and d respectively. This proves that $X[C]$ is Hausdorff.

To see that $X[C]$ is compact take an open cover $\mathcal{V}$ of the space $X[C]$. For any $x \in X[C]$ fix a set $O_x \in \mathcal{U}_x$ which is contained in some element of the family $\mathcal{V}$. For any $t \in X$, there is $U_t \in \tau(t, X)$ such that $O_{(t,a)} = O(t, U_t, A_t)$ for some $A_t \in \mathcal{A}$. The space X being compact, there is a finite set $P \subset X$ such that $X = \bigcup \{U_t : t \in P\}$. It is clear that we have $(X \backslash P) \times C \subset O = \bigcup \{O_t : t \in P\}$; besides, we have the inclusion $X[C] \backslash O \subset \bigcup \{\{t\} \times (C \backslash A_t) : t \in P\}$. Every subspace $\{t\} \times (C \backslash A_t)$ is homeomorphic to the clopen (and hence compact) subspace $C \backslash A_t$ of the compact space C. An immediate consequence is that $X[C] \backslash O$ is contained in a compact subspace of $X[C]$; so there is a finite $\mathcal{V}' \subset \mathcal{V}$ with $X[C] \backslash O \subset \bigcup \mathcal{V}'$. Since every O_x is contained in an element of $\mathcal{V}$, there is a finite $\mathcal{V}'' \subset \mathcal{V}$ such that $O \subset \bigcup \mathcal{V}''$. Thus $\mathcal{V}' \cup \mathcal{V}''$ is a finite subcover of $\mathcal{V}$, which shows that $X[C]$ is compact and hence (i) is proved.

Now, if X is Eberlein compact apply Problem 324 to find a σ-point-finite T_0-separating family $\mathcal{U}$ of cozero subsets of X. It is clear that the family $\mathcal{V} = \{U \times C : U \in \mathcal{U}\}$ is σ-point-finite and consists of cozero subsets of $X[C]$. It is easy to choose a family $\mathcal{B} = \{B_n : n \in \omega\} \subset \mathcal{C}(C) \backslash \mathcal{A}$ such that $\mathcal{B}$ is a base in $C \backslash \{a\}$. For any $n \in \omega$ the family $\mathcal{D}_n = \{\{t\} \times B_n : t \in X\}$ is disjoint; so the family $\mathcal{D} = \bigcup_{n \in \omega} \mathcal{D}_n$ is σ-disjoint and hence σ-point-finite. Since all elements of $\mathcal{D}$ are compact, the family $\mathcal{D}$ consists of cozero subsets of $X[C]$. Thus the family $\mathcal{E} = \mathcal{V} \cup \mathcal{D}$ is σ-point-finite and consists of cozero subsets of $X[C]$. It is an easy exercise that $\mathcal{E}$ is T_0-separating in $X[C]$; so $X[C]$ is Eberlein compact by Problem 324 and hence (ii) is proved.

Let $\varphi(t) = (t, a)$ for any $t \in X$; then $\varphi : X \to X' = X \times \{a\}$ is a bijection. Observe that the subspace X' is compact being closed in $X[C]$. For any point $x = (t, a) \in X'$ if $O = O(t, U, A) \in \mathcal{U}_x$ then $\varphi^{-1}(O \cap X') = U \in \tau(t, X)$. This proves continuity of φ; the space X being compact, φ is an embedding (see TFS-123); so (iii) is proved.

To finally see that (iv) holds, observe that $\pi^{-1}(U) = U \times C$ is open in $X[C]$ for any $U \in \tau(X)$; so π is continuous and Fact 1 is proved.

Given an arbitrary compact space X let $X_0 = X$ and $X_{n+1} = X_n[C]$ for any $n \in \omega$. Let $\pi_X^n : X_{n+1} \to X_n$ be the projection for any number $n \in \omega$. In the space $\tilde{X} = \prod_{n \in \omega} X_n$ consider the set $S(X) = \{(x_n)_{n \in \omega} \in \tilde{X} : \pi_X^n(x_{n+1}) = x_n \text{ for all } n \in \omega\}$ and let $p_X^n : S(X) \to X_n$ be the restriction to $S(X)$ of the natural projection of $\tilde{X}$ onto its factor X_n for any $n \in \omega$.

Fact 2. For any compact space X the set $S(X)$ is closed in $\tilde{X}$ and hence compact. Besides, $p_X^n(S(X)) = X_n$ for any $n \in \omega$.

Proof. It follows from Fact 1 that every space X_n is compact; so $\tilde{X}$ is compact as well. If $y = (y_n)_{n \in \omega} \in \tilde{X} \backslash S(X)$ then there exists $m \in \omega$ such that $z = \pi_X^m(y_{m+1}) \neq y_m$; choose disjoint $U, V \in \tau(X_m)$ such that $z \in U$ and $y_m \in V$. By continuity of π_X^m, there is $W \in \tau(y_{m+1}, X_{m+1})$ for which $\pi_X^m(W) \subset U$. It is straightforward that the set $O = \{(x_n)_{n \in \omega} \in \tilde{X} : x_{m+1} \in W$ and $x_m \in V\}$ is open in $\tilde{X}$ and $y \in O$; besides, $O \cap S(X) = \emptyset$ which proves that $S(X)$ is closed in $\tilde{X}$ and hence compact.

Now fix any number $m \in \omega$ and $x \in X_m$; we need to show that there exists a point $y = (x_n)_{n \in \omega} \in S(X)$ such that $x_m = x$. If $m = 0$ then let $x_0 = x$; if $m > 0$ then let $x_i = \pi_X^i \circ \ldots \circ \pi_X^{m-1}(x)$ whenever $0 \leq i < m$. Letting $x_m = x$ we obtain a sequence $x_0, \ldots, x_m$ such that $x_i \in X_i$ and $x_i = \pi_X^i(x_{i+1})$ for any $i < m$.

Proceeding inductively, let $x_{m+1} = (x_m, a)$ and, if a number $n \in \mathbb{N}$ and a point $x_{m+n} \in X_{m+n}$ are given let $x_{m+n+1} = (x_{m+n}, a)$. This gives us a point $y = (x_n)_{n \in \omega} \in \tilde{X}$; by our construction we have $x_n = \pi_X^n(x_{n+1})$ for any $n \in \omega$, i.e., $y \in S(X)$. Since also $p_X^m(y) = x_m = x$, we established that, for any $m \in \omega$ and $x \in X_m$ there is $y \in S(X)$ with $p_X^m(y) = x$, i.e., Fact 2 is proved.

Fact 3. If X and Y are homeomorphic compact spaces then $S(X) \simeq S(Y)$.

Proof. Fix a homeomorphism $\alpha_0 : X \to Y$. If $n \in \omega$ and we have a homeomorphism $\alpha_n : X_n \to Y_n$ then let $\alpha_{n+1} = \alpha_n \times \mathrm{id}_C$, i.e., $\alpha_{n+1}(x) = (\alpha_n(t), c)$ for any point $x = (t, c) \in X_{n+1} = X_n[C]$. It is easy to check that this inductive procedure gives us a homeomorphism $\alpha_n : X_n \to Y_n$ for any $n \in \omega$. Since $\alpha_n \circ \pi_X^n = \pi_Y^n \circ \alpha_{n+1}$ for any $n \in \omega$, the product map $\alpha = \prod_{n \in \omega} \alpha_n$ takes $S(X)$ onto $S(Y)$. An easy consequence of Fact 4 of S.271 is that any product of homeomorphisms is a homeomorphism so $\alpha | S(X)$ is a homeomorphism between $S(X)$ and $S(Y)$. Fact 3 is proved.

Fact 4. Suppose that a zero-dimensional first countable space X is *strongly homogeneous*, i.e., any non-empty clopen subset of X is homeomorphic to X. Then X is homogeneous.

Proof. If X has an isolated point then the respective singleton is homeomorphic to X, i.e., X is a singleton and hence there is nothing to prove. Therefore, if $|X| > 1$ then X has no isolated points. Given $x, y \in X$ it is easy to find local bases $\mathcal{U} = \{U_n : n \in \omega\} \subset \mathcal{C}(X)$ and $\mathcal{V} = \{V_n : n \in \omega\} \subset \mathcal{C}(X)$ at the points x and y respectively such that $U_0 = X$ and $V_0 = X$ while $U_{n+1} \subset U_n$, $U_{n+1} \neq U_n$ and $V_{n+1} \subset V_n$, $V_{n+1} \neq V_n$ for any $n \in \omega$. Let $U_n' = U_n \backslash U_{n+1}$ and $V_n' = V_n \backslash V_{n+1}$ for all $n \in \omega$. Then $\{U_n' : n \in \omega\}$ and $\{V_n' : n \in \omega\}$ are disjoint clopen covers of $X \backslash \{x\}$ and $X \backslash \{y\}$ respectively. By strong homogeneity of X there exists a homeomorphism $\varphi_n : U_n' \to V_n'$ for any $n \in \omega$.

Let $\varphi(x) = y$; if $z \in X \backslash \{x\}$ then there is a unique $n \in \omega$ such that $z \in U_n'$; let $\varphi(z) = \varphi_n(z)$. We leave to the reader a simple checking that $\varphi : X \to X$ is a homeomorphism such that $\varphi(x) = y$. Fact 4 is proved.

Fact 5. If X is a zero-dimensional compact space then

(i) for any $m \in \omega$ and any closed $F \subset X_m$, the space $(p_X^m)^{-1}(F) \subset S(X)$ is homeomorphic to $S(F)$;

(ii) if $O \subset S(X)$ is clopen in $S(X)$ then there is $m \in \omega$ and a clopen $U \subset X_m$ such that $O = (p_X^m)^{-1}(U)$.

Proof. (i) Fix any $m \in \omega$ and construct the sequence $\{F_n : n \in \omega\}$ as in the definition of $S(F)$. We have $F_1 = F[C] = F \times C \subset X_{m+1} = X_m[C]$. It is easy to check that the topology induced from X_{m+1} on F_1 coincides with the topology of F_1; a trivial induction shows that F_n is a subspace of X_{m+n}, i.e., the identity $i_n = i_{F_n} : F_n \to X_{m+n}$ is an embedding for any $n \in \omega$.

Given an arbitrary $x = (x_n)_{n \in \omega} \in S(F)$ let $y_k = \pi_X^k \circ \ldots \pi_X^{m-1}(x_0)$ for any $k < m$. If $y_{m+n} = x_n$ for any $n \in \omega$ then $y = (y_i)_{i \in \omega} \in S(X)$. Since $x_0 = y_m \in F$, the point y belongs to $G = (p_X^m)^{-1}(F)$; let $y = \varphi(x)$.

To prove that the map $\varphi : S(F) \to G$ is continuous observe first that, for any $n \in \omega$, the map $p_X^{m+n} \circ \varphi$ coincides with $i_n \circ p_F^n$; so $p_X^k \circ \varphi$ is continuous for any $k \geq m$; if $k < m$ then $p_X^k \circ \varphi = \pi_X^k \circ \ldots \pi_X^{m-1} \circ i_0 \circ p_F^0$ is also continuous; so φ is a continuous map by TFS-102. It is immediate from the definition that φ is injective; so φ is an embedding because $S(F)$ is compact by Fact 1.

Finally take any point $x = (x_n)_{n \in \omega} \in G$; then $x_m \in F$. Since $\pi_X^m(x_{m+1}) = x_m$, we have $x_{m+1} \in F \times C = F_1$; repeating this step for all $n \in \mathbb{N}$ we can show by a trivial induction that $x_{m+n} \in F_n$ for any $n \in \omega$. Thus $y = (x_{m+n} : n \in \omega) \in S(F)$ and it is immediate that $\varphi(y) = x$. This shows that φ is surjective; so $\varphi : S(F) \to G$ is a homeomorphism and hence (i) is proved.

(ii) Take an arbitrary clopen set $O \subset S(X)$. For any point $x = (x_n)_{n \in \omega} \in O$ we can find $n \in \omega$ and $U_0, \ldots, U_n \in C(X)$ such that $x_i \in U_i$ for all $i \leq n$ and the set $O'_x = \{y \in S(X) : p_X^i(y) \in U_i \text{ for all } i \leq n\}$ is contained in O. Let $\delta_i = \pi_X^i \circ \ldots \circ \pi_X^{n-1}$; we have $\delta_i(x_n) = x_i$ for any $i < n$; so

$$x_n \in U = U_n \cap \left(\bigcap\{\delta_i^{-1}(U_i) : i < n\}\right).$$

It is easy to see that $x \in (p_X^n)^{-1}(U) \subset O'_x \subset O$. This shows that

(∗) for any point $x \in O$ there exists a number $n \in \omega$ and a set $U_x \in C(X_n)$ such that $x \in (p_X^n)^{-1}(U_x) \subset O$.

The property (∗) together with compactness of the set O implies that we can choose numbers $m_1, \ldots, m_k \in \omega$ and a set $U_i \in C(X_{m_i})$ for every $i \leq k$ such that $O = \bigcup\{(p_X^{m_i})^{-1}(U_i) : i \leq k\}$. Let $m = m_1 + \ldots + m_k$ and consider the map $\mu_i = \pi_X^{m_i} \circ \pi_X^{m_i+1} \circ \ldots \circ \pi_X^{m-1}$ for any $i \leq k$. The set $U = \bigcup\{\mu_i^{-1}(U_i) : i \leq k\}$ is clopen in X_m; it follows from $\mu_i \circ p_X^m = p_X^{m_i}$ that $(p_X^m)^{-1}(\mu_i^{-1}(U_i)) = (p_X^{m_i})^{-1}(U_i)$ for any $i \leq k$. An immediate consequence is that $(p_X^m)^{-1}(U) = O$ which settles (ii) and shows that Fact 5 is proved.

Returning to our solution observe that it follows from Fact 1 that the space $E = S(C)$ is Eberlein compact (see Problems 306 and 307). Observe that the space $C_1 = C[C]$ is not metrizable because, for any $b \in C \backslash \{a\}$, the set $\{(x, b) : x \in C\}$ is uncountable and discrete. The space E maps continuously onto a non-metrizable compact space C_1 (see Fact 2); so it is not metrizable by Fact 5 of S.307.

Let us prove by induction on $n \in \omega$ that

(1) if G is a non-empty clopen subset of C_n then $(p_C^n)^{-1}(G)$ is homeomorphic to the space E.

If $n = 0$ then G is clopen in $C_0 = C$; so it is homeomorphic to C by SFFS-348. Therefore $(p_C^0)^{-1}(G)$ is homeomorphic to $S(G) \simeq S(C) = E$ by Fact 3 and Fact 5. This shows that (1) is true for $n = 0$. Now take any $U \in \mathcal{C}(C)$ with $U \neq \emptyset$ and $U \neq C$; then $U \simeq C$ and $U' = C \backslash U \simeq C$. The sets $W = (p_C^0)^{-1}(U)$ and $W' = (p_C^0)^{-1}(U')$ are clopen and disjoint in E. We saw already that $W \simeq E$ and $W' \simeq E$; since $W \cup W' = E$, this proves that

(2) $E \oplus E \simeq E$.

Now, assume that we have proved (1) for all $n \leq k$ and take a non-empty clopen set $G \subset C_{k+1}$.

Suppose first that $G \cap (C_k \times \{a\}) = \emptyset$. It follows from the definition of the topology of $C_{k+1} = C_k[C]$ that for any set $P \subset C_{k+1}$ such that $P' = \pi_C^k(P)$ is infinite, for any accumulation point x of P' in C_k, the point (x, a) is an accumulation point of P in C_{k+1}. An immediate consequence is that $D = \pi_C^k(G)$ is finite. The set $G_t = (\{t\} \times C) \cap G$ is clopen and non-empty in $\{t\} \times C$; so $G_t \simeq C$ for any $t \in D$ (we used SFFS-348 again as well as the fact that $\{t\} \times C$ is homeomorphic to C). Therefore $G = \bigcup \{G_t : t \in D\}$ is a metrizable compact space without isolated points; so we can use SFFS-348 once more to see that $G \simeq C$. Now apply Fact 3 and Fact 5 to see that $(p_C^{k+1})^{-1}(G) \simeq S(G) \simeq S(C) = E$.

If the set $H = G \cap (C_k \times \{a\})$ is non-empty, then let $H' = \pi_C^k(H)$ and take, for any $t \in H'$ a basic neighbourhood $O(t, U_t, A_t)$ of the point (t, a) in C_{k+1} such that $O(t, U_t, A_t) \subset G$. The space H being compact there exists a finite set $F' \subset H'$ such that $H \subset P = \bigcup \{O(t, U_t, A_t) : t \in F'\}$.

Observe that $P = ((H' \backslash F') \times C) \cup (\bigcup \{\{t\} \times W_t : t \in F'\})$ where $A_t \subset W_t \in \mathcal{A}$ for any $t \in F'$. For any $t \in F'$ choose a set $B_t \in \mathcal{A}$ such that $B_t \subset W_t$ and $W_t \backslash B_t \neq \emptyset$. Use SFFS-348 to construct a homeomorphism $h_t : W_t \to C$ such that $h_t|B_t = \mathrm{id}_{B_t}$ and $h_t(W_t \backslash B_t) = C \backslash B_t$.

Now, we are going to define a map $h : P \to H' \times C$ as follows: $h(t, c) = (t, c)$ whenever $(t, c) \in (H' \backslash F') \times C$; if $t \in F'$ then let $h(t, c) = (t, h_t(c))$ for any $c \in W_t$. To see that h is a homeomorphism note that $h(\{t\} \times C) = \{t\} \times C$ and $h|(\{t\} \times C)$ is a homeomorphism for any $t \in H'$. Therefore h and h^{-1} are both continuous at the points whose second coordinate is distinct from a. Besides, the map h is locally an identity at any point $x \in H' \times \{a\}$ (i.e., there is $O \in \tau(x, P)$ such that $h|O = \mathrm{id}_O$); so h and h^{-1} are both continuous at every point $x \in H' \times \{a\}$ as well.

This proves that $P \simeq H' \times C$; since H' is clopen in C_k, we can apply the induction hypothesis to see that $(p_C^k)^{-1}(H') \simeq E$. By Fact 3 and Fact 5, we have $(p_C^{k+1})^{-1}(P) \simeq (p_C^{k+1})^{-1}(H' \times C) = (p_C^k)^{-1}(H') \simeq E$. The set $G' = G \backslash P$ is clopen in C_{k+1} and $G' \cap (C_k \times \{a\}) = \emptyset$. If $G' = \emptyset$ then the proof of (1) is over. If $G' \neq \emptyset$ then $(p_C^{k+1})^{-1}(G') \simeq E$ by what we proved in the first case. Therefore

$(p_C^{k+1})^{-1}(G)$ is a union of two clopen subspaces which are both homeomorphic to E. Since $E \oplus E \simeq E$ by (2), we have $(p_C^{k+1})^{-1}(G) \simeq E$; so our induction procedure shows that (1) is proved.

An immediate consequence of (1) and Fact 5 is that $W \simeq E$ for any clopen nonempty $W \subset E$. Now, Fact 4 shows that E is homogeneous and makes our solution complete.

U.360. *Give an example of a hereditarily normal but not perfectly normal Eberlein compact space.*

Solution. The space $A(\omega_1)$ is Eberlein compact (see Fact 1 of U.340). It is not perfectly normal because the pseudocharacter at the unique non-isolated point a of $A(\omega_1)$ is equal to ω_1. If $Y \subset A(\omega_1)$ and $a \notin Y$ then Y is discrete and hence normal. If $a \in Y$ then Y is compact and hence also normal. Thus any $Y \subset A(\omega_1)$ is normal, i.e., $A(\omega_1)$ is a hereditarily normal Eberlein compact space which is not perfectly normal.

U.361. *Let X be an Eberlein compact space such that $X \times X$ is hereditarily normal. Prove that X is metrizable.*

Solution. If X is finite then there is nothing to prove; so suppose that X is infinite. By Fact 2 of T.090 the space X has a non-closed countable subspace; so we can apply Fact 1 of T.090 (letting $Y = Z = X$) to conclude that X is perfectly normal. By SFFS-001 the space X is hereditarily Lindelöf; so $c(X) = \omega$. Consequently, $w(X) = c(X) = \omega$ by Problem 343, i.e., X is metrizable.

U.362. *Prove that there exists an Eberlein compact space X such that $X^2 \backslash \Delta$ is not metacompact.*

Solution. Recall that $A(\omega_1) = \omega_1 \cup \{a\}$ where $a \notin \omega_1$ is the unique non-isolated point of $A(\omega_1)$ and a set $U \subset A(\omega_1)$ with $a \in U$ is open in $A(\omega_1)$ if and only if $A(\omega_1) \backslash U$ is finite.

Fact 1. Let $u \in K = (A(\omega_1))^\omega$ be the point with $u(n) = a$ for all $n \in \omega$. Then the space $K \backslash \{u\}$ is not metacompact.

Proof. For any $\alpha \in \omega_1$ and $n \in \omega$ let $O(\alpha, n) = \{x \in K : x(n) = \alpha\}$. It is clear that $\mathcal{O} = \{O(\alpha, n) : \alpha < \omega_1, n \in \omega\}$ is an open cover of $K \backslash \{u\}$. Assume that $\mathcal{U}$ is a point-finite open refinement of $\mathcal{O}$.

For any $\alpha < \omega_1$ and $n \in \omega$ let $u_\alpha^n(n) = \alpha$ and $u_\alpha^n(m) = a$ for any $m \neq n$. It is clear that $O(\alpha, n)$ is the unique element of $\mathcal{O}$ which contains u_α^n. Therefore there is $U(\alpha, n) \in \mathcal{U}$ such that $u_\alpha^n \in U(\alpha, n) \subset O(\alpha, n)$ for any $\alpha < \omega_1$ and $n \in \omega$. There exists $k_\alpha^n \in \omega$ and a finite set $F(\alpha, n) \subset \omega_1$ such that $k_\alpha^n > n$ and the set $H(\alpha, n) = \{x \in K : x(n) = \alpha$ and $x(m) \notin F(\alpha, n)$ for all $m \in k_\alpha^n \backslash \{n\}\}$ is contained in $U(\alpha, n)$ for all $\alpha < \omega_1$ and $n \in \omega$. Thus the family $\mathcal{H} = \{H(\alpha, n) : \alpha < \omega_1, n \in \omega\}$ is point-finite.

Let $r_0 = 0$ and choose an uncountable $\Omega_0 \subset \omega_1$ for which there is $r_1 \in \omega$ such that $k_\alpha^0 = r_1$ for all $\alpha \in \Omega_0$. Suppose that, for some $m \in \omega$ we have uncountable sets $\Omega_0, \dots, \Omega_m \subset \omega_1$ and $r_0, \dots, r_{m+1} \in \omega$ with the following properties:

(1) $\Omega_{i+1} \subset \Omega_i$ for all $i < m$;
(2) if $i \le m$ then $r_{i+1} = k_\alpha^{r_i}$ for any $\alpha \in \Omega_i$;
(3) for any $i < m$ there is a set $F_{i+1} \subset \omega_1$ such that $F(\alpha, r_{i+1}) \cap F(\beta, r_{i+1}) = F_{i+1}$ for any distinct $\alpha, \beta \in \Omega_{i+1}$.

Since the set Ω_m is uncountable, we can apply the Δ-lemma (see SFFS-038) to find an uncountable set $\Omega'_{m+1} \subset \Omega_m$ for which there exists $F_{m+1} \subset \omega_1$ such that $F(\alpha, r_{m+1}) \cap F(\beta, r_{m+1}) = F_{m+1}$ for any distinct $\alpha, \beta \in \Omega'_{m+1}$. Choose an uncountable set $\Omega_{m+1} \subset \Omega'_{m+1}$ for which there exists $r_{m+2} \in \omega$ such that $k_\alpha^{r_{m+1}} = r_{m+2}$ for any $\alpha \in \Omega_{m+1}$.

Since the properties (1)–(3) now hold if substitute m by $m+1$, our inductive procedure can be continued to construct a family $\{\Omega_n : n \in \omega\}$ of uncountable subsets of ω_1 and a sequence $\{r_i : i \in \omega\} \subset \omega$ such that (1)–(3) are fulfilled for all $m \in \omega$; let $F = \bigcup_{i\in\omega} F_{i+1}$. Observe that (2) implies that $r_i < r_{i+1}$ for all $i \in \omega$.

Take an arbitrary $\alpha_0 \in \Omega_0 \backslash F$. Proceeding inductively, assume that $m \in \omega$ and we have chosen $\alpha_0, \ldots, \alpha_m$ such that

(4) $\alpha_i \in \Omega_i \backslash F$ for any $i \le m$;
(5) if $0 \le i < j \le m$ then $\alpha_i \notin F(\alpha_j, r_j)$.

Since the family $\{F(\alpha, r_{m+1}) \backslash F : \alpha \in \Omega_{m+1} \backslash F\}$ is disjoint and uncountable by (3), there is an ordinal $\alpha_{m+1} \in \Omega_{m+1} \backslash F$ such that $\alpha_i \notin F(\alpha_{m+1}, r_{m+1}) \backslash F$ and hence $\alpha_i \notin F(\alpha_{m+1}, r_{m+1})$ for any $i \le m$. It is clear that the properties (4) and (5) still hold if we substitute m by $m+1$; so our inductive construction can be continued to obtain a sequence $\{\alpha_i : i \in \omega\}$ with the properties (4) and (5) fulfilled for all $m \in \omega$; let $L = \{r_i : i \in \omega\}$.

The set $G = \bigcup\{F(\alpha_i, r_i) : i \in \omega\}$ being countable we can fix an ordinal $\beta_i \in \Omega_i \backslash G$ for any $i \in \omega \backslash L$. Finally, let $z(r_i) = \alpha_i$ for all $i \in \omega$; if $i \in \omega \backslash L$ then let $z(i) = \beta_i$. This gives us a point $z \in K \backslash \{u\}$.

Given $n \in \omega$ take an arbitrary number $i < r_{n+1}$; if $i \ne r_k$ for any $k \le n$ then $z(i) = \beta_i \notin F(\alpha_n, r_n)$ because $\beta_i \notin G$. Now, assume that $i = r_k$ for some $k \le n$; if $k < n$ then $z(i) = \alpha_k \notin F(\alpha_n, r_n)$ by the property (5). If $k = n$ then $z(i) = \alpha_n$; recalling the definition of $H(\alpha_n, r_n)$ we can see that $z \in H(\alpha_n, r_n)$.

The number $n \in \omega$ was chosen arbitrarily; so $z \in \bigcap_{n\in\omega} H(\alpha_n, r_n)$ which is a contradiction because $\{H(\alpha_n, r_n) : n \in \omega\}$ is an infinite subfamily of the point-finite family $\mathcal{H}$. Therefore $K \backslash \{u\}$ is not metacompact and hence Fact 1 is proved.

Returning to our solution let $X = (A(\omega_1))^\omega$. Then X is Eberlein compact by Problem 307 and Fact 1 of U.340. There is a point $u \in X$ such that $X \backslash \{u\}$ is not metacompact (see Fact 1). It is evident that the correspondence $x \to (x, u)$ embeds $X \backslash \{u\}$ in $X^2 \backslash \Delta$ as a closed subspace. It is an easy exercise to show that a closed subspace of a metacompact space is metacompact; so $X^2 \backslash \Delta$ cannot be metacompact and hence our solution is complete.

U.363. *Prove that any Eberlein compact space is hereditarily σ-metacompact.*

Solution. Take any Eberlein compact space K and let $\Delta = \{(x, x) : x \in K\}$ be the diagonal of K. We can assume that $K \subset \Sigma_*(A)$ for some A (see Problem 322).

Denote by $\mathcal{Q}$ the family of all intervals $(a, b) \subset \mathbb{R}$ such that $a < b$ and $a, b \in \mathbb{Q}$. For any $J \in \mathcal{Q}$ let $O_J^a = \{x \in K : x(a) \in J\}$ for any $a \in A$. It is evident that the family $\mathcal{O} = \{O_I^a \times O_J^a : a \in A, \ I, J \in \mathcal{Q} \text{ and } \overline{I} \cap \overline{J} = \emptyset\}$ consists of open subsets of $K \times K$; it is immediate that $\bigcup \mathcal{O} = K^2 \backslash \Delta$ and $\overline{O} \cap \Delta = \emptyset$ for any $O \in \mathcal{O}$.

Fix $I, J \in \mathcal{Q}$ such that $\overline{I} \cap \overline{J} = \emptyset$ and let $\mathcal{O}(I, J) = \{O_I^a \times O_J^a : a \in A\}$; one of the sets $\overline{I}, \overline{J}$, say $\overline{J}$, does not contain zero and therefore there is $\varepsilon > 0$ such that $|t| \geq \varepsilon$ for any $t \in J$. If the family $\mathcal{O}(I, J)$ is not point-finite then there is an infinite $B \subset A$ and $z = (x, y) \in K^2$ such that $z \in \bigcap\{O_I^a \times O_J^a : a \in B\}$. This implies $y \in O_J^a$, i.e., $y(a) \in J$ and therefore $|y(a)| \geq \varepsilon$ for any $a \in B$ which is a contradiction with $y \in K \subset \Sigma_*(A)$. This proves that

(1) the family $\mathcal{O}(I, J)$ is point-finite for any $I, J \in \mathcal{Q}$ with $\overline{I} \cap \overline{J} = \emptyset$.

Since $\mathcal{O} = \bigcup\{\mathcal{O}(I, J) : I, J \in \mathcal{Q} \text{ and } \overline{I} \cap \overline{J} = \emptyset\}$, it follows from (1) that the family $\mathcal{O}$ is σ-point-finite; thus $\mathcal{O} = \bigcup_{n \in \omega} \mathcal{O}_n$ where $\mathcal{O}_n$ is point-finite for any $n \in \omega$.

Fact 1. Suppose that a space X has an open σ-point-finite cover $\mathcal{U}$ such that $\overline{U}$ is compact for any $U \in \mathcal{U}$. Then X is σ-metacompact.

Proof. Represent $\mathcal{U}$ as $\bigcup_{n \in \omega} \mathcal{U}_n$ where $\mathcal{U}_n$ is point-finite and $\mathcal{U}_n \subset \mathcal{U}_{n+1}$ for any $n \in \omega$. Take an open cover $\mathcal{V}$ of the space X and choose, for any $U \in \mathcal{U}$ a number $n(U) \in \omega$ and a family $\mathcal{S}(U) = \{V_i : i \leq n(U)\} \subset \mathcal{V}$ such that $\overline{U} \subset \bigcup \mathcal{S}(U)$; let $V_i(U) = V_i \cap U$ for any $i \leq n(U)$. For any $k, i \in \omega$ the family $\mathcal{V}(k, i) = \{V_i(U) : U \in \mathcal{U}_k \text{ and } i \leq n(U)\}$ is point-finite because so is $\mathcal{U}_k$. It is easy to check that $\mathcal{V}' = \bigcup\{\mathcal{V}(k, i) : k \in \omega\}$ is a cover of X; so $\mathcal{V}'$ is a σ-point-finite refinement of $\mathcal{V}$. Fact 1 is proved.

Returning to our solution take an arbitrary subspace $Y \subset K$ and a family $\mathcal{W} \subset \tau(Y)$ such that $Y = \bigcup \mathcal{W}$. For any $W \in \mathcal{W}$ fix a set $O_W \in \tau(K)$ such that $O_W \cap Y = W$; the set $G = \bigcup\{O_W : W \in \mathcal{W}\}$ is open in K and contains Y. For every $n \in \omega$ re-index the family $\mathcal{O}_n$ as $\{U_t \times V_t : t \in T_n\}$ so that $T_n \cap T_m = \emptyset$ if $n \neq m$; let $T = \bigcup_{n \in \omega} T_n$. Let $F = K \backslash G$ and consider the family $\mathcal{U} = \{U(S) = \bigcap\{U_t : t \in S\} : \text{ the set } S \subset T \text{ is finite and } \mathcal{V}(S) = \{V_t : t \in S\} \text{ is a minimal cover of the set } F\}$. It is easy to see that $U(S) \cap (\bigcup \mathcal{V}(S)) = \emptyset$ for any finite $S \subset T$; so $U(S) \in \mathcal{U}$ implies $\overline{U(S)} \cap F = \emptyset$.

Fix a number $m \in \omega$; we will prove that the family $\mathcal{U}(m) = \{U(S) \in \mathcal{U} : |S| \leq m \text{ and } S \subset T[m] = T_0 \cup \ldots \cup T_m\}$ is point-finite. Indeed, if this is false then there is a point $x \in K$ and an infinite family $\mathcal{S}$ of finite subsets of $T[m]$ such that $|S| \leq m$ for any $S \in \mathcal{S}$ and $x \in \bigcap\{U(S) : S \in \mathcal{S}\}$. By the countable version of the Δ-lemma (see Fact 2 of U.337) there exists a countably infinite $\mathcal{S}' \subset \mathcal{S}$, which is a Δ-system, i.e., for some $P \subset T$, we have $S \cap S' = P$ for any distinct $S, S' \in \mathcal{S}'$. Fix a faithful enumeration $\{S_i : i \in \omega\}$ of the family $\mathcal{S}'$.

Since $P \neq S_0$ and the family $\mathcal{V}(S_0)$ is a minimal cover of F, there is a point $y \in F$ such that $y \notin \bigcup \mathcal{V}(P)$. Therefore, for every $i \in \mathbb{N}$ there is $s_i \in S_i \backslash P$ such that $y \in V_{s_i}$. The family $\mathcal{S}'$ being a Δ-system, we have $s_i \neq s_j$ whenever $i \neq j$. Now it follows from $(x, y) \in \bigcap\{U_{s_i} \times V_{s_i} : i \in \mathbb{N}\}$ that the family $\mathcal{O}_0 \cup \ldots \cup \mathcal{O}_m$ is not point-finite; this contradiction proves that every $\mathcal{U}(m)$ is point-finite. It follows from $\mathcal{U} = \bigcup_{m \in \omega} \mathcal{U}(m)$ that

(2) the family $\mathcal{U}$ is σ-point-finite and $\overline{U} \cap F = \emptyset$ for any $U \in \mathcal{U}$;

Given a point $x \in K\backslash F$, for any $y \in F$ the point (x, y) does not belong to the diagonal Δ; so there is $t(y) \in T$ such that $(x, y) \in U_{t(y)} \times V_{t(y)}$ and hence $y \in V_{t(y)}$. The set F being compact, there is a finite $Z \subset Y$ such that $F \subset \bigcup\{V_{t(y)} : y \in Z\}$. Take a subset $S \subset \{t(y) : y \in Z\}$ such that $\mathcal{V}(S)$ is a minimal cover of F; then $x \in U(S)$. This, together with the property (2) shows that the family $\mathcal{U}$ is a σ-point-finite cover of $K\backslash F$ such that $\overline{U} \subset K\backslash F$ for any set $U \in \mathcal{U}$. By Fact 1, the space $G = K\backslash F$ is σ-metacompact and hence there exists a σ-point-finite refinement $\mathcal{H}$ of the open cover $\mathcal{G} = \{O_W : W \in \mathcal{W}\}$ of the space G. It is straightforward that the family $\mathcal{W}' = \{V \cap Y : V \in \mathcal{H}\} \subset \tau(Y)$ is a σ-point-finite refinement of $\mathcal{W}$. This proves that Y is σ-metacompact and completes our solution.

U.364. *Prove that, for any compact space X, the subspace $(X \times X)\backslash\Delta \subset X \times X$ is σ-metacompact if and only if X is Eberlein compact.*

Solution. If X is Eberlein compact then so is X^2 by Problem 307. Every Eberlein compact space is hereditarily σ-metacompact by Problem 363; so $X^2\backslash\Delta$ is σ-metacompact and hence we established necessity.

Now assume that X is a compact space such that $Y = X^2\backslash\Delta$ is σ-metacompact. Then there exists a σ-point-finite open cover $\mathcal{Q}$ of the set Y such that $\overline{U} \subset Y$ for any $U \in \mathcal{Q}$. This implies that $\overline{U} = \mathrm{cl}_Y(U)$ is a compact set; so Fact 1 of U.188 is applicable to find a shrinking $\{F_U : U \in \mathcal{Q}\}$ of the family $\mathcal{Q}$. It is easy to see that every F_U is compact; so there is a finite family $\mathcal{O}_U$ of open subsets of $X \times X$ such that $F_U \subset \bigcup \mathcal{O}_U \subset \overline{\bigcup \mathcal{O}_U} \subset U$ and every $O \in \mathcal{O}_U$ is *standard*, i.e., $O = O_1 \times O_2$ for some σ-compact sets $O_1, O_2 \in \tau(X)$ such that $\overline{O}_1 \cap \overline{O}_2 = \emptyset$. It is an easy exercise that the family $\mathcal{O} = \{\overline{O} : O \in \mathcal{O}_U, U \in \mathcal{Q}\}$ is σ-point-finite.

The map $\varphi : X \times X \to X \times X$ defined by the formula $\varphi(x, y) = (y, x)$ is, evidently, a homeomorphism and $\varphi(Y) = Y$ so the family $\mathcal{O} \cup \{\varphi(O) : O \in \mathcal{O}\}$ is also σ-point-finite. Thus we can choose an indexation $\{U_t \times V_t : t \in T\}$ of the family $\mathcal{O}' = \{O : O \in \mathcal{O}_U, U \in \mathcal{Q}\} \cup \{\varphi(O) : O \in \mathcal{O}_U, U \in \mathcal{Q}\}$ such that $T = \bigcup_{n\in\omega} T_n$ and $\{\overline{U}_t \times \overline{V}_t : t \in T_n\}$ is point-finite for any $n \in \omega$. In other words, the family $\mathcal{O}' = \{U_t \times V_t : t \in T\}$ has the following properties:

(1) U_t, V_t are open σ-compact subsets of X for any $t \in T$;
(2) $\overline{U}_t \cap \overline{V}_t = \emptyset$ for any $t \in T$;
(3) $T = \bigcup_{n\in\omega} T_n$ and the family $\{\overline{U}_t \times \overline{V}_t : t \in T_n\}$ is point-finite for any $n \in \omega$;
(4) $\bigcup \mathcal{O}' = Y$ and $U \times V \in \mathcal{O}'$ implies $V \times U \in \mathcal{O}'$.

Let $\kappa = d(X)$ and choose a dense set $D = \{p_\alpha : \alpha < \kappa\}$ in the space X; let $X_\alpha = \{p_\beta : \beta < \alpha\}$ for any $\alpha < \kappa$; call a family $\mathcal{F} \subset \exp X$ a *minimal cover* of X_α if $X_\alpha \subset \bigcup \mathcal{F}$ and $X_\alpha \not\subset \bigcup(\mathcal{F}\backslash\{F\})$ for any $F \in \mathcal{F}$. Given a finite $F \subset T$ we will often use the set $U_F = \bigcap\{U_t : t \in F\}$; for any $\alpha < \kappa$ let $\mathcal{U}_\alpha = \{U_F : F \subset T$ is finite and $\{V_t : t \in F\}$ is a minimal cover of $X_\alpha\}$. It is easy to see that $\mathcal{U}_\alpha$ consists of σ-compact open subsets of X for any $\alpha < \kappa$. It turns out that

(5) the family $\mathcal{U} = \bigcup\{\mathcal{U}_\alpha : \alpha < \kappa\}$ is σ-point-finite.

To prove (5) let $\mathcal{U}_{\alpha,n} = \{U_F \in \mathcal{U} : F \subset T, |F| \le n$ and $F \subset T_0 \cup \ldots \cup T_n\}$ for any $\alpha < \kappa$ and $n \in \omega$. It is evident that $\mathcal{U}_\alpha = \bigcup_{n\in\omega} \mathcal{U}_{\alpha,n}$ for any $\alpha < \kappa$; so $\mathcal{U} = \bigcup_{n\in\omega} \mathcal{W}_n$ where $\mathcal{W}_n = \bigcup\{\mathcal{U}_{\alpha,n} : \alpha < \kappa\}$ for all $n \in \omega$. Thus it suffices to show that every $\mathcal{W}_n$ is point-finite.

If this is not true for some $n \in \omega$ then there exists a family $\{F_k : k \in \omega\}$ of finite subsets of T such that

(6) there is $p \in \omega$ such that $|F_k| = p$ for all $k \in \omega$;
(7) for every $k < \omega$ there is $\beta(k) < \kappa$ such that $U_{F_k} \in \mathcal{U}_{\beta(k),n}$ and the sequence $\{\beta(k) : k \in \omega\}$ is non-decreasing;
(8) there a set $F \subset T$ such that $F_k \cap F_l = F$ for any distinct $k, l \in \omega$;
(9) there is a point $x \in X$ with $x \in \bigcap\{U_{F_k} : k \in \omega\}$.

The properties (6) and (8) can be guaranteed taking an infinite family with the property (9) and passing to an appropriate infinite subfamily applying the countable version of the Δ-lemma (see Fact 2 of U.337) and the fact that $|F_k| \le n$ for all $k \in \omega$. Once we have (6) and (8) choose a function $\varphi : \omega \to \kappa$ such that $U_{F_k} \in \mathcal{U}_{\varphi(k)}$ for any $k < \omega$. Suppose first that there is an ordinal $\alpha_0 < \kappa$ for which the set $S = \varphi^{-1}(\alpha_0)$ is infinite. Passing to $\{F_k : k \in S\}$ gives a set for which $\beta(k)$ is the same ordinal for all $k < \omega$; so the condition (7) is satisfied.

Now if $\varphi^{-1}(\alpha)$ is finite for any $\alpha < \kappa$ then use a trivial induction to get the relevant subfamily with the property (7).

Since we have $U_{F_k} \ne U_{F_l}$ for distinct $k, l \in \omega$ we have $F_k \ne F$ for all $k \in \omega$. The family $\mathcal{V}_0 = \{V_t : t \in F_0\}$ is a minimal cover of $X_{\beta(0)}$; so there is a point $y \in X_{\beta(0)} \setminus (\bigcup\{\overline{V}_t : t \in F\})$.

We have $\beta(0) \le \beta(k)$ for any $k < \omega$; so $y \in \bigcup\{\overline{V}_t : t \in F_k\}$ and hence there is $t_k \in F_k \setminus F$ for which $y \in \overline{V}_{t_k}$. It follows from the property (8) that $t_k \ne t_l$ if $k \ne l$; since $(x, y) \in \bigcap\{U_{t_k} \times \overline{V}_{t_k} : 0 < k < \omega\}$, we obtain a contradiction with the property (3). Thus the family $\mathcal{W}_n$ is point-finite for any $n \in \omega$ and (5) is proved.

Our final step is to show that

(10) the family $\mathcal{U}$ is T_0-separating in X.

Take distinct points $x_1, x_2 \in X$ and assume first that there is $\alpha < \kappa$ such that $x_i \in X_\alpha$ while $z = x_{2-i} \notin X_\alpha$. For any $y \in X_\alpha$ there is $t_y \in T$ such that $(z, y) \in U_{t_y} \times V_{t_y}$. There exists a finite $P \subset X_\alpha$ for which $X_\alpha \subset \bigcup\{V_{t_y} : y \in P\}$; so the set $F' = \{t_y : y \in P\}$ is finite and $X_\alpha \subset \bigcup\{V_t : t \in F'\}$. Take $F \subset F'$ such that $\{\overline{V}_t : t \in F\}$ is a minimal cover of X_α; then $U_F \in \mathcal{U}$ while $z \in U_F$ and $U_F \cap X_\alpha = \emptyset$ which shows that U_F separates the points x_1 and x_2.

Therefore we can assume, without loss of generality, that there is $\alpha < \kappa$ such that $x_1, x_2 \in X_\alpha$ and $\{x_1, x_2\} \cap X_\beta = \emptyset$ for any $\beta < \alpha$. Let $\mathcal{V}_1 = \{V_t : x_1 \in U_t$ and $x_2 \notin U_t\}$ and $\mathcal{V}_2 = \{V_t : x_2 \in U_t$ and $x_1 \notin U_t\}$. Observe that $x_2 \in \bigcup \mathcal{V}_1$ and $x_1 \in \bigcup \mathcal{V}_2$.

The space $K = X \setminus \bigcup(\mathcal{V}_1 \cup \mathcal{V}_2)$ is compact; for any point $z \in K$ there is $t(z) \in T$ such that $x_1 \in U_{t(z)}$ and $z \in V_{t(z)}$. Since $V_{t(z)} \notin \mathcal{V}_1 \cup \mathcal{V}_2$, we must have $x_2 \in U_{t(z)}$. As a consequence, there is a finite $F \subset T$ such that $K \subset \bigcup\{V_t : t \in F\}$ and

$\{x_1, x_2\} \subset U_F$. It follows from $\{x_1, x_2\} \subset X_\alpha$ that $U_F \cap \{p_\beta : \beta < \alpha\} \neq \emptyset$ and therefore $H = \bigcup\{\overline{V}_t : t \in F\}$ does not contain the set $\{p_\beta : \beta < \alpha\}$; let $\delta = \min\{\beta : p_\beta \notin H\}$.

The point p_δ has to belong to $X \backslash K = \bigcup(\mathcal{V}_1 \cup \mathcal{V}_2)$; so there is an element $s \in T$ for which $V_s \in \mathcal{V}_1 \cup \mathcal{V}_2$ and $p_\delta \in V_s$. Then the set U_s separates the points x_1 and x_2. Since $\{\overline{V}_s\} \cup \{\overline{V}_t : t \in F\}$ covers $X_{\delta+1}$, there is $F' \subset F$ such that the family $\{\overline{V}_s\} \cup \{\overline{V}_t : t \in F'\}$ is a minimal cover of $X_{\delta+1}$. As a consequence, $U = U_s \cap U_{F'} \in \mathcal{U}$ and U separates the points x_1 and x_2 because so does U_s while $\{x_1, x_2\} \subset U_t$ for any $t \in F'$. Thus the property (10) is proved.

It follows from (5) and (10) that $\mathcal{U}$ is a σ-point-finite T_0-separating family of open F_σ-subsets of X; so X is Eberlein compact by Problem 324; this proves sufficiency and makes our solution complete.

U.365. *Prove that a compact space X is Eberlein compact if and only if $X \times X$ is hereditarily σ-metacompact.*

Solution. If X is Eberlein compact then so is $X \times X$ by Problem 307; therefore $X \times X$ is hereditarily σ-metacompact by Problem 363. Now, if $X \times X$ is hereditarily σ-metacompact then $X^2 \backslash \Delta$ is σ-metacompact so Problem 364 is applicable to conclude that X is Eberlein compact.

U.366. *Prove that a compact space X has a closure-preserving cover by compact metrizable subspaces if and only if it embeds into a σ-product of compact metrizable spaces. In particular, if X has a closure-preserving cover by compact metrizable subspaces then it is an Eberlein compact space.*

Solution. Given a space Z and a family $\mathcal{A} \subset \exp Z$ let $\mathcal{A}|Y = \{A \cap Y : A \in \mathcal{A}\}$. If $x, y \in Z$ and $A \subset Z$ then the set A *separates the points* x and y if $A \cap \{x, y\}$ is a singleton.

Fact 1. Suppose that a non-empty space Z has a closure-preserving cover $\mathcal{C}$ by compact subspaces. Consider the collection $\mathbb{F}(\mathcal{C})$ of all maximal centered subfamilies of $\mathcal{C}$. Given $\mathcal{F} \in \mathbb{F}(\mathcal{C})$ let $P_\mathcal{F} = \bigcap \mathcal{F}$. Then the family $\mathcal{P}(\mathcal{C}) = \{P_\mathcal{F} : \mathcal{F} \in \mathbb{F}(\mathcal{C})\}$ is discrete; in particular, if Z is compact then $\mathcal{P}(\mathcal{C})$ is finite. We will call the family $\mathcal{P}(\mathcal{C})$ *the core of the family* $\mathcal{C}$.

Proof. If $\mathcal{F} \in \mathbb{F}(\mathcal{C})$ then $P_\mathcal{F} \neq \emptyset$ because $\mathcal{F}$ is non-empty and every element of $\mathcal{F}$ is compact. If $\mathcal{F}, \mathcal{G} \in \mathbb{F}(\mathcal{C})$ are distinct then $P_\mathcal{F} \cap P_\mathcal{G} = \emptyset$ because otherwise $\mathcal{F} \cup \mathcal{G}$ is centered and hence coincides with both $\mathcal{F}$ and $\mathcal{G}$. Thus, for any $x \in Z$ there is at most one $\mathcal{F} \in \mathbb{F}(\mathcal{C})$ such that $x \in P_\mathcal{F}$. The set $U = Z \backslash (\bigcup\{C \in \mathcal{C} : x \notin C\})$ is open in Z and $x \in U$. If $\mathcal{G} \in \mathbb{F}(\mathcal{C})$ and $\mathcal{G} \neq \mathcal{F}$ then $x \notin P_\mathcal{G}$; so there is $G \in \mathcal{G}$ such that $x \notin G$. But then $G \cap U = \emptyset$ and therefore $P_\mathcal{G} \cap U = \emptyset$ which shows that U intersects at most one element of the family $\mathcal{P}(\mathcal{C})$. Thus $\mathcal{P}(\mathcal{C})$ is discrete and hence Fact 1 is proved.

Fact 2. Suppose that a compact space K has a closure-preserving cover $\mathcal{C}$ by compact metrizable subspaces. Then the set $M(K) = \{x \in K : \text{there exists a second countable } U \in \tau(K) \text{ with } x \in U\}$ is open and dense in K.

Proof. It is evident that $M(K)$ is open in K; if $M(K)$ is not dense then there is a non-empty $U \in \tau(K)$ such that $M(K) \cap U = \emptyset$. It is straightforward that $\mathcal{C}|\overline{U}$ is a closure-preserving cover of $L = \overline{U}$ by compact metrizable subspaces. If $M(L) \neq \emptyset$ then there exists a second countable $V \in \tau^*(L)$. Then $V' = V \cap U$ is a non-empty open second countable subset of K, so every point of V' belongs to $M(K)$; this contradiction with $V' \cap M(K) = \emptyset$ shows that $M(L) = \emptyset$.

This proves that we can assume, without loss of generality, that $K = L$, i.e., $M(K) = \emptyset$. Given a countable subfamily $\mathcal{C}'$ of the family $\mathcal{C}$, the set $P = \bigcup \mathcal{C}'$ is closed in K (because $\mathcal{C}$ is closure-preserving) and has a countable network (being a countable union of second countable spaces). Therefore P is compact and metrizable by Fact 4 of S.307.

Now let $\mathcal{P}_0$ be the core of the family $\mathcal{C}$ and let $F_0 = \bigcup \mathcal{P}_0$; it follows from Fact 1 that F_0 can be covered by finitely many elements of $\mathcal{C}$, so F_0 is compact and second countable. Since $M(K) = \emptyset$, the interior of F_0 is empty and hence F_0 is nowhere dense in K. Take a point $x \in K \backslash F_0$ and a set $V \in \tau(x, K)$ such that $\overline{V} \cap F_0 = \emptyset$; then $U_0 = K \backslash \overline{V}$ is an open neighbourhood of F_0 such that $\text{Int}_K(K \backslash U_0) \neq \emptyset$.

Suppose that $m \in \omega$ and we have constructed compact second countable sets $F_0, \ldots, F_m$ and open sets $U_0, \ldots, U_m$ with the following properties:

(1) F_i is second countable, $U_i \in \tau(F_i, K)$ and $\text{Int}_K(K \backslash U_i) \neq \emptyset$ for any $i \leq m$;
(2) $U_i \subset U_{i+1}$ for any $i < m$;
(3) if $i < m$ and $\mathcal{P}_{i+1}$ is the core of the family $\mathcal{C}|(K \backslash U_i)$ then $F_{i+1} = \bigcup \mathcal{P}_{i+1}$.

The core $\mathcal{P}_{m+1}$ of the family $\mathcal{C}|(K \backslash U_m)$ is finite by Fact 1; therefore the set $F_{m+1} = \bigcup \mathcal{P}_{m+1}$ is compact and second countable. The property (1) shows that there is $W \in \tau^*(K)$ such that $W \subset K \backslash U_m$. It follows from $M(K) = \emptyset$ then there is a point $x \in W \backslash F_{m+1}$; take $V \in \tau(x, K)$ such that $\overline{V} \subset W \cap (K \backslash F_{m+1})$ and let $U_{m+1} = K \backslash \overline{V}$. It is evident that the properties (1)–(3) still hold if we substitute m by $m + 1$; so our inductive procedure can be continued to construct families $\{F_n : n \in \omega\}$ and $\{U_n : n \in \omega\}$ for which the properties (1)–(3) are fulfilled for all $m \in \omega$.

It follows from (1) and (3) that F_i is non-empty for all $i \in \omega$; the conditions (1)–(3) imply that the family $\{F_i : i \in \omega\}$ is disjoint. Given a point $x \in K$ the set $O = K \backslash (\bigcup \{C : C \in \mathcal{C}$ and $x \notin C\})$ is an open neighbourhood of x. Let $U_{-1} = \emptyset$ and consider first the case when $x \notin \bigcup_{n \in \omega} U_n$ and hence $x \notin \bigcup_{n \in \omega} F_n$; given $i \in \omega$ the point x does not belong to H for any $H \in \mathcal{P}_i$. Since H is the intersection of a subfamily of $\mathcal{C}|(K \backslash U_{i-1})$, there is $C \in \mathcal{C}$ such that $H \subset C$ and $x \notin C$. Consequently, $C \cap O = \emptyset$ which shows that $O \cap H = \emptyset$ for any $H \in \mathcal{P}_i$, i.e., $O \cap F_i = \emptyset$.

Now, if $x \in U_n$ for some $n \in \omega$ then it follows from (1)–(3) that U_n is a neighbourhood of x which does not meet F_i for any $i > n$. This shows that in all possible cases x has a neighbourhood which meets finitely many elements of the family $\mathcal{F} = \{F_i : i \in \omega\}$. Since $\mathcal{F}$ is disjoint, it has to be discrete which is a contradiction with compactness of K. Fact 2 is proved.

Fact 3. A space Z condenses into a σ-product of compact metrizable spaces if and only if there exists a family $C = \{C_{n,t} : n \in \omega,\ t \in T\}$ of cozero subsets of Z such that C is T_0-separating in Z and the family $C_0 = \{\bigcup_{n \in \omega} C_{n,t} : t \in T\}$ is point-finite.

Proof. Suppose that M_t is a compact metrizable space for any $t \in T$ and a point $a \in M = \prod_{t \in T} M_t$ is chosen so that there exists a condensation $\varphi : Z \to Y$ for some $Y \subset \sigma(M, a)$. Let $\pi_t : M \to M_t$ be the natural projection and fix a countable base $\mathcal{B}_t = \{B_t^n : n \in \omega\}$ in the space $M_t \backslash \{a(t)\}$ for every $t \in T$. It is clear that $U_t^n = \pi_t^{-1}(B_t^n) \cap Y$ is a cozero set in Y; so $C_{n,t} = \varphi^{-1}(B_t^n)$ is a cozero set in Z for any $t \in T$ and $n \in \omega$. The family $\{U_t^n : n \in \omega,\ t \in T\}$ is easily seen to be T_0-separating in Y; so $C = \{C_{n,t} : n \in \omega,\ t \in T\}$ is T_0-separating in Z because φ is a condensation. Let $E_t = \pi_t^{-1}(M_t \backslash \{a(t)\}) \cap Y$ for any $t \in T$. It follows from the inclusion $Y \subset \sigma(M, a)$ that $\{E_t : t \in T\}$ is point-finite. On the other hand, $E_t = \bigcup \{U_t^n : n \in \omega\}$ for any $t \in T$, so $C_0 = \{\bigcup_{n \in \omega} C_{n,t} : t \in T\} = \{\varphi^{-1}(E_t) : t \in T\}$ is point-finite as well. This proves necessity.

Now assume that $C = \{C_{n,t} : n \in \omega,\ t \in T\}$ is a T_0-separating family of cozero subsets of Z such that $C_0 = \{\bigcup_{n \in \omega} C_{n,t} : t \in T\}$ is point-finite. Fix a continuous function $f_{n,t} : Z \to I = [0, 1]$ such that $C_{n,t} = (f_{n,t})^{-1}((0, 1])$ for any $n \in \omega$ and $t \in T$. If $\varphi_t = \Delta\{f_{n,t} : n \in \omega\}$ then $M_t = I^\omega$ is a compact metrizable space; if $a_t(n) = 0$ for all $n \in \omega$ then $a_t \in M_t$ for all $t \in T$. The family C being T_0-separating, the set $\{f_{n,t} : n \in \omega,\ t \in T\}$ separates the points of Z so $\varphi = \Delta\{\varphi_t : t \in T\}$ is a condensation of Z into $M = \prod_{t \in T} M_t$. Let $a(t) = a_t$ for all $t \in T$. To see that $\varphi(Z) \subset \sigma(M, a)$ take any $z \in Z$. The family C_0 is point-finite; so there is a finite $S \subset T$ such that $f_{n,t}(z) = 0$ for any $t \in T \backslash S$ and $n \in \omega$. This implies that $\varphi_t(z) = a_t$ for all $t \in T \backslash S$ and therefore $\varphi(z) \in \sigma(M, a)$. Thus φ is the promised condensation; this settles sufficiency and shows that Fact 3 is proved.

Fact 4. Any σ-product of compact metrizable spaces has a closure-preserving cover by metrizable compact subspaces.

Proof. Suppose that M_t is compact and metrizable for any $t \in T$ and take a point $a \in M = \prod_{t \in T} M_t$. Recall that $\sigma(M, a) = \{x \in M : |\{t \in T : x(t) \neq a(t)\}| < \omega\}$; we must prove that $\sigma(M, a)$ has a closure-preserving cover by metrizable compact sets. For any $x \in \sigma(M, a)$ let $\operatorname{supp}(x) = \{t \in T : x(t) \neq a(t)\}$. If $t_1, \ldots, t_n \in T$ then $H(t_1, \ldots, t_n) = M_{t_1} \times \ldots \times M_{t_n} \times \prod\{\{a(t)\} : t \in T \backslash \{t_1, \ldots, t_n\}\}$ is a compact metrizable subset of $\sigma(M, a)$. It is clear that $\bigcup \{H(t_1, \ldots, t_n) : n \in \mathbb{N},\ t_i \in T$ for all $i \leq n\} = \sigma(M, a)$; so it suffices to establish that the family $\mathcal{H} = \{H(t_1, \ldots, t_n) : n \in \mathbb{N},\ t_i \in T$ for all $i \leq n\}$ is closure-preserving.

Assume, towards a contradiction, that there is a family $\mathcal{H}' \subset \mathcal{H}$ such that $x \in \bigcup \mathcal{H}' \backslash (\bigcup \mathcal{H}')$ for some $x \in \sigma(M, a)$; let $S = \operatorname{supp}(x)$. It follows from Problem 101 that $\sigma(M, a)$ is a Fréchet–Urysohn space; so there exist sequences $\{H_n : n \in \omega\} \subset \mathcal{H}'$ and $\{x_n : n \in \omega\} \subset \sigma(M, a)$ such that $x_n \to x$ and $x_n \in H_n$ for all $n \in \omega$. We have $H_n = H(t_1^n, \ldots, t_{k_n}^n)$ for any $n \in \omega$; it follows from $x \notin H_n$ that there is $s_n \in \operatorname{supp}(x)$ such that $s_n \notin \{t_1^n, \ldots, t_{k_n}^n\}$. The set $\operatorname{supp}(x)$ being finite there is an infinite $A \subset \omega$ and $s \in \operatorname{supp}(x)$ such that $s_n = s$ for all $n \in A$. The sequence $\{x_n : n \in A\}$ still converges to the point x; we have, $x_n(s) = a(s)$ for all $s \in A$ which implies

$x(s) = a(s)$. However, $s \in \operatorname{supp}(x)$ so $x(s) \neq a(s)$. This contradiction shows that $\mathcal{H}$ is a closure-preserving cover of $\sigma(M, a)$ by compact metrizable subspaces, i.e., Fact 4 is proved.

Returning to our solution suppose that X is a compact subspace of a σ-product Y of metrizable compact spaces. By Fact 4, the space Y has a closure-preserving cover $\mathcal{C}'$ by compact metrizable subspaces. It is clear that the family $\mathcal{C} = \{C \cap X : C \in \mathcal{C}'\}$ is a closure-preserving cover of X by compact metrizable subspaces; this proves sufficiency.

To establish necessity suppose that $\mathcal{C}$ is a closure-preserving cover of a compact space X and every element of $\mathcal{C}$ is compact and metrizable. It is easy to see that if $\mathcal{C}' \subset \mathcal{C}$ is countable then $F = \bigcup \mathcal{C}'$ is compact and metrizable. Therefore we can add F and all other countable unions of elements of $\mathcal{C}$ to $\mathcal{C}$; the resulting family will still be closure-preserving. Thus we can consider, without loss of generality, that

(4) $\bigcup \mathcal{C}' \in \mathcal{C}$ for any countable $\mathcal{C}' \subset \mathcal{C}$.

Let $X_0 = X$ and $G_0 = M(X)$; then G_0 is open and dense in X_0 by Fact 2. Suppose that α is an ordinal and we have a family $\{X_\beta : \beta < \alpha\}$ such that

(5) X_β is closed in X and non-empty for all $\beta < \alpha$;
(6) if $\beta + 1 < \alpha$ and $G_\beta = M(X_\beta)$ then $X_{\beta+1} = X_\beta \backslash G_\beta$;
(7) if $\beta < \alpha$ is a limit ordinal then $X_\beta = \bigcap_{\gamma < \beta} X_\gamma$.

If α is a limit ordinal then let $X_\alpha = \bigcap_{\beta < \alpha} X_\beta$; then the set X_α is non-empty and we have a family $\{X_\beta : \beta \leq \alpha\}$ such that the properties (5)–(7) still hold if we substitute α by $\alpha + 1$. If $\alpha = \beta + 1$ then let $X_\alpha = X_\beta \backslash G_\beta$; if X_α is empty then our construction stops. If not, then we obtain a family $\{X_\beta : \beta \leq \alpha\}$ such that the properties (5)–(7) still hold if we substitute α by $\alpha + 1$.

It follows from Fact 2 that $G_\beta \neq \emptyset$ if $X_\beta \neq \emptyset$; so $X_{\beta+1}$ is strictly contained in X_β if $X_{\beta+1} \neq \emptyset$. This shows that our construction cannot last $|X|^+$ steps and therefore there exists α such that $X_{\alpha+1} = \emptyset$ which means that X_α is locally metrizable and hence second countable. Let μ be the minimal such α; it turns out that

(8) for any $\alpha \leq \mu$ and $C \in \mathcal{C}$ there is $C' \in \mathcal{C}$ such that $C \subset C'$ and $C' \cap G_\alpha$ is open in G_α and hence in X_α.

Suppose first that P is a second countable subspace of G_α. Since G_α is locally second countable, the set P can be covered by a countable family $\mathcal{U}$ of open subsets of G_α such that every $U \in \mathcal{U}$ is second countable. An immediate consequence is that $W = \bigcup \mathcal{U}$ is separable; let D be a countable dense subset of W. The family $\mathcal{C}$ being closed under countable unions there is $E \in \mathcal{C}$ such that $D \subset E$; since E is compact, we have $W \subset \overline{D} \subset E$. Thus, for every second countable subset P of G_α there is $E \in \mathcal{C}$ such that $P \subset \operatorname{Int}_{X_\alpha}(E)$.

Now, given $C \in \mathcal{C}$ let $C_0 = C$ and use the observation of the previous paragraph to construct a sequence $\{C_n : n \in \omega\}$ such that $C_n \cap G_\alpha \subset \operatorname{Int}_{X_\alpha}(C_{n+1} \cap G_\alpha)$ for any $n \in \omega$. It is straightforward that $C' = \bigcup_{n \in \omega} C_n$ is as promised in (8).

Given sets $A, B, C \subset X$ we will use the expression $\langle A, B, C \rangle$ to say that $B \subset A$ and $A \cap C = B \cap C$. Let us show that

(9) for any $\alpha \le \mu$ there exists a closure-preserving family $\mathcal{C}_\alpha$ on X_α such that any $C \in \mathcal{C}_\alpha$ is compact for which there is $E_C \in \mathcal{C}$ with $\langle E_C, C, X_{\alpha+1} \rangle$ and $\{C \cap G_\alpha : C \in \mathcal{C}_\alpha\}$ is a disjoint family of clopen subsets of G_α.

It follows from (8) that the family $\mathcal{E} = \{C \cap X_\alpha : C \in \mathcal{C}$ and $C \cap G_\alpha$ is non-empty and open in $X_\alpha\}$ is a closure-preserving cover of X_α. The set X_α being non-empty, the family $\mathcal{E}$ is non-empty as well; take a set $F_0 \in \mathcal{E}$. Suppose that β is an ordinal and we have a family $\{F_\gamma : \gamma < \beta\} \subset \mathcal{E}$ such that $(F_\delta \cap G_\alpha)\backslash(\bigcup\{F_\gamma : \gamma < \delta\}) \ne \emptyset$ for any $\delta < \beta$.

If $G_\alpha\backslash(\bigcup\{F_\gamma : \gamma < \beta\}) = \emptyset$ then our construction stops. If not, then we can choose $F_\beta \in \mathcal{E}$ for which $(F_\beta \cap G_\alpha)\backslash(\bigcup\{F_\gamma : \gamma < \beta\}) \ne \emptyset$. It is easy to see that, for any ordinal β the family $\{(F_\delta \cap G_\alpha)\backslash(\bigcup\{F_\gamma : \gamma < \delta\}) : \delta < \beta\}$ is disjoint and consists of non-empty subsets of G_α. Therefore we cannot make $|G_\alpha|^+$ steps in our construction, i.e., there is an ordinal $\beta < |G_\alpha|^+$ such that $G \subset \bigcup\{F_\gamma : \gamma < \beta\}$. Let $H_\gamma = F_\gamma\backslash(\bigcup\{F_\delta \cap G_\alpha : \delta < \gamma\})$ for any $\gamma < \beta$. We will check that the family $\mathcal{C}_\alpha = \{H_\gamma : \gamma < \beta\}$ is as promised.

It is evident that $G \subset \bigcup \mathcal{C}_\alpha$; every set $F_\delta \cap G_\alpha$ is open in X_α which shows that each H_γ is closed in X_α being a difference of a closed set and an open set. For every $\gamma < \beta$ there is $D \in \mathcal{C}$ such that $C = F_\gamma = D \cap X_\alpha$. Since $F_\gamma\backslash H_\gamma \subset G_\alpha$, we have $D \cap X_{\alpha+1} = C \cap X_{\alpha+1}$ which shows that, for $E_C = D$, the property $\langle E_C, C, X_{\alpha+1}\rangle$ holds as well.

Now it follows from $H_\gamma \cap G_\alpha = (F_\gamma \cap G_\alpha)\backslash(\bigcup\{F_\delta : \delta < \gamma\})$ that $H_\gamma \cap G_\alpha$ is open in X_α being the difference of an open set and a closed set. It easily follows from the definition of $\mathcal{C}_\alpha$ that $\{C \cap G_\alpha : C \in \mathcal{C}_\alpha\}$ is disjoint. To finally see that $\mathcal{C}_\alpha$ is closure-preserving take a family $\mathcal{C}' \subset \mathcal{C}_\alpha$ and suppose that $x \in \overline{\bigcup \mathcal{C}'}$. If $x \in G_\alpha$ then there is a unique $C \in \mathcal{C}_\alpha$ such that $x \in C \cap G_\alpha$. The set $C \cap G_\alpha$ is a neighbourhood of x; so $C \in \mathcal{C}'$ and $x \in C$.

If $x \in X_\alpha\backslash G_\alpha = X_{\alpha+1}$ then fix, for any $C \in \mathcal{C}'$ a set $E_C \in \mathcal{C}$ for which we have $\langle E_C, C, X_{\alpha+1}\rangle$. If $\mathcal{C}'' = \{E_C : C \in \mathcal{C}'\}$ then $x \in \overline{\bigcup \mathcal{C}''}$; the family $\mathcal{C}$ being closure-preserving there exists a set $C \in \mathcal{C}'$ such that $x \in E_C$ and therefore $x \in E_C \cap X_{\alpha+1} = C \cap X_{\alpha+1}$, i.e., $x \in C$ which shows that $\mathcal{C}_\alpha$ is closure-preserving and hence (9) is proved.

Our next step is to show that

(10) the family $\mathcal{D} = \bigcup\{\mathcal{C}_\alpha : \alpha \le \mu\}$ is closure-preserving.

It suffices to prove that $\mathcal{D}_\mu = \bigcup\{\mathcal{C}_\alpha : \alpha < \mu\}$ is closure-preserving; we will do this by showing by transfinite induction that $\mathcal{D}_\beta = \bigcup\{\mathcal{C}_\alpha : \alpha < \beta\}$ is closure-preserving for any $\beta \le \mu$. This is clear for any $\beta < \omega$ because a finite union of closure-preserving families is a closure-preserving family. If $\mathcal{D}_\beta$ is closure-preserving then $\mathcal{D}_{\beta+1}$ is also closure-preserving being the inion of two closure-preserving families.

Finally assume that $\beta \le \mu$ is a limit ordinal and we proved that $\mathcal{D}_\alpha$ is closure-preserving for any $\alpha < \beta$. Fix a family $\mathcal{E}_\alpha \subset \mathcal{C}_\alpha$ for any $\alpha < \beta$ and suppose that, for the set $P = \bigcup\{\bigcup \mathcal{E}_\alpha : \alpha < \beta\}$, we have $x \in \overline{P}$ for some $x \in X$.

If $x \in G_\alpha$ for some $\alpha < \beta$ then $U = X \backslash X_{\alpha+1}$ is an open neighbourhood of x such that $(\bigcup \mathcal{E}_\gamma) \cap U = \emptyset$ for all $\gamma > \alpha$. Therefore $x \in \bigcup \{\bigcup E_\gamma : \gamma \leq \alpha\}$. The family $\mathcal{D}_{\alpha+1} \supset \bigcup \{\mathcal{E}_\gamma : \gamma \leq \alpha\}$ being closure-preserving there is $E \in \bigcup \{\mathcal{E}_\gamma : \gamma \leq \alpha\}$ with $x \in E$.

Therefore we can assume that $x \notin \bigcup \{G_\alpha : \alpha < \beta\}$ and hence $x \in X_\beta$. For any ordinal $\alpha < \beta$ and $C \in \mathcal{E}_\alpha$ we can fix a set $E_C \in C$ such that $\langle E_C, C, X_{\alpha+1} \rangle$. Then $x \in \bigcup \{\bigcup \{E_C : C \in \mathcal{E}_\alpha\} : \alpha < \beta\}$ which, together with the family C being closure-preserving, implies that there is $\alpha < \beta$ and $C \in \mathcal{E}_\alpha$ such that $x \in E_C$. The property $\langle E_C, C, X_{\alpha+1} \rangle$ implies that $E_C \cap X_{\alpha+1} = C \cap X_{\alpha+1}$ and hence $E_C \cap X_\beta = C \cap X_\beta$. Consequently, $x \in E_C \cap X_\beta = C \cap X_\beta$ whence $x \in C$; so (10) is proved.

We will show next that

(11) the space X is Corson compact.

Observe first that, for any $A \subset X$ the set $ast(A) = X \backslash (\bigcup \{D \in \mathcal{D} : D \cap A = \emptyset\})$ is an open neighbourhood of the set A. Observe that, for any ordinal $\alpha \leq \mu$ the family $\mathcal{B}_\alpha = \{C \cap G_\alpha : C \in \mathcal{C}_\alpha\}$ is disjoint and consists of clopen subsets of G_α. We claim that

(12) the family $\mathcal{H} = \{ast(B) : B \in \mathcal{B} = \bigcup \{\mathcal{B}_\alpha : \alpha \leq \mu\}\}$ is point-countable.

To see that (12) holds fix $x \in X$ and $D \in \mathcal{D}$ such that $x \in D$. The property (9) shows that D is second countable. Let $L = \{\alpha \leq \mu : D \cap G_\alpha \neq \emptyset\}$. If $\alpha, \beta \in L$ and $\alpha < \beta$ then $D \cap X_\beta \subset D \cap X_\alpha$ and $D \cap X_\beta \neq D \cap X_\alpha$. Thus $\{D \cap X_\alpha : \alpha \in L\}$ is a strictly decreasing family of compact subsets of compact metrizable space D. Therefore L is countable; for any $\alpha \in L$ the set D intersects at most countably many elements of $\mathcal{B}_\alpha$ because $c(D) \leq w(D) = \omega$. Therefore there is a countable $\mathcal{B}' \subset \mathcal{B}$ such that, for any $B \in \mathcal{B} \backslash \mathcal{B}'$, we have $D \cap B = \emptyset$ and hence $D \cap ast(B) = \emptyset$ which shows that $x \notin ast(B)$. Therefore x can only belong to some elements of the family $\{ast(B) : B \in \mathcal{B}'\}$, i.e., $\mathcal{H}$ is point-countable and hence (12) is proved.

Now fix $B \in \mathcal{B}$ and let $H = ast(B)$. The space B is locally compact and second countable; so it has a base $\mathcal{W}_B = \{W_B^n : n \in \omega\}$ such that $cl(W_B^n) \subset B$ is compact for any $n \in \omega$. The property (9) shows that B is contained in a second countable compact space; so $\overline{B}$ is second countable. The space $B' = (X \backslash H) \cup \overline{B}$ is compact and every W_B^n is an open σ-compact subspace of B'. Therefore there exists a continuous function $f : B' \to [0, 1]$ such that $W_B^n = f^{-1}((0, 1])$. If $g \in C_p(X, [0, 1])$ and $g|B' = f$ then $V_B^n = g^{-1}((0, 1])$ is a cozero set in X such that $V_B^n \subset H$ and $V_B^n \cap B = W_B^n$. This shows that

(13) for any $B \in \mathcal{B}$ there is a family $\{V_B^n : n \in \omega\}$ of cozero sets in X such that $V_B^n \subset ast(B)$ for any $n \in \omega$ and the family $\{V_B^n \cap B : n \in \omega\}$ is a base in B.

It is an easy consequence of (12) that the family $\mathcal{V} = \{V_B^n : n \in \omega, B \in \mathcal{B}\}$ is point-countable. Let x, y be distinct points of X. Since $\bigcup \{G_\alpha : \alpha \leq \mu\} = X$, there exist $\alpha, \beta \leq \mu$ such that $x \in G_\alpha$ and $y \in G_\beta$. If $\alpha \neq \beta$, say, $\alpha < \beta$ then fix $B \in \mathcal{B}_\alpha$ with $x \in B$ and observe that $y \in C$ for some $C \in \mathcal{C}_\beta$; since $C \cap G_\alpha = \emptyset$ and $B \subset G_\alpha$, we have $B \cap C = \emptyset$ and therefore $ast(B) \cap C = \emptyset$. The family $\{V_B^n \cap B : n \in \omega\}$ being a base in B there is $n \in \omega$ such that $x \in V_B^n$. However, $V_B^n \subset ast(B) \subset X \backslash C$ which shows that V_B^n separates the points x and y.

If $\alpha = \beta$ and there is $B \in \mathcal{B}_\alpha$ such that $x, y \in B$ then some V_B^n separates x and y because $\{V_B^n \cap B : n \in \omega\}$ is a base in B. If some $B \in \mathcal{B}_\alpha$ separates the points x and y then some V_B^n also separates x and y because $\{V_B^n \cap B : n \in \omega\}$ is a base in B and $V_B^n \cap G_\alpha \subset B$ for any $n \in \omega$.

Therefore the family $\mathcal{V}$ is point-countable, T_0-separating and consists of cozero subsets of X. Applying Problem 118 we conclude that X is Corson compact, i.e., (11) is proved.

To make the final step in our proof of necessity we will need the following concept: say that a family $\mathcal{A}' \subset \exp X$ is *an enlargement* of a family $\mathcal{A} \subset \exp X$ if there exists a map $\varphi : \mathcal{A} \to Q$ such that $\varphi^{-1}(q)$ is countable for any $q \in Q$ and $\mathcal{A}' = \{\bigcup \varphi^{-1}(q) : q \in Q\}$. In other words, to enlarge a family $\mathcal{A}$ we must split it in disjoint countable subfamilies and take the unions of these subfamilies. Recalling Fact 3 we can see that, to prove that X embeds in a σ-product of metrizable compact spaces, it suffices to show that the family $\mathcal{V}$ has a point-finite enlargement. To do so we will prove by induction on κ that

(14) for every cardinal $\kappa \leq \mu$ any point-countable family of cozero subsets of X of cardinality κ admits a point-finite enlargement.

If $\mathcal{A}$ is a countable family of cozero subsets of X then $\bigcup \mathcal{A}$ is the desired enlargement; so (14) is true for $\kappa \leq \omega$. Now assume that λ is a cardinal with $\omega < \lambda \leq \mu$ and we have proved (14) for any cardinal $\kappa < \lambda$.

Fix a point-countable family $\mathcal{U} = \{U_\beta : \beta < \lambda\}$ of cozero subsets of X; every U_β is σ-compact; so we can represent it as $U_\beta = \bigcup_{n \in \omega} K_{\beta,n}$ where $K_{\beta,n}$ is compact for any $n \in \omega$.

Given a non-empty closed $F \subset X$ let $A_F = \{\alpha \leq \mu : F \cap X_\alpha \neq \emptyset\}$. The family $\{F \cap X_\alpha : \alpha \in A_F\}$ is decreasing and consists of non-empty closed subsets of F; so $N(F) = \bigcap \{F \cap X_\alpha : \alpha \in A_F\} \neq \emptyset$. There is $\beta \leq \mu$ such that $N(F) \cap G_\beta \neq \emptyset$. It is clear that $\beta \in A_F$; if $\alpha > \beta$ and $F \cap X_\alpha \neq \emptyset$ then $N(F) \not\subset F \cap X_\alpha$ which is a contradiction. Therefore β is the maximal element of A_F, i.e., we have established that

(15) for any closed non-empty $F \subset X$ the set A_F has a maximal element β; let $\alpha(F) = \beta$.

If $F = \emptyset$ then let $N(F) = \emptyset$. Now take again a non-empty closed $F \subset X$ and let $\beta = \alpha(F)$. It follows from (15) that $N(F) = F \cap X_\beta$; furthermore, $N(F) \cap X_{\beta+1} = \emptyset$ and therefore $N(F) \subset G_\beta$. As a consequence, $N(F)$ is locally second countable and hence second countable. This proves that

(16) if F is a non-empty closed subset of X and $\beta = \alpha(F)$ then $N(F) = F \cap X_\beta$ is compact and metrizable.

For any $\beta_1, \ldots, \beta_n < \lambda$ let

$$Q(\beta_1, \ldots, \beta_n) = \bigcup \{N(K_{\beta_1,m_1} \cap \ldots \cap K_{\beta_n,m_n}) : m_1, \ldots, m_n \in \omega\}.$$

Every set $Q(\beta_1, \ldots, \beta_n)$ is separable being a countable union of second countable spaces.

Given $\beta < \lambda$ let $\Phi_\beta^0 = \{U_\gamma : \gamma \le \beta\}$. If we have Φ_β^n let $\Phi_\beta^{n+1} = \{U_\gamma : \gamma < \lambda$ and there exist $n \in \mathbb{N}$ and $U_{\beta_1}, \ldots, U_{\beta_n} \in \Phi_\alpha^n$ such that $U_\gamma \cap Q(\beta_1, \ldots, \beta_n) \ne \emptyset\}$. This makes it possible to construct the sequence $\{\Phi_\beta^n : n \in \omega\}$; let $\Phi_\beta = \bigcup_{n \in \omega} \Phi_\beta^n$ for any $\beta < \lambda$.

Fix an ordinal $\beta < \lambda$ and observe that $|\Phi_\beta^0| \le |\beta| < \lambda$. Assume that $m \in \omega$ and we proved that $|\Phi_\beta^m| \le |\beta| \cdot \omega$. The family $\mathcal{I} = \{(\beta_1, \ldots, \beta_n) : n \in \mathbb{N}$ and $U_{\beta_i} \in \Phi_\beta^m\}$ has cardinality at most $|\Phi_\beta^m| \cdot \omega$; besides, for any $(\beta_1, \ldots, \beta_n) \in \mathcal{I}$ the set $Q = Q(\beta_1, \ldots, \beta_n)$ is separable; so there are at most countably many elements of the family $\mathcal{U}$ which can intersect Q because $\mathcal{U}$ is point-countable. Therefore $|\Phi_\beta^{m+1}| \le |\mathcal{I}| \cdot \omega \le |\Phi_\beta^m| \cdot \omega \le |\beta| \cdot \omega$. Thus $|\Phi_\beta| \le |\beta| \cdot \omega < \lambda$ for any $\beta < \lambda$.

Now let $\Omega_\beta = \Phi_\beta \backslash (\bigcup\{\Phi_\gamma : \gamma < \beta\})$ for any $\beta < \lambda$. It is easy to see that the collection $\{\Omega_\beta : \beta < \lambda\}$ is disjoint and $\bigcup\{\Omega_\beta : \beta < \lambda\} = \mathcal{U}$. Since $|\Omega_\beta| < \lambda$, we can apply the induction hypothesis to Ω_β to find a point-finite enlargement $\mathcal{J}_\beta$ of the family Ω_β for any $\beta < \lambda$. It is clear that $\mathcal{J} = \bigcup\{\mathcal{J}_\beta : \beta < \lambda\}$ is an enlargement of the family $\mathcal{U}$.

To see that $\mathcal{J}$ is point-finite suppose not. Since every $\mathcal{J}_\beta$ is point-finite, there is a strictly increasing sequence $\{\beta_i : i \in \omega\} \subset \lambda$ and $V_i \in \mathcal{J}_{\beta_i}$ for any $i \in \omega$ such that $\bigcap_{i \in \omega} V_i \ne \emptyset$. Every V_i is a countable union of some elements of Ω_{β_i}; so we can choose $W_i \in \Omega_{\beta_i}$ such that $\bigcap_{i \in \omega} W_i \ne \emptyset$. For any $i \in \omega$ we can fix $\gamma_i < \lambda$ such that $W_i = U_{\gamma_i}$. Recalling that $U_{\gamma_i} = \bigcup_{n \in \omega} K_{\gamma_i, n}$ for any $i \in \omega$, we can choose a sequence $\{m_i : i \in \omega\} \subset \omega$ such that $\bigcap_{i \in \omega} K_{\gamma_i, m_i} \ne \emptyset$. Let $v_i = \alpha(K_{\gamma_1, m_1} \cap \ldots \cap K_{\gamma_i, m_i})$ for any $i \in \omega$.

Now fix an arbitrary natural number $i \in \omega$; we have $\beta_{i+1} > \beta_i$ which implies $U_{\gamma_{i+1}} \notin \Phi_{\beta_i}$ and therefore $U_{\gamma_{i+1}} \cap N(K_{\gamma_1, m_1} \cap \ldots \cap K_{\gamma_i, m_i}) = \emptyset$ which is the same as saying that $U_{\gamma_{i+1}} \cap K_{\gamma_1, m_1} \cap \ldots \cap K_{\gamma_i, m_i} \cap X_{v_i} = \emptyset$. An immediate consequence is that $K_{\gamma_{i+1}, m_{i+1}} \cap K_{\gamma_1, m_1} \cap \ldots \cap K_{\gamma_i, m_i} \cap X_{v_i} = \emptyset$. Recalling the definition of $\alpha(F)$ for any closed $F \subset X$ we conclude that $v_{i+1} = \alpha(K_{\gamma_{i+1}, m_{i+1}} \cap K_{\gamma_1, m_1} \cap \ldots \cap K_{\gamma_i, m_i}) < v_i$. Thus $\{v_i : i \in \omega\}$ is a strictly decreasing sequence of ordinals; this contradiction shows that $\mathcal{J}$ is a point-finite enlargement of $\mathcal{U}$; so (14) is proved.

As a consequence, the family $\mathcal{V}$ has a point-finite enlargement; so we can apply Fact 3 to see that X embeds into a σ-product of compact metrizable spaces. This settles necessity.

Finally assume that X is a compact space with a closure-preserving cover by compact metrizable subspaces. We have proved that such a space X is embeddable into a σ-product of metrizable compact spaces; so we can apply Fact 3 to find a T_0-separating family $\mathcal{W} = \{W_{t,n} : n \in \omega, t \in T\}$ of cozero subsets of X such that the family $\tilde{\mathcal{W}} = \{\bigcup_{n \in \omega} W_{t,n} : t \in T\}$ is point-finite. It is clear that the collection $\mathcal{W}_n = \{W_{t,n} : t \in T\}$ has to be point-finite for any $n \in \omega$. Since also $\mathcal{W} = \bigcup_{n \in \omega} \mathcal{W}_n$, the space X has a σ-point-finite T_0-separating family of cozero sets; so X is Eberlein compact by Problem 324. This proves that any compact space with a closure-preserving cover by compact metrizable subspaces is Eberlein compact and completes our solution.

U.367. *Construct an Eberlein compact space which does not have a closure-preserving cover by compact metrizable subspaces.*

Solution. Let $K = (A(\omega_1))^\omega$. Then K is Eberlein compact by Fact 1 of U.340 and Problem 307. If U is a non-empty open subset of K then U contains a standard open subset of $(A(\omega_1))^\omega$, i.e., there is $m \in \omega$ and $V_i \in \tau^*(A(\omega_1))$ for all $i < m$ such that $V = V_0 \times \ldots \times V_{m-1} \times (A(\omega_1))^{\omega \setminus m} \subset U$. It is easy to see that $A(\omega_1)$ embeds in V and therefore $w(U) \geq w(V) \geq w(A(\omega_1)) = \omega_1$. This proves that K has no non-empty open subspaces of countable weight. If K has a closure-preserving cover by compact metrizable subspaces then the set of points which have a second countable neighbourhood is dense in K by Fact 2 of U.366. Thus K must have non-empty open second countable subspaces. This contradiction shows that K does not have a closure-preserving cover by compact metrizable subspaces.

U.368. *Observe that every strong Eberlein compact is Eberlein compact. Prove that a metrizable compact space is strong Eberlein compact if and only if it is countable.*

Solution. Given a set A let $\sigma_0(A) = \{x \in \mathbb{D}^A : |x^{-1}(1)| < \omega\}$. If X is a strong Eberlein compact space then it embeds in $\sigma_0(A)$ for some A. It is evident that $\sigma_0(A)$ is a subspace of $\Sigma_*(A)$; so X is Eberlein compact by Problem 322. Therefore every strong Eberlein compact space is Eberlein compact.

Now assume that X is a metrizable (and hence second countable) strong Eberlein compact space; we can consider, without loss of generality, that $X \subset \sigma_0(A)$ for some infinite set A. For arbitrary elements $a_1, \ldots, a_n \in A$ and $i_1, \ldots, i_n \in \mathbb{D}$ let $[a_1, \ldots, a_n; i_1, \ldots, i_n] = \{x \in \sigma_0(A) : x(a_j) = i_j \text{ for all } j \leq n\}$. It is evident that the family $\mathcal{O} = \{[a_1, \ldots, a_n; i_1, \ldots, i_n] : n \in \mathbb{N}, a_j \in A \text{ and } i_j \in \mathbb{D} \text{ for all } j \leq n\}$ is a base in $\sigma_0(A)$; for any $O = [a_1, \ldots, a_n; i_1, \ldots, i_n] \in \mathcal{O}$ let $\text{supp}(O) = \{a_1, \ldots, a_n\}$. The family $\{O \cap X : O \in \mathcal{O}\}$ is a base in X; so we can apply Claim proved in S.088 to conclude that there is a countable $\mathcal{O}' \subset \mathcal{O}$ such that $\mathcal{B} = \{O \cap X : O \in \mathcal{O}'\}$ is a base in X.

The set $B = \bigcup\{\text{supp}(O) : O \in \mathcal{O}'\}$ is countable; let $\pi : \sigma_0(A) \to \mathbb{D}^B$ be the restriction of the natural projection of $\mathbb{D}^A$ onto its face $\mathbb{D}^B$. If x and y are distinct points of X then there is $O \in \mathcal{O}'$ such that $x \in O$ and $y \notin O$. It follows from $\text{supp}(O) \subset B$ that $\pi^{-1}(O) = O$; therefore $\pi(x) \in \pi(O)$ while $\pi(y) \notin O$. In particular, $\pi(x) \neq \pi(y)$. This proves that $\pi_0 = \pi|X$ is an injection; the space X being compact, the map π_0 embeds X in $\pi(\sigma_0(A)) \subset \sigma_0(B)$. It is an easy exercise that $\sigma_0(B)$ is countable; so X also has to be countable.

Finally, suppose that X is a countable (and hence metrizable) compact space. Observe that $\sigma_0(\omega)$ is a countable second countable space without isolated points; so we can apply SFFS-349 to see that $\sigma_0(\omega) \simeq \mathbb{Q}$. By SFFS-350 the space X is homeomorphic to a subspace of $\mathbb{Q}$; an immediate consequence is that X also embeds in $\sigma_0(\omega)$; so X is a strong Eberlein compact space.

U.369. *Prove that a compact X is strong Eberlein compact if and only if it has a point-finite T_0-separating cover by clopen sets.*

Solution. Suppose that X is a strong Eberlein compact space. We can assume, without loss of generality, that $X \subset \sigma_0(A) = \{x \in \mathbb{D}^A : |x^{-1}(1)| < \omega\}$ for some infinite set A. For any $a \in A$ let $\pi_a : \sigma_0(A) \to \mathbb{D}$ be the restriction of the natural projection of $\mathbb{D}^A$ onto is factor determined by a.

The set $U_a = \pi_a^{-1}(1) \cap X$ is clopen in X for any $a \in A$ and it is straightforward that the family $\mathcal{U} = \{U_a : a \in A\}$ is T_0-separating in X. If $x \in X$ then $x \in U_a$ if and only if $a \in x^{-1}(1)$; so it follows from $X \subset \sigma_0(A)$ that every $x \in X$ belongs to finitely many elements of $\mathcal{U}$. Thus $\mathcal{U}$ is point-finite and we have proved necessity.

Finally, assume that a family $\mathcal{U}$ is a point-finite T_0-separating cover of X such that every $U \in \mathcal{U}$ is clopen in X. Let $\chi_U(x) = 1$ if $x \in U$ and $\chi_U(x) = 0$ for all $x \in X \setminus U$; then $\chi_U : X \to \mathbb{D}$ is a continuous function for any set $U \in \mathcal{U}$. The map $\varphi = \Delta\{\chi_U : U \in \mathcal{U}\} : X \to \mathbb{D}^{\mathcal{U}}$ is continuous and injective because the family $\mathcal{U}$ is T_0-separating. The space X being compact, φ embeds X into $\mathbb{D}^{\mathcal{U}}$. If $\varphi(x)(U) = 1$ then $x \in U$; the family $\mathcal{U}$ is point-finite, so the set $\{U \in \mathcal{U} : x \in U\}$ is finite which shows that $\varphi(X) \subset \sigma_0(\mathcal{U})$. Thus X is strong Eberlein compact being homeomorphic to a subspace of $\sigma_0(\mathcal{U})$ which shows that we settled sufficiency.

U.370. *Prove that every σ-discrete compact space is scattered. Give an example of a scattered compact non-σ-discrete space.*

Solution. Suppose that X is a compact space and $X = \bigcup_{n \in \omega} X_n$ where X_n is a discrete subspace of X for any $n \in \omega$. A subspace $A \subset X$ has an isolated point if and only if $\overline{A}$ has an isolated point; so it suffices to show that every *closed* $A \subset X$ has an isolated point. Assume, towards a contradiction, that a closed $A \subset X$ is dense-in-itself and let $A_n = X_n \cap A$ for each $n \in \omega$.

Every A_n is a discrete subspaces of A; it is an easy exercise to see that any discrete subspace of a dense-in-itself space is nowhere dense so $F_n = \overline{A}_n$ is nowhere dense in A. Now, $A = \bigcup_{n \in \omega} F_n$ shows that A is of first category in itself which is a contradiction with the fact that any compact space has the Baire property (see TFS-274). This proves that any σ-discrete compact space is scattered.

Fact 1. If $A \subset \omega_1$ is a stationary set then A is not discrete as a subspace of ω_1.

Proof. Assume that A is a discrete subspace of ω_1 and consider the set $C = \overline{A} \setminus A$; it is easy to see that C is a closed subset of ω_1. The set A being uncountable (see SFFS-065), for any ordinal $\beta < \omega_1$ we can choose a strictly increasing sequence $S = \{\alpha_n : n \in \omega\} \subset A$ such that $\beta < \alpha_0$. If $\alpha = \sup\{\alpha_n : n \in \omega\}$ then $\alpha \in \overline{A} \setminus A$ and $\alpha > \beta$ which shows that C is an unbounded subset of ω_1. Since A is stationary, there is a point $\alpha \in A \cap C$; this contradiction shows that Fact 1 is proved.

Finally, consider the compact space $K = \omega_1 + 1$; then K is scattered by Fact 4 of U.074. Assume that K is σ-discrete and fix a family $\{D_n : n \in \omega\}$ of discrete subspaces of K such that $K = \bigcup_{n \in \omega} D_n$. If $D_n' = D_n \cap \omega_1$ for all $n \in \omega$ then $\omega_1 = \bigcup_{n \in \omega} D_n'$. The set ω_1 being stationary, we can apply SFFS-065 to see that there is $n \in \omega$ such that D_n' is a stationary discrete subspace of ω_1. This contradiction with Fact 1 shows that K is a scattered compact space which is not σ-discrete.

U.371. *Prove that every strong Eberlein compact space is σ-discrete and hence scattered.*

Solution. Suppose that X is strong Eberlein compact and fix a T_0-separating point-finite family $\mathcal{U}$ of clopen subsets of X (see Problem 369). For any $x \in X$ the family $\mathcal{U}(x) = \{U \in \mathcal{U} : x \in U\}$ is finite; let $n(x) = |\mathcal{U}(x)|$. Then $X = \bigcup_{m \in \omega} X_m$ where $X_m = \{x \in X : n(x) = m\}$ for any $m \in \omega$.

To see that every set X_m is a discrete subspace of X fix an arbitrary point $x \in X_m$ and let $U_x = \bigcap \mathcal{U}(x)$. If $y \neq x$ and $y \in U_x$ then $\mathcal{U}(y) \supset \mathcal{U}(x)$; since the family $\mathcal{U}$ is T_0-separating, there exists $V \in \mathcal{U}(y)$ such that $x \notin V$ and therefore $n(y) \geq m+1$. This proves that $U_x \cap X_m = \{x\}$ for any $x \in X_m$, i.e., X_m is a discrete subspace of X for any $m \in \omega$. Thus our space X is σ-discrete and hence scattered by Problem 370.

U.372. *Prove that a hereditarily metacompact scattered compact space is strong Eberlein compact.*

Solution. Given a scattered compact space Y let $I(Y)$ be the (dense) set of isolated points of Y. Now, if X is a compact scattered space let $X_0 = X$; if α is an ordinal and we have a closed set $X_\alpha \subset X$ let $X_{\alpha+1} = X_\alpha \backslash I(X_\alpha)$. If α is a limit ordinal and we have a closed set $X_\beta \subset X$ for all $\beta < \alpha$ then let $X_\alpha = \bigcap \{X_\beta : \beta < \alpha\}$.

It is immediate that if α is a limit ordinal and $X_\beta \neq \emptyset$ for all $\beta < \alpha$ then $X_\alpha \neq \emptyset$. Besides, if $X_\alpha \neq \emptyset$ then $X_{\alpha+1} \neq X_\alpha$ because X_α is a scattered space. Our construction cannot last $|X|^+$ steps; so there is an ordinal $\alpha < |X|^+$ such that $X_{\alpha+1} = \emptyset$; let $\alpha(X)$ be the minimal such α. If $\alpha = \alpha(X)$ then X_α is discrete and hence finite.

Let us prove, by induction on $\alpha(X)$, that any scattered compact hereditarily metacompact space X is strong Eberlein compact. If $\alpha(X) = 0$ then X is finite and hence strong Eberlein compact by Problem 368. Now assume that X is a scattered compact hereditarily metacompact space such that $\alpha(X) = \mu$ and we have proved that any scattered compact hereditarily metacompact space Y with $\alpha(Y) < \mu$, is strong Eberlein compact.

The set X_μ is finite and $X \backslash X_\mu$ is metacompact; so there is a point-finite open cover $\mathcal{U}$ of $X \backslash X_\mu$ such that $\overline{U} \subset X \backslash X_\mu$ for any $U \in \mathcal{U}$. Applying Fact 1 of U.188 to the cover $\mathcal{U}$ we can find a compact set $F_U \subset U$ for any $U \in \mathcal{U}$ such that $\{F_U : U \in \mathcal{U}\}$ is still a cover of $X \backslash X_\mu$. The space X is strongly zero-dimensional (see SFFS-129, SFFS-305 and SFFS-306); so we can find a clopen set $O_U \subset X$ such that $F_U \subset O_U \subset U$ for any $U \in \mathcal{U}$.

It is clear that $\{O_U : U \in \mathcal{U}\}$ is a point-finite cover of $X \backslash X_\mu$ by compact open subsets of X. Every O_U is a compact scattered hereditarily metacompact space such that $\alpha(O_U) < \mu$; so the induction hypothesis is applicable to see that O_U is strong Eberlein compact and hence there exists a point-finite clopen cover $\mathcal{G}_U$ of the space O_U which T_0-separates the points of O_U (see Problem 369). Represent X_μ as $\{x_0, \ldots, x_n\}$ and fix pairwise disjoint clopen sets $W_0, \ldots, W_n$ such that $x_i \in W_i$ for all $i \leq n$; let $\mathcal{H} = \{W_0, \ldots, W_n\}$. It is easy to check that the family $\mathcal{H} \cup (\bigcup \{\mathcal{G}_U : U \in \mathcal{U}\})$ is a point-finite cover of X which T_0-separates the points of X and consists of clopen subsets of X; so X is strong Eberlein compact by Problem 369. This completes our inductive proof and shows that any compact scattered hereditarily metacompact space is strong Eberlein compact.

U.373. *Prove that the following conditions are equivalent for any compact X:*

(i) X is σ-discrete and Corson compact;
(ii) X is scattered and Corson compact;
(iii) X is strong Eberlein compact.

Solution. Any strong Eberlein compact space is σ-discrete by Problem 371; besides, any Eberlein compact space is Corson compact; so (iii)$\Longrightarrow$(i). Any σ-discrete compact space is scattered by Problem 370; so (i)$\Longrightarrow$(ii). Now, assume that X is a scattered Corson compact space and fix any $Y \subset X$.

If $\mathcal{U} \subset \tau(Y)$ and $Y = \bigcup \mathcal{U}$ then choose $O_U \in \tau(X)$ such that $O_U \cap Y = U$ for any $U \in \mathcal{U}$. Then $O = \bigcup \{O_U : U \in \mathcal{U}\}$ is an open subset of X such that $Y \subset O$. It follows from Problem 184 that $F = X \backslash O$ is a W-set in X; apply Problem 186 to conclude that O is metacompact and hence the open cover $\{O_U : U \in \mathcal{U}\}$ of the space O has a point-finite refinement $\mathcal{V}$. It is evident that $\mathcal{U}' = \{V \cap Y : V \in \mathcal{V}\}$ is a point-finite open (in Y) refinement of the cover $\mathcal{U}$ of the space Y. Therefore Y is metacompact for any $Y \subset X$, i.e., X is hereditarily metacompact. Therefore Problem 372 is applicable to see that X is strong Eberlein compact. This proves that (ii)$\Longrightarrow$(iii) and completes our solution.

U.374. *Prove that any continuous image of a strong Eberlein compact space is a strong Eberlein compact space.*

Solution. It follows from SFFS-129 and SFFS-133 that a compact space is scattered if and only if it cannot be continuously mapped onto $\mathbb{I}$. An immediate consequence is that a continuous image of any scattered compact space is scattered. Now, assume that X is a strong Eberlein compact space and $f : X \to Y$ is a continuous onto map. By the observation above, the space Y is compact and scattered. Besides, X is Corson compact; so Y is Corson compact as well by Problem 151. Thus Y is a scattered Corson compact space; so we can apply Problem 373 to conclude that Y is strong Eberlein compact.

U.375. *Prove that any Eberlein compact space is a continuous image of a closed subset of a countable product of strong Eberlein compact spaces.*

Solution. If A is a set then $\sigma_0(A) = \{x \in \mathbb{D}^A : |x^{-1}(1)| < \omega\} \subset \mathbb{D}^A$. Let X be an Eberlein compact space. By Problem 336, there exists a zero-dimensional Eberlein compact space Y such that X is a continuous image of Y. By Problem 334, the space Y has a T_0-separating σ-point-finite family $\mathcal{U}$ which consists of clopen subsets of Y. Fix a sequence $\{\mathcal{U}_n : n \in \omega\}$ of subfamilies of $\mathcal{U}$ such that $\mathcal{U} = \bigcup_{n \in \omega} \mathcal{U}_n$ and every $\mathcal{U}_n$ is point-finite. It is easy to see that we can assume, without loss of generality, that $\mathcal{U}_n \cap \mathcal{U}_m = \emptyset$ if $n \neq m$.

Given any $U \in \mathcal{U}$ let $\chi_U : Y \to \mathbb{D}$ be the characteristic function of U, i.e., $\chi_U(x) = 1$ for all $x \in U$ and $\chi_U(x) = 0$ whenever $x \in Y \backslash U$. Then the map χ_U is continuous for any $U \in \mathcal{U}$. Therefore the diagonal product $\varphi_n = \Delta\{\chi_U : U \in \mathcal{U}_n\}$

is continuous for any $n \in \omega$. Let $Y_n = \varphi_n(Y) \subset \mathbb{D}^{\mathcal{U}_n}$; given $y \in Y_n$ fix $x \in Y$ with $\varphi_n(x) = y$ and observe that $y(U) \neq 0$ if and only if $x \in U$. The family $\mathcal{U}_n$ is point-finite and hence $\{U \in \mathcal{U}_n : y(U) \neq 0\}$ is finite, i.e., $y \in \sigma_0(\mathcal{U}_n)$. The point $y \in Y$ was taken arbitrarily; so we proved that $Y_n \subset \sigma_0(\mathcal{U}_n)$ and hence Y_n is strong Eberlein compact for any $n \in \omega$.

The map $\varphi = \Delta_{n\in\omega}\varphi_n : Y \to \prod_{n\in\omega} Y_n$ is an embedding: this follows from the fact that $\mathcal{U}$ is T_0-separating and hence the family $\{\chi_U : U \in \mathcal{U}\}$ separates the points of Y. Thus Y is a compact space which maps continuously onto X while being homeomorphic to a closed subset of a countable product of strong Eberlein compact spaces.

U.376. *Let X be a strong Eberlein compact space. Prove that the Alexandroff double of X is also strong Eberlein compact.*

Solution. Recall that the underlying set of the Alexandroff double $AD(X)$ of the space X is the set $X \times \mathbb{D}$ and the topology of $AD(X)$ is defined by declaring all points of $X_1 = X \times \{1\}$ isolated whereas a local base at a point $z = (x, 0)$ of the set $X_0 = X \times \{0\}$ is formed by the sets $(U \times \mathbb{D})\backslash\{(x, 1)\}$ where the set U runs over all open neighbourhoods of the point x in X.

The space X being strong Eberlein compact it has a point-finite T_0-separating family $\mathcal{U}$ which consists of clopen subsets of X (see Problem 369). It is straightforward that the family $\mathcal{V} = \{U \times \mathbb{D} : U \in \mathcal{U}\} \cup \{\{z\} : z \in X_1\}$ is point-finite and consists of clopen subsets of $AD(X)$. It is also easy to see that $\mathcal{V}$ is T_0-separating in $AD(X)$; so we can apply Problem 369 once more to conclude that the space $AD(X)$ is strong Eberlein compact.

U.377. *Suppose that X_t is strong Eberlein compact for each $t \in T$. Prove that the Alexandroff one-point compactification of the space $\bigoplus\{X_t : t \in T\}$ is also strong Eberlein compact.*

Solution. Let $X = \bigoplus_{t\in T} X_t$; we will identify every X_t with the respective clopen subspace of X. Take a point $a \notin X$ and let $Y = X \cup \{a\}$. Define a topology μ on the set Y as follows: a set $U \subset X$ belongs to μ if and only if $U \in \tau(X)$; a set $U \ni a$ belongs to μ if and only if $Y\backslash U$ is a compact subspace of X. The space $Y = (Y, \mu)$ is homeomorphic to the Alexandroff compactification of X.

Since X_t is strong Eberlein compact, we can apply Problem 369 to fix a point-finite cover $\mathcal{U}_t$ of the space X_t which T_0-separates the points of X_t and consists of clopen subsets of X_t for any $t \in T$. It is evident that the elements of every $\mathcal{U}_t$ are also clopen in Y. It is easy to see that the family $\mathcal{U} = \bigcup\{\mathcal{U}_t : t \in T\} \cup \{Y\}$ is a point-finite clopen cover of the space Y which T_0-separates the points of Y. Applying Problem 369 once more we conclude that the space Y is strong Eberlein compact.

U.378. *Observe that any uniform Eberlein compact is Eberlein compact. Prove that any metrizable compact space is uniform Eberlein compact.*

Solution. If X is uniform Eberlein compact then X embeds in $\Sigma_*(A)$ for some A; so X is Eberlein compact by Problem 322. Now assume that X is a metrizable compact space; if $I = [0,1] \subset \mathbb{R}$ then X is embeddable in I^ω. Let $I_n = [0, \frac{1}{n+1}]$ for any $n \in \omega$; then the space $Q = \prod_{n \in \omega} I_n \subset I^\omega \subset \mathbb{R}^\omega$ is homeomorphic to I^ω so we can consider that $X \subset Q$. Given any point $x \in Q$ we have $|x(n)| = x(n) \leq \frac{1}{n+1}$ for any $n \in \omega$; if $m \in \omega$ and $\frac{1}{m+1} < \varepsilon$ then the set $\{n \in \omega : |x(n)| \geq \varepsilon\} \subset \{0, \ldots, m\}$ is finite. This proves that $Q \subset \Sigma_*(\omega)$ and hence $X \subset \Sigma_*(\omega)$ as well.

For any $\varepsilon > 0$ fix $N(\varepsilon) \in \mathbb{N}$ such that $\frac{1}{N(\varepsilon)} < \varepsilon$. Given any point $x \in X$, if $|x(n)| = x(n) \geq \varepsilon$ then $n < N(\varepsilon)$; so the set $A(x, \varepsilon) = \{n \in \omega : |x(n)| \geq \varepsilon\}$ is contained in $\{0, \ldots, N(\varepsilon) - 1\}$ which shows that $|A(x, \varepsilon)| \leq N(\varepsilon)$ for any $x \in X$ and $\varepsilon > 0$. Therefore our embedding of X in $\Sigma_*(\omega)$ witnesses that X is a uniform Eberlein compact space.

U.379. *Observe that any closed subspace of a uniform Eberlein compact space is uniform Eberlein compact. Prove that any countable product of uniform Eberlein compact spaces is uniform Eberlein compact.*

Solution. Suppose that X is a uniform Eberlein compact space and F is a closed subspace of X. We can consider that $X \subset \Sigma_*(A)$ for some set A and there is a function $N : (0, +\infty) \to \mathbb{N}$ such that $|\{a \in A : |x(a)| \geq \varepsilon\}| \leq N(\varepsilon)$ for any $x \in X$. The embedding of F in X is at the same time an embedding of F in $\Sigma_*(A)$ and it is evident that this embedding witnesses that F is also uniform Eberlein compact. Thus every closed subspace of a uniform Eberlein compact space is uniform Eberlein compact.

Given a family $\mathcal{A}$ of subsets of a space X let $\mathrm{ord}(\mathcal{A}, x) = |\{A \in \mathcal{A} : x \in A\}|$ for any $x \in X$.

Fact 1. A compact space X is uniform Eberlein compact if and only if there is a T_0-separating family $\mathcal{U}$ of cozero subsets of X such that $\mathcal{U} = \bigcup_{n \in \omega} \mathcal{U}_n$ and, for each $n \in \omega$, there is $m(n) \in \omega$ for which $\mathrm{ord}(\mathcal{U}_n, x) \leq m(n)$ for any $x \in X$.

Proof. Suppose that X is uniform Eberlein compact and hence we can consider that there is a set A such that $X \subset \Sigma_*(A)$ and there exists a function $N : (0, +\infty) \to \mathbb{N}$ for which $|\{a \in A : |x(a)| \geq \varepsilon\}| \leq N(\varepsilon)$ for any $x \in X$. For any $a \in A$ let $\pi_a : \Sigma_*(A) \to \mathbb{R}$ be the restriction to $\Sigma_*(A)$ of the natural projection of $\mathbb{R}^A$ onto the factor determined by a.

For any $q \in Q = \mathbb{Q} \setminus \{0\}$ let $O_q = (q, +\infty)$ if $q > 0$ and $O_q = (-\infty, q)$ if $q < 0$. It is clear that $U(a, q) = \pi_a^{-1}(O_q) \cap X$ is a cozero subset of X for any $q \in Q$ and $a \in A$. Besides, the family $\mathcal{U} = \{U(a, q) : a \in A, q \in Q\}$ is T_0-separating in X. Given $q \in Q$, let $V_q = \{U(a, q) : a \in A\}$; for any $x \in X$ the set $\{a \in A : |x(a)| \geq |q|\}$ has at most $N(|q|)$ elements which implies that $|\{a \in A : x \in U(a, q)\}| \leq N(|q|)$ for any $x \in X$. Thus $\mathrm{ord}(V_q, x) \leq N(|q|)$ for any $x \in X$ and $q \in Q$.

If $\{q_n : n \in \omega\}$ is an enumeration of the set Q let $m(n) = N(|q_n|)$ and $\mathcal{U}_n = V_{q_n}$ for each $n \in \omega$. Then $\mathcal{U} = \bigcup_{n \in \omega} \mathcal{U}_n$ and $\mathrm{ord}(\mathcal{U}_n, x) \leq m(n)$ for any $x \in X$ and $n \in \omega$. This proves necessity.

For sufficiency assume that X is a compact space and there exists a T_0-separating family $\mathcal{U}$ of cozero subsets of X such that $\mathcal{U} = \bigcup_{n \in \omega} \mathcal{U}_n$ and, for any $n \in \omega$, there is $m(n) \in \omega$ such that $\mathrm{ord}(\mathcal{U}_n, x) \leq m(n)$ for any $x \in X$. For any $n \in \omega$ and $U \in \mathcal{U}_n$ fix a continuous function $f_U : X \to [0, \frac{1}{n+1}]$ such that $U = (f_U)^{-1}((0, \frac{1}{n+1}])$.

It is clear that the map $\varphi = \Delta\{f_U : U \in \mathcal{U}\} : X \to \mathbb{R}^{\mathcal{U}}$ is continuous; the family $\mathcal{U}$ being T_0-separating, the set $\{f_U : U \in \mathcal{U}\}$ separates the points of the space X so φ is an embedding. Given any $\varepsilon > 0$ there exists a number $k \in \omega$ such that $\frac{1}{k+1} < \varepsilon$; let $N(\varepsilon) = m(0) + \ldots + m(k)$. For any point $x \in X$ and $n > k$ we have $|f_U(x)| = f_U(x) \leq \frac{1}{n+1} \leq \frac{1}{k+1} < \varepsilon$ for any $U \in \mathcal{U}_n$; so the set $\{U \in \mathcal{U} : |\varphi(x)(U)| \geq \varepsilon\}$ is contained in the finite set $V = \{U \in \bigcup_{n \leq k} \mathcal{U}_n : x \in U\}$. It is immediate that we have the inequality $|V| \leq m(0) + \ldots + m(k) = N(\varepsilon)$; so $|\{U \in \mathcal{U} : |\varphi(x)(U)| \geq \varepsilon\}| \leq N(\varepsilon)$ for any point $x \in X$ which shows that the map φ embeds X in $\Sigma_*(\mathcal{U})$ and the function $N : (0, +\infty) \to \omega$ together with this embedding witness that X is uniform Eberlein compact. This settles sufficiency and shows that Fact 1 is proved.

Returning to our solution assume that X_n is uniform Eberlein compact for any $n \in \omega$; we must prove that $X = \prod_{n \in \omega} X_n$ is also uniform Eberlein compact. Let $\pi_n : X \to X_n$ be the natural projection for every $n \in \omega$. By Fact 1, we can fix, for every $n \in \omega$, a family $\mathcal{U}_n$ of cozero subsets of X_n which T_0-separates the points of X_n and $\mathcal{U}_n = \bigcup_{m \in \omega} \mathcal{U}_{mn}$ where, for each $m \in \omega$ there is $k(n, m) \in \omega$ for which $\mathrm{ord}(\mathcal{U}_{nm}, x) \leq k(n, m)$ for any $x \in X_n$.

It is straightforward that the family $W = \{\pi_n^{-1}(U) : n \in \omega, U \in \mathcal{U}_n\}$ consists of cozero subsets of X and T_0-separates the points of X. If $V_{nm} = \{\pi_n^{-1}(U) : U \in \mathcal{U}_{nm}\}$ then $\mathrm{ord}(V_{nm}, x) \leq k(n, m)$ for any $x \in X$ and $m, n \in \omega$. Choose a surjective map $\nu : \omega \to \omega \times \omega$. For any $i \in \omega$ we have $\nu(i) = (n, m) \in \omega \times \omega$; let $l(i) = k(n, m)$ and $W_i = V_{nm}$. It is evident that $W = \bigcup_{i \in \omega} W_i$. If $i \in \omega$ and $\nu(i) = (n, m)$ then $\mathrm{ord}(W_i, x) = \mathrm{ord}(V_{nm}, x) \leq k(n, m) = l(i)$ for any $x \in X$; so Fact 1 shows that the family W and the collection $\{W_i : i \in \omega\}$ witness that X is uniform Eberlein compact. This proves that any countable product of uniform Eberlein compact spaces is uniform Eberlein compact and makes our solution complete.

U.380. *Prove that if X is a uniform Eberlein compact space then it is a continuous image of a closed subspace of $(A(\kappa))^{\omega}$ for some infinite cardinal κ.*

Solution. If Z is a space and $\mathcal{U} \subset \exp Z$ then $\mathrm{ord}(\mathcal{U}, z) = |\{U \in \mathcal{U} : z \in U\}|$ for any $z \in Z$. Say that a family $\mathcal{U} \subset \exp Z$ is *of bounded order* if there is $n \in \omega$ such that $\mathrm{ord}(\mathcal{U}, z) \leq n$ for any $z \in Z$.

Denote by $\mathcal{A}$ the class of spaces which can be represented as a continuous image of a closed subspace of $(A(\kappa))^{\omega}$ for some infinite cardinal κ. Recall that $A(\kappa) = \kappa \cup \{a\}$ where $a \notin \kappa$ is the unique non-isolated point of $A(\kappa)$. Therefore we can consider that $\lambda < \kappa$ implies that $A(\lambda) \subset A(\kappa)$ and hence any closed subspace of $(A(\lambda))^{\omega}$ is also a closed subspace of $(A(\kappa))^{\omega}$. Now it is easy to see that

(1) if $Y \in \mathcal{A}$ and F is closed in Y then $F \in \mathcal{A}$;
(2) if $Y \in \mathcal{A}$ then any continuous image of Y also belongs to $\mathcal{A}$;
(3) if $Y_n \in \mathcal{A}$ for any $n \in \omega$ then $\prod_{n \in \omega} Y_n \in \mathcal{A}$.

Given a set B let $\sigma[B] = \{x \in \mathbb{D}^B : |x^{-1}(1)| < \omega\}$; we will also need the set $\sigma_n[B] = \{x \in \mathbb{D}^B : |x^{-1}(1)| \leq n\}$ for any $n \in \omega$. It is easy to see that $\sigma_n[B]$ is a compact subset of $\sigma[B]$ for any $n \in \omega$.

Fact 1. Given an infinite cardinal κ and a set B with $|B| = \kappa$, the space $\sigma_n[B]$ is a continuous image of $(A(\kappa))^n$ (and hence $\sigma_n[B] \in \mathcal{A}$) for any $n \in \mathbb{N}$.

Proof. For every $b \in B$ let $z_b(b) = 1$ and $z_b(c) = 0$ for any $c \in B \backslash \{b\}$; denote by u the function which is identically zero on B. It is an easy exercise that the set $K = \{u\} \cup \{z_b : b \in B\} = \sigma_1[B]$ is homeomorphic to $A(\kappa)$ and u is the unique non-isolated point of K. For any $x, y \in \sigma[B]$ let $x * y = x + y - xy$; then $x * y \in \sigma[B]$, the operation $*$ is associative, commutative and the map $s_n : (\sigma[B])^n \to \sigma[B]$ defined by $s_n(x_0, \ldots, x_{n-1}) = x_0 * \ldots * x_{n-1}$ for any $(x_0, \ldots, x_{n-1}) \in (\sigma[B])^n$, is continuous (see Fact 1 of U.150). It is clear that $(x_0 * \ldots * x_{n-1})^{-1}(1) = x_0^{-1}(1) \cup \ldots \cup x_{n-1}^{-1}(1)$ for any $x_0, \ldots, x_{n-1} \in \mathbb{D}^B$.

Thus $L = s_n(K^n)$ is a continuous image of K^n; if $x = (x_0, \ldots, x_{n-1}) \in K^n$ then $|x_i^{-1}(1)| \leq 1$ for every $i < n$; so $|(s_n(x))^{-1}(1)| \leq n$ which proves that $L \subset \sigma_n[B]$. Now, if $y \in \sigma_n[B]$ then $y^{-1}(1) = \{b_1, \ldots, b_k\}$ for some $k \leq n$ (if $y = u$ then $k = 0$); let $x = (z_{b_1}, \ldots, z_{b_k}, t_1, \ldots, t_{n-k})$ where $t_i = u$ for any $i \leq n - k$. It is straightforward that $s_n(x) = y$; the point $y \in \sigma_n[B]$ was taken arbitrarily, so we proved that $\sigma_n[B] = L = s_n(K^n)$ is a continuous image of $K^n \simeq (A(\kappa))^n$. It is clear that $(A(\kappa))^n \in \mathcal{A}$; so $\sigma_n[B] \in \mathcal{A}$ by (2) and hence Fact 1 is proved.

Fact 2. A zero-dimensional compact space K is uniform Eberlein compact if and only if there exists a T_0-separating family $\mathcal{U}$ of clopen subsets of K such that $\mathcal{U} = \bigcup_{n \in \omega} \mathcal{U}_n$ and every $\mathcal{U}_n$ is of bounded order.

Proof. Sufficiency follows from Fact 1 of U.379. Now, if K is uniform Eberlein compact then we can apply Fact 1 of U.379 once more to find a T_0-separating family $\mathcal{V}$ of cozero subsets of X such that $\mathcal{V} = \bigcup_{n \in \omega} \mathcal{V}_n$ and every $\mathcal{V}_n$ is of bounded order.

For each $V \in \mathcal{V}$ there is a family $\{F_V^n : n \in \omega\}$ of compact subspaces of K such that $V = \bigcup_{n \in \omega} F_V^n$. By zero-dimensionality of K there is a clopen $W_V^n \subset K$ such that $F_V^n \subset W_V^n \subset V$ for any $n \in \omega$. The family $\mathcal{W}_{i,n} = \{W_V^n : V \in \mathcal{V}_i\}$ consists of clopen subspaces of K and it is immediate that $\mathcal{W}_{i,n}$ is of bounded order for any $i, n \in \omega$. It is easy to check that the family $\mathcal{U} = \bigcup_{i,n \in \omega} \mathcal{W}_{i,n}$ is T_0-separating in K; if we enumerate the collection $\{\mathcal{W}_{i,n} : i, n \in \omega\}$ as $\{\mathcal{U}_n : n \in \omega\}$ then we obtain the desired decomposition of $\mathcal{U}$ into subfamilies of bounded order. Fact 2 is proved.

Fact 3. Any uniform Eberlein compact is a continuous image of a zero-dimensional uniform Eberlein compact space.

Proof. Let K be a uniform Eberlein compact space. By Fact 1 of U.379 the space K has a T_0-separating family $\mathcal{U}$ of open F_σ-subsets such that $\mathcal{U} = \bigcup \{\mathcal{U}_n : n \in \omega\}$ and every $\mathcal{U}_n$ is of bounded order. There is no loss of generality to assume that

$\mathcal{U}_m \cap \mathcal{U}_n = \emptyset$ whenever $m \neq n$. For every $U \in \mathcal{U}$ choose a continuous function $f_U : K \to I = [0, 1] \subset \mathbb{R}$ such that $K \backslash U = f_U^{-1}(0)$ (see Fact 1 of S.358 and Fact 1 of S.499).

For any $n \in \omega$ and $U \in \mathcal{U}_n$ let $g_U = \frac{1}{n+1} \cdot f_U$; then $\varphi_n = \Delta\{g_U : U \in \mathcal{U}_n\}$ maps K continuously into $I^{\mathcal{U}_n}$ in such a way that $\varphi_n(K) \subset [0, \frac{1}{n+1}]^{\mathcal{U}_n}$. It turns out that the map $\varphi = \Delta\{\varphi_n : n \in \omega\} : K \to \prod\{I^{\mathcal{U}_n} : n \in \omega\} = I^{\mathcal{U}}$ is an embedding. To see it take distinct $x, y \in K$. There is $n \in \omega$ and $U \in \mathcal{U}_n$ such that $U \cap \{x, y\}$ is a singleton. Then $f_U(x) \neq f_U(y)$ and hence $g_U(x) \neq g_U(y)$ which shows that $\varphi_n(x) \neq \varphi_n(y)$ and therefore $\varphi(x) \neq \varphi(y)$. This proves that φ is, indeed, an embedding being an injection; let $K' = \varphi(K)$.

For any $\varepsilon > 0$ there is $m = m(\varepsilon) \in \omega$ such that $|\{U \in \mathcal{U} : |x(U)| \geq \varepsilon\}| \leq m$ for every $x \in K'$. Indeed, fix $n \in \omega$ such that $\frac{1}{n+1} < \varepsilon$. The family $\mathcal{V} = \bigcup\{\mathcal{U}_k : k < n\}$ is of bounded order; so there is $m \in \omega$ such that $\text{ord}(\mathcal{V}, y) \leq m$ for any $y \in K$. Given $x \in K'$, there is $y \in K$ with $\varphi(y) = x$. We have $|x(U)| = g_U(y) \leq \frac{1}{k+1} \leq \frac{1}{n+1} < \varepsilon$ for any $k \geq n$ and $U \in \mathcal{U}_k$. Therefore $|\{U \in \mathcal{U} : |x(U)| \geq \varepsilon\}| \leq \text{ord}(\mathcal{V}, y) \leq m$.

To simplify the notation we will reformulate the obtained result as follows:

(4) for some set T the space K embeds in I^T in such a way that, for any $\varepsilon > 0$ there exists $m(\varepsilon) \in \omega$ for which $|\{t : x(t) \geq \varepsilon\}| \leq m(\varepsilon)$ for any $x \in K$.

Let $C = \mathbb{D}^\omega$ be the Cantor set; fix a point $d \in C$ and let $I_n = [\frac{1}{n+2}, \frac{1}{n+1}] \subset I$ for any $n \in \omega$. The space C being zero-dimensional, there exists a local base $\mathcal{O} = \{O_n : n \in \omega\}$ at the point d in C such that the set O_n is clopen in C and $O_{n+1} \subset O_n$ for any $n \in \omega$. Making the relevant changes in $\mathcal{O}$ if necessary, we can assume that $O_0 = C$ and $C_n = O_n \backslash O_{n+1} \neq \emptyset$ for any $n \in \omega$.

Since no point of C is isolated, the same is true for any non-empty clopen subset of C and hence every non-empty clopen subset of C is homeomorphic to C (see SFFS-348). This shows that C_n is homeomorphic to C and hence there is a continuous onto map $\xi_n : C_n \to I_n$ for any $n \in \omega$ (see TFS-128). Let $\xi(d) = 0$; if $x \in C \backslash \{d\}$ then there is a unique $n \in \omega$ such that $x \in C_n$; let $\xi(x) = \xi_n(x)$. It is an easy exercise that $\xi : C \to I$ is a continuous onto map such that $\xi^{-1}(0) = \{d\}$. We proved that

(5) for any $d \in \mathbb{D}^\omega$ there is a continuous onto map $\xi : \mathbb{D}^\omega \to I$ with $\xi^{-1}(0) = \{d\}$.

From now on we will consider that $C \subset I$ and $0 \in C$ (see TFS-128); apply (5) to fix a continuous onto map $\xi : C \to I$ such that $\xi^{-1}(0) = \{0\}$. Given any $\varepsilon > 0$, the set $L = C \backslash [0, \varepsilon)$ is compact so $\xi(L)$ is compact in I; since $0 \notin \xi(L)$, there is $r(\varepsilon) > 0$ such that $[0, r(\varepsilon)) \cap \xi(L) = \emptyset$. This proves that

(6) for any $\varepsilon > 0$ there is $r(\varepsilon) > 0$ such that $a \in C$ and $a \geq \varepsilon$ implies $\xi(a) \geq r(\varepsilon)$.

Let $\Phi : C^T \to I^T$ be the product of T-many copies of ξ, i.e., $\Phi(x)(t) = \xi(x(t))$ for any $x \in C^T$ and $t \in T$. The map Φ is continuous by Fact 1 of S.271 and it is easy to see that $\Phi(C^T) = I^T$. The space $Y = \Phi^{-1}(K)$ is compact being closed in the compact space $C^T \subset I^T \subset \mathbb{R}^T$.

Take any $x \in K$ and $y \in \Phi^{-1}(x)$. For any $t \in T$, if $\varepsilon > 0$ and $y(t) \geq \varepsilon$ then $x(t) = \xi(y(t)) \geq r(\varepsilon)$ by (6); so the set $S(y, \varepsilon) = \{t \in T : y(t) \geq \varepsilon\}$ is contained in $\mathrm{supp}(x, \varepsilon) = \{t \in T : x(t) \geq r(\varepsilon)\}$. Now it follows from (4) that $|S(y, \varepsilon)| \leq |\mathrm{supp}(x, \varepsilon)| \leq m(r(\varepsilon))$ for any $\varepsilon > 0$. This proves that Y is contained in $\Sigma_*(T)$ in such a way that, for any $\varepsilon > 0$ the set $S(y, \varepsilon)$ has cardinality at most $m(r(\varepsilon))$; so Y is a uniform Eberlein compact space. It is clear that $\Phi|Y$ maps Y continuously onto K. Besides, Y is zero-dimensional because so is $C^T \supset Y$ (see SFFS-301 and SFFS-302). Therefore Y is a zero-dimensional uniform Eberlein compact space which maps continuously onto K, i.e., Fact 3 is proved.

Returning to our solution take an arbitrary uniform Eberlein compact space X. By Fact 3, there is a zero-dimensional uniform Eberlein compact space Y which maps continuously onto X. Apply Fact 2 to find a T_0-separating family $\mathcal{U}$ of clopen subsets of Y such that $\mathcal{U} = \bigcup_{n \in \omega} \mathcal{U}_n$ and every $\mathcal{U}_n$ is of bounded order; fix $m(n) \in \omega$ such that $\mathrm{ord}(\mathcal{U}_n, y) \leq m(n)$ for any $y \in Y$.

For any $U \in \mathcal{U}$ let χ_U be the characteristic function of U, i.e., $\chi_U(x) = 1$ for each $x \in U$ and $\chi_U(Y \backslash U) \subset \{0\}$. Let $\varphi_n = \Delta\{\chi_U : U \in \mathcal{U}_n\}$; it is easy to check that $Y_n = \varphi_n(Y) \subset \sigma_{m(n)}[\mathcal{U}_n]$ for any $n \in \omega$. The family $\mathcal{U}$ is T_0-separating; so the map $\varphi = \Delta_{n \in \omega} \varphi_n : Y \to \prod_{n \in \omega} Y_n$ is an embedding. It follows from (1) and Fact 1 that $Y_n \in \mathcal{A}$ for every $n \in \omega$. The properties (1) and (3) show that $Y \in \mathcal{A}$; applying (2) we conclude that also $X \in \mathcal{A}$, i.e., X is a continuous image of a closed subspace of $(A(\kappa))^\omega$ for some κ and hence our solution is complete.

U.381. *Prove that any continuous image of a uniform Eberlein compact space is uniform Eberlein compact. Deduce from this fact that a space X is uniform Eberlein compact if and only if it is a continuous image of a closed subspace of $(A(\kappa))^\omega$ for some infinite cardinal κ.*

Solution. If Z is a space and $\mathcal{A} \subset \exp Z$ then $\mathrm{ord}(\mathcal{A}, z) = |\{A \in \mathcal{A} : z \in A\}|$ for any $z \in Z$.

Suppose that X is a uniform Eberlein compact space and $f : X \to Y$ is a continuous onto map. If X is metrizable then Y is also metrizable and hence it is uniform Eberlein compact (see Problem 378 and Fact 5 of S.307). Thus we can assume that X is not metrizable and hence $|X| > \omega$. For any set $A \subset X$ let $f^\#(A) = Y \backslash f(X \backslash A)$. It is easy to see that if U is an open subset of X and $V = f^\#(U)$ then V is open in Y and $f^{-1}(V) \subset U$. It was proved in Fact 3 of U.380 that there exists a zero-dimensional Eberlein compact space X' such that X is a continuous image of X'. Then Y is a continuous image of X'; so we can assume, without loss of generality, that $X' = X$, i.e., X is a zero-dimensional uniform Eberlein compact space.

Our first observation is that $C_p(Y)$ embeds in $C_p(X)$ as a closed subspace (see TFS-163); thus $C_p(Y)$ is a Lindelöf Σ-space because so is $C_p(X)$ (see Problem 020). Therefore Y is Gul'ko compact, so it is ω-monolithic and Fréchet–Urysohn (see Problem 208, and Fact 1 of U.080).

Apply Fact 2 of U.380 to find a T_0-separating family $\mathcal{U}$ of clopen subsets of X such that $\mathcal{U} = \bigcup\{\mathcal{U}_n : n \in \omega\}$ and every $\mathcal{U}_n$ is of *bounded order*, i.e., there exists

$m(n) \in \omega$ such that $\mathrm{ord}(\mathcal{U}_n, x) \leq m(n)$ for every $x \in X$. Given a family $\mathcal{A}$ of subsets of X we consider that $\bigcap \mathcal{A} = X$ if $\mathcal{A}$ is empty.

Fact 1. For any $n, l \in \omega$ there exists a number $g(n, l) \in \omega$ such that, for any family $\mathcal{N} = \{N_t : t \in T\}$ with $|N_t| \leq n$ for any $t \in T$, if $|T| \geq g(n, l)$ then there is $S \subset T$ such that $|S| \geq l$ and $\{N_t : t \in S\}$ is a Δ-system, i.e., there is a set D such that $N_s \cap N_t = D$ for any distinct $s, t \in S$.

Proof. If $n = 0$ then $D = \emptyset$ and $g(0, l) = l$ work for any $l \in \omega$. Proceeding inductively, assume that $m \in \omega$ and we have constructed $g(n, l)$ for all $n \leq m$ and $l \in \omega$. Let $g(m + 1, l) = (m + 1)lg(m, l) + 1$ for any $l \in \omega$ and fix a family $\mathcal{N} = \{N_t : t \in T\}$ such that $|T| \geq g(m + 1, l)$ and $|N_t| \leq m + 1$ for all $t \in T$. Here, two cases are possible.

(a) There is a set $S' \subset T$ such that $|S'| \geq g(m, l)$ and $M = \bigcap \{N_t : t \in S'\} \neq \emptyset$; choose a point $x \in M$. Then every element of the family $\mathcal{M} = \{N_t \setminus \{x\} : t \in S'\}$ has at most m elements; so our inductive assumption is applicable to find $S \subset S'$ and a set D' such that $|S| \geq l$ and $(N_t \setminus \{x\}) \cap (N_s \setminus \{x\}) = D'$ for any distinct $s, t \in S$. It is straightforward that, letting $D = D' \cup \{x\}$ we will have the equality $N_t \cap N_s = D$ for any distinct $t, s \in S$. Therefore $\{N_t : t \in S\}$ is a Δ-system of cardinality at least l.

(b) For any $S' \subset T$ with $|S'| \geq g(m, l)$ we have $\bigcap \{N_t : t \in S'\} = \emptyset$. Then every point of $\bigcup \mathcal{N}$ belongs to at most $(g(m, l) - 1)$-many elements of $\mathcal{N}$ and therefore every N_t meets at most $(m + 1)(g(m, l) - 1)$-many elements of $\mathcal{N}$. Take an arbitrary $t_1 \in T$; then there are at most $(m + 1)(g(m, l) - 1)$-many elements of $\mathcal{N}$ which meet the set N_{t_1}. Proceeding by induction, suppose that $k < l$ and we have disjoint sets $N_{t_1}, \ldots, N_{t_k} \in \mathcal{N}$; then there are at most $k(m + 1)(g(m, l) - 1)$-many elements of $\mathcal{N}$ which meet some N_{t_i}; since $|T| > k(m + 1)(g(m, l) - 1)$, there is $t_{k+1} \in T$ such that the sets $N_{t_0}, \ldots, N_{t_k}, N_{t_{k+1}}$ are still disjoint. Therefore we can construct a disjoint family $\mathcal{M} = \{N_{t_1}, \ldots, N_{t_l}\}$ which is a Δ-system of cardinality l for $D = \emptyset$.

This settles the case of $n = m + 1$ and hence our inductive procedure shows that Fact 1 is proved.

Returning to our solution denote by $\mathbb{F}$ the collection of all finite subfamilies of the family $\mathcal{U}$. We will also need the families $\mathcal{U}[x] = \{U \in \mathcal{U} : x \in U\}$ and $\mathcal{U}[x, n] = \{U \in \bigcup \{\mathcal{U}_k : k \leq n\} : x \in U\}$ for any $x \in X$ and $n \in \omega$. For any $\mathcal{A} \in \mathbb{F}$ the set $U(\mathcal{A}) = f^\#(\bigcup \mathcal{A})$ is open in Y; so $F(\mathcal{A}) = U(\mathcal{A}) \setminus (\bigcup \{U(\mathcal{A}') : \mathcal{A}' \subset \mathcal{A}$ and $\mathcal{A}' \neq \mathcal{A}\})$ is closed in $U(\mathcal{A})$. We claim that the collection $\{F(\mathcal{A}), U(\mathcal{A}) : \mathcal{A} \in \mathbb{F}\}$ separates the points of Y in the sense that

(1) for any distinct points $x, y \in Y$ there exists a family $\mathcal{A} \in \mathbb{F}$ such that either $x \in F(\mathcal{A}) \subset U(\mathcal{A}) \subset Y \setminus \{y\}$ or $y \in F(\mathcal{A}) \subset U(\mathcal{A}) \subset Y \setminus \{x\}$.

To prove (1) we will first establish that

(2) either there exists a point $a \in f^{-1}(x)$ such that $\mathcal{U}[b] \not\subset \mathcal{U}[a]$ for any $b \in f^{-1}(y)$ or there is $a \in f^{-1}(y)$ such that $\mathcal{U}[b] \not\subset \mathcal{U}[a]$ for any $b \in f^{-1}(x)$.

Indeed, if (2) is not true then we can find sequences $A = \{a_n : n \in \omega\} \subset f^{-1}(x)$ and $B = \{b_n : n \in \omega\} \subset f^{-1}(y)$ such that $\mathcal{U}[a_n] \supset \mathcal{U}[b_n] \supset \mathcal{U}[a_{n+1}]$ for any $n \in \omega$. This property will still hold if a we pass to any subsequences $\{a_{k_n} : n \in \omega\}$ and $\{b_{k_n} : n \in \omega\}$ of our sequences A and B. Since Y is a Fréchet–Urysohn space, we can assume, without loss of generality, that there are points $a \in f^{-1}(x)$ and $b \in f^{-1}(y)$ such that $A \to a$ and $B \to b$.

The family $\mathcal{U}$ being T_0-separating we can pick a set $U \in \mathcal{U}$ such that $a \in U$ and $b \notin U$ or vice versa. If we have the first case then the sequence A is eventually in U and B is eventually in $X \backslash U$ (here we use the fact that U is a clopen set). Therefore there is $n \in \omega$ for which $a_{n+1} \in U$ and $b_n \notin U$. This contradiction with $\mathcal{U}[a_{n+1}] \subset \mathcal{U}[b_n]$ shows that the first case is impossible. In the second case the contradiction is obtained in an analogous way; so (2) is proved.

Now apply (2) to find a point $a \in f^{-1}(x)$ such that $\mathcal{U}[b] \not\subset \mathcal{U}[a]$ for any $b \in f^{-1}(y)$ (or vice versa). Thus we can fix a set $U_b \in \mathcal{U}[b]$ such that $a \notin U_b$; the set $f^{-1}(y)$ being compact, there is a finite $B' \subset B$ for which $\bigcup \{U_b : b \in B'\} \supset f^{-1}(y)$. Since the family $\mathcal{U}' = \{U_b : b \in B'\}$ is finite, we can choose a minimal $\mathcal{A} \subset \mathcal{U}'$ for which $f^{-1}(y) \subset \bigcup \mathcal{A}$. Thus $y \in f^{\#}(\bigcup \mathcal{A}) = U(\mathcal{A})$ and it follows from minimality of $\mathcal{A}$ that $f^{-1}(y) \not\subset \bigcup \mathcal{A}'$ for any proper $\mathcal{A}' \subset \mathcal{A}$. As a consequence, $y \in F(\mathcal{A})$; it follows from $a \in f^{-1}(x) \backslash (\bigcup \mathcal{A})$ that $x \notin U(\mathcal{A})$ so the pair $(F(\mathcal{A}), U(\mathcal{A}))$ separates the points x and y. The case when $a \in f^{-1}(y)$ and $\mathcal{U}[b] \not\subset \mathcal{U}[a]$ for any $b \in f^{-1}(x)$ is considered analogously; so (1) is proved.

Let us show that the family $\mathcal{U}$ has another useful property.

(3) If F and G are disjoint closed subsets of the space X then there is $k \in \omega$ such that $\mathcal{U}[x, k] \neq \mathcal{U}[y, k]$ for any $x \in F$ and $y \in G$.

Indeed, if (3) is not true then there exist sequences $\{x_n : n \in \omega\} \subset F$ and $\{y_n : n \in \omega\} \subset G$ for which $\mathcal{U}[x_n, n] = \mathcal{U}[y_n, n]$ for any $n \in \omega$. It is easy to see that we can assume, without loss of generality, that there are points $x \in F$ and $y \in G$ such that $x_n \to x$ and $y_n \to y$. Fix $U \in \mathcal{U}$ such that $x \in U$ and $y \notin U$ (or vice versa). There is $k \in \omega$ with $U \in \mathcal{U}_k$; the sequence $\{x_n : n \in \omega\}$ is eventually in U while $\{y_n : n \in \omega\}$ is eventually outside U (here we used again the fact that U is a clopen set). Therefore there exists $n > k$ such that $x_n \in U$ and $y_n \notin U$. This, together with $U \in \mathcal{U}[x_n, n]$ implies that $\mathcal{U}[x_n, n] \neq \mathcal{U}[y_n, n]$ which is contradiction. In the symmetric case the contradiction is obtained analogously; so (3) is proved.

It follows from Fact 1 of U.337 that, for the cardinal $\kappa = |Y|$, we can construct, for any $\mathcal{A} \in \mathbb{F}$, a family $\{S(\alpha, n, \mathcal{A}) : \alpha < \kappa, n \in \omega\}$ of compact subsets of Y and a family $\{V(\alpha, n, \mathcal{A}), \alpha < \kappa, n \in \omega\}$ of open subsets of Y such that

(4) $\bigcup \{S(\alpha, n, \mathcal{A}) : \alpha < \kappa, n \in \omega\} = F(\mathcal{A})$ for any $\mathcal{A} \in \mathbb{F}$;
(5) $S(\alpha, n, \mathcal{A}) \subset V(\alpha, n, \mathcal{A}) \subset U(\mathcal{A})$ for any $\mathcal{A} \in \mathbb{F}$;
(6) $S(\alpha, n, \mathcal{A}) \cap V(\beta, m, \mathcal{A}) = \emptyset$ for any $\beta < \alpha$, $m, n \in \omega$ and $\mathcal{A} \in \mathbb{F}$.

Given an ordinal $\alpha < \kappa$ a family $\mathcal{A} \in \mathbb{F}$ and $n \in \omega$, the sets $f^{-1}(S(\alpha, n, \mathcal{A}))$ and $f^{-1}(Y \backslash V(\alpha, n, \mathcal{A}))$ are closed and disjoint in the space X; so it follows from the property (3) that there is $k = k(\alpha, n, \mathcal{A})$ such that the conditions $x \in f^{-1}(S(\alpha, n, \mathcal{A}))$ and $y \in f^{-1}(Y \backslash V(\alpha, n, \mathcal{A}))$ imply $\mathcal{U}[x, k] \neq \mathcal{U}[y, k]$. An immediate consequence is that

(7) given any $A \in \mathbb{F}$, $\alpha < \kappa$ and $n \in \omega$, if $k = k(\alpha, n, A)$ and $x \in f^{-1}(S(\alpha, n, A))$ then $\mathcal{U}[y, k] = \mathcal{U}[x, k]$ implies $f(y) \in V(\alpha, n, A)$.

Now fix $\alpha < \kappa$, $n \in \omega$ and $A \in \mathbb{F}$; if $S(\alpha, n, A) = \emptyset$ then let $G(\alpha, n, A) = \emptyset$. If the set $S(\alpha, n, A)$ is non-empty then let $k = k(\alpha, n, A)$. It is evident that the set $O(\alpha, n, A) = \bigcup\{\bigcap \mathcal{U}[x, k] : x \in f^{-1}(S(\alpha, n, A))\}$ is open in X and contains $f^{-1}(S(\alpha, n, A))$; so $W(\alpha, n, A) = f^\#(O(\alpha, n, A))$ is an open neighbourhood of the set $S(\alpha, n, A)$; by normality of Y there exists an F_σ-set $G(\alpha, n, A) \in \tau(Y)$ such that $S(\alpha, n, A) \subset G(\alpha, n, A) \subset W(\alpha, n, A) \cap V(\alpha, n, A)$. We claim that

(8) the family $\mathcal{G} = \{G(\alpha, n, A) : \alpha < \kappa, n \in \omega \text{ and } A \in \mathbb{F}\}$ is T_0-separating.

Indeed, given distinct $x, y \in Y$, we can apply (1) to find $A \in \mathbb{F}$ such that either $x \in F(A)$ and $y \notin U(A)$ or $y \in F(A)$ and $x \notin U(A)$. In the first case apply (4) to see that there are $\alpha < \kappa$ and $n \in \omega$ for which $x \in S(\alpha, n, A)$. Then $x \in G(\alpha, n, A)$ and $y \notin G(\alpha, n, A)$ because $G(\alpha, n, A) \subset V(\alpha, n, A) \subset U(A)$ while $y \notin U(A)$. The second case can be considered analogously; so (8) is proved.

The next property of $\mathcal{G}$ is crucial.

(9) the family $\mathcal{G}$ can be represented as a countable union of families of bounded order.

Consider the family

$$\mathcal{H}(m) = \{G(\alpha, n, A) : \alpha < \kappa, n \leq m, k(\alpha, n, A) \leq m,$$

$$|A| \leq m \text{ and } A \subset \mathcal{U}_0 \cup \ldots \cup \mathcal{U}_m\}$$

for any $m \in \omega$ then $\bigcup\{\mathcal{H}(m) : m \in \omega\} = \mathcal{G}$; so it suffices to show that every $\mathcal{H}(m)$ is of bounded order.

Assume, towards a contradiction, that $\mathcal{H}(m)$ is not of bounded order for some $m \in \omega$. Since every $\mathcal{U}_i$ is of bounded order, we can fix $l \in \mathbb{N}$ such that $|\mathcal{U}[x, m]| \leq l$ for any point $x \in X$. Since the family $\mathcal{H}(m)$ fails to be of bounded order, there is $y \in Y$ and a family $\mathcal{H}_0 \subset \mathcal{H}(m)$ such that $|\mathcal{H}_0| > m^3 \cdot g(m, 2^l(l + 2))$ and $y \in \bigcap \mathcal{H}_0$. An elementary computing shows that there exists a family $\mathcal{H}_1 \subset \mathcal{H}_0$ such that $|\mathcal{H}_1| \geq g(m, 2^l(l + 2))$ and there are $k, p, r \in \omega$ such that $k(\alpha, n, A) = k$, $|A| = r$ and $n = p$ whenever $G(\alpha, n, A) \in \mathcal{H}_1$.

It follows from Fact 1 that there exists $\mathcal{H}_2 \subset \mathcal{H}_1$ such that $|\mathcal{H}_2| \geq 2^l(l + 2)$ and, for some $D \subset \mathcal{U}$ we have $A \cap A' = D$ for any distinct $G(\alpha, p, A) \in \mathcal{H}_2$ and $G(\beta, p, A') \in \mathcal{H}_2$; let $\{G(\alpha_i, p, A_i) : i < 2^l(l + 2)\}$ be a faithful enumeration of some subfamily $\mathcal{H}_3 \subset \mathcal{H}_2$. Pick a point $x \in f^{-1}(y)$; then, for any $i < 2^l(l + 2)$, we have $x \in f^{-1}(W(\alpha_i, n, A_i)) \subset O(\alpha_i, n, A_i)$; so there is $x_i \in f^{-1}(S(\alpha_i, p, A_i))$ such that $x \in \bigcap \mathcal{U}[x_i, k]$ (here we used the fact that $k(\alpha_i, p, A_i) = k$ for any $i \leq 2^l(l + 2)$).

As a consequence, $\mathcal{U}[x_i, k] \subset \mathcal{U}[x, k]$ for any $i < 2^l(l + 2)$; the family $\mathcal{U}[x, k]$ has cardinality at most l so it has at most 2^l subfamilies which shows that we can find a set $E \subset 2^l(l + 2)$ such that $|E| > l + 1$ and $\mathcal{U}[x_i, k] = \mathcal{U}[x_j, k]$ for any $i, j \in E$. To finally obtain the promised contradiction consider the following cases.

(a) There are distinct elements $i, j \in E$ such that $A_i = A_j = D$. Then $\alpha_i \neq \alpha_j$, say, $\alpha_i < \alpha_j$. It follows from the property (6) that $S(\alpha_j, p, A_j) \cap V(\alpha_i, p, A_i) = \emptyset$ and hence $f(x_j) \notin V(\alpha_i, p, A_i)$. On the other hand, it follows from (7) and $U[x_i, k] = U[x_j, k]$ that $f(x_j) \in V(\alpha_i, p, A_i)$ which is a contradiction.

(b) We have $A_i \neq D$ for any $i \in E$ (to see that this case covers the complement of (a) observe that $|A_i| = |A_j| = r$ for any $i, j \in E$). Since $f(x_i) \in F(A_i)$, there is $y_i \in f^{-1}(f(x_i)) \backslash (\bigcup D)$ for any $i \in E$; let i_0 be the minimal element of E.

Recall that $|U[y_{i_0}, m]| \leq l$ while the collection $\{A_i \backslash D : i \in E \backslash \{i_0\}\}$ is disjoint and has cardinality strictly greater than l. Therefore there exists $i \in E \backslash \{i_0\}$ for which $U[y_{i_0}, m] \cap (A_i \backslash D) = \emptyset$. Since $A_i \backslash D \subset U_0 \cup \ldots \cup U_m$ and $D \cap U[y_{i_0}, m] = \emptyset$, we have $A_i \cap U[y_{i_0}] = \emptyset$ and therefore $f^{-1}(f(y_{i_0})) \not\subset \bigcup A_i$. An immediate consequence is that $f(x_{i_0}) = f(y_{i_0}) \notin U(A_i)$; on the other hand, the property (7) and the equality $U[x_{i_0}, k] = U[x_i, k]$ imply that $f(x_{i_0}) \in V(\alpha_i, p, A_i) \subset U(A_i)$. This final contradiction shows that every family $\mathcal{H}(m)$ is, indeed, of bounded order so (9) is proved.

Finally, observe that the properties (8) and (9) show that $\mathcal{G}$ is a T_0-separating family of open F_σ-subsets of Y which can be represented as a countable union of subfamilies of bounded order; so Fact 1 of U.379 implies that Y is uniform Eberlein compact and shows that any continuous image of a uniform Eberlein compact space is uniform Eberlein compact.

Finally observe that any uniform Eberlein compact space is a continuous image of a closed subspace of $(A(\kappa))^\omega$ for some infinite cardinal κ (see Problem 380). Now assume that F is a closed subspace of $(A(\kappa))^\omega$ and X is a continuous image of F. The family $\mathcal{C} = \{\{\alpha\} : \alpha \in \kappa\} \cup \{A(\kappa)\}$ is a T_0-separating clopen cover of $A(\kappa)$ and it is immediate that $\mathcal{C}$ is of bounded order (in fact, its order is ≤ 2). Therefore $A(\kappa)$ is uniform Eberlein compact; apply Problem 379 to conclude that F is also uniform Eberlein compact. We have already established that any continuous image of a uniform Eberlein compact space is uniform Eberlein compact; so X is uniform Eberlein compact. Thus, a space is uniform Eberlein compact if and only if it is a continuous image of a closed subspace of $(A(\kappa))^\omega$ for some infinite cardinal κ, i.e., our solution is now complete.

U.382. *Given an infinite set T suppose that a space $X_t \neq \emptyset$ is uniform Eberlein compact for each $t \in T$. Prove that the Alexandroff compactification of the space $\bigoplus \{X_t : t \in T\}$ is also uniform Eberlein compact.*

Solution. We will identify every X_t with the respective clopen subspace of the space $X = \bigoplus_{t \in T} X_t$; let K be the Alexandroff compactification of the space X. Then the set $K \backslash X$ is a singleton. Apply Fact 2 of U.380 to fix, for any $t \in T$, a family U_t of open F_σ-subsets of X_t which T_0-separates the points of X_t and $U_t = \bigcup_{n \in \omega} U_t^n$ where every U_t^n is *of bounded order*, i.e., there is $k_t^n \in \omega$ such that every $x \in X_t$ belongs to at most k_t^n-many elements of U_t^n.

It is easy to see that the family $W = (\bigcup_{t \in T} \mathcal{U}_t) \cup \{X_t : t \in T\}$ consists of cozero subsets of K and T_0-separates the points of K. The family $W_0 = \{X_t : t \in T\}$ is of bounded order because it is disjoint. For any $m \in \mathbb{N}$ let $W_m = \bigcup\{\mathcal{U}_t^n : k_t^n \le m\}$; it is straightforward that every point of K belongs to at most m elements of W_m and hence W_m is of bounded order. It is also evident that $W = \bigcup_{m \in \omega} W_m$; so W is a countable union of its subfamilies of bounded order. Therefore we can apply Fact 2 of U.380 again to conclude that K is uniform Eberlein compact.

U.383. *Let T be an infinite set. Suppose that $\mathcal{A}$ is an adequate family on T. Prove that the space $K_{\mathcal{A}}$ is Eberlein compact if and only if $T_{\mathcal{A}}^*$ is σ-compact.*

Solution. It was proved in Problem 170 that the space $T_{\mathcal{A}}^*$ is homeomorphic to a closed subspace of $C_p(K_{\mathcal{A}}, \mathbb{D})$; if $K_{\mathcal{A}}$ is Eberlein compact then $C_p(K_{\mathcal{A}}, \mathbb{D})$ is σ-compact by Problem 335; so $T_{\mathcal{A}}^*$ is also σ-compact.

Now, if $T_{\mathcal{A}}^*$ is σ-compact then let $u(f) = 0$ for all $f \in K_{\mathcal{A}}$; for any $t \in T$ define a function $e_t : K_{\mathcal{A}} \to \mathbb{D}$ by the equality $e_t(f) = f(t)$ for any $f \in \mathbb{K}_{\mathcal{A}}$. Then the set $Z = \{e_t : t \in T\} \cup \{u\} \subset C_p(K_{\mathcal{A}}, \mathbb{D})$ is homeomorphic to $T_{\mathcal{A}}^*$ (see Problem 170); this implies that Z is σ-compact. It is easy to see that Z separates the points of $K_{\mathcal{A}}$; so we can apply Problem 035 to conclude that $K_{\mathcal{A}}$ is Eberlein compact.

U.384. *Let T be an infinite set. Suppose that $\mathcal{A}$ is an adequate family on T. Prove that the space $K_{\mathcal{A}}$ is Eberlein compact if and only if there exists a disjoint family $\{T_i : i \in \omega\}$ such that $T = \bigcup\{T_i : i \in \omega\}$ and $x^{-1}(1) \cap T_i$ is finite for every $x \in K_{\mathcal{A}}$ and $i \in \omega$.*

Solution. Recall that $T_{\mathcal{A}}^*$ is a space with the underlying set $T \cup \{\xi\}$ for some $\xi \notin T$, all points of T are isolated in $T_{\mathcal{A}}^*$ while the local base at ξ is given by the complements of all finite unions of elements of $\mathcal{A}$. Suppose that $\{T_i : i \in \omega\}$ is a disjoint family such that $T = \bigcup_{i \in \omega} T_i$ and $x^{-1}(1) \cap T_i$ is finite for any $x \in K_{\mathcal{A}}$ and $i \in \omega$. Let $\pi_i : \mathbb{D}^T \to \mathbb{D}^{T_i}$ be the natural projection for any $i \in \omega$. The space $X_i = \pi_i(K_{\mathcal{A}})$ is compact for any $i \in \omega$ and $K_{\mathcal{A}} \subset \prod_{i \in \omega} X_i$.

If $x \in K_{\mathcal{A}}$ then it follows from $|x^{-1}(1) \cap T_i| < \omega$ that $\pi_i(x) \in \sigma(T_i) \subset \Sigma_*(T_i)$; therefore $X_i \subset \Sigma_*(T_i)$; so X_i is Eberlein compact for any $i \in \omega$ (see Problem 322). Applying Problems 306 and 307 we conclude that $K_{\mathcal{A}}$ is Eberlein compact. This settles sufficiency.

Now, if $K_{\mathcal{A}}$ is Eberlein compact then $T_{\mathcal{A}}^*$ is σ-compact by Problem 383; fix a family $\{Y_i : i \in \omega\}$ of compact subsets of $T_{\mathcal{A}}^*$ such that $T_{\mathcal{A}}^* = \bigcup_{i \in \omega} Y_i$. Choose any sequence $\mathcal{S} = \{T_i : i \in \omega\}$ of disjoint subsets of T such that $T = \bigcup_{i \in \omega} T_i$ and $T_i \subset Y_i$ for every $i \in \omega$. To see that $\mathcal{S}$ is as promised suppose that there is $i \in \omega$ and an infinite set $A \subset T_i$ such that $y(A) \subset \{1\}$ for some $y \in K_{\mathcal{A}}$. The set Y_i being compact, the point ξ must be in the closure of $A \subset y^{-1}(1)$ and hence ξ belongs to the closure of $y^{-1}(1)$. This contradicts, however, the definition of the local base at ξ because the complement of $x^{-1}(1) \in \mathcal{A}$ has to be a neighbourhood of ξ for any $x \in K_{\mathcal{A}}$. This proves that $x^{-1}(1) \cap T_i$ is finite for any $i \in \omega$ and $x \in K_{\mathcal{A}}$, i.e., we established necessity; so our solution is complete.

U.385. *Let T be an infinite set and $\mathcal{A}$ an adequate family on T. Prove that the adequate compact $K_{\mathcal{A}}$ is uniform Eberlein compact if and only if there exists a disjoint family $\{T_i : i \in \omega\}$ and a function $N : \omega \to \omega$ such that $T = \bigcup\{T_i :\in \omega\}$ and $|x^{-1}(1) \cap T_i| \leq N(i)$ for any $x \in K_{\mathcal{A}}$ and $i \in \omega$.*

Solution. Given a set B let $\sigma_n[B] = \{x \in \mathbb{D}^B : |x^{-1}(1)| \leq n\}$ for any $n \in \omega$. If $\mathcal{V}$ is a family of subsets of a space Z then let $\mathcal{V}(z) = \{V \in \mathcal{V} : z \in V\}$ and $\operatorname{ord}(\mathcal{V}, z) = |\mathcal{V}(z)|$ for any $z \in Z$. Recall that $T_{\mathcal{A}}^*$ is a space with the underlying set $T \cup \{\xi\}$ for some $\xi \notin T$, all points of T are isolated in $T_{\mathcal{A}}^*$ while the local base at ξ is given by the complements of all finite unions of elements of $\mathcal{A}$. Suppose that $\{T_i : i \in \omega\}$ is a disjoint family and $N : \omega \to \omega$ is a map such that $T = \bigcup_{i \in \omega} T_i$ and $|x^{-1}(1) \cap T_i| \leq N(i)$ for any $x \in K_{\mathcal{A}}$ and $i \in \omega$. Let $\pi_i : \mathbb{D}^T \to \mathbb{D}^{T_i}$ be the natural projection for any $i \in \omega$. The space $X_i = \pi_i(K_{\mathcal{A}})$ is compact for any $i \in \omega$ and $K_{\mathcal{A}} \subset \prod_{i \in \omega} X_i$.

If $x \in K_{\mathcal{A}}$ then it follows from $|x^{-1}(1) \cap T_i| \leq N(i)$ that $\pi_i(x) \in \sigma_{N(i)}[T_i]$; therefore $X_i \subset \sigma_{N(i)}[T_i]$ is uniform Eberlein compact (see Problem 380 and Fact 1 of U.380) for any $i \in \omega$. Applying Problem 379 we conclude that $K_{\mathcal{A}}$ is uniform Eberlein compact. This settles sufficiency.

Now, assume that $K_{\mathcal{A}}$ is uniform Eberlein compact and apply Fact 2 of U.380 to find a T_0-separating family $\mathcal{U}$ of clopen subsets of $K_{\mathcal{A}}$ such that $\mathcal{U} = \bigcup_{n \in \omega} \mathcal{U}_n$ where every $\mathcal{U}_n$ is *of bounded order*, i.e., there is $m_n \in \omega$ for which $\operatorname{ord}(\mathcal{U}_n, x) \leq m_n$ for any $x \in K_{\mathcal{A}}$. Denote by u the point of $K_{\mathcal{A}}$ such that $u(t) = 0$ for all $t \in T$. For any $n \in \omega$ consider the family $\mathcal{U}_n' = (\mathcal{U}_n \backslash \mathcal{U}_n(u)) \cup \{K_{\mathcal{A}} \backslash U : U \in \mathcal{U}_n(u)\}$. It is easy to see that every $\mathcal{U}_n'$ is of bounded order while the family $\mathcal{U}' = \bigcup_{n \in \omega} \mathcal{U}_n'$ is still T_0-separating and $u \notin \bigcup \mathcal{U}'$. Therefore we can assume, without loss of generality, that $\mathcal{U} = \mathcal{U}'$, i.e.,

(1) the point u does not belong to $\bigcup \mathcal{U}$ and hence $\bigcup \mathcal{U} = K_{\mathcal{A}} \backslash \{u\}$.

An easy consequence of the definition of an adequate family is that, for any $t \in T$, the function $u_t : T \to \mathbb{D}$ defined by $u_t(t) = 1$ and $u_t(T \backslash \{t\}) = \{0\}$ belongs to $K_{\mathcal{A}}$. Let $U_n = \bigcup \mathcal{U}_n$ for every $n \in \omega$; it follows from the property (1) that $\{u_t : t \in T\} \subset \bigcup_{n \in \omega} U_n$; so it is easy to find a disjoint family $\{Q_n : n \in \omega\}$ such that $T = \bigcup_{i \in \omega} Q_n$ and $\{e_t : t \in Q_n\} \subset U_n$ for any $n \in \omega$.

For any element $t \in T$ there is a unique number $n \in \omega$ such that $t \in Q_n$ and hence there exists a set $W_t \in \mathcal{U}_n$ with $u_t \in W_t$. Fix a finite set $E_t \subset T \backslash \{t\}$ such that $[E_t] = \{x \in K_{\mathcal{A}} : x(t) = 1 \text{ and } x(E_t) = \{0\}\} \subset W_t$. It is clear that $[E_t]$ is an open subset of $K_{\mathcal{A}}$ such that $u_t \in [E_t] \subset W_t$ for any $t \in T$ and therefore

(2) for the family $\mathcal{Q}_n = \{[E_t] : t \in Q_n\}$ we have $\operatorname{ord}(\mathcal{Q}_n, x) \leq m_n$ for any $x \in K_{\mathcal{A}}$ and $n \in \omega$.

Let $Q_n^j = \{t \in Q_n : |E_t| = j\}$ for any $n, j \in \omega$. It is clear that the family $\mathcal{Q} = \{Q_n^j : n, j \in \omega\}$ is disjoint and $T = \bigcup \mathcal{Q}$. Let us show that, enumerating $\mathcal{Q}$ as $\{T_i : i \in \omega\}$, we obtain the desired partition of T.

Given $i \in \omega$ let $\mathcal{S}_i = \{[E_t] : t \in T_i\}$; since T_i is a subset of some Q_n, the property (2) shows that there exists $l_i \in \omega$ such that $\operatorname{ord}(\mathcal{S}_i, x) \leq l_i$ for any $x \in K_{\mathcal{A}}$. Besides, there is $k_i \in \omega$ such that $|E_t| = k_i$ for any $t \in T_i$.

Fix $i \in \omega$ and suppose that there is no finite upper bound for the number of elements in the sets $x^{-1}(1) \cap T_i$ when x runs over K_A. By Fact 1 of U.381 there exists a point $x \in K_A$ for which the family $\{E_t : t \in x^{-1}(1) \cap T_i\}$ contains a Δ-system of cardinality $k_i \cdot l_i + k_i + 2l_i + 1$, i.e., there is a set $S \subset T_i \cap x^{-1}(1)$ such that $|S| \geq k_i \cdot l_i + k_i + 2l_i + 1$ and there exists $E \subset T$ such that $E_t \cap E_s = E$ for any distinct $s, t \in S$.

Since $|E| \leq k_i$, the set $S' = S \backslash E$ has at least $k_i \cdot l_i + 2l_i + 1$ elements. Take $t_1 \in S'$ arbitrarily; assume that $l \leq l_i$ and we have chosen distinct $t_1, \ldots, t_l \in S'$ in such a way that

(*) $\{t_1, \ldots, t_l\} \cap E_{t_j} = \emptyset$ for any $j \leq l$.

The set $H = \{t_1, \ldots, t_l\} \cup (\bigcup \{E_{t_j} : j \leq l\})$ has at most $k_i \cdot l + l \leq k_i \cdot l_i + l_i$ elements; so we have at least $l_i + 1 > l$ elements left in S'. This makes it possible to take distinct $s_1, \ldots, s_{l+1} \in S' \backslash H$; the family $\mathcal{E} = \{E_{s_j} \cap S' : j \leq l+1\}$ being disjoint, some element of $\mathcal{E}$ does not meet the set $\{t_1, \ldots, t_l\}$. Therefore we can choose $t_{l+1} \in S' \backslash H$ such that $E_{t_{l+1}} \cap \{t_1, \ldots, t_l\} = \emptyset$.

It is straightforward that the property (*) is now fulfilled if we substitute l by $l + 1$; so our inductive procedure shows that we can construct a set $P = \{t_1, \ldots, t_{l_i+1}\}$ such that (*) holds for all $j \leq l_i + 1$. Now, if $z(t) = 1$ for all $t \in P$ and $z(T \backslash P) = \{0\}$ then $z \in K_A$ (because $P \subset x^{-1}(1)$) and $z \in \bigcap \{[E_{t_j}] : j \leq l_i + 1\}$ which is a contradiction with the fact that $\text{ord}(S_i, z) \leq l_i$.

Thus there exists a number $N(i) \in \omega$ such that $|x^{-1}(1) \cap T_i| \leq N(i)$ for any $x \in K_A$ and $i \in \omega$; this shows that the partition $\{T_i : i \in \omega\}$ of the set T is as promised and hence our solution is complete.

U.386. *For the set $T = \omega_1 \times \omega_1$ let us introduce an order $<$ on T declaring that $(\alpha_1, \beta_1) < (\alpha_2, \beta_2)$ if and only if $\alpha_1 < \alpha_2$ and $\beta_1 > \beta_2$. Denote by $\mathcal{A}$ the family of all subsets of T which are linearly ordered by $<$ (the empty set and the one-point sets are considered to be linearly ordered). Prove that $\mathcal{A}$ is an adequate family and $X = K_A$ is a strong Eberlein compact space which is not uniform Eberlein compact.*

Solution. Given a set B let $\sigma[B] = \{x \in \mathbb{D}^B : |x^{-1}(1)| < \omega\}$. We leave it to the reader to verify that $<$ is, indeed, a partial order on T. Recall that, for any $A \subset T$ the function $\chi_A \in \mathbb{D}^T$ is defined by $\chi_A(A) \subset \{1\}$ and $\chi_A(T \backslash A) \subset \{0\}$. If $A \subset T$ is linearly ordered by $<$ then any subset of A is also linearly ordered; since we have declared the singletons to be linearly ordered, we have $\bigcup \mathcal{A} = T$. A set is linearly ordered if and only if any pair of its elements is linearly ordered; so if $A \subset T$ and all finite subsets of A belong to $\mathcal{A}$ then A is linearly ordered, i.e., $A \in \mathcal{A}$. This proves that $\mathcal{A}$ is an adequate family and hence $K = K_A = \{\chi_A : A \in \mathcal{A}\}$ is compact (see Problem 168).

Suppose that some set $A \in \mathcal{A}$ is infinite; by an evident induction it is possible to choose $(\alpha_n, \beta_n) \in A$ such that $\alpha_n < \alpha_{n+1}$ for all $n \in \omega$. Since every two elements of A are comparable, we have $\beta_{n+1} < \beta_n$ for all $n \in \omega$ which is impossible because $\omega_1 \supset \{\beta_n : n \in \omega\}$ is a well-ordered set. This contradiction shows that all elements of $\mathcal{A}$ are finite and hence $K \subset \sigma[T]$ is strong Eberlein compact.

Let us establish that

(1) if $T = \bigcup_{n \in \omega} T_n$ then there is a number $m \in \omega$ such that, for any $k \in \omega$, we can find a set $A \in \mathcal{A}$ such that $A \subset T_m$ and $|A| = k$; in particular, $x = \chi_A \in K$ and $|x^{-1}(1) \cap T_m| = |x^{-1}(1)| = |A| = k$.

Given an ordinal $\alpha < \omega_1$ let $T_n^\alpha = \{\beta < \omega_1 : (\alpha, \beta) \in T_n\}$ for any $n \in \omega$; since $\bigcup_{n \in \omega} T_n^\alpha = \omega_1$, there is $n_\alpha \in \omega$ such that $T_{n_\alpha}^\alpha$ is uncountable. Fix $m \in \omega$ such that the set $\Gamma = \{\alpha < \omega_1 : n_\alpha = m\}$ is uncountable; we claim that m is as promised.

Indeed, fix any $k \in \omega$; the set Γ being uncountable, we can find $\alpha_1, \ldots, \alpha_k \in \Gamma$ with $\alpha_1 < \ldots < \alpha_k$. Take an ordinal $\beta_k \in T_m^{\alpha_k}$ arbitrarily and assume that $1 < l \leq k$ and we have chosen $\beta_k, \beta_{k-1}, \ldots, \beta_l$ so that $\beta_i \in T_m^{\alpha_i}$ for every $i \in \{k, k-1, \ldots, l\}$ and $\beta_i < \beta_{i-1}$ whenever $i \in \{k, k-1, \ldots, l+1\}$.

The set $T_m^{\alpha_{l-1}}$ being uncountable there is $\beta_{l-1} \in T_m^{\alpha_{l-1}}$ for which $\beta_{l-1} > \beta_l$. Thus our inductive procedure shows that we can choose an ordinal $\beta_i \in T_m^{\alpha_i}$ for each $i \in \{1, \ldots, k\}$ in such a way that $\beta_1 > \beta_2 > \ldots > \beta_k$. It is straightforward that the set $A = \{(\alpha_i, \beta_i) : i \leq k\} \subset T_m$ belongs to $\mathcal{A}$ and $|A| = k$, i.e., (1) is proved.

Finally, if K is uniform Eberlein compact then there is a map $N : \omega \to \omega$ and a sequence $\{T_i : i \in \omega\}$ such that $T = \bigcup_{i \in \omega} T_i$ and $|x^{-1}(1) \cap T_i| \leq N(i)$ for all $i \in \omega$ and $x \in K$ (see Problem 385). This evident contradiction with (1) shows that K is not uniform Eberlein compact.

U.387. *(Talagrand's example) For any distinct $s, t \in \omega^\omega$, consider the number $\delta(s, t) = \min\{k \in \omega : s(k) \neq t(k)\}$. For each $n \in \omega$, let $\mathcal{A}_n = \{A \subset \omega^\omega : \text{for any distinct } s, t \in A \text{ we have } \delta(s, t) = n\}$. Prove that $\mathcal{A} = \bigcup\{\mathcal{A}_n : n \in \omega\}$ is an adequate family and $X = K_\mathcal{A}$ is a Talagrand compact space (i.e., $C_p(X)$ is K-analytic and hence X is Gul'ko compact) while X is not Eberlein compact.*

Solution. Observe first that $\emptyset \in \mathcal{A}_n$ and $\{s\} \in \mathcal{A}_n$ for any $s \in \omega^\omega$ and $n \in \omega$. Therefore every $\mathcal{A}_n$ covers ω^ω; an immediate consequence of the definition is that $A \in \mathcal{A}_n$ implies $B \in \mathcal{A}_n$ for any $B \subset A$. Now, if $A \subset \omega^\omega$ and every two-element subset of A belongs to $\mathcal{A}_n$ then $\delta(s, t) = n$ for any distinct $s, t \in A$ which shows that $A \in \mathcal{A}_n$. This proves that every $\mathcal{A}_n$ is an adequate family.

Therefore, to prove that $\mathcal{A} = \bigcup_{n \in \omega} \mathcal{A}_n$ is an adequate family, it suffices to establish that the last requirement of the definition is fulfilled. So, assume that $A \subset \omega^\omega$ and every finite $B \subset A$ belongs to $\mathcal{A}$. The only non-trivial case is when A is infinite; so we can fix distinct $s, t \in A$; let $n = \delta(s, t)$. Given any distinct $u, v \in A$, the set $B = \{s, t, u, v\}$ belongs to $\mathcal{A}$ and hence there is some $m \in \omega$ such that $B \in \mathcal{A}_m$; since s and t are distinct elements of B, we have $n = \delta(s, t) = m$ and hence $m = n$. Consequently, $\delta(u, v) = m = n$ which shows that $\delta(u, v) = n$ for any distinct $u, v \in A$; this implies that $A \in \mathcal{A}_n$ and therefore $A \in \mathcal{A}$. Thus $\mathcal{A}$ is an adequate family.

Given any $A \in \mathcal{A}$ there is $n \in \omega$ such that $A \in \mathcal{A}_n$ and hence the correspondence $s \to s|(n+1)$ is an injection of A into ω^{n+1}. This proves that every $A \in \mathcal{A}$ is countable and hence the space $X = K_\mathcal{A}$ is Corson compact (see Problem 169).

If X is Eberlein compact then there exists a disjoint family $\{T_n : n \in \omega\}$ such that $\bigcup_{n\in\omega} T_n = \omega^\omega$ and $x^{-1}(1) \cap T_n$ is finite for any $x \in X$ and $n \in \omega$ (see Problem 384).

The space ω^ω is Polish, so it has the Baire property; therefore there is $n \in \omega$ such that the closure of T_n in ω^ω contains some non-empty open subset of ω^ω. Consequently, there is $m \in \mathbb{N}$ and $h \in \omega^m$ such that the set $U_h = \{s \in \omega^\omega : s|m = h\}$ is contained in the closure of T_n in ω^ω. The set $U_{h,i} = \{s \in U_h : s(m) = i\}$ is open in U_h and non-empty; so we can fix $t_i \in T_n \cap U_{h,i}$ for any $i \in \omega$. It is straightforward that $\delta(t_i, t_j) = m$ for any distinct $i, j \in \omega$; so the set $A = \{t_i : i \in \omega\}$ belongs to $\mathcal{A}_m$. Therefore $x = \chi_A \in X$ and $x^{-1}(1) \cap T_n = A$ is infinite which is a contradiction. This proves that X is not Eberlein compact.

For any $n \in \omega$ let $X_n = K_{\mathcal{A}_n}$; we already saw that $\mathcal{A}_n$ is adequate, so X_n is a Corson compact space.

Fix any $n \in \omega$; the set ω^{n+1} being countable we can faithfully enumerate it as $\{h_i : i \in \omega\}$. Let $T_i = \{s \in \omega^\omega : s|(n+1) = h_i\}$ for any $i \in \omega$. It is evident that the family $\{T_i : i \in \omega\}$ is disjoint and $\bigcup_{i\in\omega} T_i = \omega^\omega$. Given any $x \in X_n$ there is $A \in \mathcal{A}_n$ such that $x = \chi_A$. For any distinct $s, t \in A$ we have $\delta(s, t) = n$; so $s|(n+1) \neq t|(n+1)$ and therefore the set $A \cap T_i$ cannot have more than one element for any $i \in \omega$. Thus $|x^{-1}(1) \cap T_i| = |A \cap T_i| \leq 1$ for any $i \in \omega$ and $x \in X_n$ which shows that X_n is Eberlein compact by Problem 384.

It is clear that $X = \bigcup_{n\in\omega} X_n$; so we can apply Problem 355 to conclude that $C_p(X)$ is K-analytic and hence X is Gul'ko compact.

U.388. *Given a compact space X let $\|f\| = \sup\{|f(x)| : x \in X\}$ for any function $f \in C(X)$. Prove that $(C(X), \|\cdot\|)$ is a Banach space.*

Solution. It is straightforward that $C(X)$ is a linear space. Since $|f(x)| \geq 0$ for any $x \in X$, we have $\|f\| \geq 0$ for each $f \in C(X)$. If $\|f\| = 0$ then $f(x) = 0$ for any $x \in X$ and hence $f = \mathbf{0}_X$ is the zero vector of $C(X)$. It is evident that $\|\mathbf{0}_X\| = 0$; so we proved (N1).

Now, fix $f \in C(X)$ and $\alpha \in \mathbb{R}$. The equality $\|\alpha f\| = 0 = |\alpha|\|f\|$ is evident if $\alpha = 0$; so assume that $\alpha \neq 0$. For any $x \in X$ we have $|(\alpha f)(x)| = |\alpha||f(x)| \leq |\alpha|\|f\|$ which shows that $\|\alpha f\| \leq |\alpha|\|f\|$. If $t < |\alpha|\|f\|$ then $\frac{t}{|\alpha|} < \|f\|$; so there is $x \in X$ for which $\frac{t}{|\alpha|} < |f(x)|$ and hence $\|\alpha f\| \geq |(\alpha f)(x)| > t$. Thus $\|\alpha f\| > t$ for any $t < |\alpha|\|f\|$ which proves that $\|\alpha f\| \geq |\alpha|\|f\|$ and therefore $\|\alpha f\| = |\alpha|\|f\|$, i.e., the property (N2) also holds for $\|\cdot\|$.

Finally, given any functions $f, g \in C(X)$, for any $x \in X$ we have the inequality $|f(x) + g(x)| \leq |f(x)| + |g(x)| \leq \|f\| + \|g\|$ which proves that $\|f + g\| \leq \|f\| + \|g\|$, i.e., (N3) is also fulfilled for $\|\cdot\|$ and hence $(C(X), \|\cdot\|)$ is a normed space. The metric defined by the norm $\|\cdot\|$ is complete by Problem 344; so $(C(X), \|\cdot\|)$ is a Banach space.

U.389. *(Hahn–Banach Theorem) Let M be a linear subspace of a normed space $(L, \|\cdot\|)$. Suppose that $f : M \to \mathbb{R}$ is a linear functional such that $|f(x)| \leq \|x\|$ for any $x \in M$. Prove that there exists a linear functional $F : L \to \mathbb{R}$ such that $F|M = f$ and $|F(x)| \leq \|x\|$ for all $x \in L$.*

Solution. Consider the family $\mathcal{L} = \{(N, \varphi) : N$ is a linear subspace of L with $M \subset N$ and $\varphi : N \to \mathbb{R}$ is a linear functional such that $\varphi | M = f$ and $|\varphi(x)| \leq ||x||$ for any $x \in N\}$. Given $(N, \varphi) \in \mathcal{L}$ and $(N', \varphi') \in \mathcal{L}$ say that $(N, \varphi) \prec (N', \varphi')$ if $N \subset N'$ and $\varphi \subset \varphi'$., i.e., $\varphi' | N = \varphi$. It is easy to see that $(\mathcal{L}, \prec)$ is a partially ordered set; for any $p = (N, \varphi) \in \mathcal{L}$ let $S(p) = N$ and $\Phi(p) = \varphi$.

Suppose that $\mathcal{C}$ is a chain in $(\mathcal{L}, \prec)$ and let $N = \bigcup\{S(p) : p \in \mathcal{C}\}$. The family $\{S(p) : p \in \mathcal{C}\}$ is a chain of linear subspaces of L; so N is a linear subspace of L as well. We leave to the reader the checking that $\varphi = \bigcup\{\Phi(p) : p \in \mathcal{C}\}$ is a well-defined linear functional on N. Since $\Phi(p) | M = f$ for any $p \in \mathcal{C}$, we have $\varphi | M = f$. Given $x \in N$ there is $p \in \mathcal{C}$ such that $|\varphi(x)| = |\Phi(p)(x)| \leq ||x||$; so $\varphi(x) \leq ||x||$ for any $x \in N$. This proves that $(N, \varphi) \in \mathcal{L}$; it is evident that $p \prec (N, \varphi)$ for any $p \in \mathcal{C}$; so (N, φ) is an upper bound for the chain $\mathcal{C}$.

We proved that any chain of $(\mathcal{L}, \prec)$ has an upper bound which belongs to $\mathcal{L}$; so Zorn's lemma is applicable to conclude that $(\mathcal{L}, \prec)$ has a maximal element (N, φ). If $N = L$ then letting $F = \varphi$ completes our solution.

Assume, towards a contradiction, that $N \neq L$ and fix a vector $v \in L \backslash N$. It is evident that $N' = \{x + tv : x \in N, t \in \mathbb{R}\}$ is a linear subspace of L such that $N \subset N'$ and $N \neq N'$. Observe also that, for any $w \in N'$ there are unique $x \in N$ and $t \in \mathbb{R}$ such that $w = x + tv$.

We have $||z - y|| = ||z + v - v - y|| \leq ||z + v|| + || - v - y|| = ||z + v|| + ||v + y||$ for any $y, z \in N$. This implies $|\varphi(z) - \varphi(y)| = |\varphi(z - y)| \leq ||z - y|| \leq ||z + v|| + ||v + y||$. This proves that $\varphi(z) - \varphi(y) \leq ||z + v|| + ||v + y||$ and therefore

(1) $-||v + y|| - \varphi(y) \leq ||v + z|| - \varphi(z)$ for any $y, z \in N$.

Let $a = \sup\{-||v + y|| - \varphi(y) : y \in N\}$ and $b = \inf\{||v + z|| - \varphi(z) : z \in N\}$; an immediate consequence of (1) is that $a \leq b$. Choose any number $c \in [a, b]$. For any $x \in N$ and $t \in \mathbb{R}$ let $\varphi'(x + tv) = \varphi(x) + tc$; it is straightforward that $\varphi' : N' \to \mathbb{R}$ is a linear functional such that $\varphi' | N = \varphi$.

Fix any $w \in N'$; there are unique $x \in N$ and $t \in \mathbb{R}$ such that $w = x + tv$. If $t = 0$ then $w \in N$; so we have $|\varphi'(w)| = |\varphi(w)| \leq ||w||$. Now assume that $t \neq 0$.

It follows from (1) that $-||v + y|| - \varphi(y) \leq c \leq ||v + y|| - \varphi(y)$ for any $y \in N$; in particular, for $y = \frac{1}{t}x$ we obtain the inequality $-||v + \frac{x}{t}|| \leq c + \varphi(\frac{x}{t}) \leq ||v + \frac{x}{t}||$ or, equivalently, $|c + \varphi(\frac{x}{t})| \leq ||v + \frac{x}{t}||$. An immediate consequence of the last inequality is that $|\frac{1}{t}| \cdot |\varphi(x) + ct| \leq |\frac{1}{t}| \cdot ||x + vt||$, i.e., $|\varphi'(w)| \leq ||w||$. We proved that $|\varphi'(w)| \leq ||w||$ for any $w \in N'$ and hence the pair (N', φ') belongs to $\mathcal{L}$. We also have $(N, \varphi) \prec (N', \varphi')$ and $(N, \varphi) \neq (N', \varphi')$; this contradiction with maximality of (N, φ) shows that the case of $N \neq L$ is impossible. Thus $N = L$ and the linear functional $F = \varphi$ is an extension of f such that $|F(x)| \leq ||x||$ for any $x \in L$.

U.390. *Given a normed space $(L, || \cdot ||)$ let $S = \{x \in L : ||x|| \leq 1\}$ be the unit ball of L. Prove that a linear functional $f : L \to \mathbb{R}$ is continuous if and only if there exists $k \in \mathbb{N}$ such that $|f(x)| \leq k$ for any $x \in S$.*

Solution. Denote by u the zero vector of L. If f is continuous then it is continuous at u; so it follows from $f(u) = 0$ that there is $U \in \tau(u, L)$ such that

$f(U) \subset (-1, 1)$. Since the topology of the space L is generated by the metric $d(x, y) = ||x - y||$, the open balls form a base in L and hence there is $\varepsilon > 0$ such that $B = \{x \in L : ||x - u|| = ||x|| < 2\varepsilon\} \subset U$.

Take a number $k \in \mathbb{N}$ such that $\frac{1}{\varepsilon} < k$; given any $x \in S$ we have $||x|| \le 1$ and hence $||\varepsilon x|| \le \varepsilon < 2\varepsilon$ which implies $\varepsilon x \in B \subset U$. Therefore $f(\varepsilon x) \in (-1, 1)$ which is equivalent to $|f(\varepsilon x)| = \varepsilon |f(x)| < 1$. As a consequence, $|f(x)| < \frac{1}{\varepsilon} < k$ which proves that $|f(x)| < k$ for any $x \in S$, i.e., we have established necessity.

Now assume that there is $k \in \mathbb{N}$ such that $|f(x)| \le k$ for any $x \in S$ and take any $\varepsilon > 0$. The set $B = \{x \in L : ||x|| < \frac{\varepsilon}{k+1}\}$ is an open neighbourhood of u. If $x \in B$ then $||\frac{k+1}{\varepsilon}x|| = \frac{k+1}{\varepsilon}||x|| < \frac{k+1}{\varepsilon} \cdot \frac{\varepsilon}{k+1} = 1$ which implies $\frac{k+1}{\varepsilon}x \in S$ and therefore $|f(\frac{k+1}{\varepsilon}x)| \le k$. Using once more linearity of f we obtain $\frac{k+1}{\varepsilon}|f(x)| \le k$ and hence $|f(x)| \le \frac{k}{k+1}\varepsilon < \varepsilon$. Thus B is an open neighbourhood of u such that $f(B) \subset (-\varepsilon, \varepsilon)$; so f is continuous at u. Finally, apply Fact 2 of S.496 to conclude that f is continuous.

U.391. *Given a normed space* $(L, || \cdot ||)$, *consider the sets* $S = \{x \in L : ||x|| \le 1\}$ *and* $S^* = \{f \in L^* : f(S) \subset [-1, 1]\}$. *Prove that* S^* *separates the points of* L.

Solution. Denote by $\mathbf{0}_L$ the zero vector of L and suppose that x and y are distinct points of L. Then $z = x - y \ne \mathbf{0}_L$; so $M = \{tz : t \in \mathbb{R}\}$ is a linear subspace of L such that $z \in M$. Let $f(tz) = t||z||$ for any $t \in \mathbb{R}$; it is straightforward that $f : M \to \mathbb{R}$ is a linear functional. If $u \in M$ then there is a unique number $t \in \mathbb{R}$ such that $u = tz$. Thus, $|f(u)| = |f(tz)| = |t| \cdot ||z|| = ||u||$ whence $|f(u)| = ||u||$ for any $u \in M$. Therefore we can apply Hahn–Banach Theorem (see Problem 389) to find a linear functional $F : L \to \mathbb{R}$ such that $F|M = f$ and $|F(x)| \le ||x||$ for any $x \in L$. In particular, if $x \in S$ then $|F(x)| \le ||x|| \le 1$ which shows that $F(S) \subset [-1, 1]$, i.e., $F \in S^*$ (we applied Problem 390 to convince ourselves that F is continuous). Finally observe that $F(z) = f(z) = ||z|| \ne 0$ and hence $F(z) = F(x - y) = F(x) - F(y) \ne 0$ which implies $F(x) \ne F(y)$. Thus S^* separates the points of L.

U.392. *Let* L *be a linear space without any topology. Suppose that* $\mathcal{F}$ *is a family of linear functionals on* L *which separates the points of* L. *Prove that the topology on* L *generated by* $\mathcal{F}$, *is Tychonoff and makes* L *locally convex linear topological space.*

Solution. Let $\varphi = \Delta \mathcal{F} : L \to \mathbb{R}^{\mathcal{F}}$ be the diagonal product of the family $\mathcal{F}$. Recall that φ is defined by $\varphi(x)(f) = f(x)$ for any $x \in L$ and $f \in \mathcal{F}$. Since $\mathcal{F}$ separates the points of L, the map φ is injective; it is easy to see that φ is a linear map; so φ is a linear isomorphism between L and $M = \varphi(L)$. As a consequence, M is a linear subspace of $\mathbb{R}^{\mathcal{F}}$; let μ be the topology on M induced from $\mathbb{R}^{\mathcal{F}}$. For any $f \in \mathcal{F}$ let $\pi_f : \mathbb{R}^{\mathcal{F}} \to \mathbb{R}$ be the natural projection onto the factor determined by f.

It is clear that the family $\mathcal{B} = \{\pi_f^{-1}(U) \cap M : U \in \tau(\mathbb{R}), f \in \mathcal{F}\}$ is a subbase of the space (M, μ). We have $\varphi^{-1}(\pi_f^{-1}(U) \cap M) = f^{-1}(U)$ for any function $f \in \mathcal{F}$ and $U \in \tau(\mathbb{R})$; so the family $\mathcal{U} = \{f^{-1}(U) : U \in \tau(\mathbb{R}), f \in \mathcal{F}\}$ coincides with the family $\{\varphi^{-1}(B) : B \in \mathcal{B}\}$. Since $\mathcal{U}$ is a subbase of the topology τ generated by

$\mathcal{F}$, the map φ is a linear homeomorphism between the spaces (L, τ) and (M, μ). This shows that (L, τ) is a linear topological space identifiable with (M, μ).

It is an easy exercise that any linear subspace of a locally convex space is locally convex; so (M, μ) is Tychonoff and locally convex because so is $\mathbb{R}^{\mathcal{F}}$ (see Fact 1 of T.131). The space (L, τ) being linearly homeomorphic to (M, μ), it is also a Tychonoff locally convex linear topological space.

U.393. *Let L be a linear space. Denote by L' the set of all linear functionals on L. Considering L' a subspace of $\mathbb{R}^L$, prove that L' is closed in $\mathbb{R}^L$.*

Solution. Take any $h \in \mathbb{R}^L \backslash L'$; then there are points $x, y \in L$ and $\alpha, \beta \in \mathbb{R}$ such that $p = h(\alpha x + \beta y) \neq q = \alpha h(x) + \beta h(y)$. Fix $U, V \in \tau(\mathbb{R})$ such that $p \in U$, $q \in V$ and $U \cap V = \emptyset$. The map $\delta : \mathbb{R}^2 \to \mathbb{R}$ defined by $\delta(t, s) = \alpha t + \beta s$ for any $(t, s) \in \mathbb{R}^2$, is continuous; so we can find sets $G, H \in \tau(\mathbb{R})$ such that $h(x) \in G$, $h(y) \in H$ and $\delta(G \times H) \subset V$.

The set $O = \{f \in \mathbb{R}^L : f(\alpha x + \beta y) \in U, \ f(x) \in G \text{ and } f(y) \in H\}$ is open in $\mathbb{R}^L$ and it is evident that $h \in O$. If $f \in O$ then $\alpha f(x) + \beta f(y) \in \delta(G \times H) \subset V$ while $f(\alpha x + \beta y) \in U$; since U and V are disjoint, we have $f(\alpha x + \beta y) \neq \alpha f(x) + \beta f(y)$, i.e., $f \notin L'$. We proved that any function $h \in \mathbb{R}^L \backslash L'$ has an open neighbourhood O with $O \cap L' = \emptyset$. This implies that $\mathbb{R}^L \backslash L'$ is open in $\mathbb{R}^L$ and hence L' is closed in $\mathbb{R}^L$.

U.394. *Given a normed space $(L, || \cdot ||)$, consider the sets $S = \{x \in L : ||x|| \leq 1\}$ and $S^* = \{f \in L^* : f(S) \subset [-1, 1]\}$. Prove that, for any point $x \in L$, the set $S^*(x) = \{f(x) : f \in S^*\}$ is bounded in $\mathbb{R}$.*

Solution. If x is the zero vector of the space L then $f(x) = 0$ for any $f \in S^*$; so $S^*(x) = \{0\}$ is a bounded subset of $\mathbb{R}$. Now, if x is a non-zero vector of L then $||x|| > 0$; for any $f \in S^*$ we have $|f(x)| = |f(||x|| \cdot \frac{x}{||x||})| = ||x|| \cdot |f(\frac{x}{||x||})|$. The point $y = \frac{x}{||x||}$ belongs to the set S because $||y|| = 1$; so $|f(y)| \leq 1$. Consequently, $|f(x)| = ||x|| \cdot |f(y)| \leq ||x||$ for any $f \in S^*$ which shows that $S^*(x) \subset [-||x||, ||x||]$ is a bounded subset of $\mathbb{R}$.

U.395. *Given a normed space $(L, || \cdot ||)$, consider the sets $S = \{x \in L : ||x|| \leq 1\}$ and $S^* = \{f \in L^* : f(S) \subset [-1, 1]\}$. Denote by L_w the set L with the topology generated by L^*. Observe that $S^* \subset C(L_w)$ and give S^* the topology τ induced from $C_p(L_w)$. Prove that (S^*, τ) is a compact space.*

Solution. The family $\mathcal{B} = \{f^{-1}(U) : f \in L^*, \ U \in \tau(\mathbb{R})\}$ is a subbase of L_w and hence the family $\{f^{-1}(U) : U \in \tau(\mathbb{R})\}$ is contained in $\mathcal{B}$ for any $f \in L^*$. This shows that, for any $f \in L^*$, the set $f^{-1}(U) \in \mathcal{B}$ is open in L_w for any $U \in \tau(\mathbb{R})$, i.e., the functional f is continuous on L_w. Therefore $S^* \subset L^* \subset C(L_w)$ and hence all elements of S^* are continuous on L_w.

The space (S^*, τ) being a subspace of $C_p(L_w)$, is also a subspace of $\mathbb{R}^L$; denote by K the closure of S^* in $\mathbb{R}^L$. For any $x \in L$ the set $S^*(x) = \{f(x) : f \in S^*\}$ is bounded in $\mathbb{R}$ by Problem 394; so there is $a_x > 0$ such that $S^*(x) \subset [-a_x, a_x]$. It is immediate that $S^* \subset M = \prod\{[-a_x, a_x] : x \in L\}$; so $K \subset M$ is compact being a closed subspace of a compact space M.

Furthermore, $S^* \subset L^* \subset L'$ where L' is the set of all linear (not necessarily continuous) functionals on L. Since L' is closed in $\mathbb{R}^L$ (see Problem 393), the set K is contained in L'.

Fix $f \in K$ and $x \in S$. If $|f(x)| > 1$ then $O = \{g \in \mathbb{R}^L : |g(x)| > 1\}$ is an open neighbourhood of f in $\mathbb{R}^L$; since f is in the closure of S^*, there is $g \in S^* \cap O$ which implies $|g(x)| > 1$, a contradiction with the definition of S^*. This proves that $|f(x)| \leq 1$ for any $x \in S$; so we can apply Problem 390 to see that the functional f is continuous on L. Since $f \in K$ was chosen arbitrarily, we proved that all elements of K are continuous on L, i.e., $K \subset L' \cap C(L) = L^*$. Now, it follows from $S^* = [-1, 1]^S \cap L^*$ that the set S^* is closed in L^*. An immediate consequence is that $K = S^*$ and hence (S^*, τ) is compact because so is K.

U.396. *Prove that L_w is functionally perfect for any normed space $(L, \|\cdot\|)$. As a consequence, any compact subspace of L_w is Eberlein compact.*

Solution. Let $S = \{x \in L : \|x\| \leq 1\}$ and $S^* = \{f \in L^* : f(S) \subset [-1, 1]\}$. The space S^* separates the points of L by Problem 391. Besides, $S^* \subset C(L_w)$ and S^* is compact if considered with the topology induced from $C_p(L_w)$ (see Problem 395). Therefore S^* is a compact subspace of $C_p(L_w)$ which separates the points of L_w, i.e., L_w is functionally perfect.

U.397. *Let L be a linear topological space. Given a sequence $\{x_n : n \in \omega\} \subset L$, prove that $x_n \to x$ in the weak topology on L if and only if $f(x_n) \to f(x)$ for any $f \in L^*$.*

Solution. Denote by L_w the space L with the weak topology, i.e., with the topology generated by the set L^*. The family $\mathcal{B} = \{f^{-1}(U) : f \in L^*, U \in \tau(\mathbb{R})\}$ is a subbase of L_w which shows that, for any $f \in L^*$, the set $f^{-1}(U)$ is open in L_w, i.e., f is also continuous on L_w. Since continuous maps preserve convergence of sequences, if $x_n \to x$ in the space L_w then $f(x_n) \to f(x)$ for any $f \in L^*$. This proves necessity.

Now assume that we have a sequence $\{x_n : n \in \omega\} \subset L$ and there exists a point $x \in L$ such that $f(x_n) \to f(x)$ for any $f \in L^*$. If $y_n = x_n - x$ then $f(y_n) = f(x_n) - f(x)$ for any $n \in \omega$ which shows that $f(y_n) \to 0$ for any $f \in L^*$. Let $\mathbf{0}$ be the zero vector of L and take any $U \in \tau(\mathbf{0}, L_w)$. There exist $f_1, \ldots, f_k \in L^*$ and $O_1, \ldots, O_k \in \tau(0, \mathbb{R})$ such that $V = \bigcap \{f_i^{-1}(O_i) : i \leq k\} \subset U$. There is $\varepsilon > 0$ such that $(-\varepsilon, \varepsilon) \subset O_i$ for any $i \leq k$; so $W = \bigcap \{f_i^{-1}((-\varepsilon, \varepsilon)) : i \leq k\} \subset V \subset U$.

Since $f_i(y_n) \to 0$ for every $i \leq k$, there is $m \in \omega$ such that $|f_i(y_n)| < \varepsilon$ for any $i \leq k$ and $n \geq m$. As a consequence, $y_n \in W \subset U$ for any $n \geq m$ which proves that the sequence $\{y_n : n \in \omega\}$ converges to $\mathbf{0}$ in the space L_w. Since L_w is a linear topological space (see Problem 392), we can apply Fact 1 of S.496 to see that the map $z \to z + x$ is continuous on L_w and therefore $y_n + x \to \mathbf{0} + x = x$ in the space L_w. Since $y_n + x = x_n$ for every $n \in \omega$, the sequence $\{x_n : n \in \omega\}$ converges to x and hence we have established sufficiency.

U.398. *For an arbitrary compact space K let $\|f\| = \sup\{|f(x)| : x \in K\}$ for any $f \in C(K)$. Denote by $C_w(K)$ the space $C(K)$ endowed with the weak topology of*

the normed space $(C(K), || \cdot ||)$. *Prove that* $\tau(C_w(K)) \supset \tau(C_p(K))$ *and show that,
in the case of* $K = \mathbb{I}$, *this inclusion is strict, i.e.,* $\tau(C_w(\mathbb{I})) \neq \tau(C_p(\mathbb{I}))$.

Solution. The topology μ of the space $(C(K), || \cdot ||)$ coincides with $\tau(C_u(K))$
by Problem 344. Since $\tau(C_p(K)) \subset \tau(C_u(K))$ (see TFS-086), we have
$\tau(C_p(K)) \subset \mu$.

For any $x \in K$ let $e_x(f) = f(x)$ for any $f \in C(K)$; the map $e_x : C(K) \to \mathbb{R}$ is
a linear functional on $C(K)$. It follows from the definition of the pointwise topology
on $C(K)$ that the set $e_x^{-1}(U)$ is open in $C_p(K)$ for any $x \in K$ and $U \in \tau(\mathbb{R})$; so
every e_x is continuous on $C_p(K)$ and hence on $(C(K), ||\cdot||)$. Therefore the topology
of $C_w(K)$ contains the topology generated by the family $\mathcal{E} = \{e_x : x \in K\}$. It is
straightforward that $\{e_x^{-1}(U) : x \in K, U \in \tau(\mathbb{R})\}$ is a subbase in $C_p(K)$; so
the topology generated by the family $\mathcal{E}$ coincides with $\tau(C_p(K))$. This proves that
$\tau(C_p(K)) \subset \tau(C_w(K))$.

For any $f \in C(\mathbb{I})$ let $\varphi(f) = \int_{-1}^{1} f(x)dx$; it is straightforward that $\varphi : C(\mathbb{I}) \to$
$\mathbb{R}$ is a linear functional on $C(\mathbb{I})$. Observe that,

(1) $|\varphi(f)| \leq \int_{-1}^{1} |f(x)|dx \leq \int_{-1}^{1} ||f||dx \leq 2||f||$ for any $f \in C(\mathbb{I})$.

If $S = \{f \in C(\mathbb{I}) : ||f|| \leq 1\}$ then it follows from (1) that $\varphi(S) \subset [-2, 2]$; so
the functional φ is continuous on $(C(\mathbb{I}), || \cdot ||)$ (see Problem 390).

For a fixed $n \in \omega$ let $f_n(t) = 0$ for all $t \in [-1, 0] \cup [\frac{1}{n+1}, 1]$. If $t \in [0, \frac{1}{2(n+1)}]$
then $f_n(t) = 4(n+1)^2 t$; if $t \in [\frac{1}{2(n+1)}, \frac{1}{n+1}]$ then $f_n(t) = 4(n+1)^2(\frac{1}{n+1} - t)$. It is
easy to see that $f_n \in C(\mathbb{I})$ for all $n \in \omega$ and $f_n(t) \to 0$ for any $t \in \mathbb{I}$. Let $u(t) = 0$
for all $t \in \mathbb{I}$; then $f_n \to u$ in the space $C_p(\mathbb{I})$ by TFS-143. However, $\varphi(f_n) = 1$
for all $n \in \omega$ while $\varphi(u) = 0$. Therefore the sequence $\{\varphi(f_n) : n \in \omega\}$ does not
converge to $\varphi(u)$ which, together with Problem 397, shows that $F = \{f_n : n \in \omega\}$
does not converge to u in $C_w(\mathbb{I})$. Thus we have found a sequence F which converges
to u in $C_p(\mathbb{I})$ and does not converge to u in $C_w(\mathbb{I})$. This proves that the topologies of
the spaces $C_p(\mathbb{I})$ and $C_w(\mathbb{I})$ do not coincide.

U.399. *Suppose that* K *is a compact space and let* $||f|| = \sup\{|f(x)| : x \in K\}$ *for
any* $f \in C(K)$. *Denote by* $C_w(K)$ *the space* $C(K)$ *endowed with the weak topology
of the normed space* $(C(K), ||\cdot||)$. *Prove that, for any* $X \subset C(K, \mathbb{I})$, *if* X *is compact
as a subspace of* $C_p(K)$ *then the topologies, induced on* X *from the spaces* $C_p(K)$
and $C_w(K)$, *coincide.*

Solution. Let τ_p be the topology on X induced from $C_p(K)$ and denote by τ_w the
topology on X as a subspace of $C_w(K)$. It follows from Problem 398 that $\tau_p \subset \tau_w$;
so the identity map $id_X : (X, \tau_w) \to (X, \tau_p)$ is a condensation. For any $r > 0$ let
$S_r = \{f \in C(K) : ||f|| \leq r\}$. As usual, the function u is the zero vector of $C(K)$,
i.e., $u(x) = 0$ for every $x \in K$; we will prove first that

(1) if $r > 0$, $\{f_n : n \in \omega\} \subset S_r$ and $f_n \to u$ in $C_p(K)$ then $f_n \to u$ in the
　　space $C_w(K)$.

Assume, towards a contradiction, that the sequence $F = \{f_n : n \in \omega\}$ does not converge to u in $C_w(K)$. Then there exists a linear functional $\varphi : C(K) \to \mathbb{R}$ such that φ is continuous on the space $(C(K), ||\cdot||)$ and the sequence $\varphi(F)$ does not converge to 0 (see Problem 397). Consequently, there is $\varepsilon > 0$ and an infinite $A \subset \omega$ such that the set $\{\varphi(f_n) : n \in A\}$ is contained either in $[\varepsilon, +\infty)$ or $(-\infty, -\varepsilon]$. Passing, if necessary, to the subsequence $\{f_n : n \in A\}$ and multiplying every f_n by -1 we can assume, without loss of generality, that $A = \omega$ and $\varphi(f_n) \geq \varepsilon$ for every $n \in A$.

Recall that $\text{conv}(P) = \{t_1 g_1 + \ldots + t_k g_k : k \in \mathbb{N}, g_i \in P, t_i \in [0,1]$ for all $i \leq k$ and $\sum_{i=1}^{k} t_i = 1\}$ is the convex hull of the set P; it is convex for any $P \subset C(K)$ and a set $Q \subset C(K)$ is convex if and only if $Q = \text{conv}(Q)$ (see Fact 1 of T.104). It is also evident that $\varphi(P)$ is a convex subset of $\mathbb{R}$ whenever P is a convex subset of $C(K)$.

It follows from Problem 346 that there exists a sequence $\{h_n : n \in \omega\} \subset \text{conv}(F)$ which converges uniformly to u; since φ is continuous on $(C(K), ||\cdot||)$, the sequence $H = \{\varphi(h_n) : n \in \omega\}$ converges to $\varphi(u) = 0$. In particular, there is $m \in \omega$ such that $\varphi(h_m) < \varepsilon$. Since $h_m \in \text{conv}(F)$, there are $n \in \mathbb{N}$ and $t_1, \ldots, t_n \in [0,1]$ such that $\sum_{i=1}^{n} t_i = 1$ and $h_m = t_1 f_1 + \ldots + t_n f_n$. However, $\varphi(f_i) \in R_\varepsilon = [\varepsilon, +\infty)$ for every $i \leq n$; so $\varphi(h_m) = t_1 \varphi(f_1) + \ldots + t_n \varphi(f_n) \in \text{conv}(R_\varepsilon)$. The set R_ε is convex; so $\varphi(h_m) \in R_\varepsilon$, i.e., $\varphi(h_m) \geq \varepsilon$; this contradiction shows that (1) is proved.

Our next step is to show that

(2) if $G = \{g_n : n \in \omega\} \subset X$ and $G \to g$ in (X, τ_p) then $G \to g$ in (X, τ_w).

Observe first that, the space (X, τ_p) being compact, the function g belongs to X and hence $G \cup \{g\} \subset C(K, \mathbb{I}) = S_1$. If $f_n = g_n - g$ for any $n \in \omega$ then we have the inequality $||f_n|| \leq ||g_n|| + ||g|| \leq 2$; so $F = \{f_n : n \in \omega\} \subset S_2$ and $F \to u$ in $C_p(K)$. Apply (1) to conclude that $f_n \to u$ in the space $C_w(K)$. Since $C_w(K)$ is a linear topological space by Problem 392, the operation $f \to f + g$ is a homeomorphism on $C_w(K)$ (see Fact 1 of S.496); so the sequence $\{f_n + g : n \in \omega\} = \{g_n : n \in \omega\}$ converges to $u + g = g$ in the space $C_w(K)$ and hence (2) is proved.

If $D \subset X$ is an infinite closed and discrete subset of (X, τ_w) then use compactness and Fréchet–Urysohn property of (X, τ_p) (recall that (X, τ_p) is Eberlein compact) to choose a faithfully indexed sequence $G = \{g_n : n \in \omega\} \subset D$ which τ_p-converges to some $g \in X \backslash G$. By (2), the sequence G also converges to g in (X, τ_w) which is a contradiction with the fact that $G \subset D$ is closed and discrete. This proves that $Y = (X, \tau_w)$ is countably compact. The map $id_X : Y \to (X, \tau_p)$ is a condensation; so we can apply Problem 223 to see that id_X is a homeomorphism and hence $\tau_w = \tau_p$, i.e., the topologies induced on X from the spaces $C_w(K)$ and $C_p(K)$ coincide.

U.400. *(The original definition of an Eberlein compact space) Prove that X is an Eberlein compact space if and only if it is homeomorphic to a weakly compact subset of a Banach space.*

Solution. If $(L, ||\cdot||)$ is a Banach space then denote by L_w the set L endowed with the weak topology of $(L, ||\cdot||)$. It was proved in Problem 396 that L_w is functionally perfect and hence any compact $K \subset L_w$ is Eberlein compact. Therefore, if a compact space X is homeomorphic to a weakly compact subspace of a Banach space then X is Eberlein compact.

Now assume that X is an Eberlein compact space; then there exists a compact space K such that X embeds in $C_p(K)$ (see Problem 301). The spaces $C_p(K)$ and $C_p(K, (-1, 1))$ are homeomorphic; so we can assume, without loss of generality, that $X \subset C_p(K, (-1, 1)) \subset C_p(K, \mathbb{I})$. If $||f|| = \sup\{|f(x)| : x \in K\}$ for any $f \in C(K)$ then $(C(K), ||\cdot||)$ is a Banach space by Problem 388. Denote by $C_w(K)$ the space $C(K)$ endowed with the weak topology of $(C(K), ||\cdot||)$. Then the topologies induced on X from the spaces $C_p(K)$ and $C_w(K)$ coincide by Problem 399; so X embeds in the space $C_w(K)$ which shows that X is homeomorphic to a weakly compact subset of the Banach space $(C(K), ||\cdot||)$.

U.401. *Prove that every subspace of a splittable space is splittable.*

Solution. Suppose that X is a splittable space and take any $Y \subset X$; then the restriction map $\pi : \mathbb{R}^X \to \mathbb{R}^Y$ is continuous (to see it apply TFS-152 to the sets X and Y with their respective discrete topologies). Given any $f \in \mathbb{R}^Y$ let $g(x) = f(x)$ for any $x \in Y$ and $g(x) = 0$ whenever $x \in X \backslash Y$. Then $g \in \mathbb{R}^X$ and $\pi(g) = f$; the space X being splittable, there is a countable $B \subset C_p(X)$ such that $g \in \overline{B}$ (the bar denotes the closure in $\mathbb{R}^X$). The set $A = \pi(B)$ is countable and it follows from continuity of π that $f = \pi(g) \in \overline{A}$ (here the bar stands for the closure in $\mathbb{R}^Y$). Therefore every $f \in \mathbb{R}^Y$ is in the closure of a countable subset of $C_p(Y)$, i.e., Y is splittable.

U.402. *Prove that every second countable space is splittable.*

Solution. If a space X is second countable then $d(C_p(X)) = iw(X) \le \omega$ (see TFS-174) and hence $C_p(X)$ is separable. If $A \subset C_p(X)$ is countable and dense in $C_p(X)$ then A is also dense in $\mathbb{R}^X$ because $C_p(X)$ is dense in $\mathbb{R}^X$. Thus, there is a countable set $A \subset C_p(X)$ such that $f \in \overline{A}$ for any $f \in \mathbb{R}^X$ (the bar denotes the closure in $\mathbb{R}^X$); so X is splittable.

U.403. *Prove that $\psi(X) \le \omega$ for every splittable space X.*

Solution. Suppose that a space X is splittable and fix any $a \in X$. If $g(a) = 0$ and $g(x) = 1$ for any $x \in X \backslash \{a\}$ then $g \in \mathbb{R}^X$; so there is a countable $A \subset C_p(X)$ such that $g \in \overline{A}$ where the bar denotes the closure in $\mathbb{R}^X$. If $f \in A$ then f is continuous; so $P_f = f^{-1}(f(a))$ is a G_δ-subset of X such that $a \in P_f$; consequently, $P = \bigcap\{P_f : f \in A\}$ is a G_δ-subset of X with $a \in P$. For any $x \in X \backslash \{a\}$ we have $g(x) = 1$; so the set $V = \{f \in \mathbb{R}^X : f(a) < \frac{1}{3}, f(x) > \frac{2}{3}\}$ is an open neighbourhood of the function g in $\mathbb{R}^X$. It follows from $g \in \overline{A}$ that there is $f \in A \cap V$ and hence $f(x) > \frac{2}{3}$, $f(a) < \frac{1}{3}$ which implies that $f(x) \ne f(a)$ and therefore $x \notin P_f$; it follows from $P \subset P_f$ that $x \notin P$.

This proves that $x \notin P$ for any $x \in X \setminus \{a\}$ and hence $\{a\} = P$ is a G_δ-set. The point $a \in X$ was chosen arbitrarily; so we proved that any point of X is a G_δ-set, i.e., $\psi(X) \leq \omega$.

U.404. *Prove that, if X condenses onto a splittable space, then X is splittable. In particular, any space of countable i-weight is splittable.*

Solution. Let $\varphi : X \to Y$ be a condensation of a space X onto a splittable space Y. For any $f \in \mathbb{R}^Y$ let $\varphi^*(f) = f \circ \varphi$; then $\varphi^* : \mathbb{R}^Y \to \mathbb{R}^X$ is an embedding (to see this apply TFS-163 to the sets X and Y with their respective discrete topologies). Given any $f \in \mathbb{R}^X$ the function $f_0 = f \circ \varphi^{-1}$ belongs to $\mathbb{R}^Y$ and $\varphi^*(f_0) = f$; this shows that $\varphi^*(\mathbb{R}^Y) = \mathbb{R}^X$. Besides, $\varphi^*(C_p(Y))$ is a (dense) subspace of $C_p(X)$ (this was also proved in TFS-163).

Given any $g \in \mathbb{R}^X$ there is $g_0 \in \mathbb{R}^Y$ such that $\varphi^*(g_0) = g$. The space Y being splittable, there is a countable set $B \subset C_p(Y)$ such that $g_0 \in \mathrm{cl}_{\mathbb{R}^Y}(B)$. The set $A = \varphi^*(B) \subset C_p(X)$ is countable and it follows from continuity of φ^* that $g \in \mathrm{cl}_{\mathbb{R}^X}(A)$. This proves that every $g \in \mathbb{R}^X$ belongs to the closure of a countable subset of $C_p(X)$, i.e., X is splittable. Finally apply our result and Problem 402 to conclude that $iw(X) = \omega$ implies that X is splittable.

U.405. *Give an example of a splittable space which does not condense onto a second countable space.*

Solution. Let X be a discrete space of cardinality $2^\mathfrak{c}$. Then $C_p(X) = \mathbb{R}^X$; so X is splittable. Any space that condenses onto a second countable one has cardinality at most $\mathfrak{c}$ (because any second countable space has cardinality at most $\mathfrak{c}$). Thus X is a splittable space which does not condense onto a second countable space because $|X| = 2^\mathfrak{c} > \mathfrak{c}$.

U.406. *Give an example of a metrizable space which is not splittable.*

Solution. Our main tool will be the following fact.

Fact 1. Suppose that M is a space such that $|M| \geq \mathfrak{c}$ and $|C_p(M)| \leq \mathfrak{c}$. If X is a space for which there exists a disjoint family $\{M_\alpha : \alpha < \mathfrak{c}^+\}$ of subspaces of X with $M_\alpha \simeq M$ for any $\alpha < \mathfrak{c}$ then X is not splittable.

Proof. Assume towards a contradiction, that X is splittable and fix a homeomorphism $\varphi_\alpha : M \to M_\alpha$ for any $\alpha < \mathfrak{c}^+$. It follows from $|\exp M| = 2^{|M|} \geq 2^\mathfrak{c} \geq \mathfrak{c}^+$ that we can choose a family $\{A_\alpha : \alpha < \mathfrak{c}^+\}$ of subsets of M such that $A_\alpha \neq A_\beta$ whenever $\alpha \neq \beta$; let $N_\alpha = \varphi_\alpha(A_\alpha)$ for any $\alpha < \mathfrak{c}^+$. For the set $N = \bigcup\{N_\alpha : \alpha < \mathfrak{c}^+\} \subset X$ let $h(x) = 1$ for all $x \in N$ and $h(x) = 0$ if $x \in X \setminus N$. Then $h \in \mathbb{R}^X$ so, by splittability of X, there is a set $P = \{f_n : n \in \omega\} \subset C_p(X)$ such that $h \in \overline{P}$ (the bar denotes the closure in $\mathbb{R}^X$).

The function $g_n^\alpha = (f_n|M_\alpha) \circ \varphi_\alpha$ is continuous on M for any $\alpha < \mathfrak{c}^+$ and $n \in \omega$; since $|C_p(M)| \leq \mathfrak{c}$, there are at most $\mathfrak{c}$-many distinct countable subsets of $C_p(M)$. As a consequence, there are distinct $\alpha, \beta < \mathfrak{c}^+$ such that $g_n^\alpha = g_n^\beta$ for any $n \in \omega$.

The sets A_α and A_β being distinct, we can assume, without loss of generality, that there is $t \in A_\alpha \backslash A_\beta$; let $t_\alpha = \varphi_\alpha(t)$ and $t_\beta = \varphi_\beta(t)$. Then $t_\alpha \in N$ and $t_\beta \notin N$; so the set $V = \{f \in \mathbb{R}^X : f(t_\alpha) > \frac{2}{3}, \ f(t_\beta) < \frac{1}{3}\}$ is open in $\mathbb{R}^X$ and contains h. Since $h \in \overline{P}$, there is $n \in \omega$ such that $f_n \in V$ and therefore $f_n(t_\alpha) \neq f_n(t_\beta)$. This implies $g_n^\alpha(t) = f_n(\varphi_\alpha(t)) = f_n(t_\alpha) \neq f_n(t_\beta) = g_n^\beta(t)$ and hence $g_n^\alpha \neq g_n^\beta$ which is a contradiction. This shows that X cannot be splittable and completes the proof of Fact 1.

Returning to our solution take a discrete space D of cardinality $\mathfrak{c}^+$ and let $X = \mathbb{R} \times D$. If $\{d_\alpha : \alpha < \mathfrak{c}^+\}$ is a faithful enumeration of D then the family $\{\mathbb{R} \times \{d_\alpha\} : \alpha < \mathfrak{c}^+\}$ consists of $\mathfrak{c}^+$-many disjoint homeomorphic copies of $\mathbb{R}$. Since $|\mathbb{R}| = \mathfrak{c}$ and $|C_p(\mathbb{R})| = \mathfrak{c}$, we can apply Fact 1 to see that X is not splittable. Finally observe that the product of two metrizable spaces is metrizable; so X is a metrizable non-splittable space.

U.407. *Give an example of a splittable space whose square is not splittable.*

Solution. Take a discrete space D of cardinality $\mathfrak{c}^+$ and consider the set $Y = \mathbb{R} \times D$. Choose a faithful enumeration $\{d_\alpha : \alpha < \mathfrak{c}^+\}$ of the set D and note that the family $\{\mathbb{R} \times \{d_\alpha\} : \alpha < \mathfrak{c}^+\}$ consists of $\mathfrak{c}^+$-many disjoint homeomorphic copies of $\mathbb{R}$. Since $|\mathbb{R}| = \mathfrak{c}$ and $|C_p(\mathbb{R})| = \mathfrak{c}$, we can apply Fact 1 of U.406 to see that Y is not splittable.

To see that the space $X = \mathbb{R} \oplus D$ is splittable let us identify $\mathbb{R}$ and D with the respective clopen subsets of X and take any function $h \in \mathbb{R}^X$. Since $C_p(\mathbb{R})$ is separable, there is a set $P = \{f_n : n \in \omega\} \subset C_p(\mathbb{R})$ such that $h|\mathbb{R} \in \overline{P}$ (the bar denotes the closure in $\mathbb{R}^\mathbb{R}$). For any $n \in \omega$ let $g_n(x) = h(x)$ if $x \in D$ and $g_n(x) = f_n(x)$ for all $x \in \mathbb{R}$. It is straightforward that $Q = \{g_n : n \in \omega\} \subset C_p(X)$.

Now take any $U \in \tau(h, \mathbb{R}^X)$; there is a finite $A \subset X$ and $\varepsilon > 0$ such that $W = \{f \in \mathbb{R}^X : |f(x) - h(x)| < \varepsilon$ for any $x \in A\} \subset U$. The function $h|\mathbb{R}$ being in the closure of P, there is $n \in \omega$ such that $|f_n(x) - h(x)| < \varepsilon$ for any $x \in A \cap \mathbb{R}$. Since $f_n|D = h|D$, the inequality $|f_n(x) - h(x)| < \varepsilon$ holds for all $x \in A$ which shows that $f_n \in W \cap Q \subset U \cap Q$. Therefore any neighbourhood of h in $\mathbb{R}^X$ meets Q, i.e., h belongs to the closure of Q in $\mathbb{R}^X$; since $h \in \mathbb{R}^X$ was chosen arbitrarily, we proved that the space X is splittable. However, it is easy to see that Y embeds in $X \times X$; this, together with Problem 401, implies that $X \times X$ is not splittable.

U.408. *Prove that a space X with a unique non-isolated point is splittable if and only if $\psi(X) \leq \omega$.*

Solution. Necessity follows from Problem 403; so assume that a is a unique non-isolated point of a space X and $\psi(X) = \omega$; let $Y = X \backslash \{a\}$ and choose a family $\mathcal{U} = \{U_n : n \in \omega\} \subset \tau(a, X)$ such that $\bigcap \mathcal{U} = \{a\}$ and $U_{n+1} \subset U_n$ for any $n \in \omega$.

The space X is normal (see Claim 2 of S.018); since the set $D_n = (Y \backslash U_n) \cup \{a\}$ is closed and discrete in X, any real-valued function on D_n extends to a continuous real-valued function on the space X for any $n \in \omega$.

Now, if we are given an arbitrary $h \in \mathbb{R}^X$ then, for every $n \in \omega$, there is a function $f_n \in C_p(X)$ such that $f_n|D_n = h|D_n$. Since $X = \bigcup_{n \in \omega} D_n$, for any

$x \in X$ there is $m \in \omega$ such that $f_n(x) = h(x)$ for all $n \geq m$. Thus the sequence $S = \{f_n : n \in \omega\} \subset C_p(X)$ converges to h (see TFS-143) and hence h belongs to the closure of S. Therefore X is splittable, i.e., we settled sufficiency.

U.409. *Let X be a non-discrete space. Prove that, for any $f \in \mathbb{R}^X$, there exists a countable $A \subset \mathbb{R}^X \backslash C_p(X)$ such that $f \in \overline{A}$ (the closure is taken in $\mathbb{R}^X$).*

Solution. If $f \in \mathbb{R}^X \backslash C_p(X)$ then $A = \{f\}$ does the job; so we can assume, without loss of generality, that $f \in C_p(X)$. Since X is non-discrete, there is a set $Y \subset X$ which is not open; let $h(x) = 1$ for any $x \in Y$ and $h(x) = 0$ whenever $x \in X \backslash Y$. It is evident that $h \in \mathbb{R}^X \backslash C_p(X)$; so $h_n = \frac{1}{n} h \in \mathbb{R}^X \backslash C_p(X)$; an immediate consequence is that $f_n = f + h_n \in \mathbb{R}^X \backslash C_p(X)$ for any $n \in \mathbb{N}$. Since $h_n \in [0, \frac{1}{n}]^X$ for any $n \in \mathbb{N}$, the sequence $\{f_n - f : n \in \mathbb{N}\} = \{h_n : n \in \mathbb{N}\}$ converges uniformly to zero; this implies that the sequence $A = \{f_n : n \in \mathbb{N}\} \subset \mathbb{R}^X \backslash C_p(X)$ converges to f and hence $f \in \overline{A}$.

U.410. *Let X be a splittable space. Prove that every regular uncountable cardinal is a caliber of $C_p(X)$.*

Solution. Suppose that κ is an uncountable regular cardinal and take an arbitrary family $\mathcal{U} \subset \tau^*(C_p(X))$ of cardinality κ. Choose, for every $U \in \mathcal{U}$, a set $O_U \in \tau(\mathbb{R}^X)$ such that $O_U \cap C_p(X) = U$. It is clear that $U \neq V$ implies $O_U \neq O_V$; so the family $\mathcal{O} = \{O_U : U \in \mathcal{U}\} \subset \tau^*(\mathbb{R}^X)$ has cardinality κ. The cardinal κ being a caliber of $\mathbb{R}^X$ (see SFFS-281), there is a function $h \in \mathbb{R}^X$ and a family $\mathcal{U}' \subset \mathcal{U}$ such that $|\mathcal{U}'| = \kappa$ and $h \in \bigcap\{O_U : U \in \mathcal{U}'\}$. By splittability of X there is a countable $A \subset C_p(X)$ such that $h \in \overline{A}$ (the bar denotes the closure in $\mathbb{R}^X$).

For any $U \in \mathcal{U}'$ we can take a function $f_U \in A \cap O_U$; since κ is regular and uncountable, there is $f \in A$ and a family $\mathcal{W} \subset \mathcal{U}'$ such that $|\mathcal{W}| = \kappa$ and $f_U = f$ for any $U \in \mathcal{W}$. In particular, $f \in O_U \cap C_p(X) = U$ for any $U \in \mathcal{W}$; so $\bigcap \mathcal{W} \neq \emptyset$. Since for any $\mathcal{U} \subset \tau^*(C_p(X))$ of cardinality κ we have found $\mathcal{W} \subset \mathcal{U}$ with $|\mathcal{W}| = \kappa$ and $\bigcap \mathcal{W} \neq \emptyset$, the cardinal κ is a caliber of $C_p(X)$.

U.411. *Prove that every splittable space has a small diagonal.*

Solution. If X is a splittable space then ω_1 is a caliber of $C_p(X)$ by Problem 410; so X has a small diagonal by SFFS-293.

U.412. *Prove that $C_p(X)$ is splittable if and only if it condenses onto a second countable space.*

Solution. Sufficiency is a consequence of Problem 404; now, if $C_p(X)$ is splittable then $\psi(C_p(X)) = \omega$ by Problem 403 and therefore $iw(C_p(X)) = \psi(C_p(X)) = \omega$ (see TFS-173), i.e., $C_p(X)$ condenses onto a second countable space.

U.413. *Show that an open continuous image of a splittable space can fail to be splittable.*

Solution. The space $\beta\omega$ is separable; so $iw(C_p(\beta\omega)) = d(\beta\omega) = \omega$ and hence $C_p(\beta\omega)$ is splittable (see TFS-173 and Problem 404). If $\pi : C_p(\beta\omega) \to C_p(\beta\omega \backslash \omega)$ is the restriction map then π is continuous, onto and open by TFS-152. Apply TFS-

173 once more to observe that $\psi(C_p(\beta\omega\backslash\omega)) = d(\beta\omega\backslash\omega) > \omega$ (see TFS-371); so $C_p(\beta\omega\backslash\omega)$ is not splittable by Problem 403. Thus the non-splittable space $C_p(\beta\omega\backslash\omega)$ is an open continuous image of the splittable space $C_p(\beta\omega)$.

U.414. *Let X be a space of cardinality $\leq \mathfrak{c}$. Prove that X is splittable if and only if the i-weight of X is countable.*

Solution. If $iw(X) = \omega$ then X is splittable by Problem 404; so we have sufficiency. Now assume that $|X| \leq \mathfrak{c}$ and X is splittable. The space $\mathbb{R}^X$ is separable by TFS-108; so let $H = \{h_n : n \in \omega\}$ be a dense subset of $\mathbb{R}^X$. By splittability of X, for any $n \in \omega$, there is a countable set $A_n \subset C_p(X)$ such that $h_n \in \overline{A}_n$ (the bar denotes the closure in $\mathbb{R}^X$). The set $A = \bigcup_{n\in\omega} A_n$ is countable and $H \subset \overline{A}$; so $\mathbb{R}^X = \overline{H} \subset \overline{A}$ which shows that A is dense in $\mathbb{R}^X$ and hence in $C_p(X)$.

Thus the space $C_p(X)$ is separable; so we can apply TFS-174 to conclude that $iw(X) = d(C_p(X)) \leq \omega$, i.e., i-weight of X is countable.

U.415. *Prove that a space X is splittable if and only if, for every $f \in \mathbb{D}^X$, there is a countable $A \subset C_p(X)$ such that $f \in \overline{A}$ (the closure is taken in $\mathbb{R}^X$).*

Solution. Necessity is evident; so assume that X is $\mathbb{D}$-splittable, i.e., for any $h \in \mathbb{D}^X$ there is a countable $A \subset C_p(X)$ such that $h \in \overline{A}$ (the bar denotes the closure in $\mathbb{R}^X$).

Fact 1. For any space Z the set $\mathbb{Q}^Z$ is uniformly dense in $\mathbb{R}^Z$, i.e., for any $f \in \mathbb{R}^Z$ there is a sequence $\{f_n : n \in \mathbb{N}\} \subset \mathbb{Q}^Z$ which converges uniformly to f.

Proof. Take any $f \in \mathbb{R}^Z$; denote by $\mathbb{Z}$ the set of all integers and fix $n \in \mathbb{N}$. For any $m \in \mathbb{Z}$ let $I_m = [\frac{m}{n}, \frac{m+1}{n})$. It is clear that $\{I_m : m \in \mathbb{Z}\}$ is a disjoint cover of $\mathbb{R}$; so the family $\{f^{-1}(I_m) : m \in \mathbb{Z}\}$ is a disjoint cover of Z.

For any $z \in Z$ there is a unique $m \in \mathbb{Z}$ such that $z \in f^{-1}(I_m)$; let $f_n(z) = \frac{m}{n}$. It is straightforward that f_n is a well-defined function and $f_n \in \mathbb{Q}^Z$ for any $n \in \mathbb{N}$. Besides, it follows from $z \in f^{-1}(I_m)$ that $f(z) \in I_m$ and $f_n(z) \in I_m$ and therefore $|f(z) - f_n(z)| < \frac{1}{n}$. This proves that $|f(z) - f_n(z)| < \frac{1}{n}$ for any $z \in Z$ and $n \in \mathbb{N}$; an immediate consequence is that the sequence $\{f_n : n \in \mathbb{N}\}$ converges uniformly to f, i.e., Fact 1 is proved.

Returning to our solution let $\{q_i : i \in \omega\}$ be a faithful enumeration of $\mathbb{Q}$ and take any function $h \in \mathbb{Q}^X$. For any $n \in \omega$ consider the function h_n defined by $h_n(x) = h(x)$ if $h(x) = q_i$ for some $i \leq n$; if $h(x) = q_i$ and $i > n$ then $h_n(x) = 0$. It is clear that, for any $x \in X$ if $h(x) = q_i$ then $h_n(x) = h(x)$ for all $n \geq i$; as a consequence, the sequence $\{h_n : n \in \omega\}$ converges to h (see TFS-143).

Since $h_n(X) \subset \{q_0, \dots, q_n\}$, there are $g_0^1, \dots, g_n^n \in \mathbb{D}^X$ such that $h_n = \sum_{i=0}^{n} q_i g_n^i$ for all $n \in \omega$. Fix $n \in \omega$; by $\mathbb{D}$-splittability of X, for any $i \leq n$, there is a countable set $B_i \subset C_p(X)$ such that $g_n^i \in \overline{B}_i$. The set $C_i = \{q_i f : f \in B_i\}$ is still contained in $C_p(X)$ and $q_i g_n^i \in \overline{C}_i$ because the operation of multiplication by q_i is continuous in $\mathbb{R}^X$. Let $C = \{f_0 + \ldots + f_n : f_i \in C_i \text{ for all } i \leq n\}$; recalling that the sum operation is also continuous in $\mathbb{R}^X$, we conclude that $h_n \in \overline{C}$. This proves that

(1) for any $n \in \omega$ there is a set $A_n \subset C_p(X)$ such that $h_n \in \overline{A}_n$.

Since $h_n \to h$, it follows from (1) that the set $A = \bigcup_{n\in\omega} A_n$ contains h in its closure. This shows that

(2) for every $h \in \mathbb{Q}^X$ there is a countable $A \subset C_p(X)$ such that $h \in \overline{A}$.

Finally, if $g \in \mathbb{R}^X$ then there is a sequence $\{g_n : n \in \omega\} \subset \mathbb{Q}^X$ such that $g_n \to g$ (see Fact 1). The property (2) makes it possible to find a countable $P_n \subset C_p(X)$ such that $g_n \in \overline{P}_n$ for any $n \in \omega$. It immediate that the set $P = \bigcup_{n\in\omega} P_n \subset C_p(X)$ is countable and $g \in \overline{P}$; this proves that X is splittable and settles sufficiency; so our solution is complete.

U.416. *Prove that a space X is splittable if and only if, for any $A \subset X$, there exists a continuous map $\varphi : X \to \mathbb{R}^\omega$ such that $A = \varphi^{-1}\varphi(A)$.*

Solution. To prove sufficiency fix any $h \in \mathbb{D}^X$; for the set $A = h^{-1}(1)$, there is a continuous map $\varphi : X \to \mathbb{R}^\omega$ such that $A = \varphi^{-1}(\varphi(A))$. The space $Y = \varphi(X)$ is second countable, so $C_p(Y)$ must be separable; take a countable dense $Q \subset C_p(Y)$. Then the set $P = \{f \circ \varphi : f \in Q\} \subset C_p(X)$ is countable; we will prove that $h \in \overline{P}$ (the bar denotes the closure in $\mathbb{R}^X$).

Let $A' = \varphi(A)$; given any $O \in \tau(h, \mathbb{R}^X)$, there is $\varepsilon > 0$ and a finite set $K \subset X$ such that $U = \{f \in \mathbb{R}^X : |f(x) - h(x)| < \varepsilon \text{ for all } x \in K\} \subset O$. If $K_0 = K \cap A$ then it follows from the choice of φ that $\varphi(K\backslash K_0) \cap A' = \emptyset$. The set Q being dense in $C_p(Y)$ there is a function $f \in Q$ such that $|f(y) - 1| < \varepsilon$ for all $y \in \varphi(K_0)$ and $|f(y)| < \varepsilon$ whenever $y \in \varphi(K\backslash K_0)$. Then $g = f \circ \varphi \in P$ and $|g(x) - 1| < \varepsilon$ for any $x \in K_0$ while $|g(x)| < \varepsilon$ if $x \in K\backslash K_0$. Since $h(x) = 1$ for all $x \in K_0$ and $h(x) = 0$ whenever $x \in K\backslash K_0$, we have $|h(x) - g(x)| < \varepsilon$ for all $x \in K$. An immediate consequence is that $g \in U \cap P \subset O \cap P$ and hence $h \in \overline{P}$. Therefore, for any $h \in \mathbb{D}^X$ there is a countable set $P \subset C_p(X)$ such that $h \in \overline{P}$; this shows that X is splittable (see Problem 415) and hence we proved sufficiency.

Now assume that X is splittable and take any $A \subset X$; let $h(x) = 1$ for all $x \in A$ and $h(x) = 0$ whenever $x \in X\backslash A$. There is a countable set $P \subset C_p(X)$ such that $h \in \overline{P}$. The map $\varphi = \Delta P : X \to \mathbb{R}^P$ is continuous; since P is countable, the space $\mathbb{R}^P$ is homeomorphic to a subspace of $\mathbb{R}^\omega$; so φ can be considered to be a map from X to $\mathbb{R}^\omega$. Given $x \in A$ and $y \in X\backslash A$ we have $h(x) = 1$ and $h(y) = 0$; so there is a function $f \in P$ such that $f(x) > \frac{2}{3}$ and $f(y) < \frac{1}{3}$. In particular, $f(x) \neq f(y)$ and hence $\varphi(x) \neq \varphi(y)$. This proves that $\varphi(A) \cap \varphi(X\backslash A) = \emptyset$; an immediate consequence is that $A = \varphi^{-1}(\varphi(A))$ so we settled necessity.

U.417. *Prove that any pseudocompact splittable space is compact and metrizable.*

Solution. If Z is a space and $\mathcal{A} \subset \exp Z$ then $\mathcal{A}|B = \{A \cap B : A \in \mathcal{A}\}$ for any set $B \subset Z$.

Fact 1. If Z is a pseudocompact space and $\psi(z, Z) \leq \omega$ for some point $z \in Z$ then $\chi(z, Z) \leq \omega$.

Proof. Choose a sequence $\mathcal{B} = \{U_n : n \in \omega\} \subset \tau(z, Z)$ such that $\overline{U}_{n+1} \subset U_n$ for all $n \in \omega$ and $\bigcap \mathcal{B} = \{z\}$. If $\mathcal{B}$ is not a local base at z then there is $U \in \tau(z, Z)$ for which $U_n \backslash U \neq \emptyset$ for every $n \in \omega$. Fix $V \in \tau(z, Z)$ such that $\overline{V} \subset U$; then $W_n = U_n \backslash \overline{V} \in \tau^*(Z)$ for each $n \in \omega$. It is clear that $W_{n+1} \subset W_n$ for any $n \in \omega$, so the family $\{W_n : n \in \omega\}$ is centered; by pseudocompactness of the space Z, the set $F = \bigcap_{n \in \omega} \overline{W}_n$ is non-empty. It is easy to see that $F \subset Z \backslash V$ and at the same time, $F \subset \bigcap_{n \in \omega} \overline{U}_n = \{z\} \subset V$; this contradiction shows that $\mathcal{B}$ is a countable local base at z, i.e., $\chi(z, Z) \leq \omega$ and hence Fact 1 is proved.

Fact 2. If Z is a pseudocompact first countable space then, for any $Y \subset Z$ with $|Y| \leq \mathfrak{c}$ there is $Y' \subset Z$ such that $|Y'| \leq \mathfrak{c}$, the space Y' is pseudocompact and $Y \subset Y'$.

Proof. Fix a countable local base $\mathcal{B}_x$ at any point $x \in Z$ and let $Y_0 = Y$. Suppose that $\alpha < \omega_1$ and we have constructed a family $\{Y_\beta : \beta < \alpha\}$ of subsets of Z with the following properties:

(1) $\gamma < \beta < \alpha$ implies $Y_\gamma \subset Y_\beta$;
(2) if $\gamma < \beta < \alpha$ and $\mathcal{U}$ is a countably infinite subfamily of $\bigcup\{\mathcal{B}_x : x \in Y_\gamma\}$ then there is a point $z \in Y_\beta$ such that every neighbourhood of z meets infinitely many elements of the family $\mathcal{U}|Y_{\gamma+1}$;
(3) $|Y_\beta| \leq \mathfrak{c}$ for any $\beta < \alpha$.

If α is a limit ordinal then we let $Y_\alpha = \bigcup_{\beta < \alpha} Y_\beta$; it is clear that the properties (1)–(3) still hold for all $\beta \leq \alpha$. If $\alpha = \beta + 1$ then the family $\mathbb{U}$ of all countably infinite subfamilies of $\bigcup\{\mathcal{B}_x : x \in Y_\beta\}$ has cardinality at most $\mathfrak{c}$. For any $\mathcal{U} \in \mathbb{U}$ apply pseudocompactness of Z to find a point $z(\mathcal{U})$ such that every neighbourhood of $z = z(\mathcal{U})$ meets infinitely many elements of $\mathcal{U}$. For any $B \in \mathcal{B}_z$ we can choose a countable set $N_B \subset Z$ such that, for any $U \in \mathcal{U}$, if $B \cap U \neq \emptyset$ then $B \cap U \cap N_B \neq \emptyset$; then the set $M_z = \bigcup\{N_B : B \in \mathcal{B}_z\}$ is countable.

Since $|\mathbb{U}| \leq \mathfrak{c}$, the set $Y_\alpha = \bigcup\{M_{z(\mathcal{U})} : \mathcal{U} \in \mathbb{U}\}$ has cardinality at most $\mathfrak{c}$. Our construction shows that the conditions (1)-(3) are satisfied for all $\beta \leq \alpha$. Therefore our inductive procedure can be continued to obtain a family $\{Y_\alpha : \alpha < \omega_1\}$ such that (1)–(3) are true for all $\alpha < \omega_1$.

It is evident that the set $Y' = \bigcup\{Y_\alpha : \alpha < \omega_1\} \supset Y$ has cardinality at most $\mathfrak{c}$; so we must only prove that Y' is pseudocompact. To do so take a countably infinite discrete family $\mathcal{U}' \subset \tau^*(Y')$. For any $U' \in \mathcal{U}'$ there is a point $y \in Y'$ and $B \in \mathcal{B}_y$ such that $B \cap Y' \subset U'$. This shows that we can assume, without loss of generality, that there is a set $A = \{y_n : n \in \omega\} \subset Y'$ such that, for any $n \in \omega$, a set $B_n \in \mathcal{B}_{y_n}$ is chosen in such a way that $\mathcal{U} = \{B_n \cap Y' : n \in \omega\}$.

Fix an ordinal $\beta < \omega_1$ for which $A \subset Y_\beta$. The property (2) shows that there is a point $z \in Y'$ such that every neighbourhood of z intersects infinitely many elements of $\mathcal{U}$. This, however, contradicts discreteness of $\mathcal{U}$ in Y' and proves that Y' has no discrete infinite families of non-empty open subsets, i.e., Y' is pseudocompact; so Fact 2 is proved.

Returning to our solution assume that X is a pseudocompact splittable space. We have $\psi(X) \leq \omega$ by Problem 403; so X is first countable by Fact 1. If X is not hereditarily Lindelöf then there is a right-separated $Y \subset X$ with $|Y| = \omega_1$ (see SFFS-005). Apply Fact 2 to find a pseudocompact $Y' \subset X$ such that $Y \subset Y'$ and $|Y'| \leq \mathfrak{c}$. The space Y' is also splittable (see Problem 401); so it can condensed onto a second countable space by Problem 414. Such a condensation has to be a homeomorphism (see TFS-140); so $w(Y') = \omega$ which is a contradiction with the fact that $Y \subset Y'$ is right-separated and hence $hl(Y) > \omega$.

As a consequence, the space X is hereditarily Lindelöf; apply Fact 1 of T.015 to see that $|X| \leq 2^{\psi(X) \cdot s(X)} \leq 2^{hl(X)} \leq 2^\omega = \mathfrak{c}$. Thus we can apply Problem 414 again to conclude that $iw(X) \leq \omega$ and hence X is compact and metrizable (see TFS-140), i.e., our solution is complete.

U.418. *Prove that a Lindelöf space X is splittable if and only if $iw(X) \leq \omega$.*

Solution. If $iw(X) \leq \omega$ then X is splittable by Problem 404; so we have sufficiency. Now assume that X is a splittable Lindelöf space. If $Y \subset X$ then there is a continuous map $\varphi : X \to \mathbb{R}^\omega$ such that $Y = \varphi^{-1}(\varphi(Y))$ (see Problem 416). The set $Z = \varphi(Y)$ has cardinality at most $\mathfrak{c}$ and $\varphi^{-1}(z)$ is closed in X for any $z \in Z$. The set $Y = \bigcup\{\varphi^{-1}(z) : z \in Z\}$ is a union of $\leq \mathfrak{c}$-many closed subspaces of X; so it follows from $l(X) \leq \omega$ that $l(Y) \leq \mathfrak{c}$.

The set $Y \subset X$ was taken arbitrarily; so we proved that $hl(X) \leq \mathfrak{c}$ and therefore $|X| \leq 2^{\psi(X)s(X)} \leq 2^{hl(X)} \leq 2^\mathfrak{c}$ (see Fact 1 of T.015). The space $\mathbb{R}^X$ has density at most $\mathfrak{c}$ (see TFS-108); so we can fix a dense set $P \subset \mathbb{R}^X$ with $|P| \leq \mathfrak{c}$. Since X is splittable, for any $f \in P$ we can find a countable set $A_f \subset C_p(X)$ such that $f \in \overline{A}_f$ (the bar denotes the closure in $\mathbb{R}^X$). If $A = \bigcup\{A_f : f \in P\} \subset C_p(X)$ then $|A| \leq \mathfrak{c}$ and $P \subset \overline{A}$ which shows that A is dense in $\mathbb{R}^X$ and hence in $C_p(X)$.

Thus $iw(X) = d(C_p(X)) \leq \mathfrak{c}$ and hence there exists a condensation $\varphi : X \to Y$ of the space X onto a space Y such that $w(Y) \leq \mathfrak{c}$. Observe that $\psi(X) \leq \omega$ (see Problem 403) which, together with the Lindelöf property of X implies that $l(X \setminus \{x\}) \leq \omega$; as a consequence, $l(Y \setminus \{y\}) \leq \omega$ for any $y \in Y$ and hence $\psi(Y) \leq \omega$ (see Fact 1 of U.027).

Fact 1. If Z is a space then $|Z| \leq nw(Z)^{\psi(Z)}$.

Proof. Let $\mathcal{N}$ be a network in Z of cardinality $\kappa = nw(Z)$. For any $z \in Z$ fix a family $\mathcal{B}_z \subset \tau(z, Z)$ such that $|\mathcal{B}_z| \leq \lambda = \psi(Z)$ and $\bigcap \mathcal{B}_z = \{z\}$.

For any point $z \in Z$ and any element $B \in \mathcal{B}_z$ choose a set $N(z, B) \in \mathcal{N}$ such that $z \in N(z, B) \subset B$. The family $\mathcal{N}_z = \{N(z, B) : B \in \mathcal{B}_z\}$ is contained in $\mathcal{N}$ and $|\mathcal{N}_z| \leq \lambda$. If we let $\varphi(z) = \mathcal{N}_z$ for any $z \in Z$ then φ maps Z into the family $\mathcal{A} = \{\mathcal{N}' \subset \mathcal{N} : |\mathcal{N}'| \leq \lambda\}$. It is clear that $|\mathcal{A}| \leq \kappa^\lambda$. Besides, if $x \neq y$ then there is $B \in \mathcal{B}_x$ such that $y \notin B$ and hence $N(x, B) \subset B$ cannot belong to $\mathcal{N}_y$ because all elements of $\mathcal{N}_y$ contain y. This proves that $\mathcal{N}_x \neq \mathcal{N}_y$ for any distinct $x, y \in Z$, i.e., the map φ is an injection of Z into the set $\mathcal{A}$ of cardinality $\leq \kappa^\lambda$. Therefore $|Z| \leq \kappa^\lambda$ and Fact 1 is proved.

Finally, apply Fact 1 to see that $|Y| \leq w(Y)^{\psi(Y)} \leq \mathfrak{c}^\omega = \mathfrak{c}$; since X condenses onto Y, we also have $|X| \leq \mathfrak{c}$; so $iw(X) \leq \omega$ by Problem 414. This settles necessity and makes our solution complete.

U.419. *Prove that a Lindelöf Σ-space is splittable if and only if it has a countable network.*

Solution. If $nw(X) = \omega$ then $iw(X) \leq \omega$ (see TFS-156) and hence X is splittable (see Problem 404); so we have sufficiency. Now, if X is a splittable Lindelöf Σ-space then $iw(X) \leq \omega$ by Problem 418; so $nw(X) \leq \omega$ by SFFS-266; this proves necessity.

U.420. *Prove that a Lindelöf p-space is splittable if and only if it is second countable.*

Solution. If $w(X) = \omega$ then X is splittable (see Problem 402); so we have sufficiency. Now, if X is a splittable Lindelöf p-space then $iw(X) \leq \omega$ by Problem 418, so $w(X) \leq \omega$ by SFFS-244; this proves necessity.

U.421. *Prove that any Čech-complete splittable paracompact space is metrizable.*

Solution. Call a non-empty space X *crowded* if X has no isolated points. For any $\mathcal{U} \subset \exp X$ and $A \subset X$ let $\mathcal{U}(A) = \bigcup\{U \in \mathcal{U} : U \cap A \neq \emptyset\}$ be the star of the set A with respect to $\mathcal{U}$. For any $x \in X$ we write $\mathcal{U}(x)$ instead of $\mathcal{U}(\{x\})$. A sequence $\{\mathcal{U}_n : n \in \omega\}$ of open covers of X is called *star-decreasing* if $\mathcal{U}_{n+1}$ is a star refinement of $\mathcal{U}_n$ for any $n \in \omega$; recall that this means that, for any $U \in \mathcal{U}_{n+1}$, the set $\mathcal{U}_{n+1}(U)$ is contained in some element of $\mathcal{U}_n$.

Fact 1. If X is a sequential splittable space then there is a closed $Y \subset X$ such that $|Y| \leq \mathfrak{c}$ and $K \subset Y$ whenever K is a crowded compact subspace of X.

Proof. Let $\mathcal{F}$ be a maximal disjoint family of compact crowded subspaces of X. Every $F \in \mathcal{F}$ is metrizable (see Problems 401 and 417); so we can apply Fact 4 of T.250 to see that there is $K_F \subset F$ with $K_F \simeq \mathbb{D}^\omega$. Since $|\mathbb{D}^\omega| = \mathfrak{c}$ and $|C_p(\mathbb{D}^\omega)| = \mathfrak{c}$, we can apply Fact 1 of U.406 to see that $|\mathcal{F}| \leq \mathfrak{c}$ and hence the set $Z = \bigcup \mathcal{F}$ has cardinality at most $\mathfrak{c}$.

It follows easily from sequentiality of X that $Y = \overline{Z}$ also has cardinality which does not exceed $\mathfrak{c}$. If $K \subset X$ is a crowded compact space and $K \backslash Y \neq \emptyset$ then $K \backslash Y$ is a crowded locally compact subspace of X which shows that there is a crowded compact $K' \subset K \backslash Y$. Thus $\mathcal{F}' = \mathcal{F} \cup \{K'\}$ is a disjoint family of crowded compact subspaces of X such that $\mathcal{F} \subset \mathcal{F}'$ and $\mathcal{F} \neq \mathcal{F}'$; this contradiction with maximality of $\mathcal{F}$ shows that $K \subset Y$ and hence Fact 1 is proved.

Fact 2. For any sequential splittable space X there are disjoint $A, B \subset X$ such that $A \cup B = X$ and, for any compact $K \subset X$, if $K \subset A$ or $K \subset B$ then K is countable.

Proof. By Fact 1 there exists a closed $Y \subset X$ such that $|Y| \leq \mathfrak{c}$ and any crowded compact subspace of X is contained in Y. By Problems 414 and 401 there exists a condensation $\varphi : Y \to M$ of the space Y onto a second countable space M. If X has no crowded compact subsets then every compact $K \subset X$ is countable (see Problem 417 and SFFS-353); so the sets $A = \emptyset$ and $B = X$ do the job. Now suppose that the family $\mathcal{K} = \{K : K$ is a crowded compact subspace of $X\}$ is non-empty.

If $K \in \mathcal{K}$ then $K \subset Y$ and hence $\varphi(K) \simeq K$ is a crowded compact subset of M; besides, $K, K' \in \mathcal{K}$ and $K \neq K'$ implies $\varphi(K) \neq \varphi(K')$ because φ is a bijection. Thus cardinality of $\mathcal{K}$ does not exceed the cardinality of the family of all closed subsets of M. It is an easy exercise that the family of closed subspaces of a second countable space has at most $\mathfrak{c}$ elements so $|\mathcal{K}| \leq \mathfrak{c}$; let $\{K_\alpha : \alpha < \mathfrak{c}\}$ be an enumeration of $\mathcal{K}$.

It follows from SFFS-353 that any set $K \in \mathcal{K}$ has cardinality $\mathfrak{c}$; so it is easy to construct by transfinite induction faithfully indexed disjoint subsets $\{x_\alpha : \alpha < \mathfrak{c}\}$ and $\{y_\alpha : \alpha < \mathfrak{c}\}$ of the space Y such that $\{x_\alpha, y_\alpha\} \subset K_\alpha$ for all $\alpha < \mathfrak{c}$. We claim that the sets $A = \{x_\alpha : \alpha < \mathfrak{c}\}$ and $B = X \backslash A$ are as promised. Indeed, if K' is an uncountable compact subset of X then it is metrizable by Problem 417; so it contains a crowded compact subset K; by our choice of Y we have $K \subset Y$ and hence $K = K_\alpha$ for some $\alpha < \mathfrak{c}$. Since $\{x_\alpha, y_\alpha\} \subset K_\alpha = K$, the set K (and hence K') intersects both A and B. Therefore no uncountable compact subset of X can be contained either in A or in B, i.e., Fact 2 is proved.

Fact 3. If a space X is Čech-complete then $\psi(x, X) = \chi(x, X)$ for any $x \in X$. In particular, any splittable Čech-complete space is first countable.

Proof. Fix an arbitrary point $x \in X$; we have to prove that $\chi(x, X) \leq \psi(x, X)$ so take an infinite cardinal κ and assume that $\psi(x, X) \leq \kappa$. There exists a compact subspace $K \subset X$ such that $x \in K$ and $\chi(K, X) \leq \omega$ (see TFS-263). It follows from the formula $\chi(x, K) = \psi(x, K) \leq \psi(x, X) \leq \kappa$ (see TFS-327) that we can choose a local base $\{U_\alpha : \alpha < \kappa\}$ of the space K at the point x. Let $\mathcal{V} = \{V_n : n \in \omega\}$ be a decreasing outer base of K in X. It is easy to see that the set $W_{n,\alpha} = V_n \backslash (K \backslash U_\alpha)$ is an open neighbourhood of x in X; using regularity of X fix a decreasing family $\{B_{n,\alpha} : n \in \omega\} \subset \tau(x, X)$ such that $\overline{B}_{n,\alpha} \subset W_{n,\alpha}$ for any $n \in \omega$ and $\alpha < \kappa$.

The family $\mathcal{B} = \{B_{n,\alpha} : n \in \omega, \alpha < \kappa\} \subset \tau(x, X)$ has cardinality $\leq \kappa$. To see that $\mathcal{B}$ is a local base at x in the space X take any $U \in \tau(x, X)$. There is $\alpha < \kappa$ such that $U_\alpha \subset U \cap K$; since $W_{0,\alpha} \cap K = U_\alpha$ and $\overline{B}_{0,\alpha} \subset W_{0,\alpha}$, the closed set $F = \overline{B}_{0,\alpha} \backslash U$ is disjoint from K. The family $\mathcal{V}$ being an outer base of K in X, there is $n \in \omega$ such that $V_n \cap F = \emptyset$. Then $B_{n,\alpha} \backslash U \subset (\overline{B}_{0,\alpha} \backslash U) \cap V_n = \emptyset$ and hence $B_{n,\alpha} \subset U$ which proves that $\mathcal{B}$ is a local base of X at the point x, i.e., $\chi(x, X) \leq \kappa$. This shows that $\chi(x, X) \leq \psi(x, X)$ and hence $\psi(x, X) = \chi(x, X)$ for any $x \in X$. Finally observe that any splittable space has countable pseudocharacter by Problem 403; so $\chi(X) \leq \omega$ for any splittable Čech-complete space X and hence Fact 3 is proved.

Fact 4. Let T be a countable set (which can be finite). Suppose that X is a splittable space and $\mathcal{A} = \{A_t : t \in T\}$ is a family of subsets of X. Then there exists a second countable space M and a continuous map $\varphi : X \to M$ such that $\varphi^{-1}(\varphi(A_t)) = A_t$ for any $t \in T$.

Proof. We can apply Problem 416 to see that for any element $t \in T$, there exists a continuous map $\varphi_t : X \to \mathbb{R}^\omega$ such that $\varphi_t^{-1}(\varphi_t(A_t)) = A_t$. The diagonal product $\varphi = \Delta_{t \in T}\varphi_t : X \to (\mathbb{R}^\omega)^T$ is continuous and $M = \varphi(X)$ is second countable. To see that M is as promised take any $t \in T$ and $x \in A_t$. If $\varphi(x') = \varphi(x)$ then $\varphi_t(x') = \varphi_t(x)$ and hence $x' \in A_t$ by the choice of φ_t. This proves that $\varphi^{-1}(\varphi(A_t)) = A_t$ for any $t \in T$, i.e., Fact 4 is proved.

Fact 5. Suppose that X is a paracompact Čech-complete space and fix a sequence $\mathcal{O} = \{O_n : n \in \omega\} \subset \tau(\beta X)$ such that $\bigcap_{n \in \omega} O_n = X$ and $O_{n+1} \subset O_n$ for every $n \in \omega$. Call a sequence $\{\mathcal{A}_n : n \in \omega\}$ of families of subsets of X *subordinated to* $\mathcal{O}$ if, for any $n \in \omega$ and $A \in \mathcal{A}_n$, we have $\mathrm{cl}_{\beta X}(A) \subset O_n$. Then, for any sequence $\mathcal{S} = \{\mathcal{U}_n : n \in \omega\}$ of open covers of X which is star-decreasing and subordinated to $\mathcal{O}$, the set $K_x = \bigcap_{n \in \omega} \mathcal{U}_n(x)$ is compact and the family $\{\mathcal{U}_n(x) : n \in \omega\}$ is an outer base of K_x for any $x \in X$.

Proof. Fix $x \in X$ and let $W_n = \mathcal{U}_n(x)$; observe that $W_{n+1} \subset W_n$ for any $n \in \omega$. Since $\mathcal{S}$ is star-decreasing, for any $n \in \mathbb{N}$, the set W_n is contained in an element of $\mathcal{U}_{n-1}$; so it follows from the fact that $\mathcal{S}$ is subordinated to $\mathcal{O}$ that $\mathrm{cl}_{\beta X}(W_n) \subset O_{n-1}$. Therefore $K = \bigcap\{\mathrm{cl}_{\beta X}(W_n) : n \in \mathbb{N}\} \subset \bigcap\{O_{n-1} : n \in \mathbb{N}\} = X$; it is evident that K is compact and $K_x \subset K$.

It follows from Fact 1 of S.326 that the family $\mathcal{K} = \{\mathrm{cl}_{\beta X}(W_n) : n \in \omega\}$ is a network of the set K in βX, i.e., for any $U \in \tau(K, \beta X)$ there is $n \in \omega$ such that $\mathrm{cl}_{\beta X}(W_n) \subset U$.

Given an element $V \in \mathcal{U}_{n+1}$ there is $V' \in \mathcal{U}_n$ such that $\mathcal{U}_{n+1}(V) \subset V'$ and therefore $\mathcal{U}_{n+1}(X \backslash V') \cap V = \emptyset$. An immediate consequence is that $\overline{V} \subset V'$. Thus the closure of every element of $\mathcal{U}_{n+1}$ is contained in an element of $\mathcal{U}_n$; the sequence $\mathcal{S}$ being star-decreasing, we have $\overline{W}_{n+2} \subset W_n$ for any $n \in \omega$. This proves that $K_x = \bigcap\{\overline{W}_n : n \in \omega\}$ is a closed subset of X.

Now, if $y \in K$ then $y \in \mathrm{cl}_{\beta X}(W_n) \cap X = \overline{W}_n$ for any number $n \in \omega$ and hence $y \in \bigcap_{n \in \omega} \overline{W}_n = K_x$; this proves that $K_x = K$ is compact. Since the family $\mathcal{K}$ is a network of $K = K_x$ in βX, the family $\{W_n : n \in \omega\}$ is an outer base of K_x in X; so Fact 5 is proved.

Fact 6. Suppose that X is a Čech-complete paracompact splittable space. If $\mathcal{A}$ is a countable disjoint family of subsets of X then there exists a star-decreasing sequence $\{\mathcal{U}_n : n \in \omega\}$ of open covers of X such that for any point $x \in X$, the set $K_x = \bigcap\{\mathcal{U}_n(x) : n \in \omega\}$ is compact and $x \in A \in \mathcal{A}$ implies $K_x \subset A$.

Proof. Apply Fact 4 to find a second countable space M and a continuous map $\varphi : X \to M$ such that $\varphi^{-1}(\varphi(A)) = A$ for any $A \in \mathcal{A}$. Fix a decreasing family $\mathcal{O} = \{O_n : n \in \omega\} \subset \tau(\beta X)$ such that $\bigcap \mathcal{O} = X$. Since M is metrizable, there exists a sequence $\{\mathcal{V}'_n : n \in \omega\}$ of open covers of M for which $\bigcap\{\mathcal{V}'_n(x) : n \in$

$\omega\} = \{x\}$ for any $x \in M$; let $\mathcal{V}_n = \{\varphi^{-1}(V) : V \in \mathcal{V}'_n\}$ for every $n \in \omega$. It is straightforward that

(1) $\bigcap_{n\in\omega} \mathcal{V}_n(x) = \varphi^{-1}(\varphi(x))$ for any $x \in X$; so $x \in A \in \mathcal{A}$ implies $\bigcap_{n\in\omega} \mathcal{V}_n(x) \subset A$.

For any number $n \in \omega$ there exists an open cover $\mathcal{W}_n$ of the space X such that $\text{cl}_{\beta X}(W) \subset O_n$ for any $W \in \mathcal{W}_n$. Using paracompactness of X it is easy to construct a star-decreasing sequence $\mathcal{S} = \{\mathcal{U}_n : n \in \omega\}$ of open covers of X such that every $\mathcal{U}_n$ refines both covers $\mathcal{V}_n$ and $\mathcal{W}_n$. Therefore the sequence $\mathcal{S}$ is subordinated to $\mathcal{O}$ in the sense of Fact 5; so the set K_x is compact for any $x \in X$. Besides, $K_x \subset \bigcap_{n\in\omega} \mathcal{V}_n(x)$; so the property (1) shows that $K_x \subset A$ whenever $x \in A \in \mathcal{A}$ and hence Fact 6 is proved.

Fact 7. If X is a Čech-complete paracompact space with a G_δ-diagonal then X is metrizable.

Proof. By Fact 1 of T.235 there is a sequence $\{\mathcal{V}_n : n \in \omega\}$ of open covers of X such that $\bigcap_{n\in\omega} \mathcal{V}_n(x) = \{x\}$ for any $x \in X$. By Čech-completeness of X there is a decreasing family $\mathcal{O} = \{O_n : n \in \omega\} \subset \tau(\beta X)$ such that $\bigcap \mathcal{O} = X$. It is evident that there exists an open cover $\mathcal{W}_n$ of the space X such that $\text{cl}_{\beta X}(W) \subset O_n$ for any $W \in \mathcal{W}_n$. Using paracompactness of X we can construct a star-decreasing sequence $\{\mathcal{U}_n : n \in \omega\}$ of open covers of X such that every $\mathcal{U}_n$ refines both covers $\mathcal{V}_n$ and $\mathcal{W}_n$. Then $\bigcap_{n\in\omega} \mathcal{U}_n(x) \subset \bigcap_{n\in\omega} \mathcal{V}_n(x) = \{x\}$ for any $x \in X$; this, together with the fact that the sequence $\{\mathcal{U}_n : n \in \omega\}$ is subordinated to $\mathcal{O}$ in the sense of Fact 5, implies that

(2) the family $\{\mathcal{U}_n(x) : n \in \omega\}$ is a local base in X at x for any $x \in X$.

Use once more paracompactness of X to take a locally finite refinement $\mathcal{B}_n$ of the cover $\mathcal{U}_n$ for any $n \in \omega$. It turns out that the family $\mathcal{B} = \bigcup_{n\in\omega} \mathcal{B}_n$ is a base in X. Indeed, it follows from (2) that if $x \in U \in \tau(X)$ then there is $n \in \omega$ such that $\mathcal{U}_n(x) \subset U$. Take any $B \in \mathcal{B}_n$ with $x \in B$; since $\mathcal{B}_n$ is a refinement of $\mathcal{U}_n$, we have $B \subset \mathcal{U}_n(x) \subset U$. As a consequence, $\mathcal{B}$ is a σ-locally finite base of X; so X is metrizable by TFS-221 and hence Fact 7 is proved.

Returning to our solution assume that X is a paracompact Čech-complete splittable space and fix a family $\mathcal{O} \subset \tau(\beta X)$ such that $\bigcap \mathcal{O} = X$ and $O_{n+1} \subset O_n$ for any $n \in \omega$. The space X is first countable and hence sequential by Fact 3; so we can apply Fact 2 to see that there exists a set $A \subset X$ such that a compact subspace $K \subset X$ is countable whenever $K \subset A$ or $K \subset X\backslash A$. Apply Fact 6 to find a star-decreasing sequence $\{\mathcal{U}_n : n \in \omega\}$ of open covers of X such that the set $K_x = \bigcap_{n\in\omega} \mathcal{U}_n(x)$ is compact and $x \in A$ ($x \in X\backslash A$) implies $K_x \subset A$ (or $K_x \subset X\backslash A$ respectively) for any $x \in X$.

Take any $x \in X$; observe first that $y \in K_x$ if and only if $x \in K_y$ and fix a point $y \in K_x$. If $t \in K_y$ and $n \in \omega$ then $t \in \mathcal{U}_{n+1}(y)$ and $y \in \mathcal{U}_{n+1}(x)$; so there are $U, V \in \mathcal{U}_{n+1}$ such that $\{t, y\} \subset U$ and $\{x, y\} \subset V$. Therefore $t \in \mathcal{U}_{n+1}(V)$; since $\mathcal{U}_{n+1}$ is a star refinement of $\mathcal{U}_n$, there is $W \in \mathcal{U}_n$ such that $U \cup V \subset W$ which shows that $t \in \mathcal{U}_n(x)$. This proves that $t \in \mathcal{U}_n(x)$ for any $n \in \omega$ and hence $t \in K_x$. The point $t \in K_y$ was chosen arbitrarily; so we proved that $K_y \subset K_x$. We already noted that $y \in K_x$ implies $x \in K_y$; so replacing x by y and vice versa in the previous reasoning we also obtain the inclusion $K_x \subset K_y$. In other words,

(3) if $x \in X$ and $y \in K_x$ then $K_y = K_x$.

An easy consequence of (3) is that the family $\{K_x : x \in X\}$ is a partition of X in the sense that $K_x \cap K_y \neq \emptyset$ implies $K_x = K_y$; to see it suffices to take any $z \in K_x \cap K_y$ and observe that $K_x = K_z = K_y$ by (3).

Every set K_x being countable by the choice of A, we can choose a disjoint family $\{P_n : n \in \omega\}$ of subsets of X such that $\bigcup_{n \in \omega} P_n = X$ and $|P_n \cap K_x| \leq 1$ for any $x \in X$ and $n \in \omega$. Apply Fact 6 once more to find a star-decreasing sequence $\{\mathcal{V}_n : n \in \omega\}$ of open covers of X such that the set $M_x = \bigcap_{n \in \omega} \mathcal{V}_n(x)$ is compact and $x \in P_n$ implies $M_x \subset P_n$ for any $x \in X$ and $n \in \omega$.

For any $n \in \omega$ choose an open cover $\mathcal{W}_n$ of the space X which refines both covers $\mathcal{U}_n$ and $\mathcal{V}_n$. If $x \in X$ then there is a unique $m \in \omega$ such that $x \in P_m$. We have $N_x = \bigcap_{n \in \omega} \mathcal{W}_n(x) \subset M_x \subset P_m$; besides, it follows from $N_x \subset K_x$ that $N_x \cap P_m$ has at most one point; so $N_x = N_x \cap P_m = \{x\}$. The point $x \in X$ was chosen arbitrarily; so we proved that $\bigcap_{n \in \omega} \mathcal{W}_n(x) = \{x\}$ for any $x \in X$ and hence X has a G_δ-diagonal by Fact 1 of T.235. Finally, apply Fact 7 to conclude that X is metrizable and hence our solution is complete.

U.422. *Let X be a complete metrizable dense-in-itself space. Prove that X is splittable if and only if $|X| \leq \mathfrak{c}$.*

Solution. If $|X| \leq \mathfrak{c}$ then $w(X) \leq \mathfrak{c}$; so $iw(X) \leq \omega$ by SFFS-102 and hence X is splittable by Problem 404; this gives sufficiency.

Now assume that X is a complete metrizable splittable space with $|X| > \mathfrak{c}$; apply SFFS-102 once more to see that $w(X) > \mathfrak{c}$ and hence $c(X) = w(X) > \mathfrak{c}$. Fix a disjoint family $\{U_\alpha : \alpha < \mathfrak{c}^+\} \subset \tau^*(X)$ and choose a set $V_\alpha \in \tau^*(X)$ such that $\overline{V}_\alpha \subset U_\alpha$ for any $\alpha < \mathfrak{c}^+$. Every space $\overline{V}_\alpha$ is dense-in-itself; so we can apply Fact 1 of T.045 to see that there exists a closed separable dense-in-itself subspace $F_\alpha \subset \overline{V}_\alpha$.

It is clear that every F_α is also a complete metric space. If F_α is countable then it is of first category being the union of its singletons which are nowhere dense in F_α; this contradiction proves that F_α is an uncountable Polish space and hence we can apply SFFS-353 to find a subspace $K_\alpha \subset F_\alpha$ such that $K_\alpha \simeq \mathbb{D}^\omega$ for any $\alpha < \mathfrak{c}^+$.

The family $\{K_\alpha : \alpha < \mathfrak{c}^+\}$ is disjoint and consists of homeomorphic copies of $\mathbb{D}^\omega$. Since $|\mathbb{D}^\omega| = \mathfrak{c}$ and $|C_p(\mathbb{D}^\omega)| = \mathfrak{c}$, we can apply Fact 1 of U.406 to see that X is not splittable. This contradiction shows that $|X| \leq \mathfrak{c}$ and hence we proved necessity.

U.423. *Suppose that $X = \bigcup\{X_n : n \in \omega\}$, where $X_n \subset X_{n+1}$ for each $n \in \omega$, the subspace X_n is splittable and C^*-embedded in X for every n. Prove that X is splittable.*

Solution. Take any $h \in \mathbb{D}^X$; then $h_n = h|X_n \in \mathbb{D}^{X_n}$ for any $n \in \omega$. The space X_n being splittable there is a countable set $B_n \subset C_p(X_n)$ such that h_n belongs to the closure of B_n in $\mathbb{R}^{X_n}$ for any $n \in \omega$.

Let $\eta(t) = \frac{1}{2}(|t + 2| - |t - 2|)$ for every $t \in \mathbb{R}$; then $\eta : \mathbb{R} \to [-2, 2]$ is a continuous function with $\eta(t) = t$ for any $t \in [-2, 2]$. For any $n \in \omega$ the set $C_n = \{\eta \circ f : f \in B_n\}$ is contained in $C_p^*(X_n)$ and it is straightforward that h_n is also in the closure of C_n in $\mathbb{R}^{X_n}$.

Since X_n is C^*-embedded in X, for every $f \in C_n$ there is $g_f \in C_p(X)$ such that $g_f|X_n = f$; therefore the set $A_n = \{g_f : f \in C_n\} \subset C_p(X)$ is countable for any $n \in \omega$. The set $A = \bigcup_{n \in \omega} A_n$ is also countable; let us check that $h \in \overline{A}$ (the bar denotes the closure in $\mathbb{R}^X$).

For any $U \in \tau(h, \mathbb{R}^X)$ there exists a finite set $E \subset X$ and a number $\varepsilon > 0$ such that $V = \{f \in \mathbb{R}^X : |f(x) - h(x)| < \varepsilon$ for all $x \in E\} \subset U$. Take $n \in \omega$ such that $E \subset X_n$; since h_n is in the closure of C_n, there is $f \in C_n$ such that $|h_n(x) - f(x)| < \varepsilon$ for all $x \in E$. We have $h_n(x) = h(x)$ for all $x \in E$; so $|g_f(x) - h(x)| = |f(x) - h_n(x)| < \varepsilon$ for all $x \in E$ which shows that $g_f \in A \cap V \subset A \cap U$, i.e., $A \cap U \neq \emptyset$ for any $U \in \tau(h, \mathbb{R}^X)$ and therefore $h \in \overline{A}$. Thus every $h \in \mathbb{D}^X$ is in the closure (in $\mathbb{R}^X$) of a countable subset of $C_p(X)$; so X is splittable by Problem 415.

U.424. *Prove that any normal strongly σ-discrete space is strongly splittable.*

Solution. Suppose that X is strongly σ-discrete and normal. There exists a sequence $\{X_n : n \in \omega\}$ of closed discrete subspaces of X such that $X = \bigcup_{n \in \omega} X_n$ and $X_n \subset X_{n+1}$ for any $n \in \omega$. Given any $h \in \mathbb{R}^X$, the function $h_n = h|X_n$ is continuous on X_n because X_n is discrete. By normality of X, for any $n \in \omega$, there exists $f_n \in C_p(X)$ such that $f_n|X_n = h_n$. If $x \in X$ then $x \in X_m$ for some $m \in \omega$ and therefore $f_n(x) = h_n(x) = h(x)$ for any $n \geq m$. Thus the sequence $\{f_n(x) : n \in \omega\}$ converges to $h(x)$ for any $x \in X$ which shows that the sequence $\{f_n : n \in \omega\}$ converges to h (see TFS-143) and hence X is strongly splittable.

U.425. *Give an example of a strongly σ-discrete space which is not splittable.*

Solution. Let X be a Mrowka space (see TFS-142); then X is pseudocompact, non-compact and $X = D \cup A$ where the set D is countable and dense in X while A is closed and discrete. Thus the family $\{\{d\} : d \in D\} \cup \{A\}$ witnesses that X is strongly σ-discrete. However, X is not splittable because any splittable pseudocompact space has to be compact by Problem 417.

U.426. *Show that, for any cardinal κ, there exists a normal strongly σ-discrete (and hence splittable) space X with $c(X) = \omega$ and $|X| \geq \kappa$.*

Solution. Given a set T and $t \in T$ say that a family $\mathcal{B}_t \subset \exp T$ is a *weak local base* at t if $\bigcap \mathcal{B}_t = \{t\}$ and, for any $B, B' \in \mathcal{B}_t$ there is $C \in \mathcal{B}_t$ such that $C \subset B \cap B'$. Given a set $Y \subset T$ say that a set $U \subset T$ is a *weak neighbourhood* of Y if, for any $t \in Y$ there exists $B \in \mathcal{B}_t$ with $B \subset U$.

Fact 1. Suppose that λ is a cardinal with $\lambda^\omega = \lambda$ and A is a set such that $|A| = \lambda$. Then the space $\mathbb{I}^A$ has a strongly σ-discrete dense subspace of cardinality λ.

Proof. Take a disjoint family $\mathcal{A} = \{A_\alpha : \alpha < \lambda\} \subset \exp A$ such that $|A_\alpha| = \lambda$ for any $\alpha < \lambda$. Let $\mathcal{F} = \bigcup \{\mathbb{I}^F : F$ is a non-empty finite subset of $A\}$. It is easy to see that $|\mathcal{F}| = \lambda$; so let $\{f_\alpha : \alpha < \lambda\}$ be an enumeration of $\mathcal{F}$; for any $\alpha < \lambda$ denote by D_α the domain of the function f_α.

For each $\alpha < \lambda$ define a point $x_\alpha \in \mathbb{I}^A$ as follows: $x_\alpha(t) = f_\alpha(t)$ for all $t \in D_\alpha$; if $t \in A_\alpha \backslash D_\alpha$ then $x_\alpha(t) = 1$ and $x_\alpha(t) = 0$ for all $t \in A \backslash (D_\alpha \cup A_\alpha)$. It turns out that the set $Z = \{x_\alpha : \alpha < \lambda\}$ is as promised.

To see it let $P_n = \{\alpha < \lambda : |D_\alpha| = n\}$ for every $n \in \mathbb{N}$. If $Z_n = \{x_\alpha : \alpha \in P_n\}$ for any $n \in \mathbb{N}$ then $Z = \bigcup \{Z_n : n \in \mathbb{N}\}$. Fix an arbitrary $n \in \mathbb{N}$; given $\alpha < \lambda$ choose a set $F_\alpha \subset A_\alpha \backslash D_\alpha$ with $|F_\alpha| = n + 1$ and consider the set $O_\alpha = \{x \in Z : x(t) > 0$ for any $t \in F_\alpha\}$. It is clear that $O_\alpha \in \tau(x_\alpha, Z)$; if $\beta \in P_n$ and $\beta \neq \alpha$ then there is a point $t \in F_\alpha \backslash D_\beta$ (because $|D_\beta| = n < |F_\alpha|$). It is immediate that $x_\beta(t) = 0$ and hence $x_\beta \notin O_\alpha$. This proves that every $x_\alpha \in Z$ has a neighbourhood O_α such that $O_\alpha \cap Z_n$ contains at most one point. This, of course, implies that Z_n is closed and discrete in Z and hence Z is strongly σ-discrete.

Finally take a non-empty open $U \subset \mathbb{I}^A$; there is a finite non-empty $D \subset A$ and a function $f \in \mathbb{I}^D$ such that the set $\{x \in \mathbb{I}^A : x|D = f\} \subset U$. There is $\alpha < \lambda$ such that $D = D_\alpha$ and $f = f_\alpha$; it is straightforward that $x_\alpha|D = f$ and hence $x_\alpha \in U \cap Z$. This shows that Z is dense in $\mathbb{I}^A$; it is evident that $|Z| = \lambda$; so Fact 1 is proved.

Fact 2. Given a set T suppose that $\mathcal{B}_t \subset \exp T$ is a weak local base at T for any $t \in T$. If $\tau = \{U \subset T : t \in U$ implies $B \subset U$ for some $B \in \mathcal{B}_t\}$ then τ is a T_1-topology on T which is said to be generated by the collection $\{\mathcal{B}_t : t \in T\}$ of weak local bases.

Proof. It is clear that $\emptyset \in \tau$ and $T \in \tau$. If $U, V \in \tau$ and $t \in U \cap V$ then there are $B, B' \in \mathcal{B}_t$ such that $B \subset U$ and $B' \subset V$. Take $C \in \mathcal{B}_t$ with $C \subset B \cap B'$; then $C \subset U \cap V$ and hence $U \cap V \in \tau$. It is evident that $\mathcal{U} \subset \tau$ implies $\bigcup \mathcal{U} \in \tau$; so τ is indeed a topology on T. Finally, if $t \in T$ and $z \in T \backslash \{t\}$ then there is $B \in \mathcal{B}_z$ such that $t \notin B$ and hence $B \subset T \backslash \{t\}$. This shows that $T \backslash \{t\}$ is open in (T, τ) for any $t \in T$, i.e., (T, τ) is a T_1-space and hence Fact 2 is proved.

Fact 3. Given a set T suppose that $\mathcal{B}_t \subset \exp T$ is a weak local base at t for any $t \in T$ and let τ be the topology on T generated by the collection $\{\mathcal{B}_t : t \in T\}$. Then

(i) if U and V are weak neighbourhoods of a set $Y \subset T$ then $U \cap V$ is also a weak neighbourhood of Y;

(ii) for every sequence $\{U_n : n \in \omega\} \subset \exp T$ such that U_{n+1} is a weak neighbourhood of U_n for any $n \in \omega$, the set $U = \bigcup_{n \in \omega} U_n$ is open in (T, τ).

Proof. If $x \in Y$ then there are $B, B' \in \mathcal{B}_x$ such that $B \subset U$ and $B' \subset V$. Take any $C \in \mathcal{B}_x$ with $C \subset B \cap B'$; then $C \subset U \cap V$; so $U \cap V$ is a weak neighbourhood of Y which proves (i).

As to (ii), if $x \in U$ then $x \in U_n$ for some n; by our assumption, there is $B \in \mathcal{B}_x$ such that $B \subset U_{n+1} \subset U$; so $U \in \tau$ and Fact 3 is proved.

Fact 4. Suppose that T is a set and we have a collection $\mathbb{B} = \{\mathcal{B}_t : t \in T\}$ of weak local bases and a disjoint family $\{T_n : n \in \omega\} \subset \exp T$ with the following properties:

(a) $\bigcup_{n\in\omega} T_n = T$;
(b) for any $n \in \omega$, if $t \in T_n$ then $B\backslash\{t\} \subset T_{n+1}$ for every $B \in \mathcal{B}_t$;
(c) for any $n \in \omega$ and $Y \subset T_n$ there exist $U, V \subset T$ such that U is a weak neighbourhood of Y, the set V is a weak neighbourhood of $T_n\backslash Y$ and $U \cap V = \emptyset$.

Then the set T, with the topology τ generated by the collection $\mathbb{B}$ of weak local bases, is a T_4-space.

Proof. It follows from Fact 2 that (T, τ) is a T_1-space; so it suffices to establish normality of (T, τ). Suppose that F and G are disjoint closed subspaces of (T, τ); let $F_n = F \cap T_n$, $G_n = G \cap T_n$ for all $n \in \omega$. Our first inductive step is to define the set $U_0 = F_0$; Suppose that $m \in \omega$ and we have defined sets $U_0, \ldots, U_m$ with the following properties:

(1) $F_i \subset U_i \subset T_i\backslash G_i$ for all $i \le m$;
(2) if $V_i = U_0 \cup \ldots \cup U_i$ for each $i \le m$ then V_{i+1} is a weak neighbourhood of V_i for all $i < m$;
(3) if $W_i = (T_0\backslash U_0) \cup \ldots \cup (T_i\backslash U_i)$ for each $i \le m$ then W_{i+1} is a weak neighbourhood of W_i for all $i < m$;

Apply the property (c) to the set $Y = U_m$ to find disjoint sets $P, Q \subset T$ such that $U_m \subset P$, $T_m\backslash U_m \subset Q$ while the sets P and Q are weak neighbourhoods of U_m and $T_m\backslash U_m$ respectively. Fact 3 and (b) imply that $U_m \cup ((P \cap T_{m+1})\backslash G_{m+1})$ is still a weak neighbourhood of U_m; let $U_{m+1} = (P \cap T_{m+1})\backslash G_{m+1}$. It is straightforward that the conditions (1) and (2) are still satisfied if we replace m by $m + 1$. As to (3), the set $(T_m\backslash U_m) \cup (Q \cap T_{m+1})$ is a weak neighbourhood of $T_m\backslash U_m$. Since $Q \cap T_{m+1} \subset T_{m+1}\backslash U_{m+1}$, the set $(T_m\backslash U_m) \cup (T_{m+1}\backslash U_{m+1})$ is a weak neighbourhood of $T_m\backslash U_m$. An easy application of (b) shows that (3) is also fulfilled if we replace m by $m + 1$.

Therefore our inductive procedure can be continued to construct the sequence $\{U_i : i \in \omega\}$ such that (1)–(3) hold for all $m \in \omega$. If $V = \bigcup_{i\in\omega} V_i$ and $W = \bigcup_{i\in\omega} W_i$ then it follows from (1) and (3) that $F \subset V$, $G \subset W$ and $V \cap W = \emptyset$. Besides, the properties (2) and (3) together with Fact 3 show that V and W are both open in (T, τ). Thus (T, τ) is a T_4-space and Fact 4 is proved.

Returning to our solution take a cardinal $\kappa_0 \ge \kappa$ such that $\kappa_0^\omega = \kappa_0$; apply Fact 1 to find a strongly discrete dense subset X_0 of $\mathbb{I}^{\kappa_0}$ of cardinality κ_0 and denote by μ_0 the topology induced on X_0 from $\mathbb{I}^{\kappa_0}$. If λ is a cardinal, say that a subspace $D \subset \mathbb{I}^\lambda$ is β-*discrete* if D is discrete and there is a homeomorphism $h : \beta D \to \overline{D}$ such that $h(d) = d$ for any $d \in D$.

Proceeding inductively suppose that, for some $n \in \omega$ we have constructed disjoint sets $X_0, \ldots, X_n$, cardinals $\kappa_0, \ldots, \kappa_n$ and topologies $\mu_0, \ldots, \mu_n$ with the following properties:

(4) $\kappa \le \kappa_0 < \ldots < \kappa_n$;

(5) μ_i is a topology on X_i and there is a family $\mathcal{D}_i = \{D^i_j : j \in \omega\}$ of closed
  discrete subspaces of (X_i, μ_i) such that $D^i_j \subset D^i_{j+1}$ for each $j \in \omega$ and
  $\bigcup \mathcal{D}_i = X_i$;

(6) for any $i \leq n$ the set X_i is a dense subspace of $\mathbb{I}^{\kappa_i}$ and μ_i is the topology
  induced on X_i from $\mathbb{I}^{\kappa_i}$;

(7) for any $i < n$ we have $d(\mathbb{I}^{\kappa_i+1}) \geq \kappa_i$ and there is an injection $\varphi_i : X_i \to \mathbb{I}^{\kappa_i+1}$
  such that $\varphi_i(X_i) \cap X_{i+1} = \emptyset$ while the subspace $\varphi(X_i)$ is β-discrete in $\mathbb{I}^{\kappa_i+1}$.

Take a bijection $\eta : X_n \to D$ of X_n onto a discrete space D and choose a cardinal
$\kappa_{n+1} > \kappa_n$ with $\kappa^\omega_{n+1} = \kappa_{n+1}$ and $d(\mathbb{I}^{\kappa_n+1}) \geq k_n$ such that βD embeds in $\mathbb{I}^{\kappa_n+1}$; let
$\xi : \beta D \to \mathbb{I}^{\kappa_n+1}$ be an embedding. It is easy to see that $\xi(\beta D)$ is nowhere dense
in $\mathbb{I}^{\kappa_n+1}$; so if E is a strongly σ-discrete dense subspace of $\mathbb{I}^{\kappa_n+1}$ (see Fact 1) then
$X_{n+1} = E \backslash \xi(\beta(D))$ is still strongly discrete and dense in $\mathbb{I}^{\kappa_n+1}$. Let μ_{n+1} be the
topology induced on X_{n+1} from $\mathbb{I}^{\kappa_n+1}$; the map $\varphi_n = \xi \circ \eta : X_n \to \mathbb{I}^{\kappa_n+1}$ is an
injection and $X_{n+1} \cap \varphi_n(X_n) = \emptyset$. Besides, it follows from $\varphi_n(X_n) \subset \xi(D)$ that
$\varphi_n(X_n)$ is β-discrete in $\mathbb{I}^{\kappa_n+1}$.

Thus all conditions (4)–(7) are satisfied for n replaced by $n + 1$; so our inductive
procedure can be continued to obtain the families $\{X_i : i \in \omega\}$, $\{\kappa_i : i \in \omega\}$ and
$\{\mu_i : i \in \omega\}$ such that (4)–(7) are fulfilled for all $n \in \omega$. It follows from (4) and (6)
that the family $\mathcal{X} = \{X_i : i \in \omega\}$ is disjoint; let $X = \bigcup \mathcal{X}$.

If $n \in \omega$ and $x \in X_n$ then $\mathcal{B}_x = \{\{x\} \cup ((U \cap X_{n+1}) \backslash D^{n+1}_j) : U \in$
$\tau(\varphi_n(x), \mathbb{I}^{\kappa_n+1})$ and $j \in \omega\}$ is easily seen to be a weak local base at x such that
$B \backslash \{x\} \subset X_{n+1}$ for any $B \in \mathcal{B}_x$. Let τ be the topology on X generated by the
collection $\{\mathcal{B}_x : x \in X\}$ of weak local bases.

Fix a number $n \in \omega$ and $Y \subset X_n$; since $\varphi_n(X_n)$ is β-discrete in $\mathbb{I}^{\kappa_n+1}$, the
sets $\varphi_n(Y)$ and $\varphi_n(X_n \backslash Y)$ have disjoint closures in $\mathbb{I}^{\kappa_n+1}$. Therefore there exist
disjoint sets $U', V' \in \tau(\mathbb{I}^{\kappa_n+1})$ such that $\varphi_n(Y) \subset U'$ and $\varphi_n(X_n \backslash Y) \subset V'$. It is
straightforward to check that the sets $U = Y \cup (U' \cap X_{n+1})$ and $V = (X_n \backslash Y) \cup$
$(V' \cap X_{n+1})$ are disjoint weak neighbourhoods of Y and $X_n \backslash Y$ respectively. This
makes it possible to apply Fact 4 to see that (X, τ) is a T_4-space.

It follows from (4),(6) and (7) that $|X| \geq \kappa$; the family $\mathcal{D} = \{D^i_j : i, j \in$
$\omega\}$ consists of closed discrete subspaces of X and $X = \bigcup \mathcal{D}$; so X is strongly
σ-discrete.

Finally, assume that a family $\mathcal{U} \subset \tau \backslash \{\emptyset\}$ is disjoint and uncountable. There
exists $n \in \omega$ such that $\mathcal{U}' = \{U \in \mathcal{U} : U \cap X_n \neq \emptyset\}$ is uncountable. For any
$U \in \mathcal{U}'$ fix a point $x_U \in U \cap X_n$; then there is $B_U \in \mathcal{B}_{x_U}$ with $B_U \subset U$. It is
immediate from the definition of $\mathcal{B}_{x_U}$ that $B_U \backslash \{x_U\}$ is a non-empty open subset of
(X_{n+1}, μ_{n+1}). Thus $\{B_U \backslash \{x_U\} : U \in \mathcal{U}'\}$ is an uncountable disjoint family of non-
empty open subsets of (X_{n+1}, μ_{n+1}) which contradicts (6). This shows that X is a
normal strongly σ-discrete space with $|X| \geq \kappa$ and $c(X) \leq \omega$, i.e., our solution is
complete.

U.427. *Show that there exists a splittable space which cannot be condensed onto a
first countable space.*

Solution. There exists a splittable space X such that $c(X) = \omega$ and $|X| \geq 2^{\mathfrak{c}}$ (see Problem 426). Suppose that there is a condensation $\varphi : X \to Y$ of the space X onto a space Y with $\chi(Y) \leq \omega$. Then $c(Y) \leq \omega$; so we can apply Fact 4 of U.083 to see that $w(Y) \leq \pi\chi(Y)^{c(Y)} \leq \chi(Y)^{c(Y)} \leq \omega^{\omega} = \mathfrak{c}$. Further, apply Fact 1 of U.418 to conclude that $|Y| \leq nw(Y)^{\psi(Y)} \leq w(Y)^{\psi(Y)} \leq \mathfrak{c}^{\omega} = \mathfrak{c}$; this, together with $|Y| = |X| > \mathfrak{c}$ gives a contradiction. Therefore X is a splittable space which cannot be condensed onto a first countable space.

U.428. *Assuming the Generalized Continuum Hypothesis prove that, if X is a splittable space and $A \subset X$, $|A| \leq \mathfrak{c}$ then $|\overline{A}| \leq \mathfrak{c}$.*

Solution. Let $Y = \overline{A}$; then the restriction map $\pi : C_p(Y) \to C_p(A)$ is injective; so $|C_p(Y)| \leq |C_p(A)| \leq |\mathbb{R}^A| \leq 2^{\mathfrak{c}}$. By splittability of Y (see Problem 401), every $f \in \mathbb{R}^Y$ is in the closure of a countable subset of $C_p(Y)$; so $\mathbb{R}^Y = \bigcup\{[P] : P \subset C_p(Y)$ and $|P| \leq \omega\}$ (the brackets denote the closure in $\mathbb{R}^Y$).

If a set $P \subset C_p(Y)$ is countable then $w([P]) \leq 2^{d([P])} \leq \mathfrak{c}$ (see Fact 2 of S.368) and hence we have the inequality $\psi([P]) \leq \mathfrak{c}$. Next, apply Fact 1 of U.418 to conclude that $|[P]| \leq nw([P])^{\psi([P])} \leq \mathfrak{c}^{\mathfrak{c}} = 2^{\mathfrak{c}}$. Thus $|[P]| \leq 2^{\mathfrak{c}}$ for any countable $P \subset C_p(Y)$ and therefore $|\mathbb{R}^Y| \leq 2^{\mathfrak{c}} \cdot |C_p(Y)|^{\omega} \leq 2^{\mathfrak{c}} \cdot (2^{\mathfrak{c}})^{\omega} = 2^{\mathfrak{c}}$; by Generalized Continuum Hypothesis, we have $|\mathbb{R}^Y| \leq \mathfrak{c}^+$. However, if $|Y| > \mathfrak{c}$ then $|Y| \geq \mathfrak{c}^+$ and therefore $|\mathbb{R}^Y| = 2^{|Y|} \geq 2^{\mathfrak{c}^+} > \mathfrak{c}^+$; this contradiction shows that $|\overline{A}| = |Y| \leq \mathfrak{c}$.

U.429. *Prove that any Čech-complete splittable space has a dense metrizable subspace.*

Solution. Say that a family $\mathcal{A}$ of subsets of a space X is called *strongly disjoint* if $\overline{A} \cap \overline{B} = \emptyset$ for any distinct $A, B \in \mathcal{A}$. Given families $\mathcal{A}, \mathcal{B} \subset \exp X$ say that $\mathcal{A}$ is *strongly inscribed* in $\mathcal{B}$ if, for any $A \in \mathcal{A}$ there is $B \in \mathcal{B}$ such that $\overline{A} \subset B$.

Fact 1. For any Čech-complete space X there exists a dense paracompact subspace $Y \subset X$ such that Y is a G_δ-set in X (and hence Y is Čech-complete).

Proof. Fix a family $\mathcal{O} = \{O_n : n \in \omega\} \subset \tau(\beta X)$ such that $O_{n+1} \subset O_n$ for any $n \in \omega$ and $\bigcap \mathcal{O} = X$. Let $\mathcal{G}_0 = \{\beta X\}$ and assume that, for some $n \in \omega$, we have constructed families $\mathcal{G}_0, \ldots, \mathcal{G}_n$ of non-empty open subsets of βX such that

(1) for any $i < n$, the family $\mathcal{G}_{i+1}$ is a maximal element (with respect to inclusion) in the collection of all strongly disjoint subfamilies of $\tau^*(\beta X)$ which are strongly inscribed in $\{G \cap O_i : G \in \mathcal{G}_i\}$.

Let $\mathcal{G}_{n+1} \subset \tau^*(\beta X)$ be a maximal element (by inclusion) in the collection of all strongly disjoint subfamilies of $\tau^*(\beta X)$ which are strongly inscribed in the family $\{G \cap O_n : G \in \mathcal{G}_n\}$. It is straightforward that the condition (1) is now satisfied for all $i \leq n$; so our inductive construction can go ahead to give us a sequence $\{\mathcal{G}_n : n \in \omega\}$ of families such that (1) holds for each $n \in \omega$.

It is easy to see that the set $\bigcup \mathcal{G}_n$ is dense in βX and hence $(\bigcup \mathcal{G}_n) \cap X$ is dense in X for any $n \in \omega$. Any Čech-complete space has the Baire property (see TFS-274); so $Y = \bigcap_{n\in\omega}((\bigcup \mathcal{G}_n) \cap X)$ is a dense G_δ-subspace of X. Observe that it follows from (1) that $Y = \bigcap_{n\in\omega}(\bigcup \mathcal{G}_n)$. As a consequence, the family $\mathcal{H}_n = \{G \cap Y : G \in$

$\mathcal{G}_n\}$ is a disjoint cover of Y which consists of clopen subsets of Y for any $n \in \omega$. In particular, every $\mathcal{H}_n$ is a discrete cover of Y; let $\mathcal{H} = \bigcup_{n\in\omega} \mathcal{H}_n$.

Denote by $\mathbb{S}$ the collection of all families $\{H_n : n \in \omega\}$ such that $H_n \in \mathcal{H}_n$ and $H_{n+1} \subset H_n$ for all $n \in \omega$. Given any sequence $\mathcal{S} = \{H_n : n \in \omega\} \in \mathbb{S}$ take a set $G_n \in \mathcal{G}_n$ such that $G_n \cap Y = H_n$; then $G_{n+1} \cap G_n \neq \emptyset$; so the property (1) implies that $\mathrm{cl}_{\beta X}(G_{n+1}) \subset G_n \cap O_n$ for every $n \in \omega$. Therefore $\mathrm{cl}_Y(H_{n+1}) \subset H_n$ for every $n \in \omega$ while the set $P(\mathcal{S}) = \bigcap_{n\in\omega} \mathrm{cl}_{\beta X}(G_n) = \bigcap_{n\in\omega} G_n$ is compact and contained in X; hence $P(\mathcal{S}) = \bigcap_{n\in\omega} H_n = \bigcap_{n\in\omega} \mathrm{cl}_{\beta X}(H_n)$. Since every set $\mathrm{cl}_{\beta X}(H_n)$ is compact, the family $\{\mathrm{cl}_{\beta X}(H_n) : n \in \omega\}$ is a network of βX at $P(\mathcal{S})$, i.e., for any $U \in \tau(P(\mathcal{S}), \beta X)$ there is $n \in \omega$ such that $\mathrm{cl}_{\beta X}(H_n) \subset U$ (see Fact 1 of S.326); therefore $\mathcal{S}$ is an outer base of $P(\mathcal{S})$ in Y. Besides, the family $\mathcal{K} = \{P(\mathcal{S}) : \mathcal{S} \in \mathbb{S}\}$ is disjoint and $Y = \bigcup \mathcal{K}$.

Take any open cover $\mathcal{U}$ of the space Y and denote by $\mathrm{Fin}(\mathcal{U})$ the family of all finite unions of elements of $\mathcal{U}$. The family $\mathcal{H}' = \{H \in \mathcal{H} :$ there is $W \in \mathrm{Fin}(\mathcal{U})$ such that $H \subset W\}$ is σ-discrete being a subfamily of $\mathcal{H}$. Given a point $y \in Y$ there is $\mathcal{S} = \{H_n : n \in \omega\} \in \mathbb{S}$ with $x \in K = P(\mathcal{S})$; so there exists $W \in \mathrm{Fin}(\mathcal{U})$ with $K \subset W$. The family $\mathcal{S}$ being an outer base of K there is $n \in \omega$ for which $H_n \subset W$ and therefore $H_n \in \mathcal{H}'$. Since $x \in H_n$, we proved that $\mathcal{H}'$ is a cover of Y.

For every $H \in \mathcal{H}'$ fix a finite family $\mathcal{U}_H \subset \mathcal{U}$ such that $H \subset \bigcup \mathcal{U}_H$; repeating infinitely many times some element of $\mathcal{U}_H$ we can enumerate $\mathcal{U}_H$ as $\{U_i^H : i \in \omega\}$. The family $\mathcal{V}_i = \{U_i^H \cap H : H \in \mathcal{H}'\}$ is σ-discrete for any $i \in \omega$ because so is $\mathcal{H}'$. Therefore the family $\mathcal{V} = \bigcup_{i\in\omega} \mathcal{V}_i$ is also σ-discrete. Given a point $x \in Y$ take $H \in \mathcal{H}'$ such that $x \in H$; since $H \subset \bigcup \mathcal{U}_H$, there is $i \in \omega$ with $x \in U_i^H \cap H \in \mathcal{V}$. This proves that $\mathcal{V}$ is a σ-discrete open refinement of $\mathcal{U}$ and hence Y is paracompact by TFS-230. We already saw that Y is a dense G_δ-subspace of X; so Y is Čech-complete by TFS-260. Fact 1 is proved.

Returning to our solution suppose that X is a Čech-complete splittable space. Apply Fact 1 to find a dense Čech-complete paracompact $Y \subset X$. The space Y is also splittable by Problem 401; so we can apply Problem 421 to conclude that Y is the promised metrizable dense subspace of X.

U.430. *Prove that every subspace of a weakly splittable space must be weakly splittable.*

Solution. Suppose that X is a weakly splittable space and take any $Y \subset X$; then the restriction map $\pi : \mathbb{R}^X \to \mathbb{R}^Y$ is continuous (to see it apply TFS-152 to the sets X and Y with their respective discrete topologies). Given any $f \in \mathbb{R}^Y$ let $g(x) = f(x)$ for any $x \in Y$ and $g(x) = 0$ whenever $x \in X \backslash Y$. Then $g \in \mathbb{R}^X$ and $\pi(g) = f$; the space X being weakly splittable, there is a σ-compact subspace $B \subset C_p(X)$ such that $g \in \overline{B}$ (the bar denotes the closure in $\mathbb{R}^X$). The set $A = \pi(B)$ is also σ-compact and it follows from continuity of π that $f = \pi(g) \in \overline{A}$ (here the bar stands for the closure in $\mathbb{R}^Y$). Therefore every $f \in \mathbb{R}^Y$ is in the closure of a σ-compact subset of $C_p(Y)$, i.e., Y is weakly splittable.

U.431. *Prove that, if X condenses onto a weakly splittable space then X is weakly splittable.*

Solution. Let $\varphi : X \to Y$ be a condensation of a space X onto a weakly splittable space Y. For any $f \in \mathbb{R}^Y$ let $\varphi^*(f) = f \circ \varphi$; then $\varphi^* : \mathbb{R}^Y \to \mathbb{R}^X$ is an embedding (to see this apply TFS-163 to the sets X and Y with their respective discrete topologies). Given any $f \in \mathbb{R}^X$ the function $f_0 = f \circ \varphi^{-1}$ belongs to $\mathbb{R}^Y$ and $\varphi^*(f_0) = f$; this shows that $\varphi^*(\mathbb{R}^Y) = \mathbb{R}^X$. Besides, $\varphi^*(C_p(Y))$ is a (dense) subspace of $C_p(X)$ (this was also proved in TFS-163).

Given any $g \in \mathbb{R}^X$ there is $g_0 \in \mathbb{R}^Y$ such that $\varphi^*(g_0) = g$. The space Y being weakly splittable, there is a σ-compact set $B \subset C_p(Y)$ such that $g_0 \in \mathrm{cl}_{\mathbb{R}^Y}(B)$. The set $A = \varphi^*(B) \subset C_p(X)$ is also σ-compact and it follows from continuity of φ^* that $g \in \mathrm{cl}_{\mathbb{R}^X}(A)$. This proves that every $g \in \mathbb{R}^X$ belongs to the closure of a σ-compact subset of $C_p(X)$, i.e., X is weakly splittable.

U.432. *Give an example of a weakly splittable non-splittable space.*

Solution. Let $X = A(\omega_1)$ be the one-point compactification of a discrete space of cardinality ω_1. Then X is Eberlein compact (see Fact 1 of U.340); so there is a dense σ-compact $P \subset C_p(X)$ (see Problem 035). The set P is also dense in $\mathbb{R}^X$; so $f \in \overline{P}$ for any $f \in \mathbb{R}^X$ (the bar denotes the closure in $\mathbb{R}^X$) and hence X is weakly splittable. Since X is not metrizable, it cannot be splittable by Problem 417.

U.433. *Prove that a separable weakly splittable space is splittable.*

Solution. Suppose that X is a separable weakly splittable space. Apply TFS-173 to see that $iw(C_p(X)) = \omega$ and hence every compact subspace of $C_p(X)$ is metrizable. Given any $f \in \mathbb{R}^X$ there is a σ-compact $P \subset C_p(X)$ such that $f \in \overline{P}$ (the bar denotes the closure in $\mathbb{R}^X$). The space P is a countable union of metrizable compact (and hence separable) spaces. This implies separability of P; so there is a countable $A \subset P$ with $P \subset \overline{A}$. Then $f \in \overline{A}$; so the space X is splittable.

U.434. *Let X be a space with ω_1 caliber of X. Prove that, under CH, X is splittable if and only if it is weakly splittable.*

Solution. Since every splittable space is weakly splittable, we must only prove sufficiency; so assume that X is a weakly splittable space with ω_1 a caliber of X. Apply SFFS-290 to see that the space $C_p(X)$ has a small diagonal. Since having a small diagonal is a hereditary property, every subspace of $C_p(X)$ has a small diagonal; this, together with SFFS-298 implies that every compact $K \subset C_p(X)$ is metrizable.

Given any $f \in \mathbb{R}^X$ there is a σ-compact $P \subset C_p(X)$ such that $f \in \overline{P}$ (the bar denotes the closure in $\mathbb{R}^X$). The space P is a countable union of metrizable compact (and hence separable) spaces. This implies separability of P; so there is a countable $A \subset P$ with $P \subset \overline{A}$. Then $f \in \overline{A}$; so the space X is splittable.

U.435. *Show that every functionally perfect space is weakly splittable. In particular, every Eberlein compact space is weakly splittable.*

Solution. If X is a functionally perfect space then there is a σ-compact dense set $P \subset C_p(X)$ (see Problem 301). Then P is also dense in $\mathbb{R}^X$; so any $f \in \mathbb{R}^X$ belongs to the closure of P in $\mathbb{R}^X$. Therefore X is weakly splittable.

U.436. *Prove that every metrizable space is weakly splittable.*

Solution. Observe that every metrizable space is functionally perfect (see Problem 316) and apply Problem 435.

U.437. *Let X be a weakly splittable space of cardinality $\leq \mathfrak{c}$. Prove that $C_p(X)$ is k-separable.*

Solution. The space $\mathbb{R}^X$ is separable by TFS-108; so fix a countable dense $A \subset \mathbb{R}^X$. For any $f \in A$ there is a σ-compact set $P_f \subset C_p(X)$ such that $f \in \overline{P}_f$ (the bar denotes the closure in $\mathbb{R}^X$). The set $P = \bigcup_{f \in A} P_f$ is σ-compact; since $A \subset \overline{P}$, the set P is dense in $\mathbb{R}^X$ and hence in $C_p(X)$. Therefore P is a dense σ-compact subspace of $C_p(X)$, i.e., $C_p(X)$ is k-separable.

U.438. *Prove that if X is a weak Eberlein compact space and $|X| \leq \mathfrak{c}$ then X is Eberlein compact.*

Solution. If X is a compact weakly splittable space of cardinality at most $\mathfrak{c}$ then $C_p(X)$ has a dense σ-compact subspace by Problem 437; so X is Eberlein compact.

U.439. *Let X be a weak Eberlein compact space. Prove that $w(X) = c(X)$. In particular, a weak Eberlein compact space is metrizable whenever it has the Souslin property.*

Solution. Since $c(Z) \leq w(Z)$ for any space Z, it suffices to prove the inequality $w(X) \leq c(X)$ for any weakly splittable compact space X. So, assume that X is a weakly splittable compact space with $c(X) = \kappa$. Then $w(K) \leq \kappa$ for any compact subspace $K \subset C_p(X)$ (see Problem 343). Given any point $x \in X$ let $h(x) = 1$ and $h(y) = 0$ for any $y \in X \setminus \{x\}$. Then $h \in \mathbb{R}^X$; so there is a σ-compact $P \subset C_p(X)$ such that $f \in \overline{P}$ (the bar denotes the closure in $\mathbb{R}^X$).

We have $P = \bigcup_{n \in \omega} P_n$ where every P_n is compact and hence $w(P_n) \leq \kappa$. Therefore $d(P) \leq \sup\{d(P_n) : n \in \omega\} \leq \kappa$; take a set $Q \subset P$ such that $|Q| \leq \kappa$ and $P \subset \overline{Q}$. Observe that $E_f = f^{-1}(f(x))$ is a G_δ-set in X for any $f \in Q$. Given any $y \in X \setminus \{x\}$ it follows from $h \in \overline{P} \subset \overline{Q}$ that there is $f \in Q$ for which $f(x) > \frac{2}{3}$ and $f(y) < \frac{1}{3}$; this shows that $f(x) \neq f(y)$ and hence $y \notin E_f$. As a consequence, $\{f\} = \bigcap_{f \in Q} E_f$; so the point x is a G_κ-set in X. The point $x \in X$ was chosen arbitrarily so we established that $\psi(X) \leq \kappa$ and hence $\chi(X) \leq \kappa$ (see TFS-327); it follows from TFS-329 that $|X| \leq 2^\kappa$.

Next, apply TFS-108 to see that $d(\mathbb{R}^X) \leq \kappa$; so we can fix a dense $B \subset \mathbb{R}^X$ such that $|B| \leq \kappa$. For any $f \in B$ take a σ-compact H_f such that $f \in \overline{H}_f$. The set $H = \bigcup_{f \in B} H_f$ is the union of at most κ-many compact subspaces of $C_p(X)$. Since all of them have weight $\leq \kappa$, the density of H does not exceed κ; choose a set $G \subset H$ such that $H \subset \overline{G}$ and $|G| \leq \kappa$. Observe that $B \subset \overline{H} \subset \overline{G}$; so G is dense in $\mathbb{R}^X$ and hence in $C_p(X)$. This proves that $iw(X) = d(C_p(X)) \leq \kappa$ (see TFS-174); since X is compact, we have $w(X) = iw(X) \leq \kappa$ and hence $w(X) = \kappa = c(X)$.

U.440. *Given a weak Eberlein compact space X prove that X is ω-monolithic, Fréchet–Urysohn and $C_p(X)$ is Lindelöf.*

Solution. Suppose that X is a weakly splittable compact space and take a countable $A \subset X$. The space $\overline{A}$ is also weakly splittable (see Problem 430); being separable, it must be splittable (see Problem 433); so $\overline{A}$ is metrizable and hence second countable (see Problem 417). This proves that X is ω-monolithic.

Assume towards a contradiction that $t(X) > \omega$; then we can apply TFS-328 to find a free sequence $S = \{x_\alpha : \alpha < \omega_1\} \subset X$. By ω-monolithity of X the space $Y_\alpha = \{x_\beta : \beta < \alpha\}$ is second countable for any $\alpha < \omega_1$. If $F_\alpha = \{x_\beta : \beta \geq \alpha\}$ then $F_\alpha \cap Y_\alpha = \emptyset$ for any $\alpha < \omega_1$; the family $\{F_\alpha : \alpha < \omega_1\}$ being centered there is a point $x \in \bigcap\{F_\alpha : \alpha < \omega_1\}$. If $Y = \bigcup\{Y_\alpha : \alpha < \omega_1\}$ then $x \notin Y$; on the other hand, it follows from $S \subset Y$ that $x \in \overline{Y}$.

The space $Z = \{x\} \cup Y$ has cardinality at most $\mathfrak{c}$. Since any countable subset of Y is contained in some Y_α, the space Y is countably compact. The space Z is weakly splittable by Problem 430; so we can apply Problem 437 to see that $C_p(Z)$ has a dense σ-compact subspace P.

For every $z \in Z$ let $\varphi(z)(f) = f(z)$ for any $f \in P$; then the map $\varphi : Z \to C_p(P)$ is injective and continuous (see TFS-166); let $Z' = \varphi(Z)$ and $Y' = \varphi(Y)$. The space Y' is countably compact being a continuous image of Y; so we can apply Problems 046 and 035 to see that Y' is Eberlein compact. This, together with the fact that Y' is dense in Z' and $Y' \neq Z'$, gives a contradiction. Therefore $t(X) \leq \omega$; since X is also ω-monolithic, we can apply Fact 1 of U.080 to conclude that X is Fréchet–Urysohn.

To finally show that $C_p(X)$ is Lindelöf suppose not. Then there is a closed discrete $D \subset C_p(X)$ with $|D| = \omega_1$ (see SFFS-269). For any $x \in X$ let $\mu(x)(f) = f(x)$ for every $x \in D$. Then $\mu : X \to C_p(D)$ is a continuous map; let $X' = \mu(X)$. It is clear that $w(X') = nw(X') \leq nw(C_p(D)) \leq \omega_1$; so we can find a set $A \subset X$ such that $|A| \leq \omega_1$ and $\mu(A)$ is dense in X'. If $F = \overline{A}$ then an easy consequence of the Fréchet–Urysohn property of X is the inequality $|F| \leq \mathfrak{c}$; so F is Eberlein compact by Problem 438. Since $X' = \mu(F)$, the space X' is also Eberlein compact (see Problem 337). For any $f \in C_p(X')$ let $\mu^*(f) = f \circ \mu$; then the map $\mu^* : C_p(X') \to C_p(X)$ is an embedding (see TFS-163) and $D \subset \mu^*(C_p(X'))$ by Fact 5 of U.086; so D is an uncountable closed discrete subspace of $\mu^*(C_p(X'))$. Since $\mu^*(C_p(X')) \simeq C_p(X')$, we have $ext(C_p(X')) \geq \omega_1$ which is a contradiction with the Lindelöf property of $C_p(X')$ (see Problem 150). This proves that $C_p(X)$ is Lindelöf and completes our solution.

U.441. *Give an example of a Gul'ko compact space which is not weakly splittable.*

Solution. By Problem 387, there exists a Gul'ko compact space X which fails to be Eberlein and consists of characteristic functions of some countable subsets of ω^ω. Since there are only $\mathfrak{c}$-many countable subsets of ω^ω, the cardinality of X does not exceed $\mathfrak{c}$. This, together with Problem 438, implies that X cannot be weakly splittable. Thus X is a Gul'ko compact space which is not weakly splittable.

U.442. *Prove that any subspace of a strongly splittable space is strongly splittable.*

Solution. Suppose that X is a strongly splittable space and take any $Y \subset X$; then the restriction map $\pi : \mathbb{R}^X \to \mathbb{R}^Y$ is continuous (to see it apply TFS-152 to the sets X and Y with their respective discrete topologies). Given any $f \in \mathbb{R}^Y$ let $g(x) = f(x)$ for any $x \in Y$ and $g(x) = 0$ whenever $x \in X \backslash Y$. Then $g \in \mathbb{R}^X$ and $\pi(g) = f$; the space X being strongly splittable, there exists a sequence $\{g_n : n \in \omega\} \subset C_p(X)$ which converges to g. It follows from continuity of π that the sequence $\{\pi(g_n) : n \in \omega\} \subset C_p(Y)$ converges to $\pi(g) = f$. Therefore every $f \in \mathbb{R}^Y$ is the limit of a convergent sequence from $C_p(Y)$, i.e., Y is strongly splittable.

U.443. *Prove that, if X condenses onto a strongly splittable space then X is strongly splittable.*

Solution. Let $\varphi : X \to Y$ be a condensation of a space X onto a strongly splittable space Y. For any $f \in \mathbb{R}^Y$ let $\varphi^*(f) = f \circ \varphi$; then $\varphi^* : \mathbb{R}^Y \to \mathbb{R}^X$ is an embedding (to see this apply TFS-163 to the sets X and Y with their respective discrete topologies). Given any $f \in \mathbb{R}^X$ the function $f_0 = f \circ \varphi^{-1}$ belongs to $\mathbb{R}^Y$ and $\varphi^*(f_0) = f$; this shows that $\varphi^*(\mathbb{R}^Y) = \mathbb{R}^X$. Besides, $\varphi^*(C_p(Y))$ is a (dense) subspace of $C_p(X)$ (this was also proved in TFS-163).

Given any $g \in \mathbb{R}^X$ there is $h \in \mathbb{R}^Y$ such that $\varphi^*(h) = g$. The space Y being strongly splittable, there exists a sequence $\{h_n : n \in \omega\} \subset C_p(Y)$ which converges to h. It follows from continuity of φ^* that the sequence $\{\varphi^*(h_n) : n \in \omega\} \subset C_p(X)$ converges to $\varphi^*(h) = g$. This proves that every $g \in \mathbb{R}^X$ is a limit of a sequence from $C_p(X)$, i.e., X is strongly splittable.

U.444. *Prove that, under MA+¬CH, there is a strongly splittable space which is not σ-discrete.*

Solution. Take any subspace $X \subset \mathbb{R}$ with $|X| = \omega_1$. Any σ-discrete second countable space is easily seen to be countable; so X is not σ-discrete. Observe that $C_p(X)$ is separable (because it has a countable network); so let A be a countable dense subset of $C_p(X)$.

Given a function $f \in \mathbb{R}^X$ if $Y = A \cup \{f\}$ then $f \in \overline{A}$ (the bar denotes the closure in Y) and $\chi(f, Y) \leq \chi(\mathbb{R}^X) = \omega_1 < \mathfrak{c}$; this makes it possible to apply SFFS-054 to conclude that there is a sequence $\{f_n : n \in \omega\} \subset A$ which converges to f. Therefore X is a strongly splittable space which fails to be σ-discrete.

U.445. *Prove that every subset of a strongly splittable space is a G_δ-set.*

Solution. Suppose that X is a strongly splittable space and fix a set $A \subset X$. Let f be the characteristic function of A, i.e., $f(x) = 1$ if $x \in A$ and $f(x) = 0$ for all $x \in X \backslash A$. By strong splittability of X there is a sequence $\{f_n : n \in \mathbb{N}\} \subset C_p(X)$ which converges to f. For any $m, n \in \mathbb{N}$, the set $F_{nm} = \{x \in X : f_n(x) \geq 1 - \frac{1}{m+1}\}$ is closed in X; so the set $P_{nm} = \bigcap \{F_{km} : k \geq n\}$ is closed in X as well.

Given a point $x \in X \backslash A$ there is $l \in \mathbb{N}$ such that $f_k(x) < \frac{1}{2}$ for all $k \geq l$; if $x \in P_{nm}$ then $f_k(x) \geq 1 - \frac{1}{m+1} \geq \frac{1}{2}$ for all $k \geq n$ which gives a contradiction for $k = n + l$. Therefore, no point of P_{nm} belongs to $X \backslash A$, i.e., $P_{nm} \subset A$ for all $n, m \in \mathbb{N}$. Therefore $P = \bigcup \{P_{nm} : n, m \in \mathbb{N}\}$ is an F_σ-set contained in A.

For any $x \in A$ the sequence $\{f_n(x)\}$ converges to 1 so, for any $m \in \mathbb{N}$, there is $n \in \mathbb{N}$ such that $f_k(x) > 1 - \frac{1}{m+1}$ for all $k \geq n$. This shows that $x \in P_{nm}$; since the point $x \in A$ was chosen arbitrarily, we proved that $A = P$ and hence A is an F_σ-subset of X. Thus every subset of X is an F_σ-set which is equivalent to saying that every subset of X is a G_δ-set.

U.446. *Show that there exists a space X in which every subset is a G_δ-set while X is not splittable.*

Solution. In Problem 425 it was proved that there exists a strongly σ-discrete space X which is not splittable. It is evident that any subset of X is also a countable union of closed discrete subspaces of X; in particular, every subset of X is an F_σ-set. Passing to complements it is easy to see that every subset of X is a G_δ-set; thus X is a non-splittable space in which every subset is a G_δ-set.

U.447. *Let X be a normal space in which every subset is G_δ. Prove that X is strongly splittable.*

Solution. It is immediate that any $A \subset X$ is also an F_σ-set; consider the set

$$B_1 = \{f \in \mathbb{R}^X : \text{there is a sequence } \{f_n : n \in \omega\} \subset C_p(X) \text{ which converges to } f\}.$$

We must prove that $B_1 = \mathbb{R}^X$; let us first establish that

(1) if $f \in \mathbb{R}^X$ and $f(X)$ is countable then $f \in B_1$.

To prove the property (1) let $M = f(X)$; since M is countable, there is a sequence $\{M_n : n \in \omega\}$ of finite subsets of M such that $M_n \subset M_{n+1}$ for every $n \in \omega$ and $M = \bigcup_{n \in \omega} M_n$. For each $r \in M$ there is a family $\{F_n^r : n \in \omega\}$ of closed subsets of X such that $F_n^r \subset F_{n+1}^r$ for all $n \in \omega$ and $f^{-1}(r) = \bigcup\{F_n^r : n \in \omega\}$. For any $n \in \omega$ the family $\mathcal{F}_n = \{F_n^r : r \in M_n\}$ is finite, disjoint and consists of closed subspaces of X; so we can use normality of X to find a function $f_n \in C_p(X)$ such that $f(x) = r$ whenever $r \in M_n$ and $x \in F_n^r$.

Observe that, for any $x \in X$, there is $k \in \omega$ such that $r = f(x) \in M_k$. It follows from $x \in f^{-1}(r)$ that there is $m \in \omega$ such that $x \in F_m^r$ and hence $x \in F_n^r$ for all $n \geq m + k$. Since $F_n^r \in \mathcal{F}_n$ for all $n \geq m + k$, we have $f_n(x) = r = f(x)$ for all $n \geq m + k$. In particular, the sequence $\{f_n(x)\}$ converges to $f(x)$ for any $x \in X$ and hence $f_n \to f$ (see TFS-143), i.e., (1) is proved.

An immediate consequence of (1) is the inclusion $\mathbb{Q}^X \subset B_1$. Furthermore, $\mathbb{Q}^X$ is uniformly dense in $\mathbb{R}^X$, i.e., for any $f \in \mathbb{R}^X$ there is a sequence $\{f_n : n \in \omega\} \subset \mathbb{Q}^X$ with $f_n \rightrightarrows f$ (see Fact 1 of U.415). Therefore B_1 is also uniformly dense in $\mathbb{R}^X$; however, B_1 is uniformly closed in $\mathbb{R}^X$, i.e., for any sequence $S = \{f_n : n \in \omega\} \subset B_1$, if S converges uniformly to f then $f \in B_1$ (see Fact 2 of T.379). This shows that $B_1 = \mathbb{R}^X$ and completes our solution.

U.448. *Suppose that $Y \subset X$ and $\varphi : C_p(Y) \to C_p(X)$ is a continuous (linear) extender. For $I = \{ f \in C_p(X) : f(Y) \subset \{0\} \}$, define a map $\xi : C_p(Y) \times I \to C_p(X)$ by the formula $\xi(f, g) = \varphi(f) + g$ for any $(f, g) \in C_p(Y) \times I$. Prove that ξ is a (linear) embedding and hence $C_p(Y)$ embeds in $C_p(X)$ as a closed (linear) subspace.*

Solution. For any functions $f, g \in C_p(X)$ let $s(f, g) = f + g$; by TFS-115 the mapping $s : C_p(X) \times C_p(X) \to C_p(X)$ is continuous. Now let $\mu(f, g) = (\varphi(f), g)$ for any pair $(f, g) \in C_p(Y) \times C_p(X)$; the map $\mu : C_p(Y) \times C_p(X) \to (C_p(X))^2$ is continuous being the product of two continuous maps. It is straightforward that $\xi = (s \circ \mu) | (C_p(Y) \times I)$; so the map ξ is continuous.

The restriction map $\pi : C_p(X) \to C_p(Y)$ is continuous (see TFS-152); for any function $f \in C_p(X)$ let $\delta(f) = f - \varphi(\pi(f))$. It is clear that $\delta : C_p(X) \to C_p(X)$ is a continuous map as well and it is immediate that $\delta(C_p(X)) \subset I$; so we can consider that $\delta : C_p(X) \to I$. Now let $\nu(f) = (\pi(f), \delta(f))$ for any $f \in C_p(X)$; then $\nu : C_p(X) \to C_p(Y) \times I$ is a continuous map being the diagonal product of two continuous maps.

Finally, observe that both maps $\nu \circ \xi$ and $\xi \circ \nu$ are identities on the spaces $C_p(Y) \times I$ and $C_p(X)$ respectively; so ξ and ν are homeomorphisms. It φ is a linear map then it is immediate that both maps ξ and ν are linear; so ξ is a linear homeomorphism between the spaces $C_p(Y) \times I$ and $C_p(X)$.

Any factor of a product of (linear) spaces embeds in that product as a closed (linear) subspace; so $C_p(Y)$ embeds in $C_p(X)$ as a closed subspace which is linear if the map φ is linear.

U.449. *Given a space X define a map $e : X \to C_p(C_p(X))$ by $e(x)(f) = f(x)$ for any $x \in X$ and $f \in C_p(X)$. If X is a subspace of a space Y prove that X is t-embedded in Y if and only if there exists a continuous map $\varphi : Y \to C_p(C_p(X))$ such that $\varphi | X = e$. Deduce from this fact that $e(X)$ is t-embedded in $C_p(C_p(X))$ and hence X is homeomorphic to a t-embedded subspace of $C_p(C_p(X))$.*

Solution. If the space X is t-embedded in a space Y then fix a continuous extender $u : C_p(X) \to C_p(Y)$ and let $\pi : C_p(Y) \to C_p(X)$ be the restriction map. We will also need the canonical map $d : Y \to C_p(C_p(Y))$ defined by $d(y)(f) = f(y)$ for any $f \in C_p(Y)$ and $y \in Y$. Observe first that the map $u : C_p(X) \to D = u(C_p(X))$ is a homeomorphism because $\pi | D$ is the continuous inverse of u. Therefore the dual map $u^* : C_p(D) \to C_p(C_p(X))$ defined by $u^*(\xi) = \xi \circ u$ for any $\xi \in C_p(D)$, is a homeomorphism as well.

It follows from continuity of the restriction map $p : C_p(C_p(Y)) \to C_p(D)$, that the map $u^* \circ p \circ d : Y \to C_p(C_p(X))$ is also continuous. If $x \in X$ then

$$u^*(p(d(x)))(f) = ((p(d(x))) \circ u)(f) = d(x)(u(f)) = u(f)(x) = f(x) = e(x)(f)$$

for any function $f \in C_p(X)$ which shows that $u^*(p(d(x))) = e(x)$ for any $x \in X$, i.e., $(u^* \circ p \circ d) | X = e$ and therefore the map $\varphi = u^* \circ p \circ d$ is as required. This proves necessity.

Now assume that there exists a continuous map $\varphi : Y \to C_p(C_p(X))$ such that $\varphi|X = e$. Given a function $f \in C_p(X)$ let $u(f)(y) = \varphi(y)(f)$ for any $y \in Y$; it is straightforward that $u : C_p(X) \to C_p(Y)$. For any point $x \in X$ we have $u(f)(x) = \varphi(x)(f) = e(x)(f) = f(x)$ for every $f \in C_p(X)$ which shows that u is an extender.

To see that u is continuous, for any $y \in Y$, denote by $\pi_y : C_p(Y) \to \mathbb{R}$ the projection of $C_p(Y)$ to the factor of $\mathbb{R}^Y$ defined by the point y; recall that $\pi_y(f) = f(y)$ for any $f \in C_p(Y)$. Then $(\pi_y \circ u)(f) = u(f)(y) = \varphi(y)(f)$ for any $f \in C_p(X)$ which shows that $\pi_y \circ u = \varphi(y)$ is a continuous map on $C_p(X)$ for any $y \in Y$; applying TFS-102 we conclude that the map u is continuous; so u is a continuous extender from X to Y. Thus X is t-embedded in Y; this settles sufficiency.

Finally observe that e is an embedding of X in $C_p(C_p(X))$ by TFS-167 so, if we identify X and $e(X)$ then the identity map of $C_p(C_p(X))$ onto itself witnesses the fact that $e(X)$ is t-embedded in $C_p(C_p(X))$.

U.450. *Prove that, for any space X every t-embedded subspace of X must be closed in X.*

Solution. Suppose that Y is t-embedded in the space X and take a continuous extender $\varphi : C_p(Y) \to C_p(X)$; let $F = \overline{Y}$. The restriction map $\pi : C_p(X) \to C_p(F)$ is continuous; so the map $\mu = \pi \circ \varphi$ is continuous as well and it is straightforward that $\mu : C_p(Y) \to C_p(F)$ is an extender. If $p : C_p(F) \to C_p(Y)$ is the restriction map then μ and p are mutually inverse continuous maps; so p is a homeomorphism. Applying TFS-152 we conclude that $Y = F$ and hence Y is closed in X.

U.451. *Prove that $\beta\omega \setminus \omega$ is not t-embedded in $\beta\omega$.*

Solution. The space $\beta\omega$ is separable while $d(\beta\omega \setminus \omega) > \omega$ (see TFS-371). Therefore $iw(C_p(\beta\omega)) = \omega$ and $iw(C_p(\beta\omega \setminus \omega)) > \omega$ (see TFS-173); i-weight being hereditary, the space $C_p(\beta\omega \setminus \omega)$ is not embeddable in $C_p(\beta\omega)$. Finally, apply Problem 448 to conclude that $\beta\omega \setminus \omega$ is not t-embedded in $\beta\omega$.

U.452. *Suppose that Y is t-embedded in a space X. Prove that $p(Y) \leq p(X)$ and $d(Y) \leq d(X)$.*

Solution. For any space Z the cardinal $a(Z) = \sup\{\kappa : A(\kappa)$ embeds in $Z\}$ is called *the Alexandroff number of Z*. It follows from Problem 448 that $C_p(Y)$ embeds in $C_p(X)$; the Alexandroff number being a hereditary cardinal function, we have $p(Y) = a(C_p(Y)) \leq a(C_p(X)) = p(X)$ (see TFS-178). Since i-weight is also hereditary, we have $d(Y) = iw(C_p(Y)) \leq iw(C_p(X)) = d(X)$ (see TFS-173); this proves that $p(Y) \leq p(X)$ and $d(Y) \leq d(X)$.

U.453. *Suppose that Y is t-embedded in X and a regular cardinal κ is a caliber of X. Prove that κ is a caliber of Y.*

Solution. It follows from Problem 448 that $C_p(Y)$ embeds in $C_p(X)$. The diagonal of $C_p(X)$ is κ-small by SFFS-290; since having a κ-small diagonal is a hereditary property, the space $C_p(Y)$ also has a κ-small diagonal; so we can apply SFFS-290 again to conclude that κ is also a caliber of Y.

U.454. *Prove that any closed subspace of a t-extendial space is t-extendial.*

Solution. Suppose that X is a t-extendial space and Y is a closed subspace of X. If F is closed in Y then F is also closed in X; so there exists a continuous extender $\varphi : C_p(F) \to C_p(X)$. The restriction map $\pi : C_p(X) \to C_p(Y)$ being continuous, the map $\eta = \pi \circ \varphi$ is continuous as well and it is evident that $\eta : C_p(F) \to C_p(Y)$ is an extender. Thus, for every closed $F \subset Y$, there exists a continuous extender $\eta : C_p(F) \to C_p(Y)$, i.e., F is t-embedded in Y. This proves that Y is t-extendial.

U.455. *Let X be a t-extendial space. Prove that $s(X) = p(X)$.*

Solution. It was proved in TFS-179 that $p(X) \leq s(X)$; to show that $s(X) \leq p(X)$ take an arbitrary discrete subspace $D \subset X$. The set $Y = \overline{D}$ is t-embedded in X; so $p(Y) \leq p(X)$ by Problem 452. The family $\{\{d\} : d \in D\}$ is disjoint and consists of non-empty open subsets of Y. Therefore $|D| \leq p(Y) \leq p(X)$; so $|D| \leq p(X)$ for any discrete subspace $D \subset X$. This proves that $s(X) \leq p(X)$ and hence $s(X) = p(X)$.

U.456. *Let X be a t-extendial Baire space. Prove that $s(X) = c(X)$. In particular, if X is a pseudocompact t-extendial space, then $c(X) = s(X)$.*

Solution. It follows from TFS-282 and Problem 455 that $s(X) = p(X) = c(X)$. If X is a pseudocompact space then X has the Baire property (see TFS-274); so $s(X) = c(X)$ as well.

U.457. *Prove that, for any compact t-extendial space X, we have $t(X) \leq c(X)$.*

Solution. Given a free sequence $S \subset X$, it is an easy exercise that S is a discrete subspace of X; as a consequence, $|S| \leq s(X)$. Since $t(X)$ is a supremum of cardinalities of free sequences of X (see TFS-328), we have $t(X) \leq s(X)$. Finally, apply Problem 456 to conclude that $t(X) \leq s(X) = c(X)$.

U.458. *Prove that, under MA$+\neg$CH, if $X \times X$ is a t-extendial compact space and $c(X) \leq \omega$ then X is metrizable.*

Solution. Apply SFFS-050 to see that $c(X \times X) \leq \omega$ and hence $s(X \times X) \leq \omega$ by Problem 456; applying SFFS-062 we conclude that X is metrizable.

U.459. *Suppose that X is a t-extendial Čech-complete space such that ω_1 is a caliber of X. Prove that X is hereditary separable.*

Solution. As ω_1 is a caliber of X, we have $p(X) = \omega$ and hence $s(X) = \omega$ by Problem 455. If $K \subset X$ is compact and $S \subset K$ is a free sequence in K then S is a discrete subspace of K, so $|S| \leq s(K) \leq s(X) = \omega$; this, together with TFS-328, implies that $t(K) \leq \omega$ for any compact $K \subset X$.

If $A \subset X$ is not closed then there is a compact $K \subset X$ such that $K \cap A$ is not closed in K (see Fact 1 of T.210); since $t(K) = \omega$, there is a countable $B \subset A \cap K$ such that $\overline{B} \backslash A \neq \emptyset$. This proves that $t(X) \leq \omega$ (see Lemma of S.162).

If X is not hereditarily separable then there exists a set $Y = \{x_\alpha : \alpha < \omega_1\} \subset X$ which is left-separated, i.e., $x_\alpha \notin \overline{\{x_\beta : \beta < \alpha\}}$ for any $\alpha < \omega_1$ (see SFFS-004). The set $F_\alpha = \overline{\{x_\beta : \beta < \alpha\}}$ is closed in X for any $\alpha < \omega_1$; since $t(X) = \omega$, the set $F = \bigcup \{F_\alpha : \alpha < \omega_1\}$ is closed in X. The set Y being left-separated, we have $U_\alpha = F \backslash F_\alpha \in \tau^*(F)$ for any $\alpha < \omega_1$. The ω_1-sequence $\{F_\alpha : \alpha < \omega_1\}$ is increasing; so the family $\{U_\alpha : \alpha < \omega_1\} \subset \tau^*(F)$ is point-countable. However, F is t-embedded in X, so ω_1 is a caliber of F by Problem 453; this contradiction shows that X is hereditarily separable.

U.460. *Assuming MA+¬CH prove that any t-extendial Čech-complete space with the Souslin property is hereditarily separable.*

Solution. Say that a space X is *inadequate* if X is t-extendial, Čech-complete, $c(X) = \omega$ and $hd(X) > \omega$; we must prove that inadequate spaces do not exist. Assume towards a contradiction that X is an inadequate space and apply Problem 456 to see that $s(X) = c(X) = \omega$. If $K \subset X$ is compact and $S \subset K$ is a free sequence in K then S is a discrete subspace of K, so $|S| \leq s(K) \leq s(X) = \omega$; this, together with TFS-328, implies that $t(K) \leq \omega$ for any compact $K \subset X$.

If $A \subset X$ is not closed then there is a compact $K \subset X$ such that $K \cap A$ is not closed in K (see Fact 1 of T.210); since $t(K) = \omega$, there is a countable $B \subset A \cap K$ such that $\overline{B} \backslash A \neq \emptyset$. This proves that $t(X) \leq \omega$ (see Lemma of S.162). Therefore

(1) $t(X) = s(X) = \omega$ for any inadequate space X.

Now assume that X is inadequate and any $U \in \tau^*(X)$ contains a non-empty separable open subspace. Take a maximal disjoint family $\mathcal{U}$ of non-empty separable open subspaces of X. It is easy to see that $\bigcup \mathcal{U}$ is dense in X; it follows from $c(X) = \omega$ that $|\mathcal{U}| \leq \omega$; so $\bigcup \mathcal{U}$ is separable and hence X is separable as well. An easy consequence is that ω_1 is a caliber of X; so X is hereditarily separable by Problem 459. This contradiction shows that

(2) for every inadequate space X there exists a set $U \in \tau^*(X)$ which is nowhere separable, i.e., the closure of any countable subset of U is nowhere dense in U.

Take again an inadequate space X; by (2), there exists a set $U \in \tau^*(X)$ which is nowhere separable. It is clear that $F = \overline{U}$ is nowhere separable as well. Besides, F is Čech-complete, extendial and $c(F) = \omega$ (see TFS-260 and Problem 454); therefore F is inadequate. As a consequence,

(3) there exists an inadequate nowhere separable space.

Apply (3) to choose an inadequate nowhere separable space X. Given a closed separable $F \subset X$ take a countable dense $A \subset F$; since F is nowhere dense, it follows from (1) that for any $a \in A$ there is a countable $P_a \subset X \backslash F$ such that $a \in \overline{P_a}$. The set $P = \bigcup \{P_a : a \in A\}$ is closed and separable; it is easy to see that $F \subset P$ and F is nowhere dense in P. This proves that

(4) for any closed separable $F \subset X$ there is a closed separable $P \subset X$ such that $F \subset P$ and F is nowhere dense in P.

Apply (4) to construct, by a transfinite induction, an ω_1-sequence $\{F_\alpha : \alpha < \omega_1\}$ of closed separable subspaces of X such that $\alpha < \beta < \omega_1$ implies that $F_\alpha \subset F_\beta$ and F_α is nowhere dense in F_β. It follows from (1) that the set $F = \bigcup\{F_\alpha : \alpha < \omega_1\}$ is closed in X. Besides, F is Čech-complete and $c(F) \leq s(F) = \omega$. Fix a countable family Q of compact subsets of $\beta F \backslash F$ such that $\bigcup Q = \beta F \backslash F$. It is evident that every element of Q is nowhere dense in βF. Since F_α is nowhere dense in F, the set $K_\alpha = \text{cl}_{\beta F}(F_\alpha)$ is nowhere dense in βF for any $\alpha < \omega_1$.

The family $\mathcal{A} = \{K_\alpha : \alpha < \omega_1\} \cup Q$ consists of nowhere dense subsets of βF and $\bigcup \mathcal{A} = \beta F$; since $|\mathcal{A}| = \omega_1 < \mathfrak{c}$ and $c(\beta F) = c(F) = \omega$, we obtain a contradiction with the topological version of Martin's Axiom (see SFFS-058). Therefore inadequate spaces do not exist and hence our solution is complete.

U.461. *Prove that a t-extendial compact space cannot be mapped onto $\mathbb{I}^{\omega_1}$.*

Solution. Suppose that a space X is t-extendial, compact and there exists a continuous onto map $f : X \to \mathbb{I}^{\omega_1}$. Apply TFS-366 to find a closed $F \subset X$ such that $f(F) = \mathbb{I}^{\omega_1}$ and the map $g = f|F$ is irreducible. Since $\mathbb{I}^{\omega_1}$ is separable, so is F by Problem 130; it is easy to see that this implies that ω_1 is a caliber of F. By Problem 454, the space F has to be t-extendial; so $hd(F) = \omega$ by Problem 459. Any continuous image of a hereditarily separable space is hereditarily separable so $\mathbb{I}^{\omega_1}$ is hereditarily separable which is easily seen to be false. This contradiction shows that X cannot be mapped onto $\mathbb{I}^{\omega_1}$.

U.462. *Prove that the set $\{x \in X : \pi\chi(x, X) \leq \omega\}$ is dense in any t-extendial compact space X.*

Solution. Suppose that the set $P = \{x \in X : \pi\chi(x, X) \leq \omega\}$ is not dense in X. The set $U = X \backslash \overline{P}$ is non-empty, open in X and $\pi\chi(x, U) \geq \omega_1$ for any $x \in U$. Take $V \in \tau^*(X)$ such that $F = \overline{V} \subset U$; it is straightforward that $\pi\chi(x, F) \geq \omega_1$ for any $x \in F$. Apply Fact 1 of U.086 to see that F can be mapped continuously onto $\mathbb{I}^{\omega_1}$; however, F is t-extendial by Problem 454, so F cannot be mapped onto $\mathbb{I}^{\omega_1}$ by Problem 461. This contradiction shows that the set P is dense in X.

U.463. *Give an example of a countable space which is not t-extendial.*

Solution. Given a space X we denote by $B_1(X)$ the first Baire class functions on X, i.e., $B_1(X) = \{f \in \mathbb{R}^X : \text{there exists a sequence } \{f_n : n \in \omega\} \subset C_p(X)$ such that $f_n \to f\}$. If $\mathcal{A} \subset \exp X$ then a family $\mathcal{B} \subset \exp X$ is *inscribed* in $\mathcal{A}$ if, for any $B \in \mathcal{B}$ there is $A \in \mathcal{A}$ such that $B \subset A$.

Fact 1. There exists a surjective map $\xi : \mathbb{D}^\omega \to \omega^\omega$ such that $\xi^{-1}(U)$ is an F_σ-set in $\mathbb{D}^\omega$ for any open $U \subset \omega^\omega$.

Proof. Let $Q = \{x_n : n \in \omega\}$ be a faithfully indexed dense subset of $\mathbb{D}^\omega$. The set $I = \mathbb{D}^\omega \backslash Q$ is dense in $\mathbb{D}^\omega$ and Polish being a G_δ-subset of $\mathbb{D}^\omega$. If $K \subset I$ is compact then it is nowhere dense in $\mathbb{D}^\omega$ and hence in I. Besides, I is zero-dimensional because so is $\mathbb{D}^\omega$. This makes it possible to apply SFFS-347 to see that I is homeomorphic to ω^ω; therefore it suffices to construct a surjective map $\xi : \mathbb{D}^\omega \to I$ such that $\xi^{-1}(U)$ is an F_σ-set in $\mathbb{D}^\omega$ for any $U \in \tau(I)$.

To do so denote by $\mathcal{C}$ the family of all non-empty clopen subsets of $\mathbb{D}^\omega$ and take a metric ρ on $\mathbb{D}^\omega$ which generates the topology of $\mathbb{D}^\omega$. Use zero-dimensionality of $\mathbb{D}^\omega$ to find a disjoint cover $\mathcal{U}_0 \subset \mathcal{C}$ of the set $\mathbb{D}^\omega\setminus\{x_0\}$ such that $\text{diam}_\rho(U) \leq 1$ for all $U \in \mathcal{U}_0$.

Suppose that $n \in \omega$ and we have families $\mathcal{U}_0, \ldots, \mathcal{U}_n$ with the following properties:

(1) $\mathcal{U}_i$ is disjoint, $\bigcup \mathcal{U}_i = \mathbb{D}^\omega\setminus\{x_0, \ldots, x_i\}$ and $\mathcal{U}_i \subset \mathcal{C}$ for any $i \leq n$;
(2) if $i \leq n$ then $\text{diam}_\rho(U) \leq 2^{-i}$ for any $U \in \mathcal{U}_i$;
(3) $\mathcal{U}_{i+1}$ is inscribed in $\mathcal{U}_i$ for all $i < n$.

Every $U \in \mathcal{U}_n$ is compact and zero-dimensional; so we can find a disjoint family $\mathcal{A}_U \subset \mathcal{C}$ such that $\bigcup \mathcal{A}_U = U\setminus\{x_{n+1}\}$ and $\text{diam}_\rho(V) \leq 2^{-n-1}$ for any $V \in \mathcal{A}_U$. It is straightforward that, for the family $\mathcal{U}_{n+1} = \bigcup\{\mathcal{A}_U : U \in \mathcal{U}_n\}$, the properties (1)–(3) are fulfilled if we substitute n by $n+1$. Therefore we can proceed inductively to construct a sequence $\{\mathcal{U}_i : i \in \omega\}$ for which (1)–(3) are satisfied for all $n \in \omega$; let $\mathcal{U} = \bigcup_{i \in \omega} \mathcal{U}_i$. It follows from (1) that $I \subset \bigcup \mathcal{U}_i$ for all $i \in \omega$; so (2) implies that $\mathcal{U}$ contains a local base at any point of I. Therefore the family $\mathcal{V} = \{U \cap I : U \in \mathcal{U}\}$ is a base of I.

Fix $n \in \omega$; it follows from (1) and (3) that there exist $U_0, \ldots, U_{n-1} \in \mathcal{U}$ such that $x_n \in U_i \in \mathcal{U}_i$ for every $i < n$ and $x_n \notin U$ for all $U \in \mathcal{U}\setminus\{U_0, \ldots, U_{n-1}\}$. The property (3) shows that $U_0 \supset \ldots \supset U_{n-1}$; take a point $y_n \in U_{n-1} \cap I$. After we accomplish the choice of y_n for all $n \in \omega$ it is easy to see that

(4) for any $n \in \omega$, if $x_n \in U \in \mathcal{U}$ then $y_n \in U \cap I$.

Now let $\xi(x) = x$ if $x \in I$ and $\xi(x_n) = y_n$ for any $n \in \omega$. This gives a surjective map $\xi : \mathbb{D}^\omega \to I$. Given any $V \in \mathcal{V}$ there is $U \in \mathcal{U}$ such that $V = U \cap I$ and hence $\overline{V} = U$. If $x \in V$ then $\xi(x) = x \in V$; if $x \in U\setminus V$ then $x = x_n$ for some $n \in \omega$ and hence $y_n = \xi(x_n) = \xi(x) \in U \cap I = V$ by (4); this proves that $\xi(U) \subset V$ and therefore $U \subset \xi^{-1}(V)$. Thus $\xi^{-1}(V) = U \cup Q'$ for some $Q' \subset Q$ which, together with compactness of U, shows that

(5) $\xi^{-1}(V)$ is an F_σ-subset of $\mathbb{D}^\omega$ for any $V \in \mathcal{V}$.

Finally take any set $U \in \tau(I)$; since $\mathcal{V}$ is a base in I, there is $\mathcal{V}' \subset \mathcal{V}$ such that $\bigcup \mathcal{V}' = U$. It follows from the property (5) and countability of the family $\mathcal{V}'$ that $\xi^{-1}(U) = \bigcup\{\xi^{-1}(V) : V \in \mathcal{V}'\}$ is an F_σ-subset of $\mathbb{D}^\omega$; so Fact 1 is proved.

Fact 2. The space $C_p(\omega^\omega)$ embeds linearly in $B_1(\mathbb{D}^\omega)$.

Proof. Apply Fact 1 to choose a surjective map $\xi : \mathbb{D}^\omega \to \omega^\omega$ such that $\xi^{-1}(U)$ is an F_σ-subset of $\mathbb{D}^\omega$ for any $U \in \tau(\omega^\omega)$. If D is the set $\mathbb{D}^\omega$ with the discrete topology then the map $\xi : D \to \omega^\omega$ is continuous; so the dual map $\xi^* : C_p(\omega^\omega) \to \mathbb{R}^D$ is a linear embedding (see TFS-163). For any $f \in C_p(\omega^\omega)$ and $O \in \tau(\mathbb{R})$ the set $U = f^{-1}(O)$ is open in ω^ω; so $(\xi^*(f))^{-1}(O) = (f \circ \xi)^{-1}(O) = \xi^{-1}(U)$ is an F_σ-subset of $\mathbb{D}^\omega$. This makes it possible to apply SFFS-379 to conclude that $\xi^*(f) \in B_1(\mathbb{D}^\omega)$ for any $f \in C_p(\omega^\omega)$, i.e., ξ^* embeds $C_p(\omega^\omega)$ linearly in $B_1(\mathbb{D}^\omega)$; so Fact 2 is proved.

Returning to our solution apply SFFS-372 to choose a countable space P with a unique non-isolated point such that $C_p(P) \in \mathbb{B}(\mathbb{R}^P) \setminus \Sigma_\omega^0(\mathbb{R}^P)$ and, in particular, $C_p(P)$ is a Borel subset of $\mathbb{R}^P$ which does not belong to $\Sigma_\alpha^0(\mathbb{R}^P)$ for any $\alpha < \omega$. Denote by w the unique non-isolated point of P. Since $C_p(P)$ is analytic by SFFS-334, we can apply SFFS-370 to conclude that P embeds in $C_p(\omega^\omega)$; by Fact 2, the space P also embeds in $B_1(\mathbb{D}^\omega)$; so we consider that $P = \{w\} \cup \{g_n : n \in \omega\} \subset B_1(\mathbb{D}^\omega)$ and the enumeration of the set P is faithful. Let $u(x) = 0$ for all $x \in \mathbb{D}^\omega$; if $f_n = g_n - w$ for all $n \in \omega$ then the space $F = \{u\} \cup \{f_n : n \in \omega\} \subset B_1(\mathbb{D}^\omega)$ is homeomorphic to P.

The subspace $\{f_n : n \in \omega\} \subset B_1(\mathbb{D}^\omega)$ being discrete, there exists a disjoint family $\mathcal{O} = \{O_n : n \in \omega\} \subset \tau(B_1(\mathbb{D}^\omega))$ such that $u \notin \bigcup \mathcal{O}$ and $f_n \in O_n$ for all $n \in \omega$ (see Fact 1 of S.369). Since every f_n is a limit of a non-trivial sequence from $C_p(\mathbb{D}^\omega)$, we can fix a sequence $S_n \subset (O_n \cap C_p(\mathbb{D}^\omega)) \setminus \{f_n\}$ which converges to f_n; let $S = \bigcup_{n \in \omega} S_n$ and choose a faithful enumeration $\{f_n^k : k \in \omega\}$ of the sequence S_n for every $n \in \omega$. The space $X = S \cup F$ is countable; since all points of S are isolated in X, the set F is closed in X. Observe also that

(6) for any $n \in \omega$, the family $\{\{f_n\} \cup (S_n \setminus K) : K$ is a finite subset of $S_n\}$ is a local base of the space X at the point f_n.

The following property of X is crucial:

(7) a function $\varphi \in \mathbb{R}^X$ is continuous on X if and only if $\varphi|(\{u\} \cup S)$ is continuous and φ is continuous at f_n for every $n \in \omega$.

Only sufficiency must be proved in (7); so assume that a function $\varphi \in \mathbb{R}^X$ is continuous on the space $\{u\} \cup S$ and at every f_n. The points of S being isolated in X we must only establish continuity of φ at the point u; so take any $\varepsilon > 0$. Since φ is continuous at u in $S' = \{u\} \cup S$, there is $U \in \tau(u, X)$ such that $|\varphi(f) - \varphi(u)| < \frac{\varepsilon}{2}$ for any $f \in U \cap S'$. If $f \in U \cap (F \setminus \{u\})$ then $f = f_n$ for some number $n \in \omega$ and hence $\varphi(f_n^k) \to \varphi(f)$ when $k \to \infty$. There is $p \in \omega$ such that $f_n^k \in U$ and hence $|\varphi(f_n^k) - \varphi(u)| < \frac{\varepsilon}{2}$ for all $k \geq p$; so it follows from continuity of the map φ at f_n that $|\varphi(f_n) - \varphi(u)| \leq \frac{\varepsilon}{2} < \varepsilon$. This proves that $|\varphi(f) - \varphi(u)| < \varepsilon$ for any $f \in U$, i.e., U witnesses that φ is continuous at the point u in X and hence (7) is proved.

Given $k, m, n \in \omega$ the set $C(k, m, n) = \{\varphi \in \mathbb{R}^X : |\varphi(f_n^i) - \varphi(f_n)| \leq 2^{-m}$ for any $i \geq k\}$ is closed in $\mathbb{R}^X$; so the set $D(m, n) = \bigcup_{k \in \omega} C(k, m, n)$ is an F_σ-subset of $\mathbb{R}^X$. It is easy to see, applying (6), that a function $\varphi \in \mathbb{R}^X$ is continuous at f_n if and only if $\varphi \in E(n) = \bigcap_{m \in \omega} D(m, n)$. Observe that $E = \bigcap_{n \in \omega} E(n)$ is an $F_{\sigma\delta}$-subset of $\mathbb{R}^X$ and a function $\varphi \in \mathbb{R}^X$ is continuous at all points of $F \setminus \{u\}$ if and only if $\varphi \in E$.

Now, it follows from the property (7) that $C_p(X) = E \cap (C_p(S') \times \mathbb{R}^{X \setminus S'})$. Since $S' \subset C_p(\mathbb{D}^\omega)$, we can apply SFFS-375 to see that $C_p(S')$ is an $F_{\sigma\delta}$-subset of $\mathbb{R}^{S'}$ and hence $C_p(S') \times \mathbb{R}^{X \setminus S'}$ is an $F_{\sigma\delta}$-subset of $\mathbb{R}^X$. This, together with Fact 1 of T.341, implies that

(8) $C_p(X)$ is an $F_{\sigma\delta}$-subset of $\mathbb{R}^X$, i.e., $C_p(X) \in \Pi_2^0(\mathbb{R}^X)$.

Assume that the set F is t-embedded in X and hence $C_p(F)$ embeds in $C_p(X)$ as a closed subspace (see Problem 448). The respective copy of $C_p(F)$ is, therefore, the intersection of $C_p(X)$ and a closed subspace of $\mathbb{R}^X$ which shows, together with (8), that $C_p(F)$ embeds in $\mathbb{R}^X$ as an $F_{\sigma\delta}$-subspace and hence the mentioned copy of $C_p(F)$ belongs to $\Sigma_3^0(\mathbb{R}^X)$.

The space $C_p(F)$ being homeomorphic to $C_p(P)$ we conclude that $C_p(P)$ embeds in $\mathbb{R}^X$ as a subset of class $\Sigma_3^0(\mathbb{R}^X)$. Recalling that $C_p(P) \in \mathbb{B}(\mathbb{R}^P)\backslash\Sigma_\omega^0(\mathbb{R}^P)$, we discover a contradiction with Fact 3 of T.333. Thus the closed subset F of the space X is not t-embedded in X; this shows that X is an example of a countable space which is not t-extendial.

U.464. *Prove that any strongly discrete subspace $A \subset X$ is l-embedded in X.*

Solution. This was proved (in other terminology) in Fact 5 of T.132.

U.465. *Prove that, if Y is a retract of X, then Y is l-embedded in X.*

Solution. Let $r : X \to Y$ be a retraction; then it dual map $r^* : C_p(Y) \to C_p(X)$ is linear and continuous (see TFS-163). For any $f \in C_p(Y)$ and $x \in Y$ we have $r^*(f)(x) = f(r(x)) = f(x)$ which shows that $r^*(f)|Y = f$. Therefore r^* is the required linear extender and hence Y is l-embedded in X.

U.466. *Given a space X define a map $e : X \to C_p(C_p(X))$ by $e(x)(f) = f(x)$ for any $x \in X$ and $f \in C_p(X)$. Observe that $e(X) \subset L_p(X)$; prove that $e(X)$ is l-embedded in $L_p(X)$ and hence any space X is homeomorphic to an l-embedded subspace of $L_p(X)$.*

Solution. The map e is an embedding by TFS-167 and $e(X) \subset L_p(X)$ by TFS-196. For any $f \in C_p(e(X))$ let $e^*(f) = f \circ e$; then $e^* : C_p(e(X)) \to C_p(X)$ is a homeomorphism by TFS-163 and it is straightforward that e^* is a linear map. Given any function $f \in C_p(X)$ and $\xi \in L_p(X)$ let $u(f)(\xi) = \xi(f)$; this defines a continuous map $u : C_p(X) \to C_p(L_p(X))$ (see TFS-166). An immediate consequence is that the map $\varphi = u \circ e^* : C_p(e(X)) \to C_p(L_p(X))$ is continuous.

Next take $f, g \in C_p(X)$ and $\alpha, \beta \in \mathbb{R}$. Given $\xi \in L_p(X)$ apply linearity of ξ to see that $u(\alpha f + \beta g)(\xi) = \xi(\alpha f + \beta g) = \alpha\xi(f) + \beta\xi(g) = (\alpha u(f) + \beta u(g))(\xi)$ which shows that $u(\alpha f + \beta g) = \alpha u(f) + \beta u(g)$, i.e., $u : C_p(X) \to C_p(L_p(X))$ is a linear map. Therefore φ is linear being the composition of two linear maps.

Finally fix any $y \in e(X)$, $f \in C_p(e(X))$ and $x \in X$ with $e(x) = y$; then $\varphi(f)(y) = u(e^*(f))(y) = u(f \circ e)(y) = y(f \circ e) = e(x)(f \circ e) = f(e(x)) = f(y)$ which proves that $\varphi(f)(y) = f(y)$ for any $y \in e(X)$, i.e., $\varphi(f)|e(X) = f$ and hence e is a linear continuous extender. Thus $e(X)$ is a homeomorphic copy of X which is l-embedded in $L_p(X)$.

U.467. *Given a space X define a map $e : X \to C_p(C_p(X))$ by $e(x)(f) = f(x)$ for any $x \in X$ and $f \in C_p(X)$. Observe that $e(X) \subset L_p(X)$; so we can consider that $e : X \to L_p(X)$. If X is a subspace of a space Y prove that X is l-embedded in Y if and only if there exists a continuous map $\varphi : Y \to L_p(X)$ such that $\varphi|X = e$.*

Solution. The map e is an embedding by TFS-167 and $e(X) \subset L_p(X)$ by TFS-196; so we can and will consider that $e : X \to L_p(X)$. If X is l-embedded in Y then fix a linear continuous extender $u : C_p(X) \to C_p(Y)$ and let $\pi : C_p(Y) \to C_p(X)$ be the restriction map; it is clear that π is also linear and continuous. We will also need the canonical map $d : Y \to C_p(C_p(Y))$ defined by $d(y)(f) = f(y)$ for any $f \in C_p(Y)$ and $y \in Y$. Observe first that the map $u : C_p(X) \to D = u(C_p(X))$ is a linear homeomorphism because $\pi|D$ is the continuous linear inverse of u. This implies that the dual map $u^* : C_p(D) \to C_p(C_p(X))$ defined by $u^*(\xi) = \xi \circ u$ for any $\xi \in C_p(D)$, is a linear homeomorphism as well.

Since the restriction map $p : C_p(C_p(Y)) \to C_p(D)$ is continuous, the map $\varphi = u^* \circ p \circ d : Y \to C_p(C_p(X))$ is continuous as well. Take an arbitrary $y \in Y$; then $d(y)$ is a linear continuous functional on $C_p(Y)$ (see TFS-196); so $p(d(y)) = d(y)|D$ is a continuous linear functional on D. The map u being linear and continuous, the equality $\varphi(y) = u^*(p(d(y))) = p(d(y)) \circ u$ shows that $\varphi(y)$ is a linear continuous functional on $C_p(X)$, i.e., $\varphi(y) \in L_p(X)$. Thus we can consider that $\varphi : Y \to L_p(X)$.

If $x \in X$ then

$$u^*(p(d(x)))(f) = ((p(d(x)) \circ u)(f) = d(x)(u(f)) = u(f)(x) = f(x) = e(x)(f)$$

for any $f \in C_p(X)$ which shows that $\varphi(x) = e(x)$ for any $x \in X$, i.e., $\varphi|X = e$ and therefore the map φ is as required. This proves necessity.

Now assume that there exists a continuous map $\varphi : Y \to L_p(X)$ such that $\varphi|X = e$. Given a function $f \in C_p(X)$ let $u(f)(y) = \varphi(y)(f)$ for any $y \in Y$; it is straightforward that $u : C_p(X) \to C_p(Y)$. For any point $x \in X$ we have $u(f)(x) = \varphi(x)(f) = e(x)(f) = f(x)$ for every $f \in C_p(X)$ which shows that u is an extender. Given functions $f, g \in C_p(X)$ and $\alpha, \beta \in \mathbb{R}$ fix $y \in Y$; recalling that $\varphi(y)$ is a linear functional on the space $C_p(X)$ we conclude that

$$u(\alpha f + \beta g)(y) = \varphi(y)(\alpha f + \beta g) = \alpha\varphi(y)(f) + \beta\varphi(y)(g) = \alpha u(f)(y) + \beta u(g)(y).$$

Since the point $y \in Y$ was chosen arbitrarily, we established that $u(\alpha f + \beta g) = \alpha u(f) + \beta u(g)$, i.e., u is a linear map.

To see that u is continuous, for any $y \in Y$, denote by $\pi_y : C_p(Y) \to \mathbb{R}$ the projection of the space $C_p(Y)$ to the factor of $\mathbb{R}^Y$ defined by the point y; recall that $\pi_y(f) = f(y)$ for any function $f \in C_p(Y)$. Then $(\pi_y \circ u)(f) = u(f)(y) = \varphi(y)(f)$ for any $f \in C_p(X)$ which shows that $\pi_y \circ u = \varphi(y)$ is a continuous map on $C_p(X)$ for any $y \in Y$; applying TFS-102 we infer that the map u is continuous. Therefore $u : C_p(X) \to C_p(Y)$ is a continuous linear extender, i.e., X is l-embedded in Y; this settles sufficiency.

U.468. *Prove that any closed subspace of an extendial space is extendial.*

Solution. Suppose that X is an extendial space and Y is a closed subspace of X. If F is closed in Y then F is also closed in X; so there exists a continuous linear extender $\varphi : C_p(F) \to C_p(X)$. The restriction map $\pi : C_p(X) \to C_p(Y)$ being linear and continuous, the map $\eta = \pi \circ \varphi$ is linear and continuous as well and it is evident that $\eta : C_p(F) \to C_p(Y)$ is an extender. Thus, for every closed $F \subset Y$, there exists a continuous linear extender $\eta : C_p(F) \to C_p(Y)$, i.e., F is l-embedded in Y. This proves that Y is extendial.

U.469. *Prove that every metrizable space is extendial.*

Solution. This was proved in Fact 1 of U.062.

U.470. *Prove that, for any zero-dimensional linearly ordered compact space X any closed $F \subset X$ is a retract of X; in particular, the space X is extendial.*

Solution. Let $<$ be a linear order that generates the topology of X. We will apply the usual notation for the intervals in X. Thus for any $a, b \in X$ we let $(a, b) = \{x \in X : a < x < b\}$; besides, $[a, b) = \{x \in X : a \leq x < b\}$ and $(a, b] = \{x \in X : a < x \leq b\}$. If $a \in X$ then $(\leftarrow, a) = \{x \in X : x < a\}$ and $(a, \rightarrow) = \{x \in X : a < x\}$; furthermore, $(\leftarrow, a] = \{x \in X : x \leq a\}$ and $[a, \rightarrow) = \{x \in X : a \leq x\}$.

By TFS-305 there exist $a, b \in X$ such that $a = \min F$ and $b = \max F$. Given any point $x \in [a, b] \backslash F$ it is evident that both sets $L_x = (\leftarrow, x) \cap F = (\leftarrow, x] \cap F$ and $R_x = (x, \rightarrow) \cap F = [x, \rightarrow) \cap F$ are compact; so we can take $a_x = \max L_x$ and $b_x = \min R_x$. It is clear that $x \in (a_x, b_x) \subset X \backslash F$ for any $x \in [a, b] \backslash F$. Observe also that $y \in (a_x, b_x)$ implies $a_y = a_x$ and $b_y = b_x$. An immediate consequence is that,

(1) for any $x, y \in [a, b] \backslash F$ either $(a_x, b_x) \cap (a_y, b_y) = \emptyset$ or $(a_x, b_x) = (a_y, b_y)$.

It follows from (1) that we can find a set $A \subset X$ such that $(a_x, b_x) \cap (a_y, b_y) = \emptyset$ for any distinct $x, y \in A$ and, for any $x \in [a, b] \backslash F$ there exists a point $y \in A$ such that $(a_x, b_x) = (a_y, b_y)$.

Given $x \in A$ there exists a clopen set $U \in \tau(x, X)$ such that $U \subset (a_x, b_x)$; let $e_x = \min U$ (apply TFS-305 again to see that e_x is well defined). It is easy to see that the sets (a_x, e_x) and $[e_x, b_x)$ are open in X for any $x \in A$.

We are now ready to define a retraction $r : X \to F$. It is obligatory that $r(x) = x$ for any $x \in F$; we let $r(x) = a$ for all $x < a$ and $r(x) = b$ for every $x > b$. If $x \in [a, b] \backslash F$ then there is a unique $y \in A$ such that $x \in (a_y, b_y)$. If $x < e_y$ then let $r(x) = a_x$; if $x \geq e_x$ then let $r(x) = b_x$. This defines a map $r : X \to F$ such that $r(x) = x$ for any $x \in F$.

To see that r is continuous observe that, for any $x \in X \backslash F$ there is $U \in \tau(x, X)$ such that $r(U) = \{r(x)\}$; an immediate consequence is that r is continuous at all points of $X \backslash F$. Now suppose that $x, y \in F$ and $x \leq y$. If $z \in [x, y] \backslash F$ then there is $p \in A$ for which $z \in (a_p, b_p)$; by the definition of a_p and b_p we have $x \leq a_p \leq b_p \leq y$ and therefore $r(z) \in \{a_p, b_p\} \subset [x, y]$. This proves that

(2) for any $x, y \in F$ with $x \leq y$ we have $r([x, y]) \subset [x, y]$.

Finally, fix a point $x \in F$ and a neighbourhood $U \in \tau(x, X)$; making the set U smaller if necessary, we can assume that U is an interval. Let $U_0 = (\leftarrow, x) \cap U$ and $U_1 = (x, \rightarrow) \cap U$. We have four cases:

Case 1. There are $y, z \in F \cap U$ such that $x \in (y, z)$. Since U is an interval, we have $(y, z) \subset U$ and hence $[y, z] \subset U$. The property (2) implies that $r((y, z)) \subset [y, z] \subset U$; so (y, z) is a neighbourhood of x which witnesses continuity of r at x.

Case 2. There is $y \in U_0 \cap F$ and $U_1 \cap F = \emptyset$. If $U_1 \neq \emptyset$ then there exists a point $q \in A$ such that $(a_q, b_q) \cap U_1 \neq \emptyset$; then $a_q = x$ and hence there exists $d \in (x, e_q]$ such that $(x, d) \subset U_1$. Apply the property (2) again to see that $r((y, x]) \subset [y, x] \subset U$ and $r((x, d)) \subset r((x, e_q)) = \{x\} \subset U$. Therefore (y, d) is a neighbourhood of x which witnesses continuity of r at x. If $U_1 = \emptyset$ then $(y, x]$ is a neighbourhood of x with $r((y, x]) \subset U$; so r is continuous at the point x.

Case 3. There is $z \in U_1 \cap F$ and $U_0 \cap F = \emptyset$. If $U_0 \neq \emptyset$ then there exists a point $q \in A$ such that $(a_q, b_q) \cap U_0 \neq \emptyset$; then $b_q = x$ and hence there exists $d \in [e_q, x)$ such that $(d, x) \subset U_0$. Apply the property (2) again to see that $r([x, z)) \subset [x, z] \subset U$ and $r((d, x)) \subset r((e_q, x)) = \{x\} \subset U$. Therefore (d, z) is a neighbourhood of x which witnesses continuity of r at x. If $U_0 = \emptyset$ then $[x, z)$ is a neighbourhood of x with $r([x, z)) \subset U$; this shows that r is continuous at the point x.

Case 4. $U_0 \cap F = \emptyset$ and $U_1 \cap F = \emptyset$. We actually have four subcases here.

Case 4.1. $U_0 = \emptyset$ and $U_1 = \emptyset$. Then $\{x\} \in \tau(X)$; so r is continuous at x.

Case 4.2. $U_0 \neq \emptyset$ and $U_1 \neq \emptyset$. Then there exist $p, q \in A$ such that $(a_p, b_p) \cap U_0 \neq \emptyset$ and $(a_q, b_q) \cap U_1 \neq \emptyset$. Consequently, $b_p = a_q = x$ and we can choose a point $c \in U_0$ and $d \in U_1$ such that $(c, d) \subset (e_p, e_q)$. Then $x \in (c, d)$ and $r((c, d)) \subset r((e_p, e_q)) = \{x\} \subset U$; so r is continuous at x.

Case 4.3. $U_0 \neq \emptyset$ and $U_1 = \emptyset$. Then there exists $p \in A$ such that $(a_p, b_p) \cap U_0 \neq \emptyset$. Consequently, $b_p = x$ and we can choose $c \in U_0$ such that $(c, x) \subset (e_p, x)$. Then $(c, x] \in \tau(x, X)$ and $r((c, x]) \subset r((e_p, x]) = \{x\} \subset U$; so r is continuous at x.

Case 4.4. $U_0 = \emptyset$ and $U_1 \neq \emptyset$. Then there exists $q \in A$ such that $(a_q, b_q) \cap U_1 \neq \emptyset$. Consequently, $a_q = x$ and we can choose $d \in U_1$ such that $(x, d) \subset (x, e_q)$. Then $[x, d) \in \tau(x, X)$ and $r([x, d)) \subset r([x, e_q)) = \{x\} \subset U$; so r is continuous at x.

Thus $r : X \to F$ is a continuous retraction which, together with Problem 465, implies that F is l-embedded in X. This shows that every closed $F \subset X$ is l-embedded in X, i.e., the space X is extendial and hence our solution is complete.

U.471. *Give an example of a perfectly normal, hereditarily separable extendial compact space which is not metrizable.*

Solution. Consider the set $T = ((0, 1] \times \{0\}) \cup ([0, 1) \times \{1\}) \subset \mathbb{R}^2$.

If $z = (t, 0) \in T$, then $\mathcal{B}_z = \{((a, t] \times \{0\}) \cup ((a, t) \times \{1\}) : 0 < a < t\}$.

Now if $z = (t, 1) \in T$, let $\mathcal{B}_z = \{([t, a) \times \{1\}) \cup ((t, a) \times \{0\}) : t < a < 1\}$. Let τ be the topology generated by the families $\{\mathcal{B}_z : z \in T\}$ as local bases. The space $X = (T, \tau)$ is called *two arrows (or double arrow) space*.

Let $\pi : X \to [0,1]$ be the projection in the plane, i.e., $\pi((r,i)) = r$ for any point $(r,i) \in X$. Given distinct $x, y \in X$ say that $x < y$ if $\pi(x) < \pi(y)$. If $\pi(x) = \pi(y) = r$ then $x < y$ if $x = (r,0)$ and $y = (r,1)$. It is an easy exercise that $<$ is a linear order on X; it is called *the lexicographical order* on X. To distinguish the intervals of the order $<$ from the pairs or reals which represent the elements of X the intervals will have the index lo, i.e., given $x, y \in X$ with $x < y$ we let $(x,y)_{lo} = \{z \in X : x < z < y\}$ and $[x,y)_{lo} = \{z \in X : x \le z < y\}$.

Fact 1. The topology of the double arrow space is zero-dimensional and generated by its lexicographical order. In particular, the double arrow space is a linearly ordered zero-dimensional perfectly normal, hereditarily separable non-metrizable compact space.

Proof. It was proved in TFS-384 that X is a perfectly normal, hereditarily separable non-metrizable compact space. Zero-dimensionality of X follows from the fact that, for every $z \in X$, all elements of $\mathcal{B}_z$ are clopen in X. Let μ be the topology on T generated by the lexicographical order.

If $x = (t,0) \in X$ and $U = ((a,t] \times \{0\}) \cup ((a,t) \times \{1\}) \in \mathcal{B}_x$ then, for the points $p = (a,1)$ and $q = (t,1)$ we have $(p,q)_{lo} = U$ which shows that $U \in \mu$ for any $U \in \mathcal{B}_x$. Analogously, if $x = (t,1)$ and $U = ([t,a) \times \{1\}) \cup ((t,a) \times \{0\}) \in \mathcal{B}_x$ then, for the points $p = (t,0)$ and $q = (a,0)$ we have $(p,q)_{lo} = U$; so $U \in \mu$. This proves that $\mathcal{B} = \bigcup\{\mathcal{B}_x : x \in X\} \subset \mu$ and hence $\tau(X) \subset \mu$.

Now assume that $x = (t,0) \in X$ and $0 < t < 1$. The intervals $(p,q)_{lo}$ such that $p < x < q$ form a local base in (T,μ) at x; fix $p, q \in X$ with $p < x < q$. Let $q' = (t,1)$ and $a = \pi(p)$; if $p' = (a,1)$ then we have the inequalities $p \le p' < x < q' \le q$ and $(p',q')_{lo} = ((a,t] \times \{0\}) \cup ((a,t) \times \{1\}) \in \mathcal{B}_x$.

Thus, for any $p, q \in X$ such that $x \in (p,q)_{lo}$ there is $U \in \mathcal{B}_x$ with $U \subset (p,q)_{lo}$. The same is true if $x = (t,1)$ for some t with $0 < t < 1$: the proof is analogous.

Now, if $x = (0,1) \in X$ then x is the minimal element of the space X and hence the family $\{[x,p)_{lo} : p \in X \text{ and } x < p\}$ is a local base of (T,μ) at x. Take any $p > x$; if $a = \pi(p)$ and $p' = (a,0)$ then $x < p' \le p$; so $[x,p')_{lo} \subset [x,p)_{lo}$ and $[x,p')_{lo} = ([0,a) \times \{1\}) \cup ((0,a) \times \{0\}) \in \mathcal{B}_x$. The proof for the point $x = (1,0) \in X$ being analogous, we have established that

(1) for any $x \in X$ and $V \in \mu$ with $x \in V$ there is $U \in \mathcal{B}_x$ such that $U \subset V$.

The property (1) says that $\mathcal{B}$ is a base for μ and hence $\mu \subset \tau(X)$. We already showed that $\tau(X) \subset \mu$; so $\mu = \tau(X)$, i.e., the topology of X is generated by the lexicographical order of X. Fact 1 is proved.

Returning to our solution observe that the double arrow space X is compact, zero-dimensional and linearly ordered by Fact 1; therefore X is extendial by Problem 470. It also follows from Fact 1 that X is perfectly normal, hereditarily separable and non-metrizable; so X is the required example and hence our solution is complete.

U.472. *Give an example of an extendial compact space X such that $X \times X$ is not t-extendial.*

Solution. Let $T = ((0, 1] \times \{0\}) \cup ([0, 1) \times \{1\}) \subset \mathbb{R}^2$. If $z = (t, 0) \in T$, then

$$\mathcal{B}_z = \{((a, t] \times \{0\}) \cup ((a, t) \times \{1\}) : 0 < a < t\}.$$

Now if $z = (t, 1) \in T$, let $\mathcal{B}_z = \{([t, a) \times \{1\}) \cup ((t, a) \times \{0\}) : t < a < 1\}$. Let τ be the topology generated by the families $\{\mathcal{B}_z : z \in T\}$ as local bases. The space $X = (T, \tau)$ is called *two arrows (or double arrow) space*. By Fact 1 of U.471 the space X is compact, linearly ordered and zero-dimensional. Therefore X is extendial by Problem 470.

Let $T_1 = [0, 1) \times \{1\}$; given any $x = (t, 1) \in T_1$ the family $\mathcal{C}_x = \{U \cap T_1 : U \in \mathcal{B}_x\}$ is a local base at x in the space T_1 with the topology induced from X. It is easy to see that $\mathcal{C}_x = \{[t, a) \times \{1\} : t < a < 1\}$. Let $\pi : T_1 \to [0, 1)$ be the projection in the plane, i.e., $\pi((r, 1)) = r$ for any $(r, 1) \in T_1$. The map π is a bijection and it is easy to see that it is a homeomorphism if we consider $[0, 1)$ with the topology μ induced from the Sorgenfrey line (see TFS-165). Denote by S the space $([0, 1), \mu)$; then the map π^{-1} embeds S in X.

Consider the set $D = \{(t, 1-t) : t \in (0, 1)\} \subset S \times S$; for any $d = (t, 1 - t) \in D$ take numbers a, b with $t < a < 1$ and $1 - t < b < 1$; then the set $U = [t, a) \times [1-t, b)$ is open in $S \times S$ and $U \cap D = \{d\}$ which shows that the set D is discrete. Therefore $s(S \times S) \geq |D| > \omega$; we already saw that S embeds in X; so $S \times S$ embeds in $X \times X$ and therefore $s(X \times X) > \omega$. It was proved in TFS-384 that X is hereditarily separable and hence separable. This implies that $X \times X$ is also separable; so $p(X \times X) = \omega$. If $X \times X$ is t-extendial then we can apply Problem 455 to see that $s(X \times X) = \omega$ which is a contradiction. Thus X is an example of an extendial compact space such that $X \times X$ is not t-extendial.

U.473. *Show that there exist extendial compact spaces of uncountable tightness.*

Solution. The space $X = \omega_1 + 1$ is compact, linearly ordered and zero-dimensional; so it is extendial by Problem 470. The point $\omega_1 \in X$ is in the closure of the set $A = X \setminus \{\omega_1\}$ but no countable subset of A contains ω_1 in its closure. Thus X is a compact extendial space of uncountable tightness.

U.474. *Give an example of a non-linearly orderable extendial compact space.*

Solution. Let $X = A(\omega_1)$; then $X = \omega_1 \cup \{a\}$ is a compact space with the unique non-isolated point a. To see that X is extendial suppose that F is a closed subset of X. If $a \notin F$ then F is finite and open in X. It is an easy exercise that any clopen subset of X is a retract of X; so F is a retract of X.

Now, if $a \in F$ then let $f(x) = x$ for $x \in F$ and $f(x) = a$ for all $x \in X \setminus F$. Then $f : X \to F$ and $f(x) = x$ for any $x \in F$. To see that f is continuous we must only establish continuity at the point a, so take any $U \in \tau(a, X)$; there is a finite $K \subset \omega_1$ such that $U = X \setminus K$. If $x \in F \cap U$ then $f(x) = x \in U$; if $x \in U \setminus F$ then $f(x) = a \in U$ which shows that $f(U) \subset U$. Since $f(a) = a$, the set U witnesses continuity of f at the point a. Thus f is a retraction and hence we proved that every closed $F \subset X$ is a retract of X. This shows that X is extendial (see Problem 465).

If the space X is linearly orderable then it follows from $t(X) = \omega$ (see TFS-129) that $\chi(X) \leq \omega$ (see TFS-303) which is false because $\psi(a, X) = \omega_1$. This contradiction shows that X is an extendial non-linearly orderable compact space.

U.475. *Prove that every t-extral (and hence every extral) space X is compact and every uncountable regular cardinal is a caliber of X.*

Solution. Let $\kappa = w(X)$; there is no loss of generality to consider that $\kappa \geq \omega$. Choose disjoint sets $P, Q \subset \kappa^+$ such that $|P| = |Q| = \kappa^+$ and $P \cup Q = \kappa^+$. In the Tychonoff cube $\mathbb{I}^{\kappa^+}$ consider the subspace $E = \{x \in \mathbb{I}^{\kappa^+} : x(\alpha) = 1$ for any $\alpha \in P\}$. It is evident that $E \simeq \mathbb{I}^Q \simeq \mathbb{I}^{\kappa^+}$; so the space X can be embedded in E and hence we can assume, without loss of generality, that $X \subset E$.

For any $A \subset \kappa^+$ let $u_A \in \mathbb{I}^A$ be the "zero point" of $\mathbb{I}^A$ defined by $u_A(\alpha) = 0$ for all $\alpha \in A$. The subspace $S = \{x \in \mathbb{I}^{\kappa^+} : |x^{-1}(\mathbb{I} \setminus \{0\})| \leq \kappa\} \subset \mathbb{I}^{\kappa^+}$ is dense in $\mathbb{I}^{\kappa^+}$ and has the following property:

(1) if $Y \subset S$ and $|Y| \leq \kappa$ then $\text{cl}_S(Y)$ is compact.

Indeed, the set $C = \bigcup\{y^{-1}(\mathbb{I} \setminus \{0\}) : y \in Y\}$ has cardinality at most κ; so the set $F = \mathbb{I}^C \times \{u_{\kappa^+ \setminus C}\}$ is contained in S. It is straightforward that $Y \subset F$ and hence $\text{cl}_S(Y) = \text{cl}_F(Y)$ is a compact space because so is F; this settles (1).

For any $x \in X$ let $e(x)(f) = f(x)$ for any $f \in C_p(X)$; this defines an embedding $e : X \to C_p(C_p(X))$ (see TFS-167). Let $T = S \cup X$; then X is closed in T because E is a closed subset of $\mathbb{I}^{\kappa^+}$ such that $E \cap S = \emptyset$ and therefore $X = E \cap T$. The space X being t-extral there exists a continuous map $\varphi : T \to C_p(C_p(X))$ such that $\varphi|X = e$ (see Problem 449); let $T' = \varphi(T)$. Since S is dense in T, the set $S' = \varphi(S)$ is dense in T'; it follows from $nw(S') \leq nw(C_p(C_p(X))) = nw(X) \leq \kappa$ that we can find a dense $Y' \subset S'$ with $|Y'| \leq \kappa$.

Choose $Y \subset S$ such that $|Y| \leq \kappa$ and $\varphi(Y) = Y'$; then $K = \text{cl}_S(Y)$ is compact by (1) and the compact set $K' = \varphi(K)$ is dense in T' because $K' \supset Y'$ and Y' is dense in T'. An immediate consequence is that $K' = T'$, i.e., the space T' is compact. The set $e(X) \subset T'$ is closed in $C_p(C_p(X))$ (this was also proved in TFS-167), so it has to be closed in T'; this implies that $e(X)$ is compact and hence X is compact as well being homeomorphic to $e(X)$.

Finally take an uncountable regular cardinal λ. The space X can be embedded in $\mathbb{I}^\kappa$; so we can assume that $X \subset \mathbb{I}^\kappa$. Since X compact, it is closed and hence t-embedded in $\mathbb{I}^\kappa$. The cardinal λ being a caliber of $\mathbb{I}^\kappa$ (see SFFS-282) it has to be a caliber of X by Problem 453. Thus every uncountable regular cardinal is a caliber of X.

U.476. *Prove that X is t-extral if and only if it can be t-embedded in $\mathbb{I}^\kappa$ for some cardinal κ.*

Solution. If X is t-extral then it is compact by Problem 475 so, for $\kappa = w(X)$, it can be embedded in $\mathbb{I}^\kappa$ as a closed subspace. It follows from t-extrality of X that this embedding is automatically a t-embedding in $\mathbb{I}^\kappa$. This proves necessity.

Now assume that X is t-embedded in $\mathbb{I}^\lambda$ for some cardinal λ. Then X is compact because it has to be closed in $\mathbb{I}^\lambda$ (see Problem 450). Suppose that X is embedded in some space Y; denote by X' the respective copy of X in Y and fix a homeomorphism $\xi : X' \to X$. Then $\xi^{-1} : X \to X'$ is a homeomorphism as well an hence its dual map $\mu = (\xi^{-1})^* : C_p(X') \to C_p(X)$ defined by $\mu(f) = f \circ \xi^{-1}$ for any $f \in C_p(X')$ is also a homeomorphism (see TFS-163).

For any $\alpha \in \lambda$ let $\pi_\alpha : \mathbb{I}^\lambda \to \mathbb{I}$ be the natural projection of $\mathbb{I}^\lambda$ onto its factor determined by α. Any compact subspace of Y is C-embedded in Y by Fact 1 of T.218; so X' is C-embedded in Y. Therefore there exists a continuous function $\Phi_\alpha : Y \to \mathbb{I}$ such that $\pi_\alpha \circ \xi = \Phi_\alpha$ for all $\alpha < \lambda$. Then $\Phi = \Delta_{\alpha<\lambda}\Phi_\alpha : Y \to \mathbb{I}^\lambda$ and $\Phi|X' = \xi$; besides, the set X is contained in $Y' = \Phi(Y)$.

Since the space X is t-embedded in $\mathbb{I}^\lambda$, there exists a continuous extender $\varphi : C_p(X) \to C_p(\mathbb{I}^\lambda)$. The restriction map $\pi : C_p(\mathbb{I}^\lambda) \to C_p(Y')$ being continuous, the map $\varphi' = \pi \circ \varphi : C_p(X) \to C_p(Y')$ is continuous as well and it is straightforward that φ' is an extender.

Finally let $\Phi^* : C_p(Y') \to C_p(Y)$ be the dual map of Φ defined by the equality $\Phi^*(f) = f \circ \Phi$ for any $f \in C_p(Y')$. Since Φ^* is continuous by TFS-163, the map $\eta = \Phi^* \circ \varphi' \circ \mu : C_p(X') \to C_p(Y)$ is also continuous; let us check that η is an extender.

Fix $f' \in C_p(X')$, $x' \in X'$ and let $f = \mu(f')$, $x = \xi(x')$. Since φ' is an extender, we have $\varphi'(f)(x) = f(x)$. Now, $\Phi^*(\varphi'(f))(x') = (\varphi'(f))(\Phi(x')) = f(x)$ because $\Phi(x') = \xi(x') = x$. Thus $\eta(f')(x') = f(x) = f'(\xi^{-1}(x)) = f'(x')$; since $x' \in X'$ was chosen arbitrarily, we proved that $\eta(f')|X' = f'$, i.e., η is a continuous extender and hence X' is t-embedded in Y. This shows that X is t-extral and settles sufficiency.

U.477. *Prove that any retract of a t-extral space is a t-extral space.*

Solution. Suppose that X is a t-extral space and $r : X \to Y$ is a retraction; we can consider that $X \subset \mathbb{I}^\kappa$ for some cardinal κ. It follows from t-extrality of X that there exists a continuous extender $\varphi : C_p(X) \to C_p(\mathbb{I}^\kappa)$.

The dual map $r^* : C_p(Y) \to C_p(X)$ of the retraction r, is continuous (see TFS-163); so the map $\eta = \varphi \circ r^* : C_p(Y) \to C_p(\mathbb{I}^\kappa)$ is continuous as well.

Given any function $f \in C_p(Y)$ and $y \in Y$ the equality $r(y) = y$ implies that $\eta(f)(y) = \varphi(f \circ r)(y) = (f \circ r)(y) = f(r(y)) = f(y)$ (we also applied the fact that φ is an extender and hence $\varphi(f \circ r)|X = f \circ r$). Therefore $\eta(f)(y) = f(y)$ for any $y \in Y$, i.e., $\eta(f)|Y = f$ which shows that η is a continuous extender. As a result, the space Y is t-embedded in $\mathbb{I}^\kappa$; so it is t-extral by Problem 476.

U.478. *Let X be a t-extral space such that $w(X) \leq \mathfrak{c}$. Prove that X is separable.*

Solution. It follows from $w(X) \leq \mathfrak{c}$ that the space X is embeddable in $\mathbb{I}^\mathfrak{c}$; so we can consider that $X \subset \mathbb{I}^\mathfrak{c}$. Since X is t-extral, it is t-embedded in $\mathbb{I}^\mathfrak{c}$ and hence $C_p(X)$ embeds in $C_p(\mathbb{I}^\mathfrak{c})$ (see Problem 448). The space $\mathbb{I}^\mathfrak{c}$ being separable (see TFS-108), we have $iw(C_p(X)) \leq iw(C_p(\mathbb{I}^\mathfrak{c})) = d(\mathbb{I}^\mathfrak{c}) = \omega$ which, together with TFS-173, shows that X is separable.

U.479. *Suppose that an ω-monolithic space X is t-extral and has countable tightness. Prove that X is metrizable.*

Solution. The space X being compact (see Problem 475), it follows from $t(X) = \omega$ that there exists a point-countable π-base $\mathcal{B}$ in X. Since ω_1 is a caliber of X (this was also proved in Problem 475), the family $\mathcal{B}$ must be countable which shows that $\pi w(X) \leq \omega$. Thus X is separable; so it follows from ω-monolithity of X that $w(X) = nw(X) = \omega$ (see Fact 4 of S.307) and hence X is metrizable.

U.480. *Prove that a t-extral space X is metrizable whenever $C_p(X)$ is Lindelöf.*

Solution. We can consider that $X \subset \mathbb{I}^\kappa$ for some infinite cardinal κ. It follows from t-extrality of X that X is t-embedded in $\mathbb{I}^\kappa$ and therefore $C_p(X)$ embeds in $C_p(\mathbb{I}^\kappa)$ (see Problem 448). The compact space $\mathbb{I}^\kappa$ being dyadic (see Fact 2 of T.298), we can apply Problem 086 to conclude that $nw(C_p(X)) = \omega$ and hence $w(X) = nw(X) = \omega$ (see Fact 4 of S.307); so X is metrizable.

U.481. *Give an example of a t-extral space which is not extral.*

Solution. Given a space Z and $f \in C^*(Z)$ let $\|f\|_Z = \sup\{|f(x)| : x \in Z\}$; if the space Z is clear we will write $\|f\|$ instead of $\|f\|_Z$. For a subspace $Y \subset Z$ a map $e : \tau(Y) \to \tau(Z)$ is *an extender of open sets* if $e(U) \cap Y = U$ for any $U \in \tau(Y)$. Given a natural $n \geq 2$ a subspace $Y \subset Z$ is K_n-*embedded* in Z if there exists an extender of open sets $e : \tau(Y) \to \tau(Z)$ such that, for any disjoint family $\{U_0, \ldots, U_{n-1}\} \subset \tau(Y)$ we have $e(U_0) \cap \ldots \cap e(U_{n-1}) = \emptyset$; in this case the map e is called *a K_n-extender*.

If X is a space and $f_0, \ldots, f_{n-1}$ are functions on X then, as usual, the functions $m = \min(f_0, \ldots, f_{n-1})$ and $M = \max(f_0, \ldots, f_{n-1})$ are defined by the equalities $m(x) = \min\{f_i(x) : i < n\}$ and $M(x) = \max\{f_i(x) : i < n\}$ for any $x \in X$.

Fact 1. Given compact spaces X and Y suppose that $\varphi : C_p(X) \to C_p(Y)$ is a continuous linear map. Then the map $\varphi : C_u(X) \to C_u(Y)$ is continuous as well.

Proof. Recall that the topology of $C_u(X)$ is generated by the metric d defined by the equality $d(f, g) = \|f - g\|_X$ for any $f, g \in C_p(X)$ (see Problem 344); this metric is complete; so $C_u(X)$ is a Čech-complete space by TFS-269.

It follows from $\tau(C_p(X)) \subset \tau(C_u(X))$ that the map $\varphi : C_u(X) \to C_p(Y)$ is continuous; so we can apply Problem 347 to see that $\varphi : C_u(X) \to C_u(Y)$ is continuous at some point $f \in C_u(X)$. For any function $g \in C_u(X)$ let $L(g) = g - f$; then $L : C_u(X) \to C_u(X)$ is a homeomorphism (see Fact 2 of T.241). If u is the function which is identically zero on X then $L(f) = u$. By linearity of φ, the function $v = \varphi(u)$ is identically zero on Y. Take an arbitrary set $V \in \tau(v, C_u(Y))$. By Fact 2 of T.104, there is $W \in \tau(f, C_u(X))$ such that $\varphi(g) - \varphi(f) \in V$ for any $g \in W$. Since L is a homeomorphism and $U = L(W) \ni u$, the set U is an open neighbourhood of u in $C_u(X)$.

For any $h \in U$, there is $g \in W$ with $L(g) = g - f = h$; recall that v is identically zero on Y; so $\varphi(h) - \varphi(u) = \varphi(g - f) - v = \varphi(g) - \varphi(f) \in V$. Therefore we can apply Fact 2 of T.104 again to see that φ is continuous at u. Finally apply Fact 1 of T.105 and Fact 2 of S.496 to conclude that $\varphi : C_u(X) \to C_u(Y)$ is continuous and hence Fact 1 is proved.

Fact 2. Given a compact space X if a set $Y \subset X$ is l-embedded in X then there exists a natural $n \geq 2$ such that Y is K_n-embedded in X.

Proof. The set Y is compact being closed in X (see Problem 450); fix a linear continuous extender $\varphi : C_p(Y) \to C_p(X)$. For any $r > 0$ let $B_r = \{f \in C_p(Y) : \|f\|_Y < r\}$ be the ball of radius r centered at zero; the map $\varphi : C_u(Y) \to C_u(X)$ is continuous by Fact 1; so there is $r > 0$ such that $\|\varphi(f)\|_X < 1$ for any $f \in B_r$.

Let $S = \{f \in C_p(Y) : \|f\|_Y \leq 1\}$; for any $f \in S$, the point $g = \frac{r}{2}f$ belongs to B_r and therefore $\frac{r}{2}\|\varphi(f)\|_X = \|\varphi(g)\|_X < 1$; this proves that $\|\varphi(f)\|_X \leq \frac{2}{r}$ for any $f \in S$. Choose a natural $n \geq \max\{2, \frac{4}{r}\}$; then

(1) $\|\varphi(f)\|_X \leq \frac{n}{2}$ for any $f \in S$.

Given a set $U \in \tau(Y)$ let $S_U = \{f \in S : f(Y\backslash U) \subset \{0\}\}$; then the set $e(U) = \bigcup\{\varphi(f)^{-1}((1 - \frac{1}{n}, +\infty)) : f \in S_U\}$ is open in X and $e(U) \cap Y \subset U$. If $U = \emptyset$ then the unique element of S_U is the function which is identically zero on Y. The map φ being linear, the function $\varphi(f)$ is identically zero on X; so $e(U) = \emptyset$ and hence $e(U) \cap Y = U$. If $U \neq \emptyset$ then, for any $x \in U$ there is $f \in S_U$ such that $f(x) = 1$. The function $\varphi(f)$ extends f; so $\varphi(f)(x) = 1$ and $\varphi(f)(Y\backslash U) \subset \{0\}$ which shows that $x \in \varphi(f)^{-1}((1 - \frac{1}{n}, +\infty)) \subset e(U)$; this proves that $e(U) \cap Y = U$; so e is an extender of open sets.

Suppose that $\mathcal{U} = \{U_0, \ldots, U_{n-1}\}$ is a disjoint family of open subsets of Y such that $E = e(U_0) \cap \ldots \cap e(U_{n-1}) \neq \emptyset$. Fix a point $y \in E$; there are functions $f_0, \ldots, f_{n-1} \in S$ such that $f_i \in S_{U_i}$ and $\varphi(f_i)(y) > 1 - \frac{1}{n}$ for any $i < n$. It follows from disjointness of $\mathcal{U}$ that $f = f_0 + \ldots + f_{n-1} \in S$ which, together with (1), implies that $\|\varphi(f)\|_X \leq \frac{n}{2}$ and therefore $|\varphi(f)(y)| \leq \frac{n}{2} \leq n - 1$. The map φ being linear, we have $|\varphi(f)(y)| = |\sum_{i<n} \varphi(f_i)(y)| > n(1 - \frac{1}{n}) = n - 1$. This contradiction shows that e is, indeed, a K_n-extender; so Fact 2 is proved.

Recall that, for any infinite cardinal κ, the space $A(\kappa) = \kappa \cup \{a\}$ is the one-point compactification of the discrete space $D(\kappa)$ with the underlying set κ and $a \notin \kappa$ is the unique non-isolated point of $A(\kappa)$. We will call the space $A(\kappa)$ *supersequence with the limit a*.

Fact 3. Suppose that A is a set and $p \in \mathbb{I}^A$. Assume that, for some $n \in \mathbb{N}$, we have subspaces $K_0, \ldots, K_{n-1} \subset \mathbb{I}^A$ such that every K_i is homeomorphic to $A(\omega_1)$ and p is the unique non-isolated point of K_i; let $D_i = F_i\backslash\{p\}$. Then, for any family $\{U_0, \ldots, U_{n-1}\} \subset \tau(\mathbb{I}^A)$ such that $D_i \subset U_i$ for all $i < n$, we have $U_0 \cap \ldots \cap U_{n-1} \neq \emptyset$.

Proof. Given a set $B \subset A$ the map $\pi_B : \mathbb{I}^A \to \mathbb{I}^B$ is the projection of $\mathbb{I}^A$ onto its face $\mathbb{I}^B$. It suffices to prove by induction on n, that, in our situation, the point p belongs to the closure of the set $U_0 \cap \ldots \cap U_{n-1}$. The case of $n = 1$ being evident, assume that $n \in \mathbb{N}$ and the respective statement S_k was proved for any $k \leq n$.

To see that the statement $\mathcal{S}_{n+1}$ is true take any supersequences $K_0, \ldots, K_n$ with the common limit p and suppose that we are given sets $U_0, \ldots, U_n \in \tau(\mathbb{I}^A)$ such that $U_i \cap K_i = D_i = K_i \backslash \{p\}$ for any $i \leq n$. For the set $V = U_0 \cap \ldots \cap U_{n-1}$ the induction hypothesis shows that $p \in \overline{V}$. There exists a countable $B \subset A$ such that $\pi_B^{-1}(\pi_B(\overline{V})) = \overline{V}$ (see Fact 6 of T.298).

The space $\mathbb{I}^B$ being second countable, $\pi_B^{-1}\pi_B(p)$ is a G_δ-subset of K_n which contains the point p. An immediate consequence is that $|K_n \backslash \pi_B^{-1}\pi_B(p)| = \omega$ and hence p belongs to the closure of the set $E = (K_n \cap \pi_B^{-1}\pi_B(p)) \backslash \{p\}$. It follows from $\pi_B(p) \in \pi_B(\overline{V})$, that $E \subset \overline{V}$ and therefore $E \subset \overline{U_n \cap V}$. Recalling that we have $U_n \cap V = U = U_0 \cap \ldots \cap U_n$ we conclude that $p \in \overline{U}$; this accomplishes the induction step and shows that Fact 3 is proved.

Fact 4. Given a space X suppose that $F \subset X$ is C^*-embedded in X and $f : F \to \mathbb{I}^A$ is a continuous map for some A. Then there exists a continuous map $g : X \to \mathbb{I}^A$ such that $g|F = f$. In particular, if the space X is normal and F is closed in X then any continuous map from F to a Tychonoff cube can be continuously extended to the whole space X.

Proof. For any $a \in A$ let $\pi_a : \mathbb{I}^A \to \mathbb{I}$ be the natural projection of $\mathbb{I}^A$ onto its factor determined by a; then $g_a = \pi_a \circ f : F \to \mathbb{I}$ is continuous. Since F is C^*-embedded in X, there is a continuous function $g_a : X \to \mathbb{I}$ such that $g_a|F = f_a$ for any $a \in A$. Let $g = \Delta\{g_a : a \in A\} : X \to \mathbb{I}^A$ be the diagonal product of the family $\{g_a : a \in A\}$ (recall that g is defined by $g(x)(a) = g_a(x)$ for any $a \in A$). The map g is continuous; given $x \in F$ we have $g(x)(a) = g_a(x) = f_a(x) = f(x)(a)$ for any $a \in A$ and hence $g(x) = f(x)$. Thus $g|F = f$ and Fact 4 is proved.

Fact 5. Suppose that Y is a compact space and $K \subset Y$ is a closed subspace of Y such that $K = K_0 \cup \ldots \cup K_{n-1}$, every K_i is homeomorphic to some Tychonoff cube $\mathbb{I}^{\kappa_i}$ and there is a point $a \in Y$ such that $K_i \cap K_j = \{a\}$ for any distinct $i, j < n$. Then there exists a continuous extender $\varphi : C_p(K) \to C_p(Y)$ such that $\|\varphi(f)\|_Y \leq 5\|f\|_K$ for any $f \in C_p(K)$.

Proof. Given a function $f \in C_p(Y)$ let $f^+ = \frac{f+|f|}{2}$ and $f^- = \frac{f-|f|}{2}$; define maps $\xi_+, \xi_- : C_p(Y) \to C_p(Y)$ by $\xi_+(f) = f^+$ and $\xi_-(f) = f^-$ for any $f \in C_p(Y)$. It is an easy exercise that both maps ξ_+ and ξ_- are continuous. It is straightforward that $f^+(x) = f(x)$ if $f(x) \geq 0$ and $f^+(x) = 0$ if $f(x) < 0$. Analogously, $f_-(x) = f(x)$ if $f(x) \leq 0$ while $f_-(x) = 0$ if $f(x) > 0$. We also have the inequalities $\|f^+\|_Y \leq \|f\|_Y$ and $\|f^-\|_Y \leq \|f\|_Y$. Let $r_i : K \to K_i$ be the map defined by $r_i(x) = a$ for all $x \in K \backslash K_i$ and $r(x) = x$ whenever $x \in K_i$; it is clear that r_i is a retraction for any $i < n$.

Apply Fact 4 to find a continuous map $s_i : Y \to K_i$ such that $s_i|K = r_i$ for any $i < n$. Let $\mu_i(f) = (f|K_i) \circ s_i$ for any $i < n$ and $f \in C_p(K)$. It is straightforward that every $\mu_i : C_p(K) \to C_p(Y)$ is a continuous map such that $\|\mu_i(f)\|_Y \leq \|f\|_K$ and $\mu_i(f)|K_i = f|K_i$ for any $f \in C_p(K)$. Let $v_0(f) = \max(\mu_0(f)^+, \ldots, \mu_{n-1}(f)^+)$ and $v_1(f) = \min(\mu_0(f)^-, \ldots, \mu_{n-1}(f)^-)$ for any $f \in C_p(K)$.

We leave it to the reader to verify that then map $\nu_i : C_p(K) \to C_p(Y)$ is continuous and $||\nu_i(f)||_Y \leq ||f||_K$ for any $f \in C_p(K)$ and $i \in \mathbb{D}$. Therefore the map $\nu = \nu_0 + \nu_1 : C_p(K) \to C_p(Y)$ is continuous and $||\nu(f)||_Y \leq 2||f||_K$ for any $f \in C_p(K)$.

Let $\varphi(f) = \nu(f - f(a)) + f(a)$ for any $f \in C_p(K)$. It follows easily from continuity of ν that the map $\varphi : C_p(K) \to C_p(Y)$ is continuous. For any $f \in C_p(K)$ we have $||f - f(a)||_K \leq 2||f||_K$; so $||\nu(f - f(a))||_Y \leq 2||f - f(a)||_K \leq 4||f||_K$. Therefore $||\varphi(f)||_Y \leq ||\nu(f - f(a))||_Y + |f(a)| \leq 5||f||_K$ for each $f \in C_p(K)$.

To finally see that the map φ is an extender take any $f \in C_p(K)$ and $x \in K$; let $g = f - f(a)$ and fix a number $i < n$ with $x \in K_i$. Then $\mu_j(g)(x) = 0$ and hence $\mu_j(g)^+(x) = \mu_j(g)^-(x) = 0$ for all $j \neq i$. Suppose first that $g(x) > 0$; then $\mu_i(g)^+(x) = g(x)$ which implies $\nu_0(g)(x) = g(x)$ and $\nu_1(g)(x) = 0$. As a consequence, $\nu(g)(x) = g(x)$ and therefore $\varphi(f)(x) = g(x) + f(a) = f(x)$. Analogously, if $g(x) \leq 0$ then $\mu_i(g)^-(x) = g(x)$ which implies $\nu_1(g)(x) = g(x)$ and $\nu_0(g)(x) = 0$. Consequently, $\nu(g)(x) = g(x)$ and therefore $\varphi(f)(x) = g(x) + f(a) = f(x)$. We conclude that the extender φ has all promised properties; so Fact 5 is proved.

Fact 6. Let $\mathcal{U} = \{U_n : n \in \omega\}$ be disjoint family of non-empty open subsets of a space Z. Assume that $f_n \in C^*(Z)$ and $f_n(Z \backslash U_n) \subset \{0\}$ for all $n \in \omega$. If, additionally, the sequence $\{||f_n|| : n \in \omega\}$ converges to zero then the function $f = \sum_{n \in \omega} f_n$ is continuous on Z.

Proof. Let $r_n = ||f_n||$ and $g_n = f_0 + \ldots + f_n$ for any $n \in \omega$. Given $\varepsilon > 0$ there is $m \in \omega$ such that $r_n < \frac{\varepsilon}{2}$ for all $n \geq m$. The family $\mathcal{U}$ being disjoint, for any set $A \subset \omega$, we have the inequality $|| \sum\{f_i : i \in A\}|| \leq \sup\{r_i : i \in A\}$. In particular, $||f - g_n|| \leq \sup\{r_n : n \geq m\} \leq \frac{\varepsilon}{2} < \varepsilon$ which shows that the sequence $\{g_n : n \in \omega\}$ of continuous functions on Z converges uniformly to f. Therefore f is continuous (see TFS-029); so Fact 6 is proved.

Returning to our solution, fix a number $n \in \mathbb{N}$ and let $I_n = [0, \frac{1}{n+1}] \subset \mathbb{I}$. It is easy to find a disjoint family $\mathcal{A}_n = \{A_0^n, \ldots, A_{n-1}^n\}$ of subsets of ω_1 such that $|A_i^n| = \omega_1$ for any $i < n$ and $\bigcup \mathcal{A}_n = \omega_1$. The set $K_i^n = \{x \in (I_n)^{\omega_1} : x(\alpha) = 0$ for any $\alpha \in \omega_1 \backslash A_i^n\}$ is homeomorphic to $\mathbb{I}^{\omega_1}$ for any $i < n$. Let $u \in \mathbb{I}^{\omega_1}$ be the zero point of $\mathbb{I}^{\omega_1}$, i.e., $u(\alpha) = 0$ for all $\alpha \in \omega_1$. It is evident that $K_i^n \cap K_j^n = \{u\}$ for any $i \neq j$ (observe that we consider that $I_n \subset \mathbb{I}$ and hence $(I_n)^{\omega_1} \subset \mathbb{I}^{\omega_1}$); let $K_n = K_0^n \cup \ldots \cup K_{n-1}^n$. It is easy to find a discrete subspace $D_i \subset K_i^n \backslash \{u\}$ such that $E_i = D_i \cup \{u\}$ is homeomorphic to $A(\omega_1)$ for any $i < n$.

In the space $\mathbb{I} \times \mathbb{I}^{\omega_1}$ consider the set $X = \{(0, u)\} \cup (\bigcup\{\{\frac{1}{n}\} \times K_n : n \in \mathbb{N}\})$; it is straightforward that any $U \in \tau((0, u), \mathbb{I} \times \mathbb{I}^{\omega_1})$ contains the set $L_n = \{\frac{1}{n}\} \times K_n$ for all but finitely many $n \in \mathbb{N}$ so the space X is compact. Let $v_0 = (0, u)$ and $v_n = (\frac{1}{n}, u)$ for any $n \in \mathbb{N}$.

If X is l-embedded in $\mathbb{I} \times \mathbb{I}^{\omega_1}$ then there is a natural $n \geq 2$ such that X is K_n-embedded in $\mathbb{I} \times \mathbb{I}^{\omega_1}$ (see Fact 2); let $e : \tau(X) \to \tau(\mathbb{I} \times \mathbb{I}^{\omega_1})$ be the respective

K_n-extender. The set $U_i = \{\frac{1}{n}\} \times (K_i^n \backslash \{u\})$ is open in X for any $i < n$ and the family $\{U_i : i < n\}$ is disjoint. Consequently, $\bigcap_{i<n} e(U_i) = \emptyset$.

On the other hand, $E_0, \ldots, E_{n-1}$ are supersequences with the common limit v_n, such that $e(U_i) \supset U_i \supset D_i$ for every $i < n$; so we can apply Fact 3 to see that $\bigcap_{i<n} e(U_i) \neq \emptyset$ which is a contradiction. Thus X is not l-embedded in $\mathbb{I} \times \mathbb{I}^{\omega_1}$ and hence the space X is not extral.

Now take a disjoint family $\mathcal{O} = \{O_n : n \in \mathbb{N}\}$ of open subsets of the space $Q = \mathbb{I} \times \mathbb{I}^{\omega_1}$ such that $v_0 \notin \bigcup \mathcal{O}$ and $L_n \subset O_n$ for each $n \in \mathbb{N}$; choose a continuous function $h_n : Q \to [0,1]$ such that $h_n(L_n) = 1$ and $h_n(x) = 0$ for every $x \in Q \backslash O_n$. Every space L_n satisfies the assumptions of Fact 5; so we can take a continuous extender $\xi_n : C_p(L_n) \to C_p(Q)$ such that $\|\xi_n(f)\|_Q \le 5\|f\|_{L_n}$ for any $f \in C_p(L_n)$. Given $n \in \mathbb{N}$, let $\varphi_n(f) = \xi_n(f) \cdot h_n$ for any $f \in C_p(L_n)$. It is easy to see that

(2) every $\varphi_n : C_p(L_n) \to C_p(Q)$ is a continuous extender with $\varphi_n(f)(Q \backslash O_n) = \{0\}$ and $\|\varphi_n(f)\|_Q \le 5\|f\|_{L_n}$ for any $f \in C_p(L_n)$.

Fix a $f \in C_p(X)$ and let $\varphi(f) = \sum \{\varphi_n((f - f(v_0))|L_n) : n \in \mathbb{N}\} + f(v_0)$. Then $\varphi(f)$ is a real-valued function on Q. If $g_n = (f - f(v_0))|L_n$ for any $n \in \mathbb{N}$ then the sequence $\{\|g_n\|_{L_n} : n \in \mathbb{N}\}$ converges to zero and hence so does the sequence $\{\|\varphi_n(g_n)\|_Q : n \in \mathbb{N}\}$ by the property (2). This makes it possible to apply Fact 6 to see that $\varphi(f)$ is a continuous function on Q and hence $\varphi : C_p(X) \to C_p(Q)$.

It follows from the property (2) that $\varphi_n(g_n)(v_0) = 0$ for all $n \in \omega$ and therefore $\varphi(f)(v_0) = f(v_0)$. If $x \in X \backslash \{v_0\}$ then there is a unique $n \in \mathbb{N}$ such that $x \in L_n$; the map φ_n being an extender, we have $\varphi_n((f - f(v_0))|L_n)(x) = f(x) - f(v_0)$. Besides, $\varphi_m(g_m)(x) = 0$ for any $m \neq n$; so $\varphi(f)(x) = f(x) - f(v_0) + f(v_0) = f(x)$. Thus $\varphi(f)(x) = f(x)$ for any $x \in X$; so φ is an extender.

To see that φ is continuous take any $x \in Q$ and let $\pi_x : C_p(Q) \to \mathbb{R}$ be the projection of $C_p(Q)$ onto the factor of $\mathbb{R}^Q$ determined by x. Recall that $\pi_x(f) = f(x)$ for any $f \in C_p(Q)$. If $x \in X$ then $(\pi_x \circ \varphi)(f) = f(x) = \pi_x(f)$ because φ is an extender so $\pi_x \circ \varphi$ coincides with the projection of $C_p(X)$ onto the factor of $\mathbb{R}^X$ determined by the point x and hence $\pi_x \circ \varphi$ is continuous.

If $x \in Q \backslash (\bigcup \mathcal{O})$ then $\pi_x \circ \varphi$ is continuous being identically zero on the space $C_p(X)$. Now, if $x \in \bigcup \mathcal{O}$ then there is a unique $n \in \mathbb{N}$ such that $x \in O_n$ and hence $\varphi_m(f - f(v_0))(x) = 0$ for any $m \neq n$. Thus $\varphi(f)(x) = \varphi_n(f - f(v_0))(x) + f(v_0)$ for any $f \in C_p(X)$; so the map $\pi_x \circ \varphi$ is continuous because so is φ_n.

This shows that $\pi_x \circ \varphi$ is a continuous map for any $x \in Q$ and hence φ is a continuous extender (see TFS-102) and, in particular, X is t-embedded in Q. The space Q being homeomorphic to $\mathbb{I}^{\omega_1}$ we can apply Problem 476 to conclude that X is t-extral. Therefore X is an example of a t-extral space which is not extral and hence our solution is complete.

U.482. *Prove that every metrizable compact space is extral.*

Solution. If X is a metrizable compact space and $X \subset Y$ then X is l-embedded in Y: this was proved in Fact 1 of U.216. Therefore X is extral.

U.483. *Prove that X is extral if and only if it can be l-embedded in $\mathbb{I}^\kappa$ for some cardinal κ.*

Solution. If X is extral then it is compact by Problem 475 so, for $\kappa = w(X)$, it can be embedded in $\mathbb{I}^\kappa$ as a closed subspace. It follows from extrality of X that this embedding is automatically an l-embedding in $\mathbb{I}^\kappa$. This proves necessity.

Now assume that X is l-embedded in $\mathbb{I}^\lambda$ for some cardinal λ. Then X is compact because it has to be closed in $\mathbb{I}^\lambda$ (see Problem 450). Suppose that X is embedded in some space Y; denote by X' the respective copy of X in Y and fix a homeomorphism $\xi : X' \to X$. Then $\xi^{-1} : X \to X'$ is a homeomorphism as well an hence its dual map $\mu = (\xi^{-1})^* : C_p(X') \to C_p(X)$ defined by $\mu(f) = f \circ \xi^{-1}$ for any $f \in C_p(X')$ is a linear homeomorphism (see TFS-163).

Any compact subspace of Y is C-embedded in Y by Fact 1 of T.218; so X' is C-embedded in Y. Therefore there exists a continuous function $\Phi : Y \to \mathbb{I}^\lambda$ such that $\Phi|X' = \xi$ (see Fact 4 of U.481); besides, the set X is contained in $Y' = \Phi(Y)$.

Since the space X is l-embedded in $\mathbb{I}^\lambda$, there exists a linear continuous extender $\varphi : C_p(X) \to C_p(\mathbb{I}^\lambda)$. The restriction map $\pi : C_p(\mathbb{I}^\lambda) \to C_p(Y')$ being linear and continuous, the map $\varphi' = \pi \circ \varphi : C_p(X) \to C_p(Y')$ is linear and continuous as well; it is straightforward that φ' is an extender.

Finally let $\Phi^* : C_p(Y') \to C_p(Y)$ be the dual map of Φ defined by the equality $\Phi^*(f) = f \circ \Phi$ for any $f \in C_p(Y')$. Since Φ^* is linear and continuous by TFS-163, the map $\eta = \Phi^* \circ \varphi' \circ \mu : C_p(X') \to C_p(Y)$ is also linear and continuous; let us check that η is an extender.

Fix $f' \in C_p(X')$, $x' \in X'$ and let $f = \mu(f')$, $x = \xi(x')$. Since φ' is an extender, we have $\varphi'(f)(x) = f(x)$. Now, $\Phi^*(\varphi'(f))(x') = (\varphi'(f))(\Phi(x')) = f(x)$ because $\Phi(x') = \xi(x') = x$. Thus $\eta(f')(x') = f(x) = f'(\xi^{-1}(x)) = f'(x')$; since $x' \in X'$ was chosen arbitrarily, we proved that $\eta(f')|X' = f'$, i.e., η is a linear continuous extender and hence X' is l-embedded in Y. This shows that X is l-extral and settles sufficiency.

U.484. *Prove that any retract of an extral space is an extral space.*

Solution. Suppose that X is an extral space and $r : X \to Y$ is a retraction; we can consider that $X \subset \mathbb{I}^\kappa$ for some cardinal κ. It follows from extrality of the space X that there exists a linear continuous extender $\varphi : C_p(X) \to C_p(\mathbb{I}^\kappa)$. The dual map $r^* : C_p(Y) \to C_p(X)$ of the retraction r is linear and continuous (see TFS-163); so the map $\eta = \varphi \circ r^* : C_p(Y) \to C_p(\mathbb{I}^\kappa)$ is linear and continuous as well.

Given any function $f \in C_p(Y)$ and $y \in Y$ the equality $r(y) = y$ implies that $\eta(f)(y) = \varphi(f \circ r)(y) = (f \circ r)(y) = f(r(y)) = f(y)$ (we also applied the fact that φ is an extender and hence $\varphi(f \circ r)|X = f \circ r$). Therefore $\eta(f)(y) = f(y)$ for any $y \in Y$, i.e., $\eta(f)|Y = f$ which shows that η is a linear continuous extender. As a result, the space Y is l-embedded in $\mathbb{I}^\kappa$; so it is extral by Problem 483.

U.485. *Given an extral space X and an infinite cardinal κ prove that $w(X) > \kappa$ implies that $\mathbb{D}^{\kappa^+}$ embeds in X.*

Solution. If we are given sets $A, B \subset \mathbb{R}$ then $A \cdot B = \{ab : a \in A$ and $b \in B\}$; if, additionally, $A \subset \mathbb{R}\backslash\{0\}$ then $A^{-1} = \{r^{-1} : r \in A\}$. Given a space Z let $e_z(f) = f(z)$ for any $z \in Z$ and $f \in C_p(Z)$. Then $e_z : C_p(Z) \to \mathbb{R}$ is a linear continuous functional (see TFS-196) and the map $e : Z \to C_p(C_p(Z))$ defined by $e(z) = e_z$ for each $z \in Z$, is a closed embedding (see TFS-167). This makes it possible to identify every $z \in Z$ with e_z and consider that Z is a closed subspace of $C_p(C_p(Z))$. Then $Z \subset L_p(Z)$ and Z is a closed Hamel basis in $L_p(Z)$ (see Fact 5 of S.489). For any sets $A \subset \mathbb{R}$ and $H \subset L_p(Z)$ let $A \cdot H = \{a \cdot u : a \in A$ and $u \in H\}$.

If $\mathbf{0} \in L_p(Z)$ is the zero functional then we let $l(\mathbf{0}) = 0$. If $u \in L_p(Z)\backslash\{\mathbf{0}\}$ then there exist $\lambda_1, \dots, \lambda_n \in \mathbb{R}\backslash\{0\}$ and distinct points $z_1, \dots, z_n \in Z$ such that $u = \lambda_1 z_1 + \dots + \lambda_n z_n$; let $l(z) = n$. This representation of u will be called *canonical*; the canonical representation is unique up to a permutation in the sense that, for any distinct $y_1, \dots, y_k \in Z$ and $\mu_1, \dots, \mu_k \in \mathbb{R}\backslash\{0\}$ such that $u = \mu_1 y_1 + \dots + \mu_k y_k$ we have $n = k$ and there exists a permutation $\{i_1, \dots, i_n\}$ of the set $\{1, \dots, n\}$ such that $y_{i_j} = z_j$ and $\mu_{i_j} = \lambda_j$ for any $j \in \{1, \dots, n\}$.

Fact 1. For any space Z and $n \in \omega$ the set $L_p^n(Z) = \{u \in L_p(Z) : l(u) \leq n\}$ is closed in $L_p(Z)$.

Proof. Take any $u \in L_p(Z)\backslash L_p^n(Z)$; then there is $k > n$, distinct points $z_1, \dots, z_k \in Z$ and $\lambda_1, \dots, \lambda_k \in \mathbb{R}\backslash\{0\}$ such that $u = \lambda_1 z_1 + \dots + \lambda_k z_k$.

Choose disjoint sets $U_1, \dots, U_k \in \tau(Z)$ and functions $f_1, \dots, f_k \in C_p(Z)$ such that $z_i \in U_i$ while $f_i(z_i) = 1$ and $f_i(Z\backslash U_i) \subset \{0\}$ for every $i \leq k$. The set $O_i = \{v \in L_p(Z) : v(f_i) \neq 0\}$ is open in $L_p(Z)$ for any $i \leq k$; let $O = \bigcap_{i \leq k} O_i$. Observe that $u(f_i) = \lambda_i$ for any $i \leq k$; so $O \in \tau(u, L_p(Z))$. Given any point $v \in O$ let $v = \mu_1 y_1 + \dots + \mu_m y_m$ be a canonical representation of v. It follows from $v(f_i) = \mu_1 f_i(y_1) + \dots + \mu_m f_i(y_m) \neq 0$ that there is $r_i \leq m$ such that $f_i(y_{r_i}) \neq 0$ and therefore $y_{r_i} \in U_i$ for any $i \leq k$. The family $\{U_1, \dots, U_k\}$ being disjoint the points y_{r_i} and y_{r_j} are distinct if $i \neq j$. An immediate consequence is that $m \geq k > n$; this proves that $l(v) > n$ for any $v \in O$, so $O \cap L_p^n(Z) = \emptyset$.

Thus every point $u \in L_p(Z)\backslash L_p^n(Z)$ has a neighbourhood which does not meet $L_p^n(Z)$ and hence $L_p^n(Z)$ is closed in $L_p(Z)$. Fact 1 is proved.

Fact 2. Given an arbitrary space Z suppose that, for some number $n \in \mathbb{N}$, we are given a disjoint family $\mathcal{U} = \{U_1, \dots, U_n\} \subset \tau^*(Z)$ and sets $O_1, \dots, O_n \in \tau^*(\mathbb{R}\backslash\{0\})$. Then $G = O_1 \cdot U_1 + \dots + O_n \cdot U_n \subset L_p^n(Z)\backslash L_p^{n-1}(Z)$ and G is open in $L_p^n(Z)$.

Proof. Take any element $u \in G$. By definition of G there exist $y_1, \dots, y_n \in Z$ and $\mu_1, \dots, \mu_n \in \mathbb{R}$ such that $y_i \in U_i$, $\mu_i \in O_i$ for all $i \leq n$ and $u = \mu_1 y_1 + \dots + \mu_n y_n$. We have $y_i \neq y_j$ whenever $i \neq j$ because the family $\mathcal{U}$ is disjoint. It follows from $\mu_i \in O_i$ that $\mu_i \neq 0$ for every $i \leq n$; so the equality $u = \mu_1 y_1 + \dots + \mu_n y_n$ gives a canonical representation of u. This shows that $l(u) = n$ and hence $u \in L_p^n(Z)\backslash L_p^{n-1}(Z)$. The point $u \in G$ was chosen arbitrarily; so we proved that $G \subset L_p^n(Z)\backslash L_p^{n-1}(Z)$.

To see that G is open in $L_p^n(Z)$ take any $u \in G$. There are $\lambda_i \in O_i$ and $z_i \in U_i$ for any $i \leq n$ such that $u = \lambda_1 z_1 + \ldots + \lambda_n z_n$; as before, it is easy to see that the last equality is the canonical representation of u. For any $i \leq n$ fix a function $f_i \in C_p(Z)$ such that $f_i(z_i) = \lambda_i^{-1}$ and $f_i(Z \backslash U_i) \subset \{0\}$.

The multiplication and taking the inverse are continuous operations in $\mathbb{R}$; so it follows from the equality $1 \cdot (\lambda_i^{-1})^{-1} = \lambda_i$ that

(1) for every $i \leq n$, there exist sets $P_i, Q_i \in \tau(\mathbb{R}\backslash\{0\})$ such that $1 \in P_i$, $\lambda_i^{-1} \in Q_i$ and $P_i \cdot (Q_i)^{-1} \subset O_i$.

It follows from (1) that $z_i \in W_i = f_i^{-1}(Q_i)$; so we can find a function $g_i \in C_p(Z)$ such that $g_i(z_i) = 1$ and $g_i(Z \backslash W_i) \subset \{0\}$ for all $i \leq n$. The set $H_i = \{v \in L_p(Z) : v(f_i) \in P_i \text{ and } v(g_i) \neq 0\}$ is open in $L_p(Z)$; furthermore, $u(f_i) = \lambda_i f_i(z_i) = 1$ and $u(g_i) = \lambda_i g_i(z_i) = \lambda_i \neq 0$ (we used the fact that $\mathcal{U}$ is disjoint and $W_i \subset U_i$); so $u \in H_i$ for every $i \leq n$. Therefore $H = \bigcap_{i \leq n} H_i \in \tau(u, L_p(Z))$.

Suppose that $v \in H \cap L_p^n(Z)$ and hence $l(v) = k \leq n$; let $\mu_1 y_1 + \ldots + \mu_k y_k$ be the canonical representation of v. Since $v(f_i) = \mu_1 f_i(y_1) + \ldots + \mu_k f_i(y_k) \neq 0$, there is $m_i \leq k$ such that $f_i(y_{m_i}) \neq 0$ and hence $y_{m_i} \in U_i$. The family $\mathcal{U}$ being disjoint, the points $y_{m_1}, \ldots, y_{m_n}$ have to be distinct; so $k = n$ and we can change the order of summation if necessary to be able to assume, without loss of generality, that $v = \mu_1 y_1 + \ldots + \mu_n y_n$ and $y_i \in U_i$ for all $i \leq n$.

Now, $v(g_i) = \mu_i g_i(y_i) \neq 0$ implies that $g_i(y_i) \neq 0$ and hence $y_i \in W_i$; an immediate consequence is that $f_i(y_i) \in Q_i$. Since we also have $v(f_i) = \mu_i f(y_i) \in P_i$, we conclude that $\mu_i \in P_i \cdot Q_i^{-1} \subset O_i$ for any $i \leq n$. Thus $v \in G$ which shows that any point $u \in G$ has a neighbourhood H in $L_p(Z)$ such that $H \cap L_p^n(Z) \subset G$; an immediate consequence is that G is open in $L_p^n(Z)$ and hence Fact 2 is proved.

Fact 3. Suppose that Z is a space and we have distinct points $z_1, \ldots, z_n \in Z$. Then for any $\lambda_1, \ldots, \lambda_n \in \mathbb{R}\backslash\{0\}$ the family $\mathcal{B} = \{O_1 \cdot U_1 + \ldots + O_n \cdot U_n : \text{the sets } U_1, \ldots, U_n \text{ are disjoint while } O_i \in \tau(\lambda_i, \mathbb{R}\backslash\{0\}) \text{ and } U_i \in \tau(z_i, Z) \text{ for any } i \leq n\}$ is a local base of the point $u = \lambda_1 z_1 + \ldots + \lambda_n z_n$ in $L_p^n(Z)$.

Proof. All elements of $\mathcal{B}$ are open in $L_p^n(Z)$ by Fact 2 and it is evident that $u \in \bigcap \mathcal{B}$. Consider the map $\varphi_n : \mathbb{R}^n \times Z^n \to L_p^n(Z)$ defined by $\varphi_n(\mu, y) = \mu_1 y_1 + \ldots + \mu_n y_n$ for any $\mu = (\mu_1, \ldots, \mu_n) \in \mathbb{R}^n$ and $y = (y_1, \ldots, y_n) \in Z^n$. Since $L_p^n(Z)$ is a subspace of $C_p(C_p(Z))$ where the sum and the multiplication by a number are continuous operations, the map φ_n is continuous.

Thus for any set $G \in \tau(u, L_p^n(Z))$ we can choose, for any number $i \leq n$, some $Q_i \in \tau(\lambda_i, \mathbb{R})$ and $V_i \in \tau(z_i, Z)$ such that $\varphi_n(Q, V) \subset G$ where $Q = Q_1 \times \ldots \times Q_n$ and $V = V_1 \times \ldots \times V_n$. It is clear that there exists a disjoint family $\{U_1, \ldots, U_n\} \subset \tau(Z)$ and a collection $\{O_1, \ldots, O_n\} \subset \tau(\mathbb{R}\backslash\{0\})$ such that $z_i \in U_i \subset V_i$ and $\lambda_i \in O_i \subset Q_i$ for any $i \leq n$. Then, for the sets $O = O_1 \times \ldots \times O_n$ and $U = U_1 \times \ldots \times U_n$ we have $H = O_1 \cdot U_1 + \ldots + O_n \cdot U_n = \varphi_n(O, U) \subset \varphi_n(Q, V) \subset G$ which implies that $u \in H \subset G$; since $H \in \mathcal{B}$, this shows that $\mathcal{B}$ is a local base at u in the space $L_p^n(Z)$; so Fact 3 is proved.

Fact 4. Given a space Z and $n \in \mathbb{N}$ consider the map $\varphi_n : \mathbb{R}^n \times Z^n \to L_p^n(Z)$ defined by $\varphi_n(\mu, y) = \mu_1 y_1 + \ldots + \mu_n y_n$ for any points $\mu = (\mu_1, \ldots, \mu_n) \in \mathbb{R}^n$ and $y = (y_1, \ldots, y_n) \in Z^n$. Say that a set $Q \subset \mathbb{R}^n \times Z^n$ is *canonical* if there exist $O_1, \ldots, O_n \in \tau^*(\mathbb{R}\backslash\{0\})$ and disjoint $U_1, \ldots, U_n \in \tau^*(Z)$ such that, for the sets $O = O_1 \times \ldots \times O_n$ and $U = U_1 \times \ldots \times U_n$, we have $Q = O \times U$. Then the map $\varphi_n | Q$ is an embedding for any canonical set $Q \subset \mathbb{R}^n \times Z^n$. As a consequence, for every $u \in L_p^n(Z) \backslash L_p^{n-1}(Z)$ the family $\mathcal{B} = \{G \in \tau(u, L_p^n(Z)): G$ is homeomorphic to a canonical subset of $\mathbb{R}^n \times Z^n\}$ is a local base at the point u in $L_p^n(Z)$.

Proof. Fix an arbitrary canonical set $Q = O \times U$ where $O = O_1 \times \ldots \times O_n$ for some sets $O_1, \ldots, O_n \in \tau^*(\mathbb{R}\backslash\{0\})$ and $U = U_1 \times \ldots \times U_n$ for some disjoint collection $\mathcal{U} = \{U_1, \ldots, U_n\} \subset \tau^*(Z)$. Since $L_p^n(Z)$ is a subspace of $C_p(C_p(Z))$ where the sum and the multiplication by a number are continuous operations, the map φ_n is continuous.

Suppose that $u, v \in Q$ and $\varphi_n(u) = \varphi_n(v)$. For any $i \leq n$ there are points $y_i, z_i \in U_i$ and $\lambda_i, \mu_i \in O_i$ such that $u = (\lambda, z)$ and $v = (\mu, y)$ where $\lambda = (\lambda_1, \ldots, \lambda_n)$ and $\mu = (\mu_1, \ldots, \mu_n)$ while $z = (z_1, \ldots, z_n)$ and $y = (y_1, \ldots, y_n)$. The set Z being a Hamel basis in $L_p(Z)$ (see Fact 5 of S.489), it follows from the equality $\sum_{i \leq n} \mu_i y_i = \sum_{i \leq n} \lambda_i z_i$ that, for any $i \leq n$ there is $j_i \leq n$ such that $y_i = z_{j_i}$. Since $\mathcal{U}$ is disjoint, we have $j_i = i$ and hence $y_i = z_i$ for any $i \leq n$. Thus we have the equality $\sum_{i \leq n} (\mu_i - \lambda_i) y_i = \mathbf{0}$ which, together with linear independence of the set $\{y_1, \ldots, y_n\}$ shows that $\mu_i = \lambda_i$ for all $i \leq n$ and hence $u = v$. Therefore the map $\xi = \varphi_n | Q$ is an injection; let $P = \xi(Q)$.

It is easy to see that canonical subsets of Q form a base $\mathcal{Q}$ in the space Q; besides, the set $\xi(Q') = \varphi_n(Q')$ is open in $L_p^n(Z)$ and hence in P (see Fact 2) for any $Q' \in \mathcal{Q}$. An immediate consequence is that the map ξ is a homeomorphism being open and bijective. This shows that $\varphi_n | Q$ is an embedding.

Finally, take any point $u = L_p^n(Z) \backslash L_p^{n-1}(Z)$; there are distinct $z_1, \ldots, z_n \in Z$ and $\lambda_1, \ldots, \lambda_n \in \mathbb{R}\backslash\{0\}$ such that $u = \lambda_1 z_1 + \ldots + \lambda_n z_n$. Then $z = (z_1, \ldots, z_n) \in Z^n$ and $\lambda = (\lambda_1, \ldots, \lambda_n) \in \mathbb{R}^n$ while $\varphi_n(\lambda, z) = u$. Choose disjoint sets $U_1, \ldots, U_n \in \tau(Z)$ such that $z_i \in U_i$ for all $i \leq n$ and let $U = U_1 \times \ldots \times U_n$. The set $Q = (\mathbb{R}\backslash\{0\})^n \times U$ is canonical; so $\varphi_n(Q) \in \tau(u, L_p^n(Z))$. Since canonical subsets of Q constitute a local base at the point (λ, z) in $\mathbb{R}^n \times Z^n$, their (homeomorphic) images under φ_n constitute a local base in $L_p^n(Z)$ at the point u. Fact 4 is proved.

Returning to our solution suppose that X is an extral space with $w(X) > \kappa$. We can assume that X is a subspace of $\mathbb{I}^\nu$ for some cardinal ν. Thus X is l-embedded in $\mathbb{I}^\nu$ and hence there exists a continuous map $\varphi : \mathbb{I}^\nu \to L_p(X)$ such that $\varphi(x) = x$ for any $x \in X$ (see Problem 467). The compact space $K = \varphi(\mathbb{I}^\nu)$ is dyadic (see Fact 2 of T.298) and $X \subset K$; so $w(K) \geq w(X) > \kappa$. This makes it possible to apply Fact 3 of U.086 to conclude that there is a set $E \subset K$ such that $E \simeq \mathbb{D}^{\kappa^+}$.

The set $E_n = E \cap L_p^n(X)$ is closed in E for any $n \in \omega$ (see Fact 1) and $E = \bigcup_{n \in \omega} E_n$. By the Baire property of E, there is a number $n \in \omega$ such that $\text{Int}_E(E_n) \neq \emptyset$. Let $m = \min\{n \in \omega : \text{Int}_E(E_n) \neq \emptyset\}$. Then there exists a set $U \in \tau^*(E)$ with $U \subset L_p^m(X) \backslash L_p^{m-1}(X)$. It is easy to see that $\mathbb{D}^{\kappa^+}$ embeds into any non-empty open subset of $\mathbb{D}^{\kappa^+}$, so E embeds in the space $L_p^m(X)\backslash L_p^{m-1}(X)$; thus we can consider that $E \subset L_p^m(X)\backslash L_p^{m-1}(X)$. Take a point $u \in E$ and apply Fact 4 to see that there is $W \in \tau(u, L_p^m(X))$ which is homeomorphic to a subspace of $\mathbb{R}^m \times X^m$. The set $W' = W \cap E$ is non-empty and open in E; so E embeds in W' and hence in $\mathbb{R}^m \times X^m$.

To simplify the notation we are going to consider that E is a subspace of the space $\mathbb{R}^m \times X^m$; let $\pi : \mathbb{R}^m \times X^m \to \mathbb{R}^m$ and $q : \mathbb{R}^m \times X^m \to X^m$ be the projections. Since $E \subset \pi(E) \times q(E)$ and $w(\pi(E)) = \omega \leq \kappa$, we must have $w(q(E)) > \kappa$. Thus $L = q(E)$ is a dyadic compact space with $w(L) > \kappa$. Let $p_i : X^m \to X$ be the natural projection of X^m onto its i-th factor for any $i \leq m$. We have $L \subset p_1(L) \times \ldots \times p_m(L)$; so all spaces $p_i(L)$ cannot have weight $\leq \kappa$. Thus there is $i \leq m$ for which $w(p_i(L)) > \kappa$; the space $p_i(L)$ being dyadic, we can apply Fact 3 of U.086 again to see that $\mathbb{D}^{\kappa^+}$ embeds in $p_i(L)$. Of course, the same map embeds $\mathbb{D}^{\kappa^+}$ in X; so our solution is complete.

U.486. *Let X be an extral space. Prove that $w(X) = t(X) = \chi(X)$.*

Solution. Since the inequalities $t(Z) \leq \chi(Z) \leq w(Z)$ hold for any space Z, it suffices to show that $w(X) \leq t(X)$; let $\kappa = t(X)$. If $w(X) > \kappa$ then $\mathbb{D}^{\kappa^+}$ embeds in X by Problem 485 and hence $t(\mathbb{D}^{\kappa^+}) \leq \kappa$. The space $Y = \kappa^+ + 1$ is zero-dimensional and $w(Y) \leq \kappa^+$; by Fact 2 of U.003, there is an embedding of Y in $\mathbb{D}^{\kappa^+}$. On the one hand, the point $\kappa^+ \in Y$ is non-isolated in Y and, for any $A \subset Y\backslash\{\kappa^+\}$ with $|A| \leq \kappa$, we have $\kappa^+ \notin \overline{A}$. This implies $t(Y) \geq \kappa^+$. The space Y being embeddable in $\mathbb{D}^{\kappa^+}$, we must have $t(Y) \leq t(\mathbb{D}^{\kappa^+}) \leq \kappa$; this contradiction shows that $w(X) \leq \kappa = t(X)$ and hence $w(X) = \chi(X) = t(X)$.

U.487. *Assuming that $\mathfrak{c} < 2^{\omega_1}$ prove that any extral space X with $|X| \leq \mathfrak{c}$ is metrizable.*

Solution. If X is not metrizable then $w(X) > \omega$ and hence $\mathbb{D}^{\omega_1}$ embeds in X by Problem 485. Therefore $\mathfrak{c} < 2^{\omega_1} = |\mathbb{D}^{\omega_1}| \leq |X| \leq \mathfrak{c}$ which is a contradiction.

U.488. *Prove that every extral t-extendial space is metrizable.*

Solution. The space X is compact and ω_1 is a caliber of X by Problem 475. It follows from Problem 459 that X is hereditarily separable. If X is not metrizable then $w(X) > \omega$ and hence $\mathbb{D}^{\omega_1}$ embeds in X by Problem 485. Therefore $\mathbb{D}^{\omega_1}$ is hereditarily separable which is easily seen to be a contradiction.

U.489. *Suppose that every closed subspace of a space X is extral. Prove that X is metrizable.*

Solution. Since X is closed in X, the space X is extral. Given a closed $F \subset X$ it follows from extrality of F that F is l-embedded in X; this proves that X is l-extendial and hence t-extendial; so we can apply Problem 488 to conclude that X is metrizable.

U.490. *Prove that any extral linearly orderable space is metrizable.*

Solution. The main tool of this solution is the following fact.

Fact 1. The space $A(\omega_1)$ does not embed in a linearly ordered space.

Proof. Recall that $A(\omega_1) = \omega_1 \cup \{a\}$ is a compact space and a is the unique non-isolated point of $A(\omega_1)$. Suppose that $(X, \prec)$ is a linearly ordered space and $A(\omega_1)$ embeds in X; to simplify the notation, we will assume that $A(\omega_1) \subset X$. For any ordinal $\alpha \in \omega_1$ either $\alpha \prec a$ or $a \prec \alpha$. Thus, for the sets $L_a = \{y \in X : y \prec a\}$ and $R_a = \{y \in X : a \prec y\}$ either $|\omega_1 \cap L_a| > \omega$ or $|\omega_1 \cap R_a| > \omega$.

Suppose first that $C = \omega_1 \cap L_a$ is uncountable and choose a countably infinite $D \subset C$. In the space $A(\omega_1)$ every infinite subset contains a in its closure; so $a \in \overline{D}$. For any $d \in D$ the set $(d, a] = \{y \in X : d \prec y \preceq a\}$ is open in $L_a \cup \{a\}$; so $(d, a] \cap (C \cup \{a\})$ is open in $C \cup \{a\}$. The space $C \cup \{a\}$ being compact with the unique non-isolated point a, the set $C \backslash (d, a]$ must be finite for any $d \in D$.

On the other hand, for any $c \in C$ the set $(c, a]$ is an open neighbourhood of a in $L_a \cup \{a\}$; so it follows from $a \in \overline{D}$ that there is $d \in D \cap (c, a]$; this, evidently, implies that $c \notin (d, a]$. Consequently, $C \cap (\bigcap\{(d, a] : d \in D\}) = \emptyset$ and hence $C \subset \bigcup\{C \backslash (d, a] : d \in D\}$; so the set C is countable which is a contradiction.

In the case of $|\omega_1 \cap R_a| > \omega$ the contradiction is obtained analogously; so $A(\omega_1)$ cannot be embedded in X and hence Fact 1 is proved.

Returning to our solution assume that X is an extral linearly ordered space. If X is not metrizable then $w(X) > \omega$ and hence $\mathbb{D}^{\omega_1}$ embeds in X by Problem 485. It is an easy exercise that $A(\omega_1)$ embeds in $\mathbb{D}^{\omega_1}$; so $A(\omega_1)$ embeds in our linearly ordered space X which is a contradiction with Fact 1. Therefore X has to be metrizable.

U.491. *Give an example of an extral space which is not dyadic.*

Solution. If Z is a space and $f \in C^*(Z)$ then $||f||_Z = \sup\{|f(z)| : z \in Z\}$. Recall that $AD(Z)$ is the *Alexandroff double* of Z whose underlying set is $Z \times \mathbb{D}$; for any $z \in Z$ let $u_0(z) = (z, 0)$ and $u_1(z) = (z, 1)$. Then $AD(Z) = u_0(Z) \cup u_1(Z)$ and the points of $u_1(Z)$ are isolated while, for any $a = (z, 0) \in AD(Z)$, the base at a is formed by the sets $u_0(V) \cup (u_1(V) \backslash \{u_1(z)\})$ where V runs over the open neighbourhoods of z.

The space $Y = AD(\mathbb{I}^{\omega_1})$ is compact (see TFS-364); the projection $\pi : Y \to \mathbb{I}^{\omega_1}$ defined by $\pi(x, i) = x$ for any $x \in \mathbb{I}^{\omega_1}$ and $i \in \mathbb{D}$ is easily seen to be continuous. Let $Y_i = \mathbb{I}^{\omega_1} \times \{i\}$ for every $i \in \mathbb{D}$; it is evident that $\pi | Y_0 : Y_0 \to \mathbb{I}^{\omega_1}$ is a homeomorphism and Y_1 is discrete. As above, let $u_i : \mathbb{I}^{\omega_1} \to Y$ be defined by $u_i(x) = (x, i)$ for any $i \in \mathbb{D}$ and $x \in \mathbb{I}^{\omega_1}$. Then $u_0 : \mathbb{I}^{\omega_1} \to Y_0$ is the homeomorphism inverse to $\pi | Y_0$.

Apply TFS-108 to fix a countable dense set $Q \subset \mathbb{I}^{\omega_1}$ and let $Q_1 = Q \times \{1\} \subset Y_1$. The set $X = Y_0 \cup Q_1$ is closed in Y its complement being a subspace of the set of isolated points of Y; therefore X is compact. Besides, X is not metrizable because $Y_0 \simeq \mathbb{I}^{\omega_1}$ is a non-metrizable subspace of X. Observe that Q_1 is dense in X; so X has a dense set of points of countable character which, together with TFS-360, shows that X is not dyadic.

Consider the map $r : X \to Y_0$ defined by $r(x, i) = (x, 0)$ for any $(x, i) \in X$. The equality $r = u_0 \circ \pi$ implies that r is continuous. Since $r(y) = y$ for any $y \in Y_0$, the map r is a retraction. Let $\{q_n : n \in \omega\}$ be a faithful enumeration of the set Q_1. To show that X is extral assume that $X \subset Z$ for some space Z.

It is easy to see that $E = \{f \in C_p(X) : f(Y_0) = \{0\}\}$ is a closed linear subspace of $C_p(X)$. Our first step is to show that

(1) there exists a linear continuous extender $\varphi_0 : E \to C_p(Z)$.

Apply Fact 1 of S.369 to find a disjoint family $\mathcal{U} = \{U_n : n \in \omega\} \subset \tau(Z)$ such that $q_n \in U_n$ and $U_n \cap Y_0 = \emptyset$; fix a continuous function $g_n : Z \to [0, 1]$ such that $g_n(q_n) = 1$ and $g_n(Z \backslash U_n) = \{0\}$ for any $n \in \omega$. For any $f \in E$ and $\varepsilon > 0$, the set $f^{-1}(\mathbb{R} \backslash (-\varepsilon, \varepsilon))$ is closed in X and contained in Q_1; therefore it is finite being compact and discrete. This shows that

(2) the sequence $\{f(q_n) : n \in \omega\}$ converges to zero for any $f \in E$.

For any $f \in E$ let $\varphi_0(f) = \sum_{n \in \omega} f(q_n) \cdot g_n$; then $\|f(q_n) \cdot g_n\|_Z \leq |f(q_n)|$ for every $n \in \omega$; so it follows from (2) that we can apply Fact 6 of U.481 to see that $\varphi_0(f)$ is continuous on Z. Thus $\varphi_0 : E \to C_p(Z)$ and it is straightforward that φ_0 is a linear map. Given $f \in C_p(X)$ it follows from disjointness of $\mathcal{U}$ that $\varphi_0(f)(q_n) = f(q_n) \cdot g(q_n) = f(q_n)$ for any $n \in \omega$; since $\bigcup \mathcal{U}$ does not meet Y_0, we have $\varphi_0(f)(x) = 0$ for any $x \in Y_0$. Therefore $\varphi_0(f)|X = f$ for any $f \in E$, i.e., φ_0 is a linear extender.

For any $z \in Z$ let $\mu_z : C_p(Z) \to \mathbb{R}$ be the natural projection onto the factor of the product $\mathbb{R}^Z$ determined by z. If $z \notin \bigcup \mathcal{U}$ then $(\mu_z \circ \varphi_0)(f) = 0$ for any $f \in E$; so the map $\mu_z \circ \varphi_0$ is continuous. If $z \in \bigcup \mathcal{U}$ then there is $n \in \omega$ such that $z \in U_n$ and hence $(\mu_z \circ \varphi_0)(f) = f(q_n) \cdot g_n(z)$ for any $f \in E$ which shows that $\mu_z \circ \varphi_0$ is continuous being the product of a constant $g_n(z)$ and the evaluation function at q_n (see TFS-166). Therefore we proved that $\mu_z \circ \varphi$ is continuous for any $z \in Z$; as a consequence, φ_0 is continuous by TFS-102 and (1) is proved.

The space X is compact; so it is C-embedded in Z by Fact 1 of T.218 and hence we can apply Fact 4 of U.481 to see that there exists a continuous map $s : Z \to Y_0$ such that $s|X = r$; it is clear that s is a retraction of Z onto Y_0. Take any function $f \in C_p(X)$; then $f(x) = f(r(x))$ for any $x \in Y_0$ and hence $v(f) = f - f \circ r \in E$. Let $e(f) = \varphi_0(v(f)) + f \circ s$; it is immediate that $e(f)$ is continuous on Z for any $f \in C_p(X)$. The map φ_0 being linear it is straightforward that e is linear as well; let us check that e is an extender.

If $x \in X$ then $e(f)(x) = v(f)(x) + f(s(x))$ because the map φ_0 is an extender. Consequently, $e(f)(x) = f(x) - f(r(x)) + f(s(x)) = f(x)$; this follows from $s|X = r$. Therefore $e(f)(x) = f(x)$ for any $x \in X$, i.e., e is, indeed, an extender. Finally, to see that e is continuous, observe that $v : C_p(X) \to C_p(X)$ is continuous (this is an easy exercise which we leave to the reader) and the map $\gamma : C_p(X) \to C_p(Z)$ defined by $\gamma(f) = f \circ s$ is continuous being the composition of the restriction to Y_0 and the dual map of s (see TFS-163). Thus e is a continuous linear extender and therefore X is l-embedded in Z. The space Z was chosen arbitrarily; so we established that X is an extral non-dyadic space and hence our solution is complete.

U.492. *Give an example of an extral space X such that some continuous image of X is not extral.*

Solution. In the Tychonoff cube $\mathbb{I}^{\omega_1}$ let a be the point whose all coordinates are equal to zero. If $a_n(\alpha) = 2^{-n}$ for any $\alpha < \omega_1$ then $a_n \in \mathbb{I}^{\omega_1}$ for any $n \in \omega$ and the sequence $\{a_n : n \in \omega\}$ converges to a; let $S = \{a\} \cup \{a_n : n \in \omega\}$. The set S is compact and hence closed in the normal space $\mathbb{I}^{\omega_1}$; let X be the space obtained from $\mathbb{I}^{\omega_1}$ by collapsing S to a point (see Fact 2 of T.245). Recall that $X = (\mathbb{I}^{\omega_1} \backslash S) \cup \{S\}$; let $\pi(x) = x$ for any $x \in \mathbb{I}^{\omega_1} \backslash S$ and $\pi(x) = S$ for any $x \in S$. Then $\pi : \mathbb{I}^{\omega_1} \to X$ is a continuous onto map (this was also proved in Fact 2 of T.245); in particular, X is a continuous image of $\mathbb{I}^{\omega_1}$. To avoid confusion denote by s the point of X represented by the set S; then $\pi(x) = s$ for any $x \in S$.

Given a number $n \in \omega$ and an ordinal α such that $0 < \alpha < \omega_1$ let $p_\alpha^n(\alpha) = 0$ and $p_\alpha^n(\beta) = 2^{-n}$ for any $\beta \neq \alpha$. It is clear that $p_\alpha^n \in \mathbb{I}^{\omega_1}$ whenever $n \in \omega$ and $0 < \alpha < \omega_1$; besides, the subspace $L_n = \{a_n\} \cup \{p_\alpha^n : 0 < \alpha < \omega_1\}$ is homeomorphic to one-point compactification $A(\omega_1)$ of a discrete space of cardinality ω_1 while a_n is the unique non-isolated point of L_n and $L_n \cap S = \{a_n\}$ for any $n \in \omega$.

Fix a disjoint family $\{O_n : n \in \omega\}$ of open subsets of $\mathbb{I} \backslash \{0\}$ such that $2^{-n} \in O_n$ for any $n \in \omega$. The set $U_n = \{x \in \mathbb{I}^{\omega_1} : x \neq a_n$ and $x(0) \in O_n\}$ is open in $\mathbb{I}^{\omega_1}$ while $U_n \cap S = \emptyset$ and $L_n \backslash \{a_n\} \subset U_n$ for any $n \in \omega$. It is easy to see that the family $\mathcal{U} = \{U_n : n \in \omega\}$ is disjoint.

The map π is injective on the compact space L_n; therefore $K_n = \pi(L_n)$ is still homeomorphic to $A(\omega_1)$ and s is the unique non-isolated point of K_n for every $n \in \omega$. The set $V_n = \pi(U_n)$ is open in X and $K_n \backslash \{s\} \subset V_n$ for any $n \in \omega$. Since π is injective on the set $\mathbb{I}^{\omega_1} \backslash S \supset \bigcup \mathcal{U}$, the family $\mathcal{V} = \{V_n : n \in \omega\}$ is disjoint.

Now assume that X is an extral space; as any Tychonoff space it can be embedded in a Tychonoff cube $\mathbb{I}^\kappa$ for some cardinal κ; so let us consider that $X \subset \mathbb{I}^\kappa$. It follows from extrality of X that X is l-embedded in $\mathbb{I}^\kappa$. By Fact 2 of U.481, there exists $n \in \omega$ for which the set X is K_n-embedded in $\mathbb{I}^\kappa$, i.e., we can find a map $e : \tau(X) \to \tau(\mathbb{I}^\kappa)$ such that $e(W) \cap X = W$ for any $W \in \tau(X)$ and $\bigcap_{i<n} e(W_i) = \emptyset$ for any disjoint family $\{W_0, \ldots, W_{n-1}\} \subset \tau(X)$.

The family $\mathcal{V}$ being disjoint, we have $e(V_0) \cap \ldots \cap e(V_{n-1}) = \emptyset$. However, it follows from $K_i \backslash \{s\} \subset V_i$ that $K_i \backslash \{s\} \subset e(V_i)$ for any $i < n$; so we can apply Fact 3 of U.481 to conclude that $e(V_0) \cap \ldots \cap e(V_{n-1}) \neq \emptyset$. This contradiction shows that the space X is not l-embedded in $\mathbb{I}^\kappa$ and hence X is not extral. Since $\mathbb{I}^{\omega_1}$ is extral (see Problem 483), our space X is an example of a continuous non-extral image of an extral space.

U.493. *Prove that any zero-dimensional extral space is metrizable.*

Solution. If Z is a space say that a subspace $A \subset Z$ is *non-trivial* if $A \neq \emptyset$ and $A \neq Z$. A space is called *connected* if it has no non-trivial clopen subspaces. Given a space Z let $e_z(f) = f(z)$ for any $z \in Z$ and $f \in C_p(Z)$. Then $e_z : C_p(Z) \to \mathbb{R}$ is a linear continuous functional (see TFS-196) and the map $e : Z \to C_p(C_p(Z))$ defined by $e(z) = e_z$ for each $z \in Z$, is a closed embedding (see TFS-167). This makes it possible to identify every $z \in Z$ with e_z and consider that Z is a closed subspace of $C_p(C_p(Z))$. Then $Z \subset L_p(Z)$ and Z is a closed Hamel basis in $L_p(Z)$ (see Fact 5 of S.489).

If $\mathbf{0} \in L_p(Z)$ is the zero functional then we let $l(\mathbf{0}) = 0$. If $u \in L_p(Z)\backslash\{\mathbf{0}\}$ then there exist $\lambda_1, \ldots, \lambda_n \in \mathbb{R}\backslash\{0\}$ and distinct points $z_1, \ldots, z_n \in Z$ such that $u = \lambda_1 z_1 + \ldots + \lambda_n z_n$; let $l(z) = n$. This representation of u will be called *canonical*; the canonical representation is unique up to a permutation in the sense that, for any distinct $y_1, \ldots, y_k \in Z$ and $\mu_1, \ldots, \mu_k \in \mathbb{R}\backslash\{0\}$ such that $u = \mu_1 y_1 + \ldots + \mu_k y_k$ we have $n = k$ and there exists a permutation $\{i_1, \ldots, i_n\}$ of the set $\{1, \ldots, n\}$ such that $y_{i_j} = z_j$ and $\mu_{i_j} = \lambda_j$ for any $j \in \{1, \ldots, n\}$. Let $L_p^n(Z) = \{u \in L_p(Z) : l(u) \leq n\}$ for any $n \in \omega$.

Fact 1. For any set A both spaces $\mathbb{R}^A$ and $\mathbb{I}^A$ are connected.

Proof. The set $\sigma(A) = \{x \in \mathbb{R}^A : x^{-1}(\mathbb{R}\backslash\{0\})$ is finite$\}$ is dense in $\mathbb{R}^A$ and connected by Fact 2 of T.312. Therefore $\mathbb{R}^A$ is connected by Fact 1 of T.312. The space $(-1, 1)^A$ is dense in $\mathbb{I}^A$ and homeomorphic to $\mathbb{R}^A$; thus $\mathbb{I}^A$ has a dense connected subspace; so we can apply Fact 1 of T.312 again to conclude that $\mathbb{I}^A$ is also connected and finish the proof of Fact 1.

Fact 2. Any continuous image of a connected space is connected. As a consequence, if X is connected, Y is zero-dimensional and $f : X \to Y$ is a continuous map then $f(X)$ is a singleton.

Proof. Suppose that X is a connected space and $f : X \to Z$ is a continuous onto map. If Z is not connected then there is a non-trivial clopen $U \subset Z$; it is straightforward that $f^{-1}(U)$ is a non-trivial clopen subset of X which is a contradiction with connectedness of X. Therefore Z has to be connected.

Finally, assume that X is connected, Y is zero-dimensional and $f : X \to Y$ is a continuous map. The space $f(X)$ is connected and zero-dimensional (see SFFS-301); so $|f(X)| > 1$ gives a contradiction with Fact 1 of T.312 (where it was proved that a connected space which has more than one point cannot be zero-dimensional). Therefore $f(X)$ has to be a singleton and hence Fact 2 is proved.

Returning to our solution suppose that X is a zero-dimensional extral space; we can assume, without loss of generality, that $X \subset \mathbb{I}^\kappa$ for some cardinal κ and therefore X is l-embedded in $\mathbb{I}^\kappa$. Assume that X is not metrizable and apply Problem 467 to take a continuous map $\varphi : \mathbb{I}^\kappa \to L_p(X)$ such that $\varphi(x) = x$ for any $x \in X$. Say that a space Z is *thick* if Z is a non-metrizable image of a Tychonoff cube. It follows from Fact 2 of T.298 that every thick space is dyadic.

The space $K = \varphi(\mathbb{I}^\kappa)$ is thick because X is a non-metrizable subspace of K. It follows from TFS-360 that the set $P = \{x \in K : \chi(x, K) \leq \omega\}$ cannot be dense in K; so take $U \in \tau^*(K)$ such that $U \cap P = \emptyset$. If $P_n = U \cap L_p^n(X)$ then P_n is closed in U (see Fact 1 of U.485) for any $n \in \omega$ and $U = \bigcup_{n\in\omega} P_n$. The space U being locally compact, it has the Baire property; so some P_n has a non-empty interior in U. If $m = \min\{n \in \omega : \text{Int}_U(P_n) \neq \emptyset\}$ then there is $V \in \tau^*(U)$ such that $V \subset S_m = L_p^m(X)\backslash L_p^{m-1}(X)$.

Choose $W \in \tau^*(K)$ such that $\overline{W} \subset V$; it is evident that the compact space $F = \overline{W}$ is not metrizable. Let $F_1 = \varphi^{-1}(F)$ and suppose first that for every $x \in F_1$ there is $O_x \in \tau(x, \mathbb{I}^\kappa)$ such that $nw(\varphi(O_x)) \leq \omega$. The set F_1 being compact, there is a finite $A \subset F_1$ such that $F_1 \subset \bigcup\{O_x : x \in A\}$. Then $F \subset \bigcup\{\varphi(O_x) : x \in A\}$ and therefore $nw(F) \leq \omega$ (it is an easy exercise that a finite union of spaces with a countable network has a countable network); so F is metrizable (see Fact 4 of S.307) which is a contradiction.

So, fix a point $x \in F_1$ such that $nw(\varphi(O)) > \omega$ for any $O \in \tau(x, \mathbb{I}^\kappa)$ and let $y = \varphi(x)$. Since $y \in S_m$, we can apply Fact 4 of U.485 to find a set $G \in \tau(y, S_m)$ which is homeomorphic to a subspace of $\mathbb{R}^m \times X^m$. It follows from $V \subset S_m$ and $V \in \tau(y, K)$ that $G' = G \cap V \in \tau(y, K)$. By continuity of φ there exists a set $O \in \tau(x, \mathbb{I}^\kappa)$ such that $\overline{\varphi(O)} \subset G'$ and $O = \prod\{O_\alpha : \alpha < \kappa\}$ where $O_\alpha = (a_\alpha, b_\alpha) \cap \mathbb{I}$ for some $a_\alpha, b_\alpha \in \mathbb{R}$ with $a_\alpha < b_\alpha$ (and only finitely many O_α's are distinct from $\mathbb{I}$, but we won't need that). Then $Q_\alpha = \overline{O}_\alpha$ is a non-trivial closed subinterval of $\mathbb{I}$ for any $\alpha < \kappa$ and hence the set $Q = \prod\{Q_\alpha : \alpha < \kappa\}$ is homeomorphic to $\mathbb{I}^\kappa$. The set O being dense in Q, we have $\varphi(Q) \subset \overline{\varphi(O)} \subset G'$.

Observe that the compact space $E = \varphi(Q)$ is not metrizable because $\varphi(O)$ is a non-metrizable subspace of E; being a subspace of G' the space E is homeomorphic to some $E' \subset \mathbb{R}^m \times X^m$. Let $q_0 : E' \to \mathbb{R}^m$ and $q_1 : E' \to X^m$ be the restrictions of the projections of the product $\mathbb{R}^m \times X^m$ onto its factors $\mathbb{R}^m$ and X^m respectively. The space E' is connected being homeomorphic to a continuous image E of a Tychonoff cube (see Fact 1 and Fact 2); the space X^m is zero-dimensional by SFFS-302; so $q_1(E')$ is singleton by Fact 2. Now, it follows from $E' \subset q_0(E') \times q_1(E')$ that $w(E') \leq \omega$; this contradiction with the set E' being thick shows that the space X has to be metrizable and hence our solution is complete.

U.494. *Prove that any continuous image of a t-dyadic space is a t-dyadic space.*

Solution. Given an arbitrary space Z let $e_z(f) = f(z)$ for any $z \in Z$ and $f \in C_p(Z)$. Then $e_z : C_p(Z) \to \mathbb{R}$ is a linear continuous functional (see TFS-196) and the map $e : Z \to C_p(C_p(Z))$ defined by $e(z) = e_z$ for each $z \in Z$, is a closed embedding (see TFS-167). This makes it possible to identify every $z \in Z$ with e_z and consider that Z is a closed subspace of $C_p(C_p(Z))$.

Suppose that X is a t-dyadic space and hence we can consider that there is a dyadic space D such that X is t-embedded in D. By Problem 449 there exists a continuous map $\varphi : D \to C_p(C_p(X))$ such that $\varphi(x) = x$ for any $x \in X$; then $D' = \varphi(D) \subset C_p(C_p(X))$ is a dyadic space.

Now, if we have a continuous onto map $q : X \to Y$ of the space X onto a space Y then there exists a continuous map $\varphi_q : C_p(C_p(X)) \to C_p(C_p(Y))$ such that $\varphi_q(x) = q(x)$ for any $x \in X$ (see SFFS-467). The space $E = \varphi_q(D') \subset C_p(C_p(Y))$ is dyadic and it follows from $X \subset D'$ that $Y \subset E$. Letting $\delta(x) = x$ for any $x \in E$ we obtain a continuous map $\delta : E \to C_p(C_p(Y))$ which shows that Problem 449 is applicable to conclude that Y is t-embedded in E. The space E being dyadic, Y is a t-dyadic space.

U.495. *Prove that any continuous image of an l-dyadic space is an l-dyadic space.*

Solution. Given an arbitrary space Z let $e_z(f) = f(z)$ for any $z \in Z$ and $f \in C_p(Z)$. Then $e_z : C_p(Z) \to \mathbb{R}$ is a linear continuous functional (see TFS-196) and the map $e : Z \to C_p(C_p(Z))$ defined by $e(z) = e_z$ for each $z \in Z$, is a closed embedding (see TFS-167). This makes it possible to identify every $z \in Z$ with e_z and consider that Z is a closed subspace of $C_p(C_p(Z))$ which is, at the same time, a closed Hamel basis in $L_p(Z)$ (see Fact 5 of S.489).

Suppose that X is an l-dyadic space and hence we can consider that there is a dyadic space D such that X is l-embedded in D. By Problem 467 there exists a continuous map $\varphi : D \to L_p(X)$ such that $\varphi(x) = x$ for any $x \in X$; then $D' = \varphi(D) \subset L_p(X)$ is a dyadic space.

Now, if we have a continuous onto map $q : X \to Y$ of the space X onto a space Y then there exists a linear continuous map $\varphi_q : C_p(C_p(X)) \to C_p(C_p(Y))$ such that $\varphi_q(x) = q(x)$ for any $x \in X$ (see SFFS-467). Take any $u \in L_p(X)$; since X is a basis in $L_p(X)$, there exist $x_1, \ldots, x_n \in X$ and $\lambda_1, \ldots, \lambda_n \in \mathbb{R}$ such that $u = \lambda_1 x_1 + \ldots + \lambda_n x_n$. Then $\varphi_q(u) = \lambda_1 q(x_1) + \ldots + \lambda_n q(x_n) \in L_p(Y)$. This proves that $\varphi_q(L_p(X)) \subset L_p(Y)$.

The space $E = \varphi_q(D') \subset \varphi_q(L_p(X)) \subset L_p(Y)$ is dyadic and it follows from $X \subset D'$ that $Y \subset E$. Letting $\delta(x) = x$ for any point $x \in E$ we obtain a continuous map $\delta : E \to L_p(Y)$ which shows that Problem 467 is applicable to conclude that Y is l-embedded in E. The space E being dyadic, Y is an l-dyadic space.

U.496. *Given an l-dyadic space X prove that, for any infinite cardinal κ such that $\kappa < w(X)$, the space $\mathbb{D}^{\kappa^+}$ embeds in X.*

Solution. Given an arbitrary space Z let $e_z(f) = f(z)$ for any $z \in Z$ and $f \in C_p(Z)$. Then $e_z : C_p(Z) \to \mathbb{R}$ is a linear continuous functional (see TFS-196) and the map $e : Z \to C_p(C_p(Z))$ defined by $e(z) = e_z$ for each $z \in Z$, is a closed embedding (see TFS-167). This makes it possible to identify every $z \in Z$ with e_z and consider that Z is a closed subspace of $C_p(C_p(Z))$. Then $Z \subset L_p(Z)$ and Z is a closed Hamel basis in $L_p(Z)$ (see Fact 5 of S.489).

If $\mathbf{0} \in L_p(Z)$ is the zero functional then we let $l(\mathbf{0}) = 0$. If $u \in L_p(Z) \backslash \{\mathbf{0}\}$ then there exist $\lambda_1, \ldots, \lambda_n \in \mathbb{R} \backslash \{0\}$ and distinct points $z_1, \ldots, z_n \in Z$ such that $u = \lambda_1 z_1 + \ldots + \lambda_n z_n$; let $l(z) = n$. This representation of u will be called *canonical*; the canonical representation is unique up to a permutation in the sense that, for any distinct $y_1, \ldots, y_k \in Z$ and $\mu_1, \ldots, \mu_k \in \mathbb{R} \backslash \{0\}$ such that $u = \mu_1 y_1 + \ldots + \mu_k y_k$ we have $n = k$ and there exists a permutation $\{i_1, \ldots, i_n\}$ of the set $\{1, \ldots, n\}$ such that $y_{i_j} = z_j$ and $\mu_{i_j} = \lambda_j$ for any $j \in \{1, \ldots, n\}$. Let $L_p^n(Z) = \{u \in L_p(Z) : l(u) \le n\}$ for any $n \in \omega$.

Suppose that X is an l-dyadic space with $w(X) > \kappa$ and fix a dyadic space D such that X is l-embedded in D. There exists a continuous map $\varphi : D \to L_p(X)$ such that $\varphi(x) = x$ for any $x \in X$ (see Problem 467). The compact space $K = \varphi(D)$ is dyadic and $X \subset K$; so $w(K) \ge w(X) > \kappa$. This makes it possible to apply Fact 3 of U.086 to conclude that there is a set $E \subset K$ such that $E \simeq \mathbb{D}^{\kappa^+}$.

The set $E_n = E \cap L_p^n(X)$ is closed in E for any $n \in \omega$ (see Fact 1 of U.485) and $E = \bigcup_{n \in \omega} E_n$. By the Baire property of the space E, there is $n \in \omega$ such that $\text{Int}_E(E_n) \neq \emptyset$. Let $m = \min\{n \in \omega : \text{Int}_E(E_n) \neq \emptyset\}$. Then there is a set $U \in \tau^*(E)$ with $U \subset L_p^m(X)\backslash L_p^{m-1}(X)$. It is easy to see that $\mathbb{D}^{\kappa^+}$ embeds into any non-empty open subset of $\mathbb{D}^{\kappa^+}$, so E embeds in $L_p^m(X)\backslash L_p^{m-1}(X)$; thus we can consider that $E \subset L_p^m(X)\backslash L_p^{m-1}(X)$. Take a point $u \in E$ and apply Fact 4 of U.485 to see that there is $W \in \tau(u, L_p^m(X))$ which is homeomorphic to a subspace of $\mathbb{R}^m \times X^m$. The set $W' = W \cap E$ is non-empty and open in E; so E embeds in W' and hence in $\mathbb{R}^m \times X^m$.

To simplify the notation we will consider that E is a subspace of $\mathbb{R}^m \times X^m$; let $\pi : \mathbb{R}^m \times X^m \to \mathbb{R}^m$ and $q : \mathbb{R}^m \times X^m \to X^m$ be the respective natural projections. Since $E \subset \pi(E) \times q(E)$ and $w(\pi(E)) = \omega \leq \kappa$, we must have $w(q(E)) > \kappa$. Thus $L = q(E)$ is a dyadic compact space with $w(L) > \kappa$. Let $p_i : X^m \to X$ be the natural projection of X^m onto its i-th factor for any $i \leq m$. We have $L \subset p_1(L) \times \ldots \times p_m(L)$; so all spaces $p_i(L)$ cannot have weight $\leq \kappa$. Thus there is $i \leq m$ for which $w(p_i(L)) > \kappa$; the space $p_i(L)$ being dyadic, we can apply Fact 3 of U.086 again to see that $\mathbb{D}^{\kappa^+}$ embeds in $p_i(L)$. Of course, the same map embeds $\mathbb{D}^{\kappa^+}$ in X; so our solution is complete.

U.497. *Prove that, if βX is an l-dyadic space then X is pseudocompact.*

Solution. If the space X is not pseudocompact then there exists a discrete family $\{U_n : n \in \omega\} \subset \tau^*(X)$; choose a point $x_n \in U_n$ for any $n \in \omega$ and let $\{q_n : n \in \omega\}$ be an enumeration of $\mathbb{Q}$. The space $D = \{x_n : n \in \omega\}$ is discrete so the map $f : D \to \mathbb{R}$ defined by $f(x_n) = q_n$ for every $n \in \omega$, is continuous. It follows from Fact 5 of T.132 that D is C-embedded in X; so there is a continuous map $g : X \to \mathbb{R}$ such that $g|D = f$ and hence $g(X) \supset \mathbb{Q}$ is dense in $\mathbb{R}$. We can consider that $g : X \to \beta\mathbb{R}$ and therefore there exists a continuous map $h : \beta X \to \beta\mathbb{R}$ such that $h|X = g$; as a consequence, the set $h(\beta X) \supset g(X)$ is dense in $\beta\mathbb{R}$, which, together with compactness of $h(\beta X)$ implies that $h(\beta X) = \beta\mathbb{R}$. This proves that $\beta\mathbb{R}$ is a continuous image of βX and hence $\beta\mathbb{R}$ is l-dyadic by Problem 494.

Fact 1. If Z is a space and Y is a dense locally compact subspace of Z then Y is open in Z.

Proof. Given a point $y \in Y$ there is $U \in \tau(y, Y)$ such that the set $K = \text{cl}_Y(U)$ is compact and hence closed in Z. Take a set $V \in \tau(Z)$ such that $V \cap Y = U$; then U is dense in V and hence $V \subset \text{cl}_Z(V) = \text{cl}_Z(U) \subset K \subset Y$ which shows that $V \subset Y$ and hence $U = V$, i.e., U is open in Z. This proves that, for any $y \in Y$ there is $U \in \tau(y, Z)$ with $U \subset Y$; thus Y is open in Z and hence Fact 1 is proved.

Fact 2. The space $\beta\mathbb{R}$ is not metrizable and there are no non-trivial convergent sequences in $\beta\mathbb{R}\backslash\mathbb{R}$.

Proof. The space ω is closed in $\mathbb{R}$; by normality of $\mathbb{R}$ the closure of ω in $\beta\mathbb{R}$ is homeomorphic to $\beta\omega$ (see Fact 2 of S.451); so $\beta\omega$ is a non-metrizable subspace of $\beta\mathbb{R}$ (apply TFS-368 to see that $\beta\omega$ is not metrizable). Thus $\beta\mathbb{R}$ is not metrizable either.

To prove the second statement of our Fact suppose that we have a faithfully indexed sequence $S = \{a_n : n \in \omega\} \subset \beta\mathbb{R}\backslash\mathbb{R}$ which converges to some $a \in \beta\mathbb{R}\backslash S$. The space $\mathbb{R}$ being locally compact, it is open in $\beta\mathbb{R}$ by Fact 1; so $\beta\mathbb{R}\backslash\mathbb{R}$ is closed in $\beta\mathbb{R}$ and hence $a \in \beta\mathbb{R}\backslash\mathbb{R}$. It is easy to see that S is a discrete subspace of $\beta\mathbb{R}$; so there exists a disjoint family $\{O_n : n \in \omega\} \subset \tau(\beta\mathbb{R})$ such that $a_n \in O_n$ for any $n \in \omega$ (see Fact 1 of S.369).

The space $L_n = [-n, n] \subset \mathbb{R}$ is compact and $a_n \notin L_n$; so we can choose a set $W_n \in \tau(a_n, \beta\mathbb{R})$ such that $\overline{W}_n \subset O_n\backslash L_n$; then $G_n = W_n \cap \mathbb{R}$ is a non-empty open subset of $\mathbb{R}$ for any $n \in \omega$.

Fix a point $r \in \mathbb{R}$ and take $n \in \omega$ with $|r| < n$. We have $G_m \cap [-m, m] = \emptyset$ and hence $G_m \cap [-n, n] = \emptyset$ for any $m \geq n$. Therefore the set $W = (-n, n) \in \tau(r, \mathbb{R})$ meets at most n elements of the family $\mathcal{G} = \{G_i : i \in \omega\}$. Since the family $\mathcal{G}' = \{\text{cl}_{\mathbb{R}}(G_i) : i < n\}$ is disjoint, there is $W' \in \tau(r, \mathbb{R})$ such that W' meets at most one element of $\mathcal{G}'$. It is immediate that the set $W \cap W' \in \tau(r, \mathbb{R})$ meets at most one element of $\mathcal{G}$; so the family $\mathcal{G}$ is discrete in $\mathbb{R}$. As a consequence, the sets $A = \text{cl}_{\mathbb{R}}(\bigcup\{G_{2i} : i \in \omega\})$ and $B = \text{cl}_{\mathbb{R}}(\bigcup\{G_{2i+1} : i \in \omega\})$ are disjoint and closed in $\mathbb{R}$. By normality of $\mathbb{R}$ there is a continuous function $\varphi : \mathbb{R} \to [0, 1]$ such that $\varphi(A) = \{0\}$ and $\varphi(B) = \{1\}$.

There exists a continuous function $\Phi : \beta\mathbb{R} \to [0, 1]$ such that $\Phi|\mathbb{R} = \varphi$. The set G_n is dense in W_n; so $a_n \in \overline{G}_n$ for any $n \in \omega$. Besides, $\Phi(G_{2i}) = \{0\}$ because $G_{2i} \subset A$; this, together with continuity of Φ, implies that $\Phi(a_{2i}) = 0$ for any $i \in \omega$. Analogously, it follows from $G_{2i+1} \subset B$ that $\Phi(G_{2i+1}) = \{1\}$ and therefore $\Phi(a_{2i+1}) = 1$ for any $i \in \omega$. Thus the sequence $\{\Phi(a_n) : n \in \omega\}$ is not convergent which, together with $a_n \to a$, contradicts continuity of Φ. Thus, there are no non-trivial convergent sequences in $\beta\mathbb{R}\backslash\mathbb{R}$ so Fact 2 is proved.

Returning to our solution observe that l-dyadicity and non-metrizability of $\beta\mathbb{R}$ (see Fact 2) imply that $\mathbb{D}^{\omega_1}$ embeds in $\beta\mathbb{R}$ (see Problem 496); so take a set $E \subset \beta\mathbb{R}$ with $E \simeq \mathbb{D}^{\omega_1}$. The set $\mathbb{R}$ is dense and open in $\beta\mathbb{R}$; so $\mathbb{R} \cap E$ is a second countable open subset of E. It is easy to see that E embeds in any $U \in \tau^*(E)$; so there are no non-empty open second countable subsets of E; this proves that $\mathbb{R} \cap E = \emptyset$ and hence $E \subset \beta\mathbb{R}\backslash\mathbb{R}$. It is an easy exercise that $\mathbb{D}^{\omega_1}$ has non-trivial convergent sequences; therefore there are non-trivial convergent sequences in $\beta\mathbb{R}\backslash\mathbb{R}$; this contradiction with Fact 2 shows that $\beta\mathbb{R}$ is not l-dyadic. This final contradiction (recall that we proved that non-pseudocompactness of X implies that $\beta\mathbb{R}$ is l-dyadic) shows that X is pseudocompact and completes our solution.

U.498. *Prove that any hereditarily normal l-dyadic space is metrizable.*

Solution. Suppose that X is a hereditarily normal l-dyadic space. If X is not metrizable then $\mathbb{D}^{\omega_1}$ embeds in X (see Problem 496) and hence $\mathbb{D}^{\omega_1}$ is hereditarily normal. Since $\mathbb{D}^{\omega_1}$ is also dyadic, we can apply TFS-361 to conclude that $\mathbb{D}^{\omega_1}$ is metrizable. This contradiction shows that X is metrizable.

U.499. *Prove that any radial l-dyadic space is metrizable.*

Solution. Suppose that X is a radial l-dyadic space. If X is not metrizable then $\mathbb{D}^{\omega_1}$ embeds in X (see Problem 496) and hence $\mathbb{D}^{\omega_1}$ is radial. Since $\mathbb{D}^{\omega_1}$ is also dyadic, we can apply Problem 069 to conclude that $\mathbb{D}^{\omega_1}$ is metrizable. This contradiction shows that X is metrizable.

U.500. *Give an example of an l-dyadic space which is not extral.*

Solution. It was proved in Problem 492 that there exists an extral space X such that, for some continuous onto map $f : X \to Y$, the space Y is not extral. The space X is, evidently, l-dyadic and hence so is Y by Problem 495. Therefore Y is an example of an l-dyadic space which is not extral.

Chapter 3
Bonus Results: Some Hidden Statements

The reader has, evidently, noticed that an essential percentage of the problems of the main text is formed by purely topological statements some of which are quite famous and difficult theorems. A common saying among C_p-theorists is that any result on C_p-theory contains only 20% of C_p-theory; the rest is general topology.

It is evident that the author could not foresee all topology which would be needed for the development of C_p-theory; so a lot of material had to be dealt with in the form of auxiliary assertions. After accumulating more than seven hundred such assertions, the author decided that some deserve to be formulated together to give a "big picture" of the additional material that can be found in solutions of problems.

This section presents 100 topological statements which were proved in the solutions of problems without being formulated in the main text. In these formulations the main principle is to make them clear for an average topologist. A student could lack the knowledge of some concepts of the formulation; so the index of this book can be used to find the definitions of the necessary notions.

After every statement we indicate the exact place (in this book) where it was proved. We did not include any facts from C_p-theory because more general statements are proved sooner or later in the main text.

The author considers that most of the results that follow are very useful and have many applications in topology. Some of them are folkloric statements and quite a few are published theorems, sometimes famous ones. For example, Fact 2 of U.086 is a famous result of Efimov (1977), Fact 1 of U.071 is a result of Arhangel'skii (1978b). Fact 1 of U.190 is a theorem of Yakovlev (1980) and Fact 4 of U.083 is a result of Shapirovsky (1974).

To help the reader find a result he/she might need, we have classified the material of this section according to the following topics: *standard spaces, metrizable spaces, compact spaces and their generalizations, properties of continuous maps, covering properties, normality and open families, completeness and convergence properties, ordered, zero-dimensional and product spaces, and cardinal invariants and set*

© Springer International Publishing Switzerland 2015
V.V. Tkachuk, *A Cp-Theory Problem Book*, Problem Books
in Mathematics, DOI 10.1007/978-3-319-16092-4_3

theory. The author hopes that once we understand in which subsection a result should be, then it will be easy to find it.

3.1 Standard Spaces

By *standard spaces* we mean the real line, its subspaces and it powers, Tychonoff and Cantor cubes as well as ordinals together with the Alexandroff and Stone–Čech compactifications of discrete spaces.

U.074. Fact 1. *Given a regular uncountable cardinal κ suppose that $\lambda < \kappa$ and $C_\alpha \subset \kappa$ is a club for any $\alpha < \lambda$. Then $C = \bigcap\{C_\alpha : \alpha < \lambda\}$ is also a club.*

U.074. Fact 2. *Let κ be an uncountable regular cardinal. Then*

 (i) if $A \subset \kappa$ is stationary then $|A| = \kappa$;
 (ii) if $A \subset B \subset \kappa$ and A is stationary then B is also stationary;
 (iii) if $A \subset \kappa$ is stationary and $C \subset \kappa$ is a club then $A \cap C$ is stationary;
 (iv) given a cardinal $\lambda < \kappa$ suppose that $A_\alpha \subset \kappa$ for all $\alpha < \lambda$ and $\bigcup\{A_\alpha : \alpha < \lambda\}$
 is stationary. Then A_α is stationary for some $\alpha < \lambda$.

U.074. Fact 3. *Suppose that κ is a regular uncountable cardinal and A is a stationary subset of κ. Assume that $f : A \to \kappa$ and $f(\alpha) < \alpha$ for any $\alpha \in A$. Then there is $\beta < \kappa$ such that the set $\{\alpha \in A : f(\alpha) = \beta\}$ is stationary.*

U.074. Fact 4. *For any ordinal ξ the space $\xi + 1$ is compact and scattered and hence $C_p(\xi + 1)$ is a Fréchet–Urysohn space.*

U.074. Fact 5. *If ξ is any ordinal then any closed non-empty $F \subset \xi$ is a retract of ξ, i.e., there exists a continuous map $r : \xi \to F$ such that $r(\alpha) = \alpha$ for any $\alpha \in F$.*

U.074. Fact 6. *If ξ is an ordinal such that $cf(\xi) > \omega$ then, for any second countable space M and a continuous map $f : \xi \to M$ there is $z \in M$ and $\eta < \xi$ such that $f(\alpha) = z$ for any $\alpha \in [\eta, \xi)$.*

U.074. Fact 8. *Given an ordinal ξ assume that $\kappa = cf(\xi) \geq \omega$ and $\mu \in \xi$. Then there exists a map $f : \kappa \to [\mu, \xi)$ such that $\alpha < \beta < \kappa$ implies $f(\alpha) < f(\beta)$, the set $F = f[\kappa]$ is closed in ξ and $f : \kappa \to F$ is a homeomorphism. In particular, κ embeds in $[\mu, \xi)$ as a closed subspace.*

U.074. Fact 9. *For any ordinal α there exists a unique $n(\alpha) \in \omega$ and a unique limit ordinal $\mu(\alpha)$ such that $\alpha = \mu(\alpha) + n(\alpha)$.*

U.086. Fact 2. *If κ is a regular uncountable cardinal and $\varphi : \mathbb{D}^\kappa \to \mathbb{I}^\kappa$ is a continuous onto map then there is a closed $F \subset \mathbb{D}^\kappa$ such that $F \simeq \mathbb{D}^\kappa$ and $\varphi|F$ is injective.*

U.174. Fact 2. *Given an infinite set A and $n \in \omega$ let $\sigma_n(A) = \{x \in \mathbb{D}^A : |x^{-1}(1)| \leq n\}$. Then $\sigma_n(A)$ is a scattered compact space.*

U.176. Fact 1. *Suppose that $A \subset \omega_1$ is a stationary set such that $\omega_1 \backslash A$ is also stationary and let $T(A) = \{F \subset A : F$ is closed in $\omega_1\}$. Then all elements of $T(A)$ are countable and hence compact; given $F, G \in T(A)$ say that $F \leq G$ if F is an initial segment of G, i.e., for the ordinal $\alpha = \max(F)$, we have*

$G \cap (\alpha + 1) = F$. Then $(T(A), \leq)$ is a tree which has no uncountable chains and no dense σ-antichains.

U.176. Fact 2. *Given an infinite set T consider the set $[P, Q] = \{x \in \mathbb{D}^T : x(P) \subset \{1\}$ and $x(Q) \subset \{0\}\}$ for any disjoint finite sets $P, Q \subset T$. Suppose additionally that we have a family $\mathcal{U} = \{[P_a, Q_a] : a \in A\}$ such that the set A is infinite and $\sup\{|P_a \cup Q_a| : a \in A\} < \omega$. Then $\mathcal{U}$ is not disjoint.*

U.342. Fact 4. *Given an infinite cardinal κ let $S_n = \{x \in \mathbb{D}^\kappa : |x^{-1}(1)| \leq n\}$ for any $n \in \omega$. Then S_n is a continuous image of $(A(\kappa))^n$ for any $n \in \omega$.*

U.342. Fact 5. *If $\kappa \leq \mathfrak{c}$ is an infinite cardinal then $A(\kappa)$ is weakly metrizably fibered. As a consequence, if A is a set such that $|A| \leq \mathfrak{c}$ then any compact subspace of $\sigma_0(A)$ is weakly metrizably fibered.*

U.362. Fact 1. *Let $u \in K = (A(\omega_1))^\omega$ be the point with $u(n) = a$ for all $n \in \omega$. Then the space $K \backslash \{u\}$ is not metacompact.*

U.370. Fact 1. *If $A \subset \omega_1$ is a stationary set then A is not discrete as a subspace of ω_1.*

U.415. Fact 1. *For any space Z the set $\mathbb{Q}^Z$ is uniformly dense in $\mathbb{R}^Z$, i.e., for any $f \in \mathbb{R}^Z$ there is a sequence $\{f_n : n \in \mathbb{N}\} \subset \mathbb{Q}^Z$ which converges uniformly to f.*

U.426. Fact 1. *Suppose that λ is a cardinal with $\lambda^\omega = \lambda$ and A is a set such that $|A| = \lambda$. Then the space $\mathbb{I}^A$ has a strongly σ-discrete dense subspace of cardinality λ.*

U.463. Fact 1. *There exists a surjective map $\xi : \mathbb{D}^\omega \rightarrow \omega^\omega$ such that $\xi^{-1}(U)$ is an F_σ-set in $\mathbb{D}^\omega$ for any open $U \subset \omega^\omega$.*

U.471. Fact 1. *The topology of the double arrow space is zero-dimensional and generated by its lexicographical order. In particular, the double arrow space is a linearly ordered zero-dimensional perfectly normal, hereditarily separable nonmetrizable compact space.*

U.481. Fact 3. *Suppose that A is a set and $p \in \mathbb{I}^A$. Assume that, for some $n \in \mathbb{N}$, we have subspaces $K_0, \ldots, K_{n-1} \subset \mathbb{I}^A$ such that every K_i is homeomorphic to $A(\omega_1)$ and p is the unique non-isolated point of K_i; let $D_i = F_i \backslash \{p\}$. Then, for any family $\{U_0, \ldots, U_{n-1}\} \subset \tau(\mathbb{I}^A)$ such that $D_i \subset U_i$ for all $i < n$, we have $U_0 \cap \ldots \cap U_{n-1} \neq \emptyset$.*

U.490. Fact 1. *The space $A(\omega_1)$ does not embed in a linearly ordered space.*

U.493. Fact 1. *For any set A both spaces $\mathbb{R}^A$ and $\mathbb{I}^A$ are connected.*

U.497. Fact 2. *The space $\beta\mathbb{R}$ is not metrizable and there are no non-trivial convergent sequences in $\beta\mathbb{R} \backslash \mathbb{R}$.*

3.2 Metrizable Spaces

The results of this Section deal with metrics, pseudometrics or metrizable spaces in some way. We almost always assume the Tychonoff separation axiom; so our second countable spaces are metrizable and hence present here too.

U.050. Fact 1. *Given a metric space (M, ρ) a family $\mathcal{C} \subset \tau(M)$ is a base in M if and only if, for any $\varepsilon > 0$ there is a collection $\mathcal{C}' \subset \mathcal{C}$ such that $\bigcup \mathcal{C}' = M$ and $mesh(\mathcal{C}') \leq \varepsilon$.*

U.050. Fact 2. *Suppose that Z is an arbitrary space. Then*

(1) for any pseudometrics d_1 and d_2 on the space Z, the function $d = d_1 + d_2$ is a pseudometric on the space Z;

(2) if d is a pseudometric on the space Z then $a \cdot d$ is a pseudometric on Z for any $a > 0$;

(3) for any pseudometrics d_1 and d_2 on the space Z, the function $d = \max\{d_1, d_2\}$ is a pseudometric on the space Z;

(4) if d_1 is a pseudometric on the space Z and $a > 0$ then the function $d : Z \times Z \to \mathbb{R}$ defined by $d(x, y) = \min\{d_1(x, y), a\}$ for all $x, y \in Z$ is a pseudometric on Z;

(5) if, for any $i \in \omega$, a function d_i is a pseudometric on the space Z and $d_i(x, y) \leq 1$ for any $x, y \in Z$ then $d = \sum_{i \in \omega} 2^{-i} \cdot d_i$ is a pseudometric on Z;

(6) if $f : Z \to \mathbb{R}$ is a continuous function then the function $d : Z \times Z \to \mathbb{R}$ defined by $d(x, y) = |f(x) - f(y)|$ for any $x, y \in Z$ is a pseudometric on the space Z;

(7) if d_1 is a metric and d_2 is a pseudometric on the space Z then $d = d_1 + d_2$ is a metric on the space Z.

U.062. Fact 1. *Let A be a non-empty closed subspace of a metrizable space M. Then there exists a continuous linear map $e : C_p(A) \to C_p(M)$ such that $e(f)|A = f$ for any $f \in C_p(A)$.*

U.062. Fact 2. *Suppose that M is a metrizable space and $A \subset M$ is a non-empty closed subset of M; let $I_A = \{f \in C_p(M) : f|A \equiv 0\}$. Then there exists a linear homeomorphism between $C_p(M)$ and $C_p(A) \times I_A$ and, in particular, $C_p(A)$ embeds in $C_p(M)$ as a closed linear subspace.*

U.094. Fact 1. *For any second countable space Z there is a countable space T such that Z embeds in $C_p(T)$ as a closed subspace.*

U.138. Fact 1. *If a space has a dense metrizable subspace then it has a σ-disjoint π-base. For first countable spaces the converse is also true, i.e., a first countable space Z has a dense metrizable subspace if and only if it has a σ-disjoint π-base.*

U.318. Fact 1. *A space Z can be condensed onto a metrizable space if and only if it has a G_δ-diagonal sequence $\{\mathcal{U}_n : n \in \omega\}$ such that $\mathcal{U}_{n+1}$ is a star refinement of $\mathcal{U}_n$ for any $n \in \omega$.*

U.347. Fact 1. *Suppose that Z is a space, (M, ρ) is a metric space and we have a map $f : Z \to M$. Then*

(a) for any $\varepsilon > 0$ the set $O_\varepsilon = \{z \in Z : osc(f, z) < \varepsilon\}$ is open in Z;
(b) the map f is continuous at a point $z \in Z$ if and only if $osc(f, z) = 0$.

3.3 Compact Spaces and Their Generalizations

This Section contains some statements on compact, countably compact and pseudo-compact spaces.

U.039. Fact 1. *Any perfect preimage of a countably compact space is countably compact.*

U.071. Fact 1. *If $MA+\neg CH$ holds then any compact space X of weight at most ω_1 is pseudoradial. In particular, $\mathbb{D}^{\omega_1}$ is pseudoradial under $MA+\neg CH$.*

U.072. Fact 1. *Let λ be an infinite cardinal. If X is a dyadic space such that the set $C = \{x \in X : \pi\chi(x, X) \leq \lambda\}$ is dense in X then $w(X) \leq \lambda$. In particular, if X has a dense set of points of countable π-character then X is metrizable.*

U.072. Fact 2. *Under CH every pseudoradial dyadic space is metrizable.*

U.077. Fact 1. *If K is an infinite compact space then $|\mathcal{C}(K)| \leq w(K)$. In particular, for any metrizable compact K the family of all clopen subsets of K is countable.*

U.080. Fact 1. *If K is a compact ω-monolithic space of countable tightness then K is Fréchet–Urysohn and has a dense set of points of countable character. This is true in ZFC, i.e., no additional axioms are needed for the proof of this Fact.*

U.086. Fact 1. *Given a regular uncountable cardinal κ if K is a compact space such that $\pi\chi(x, K) \geq \kappa$ for any $x \in K$ then there is a closed $P \subset K$ which maps continuously onto $\mathbb{D}^\kappa$ and hence K maps continuously onto $\mathbb{I}^\kappa$.*

U.086. Fact 3. *If K is a dyadic space and $w(K) > \kappa$ for some infinite cardinal κ then $\mathbb{D}^{\kappa^+}$ embeds in K.*

U.104. Fact 1. *Suppose that K is a non-empty compact space with no points of countable character. Then K cannot be represented as a union of $\leq \omega_1$-many cosmic subspaces.*

U.104. Fact 1. *Suppose that K is a non-empty compact space with no points of countable character. Then K cannot be represented as a union of $\leq \omega_1$-many cosmic subspaces.*

U.174. Fact 1. *Let X be a countably compact σ-discrete space, i.e., $X = \bigcup_{n\in\omega} X_n$ where each X_n is a discrete subspace of X. Then X is scattered.*

U.185. Fact 1. *Suppose that Z is a compact space, F is non-empty and closed in Z and, additionally, there is a point-countable open cover $\mathcal{U}$ of the set $Z \setminus F$ such that $\overline{U} \subset Z \setminus F$ for any $U \in \mathcal{U}$. Then F is a W-set in Z.*

U.222. Fact 1. *Suppose that K is a compact space and some $x \in K$ is not a π-point. Then $K \setminus \{x\}$ is pseudocompact and K is canonically homeomorphic to $\beta(K \setminus \{x\})$, i.e., there exists a homeomorphism $\varphi : \beta(K \setminus \{x\}) \to K$ such that $\varphi(y) = y$ for any $y \in K \setminus \{x\}$.*

U.337. Fact 1. *Let K be an infinite compact space with $|K| = \kappa$. Suppose that U is an open subset of K and F is a closed subset of the subspace U. Then there exist families $\{S(\alpha, n) : \alpha < \kappa, n \in \omega\}$ and $\{V(\alpha, n) : \alpha < \kappa, n \in \omega\}$ with the following properties:*

(i) the set $S(\alpha, n)$ is compact and $S(\alpha, n) \subset F$ for any $\alpha < \kappa$ and $n \in \omega$;
(ii) $V(\alpha, n)$ is open in K and $S(\alpha, n) \subset V(\alpha, n) \subset U$ for any $\alpha < \kappa$ and $n \in \omega$;
(iii) $\bigcup\{S(\alpha, n) : \alpha < \kappa, n \in \omega\} = F$;
(iv) if $\alpha < \beta < \kappa$ then $S(\beta, n) \cap V(\alpha, m) = \emptyset$ for any $m, n \in \omega$.

U.342. Fact 1. *Suppose that K is compact and $K = \bigcup\{K_n : n \in \omega\}$ where K_n is weakly metrizably fibered and closed in K for any $n \in \omega$. Then K is also weakly metrizably fibered.*

U.342. Fact 2. *If K_n is a compact weakly metrizably fibered space for any $n \in \omega$ then $K = \prod_{n \in \omega} K_n$ is also weakly metrizably fibered.*

U.342. Fact 3. *Any closed subspace and any continuous image of a compact weakly metrizably fibered space is weakly metrizably fibered.*

U.417. Fact 1. *If Z is a pseudocompact space and $\psi(z, Z) \leq \omega$ for some $z \in Z$ then $\chi(z, Z) \leq \omega$.*

U.417. Fact 2. *If Z is a pseudocompact first countable space then, for any $Y \subset Z$ with $|Y| \leq \mathfrak{c}$ there is $Y' \subset Z$ such that $|Y'| \leq \mathfrak{c}$, the space Y' is pseudocompact and $Y \subset Y'$.*

U.481. Fact 1. *Given compact spaces X and Y suppose that $\varphi : C_p(X) \to C_p(Y)$ is a continuous linear map. Then the map $\varphi : C_u(X) \to C_u(Y)$ is continuous as well.*

U.497. Fact 1. *If Z is a space and Y is a dense locally compact subspace of Z then Y is open in Z.*

3.4 Properties of Continuous Maps

We consider the most common classes of continuous maps: open, closed, perfect and quotient. The respective results basically deal with preservation of topological properties by direct and inverse images.

U.074. Fact 11. *Given spaces Y and Z assume that $f : Y \to Z$ is a continuous map such that there is $P \subset Y$ for which $f(P) = Z$ and $f|P : P \to Z$ is a quotient map. Then f is quotient. In particular, any retraction is a quotient map.*

U.077. Fact 2. *Given spaces Z, T and a continuous map $f : Z \to T$, for any $B \subset T$, the set $G(f, B) = \{(z, f(z)) : z \in f^{-1}(B)\} \subset Z \times B$ is closed in $Z \times B$.*

U.093. Fact 2. *Suppose that $\varphi_t : E_t \to M_t$ is a compact-valued upper semicontinuous onto map for any $t \in T$. Let $E = \prod_{t \in T} E_t$, $M = \prod_{t \in T} M_t$ and define a multi-valued map $\varphi = \prod_{t \in T} \varphi_t : E \to M$ by $\varphi(x) = \prod_{t \in T} \varphi_t(x(t))$ for any $x \in E$. Then $\varphi : E \to M$ is a compact-valued upper semicontinuous onto map.*

U.481. Fact 4. *Given a space X suppose that $F \subset X$ is C^*-embedded in X and we have a continuous map $f : F \to \mathbb{I}^A$ for some A. Then there exists a continuous map $g : X \to \mathbb{I}^A$ such that $g|F = f$. In particular, if the space X is normal and F is closed in X then any continuous map from F to a Tychonoff cube can be continuously extended to the whole space X.*

U.481. Fact 6. *Let $\mathcal{U} = \{U_n : n \in \omega\}$ be disjoint family of non-empty open subsets of a space Z. Assume that $f_n \in C^*(Z)$ and $f_n(Z \backslash U_n) \subset \{0\}$ for all $n \in \omega$. If, additionally, the sequence $\{\|f_n\| : n \in \omega\}$ converges to zero then the function $f = \sum_{n \in \omega} f_n$ is continuous on Z.*

U.493. Fact 2. *Any continuous image of a connected space is connected. As a consequence, if X is connected, Y is zero-dimensional and $f : X \to Y$ is a continuous map then $f(X)$ is a singleton.*

3.5   Covering Properties, Normality and Open Families

This section contains results on the covering properties which are traditionally considered not to be related to compactness, such as the Lindelöf property, paracompactness and its derivatives.

U.082. Fact 1. *If Z is a space and $l(Z) \leq \kappa$ for some infinite cardinal κ then any indexed set $Y = \{y_\alpha : \alpha < \kappa^+\} \subset Z$ has a complete accumulation point, i.e., there is $z \in Z$ such that $|\{\alpha < \kappa^+ : y_\alpha \in U\}| = \kappa^+$ for any $U \in \tau(z, Z)$.*

U.093. Fact 3. *If Z^ω is Lindelöf then $Z^\omega \times T$ is also Lindelöf for any K-analytic space T.*

U.095. Fact 1. *Assume that we have an uncountable space Z such that $w(Z) \leq \mathfrak{c}$ and there is a countable $Q \subset Z$ such that Z is concentrated around Q, i.e., $|Z \backslash U| \leq \omega$ for any $U \in \tau(Q, Z)$. Then the Continuum Hypothesis (CH) implies that there is an uncountable $T \subset Z$ such that $Q \subset T$ and T^n is Lindelöf for any $n \in \mathbb{N}$.*

U.102. Fact 1. *Given a space Z, any σ-locally finite open cover of Z has a locally finite refinement.*

U.102. Fact 2. *Suppose that Z is a space and $\mathcal{F}$ is a discrete family of closed subsets of Z. If there exists a locally finite closed cover $\mathcal{C}$ of the space Z such that every $C \in \mathcal{C}$ meets at most one element of $\mathcal{F}$ then the family $\mathcal{F}$ is open-separated, i.e., for any $F \in \mathcal{F}$ we can choose $O_F \in \tau(F, Z)$ such that the family $\{O_F : F \in \mathcal{F}\}$ is disjoint.*

U.175. Fact 1. *If Z is an uncountable space which is a continuous image of a Lindelöf k-space then there is an infinite compact $K \subset Z$.*

U.177. Fact 1. *In a Lindelöf space Z every uncountable $A \subset Z$ has a condensation point, i.e., there is $z_0 \in Z$ for which $|A \cap U| > \omega$ for any $U \in \tau(z_0, Z)$. In addition, if Z is a space with $l(Z) \leq \omega_1$ then Z is Lindelöf if and only if every uncountable $A \subset Z$ has a condensation point.*

U.177. Fact 2. *If Z is a space with strong condensation property and $l(Z^\omega) \leq \omega_1$ then Z^ω is Lindelöf.*

U.188. Fact 1. *Given a space Z suppose that $\mathcal{U}$ is an open cover of Z such that $\overline{U}$ is Lindelöf for any $U \in \mathcal{U}$. Then $\mathcal{U}$ can be shrunk, i.e., for any $U \in \mathcal{U}$ there is a closed set $F_U \subset U$ such that $\{F_U : U \in \mathcal{U}\}$ is a cover of Z.*

U.193. Fact 3. *A space Z is hereditarily normal if and only if any pair of separated subsets of Z are open-separated.*

U.271. Fact 1. *Given a space Z and a family $\mathcal{U} \subset \tau^*(Z)$ there is a discrete $D \subset Z$ such that $\bigcup \{U \in \mathcal{U} : D \cap U \neq \emptyset\} = \bigcup \mathcal{U}$.*

U.271. Fact 2. *Suppose that λ is an infinite cardinal, Z is a space and $\mathcal{B} \subset \tau^*(Z)$ is a family with $\mathrm{ord}(z, \mathcal{B}) < \lambda$ for any $z \in Z$. Then there exists a family $\{D_\alpha : \alpha < \lambda\}$ of discrete subspaces of Z such that, for the set $D = \bigcup\{D_\alpha : \alpha < \lambda\}$, we have $D \cap B \neq \emptyset$ for any $B \in \mathcal{B}$. In particular, if $\mathcal{B}$ is point-finite then there is a σ-discrete subset of Z which is dense in $\mathcal{B}$.*

U.284. Fact 1. *Suppose that Z is a Lindelöf Σ-space, $Y \subset Z$ and there is a countable family $\mathcal{A}$ of Lindelöf Σ-subspaces of Z that separates Y from $Z \backslash Y$. Then Y is a Lindelöf Σ-space.*

U.285. Fact 5. *Suppose that X is a Lindelöf Σ-space and $\mathcal{F}$ is a fixed countable network with respect to a compact cover $\mathcal{C}$ of the space X. Assume additionally that $\mathcal{F}$ is closed under finite intersections and $f : X \to Y$ is a continuous onto map. If $A \subset X$ is a set such that $f(A \cap F)$ is dense in $f(F)$ for any $F \in \mathcal{F}$ then $f(\overline{A}) = Y$.*

U.363. Fact 1. *Suppose that a space X has an open σ-point-finite cover $\mathcal{U}$ such that $\overline{U}$ is compact for any $U \in \mathcal{U}$. Then X is σ-metacompact.*

3.6   Completeness and Convergence Properties

This Section deals mainly with Čech-complete spaces. Some results on convergence properties are presented as well.

U.061. Fact 1. *If Z is a sequential space, $A \subset Z$ and $z \in \overline{A} \backslash A$ then z has a countable π-network in Z which consists of infinite subsets of A.*

U.074. Fact 10. *For any space Z, if $C_p(Z)$ is Fréchet–Urysohn and $Y \neq \emptyset$ is an F_σ-subset of Z then $C_p(Y)$ is also Fréchet–Urysohn.*

U.421. Fact 3. *If X is Čech-complete and $\psi(x, X) \leq \omega$ for some $x \in X$ then $\chi(x, X) \leq \omega$. In particular, any splittable Čech-complete space is first countable.*

U.421. Fact 7. *If X is a Čech-complete paracompact space with a G_δ-diagonal then X is metrizable.*

U.429. Fact 1. *For any Čech-complete space X there exists a dense paracompact subspace $Y \subset X$ such that Y is a G_δ-set in X (and hence Y is Čech-complete).*

3.7 Ordered, Zero-Dimensional and Product Spaces

The space $C_p(X)$ being dense in $\mathbb{R}^X$, the results on topological products form a fundamental part of C_p-theory. The main line here is to classify spaces which could be embedded in (or expressed as a continuous image of) a nice subspace of a product.

U.003. Fact 1. *If dim $Z_t = 0$ for any $t \in T$ and $Z = \bigoplus\{Z_t : t \in T\}$ then dim $Z = 0$.*

U.067. Fact 1. *For an infinite cardinal κ let $\preceq$ be the lexicographic order on the set $D_\kappa = \kappa \times \mathbb{Z}$, i.e., for any $a, b \in D_\kappa$ such that $a = (\alpha, n)$, $b = (\beta, m)$ let $a \prec b$ if $\alpha < \beta$; if $\beta < \alpha$ then we let $b \prec a$. Now if $\alpha = \beta$ then $a \preceq b$ if $n \leq m$ and $b \preceq a$ if $m \leq n$. Then $\preceq$ is a linear order on D_κ and the space $(D_\kappa, \tau(\preceq))$ is discrete. Besides, if $\kappa > \omega$ then $|\{a \in D_\kappa : a \preceq b\}| < \kappa$ for any $b \in D_\kappa$. In particular, any discrete space X is linearly orderable.*

U.067. Fact 2. *Suppose that, for every $t \in T$, the topology of a space X_t can be generated by a linear order $\preceq_t$ which has a maximal and a minimal element. Then the space $X = \bigoplus\{X_t : t \in T\}$ is linearly orderable.*

U.067. Fact 3. *Suppose that $X = \{x\} \cup \{x_\alpha : \alpha < \kappa\}$ where κ is an infinite regular cardinal, the enumeration of X is faithful and x is the unique non-isolated point of X. For every $\alpha < \kappa$ let $O_\alpha = \{x\} \cup \{x_\beta : \beta \geq \alpha\}$. If the family $\{O_\alpha : \alpha < \kappa\}$ is a local base at x in X then there is a linear order $\preceq$ on X such that $\tau(\preceq) = \tau(X)$, the point x is the maximal element of $(X, \preceq)$ and x_0 is its minimal element.*

U.104. Fact 2. *Suppose that N_t is a cosmic space for each $t \in T$ and take any point $u \in N = \prod\{N_t : t \in T\}$. If $|T| \leq \omega_1$ then $\Sigma(N, u)$ is a union of $\leq \omega_1$-many cosmic spaces.*

U.190. Fact 1. *Any subspace of a σ-product of second countable spaces is metacompact.*

U.359. Fact 4. *Suppose that a zero-dimensional first countable X is strongly homogeneous, i.e., any non-empty clopen subset of X is homeomorphic to X. Then X is homogeneous.*

U.366. Fact 4. *Any σ-product of compact metrizable spaces must have a closure-preserving cover by metrizable compact subspaces.*

3.8   Cardinal Invariants and Set Theory

To classify function spaces using cardinal invariants often gives crucial information. This Section includes both basic, simple results on the topic as well as very difficult classical theorems.

U.003. Fact 2. *Given an infinite cardinal κ a space Z is zero-dimensional and $w(Z) \leq \kappa$ if and only if Z is homeomorphic to a subspace of $\mathbb{D}^{\kappa}$.*

U.027. Fact 1. *For any space T and a closed $F \subset T$ we have $\psi(F, T) \leq l(T \setminus F)$. In particular, $\psi(t, T) \leq l(T \setminus \{t\})$ for any $t \in T$.*

U.074. Fact 7. *For any space Z we have $|C_p(Z)| \leq w(Z)^{l(Z)}$.*

U.083. Fact 1. *Given a space Z and an infinite cardinal κ suppose that $\psi(Z) \leq 2^{\kappa}$ and $l(Z) \cdot t(Z) \leq \kappa$. Assume additionally that $|\overline{A}| \leq 2^{\kappa}$ for any $A \subset Z$ with $|A| \leq \kappa$. Then $|Z| \leq 2^{\kappa}$.*

U.083. Fact 2. *Given a space Z let $\mathcal{R}(Z)$ be the family of all regular open subsets of Z, i.e., $\mathcal{R}(Z) = \{U \in \tau(Z) : U = Int(\overline{U})\}$. Then $|\mathcal{R}(Z)| \leq \pi w(Z)^{c(Z)}$.*

U.083. Fact 3. *For any space Z we have $\pi w(Z) \leq \pi \chi(Z) \cdot d(Z)$.*

U.083. Fact 4. *For any space Z we have $w(Z) \leq \pi \chi(Z)^{c(Z)}$.*

U.127. Fact 1. *If a space Z is κ-monolithic and $s(Z) \leq \kappa$ for some infinite cardinal κ then $hl(Z) \leq \kappa$.*

U.274. Fact 1. *If Z is a space with a unique non-isolated point then $Z \oplus \{t\} \simeq Z$ for any $t \notin Z$.*

U.337. Fact 2. *If n is a natural number and T is an infinite set then for any family $\mathcal{N} = \{N_t : t \in T\}$ such that $|N_t| \leq n$ for any $t \in T$, there is a set D and an infinite $T' \subset T$ such that $N_s \cap N_t = D$ for any distinct $s, t \in T'$.*

U.381. Fact 1. *For any $n, l \in \omega$ there exists a number $g(n, l) \in \omega$ such that, for any family $\mathcal{N} = \{N_t : t \in T\}$ with $|N_t| \leq n$ for any $t \in T$, if $|T| \geq g(n, l)$ then there is $S \subset T$ such that $|S| \geq l$ and $\{N_t : t \in S\}$ is a Δ-system, i.e., there is a set D such that $N_s \cap N_t = D$ for any distinct $s, t \in S$.*

U.418. Fact 1. *If Z is a space then $|Z| \leq nw(Z)^{\psi(Z)}$.*

Chapter 4
Open Problems

The unsolved problems form an incentive for the development of any area of mathematics. Since this book has an ambitious purpose to embrace all or almost all modern C_p-theory, it was impossible to avoid dealing with open questions.

In this book, we have a wide selection of unsolved problems of C_p-theory. Of course, "unsolved" means "unsolved to the best of the knowledge of the author". I give a classification by topics but there is no mention whatsoever of whether the given problem is difficult or not. One good parameter is the year of publication but sometimes the problem is not solved for many years because of lack of interest or effort and not because it is too difficult.

I believe that almost all unsolved problems of importance in C_p-theory are present in the volumes of this book taken together. The reader understands, of course, that there is a big difference between the textbook material of the first three chapters and open questions to which an author must be assigned. I decided that it was my obligation to make this assignment and did my best to be frowned at (or hated!) by the least possible number of potential authors of open problems.

This volume contains 100 unsolved problems which are classified by topics presented in 8 sections the names of which outline what the given group of problems is about. At the beginning of each subsection we define the notions *which are not defined in the main text*. Each published problem has a reference to the respective paper or book. If it is unpublished, then my opinion on who is the author is expressed. The last part of each problem is a very brief explanation of its motivation and/or comments referring to the problems of the main text or some papers for additional information. If the paper is published and the background material is presented in the main text, we mention the respective exercises. If the main text contains no background we refer the reader to the original paper. If no paper is mentioned in the motivation part, then the reader must consult the paper/book in which the unsolved problem was published.

© Springer International Publishing Switzerland 2015
V.V. Tkachuk, *A Cp-Theory Problem Book*, Problem Books
in Mathematics, DOI 10.1007/978-3-319-16092-4_4

To do my best to assign the right author to every problem I implemented the following simple principles:

1. If the unsolved problem is published, then I cite the publication and consider myself not to be involved in the decision about who is the author. Some problems are published many times and I have generally preferred to cite the articles in journals/books which are more available for the Western reader. Thus it may happen that I do not cite the earliest paper where the problem was formulated. Of course, I mention it explicitly, if the author of the publication attributes the problem to someone else.
2. If, to the best of my knowledge, the problem is unpublished then I mention the author according to my personal records. The information I have is based upon my personal acquaintance and communication with practically all specialists in C_p-theory. I am aware that it is a weak point and it might happen that the problem I attributed to someone, was published (or invented) by another person. However, I did an extensive work ploughing through the literature to make sure that this does not happen.

4.1 Sokolov Spaces and Corson Compact Spaces

A space X is *splittable over a class* $\mathcal{P}$ if for any set $A \subset X$ there exists a continuous map $f : X \to Y$ such that $Y \in \mathcal{P}$ and $f^{-1}f(A) = A$. This notion generalizes the class of spaces condensable into a space from $\mathcal{P}$. Observe that a space X is splittable in the sense of Section 1.5 if and only if it is splittable over $\mathbb{R}^{\omega}$ (see Problem 416). If X is a space, then the *ω-modification of X* is the set X with the topology generated by all G_{δ}-subsets of the space X.

4.1.1 Suppose that X is a Sokolov compact space and $p(C_p(X)) = \omega$. Must X be metrizable?

> **Published in** Tkachuk (2005c)
> **Motivated by** the fact that this is true if X is Corson compact (Problem 287)

4.1.2 Suppose that X is a Sokolov compact space. Does $C_p(X)$ condense into a Σ-product of real lines?

> **Published in** Tkachuk (2005c)
> **Motivated by** the fact that this is true if X is Corson compact (Problem 286)

4.1.3 Suppose that K is a scattered compact space and let X be the ω-modification of K. Must the space X be Sokolov?

> **Author** V.V. Tkachuk

4.1.4 Suppose that K is a scattered compact space of finite dispersion index and let X be the ω-modification of K. Must the space X be Sokolov?

> **Author** V.V. Tkachuk

4.1.5 Suppose that X is a Corson compact space such that $C_p(X)$ is hereditarily σ-metacompact. Is it true that X is metrizable?

> **Author** V.V. Tkachuk
> **Motivated by** the fact that if $C_p(X)$ is hereditarily Lindelöf then X is metrizable

4.1.6 Let $\mathcal{P}$ be a class of all Corson compacta. Suppose that X is a space such that each compact subspace of $C_p(X)$ belongs to $\mathcal{P}$. Is it true that, for any compact K, all compact subspaces of $C_p(X \times K)$ belong to $\mathcal{P}$?

> **Published in** Arhangel'skii (1998b)
> **Related to** some questions on Grothendieck spaces

4.1.7 Recall that $L(\kappa)$ is the one-point Lindelöfication of the discrete space of cardinality κ. Suppose that X is compact and $C_p(X)$ is a continuous image of a closed subspace of $L(\kappa)^{\omega} \times K$ for some uncountable cardinal κ and a compact space K. Is it true that X is Corson compact?

Published in Arhangel'skii (1989a)

Motivated by the fact that it is true if K is the one-point space.

4.1.8 Suppose that a compact space X is splittable over the class of Corson compact spaces. Is it true in ZFC that X is Corson compact?

Authors D. Jardón, V.V. Tkachuk

4.1.9 Suppose that a compact space X is splittable over the class of Corson compact spaces. Is it true in ZFC that $w(X) = d(X)$?

Authors D. Jardón, V.V. Tkachuk

4.1.10 Suppose that a compact space X is splittable over the class of Corson compact spaces. Is it true in ZFC that any continuous image of X is also splittable over the class of Corson compacta?

Authors D. Jardón, V.V. Tkachuk

4.1.11 Suppose that a compact space X is splittable over the class of Corson compact spaces. Is it true in ZFC that $X \times X$ is also splittable over the class of Corson compacta?

Authors D. Jardón, V.V. Tkachuk

4.1.12 Suppose that X is a compact space such that $\overline{A}$ is Corson compact for any $A \subset X$ with $|A| \leq \mathfrak{c}$. Must X be Corson compact?

Authors D. Jardón, V.V. Tkachuk

4.2 Gul'ko Compact Spaces

A space X is *splittable over a class* $\mathcal{P}$ if for any set $A \subset X$ there exists a continuous map $f : X \to Y$ such that $Y \in \mathcal{P}$ and $f^{-1}f(A) = A$. This notion generalizes the class of spaces condensable into a space from $\mathcal{P}$. Observe that a space X is splittable in the sense of Section 1.5 if and only if it is splittable over $\mathbb{R}^\omega$ (see Problem 416).

4.2.1 Suppose that a compact space X is splittable over the class of Gul'ko compact spaces. Is it true in ZFC that X is Gul'ko compact?

Authors D. Jardón, V.V. Tkachuk

4.2.2 Suppose that a compact space X is splittable over the class of Gul'ko compact spaces. Is it true in ZFC that any continuous image of X is also splittable over the class of Gul'ko compacta?

Authors D. Jardón, V.V. Tkachuk

4.2.3 Suppose that a compact space X is splittable over the class of Gul'ko compact spaces. Is it true in ZFC that $X \times X$ is also splittable over the class of Gul'ko compacta?

Authors D. Jardón, V.V. Tkachuk

4.2.4 Suppose that a compact space X is splittable over the class of Gul'ko compact spaces. Is it true in ZFC that X has a dense metrizable subspace?

Authors D. Jardón, V.V. Tkachuk

4.2.5 Suppose that X is a compact space such that $\overline{A}$ is Gul'ko compact for any $A \subset X$ with $|A| \leq \mathfrak{c}$. Must X be Gul'ko compact?

Authors D. Jardón, V.V. Tkachuk

4.2.6 Suppose that X is a compact space such that $\overline{A}$ is a Lindelöf Σ-space for any $A \subset C_p(X)$ with $|A| \leq \mathfrak{c}$. Must X be Gul'ko compact?

Authors D. Jardón, V.V. Tkachuk

4.3 Eberlein Compact Spaces

A space X is *splittable over a class* $\mathcal{P}$ if for any set $A \subset X$ there exists a continuous map $f : X \to Y$ such that $Y \in \mathcal{P}$ and $f^{-1}f(A) = A$. This notion generalizes the class of spaces condensable into a space from $\mathcal{P}$. Observe that a space X is splittable in the sense of Section 1.5 if and only if it is splittable over $\mathbb{R}^{\omega}$ (see Problem 416).

4.3.1 Suppose that X has a dense σ-bounded subspace. Is it true that every countably compact subspace of $C_p(X)$ is an Eberlein compact?

Published in Arhangel'skii (1989a)
Comment this is true if X has a dense σ-pseudocompact subspace.

4.3.2 Given a cardinal κ, does there exist an Eberlein compact X such that every Eberlein compact of weight $\leq \kappa$ is a continuous image of X?

Published in Benyamini et al. (1977)

4.3.3 Given a cardinal κ, does there exist a uniform Eberlein compact X such that every uniform Eberlein compact of weight $\leq \kappa$ is a continuous image of X?

Published in Benyamini et al. (1977)

4.3.4 Suppose that a compact space X is splittable over the class of Eberlein compact spaces. Is it true in ZFC that X is Eberlein compact?

Authors D. Jardón, V.V. Tkachuk

4.3.5 Suppose that a compact space X is splittable over the class of Eberlein compact spaces. Is it true in ZFC that X has a dense metrizable subspace?

Authors D. Jardón, V.V. Tkachuk

4.3.6 Suppose that a compact space X is splittable over the class of Eberlein compact spaces. Is it true in ZFC that any continuous image of X is also splittable over the class of Eberlein compacta?

Authors D. Jardón, V.V. Tkachuk

4.3.7 Suppose that a compact space X is splittable over the class of Eberlein compact spaces. Is it true in ZFC that $X \times X$ is also splittable over the class of Eberlein compacta?

Authors D. Jardón, V.V. Tkachuk

4.3.8 Suppose that X is a compact space such that $\overline{A}$ is Eberlein compact for any $A \subset X$ with $|A| \leq \mathfrak{c}$. Must X be Eberlein compact?

Authors D. Jardón, V.V. Tkachuk

4.3.9 Suppose that a pseudocompact space X is splittable over the class of Eberlein compact spaces. Must X be Eberlein compact?

Authors D. Jardón, V.V. Tkachuk

4.3.10 Suppose that a countably compact space X is splittable over the class of Eberlein compact spaces. Must X be Eberlein compact?

Authors D. Jardón, V.V. Tkachuk

4.3.11 Suppose that X is an Eberlein compact space such that $C_p(X)$ is hereditarily σ-metacompact. Is it true that X is metrizable?

Author V.V. Tkachuk
Motivated by the fact that if $C_p(X)$ is hereditarily Lindelöf then X is metrizable

4.3.12 Let $\mathcal{P}$ be a class of Eberlein compacta. Suppose that X is a space such that each compact subspace of $C_p(X)$ belongs to $\mathcal{P}$. Is it true that, for any compact K, all compact subspaces of $C_p(X \times K)$ belong to $\mathcal{P}$?

Published in Arhangel'skii (1998b)
Related to some questions on Grothendieck spaces

4.4   The Lindelöf Σ-Property in $C_p(X)$

There is no need to explain why the Lindelöf Σ-property is of vital importance in general topology and C_p-theory. However, after decades of very hard work on the subject, there is still no acceptable characterization of the Lindelöf Σ-property in $C_p(X)$. As a consequence, many basic questions solved in topology remain unanswered for the spaces $C_p(X)$.

4.4.1 Suppose that $\upsilon(C_p(X))$ is a Lindelöf Σ-space. Is it true that the space $\upsilon(C_pC_p(X))$ is Lindelöf Σ?

Published in Arhangel'skii (1992b)
Comment the answer is "yes" if $C_p(X)$ is a Lindelöf Σ-space or X is normal (Problems 246–247)

4.4.2 Suppose that $C_p(X, \mathbb{I})$ is a Lindelöf Σ-space. Let Y be the set of non-isolated points of X. Is it true that the closure of Y in υX is a Lindelöf Σ-space?

Published in Okunev (1993a)
Motivated by the fact that υX is a Lindelöf Σ-space if $C_p(X)$ is a Lindelöf Σ-space (Problem 206) and $C_p(X, \mathbb{I})$ is compact (and hence Lindelöf Σ) if X is discrete.

4.4.3 Let X be a space of countable spread such that $C_p(C_p(X))$ is a Lindelöf Σ- space. Is it true that X has a countable network?

Published in Tkachuk (2001)
Motivated by the positive answer if $C_p(X)$ is a Lindelöf Σ- space (Problem 275).

4.4.4 Is it true that, for any compact space X, the space $C_p(X)$ has a T_0-separating weakly σ-point-finite family of cozero sets? What happens in case when X is a Lindelöf Σ- space?

Published in Tkachuk (2001)
Motivated by the fact that this is true for any Lindelöf Σ- space with a unique non-isolated point (Problem 274).

4.4.5 Suppose that X is a product of separable metric spaces. Is it true that every Lindelöf subspace of $C_p(X)$ has a countable network?

Published in Tkachuk (2001)
Comment this is true for Lindelöf Σ- subspaces of $C_p(X)$ (Problem 254).

4.4.6 Is it consistent with ZFC that for any hereditarily Lindelöf space X every Lindelöf Σ- subspace of $C_p(X)$ has a countable network?

Published in Tkachuk (2001)
Motivated by the fact that if $C_p(X)$ is a Lindelöf Σ- space then it has a countable network (Problem 275).

4.4.7 Let $C_p(X)$ be a Lindelöf Σ-space and assume that $\psi(X) \leq \omega$ (or even $\chi(X) \leq \omega$). Is it true that $|X| \leq \mathfrak{c}$?

Published in Tkachuk (2005c)

Motivated by the fact that this is true if X is a closed subspace of a Σ-product of real lines (Problem 285)

4.5   The Lindelöf Property in X and $C_p(X)$

One of the basic results here is the fact that $\omega_1 + 1$ does not embed in $C_p(Y)$ for any Lindelöf space Y. A natural step forward would be to prove the same for any compact space of uncountable tightness. However, this is only known under PFA and one of the most important questions of the topic is whether this statement is true without any additional axioms.

4.5.1 Suppose that X is a compact space embeddable in $C_p(Y)$ for a Lindelöf space Y. Is it true in ZFC that $t(X) = \omega$?

Published in Arhangel'skii (1989a)
Related to Problems 085–089

4.5.2 Suppose that X is a compact space embeddable in $C_p(Y)$ for a Lindelöf space Y. Is it true that X is sequential under MA$+\neg$CH?

Published in Arhangel'skii (1989a)
Related to Problems 085–089
Comment the answer is "yes" under PFA (Arhangel'skii (1989a))

4.5.3 Suppose that X is a compact space embeddable in $C_p(Y)$ for a Lindelöf space Y. Let X_1 be a continuous image of X. Is it true that X_1 is embeddable in $C_p(Z)$ for some Lindelöf Z?

Published in Arhangel'skii (1989a)
Comment if the answer is positive then every compact space embeddable in $C_p(Y)$ for a Lindelöf space Y, has countable tightness (see Problems 085–089)

4.5.4 Assuming MA$+\neg$CH suppose that X is a separable compact subspace of $C_p(Y)$ for a Lindelöf space Y. Must X be metrizable?

Published in Arhangel'skii (1989a)
Related to Problems 080–081, 097–098
Motivated by the fact that the answer is "yes" if every finite power of Y is Lindelöf

4.5.5 Suppose that X is a separable compact subspace of $C_p(Y)$ for a Lindelöf space Y. Must X be hereditarily separable?

Published in Arhangel'skii (1998b)
Related to Problems 097–098

4.5.6 Suppose that X is a homogeneous compact space and $X \subset C_p(Y)$ for some Lindelöf space Y. Is it true that $|X| \leq \mathfrak{c}$?

Published in Arhangel'skii (1989a)
Motivated by the fact that it is true under PFA because $|X| \leq \mathfrak{c}$ for any sequential homogeneous compact X (Arhangel'skii (1978b))
Related to Problems 085–089

4.5.7 Let X be a space of countable extent. Is it consistently true (say, under MA+¬CH) that every compact subspace of $C_p(X)$ is Fréchet–Urysohn?

Published in Arhangel'skii (1997)
Related to Problems 085–089

4.5.8 Let X be a Lindelöf space. Is it consistent with ZFC that every compact subspace of $C_p(X)$ is Fréchet–Urysohn?

Published in Arhangel'skii (1998b)
Related to Problems 085–089

4.5.9 Let X be a Lindelöf k-space. Is it true in ZFC that every compact subspace of $C_p(X)$ is Fréchet–Urysohn? What happens if X is a Lindelöf first countable space?

Published in Arhangel'skii (1997)
Related to Problems 085–089

4.5.10 Let X be a Lindelöf space. Is it true that every countably compact subspace of $C_p(X)$ has countable tightness?

Published in Arhangel'skii (1997)
Related to Problems 085–089

4.5.11 Suppose that $C_p(X)$ is Lindelöf. Is it true that $C_p(X) \times C_p(X)$ is Lindelöf?

Published in Arhangel'skii (1988a), (1989a), (1990a), (1992b)
Related to Problem 032

4.5.12 Suppose that X is a compact space and $C_p(X)$ is Lindelöf. Is it true that $C_p(X) \times C_p(X)$ is Lindelöf?

Published in Arhangel'skii (1988a), [1989a], [1990a], [1992b]
Related to Problem 032

4.5.13 Is every Lindelöf space embeddable in a Lindelöf $C_p(X)$ (as a closed subspace)?

Published in Arhangel'skii (1992b)

4.5.14 Let X be a (compact) space such that $C_p(X)$ is Lindelöf. Is it true that there exists a (compact) zero-dimensional space Y such that X is a continuous image of Y and $C_p(Y)$ is Lindelöf?

Published in Arhangel'skii (1989a)
Related to Problem 032

4.5.15 Suppose that $C_p(X)$ is Lindelöf. Is it true that $t(L_p(X)) = \omega$?

Published in Arhangel'skii (1989a)

4.5.16 Suppose that $t(L_p(X)) = \omega$. Is it true that $C_p(X)$ is Lindelöf?

Author A.V. Arhangel'skii
Comment In many occasions Arhangel'skii formulated this problem personally.

4.5.17 Let X be a compact space such that $C_p(C_p(X))$ is Lindelöf. Is it true that $C_p(X)$ is Lindelöf? Must X be ω-monolithic?

Published in Arhangel'skii (1990a)
Comment if $C_p(C_p(X))$ is a Lindelöf Σ-space then so is $C_p(X)$ (Problem 221)

4.5.18 Suppose that $C_p(X)$ is Lindelöf and K is a compact subspace of $C_p(X)$. Is it true that $t(K) = \omega$?

Published in Arhangel'skii (1990a) (attributed to Okunev)
Motivated by the fact that it is true if $C_p(X)$ is a Lindelöf Σ-space
Comment this is true under PFA (Arhangel'skii (1998b))

4.5.20 Let X be a monolithic compact space of countable tightness. Suppose that every linearly orderable compact subspace of X is metrizable. Is it true in ZFC that $C_p(X)$ must be Lindelöf?

Published in Arhangel'skii (1992b)
Motivated by the fact that, in a Corson compact space, every closed linearly orderable subspace is metrizable (Problem 082)

4.6 Extral and Extendial Spaces

Every embedding of an absolute retract is an l-embedding; so the notion of an extral space is a generalization of the concept of an absolute retract. This generalization turned out to be quite ample, because the class of extral spaces contains all compact metrizable spaces. Another non-trivial result is that extendial spaces form a larger class than the class of metrizable spaces. Being extral or extendial imposes very strong restrictions on a space; since these restrictions often imply metrizability, a basic question about a given extral/extendial space is whether it is metrizable.

4.6.1 Suppose that X is a space which is t-extral and t-extendial. Must X be metrizable?

> **Published in** Arhangel'skii (1998b)
> **Motivated by** the theorem which says that every extral t-extendial space must be metrizable (Problem 488)

4.6.2 Is it true that $\mathbb{D}^{\omega_1}$ is t-embedded in $[0, 1]^{\omega_1}$?

> **Published in** Arhangel'skii (1998b)
> **Motivated by** the fact that $\mathbb{D}^\kappa$ is not l-embeddable in $[0, 1]^\kappa$ for any uncountable cardinal κ (Problems 483 and 493)

4.6.3 Is it true that $C_p(\mathbb{D}^{\omega_1})$ can be embedded in $C_p([0, 1]^{\omega_1})$?

> **Published in** Arhangel'skii (1998b)
> **Motivated by** the fact that there is no linear embedding of $C_p(\mathbb{D}^\kappa)$ in $C_p([0, 1]^\kappa)$ for any uncountable cardinal κ.

4.6.4 Suppose that $\varphi : C_p(\mathbb{D}^{\omega_1}) \to C_p([0, 1]^{\omega_1})$ is a continuous map. Is it true that $\varphi(C_p(\mathbb{D}^{\omega_1}))$ has a countable network?

> **Published in** Arhangel'skii (1998a)
> **Motivated by** the fact that if $\varphi : C_p(\mathbb{D}^{\omega_1}) \to C_p([0, 1]^{\omega_1})$ is a linear continuous map then $nw(\varphi(C_p(\mathbb{D}^{\omega_1}))) = \omega$.

4.6.5 Can every compact space be t-embedded (or l-embedded) in a homogeneous compact space? In particular, given an uncountable cardinal κ, does there exist a t-embedding of the space $A(\kappa)$ in a homogeneous compact space?

> **Published in** Arhangel'skii (1998b)
> **Motivated by** the fact that each metrizable compact space is l-embedded in any compact space containing it as a subspace.

4.6.6 Is it true that, for every compact X, the space $C_p(X)$ can be (linearly) embedded in $C_p(Y)$ for some homogeneous compact space Y?

> **Published in** Arhangel'skii (1998b)

Motivated by the fact that each metrizable compact space is l-embedded in any compact space containing it as a subspace.

4.6.7 Let κ be an infinite cardinal number. Is it true that there exists a compact space $U(\kappa)$ of weight κ such that every compact space of weight not exceeding κ can be l-embedded (or t-embedded) in $U(\kappa)$?

Published in Arhangel'skii (1998b)
Motivated by the fact that this is true for $\kappa = \omega$.

4.6.8 Let κ be an infinite cardinal number. Is it true in ZFC that there exists a compact space $W(\kappa)$ of weight κ such that, for every compact space X with $w(X) \leq \kappa$, there exists a (linear) embedding of $C_p(X)$ in $C_p(W(\kappa))$?

Published in Arhangel'skii (1998b)
Motivated by the fact that this is true for $\kappa = \omega$.
Comment this is true under GCH (Esenin-Vol'pin (1949))

4.6.9 Is $\mathbb{D}^{\omega_1}$ a t-extral space?

Published in Arhangel'skii (1992b)
Motivated by the fact that $\mathbb{D}^{\omega_1}$ is not extral (Problem 493)

4.6.10 Let X be a non-metrizable t-extral space. Is it true that $\mathbb{D}^{\omega_1}$ (or $[0, 1]^{\omega_1}$) embeds in X?

Published in Arhangel'skii (1992b)
Motivated by the analogous theorem for l-dyadic compact spaces (Problem 496).

4.6.11 Let X be a t-extral space of countable tightness. Is it true that X is metrizable?

Published in Arhangel'skii (1992b)
Motivated by the analogous theorem for l-dyadic compact spaces (Problem 496).

4.6.12 Let X be an ω-monolithic t-extral space. Is it true that X is metrizable?

Published in Arhangel'skii (1992b)
Motivated by the fact that any Corson t-extral space is metrizable (Problems 141 and 480).

4.6.13 Is it true that any continuous image of an extendial (t-extendial) compact space is extendial (t-extendial respectively)?

Published in Arhangel'skii and Choban (1992)
Motivated by the fact that any continuous image of a t-extendial compact space has some strong extension properties described in Arhangel'skii and Choban (1992)

4.6.14 Let X be a separable compact space such that $X \times X$ is t-extendial. Is it true in ZFC that X is metrizable?

> **Published in** Arhangel'skii and Choban (1989)
> **Motivated by** the fact that it is true under MA$+\neg$CH (Problem 458).

4.7   Point-Finite Cellularity and Calibers

Every space $C_p(X)$ has the Souslin property; however ω_1 need not be a caliber of $C_p(X)$ and point-finite cellularity of $C_p(X)$ is not necessarily countable. This fact triggered an extensive study of calibers and point-finite cellularity in C_p-theory. Most of the open questions in this area are consequences of too little knowledge we have about the spaces X such that $C_p(X)$ has countable point-finite cellularity.

4.7.1 Is it true that $p(C_p(X)) = p((C_p(X))^\kappa)$ for any cardinal κ?

> **Published in** Tkachuk (1984c)
> **Motivated by** the fact that $p(C_p(X)) = p((C_p(X))^c)$ and the positive answer for the case when $p(C_p(X)) = \omega$ (Tkachuk (1984c))

4.7.2 Is it consistent with ZFC that any perfectly normal compact space X with $p(C_p(X)) = \omega$ is metrizable?

> **Published in** Okunev and Tkachuk (2001)
> **Related to** Problem 287

4.7.3 Is there a ZFC example of a non-metrizable first countable compact space X such that $p(C_p(X)) = \omega$?

> **Published in** Okunev and Tkachuk (2001)
> **Related to** Problem 287

4.7.4 Let X be the Souslin continuum. Is it true that $p(C_p(X)) = \omega$?

> **Published in** Okunev and Tkachuk (2001)
> **Related to** Problem 287

4.7.5 Suppose that $p(C_p(X)) = \omega$. Is it true that $p(C_p(\upsilon X)) = \omega$?

> **Published in** Okunev and Tkachuk (2001)
> **Related to** Problems 228–232

4.7.6 Let X be a monolithic compact space with $p(C_p(X)) = \omega$. Must X be metrizable?

> **Published in** Arhangel'skii (1992b)
> **Related to** Problem 287

4.7.7 Suppose that ω_1 is a caliber of X. Is it true that all countably compact subspaces of $C_p(X)$ are metrizable?

> **Published in** Tkachuk (1988)
> **Motivated by** the fact that in some models of ZFC all compact subsets of $C_p(X)$ are metrizable (Problems SFFS-294–299) while this is not true (in ZFC) for pseudocompact subspaces of $C_p(X)$.

4.8 Grothendieck Spaces

A subset A of a space X is called *countably compact in* X if every infinite subset of A has an accumulation point in X. Call X *a g-space* if $\overline{A}$ is compact for every $A \subset X$, which is countably compact in X. A space X is called *Grothendieck* if $C_p(X)$ is a hereditarily g-space. If $C_p(X)$ is a g-space then X is called *weakly Grothendieck*.

The classes described above were introduced by Arhangel'skii to generalize the theorem of Grothendieck for compact spaces. Observe that, in the new terminology, every compact space is Grothendieck.

4.8.1 Assume that X is a k-space (a space of countable tightness) with $l^*(X) = \omega$. Is then X a Grothendieck space?

Published in Arhangel'skii (1998b)
Motivated by the fact that every Lindelöf Σ-space is Grothendieck.

4.8.2 Let X be a Lindelöf first countable space. Is then X a Grothendieck space?

Published in Arhangel'skii (1998b)
Motivated by the fact that every Lindelöf Σ-space is Grothendieck.

4.8.3 Is the product of two Grothendieck spaces a (weakly) Grothendieck space?

Published in Arhangel'skii (1998b)
Motivated by the fact that the product of two weakly Grothendieck spaces is not necessarily weakly Grothendieck.

4.8.4 Let X be a (weakly) Grothendieck space. Is it true that $X \times K$ is a (weakly) Grothendieck space for any compact space K? What happens if $K = \mathbb{I}$?

Published in Arhangel'skii (1998b)
Motivated by the fact that the product of two weakly Grothendieck spaces is not necessarily weakly Grothendieck.

4.8.5 Let X and Y be Grothendieck spaces. Is it true that $X \oplus Y$ is a Grothendieck space?

Published in Arhangel'skii (1998b)
Motivated by the fact that the discrete union of any family of weakly Grothendieck spaces is weakly Grothendieck.

4.8.6 Let X and Y be Grothendieck spaces. Suppose that K is a compact subspace of $C_p(X)$ and L is a compact subspace of $C_p(Y)$. Must the space $K \times L$ be Fréchet–Urysohn?

Published in Arhangel'skii (1998b)
Motivated by the fact that K and L are Fréchet–Urysohn and this property is not preserved by products even in compact spaces.

4.8.7 Is the product of a Grothendieck space and a countable space a (weakly) Grothendieck space?

Published in Arhangel'skii (1998b)
Motivated by the fact that any countable space is Grothendieck and it is not known whether the product of two Grothendieck spaces is Grothendieck.

4.8.8 Must a perfect preimage of a (weakly) Grothendieck space be a (weakly) Grothendieck space?

Published in Arhangel'skii (1998b)
Motivated by the fact that very little is known about the categorical behavior of (weakly) Grothendieck spaces.

4.8.9 Is every hereditarily Grothendieck space hereditarily separable?

Published in Arhangel'skii (1998b)
Motivated by the fact that any hereditarily separable space is hereditarily Grothendieck and any hereditarily Grothendieck space has countable spread.

4.8.10 Is any perfectly normal compact space hereditarily Grothendieck?

Published in Arhangel'skii (1998b)
Comment any hereditarily separable space is hereditarily Grothendieck.

4.8.11 Suppose that X is a Lindelöf ω-stable space of countable tightness. Is it true that every continuous image of X is a Grothendieck space?

Published in Arhangel'skii (1997)
Motivated by the fact that this is true under PFA.

4.8.12 A space Z is *isocompact* if any countably compact closed subspace of Z is compact. Is it true that $C_p(X)$ is isocompact if and only if it is a g-space?

Published in Arhangel'skii (1997)
Motivated by the fact that any g-space is isocompact

4.9 Raznoie (Unclassified Questions)

It is usually impossible to completely classify a complex data set such as the open problems in C_p-theory. This last group of problems contains the open questions which do not fit into any of the previous Sections.

4.9.1 Suppose that every compact subspace of $C_p(X)$ has countable tightness (is Fréchet–Urysohn). Is it true that, for any compact K, all compact subspaces of $C_p(X \times K)$ have countable tightness (are Fréchet–Urysohn)?

Published in Arhangel'skii (1998b)
Related to some questions on Grothendieck spaces

4.9.2 Is the countable Fréchet–Urysohn fan embeddable in $C_p(X)$ for a Hurewicz space X?

Published in Arhangel'skii (1989a)
Motivated by the fact that this fan is not embeddable in $C_p(X)$ if all finite powers of X are Hurewicz (Problem 057).

4.9.3 Let X be a hereditarily Lindelöf space. Is it true that every compact subspace of $C_p(X)$ has countable tightness?

Published in Arhangel'skii (1998b)
Motivated by the fact this is true under PFA even for Lindelöf X (Problem 089)

4.9.4 Let X be a space of countable spread. Is it true that every compact subspace of $C_p(X)$ has countable tightness?

Published in Arhangel'skii (1998b)
Motivated by the fact this is true under PFA if X is Lindelöf (Problem 089)

4.9.5 Let X be a dyadic compact space such that $C_p(X)$ is (hereditarily) subparacompact. Is it true that X is metrizable?

Author V.V. Tkachuk
Motivated by the fact that if $C_p(X)$ is Lindelöf then X is metrizable (Problems 086–087)

4.9.6 Let X be a zero-dimensional compact space such that $C_p(X, \mathbb{D})$ is normal. Must the space $C_p(X, \mathbb{D})$ be Lindelöf?

Author V.V. Tkachuk
Related to Problem 032

4.9.7 Let X be a zero-dimensional (compact) space such that $C_p(X, \mathbb{D})$ is normal. Must the space $C_p(X, \mathbb{D}) \times C_p(X, \mathbb{D})$ be normal?

Author V.V. Tkachuk
Related to Problem 032

4.9.8 Let X be a zero-dimensional (compact) space such that $C_p(X, \mathbb{D})$ is Lindelöf. Must the space $C_p(X, \mathbb{D}) \times C_p(X, \mathbb{D})$ be Lindelöf (or metacompact or subparacompact)?

Author V.V. Tkachuk
Related to Problem 032

4.9.9 Let X be a zero-dimensional (compact) space such that $C_p(X, \mathbb{D})$ is metacompact. Must the space $C_p(X, \mathbb{D}) \times C_p(X, \mathbb{D})$ be metacompact?

Author V.V. Tkachuk
Related to Problem 032

4.9.10 Let X be a compact monolithic space. Is it true that $C_p(X)$ can be condensed onto a subspace of a Σ-product of real lines?

Published in Arhangel'skii (1992b)
Related to Problems 285 and 286

4.9.11 Is it possible to embed $C_p(\mathbb{I})$ in a pseudoradial space?

Published in Arhangel'skii (1992b)
Motivated by the fact that $C_p(\mathbb{I})$ is not embeddable in any sequential space (Problem 062).

Bibliography

The bibliography of this book is intended to reflect the state of the art of modern C_p-theory; besides, it is obligatory to mention the work of all authors whose results, in one form or another, are cited here. The bibliographic selection for this volume has 400 items to solve the proportional part of the task.

ALAS, O.T.
[1978]   *Normal and function spaces,* Topology, Vol. II, Colloq. Math. Soc. János Bolyai, 23, North-Holland, Amsterdam, 1980, 29–33.

ALAS, O.T., GARCIA-FERREIRA, S., TOMITA, A.H.
[1999]   *The extraresolvability of some function spaces,* Glas. Mat. Ser. III **34(54):1**(1999), 23–35.

ALAS, O.T., TAMARIZ-MASCARÚA, A.
[2006]   *On the Čech number of $C_p(X)$ II,* Questions Answers Gen. Topology **24:1**(2006), 31–49.

ALSTER, K.
[1979]   *Some remarks on Eberlein compacta,* Fund. Math., **104:1**(1979), 43–46.

ALSTER, K., POL, R.
[1980]   *On function spaces of compact subspaces of Σ-products of the real line,* Fund. Math., **107:2**(1980), 135–143.

AMIR, D., LINDENSTRAUSS, J.
[1968]   *The structure of weakly compact sets in Banach spaces,* Annals Math., **88:1**(1968), 35–46.

ANDERSON, B.D.
[1973]   *Projections and extension maps in $C(T)$,* Illinois J. Math., **17**(1973), 513–517.

ARGYROS, S., MERCOURAKIS, S., NEGREPONTIS, S.
[1983]   *Analytic properties of Corson compact spaces,* General Topology and Its Relations to Modern Analysis and Algebra, 5. Berlin, 1983, 12–24.

ARGYROS, S., NEGREPONTIS, S.
[1983]   *On weakly K-countably determined spaces of continuous functions,* Proc. Amer. Math. Soc., **87:4**(1983), 731–736.

© Springer International Publishing Switzerland 2015
V.V. Tkachuk, *A Cp-Theory Problem Book*, Problem Books
in Mathematics, DOI 10.1007/978-3-319-16092-4

ARHANGEL'SKII, A.V.

[1959]   *An addition theorem for the weight of sets lying in bicompacta (in Russian),* DAN
            SSSR, **126**(1959), 239–241.

[1976]   *On some topological spaces occurring in functional analysis (in Russian),* Uspehi
            Mat. Nauk, **31:5**(1976), 17–32.

[1978a]   *On spaces of continuous functions with the topology of pointwise convergence (in
            Russian),* Doklady AN SSSR, **240:3**(1978), 506–508.

[1978b]   *The structure and classification of topological spaces and cardinal invariants (in
            Russian),* Uspehi Mat. Nauk, **33:6**(1978), 29–84.

[1981]   *Classes of topological groups (in Russian),* Uspehi Mat. Nauk, **36:3**(1981),
            127–146.

[1982c]   *Factorization theorems and function spaces: stability and monolithity (in Rus-
            sian),* Doklady AN SSSR, **265:5**(1982), 1039–1043.

[1983c]   *Functional tightness, ϱ-spaces and τ-embeddings,* Comment. Math. Univ. Car-
            olinae, **24:1**(1983), 105–120.

[1984b]   *Continuous mappings, factorization theorems and function spaces (in Russian),*
            Trudy Mosk. Mat. Obsch., **47**(1984), 3–21.

[1985]   *Function spaces with the topology of pointwise convergence (in Russian),* General
            Topology: Function Spaces and Dimension (in Russian), Moscow University
            P.H., 1985, 3–66.

[1986]   *Hurewicz spaces, analytic sets and fan tightness of function spaces (in Russian),*
            Doklady AN SSSR, **287:3**(1986), 525–528.

[1987]   *A survey of C_p-theory,* Questions and Answers in General Topology, **5**(1987),
            1–109.

[1988a]   *Some results and problems in $C_p(X)$-theory.* General Topology and Its Relations
            to Modern Analysis and Algebra, VI, Heldermann Verlag, Berlin, 1988, 11–31.

[1989a]   *Topological Function Spaces (in Russian),* Moscow University P.H., Moscow,
            1989.

[1989c]   *On iterated function spaces,* Bull. Acad. Sci. Georgian SSR, **134:3**(1989),
            481–483.

[1990a]   *Problems in C_p-theory,* in: Open Problems in Topology, North Holland, Amster-
            dam, 1990, 603–615.

[1990b]   *On the Lindelöf degree of topological spaces of functions, and on embeddings
            into $C_p(X)$,* Moscow University Math. Bull., **45:5**(1990), 43–45.

[1992a]   *Topological Function Spaces (translated from Russian),* Kluwer Academic Pub-
            lishers, Dordrecht, 1992.

[1992b]   *C_p-theory,* in: Recent Progress in General Topology, North Holland, Amsterdam,
            1992, 1–56.

[1995a]   *Spaces of mappings and rings of continuous functions,* in: General Topology, 3,
            Encyclopedia Math. Sci., Springer, Berlin, **51**(1995), 73–156.

[1995b]   *A generic theorem in the theory of cardinal invariants of topological spaces,*
            Comment. Math. Univ. Carolinae, **36:2**(1995), 303–325.

[1996a]   *On Lindelöf property and spread in C_p-theory,* Topol. and Its Appl., **74:**
            (1-3)(1996), 83–90.

[1996b]   *On spread and condensations,* Proc. Amer. Math. Soc., **124:11**(1996),
            3519–3527.

[1997]   *On a theorem of Grothendieck in C_p-theory,* Topology Appl. **80**(1997), 21–41.

[1998a]   *Some observations on C_p-theory and bibliography,* Topology Appl., **89**(1998),
            203–221.

[1998b]   *Embeddings in C_p-spaces,* Topology Appl., **85**(1998), 9–33.

[2000a]   *On condensations of C_p-spaces onto compacta,* Proc. Amer. Math. Soc., **128:6**(2000), 1881–1883.

[2000b]   *Projective σ-compactness, ω_1-caliber and C_p-spaces,* Topology Appl., **104**(2000), 13–26.

ARHANGEL'SKII, A.V., CALBRIX, J.
[1999]   *A characterization of σ-compactness of a cosmic space x by means of subspaces of $\mathbb{R}^x$,* Proc. Amer. Math. Soc., **127:8**(1999), 2497–2504.

ARHANGEL'SKII, A.V., CHOBAN M.M.
[1988]   *The extension property of Tychonoff spaces and generalized retracts,* Comptes Rendus Acad. Bulg. Sci., **41:12**(1988), 5–7.

[1989] $C_p(X)$ *and some other functors in general topology. Continuous extenders,* Categorical Topology, World Scientific, London, 1989, 432–445.

[1990]   *On the position of a subspace in the whole space,* Comptes Rendus Acad. Bulg. Sci., **43:4**(1990), 13–15.

[1992]   *Extenders of Kuratowski–van Douwen and classes of spaces,* Comptes Rendus Acad. Bulg. Sci., **45:1**(1992), 5–7.

[1996]   *On continuous mappings of C_p-spaces and extenders,* Proc. Steklov Institute Math., **212**(1996), 28–31.

ARHANGEL'SKII, A.V., OKUNEV, O.G.
[1985]   *Characterization of properties of spaces by properties of their continuous images (in Russian)* Vestnik Mosk. Univ., Math., Mech., **40:5**(1985), 28–30.

ARHANGEL'SKII, A.V., PAVLOV, O.I.
[2002]   *A note on condensations of $C_p(X)$ onto compacta,* Comment. Math. Univ. Carolinae, **43:3**(2002), 485–492.

ARHANGEL'SKII, A.V., PONOMAREV, V.I.
[1974]   *Basics of General Topology in Problems and Exercises (in Russian),* Nauka, Moscow, 1974.

ARHANGEL'SKII, A.V., SHAKHMATOV, D.B.
[1988]   *On pointwise approximation of arbitrary functions by countable families of continuous functions (in Russian),* Trudy Sem. Petrovsky, **13**(1988), 206–227.

ARHANGEL'SKII, A.V., SZEPTYCKI, P.J.
[1997]   *Tightness in compact subspaces of c_p-spaces,* Houston J. Math., **23:1**(1997), 1–7.

ARHANGEL'SKII, A.V., TKACHUK, V.V.
[1985]   *Function Spaces and Topological Invariants (in Russian),* Moscow University P.H., Moscow, 1985.

[1986]   *Calibers and point-finite cellularity of the spaces $C_p(X)$ and some questions of S. Gul'ko and M. Hušek,* Topology Appl., **23:1**(1986), 65–74.

ARHANGEL'SKII, A.V., USPENSKIJ, V.V.
[1986]   *On the cardinality of Lindelöf subspaces of function spaces,* Comment. Math. Univ. Carolinae, **27:4**(1986), 673–676.

ASANOV, M.O.
[1979]   *On cardinal invariants of spaces of continuous functions (in Russian),* Modern Topology and Set Theory (in Russian), Izhevsk, 1979, N 2, 8–12.

[1980]   *On spaces of continuous maps,* Izvestiia Vuzov, Mat., 1980, N 4, 6–10.

[1983]   *About the space of continuous functions,* Colloq. Math. Soc. Janos Bolyai, **41**(1983), 31–34.

ASANOV, M.O., SHAMGUNOV, N.K.
[1981]   *The topological proof of the Nachbin–Shirota's theorem,* Comment. Math. Univ.
          Carolinae, **24:4**(1983), 693–699.

ASANOV, M.O., VELICHKO, M.V.
[1981]   *Compact sets in $C_p(X)$ (in Russian),* Comment. Math. Univ. Carolinae,
          22:2(1981), 255–266.

BANACH, T., CAUTY, R.
[1997]   *Universalité forte pour les sous-ensembles totalement bornés. Applications aux
          espaces $C_p(X)$,* Colloq. Math., **73**(1997), 25–33.

BANDLOW, I.
[1991]   *A characterization of Corson compact spaces,* Commentationes Math. Univ.
          Carolinae, **32:3**(1991), 545–550.
[1994]   *On function spaces of Corson–compact spaces,* Comment. Math. Univ. Caroli-
          nae, **35:2**(1994), 347–356.

BATUROV D.P.
[1987]   *On subspaces of function spaces (in Russian),* Vestnik Moskovsk. Univ., Math.,
          Mech., **42:4**(1987), 66–69.
[1988]   *Normality of dense subsets of function spaces,* Vestnik Moskovsk. Univ., Math.,
          Mech., **43:4**(1988), 63–65.
[1990a]   *Normality in dense subspaces of products,* Topology Appl., **36**(1990), 111–116.
[1990b]   *Some properties of the weak topology of Banach spaces,* Vestnik Mosk. Univ.,
          Math., Mech., **45:6**(1990), 68–70.

BELL, M., MARCISZEWSKI, W.
[2004]   *Function spaces on t-Corson compacta and tightness of polyadic spaces,* Czech.
          Math. J., **54(129)**(2004), 899–914.

BELLA, A., YASCHENKO, I.V.
[1999]   *On AP and WAP spaces,* Comment. Math. Univ. Carolinae, **40:3**(1999), 531–536.

BENYAMINI, Y., RUDIN, M.E., WAGE, M.
[1977]   *Continuous images of weakly compact subsets of Banach spaces,* Pacific J. Math-
          ematics, **70:2**(1977), 309–324.

BENYAMINI, Y., STARBIRD, T.
[1976]   *Embedding weakly compact sets into Hilbert space,* Israel J. Math., **23**(1976),
          137–141.

BESSAGA, C., PELCZINSKI, A.
[1960]   *Spaces of continuous functions, 4,* Studia Math., **19**(1960), 53–62.
[1975]   *Selected Topics in Infinite-Dimensional Topology,* PWN, Warszawa, 1975.

BLASKO, J.L.
[1977]   *On μ-spaces and k_R-spaces,* Proc. Amer. Math. Soc., **67:1**(1977), 179–186.
[1990]   *The G_δ-topology and κ-analytic spaces without perfect compact sets,* Colloq.
          Math., **58**(1990), 189–199.

BORGES, C.J.
[1966a]   *On stratifiable spaces,* Pacific J. Math., **17:1**(1966), 1–16.
[1966b]   *On function spaces of stratifiable spaces and compact spaces,* Proc. Amer. Math.
          Soc., **17**(1966), 1074–1078.

BOURGAIN, J.
[1977]   Compact sets of the first Baire class, Bull. Soc. Math. Belg., **29:2**(1977), 135–143.
[1978]   Some remarks on compact sets of first Baire class, Bull. Soc. Math. Belg., **30**(1978), 3–10.

BOURGAIN, J., FREMLIN, D.H., TALAGRAND, M.
[1978]   Pointwise compact sets of Baire-measurable functions, Amer. J. Math., **100**(1978), 845–886.

BOURGAIN, J., TALAGRAND, M.
[1980]   Compacité extremal, Proc. Amer. Math. Soc., **80**(1980), 68–70.

BOUZIAD, A., CALBRIX, J.
[1995]   Images usco-compactes des espaces Čech-complets de Lindelöf, C. R. Acad. Sci. Paris Sér. I Math. **320:7**(1995), 839–842.

BURKE, D.K.
[1984]   Covering properties, Handbook of Set–Theoretic Topology, edited by K. Kunen and J.E. Vaughan, Elsevier Science Publishers B.V., 1984, 347–422.

BUZYAKOVA, R.Z.
[2006a]   Spaces of continuous step functions over LOTS, Fund. Math., **192**(2006), 25–35.
[2006b]   Spaces of continuous characteristic functions, Comment. Math. Universitatis Carolinae, **47:4**(2006), 599–608.
[2007]   Function spaces over GO spaces, Topology Appl., **154:4**(2007), 917–924.

CALBRIX, J.
[1985a]   Classes de Baire et espaces d'applications continues, Comptes Rendus Acad. Sci. Paris, Ser I, **301:16**(1985), 759–762.
[1985b]   Espaces K_σ et espaces des applications continues, Bulletin Soc. Mathematique de France, **113**(1985), 183–203.
[1987]   Filtres sur les entiers et ensembles analytiques, Comptes Rendus Acad. Sci. Paris, Ser I, **305**(1987), 109–111.
[1988]   Filtres Boreliens sur l'ensemble des entiers et espaces des applications continues, Rev. Roumaine Math. Pures et Appl., **33**(1988), 655–661.
[1996]   k-spaces and Borel filters on the set of integers (in French), Trans. Amer. Math. Soc., **348**(1996), 2085–2090.

CASARRUBIAS–SEGURA, F.
[1999]   Realcompactness and monolithity are finitely additive in $C_p(X)$, Topology Proc., **24**(1999), 89–102.
[2001]   On compact weaker topologies in function spaces, Topology and Its Applications, **115**(2001), 291–298.

CASCALES, B.
[1987]   On K-analytic locally convex spaces, Arch. Math., **49**(1987), 232–244.

CASCALES, B., NAMIOKA, I.
[2003]   The Lindelöf property and σ-fragmentability, Fund. Math., **180**(2003), 161–183.

CASCALES, B., ORIHUELA, J.
[1987]   On compactness in locally convex spaces, Math. Z., **195**(1987), 365–381
[1988]   On pointwise and weak compactness in spaces of continuous functions, Bull. Soc. Math. Belg., **40:2**(1988), 331–352.
[1991a]   A sequential property of set-valued maps, J. Math. Anal. and Appl., **156:1**(1991), 86–100.
[1991b]   Countably determined locally convex spaces, Portugal. Math. **48:1**(1991), 75–89.

CAUTY, R.

[1974]   *Rétractions dans les espaces stratifiables,* Bull. Soc. Math. France **102**(1974), 129–149.

[1991]   *L'espace de fonctions continues d'un espace metrique denombrable,* Proc. Amer. Math. Soc., **113**(1991), 493–501.

[1998]   *La classe borélienne ne détermine pas le type topologique de $C_p(X)$,* Serdica Math. J. **24:3-4**(1998), 307–318.

CAUTY, R., DOBROWOLSKI, T., MARCISZEWSKI, W.

[1993]   *A contribution to the topological classification of the spaces $C_p(X)$,* Fundam. Math., **142**(1993), 269–301.

CHOBAN, M.M.

[1998a]   *General theorems on functional equivalence of topological spaces,* Topol. Appl., **89**(1998), 223–239.

[1998b]   *Isomorphism problems for the Baire function spaces of topological spaces,* Serdica Math. J. **24:1**(1998), 5–20.

[1998c]   *Isomorphism of functional spaces,* Math. Balkanica (N.S.) **12:1-2**(1998), 59–91.

[2001]   *Functional equivalence of topological spaces,* Topology Appl., **111**(2001), 105–134.

[2005]   *On some problems of descriptive set theory in topological spaces,* Russian Math. Surveys **60:4**(2005), 699–719.

CHRISTENSEN, J.P.R.

[1974]   *Topology and Borel Structure,* North Holland P.C., Amsterdam, 1974.

[1981]   *Joint continuity of separably continuous functions,* Proceedings of Amer. Math. Soc., **82:3**(1981), 455–461.

CHRISTENSEN, J.P.R., KENDEROV, P.S.

[1984]   *Dense strong continuity of mappings and the Radon–Nykodym property,* Math. Scand., **54:1**(1984), 70–78.

CIESIELSKI, K.

[1993]   *Linear subspace of $\mathbb{R}^\lambda$ without dense totally disconnected subsets,* Fund. Math. **142** (1993), 85–88.

COMFORT, W.W., FENG, L.

[1993]   *The union of resolvable spaces is resolvable,* Math. Japon. **38:3**(1993), 413–414.

COMFORT, W.W., HAGER, A.W.

[1970a]   *Estimates for the number of real-valued continuous functions,* Trans. Amer. Math. Soc., **150**(1970), 619–631.

[1970b]   *Dense subspaces of some spaces of continuous functions,* Math. Z. **114**(1970), 373–389.

[1970c]   *Estimates for the number of real-valued continuous functions,* Trans. Amer. Math. Soc. **150**(1970), 619–631.

COMFORT, W.W., NEGREPONTIS, S.A.

[1982]   *Chain Conditions in Topology,* Cambridge Tracts in Mathematics, **79**, New York, 1982.

CORSON, H.H.

[1959]   *Normality in subsets of product spaces,* American J. Math., **81:3**(1959), 785–796.

[1961]   *The weak topology of a Banach space,* Trans. Amer. Math. Soc., **101:1**(1961), 1–15.

CORSON, H.H., LINDENSTRAUSS, J.
[1966a]   *On function spaces which are Lindelöf spaces*, Trans. Amer. Math. Soc., **121:2**(1966), 476–491.
[1966b]   *On weakly compact subsets of Banach spaces*, Proc. Amer. Math. Soc., **17:2**(1966), 407–412.

DEBS, G.
[1985]   *Espaces K-analytiques et espaces de Baire de fonctions continues*, Mathematika, **32**(1985), 218–228.

DIJKSTRA, J., GRILLOT, T., LUTZER, D., VAN MILL, J.
[1985]   *Function spaces of low Borel complexity*, Proc. Amer. Math. Soc., **94:4**(1985), 703–710.

DIJKSTRA, J., MOGILSKI, J.
[1996] $C_p(X)$-*representation of certain Borel absorbers*, Topology Proc., **16**(1991), 29–39.

DIMOV, G.
[1987]   *Espaces d'Eberlein et espaces de type voisins*, Comptes Rendus Acad. Sci., Paris, Ser. I, **304:9**(1987), 233–235.
[1988]   *Baire subspaces of* $c_0(\Gamma)$ *have dense* G_δ *metrizable subsets*, Rend. Circ. Mat. Palermo (2) Suppl. **18**(1988), 275–285.

DIMOV, G., TIRONI, G.
[1987]   *Some remarks on almost radiality in function spaces*, Acta Univ. Carolin. Math. Phys. **28:2**(1987), 49–58.

DOUWEN, E.K. VAN
[1975a]   *Simultaneous extension of continuous functions*, PhD Dissertation, **99**(1975), Amsterdam, Free University.
[1975b]   *Simultaneous linear extensions of continuous functions*, General Topology Appl., **5**(1975), 297–319.
[1984]   *The integers and topology*, Handbook of Set–Theoretic Topology, K. Kunen and J.E. Vaughan, editors, Elsevier Science Publishers B.V., 1984, 111–167.

DOUWEN, E.K. POL, R.
[1977]   *Countable spaces without extension properties*, Bull. Polon. Acad. Sci., Math., **25**(1977), 987–991.

DOW, A.
[2005a]   *Closures of discrete sets in compact spaces*, Studia Sci. Math. Hungar. **42:2**(2005), 227–234.
[2005b]   *Property D and pseudonormality in first countable spaces*, Comment. Math. Univ. Carolin. **46:2**(2005), 369–372.

DOW, A., JUNNILA H., PELANT, J.
[1997]   *Weak covering properties of weak topologies*, Proceedings of London Math. Soc., **(3)75:2**(1997), 349–368.
[2006]   *Coverings, networks and weak topologies*, Mathematika **53:2**(2006), 287–320.

DOW, A., PAVLOV, O.
[2006]   *More about spaces with a small diagonal*, Fund. Math. **191:1**(2006), 67–80.
[2007]   *Perfect preimages and small diagonal*, Topology Proc. **31:1**(2007), 89–95.

DOW, A., SIMON, P.
[2006]   *Spaces of continuous functions over a* ψ-*space*, Topology Appl. **153:13**(2006), 2260–2271.

DUGUNDJI, J.
[1951]   An extension of Tietze's theorem, Pacific J. Math., 1(1951), 353–367.

EBERLEIN, W.F.
[1947]   Weak compactness in Banach spaces, I, Proc. Nat. Acad. Sci. (USA), 33(1947),
          51–53.

EFIMOV, B.A.
[1977]   Mappings and imbeddings of dyadic spaces, I, Math. USSR Sbornik, 32:1(1977),
          45–57.

ENGELKING, R.
[1977]   General Topology, PWN, Warszawa, 1977.

ESENIN-VOL'PIN, A.S.
[1949]   On the existence of a universal bicompactum of arbitrary weight (in Russian).
          Dokl. Acad. Nauk SSSR, 68(1949), 649–652.

FREMLIN, D.H.
[1977]   K-analytic spaces with metrizable compacta, Mathematika, 24(1977), 257–261.
[1994]   Sequential convergence in $C_p(X)$, Comment. Math. Univ. Carolin., 35:2(1994),
          371–382.

GARTSIDE, P.
[1997]   Cardinal invariants of monotonically normal spaces, Topology Appl.
          77:3(1997), 303–314.
[1998]   Nonstratifiability of topological vector spaces, Topology Appl. 86:2(1998),
          133–140.

GARTSIDE, P., FENG, Z.
[2007]   More stratifiable function spaces, Topology Appl. 154:12(2007), 2457–2461.

GERLITS, J.
[1983]   Some properties of $C(X)$, II, Topology Appl., 15:3(1983), 255–262.

GERLITS, J., NAGY, ZS.
[1982]   Some properties of $C(X)$, I, Topology Appl., 14:2(1982), 151–161.

GERLITS, J., JUHÁSZ, I., SZENTMIKLÓSSY, Z.
[2005]   Two improvements on Tkačenko's addition theorem, Comment. Math. Univ.
          Carolin. 46:4(2005), 705–710.

GERLITS, J., NAGY, ZS., SZENTMIKLOSSY, Z.
[1988]   Some convergence properties in function spaces, in: General Topology and Its
          Relation to Modern Analysis and Algebra, Heldermann, Berlin, 1988, 211–222.

GILLMAN, L., JERISON, M.
[1960]   Rings of Continuous Functions, D. van Nostrand Company Inc., Princeton, 1960.

GRAEV, M.I.
[1950]   Theory of topological groups, I (in Russian), Uspehi Mat. Nauk, 5:2(1950), 3–56.

GROTHENDIECK, A.
[1952]   Critères de compacité dans les espaces fonctionnels génereaux, Amer. J. Math.,
          74(1952), 168–186.
[1953]   Sur les applications linéaires faiblement compactes d'espaces du type $C(K)$,
          Canadian J. Math., 5:2(1953), 129–173.

GRUENHAGE, G.

[1976]   *Infinite games and generalizations of first-countable spaces*, General Topology and Appl. **6:3**(1976), 339–352.

[1984a]   *Covering properties of $X^2 \backslash \Delta$, w-sets and compact subspaces of Σ-products*, Topology Appl., **17:3**(1984), 287–304.

[1984b]   *Generalized metric spaces*, Handbook of Set-Theoretic Topology, North-Holland, Amsterdam, 1984, 423–501.

[1986a]   *Games, covering properties and Eberlein compacta*, Topology Appl., **23:3**(1986a), 291–298.

[1986b]   *On a Corson compact space of Todorcevic*, Fund. Math., **126**(1986), 261–268.

[1987]   *A note on Gul'ko compact spaces*, Proc. Amer. Math. Soc., **100**(1987), 371–376.

[1997]   *A non-metrizable space whose countable power is σ-metrizable*, Proc. Amer. Math. Soc. **125:6**(1997), 1881–1883.

[1998]   *Dugundji extenders and retracts of generalized ordered spaces*, Fundam. Math., **158**(1998), 147–164.

[2002]   *Spaces having a small diagonal*, Topology Appl., **122**(2002), 183–200.

[2006a]   *A note on D-spaces*, Topology Appl., **153**(2006), 2229–2240.

[2006b]   *The story of a topological game*, Rocky Mountain J. Math. **36:6**(2006), 1885–1914.

[2008]   *Monotonically compact and monotonically Lindelöf spaces*, Questions and Answers Gen. Topology **26:2**(2008), 121–130.

GRUENHAGE, G., MA, D.K.

[1997]   *Bairness of $C_k(X)$ for locally compact X*, Topology Appl., **80**(1997), 131–139.

GRUENHAGE, G., MICHAEL, E.

[1983]   *A result on shrinkable open covers*, Topology Proc., **8:1**(1983), 37–43.

GRUENHAGE, G., TAMANO, K.

[2005]   *If X is σ-compact Polish, then $C_k(X)$ has a σ-closure-preserving base*, Topology Appl. **151:1-3**(2005), 99–106.

GUL'KO, S.P.

[1977]   *On properties of subsets lying in Σ-products (in Russian)*. Doklady AN SSSR, **237:3**(1977), 505–507.

[1979]   *On the structure of spaces of continuous functions and their hereditary paracompactness (in Russian)*. Uspehi Matem. Nauk, **34:6**(1979), 33–40.

[1981]   *On properties of function spaces*. Seminar Gen. Topol., Moscow University P.H., Moscow, 1981, 8–41.

GUL'KO, S.P., SOKOLOV, G.A.

[1998]   *P-points in $\mathbb{N}^*$ and the spaces of continuous functions*, Topology Appl., **85**(1998), 137–142.

[2000]   *Compact spaces of separately continuous functions in two variables*, Topology and Its Appl. **107:1-2**(2000), 89–96.

HAGER, A.W.

[1969]   *Approximation of real continuous functions on Lindelöf spaces*, Proc. Amer. Math. Soc., **22**(1969), 156–163.

HAGLER, J.

[1975]   *On the structure of S and $C(S)$ for S dyadic*, Trans. Amer. Math. Soc., **214**(1975), 415–428.

HAO-XUAN, Z.

[1982]   *On the small diagonals*, Topology Appl., **13:3**(1982), 283–293.

HAYDON, R.G.
   [1990]   *A counterexample to several questions about scattered compact spaces,* Bull.
             London Math. Soc., **22**(1990), 261–268.

HEATH, R.W., LUTZER, D.J.
   [1974a]   *The Dugundji extension theorem and collectionwise normality,* Bull. Acad. Polon.
             Sci., Ser. Math., **22**(1974), 827–830.
   [1974b]   *Dugundji extension theorem for linearly ordered spaces,* Pacific J. Math.,
             55(1974), 419–425.

HEATH, R.W., LUTZER, D.J., ZENOR, P.L.
   [1975]   *On continuous extenders,* Studies in Topology, Academic Press, New York, 1975,
             203–213.

HEWITT, E.
   [1948]   *Rings of real-valued continuous functions, I,* Trans. Amer. Math. Soc.,
             64:1(1948), 45–99.

HODEL, R.
   [1984]   *Cardinal Functions I,* in: Handbook of Set-Theoretic Topology, Ed. by K. Kunen
             and J.E. Vaughan, Elsevier Science Publishers B.V., 1984, 1–61.

HUŠEK, M.
   [1972]   *Realcompactness of function spaces and* $\upsilon(P \times Q)$, General Topology and Appl.
             2(1972), 165–179.
   [1977]   *Topological spaces without κ-accessible diagonal,* Comment. Math. Univ. Car-
             olinae, **18:4**(1977), 777–788.
   [1979]   *Mappings from products. Topological structures, II,* Math. Centre Tracts, Ams-
             terdam, **115**(1979), 131–145.
   [1997a]   *Productivity of some classes of topological linear spaces,* Topology Appl. **80:1-
             2**(1997), 141–154.
   [2005] $C_p(X)$ *in coreflective classes of locally convex spaces,* Topology and Its Appli-
             cations, **146/147**(2005), 267–278.

IVANOV, A.V.
   [1978]   *On bicompacta all finite powers of which are hereditarily separable,* Soviet
             Math., Doklady, **19:6**(1978), 1470–1473.

JARDÓN, D., TKACHUK, V.V.
   [2002]   *Ultracompleteness in Eberlein–Grothendieck spaces,* Boletín de la Sociedad Mat.
             Mex., **10:3**(2004), 209–218.

JUHÁSZ, I.
   [1971]   *Cardinal functions in topology,* Mathematical Centre Tracts, **34**, Amsterdam,
             1971.
   [1980]   *Cardinal Functions in Topology—Ten Years Later,* Mathematical Centre Tracts,
             North Holland P.C., Amsterdam, 1980.
   [1991]   *Cardinal functions,* Recent Progress in General Topology, North-Holland, Ams-
             terdam, 1992, 417–441.
   [1992]   *The cardinality and weight-spectrum of a compact space,* Recent Devel. Gen.
             Topol. and Appl., Math. Research Berlin **67**(1992), 170–175.

JUHÁSZ, I., MILL, J. VAN
   [1981]   *Countably compact spaces all countable subsets of which are scattered,* Com-
             ment. Math. Univ. Carolin. **22:4**(1981), 851–855.

JUHÁSZ, I., SOUKUP, L., SZENTMIKLÓSSY, Z.
[2007]   *First countable spaces without point-countable π-bases,* Fund. Math.
           196:2(2007), 139–149.

JUHÁSZ, I., SZENTMIKLÓSSY, Z.
[1992]   *Convergent free sequences in compact spaces,* Proc. Amer. Math. Soc.,
           116:4(1992), 1153–1160.
[1995]   *Spaces with no smaller normal or compact topologies,* 1993), Bolyai Soc. Math.
           Stud., **4**(1995), 267–274.
[2002]   *Calibers, free sequences and density,* Topology Appl. **119:3**(2002), 315–324.
[2008]   *On d-separability of powers and $C_p(X)$,* Topology Appl. **155:4**(2008), 277–281.

JUHÁSZ, I., SZENTMIKLO'SSY, Z., SZYMANSKI, A.
[2007]   *Eberlein spaces of finite metrizability number,* Comment. Mathem. Univ. Caroli-
           nae **48:2**(2007), 291–301.

JUST, W., SIPACHEVA, O.V., SZEPTYCKI, P.J.
[1996]   *Non-normal spaces $C_p(X)$ with countable extent,* Proceedings Amer. Math. Soc.,
           124:4(1996), 1227–1235.

KALAMIDAS, N.D.
[1985]   *Functional properties of $C(X)$ and chain conditions on X,* Bull. Soc. Math.
           Grèce **26**(1985), 53–64.
[1992]   *Chain condition and continuous mappings on $C_p(X)$,* Rendiconti Sem. Mat.
           Univ. Padova, **87**(1992), 19–27.

KALAMIDAS, N.D., SPILIOPOULOS, G.D.
[1992]   *Compact sets in $C_p(X)$ and calibers,* Canadian Math. Bull., **35:4**(1992),
           497–502.

KAUL, S.K.
[1969]   *Compact subsets in function spaces,* Canadian Math. Bull., **12**(1969), 461–466.

KENDEROV, P.S.
[1980]   *Dense strong continuity of pointwise continuous mappings,* Pacif. J. Math.,
           89(1980), 111–130.
[1987]   *C(T) is weak Asplund for every Gul'ko compact T,* Comptes Rendus Acad. Bulg.
           Sci., **40:2**(1987), 17–19.

KOČINAC, L.D., SCHEEPERS, M.
[1999]   *Function spaces and strong measure zero sets,* Acta Math. Hungar. **82:4**(1999),
           341–351.
[2002]   *Function spaces and a property of Reznichenko,* Topology Appl. **123:1**(2002),
           135–143.

KRIVORUCHKO, A.I.
[1972]   *On the cardinality of the set of continuous functions,* Soviet Math., Dokl.,
           13(1972), 1364–1367.

[1973]   *On cardinal invariants of spaces and mappings,* Soviet Math., Doklady,
           14(1973), 1642–1647.
[1975]   *The cardinality and density of spaces of mappings,* Soviet Math., Doklady,
           16(1975), 281–285.

KUBIS, W., LEIDERMAN, A.
[2004]   *Semi-Eberlein spaces,* Topology Proc. **28:2**(2004), 603–616.

KUNEN, K.
[1980]   *Set Theory. An Introduction to Independence Proofs,* Studies Logic Found.
          Mathematics, **102**(1980), North Holland P.C., Amsterdam, 1980
[1981]   *A compact L-space under CH,* Topology Appl., **12**(1981), 283–287.
[1998]   *Bohr topologies and partition theorems for vector spaces,* Topology Appl. **90:**
          1-3(1998), 97–107.

KUNEN, K., MILL, J. VAN
[1995]   *Measures on Corson compact spaces,* Fund. Math., **147**(1995), 61–72.

KUNEN, K., DE LA VEGA, R.
[2004]   *A compact homogeneous S-space,* Topology Appl., **136**(2004), 123–127.

KURATOWSKI, C.
[1966]   *Topology, vol. 1,* Academic Press Inc., London, 1966.

LEIDERMAN, A.G.
[1985]   *On dense metrizable subspaces of Corson compact spaces.* Matem. Zametki,
          38:3(1985), 440–449.

LEIDERMAN, A.G., SOKOLOV, G.A.
[1984]   *Adequate families of sets and Corson compacts. Commentat. Math. Univ. Caroli-*
          nae, **25:2**(1984), 233–246.

LUTZER, D.J., MCCOY, R.A.
[1980]   *Category in function spaces I,* Pacific J. Math., **90:1**(1980), 145–168.

LUTZER, D.J., MILL, J. VAN, POL, R.
[1985]   *Descriptive complexity of function spaces,* Transactions of the Amer. Math. Soc.,
          291(1985), 121–128.

LUTZER, D. J., MILL, J. VAN, TKACHUK, V.V.
[2008]   *Amsterdam properties of $C_p(X)$ imply discreteness of X,* Canadian Math. Bull.
          51:4(2008), 570–578.

MALYKHIN, V.I.
[1987]   *Spaces of continuous functions in simplest generic extensions,* Math. Notes,
          41(1987), 301–304.
[1994]   *A non-hereditarily separable space with separable closed subspaces,* Q & A in
          General Topology, **12**(1994), 209–214.
[1998]   *On subspaces of sequential spaces (in Russian),* Matem. Zametki, **64:3**(1998),
          407–413.
[1999]   *$C_p(I)$ is not subsequential,* Comment. Math. Univ. Carolinae, **40:4**(1999),
          785–788.

MALYKHIN, V.I., SHAKHMATOV, D.B.
[1992]   *Cartesian products of Fréchet topological groups and function spaces,* Acta
          Math. Hungarica, **60**(1992), 207–215.

MARCISZEWSKI, W.
[1983]   *A pre-Hilbert space without any continuous map onto its own square,* Bull. Acad.
          Polon. Sci., **31:(9-12)**(1983), 393–397.
[1988a]   *A remark on the space of functions of first Baire class,* Bull. Polish Acad. Sci.,
          Math., **36:(1-2)**(1997), 65–67.

[1993]   *On analytic and coanalytic function spaces* $C_p(X)$, Topology and Its Appl., **50**(1993), 241–248.

[1995a]   *On universal Borel and projective filters*, Bull. Acad. Polon. Sci., Math., **43:1**(1995), 41–45.

[1995b]   *A countable X having a closed subspace A with* $C_p(A)$ *not a factor of* $C_p(X)$, Topology Appl., **64**(1995), 141–147.

[1997a]   *A function space* $C_p(X)$ *not linearly homeomorphic to* $C_p(X) \times R$, Fundamenta Math., **153:2**(1997), 125–140.

[1997b]   *On hereditary Baire products*, Bull. Polish Acad. Sci., Math., **45:3**(1997), 247–250.

[1998a]   *P-filters and hereditary Baire function spaces*, Topology Appl., **89**(1998), 241–247.

[1998b]   *Some recent results on function spaces* $C_p(X)$, Recent Progress in Function Spaces., Quad. Mat. **3**(1998), Aracne, Rome, 221–239.

[2002]   *Function Spaces*, in: Recent Progress in General Topology II, Ed. by M. Hušek and J. van Mill, Elsevier Sci. B.V., Amsterdam, 2002, 345–369.

[2003a]   *A function space* $C_p(X)$ *without a condensation onto a* σ-*compact space*, Proc. Amer. Math. Society, **131:6**(2003), 1965–1969.

MARCISZEWSKI, W., PELANT, J.
[1995]   *Absolute Borel sets and function spaces*, Trans. Amer. Math. Soc., **349:9**(1997), 3585–3596.

MCCOY, R.A.
[1975]   *First category function spaces under the topology of pointwise convergence*, Proc. Amer. Math. Soc., **50**(1975), 431–434.

[1978a]   *Characterization of pseudocompactness by the topology of uniform convergence on function spaces*, J. Austral. Math. Soc., **26**(1978), 251–256.

[1978b]   *Submetrizable spaces and almost* σ-*compact function spaces*, Proc. Amer. Math. Soc., **71**(1978), 138–142.

[1978c]   *Second countable and separable function spaces*, Amer. Math. Monthly, **85:6**(1978), 487–489.

[1980a]   *Countability properties of function spaces*, Rocky Mountain J. Math., **10**(1980), 717–730.

[1980b]   *A K-space function space*, Int. J. Math. Sci., **3**(1980), 701–711.

[1986]   *Fine topology on function spaces*, Internat. J. Math. Math. Sci. **9:3**(1986), 417–424.

MCCOY, R.A., NTANTU, I.
[1986]   *Completeness properties of function spaces*, Topology Appl., **22:2**(1986), 191–206.

[1988]   *Topological Properties of Spaces of Continuous Functions*, Lecture Notes in Math., 1315, Springer, Berlin, 1988.

MICHAEL, E.
[1966] $\aleph_0$-*spaces*, J. Math. and Mech., **15:6**(1966), 983–1002.

[1973]   *On k-spaces,* k_R-*spaces and* $k(X)$, Pacific J. Math., **47:2**(1973), 487–498.

[1977] $\aleph_0$-*spaces and a function space theorem of R. Pol*, Indiana Univ. Math. J., **26**(1977), 299–306.

MICHAEL, E., RUDIN, M.E.
[1977]   *A note on Eberlein compacta*. Pacific J. Math., **72:2**(1977), 487–495.

MILL, J. VAN
[1984]   *An introduction to* $\beta\omega$, Handbook of Set-Theoretic Topology, North-Holland, Amsterdam, 1984, 503–567.

[1989]   *Infinite-Dimensional Topology.* Prerequisites and Introduction, North Holland, Amsterdam, 1989.

[1999] $C_p(X)$ *is not* $G_{\delta\sigma}$: *a simple proof,* Bull. Polon. Acad. Sci., Ser. Math., **47**(1999), 319–323.

[2002]   *The Infinite-Dimensional Topology of Function Spaces,* North Holland Math. Library **64**, Elsevier, Amsterdam, 2002.

MORISHITA, K.

[1992a]   *The minimal support for a continuous functional on a function space,* Proc. Amer. Math. Soc. **114:2**(1992), 585–587.

[1992b]   *The minimal support for a continuous functional on a function space. II,* Tsukuba J. Math. **16:2**(1992), 495–501.

NAKHMANSON, L. B.

[1982]   *On continuous images of* σ-*products (in Russian),* Topology and Set Theory, Udmurtia Universty P.H., Izhevsk, 1982, 11–15.

[1984]   *The Souslin number and calibers of the ring of continuous functions (in Russian),* Izv. Vuzov, Matematika, 1984, N 3, 49–55.

[1985]   *On Lindelöf property of function spaces (in Russian),* Mappings and Extensions of Topological Spaces, Udmurtia University P.H., Ustinov, 1985, 8–12.

NAMIOKA, I.

[1974]   *Separate continuity and joint continuity.* Pacific J. Math., **51:2**(1974), 515–531.

NOBLE, N.

[1969a]   *Products with closed projections,* Trans Amer. Math. Soc., **140**(1969), 381–391.

[1969b]   *Ascoli theorems and the exponential map,* Trans Amer. Math. Soc., **143**(1969), 393–411.

[1971]   *Products with closed projections II,* Trans Amer. Math. Soc., **160**(1971), 169–183.

[1974]   *The density character of function spaces,* Proc. Amer. Math. Soc., **42:1**(1974), 228–233.

NOBLE, N., ULMER, M.

[1972]   *Factoring functions on Cartesian products,* Trans Amer. Math. Soc., **163**(1972), 329–339.

OKUNEV, O.G.

[1984]   *Hewitt extensions and function spaces (in Russian),* Cardinal Invariants and Mappings of Topological Spaces (in Russian), Izhvsk, 1984, 77–78.

[1985]   *Spaces of functions in the topology of pointwise convergence: Hewitt extension and* τ-*continuous functions,* Moscow Univ. Math. Bull., **40:4**(1985), 84–87.

[1993a]   *On Lindelöf* Σ-*spaces of functions in the pointwise topology,* Topology and Its Appl., **49**(1993), 149–166.

[1993b]   *On analyticity in cosmic spaces,* Comment. Math. Univ. Carolinae, **34**(1993), 185–190.

[1995a]   *On Lindelöf sets of continuous functions,* Topology Appl., **63**(1995), 91–96.

[1996]   *A remark on the tightness of products,* Comment. Math. Univ. Carolinae, **37:2**(1996), 397–399.

[1997b]   *On the Lindelöf property and tightness of products,* Topology Proc., **22**(1997), 363–371.

[2002]   *Tightness of compact spaces is preserved by the t-equivalence relation,* Comment. Math. Univ. Carolin. **43:2**(2002), 335–342.

[2005a]   *Fréchet property in compact spaces is not preserved by M-equivalence,* Comment. Math. Univ. Carolin. **46:4**(2005), 747–749.

[2005b]   *A σ-compact space without uncountable free sequences can have arbitrary tightness,* Questions Answers Gen. Topology **23:2**(2005), 107–108.

OKUNEV, O.G., SHAKHMATOV, D.B.
[1987]   *The Baire property and linear isomorphisms of continuous function spaces (in Russian),* Topological Structures and Their Maps, Latvian State University P.H., Riga, 1987, 89–92.

OKUNEV, O.G., TAMANO, K.
[1996]   *Lindelöf powers and products of function spaces,* Proceedings of Amer. Math. Soc., **124:9**(1996), 2905–2916.

OKUNEV, O.G., TAMARIZ-MASCARÚA, A.
[2004]   *On the Čech number of $C_p(X)$,* Topology Appl., **137**(2004), 237–249.

OKUNEV, O.G., TKACHUK, V.V.
[2001]   *Lindelöf Σ-property in $C_p(X)$ and $p(C_p(X)) = \omega$ do not imply countable network weight in X,* Acta Mathematica Hungarica, **90:(1–2)**(2001), 119–132.

[2002]   *Density properties and points of uncountable order for families of open sets in function spaces,* Topology Appl., **122**(2002), 397–406.

PASYNKOV, B.A.
[1967]   *On open mappings,* Soviet Math. Dokl., **8**(1967), 853–856.

PELANT, J.
[1988]   *A remark on spaces of bounded continuous functions,* Indag. Math., **91**(1988), 335–338.

POL, R.
[1972]   *On the position of the set of monotone mappings in function spaces,* Fund. Math., **75**(1972), 75–84.

[1974]   *Normality in function spaces,* Fund. Math., **84:2**(1974), 145–155.

[1977]   *Concerning function spaces on separable compact spaces,* Bull. Acad. Polon. Sci., Sér. Math., Astron. et Phys., **25:10**(1977), 993–997.

[1978]   *The Lindelöf property and its analogue in function spaces with weak topology,* Topology. 4-th Colloq. Budapest, **2**(1978), Amsterdam, 1980, 965–969.

[1979]   *A function space $c(x)$ which is weakly Lindelöf but not weakly compactly generated,* Studia Math., **64:3**(1979), 279–285.

[1980a]   *A theorem on the weak topology of $c(x)$ for compact scattered x,* Fundam. Math., **106:2**(1980), 135–140.

[1980b]   *On a question of H.H. Corson and some related matters,* Fund. Math., **109:2**(1980), 143–154.

[1982]   *Note on the spaces $P(S)$ of regular probability measures whose topology is determined by countable subsets,* Pacific J. Math., **100**(1982), 185–201.

[1984a]   *An infinite-dimensional pre-Hilbert space not homeomorphic to its own square,* Proc. Amer. Math. Soc., **90:3**(1984), 450–454.

[1984b]   *On pointwise and weak topology in function spaces,* Univ. Warszawski, Inst. Matematiki, Preprint 4/84, Warszawa, 1984.

[1986]   *Note on compact sets of first Baire class functions,* Proceedings Amer. Math. Soc., **96:1**(1986), 152–154.

[1989]   *Note on pointwise convergence of sequences of analytic sets,* Mathem., **36**(1989), 290–300.

[1995]   *For a metrizable X, $C_p(X)$ need not be linearly homeomorphic to $C_p(X) \times C_p(X)$,* Mathematika, **42**(1995), 49–55.

PONTRIAGIN, L.S.
   [1984]   *Continuous Groups (in Russian)*, Nauka, Moscow, 1984.

PYTKEEV, E.G.
   [1976]   *Upper bounds of topologies*, Math. Notes, **20**:4(1976), 831–837.
   [1982a]   *On the tightness of spaces of continuous functions (in Russian)*, Uspehi Mat. Nauk, **37**:1(1982), 157–158.
   [1982b]   *Sequentiality of spaces of continuous functions (in Russian)*, Uspehi Matematich-eskih Nauk, **37**:5(1982), 197–198.
   [1985]   *The Baire property of spaces of continuous functions, (in Russian)*, Matem. Zametki, **38**:5(1985), 726–740.
   [1990]   *A note on Baire isomorphism*, Comment. Math. Univ. Carolin. **31**:1(1990), 109–112.
   [1992a]   *On Fréchet–Urysohn property of spaces of continuous functions, (in Russian)*, Trudy Math. Inst. RAN, **193**(1992), 156–161.
   [1992b]   *Spaces of functions of the first Baire class over K-analytic spaces (in Russian)*, Mat. Zametki **52**:3(1992), 108–116.
   [2003]   *Baire functions and spaces of Baire functions*, Journal Math. Sci. (N.Y.) **136**:5(2006), 4131–4155

PYTKEEV, E.G., YAKOVLEV, N.N.
   [1980]   *On bicompacta which are unions of spaces defined by means of coverings*, Comment. Math. Univ. Carolinae, **21**:2(1980), 247–261.

RAJAGOPALAN, M., WHEELER, R.F.
   [1976]   *Sequential compactness of X implies a completeness property for C(X)*, Canadian J. Math., **28**(1976), 207–210.

REZNICHENKO, E.A.
   [1987]   *Functional and weak functional tightness (in Russian)*, Topological Structures and Their Maps, Latvian State University P.H., Riga, 1987, 105–110.
   [1989b]   *A pseudocompact space in which only the subsets of not full cardinality are not closed and not discrete*, Moscow Univ. Math. Bull. **44**:6(1989), 70–71.
   [1990a]   *Normality and collectionwise normality in function spaces*, Moscow Univ. Math. Bull., **45**:6(1990), 25–26.
   [2008]   *Stratifiability of $C_k(X)$ for a class of separable metrizable X*, Topology Appl. **155**:17-18(2008), 2060–2062.

RUDIN, M.E.
   [1956]   *A note on certain function spaces*, Arch. Math., **7**(1956), 469–470.

RUDIN, W.
   [1973]   *Functional Analysis*, McGraw-Hill Book Company, New York, 1973.

SAKAI, M.
   [1988a]   *On supertightness and function spaces*, Commentationes Math. Univ. Carolinae, **29**:2(1988), 249–251.
   [1988b]   *Property C'' and function spaces*, Proc. Amer. Math. Soc., **104**:3(1988), 917–919.
   [1992]   *On embeddings into $C_p(X)$ where X is Lindelöf*, Comment. Math. Univ. Carolinae, **33**:1(1992), 165–171.
   [1995]   *Embeddings of κ-metrizable spaces into function spaces*, Topol. Appl., **65**(1995), 155–165.
   [2000]   *Variations on tightness in function spaces*, Topology Appl., **101**(2000), 273–280.
   [2003]   *The Pytkeev property and the Reznichenko property in function spaces*, Note di Matem. **22**:2(2003/04), 43–52

[2006]   *Special subsets of reals characterizing local properties of function spaces,*
           Selection Principles and Covering Properties in Topology, Dept. Math., Seconda
           Univ. Napoli, Caserta, Quad. Mat., **18**(2006), 195–225,
[2007]   *The sequence selection properties of* $C_p(X)$, Topology Appl., **154**(2007),
           552–560.
[2008]   *Function spaces with a countable cs*-network at a point,* Topology and Its Appl.,
           156:1(2008), 117–123.

SHAKHMATOV, D.B.
[1986]   *A pseudocompact Tychonoff space all countable subsets of which are closed and*
           C*-embedded, Topology Appl., **22:2**(1986), 139–144.

SHAPIROVSKY, B.E.
[1974]   *Canonical sets and character. Density and weight in compact spaces,* Soviet
           Math. Dokl., **15**(1974), 1282–1287.
[1978]   *Special types of embeddings in Tychonoff cubes, subspaces of* Σ*-products and*
           cardinal invariants, Colloquia Mathematica Soc. Janos Bolyai, **23**(1978), 1055–
           1086.
[1981]   *Cardinal invariants in bicompacta (in Russian),* Seminar on General Topology,
           Moscow University P.H., Moscow, 1981, 162–187.

SIPACHEVA, O.V.
[1988]   *Lindelöf* Σ*-spaces of functions and the Souslin number,* Moscow Univ. Math.
           Bull., **43:3**(1988), 21–24.
[1989]   *On Lindelöf subspaces of function spaces over linearly ordered separable*
           compacta, General Topology, Spaces and Mappings, Mosk. Univ., Moscow,
           1989, 143–148.
[1992]   *On surlindelöf compacta (in Russian),* General Topology. Spaces, Mappings and
           Functors, Moscow University P.H., Moscow, 1992, 132–140.

SOKOLOV, G.A.
[1986]   *On Lindelöf spaces of continuous functions (in Russian),* Mat. Zametki,
           39:6(1986), 887–894.
[1993]   *Lindelöf property and the iterated continuous function spaces,* Fundam. Math.,
           143(1993), 87–95.

STONE, A.H.
[1963]   *A note on paracompactness and normality of mapping spaces,* Proc. Amer. Math.
           Soc., **14**(1963), 81–83.

TALAGRAND, M.
[1979b]   *Sur la K-analyticité des certains espaces d'operadeurs,* Israel J. Math., **32**(1979),
           124–130.
[1984]   *A new countably determined Banach space,* Israel J. Math., **47:1**(1984), 75–80.
[1985]   *Espaces de Baire et espaces de Namioka,* Math. Ann., **270**(1985), 159–164.

TAMANO, K., TODORCEVIC, S.
[2005]   *Cosmic spaces which are not* μ*-spaces among function spaces with the topology*
           of pointwise convergence, Topology Appl. **146/147**(2005), 611–616.

TAMARIZ–MASCARÚA, A.
[1996]   *Countable product of function spaces having p-Frechet-Urysohn like properties,*
           Tsukuba J. Math. **20:2**(1996), 291–319.
[1998] α*-pseudocompactness in* C_p*-spaces,* Topology Proc., **23**(1998), 349–362.
[2006]   *Continuous selections on spaces of continuous functions,* Comment. Math. Univ.
           Carolin. **47:4**(2006), 641–660.

TANI, T.
[1986]   *On the tightness of $C_p(X)$*, Memoirs Numazu College Technology, **21**(1986), 217–220.

TKACHENKO, M.G.
[1978]   *On the behaviour of cardinal invariants under the union of chains of spaces (in Russian)*, Vestnik Mosk. Univ., Math., Mech., **33:4**(1978), 50–58.
[1979]   *On continuous images of dense subspaces of topological products (in Russian)*, Uspehi Mat. Nauk, **34:6**(1979), 199–202.
[1982]   *On continuous images of dense subspaces of Σ-products of compacta (in Russian)*, Sibirsk. Math. J., **23:3**(1982), 198–207.
[1985]   *On continuous images of spaces of functions*, Sibirsk. Mat. Zhurnal, **26:5**(1985), 159–167.

TKACHUK, V.V.
[1984b]   *Characterization of the Baire property in $C_p(X)$ in terms of the properties of the space X (in Russian)*, Cardinal Invariants and Mappings of Topological Spaces (in Russian), Izhevsk, 1984, 76–77.
[1984c]   *On a supertopological cardinal invariant*. Vestn. Mosk. Univ., Matem., Mech., **39:4**(1984), 26–29.
[1986a]   *The spaces $C_p(X)$: decomposition into a countable union of bounded subspaces and completeness properties*, Topology Appl., **22:3**(1986), 241–254.
[1986b]   *Approximation of $\mathbf{R}^X$ with countable subsets of $C_p(X)$ and calibers of the space $C_p(X)$*, Comment. Math. Univ. Carolinae, **27:2**(1986), 267–276.
[1987a]   *The smallest subring of the ring $C_p(C_p(X))$ which contains $X \cup \{1\}$ is dense in $C_p(C_p(X))$ (in Russian)*, Vestnik Mosk. Univ., Math., Mech., **42:1**(1987), 20–22.
[1987b]   *Spaces that are projective with respect to classes of mappings*, Trans. Moscow Math. Soc., **50**(1987), 139–156.
[1988]   *Calibers of spaces of functions and metrization problem for compact subsets of $C_p(X)$ (in Russian)*, Vestnik Mosk. Univ., Matem., Mech., **43:3**(1988), 21–24.
[1991]   *Methods of the theory of cardinal invariants and the theory of mappings applied to the spaces of functions (in Russian)*, Sibirsk. Mat. Zhurnal, **32:1**(1991), 116–130.
[1994]   *Decomposition of $C_p(X)$ into a countable union of subspaces with "good" properties implies "good" properties of $C_p(X)$*, Trans. Moscow Math. Soc., **55**(1994), 239–248.
[1995]   *What if $C_p(X)$ is perfectly normal?* Topology Appl., **65**(1995), 57–67.
[1998]   *Mapping metric spaces and their products onto $C_p(X)$*, New Zealand J. Math., **27:1**(1998), 113–122.
[2000]   *Behaviour of the Lindelöf Σ-property in iterated function spaces*, Topology Appl., **107:3-4**(2000), 297–305.
[2001]   *Lindelöf Σ-property in $C_p(X)$ together with countable spread of X implies X is cosmic*, New Zealand J. Math., **30**(2001), 93–101.
[2003]   *Properties of function spaces reflected by uniformly dense subspaces*, Topology Appl., **132**(2003), 183–193.
[2005a]   *A space $C_p(X)$ is dominated by irrationals if and only if it is K-analytic*, Acta Math. Hungarica, **107:4**(2005), 253–265.
[2005c]   *A nice class extracted from C_p-theory*, Comment. Math. Univ. Carolinae, **46:3**(2005), 503–513.
[2007a]   *Condensing function spaces into Σ-products of real lines*, Houston Journal of Math., **33:1**(2007), 209–228.
[2007b]   *Twenty questions on metacompactness in function spaces*, Open Problems in Topology II, ed. by E. Pearl, Elsevier B.V., Amsterdam, 2007, 595–598.

[2007c]   *A selection of recent results and problems in C_p-theory*, Topology and Its Appl.,
            154:12(2007), 2465–2493.
[2009]   *Condensations of $C_p(X)$ onto σ-compact spaces*, Appl. Gen. Topology,
            10:1(2009), 39–48.

TKACHUK, V.V., SHAKHMATOV, D.B.
[1986]   *When the space $C_p(X)$ is σ-countably compact? (in Russian)*, Vestnik Mosk.
            Univ., Math., Mech., **41:1**(1986), 70–72.

TKACHUK, V.V., YASCHENKO, I.V.
[2001]   *Almost closed sets and topologies they determine*, Comment. Math. Univ.
            Carolinae, **42:2**(2001), 395–405.

TODORCEVIC, S.
[1989]   *Partition Problems in Topology*, Contemporary Mathematics, American Mathe-
            matical Society, **84**(1989). Providence, Rhode Island, 1989.
[1993]   *Some applications of S and L combinatorics*, Ann. New York Acad. Sci.,
            705(1993), 130–167.
[2000]   *Chain-condition methods in topology*, Topology Appl. **101:1**(2000), 45–82.

TYCHONOFF A.N.
[1935]   *Über einer Funktionenraum*, Math. Ann., **111**(1935), 762–766.

USPENSKIJ, V.V.
[1978]   *On embeddings in functional spaces (in Russian)*, Doklady Acad. Nauk SSSR,
            242:3(1978), 545–546.
[1982a]   *On frequency spectrum of functional spaces (in Russian)*, Vestnik Mosk. Univ.,
            Math., Mech., **37:1**(1982), 31–35.
[1982b]   *A characterization of compactness in terms of the uniform structure in function
            space (in Russian)*, Uspehi Mat. Nauk, **37:4**(1982), 183–184.
[1983b]   *A characterization of realcompactness in terms of the topology of pointwise con-
            vergence on the function space*, Comment. Math. Univ. Carolinae, **24:1**(1983),
            121–126.

VALOV, V.M.
[1986]   *Some properties of $C_p(X)$*, Comment. Math. Univ. Carolinae, **27:4**(1986),
            665–672.
[1997a]   *Function spaces*, Topology Appl. **81:1**(1997), 1–22.
[1999]   *Spaces of bounded functions*, Houston J. Math. **25:3**(1999), 501–521.

VALOV, V., VUMA, D,
[1996]   *Lindelöf degree and function spaces*, Papers in honour of Bernhard
            Banaschewski, Kluwer Acad. Publ., Dordrecht, 2000, 475–483.
[1998]   *Function spaces and Dieudonné completeness*, Quaest. Math. **21:3-4**(1998),
            303–309.

VELICHKO, N.V.
[1981]   *On weak topology of spaces of continuous functions (in Russian)*, Matematich.
            Zametki, **30:5**(1981), 703–712.
[1982]   *Regarding the theory of spaces of continuous functions (in Russian)*, Uspehi
            Matem. Nauk, **37:4**(1982), 149–150.
[1985]   *Networks in spaces of mappings (in Russian)*, Mappings and Extensions of
            Topological Spaces, Udmurtia University P.H., Ustinov, 1985, 3–6.
[1995b]   *On normality in function spaces*, Math. Notes, **56:5-6**(1995), 1116–1124.
[2001]   *On subspaces of functional spaces*, Proc. Steklov Inst. Math. **2**(2001), 234–240.
[2002]   *Remarks on C_p-theory*, Proc. Steklov Inst. Math., **2**(2002), 190–192.

VIDOSSICH, G.
  [1969a]   *On topological spaces whose function space is of second category,* Invent. Math.,
            8:2(1969), 111–113.
  [1969b]   *A remark on the density character of function spaces,* Proc. Amer. Math. Soc.,
            22(1969), 618–619.
  [1970]   *Characterizing separability of function spaces,* Invent. Math., **10:3**(1970),
            205–208.
  [1972]   *On compactness in function spaces,* Proc. Amer. Math. Soc., **33**(1972), 594–598.

WHITE, H.E., JR.
  [1978]   *First countable spaces that have special pseudo-bases,* Canadian Mathematical
            Bull., **21:1**(1978), 103–112.

YAKOVLEV, N.N.
  [1980]   *On bicompacta in Σ-products and related spaces.* Comment. Math. Univ. Caroli-
            nae, **21:2**(1980), 263–283.

YASCHENKO, I.V.
  [1989]   *Baire functions as restrictions of continuous ones (in Russian),* Vestnik Mosk.
            Univ., Math., Mech., **44:6**(1989), 80–82.
  [1991]   *On rings of Baire functions (in Russian),* Vestnik Mosk. Univ., Math., **46:2**(1991),
            88.
  [1992a]   *On the extent of function spaces,* Comptes Rendus Acad. Bulg. Sci., **45:1**(1992),
            9–10.
  [1992b]   *Embeddings in R^n and in R^ω and splittability,* Vestnik Mosk. Univ., Math., Mech.,
            47:2(1992), 107.
  [1992c]   *On fixed points of mappings of topological spaces,* Vestnik Mosk. Univ., Math.,
            Mech., **47:5**(1992), 93.
  [1992d]   *Cardinality of discrete families of open sets and one-to-one continuous mappings,*
            Questions and Answers in General Topology, **2**(1992), 24–26.
  [1994]   *On the monotone normality of functional spaces,* Moscow University Math. Bull.,
            49:3(1994), 62–63.

ZENOR, PH.
  [1980]   *Hereditary m-separability and hereditary m-Lindelöf property in product spaces
            and function spaces,* Fund. Math., **106**(1980), 175–180.

List of Special Symbols

For every symbol of this list we refer the reader to a place where it was defined. There could be many such places but we only mention one here. Note that a symbol is often defined in the first volume of this book entitled "Topology and Function Spaces"; we denote it by *TFS*. If it is defined in the second volume of this book entitled "Special Features of Function Spaces", we denote it by *SFFS*. We *never* use page numbers; instead, we have the following types of references:

(a) *A reference to an introductory part of a section.*
   For example,
   $C_{p,n}(X)$ ⋯⋯⋯⋯⋯⋯⋯⋯⋯⋯⋯⋯⋯ **1.2**
   says that $C_{p,n}(X)$ is defined in the Introductory Part of Section 1.2.
   Of course,
   $C_p(X)$ ⋯⋯⋯⋯⋯⋯⋯⋯⋯⋯⋯ **TFS-1.1**
   shows that $C_p(X)$ is defined in the Introductory Part of Section 1.1 of the book TFS.
   Analogously,
   $\chi(A, X)$ ⋯⋯⋯⋯⋯⋯⋯⋯⋯⋯ **SFFS-1.2**
   says that $\chi(A, X)$ was defined in the Introductory Part of Section 1.2 of the book SFFS.
(b) *A reference to a problem.*
   For example,
   $X[A]$ ⋯⋯⋯⋯⋯⋯⋯⋯⋯⋯⋯⋯⋯ **090**
   says that $X[A]$ is defined in Problem 090 of this book;
   and, naturally,
   $C_u(X)$ ⋯⋯⋯⋯⋯⋯⋯⋯⋯⋯⋯ **TFS-084**
   means that the expression $C_u(X)$ is defined in Problem 084 of the book TFS.

© Springer International Publishing Switzerland 2015
V.V. Tkachuk, *A Cp-Theory Problem Book*, Problem Books
in Mathematics, DOI 10.1007/978-3-319-16092-4

(c) *A reference to a solution.*

For example,

$O(f, K, \varepsilon)$ ································· S.321

says that the definition of $O(f, K, \varepsilon)$ can be found in the Solution of Problem 321 of the book TFS.

Analogously,

$\Delta_n(Z)$ ································· T.019

says that the definition of $\Delta_n(Z)$ can be found in the Solution of Problem 019 of the book SFFS.

The expression,

$B_1(X)$ ································· U.463

says that the definition of $B_1(X)$ can be found in the Solution of Problem 463 of this volume.

Every problem is short so it won't be difficult to find a reference in it. An introductory part *is never longer than two pages*; so, hopefully, it is not hard to find a reference in it either. Please, keep in mind that a solution of a problem can be pretty long but its definitions *are always given in the beginning.*

The symbols are arranged in alphabetical order; this makes it easy to find the expressions $B(x, r)$ and βX but it is not immediate what to do if we are looking for $\bigoplus_{t \in T} X_t$. I hope that the placement of the expressions which start with Greek letters or mathematical symbols is intuitive enough to be of help to the reader. Even if it is not, then there are only three pages to plough through. The alphabetic order is *by line* and not by column. For example, the first three lines contain symbols which start with "A" or something similar and lines 3–5 are for the expressions beginning with "B", "β" or "$\mathbb{B}$".

$A(\kappa)$ ··············· **TFS-1.2**		$a(X)$ ··············· **TFS-1.5**	
$AD(X)$ ··············· **TFS-1.4**		$\bigwedge \mathcal{A}$ ··············· **T.300**	
$\mathcal{A}|Y$ ··············· **T.092**		$\bigvee \mathcal{A}$ ··············· **U.031**	
$B_1(X)$ ··············· **U.463**		$B_d(x, r)$ ··············· **TFS-1.3**	
$B(x, r)$ ··············· **TFS-1.3**		$(B1)$–$(B2)$ ··············· **TFS-006**	
βX ··············· **TFS-1.3**		$\mathbb{B}(X)$ ··············· **SFFS-1.4**	
$\mathrm{cl}_X(A)$ ··············· **TFS-1.1**		$\mathrm{cl}_\tau(A)$ ··············· **TFS-1.1**	
$C(X)$ ··············· **TFS-1.1**		$C^*(X)$ ··············· **TFS-1.1**	
$C(X, Y)$ ··············· **TFS-1.1**		$C_p(X, Y)$ ··············· **TFS-1.1**	
$C_u(X)$ ··············· **TFS-084**		$C_p(Y|X)$ ··············· **TFS-1.5**	

Index

© Springer International Publishing Switzerland 2015
V.V. Tkachuk, *A Cp-Theory Problem Book*, Problem Books
in Mathematics, DOI 10.1007/978-3-319-16092-4

Printed in the United States
By Bookmasters